MEMBRANE TRANSPORT PROCESSES IN ORGANIZED SYSTEMS

MEMBRANE TRANSPORT PROCESSES IN ORGANIZED SYSTEMS

Edited by

Thomas E. Andreoli, M.D.
University of Texas Medical School
Houston, Texas

Joseph F. Hoffman, Ph.D.
Yale University School of Medicine
New Haven, Connecticut

Darrell D. Fanestil, M.D.
University of California, San Diego
La Jolla, California

and

Stanley G. Schultz, M.D.
University of Texas Medical School
Houston Texas

PLENUM MEDICAL BOOK COMPANY
New York and London

Library of Congress Cataloging in Publication Data

Physiology of membrane disorders. Selections.
Membrane transport processes in organized systems.

"This volume is a reprint with minor modifications of parts IV and V of Physiology of membrane disorders, second edition, published by Plenum Medical Book Company in 1986"—T.p. verso.
Includes bibliographies and index.
1. Epithelium—Congresses. 2. Biological transport—Congresses. 3. Membranes (Biology)—Congresses. 4. Cell membranes—Congresses. I. Andreoli, Thomas E., 1935- . II. Title. [DNLM: 1. Biological Transport. 2. Ions. 3. Membranes—physiology. QS 532.5.M3 P5782ma]
QP88.4.P4825 1987 599′.0875 87-18655
ISBN 0-306-42698-6

This volume is a reprint with minor modifications of Parts IV and V of *Physiology of Membrane Disorders, Second Edition,* published by Plenum Medical Book Company in 1986.

233 Spring Street, New York, N.Y. 10013

Plenum Medical Book Company is an imprint of Plenum Publishing Corporation

Printed in the United States of America

Contributors

RICHARD W. ALDRICH, PH.D.
Postdoctoral Fellow
Section of Molecular Neurobiology
Yale University School of Medicine
New Haven, Connecticut 06510

THOMAS E. ANDREOLI, M.D.
Edward Randall III Professor and Chairman
Department of Internal Medicine
Professor
Department of Physiology and Cell Biology
University of Texas Medical School
Houston, Texas 77225

PETER S. ARONSON, M.D.
Associate Professor of Medicine
and Physiology
Departments of Physiology and Internal Medicine
Yale University School of Medicine
New Haven, Connecticut 06510

P. DARWIN BELL, PH.D.
Assistant Professor of Physiology
and Biophysics
Nephrology Research and Training Center and
Department of Physiology and Biophysics
University of Alabama School of Medicine
Birmingham, Alabama 35294

FRANCISCO BEZANILLA, PH.D.
Professor
Department of Physiology
Ahmanson Laboratory of Neurobiology, and
the Jerry Lewis Neuromuscular Research Center,
University of California Medical School
Los Angeles, California 90024

WALTER F. BORON, M.D., PH.D.
Assistant Professor
Department of Physiology
Yale University School of Medicine
New Haven, Connecticut 06510

JAMES L. BOYER, M.D.
Professor of Medicine
Director, Liver Study Unit
Department of Medicine
Yale University School of Medicine
New Haven, Connecticut 06510

LINDA S. COSTANZO, PH.D.
Assistant Professor
Department of Physiology and Biophysics
Medical College of Virginia
Richmond, Virginia 23289

JOHN M. DIETSCHY, M.D.
Professor of Medicine
Department of Internal Medicine
Southwestern Medical School
University of Texas Health Science Center
Dallas, Texas 75235

VINCENT E. DIONNE
Division of Pharmacology
Department of Medicine
University of California at San Diego
La Jolla, California 92093

ANDREW P. EVAN, PH.D.
Professor of Anatomy
Department of Anatomy
Indiana University Medical Center
Indianapolis, Indiana 46223

JOHN G. FORTE, PH.D.
Professor of Physiology
Department of Physiology–Anatomy
University of California
Berkeley, California 94720

GERHARD GIEBISCH, M.D.
Sterling Professor of Physiology
Department of Physiology
Yale University School of Medicine
New Haven, Connecticut 06510

EDWARD HAWROT, PH.D.
Assistant Professor of Pharmacology
Department of Pharmacology
Yale University School of Medicine
New Haven, Connecticut 06510

STEVEN C. HEBERT, M.D.
Assistant Professor of Medicine
Division of Nephrology
University of Texas Medical School
Houston, Texas 77225
Present address
Harvard Medical School Renal Division
Brigham and Women's Hospital
Boston, Massachusetts 02115

PETER HESS, M.D.
Research Associate
Department of Physiology
Yale University School of Medicine
New Haven, Connecticut 06510

H. RONALD KABACK, M.D.
Head, Laboratory of Membrane Biochemistry
Roche Institute of Molecular Biology
Roche Research Center
Nutley, New Jersey 07110

MARK A. KNEPPER, M.D., PH.D.
Medical Staff Fellow
National Heart, Lung and Blood Institute
National Institutes of Health
Bethesda, Maryland 20205

HANS CHRISTOPH LÜTTGAU, M.D.
Professor of Physiology
Department of Cell Physiology
Ruhr University
Bochum, West Germany

TERRY E. MACHEN, PH.D.
Associate Professor of Physiology
Department of Physiology—Anatomy
University of California
Berkeley, California 94720

L. GABRIEL NAVAR, PH.D.
Professor of Physiology and Biophysics
Nephrology Research and Training Center
and Department of Physiology and Biophysics
University of Alabama School of Medicine
Birmingham, Alabama 35294

DON W. POWELL, M.D.
Professor and Chief
Department of Medicine
University of North Carolina School
of Medicine
Chapel Hill, North Carolina 27514

GARY RUDNICK, PH.D.
Associate Professor of Pharmacology
Department of Pharmacology
Yale University School of Medicine
New Haven, Connecticut 06510

STANLEY G. SCHULTZ, M.D.
Professor and Chairman
Department of Physiology and Cell Biology
University of Texas Medical School
Houston, Texas 77225

GEORGE DIMITRIE STEPHENSON
Department of Zoology
La Trobe University
Melbourne, Australia

JOHN L. STEPHENSON, M.D.
Professor of Biomathematics
Department of Physiology
Cornell University Medical College
New York, New York 10021

CHARLES F. STEVENS, M.D., PH.D.
Professor of Physiology
Section of Molecular Neurobiology
Yale University School of Medicine
New Haven, Connecticut 06510

RICHARD W. TSIEN, D. PHIL.
Professor of Physiology
Department of Physiology
Yale University School of Medicine
New Haven, Connecticut 06510

MICHAEL J. WELSH, M.D.
Assistant Professor
Laboratory of Epithelial Transport and Pulmonary Division
Department of Internal Medicine
University of Iowa College of Medicine
Iowa City, Iowa 52242

HENRIK WESTERGAARD, M.D.
Assistant Professor of Medicine
Department of Internal Medicine
Southwestern Medical School
University of Texas Health Science Center
Dallas, Texas 75235

MICHAEL M. WHITE, PH.D.
Postdoctoral Researcher
Department of Physiology
Ahmanson Laboratory of Neurobiology, and
the Jerry Lewis Neuromuscular Research Center,
University of California Medical School
Los Angeles, California 90024

ERICH E. WINDHAGER, M.D.
Professor and Chairman
Department of Physiology
Cornell University Medical College
New York, New York 10021

Preface

Membrane Transport Processes in Organized Systems is a softcover book containing portions of *Physiology of Membrane Disorders* (*Second Edition*). The parent volume contains six major sections. This text encompasses the fourth and fifth sections: Transport Events in Single Cells and Transport in Epithelia: Vectorial Transport through Parallel Arrays.

We hope that this smaller volume, which deals with transport processes in single cells and in organized epithelia, will be helpful to individuals interested in general physiology, transport in single cells and epithelia, and the methods for studying those transport processes.

THOMAS E. ANDREOLI
JOSEPH F. HOFFMAN
DARRELL D. FANESTIL
STANLEY G. SCHULTZ

Preface to the Second Edition

The second edition of *Physiology of Membrane Disorders* represents an extensive revision and a considerable expansion of the first edition. Yet the purpose of the second edition is identical to that of its predecessor, namely, to provide a rational analysis of membrane transport processes in individual membranes, cells, tissues, and organs, which in turn serves as a frame of reference for rationalizing disorders in which derangements of membrane transport processes play a cardinal role in the clinical expression of disease.

As in the first edition, this book is divided into a number of individual, but closely related, sections. Part V represents a new section where the problem of transport across epithelia is treated in some detail. Finally, Part VI, which analyzes clinical derangements, has been enlarged appreciably.

THE EDITORS

Contents

PART I: Transport Events in Single Cells

CHAPTER 1: **Active Transport in *Escherichia coli:* From Membrane to Molecule**

H. RONALD KABACK

1. Introduction ... 3
2. Membrane Vesicles and Active Transport: General Aspects ... 3
3. Energetics of Active Transport ... 5
4. Active Transport at the Molecular Level: The β-Galactoside Transport System ... 10
5. Summary ... 19
References ... 19

CHAPTER 2: **Acidification of Intracellular Organelles: Mechanism and Function**

GARY RUDNICK

1. Introduction ... 25
2. Evidence for Acid Interior ... 25
3. Generation of ΔpH ... 27
4. Uses of ΔpH ... 31
5. The Nature of the ATPase ... 34
6. Conclusion ... 35
References ... 35

CHAPTER 3: **Intracellular pH Regulation**

WALTER F. BORON

1. Introduction ... 39
2. Measurement of Intracellular pH ... 39
3. Cellular Buffering Processes ... 40
4. Effect of Externally Applied Weak Acids and Bases ... 42
5. Ion-Transport Systems ... 44
References ... 48

CHAPTER 4: **Properties of Ionic Channels in Excitable Membranes**

FRANCISCO BEZANILLA and MICHAEL M. WHITE

1. Introduction ... 53
2. How Do You Get a Resting Potential? ... 53
3. How Do You Change the Membrane Potential? ... 54
4. Ionic Channels ... 55
5. The Two-State Model ... 55

6. Real Channels Have More Than Two States 57
7. Na^+ Channels 58
8. K^+ Channels 61
9. Summary 63
References 63

CHAPTER 5: **Ion Movements in Skeletal Muscle in Relation to the Activation of Contraction**

HANS CHRISTOPH LÜTTGAU and GEORGE DIMITRIE STEPHENSON

1. Introduction 65
2. The Ultrastructure of the Tubular System in Skeletal Muscle Fibers 65
3. Electrical Properties of the Surface and Tubular Membrane 66
4. Inward Spread of Excitation 69
5. Cellular Ca^{2+} Movements Related to the Activation of Contraction 73
6. Summary 79
References 79

CHAPTER 6: **Excitable Tissues: The Heart**

RICHARD W. TSIEN and PETER HESS

1. Introduction 85
2. Multicellular Structure of the Heart 85
3. Electrical Activity in Different Regions of the Heart 88
4. Na^+ Channels and Excitability 89
5. Ca^{2+} Channels and Slow Responses 92
6. K^+ Channels Support the Resting Potential and Action Potential Repolarization 96
7. Inward Currents and Pacemaker Activity 97
8. Adrenergic and Cholinergic Modulation of Cardiac Activity 99
9. Summary 101
References 102

CHAPTER 7: **Ion Transport through Ligand-Gated Channels**

RICHARD W. ALDRICH, VINCENT E. DIONNE, EDWARD HAWROT, and CHARLES F. STEVENS

1. Introduction and Overview 107
2. Structure of the Nicotinic AChR 109
3. Immunological Approaches to the Study of the Nicotinic AChR 112
4. Biogenesis, Membrane Localization, and Regulation 114
5. Dose–Response 119
6. Kinetics of Channel Gating 120
7. AChR Cation Selectivity and Permeation 123
8. Ligand-Gated Channels Other Than the AChR 124
9. An Emerging View of Transmitter-Activated Channels 126
References 127

PART II: Transport in Epithelia: Vectorial Transport through Parallel Arrays

CHAPTER 8: **Cellular Models of Epithelial Ion Transport**

STANLEY G. SCHULTZ

1. Introduction 135
2. Models of Sodium- and Chloride-Absorbing Epithelial Cells 136
3. A Model for Active Chloride Secretion by Epithelial Cells 143
4. Summary 144
References 144

CHAPTER 9: Ion Transport by Gastric Mucosa

JOHN G. FORTE and TERRY E. MACHEN

1. Introduction ... 151
2. Organization of Gastric Epithelial Cells ... 151
3. Stimulus–Secretion Coupling in Oxyntic Cells ... 153
4. Metabolism and Energetics Associated with Gastric HCl Secretion ... 155
5. Studies with Isolated Cell Fractions and Membranes ... 156
6. Electrophysiological and Tracer Flux Studies of Gastric Ion Transport ... 161
7. Summary ... 169
References ... 170

CHAPTER 10: Ion and Water Transport in the Intestine

DON W. POWELL

1. Introduction ... 175
2. Models of Intestinal Na^+, Cl^-, and H_2O Transport ... 175
3. Intestinal Na^+ and Cl^- Absorption ... 177
4. Intestinal Na^+ and Cl^- Secretion ... 183
5. HCO_3^-, Short-Chain Fatty Acid, and K^+ Transport ... 184
6. Shunt Pathway and Water Transport ... 187
7. Control of Intestinal Electrolyte Transport ... 188
8. Summary and Conclusions ... 199
References ... 199

CHAPTER 11: The Uptake of Lipids into the Intestinal Mucosa

HENRIK WESTERGAARD and JOHN M. DIETSCHY

1. Introduction ... 213
2. Chemical Species of Lipids That Are Involved during Fat Absorption ... 213
3. The Barriers to Lipid Absorption in the Intestine ... 214
4. Characteristics of the Intestinal Microvillus Membrane Barrier to Lipid Absorption ... 216
5. Characteristics of the Intestinal Unstirred Water Layer Barrier to Lipid Absorption ... 217
6. Characteristics of Fatty Acid and Cholesterol Absorption in the Intestine ... 219
7. Role of Bile Acid Micelles in Facilitating Lipid Absorption in the Intestine ... 221
8. Nonpolar Lipids ... 222
9. Summary Description of the Process of Lipid Uptake ... 223
References ... 223

CHAPTER 12: Mechanisms of Bile Secretion and Hepatic Transport

JAMES L. BOYER

1. Introduction ... 225
2. Structural Determinants of Bile Secretory Function ... 225
3. Mechanisms of Hepatocellular Water and Electrolyte Secretion ... 230
4. Other Primary Driving Forces for Canalicular Bile Secretion (Bile Acid-Independent Secretion) ... 234
5. Model for Hepatocyte Water and Electrolyte Secretion ... 235
6. Physiological Modifiers of Hepatocyte Bile Formation ... 236
7. Organic Anion Solute Transport ... 238
8. Lipid Excretion in Bile ... 240
9. Proteins in Bile ... 242
10. Miscellaneous Substances Found in Bile ... 243
11. Bile Duct Function ... 243
12. Summary ... 244
References ... 244

CHAPTER 13: The Regulation of Glomerular Filtration Rate in Mammalian Kidneys

L. GABRIEL NAVAR, P. DARWIN BELL, and ANDREW P. EVAN

1. Introduction ... 253
2. Ultrastructural Considerations ... 254

3. Characteristics of the Filtration Process ... 258
4. Quantitative Description of Glomerular Dynamics ... 264
5. Physiological Regulation of Glomerular Filtration Rate ... 267
6. Intrarenal Distribution of Glomerular Filtration Rate ... 276
7. Summary ... 277
References ... 277

CHAPTER 14: **The Proximal Nephron**

GERHARD GIEBISCH and PETER S. ARONSON

1. General Properties of the Proximal Nephron ... 285
2. Distribution of Transport Functions along the Proximal Tubule ... 286
3. Transepithelial Potentials and Passive Permeabilities ... 288
4. NaCl and $NaHCO_3$ Transport ... 293
5. Solute–Solvent Coupling–Role of the Intercellular Shunt Pathway ... 304
References ... 309

CHAPTER 15: **The Effects of ADH on Salt and Water Transport in the Mammalian Nephron: The Collecting Duct and Thick Ascending Limb of Henle**

STEVEN C. HEBERT and THOMAS E. ANDREOLI

1. Introduction ... 317
2. Intracellular Mediators of ADH Action ... 317
3. The Medullary Thick Ascending Limb ... 318
4. The Collecting Tubule ... 320
5. Homology of Hormone Action ... 323
6. Modulation of the ADH Response ... 324
7. Summary: Integration of ADH Action on Urinary Concentration ... 325
References ... 325

CHAPTER 16: **Urinary Concentrating and Diluting Processes**

MARK A. KNEPPER and JOHN L. STEPHENSON

1. Introduction ... 329
2. Renal Structure ... 329
3. Basic Concepts ... 331
4. Handling of Individual Solutes in the Medulla ... 334
5. Properties of the Thin Limbs of Henle's Loops ... 336
6. Concentration in the Inner Medulla ... 337
7. Summary ... 340
References ... 340

CHAPTER 17: **Transport Functions of the Distal Convoluted Tubule**

LINDA S. COSTANZO and ERICH E. WINDHAGER

1. Introduction ... 343
2. Structural Heterogeneity ... 343
3. Transepithelial Net Transport of Solutes and Water ... 344
4. Electrophysiological Considerations ... 355
5. Mechanisms of Transport ... 357
6. Summary ... 360
References ... 360

CHAPTER 18: **The Respiratory Epithelium**

MICHAEL J. WELSH

1. Introduction ... 367
2. The Tracheal Epithelium ... 367
3. The Bronchial Epithelium ... 374

4. The Alveolar Epithelium 375
5. The Fetal Lung 377
6. Summary 378
References 379

Index 383

PART I

Transport Events in Single Cells

CHAPTER 1

Active Transport in *Escherichia coli*

From Membrane to Molecule

H. Ronald Kaback

1. Introduction

Although the 1970s is regarded as the era of molecular genetics, when exciting breakthroughs made possible the isolation, cloning, and sequencing of genetic material from viruses to man, another revolution in our concepts of energy transduction in biological membranes also occurred over the same period of time, but without the same drama. Thus, in much the same way that the Crick–Watson double helix provided the backbone for many advances in molecular biology, the chemiosmotic hypothesis, formulated and refined by Peter Mitchell during the 1960s,[1–5] is now the conceptual framework for a wide array of bioenergetic phenomena from photophosphorylation to the uptake and storage of neurogenic amines in the adrenal medulla. Curiously, however, the far-reaching importance of the chemiosmotic concept and the experimental evidence supporting its validity have gone relatively unnoticed because: (1) the chemiosmotic hypothesis was formulated initially to explain oxidative phosphorylation and is still strongly identified with this traditionally controversial field; (2) few biochemists are comfortable with ephemeral entities such as electrochemical ion gradients; and (3) various disciplines within the area of bioenergetics use different terminologies to describe similar phenomena. By providing an overview of active transport, this volume will help to abolish barriers, and it is with that notion in mind that this contribution is intended.

It should be emphasized, however, that the following is not a general review, but is concerned primarily with active transport in cytoplasmic membrane vesicles isolated from *Escherichia coli* and in proteoliposomes reconstituted with purified components from the membrane of this organism. As discussed in other chapters, such studies are highly relevant to other systems, as evidenced by the profusion of similar experimental systems that have been developed from other cells, organelles, and epithelia. Furthermore, the discussion is pertinent not only to active transport, but in a broad sense, to the general problem of energy transduction in biological membranes. As opposed to mitochondrial or chloroplast membranes, for example, whose primary function is to convert respiratory energy or light, respectively, into chemical energy (i.e., ATP), respiratory energy in *E. coli* membrane vesicles is converted into work in the form of solute concentration against an electrochemical or osmotic gradient.

Abbreviations used in this chapter: RSO, right-side-out; ISO, inside-out; FAD, flavin-adenine dinucleotide; D-LDH, D-lactate dehydrogenase; PMS, phenazine methosulfate; TMPD, *N,N,N′,N′*-tetramethylphenylenediamine; DAD, diaminodurene; CCCP, carbonylcyanide-*m*-chlorophenylhydrazone; DCCD, *N,N′*-dicyclohexylcarbodiimide; DDA, dimethyldibenzylammonium; TPB^-, tetraphenylboron; $TPMP^+$, triphenylmethylphosphonium; TPP^+, tetraphenylphosphonium; SCN^-, thiocyanate; DMO, 5,5′-dimethyloxazolidine-2,4-dione; $\Delta\tilde{\mu}_H$, the proton electrochemical gradient across the membrane; $\Delta\psi$, membrane potential; ΔpH, the pH gradient across the membrane; octylglucoside, octyl-β-D-glucopyranoside; NPG, *p*-nitrophenyl-α-D-galactopyranoside; TMG, methyl-1-thio-β-D-galactopyranoside; Q_1H_2, ubiquinol-1; *p*-CMBS, *p*-chloromercuribenzene-sulfonate; DEPC, diethylpyrocarbonate; HPLC, high-performance liquid chromatography; SP-RIA, solid-phase radioimmunoassay; *p*CMBS, *p*-chloromercuribenzenesulfonate.

H. Ronald Kaback • Roche Institute of Molecular Biology, Roche Research Center, Nutley, New Jersey 07110.

2. Membrane Vesicles and Active Transport: General Aspects

Preliminary evidence reported in the 1960s[6,7] suggested that cytoplasmic membrane vesicles from *E. coli* would provide a useful model system for studying active transport, and this early promise has been more than fulfilled. Thus, numerous studies demonstrate that vesicles prepared from *E. coli* as well as many other bacteria, eukaryotic cells, intracellular organelles, and epithelia, catalyze the accumulation of many different solutes under appropriate experimental conditions. Furthermore, in some instances, initial rates of transport are comparable to those of the intact cell,[8,9] and the vesicles accumulate many solutes to concentrations markedly in excess of those in the external medium.[10] Remarkably, moreover, it has been demonstrated with *E. coli* that essentially each vesicle in the preparations is functional for active transport.[11] Early progress with the system was slow primarily because of preconceived ideas regarding the physical nature of the vesicles and the energetics of active transport. It was generally thought that the vesicles, by the very nature of their preparation, had to be "leaky," and in addition, that high-energy phosphate bond energy would be directly involved in active transport. Considerable time and effort were expended before these suppositions were dispelled.

Right-side-out (RSO) bacterial membrane vesicles are prepared by lysis of osmotically sensitized cells (i.e., protoplasts or spheroplasts), and they consist of osmotically intact, unit-membrane-bound sacs that are 0.5 to 1.0 μm in diameter.[12] Actually, RSO vesicles are probably more aptly described as "ghosts,"[7] Since a single structure is obtained from each cell if care is taken to avoid excessive mechanical stress. In any event, the vesicles are devoid of internal structure, their metabolic activities are restricted to those provided by the enzymes of the membrane itself, and numerous observations demonstrate clearly that the vesicle membrane retains the same polarity and configuration as the membrane in the intact cell (Refs. 13–15: in addition, see Ref. 16).

Alternatively, by subjecting cells to relatively low shear forces in a French pressure cell, inside-out (ISO) vesicles can be prepared.[17,18] Although ISO vesicles are about 10 times smaller than RSO vesicles and the yield from the preparation is relatively low, it is apparent that these vesicles have a polarity opposite to that of the intact cell (see Ref. 19 for a review).

Transport by membrane vesicles *per se* is practically nil, and the energy source for uptake of a particular substrate can be determined by studying which compounds or experimental manipulations drive accumulation. Moreover, metabolic conversion of the transport substrate and the energy source is minimal. These properties of the vesicles constitute a considerable advantage over intact cells with respect to the study of active transport, as the system provides clear definition of the reactions involved in the transport process.

Generally, the transport systems elucidated in RSO vesicles from *E. coli* fall into three main categories: (1) group translocation in which a covalent change is exerted upon the transported molecule so that the reaction itself results in passage of the solute through the diffusion barrier; (2) "classic" active transport in which solute is accumulated against an electrochemical or osmotic gradient; and (3) passive diffusion of certain weak acids and lipophilic ions, followed by equilibration with the pH gradient (ΔpH) and the electrical potential ($\Delta\psi$), respectively, across the membrane.

In *E. coli* membrane vesicles, uptake of D-glucose, D-fructose, D-mannose, and some other carbohydrates occurs by vectorial phosphorylation via the phosphoenolpyruvate–phosphotransferase system (PTS) (a group translocation mechanism[20]). The PTS, described originally in 1964 by Roseman and his colleagues[21] and subsequently studied in considerable detail,[22] catalyzes the transfer of phosphate from phosphoenolpyruvate through a sequence of specific proteins to the appropriate carbohydrates in such a manner that the carbohydrate is translocated across the membrane and accumulated as a result of phosphorylation. Although the system is crucial to the bacteria that employ it and the PTS plays a central role in the regulation of carbohydrate metabolism in *E. coli* and *Salmonella typhimurium*,[23] it is not ubiquitous among bacteria and has not been found thus far in organisms phylogenetically higher than bacteria. Evidence for other group translocation reactions has also been reported in vesicles from *E. coli* and other bacteria.[24]

As opposed to group translocation mechanisms, transport of a variety of other solutes (sugars, amino acids, organic acids, and ions) by *E. coli* membrane vesicles occurs by active transport.[24] These transport systems are coupled to the oxidation of D-lactate to pyruvate, catalyzed by a flavin-adenine dinucleotide (FAD)-linked, membrane-bound D-lactate dehydrogenase (D-LDH) which has been purified to homogeneity[25,26] and synthesized *in vitro*[27] from a hybrid plasmid containing the *dld* gene.[28] Electrons derived from D-lactate are passed to oxygen via a membrane-bound respiratory chain, and in this sequence of reactions, respiratory energy is converted into work in the form of active transport. Although other oxidizable substrates such as L-lactate, succinate, α-gycerophosphate, and NADH also stimulate transport to some extent, they are not nearly as effective as D-lactate unless ubiquinone is added to the vesicles[16,29] or the vesicles are prepared from cells induced for the *o*- and/or *d*-type cytochrome oxidases (unpublished data). Active transport in the vesicle system is also driven by nonphysiological electron carriers such as reduced phenazine methosulfate (PMS),[30] pyocyanine,[16] *N,N,N′,N′*-tetramethylphenylenediamine (TMPD) or diaminodurene (DAD) [TMPD and DAD are effective only when the *o*- and/or *d*-type cytochrome oxidases are present (unpublished data)]. Some of these electron carriers, particularly reduced PMS, drive transport more effectively than physiological electron donors such as D-lactate.

Active transport in *E. coli* membrane vesicles also functions in the absence of oxygen when the appropriate anaerobic electron transfer systems are present.[31] Lactose and amino acid transport under anaerobic conditions can be coupled to the oxidation of α-glycerophosphate with fumarate as electron acceptor or to oxidation of formate utilizing nitrate as electron acceptor. Both of these anaerobic electron transfer systems are induced by growth of the organism under appropriate conditions, and components of both systems are loosely bound to the membrane, which necessitates the use of a modified procedure for vesicle preparation.[32]

Although there is a convincing body of evidence demonstrating that high-energy phosphate bond energy is not directly involved in active transport in this system,[24,33] it has been shown recently[34] that internally generated ATP drives active transport in RSO membrane vesicles from *S. typhimurium* and, under certain conditions, *E. coli*. In these experiments, *S. typhimurium* induced for phosphoglycerate transport[35] were loaded with pyruvate kinase and ADP by lysing spheroplasts under appropriate conditions. Vesicles so prepared catalyze active transport of proline and serine in the presence of phosphoenolpyruvate, and this activity is blocked by the protonophore carbonylcyanide-*m*-chlorophenylhydrazone (CCCP) and by the H^+-ATPase inhibitor *N,N′*-dicyclohexylcarbodiimide (DCCD), but not by anoxia or cyanide. In contrast, respiration-driven active transport is abolished by CCCP and by anoxia or cyanide but not by DCCD. Moreover, phosphoenolpyruvate does not drive transport effectively in vesicles that lack the phosphoglycerate transport system. The results are consistent with an overall mechanism in which phosphoenolpyruvate gains access to the interior of the vesicles by means of the phosphoglycerate transporter and is then acted on by pyruvate kinase to phosphorylate ADP. ATP formed inside of the vesicles is then hydrolyzed by the H^+-ATPase to drive transport. By using the plasmid pBR322 as vector and *E. coli* as host, a fragment of *S. typhimurium* DNA encoding the phosphoglycerate transporter was cloned, and RSO vesicles prepared from the host in the presence of pyruvate kinase and ADP also catalyze ATP-dependent active transport.

Finally and importantly, active transport in RSO vesicles is driven by artificially imposed K^+ gradients ($K^+_{in} \rightarrow K^+_{out}$) in the presence of the K^+-specific ionophore valinomycin.[36–39] When membrane vesicles prepared in K^+-containing buffers are diluted into media lacking the cation and valinomycin is added, efflux of K^+ creates a diffusion potential ($\Delta\psi$, interior negative) across the membrane that is able to drive substrate accumulation. Alternatively, artificially imposed ΔpHs (interior alkaline) produce similar effects.[38] The finding that substrate accumulation occurs under these conditions has important implications with respect to the energetics of respiration- and ATP-driven active transport.

3. Energetics of Active Transport

3.1. The Chemiosmotic Hypothesis

Since it is behond the scope of this chapter to review the chemiosmotic hypothesis and all of its ramifications extensively, for additional discussion, the reader is referred to the "gray books" of Mitchell,[3,4] a few particularly lucid reviews,[40–43] and a recent compendium[44] commemorating Mitchell's 60th birthday.

In its most general form (Fig. 1), the chemiosmotic concept postulates that the immediate driving force for many processes in energy-coupling membranes is a proton electrochemical gradient ($\Delta\tilde{\mu}_H$) composed of electrical and chemical parameters according to the following relationship:

$$\Delta\tilde{\mu}_H/F = \Delta\psi - 2.3RT/F\ \Delta\text{pH}$$

where $\Delta\psi$ represents the electrical potential across the membrane and ΔpH is the chemical difference in proton concentration across the membrane (R is the gas constant, T the absolute temperature, F the Faraday constant, and $2.3RT/F = 58.8$ at room temperature).

Accordingly, the basic energy-yielding processes of the cell—respiration or absorption of light—generate $\Delta\tilde{\mu}_H$, and the energy stored therein is utilized to drive a number of seemingly unrelated phenomena such as the formation of ATP from ADP and inorganic phosphate, active transport, and transhydrogenation of NADP by NADH. More recently, it has become apparent that $\Delta\tilde{\mu}_H$ or one of its components is involved in many other diverse phenomena such as bacterial motility,[45] nitrogen fixation,[46] transfer of genetic information,[47–52] sensitivity and resistance to certain antibiotics,[53,54] cellulose synthesis,[55] and processing of secreted proteins.[56–58] Importantly, many processes driven by $\Delta\tilde{\mu}_H$ are reversible. Thus, hydrolysis of ATP via the H^+-ATPase leads to the generation of $\Delta\tilde{\mu}_H$. Similarly, transport of solutes down a concentration gradient (i.e., the reverse of active transport) can also generate $\Delta\mu_H$.[59] Clearly, therefore, the "common currency of energy exchange," particularly in the bacterial cell, is *not* ATP, but $\Delta\tilde{\mu}_H$.

The molecular mechanism of H^+ translocation is unsolved for any of the processes described. However, since respiration-driven $\Delta\tilde{\mu}_H$ generation is particularly relevant to bacterial active transport and because it is not immediately obvious why respiration should give rise to $\Delta\tilde{\mu}_H$, at least a cursory discussion is important to put the problem into perspective. Although the concept is controversial, Mitchell[3,4] has postulated that "loops" in the respiratory chain may be an important means of H^+ translocation across the membrane during electron flow (Fig. 2). According to this elegantly simple notion, the electron and H^+ carriers that comprise the membrane-bound respiratory chain are disposed alternatively and asymmetrically across the membrane in such a manner that the first component of a "coupling site" on the inner surface of the membrane (e.g., a reduced flavoprotein that transfers both electrons and H^+) passes two H^+ and two electrons to a second component on the outer surface of the membrane. However, the second intermediate ac-

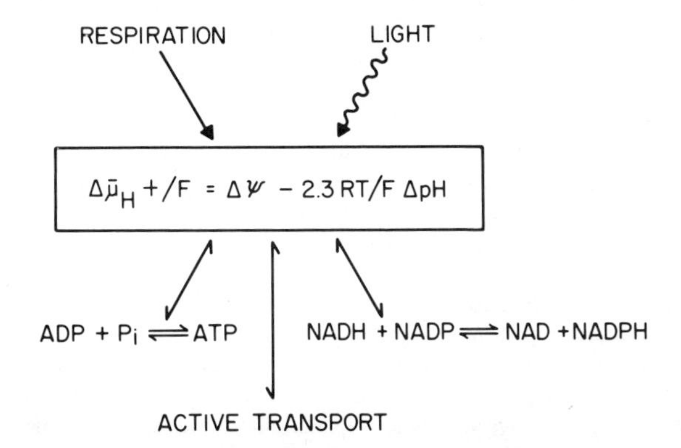

Fig. 1. Generalized chemiosmotic hypothesis. Respiration or absorption of light leads to the generation of a proton electrochemical gradient that provides the immediate driving force for a number of seemingly unrelated phenomena including oxidative- and photophosphorylation, active transport, and transhydrogenation.

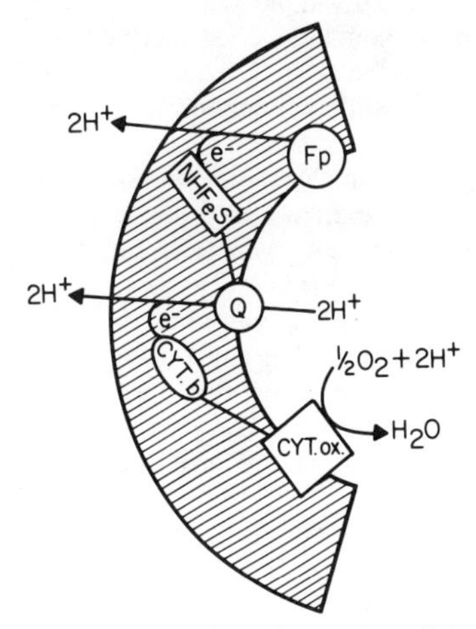

Fig. 2. Conceptual representation of two "loops" in the respiratory chain. Electrons and protons are transferred from flavoprotein (e.g., a dehydrogenase) on the inner surface of the membrane to a nonheme iron–sulfur center (NHFeS) oriented toward the outer surface of the membrane. Since the NHFeS center accepts electrons (e^-), but not H^+, 2 H^+ are released into the medium on the external surface of the membrane. Electrons are then transferred to quinone (Q) on the inner surface of the membrane, and the Q abstracts 2 H^+ from the medium bathing the inner surface of the membrane. Electrons are then transferred to cytochrome b with release of 2 H^+ into the external medium and finally to a terminal oxidase that reduces oxygen to water on the inner surface of the membrane.

cepts electrons only (e.g., a nonheme iron protein), and the two H^+ are lost to the medium on the side of the membrane opposite from where they originated. Electrons from the second component on the outer surface of the membrane are then passed to a third component on the inner surface, and this component (e.g., quinone) can accept H^+ in addition to electrons. Thus, upon accepting electrons from the second component, two H^+ are taken up from the medium bathing the inner surface of the membrane. The process is then repeated with different redox intermediates, and the overall result is that H^+ move from one side of the membrane to the other with a net flux of electrons in the opposite direction. In addition, Mitchell[60] has also postulated the existence of a "protonmotive ubiquinone cycle" in order to explain the absence of appropriate H^+ carriers after cytochrome *b* in the mitochondrial respiratory chain.

Given this type of scheme, it is apparent that the "protonmotive stoichiometry" (i.e., the stoichiometry between H^+ extrusion and electron transfer) in the respiratory chain cannot be greater than 2, since the known H^+ carriers do not react with more than 2 H^+ during a single turnover. Although Mitchell's laboratory and others (see Ref. 61) have reported stoichiometries of 2, the findings have been challenged by a number of other workers who report values of 3 or 4.[62–69] Furthermore, with *E. coli* membrane vesicles, inhibition of electron transfer at the end of the respiratory chain leads to rapid collapse of $\Delta\bar{\mu}_H$, while inhibition at the dehydrogenase level causes only slow collapse.[8,39,70] In addition, oxidation of either D-lactate[71] or D-alanine[72,73] by the appropriate dehydrogenase leads to generation of $\Delta\bar{\mu}_H$ even when the enzymes are bound to the wrong surface of the vesicle membrane (i.e., the outside). Thus, it is unlikely that either of these dehydrogenases comprises part of a loop. Finally, purified mitochondrial cytochrome oxidase (see Ref. 74) and terminal oxidases purified from *Paracoccus denitrificans*,[75] the thermophile PS-3,[76] and *E. coli*[77,78] all appear to generate $\Delta\bar{\mu}_H$ in the absence of H^+ carriers. At present, therefore, although the loop mechanism remains highly attractive in its simplicity, it seems unlikely that this is the only mechanism for respiration-dependent vectorial H^+ translocation.

The salient features of the chemiosmotic hypothesis with respect to active transport in *E. coli* are presented schematically in Fig. 3. As shown, $\Delta\bar{\mu}_H$ generated either by respiration or through the action of the H^+-ATPase acts to drive transport via different substrate-specific porter or carrier proteins that extend through the membrane binding substrates on the external surface of the membrane and releasing them on the internal surface. In addition, it is postulated that $\Delta\bar{\mu}_H$ drives transport by several different mechanisms depending on the nature of the substrate in the following manner:

1. Transport of cationic substrates such as lysine or K^+ occurs by *uniport*, a mechanism dependent specifically on $\Delta\psi$. Accordingly, a transmembrane uniporter facilitates electrophoresis of external cationic substrate in response to the internally negative $\Delta\psi$.

2. Transport of acids such as lactate or succinate, in contrast, is coupled to the interior alkaline ΔpH across the membrane. In this manner, the protonated form of the acid is translocated across the membrane, and since the internal space is alkaline relative to the external medium, the acid dissociates, H^+ is pumped out, and the anion accumulates. Formally, this is classified as a *symport* mechanism because H^+ is translocated with substrate. However, it is important to make the distinction between this type of symport mechanism and that postulated for

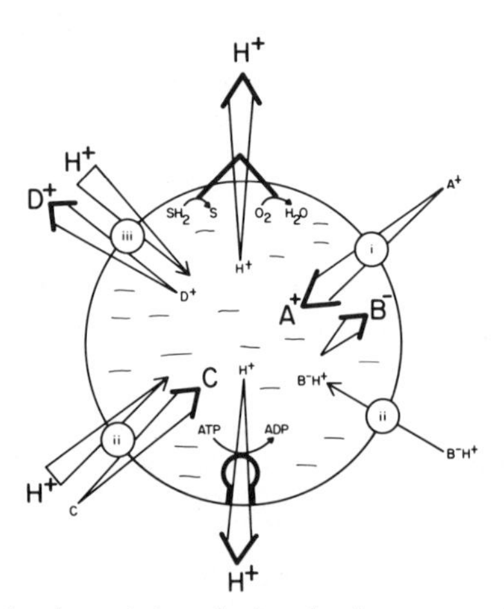

Fig. 3. The chemiosmotic hypothesis and active transport. The proton electrochemical gradient (interior negative and alkaline) generated by respiration or ATP hydrolysis drives influx of cationic substrates (A^+) by uniport (i), acidic (B^-H^+) and neutral substrates (C) by symport (ii), and efflux of cations like Na^+ or Ca^{2+} (D^+) by antiport (iii). See text for further discussion.

neutral substrates (e.g., sugars and neutral amino acids) in which H^+ is cotransported on the porter rather than on the substrate. In this case, the porter is presumed to bind substrate and H^+ independently and couples the energy released from the downhill translocation of H^+ in response to $\Delta\bar{\mu}_H$ (interior negative and alkaline) to the uphill transport of substrate against a concentration gradient.

3. Finally, solutes such as Na^+ or Ca^{2+} which are pumped out of the cell are transported by *antiport*. By this means, a transmembrane, substrate-specific antiporter couples the downhill translocation of H^+ to the *efflux* of substrate against a concentration gradient.

Although these considerations are remarkably simple and lead to straightforward predictions regarding the coupling between $\Delta\bar{\mu}_H$ and its components and specific transport substrates, it should be emphasized that the basic predictions become more complicated if the stoichiometry between H^+ and substrate translocation is greater than unity. For example if H^+:Na^+ antiport has a 1:1 stoichiometry, it is apparent that the process will be electrically neutral and respond specifically to the ΔpH across the membrane. On the other hand, if the stoichiometry is 2 H^+ : 1 Na^+, the mechanism becomes electrogenic and will respond to ΔpH and/or $\Delta\psi$. Such considerations will be discussed in more detail below.

Another notable aspect of the proposed mechanisms is their reversibility. Accordingly, *efflux* of cations under nonenergized conditions should lead to generation of $\Delta\psi$ (interior negative), while *efflux* of acids should generate a ΔpH (interior alkaline) and *efflux* of sugars or neutral amino acids should give rise to $\Delta\bar{\mu}_H$. In contrast, Na^+ or Ca^{2+} *influx* should generate ΔpH and/or $\Delta\psi$ depending on the stoichiometry of the antiport reaction. As shown by Konings *et al.*,[79,80] these reverse reactions may have more than academic importance. Thus, carrier-mediated efflux of glycolytically generated lactic acid has been shown to provide a significant increase in growth yield under certain conditions due to generation of $\Delta\bar{\mu}_H$.

Clearly, the chemiosmotic formulation stipulates that transport is the result of an indirect process whereby respiration or ATP hydrolysis is coupled to active transport through the mediation of $\Delta\bar{\mu}_H$. As such, the experimental approach to the problem at this level of resolution is dependent upon the development of techniques that allow measurement of $\Delta\bar{\mu}_H$ in a system that is not readily amenable to a direct electrophysiological approach and upon the demonstration that changes in $\Delta\bar{\mu}_H$ and its components correlate with changes in the transport of appropriate substrates.

3.2. Determination of $\Delta\psi$

In 1971, Grinius *et al.*[81] initiated the use of lipophilic ions to measure the polarity of $\Delta\psi$ in biological systems too small for the introduction of microelectrodes. The ions are constructed in such a fashion as to be sufficiently lipophilic to enter the hydrophobic core of the membrane, and apparently, they must also be able to delocalize their charge in order to allow passive equilibration with the electrical potential across the membrane.[82] There is convincing evidence that $\Delta\psi$ can be measured quantitatively in intact *E. coli,* as well as RSO membrane vesicles. Thus, although tetraphenylphosphonium (TPP^+), tetraphenylarsonium, and Rb^+ (in the presence of valinomycin) equilibrate with $\Delta\psi$ fastest, these cations and in addition triphenylmethylphosphonium, triphenylmethylarsonium, triphenylmethylammonium, and dimethyldibenzyalammonium (in the presence of tetraphenylborate)[39,83] all accumulate to the same steady-state level over a wide range of concentrations. Moreover, the magnitude of $\Delta\psi$ determined from the steady-state levels of accumulation of these cations is virtually identical to that obtained from fluorescence quenching studies with 3,3′-diisopropylthiodicarbocyanine.[84]

Most importantly, using giant cells of *E. coli* induced by growth in the presence of 6-amidinopenicillanic acid, Felle *et al.*[85] measured $\Delta\psi$ by two completely independent techniques: *directly* with intracellular microelectrodes and *indirectly* from the steady-state distribution of [^{3}H]-TPP^+. Under a variety of conditions, the two methods yield values that agree very closely. With both techniques, $\Delta\psi$ (interior negative) approximates −85 mV at pH 5.0 and −142 mV at pH 8.0, with an average slope of −22 mV/pH unit over the range of pH 5.0–7.0. In a parallel study of membrane vesicles using TPP^+ distribution alone as a measure of $\Delta\psi$, values of about −90 mV at pH 5.0 and −110 mV at pH 7.5–8.0 with an average slope of −6 mV/pH unit were obtained. Although the difference in slopes between intact cells and RSO vesicles is yet to be understood, the results lend firm support to the conclusion that distribution studies with lipophilic cations in *E. coli* provide an excellent quantitative measure of $\Delta\psi$. Parenthetically, it is noteworthy that distribution studies with [^{3}H]-TPP^+ also yield $\Delta\psi$ values similar to those obtained electrophysiologically in neuroblastoma/glioma NG108-15 cells,[86,87] chick embryo heart cells (L. J. Elsas, J. H. Krick, J. Serravezzo, and R. L. DeHaan, unpublished experiments), and J774 macrophages.[88]

Since ISO vesicles have the opposite polarity from intact cells and RSO vesicles, $\Delta\psi$s generated by respiration or ATP hydrolysis should be interior positive, necessitating the use of lipophilic anions. Although studied in less detail than the lipophilic cations in the *E. coli* system, thiocyanate (SCN^-) distribution studies with ISO vesicles[19] provide a strong indication that respiration or ATP hydrolysis generates a $\Delta\psi$ (interior positive) of similar magnitude to that observed in RSO vesicles. The results, in addition to being interesting in their own right, provide additional support for the quantitative validity of the distribution measurements.

3.3. Determination of ΔpH

The basic principle behind the measurement of ΔpH is clear-cut. If the internal compartment under consideration is alkaline relative to the external medium, as it is in intact *E. coli* or RSO vesicles, a permeant weak acid is used, the most popular being 5,5′-dimethyloxazolidine-2,4-dione (DMO).[89] The protonated form of the acid is passively permeant, but once the acid reaches the internal space which is alkaline relative to the outside, the acid dissociates and the anion, which is impermeant, accumulates. Conversely, if the internal compartment is acid relative to the outside, as in ISO vesicles, permeant weak bases such as methylamine are used. In this case, the unprotonated species is passively permeant, and it is protonated to form a positively charged, impermeant species in the internally acid environment. In general, if the pK of the probe utilized is 2 pH units or more removed from internal pH, ΔpH can be calculated by substitution of the steady-state distribution ratio into the Nernst equation (i.e., $\Delta\text{pH} = 2.3RT/F$ log distribution ratio). On the other hand, if the pK of the probe is within 2 pH units of internal pH, a slightly more complicated calculation should be used to obtain accurate values (see Ref. 90). As it turns out, however, although the principles described are straightforward, determination of ΔpH is extremely dependent on the methods used to determine uptake of the probes. Thus, for many years, it was not possible to demonstrate that *E. coli* membrane vesicles generate a significant ΔpH (see Ref. 91).

In 1976, Padan *et al.*[92] made the critical observation that *E. coli* rigidly maintains internal pH at pH 7.5–7.8 and demonstrated that the magnitude of ΔpH is very dependent upon external pH, exhibiting a maximal value of about 2 pH units (i.e., −120 mV, interior alkaline) at pH 5.5–6.0 which decreases to zero at about pH 7.6 and then reverses about pH 7.6 such that the interior of the cell becomes acid relative to the external medium. In addition, Rottenberg[93] demonstrated that acetate may be utilized to determine ΔpH in mitochondria and suggested that this weak acid might be more useful than DMO in the vesicle system because it is less permeant. Initial experiments with RSO vesicles using standard filtration assays revealed a small amount of acetate uptake that was sensitive to external pH, but the amount of acetate taken up was so small that the putative ΔpH appeared to be thermodynamically insignificant. It was not until flow dialysis, a technique introduced by Colowick and Womack[94] to measure ligand binding, was utilized[10,95–98,101] that the problem was solved. By this means, the concentration of any solute in the medium bathing the vesicles, particularly weak acids, can be monitored easily and continuously without experimental manipulations that cause artifactual loss of the accumulated probe.

Using flow dialysis and a variety of permeant weak acids (i.e., acetate, propionate, butyrate, DMO, benzoate, or acetylsalicylate), it is readily demonstrable that RSO vesicles from *E. coli* generate a ΔpH (interior alkaline) of about −120 mV (i.e., 2 pH units) at pH 5.5 in the presence of reduced PMS or D-lactate,[10,90,95–98] a value remarkably similar to that reported by Padan *et al.*[92] for intact cells. Furthermore, similar experiments with ISO vesicles using methylamine as a probe show that these vesicles establish a ΔpH of similar magnitude but opposite polarity (i.e., interior acid) under appropriate conditions.[19] Finally, Navon *et al.*[99] and Ogawa *et al.*[100] have provided strong confirmation for the quantitative nature of the measurements. These workers utilized high-resolution ^{31}P NMR spectroscopy to measure ΔpH in intact *E. coli,* and the results are

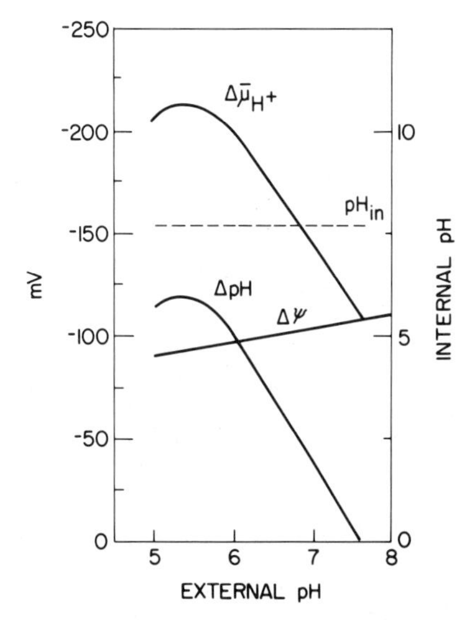

Fig. 4. Effect of external pH on ΔpH (interior alkaline), $\Delta\psi$ (interior negative), $\Delta\bar{\mu}_H$ (interior negative and alkaline), and internal pH. The curves shown are an idealized representation of data presented in Refs. 95–98.

very similar to those obtained from distribution studies with permeant weak acids in RSO membrane vesicles and intact *E. coli.*

3.4. Effect of External pH on $\Delta\bar{\mu}_H$ and Its Components

As reported originally by Padan *et al.*[92,101] with intact *E. coli,* ΔpH (interior alkaline) varies dramatically with external pH in RSO membrane vesicles[10,90,95–98] (Fig. 4). From pH 5.0 to 5.5, ΔpH is almost constant at −115 to −120 mV, decreases markedly above pH 5.5, and is negligible at about pH 7.5–7.8. Although not studied in detail, preliminary evidence suggests that RSO vesicles, like intact cells, may also acidify the internal compartment at pH 8.0–8.5 (unpublished). In contrast, $\Delta\psi$ (interior negative) is about −90 mV at pH 5.0 and increases to about −110 mV at pH 7.5–8.0. As a result of these variations, $\Delta\bar{\mu}_H$ exhibits a maximum of about −220 mV at pH 5.5 and a minimum of about −110 mV at pH 7.5–8.0. With the exception that $\Delta\psi$ increases to a lesser extent with external pH, the results are both qualitatively and quantitatively similar to those obtained with intact cells.

Clearly, the variation in ΔpH with external pH results from the propensity of the system to maintain internal pH at pH 7.5–7.8 (Fig. 4), an observation that is conceptually important. That is, since the internal space is very small relative to the external medium, the system does not have to extrude many protons in order to establish ΔpH.

Although the mechanism responsible for collapse of ΔpH from pH 5.5 to 7.5 is unknown,[19,95,96] a simple explanation that could account for the data is the operation of an antiport mechanism with an alkaline pH optimum. By this means, as external pH increases beyond pH 5.5, exchange of internal Na^+ and/or K^+ for external H^+ would occur at an increasingly rapid rate, thereby collapsing ΔpH with a compensatory increase in $\Delta\psi$ (see below). In an effort to investigate this possibility, a detailed series of experiments was carried out with ISO vesicles.[19] As opposed to RSO vesicles, which extrude Na^+,[98,102] ISO vesicles catalyze Na^+ accumulation when a $\Delta\bar{\mu}_H$ (interior positive and/or acid) is present across the membrane, and under no circumstances is K^+ or Rb^+ accumulation observed. However, the properties of the H^+/Na^+ antiport activity make it difficult to conclude that this mechanism, in itself, can account for the phenomenon. Thus, the concentration gradient of Na^+ established by ISO vesicles is constant with pH from pH 5.7 to 8.0, Na^+ accumulation is driven by either ΔpH (interior acid) or $\Delta\psi$ (interior positive), and Na^+ accumulation does not lead to dissipation of ΔpH over the pH range studied. Finally and remarkably, the pH profile for ΔpH in ISO vesicles is essentially the mirror image of that observed in RSO vesicles. ΔpH is maximal at pH 7.0–7.5 and *decreases* at relatively acid pH values which makes it even more difficult to envisage how H^+/Na^+ antiport, by functioning at alkaline pH specifically, can be responsible for the collapse of pH from pH 5.5 to 7.5. In summary, therefore, the mechanism responsible for the maintenance of internal pH at relatively acid pH values remains a fascinating, but unsolved problem.

In contrast, convincing evidence has been obtained recently indicating that the ability of cells to establish an internally acid ΔpH above pH 8.0–8.5 is due specifically to electrogenic H^+/Na^+ antiport activity. Clearly defined mutants of *E. coli*[103,104] and *Bacillus alkalophilus*[105,106] have been isolated that do not grow at alkaline pH, do not establish ΔpH (interior acid) at alkaline pH, and do not catalyze H^+/Na^+ antiport activity.

3.5. Effect of Ionophores on $\Delta\bar{\mu}_H$

As demonstrated with RSO[10,90,95–98] and ISO vesicles,[19] $\Delta\psi$ and ΔpH can be altered reciprocally with little or no change in $\Delta\bar{\mu}_H$ and no change in respiration. In RSO vesicles, for example, $\Delta\psi$ (interior negative) decreases markedly as increasing concentrations of valinomycin are added in the presence of K^+ at pH 5.5 or 7.5 (Fig. 5) because valinomycin-mediated K^+ influx occurs at a rate that approximates electrogenic H^+ efflux. Remarkably, with the decrease in $\Delta\psi$ at pH 5.5, ΔpH (interior alkaline) increases for reasons that will be

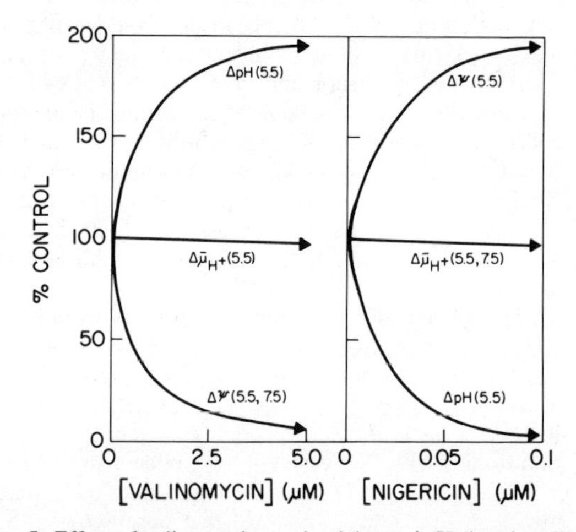

Fig. 5. Effect of valinomycin or nigericin on ΔpH (interior alkaline), $\Delta\psi$ (interior negative), and $\Delta\bar{\mu}_H$ (interior negative and alkaline at pH 5.5 and 7.5). The curves shown are an idealized representation of data presented in Refs. 95–98.

discussed. As a result of these reciprocal alterations, $\Delta\bar{\mu}_H$ is either unaffected or decreases slightly.

In contrast, nigericin, an ionophore that catalyzes H^+/K^+ exchange preferentially, induces effects that are opposite to those of valinomycin at pH 5.5 (Fig. 5). Thus, with increasing concentrations of nigericin, ΔpH decreases with an increase in $\Delta\psi$ and $\Delta\bar{\mu}_H$ remains constant or decreases slightly. Importantly, at pH 7.5, where there is no ΔpH, the ionophore has no effect on $\Delta\psi$.

Finally, increasing concentrations of the protonophore CCCP collapse both $\Delta\psi$ and ΔpH, thereby leading to dissipation of $\Delta\bar{\mu}_H$.[95] Notably, moreover, CCCP inhibits more effectively at pH 5.5 than at pH 7.5, an effect that is presumably due to the pK of the protonophore itself. The observation that CCCP abolishes both ΔpH and $\Delta\psi$ indicates that protons are the electrogenic species.

As opposed to mitochondria, where these ionophores cause an increase in respiration (i.e., the system exhibits respiratory control), no such effect is observed with *E. coli* membrane vesicles which led initially to some confusion regarding the effects of valinomycin and nigericin on $\Delta\psi$ and ΔpH. The phenomena can be rationalized, however, by considering the interrelationship between these components of $\Delta\bar{\mu}_H$.[19] At steady state, under conditions where both $\Delta\psi$ and ΔpH are present across the membrane, the magnitude of $\Delta\bar{\mu}_H$ is determined by the efficiency of the H^+ pump and the back leak of H^+ through the membrane, and each parameter of $\Delta\bar{\mu}_H$ limits the magnitude of the other. With RSO vesicles, $\Delta\psi$ (interior negative) will act to draw H^+ toward the interior of the vesicles and thus limit the pH gradient established. Similarly, ΔpH (interior alkaline) will limit $\Delta\psi$ (interior negative) because H^+ will tend to diffuse into the vesicles down the concentration gradient, thus decreasing the net extrusion of positive charge and decreasing $\Delta\psi$. Therefore, by dissipating either $\Delta\psi$ or ΔpH, a force that is limiting for the other parameter is removed, allowing it to increase without a corresponding increase in the rate of H^+ extrusion. Clearly, the same explanation in reverse applies to ISO vesicles.

3.6. Relationship between $\Delta\bar{\mu}_H$ and Active Transport

In addition to providing direct support for one of the major contentions of the chemiosmotic hypothesis, that respiration or ATP hydrolysis leads to the generation of a transmembrane $\Delta\bar{\mu}_H$, the studies discussed above establish a powerful experimental framework within which to test more specific predictions.

3.6.1. Effect of Substrate Accumulation on ΔpH and $\Delta\psi$

If substrate accumulation is coupled to $\Delta\bar{\mu}_H$, and if passive accumulation of weak acids and lipophilic cations in RSO vesicles reflects the individual components of $\Delta\bar{\mu}_H$, it follows that weak acid and/or lipophilic cation accumulation should be diminished in the presence of transport substrates such as lactose or glucose-6-P that are accumulated in relatively large amounts by the vesicles. Both predictions have been borne out experimentally. When vesicles containing the *lac* carrier protein or the glucose-6-P porter are allowed to accumulate acetate in the presence of reduced PMS and valinomycin, addition of lactose or glucose-6-P causes release of about 50% of the accumulated acetate.[10] In contrast, with vesicles that contain neither of these transporters, acetate accumulation is unaffected by either substrate. Similarly, addition of lactose to vesicles that contain the *lac* carrier causes a decrease in $TPMP^+$ accumulation, and no effect is observed with vesicles devoid of the carrier.[39] Clearly, a good explanation for these effects is that the substrates are accumulated in symport with H^+, as suggested by Mitchell.[2] It should also be emphasized that downhill transport of lactose and certain other sugars in deenergized cells[107–113] or RSO vesicles[114,115] occurs with alkalinization of the external medium.

3.6.2. Effect of Valinomycin and Nigericin on Substrate Accumulation

Before proceeding, a few important points will be reemphasized: (1) $\Delta\bar{\mu}_H$ is maximal at pH 5.5 where approximately one-half of the total driving force is ΔpH and the other half is $\Delta\psi$. (2) At pH 7.5, $\Delta\bar{\mu}_H$ is reduced by about one-half and consists solely of a $\Delta\psi$ component. (3) At pH 5.5, ΔpH and $\Delta\psi$ can be manipulated reciprocally with little or no effect on $\Delta\bar{\mu}_H$. (4) Nigericin has no effect on $\Delta\psi$ at pH 7.5 and therefore no effect on $\Delta\bar{\mu}_H$ at this pH.

With these observations as a framework, Ramos and Kaback[10] carried out a series of experiments in which the effects of valinomycin and nigericin on the steady-state levels of accumulation of 14 individual substrates at pH 5.5 and 7.5 were tested in a manner similar to that discussed in relation to Fig. 5. Direct quantitative correlations between variations in the accumulation of a particular substrate and variations in $\Delta\bar{\mu}_H$, $\Delta\psi$, or ΔpH are observed in only a few instances at pH 5.5. Nevertheless, certain qualitative statements are justified when the experiments are considered as a whole: (1) Accumulation of lactose, proline, tyrosine, serine, glycine, leucine, and, surprisingly, lysine (a cation), glutamate (an acidic amino acid), and succinate (a dicarboxylic acid) at pH 5.5 responds to increasing concentrations of valinomycin or nigericin in a manner that correlates reasonably well with the effect of these ionophores on $\Delta\bar{\mu}_H$. That is, accumulation of each substrate is progressively and mildly inhibited or relatively unaffected by increasing concentrations of either inophore. (2) Accumulation of glucose-6-P, lactate, gluconate, and glucuronate at pH 5.5 is stimulated by valinomycin and inhibited by nigericin in a manner clearly analogous to the effects of the ionophores on ΔpH. (3) Regardless of whether the accumulation of a particular substrate is stimulated, inhibited, or unaffected by valinomycin or nigericin at pH 5.5, in each and every case, valinomycin causes marked inhibition of accumulation at pH 7.5 and nigericin has no effect whatsoever at this external pH. Generally, therefore, at pH 5.5, the transport systems fall into two categories: those driven preferentially by $\Delta\bar{\mu}_H$ and those driven preferentially by ΔpH. Moreover, all of the systems, including those driven by ΔpH at pH 5.5, are driven by $\Delta\psi$ at pH 7.5 where this parameter represents the totality of $\Delta\bar{\mu}_H$.

Since the coupling between accumulation of a particular substrate and $\Delta\bar{\mu}_H$, ΔpH, and $\Delta\psi$ varies with the external pH, it is not surprising that quantitative correlations are observed in only a few instances at pH 5.5. With tyrosine, leucine, lysine, and succinate, for example, there is reasonably good correlation between the effects of valinomycin and nigericin on $\Delta\bar{\mu}_H$ and the effects of the ionophores on accumulation of these substrates. However, lactose and glycine accumulation are also coupled preferentially to $\Delta\bar{\mu}_H$ at pH 5.5, but there is an apparent bias toward $\Delta\psi$ in both instances, since valinomycin inhibits lactose and glycine accumulation more effectively than it dissipates

$\Delta\bar{\mu}_H$ at pH 5.5. Similarly, although accumulation of glucose-6-P, lactate, gluconate, and glucuronate is coupled to ΔpH at pH 5.5, in only one case (i.e., lactate) is accumulation in full equilibrium with ΔpH, and nigericin does not inhibit accumulation of these acids as effectively as it dissipates ΔpH at pH 5.5. Finally, the reader is referred to the studies of Robertson *et al.*(116) in which the kinetics of many of these transport systems were investigated under similar conditions. Basically, the results support the conclusions drawn from studying steady-state levels of accumulation (e.g., initial rates of glutamate and lysine transport are driven by $\Delta\bar{\mu}_H$); however, certain subtleties are also revealed (e.g., succinate transport is biased kinetically toward ΔpH).

3.6.3. H^+ : Substrate Stoichiometry

Clearly, the overall impact of the results discussed is that $\Delta\bar{\mu}_H$ is the immediate driving force for active transport. However, at the same time, the experiments reveal certain details that are not fully explained by the chemiosmotic hypothesis as it is formally presented. For instance, according to dogma, accumulation of organic acids is obligatorily dependent on the relative alkalinity of the internal pH, and should not be driven by $\Delta\psi$ (interior negative). Since there is no ΔpH across the membrane at pH 7.5–7.8 in either intact cells or RSO vesicles, however, the putative mechanism cannot account for acid accumulation at high external pH. Furthermore, LeBlanc *et al.*,(117) using artificially imposed pH gradients (interior alkaline) and diffusion potentials (interior negative), have provided strong support for the argument that the glucose-6-P and glucuronate porters catalyze electrically neutral reactions at acid pH and electrogenic reactions at alkaline pH. In addition, when the steady-state levels of accumulation of certain substrates that are coupled to $\Delta\bar{\mu}_H$ at pH 5.5 (i.e., lactose, proline, lysine, and succinate) are examined as a function of pH, it appears that $\Delta\bar{\mu}_H$ is insufficient to account for the magnitude of the concentration gradients at alkaline pH if the stoichiometry between H^+ and substrate is 1 : 1.(10,96)

There is at least one simple explanation for these observations within the bounds of the chemiosmotic hypothesis.(96,118) Possibly, the stoichiometry between H^+ and substrate varies as a function of external pH in such a manner that it is 1 : 1 at pH 5.5, but increases to higher values as external pH increases. If, for example, the stoichiometry between H^+ and proline or lactose were 2 : 1 at pH 7.5 rather than 1 : 1, the concentration gradients would be thermodynamically compatible with $\Delta\bar{\mu}_H$ at pH 7.5 (i.e., if the stoichiometry is 2 : 1, the concentration gradient varies as the square of the charge gradient(5,118)). In a similar vein, it is conceivable that, at pH 5.5, transport of certain organic acid occurs by a formal chemiosomotic mechanism (i.e., 1 H^+—2, if the acid is glucose-6-P—is take up per mole of undissociated acid), while at pH 7.5, 2 or more H^+ are taken up per mole of acid, 1 (2 with glucose-6-P) in association with the substrate itself and 1 in association with the porter molecule. By this means, the transport of glucose-6-P, lactate, glucuronate, and gluconate at pH 7.5 would become electrogenic, having become symport mechanisms.

Subsequent to the reports of Ramos and Kaback(10,96) which lend support to these ideas, studies with intact cells(85,101,119) cast doubt on the contention that there is a discrepancy between the steady-state level of lactose accumulation and $\Delta\bar{\mu}_H$ at alkaline pH. Specifically, it was demonstrated that in intact cells, as opposed to RSO vesicles, $\Delta\psi$ increases markedly with pH in such a manner as to compensate for the decrease in ΔpH. Thus, $\Delta\bar{\mu}_H$ in intact cells does not decrease as drastically with increasing pH as observed in vesicles and the steady-state level of lactose accumulation at high pH can be accommodated without a change in H^+ : lactose stoichiometry. In addition, more direct studies of H^+/lactose symport in deenergized cells are not indicative of a change in stoichiometry at high pH.(101,119)

On the other hand, numerous studies with both intact cells and RSO vesicles demonstrate that ΔpH is absent at pH 7.5–7.8. Thus, it is difficult to explain how the transport of certain organic acids can be coupled to ΔpH at acid pH and to $\Delta\psi$ at alkaline pH without invoking a pH-dependent increase in H^+ : substrate stoichiometry,(10,96,117,118) and direct measurements in intact cells supporting this notion have been reported.(120)

3.7. Na^+-Dependent Transport

Although the majority of bacterial transport systems are probably of the H^+ symport type, several instances have been reported in which the transport of a specific solute is dependent upon the presence of Na^+ or Li^+ (see Refs. 98, 105, 106 for reviews). Moreover, some of these studies, in particular those of Stock and Roseman,(121) Lanyi *et al.*,(122) and Krulwich,(105,106) indicate that symport mechanisms may be operative. Since the basic energy-yielding process in most bacteria is thought to be H^+ extrusion and bacteria do not possess a primary Na^+ pump, the existence of such transport systems presents certain obvious problems.

RSO membrane vesicles isolated from *S. typhimurium* grown in the presence of melibiose catalyze methyl-1-thio-β-D-galactopyranoside (TMG) transport in the presence of Na^+ or Li^+.(98,123) TMG-dependent Na^+ uptake is also observed when a K^+ diffusion potential (interior negative) is imposed across the vesicle membrane. Cation-dependent TMG accumulation varies with the $\Delta\bar{\mu}_H$ generated as a result of D-lactate or reduced PMS oxidation, and the vesicles catalyze Na^+ efflux in a manner that is consistent with the operation of a $H^+/Na^+(Li^+)$ antiport mechanism. The results with this system are consistent with a model (Fig. 6) in which $TMG/Na^+(Li^+)$ symport is driven by $\Delta\bar{\mu}_H$ which functions to maintain low intravesicular Na^+ and Li^+ concentrations through $H^+/Na^+(Li^+)$ antiport. Similar mechanisms have been suggested for light- and respiration-dependent amino acid transport in vesicles from *Halobacterium halobium*(122) and *B. alkalophilus*.(105,106)

It is also particularly interesting that recent studies with H^+/Na^+ antiport mutants in *B. alkalophilus*(105,106,124) and *E. coli*(103,104) demonstrate that such cells, in addition to being unable to grow at alkaline pH, exhibit pleiotropic defects in Na^+-dependent substrate translocation. Based on the observations, it has been suggested that a number of Na^+/substrate symport systems and the H^+/Na^+ antiporter may share a common Na^+-translocating subunit.

4. Active Transport at the Molecular Level: The β-Galactoside Transport System

The β-galactoside or *lac* transport system in *E. coli* is the most extensively studied of bacterial transport systems. It was described originally in 1955 by Cohen and Rickenberg(125–129) and is part of the well-known *lac* operon which enables the

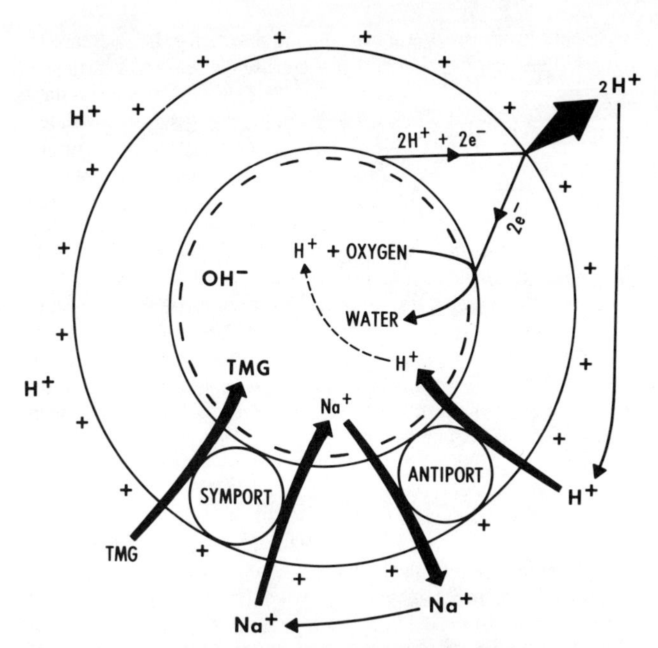

Fig. 6. Schematic representation of Na^+-dependent TMG accumulation in *S. typhimurium* G-30 membrane vesicles. From Tokuda and Kaback.[98]

organism to utilize the disaccharide lactose. In addition to its regulatory loci, the *lac* operon contains three structural genes: (1) the *z* gene encoding β-galactosidase, a cytosolic enzyme that cleaves lactose once it enters the cell; (2) the *y* gene encoding the *lac* carrier protein or *lac* permease which catalyzes transport of lactose through the plasma membrane of the cell; and (3) the *a* gene encoding thiogalactoside transacetylase, an enzyme that catalyzes the acetylation of thiogalactosides with acetyl-CoA as the acetyl donor and has no known physiological function.

In 1963, Mitchell[2] postulated explicitly that lactose transport occurs in symport with H^+ (Fig. 3) and that $\Delta\tilde{\mu}_H$ is the immediate driving force for accumulation against a concentration gradient. Subsequently, West[110] and West and Mitchell[111,112] demonstrated that addition of lactose to deenergized cells causes alkalinization of the external medium, thus providing the first evidence for H^+/lactose symport in *E. coli*. During the next 5 or 6 years, as discussed above, a wealth of evidence was reported demonstrating virtually unequivocally that the *lac* carrier protein catalyzes H^+/β-galactoside symport, and more recently, the focus of the field has shifted to a more molecular, mechanistic approach.

4.1. Purification of Functional *lac* Carrier Protein

Although the kinetics,[38,59,116,130] substrate specificity,[131] and genetics[132] of the β-galactoside transport system had been studied intensively, and the *lac y* gene product was shown to be a membrane protein,[133] relatively little progress was made with respect to purification primarily because all attempts to solubilize the protein in a functional state were unsuccessful (see Ref. 134). In 1978,[135] however, the *lac y* gene was cloned into a recombinant plasmid, allowing amplification of the carrier,[136] as well as the elucidation of its nucleotide sequence and the amino acid sequence of the *lac* carrier protein,[137] and its synthesis *in vitro*.[138] Shortly thereafter, Newman and Wilson[139] solubilized the carrier in octyl-β-D-glucopyranoside (octylglucoside) and successfully reconstituted lactose transport activity in proteoliposomes by using the octylglucoside dilution technique described by Racker *et al.*[140] Almost simultaneously, it was demonstrated that *p*-nitrophenyl-α-D-galactopyranoside (NPG) is a highly specific photoaffinity label for the *lac* carrier protein.[141] By using a strain of *E. coli* with amplified levels of the *lac y* gene, [^{3}H]-NPG to photolabel the carrier specifically and thus follow its distribution during purification and the transport activity of proteoliposomes reconstituted with the *lac* carrier, the product of the *lac y* gene was purified to homogeneity in a completely functional state.[142,143]

The pie diagram presented in Fig. 7 summarizes a purification scheme in which *E. coli* membranes are first sequentially extracted with high concentrations of urea and cholate to effect about a 3-fold purification of the carrier *in situ*. Both of these operations are based upon earlier studies[134,144] demonstrating that the treatment of RSO membrane vesicles with these reagents extracts considerable amounts of protein from the membrane with little or no effect on the *lac* carrier. Subsequent extraction with octylglucoside in the presence of *E. coli* phospholipids solubilizes most of the carrier, but only about 15% of the remaining protein, leading to an additional 4-fold enrichment and a 12-fold purification relative to the original membrane. The octylglucoside extract is then subjected to DEAE–Sepharose column chromatography under isocratic conditions at pH 6.0. Transport activity and most of the protein-associated photolabel is eluted in a symmetrical peak slightly behind the void volume of the column. Overall, the procedure results in a 35-fold purification relative to the crude membrane fraction and a yield of about 50%, based on the recovery of the photolabel.

Since photolabeling studies with [^{3}H]-NPG indicate that

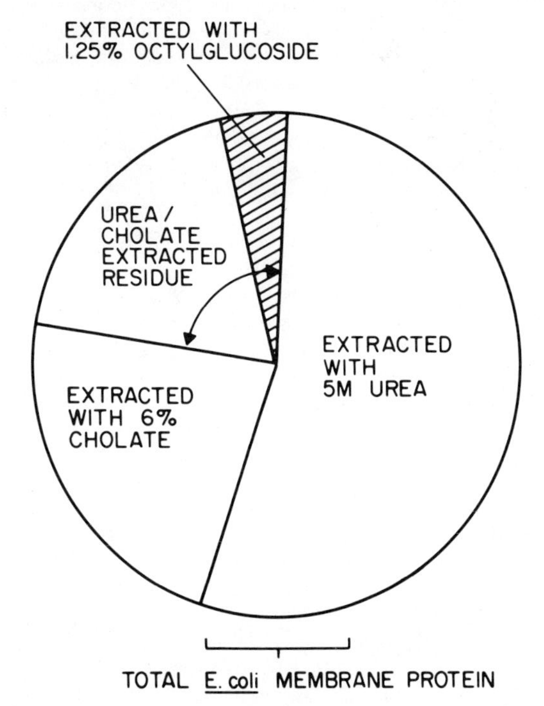

Fig. 7. Pie slice schematic showing fractionation resulting from differential solubilization of *E. coli* membrane proteins.

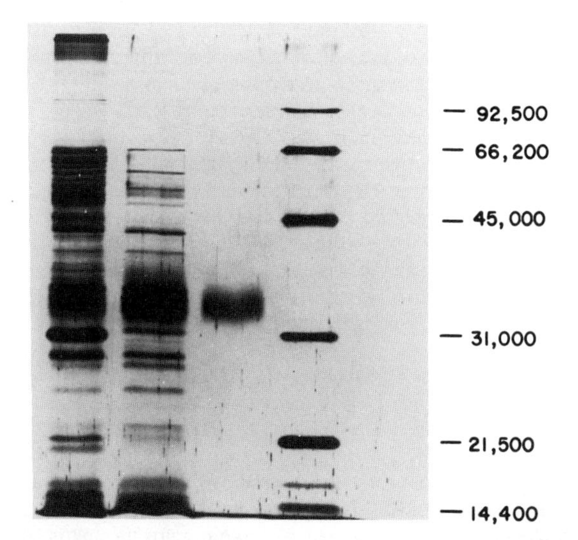

Fig. 8. SDS–PAGE of various fractions obtained during purification of the *lac* carrier. From the left, *lane* 1, urea/cholate-extracted membranes, 4 μg; *lane* 2, octylglucoside extract, 4 μg; *lane* 3, pooled DEAE fractions, 0.8 μg; *lane* 4, molecular weight standards, 2 μg. Gels were silver stained. From Newman *et al.*[143]

3% of the protein in the membrane of the amplified strain is *lac* carrier, a 35-fold enrichment of the photolabeled material suggests that a high degree of purification was achieved. This is confirmed by SDS–PAGE of the purified material (Fig. 8). Clearly, the pooled DEAE–Sepharose fractions (third lane from the left) yield a single broad band* with an M_r of about 33K, which is in close agreement with published values for the molecular weight of the carrier as determined by SDS–PAGE.[135,145] When membranes are prepared from cells that were not induced, the band corresponding to the purified carrier is only a minor constituent of the octylglucoside extract of urea/cholate-treated membranes, thus demonstrating that the purified protein exhibits an important property expected of the product of the *lac y* gene in the recombinant plasmid.[136] Importantly, the amino acid composition of the purified protein closely matches the composition predicted from the DNA sequence of the *lac y* gene.[137] This result indicates that the functional *lac* carrier has a molecular weight similar to the value predicted[137,143] from the DNA sequence. Furthermore, N-terminal sequencing of the first 13 amino acids of the purified *lac* carrier yields results that are in complete agreement with the DNA sequence, providing additional evidence for the high degree of purity of the preparation.

It is not known why the carrier yields a spuriously low molecular weight on SDS–PAGE, although the high content of hydrophobic amino acids in the protein suggests that this phenomenon may be due to unusually high binding of SDS. Furthermore, it is not known why the carrier migrates as a broad band. However, it should be emphasized that when the protein is subjected to SDS–PAGE at increasing concentrations of polyacrylamide and the data are treated quantitatively,[146,147] an M_r of about 46K is obtained. A similar value is also obtained by gel permeation chromatography on Sephacryl S-300 in hexamethylphosphoric triamide.[148]

*At higher protein concentrations, a less intense band is also observed at about M_r 6K. Since this band is observed after photoaffinity labeling with NPG and reacts with antibody prepared against purified *lac* carrier, it is probably an aggregate of the *lac* carrier protein.

4.2. Morphology and Ion Permeability of Proteoliposomes Reconstituted with *lac* Carrier Protein

Proteoliposomes prepared by octylglucoside dilution followed by freeze–thaw/sonication are unilamellar vesicles about 100 nm in diameter that exhibit no internal structure.[149] Relatively low-magnification electron microscopy of platinum/carbon replicas of freeze-fractured proteoliposomes containing purified *lac* carrier confirm the unilamellar nature of the preparation (Fig. 9A). Higher magnification reveals that both convex and concave fracture surfaces exhibit a relatively uniform distribution of particles that are 70 Å in diameter (Fig. 9B). Since particles, but no pits, are observed on both surfaces of the membranes, it seems likely that the *lac* carrier has equal affinity for the phospholipids in each leaflet of the bilayer. Given the mass of the *lac* carrier,[137,143] a size of 70 Å suggests that the particles may contain one to two polypeptides, depending on the degree to which the metal shadowing increases the observed particle diameter.

When proteoliposomes containing purified *lac* carrier are equilibrated with $^{86}Rb^+$, treated with valinomycin, and diluted 200-fold into sodium phosphate, efflux of the cation occurs very slowly, and at 20 min the proteoliposomes still retain at least 80% of the label (Fig. 10). On addition of CCCP, which increases permeability to H^+ specifically, a marked increase in the rate of Rb^+ efflux is evident. If the same experiments are performed in the absence of valinomycin, Rb^+ efflux is almost negligible, and addition of CCCP has no significant effect. The observations demonstrate, albeit indirectly, that the proteoliposomes are highly impermeabile to the ions present in the reaction mixture (i.e., H^+, Rb^+, Na^+, Cl^-, and P_i). Thus, the slow rate of Rb^+ efflux observed in the presence of valinomycin is caused by the generation of $\Delta\psi$ (interior negative) that is maintained for long periods of time because of the impermeability of the proteoliposomes to counterions. Addition of CCCP, on the other hand, provides a pathway for H^+ and results in dissipation of $\Delta\psi$ with rapid downhill movement of Rb^+.

Clearly, these proteoliposomes are almost ideally suited for studies of H^+/solute symport. Morphologically, the preparation consists of a population of unilamellar, closed, unit-membrane-bound sacs that are relatively uniform in diameter and contain no internal structure, findings that correlate nicely with the pseudo-first-order efflux and exchange kinetics observed for Rb^+ and lactose.[149] Furthermore, it is apparent that the proteoliposomes are passively impermeable to many ions, a property that is highly advantageous. Thus, certain aspects of H^+/lactose symport that were impossible to document with RSO vesicles (e.g., stimulation of efflux by ionophores; see below) are readily elucidated with the reconstituted system. Generally, proteoliposomes reconstituted with the *lac* carrier exhibit all of the phenomena described in RSO membrane vesicles, but the results are significantly more clear-cut and provide firmer support for certain ideas concerning reaction mechanisms.

4.3. A Single Polypeptide Is Required for Lactose Transport

Although it is readily apparent that the *lac* carrier protein purified to apparently homogeneity catalyzes counterflow, H^+

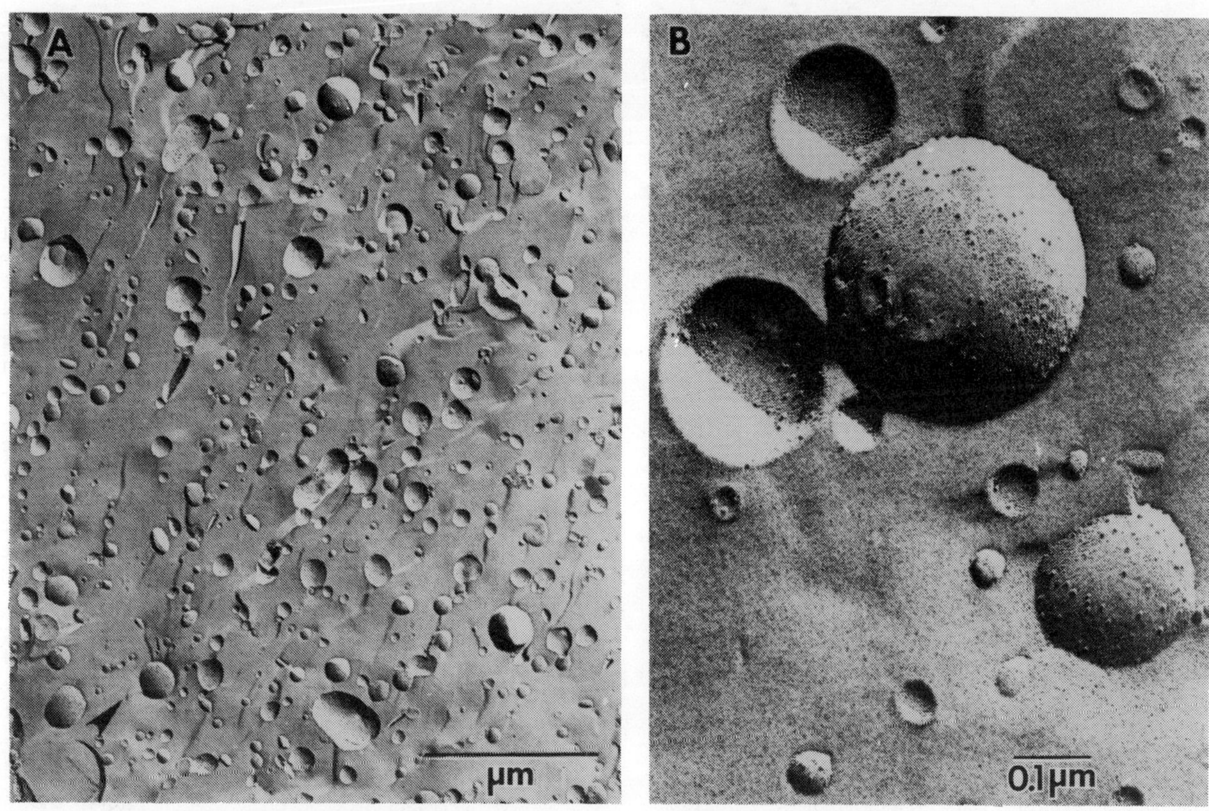

Fig. 9. Freeze–fracture electron microscopy of proteoliposomes reconstituted with purified *lac* carrier. Platinum/carbon replicas of freeze-fractured proteoliposomes prepared by octylglucoside dilution followed by freeze–thaw/sonication. The study was performed by Joseph Costello in the Department of Anatomy, Duke University Medical Center.

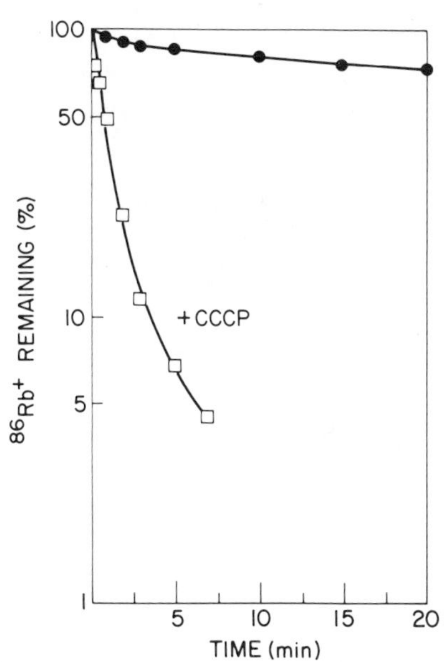

Fig. 10. $^{86}Rb^{+}$ efflux from proteoliposomes. Proteoliposomes containing purified *lac* carrier were treated with 20 μM valinomycin and equilibrated with ^{86}RbCl. Aliquots were then diluted 200-fold into appropriate buffer at 25°C. At the times indicated, samples were assayed by filtration. ●, control; □, plus 20 μM CCCP. See Garcia *et al.*[164] for experimental details.

influx and efflux in response to appropriately directed lactose concentration gradients, and $\Delta\tilde{\mu}_{H}$-driven lactose accumulation against a concentration gradient, evidence has been presented that was interpreted to indicate that active lactose transport may require more than a single polypeptide.[150–153] Furthermore, Wright *et al.*,[154] using *lac* carrier partially purified and reconstituted by techniques different from those described,[142,143] were able to elicit counterflow activity, but were unable initially to demonstrate $\Delta\psi$- or ΔpH-driven lactose accumulation.

For these reasons, careful kinetic experiments were performed on proteoliposomes reconstituted with purified *lac* carrier. Turnover numbers were calculated for the carrier operating in various modes of translocation and compared to those calculated from published V_{max} values for RSO membrane vesicles (Table I). As shown, both the turnover number of the *lac* carrier, as well as its apparent K_m for lactose, are virtually identical in proteoliposomes and membrane vesicles with respect to $\Delta\psi$-driven lactose accumulation, counterflow, facilitated diffusion (i.e., lactose influx under nonenergized conditions), and efflux.

In addition, Matsushita *et al.*[78] have demonstrated that proteoliposomes simultaneously reconstituted with a purified *o*-type cytochrome oxidase and the *lac* carrier protein catalyze electron transfer-driven active lactose accumulation. The *o*-type cytochrome oxidase was purified from a mutant of *E. coli* defective in cytochrome *d* oxidase by extraction with octylglucoside after sequential treatment of membranes with urea and cholate. The oxidase was then purified to homogeneity by DEAE–Sepharose chromatography. The purified oxidase contains four polypeptides (M_r 55K, 35K, 21K, and 14K) and two *b*

Table I. Comparison of Turnover Numbers for the *lac* Carrier Protein: ML 308-225 Membrane Vesicles versus Proteoliposomes Reconstituted with Purified Carrier

Reaction	Turnover numbers (seconds)	
	Membrane vesicles[a]	Proteoliposomes
$\Delta\Psi$-driven influx ($\Delta\Psi$ = 100 mV)	16 (K_m = 0.2 mM)	18 (K_m = 0.5 mM)
Counterflow	16–39 (K_m = 0.45 mM)	28 (K_m = 0.65 mM)
Facilitated diffusion	8–15.5 (K_m = 20 mM)	13 (K_m = 7 mM)
Efflux	8 (K_m = 2.1 mM)	6 (K_m = 2.0 mM)

[a]Determination of the amount of *lac* carrier protein in ML 308-225 membrane vesicles is based on photolabeling experiments with [^{3}H]-NPG which indicate that the carrier represents about 0.5% of the membrane protein.

-type cytochromes (b_{558} and b_{563}), and catalyzes the oxidation of ubiquinol-1 (Q_1H_2) and other electron donors with specific activities 20- to 30-fold higher than crude membranes. Proteoliposomes were reconstituted simultaneously with the purified oxidase and *lac* carrier protein by octylglucoside dilution followed by freeze–thaw/sonication. The reconstituted system generates a $\Delta\bar{\mu}_H$ (interior negative and alkaline) with Q_1H_2 as electron donor, and magnitude of the $\Delta\bar{\mu}_H$ is dependent on the concentration of the oxidase in the proteoliposomes. As shown in Fig. 11, in the presence of Q_1H_2, the proteoliposomes accumulate lactose against a concentration gradient, and the phenomenon is completely abolished by addition of valinomycin and nigericin. Since uptake in the absence of Q_1H_2 or in the presence of valinomycin and nigericin represents equilibrium with the medium, it is apparent that the steady-state level of lactose accumulation observed during oxidase turnover represents a concentration gradient of at least 10-fold. Moreover, by comparing lactose transport induced by Q_1H_2 to that induced by valinomycin-mediated K^+ diffusion potentials ($K_{in}^+ \rightarrow K_{out}^+$) and quantitating the magnitude of the $\Delta\psi$s generated under each condition, it is clear that the lactose transport activity observed is commensurate with the magnitude of $\Delta\bar{\mu}_H$.

It is highly likely therefore that only a single polypeptide species, the product of the *lac y* gene, is necessary for each of the reactions catalyzed by the *lac* transport system in the *E. coli* membrane, including active transport energized by electron transfer. Furthermore, the double-reconstitution experiment provides yet another strong line of evidence—this time on a molecular level—supporting the concept that active transport is driven by a *transmembrane* $\Delta\bar{\mu}_H$.

4.4. Mechanistic Studies

Studies with RSO membrane vesicles[38,59] demonstrate that carrier-mediated lactose efflux down a concentration gradient occurs in symport with H^+ and suggest that the translocation reaction is limited either by deprotonation of the carrier on the outer surface of the membrane or by a step corresponding to the return of the unloaded carrier to the inner surface of the membrane. In addition, the observations[38,59] led to the following conclusions (Fig. 12): (1) efflux occurs by an ordered mechanism in which lactose is released first from the carrier, followed by loss of the symported H^+; (2) the carrier recycles in the protonated form during exchange and counterflow; and (3) reac-

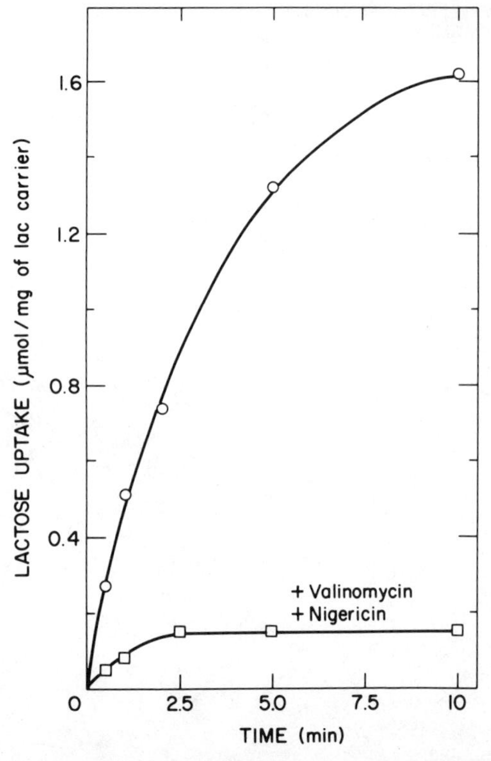

Fig. 11. Electron transfer-driven lactose accumulation by proteoliposomes simultaneously reconstituted with purified *o*-type cytochrome oxidase and purified *lac* carrier protein. Proteoliposomes containing purified cytochrome oxidase and *lac* carrier at a molar ratio of 1.3 to 1.0 were prepared by octylglucoside dilution followed by freeze–thaw/sonication. Lactose transport was assayed by filtration in the absence (□) and presence (○) of 16 μM ubiquinol-1 (Q_1H_2). When 20 μM valinomycin and 0.5 μM nigericin were added in addition to Q_1H_2, data identical to those observed in the absence of Q_1H_w (□) were obtained.

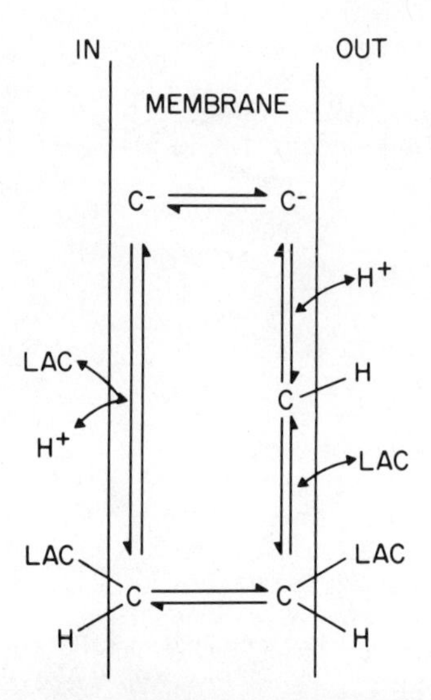

Fig. 12. Schematic representation of reactions involved in lactose efflux. *C* represents the *lac* carrier protein. The order of substrate binding at the inner surface of the membrane is not implied. From Kaczorowski and Kaback.[59]

tions catalyzed by the unloaded carrier involve net movement of negative charge. Recent experiments with proteoliposomes reconstituted with purified *lac* carrier protein provide convincing support for many of these ideas.[149,155]

Transient accumulation of Rb^+ during lactose efflux in the presence of valinomycin argues strongly for the coupled translocation of a charged species with lactose which leads to the generation of $\Delta\psi$ (interior negative). In addition to the phenomenon itself, the process is abolished by CCCP, and efflux-induced Rb^+ uptake is blocked by *p*-chloromercuribenzenesulfonate (*p*CMBS), a sulfhydryl reagent that completely inactivates the *lac* carrier protein. Furthermore, the rate of lactose efflux is enhanced by ionophores that collapse $\Delta\psi$, and artificial imposition of $\Delta\psi$ (interior negative) dramatically slows the rate of efflux with no significant change in apparent K_m.

The maximal rate of efflux is pH dependent, increasing more than 100-fold from pH 5.5 to pH 9.5 in a sigmodial fashion with a midpoint at about pH 8.3. In contrast, experiments performed under identical conditions with equimolar lactose in the external medium (i.e., under exchange conditions) demonstrate that the exchange reaction is insensitive to pH and very fast relative to efflux, particularly at relatively acid pH values (below pH 7.5). Therefore, the rate-determining step for efflux must involve either deprotonation of the carrier on the external surface of the membrane or the reaction corresponding to return of the unloaded carrier to the inner surface of the membrane, as these are the only steps by which efflux and exchange differ (Fig. 12). Assuming that loss of lactose and H^+ from the carrier is necessary for reinitiation of an efflux cycle, external pH would influence the rate of turnover in either of two ways. First, deprotonation could be slow and thereby limit the overall rate of efflux in a pH-dependent manner. Although H^+ transfers between accessible amino acid residues and water in soluble enzymes are usually fast,[156] little is known about such reactions with hydrophobic membrane proteins. Alternatively, pH could alter the equilibrium between protonated and unprotonated forms of the *lac* carrier, favoring the unprotonated form at more alkaline pH. Since it is assumed that only the deprotonated form of the carrier can recycle, the rate of efflux would be at least partially controlled by external pH, and the rate-determining step might then involve "movement" of the unloaded carrier to the inner surface of the membrane. The observation that the rate of efflux increases with pH is consistent with either possibility. In contrast, if deprotonation of the carrier is not obligatory for exchange, H^+ might remain bound to the carrier during this mode of translocation, rendering exchange insensitive to pH. If efflux is an ordered mechanism in which the carrier releases lactose first, followed by loss of H^+ (Fig. 12), deprotonation and/or return of the unloaded carrier could be slow and appear as the limiting step for efflux.

Counterflow experiments conducted at various pH values reveal that external lactose affects H^+ loss from the carrier and therefore provide strong support for the ordered efflux mechanism shown in Fig. 12. When external lactose is saturating, counterflow is unaffected by pH; moreover, transient formation of $\Delta\psi$ observed during lactose efflux is abolished under these conditions. The results can be interpreted in the following way. On initiation of efflux, lactose and H^+ bind to the carrier on the inner surface of the membrane (in an unspecified order) and are translocated to the outer surface. Lactose is released from the carrier, but in the presence of excess labeled substrate, rebinding and influx occur rapidly before deprotonation occurs. Under these conditions, therefore, H^+ release is infrequent and pH has no effect on the overall phenomenon. When external [^{14}C]lactose is limiting, however, rebinding of labeled substrate is less frequent, allowing deprotonation and return of the unloaded carrier. Moreover, as pH is increased, deprotonation and return of the unloaded carrier are enhanced, resulting in further diminution of counterflow. Inhibition of efflux-generated $\Delta\psi$ formation by external lactose is also readily explained by this scheme. When lactose is present externally at saturating concentrations, release of lactose from the carrier and rebinding of substrate occur rapidly before deprotonation can occur, and the ability of the system to generate $\Delta\psi$ is abolished.

Given the indication that H^+ loss may be a limiting step for efflux, it is apparent that one means by which to further investigate the suggested mechanism is to search for a solvent deuterium isotope effect.[38,155] At equivalent pH and pD (i.e., pD = pH + 0.4),[156] the rate of lactose facilitated diffusion (influx as well as efflux) is approximately 3–4 times slower in deuterated medium (with over 95% of the protium replaced with deuterium) relative to control conditions in protium, while the rate of exchange is identical in the presence of deuterium or protium. Furthermore, during counterflow with the external lactose concentration below the K_m of the carrier, the magnitude of the overshoot is 2–3 times *greater* in deuterium relative to protium. With respect to the kinetic model (see Fig. 12), high external lactose concentrations prevent deprotonation of the carrier (the C-H form), and it recycles across the membrane in the fully loaded state, catalyzing 1 : 1 exchange of internal unlabeled lactose with external [^{14}C]lactose. When the external lactose concentration is well below the K_m, however, D_2O increases the coupling efficiency for counterflow, particularly at higher pH (pD) values. Under these conditions, the C-H or C-D form of the carrier partitions between two pathways, one involving loss of protium or deuterium which results in net efflux and the other involving rebinding of lactose prior to loss of protium or deuterium which results in exchange (i.e., counterflow). The former pathway is favored at high pH (pD) values, but the C-D form of the carrier is deprotonated at a slower rate than the C-H form, favoring binding of external [^{14}C]lactose. Consequently, the frequency with which the carrier returns to the inner surface of the membrane in the loaded versus the unloaded form is enhanced in the presence of D_2O, and the effect is most pronounced at alkaline pH. At relatively acid pH (pD), where the coupling efficiency is already essentially 1 : 1, the D_2O effect is masked.

Remarkably, $\Delta\psi$-driven lactose accumulation exhibits essentially no solvent deuterium isotope effect. Based on the observations as a whole, it is reasonable to suggest that under conditions where carrier turnover is driven by a lactose concentration gradient, the rate of translocation is determined by a step(s) involving protonation, deprotonation, or the subsequent step (i.e., a reaction corresponding to the return of the unloaded carrier). In contrast, when there is a driving force on H^+ (i.e., in the presence of $\Delta\tilde{\mu}_H$), this step(s) is no longer rate-determining. As a cautionary note, it should be emphasized that the solvent deuterium isotope effects described cannot be attributed definitively to a true kinetic isotope effect as opposed to a pK_a effect (i.e., deuterium increases the pK_a values of various functional groups from 0.4 to 0.7 pH unit) because the isotope effect on efflux disappears at pH 9.0 and above. On the other hand, at these alkaline pHs, the rate of efflux approaches the rate of exchange, suggesting that the rate-determining step for efflux may change at high pH. In any event, it seems evident that

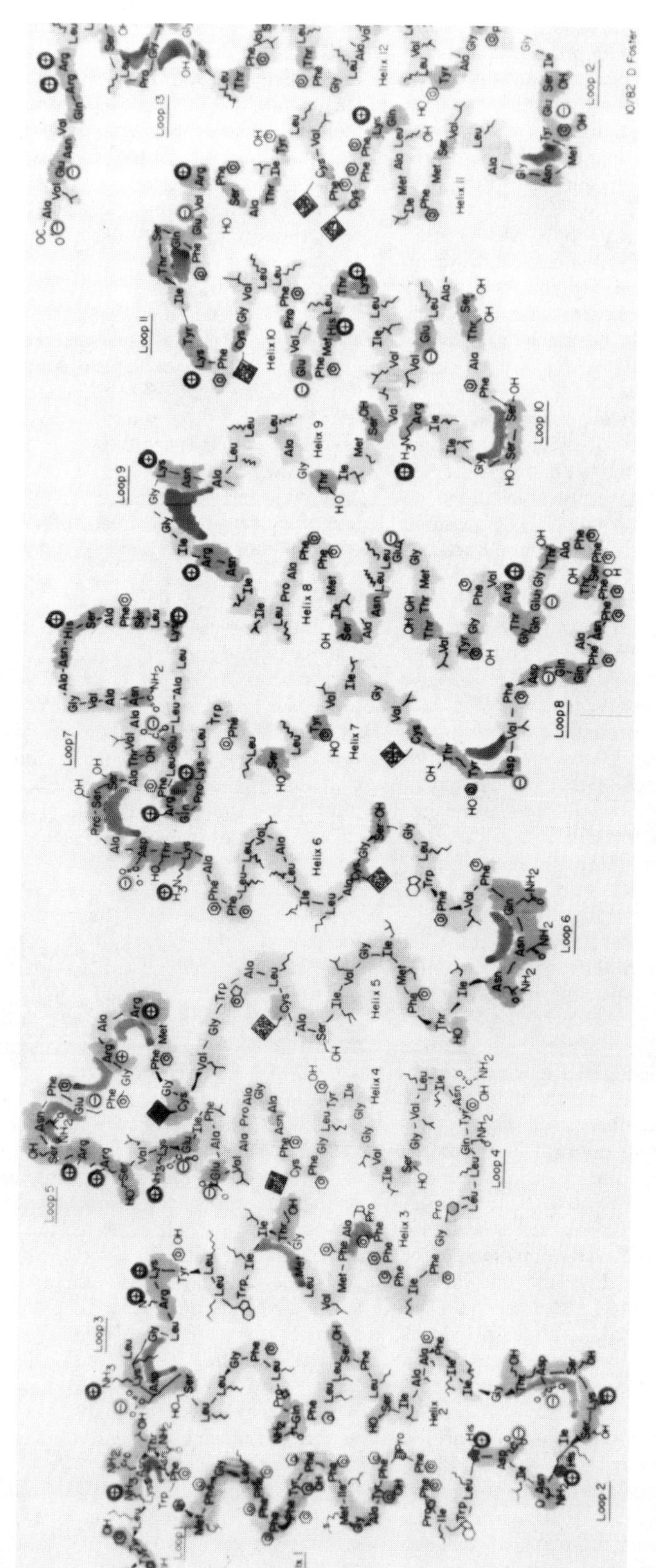

Fig. 13. Secondary structure model of the *lac* carrier protein based on the hydropathic profile of the protein.[158] Light gray shading is used for hydrophobic residues, medium gray shading for hydrophilic and charged residues, and dark gray shading for β turns. The orientation of the protein with respect to the membrane is unknown.

different steps are limiting when carrier turnover is driven by $\Delta\tilde{\mu}_H$ or by a solute concentration gradient.

4.5. A Secondary Structure Model for the *lac* Carrier Protein

Circular dichroic measurements on purified *lac* carrier indicate that 85 ± 5% of the amino acid residues are arranged in helical secondary structures whether the protein is solubilized in octylglucoside or reconstituted into proteoliposomes.[157] This finding led to a systematic examination of primary structure as determined from the DNA sequence of the *lac y* gene.[137] When the hydrophilicity and hydrophobicity (i.e., hydropathy) of the protein is evaluated along the amino acid sequence according to the method of Kyte and Doolittle,[158] it is apparent that the carrier contains a number of relatively long hydrophobic regions punctuated by shorter hydrophilic regions. In light of the circular dichroism data, this finding suggests strongly that most, if not all, of these segments are α-helical. Furthermore, since the segments are markedly hydrophobic, it seems likely that they are embedded in the lipid layer. About 12 of the longest hydrophobic segments exhibit a mean length of 24 ± 4 amino acid residues, and they comprise approximately 70% of the length of the polypeptide. The mean length of these segments correlates remarkably well with the mean lengths calculated for similar domains found in four other integral membrane proteins involved in H^+ translocation (i.e., bacteriorhodopsin and the three subunits of the F_0 portion of the H^+-ATPase). A 24-residue α-helical peptide would be expected to be a maximum of 36 Å in length, a distance that corresponds roughly to the thickness of the hydrophobic core of the membrane.

According to the secondary structure predictions of Chou and Fasman,[159] eighteen regions of the *lac* carrier contain reverse turns (180° reversals). Fifteen (83%) of the putative turns fall within hydrophilic regions between the hydrophobic segments postulated to traverse the bilayer.

Based on these considerations, the model shown in Fig. 13 is proposed. The *lac* carrier is postulated to consist of 12 α-helical segments that traverse the membrane in a zigzag fashion as suggested for bacteriorhodopsin.[160,161]

Experiments utilizing proteolytic enzymes and site-directed polyclonal antibodies (i.e., antibodies directed against synthetic peptides corresponding in sequence to specified regions of the permease) provide preliminary support for the model. Thus, photoaffinity-labeled permease in RSO or ISO membrane vesicles is accessible to chymotrypsin, trypsin, or papain, demonstrating that the permease extends through the bilayer.[162] Moreover, antibodies directed against peptides corresponding to the C-terminus[163–165] and hydrophilic segments 5 and 7 (Fig. 13) bind preferentially to ISO vesicles relative to RSO vesicles, indicating that each of these portions of the permease is present on the same side of the membrane, the cytoplasmic surface. On the other hand, antibodies directed against a number of other hydrophilic segments do not bind to vesicles of either orientation, although the antibodies react with the permease after immunoblotting. Presumably, these portions of the protein are either buried in the membrane or inaccessible within the tertiary structure of the native polypeptide.

4.6. Immunological Reagents as Structure/Function Probes

Monoclonal antibodies (Mabs) directed against purified *lac* permease have been prepared by somatic cell fusion of mouse myeloma cells with splenocytes from an immunized mouse. Several clones produce antibodies that react with purified permease as demonstrated by solid-phase radioimmunoassay and by immunoblotting. The effects of the Mabs on lactose transport were studied in RSO membrane vesicles and in proteoliposomes reconstituted with purified permease.[166] Out of more than 60 Mabs tested, only one, designated 4B1, inhibits transport. Furthermore, the nature of the inhibition is highly specific in that 4B1 inhibits only those transport reactions that involve net H^+ translocation (i.e., active transport, carrier-mediated influx and efflux under nonenergized conditions and lactose-induced H^+ influx). In contrast, 4B1 has little effect on equilibrium exchange and no effect on generation of $\Delta\tilde{\mu}_{H^+}$ or on the ability of the permease to bind a high-affinity ligand. Clearly, therefore, 4B1 alters the relationship between lactose and H^+ translocation at the level of the permease. By studying entrance counterflow with external ^{14}C-lactose at saturating and subsaturating concentrations, it is apparent that 4B1 mimics the effects of deuterium oxide,[155] and the results suggest that the Mab either inhibits the rate of deprotonation or alters the equilibrium between protonated and deprotonated forms of the permease. Monovalent Fab fragments prepared from 4B1 inhibit transport in a manner that is similar qualitatively to that of the intact IgG. However, 4B1 IgG is approximately twice as effective as the Fab fragments on a molar basis, suggesting that the intact IgG binds bivalently, while the Fab fragments bind 1 : 1. Support for this conclusion is provided by binding experiments with radiolabeled 4B1 and 4B1 Fab fragments.[167]

Radioiodinated 4B1 and 5F7 (another Mab obtained from the same fusion that does not inhibit transport) bind to distinct, nonoverlapping epitopes in the permease.[167] By using immunofluorescence microscopy and radioiodinated IgGs and Fab fragments, it is apparent that both Mabs bind to spheroplasts and to RSO membrane vesicles, but only to a small extent to ISO vesicles. Therefore, as opposed to C-terminus and hydrophilic segments 5 and 7 which are on the cytoplasmic surface of the membrane (see above), the 4B1 and 5F7 epitopes are on the external (i.e., periplasmic) surface of the membrane. In RSO vesicles, ^{125}I-4B1 binds with a stoichiometry of 1 mol of antibody per 2 mol of permease, while ^{125}I-4B1 Fab fragments bind 1 : 1. Importantly, intact 4B1 and its Fab fragments bind to proteoliposomes reconstituted with purified *lac* permease with a stoichiometry very similar to that observed in RSO membrane vesicles. Thus, with respect to the 4B1 epitope, the orientation of the permease in the reconstituted system appears to be similar to that in the bacterial cytoplasmic membrane.

In contradistinction to these results, recent experiments with site-directed polyclonal antibodies directed against the C-terminus of the permease indicate that the reconstituted system may not be entirely representative of the native membrane.[165,168] ^{125}I-Anti-C-terminal Fab fragments, which bind to ISO vesicles relatively exclusively, bind to proteoliposomes containing the permease, indicating that a significant percentage of the C-terminus is on the outside of the membrane in the reconstituted system (i.e., on the wrong side of the membrane). In addition, treatment of reconstituted proteoliposomes with carboxypeptidase partially degrades the ultimate C-terminus with no effect on the 4B1 epitope or on transport activity. Finally, antibody to hydrophilic segment 7, which also binds to ISO vesicles, does not bind to proteoliposomes, indicating that this portion of the molecule—like the 4B1 epitope—maintains the proper orientation after reconstitution. Therefore, although the obvious possibility that reconstituted permease molecules are scrambled has been considered, the data are more consistent with the notion that a portion of the permease molecules undergoes intramolecular dislocation of the C-terminus during reconstitution with no effect on catalytic activity. In this context, it is also significant that 4B1 causes enhanced binding of Mab 4A10R in the reconstituted system. Mab 4A10R is directed against a cytoplasmically disposed epitope that is partially related to the C-terminus of the permease (i.e., in native membrane, the epitopes for 4A10R and 4B1 are on opposite sides of the membrane).[169] Since it is unlikely that this effect of 4B1 is intermolecular, a more reasonable interpretation is that these conformationally-coupled epitopes which are normally on opposite sides of the native membrane are present on the external surface of the proteoliposomes within the same permease molecules.

RSO vesicles from *E. coli* ML 308–22, a mutant "uncoupled" for β-galactoside/H^+ symport;[170] are specifically defective in the ability to catalyze accumulation of methyl 1-thio-β-D-galactopyranoside (TMG) in the presence of $\Delta\bar{\mu}_{H^+}$.[169] Furthermore, the rate of carrier-mediated efflux under nonenergized conditions is slow and unaffected by ambient pH from pH 5.5 to 7.5, and TMG-induced H^+ influx is only about 15% of that observed in vesicles containing wild type *lac* permease. Alternatively, ML 308–22 vesicles bind *p*-nitrophenyl-α-D-galactopyranoside and Mab 4B1 to the same extent as wild type vesicles and catalyze facilitated diffusion and equilibrium exchange as well as wild type vesicles. When entrance counterflow is studied with external substrate at saturating and subsaturating concentrations, it is apparent that the mutation, like Mab 4B1, also stimulates the effects of deuterium oxide. That is, the mutation has no effect on the rate or extent of counterflow when external substrate is saturating, but stimulates the efficiency of counterflow when external substrate is below the apparent K_m. Moreover, although replacement of protium with deuterium stimulates counterflow in wild type vesicles when external substrate is limiting, the isotope has no effect on the mutant vesicles under the same conditions. It is suggested that the mutation in ML 308–22 results in a *lac* permease with a higher pK_a, thereby either limiting the rate of deprotonation or altering the equilibrium between protonated and deprotonated forms of the permease. Although Mab 4B1 binds similarly to wild type and mutant RSO vesicles, Mab 4A10R binds to ISO vesicles from the mutant only 30% as well as it binds to the same preparation from the wild type. Furthermore, antibodies against hydrophilic domains 5 and 7 bind 3-fold better and one-fifth as well, respectively, to ISO vesicles from the mutant relative to ISO vesicles from the wild type. Clearly, therefore, these immunological probes are able to discriminate between wild type and "uncoupled" permease molecules. In addition, the results suggest that mutation causes a significant alteration in the conformation of the permease.

4.7. Subunit Structure

Studies with RSO membrane vesicles demonstrate that in addition to acting thermodynamically as the driving force for active lactose transport, $\Delta\bar{\mu}_{H^+}$ alters the distribution of the permease between two markedly different kinetic pathways. In the absence of $\Delta\bar{\mu}_{H^+}$, transport exhibits an apparent K_m of about 20 mM for lactose, and when $\Delta\bar{\mu}_{H^+}$ is applied, the apparent K_m decreases by about 100-fold to 0.2 mM. Furthermore, the distribution of the permease between these two pathways varies as the square of ΔpH or Δψ. Based on these observations, it was suggested very tentatively that the permease might exist in two forms, monomer and dimer, that the monomer catalyzes facilitated diffusion (high apparent K_m) and the dimer active transport (low apparent K_m), and finally, that $\Delta\bar{\mu}_{H^+}$ causes an alteration in subunit interactions.

Although the permease is monomeric when solubilized in dodecylmaltoside,[171] use of radiation inactivation analysis indicates that the situation may be more complex when the permease is in the membrane.[172] In these experiments, vesicles containing the permease are frozen rapidly in liquid N_2 before and after energization and subjected to a high-intensity electron beam for various periods of time. Since the vesicles become very permeable after short periods of irradiation, it is necessary to extract and reconstitute the permease in order to assay activity. Thus, after irradiation, the samples are extracted with octylglucoside, reconstituted into proteoliposomes and tested for activity. Under all conditions, the decrease in activity exhibits pseudo-first-order kinetics as a function of radiation dosage, allowing straightforward application of target theory for determination of functional molecular size. When permease activity solubilized from nonenergized vesicles is assayed under these conditions, the results obtained yield a functional molecular weight of 45–50 Kdal, a value similar to the molecular weight of the permease as determined by other means. Importantly, moreover, similar values are obtained when the octylglucoside extract is irradiated, and target volumes observed for D-lactate dehydrogenase (D-LDH) and the H^+-ATPase complex in the same vesicles are in reasonable agreement with the known molecular weights of these enzymes. Strikingly, when the same procedures are carried out with vesicles that are energized with appropriate electron donors prior to freezing and irradiation, a functional molecular weight of 85–100 Kdal is obtained for the permease with no change in the target size of D-LDH. In contrast, when the vesicles are energized in the presence of a potent protonophore which collapses $\Delta\bar{\mu}_{H^+}$, the target size of the permease returns to 45–50 Kdal. In addition, genetic studies[173] indicating that certain *lac y* mutations may be dominant are also consistent with the idea that oligomerization may be important for *lac* permease function.

4.8. Oligonucleotide-Directed, Site-Specific Mutagenesis

Based on substrate protection against N-ethylmaleimide (NEM) inactivation, Fox and Kennedy[174] postulated that there is an essential sulfhydryl group in the *lac* permease located at or near the active site, and Cys_{148} has been shown to be the critical residue.[175] Although chemical modification of specific amino acid residues in a protein can provide important information, there are obvious drawbacks to this approach. Recently, site-directed mutagenesis has been utilized to introduce single amino acid changes into certain proteins,[176] and this strategy has been used to evaluate the role of Cys_{148} in the *lac* permease.[177]

By cloning the *lac y* gene into single-stranded M13 phage DNA and utilizing a synthetic deoxyoligonucleotide primer 21 bases in length that is complementary to the *lac y* template with the exception of a single mismatch, Cys_{148} in the permease is converted into a glycine residue. Cells bearing the mutated *lac y* gene exhibit initial rates of lactose transport that are about 4-fold lower than cells bearing the wild type gene on a recombinant plasmid. Furthermore, transport activity is less sensitive to inactivation by NEM, and strikingly, galactosyl 1-thio-β-D-galactopyranoside affords no protection whatsoever against inactivation. The findings suggest that although Cys_{148} is essential for substrate protection against sulfhydryl inactivation, it is not obligatory for lactose : H^+ symport and that another sulfhydryl group elsewhere within the *lac* permease may be required for full activity.

More recently, oligonucleotide-directed, site-specific mutagenesis has been utilized to introduce other alterations into the *lac* permease. For example, (1) Gln_{60} has been replaced with a glutamic acid residue, thus introducing a negative charge into the second putative α-helix of the permease; (2) His_{35} and His_{39} have been simultaneously replaced with arginine residues; and (3) a segment of the polypeptide from Met_{372} to Pro_{405} has been deleted, thereby excising the last transmembrane α-helical segment from the permease (Fig. 6). Currently, the effects of these alterations on transport activity and on the disposition of various epitopes in the membrane are being examined.

5. Summary

The immediate driving force for active transport of a wide range of substrates in plasma membrane vesicles isolated from *E. coli* is a proton electrochemical gradient ($\Delta\bar{\mu}_{H^+}$, interior negative and alkaline). Thus, in accordance with the chemisomotic hypothesis of Mitchell, active transport is the result of an indirect process whereby respiration or ATP hydrolysis drives substrate accumulation through the mediation of $\Delta\bar{\mu}_{H^+}$ or one of its components. As described, the experimental approach to the problem at this level of resolution is dependent upon the development of techniques that allow quantitation of $\Delta\bar{\mu}_{H^+}$ in a system that is not easily amenable to a direct electrophysiological approach and upon the demonstration that variations in $\Delta\bar{\mu}_{H^+}$ and its components correlate with changes in the transport of appropriate substrates.

Recent advances with the β-galactoside transport system have extended the problem to a molecular level. The *lac* carrier protein (*lac* permease) which is encoded by the *lac y* gene catalyzes the simultaneous translocation of substrate with H^+ (H^+/substrate symport). By using a strain of *E. coli* with multiple copies of the *lac y* gene, a highly specific photoaffinity label for the *lac* carrier protein, and reconstitution of transport activity in proteoliposome, the *lac* carrier protein has been purified to homogeneity in a functional state. Proteoliposomes reconstituted with purified carrier catalyze all of the translocation reactions typical of the β-galactoside transport system in intact cells and isolated membrane vesicles, and the turnover number of the purified, reconstituted carrier is similar to that observed in membrane vesicles, as is the K_m for lactose. Furthermore, proteoliposomes reconstituted simultaneously with an *o*-type cytochrome oxidase purified to homogeneity and the *lac* carrier generate $\Delta\bar{\mu}_{H^+}$ and catalyze electron transfer-driven lactose accumulation. It is apparent therefore that a single polypeptide species, the product of the *lac y* gene, is responsible for each of the reactions catalyzed by the β-galactoside transport system, including active transport energized by electron transfer.

Mechanistic studies carried out with RSO membrane vesicles and with reconstituted proteoliposomes demonstrate that carrier-mediated lactose efflux down a concentration gradient is an ordered symport reaction in which lactose is released first, followed by loss of the proton. The data also suggest that either deprotonation *per se* or a reaction corresponding to return of the unloaded carrier is rate-determining for efflux. Further experiments with deuterium oxide support these conclusions and indicate that different steps are limiting when carrier turnover is driven by $\Delta\bar{\mu}_{H^+}$ or by a solute concentration gradient.

Based on circular dichroic measurements on purified *lac* carrier and on its hydropathic profile as determined from the amino acid sequence, a secondary structure model is proposed in which the carrier is postulated to consist of 12 α-helical segments that traverse the membrane in a zigzag fashion as suggested for bacteriorhodopsin. Proteolysis experiments with RSO and ISO membrane vesicles demonstrate that the *lac* carrier spans the bilayer and binding studies with site-directed polyclonal antibodies demonstrate that the C-terminus, hydrophilic segment 5 and hydrophilic segment 7 are on the cytoplasmic surface of the bacterial membrane.

The preparation and characterization of monoclonal antibodies against the *lac* carrier protein are described. The antibodies are highly specific for the *lac* carrier, and at least three are directed against independent epitopes, two of which are located on the external surface of the membrane. One of the antibodies and its Fab fragments inhibit the ability of the *lac* carrier to catalyze H^+/lactose symport with no effect on binding and no effect on exchange. Importantly, this antibody and its Fab fragments bind to proteoliposomes reconstituted with purified *lac* carrier with a stoichiometry very similar to that observed in RSO membrane vesicles, demonstrating that the orientation of the epitope in the reconstituted system is similar to that in the bacterial cytoplasmic membrane. In contrast, experiments with site-directed polyclonal antibodies directed against the C-terminus bind significantly to proteoliposomes, and treatment of reconstituted proteoliposomes with carboxypeptidase partially degrades the ultimate C-terminus with no effect on the epitope for the monoclonal antibody and no effect on transport activity. These and other results are consistent with the suggestion that a portion of the carrier molecules undergoes intramolecular dislocation of the C-terminus during reconstitution with no effect on catalytic activity.

Detailed kinetic studies show that $\Delta\bar{\mu}_{H^+}$ causes a dramatic decrease in the K_m of the *lac* carrier for substrate with little effect on V_{max}, and the distribution of the carrier between the high- and low-K_m pathways varies with the square of $\Delta\bar{\mu}_{H^+}$. Remarkably, electron inactivation analysis is consistent with the notion that $\Delta\bar{\mu}_{H^+}$ may induce an alteration in subunit interactions.

Finally, the use of oligonucleotide-directed, site-specific mutagenesis as a structure/function probe for the *lac* carrier protein is described.

References

1. Mitchell, P. 1961. Coupling of phosphorylation to electron hydrogen transfer by a chemiosmotic type of mechanism. *Nature (London)* **191**:144.
2. Mitchell, P. 1963. Molecule, group, and electron translocation through natural membranes. *Biochem. Soc. Symp.* **22**:142.
3. Mitchell, P. 1966. *Chemiosmotic Coupling in Oxidative and Photophosphorylation.* Glynn Research Ltd., Bodmin, England.
4. Mitchell, P. 1966. *Chemiosmotic Coupling and Energy Transduction.* Glynn Research Ltd., Bodmin, England.
5. Mitchell, P. 1973. Performance and conservation of osmotic work by proton-coupled solute porter systems. *J. Bioenerg.* **4**:63.
6. Kaback, H. R. 1960. Uptake of amino acids by "ghosts" of mutant strains of *Escherichia coli. Fed. Proc.* **19**:130.
7. Kaback, H. R., and E. R. Stadtman. 1966. Proline uptake by an isolated cytoplasmic membrane preparation of *Escherichia coli. Proc. Natl. Acad. Sci. USA* **55**:920.
8. Lombardi, F. J., and H. R. Kaback. 1972. Mechanisms of active transport in isolated bacterial membrane vesicles. *J. Biol. Chem.* **247**:7844.

9. Short, S. A., D. C. White, and H. R. Kaback. 1972. Mechanisms of active transport in isolated bacterial membrane vesicles. *J. Biol. Chem.* **247**:7452.
10. Ramos, S., and H. R. Kaback. 1977. The relationship between the electrochemical proton gradient and active transport in *E. coli* membrane vesicles. *Biochemistry* **16**:854.
11. Short, S. A., H. R. Kaback, and L. D. Kohn. 1974. D-Lactate dehydrogenase binding in *E. coli dld*$^-$ membrane vesicles reconstituted for active transport. *Proc. Natl. Acad. Sci. USA* **71**:1461.
12. Kaback, H. R. 1971. Bacterial membranes. *Methods Enzymol.* **22**:99.
13. Owen, P., and H. R. Kaback. 1978. Molecular structure of membrane vesicles from *E. coli. Proc. Natl. Acad. Sci. USA* **75**:3148.
14. Owen, P., and H. R. Kaback. 1979. Immunochemical analysis of membrane vesicles from *E. coli. Biochemistry* **18**:1413.
15. Owen, P., and H. R. Kaback. 1979. Antigenic architecture of membrane vesicles from *E. coli. Biochemistry* **18**:1422.
16. Stroobant, P., and H. R. Kaback. 1975. Ubiquinone-mediated coupling of NADH dehydrogenase to active transport in membrane vesicles from *E. coli. Proc. Natl. Acad. Sci. USA* **72**:3970.
17. Hertzberg, E., and P. C. Hinkle. 1974. Oxidative phosphorylation and proton translocation in membrane vesicles prepared from *E. coli. Biochem. Biophys. Res. Commun.* **58**:178.
18. Rosen, B. P., and J. S. McClees. 1974. Active transport of calcium in inverted membrane vesicles of *E. coli. Proc. Natl. Acad. Sci. USA* **71**:5042.
19. Reenstra, W. W., L. Patel, H. Rottenberg, and H. R. Kaback. 1980. Electrochemical proton gradient in inverted membrane vesicles from *E. coli. Biochemistry* **19**:1.
20. Kaback, H. R. 1970. Transport. *Annu. Rev. Biochem.* **39**:561.
21. Kundig, W., S. Ghosh, and S. Roseman. 1964. Phosphate bound to histidine in a protein as an intermediate in a novel phosphotransferase system. *Proc. Natl. Acad. Sci. USA* **52**:1067.
22. Dills, S. S., A. Apperson, M. R. Schmidt, and M. H. Saier, Jr. 1980. Carbohydrate transport in bacteria. *Microbiol. Rev.* **44**:385.
23. Saier, M. H. 1982. The bacterial phototransferase system in regulation of carbohydrate permease synthesis and activity. In: *Membranes and Transport,* Volume 2. A. Martonosi, ed. Plenum Press, New York. p. 27.
24. Kaback, H. R. 1974. Transport studies in bacterial membrane vesicles. *Science* **186**:882.
25. Futai, M. 1973. Membrane D-lactate dehydrogenase from *Escherichia coli*: Purification and properties. *Biochemistry* **12**:2468.
26. Kohn, L. D., and H. R. Kaback. 1973. Mechanisms of active transport in isolated bacterial membrane vesicles. XV. Purification and properties of the membrane-bound D-lactate dehydrogenase from *E. coli. J. Biol. Chem.* **248**:7012.
27. Santos, E., H.-F. Kung, I. G. Ylung, and H. R. Kaback. 1982. *In vitro* synthesis of the membrane-bound D-lactate dehydrogenase of *Escherichia coli. Biochemistry* **21**:2085.
28. Young, I. G., A. Jaworowski, and M. Poulis. 1982. Proton electrochemical gradient in *Escherichia coli* cells and its relation to active transport of lactose. *Biochemistry* **21**:2092.
29. Stroobant, P., and H. R. Kaback. 1979. Reconstitution of ubiquinone-linked function in membrane vesicles from a double quinone mutant of *Escherichia coli. Biochemistry* **18**:226.
30. Konings, W. N., E. M. Barnes, Jr., and H. R. Kaback. 1971. Mechanisms of active transport in isolated membrane vesicles. *J. Biol. Chem.* **246**:5857.
31. Konings, W. N., and J. Boonstra. 1976. Anaerobic electron transfer and active transport in bacteria. *Curr. Top. Membr. Transp* **9**:177.
32. Konings, W. N., and H. R. Kaback. 1973. Anaerobic transport in *Escherichia coli* membrane vesicles. *Proc. Natl. Acad. Sci. USA* **70**:3376.
33. Kaback, H. R. 1976. Molecular biology and energetics of membrane transport. *J. Cell. Physiol.* **89**:575.
34. Hugenholtz, J., J.-S. Hong, and H. R. Kaback. 1981. ATP-driven active transport in right-side-out bacterial membrane vesicles. *Proc. Natl. Acad. Sci. USA* **78**:3446.
35. Saier, M. H., Jr., D. L. Wentzel, B. U. Feucht, and J. J. Justice. 1975. A transport system for phosphoenolpyruvate, 2-phosphoglycerate, and 3-phosphoglycerate in *Salmonella typhimurium. J. Biol. Chem.* **250**:5089.
36. Hirata, H., K. H. Altendorf, and F. M. Harold. 1973. Role of an electrical potential in the coupling of metabolic energy to active transport by membrane vesicles of *Escherichia coli. Proc. Natl. Acad. Sci. USA* **70**:1804.
37. Hirata, H., K. H. Altendorf, and F. M. Harold. 1974. Energy coupling in membrane vesicles of Escherichia coli. *J. Biol. Chem.* **249**:2939.
38. Kaczorowski, G. J., D. E. Robertson, and H. R. Kaback. 1979. Mechanism of lactose translocation in membrane vesicles from *Escherichia coli.* 2. Effect of imposed $\Delta\psi$, ΔpH and $\Delta\bar{\mu}_{H^+}$. *Biochemistry* **18**:3697.
39. Schuldiner, S., and H. R. Kaback. 1975. Membrane potential and active transport in membrane vesicles from *Escherichia coli. Biochemistry* **14**:5451.
40. Greville, G. D. 1969. Scrutiny of Mitchell's chemiosmotic hypothesis of respiratory chain and photosynthetic phosphorylation. *Curr. Top. Bioenerg.* **3**:1.
41. Harold, F. M. 1972. Conservation and transformation of energy by bacterial membranes. *Bacteriol. Rev.* **36**:172.
42. Harold, F. M. 1978. Vectorial metabolism. In: *The Bacteria,* Volume 6 I. C. Gunsalus, L. N. Ornston, and T. R. Sokatch, eds. Academic Press, New York. p. 463.
43. Hinkle, P. C., and R. E. McCarty. 1978. How cells make ATP. *Sci. Am.* **238**:104.
44. Skulachev, V. P., and P. C. Hinkle, eds. 1981. *Chemiosmotic Proton Circuits in Biological Membranes.* Addison-Wesley, Reading, Mass.
45. Doetsch, R. N., and R. D. Sjoblad. 1980. Flagellar structure and function in eubacteria. *Annu. Rev. Microbiol.* **34**:69.
46. Laane, C., W. Krone, W. Konings, H. Haaker, and C. Veeger. 1980. Short-term effect of ammonium chloride on nitrogen fixation by *Azotobacter vinelandii* and by bacteroids of *Rhizobium leguminosarum. Eur. J. Biochem.* **103**:39.
47. Grinius, L., and J. Bervinskiene. 1976. Studies on DNA transport during bacterial conjugation: Role of protonmotive force-generating H^+-ATPase and respiratory chain. *FEBS Lett.* **72**:151.
48. Grinius, L. 1980. Nucleic acid transport driven by ion gradient across cell membrane. *FEBS Lett.* **113**:1.
49. Kalasauskaite, E., and L. Grinius. 1979. The role of energy-yielding ATPase and respiratory chain at early stages of bacteriophage T4 infection. *FEBS Lett.* **99**:297.
50. Labedan, G., and E. B. Goldberg. 1979. Requirement for membrane potential in injection of phage T4 DNA. *Proc. Natl. Acad. Sci. USA* **76**:4669.
51. Santos, E., and H. R. Kaback. 1981. Involvement of the proton electrochemical gradient in genetic transformation in *Escherichia coli. Biochem. Biophys. Res. Commun.* **99**:1153.
52. Wagner, E. F., H. Ponta, and M. Schweiger. 1980. Development of *Escherichia coli* virus T1: The role of the proton-motive force. *J. Biol. Chem.* **255**:534.
53. Mates, S., E. S. Eisenberg, L. J. Mandel, L. Patel, H. R. Kaback, and M. H. Miller. 1982. Membrane potential and gentamicin uptake in *Staphylococcus aureus. Proc. Natl. Acad. Sci. USA* **79**:6693.
54. Loftfield, R. B., E. H. Eigner, A. Pastuszyn, T. N. E. Lovgren, and H. Jakubowski. 1980. Conformational changes during enzyme catalysis: Role of water in the transition state. *Proc. Natl. Acad. Sci. USA* **77**:3374.
55. Delmer, D. P., M. Benziman, and E. Padan. 1982. Requirement for a membrane potential for cellulose synthesis in intact cells of *Acetoabacter xylinum. Proc. Natl. Acad. Sci. USA* **79**:5282.
56. Daniels, C. J., D. G. Bole, S. C. Quay, and D. L. Oxender. 1981. Role for membrane potential in the secretion of protein into the periplasm of *Escherichia coli. Proc. Natl. Acad. Sci. USA* **78** :5396.
57. Date, T., C. Zwizniski, S. Ludmerer, and W. Wickner. 1980.

Mechanisms of membrane assembly. Effects of energy poisons on the conversion of soluble M13-coliphage procoat to membrane-bound coat protein. *Proc. Natl. Acad. Sci. USA* **77**:827.

58. Enequist, H. G., T. R. Hirst, S. J. S. Hardy, S. Harayama, and L. L. Randall. 1981. Energy is required for maturation of exported proteins in *Escherichia coli*. *Eur. J. Biochem.* **116**:227.
59. Kaczorowski, G. J., and H. R. Kaback. 1979. Mechanism of lactose translocation in membrane vesicles from *Escherichia coli*. 1. Effect of pH on efflux, exchange and counterflow. *Biochemistry* **18**:3691.
60. Mitchell, P. 1976. Possible molecular mechanisms of the protonmotive function of cytochrome systems. *J. Theor. Biol.* **62**:327.
61. Mitchell, P., and J. Moyle. 1979. Respiratory chain protonmotive stoichiometry. *Biochem. Soc. Trans.* **7**:887.
62. Brand, M. D., B. Reynafarje, and A. L. Lehninger. 1976. Re-evaluation of the H^+/site ratio of mitochondrial electron transport with oxygen pulse technique. *J. Biol. Chem.* **251**:5670.
63. Lawford, H. G. 1977. Energy transduction in the mitochondrial-like bacterium *Paracoccus denitrifficans* during carbon- or sulphate-limited aerobic growth in continuous culture. *Can. J. Biochem.* **56**:13.
64. Papa, S., F. Guerrieri, M. Lorusso, G. Izzo, D. Boffoli, and R. Stefanelli. 1970. Reversible effects of chaotropic agents on the proton permeability of *Escherichia coli* membrane vesicles. *FEBS Symp.* **45**:37.
65. Reynafarje, B., M. D. Brand, and A. L. Lehninger. 1976. Evaluation of the H^+/site ratio of mitochondrial electron transport from rate measurements. *J. Biol. Chem.* **251**:7442.
66. Reynafarje, B., and A. L. Lehninger. 1978. The K^+/site and H^+/site stoichiometry of mitochondrial electron transport. *J. Biol. Chem.* **253**:6331.
67. Sigel, E., and E. Carafoli. 1978. The proton pump of cytochrome *c* oxidase and its stoichiometry. *Eur. J. Biochem.* **89**:119.
68. Wikström, M., and H. T. Saari. 1977. The mechanism of energy conservation and transduction by mitochondrial cytochrome *c* oxidase. *Biochim. Biophys. Acta* **462**:347.
69. Wikström, M., and K. Krab. 1978. Cytochrome *c* oxidase is a proton pump. *FEBS Lett.* **91**:8.
70. Short, S. A., H. R. Kaback, and L. D. Kohn. 1975. Localization of D-lactate dehydrogenase in native and reconstituted *Escherichia coli* membrane vesicles. *J. Biol. Chem.* **250**:4291.
71. Kaback, H. R., and E. M. Barnes, Jr. 1971. Mechanisms of active transport in isolated membrane vesicles. *J. Biol. Chem.* **246**:5523.
72. Haldar, K., P. J. Olsiewski, C. Walsh, G. J. Kaczorowski, A. Bhaduri, and H. R. Kaback. 1982. Simultaneous reconstitution of *Escherichia coli* membrane vesicles with D-lactate and D-amino acid dehydrogenases. *Biochemistry* **21**:4590.
73. Olsiewski, P. J., G. Kaczorowski, C. T. Walsh, and H. R. Kaback. 1981. Reconstitution of *Escherichia coli* membrane vesicles with D-amino acid dehydrogenase. *Biochemistry* **20**:6272.
74. Wikström, M., and K. Krab. 1979. Proton-pumping cytochrome *c* oxidase. *Biochim. Biophys. Acta* **549**:177.
75. Solioz, M., E. Carafoli, and B. Ludwig. 1982. The cytochrome *c* oxidase of *Paracoccus denitrificans* pumps protons in a reconstituted system. *J. Biol. Chem.* **257**:1579.
76. Sone, N., and P. C. Hinkle. 1982. Proton transport of cytochrome *c* oxidase from the thermophilic bacterium PS3 reconstituted in liposomes. *J. Biol. Chem.* **257**:12600.
77. Kita, K., M. Kasahara, and Y. Anraku. 1982. Formation of a membrane potential by reconstituted liposomes made with cytochrome b_{562}-*o* complex, a terminal oxidase of *Escherichia coli* K12. *J. Biol. Chem.* **257**:7933.
78. Matsushita, K., L. Patel, R. B. Gennis, and H. R. Kaback. 1983. Reconstitution of active transport in proteoliposomes containing cytochrome *o* oxidase and *lac* carrier protein purified from *Escherichia coli*. *Proc. Natl. Acad. Sci. USA* **80**:4889.
79. Michels, P. A. M., J. P. J. Michels, J. Boonstra, and W. N. Konings. 1979. Generation of an electrochemical proton gradient in bacteria by the excretion of metabolic end products. *FEMS Microbiol. Lett.* **5**:357.
80. Otto, R., A. S. M. Sonenberg, H. Veldkamp, and W. N. Konings. 1980. Generation of an electrochemical proton gradient in *Streptococcus cremoris* by lactate efflux. *Proc. Natl. Acad. Sci. USA* **77**:5502.
81. Grinius, L. L., A. A. Jasaitis, Y. P. Kadziauskas, E. A. Liberman, V. P. Skulachev, L. M. Topali, L. M. Tsofina, and M. A. Vladimirova. 1971. Conversion of biomembrane-produced energy into electric form. I. Submitochondrial particles. *Biochim. Biophys. Acta* **216**:1.
82. Haydon, D. A., and S. B. Hladky. 1972. Ion transport across thin lipid-membranes: Critical discussion of mechanisms in selected systems. *Q. Rev. Biophys.* **5**:187.
83. Lombardi, F. J., J. P. Reeves, S. A. Short, and H. R. Kaback. 1974. Evaluation of the chemiosmotic interpretation of active transport in bacterial membrane vesicles. *Ann. NY Acad. Sci.* **227**:312.
84. Waggoner, A. J. 1979. The use of cyanine dyes for the determination of membrane potentials in cells, organelles, and vesicles. *Methods Enzymol.* **LV**:689.
85. Felle, H., J. S. Porter, C. L. Slayman, and H. R. Kaback. 1980. Quantitative measurements of membrane potential in *Escherichia coli*. *Biochemistry* **19**:3585.
86. Lichtshtein, D., H. R. Kaback, and A. J. Blume. 1979. Use of a lipophilic cation for determination of membrane potential in neuroblastomaglioma hybrid cell suspensions. *Proc. Natl. Acad. Sci. USA* **76**:650.
87. Lichtshtein, D., K. Dunlop, H. R. Kaback, and A. J. Blume. 1979. Mechanism of monensin-induced hyperpolarization of neuroblastoma-glioma hybrid NG108-15. *Proc. Natl. Acad. Sci. USA* **76**:2580.
88. Young, J. D.-E., J. C. Unkeless, H. R. Kaback, and Z. A. Cohn. 1983. Macrophage membrane potential changes associated with $\gamma 2b/\gamma 1$ Fc receptor-ligand binding. *Proc. Natl. Acad. Sci. USA* **80**:1357.
89. Waddel, W. J. and T. C. Butler. 1959. Calculation of intracellular pH from distribution of 5,5′-dimethyl-2,4-oxazolidinedione (DMO): Application to skeletal muscle of the dog. *J. Clin. Invest.* **38**:720.
90. Ramos, S., S. Schuldiner, and H. R. Kaback. 1979. The use of flow dialysis for determinations of ΔpH and active transport. *Methods Enzymol.* **55**:680.
91. Kaback, H. R. 1972. Transport across isolated bacterial cytoplasmic membranes. *Biochim. Biophys. Acta* **265**:367.
92. Padan, E., D. Zilberstein, and H. Rottenberg. 1976. The proton electrochemical gradient in *Escherichia coli* cells. *Eur. J. Biochem.* **63**:533.
93. Rottenberg, H. 1975. Measurement of transmembrane electrochemical proton gradients. *J. Bioenerg.* **7**:61.
94. Colowick, S. P., and F. C. Womack. 1969. Binding of diffusible molecules by macromolecules: Rapid measurement by rate of dialysis. *J. Biol. Chem.* **244**:774.
95. Ramos, S., S. Schuldiner, and H. R. Kaback. 1976. The electrochemical gradient of protons and its relationship to active transport in *Escherichia coli* membrane vesicles. *Proc. Natl. Acad. Sci. USA* **73**:1892.
96. Ramos, S., and H. R. Kaback. 1977. The electrochemical proton gradient in *Escherichia coli* membrane vesicles. *Biochemistry* **16**:848.
97. Ramos, S., and H. R. Kaback. 1977. pH-dependent changes in proton:substrate stoichiometries during active transport in *Escherichia coli* membrane vesicles. *Biochemistry* **16**:4271.
98. Tokuda, H., and H. R. Kaback. 1977. Sodium-dependent methyl 1-thio-β-D-galactopyranoside transport in membrane vesicles isolated from *Salmonella typhimurium*. *Biochemistry* **16**:2130.
99. Navon, G., S. Ogawa, R. G. Schulman, and T. Yamane. 1977. High resolution ^{31}P nuclear magnetic resonance studies of metabolism in aerobic *Escherichia coli* cells. *Proc. Natl. Acad. Sci. USA* **74**:888.
100. Ogawa, S., R. G. Schulman, P. Glynn, T. Yamane, and G. Navon. 1978. On the measurement of pH in *Escherichia coli* by

^{31}P nuclear magnetic resonance. *Biochim. Biophys. Acta* **502**:45.

101. Zilberstein, D., S. Schuldiner, and E. Padan. 1979. Proton electrochemical gradient in *Escherichia coli* cells and its relation to active transport of lactose. *Biochemistry* **18**:669.
102. Schuldiner, S., and H. Fishkes. 1978. Sodium-proton antiport in isolated membrane vesicles of *Escherichia coli. Biochemistry* **17**:706.
103. Schuldiner, S., and E. Padan. 1982. How does *Escherichia coli* regulate internal pH? In: *Membranes and Transport,* Volume 2. A. Martonosi, ed. Plenum Press, New York. p. 65.
104. Zilberstein, D., E. Padan, and S. Schuldiner. 1980. A single locus in *Escherichia coli* governs growth in alkaline pH and on carbon sources whose transport is sodium dependent. *FEBS Lett.* **116**:177.
105. Krulwich, T. A. 1982. Bioenergetic problems of alkalophilic bacteria. In: *Membranes and Transport,* Volume 2. A. Martonosi, ed. Plenum Press, New York. p. 75.
106. Krulwich, T. A. 1983. Sodium/proton antiporters. *Biochim. Biophys. Acta* **726**:245.
107. Henderson, P. J. F., R. A. Giddens, and M. C. Jones-Mortimer. 1977. Transport of galactose, glucose and their molecular analogues by *Escherichia coli* K12. *Biochem. J.* **162**:309.
108. Kashket, E. R., and T. H. Wilson. 1973. Proton-coupled accumulation of galactoside in *Streptococcus lactis* 7962. *Proc. Natl. Acad. Sci. USA* **70**:2866.
109. Lam, V. M. S., K. R. Daruwalla, P. J. F. Henderson, and M. C. Jones-Mortimer. Proton-linked D-xylose transport in *Escherichia coli. J. Bacteriol.* **143**:396.
110. West, I. C. 1970. Lactose transport coupled to proton movements in *Escherichia coli. Biochem. Biophys. Res. Commun.* **41**:655.
111. West, I. C., and P. Mitchell. 1972. Proton-coupled β-galactoside translocation in nonmetabolizing *Escherichia coli. J. Bioenerg.* **3**:445.
112. West, I. C., and P. Mitchell. 1973. Stoichiometry of lactose-protein symport across the plasma membrane of *Escherichia coli. Biochem. J.* **132**:587.
113. West, I. C., and T. H. Wilson. 1973. Galactoside transport dissociated from proton movement in mutants in *Escherichia coli. Biochem. Biophys. Res. Commun.* **50**:551.
114. Daruwalla, K. R., A. T. Paxton, and P. J. F. Henderson. 1981. Energization of the transport systems for arabinose and comparison with galactose transport in *Escherichia coli. Biochem. J.* **200**:611.
115. Patel, L., M. L. Garcia, and H. R. Kaback. 1982. Direct measurement of lactose/proton symport in *Escherichia coli* membrane vesicles: Further evidence for the involvement of histidine residue(s). *Biochemistry* **21**:5805.
116. Robertson, D. E., G. J. Kaczorowski, M. L. Garcia, and H. R. Kaback. 1980. Active transport in membrane vesicles from *Escherichia coli*: The electrochemical proton gradient alters the distribution by the *lac* carrier between two different kinetic states. *Biochemistry* **19**:5692.
117. LeBlanc, G., G. Rimon, and H. R. Kaback. 1980. Glucose 6-phosphate transport in membrane vesicles isolated from *Escherichia coli*: Effect of imposed electrical potential and pH gradient. *Biochemistry* **19**:2522.
118. Rottenberg, H. 1976. The driving force for proton(s)/metabolite contransport in bacterial cells. *FEBS Lett.* **66**:159.
119. Booth, I. R., W. J. Mitchell, and W. A. Hamilton. 1979. Quantitative analysis of proton-linked transport systems: The lactose permease of *Escherichia coli. Biochem. J.* **182**:687.
120. Taylor, D. J., and R. C. Essenberg. 1979. *Proc. 11th Int. Congr. Biochem.* (abstract) p. 460.
121. Stock, J., and S. Roseman. 1971. A sodium-dependent sugar cotransport system in bacteria. *Biochem. Biophys. Res. Commun.* **44**:132.
122. Lanyi, J. K., R. Renthal, and R. E. MacDonald. 1976. Light-induced glutamate transport in *Halobacterium halobium* envelope vesicles. II. Evidence that the driving force is a light-dependent sodium gradient. *Biochemistry* **15**:1603.
123. Cohn, D., G. J. Kaczorowski, and H. R. Kaback. 1981. Effect of the proton electrochemical gradient on maleimide inactivation of active transport in *Escherichia coli* membrane vesicles. *Biochemistry* **20**:3308.
124. Cuffanti, A. A., D. E. Cohn, H. R. Kaback, and T. A. Krulwich. 1981. Relationship between the Na^+/H^+ antiporter and Na^+/substrate symport in *Bacillus alcalophilus. Proc. Natl. Acad. Sci. USA* **78**:1481.
125. Cohen, G. N., and H. V. Rickenberg. 1955. Study of the fixation of an inducer of β-galactosidase by *Escherichia coli. Compt. Rendu* **240**:466.
126. Cohen, G. N., and J. Monad. 1957. Bacterial permeases. *Bacteriol. Rev.* **21**:169.
127. Kepes, A., and G. N. Cohen. 1962. Permeation. In: *The Bacteria,* Volume 4. I. C. Gunsalus, and R. Stanier, eds. Academic Press, New York. p. 179.
128. Kepes, A. 1971. β-Galactoside permease of *Escherichia coli. J. Membr. Biol.* **4**:87.
129. Rickenberg, H. V., G. N. Cohen, G. Buttin, and J. Monod. 1956. Galactoside-permease of *Escherichia coli. Ann. Inst. Pasteur* **91**:829.
130. Ghazi, A., and E. Shechter. 1981. Lactose transport in *Escherichia coli* cells: Dependence of kinetic parameters on the transmembrane electrical potential difference. *Biochim. Biophys. Acta* **645**:305.
131. Sandermann, H., Jr. 1977. β-D-Galactoside transport in *Escherichia coli*: Substrate recognition. *Eur. J. Biochem.* **80**:507.
132. Hobson, A. C., D. Gho, and B. Müller-Hill. 1977. Isolation, genetic analysis, and characterization of *Escherichia coli* mutants with defects in the *lac y* gene. *J. Bacteriol.* **131**:830.
133. Kennedy, E. P. 1970. The lactose permease system of *Escherichia coli*. In: *The Lactose Operon*. J. R. Beckwith and D. Zipser, eds. Cold Spring Harbor Laboratory, Cold Spring Harbor, N.Y. p. 49.
134. Padan, E., S. Schuldiner, and H. R. Kaback. 1979. Reconstitution of *lac* carrier function in cholate-extracted membranes from *Escherichia coli. Biochem. Biophys. Res. Commun.* **91**:854.
135. Teather, R. M., B. Müller-Hill, V. Abrutsch, G. Aichele, and P. Overath. 1978. Amplification of lactose carrier protein in *Escherichia coli* using a plasmid vector. *Mol. Gen. Genet.* **159**:239.
136. Teather, R. M., J. Bramhall, I. Riede, J. K. Wright, M. Fürst, G. Aichele, V. Wilhelm, and P. Overath. 1980. Lactose carrier protein of *Escherichia coli*: Structure and expression of plasmids carrying the *y* gene of the *lac* operon. *Eur. J. Biochem.* **108**: 223.
137. Büchel, D. E., B. Gronenborn, and B. Müller-Hill. 1980. Sequence of the lactose permease gene. *Nature (London)* **283**:541.
138. Ehring, R., K. Beyreuther, J. K. Wright, and P. Overath. 1980. *In vitro* and *in vivo* products of *Escherichia coli* lactose permease gene are identical. *Nature (London)* **283**:537.
139. Newman, M. J., and T. H. Wilson. 1980. Solubilization and reconstitution of the lactose transport system from *Escherichia coli. J. Biol. Chem.* **255**:10583.
140. Racker, E., B. Violand, S. O'Neal, M. Alfonzo, and J. Telford. 1979. Reconstitution, a way of biochemical research; some new approaches to membrane-bound enzymes. *Arch. Biochem. Biophys.* **198**:470.
141. Kaczorowski, G. J., G. LeBlanc, and H. R. Kaback. 1980. Specific labeling of the *lac* carrier protein in membrane vesicles of *Escherichia coli* by a photoaffinity reagent. *Proc. Natl. Acad. Sci. USA* **77**:6319.
142. Foster, D. L., M. L. Garcia, M. J. Newman, L. Patel, and H. R. Kaback. 1982. Lactose-proton symport by purified *lac* carrier protein. *Biochemistry* **21**:5634.
143. Newman, M. J., D. Foster, T. H. Wilson, and H. R. Kaback. 1981. Purification and reconstitution of functional lactose carrier from *Escherichia coli. J. Biol. Chem.* **256**:11804.
144. Patel, L., S. Schuldiner, and H. R. Kaback. 1975. Reversible effects of chaotropic agents on the proton permeability of *Escherichia coli* membrane vesicles. *Proc. Natl. Acad. Sci. USA* **72**:3387.

145. Jones, T. H. D., and E. P. Kennedy. 1969. Characterization of the membrane protein component of the lactose transport system of *Escherichia coli. J. Biol. Chem.* **244**:5981.
146. Banker, G. A., and C. W. Cotman. 1972. Measurement of free electrophoretic mobility and retardation coefficient of protein-sodium dodecyl sulfate complexes by gel electrophoresis. *J. Biol. Chem.* **247**:5856.
147. Neville, D., M., Jr. 1971. Molecular weight determination of protein-dodecyl sulfate complexes by gel electrophoresis in a discontinuous buffer system. *J. Biol. Chem.* **246**:6328.
148. König, B., and H. Sandermann, Jr. 1982. β-Galactoside transport in *Escherichia coli*: M_r determination of the transport protein in organic solvent. *FEBS Lett.* **147**:31.
149. Garcia, M. L., P. Viitanen, D. L. Foster, and H. R. Kaback. 1983. Mechanism of lactose translocation in proteoliposomes reconstituted with *lac* carrier protein purified from *Escherichia coli.* I. Effect of pH and imposed membrane potential on efflux, exchange and counterflow. *Biochemistry* **22**:2524.
150. Hong, J.-S. 1977. An *ecf* mutation in *Escherichia coli* pleiotropically affecting energy coupling in active transport but not generation or maintenance of membrane potential. *J. Biol. Chem.* **252**:8582.
151. Plate, C. A., and J. L. Suit. 1981. The *eup* genetic locus of *Escherichia coli* and its role in H^+/solute symport. *J. Biol. Chem.* **256**:12974.
152. Villarejo, M., and C. Ping. 1978. Localization of the lactose permease protein(s) in the *E. coli* envelope. *Biochem. Biophys. Res. Commun.* **82**:935.
153. Villarejo, M. 1980. Evidence for the two *lac y* gene derived protein products in the *Escherichia coli* membrane. *Biochem. Biophys. Res. Commun.* **93**:16.
154. Wright, J. K., H. Schwarz, E. Straub, P. Overath, B. Bieseler and K. Beyreuther. 1982. Lactose carrier protein of *Escherichia coli.* Reconstitution of galactoside binding and countertransport. *Eur. J. Biochem.* **124**:545.
155. Viitanen, P., M. L. Garcia, D. L. Foster, G. J. Kaczorowski, and H. R. Kaback. 1983. Mechanism of lactose translocation in proteoliposomes reconstituted with *lac* carrier protein purified from *Escherichia coli.* 2. Deuterium solvent isotope effects. *Biochemistry* **22**:2531.
156. Jencks, W. P. 1969. *Catalysis in Chemistry and Enzymology.* McGraw-Hill, New York.
157. Foster, D. L., M. Boublik, and H. R. Kaback. 1983. Structure of the *lac* carrier protein of *Escherichia coli. J. Biol. Chem.* **258**:31.
158. Kyte, J., and R. F. Doolittle. 1982. A simple method for displaying the hydropathic character of a protein. *J. Mol. Biol.* **157**:105.
159. Chou, P. Y., and G. D. Fasman. 1974. Prediction of protein conformation. *Biochemistry* **13**:222.
160. Engleman, D. M., P. Henderson, A. D. McLachlan, and B. A. Wallace. 1980. Path of the polypeptide in bacteriorhodopsin. *Proc. Natl. Acad. Sci. USA* **77**:2023.
161. Henderson, R. and P. N. T. Unwin. 1975. Three-dimensional model of purple membrane obtained by electron microscopy. *Nature (London)* **257**:28.
162. Goldkorn, T., G. Rimon, and H. R. Kaback. 1983. Topology of the *lac* carrier protein in the membrane of *Escherichia coli. Proc. Natl. Acad. Sci. USA* **80**:3322.
163. Seckler, R., J. K. Wright, and P. Overath. 1983. Peptide-specific antibody locates the COOH terminus of the lactose carrier of *Escherichia coli* on the cytoplasmic side of the plasma membrane. *J. Biol. Chem.* **258**:10817.
164. Carrasco, N., D. Herzlinger, S. DeChiara, W. Danho, T. F. Gabriel, and H. R. Kaback. 1984. Topology of the *lac* carrier protein in the membrane of *Escherichia coli. Biophys. J.* **45**:83a.
165. Carrasco, N., D. Herzlinger, R. Mitchell, S. DeChiara, W. Danho, T. F. Gabriel, and H. R. Kaback. 1984. Intramolecular dislocation of the C-terminus of the *lac* carrier protein in reconstituted proteoliposomes. *Proc. Natl. Acad. Sci. USA* **81**:4672.
166. Carrasco, N., P. Viitanen, D. Herzlinger, and H. R. Kaback. 1984. Monoclonal antibodies against the *lac* carrier protein from *Escherichia coli.* I. Functional studies. *Biochemistry* **23**:3681.
167. Herzlinger, D., P. Viitanen, N. Carrasco, and H. R. Kaback. 1984. Monoclonal antibodies against the *lac* carrier protein from *Escherichia coli.* II. Binding studies with membrane vesicles and proteoliposomes reconstituted with purified *lac* carrier protein. *Biochemistry* **23**:3688.
168. Seckler, R., and J. K. Wright. 1984. Sidedness of native membrane vesicles of *Escherichia coli* and orientation of the reconstituted lactose : H^+ carrier. *Eur. J. Biochem.* **142**:269.
169. Herzlinger, D., N. Carrasco, and H. R. Kaback. 1984. Functional and immunochemical characterization of a mutant of *Escherichia coli* energy-uncoupled for lactose transport. *Biochemistry,* in press.
170. Wong, P. T. S., E. R. Kashket, and T. H. Wilson. 1970. Energy-coupling in the lactose transport system of *Escherichia coli. Proc. Natl. Acad. Sci. USA* **65**:63.
171. Wright, J. K., U. Weigel, A. Lustig, H. Bocklage, M. Mieschendahl, B. Müller-Hill, and P. Overath. 1983. Does the lactose carrier of *Escherichia coli* function as a monomer? *FEBS Lett.* **162**:11.
172. Goldkorn, T., G. Rimon, and H. R. Kaback. 1983. Topology of the *lac* carrier protein in the membrane of *Escherichia coli. Proc. Natl. Acad. Sci. USA* **80**:3322.
173. Mieschendahl, M., D. Büchel, H. Bocklage, B. Müller-Hill. 1981. Mutations in the *lac y* gene of *Escherichia coli* define functional organization of lactose permease. *Proc. Natl. Acad. Sci. USA* **78**:7652.
174. Fox, C. F., and E. P. Kennedy. 1965. Specific labeling and partial purification of the M protein, a component of the β-galactoside transport system of *Escherichia coli. Proc. Natl. Acad. Sci. USA* **54**:891.
175. Beyreuther, K., B. Bieseler, R. Ehring, and B. Müller-Hill. 1981. Identification of internal residues of lactose permease of *Escherichia coli* by radiolabel of peptide mixtures. In: *Methods in Protein Sequence Analysis.* M. Elzina, ed. Humana Press, Clifton, N.J. p. 139.
176. Zoller, M. J., and M. Smith. 1983. Oligonucleotide-directed mutagenesis of DNA fragments cloned into M13 vectors. *Methods Enzymol.* **100**:468.
177. Trumble, W. R., P. V. Viitanen, H. K. Sarkar, M. S. Poonian, and H. R. Kaback. 1984. Site-directed mutagenesis of Cys_{148} in the *lac* carrier protein of *Escherichia coli. Biochem. Biophys. Res. Commun.* **119**:860.

CHAPTER 2

Acidification of Intracellular Organelles

Mechanism and Function

Gary Rudnick

1. Introduction

The interior of many intracellular organelles is maintained at a pH lower than that of the cytoplasm. These organelles include a wide variety of secretory granules, such as synaptic vesicles, chromaffin granules, and peptide storage granules, as well as lysosomes and plant vacuoles. This chapter will discuss the evidence that these organelles are acidic, and the ways in which the acidic interior is used by the cell. It also attempts to review the mechanisms which might account for maintenance of a pH difference (ΔpH) across the organelle membrane, including the proposal that a unique ATPase H^+ pump resides in the organelle membrane. It is beyond the scope of this chapter to review H^+ movements across energy-transducing membranes, such as submitochondrial particles and chloroplasts, which also acidify their interiors in a process directly connected with conversion of metabolic or light energy into ATP. In general, we can separate proton movements across intracellular membranes into two categories, those which are coupled to ATP production, as in mitochondria and chloroplasts, and those in which metabolic energy is used to acidify the interior of organelles for a specialized purpose distinct from energy production. The two processes are not mutually exclusive, since the processes of ATP synthesis and hydrolysis are innate properties of both pathways. However, there probably are physical distinctions between the machinery involved in the two systems.

This chapter will focus on the mechanism of organelle acidification, and the function of the resulting ΔpH across the organelle membrane. We will consider two leading pathways for acidification, a diffusion potential for H^+ and an ATP-driven H^+ pump. We will also discuss what is known about the processes which may depend on the ΔpH. These processes include storage of amines, neurotransmitters, and other metabolites, as well as the recycling of plasma membrane components and entry of viruses and other pathological agents. This chapter was completed in December 1982 and reviews the literature up to that date.

Abbreviations used in this chapter: ΔpH, transmembrane pH difference; ΔΨ, transmembrane electrical potential; $\Delta\tilde{\mu}_{H}$, transmembrane electrochemical potential of H^+; FCCP, carbonylcyanide-*p*-trifluoromethoxyphenylhydrazone; CCCP, carbonylcyanide-*m*-chlorophenylhydrazone; DCCD, *N,N'*-dicyclohexylcarbodiimide; FD, fluorescein isothiocyanate-labeled dextran; $TPMP^+$, triphenylmethylphosphonium; ANS, 1-anilino-naphthalene sulfonate; Nbf-Cl, 4-chloro-7-nitrobenzofuran.

Gary Rudnick • Department of Pharmacology, Yale University School of Medicine, New Haven, Connecticut 06510.

2. Evidence for Acid Interior

2.1. Early Indications

The first evidence that cells contained acidic compartments was obtained by Metchnikoff[1] who observed that protozoans change the color of ingested litmus particles from blue to red. Similar studies using other microorganisms stained with a variety of pH-sensitive dyes showed that the pH of the acid compartment (later confirmed to be lysosomes) was approximately 4.5.[2–7] In one of the more recent of these studies, Jensen and Bainton[8] used yeast stained with four dyes to estimate lysosomal pH in human polymorphonuclear neutrophilic leukocytes. They found that within 3 min, the pH of the compartment containing the ingested particles dropped to approximately 6.5, and then slowly drifted down to approximately 4.5 within 7 to 15 min. All of these measurements of pH were based on visual microscopie examination of the cells and represent only a rough estimate of the internal pH.

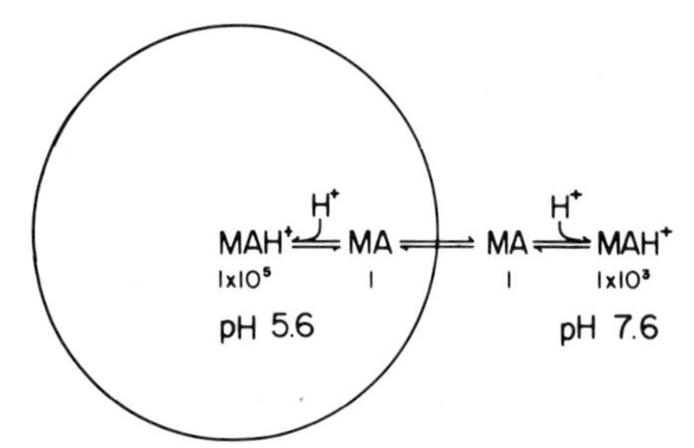

Fig. 1. Accumulation of methylamine in an acidic vesicle. MA, the neutral free base of methylamine; MAH^+, the methylammonium cation. The numbers indicate the concentration of each species, relative to the neutral form.

2.2. Technqiues of Measurement

2.2.1. Methylamine Distribution

2.2.1a. Lysosomes. In 1973, Goldman and Rottenberg[9] and Reijngoud and Tager[10] independently measured the internal pH of isolated lysosomes by monitoring the distribution of radioactive methylamine. This technique, which had been developed for use in chloroplasts by Rottenberg *et al.*,[11] takes advantage of the markedly different rates at which the neutral methylamine free base and the charged methylammonium cation move across lipid bilayers. The neutral form is so much more permeant than the cation that we can regard it as the only permeant species. Consider the situation shown in Fig. 1. A cell or vesicle with an internal pH of 5.6 is bathed in medium of pH 7.6. When a tracer amount of labeled methylamine is added to the medium, its neutral form rapidly equilibrates across the membrane so that the internal and external concentrations are equal. At the same time, the neutral form is in rapid equilibrium with the protonated cation. On each side of the membrane, the equilibrium between the neutral and the charged species depends on the pH. A larger fraction of the internal methylamine will be protonated at the lower internal pH, so the total methylamine concentration will be higher inside than outside.

The extent of methylamine accumulation gives a reasonable estimate of ΔpH. The pK of methylamine is 10.6. At pH 7.6, therefore, the ratio between the neutral and protonated forms is 1 to 1000, while at pH 5.6 it is 1 to 100,000. It we arbitrarily set the concentration of free base at 1 on both sides of the membrane, the ratio of total methylamine concentration will be 100,001/1001 or 99.9, which is a very good estimate of the 100-fold H^+ concentration ratio. The applicability of this method depends on two properties of the weak base. First, the cationic form must be essentially impermeant, so that only the neutral form crosses the membrane. Second, the pK of the base must be high enough that it is essentially all in the cationic form at both the internal and the external pH. When these conditions are fulfilled, the weak base distribution almost precisely follows that of H^+ ions. Even if the base is not almost completely ionized, the technique may still be used, but a correction must be applied to account for the un-ionized fraction.[12] The method can be used with weak acids as well. As expected, these will accumulate in cells or vesicles whose internal pH is higher than that of the medium. In fact, the use of weak acids as ΔpH indicators preceded that of weak bases, and was pioneered by the work of Waddell and Butler.[13]

When Reijngoud and Tager[10] used the methylamine distribution technique to measure the internal pH of lysosomes, they found an acid interior, as expected from earlier measurements with phagocytosed dye particles.[8] Using methylamine, however, the internal pH could be measured more accurately, and was found to be approximately 6 at a medium pH of 7. The ΔpH became smaller as the external pH was lowered, and approached zero at pH 5. Independently, Goldman and Rottenberg[9] found an internal pH approximately 1.6 pH units lower than that of the medium. The measurement has been repeated many times by many workers using both lysosome-rich fractions and purified, Triton WR-1339-loaded lysosomes, and an acid interior is universally observed (see Reijngoud and Tager[14] for a review). Lysosomal vacuoles in plants also accumulate methylamine.[15] Rubber tree sap contains vacuoles which have characteristics of both lysosomes and storage organelles. These vacuoles can be isolated in high purity without the necessity of cell lysis, and from the methylamine distribution, Marin *et al.*[15] calculated an internal pH about 1 unit lower than the medium.

2.2.1b. Secretory Granules. Using the same methylamine distribution technique, Pollard *et al.*[16] and Johnson and Scarpa[17] independently measured the internal pH of isolated adrenal medullary chromaffin granules. Both groups found an acid intragranular pH. Johnson and Scarpa determined that only at an external pH between 5.0 and 5.5 or below was there no accumulation of methylamine. Pollard *et al.*[16] found that only above a medium pH of 6.25 did methylamine concentrate within the granules. Both groups observed that estimates of internal pH were relatively independent of changes in external pH. In addition, Johnson and Scarpa[18] demonstrated dissipation of the ΔpH upon addition of ionophores [either nigericin or a combination of carbonylcyanide-*p*-trifluoromethoxyphenylhydrazone (FCCP) and valinomycin]. Nigericin acts by exchanging one external cation for one internal cation. It exchanges K^+, Na^+, and H^+, but catalyzes net cation movement only very poorly. Consequently, nigericin-induced H^+ efflux required external K^+ and was accompanied by K^+ influx. Valinomycin and FCCP each act differently from nigericin. Both ionophores are relatively specific for the cation transported—valinomycin moves K^+ and FCCP moves H^+. Furthermore, each of these compounds catalyzes net cation (and therefore charge) movement. Together, they act like nigericin in this system since FCCP-catalyzed H^+ efflux down its concentration gradient generates a membrane potential (ΔΨ) (interior negative) which is dissipated by valinomycin-mediated K^+ influx.

Serotonin storage organelles (dense granules) isolated from platelets also accumulate methylamine.[19] At an external pH of 6.85 the granule interior is over 1 pH unit lower, and changes relatively little in response to external pH changes. As with chromaffin granules, the ΔpH is collapsed by the presence of nigericin and external K^+, or by addition of high concentrations of weak bases such as ammonia or serotonin. Wilkins and Salganicoff[20] also measured the internal pH of platelet dense granules using methylamine, and reached the same conclusions, namely, that the granule pH was below 6, and was increased by ammonia or nigericin and K^+ addition.

Neurosecretory granules also maintain acid interiors. Secretory granules containing the peptide hormones oxytocin and vasopressin have been isolated from bovine neurohypophyses, and found to be acidic. Russell and Holz[21] used methylamine accumulation to measure an internal pH of approximately 5.5 which was relatively constant with changing external pH. Addition of either ammonia or nigericin and K^+ increased the granule pH. Scherman and Nordmann,[22] using the methylamine distribution technique, also found the interior of neurohypophyseal granules to be acidic, although not as acidic as Russell and Holz had reported.

2.2.2. Optical Probes

2.2.2a. Fluorescent Polymers. Ohkuma and Poole[23] measured intralysosomal pH with a variation of the dye ingestion technique using dextran derivatized with fluorescein isothiocyanate. This fluorescent macromolecule is ingested by mouse peritoneal macrophages and accumulates in lysosomes. Fluorescein-labeled dextran (FD) has a pH-dependent excitation spectrum, with maximal fluorescence above pH 8, and relatively little fluorescence below pH 4. Between these two extremes, fluorescence is a titratable function of pH, with half-maximal fluorescence around pH 6. Thus, FD fluorescence decreases dramatically as the macromolecule moves from the external medium to the lysosome. Ohkuma and Poole calibrated the pH-dependent fluorescence changes using free FD in solution, and thus were able to quantitate intralysosomal pH in these cells. From their estimates, intralysosomal pH was consistently 4.5 ± 0.1. The technique has recently been extended to measure the internal pH of lysosomes isolated from livers of rats injected intraperitoneally with FD. From the fluorescence spectrum of these isolated lysosomes, Ohkuma *et al.*[71] calculated an intralysosomal pH of approximately 5.

2.2.2b. Fluorescent Weak Bases. Salama *et al.*[25] used fluorescence quenching to measure the internal pH of chromaffin granules. They took advantage of the fact that two fluorescent dyes, 9-aminoacridine and atebrin, accumulate as weak bases inside acid compartments in the same way that methylamine does. When inside the vesicle or granule, the fluorescence of these dyes is considerably less, probably due to concentration quenching.[26] Thus, the greater the ΔpH (acid inside), the greater the dye accumulation, and the less total fluorescence is observed. The two probes both indicate an acid interior which becomes more alkaline on addition of ammonia or nigericin (in the presence of K^+). Salama *et al.*[25] calculated an internal pH of less than 5 using 9-aminoacridine and approximately 5.6 using atebrin. Mast cell granules also accumulate 9-aminoacridine.[27] Under the fluorescence microscope, this accumulation was observed as an increase in fluorescence of granules *in situ*. Addition of ammonia or decreasing external pH diminishes both ΔpH and also granule fluorescence. The internal granule acidity was confirmed using methylamine distribution. Using a similar dye, acridine orange, Dell'Antone[28] measured a ΔpH (acid inside) in isolated lysosomes, which was dissipated by nigericin.

2.2.2c. Endogenous Chromophore. Using the pH-dependent absorbance properties of endogenous anthocyanin, Lin *et al.*[29] estimated the pH of vacuoles in tulip and amaryllis petals. This chromophore is localized exclusively to the vacuoles and changes its absorbance spectrum with pH. By comparing the absorbance spectrum of anthocyanin in intact petals and isolated vacuoles with the spectrum of free anthocyanin at various pH values, they calculated that isolated vacuoles in pH 8 buffer have an internal pH of 7 to 7.3, while vacuoles *in situ* have an internal pH close to 4.

2.3. Direct Measurements

The cell-sap extracts of many plant cells are quite acidic. Since most of the intracellular volume usually consists of vacuoles, this has been taken as indicative of low vacuolar pH[30] These values fall in the range of 5–6[31–33] although some cell saps are as low as pH 1.[34] The marine brown alga *Desmarestia*, for example, has a sap pH of 1–2 and contains high concentrations of SO_4^{2-}, probably as sulfuric acid.[35] The vacuoles stain heavily with brilliant cresyl blue, which turns purple inside the vacuoles, indicating a pH of less than 1 or greater than 7.5. Since the cells release copious amounts of acid in the presence of inhibitors of energy metabolism or 2,4-dinitrophenol, Eppley and Bovell[35] suggested that the purple color was due to an extremely low pH inside the vacuoles.

2.4. Nuclear Magnetic Resonance

The fact that chromaffin granules contain a high concentration of ATP provided Casey *et al.*[36] with an endogenous probe for measuring the internal pH by NMR. The ATP γ-phosphate gives a ^{31}P NMR peak which shifts in frequency as the degree of protonation changes (see Schulman *et al.*[37] for review). Since the pK of this phosphate is approximately 6, its resonance peak reflects pH changes in the pH range expected for the granule interior. Using this γ-phosphate resonance, Ritchie[38] and Njus *et al.*[39] measured an internal pH of 5.5 to 5.7, which compares well with estimates of 5.3 to 5.7 from the same laboratory using methylamine distribution,[36] and with Pollard *et al.*[40] using a similar NMR technique.

3. Generation of ΔpH

How could the interior of a lysosome, vacuole, or secretory granule maintain its acidic pH? The two leading proposals have been a diffusion potential for H^+ and a "pump" which actively transports H^+ into the organelle. Let us consider how each of these mechanisms could account for low intraorganelle pH.

3.1. A Diffusion Potential for H^+

In the case of a diffusion potential, the primary driving force is an impermeant intraorganelle polyanion such as a protein or ATP (Fig. 2). H^+ ions enter the organelle to compensate for the internal negative charges and lower the internal pH. However, although this explanation accounts for the internal pH of a mature organelle, it does not tell us how the organelle was synthesized without a counterion to the protein or ATP, or how any original counterion was removed. One possibility is that a cation, such as Na^+, was present when the organelle was synthesized. If the membrane is sufficiently permeable, Na^+ would leak out of the organelle, carrying positive charge, and then be pumped out of the cell by the plasma membrane Na^+,K^+-ATPase. The resulting membrane potential ($\Delta\Psi$, interior negative) serves as a driving force for H^+ influx across the organelle membrane. The membrane must be selectively permeable to H^+ rather than K^+ (the predominant intracellular cation) if the potential is to result in internal acidification. At equilibrium, the H^+ gradient (in > out) will exactly balance the membrane potential (interior negative) and the electrochemical potential for H^+ will be zero.

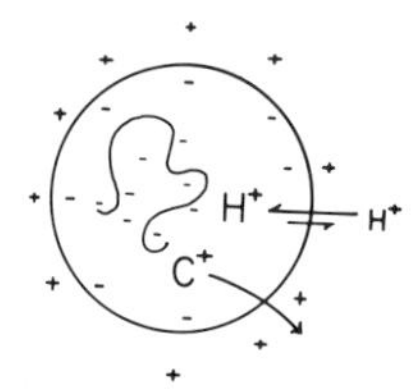

Fig. 2. A diffusion potential for H^+ in a vesicle containing fixed negative charge. H^+ accumulates inside the vesicle in response to the membrane potential (interior negative) generated by efflux of the original counterion, C^+.

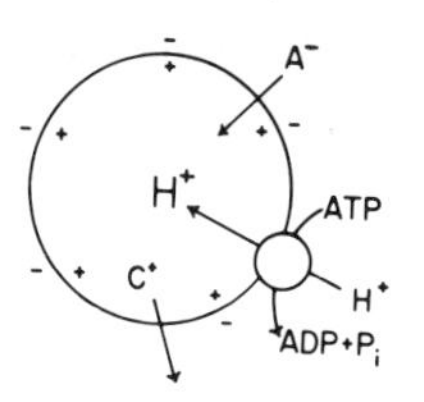

Fig. 3. Acidification of a vesicle by an H^+-pumping ATPase. H^+ influx generates a membrane potential (interior positive) which drives passive anion (A^-) influx and cation (C^+) efflux.

3.1.1. Lysosomes

A diffusion potential for H^+ was invoked by both Goldman and Rottenberg[9] and Reijngoud and Tager[10] to explain the acid interior of isolated lysosomes when they first measured lysosomal pH with methylamine. Both groups[9,41] observed that external K^+ lowered the measured ΔpH, as if K^+ could displace H^+ from the lysosome interior. Goldman and Rottenberg[9] also noted that at any given K^+ concentration, the H^+ gradient was exactly equal to the K^+ gradient across the lysosomal membrane. These results are consistent with a Donnan potential (interior negative) in a membrane with no cation selectivity. Although they account for the acidity of isolated lysosomes, cytoplasmic K^+ would be expected to dissipate the ΔpH. The experiments, however, were performed at 0°C, where the lysosomal membrane is much more permeable to K^+ than it is at 25 or 37°C.[14] Henning[42] observed the same phenomenon with lysosomes incubated at 2°C. A variety of external cations dissipated ΔpH (measured using methylamine) with the following order of effectiveness: $Cs^+ > Rb^+ > K^+ > Na^+ > Li^+ > Mg^{2+} \sim Ca^{2+}$.

3.1.2. Chromaffin Granules

Although the chromaffin granule membrane is relatively permeable to small anions such as Cl^-,[43] the granule contents consist mainly of impermeant anions such as ATP and protein.[44] The predominant cation in isolated granules is H^+, even in the absence of ATP. The driving force keeping H^+ in the granule is apparently the fixed negative charge inside. We might ask why other ions, such as K^+ or Na^+, do not take the place of H^+. Apparently, the granule membrane is relatively permeable to H^+,[45] but not to other small cations.[18] Holz[45] has suggested that the granule membrane acts like an H^+ electrode, so that the ΔpH is always exactly balanced by an equal and opposite ΔΨ.

Measurements of ΔΨ in isolated chromaffin granules have used the lipophilic ions SCN^- and triphenylmethylphosphonium ($TPMP^+$). These ions distribute across the membrane in response to ΔΨ, and from their accumulation, one can estimate ΔΨ. Chromaffin granules incubated at neutral pH excluded SCN^- from their interior, indicating a potential which is negative inside.[16,46] Holz[45] made more accurate measurements of ΔΨ using $TPMP^+$, whose distribution is easier to measure since it accumulates in vesicles whose interiors are negative. He measured a potential, of −80 mV[45] which decreased if the external pH was lowered to dissipate ΔpH. At the same time, SCN^- is excluded less strongly,[16] indicating that ΔpH and ΔΨ are in equilibrium.

3.2. H⁺ Pump

If an H^+ pump acidifies the organelle, the primary driving force for acidification is the energy source for the pump, commonly believed to be ATP. By a mechanism which is not well understood, the free energy of ATP hydrolysis is converted to electrogenic movement of H^+ from the cytoplasm to the lumen. This movement generates an electrochemical H^+ gradient ($\Delta\tilde{\mu}_H$) consisting of two components, ΔΨ (interior positive) and a ΔpH (interior acid) (Fig. 3). Initially, ΔΨ will predominate (since translocation of relatively few H^+ ions creates a sizable potential), but as anions enter or cations leave by leak pathways, more H^+ ions are pumped and ΔpH becomes paramount. If organelles are isolated in the absence of ATP, decay of ΔpH will depend on the membrane permeability to H^+ as well as other ions. If H^+ ions leak more rapidly than other ions, they will reach an equilibrium where the H^+ gradient (in > out) will exactly balance ΔΨ and $\Delta\tilde{\mu}_H$ will be zero, just as in the case described above for an H^+ diffusion potential.

Since both mechanisms of organelle acidification can result in the same H^+ equilibrium, it is obviously difficult to distinguish between the two in isolated organelles. Only in the presence of ATP would we expect the two mechanisms to show a difference in $\Delta\tilde{\mu}_H$, and that difference would be manifested primarily in ΔΨ, which would reverse from negative inside in the absence of ATP to positive inside in the presence of ATP. As we shall see, the evidence indicates that ATP-dependent acidification predominates intracellularly, even though the ΔpH of isolated organelles in the absence of ATP is primarily due to a diffusion potential. For a review of the evidence favoring an electrogenic ATPase as the driving force for lysosomal acidification, see Reeves.[47]

3.2.1. Chromaffin Granules

Because chromaffin granules were known to accumulate catecholamines in the presence of ATP,[48,49] investigators searching for the mechanism of acidification naturally examined the ATPase activity of the granule membrane. Phillips[50,51] demonstrated that membrane vesicles or "ghosts" derived from chromaffin granules by osmotic lysis also used ATP as an energy source for catecholamine transport. Most of the evidence now suggests that this activity represents an electrogenic, ATP-driven H^+ pump (for a review, see Njus *et al.*[52]). Bashford *et al.*[53] first measured generation of ΔΨ across the granule membrane with ANS, a probe whose fluorescence was known to increase when ΔΨ (positive inside) was generated by submitochondrial particles.[54] In the presence of ATP, chromaffin granules also enhanced ANS fluorescence (suggesting generation of ΔΨ) and inhibitors of the ATPase, such as *N*-ethylmaleimide, blocked the increase. Addition of S-13, another H^+ ionophore like FCCP, also prevented the increase by allowing H^+ ions to leak out of the granule.

Later, these results were confirmed by Pollard *et al.*,[16] Holz,[46] and by Johnson and Scarpa[55] who used the distribution of radioactive SCN^- to measure potential. Measurements with SCN^- indicate that during ATP hydrolysis, the granule lumen becomes approximately 50 mV more positive than the external medium.[16,46,55] Using $TPMP^+$, both Holz[45] and Johnson and Scarpa[55] determined that in the absence of ATP, the resting potential across the chromaffin granule membrane is 0 to −80 mV (interior negative). Thus, ATP increases $\Delta\Psi$ by 50 to 120 mV, indicating that some ion is pumped across the membrane. Either an anion is pumped out or a cation is pumped in.

At the same time, ΔpH increases slightly, if at all.[36,46] In the absence of ATP, the granule pH is approximately 5.5, and ΔpH almost exactly balances $\Delta\Psi$.[45] H^+ ions are at their electrochemical equilibrium. Reversal of $\Delta\Psi$ by the ATPase imposes a strong driving force on H^+. If any ion other than H^+ were pumped, $\Delta\Psi$ and ΔpH would act together to push H^+ out of the granule and dissipate ΔpH. The fact that ΔpH does not decrease, and actually may increase, indicates that H^+ is actively transported into the granule.

ATP has also been implicated in release of amines from chromaffin granules. Poisner and Trifaro[56] and Hoffman *et al.*[57] first observed ATP-dependent catecholamine release and suggested that ATP also acted as a trigger for secretion. While it is likely that energy derived from ATP hydrolysis is required for exocytosis, ATP-induced release from isolated granules probably reflects swelling induced by passive anion flux. Chromaffin granules are osmotically fragile, and if they swell past a critical point, they lyse, releasing catecholamines into the medium. Similar lysis was first observed in the absence of ATP by Dolais-Kitabgi and Perlman,[58] who used valinomycin to dramatically increase the granule K^+ permeability. As K^+ enters, it carries positive charge and generates a $\Delta\Psi$ (interior positive) across the granule membrane. Influx of very few ions is needed before the resulting potential is large enough to stop further K^+ entry. The potential, however, acts as a driving force pushing in external anions. Dolais-Kitabgi and Perlman found that if K^+ was added as the salt of an anion to which the granule membrane was permeable, swelling and lysis resulted, with concomitant release of catecholamines. The resulting order of permeability was SCN^-, I^-, $Br^- > Cl^- >$ acetate$^-$, F^-, isethionate$^-$. Diluting valinomycin-treated granules into isosmotic KCl is like diluting them into distilled water, since the salt equilibrates across the membrane almost as fast as water does.

Just as a K^+ diffusion potential drives anion influx, the $\Delta\Psi$ (interior positive) induced by ATP hydrolysis also leads to anion accumulation, swelling, and lysis. Casey *et al.*[43] demonstrated that external anions stimulate ATP-induced catecholamine release with a rank order similar to the one found by Dolais-Kitabgi and Perlman.[58] Instead of KCl influx driven by a concentration gradient, now HCl influx is driven by ATP hydrolysis. Since the primary function of the ATPase is to pump H^+ into the granule, addition of H^+ ionophores (uncouplers) prevents lysis by letting protons leak out of the vesicle, thereby preventing generation of $\Delta\Psi$. Casey *et al.*[43] demonstrated that catecholamine release was indeed due to lysis by measuring protein release and loss of turbidity which paralleled catecholamine release. Increasing the medium osmolarity with sucrose inhibited in parallel the loss of turbidity, and protein and catecholamine release. Thus, lysis was due to osmotic swelling. If lysis is inhibited by increased medium osmolarity or limiting anion concentrations, ATP significantly acidifies the granule interior. Casey *et al.*[36] measured this acidification using both methylamine distribution and ^{31}P NMR. The ATP-dependent drop in internal pH requires a permeant anion like Cl^-, confirming the proposal that ATP hydrolysis leads to net HCl influx.

The most convincing evidence for an electrogenic chromaffin granule H^+-ATPase comes from studies using resealed granule membrane vesicles or "ghosts." Johnson *et al.*[59] demonstrated that vesicles substantially free of content became acid and positive inside in the presence of ATP. FCCP blocked the changes, which were monitored by accumulation of methylamine and SCN^-. In the absence of a permeant anion, ATP addition generates a $\Delta\Psi$ (interior positive) across the vesicle membrane, without much acidification of the interior. Influx of relatively few H^+ ions is sufficient to create a potential which prevents further net H^+ pumping. As Cl^- is added to the medium, ΔpH grows at the expense of $\Delta\Psi$. At high Cl^- concentrations, each H^+ pumped into the vesicle is accompanied by a counterion, and no $\Delta\Psi$ is generated. In the absence of a potential, however, many more H^+ ions can be pumped in, and a large ΔpH develops. At intermediate Cl^- concentrations, both $\Delta\Psi$ and ΔpH are found.

3.2.2. Platelet Dense Granules

As early as 1972, Heinrich *et al.*[60] found that platelet granules contain ATPase activity. It was not until recently that Wilkins and Salganicoff[20] demonstrated that the ATPase caused acidification of granule membrane vesicles. In the presence of ATP, they found the vesicles to be approximately 1.3 pH units lower than the external medium (7.0). Both NH_4Cl and nigericin partially collapsed the ΔpH, which was generated only in vesicles incubated at 37°C, not at 0°C. Although Wilkins and Salganicoff did not measure $\Delta\Psi$, Carty *et al.*[61] used SCN^- and methylamine to estimate $\Delta\Psi$ and ΔpH in intact platelet granules. They found that ATP addition generated a $\Delta\Psi$ of approximately 25 mV (interior positive) but had little effect on ΔpH. Since these measurements were made in the absence of permeant cations, we would expect that, by analogy with chromaffin granules, ATP-driven H^+ influx would increase $\Delta\Psi$ but not ΔpH. Our laboratory[62] also measured both $\Delta\Psi$ and ΔpH, using a preparation of platelet granule membrane vesicles. We found that ATP acidified the vesicle interior only in the presence of a permeant anion such as Cl^-, where no $\Delta\Psi$ developed. In the absence of Cl^-, ATP hydrolysis generated a $\Delta\Psi$ (interior positive) but not ΔpH. This is entirely consistent with an electrogenic H^+ pump generating a potential which drives Cl^- entry, leading to net HCl influx only in the presence of Cl^-. Thus, H^+ transport across both the chromaffin granule and platelet granule membrane seems to be the result of the same electrogenic H^+-pumping ATPase.

3.2.3. Neurohypophyseal Granules

Russell and Holz[21] measured the effect of ATP-Mg^{2+} on oxytocin- and vasopressin-containing granules isolated from the neural lobe of bovine pituitaries (neurohypophysis). The resting potential (measured with SCN^-) of these acidic granules is negative inside, presumably due to an H^+ diffusion potential. Consequently, FCCP fails to alter the resting $\Delta\Psi$ or ΔpH. ATP reverses $\Delta\Psi$ to over 50 mV (interior positive) in medium free of permeant anions, and this reversal is prevented by FCCP. In the absence of permeant anions, the resulting $\Delta\tilde{\mu}_H$ limits the net influx of H^+, and ΔpH (measured with methylamine) remains unchanged.

Scherman *et al.*[63] also measured ATP-dependent H^+ influx in bovine neurohypophyseal granules. ATP induced an acidification of almost 0.5 pH unit and a $\Delta\Psi$ of up to 15 mV. $\Delta\Psi$ was measured both by SCN^- accumulation and using a fluorescent dye, bis(3-phenyl-5-oxoisoxazol-4-yl)pentamethine oxonol (OX-5). This lipophilic anion accumulates inside vesicles whose interior is positive with respect to the medium. By measuring the fluorescence quenching which accompanies uptake, Scherman *et al.* followed ATP-generated $\Delta\Psi$. Permeant anions such as SCN^- and Cl^- have opposite effects on ATP-induced $\Delta\Psi$ and ΔpH. $\Delta\Psi$ is greatest in sucrose medium but is collapsed by SCN^- and, at higher concentrations, by Cl^-. These anions presumably follow H^+, flowing down the $\Delta\Psi$ generated by H^+ influx, and dissipating the potential. As a consequence, more net H^+ is pumped by the ATPase, and ΔpH increases.

3.2.4. Lysosomes

The first evidence that lysosomes possess an H^+ pump came from studies with the so-called "lysosomotropic agents"[64] such as chloroquine. These compounds are all weak bases which accumulate inside lysosomes in response to ΔpH in the same way that methylamine accumulates in acid vesicles. Wibo and Poole[65] determined that accumulation of chloroquine by rat fibroblasts represented uptake into lysosomes, and that intralysosomal chloroquine reached high concentrations. They calculated that the amount of chloroquine taken up far exceeded the buffering capacity of the lysosome, and concluded that some active process in addition to a diffusion potential was responsible for the accumulation.

Ohkuma and Poole[23] extended these studies by measuring intralysosomal pH (using FD) in the presence and absence of energy metabolism inhibitors and lysosomotropic agents. They found that the intralysosomal pH of 4.5 in resting macrophages was increased by ammonium chloride, methylamine, tributylamine, and amantidine, all weak bases which would be expected to accumulate within lysosomes. ΔpH was also collapsed by inhibitors of energy metabolism. When both 2-deoxyglucose and azide were added to the medium, lysosomal pH increased, but neither inhibitor alone dissipated ΔpH. The results suggest that the energy source for acidification is ATP produced by glycolysis and oxidative phosphorylation.

When lysosomes accumulate some weak bases, they also swell to form large intracellular vacuoles.[24] Poole and Ohkuma[66] compared the vacuolization and accumulation to the increased lysosomal pH. They found a burst of methylamine accumulation within the first few minutes, as the lysosomal pH rose from 4.6 to 6.5, followed by continued accumulation, with concomitant osmotic lysosomal swelling but no further pH change. Again they concluded that continued accumulation after rapid establishment of the equilibrium represented an active H^+ transport process.

Schneider[67,68] first measured ATPase activity of rat liver lysosomes and suggested that the active site of the ATPase resided on the external (cytoplasmic) face of the lysosomal membrane. Using the methylamine distribution technique, he found that ATP addition decreased lysosomal pH to 5.5.[68] From these observations, he concluded that ATP hydrolysis was coupled to H^+ pumping into the lysosome. Consistent with this proposal is the fact that lysosomal ATPase activity is increased by agents such as CCCP, an H^+ ionophore like FCCP. Any H^+ ions pumped into the lysosome will readily leak out in the presence of CCCP and the pump will not be slowed down by a buildup of internal positive charge and acidity. Another interesting observation was that ANS fluorescence was enhanced by the presence of ATP.[68] In chromaffin granules, a similar enhancement was taken as evidence that ATP generated a $\Delta\Psi$ (interior positive) as expected for electrogenic proton influx.[69] In the presence of Cl^-, little $\Delta\Psi$ was observed,[70] but ATP addition generated a ΔpH of over 1 unit. Both ATPase activity and ATP-dependent ΔpH generation are sensitive to inhibitors of erythrocyte anion transport. This phenomenon may indicate that continued H^+ pumping requires dissipation of $\Delta\Psi$ by mediated anion influx.

Ohkuma *et al.*[71] also measured ATP-dependent acidification of isolated lysosomes. Liver lysosomes isolated from FD-injected rats contain endocytosed FD which reports the intralysosomal pH by virtue of its pH-dependent fluorescence spectrum. The fluorescence of freshly isolated crude lysosomal fraction or purified Triton WR 1339-containing lysosomes is quenched, presumably by an acidic internal pH. This suggestion was confirmed by the finding that addition of nigericin or valinomycin + FCCP increases the fluorescence of lysosomes suspended in KCl medium, but either FCCP or valinomycin alone has little effect. The ionophores catalyze an exchange of internal H^+ for external K^+ which alkalinizes the lysosome interior. Disrupting the lysosomal membrane with detergent also increases fluorescence as FD is exposed to the external medium at pH 7.0.

Upon addition of ATP to FD-containing lysosomes, the fluorescence decreases further, suggesting acidification of the lysosome interior by approximately 0.5 unit. The reaction requires Mg^{2+} and involves hydrolysis of ATP, since nonhydrolyzable analogs fail to replace ATP, and acidification is enhanced by an ATP-regenerating system.[71] Although FCCP has little effect on the resting pH, this H^+ ionophore rapidly reverses the ATP-dependent acidification. We can easily understand this apparent discrepancy by considering the driving forces for ΔpH in the two situations. Before addition of ATP, the ΔpH results predominantly from a diffusion potential for H^+, as postulated by Reijngoud and Tager.[14] Since H^+ is close to electrochemical equilibrium, no large H^+ fluxes result from a further increase in the membrane's H^+ conductance. ATP-driven H^+ influx, however, creates a $\Delta\tilde{\mu}_H$ (interior positive and acid) which tends to push H^+ back out. Addition of FCCP increases the H^+ conductance of the membrane to a point where H^+ leaks out faster than the ATPase can pump it in, and the internal pH returns to its resting value. Permeant anions such as SCN^- and Cl^- enhance ATP-dependent lysosome acidification, presumably by dissipating $\Delta\Psi$ and allowing more net H^+ pumping.[71]

3.2.5. Plant Vacuoles

Intracellular vacuoles isolated from a variety of plants also contain an ATPase activity. ATP induced acidification of yeast vacuole membrane vesicles by two criteria: The fluorescence of 9-aminoacridine[72] and of quinicrine, another weakly basic dye,[73] was quenched by ATP and the quenching was reversed by nigericin or SF6847, an H^+ ionophore. Methylamine was also accumulated in the vesicles in response to ATP. At the same time, SCN^- accumulation demonstrated generation of a $\Delta\Psi$ (interior positive) which was also dissipated by H^+ ionophores.[73]

Marin *et al.*[15] used distribution of methylamine and TPP^+ to measure ΔpH and $\Delta\Psi$ in vacuoles isolated from rubber tree sap. ATP lowered intravacuolar pH by 0.5 unit and made the vacuole interior more positive by 60 mV, reversing a preexisting $\Delta\Psi$ from −120 mV to −60 mV. These changes were interpreted as, and are consistent with, an ATP-driven H^+ pump in the vacuole membrane. Cretin[74] confirmed that the vacuole contents are acidic and noted that vesicles labeled with $^{86}Rb^+$ lost Rb^+ as ATP acidified the interior. Moreover, efflux was stimulated by the K^+ ionophore valinomycin. This observation supports the proposal that the ATPase pumps H^+ electrogenically, creating a $\Delta\Psi$ (interior positive) which, in this case, leads to Rb^+ efflux.

4. Uses of ΔpH

4.1. Lysosomal Protein Degradation

Many lysosomal hydrolytic enzymes have acid pH optima. This led Coffey and de Duve[75] to propose that the lysosome actively maintained an acidic internal pH, and led Barrett[76] to prophetically state that "we may well expect that an H^+ pumping ATPase will be detected in these organelles." Indeed, the rate of protein hydrolysis in lysosomes reflects internal pH, with an increase in rate at lower pH. Reijngoud *et al.*[77] compared the intralysosomal pH measured by methylamine distribution with the pH estimated from the hydrolysis rate of ^{125}I-labeled bovine serum albumin which had been endocytosed prior to lysosome isolation. By comparing the hydrolysis rate with that of disrupted lysosomes at various pH values, they calculated an internal pH in good agreement with the methylamine distribution. Mego *et al.*[78] found that addition of ATP increased the rate of protein hydrolysis. This acceleration was blocked by nigericin and 2,4-dinitrophenol (another H^+ ionophore),[79] suggesting that ATP hydrolysis led to acidification of the lysosomal contents. It is significant that nigericin inhibited hydrolysis even in the absence of ATP (presumably by exchanging external cations for internal H^+), while dinitrophenol inhibited only the acceleration due to ATP. This situation is similar to the one described above for isolated lysosomes.[71]

One potential model system for lysosomal protein degradation is the hydrolysis of amino acid esters. Reeves[80] observed that leucine methyl ester accumulated within lysosomes as free leucine, which leaked out very slowly. He concluded that the neutral, free base form of the methyl ester crossed the membrane by nonionic diffusion, and was hydrolyzed by internal proteases. Accumulation was blocked by protease inhibitors and by lowering external pH to decrease the amount of leucine methyl ester free base. Addition of Mg^{2+}-ATP causes the rate of leucine accumulation to increase up to a maximum of 1.4- to 1.6-fold greater than control rates. Reeves and Reames[81] proposed that the increase is due to the ATP-driven H^+ pump on the lysosomal membrane. By acidifying the lysosomal contents, ATP accelerates ester hydrolysis either by increasing steady-state methyl ester concentration inside the lysosome (like methylamine, leucine methyl ester is a weak base which will distribute according to ΔpH) or by increasing proteolytic enzyme activity. As expected for an electrogenic H^+ pump, the ATP-induced acceleration is blocked by CCCP as well as agents (nigericin, chloroquine) which raise intralysosomal pH.

4.2. Receptor-Mediated Endocytosis

Many nutrients, hormones, enzymes, and other macromolecules enter animal cells by binding to receptors on the plasma membrane which are subsequently internalized by a process of endocytosis.[82] Although the ligands are delivered with high efficiency to an intracellular location (usually the lysosome), the receptors often return to the cell surface free of ligand, and participate in many cycles of endocytosis and return during their lifetime.[83] The recycling process requires release of internalized ligand prior to reutilization of the receptor, and recent studies suggest that acidification of an intracellular compartment triggers receptor–ligand dissociation.

Of the ligands studied, the majority bind to their receptor best at neutral pH, and dissociate rapidly at pH 5 or below. Examples include insulin,[84] β-galactosidase,[85] asialoglycoproteins,[86] transferrin,[87] β-glucuronidase,[88] and mannosylated bovine serum albumin.[89] Addition of lysosomotropic agents capable of increasing the internal pH of acidic intracellular organelles inhibits ligand accumulation, presumably by preventing intracellular ligand release. Since many of the receptors themselves are rapidly inactivated by intralysosomal enzymes, another acidic compartment between the cell surface and the lysosome is likely to serve as the site of ligand–receptor dissociation.

Tycko and Maxfield[90] recently provided evidence that the primary endocytic vesicle (endosome) contents are acidified rapidly following their separation from the extracellular medium. Using fluorescein-labeled α_2-macroglobulin (F-α_2M), they showed that within 15 min of exposing fibroblasts to F-α_2M, the fluorescein label is in a compartment of pH 5. If the cells were kept at 4°C, where endocytosis is blocked, the fluorescence reflected the medium pH of 7.4. If cells which had endocytosed F-α_2M were fixed in formaldehyde and incubated with chloroquine, the probe also indicated a neutral pH. To determine if the low-pH compartment was lysosomal, Tycko and Maxfield compared the localization of labeled α_2M with that of acid phosphatase, a lysosomal marker, and found no overlap between the fluorescent vesicles and lysosomes. Furthermore, no proteolytic degradation of the fluorescent macromolecule is observed during the time required for acidification of F-α_2M. Similar results were obtained by Geisow *et al.*[91] using fluorescein-labeled yeast cells phagocytosed by macrophages. In this case, a transient alkalinization preceded acidification of the phagosome. Inhibitors of phagosome–lysosome fusion failed to prevent acidification, again suggesting that it occurred in a prelysosomal compartment.

Further studies in our own laboratory[92] indicate that endosomes are acidified by the same ATP-dependent process responsible for lysosomal acidification. Macrophages and fibroblasts exposed to FD for brief periods contain the label exclusively in a low-density vesicular fraction (endosomes) from which label is chased, with time, into lysosomes. The interior of both FD-containing endosomes and lysosomes becomes acidic in the presence of ATP and a permeant anion such as Cl^-. In both organelles, ΔpH is dissipated by addition of FCCP or nigericin (in the presence of external K^+) or by NH_4Cl. These results suggest that the lysosomal ATPase is present in endosomal vesicles at very early times.

In addition to physiological ligands whose intracellular accumulation depends on endosomal acidification, a variety of pathogens require an acidic internal compartment for their activity. These include a variety of envelope viruses such as Semliki forest virus and polypeptide endotoxins like diptheria toxin. In both cases, infection or intoxication is blocked either by lysosomotropic weak bases or by ionophores such as monensin and nigericin which are known to increase lysosomal pH. Moreover, virus or toxin bound to the cell surface can act if the medium is briefly acidified, even in the presence of monensin or a weak base.[93–95]

4.3. Urinary Acidification

Urinary acidification in the turtle bladder is thought to represent the same process by which mammalian urine is acidified in the renal collecting tubule. In the turtle bladder, specialized cells containing acidic organelles pump H^+ from the cell to the bladder lumen in response to stimulation by CO_2[96] When these organelles are labeled with FD by endocytosis, the probe indicates an internal pH of 5.2. Concomitant with the increase in H^+ pumping, previously accumulated FD is released by a process of exocytosis, which is responsible for inserting the H^+ pumps into the plasma membrane. Fractionation of the cells yields vesicles containing an H^+-ATPase and endocytosed ligands.[97] The vesicles separate on density gradients from both mitochondria and lysosomes, suggesting a specialized organelle whose function is to reversibly insert H^+ pumps into the plasma membrane when the cell is stimulated.

4.4. Biogenic Amine Storage

4.4.1. Chromaffin and Platelet Granules

In retrospect, the observation that ATP generates a significant $\Delta\tilde{\mu}_H$ across the chromaffin granule membrane suggests an obvious source of energy for concentrative transport of catecholamines. Unfortunately, little was known about the bioenergetics of chromaffin granules when amine transport was first studied by Falck *et al.*[98] and Blaschko *et al.*[99] It was not until 6 years later that Kirshner[48] and Carlsson *et al.*,[49] demonstrated ATP-driven catecholamine uptake, and not until the past 5 years that the coupling between ATP hydrolysis and biogenic amine transport has been understood in detail.

As described above, the chromaffin granule ATPase is an H^+ pump which creates both $\Delta\Psi$ (interior positive) and ΔpH (interior acid) across the granule membrane by electrogenic H^+ influx. We now know that both components of $\Delta\tilde{\mu}_H$ contribute to biogenic amine accumulation. The evidence for chemiosmotic coupling between ATP hydrolysis and amine transport consists of three major observations: (1) ATP hydrolysis generates both components ($\Delta\Psi$ and ΔpH) of the electrochemical H^+ gradient; (2) artificial imposition of $\Delta\Psi$, ΔpH, or both drives biogenic amine accumulation; and (3) agents which dissipate $\Delta\Psi$ or ΔpH uncouple the ATPase from amine transport.

Generation of ΔpH and $\Delta\Psi$ by the chromaffin granule ATPase was described above (Section 3). By many criteria, ATP hydrolysis pumps H^+ into intact chromaffin granules and granule-derived membrane vesicles. In the absence of permeant anions in the medium, H^+ influx generates $\Delta\Psi$ (interior positive) which limits further net H^+ pumping. Permeant anions in the medium follow H^+ into the granule, dissipating the developing $\Delta\Psi$. This in turn allows more net H^+ pumping, with the result that a significant ΔpH develops.

The first evidence that artificially imposed ΔpH could drive biogenic amine accumulation came in 1978 when Schuldiner *et al.*,[100] Phillips,[101] and Johnson *et al.*[102] demonstrated that a pH jump drives biogenic amine transport into membrane vesicles and intact granules. In the case of intact granules, raising the external pH with base increased the rate at which dopamine and norepinephrine entered the granule, but the maximal level of accumulation was less than that observed under control conditions.[102] Moreover, the possibility of exchange with endogenous granule catecholamine was not excluded. Membrane vesicles, in contrast, have little or no endogenous energy sources for transport, and do not accumulate biogenic amines in the absence of ATP. If the vesicles are equilibrated at pH 6.5, however, and then diluted into media of pH 8.5, substrate rapidly accumulates and then (as the pH gradient decays) exits the vesicles.[100,101] The same pH jump fails to drive transport in the presence of reserpine, an inhibitor of the biogenic amine transporter. Thus, ΔpH energized catecholamine accumulation in the absence of other driving forces.

At about the same time, evidence was accumulating that ΔpH was not the only force driving catecholamine transport. Casey *et al.*[36] noted that in the absence of permeant anions, ATP hydrolysis failed to generate a ΔpH measurable by either ^{31}P NMR or methylamine distribution, but increased the rate and extent of norepinephrine accumulation. They suggested that amine transport may be coupled to the $\Delta\Psi$ generated by the ATPase. Holz[46] confirmed this finding by simultaneously measuring ATP-dependent changes in ΔpH, $\Delta\Psi$, and catecholamine distribution. He found that in the absence of a permeant anion, $\Delta\Psi$ and catecholamine accumulation increased in parallel, but ΔpH was unchanged. Moreover, FCCP dissipated the catecholamine gradient and $\Delta\Psi$ to the same extent. Holz concluded that amine transport across the chromaffin granule membrane was coupled to $\Delta\Psi$, and that positive charge must leave the granule as each catecholamine molecule enters.

Njus and Radda[103] demonstrated that in chromaffin granule membrane vesicles, imposition of a $\Delta\Psi$ (interior positive) by a K^+ gradient (out > in) and valinomycin transiently increases both epinephrine accumulation (which was blocked by reserpine) and ANS fluorescence. Kanner *et al.*[104] demonstrated that serotonin transport driven by a ΔpH (interior acid) across the vesicle membrane is further stimulated by imposition of $\Delta\Psi$ (interior positive) and inhibited by potentials of the opposite polarity. Both groups concluded that biogenic amine accumulation is coupled to $\Delta\Psi$ as well as ΔpH. Johnson *et al.*[59] reached similar conclusions by energizing chromaffin granule membrane vesicles with ATP, and varying the contribution of $\Delta\Psi$ and ΔpH by changing the Cl^- concentration.

For biogenic amine transport to be coupled to both $\Delta\Psi$ (interior positive) and ΔpH (interior acid), H^+ ions must leave the granule for each amine molecule which enters (Fig. 4). The transporter must exchange a neutral amine molecule with at least one H^+, or a protonated, cationic amine molecule with at least two H^+ ions. Either stoichiometry will lead to H^+ and positive charge movement in the direction opposite to biogenic amine movement. The exact number of H^+ which exchange for each amine molecule has been estimated from the response of catecholamine gradients to changes in $\Delta\Psi$ and ΔpH. Johnson *et al.*[105] and Knoth *et al.*[106] both found that for each amine molecule entering the membrane vesicle, two H^+ and one net

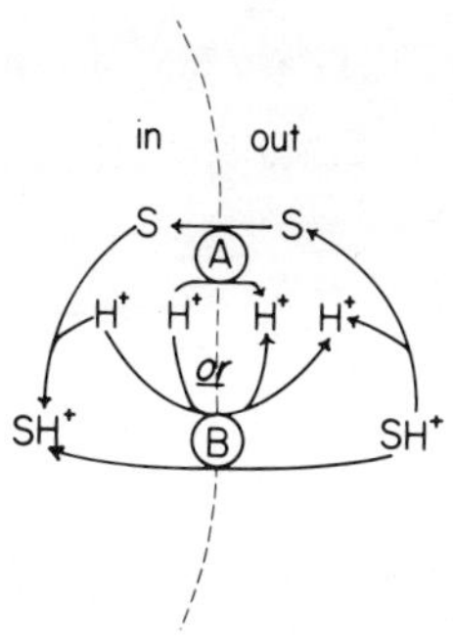

Fig. 4. Two mechanisms for accumulation of protonated biogenic amine (SH^+) into secretory granules. Either the neutral form is exchanged for H^+ (A), or the cationic form is exchanged for two H^+ ions (B). The stoichiometry is the same by either pathway.

positive charge appear in the medium. This is consistent with exchange of one H^+ with one neutral amine or exchange of two H^+ for one protonated cation. In membrane vesicles prepared from platelet dense granules, Fishkes and Rudnick[62] also found coupling of serotonin transport to both $\Delta\Psi$ and ΔpH. They obtained a stoichiometry of two to three H^+ exchanging for each protonated serotonin cation.

4.4.2 Synaptic Vesicles

Biogenic amines stored in synaptic vesicles are apparently also accumulated by the same mechanism. Norepinephrine is accumulated by preparations of rat brain synaptic vesicles in the presence of ATP.[107] The process is blocked by reserpine but not desipramine, an inhibitor of the plasma membrane transport system for norepinephrine. Ionophores expected to dissipate transmembrane pH gradients, such as nigericin and S-13, inhibit transport and cause previously accumulated norepinephrine to leak out rapidly. Maron *et al.*[108] found similar results for serotonin transport. Rat brain synaptic vesicles take up serotonin from the medium, the process is dramatically stimulated by ATP, sensitive to reserpine, and reversed by nigericin. In contrast, synaptosomal serotonin transport requires Na^+ and is blocked by imipramine, an inhibitor of serotonin movement across plasma membrane. Nerve endings apparently possess two amine transport systems, one at the level of the synaptic vesicle membrane which requires ATP for generation of ΔpH, and one at the plasma membrane which is required for amine entry into the cell or synaptosome. The existence of two serotonin transport systems (plasma membrane and granule) in the same cell has recently been established in isolated blood platelets.[109]

4.5. Acetylcholine Storage

Acetylcholine accumulation and storage within synaptic vesicles may also be coupled to ATP-dependent H^+ influx. Carpenter and Parsons[110] first described accumulation of both choline and acetylcholine within synaptic vesicles isolated from the electric organ of *Torpedo californica*. To energize acetylcholine influx, vesicles containing high concentrations of alkali cations were diluted into low-Na^+ medium in the presence of gramicidin. The vesicles develop a $\Delta\Psi$ (interior negative) which is in the opposite direction to the $\Delta\Psi$ generated by ATP in chromaffin granules. Gramicidin, however, will also exchange internal Na^+ or K^+ for external H^+, and thus develop a ΔpH (acid inside). Carpenter and Parsons considered only $\Delta\Psi$ as a driving force for accumulation, and suggested an electrogenic transport mechanism where the choline and acetylcholine cations were driven into the vesicle by $\Delta\Psi$.

Michaelson *et al.*[111] examined conditions under which endogenous acetylcholine would leave the vesicles and concluded that both $\Delta\Psi$ and ΔpH were coupled to acetylcholine accumulation. Vesicles treated with nigericin rapidly lost acetylcholine in a process dependent on extravesicular K^+, suggesting that ΔpH (interior acid) provided some driving force for acetylcholine storage. Rebois *et al.*[112] failed to observe nigericin-induced efflux from acetylcholine storage vesicles in PC-12, a clonal pheochromocytoma cell line, but nigericin markedly inhibited incorporation of labeled choline into vesicular acetylcholine by intact cells. They did not distinguish between inhibition of the transport process and impairment of vesicle synthesis, but demonstrated that nigericin had no effect on the cytoplasmic synthesis of acetylcholine from choline.[113]

In *Torpedo* synaptic vesicles, ATP also drives influx of acetylcholine. Parsons and Koenigsberger[114] found that in the presence of bicarbonate, ATP increased the acetylcholine gradient across the vesicle membrane threefold. ATP-dependent accumulation is blocked by H^+ ionophores, such as S-13, at the same concentration at which they increase ATP hydrolysis.[115] This observation is consistent with an ATPase which generates a $\Delta\Psi$ or ΔpH or both across the membrane. The $\Delta\Psi$ or ΔpH acts both to limit the rate of ATP hydrolysis and to drive acetylcholine transport. Furthermore, Anderson *et al.*[116] have shown that both nigericin and ammonia inhibit ATP-dependent acetylcholine accumulation. Although neither agent dissipates $\Delta\Psi$, both dissipate ΔpH.

4.6. Vacuolar Solute Storage

Cytoplasmic vacuoles from many plant cells serve as storage depots for a variety of metabolites.[117,118] These substances include sugars, basic amino acids, and divalent cations. Although the mechanism of transport and storage remains uncertain, the overall process is similar in two respects to biogenic amine transport into secretory granules. Specific transport systems appear to mediate solute transport into the vacuole and the driving force seems to involve ATP-dependent H^+ transport.

Evidence for a specific transporter in the vacuolar membrane is strongest for sugar accumulation by vacuoles isolated from red beet cells. Leigh *et al.*[119] determined that of the sucrose found in beet root cells, at least 95% was localized in the vacuole. Isolated vacuoles will accumulate added sucrose, partly by exchange with internal sucrose. The rate at which exogenously added sucrose is transported saturates with the sucrose concentration, and other sugars, such as raffinose, glucose, and fructose, inhibit transport competitively.[120] Both saturability and competitive inhibition indicate a specific transporter for sucrose accumulation.

Net accumulation of sucrose in beet vacuoles requires an energy source, which is probably ATP. Isolated vacuoles contain an ATPase which is stimulated by breaking the vacuole membrane, and ATP stimulates sucrose accumulation.[121] Furthermore, sucrose addition leads to acidification of the medium.[122] If ATP hydrolysis pumped H^+ into the vacuole, it would generate a $\Delta\tilde{\mu}_H$ (interior positive and acidic) which might slow down the hydrolytic rate (hence the stimulation of ATPase activity upon vacuole lysis). The same $\Delta\tilde{\mu}_H$ could drive sugar accumulation by sugar/H^+ exchange. Addition of a large pulse of sugar would be expected to push H^+ out of the vacuole and to acidify the medium. Vacuoles isolated from *Pisum sativum* mesophil also accumulate sugar in response to ATP.[123] In this case 3-*O*-methylglucose is the preferred substrate, and H^+ ionophores such as SF6847 inhibit, presumably by dissipating the $\Delta\tilde{\mu}_H$ generated by the ATPase.

Yeast vacuoles accumulate basic amino acids by a specific transport system.[124] In the absence of ATP, added arginine exchanges with endogenous amino acids. This transport process is saturable and inhibited competitively by other basic amino acids. Arginine accumulation is stimulated by ATP hydrolysis, and at the same time the vacuole interior is acidified.[72] Ionophores expected to dissipate a $\Delta\tilde{\mu}_H$ across the vacuole membrane inhibit transport. Again, the results suggest that ATP hydrolysis pumps H^+ into the vacuole, and external solute then exchanges with internal H^+.

Lutoids isolated from rubber tree latex accumulate a variety of solutes.[118] As discussed in Section 3, these organelles contain an ATPase which pumps H^+ inwardly, generating ΔpH (interior acid) and $\Delta\Psi$ (interior positive). ATP hydrolysis also leads to citrate and amino acid accumulation by these vacuoles.[125–128] Citrate accumulation is also stimulated by a ΔpH imposed using nigericin and K^+ or a pH jump. Again, the results are in agreement with ATP hydrolysis driving solute accumulation through an intermediate H^+ gradient.

5. The Nature of the ATPase

The observation that both secretory granules and lysosomes contain electrogenic H^+-pumping ATPases which acidify the organelle interior raises questions about the similarity of these enzymes with each other and with other known ATP-driven pumps. Although there is no clear consensus, there is evidence that lysosomal and granular ATPases may represent a new class of H^+ pump. On the other hand, there is evidence that chromaffin granules also contain an ATPase which resembles the F_1F_0 complex of mitochondria. The mitochondrial enzyme also pumps H^+ away from the face of the membrane on which ATP is hydrolyzed.

5.1. Inhibitor Studies

One approach to the ATPase identity is to measure its sensitivity to known inhibitors of other ATPases. One likely candidate is the class of ATPase containing Na^+,K^+-ATPase, all of which are thought to proceed through formation of a phosphoenzyme intermediate.[129] This includes at least two H^+ pumps, one found in the plasma membrane of yeast, fungi, and higher plants,[130] and an H^+,K^+-ATPase found in gastric parietal cells.[131]

It is easy to dismiss the possibility that lysosomes or secretory granules acidify their interiors by a phosphoenzyme-type ATPase. All of these enzymes are strongly inhibited by vanadate,[129] which has no effect on ATPase or ATP-dependent acidification in chromaffin granules,[132] platelet granules,[133] lysosomes,[71] endosomes,[92] or vacuoles from yeast[73] or rubber tree[127] (see Table I). In the few cases (platelet granules and lysosomes) where it has been tested, ouabain also is without effect.[20,71] Quercetin, however, an inhibitor of many ATPases including the phosphoenzyme types, inhibits acidification in both lysosomes[71] and chromaffin granules[134] (see Table I).

Table I. Effect of ATPase Inhibitors on Acidification of Organelles[a]

Inhibitor class	Secretory granules	Endosomes and lysosomes	Plant vacuoles
Phosphoenzyme ATPases			
Vanadate	−	−	−
Ouabain	−	−	nd
Mitochondrial ATPase			
Azide	−	−	−
Oligomycin	−	−	−
Efrapeptin	−	−	nd
Aurovertin	−	nd	nd
DCCD	+−	−	+
Nonspecific			
Quercetin	+	+	nd
Nbf-Cl	+	nd	nd
Trialkyltin	+	nd	nd
N-ethylmaleimide	+	+	nd

[a] +, inhibition; −, no effect; nd, not determined. See text for references.

Using inhibitors of another potential candidate, the F_1F_0 ATPase of mitochondria, chloroplasts, and bacteria, the picture is not as clear. Some inhibitors of the mitochondrial enzyme also block ATPase in secretory granules and plant vacuoles. For example, DCCD inhibits the ATPase of yeast[73] and *Neurospora*[135] vacuoles, as well as ATPase activity and ATP-dependent amine transport in chromaffin[69,139] and platelet[20] granules and synaptic vesicles[107] (see Table I). In mitochondria, DCCD is known to react covalently with the F_0 component of the ATPase complex, preventing H^+ translocation and thereby stopping further ATP hydrolysis.[137] Inhibition of secretory granule and vacuole ATPase by DCCD thus suggests participation of an F_1F_0 complex in these organelles. However, DCCD is not a universal inhibitor of ATPase activity in lysosomes and secretory granules. Acidification of isolated lysosomes is not blocked,[71] and low concentrations fail to inhibit ATPase activity or H^+ pumping in purified chromaffin granule membranes, even though the same concentration would completely inhibit mitochondrial ATPase.[138] In endosomes[92] and platelet granule membrane vesicles,[133] DCCD inhibits ATP-dependent acidification and amine transport only at concentrations many times higher than those required to completely block mitochondrial ATPase measured under the same conditions.

Many other mitochondrial ATPase inhibitors fail to inhibit the ATPase of lysosomes or secretory granules. Oligomycin, efrapeptin, aurovertin, and azide have universally been found without effect on ATPase activity, ATP-dependent acidification, or ATP-dependent amine transport in lysosomes, endosomes, plant vacuoles, or secretory granules[20,63,69,71,73,92,107,113,133–136] (see Table I). Other inhibitors, such as trialkyltin compounds, quercetin, and 4-chloro-7-nitrobenzofuran (Nbf-Cl), seem to inhibit ATPase both from these sources and from mitochondria.[71,107,132–134] In at least one case, however, secretory granule sensitivity is not the same as in mitochondria. Tributyltin inhibits ATP-dependent serotonin transport into platelet granule membrane vesicles at concentrations well below those required to block mitochondrial ATPase.[133] *N*-ethylmaleimide, a well-known inhibitor of chromaffin granule ATPase,[69] also inhibits acidification of platelet granules, lysosomes, and endosomes.[71,92,133]

From these inhibitor studies, a clear difference arises between the F_1F_0-type ATPase found in mitochondria and the ATPase of lysosomes and secretory granules. In those few studies where the inhibitor sensitivity of an acidic organelle has been compared with that of mitochondria from the same tissue or species, obvious differences appear. Moreover, the inhibitor sensitivity of chloroplast, mitochondrial, and bacterial ATPases from various sources is similar, while ATP-dependent activities in mammalian and plant lysosomes and endosomes and secretory granules show a distinctly different pattern of inhibition. In the latter case, high concentrations of DCCD, trialkyltin compounds, quercetin, *N*-ethylmaleimide, and Nbf-Cl inhibit, but azide, efrapeptin, oligomycin, and aurovertin are without effect.

5.2. Enzyme Purification

In spite of these obvious differences between mitochondrial and secretory granule ATPases, some evidence points to the presence of an F_1 ATPase in chromaffin granule preparations. Apps and Schatz[139] found that antibodies directed against mitochondrial ATPase cross-reacted with an ATPase extracted from preparations of chromaffin granules. The same antibodies inhibited ATP-dependent serotonin transport into intact chromaffin granules, suggesting that the F_1-like enzyme was responsible for H^+ pumping. Subsequently, Sutton and Apps[140] isolated a protein labeled with DCCD from chromaffin granule membranes whose ATPase was inhibited with that reagent. Although this labeled protein is not identical to the F_0 component of mitochondrial ATPase (also labeled with DCCD), it might fulfill the same function, as a channel through which H^+ is pumped, in chromaffin granule membranes.

The role of an F_1F_0 ATPase in acidifying the chromaffin granule contents has been challenged recently by Cidon and Nelson[138] who found that highly purified chromaffin granule membranes have little or no antigen recognized by antibodies raised against the β subunit of yeast mitochondrial F_1 ATPase. These antibodies cross-react with the β subunit from all known F_1 ATPases. In spite of lacking the F_1 marker, these membranes retain ATP-dependent H^+ pumping. Chung *et al.*[141] solubilized ATPase activity from lysosomes and separated the activity from mitochondrial ATPase. Both Cidon and Nelson[138] and Apps *et al.*[132] found two ATPases associated with chromaffin granule membranes, one of which is similar to mitochondrial F_1. Whether or not this enzyme is the true granule (and lysosomal) H^+ pump, it is likely to be different from most F_1 ATPases.

6. Conclusion

A variety of intracellular organelles from plant and animal cells and microorganisms contain a membrane-bound ATPase distinct both from mitochondrial F_1F_0 ATPase and from the phosphoenzyme ATPases such as Na^+,K^+-ATPase. The enzyme hydrolyzes ATP on the cytoplasmic surface of the organelle, and uses the free energy of hydrolysis to actively transport H^+ from the cytoplasm to the organelle interior. This process generates a membrane potential ($\Delta\psi$, interior positive) which may be dissipated by passive cation efflux or anion influx. This in turn leads to further H^+ pumping and generation of a pH difference (ΔpH, interior acid).

The ΔpH and $\Delta\psi$ generated by this ATPase are utilized for a variety of functions in different organelles. Acidification of endosomes is used to dissociate ligand–receptor complexes and facilitate receptor recycling. Acidification of lysosomes accelerates the action of hydrolytic enzymes. In turtle bladder and mammalian kidney, organelles containing this ATPase fuse reversibly with the plasma membrane, leading to urinary acidification. Finally, the accumulation and storage of neurotransmitters, hormones, and metabolites in secretory granules and vacuoles requirs ΔpH and, at least in the case of biogenic amines, also $\Delta\psi$ to drive transport processes.

References

1. Metchnikoff, E. 1893. *Lectures on the Comparative Pathology of Inflammation*. Kegan, Paul, Trench, Trübner, London.
2. Rous, P. 1925. The relative reaction within living mammalian tissues. I. General features of vital staining with litmus. *J. Exp. Med.* **41**:379–397.
3. Rous, P. 1925. The relative reaction within living mammalian tissues. II. On the mobilization of acid material within cells, and the reaction as influenced by cell state. *J. Exp. Med.* **41**:399–411.
4. Pulcher, C. 1927. Le variazioni del pH nella fagocitosi. Nota II. *Boll. Soc. Ital. Biol. Sper.* **2**:722.
5. Sprick, M. G. 1956. Phagocytosis of *M. tuberculosis* and *M. smegmatis* stained with indicator dyes. *Am. Rev. Tuberc. Pulm. Dis.* **74**:552–565.
6. Pavlov, E. P., and V. N. Soloviev. 1967. Changes of cytoplasm in phagocytosis of microbes stained with indicator dyes. *Biul. Eksp. Biol. Med.* **4**:78.
7. Mandell, G. L. 1970. Intraphagolysosomal pH of human polymorphonuclear neutrophils. *Proc. Soc. Exp. Biol. Med.* **134**:447–449.
8. Jensen, R. G., and D. F. Bainton. 1973. Temporal changes in ΔpH within the phagocytic vacuole of the polymorphonuclear neutrophilic leukocyte. *J. Cell Biol.* **56**:379–388.
9. Goldman, R., and H. Rottenberg. 1973. Ion distribution in lysosomal suspensions. *FEBS Lett.* **33**:233–238.
10. Reijngoud, D.-J., and J. M. Tager. 1973. Measurement of intralysosomal pH. *Biochim. Biophys. Acta* **297**:174–178.
11. Rottenberg, H., T. Grunwald, and M. Avron. 1972. Determination of pH in chloroplasts. I. Distribution of (^{14}C)methylamine. *Eur. J. Biochem.* **25**:54–63.
12. Schuldiner, S., H. Rottenberg, and M. Avron. 1972. Determination of ΔpH in chloroplasts. 2. Fluorescent amines as a probe for the determination of ΔpH in chloroplasts. *Eur. J. Biochem.* **25**:64–70.
13. Waddell, W. J., and T. C. Butler. 1959. Calculation of intracellular pH from the distribution of 5,5-dimethyl-2,4-oxazolidinedione (DMO): Application to skeletal muscle of the dog. *J. Clin. Invest.* **38**:720–729.
14. Reijngoud, D.-J., and J. M. Tager. 1977. The permeability properties of the lysosomal membrane. *Biochim. Biophys. Acta* **472**:419–449.
15. Marin, B., M. Marin-Lanza, and E. Komor. 1981. The protonmotive potential difference across the vacuo lysosomal membrane of *Hevea brasiliensis* (rubber tree) and its modification by a membrane-bound adenosine triphosphatase. *Biochem. J.* **198**: 365–372.
16. Pollard, H. B., O. Zinder, P. G. Hoffman, and O. Nikodejevic. 1976. Regulation of the transmembrane potential of isolated chromaffin granules by ATP, ATP analogs, and external pH. *J. Biol. Chem.* **251**:4544–4550.
17. Johnson, R. G., and A. Scarpa. 1976. Internal pH of isolated chromaffin vesicles. *J. Biol. Chem.* **251**:2189–2191.
18. Johnson, R. G., and A. Scarpa. 1976. Ion permeability of isolated chromaffin granules. *J. Gen. Physiol.* **68**:601–631.
19. Johnson, R. G., A. Scarpa, and L. Salganicoff. 1978. The internal pH of isolated serotonin containing granules of pig platelets. *J. Biol. Chem.* **253**:7061–7068.

20. Wilkins, J. A., and L. Salganicoff. 1981. Participation of a transmembrane proton gradient in 5-hydroxytryptamine transport by platelet dense granules and dense granule ghosts. *Biochem. J.* **198**:113–123.
21. Russell, J. T., and R. W. Holz. 1981. Measurement of ΔpH and membrane potential in isolated neurosecretory vesicles from bovine neurohypophyses. *J. Biol. Chem.* **256**:5950–5953.
22. Scherman, D., and J. J. Nordmann. 1982. Internal pH of isolated newly formed and aged neurohypophysial granules. *Proc. Natl. Acad. Sci. USA* **79**:476–479.
23. Ohkuma, S., and B. Poole. 1978. Fluorescence probe measurement of the intralysosomal pH in living cells and the perturbation of pH by various agents. *Proc. Natl. Acad. Sci. USA* **75**:3327–3331.
24. Ohkuma, S., and B. Poole. 1981. Cytoplasmic vacuolation of mouse peritoneal macrophages and the uptake into lysosomes of weakly basic substances. *J. Cell Biol.* **90**:656–664.
25. Salama, G., R. G. Johnson, and A. Scarpa. 1980. Spectrophotometric measurements of transmembrane potential and pH gradients in chromaffin granules. *J. Gen. Physiol.* **75**:109–140.
26. Rottenberg, H. 1979. The measurement of membrane potential and ΔpH in cells, organelles, and vesicles. *Methods Enzymol.* **55**:547–569.
27. Johnson, R. G., S. E. Carty, B. J. Fingerhood, and A. Scarpa. 1980. The internal pH of mast cell granules. *FEBS Lett.* **120**:75–79.
28. Dell'Antone, P. 1979. Evidence for an ATP-driven 'proton pump' in rat liver lysosomes by basic dyes uptake. *Biochem. Biophys. Res. Commun.* **86**:180–189.
29. Lin, W., G. J. Wagner, H. W. Siegelman, and G. Hind. 1977. Membrane-bound ATPase of intact vacuoles and tonoplasts isolated from mature plant tissue. *Biochim. Biophys. Acta* **465**:110–117.
30. Smith, F. A., and J. A. Raven. 1976. H^+ transport and regulation of cell pH. In: *Encyclopedia of Plant Physiology,* Volume 2. O. Luttge and M. Pittman, eds. Springer, Berlin. pp. 317–346.
31. Hurd-Karrer, A. M. 1939. Hydrogen ion concentration of leaf-juice in relation to environment and plant species. *Am. J. Bot.* **26**:834–846.
32. Small, J. 1946. *pH and Plants.* Ballière, Tindall & Cox, London.
33. Drawert, H. 1955. Der pH-Wert des Zellsaftes. In: *Encyclopedia of Plant Physiology,* Volume 1. W. Ruhland, ed. Springer, Berlin. pp. 627–648.
34. Ranson, S. L. 1965. The plant acids. In: *Plant Biochemistry.* J. Bonner and J. E. Varner, eds. Academic Press, New York. pp. 493–525.
35. Eppley, R. W., and C. R. Bovell. 1968. Sulfuric acid in *Desmarestia. Biol. Bull.* **115**:101–106.
36. Casey, R. P., D. Njus, G. K. Radda, and P. A. Sehr. 1977. Active proton uptake by chromaffin granules: Observation by amine distribution and phosphorus-31 nuclear magnetic resonance techniques. *Biochemistry* **16**:972–977.
37. Schulman, R. G., T. P. Brown, K. Ugurbil, S. Ogawa, S. M. Cohen, and J. A. den Hollander. 1979. Cellular applications of ^{31}P and ^{13}C nuclear magnetic resonance. *Science* **205**:160–166.
38. Ritchie, G. A. 1975. Ph.D. thesis. University of Oxford.
39. Njus, D., P. A. Sehr, G. K. Radda, G. A. Ritchie, and P. J. Seeley. 1978. Phosphorus-31 nuclear magnetic resonance studies of active proton translocation in chromaffin granules. *Biochemistry* **17**:4337–4343.
40. Pollard, H. B., H. Shindo, C. E. Creutz, C. J. Pazoles, and J. S. Cohen. 1979. Internal pH and state of ATP in adrenergic chromaffin granules determined by ^{31}P nuclear magnetic resonance spectroscopy. *J. Biol. Chem.* **254**:1170–1177.
41. Reijngoud, D.-J., and J. M. Tager. 1975. Effect of ionophores and temperature on intralysosomal pH. *FEBS Lett.* **54**:76–79.
42. Henning, R. 1975. pH gradient across the lysosomal membrane generated by selective cation permeability and Donnan equilibrium. *Biochim. Biophys. Acta* **401**:307–316.
43. Casey, R. P., D. Njus, G. K. Radda, and P. A. Sehr. 1976. Adenosine triphosphate-evoked catecholamine release in chromaffin granules: Osmotic lysis as a consequence of proton translocation. *Biochem. J.* **158**:583–588.
44. Winkler, H. 1976. The composition of adrenal chromaffin granules: An assessment of controversial results. *Neuroscience* **1**:65–80.
45. Holz, R. W. 1979. Measurement of membrane potential of chromaffin granules by the accumulation of triphenylmethylphosphonium cations. *J. Biol. Chem.* **254**:6703–6709.
46. Holz, R. W. 1978. Evidence that catecholamine transport into chromaffin vesicles is coupled to vesicle membrane potential. *Proc. Natl. Acad. Sci. USA* **75**:5190–5194.
47. Reeves, J. P. 1984. The mechanism of lysosomal acidification. In: *Lysosomes in Biology and Pathology,* Volume 7. J. Dingle, R. Dean, and W. Sly, eds. Elsevier, Amsterdam. pp. 175–199.
48. Kirshner, N. 1962. Uptake of catecholamines by a particulate fraction of the adrenal medulla. *J. Biol. Chem.* **237**:2311–2317.
49. Carlsson, A., N. A. Hillarp, and B. Waldeck. 1963. Analysis of the Mg^{++}-ATP dependent storage mechanism in the amine granules of the adrenal medulla. *Acta Physiol. Scand.* **59**(Suppl. 215): 1–38.
50. Phillips, J. H. 1974. Transport of catecholamines by resealed chromaffin-granule 'ghosts.' *Biochem. J.* **144**:311–318.
51. Phillips, J. H. 1974. Steady-state kinetics of catecholamine transport by chromaffin-granule 'ghosts.' *Biochem. J.* **144**:319–325.
52. Njus, D., J. Knoth, and M. Zallakian. 1980. Proton-linked transport in chromaffin granules. *Curr. Top. Bioenerg.* **11**:107–147.
53. Bashford, C. L., G. K. Radda, and G. A. Ritchie. 1975. Energy-linked activities of the chromaffin granule membrane. *FEBS Lett.* **50**:21–24.
54. Azzi, A. 1969. Redistribution of the electrical charge of the mitochondrial membrane during energy conservation. *Biochem. Biophys. Res. Commun.* **37**:254–260.
55. Johnson, R. G., and A. Scarpa. 1979. Protonmotive force and catecholamine transport in isolated chromaffin granules. *J. Biol. Chem.* **254**:3750–3760.
56. Poisner, A. M., and J. M. Trifaro. 1967. The role of ATP and ATPase in the release of catecholamines from the adrenal medulla. I. ATP-evoked release of catecholamines, ATP, and protein from isolated chromaffin granules. *Mol. Pharmacol.* **3**:561–571.
57. Hoffman, P. G., O. Zinder, O. Nikodejevik, and H. B. Pollard. 1976. ATP-stimulated transmitter release and cyclic AMP synthesis in isolated chromaffin granules. *J. Supramol. Struct.* **4**:181–184.
58. Dolais-Kitabgi, J., and R. L. Perlman. 1975. The stimulation of catecholamine release from chromaffin granules by valinomycin. *Mol. Pharmacol.* **11**:745–750.
59. Johnson, R. G., D. Pfister, S. E. Carty, and A. Scarpa. 1979. Biological amine transport in chromaffin ghosts: Coupling to the transmembrane proton and potential gradients. *J. Biol. Chem.* **254**: 10963–10972.
60. Heinrich, P., M. Da Prada, and A. Pletscher. 1972. Magnesium-dependent ATP-ase in membranes of 5-hydroxytryptamine storage organelles. *Biochem. Biophys. Res. Commun.* **46**:1769–1775.
61. Carty, S. E., R. G. Johnson, and A. Scarpa. 1981. Serotonin transport in isolated platelet granules: Coupling to the electrochemical proton gradient. *J. Biol. Chem.* **256**:11244–11250.
62. Fishkes, H., and G. Rudnick. 1982. Bioenergetics of serotonin transport by membrane vesicles derived from platelet dense granules. *J. Biol. Chem.* **257**:5671–5677.
63. Scherman, D., J. Nordmann, and J.-P. Henry. 1982. Existence of an adenosine 5′-triphosphate dependent proton translocase in bovine neurosecretory granule membrane. *Biochemistry* **21**:687–694.
64. de Duve, C., T. de Barsy, B. Poole, A. Trouet, P. Tulkens, and F. Van Hoof. 1974. Lysosomotropic agents. *Biochem. Pharmacol.* **23**:2495–2531.
65. Wibo, M., and B. Poole. 1974. Protein degradation in cultured

cells. II. The uptake of chloroquine by rat fibroblasts and the inhibition of cellular protein degradation and cathepsin B. *J. Cell Biol.* **63**:430–440.

66. Poole, B., and S. Ohkuma. 1981. Effect of weak bases on the intralysosomal pH in mouse peritoneal macrophages. *J. Cell Biol.* **90**:665–669.
67. Schneider, D. L. 1974. A membranous ATPase unique to lysosomes. *Biochem. Biophys. Res. Commun.* **61**:882–888.
68. Schneider, D. L., and E. Cornell. 1978. Evidence for a proton pump in rat liver lysosomes. In: *Protein Turnover and Lysosome Function.* H. L. Segal and D. Doyl, eds. Academic Press, New York. pp. 59–66.
69. Bashford, C. L., R. P. Casey, G. K. Radda, and G. A. Ritchie. 1976. Energy-coupling in adrenal chromaffin granules. *Neuroscience* **1**:399–412.
70. Schneider, D. L. 1981. ATP-dependent acidification of intact and disrupted lysosomes: Evidence for an ATP-driven proton pump. *J. Biol. Chem.* **256**:3858–3864.
71. Ohkuma, S., Y. Moriyama, and T. Takano. 1982. Identification and characterization of a proton pump on lysosomes by fluorescein isothiocyanate–dextran fluorescence. *Proc. Natl. Acad. Sci. USA* **79**:2758–2762.
72. Ohsumi, Y., and Y. Anraku. 1981. Active transport of basic amino acids driven by a proton motive force in vacuolar membrane vesicles of *Saccharomyces cerevisiae. J. Biol. Chem.* **256**:2079–2082.
73. Kakinuma, Y., Y. Ohsumi, and Y. Anraku. 1981. Properties of H^+-translocating adenosine triphosphatase in vacuolar membranes of *Saccharomyces cerevisiae. J. Biol. Chem.* **256**:10859–10863.
74. Cretin, H. 1982. The proton gradient across the vacuo-lysosomal membrane of lutoids from the latex of *Hevea brasiliensis.* I. Further evidence for a proton-translocating ATPase on the vacuo-lysosomal membrane of intact lutoids. *J. Membr. Biol.* **65**:175–184.
75. Coffey, J. W., and C. de Duve. 1968. Digestive activity of lysosomes. I. The digestion of proteins by extracts of rat liver lysosomes. *J. Biol. Chem.* **243**:3255–3263.
76. Barrett, A. J. 1969. Lysosomal enzymes. In: *Lysosomes in Biology and Pathology,* Volume 2. J. T. Dingle and H. B. Dell, eds. Wiley–Interscience, New York. pp. 245–312.
77. Reijngoud, D.-J., P. S. Oud, J. Kas, and J. M. Tager. 1976. Relationship between medium pH and that of the lysosomal matrix as studied by two independent methods. *Biochem. Biophys. Acta* **448**:290–302.
78. Mego, J. L., R. M. Farb, and J. Barnes. 1972. An ATP-dependent stabilization of proteolytic activity in heterolysosomes. *Biochem. J.* **128**:763–769.
79. Mego, J. L. 1975. Further evidence for a proton pump in mouse kidney phagolysosomes: Effect of nigericin and 2,4-dinitrophenol on the stimulation of intralysosomal proteolysis by ATP. *Biochem. Biophys. Res. Commun.* **67**:571–575.
80. Reeves, J. P. 1979. Accumulation of amino acids by lysosomes incubated with amino acid methyl esters. *J. Biol. Chem.* **254**:8914–8921.
81. Reeves, J. P., and T. Reames. 1981. ATP stimulates amino acid accumulation by lysosomes incubated with amino acid methyl esters. *J. Biol. Chem.* **256**:6047–6053.
82. Goldstein, J. L., R. G. W., Anderson, and M. S. Brown. 1979. Coated pits, coated vesicles, and receptor mediated endocytosis. *Nature* (*London*) **279**:679–685.
83. Steinman, R. M., I. S. Mellman, W. A. Muller, and Z. A. Cohn. 1983. Endocytosis and the recycling of plasma membrane. *J. Cell Biol.* **96**:1–27.
84. Posner, B. I., Z. Josefsberg, and J. J. M. Bergeron. 1978. Intracellular polypeptide hormone receptors: Characterization of insulin binding sites in Golgi fractions from the liver of female rats. *J. Biol. Chem.* **253**:4067–4073.
85. Sahagian, G. G., J. Distler, and G. W. Jourdian. 1981. Characterization of a membrane-associated receptor from bovine liver that binds phosphomannosyl residues of bovine testicular β-galactosidase. *Proc. Natl. Acad. Sci. USA* **78**:4289–4293.
86. Pricer, W. E., Jr., and G. Ashwell, 1971. The binding of desialylated glycoproteins by plasma membranes of rat liver. *J. Biol. Chem.* **246**:4825–4833.
87. Karin, M., and B. Mintz. 1981. Receptor-mediated endocytosis of transferrin in developmentally totipotent mouse teratocarcinoma stem cells. *J. Biol. Chem.* **256**:3245–3252.
88. Gonzales-Noreiga, V., J. H. Grubb, V. Talkad, and W. S. Sly. 1980. Chloroquine inhibits lysosomal enzyme pinocytosis and enhances lysosomal enzyme secretion by impairing receptor recycling. *J. Cell Biol.* **85**:839–852.
89. Tietze, C., P. Schlesinger, and P. Stahl. 1982. Mannose-specific endocytosis receptor of alveolar macrophages: Demonstration of two functionally distinct intracellular pools of receptor and their roles in receptor recycling. *J. Cell Biol.* **92**:417–424.
90. Tycko, B., and F. R. Maxfield. 1982. Rapid acidification of endocytic vesicles containing $alpha_2$-macroglobulin. *Cell* **28**:643–651.
91. Geisow, M. J., P. D'Arcy Hart, and M. R. Young. 1981. Temporal changes of lysosome and phagosome pH during phagolysosome formation in macrophages: Studies by fluorescence spectroscopy. *J. Cell Biol.* **89**:645–652.
92. Galloway, C. J., G. E. Dean, M. Marsh, G. Rudnick, and I. Mellman. 1983. Acidification of macrophage and fibroblast endocytic vacuoles *in vitro. Proc. Natl. Acad. Sci. USA* **80**:3334–3338.
93. Sandvig, K., and S. Olsnes. 1980. Diptheria toxin entry into cells is facilitated by low pH. *J. Cell Biol.* **87**:828–832.
94. Draper, R. K., and M. I. Simon. 1980. The entry of diptheria toxin into the mammalian cell cytoplasm: Evidence for lysosomal involvement. *J. Cell Biol.* **87**:849–854.
95. White, J., J. Kartenbeck, and A. Helenius. 1980. Fusion of Semliki forest virus with the plasma membrane can be induced by low pH. *J. Cell Biol.* **87**:264–272.
96. Gluck, S., C. Cannon, and Q. Al-Awqati. 1982. Exocytosis regulates urinary acidification in turtle bladder by rapid insertion of H^+ pumps into the luminal membrane. *Proc. Natl. Acad. Sci. USA* **79**:4327–4331.
97. Gluck, S., S. Kelly, and Q. Al-Awqati. 1982. The proton translocating ATPase responsible for urinary acidification. *J. Biol. Chem.* **257**:9230–9233.
98. Falck, B., N.-Å. Hillarp, and B. Hogberg. 1956. Content and intracellular distribution of adenosine triphosphate in cow adrenal medulla. *Acta Physiol. Scand.* **36**:360–376.
99. Blaschko, H., G. V. R. Born, A. D'Iorio, and N. R. Eade, 1956. Observations on the distribution of catecholamines and adenosine triphosphate in the bovine adrenal medulla. *J. Physiol.* (*London*) **133**:548–557.
100. Schuldiner, S., H. Fishkes, and B. I. Kanner. 1978. Active adrenaline transport in chromaffin membrane vesicles is driven by a transmembranous pH gradient. *Proc. Natl. Acad. Sci. USA* **75**:3713–3716.
101. Phillips, J. H. 1978. 5-Hydroxytryptamine transport by the chromaffin granule membrane. *Biochem. J.* **170**:673–679.
102. Johnson, R. G., N. J. Carlson, and A. Scarpa. 1978. ΔpH and catecholamine distribution in isolated chromaffin granules. *J. Biol. Chem.* **253**:1512–1521.
103. Njus, D., and G. K. Radda. 1979. A potassium ion diffusion potential causes adrenaline uptake in chromaffin granule 'ghosts.' *Biochem. J.* **180**:579–585.
104. Kanner, B. I., I. Sharon, R. Maron, and S. Schuldiner. 1980. Electrogenic transport of biogenic amines in chromaffin granule membrane vesicles. *FEBS Lett.* **111**:83–86.
105. Johnson, R. G., S. E. Carty, and A. Scarpa. 1981. Proton : substrate stoichiometries during active transport of biogenic amines in chromaffin ghosts. *J. Biol. Chem.* **256**:5773–5780.
106. Knoth, J., M. Zallakian, and D. Njus. 1981. Stoichiometry of H^+-linked dopamine transport in chromaffin granule ghosts. *Biochemistry* **20**:6625–6629.
107. Toll, L., and B. D. Howard. 1978. Role of Mg^{2+}-ATPase and a

pH gradient in the storage of catecholamines in synaptic vesicles. *Biochemistry* **17**:2517–2523.

108. Maron, R., B. I. Kanner, and S. Schuldiner. 1979. The role of a transmembrane pH gradient in 5-hydroxytryptamine uptake by synaptic vesicles from rat brain. *FEBS Lett.* **98**:237–240.
109. Rudnick, G., H. Fishkes, P. J. Nelson, and S. Schuldiner. 1980. Evidence for two distinct serotonin transport systems in platelets. *J. Biol. Chem.* **255**:3638–3641.
110. Carpenter, R. S., and S. M. Parsons. 1978. Electrogenic behavior of synaptic vesicles from *Torpedo californica*. *J. Biol. Chem.* **253**:326–329.
111. Michaelson, D. M., I. Pinchasi, I. Angel, I. Ophir, M. Sokolovsky, and G. Rudnick. 1979. The energetics and calcium dependency of acetylcholine release from *torpedo* synaptic vesicles. In: *Molecular Mechanisms of Biological Recognition.* M. Balaban, ed. Elsevier/North-Holland, Amsterdam. pp. 361–371.
112. Rebois, R. V., E. E. Reynolds, L. Toll, and B. D. Howard. 1980. Storage of dopamine and acetylcholine in granules of PC12, a clonal pheochromocytoma cell line. *Biochemistry* **19**:1240–1248.
113. Toll, L., and B. D. Howard. 1980. Evidence that an ATPase and a proton-motive force function in the transport of acetylcholine into storage vesicles. *J. Biol. Chem.* **255**:1787–1789.
114. Parsons, S. M., and R. Koenigsberger. 1980. Specific stimulated uptake of acetylcholine by *Torpedo* electric organ synaptic vesicles. *Proc. Natl. Acad. Sci. USA* **77**:6234–6238.
115. Anderson, D. C., S. C. King, and S. M. Parsons. 1981. Uncoupling of acetylcholine uptake from the *Torpedo* cholinergic synaptic vesicle ATPase. *Biochem. Biophys. Res. Commun.* **103**:422–428.
116. Anderson, D. C., S. C. King, and S. M. Parsons. 1982. Proton gradient linkage to active uptake of [^{3}H]acetylcholine by *Torpedo* electric organ synaptic vesicles. *Biochemistry* **21**:3037–3043.
117. Marty, F., D. Branton, and R. A. Leigh. 1980. Plant vacuoles. In: *The Biochemistry of Plants,* Volume 1. N. E. Tolbert, ed. Academic Press, New York. pp. 625–658.
118. D'Auzac, J., H. Cretin, B. Marin, and C. Lioret. 1982. A plant vacuolar system: The lutoids from *Hevea brasiliensis* latex. *Physiol. Veg.* **20**:311–331.
119. Leigh, R. A., T. Rees, W. A. Fuller, and J. Banfield. 1979. The location of acid invertase and sucrose in the vacuoles of storage roots of beet root (*Beta vulgaris*). *Biochem. J.* **178**:539–547.
120. Willenbrink, J., and S. Doll. 1979. Characteristics of the sucrose uptake system of vacuoles isolated from red beet tissue: Kinetics and specificity of the sucrose uptake system. *Planta* **147**:159–162.
121. Doll, S., F. Rodier, and J. Willenbrink. 1979. Accumulation of sucrose in vacuoles isolated from red beet tissue. *Planta* **144**:407–411.
122. Doll, S., V. Effelsberg, and J. Willenbrink. 1982. Characteristics of the sucrose uptake system of vacuoles isolated from red beet tissue: Relationships between ion gradients and sucrose transport. In: *Plasmalemma and Tonoplast: Their Functions in the Plant Cell.* D. Marme, E. Marre, and R. Hertel, eds. Elsevier, Amsterdam. pp. 217–224.
123. Guy, M., L. Reinhold, and D. Michaeli. 1979. Evidence for a sugar transport mechanism in isolated vacuoles. *Plant Physiol.* **64**:61–64.
124. Boller, T., M. Durr, and A. Wiemken. 1975. Characterization of a specific transport system for arginine in isolated yeast vacuoles. *Eur. J. Biochem.* **54**:81–91.
125. Lambert, C. 1975. Influence de l'ATP sur le pH intralutoidique et sur la penetration due citrate dans les lutoides du latex d'*Hevea brasiliensis*. *C.R. Acad. Sci. Ser. D* **281**:1705–1708.
126. Marin, B., J. A. C. Smith, and U. Luttge. 1981. The electrochemical proton gradient and its influence on citrate uptake in tonoplast vesicles of *Hevea brasiliensis*. *Planta* **153**:486–493.
127. Marin, B., and F. Blasco. 1982. Further evidence for the proton pumping work of tonoplast ATPase from *Hevea* latex vacuole. *Biochem. Biophys. Res. Commun.* **105**:354–361.
128. Marin, B., H. Cretin, and J. D'Auzac. 1982. Energization of solute transport and accumulation at the tonoplast in *Hevea* latex. *Physiol. Veg.* **20**:333–346.
129. Kyte, J. 1981. Molecular considerations relevant to the mechanism of active transport. *Nature (London)* **292**:201–204.
130. Bowman, B., and C. W. Slayman. 1979. The effects of vanadate on the plasma membrane ATPase of *Neurospora crassa*. *J. Biol. Chem.* **254**:2928–2934.
131. Sachs, G., H. H. Chang, E. Rabon, R. Schackman, M. Lewin, and G. Saccomani. 1976. A nonelectrogenic H^+ pump in plasma membranes of hog stomach. *J. Biol. Chem.* **251**:7690–7698.
132. Apps, D. K., J. G. Pryde, and R. Sutton. 1982. The H^+-translocating adenosine triphosphatase of chromaffin granule membranes. *Ann. N.Y. Acad. Sci.* **402**:134–145.
133. Dean, G. E., H. Fishkes, P. J. Nelson, and G. Rudnick. 1984. The hydrogen ion pumping adenosine triphosphatase of platelet dense granule membrane. Differences from F_1F_0 and phosphoenzyme-type ATPases. *J. Biol. Chem.* **259**:9569–9574.
134. Apps, D. K., and L. A. Glover. 1978. Isolation and characterization of magnesium adenosine triphosphatase from the chromaffin granule membrane. *FEBS Lett.* **85**:254–258.
135. Bowman, E. J., and B. J. Bowman. 1982. Identification and properties of an ATPase in vacuolar membranes of *Neurospora crassa*. *J. Bacteriol.* **151**:1326–1337.
136. Giraudat, J., M. P. Roisin, and J. P. Henry. 1980. Solubilization and reconstitution of the adenosine 5′-triphosphatase dependent proton translocase of bovine chromaffin granule membrane. *Biochemistry* **19**:4499–4505.
137. Racker, E. 1972. Reconstitution of oxidative phosphorylation and vesicles with respiratory control. In: *Membrane Research.* C. F. Fox, ed. Academic Press, New York. pp. 97–114.
138. Cidon, S., and N. Nelson. 1983. A novel ATPase in the chromaffin-granule membrane. *J. Biol. Chem.* **258**:2892–2898.
139. Apps, D. K., and G. Schatz. 1979. An adenosine triphosphatase isolated from chromaffin granule membranes is closely similar to F_1-adenosine triphosphatase of mitochondria. *Eur. J. Biochem.* **100**:411–419.
149. Sutton, R., and D. K. Apps. 1981. Isolation of a DCCD-binding protein from bovine chromaffin-granule membranes. *FEBS Lett.* **130**:103–106.
141. Chung, C. H., R. L. Elliott, and J. L. Mego. 1980. Lysosomal membrane adenosine triphosphatase; solubilization and partial characterization. *Arch. Biochem. Biophys.* **203**:251–259.

CHAPTER 3

Intracellular pH Regulation

Walter F. Boron

1. Introduction

Because of the vast number of cellular processes sensitive to changes in pH, the control of intracellular pH (pH_i) is of vital importance both for the individual cell and for the organism as a whole. The fundamental problem that pH_i-regulating mechanisms must address is the chronic tendency toward intracellular acidification. Depending on the conditions of incubation, a chronic intracellular acid load can be imposed by cellular metabolism. However, a nearly universal source of chronic acid loading are the fluxes across the cell membrane of H^+ and of ionized weak acids and bases. Consider a cell having a transmembrane voltage (V_m) of -60 mV (cell negative) and an extracellular pH (pH_o) of 7.4. The Nernst equation predicts that pH_i would be ~ 6.4 if H^+ were in electrochemical equilibrium across the cell membrane. Because the actual pH_i is nearly a full pH unit higher, there is a substantial gradient favoring the influx of H^+, and one of equal magnitude favoring the efflux of OH^-. It can be shown (see Section 4.1) that the anionic, conjugate weak base (e.g., HCO_3^-) of any neutral weak acid (e.g., CO_2) is influenced by the same electrochemical gradient as that for OH^-, provided the neutral weak acid is equilibrated across the cell membrane. Similarly, the electrochemical gradient for any cationic conjugate weak acid (e.g., NH_4^+) of a neutral weak base (e.g., NH_3) is the same as that for H^+, provided the neutral weak base is in equilibrium across the cell membrane. Thus, the passive fluxes of H^+, and of ionized weak acids or bases will all produce a chronic intracellular acid load. The gradual fall of pH_i toward its equilibrium value (i.e., ~ 6.4 in this example) can be forestalled only by an active-transport process that extrudes acid from the cell at a rate equal to the total rate of acid accumulation.

In this chapter, we shall consider first the major techniques available for measuring pH_i. Cellular buffering, which is responsible for minimizing the pH_i changes produced by acute acid and alkali loads, will be discussed next. This is followed by an analysis of the effects on pH_i of externally applied weak acids and bases. Finally, we will examine the systems that transport H^+ and/or HCO_3^- across the cell membrane.

The reader is directed to several recent reviews on the subject of intracellular pH.[12,26,79]

Walter F. Boron • Department of Physiology, Yale University School of Medicine, New Haven, Connecticut 06510.

2. Measurement of Intracellular pH

Although other techniques are available, we shall consider four general approaches that are in greatest use, and considered to be of greatest reliability.

2.1. Distribution of Weak Acids and Bases

Cell membranes are generally far more permeable to neutral weak acids (HA, where $HA = H^+ + A^-$) and bases (B, where $B + H^+ = HB^+$) than to their ionized conjugates (i.e., A^- and HB^+). Thus, if a cell is exposed to a neutral weak acid or base whose ionized conjugate is negligibly permeant, a time will be reached when the intracellular concentration of the neutral species is nearly the same as the extracellular concentration:

$$[HA]_i = [HA]_o \quad \text{and} \quad [B]_i = [B]_o \qquad (1a,b)$$

As described in more detail in Section 4, the influx of HA lowers pH_i and that of B raises it. However, if $[HA]_o$ and $[B]_o$ are low, the perturbation of pH_i is minimal. Given the acid dissociation constants:

$$K_{HA} = \frac{[H^+][A^-]}{[HA]} \quad \text{and} \quad K_B = \frac{[H^+][B]}{[HB^+]} \qquad (2a,b)$$

and assuming that these constants are identical in the intra- and extracellular fluids, it follows that

$$\frac{[H^+]_i}{[H^+]_o} = \frac{[A^-]_o}{[A^-]_i} \quad \text{and} \quad \frac{[H^+]_i}{[H^+]_o} = \frac{[HB^+]_i}{[HB^+]_o} \qquad (3a,b)$$

Thus, $[H^+]_i$ can be computed from the known extracellular parameters and either $[A^-]_i$ or $[HB^+]_i$. In practice, it is impossible to measure uniquely either of these last two parameters.

Simple rearrangements of Eqs. (3a, b), however, yield an analogous expression for $[H^+]_i$ in terms of total intracellular acid ($[TA]_i = [HA]_i + [A^+]_i$) or base ($[TB]_i = [B]_i + [HB^+]_i$). $[TA]_i$ and $[TB]_i$ are easily measured by employing radioisotopically labeled compounds. Inasmuch as the sample of intracellular fluid is likely to be contaminated by extracellular fluid, the sample is also assayed for a marker (e.g., inulin or mannitol) confined to the extracellular space. The water content is measured by drying the sample, or by assaying the samples's content of a compound (e.g., tritiated water) assumed to be uniformly distributed throughout all intra- and extracellular water.

The most extensively used pH indicator compounds are the weak acid 5,5-dimethyl-2,4-oxazolidine-dione (DMO) and the weak base methylamine (MA). The weak acid/base method measures of a mean pH of all intracellular compartments into which the indicator distributes. The technique is easy and relatively inexpensive. It can be used with even the smallest cells or cell fragments, and in at least the case of DMO, gives pH_i values that are within 0.1 of that measured with pH-sensitive microelectrodes.[22] Although it had been feared that the anionic form of DMO might be sufficiently permeant to compromise the accuracy of the method, more recent studies have shown that the permeability to the DMO anion is low enough to be negligible.[55] The method's major disadvantage is that continuous pH_i measurements are impossible.

2.2. pH-Sensitive Microelectrodes

When a pH-sensitive electrode is placed in the same solution as an indifferent (i.e., "reference") electrode, the voltage difference between the two electrodes (E_x) varies linearly with the solution's pH (pH_x):

$$pH_x = pH_s + (E_x - E_s) \cdot \frac{F}{RT(\ln 10)} \tag{4}$$

where E_s is the voltage measured with the electrodes in a standard solution of pH_s, and $F/RT(\ln 10)$ is the theoretical slope. In practice, the slope (~ 58 mV/pH unit at 22°C) is determined empirically by calibrating the electrodes in two or more standards. In the past, the only reliable pH sensor was glass. Two designs are suitable for intracellular use. Hinke's[51] exposed-tip microelectrode can only be used in large cells such as squid axons. Thomas'[89] recessed-tip microelectrode can be used in much smaller cells, having been employed in cells as small as amphibian renal tubules.[15] Electrodes of these two designs are very reliable and easy to use, but difficult to make, Recently, a resin pH sensor[5] has come into widespread use in microelectrodes. Such liquid-ion-exchange electrodes are easy to make, but less reliable than the glass ones. The microelectrode approach always requires that a single cell or two identical cells be impaled by both pH and reference electrodes. Double-barreled devices have been described for both recessed-tip[36] and liquid-ion-exchange electrodes that incorporate both pH and reference electrodes into a single shaft. The electrodes continuously measure the pH of the cytoplasm (e.g., that compartment in direct contact with the cell membrane), and do so with excellent resolution (~ 0.01 pH unit). Although the microelectrode approach is the method of choice in cells large and resilient enough to withstand long-lasting impalements, it is technically difficult.

2.3. pH-Sensitive Dyes

Compounds with pH-sensitive absorbance, fluorescence-excitation or fluorescence-emission spectra are convenient probes for monitoring pH in even small cells or cell fragments. The pH resolution can, under optimal conditions, approach 0.0001 pH unit, and the temporal response is nearly instantaneous. The absorbance method is considered most reliable, but requires a combination of millimolar intracellular dye concentrations and/or cytoplasmic thicknesses of a few micrometers. This approach has been used with whole cuvettes of cells,[88] small (e.g., 5 to 10) groups of cells in isolated renal tubules,[31,32] or single cells in culture.[30] Fluorescence methods, while more prone to interference from the dye's environment, are extremely sensitive, and are the methods of choice when the dye concentration or the cell thickness is low. Fluorescence techniques have been applied to large populations of cells in cuvettes[29,68,101,81] and single amebas.[50] Certain dyes, such as phenol red, must be injected into the cell (see Ref. 8). Others, such as the fluorescein derivatives, are available as permeant ester derivatives. Upon entering the cell, the derivative is hydrolyzed by native esterases, releasing the dye, which is relatively impermeant.[77,89] A third method for introducing a dye is to expose cells to a dextran-linked derivative of the dye. The dextran-linked dye is phagocytized by the cells, and then released from intracellular vesicles into the cytoplasm by a brief osmotic shock.[29,81] A convenient method for intracellular calibration is to equalize pH_o and pH_i after treating with high $[K^+]_o$ and the K^+/H^+ exchanger nigericin.[89] The use of the absorbance approach with a fluorescein derivative has been verified against pH-sensitive microelectrodes in salamander proximal tubules.[31]

2.4. Nuclear Magnetic Resonance

^{31}P-NMR can provide intracellular quantitation of a variety of phosphorus-containing compounds, such as ATP, phosphocreatine, and inorganic phosphate (for reviews see Refs. 48, 49). Because the position of the inorganic phosphate peak is pH sensitive, the technique simultaneously reports pH_i. Although NMR is expensive and requires special skills, it is noninvasive and capable of continuous monitoring. The temporal resolution and pH sensitivity are considerably less than of either microelectrodes or dyes. Nevertheless, NMR can be applied to virtually any sample for which the mass of ^{31}P is sufficient. Perhaps its most unique advantage is that NMR can be applied to whole organs and even intact animals. Three-dimensional NMR imaging, analogous to X-ray computerized tomography, is currently being employed, and three-dimensional pH mapping may soon become a reality. pH_i measurements have also been made with ^{19}F-NMR, exploiting the position of the ^{19}F peak in fluorine-substituted amino acids.[40]

3. Cellular Buffering Processes

3.1. Definitions

A pH buffer, broadly defined, is any system that moderates changes of pH by reversibly consuming or releasing H^+. We shall focus on the buffering of the cytoplasmic compartment, the fluid in direct contact with the inner surface of the cell membrane, and the outer surface of various organelles. *Buffering power* (β) is defined as

$$\beta = \frac{dB}{dpH} \quad (5)$$

where dpH is the pH change produced by the addition of an amount dB of strong base (or an amount $-d$B of strong acid). β is always positive, and is usually expressed in the units "millimolar" (inasmuch as pH is dimensionless). Note that the definition of cellular buffering does not include processes, such as acid extrusion, that transport acid or base across the cell membrane.

The procedure for measuring β is to acutely load the cell with a known amount of alkali or acid, and then monitor the change in pH_i that is due solely to the addition of the acid or base. Several methods are available for applying a defined alkali or acid load to the cytoplasm, including direct microinjection of the titrant, as well as exposure of the cell to a permeant weak acid or base. Regardless of the method, the data are generally complicated by secondary changes in pH_i. As idealized in Fig. 1, an acute acid load generally produces not only a rapid fall of pH_i (segment *ab*), but also a slower pH_i recovery (*bc*). This recovery is produced by a pH_i-regulating system that extrudes acid from the cell (see Section 5). Although acid extrusion is not included in the definition of cytoplasmic buffering, it nevertheless obscures the true pH_i change caused by the addition of the titrant [i.e., the value inserted into Eq. (5)]. It is obvious that if the measurement of pH_i is delayed too long after the imposition of the acid load (e.g., at the times indicated by *b1, b2, b3*, or *c*), or if the time resolution of the pH_i-measuring technique is poor, the measured pH_i change will be smaller than the one which would have been measured at point *b*. However, even at *b*, the pH_i change is somewhat smaller than it would have been had acid extrusion been blocked from the instant the acid load had been applied. Thus, an accurate determination of β requires the elimination of all pH-related transport processes at the cell membrane.

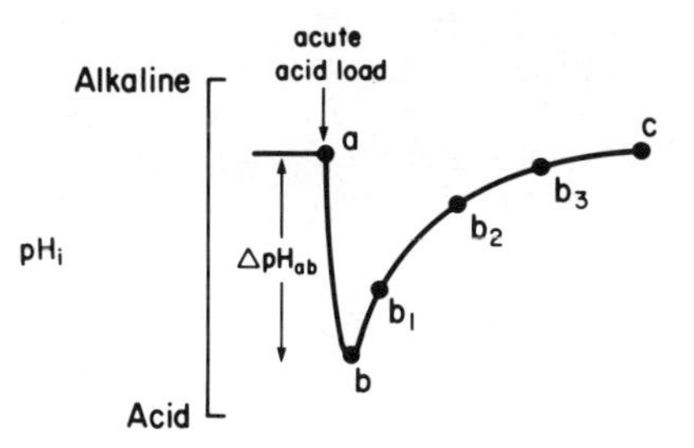

Fig. 1. Response of an idealized cell to an acute intracellular acid load. At point *a*, the cell is abruptly loaded with acid, causing pH_i to fall (segment *ab*). The intracellular buffering power (β_i) is the negative of strong acid added (given in mM) divided by the magnitude of the pH_i decrease that can be ascribed exclusively to the acid load. However, even as pH_i is declining, the cell's pH_i-regulating system responds by extruding acid from the cell, thereby minimizing the pH_i decrease. Thus, if the pH_i decrease is taken as ΔpH_{ab}, the calculated β_i will be erroneously high. If the pH_i measurement is delayed until times corresponding to b_1, b_2, or b_3, the ΔpH_i will be even smaller, and the erroneously calculated buffering power, even higher. If, in the extreme, pH_i is allowed to recover completely (*c*), then $\Delta pH_i = 0$, and the apparent buffering power would appear to be infinite. An accurate determination of β_i therefore requires the blockade of all transport processes that influence pH_i. Modified from Ref. 12. Reproduced with permission of Springer-Verlag.

The role of cytoplasmic buffering processes in pH_i regulation is to moderate the pH_i changes produced by acute acid and alkali loads. As pH_i gradually returns toward its initial value following such a load (*bc* in Fig. 1), the buffer reactions reverse direction, and the buffers are ultimately returned to their initial state. The return of pH_i, however, is not due to the buffer reactions, but rather to a change in the rate of pH-related ion transport across the cell membrane. In the steady state, when the rate of acid loading is balanced by that of acid extrusion, there is no net change in the status of cytoplasmic buffers.

3.2. Mechanisms of Cytoplasmic Buffering

Total intracellular buffering power (β_i^T) is the algebraic sum of the buffering powers of physicochemical processes (i.e., weak acid/base equilibria), as well as biochemical, and organellar buffering processes.

3.2.1. Physicochemical Buffering

H^+ reacts with a weak base (B) of valence n to produce the conjugate weak acid (HB) of valence $n + 1$:

$$H^+ + B^n = HB^{n+1} \quad (6)$$

According to the law of mass action, this equilibrium is described by the apparent acid dissociation constant K'_a:

$$K'_a = \frac{[H^+][B^n]}{[HB^{n+1}]} \quad (7)$$

If the total amount of the buffer in the system is fixed, then Eqs. (5) and (7) can be used to derive an expression for β in terms of $[H^+]$, K'_a, and total buffer concentration, TB ($[TB] = [B^n] + [HB^{n+1}]$):

$$\beta = (\ln 10)\frac{K'_a[H^+]}{(K'_a + [H^+])^2}[TB] \quad (8)$$

This was first shown, in a slightly different form, by Koppel and Spiro.[60,78] Later, Michaelis[64] and then Van Slyke[95] rewrote the equation in its present form. Koppel and Spiro also demonstrated that, regardless of the buffer, the maximal buffering power (β_{max}) occurs when $[H^+]$ equals K'_a, such that

$$\beta_{max} = \frac{(\ln 10)[TB]}{4} = 0.58[TB] \quad (9)$$

When more than one buffer is present, the total buffering power is merely the sum of the individual buffering powers.

The preceding analysis is applicable only to buffers in a *closed system* (i.e., constant [TB]). For certain buffers, however, the cell behaves as an *open system*. Consider the case of CO_2/HCO_3^-. The intracellular addition of H^+ causes the equilibria:

$$H^+ + HCO_3^- = H_2CO_3 = CO_2 + H_2O \quad (10)$$

to shift to the right. If CO_2 were impermeant, the extent of this right shift (i.e., the amount of H^+ buffered by HCO_3^-) would be limited by the accumulation of intracellular CO_2. Because the cell membrane is highly permeable to CO_2, however, the newly generated CO_2 leaves the cell (so that $[CO_2]_i$ is unchanged) and allows the equilibria to shift much further to the right. Conversely, the addition of OH^- to the cell causes the CO_2 equilibria to shift to the left. This left shift, and thus the buffering power, is greatly enhanced when CO_2 is allowed to freely enter the cell. The open-system CO_2 buffering power can be derived from Eqs. (5) and (7) (see Ref. 79):

$$\beta_{CO_2} = (\ln 10)\,[HCO_3^-] \tag{11}$$

assuming $[CO_2]_i$ to be constant. For normal mammalian skeletal muscle, the open-system CO_2/HCO_3^- buffering power is ~ 29 mM, as predicted by Eq. (11). On the other hand, if the cells were a closed system for CO_2/HCO_3^-, the CO_2/HCO_3^- buffering power would be only ~ 2.4 mM, as dictated by Eq. (8). This 26.6 mM discrepancy is a substantial fraction of the ~ 43 mM buffering power of all non-CO_2 buffers in muscle,[3] and demonstrates the importance of the open-system CO_2/HCO_3^- buffer pair. A similar open-system analysis can be applied to any monoprotic weak acid or base whose neutral form is highly permeant. In each case, the open system buffering power is the product of (ln 10) and the intracellular concentration of the ionized form of the buffer.

Buffers for which the cell behaves as a closed system have been described as *intrinsic*,[10] because their buffering powers depend only on factors internal to the cell. The total intrinsic buffering power (β^I) includes not only the buffering power of closed-system buffers, but also of biochemical and organellar buffers (see below). These three are grouped together because currently available techniques for directly measuring intracellular buffering power (see Refs. 13, 79) cannot distinguish among them. Open-system buffers have been described as *extrinsic*,[10] because their buffering powers depend largely on an extracellular factor, namely, the concentration of the buffer's neutral form. Extrinsic buffering power is generally not measured, but calculated from Eq. (11) or an analogous expression. The total physicochemical buffering power of a cell is the sum of the buffering powers (either open- or closed-system) of all physicochemical buffers.

3.2.2. Biochemical Buffering

Certain biochemical reactions and reaction sequences result in the net consumption or release of H^+ (see Ref. 79), and may thus contribute to the overall buffering of acute acid or alkali loads. For example, the conversion of one lactic acid molecule to glucose (which is non-ionizable) or to CO_2 (which is freely diffusible) results in the consumption of one H^+. When the cells of the rat brain are acutely acid loaded by rasing pCO_2, there is a fall in the concentrations of several acidic metabolic intermediates and a corresponding rise in the levels of glucose and glucose-6-phosphate.[45] This is consistent with the inhibition of glycolysis by low pH_i, possibly at the markedly pH-sensitive phosphofructokinase reaction.[93] Conversely, when rat brain cells are actutely alkali loaded by lowering pCO_2, the levels of lactic and pyruvic acids fall.[61] Thus, the biochemical apparatus of the cells responds appropriately, consuming H^+ in response to acid loads, and releasing H^+ in response to alkali loads. In rat brain, such biochemical buffers may provide about half of the buffering due to all non-CO_2, physicochemical buffers combined.[86] Biochemical buffers are included in the definition of intrinsic buffers, along with closed-system physicochemical buffers (see above) and organellar buffers (see below).

3.2.3. Organellar Buffering

Transport systems for H^+ have been described or implicated in a variety of intracellular organelles (see Ref. 79). In addition, all organellar membranes probably have a finite permeability to H^+ as well as to ionized weak acids and bases. One would thus expect that acute, cytoplasmic acid or alkali loads would result in the net transfer of H^+ equivalents across organellar membranes, thereby permitting the intra-organellar space to participate in the buffering of the cytoplasm. The contribution of organelles to total buffering is, however, unknown. Organellar buffers are included under the definition of intrinsic buffers, along with closed-system physicochemical buffers and biochemical buffers.

4. Effect of Externally Applied Weak Acids and Bases

4.1. Neutral Weak Acids and Their Anionic, Conjugate Weak Bases

A solution containing a neutral, monoprotic weak acid (HA) always contains the conjugate weak base (A^-). The cell membrane, however, is generally far more permeable to the former than to the latter, so that, at least in the short term, fluxes of HA usually have a predominant effect on pH_i.

4.1.1. Flux of the Neutral Weak Acid

When $[HA]_o$ is raised, there is a net flux of HA into the cell, and a fall of pH_i as the entering HA dissociates:

$$HA = H^+ + A^- \tag{12}$$

This process continues until $[HA]_i = [HA]_o$, after which time there should be no further change in pH_i, provided there is no net movement across the membrane of pH-related ions (e.g., H^+ and A^-). The intracellular acidifying effect of the weak acid CO_2, first demonstrated by Jacobs[52] in 1920, is illustrated in Fig. 2 for barnacle muscle. Successive exposures to solutions containing 1, 2, and 5% CO_2 (all at a $[HCO_3^-]_o$ of 10 mM) cause progressively greater declines of pH_i, as CO_2 enters the cell and undergoes the following reactions:

$$CO_2 + H_2O = H_2CO_3 = H^+ + HCO_3^- \tag{13}$$

Note that, for each CO_2 exposure, the initial decline of pH_i is followed by a period of relative stability (the "plateau phase"). The magnitude of the CO_2-induced acidification depends not only on $[CO_2]_o$, but on β_i^T and the initial pH_i as well. The pH_i dependence derives from the law of mass action, and causes the magnitude of the pH_i change to increase as the initial pH_i rises. Intracellular acidifications have been observed with a number of other weak acids, including lactic acid,[77] DMO,[55] and several fatty acids.[37]

4.1.2. Flux of the Anionic, Conjugate Weak Base

After HA has equilibrated across the membrane and produced the rapid intracellular acidification described above, the

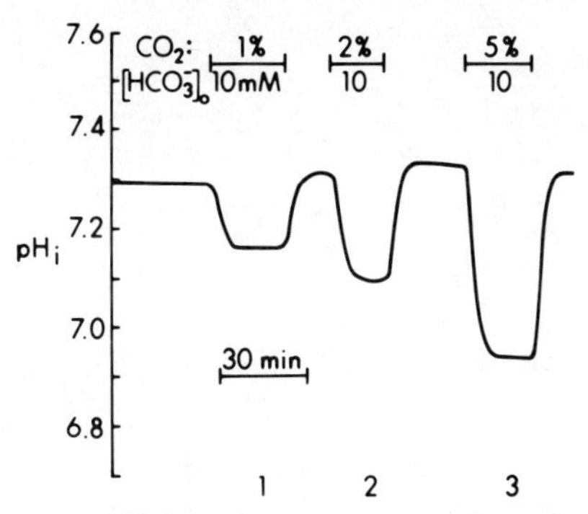

Fig. 2. Effect of exposing a cell to CO_2. The pH_i of a giant-barnacle muscle fiber was measured with a Hinke-style microelectrode.[51] On three separate occasions, the Hepes buffer was replaced by 10 mM HCO_3^- and a variable amount of CO_2. The exposures to 1% (pH_o = 7.5), 2% (pH_o = 7.2), and 5% CO_2 (pH_o = 6.8) caused progressively larger decreases in pH_i as the CO_2 enters the cell, combines with water to form H_2CO_3, and then dissociates to form H^+ and HCO_3^-. There was no significant pH_i recovery during the time allowed for CO_2 exposures, probably because the relatively low pH_o inhibited the pH_i-regulating system. From Ref. 10. Reproduced with permission of the American Physiological Society.

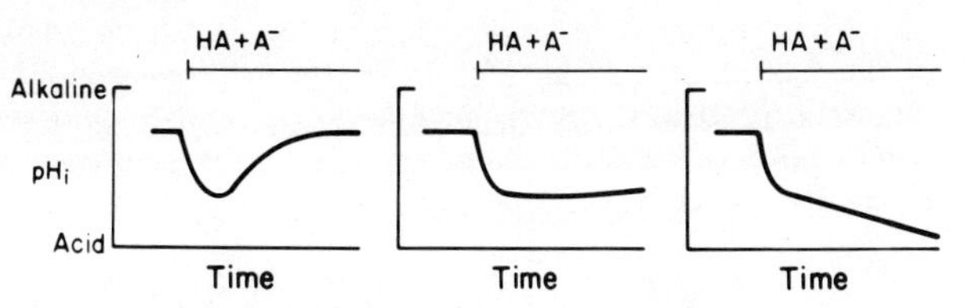

Fig. 3. Three idealized responses of pH_i to the application of a permeant weak acid (HA = $H^+ + A^-$). In all cases, the weak acid first causes pH_i to decline rapidly, due to the influx and dissociation of HA. The subsequent course of pH_i, during a period termed the plateau phase, is determined by the balance between acid extrusion (due to the cell's pH_i-regulating system) and acid loading (due to the passive fluxes of ions and cell metabolism). In the first example (left), this initial pH_i decline is followed by a recovery, as the cell's acid-extrusion rate exceeds the acid-loading rate. If the acid-extrusion rate equals the acid-loading rate (center), pH_i is stable during the plateau phase. Finally, if the acid-loading rate exceeds the acid-extrusion rate (right), pH_i falls during the plateau phase. From Ref. 12. Reproduced with permission of Springer-Verlag.

fate of pH_i during the plateau phase is determined by the interaction among at least three factors: (1) the passive flux of A^- across the cell membrane, (2) the metabolic production of acid (if any), and (3) the extrusion of acid by the pH_i-regulating mechanism. The direction of the expected A^- flux is easily determined. If $[HA]_i = [HA]_o$, the A^- distribution ratio is the inverse of that for H^+:

$$[H^+]_i\,[A^-]_i = K\,[HA]_i = K\,[HA]_o = [H^+]_o\,[A^-] \quad (14)$$

or:

$$\frac{[H^+]_i}{[H^+]_o} = \frac{[A^-]_o}{[A^-]_i} \quad (14a)$$

That is, the electrochemical gradient for A^- is equal but opposite to that for H^+. Inasmuch as the H^+ electrochemical gradient normally favors the net efflux of H^+, that for A^- favors the net efflux of this ion. Although the expected passive fluxes of H^+ and A^- are in opposite directions, both tend to produce a fall in pH_i. The metabolic production of acid, if any, might be expected to be reduced by the HA-induced decline of pH_i, and to be further modulated by possible direct effects of HA and/or A^- on metabolic pathways. Finally, acid extrusion (see below) is stimulated by the HA-induced fall in pH_i, but inhibited by concomitant decreases in pH_o, if present.

The interaction of the aforementioned three factors can produce any of three possible patterns of pH_i changes during the plateau phase. In the first (Fig. 3, left), acid extrusion (factor 3) exceeds acid loading (factors 1 and 2), and pH_i recovers toward its initial value. This is observed when snail neurons[91] and several other preparations are exposed to elevated pCO_2 at normal or elevated pH_o. The pattern of pH_i recovery also would have been observed in the experiments of Fig. 2, had pH_o been sufficiently high (e.g., pH 8.0). In the second pattern (Fig. 3, middle), acid loading is exactly balanced by acid extrusion, and pH_i is stable. An example is the response of the salamander renal proximal tubule to CO_2/HCO_3^- Ringer at constant pH_o.[16] The stimulation of acid extrusion (factor 3) by the CO_2-induced fall in pH_i is offset by a large steady-state efflux of HCO_3^- or an equivalent ion (factor 1). The pattern of a stable plateau phase is also seen in the three CO_2 pulses of Fig. 2. In that experiment, acid extrusion is inhibited by the relatively low pH_o, and the rate of HCO_3^- efflux is probably correspondingly low. The third pattern (Fig. 3, right) is a plateau-phase pH_i decline, which occurs when acid loading (factors 1 and 2) outstrips acid extrusion (factor 3). This pattern has been observed when barnacle muscle fibers are exposed to the weak acid DMO, while the acid-extrusion mechanism is blocked by lowering pH_o.[55]

4.2. Neutral Weak Bases and Their Cationic, Conjugate Weak Acids

Solutions of a neutral, monoprotic weak base (B) always contain the conjugate weak acid (HB^+). Because of the differential permeability of the cell membrane to these two buffer species, the fluxes of B usually have a dominant effect on pH_i, at least in the short term.

4.2.1. Flux of the Neutral Weak Base

Raising $[B]_o$ produces a relatively rapid intracellular alkalinization as B enters the cell and combines with H^+:

$$B + H^+ = HB^+ \quad (15)$$

The alkalinization halts when $[B]_i = [B]_o$, after which time pH_i would be expected to stabilize, provided there is no net movement of pH_i-related ions (e.g., H^+ and HB^+) across the membrane. The magnitude of the alkalinization is inversely proportional to β_i^T and directly related to $[B]_o$. Furthermore, because of the relation between pH_i and the pK governing Eq. (15), the B-induced alkalinization increases as the initial pH_i is reduced. The intracellular alkalinizing effects of a neutral weak base were first documented for NH_3 by Jacobs in 1922.[53] Similar effects have since been demonstrated for a variety of other weak bases.

4.2.2. Flux of the Cationic, Conjugate Weak Acid

After the initial alkalinization caused by the application of a weak base B, the subsequent course of pH_i is determined by at least three factors: (1) the passive flux of HB^+, (2) the metabolic production of acid (if any), and (3) acid extrusion and other transport processes. An argument analogous to the one developed above for the A^- gradient shows that the HB^+ gradient is the same as that for H^+, provided $[B]_i = [B]_o$:

$$\frac{[H^+]_0}{[HB^+]_0} = \frac{K}{[B]_0} = \frac{K}{[B]_i} = \frac{[H^+]_i}{[HB^+]_i} \tag{16}$$

or

$$\frac{[H^+]_i}{[H^+]_0} = \frac{[HB^+]_i}{[HB^+]_0} \tag{16a}$$

where K is the apparent acid dissociation constant. Normally, the H^+ electrochemical gradient favors the influx of H^+, so that HB^+ also tends to enter the cell. Figure 4 illustrates an experiment in which a barnacle muscle fiber was exposed to Ringer containing NH_3/NH_4^+. The initial rise of pH_i (segment ab) is caused by the influx of NH_3 (see above), whereas the subsequent pH_i fall (*bc*) is caused by the net influx of NH_4^+, a small fraction of which dissociates intracellularly into H^+ and NH_3 (see Fig. 4B). When the external NH_3/NH_4^+ is removed, virtually all intracellular NH_4^+ (including that which had previously accumulated as the result of NH_4^+ influx) gives up its H^+ and exits as NH_3. Because the H^+ is trapped inside the cell, pH_i falls rapidly (*cd*) and undershoots its initial value. Both the plateau-phase acidification (*bc*) and the pH_i undershoot (compare *a* and *d*) are manifestations of NH_4^+ influx. First demonstrated in squid axons,[17,18] the pattern of pH_i changes shown in Fig. 4 has since been observed in a wide variety of cells. When the NH_4^+ electrochemical gradient is reversed in barnacle muscle by depolarizing the cell, pH_i rises during the plateau-phase, and returns to a higher-than-normal value after removal of the NH_3/NH_4^+.[10]

There is evidence that, at least in some preparations, the plateau-phase acidification in an NH_3/NH_4^+ experiment is augmented by ion-transport systems. Aickin and Thomas[3] have shown that, in low-K^+ Ringer, the Na^+, K^+ pump of mouse soleus muscle can mediate the uptake of NH_4^+. In sheep cardiac Purkinje fibers, Vaughan-Jones[97] has shown that a portion of the plateau-phase acidification is due to HCO_3^- efflux mediated by a $Cl^-HCO_3^-$ exchanger. Regardless of the mechanism, the uptake of NH_4^+ is ultimately reflected in a substantial pH_i undershoot when the external NH_3/NH_4^+ is removed. The practical benefit of this undershoot, which has proven a powerful tool in studying the mechanism of acid extrusion, is that the cell is acutely acid loaded.

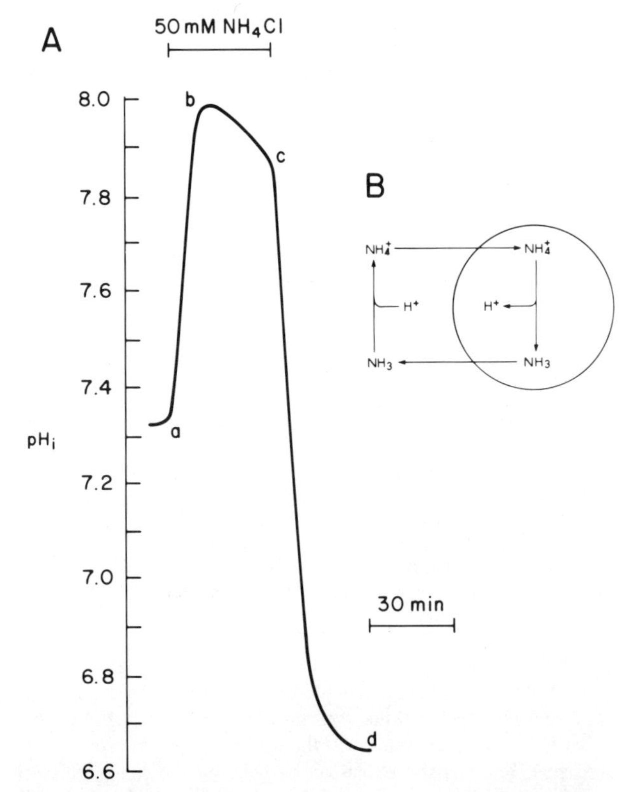

Fig. 4. The effect on pH_i of exposing a cell to a solution containing NH_3 and NH_4^+. The pH_i of a giant-barnacle muscle fiber was measured with a Hinke-style microelectrode.[51] At point *a*, 50 mM Na^+ was replaced with 50 mM NH_4^+ at constant pH_o. This causes a rapid rise in pH_i, due to the influx of NH_3, which is protonated to form NH_4^+ once inside the cell. The entry of NH_4^+ occurs simultaneously, and tends to lower pH_i as a small fraction of the entering NH_4^+ dissociated to form H^+ and NH_3. However, this acidifying influence is not obvious until *b*, when the net influx of NH_3 slows. The influx of NH_4^+ is clearly evident during the plateau phase (*bc*), when the generation of H^+ and NH_3 from incoming NH_4^+ now causes pH_i to fall and NH_3 to exit the cell (see inset, B). When external NH_3 and NH_4^+ are removed, nearly all intracellular NH_4^+ dissociates into H^+ and NH_3. The latter rapidly leaves the cells, leaving the former orphaned inside. This release of H^+ causes a larger undershoot of pH_i (*cd*). From Ref. 12. Reproduced with permission of Springer-Verlag.

5. Ion-Transport Systems

Five groups of systems have been identified that transport either H^+ and/or HCO_3^- across the cell membrane of animal cells: (1) a Na^+/HCO_3^-–Cl^-/H^+ exchanger (or its equivalent), (2) a Na^+/H^+ exchanger, (3) a Cl^-/HCO_3^- exchanger, (4) the K^+/H^+ exchange pump of the stomach, and (5) the electrogenic H^+ pump of tight epithelia. The first two transporters are clearly pH_i regulators: they respond to intracellular acid loads by increasing their acid-extrusion rate. In the steady state, these transporters counter the acidifying influence of passive H^+/HCO_3^- fluxes and of metabolism, and prevent pH_i from falling to very low levels (see Introduction). The Cl^-/HCO_3^- exchanger probably functions as an acid loader under normal conditions. The K^+-H^+ and H^+ pumps extrude H^+ against very large electrochemical gradients from specialized epithelial cells. Inasmuch as these last two transport systems have yet to be studied with respect to their effects of pH_i, they will not be further discussed in this chapter. The reader is directed to reviews on the gastric K^+/H^+ pump[83] and electrogenic H^+ pump.[87]

In addition to the aforementioned transporters, the Na^+/Ca^{2+} exchanger is also capable of producing pH_i changes, at least in cardiac Purkinje fibers. Meech and Thomas[62,63] have elegantly demonstrated that Ca^{++} injected into snail neurons causes a fall in pH_i, part of which is due to Ca^{++}/H^+ exchange by the mitochondria. A sudden infusion of Ca^{++} is also expected to lower pH_i by causing a release of H^+ from certain buffers for which Ca^{++} and H^+ compete. In experiments on cardiac Purkinje fibers, Vaughan-Jones *et al.*[98] provided evidence that Ca^{++} influx mediated by the Na-Ca^{++} exchanger can indeed produce a fall in pH_i.

5.1. Na^+/HCO_3^-–Cl^-/H^+ Exchange

Several invertebrate cells, including the squid giant axon,[14,18,23,81] snail neuron,[90–92] giant-barnacle muscle fiber,[10,20,21,24] and crayfish neuron,[65] regulate their pH_i by means of a transporter which appears to exchange external Na^+ and HCO_3^- for internal Cl^- and H^+. Recent work also indicates that a similar transport system may coexist with the Na^+/H^+ exchanger in the A431 epidermoid-carcinoma cell line.[80] Although we shall refer to this transporter as a Na^+/HCO_3^-–Cl^-/H^+ exchanger, the four models of Fig. 5 all have the same ion requirements and are thermodynamically equivalent. The transporters in these four preparations bear a strong superficial resemblance to one another, but are functionally distinctive and may be fundamentally different.

5.1.1. pH_i and pH_o Dependence

When cells are acutely acid loaded, the sudden decline of pH_i is followed by a spontaneous pH_i recovery along an approximately exponential time course. The acid-extrusion rate (J_H), defined as the equivalent H^+ efflux, is the product of the pH_i-recovery rate and β_i^T. J_H is greatest at low pH_i values[92] and approaches zero as pH_i returns toward normal.[20,24] As discussed below, this pH_i dependence may be due to an intracellular, allosteric site for H^+, in addition to any substrate effect of internal H^+. The transporter thus functions as a graded pH stat, extruding acid at a rate proportional to the degree of intracellular acid loading. The Na^+/HCO_3^--Cl^-/H^+ exchanger is also inhibited by decreases in pH_o at constant $[HCO_3^-]_o$.[18,20]

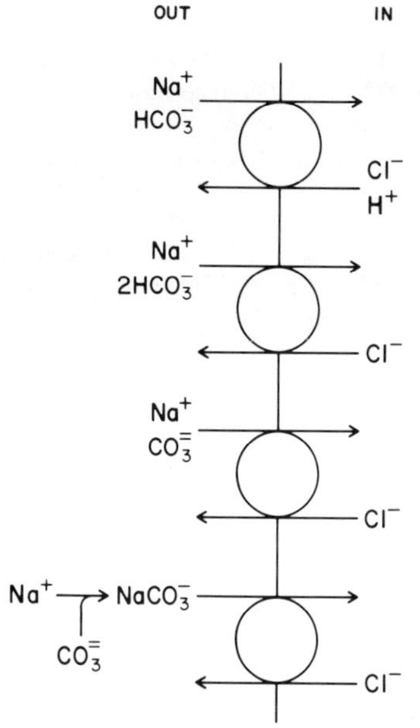

Fig. 5. Four models of the Na^+/HCO_3^-–Cl^-–H^+ pH_i regulator. The top model has external Na^+ and HCO_3^- exchanging for internal Cl^- and H^+. In the second, the exit of the H^+ is replaced by the entry of a second HCO_3^-. In the third, the entry of two HCO_3^- is replaced by the entry of a single CO_3^{2-}. Finally, in the fourth model, the separate entry of Na^+ and CO_3^{2-} is replaced by the influx of a single $NaCO_3^-$ ion pair. From Ref. 23. Reproduced with permission of the Rockefeller University Press.

5.1.2. Dependence on Na^+, HCO_3^-, and Cl^-

This transporter has an absolute requirement for external Na^+,[92] external HCO_3^-,[18] and internal Cl^-.[81] If any one of these substrates is absent, transport haults. The stoichiometry of the squid axon's transporter has been determined using pH-sensitive microelectrodes to measure J_H and radioisotopes to assay net fluxes of Na^+ and Cl^-.[23] The results confirm the predictions of each of the four models of Fig. 5: one Na^+ enters the cell for each Cl^- leaving, and for every two protons neutralized inside the cell. Ion-sensitive microelectrode experiments[92] had previously suggested that acid extrusion is accompanied by a net influx of Na^+, and a net efflux of Cl^-. The quantitative dependence of J_H on each of these three substrates is well described by Michaelis–Menten kinetics. Under standard conditions ($pH_o = 8.0$), the apparent K_m (K'_m) for external Na^+ is approximately 60–80 mM, and K'_m for external HCO_3^- is approximately 2–4 mM in both barnacle muscle[21] and squid axons.[14,23] The K'_m for internal Cl^- in squid axons is $\sim$ 84 mM.[23] Studies in which the external Na^+ dependence of J_H is examined at different, fixed $[HCO_3^-]_o$ levels, and the external HCO_3^- dependence at different, fixed $[Na^+]_o$ levels can shed light on possible mechanisms of transport. For example, the $NaCO_3^-$ ion-pair model (Fig. 5, bottom) predicts that J_H should not be a unique function of $[Na^+]_o$ or $[HCO_3^-]_o$ *per se*, but of $[NaCO_3^-]_o$. The concentration of this ion pair is in turn proportional to the product of $[Na^+]_o$ and $[HCO_3^-]_o$. That is, when J_H is plotted as a function of $[NaCO_3^-]_o$, all values should fall on the same Michaelis–Menten curve, regardless of the $[Na^+]_o$ or $[HCO_3^-]_o$. These predictions have been examined in barnacle muscle and squid axons. In the former,[21] external Na^+ dependence data obtained at a $[HCO_3^-]_o$ of 10 mM falls on a J_H vs. $[NaCO_3^-]_o$ curve quite distinct from the one describing analogous data obtained at an $[HCO_3^-]_o$ of 2.5 mM. More recently (T. J. Wilding and A. Roos, personal communication), the converse experiment has been performed in barnacle muscle. It was found that two plots of J_H vs. $[HCO_3^-]_o$, obtained at two $[Na^+]_o$ values, fall on two distinct curves. Thus, the $NaCO_3^-$ ion-pair model cannot account for the kinetic data in barnacle muscle. The situation, however, is quite different in squid axons. In this preparation, the external Na^+ dependence has been examined at three values of $[HCO_3^-]_o$, and the external HCO_3^- dependence, at three levels of $[Na^+]_o$. In each of the six cases, the data fall on the same J_H vs. $[NaCO_3^-]$ curve, consistent with the $NaCO_3^-$ ion-pair model. Although these last results do not prove that external $NaCO_3^-$ is the true substrate for the squid axon, they are suggestive of fundamental differences between squid axon and barnacle muscle.

5.1.3. Energetics

A simple analysis shows that there is normally sufficient energy in the electrochemical gradients of Na^+, HCO_3^-, Cl^-, and H^+ to support acid extrusion at pH_i values of 8.0 and above. This raises two questions: In the first place, is acid extrusion indeed driven only by ion gradients, or does ATP hydrolysis also contribute? Second, why does J_H fall toward zero at a pH_i of $\sim$ 7.4, even though there is enough energy in the ion gradients to drive pH_i much higher? Regarding the question of the energy source for transport, it is known that the squid axon's pH_i regulator is blocked by cyanide or DNP,[18] but that this blockade is overcome by the addition of exogenous ATP.[81] Further support for ATP as an energy source is the observation that the squid axon's pH_i regulator has thus far been impossible to reverse (Boron, unpublished), even with the gradients for each of the transported ions reversed simultaneously. These observations by no means prove that ATP hydrolysis fuels acid extrusion in squid axons. ATP could be required for a permissive phosphorylation. Furthermore, the failure to reverse transport could reflect kinetic impediments rather than thermodynamic ones. Thus, an energy-providing role for ATP in squid axons remains speculative. On the other hand, the pH_i regulator is readily reversed in both barnacle muscle[82] and snail neurons.[42,43] Also, sufficient metabolic inhibition to block the Na^+/K^+ pump in snail neuron fails to significantly slow the pH_i-regulating system.[90] These observations strongly favor a transport system driven soley by the gradients of the transported ions, at least in barnacle muscle and snail neurons.

As far as the pH_i dependence of acid extrusion is concerned, there is strong evidence that the pH_i sensitivity of J_H is more complex than would be expected if intracellular H^+ were merely a substrate. In barnacle muscle, the pH_i-regulating mechanism mediates Na^+ efflux and Cl^- influx, in addition to the expected Na^+ influx and Cl^- efflux. All four of these unidirectional fluxes are stimulated at low pH_i values.[24,82] If the Na^+ efflux and Cl^- influx represented the microscopic reversibility of the transporter, then low pH_i should inhibit, rather than stimulate them. The data are thus consistent with an intracellular modifier site which, when protonated, enhances transport in all modes.

5.1.4. Pharmacology

The Na^+/HCO_3^-–Cl^-/H^+ exchanger is irreversibly blocked by the stilbene derivatives 4-isothiocyano-4′-acetamido-2,2′-stilbene disulfonate (SITS) or 4,4′-diisothiocyano-2,2′-stilbene disulfonate (DIDS). Inasmuch as these compounds are amino-reactive agents, the mechanism of the irreversible inhibition is presumably the reaction of the stilbenes' isothiocyano moieties with a free amine on the transporter, forming an *N,N*-disubstituted thiourea. However, even stilbenes without amino-reactive groups can inhibit transport, though only reversibly. An example is 4,4′-dinitro-2,2′-stilbene disulfonate (DNDS), which inhibits acid extrusion by ~ 80% at a concentration of 1 mM.[23] These results are similar to those of analogous experiments on the erythrocyte's Cl^-/HCO_3^- exchanger (for review, see Refs. 27, 28 and Chapter 12). Work on that exchanger suggests that, for all the aforementioned stilbenes, there is a rapid and reversible interaction between cationic groups on the transporter and sulfonate groups on the stilbene. Those stilbenes also possessing amino-reactive groups (e.g., SITS and DIDS) eventually undergo the slower and irreversible covalent reaction. It is noteworthy that, in barnacle muscle, two nonstilbene, amino-reactive agents also inhibit the Na^+/HCO_3^-–Cl^-/H^+ system, *p*-isothiocyanatobenzenesulfonate and pyridoxal phosphate.[10]

5.2. Na^+/H^+ Exchange

Whereas the Na^+/HCO_3^-–Cl^-/H^+ exchanger is the primary pH_i regulator for the invertebrate cells thus far examined, the Na^+/H^+ exchanger fills this niche for vertebrate cells. The Na^+/H^+ exchanger has been identified in a wide variety of preparations, including skeletal muscle,[1,2,4] cardiac tissue,[38,72] renal tubules,[15,32,56,71] gallbladder,[102] and several cultured cell lines.[66,68,74,100]

5.2.1. pH_i and pH_o Dependence

Na^+/H^+ exchangers have a pH_i and pH_o dependence similar to that of $Na^+/HCO_3^-/Cl^-/H^+$ exchangers. The behavior of pH_i following an acute intracellular acid load has been examined in several vertebrate preparations, including mouse skeletal muscle,[4] renal tubules,[15,32] cardiac tissue,[38,72] and cultured cells.[67,80] After its abrupt fall due to the acid load, pH_i recovers along an approximately exponential time course. Assuming that both the acid loading rate (J_L) and β_i^T are invariant of pH_i, then the exponential shape of the pH_i recovery implies that the Na^+/H^+ exchange rate (J_H) varies linearly with pH_i, increasing at lower pH_i values. When the pH_i recovery is complete and pH_i is once again in a steady state, J_H is by definition equal to J_L. J_H does not necessarily equal zero, however. In the special case in which J_L is zero, then the asymptotic value pH_i approaches is termed the threshold pH_i,[11] and designated pH_i'. If the above assumptions are valid, J_H is zero at pH_i values exceeding pH_i', and linearly increases as pH_i falls below pH_i' (see Fig. 6). In practice, however, J_L is rarely zero, and the steady-state pH_i must therefore be lower than pH_i'. In the example of Fig. 6, J_2 represents the normal J_L, and line B, the normal dependence of J_H on pH_i. In the steady state, $J_H = J_L = J_2$ and $pH_i = pH_b$. When the cell is acutely loaded, pH_i recovers along line B, and pH_i will stabilize at the pseudothreshold of pH_b, provided J_L is fixed at J_2. Note that as long as the acid-loading rate is fixed, the rate constant of the exponential pH_i recovery (proportional to the slope of line B) is the same as if J_L were zero. The only effect of raising J_L to a fixed, nonzero value is that the steady-state pH_i (i.e., pseudothreshold) is lower than the true threshold. A practical way of determining the magnitude of J_L is to rapidly block acid extrusion (but not reverse it or alter other pH-related transport rates). J_L is calculated from the initial rate of the pH_i decline.

The only preparation in which pH_i is regulated by Na^+/H^+ exchange, and in which J_L has been demonstrated to be very low, is the salamander proximal tubule incubated at 22°C in an

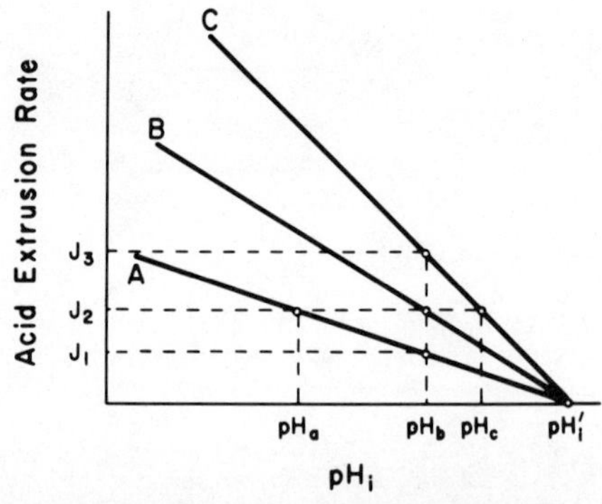

Fig. 6. Idealized dependence of a pH_i-regulating system on pH_i and other kinetic parameters. Lines A, B, and C denote the dependence of acid extrusion rate (J) on pH_i under three different kinetic conditions. In all cases, J is maximal at low pH_i values and falls to zero at values above the threshold (pH_i'). From Ref. 79. Reproduced with permission of the American Physiological Society.

HCO_3^--free Ringer.[15] When these cells are exposed to high levels of the Na^+/H^+ exchange inhibitor amiloride (see below), there is no fall in pH_i; indeed, pH_i rises slightly because amiloride is a weak base. Thus, in the absence of CO_2/HCO_3^-, J_L and J_H are both very low, and the steady-state pH_i (~ 7.45) is probably only slightly below the threshold. In the presence of CO_2/HCO_3^-, however, the continuous exit of HCO_3^- across the cell's basolateral (i.e., "blood-side") membrane imposes a substantial intracellular acid load, and causes the steady-state pH_i to fall by ~ 0.15. In the scheme of Fig. 6, introducing the HCO_3^- efflux raises J_L from zero to J_2, and causes the steady-state pH_i to fall from ~ pH_i' to pH_b. The exponential pH_i-recovery for this preparation in the absence of CO_2/HCO_3^- implies that the linear extrapolation of the J_H vs. pH_i relation should yield the pH_i at which J_H is truly zero (i.e., the threshold for the Na^+/H^+ exchanger). The present pH_i and buffering data are not of sufficiently high quality to preclude the possibility that the dependence of J_H on pH_i is curvilinear at high pH_i values that J_H remains slightly above zero at pH_i values above the apparent threshold.

As in the case for the $Na^+/HCO_3^--Cl^-/H^+$ exchanger, the basis for the Na^+/H^+ exchanger's apparent threshold pH_i is kinetic rather than thermodynamic. Taking the salamander proximal tubule as an example, the values of internal and external Na^+ activity and of pH_o predict that there should be sufficient energy in the Na^+ gradient to drive pH_i to 8.0 or above. This is more than 0.5 higher than the estimated pH_i'. The apparent threshold of the Na^+/H^+ exchanger therefore reflects kinetic factors. One reason for the stimulation of Na^+/H^+ exchange at low pH_i is probably increased substrate (i.e., intracellular H^+). However, this is unlikely to be the complete explanation, inasmuch as Michaelis–Menten kinetics predict that the dependence of J_H on pH_i should be sigmoidal, not linear. A second explanation was provided in experiments analogous to earlier studies on the $Na^+/HCO_3^--Cl^-/H^+$ system. Working with membrane vesicles derived from the renal cortex (primarily from proximal-tubule luminal membranes), Aronson *et al.*[6] showed that the Na^+ efflux mediated by the Na^+/H^+ exchanger is stimulated by low pH_i. If H^+ were merely competing with Na^+ for an internal binding site, then the Na^+ efflux should have fallen at low pH_i. Thus, the H^+ probably acts at an intracellular modifier site in addition to serving as a substrate for transport.

Low pH_o has been shown to inhibit Na^+/H^+ exchange in intact cells.[15] A detailed analysis of the effects of pH_o on Na^+/H^+ exchange kinetics has been made in membrane vesicles.[7] The rate of Na^+/H^+ exchange (measured as a Na^+ influx) is inhibited by low pH_o and enhanced by high pH_o, the apparent pK for the effect is ~ 7.4.

5.2.2. Involvement of Na^+

Aickin and Thomas[4] were the first to show that the pH_i recovery that follows an acute acid load in vertebrates is dependent on external Na^+. This has been confirmed in a variety of preparations.[15,32,38,67,72] Not only is the pH_i recovery dependent upon the presence of external Na^+, but it is presumably accompanied by a net influx of Na^+, as demonstrated by a rise of intracellular Na^+ activity,[4,15] and increased influx of radiolabeled Na^+.[46,66,74] Studies on membrane vesicles[57] have shown that the Na^+/H^+ exchanger mediates an isotopic efflux of Na^+ in addition to the expected influx. The apparent K_m for external Na^+ for the forward-running Na^+/H^+ exchanger (assayed either as a pH_i recovery or as a Na^+ influx) varies widely, depending on the preparation and the conditions of incubation. For example, the K_m is ~ 6 mM in rabbit renal cortical vesicles at pH 7.5,[58] 28 mM in MDCK cells at pH 9,[74] and 95 mM in MDCK cells at pH 6.[74] Li^+ and apparently NH_4^+[56,58] can partially replace Na^+ on the Na^+/H^+ exchanger, but K^+ and Cs^+ cannot.[56]

5.2.3. Energetics

There have been no direct, rigorous measurements of Na^+/H^+ exchange stoichiometry. In one study on intact cells,[15] measurements of pH_i and intracellular Na^+ activity changes during pH_i recovery from acute loads indicated that the H^+/Na^+ ratio is approximately 0.8, consistent with a 1 : 1 exchange. In another study on vesicles,[59] an imposed Na^+ gradient was used to drive net Na^+/H^+ exchange either forward or backward. Net transport fell to zero when the transmembrane Na^+ and H^+ gradients were approximately equal, as predicted if Na^+/H^+ exchange is driven solely by ion gradients, and the stoichiometry is 1 : 1. An ATP requirement for Na^+/H^+ exchange has never been described and, indeed, would be inconsistent with the preceding null-point experiment. Apparent reversal of Na^+/H^+ exchange has also been demonstrated in intact cells: removal of external Na^+ leads to a fall of pH_i that is inhibited by amiloride.[72] Such results must be interpreted with caution, however. The application of amiloride can temporarily slow a pH_i decline merely by entering the cell as a neutral weak base, and combining with intracellular H^+. Furthermore, a pH_i decrease elicited by removing external Na^+ might be augmented by reversal of Na^+/Ca^{++} exchange.

5.2.4. Pharmacology

The pharmacologic *sine qua non* of Na^+/H^+ exchange is sensitivity to the diuretic amiloride, first described in mouse soleus muscle,[4] and later demonstrated in numerous preparations. The drug is apparently competitive with Na^+,[57] although some investigators suggest that the inhibition is mixed.[77] The apparent K_I for amiloride is generally in the range of 10 to 30 μM (see Refs. 57, 75). Recently, several amiloride analogs have been described. One is more than 100 times more potent then amiloride, when Na^+/H^+ exchange is assayed as Na^+ influx in acid-loaded cells.[99] Na^+/H^+ exchange is also inhibited by harmaline.[56]

In intact cells, Na^+/H^+ exchange is insensitive to the stilbene derivative SITS,[15,75] and to the nominal removal of Cl^-[15,75] or HCO_3^-.[15] The pH_i-regulating mechanisms of a variety of mammalian preparations are similarly insensitive to ouabain.[4,41,77]

5.2.5. Hormones and Growth Factors

There is considerable evidence, based on DMO measurements of steady-state pH_i, that AMP may influence intracellular acid–base homeostasis in rat cardiac and skeletal muscle. When rat cardiac muscle is perfused *in vivo*, its ability to stabilize pH_i in the face of a CO_2-induced acid load is augmented by norepinephrine,[73] dibutyryl cAMP,[44] and glucagon,[44] but diminished by β-adrenergic blockade.[73] The infusion into the whole animal of a β-adrenergic blocker similarly reduced the ability of skeletal muscle to keep pH_i at a normal level during respiratory or metabolic acidosis.[33] Although these results suggest that intracellular cAMP may enhance acid extrusion (i.e., Na^+/H^+ exchange), an inhibitory effect on one or more acid-loading processes (e.g., HCO_3^- efflux) cannot presently be ruled out.

Work on frog skeletal muscle[69,70] suggests that Na^+/H^+ exchange may be stimulated by insulin. Application of this hormone to quiescent muscle results in a small (0.1–0.2) rise in pH_i and an increased uptake of Na^+. Parathyroid hormone (PTH) and several chronic electrolyte disturbances apparently influence Na^+/H^+ exchange in the rabbit proximal tubule. Acid secretion by the proximal tubule is stimulated by chronic hypercapnia,[34] and inhibited by PTH.[39] Na^+/H^+ exchange by cortical vesicles (primarily from the proximal tubule) is enhanced when the donor animals are made chronically metabolically acidotic,[35,94] are chronically K^+ depleted,[84] or are parathyroidectomized.[35] Furthermore, it is inhibited by PTH and cAMP.[54]

A number of workers[67,80,85] have shown that the application of a growth factor to quiescent, serum-starved cells in culture causes pH_i to rise by 0.1–0.2. The pH_i increase is produced by the Na^+/H^+ exchanger, and may involve an alkaline shift of the pH_i threshold for activation of the Na^+/H^+ exchanger.[67] It was previously known that such growth factors restore the ability of such cells to divide ~ 24 hr after the application of the growth factor. It is thus possible that the growth-factor-stimulated rise of pH_i, which occurs several minutes after the application of the growth factor, plays a permissive or even a triggering role in the subsequent cell division. A role for intracellular alkalinization in the activation of quiescent cells has been established in *Artemia* blastocytes.[25]

5.3. Cl^-/HCO_3^-, Exchange and Other HCO_3^- Pathways

Cl^-/HCO_3^- exchange has been well documented in erythrocytes, where this exchange is responsible for the "chloride shift" that enables the erythrocytes to transport large amounts of CO_2 from the periphery to the lungs. The reader is directed to several excellent reviews on the subject, which detail kinetics and pharmacology of this transporter (see Refs. 27, 28, 60). Cl^-/HCO_3^- exchangers have also been identified in cardiac Purkinje fibers[96] and the kidney PK-1 cell line,[30] and have been implicated in mammalian skeletal muscle[4] and in several epithelia. It is not known whether these latter Cl^-/HCO_3^- exchangers are identical either among themselves, or to the erythrocyte's exchanger.

If the transporter exchanges one Cl^- for one HCO_3^-, then it is easy to show that it must normally operate in the direction of net HCO_3^- efflux, for cells other than erythrocytes. For example, pH_i is usually only slightly less than pH_o under normal conditions, so that the transmembrane HCO_3^- gradient approximates unity. In contrast, $[Cl^-]_o$ is generally far higher than $[Cl^-]_i$. Thus, the inward Cl^- gradient should drive HCO_3^- out of the cell and lower pH_i. Such an intracellular acid load could be turned to the organism's advantage if the Cl^-/HCO_3^- exchanger were confined to the basolateral (i.e., "blood-side") membrane of acid-secreting epithelia, and an acid-extruding mechanism for regulating pH_i were present at the apical (i.e., facing the environment) membrane. The basolateral Cl^-/HCO_3^- exchanger would cause the epithelial cell to be acid loaded across the basolateral membrane, whereas the apical acid-extrusion mechanism would secrete this acid toward the environment. The net effect would be secretion of acid from one side of the epithelial cell to the other.

The teleological role of Cl^-/HCO_3^- exchange in non-epithelial cells is unclear. The continuous acid load translates to a metabolic expense, inasmuch as the efflux of HCO_3^- must be countered by Na^+/H^+ exchange. However, this may be the price the cell pays for being able to extrude acute alkali loads. Indeed, Vaughan-Jones[97] has shown that when cardiac Purkinje fibers are acutely alkali loaded, Cl^-/HCO_3^- exchange helps return pH_i toward its initial level. Cl^-/HCO_3^- exchange may also play a role in intracellular Cl^- homeostasis, with resultant disturbances in pH_i being corrected by the Na^+/H^+ exchanger.

In addition to electroneutral Cl^-/HCO_3^- exchange, two conductive pathways for HCO_3^- (or a related species) have been suggested. The first, a conductive pathway for HCO_3^- or OH^- at the basolateral membrane of the rat proximal tubule, was postulated by Fromter *et al.*[47] Changes in basolateral $[HCO_3^-]$ produce abrupt shifts in V_m that are consistent with a simple HCO_3^- conductance. These V_m changes are blocked by the carbonic anhydrase inhibitor acetazolamide.[47] Interestingly, similar V_m transients are also seen in the nominal absence of CO_2/HCO_3^-, when the basolateral levels of glycodiazide or butyrate are suddenly altered. Furthermore, acetazolamide blocks these V_m transients even though CO_2/HCO_3^- is nominally absent. The authors have suggested that the ionic species actually traversing the membrane may be OH^-. According to this view, neutral weak acids (e.g., CO_2, glycodiazine, and butyric acid) would diffuse through the membrane in parallel to the OH^-, making it appear that the anionic, conjugate weak base (e.g., HCO_3^-, glycodiazide anion, and butyrate) was the permeating species. Presumably, acetazolamide would directly block the OH^- channel.

The second, proposed conductive pathway for HCO_3^- is a transporter that mediates the tightly coupled movement of HCO_3^- (or an equivalent species) and Na^+ across the basolateral membrane of the salamander proximal tubule.[16] A reduction of basolateral $[HCO_3^-]$ or $[Na^+]$ leads to a rapid fall of pH_i and a spiking depolarization of the basolateral membrane. Returning $[HCO_3^-]$ or $[Na^+]$ to its original level produces a rapid recovery of pH_i and a spiking basolateral hyperpolarization. Not only does a reduction of basolateral $[Na^+]$ lower pH_i (i.e., $[HCO_3^-]_i$), but a reduction of basolateral $[HCO_3^-]$ lowers intracellular Na^+ activity. These changes are all blocked by SITS, suggesting that HCO_3^- transport is tightly coupled by Na^+ in an electrogenic fashion. The simplest stoichiometry would be 1 $Na^+ : 2\,HCO_3^-$. Cl^- appears not to be involved. Whether transport is driven by a Na^+ gradient or a HCO_3^- gradient, it is inhibited to a similar extent ($K_I = 5\ \mu M$) by acetazolamide.[19] Biagi and Sohtell[9] have reported, in the rabbit proximal straight tubule, V_m changes similar to those observed in the salamander. In particular, a reduction in basolateral $[Na^+]$ produced a spiking depolarization. These last data suggest that the proposed electrogenic Na^+/HCO_3^- transporter may not be confined to amphibians.

ACKNOWLEDGMENT. This work was supported by a Research Career Development Award (K01-AM-01022). The author is a Searle Scholar.

References

1. Abercrombie, R. F., R. W. Putnam, and A. Roos. 1983. The intracellular pH of frog skeletal muscle: Its regulation in isotonic solution. *J. Physiol. (London)* **345**:175–187.
2. Abercrombie, R. F., and A. Roos. 1983. The intracellular pH of

frog skeletal muscle: Its regulation in hypertonic solutions. *J. Physiol. (London)* **345**:189–204.

3. Aickin, C. C., and R. C. Thomas. 1977. Microelectrode measurement of the intracellular pH and buffering power of mouse soleus muscle fibres. *J. Physiol. (London)* **267**:791–810.
4. Aickin, C. C., and R. C. Thomas. 1977. An investigation of the ionic mechanism of intracellular pH regulation in mouse soleus muscle fibres. *J. Physiol. (London)* **273**:295–316.
5. Ammann, D., F. Lanter, R. A. Steiner, P. Schulthess, Y. Shijo, and W. Simon. 1981. Neutral carrier based hydrogen ion selective microelectrode for extra- and intracellular studies. *Anal. Chem.* **53**:2267–2269.
6. Aronson, P. S., J. Nee, and M. A. Suhm. 1982. Modifier role of internal H^+ in activating the Na^+–H^+ exchanger in renal microvillus membrane vesicles. *Nature (London)* **229**:161–163.
7. Aronson, P. A., M. A. Suhm, and J. Nee. 1983. Interaction of external H^+ with the Na^+–H^+ exchanger in renal microvillus membrane vesicles. *J. Biol, Chem.* **258**:6767–6771.
8. Baylor, S. M., W. K. Chandler, and M. W. Marshall. 1982. Optical measurements of intracellular pH and magnesium in frog skeletal muscle fibres. *J. Physiol. (London)* **331**:105–137.
9. Biagi, B., and M. Sohtell. 1984. Bicarbonate voltage transients in the rabbit proximal tubule. *Kidney Int.* **25**:271a.
10. Boron, W. F. 1977. Intracellular pH transients in giant barnacle muscle fibers. *Am. J. Physiol.* **233**:C61–C73.
11. Boron, W. F. 1980. Intracellular pH regulation. *Curr. Top. Membr. Transp.* **13**:3–22.
12. Boron, W. F. 1983. Transport of H^+ and of ionic weak acids and bases. *J. Membr. Biol.* **72**:1–16.
13. Boron, W. F. 1984. Control of intracellular pH. In: *The Kidney: Physiology and Pathology*. D. W. Seldin and G. Giebisch, eds. Raven Press, New York. Pp. 1417–1439.
14. Boron, W. F. 1985. Intracellular pH-regulating mechanism of the squid axon: Relation between the external Na^+ and HCO_3^- dependencies. *J. Gen. Physiol.* **85**:325–345.
15. Boron, W. F., and E. L. Boulpaep. 1983. Intracellular pH regulation in salamander proximal tubules: Na–H exchange. *J. Gen. Physiol.* **81**:29–52,
16. Boron, W. F., and E. L. Boulpaep. 1983. Intracellular pH regulation in salamander proximal tubules: Basolateral HCO_3^- transport. *J. Gen. Physiol.* **81**:53–94.
17. Boron, W. F., and P. DeWeer. 1976. Intracellular pH transients in squid giant axons caused by CO_2, NH_3, and metabolic inhibitors. *J. Gen. Physiol.* **67**:91–112.
18. Boron, W. F., and P. DeWeer. 1976. Active proton transport stimulated by CO_2/HCO_3^-, blocked by cyanide. *Nature (London)* **259**:240–241.
19. Boron, W. F., and P. Fong. 1983. Effect of carbonic anhydrase inhibitors on basolateral HCO_3^- transport in salamander proximal tubules. *Kidney Int.* **23**:230.
20. Boron, W. F., W. C. McCormick, and A. Roos. 1979. pH regulation in barnacle muscle fibers: Dependence on intracellular and extracellular pH. *Am. J. Physiol.* **237**:C185–C193.
21. Boron, W. F., McCormick, W. C., and A. Roos. 1981. pH regulation in barnacle muscle fibers: Dependence on extracellular sodium and bicarbonate. *Am. J. Physiol.* **240**:C80–C89.
22. Boron, W. F., and A. Roos. 1976. Comparison of microelectrode, DMO, and methylamine methods for measuring intracellular pH. *Am. J. Physiol.* **231**:799–809.
23. Boron, W. F., and J. M. Russell. 1983. Stoichiometry and ion dependencies of the intracellular pH-regulating mechanism in squid giant axons. *J. Gen. Physiol.* **81**:373–399.
24. Boron, W. F., J. M. Russell, M. S. Brodwick, D. W. Keifer, and A. Roos. 1978. Influence of cyclic AMP on intracellular pH regulation and chloride fluxes in barnacle muscle fibres. *Nature (London)* **276**:511–513.
25. Busa, W. B., and J. H. Crowe. 1983. Intracellular pH regulates transitions between dormancy and development of brine shrimp (*Artemia salina*) embryos. *Science* **221**:366–368.
26. Busa, W. B., and R. Nuccitelli. 1984. Metabolic regulation via intracellular pH. *Am. J. Physiol.* **246**:R409–R438.
27. Cabantchik, Z. I., P. A. Knauf, and A. Rothstein. 1978. The anion transport system of the red blood cell: The role of membrane protein evaluated by the use of 'probes.' *Biochim. Biophys. Acta* **515**:239–302.
28. Cabantchik, Z. I. and A. Rothstein. 1972. The nature of the membrane sites controlling anion permeability of human red blood cells as determined by studies with disulfonic stilbene derivatives. *J. Membr. Biol.* **10**:311–328.
29. Cassel, D., P. Rothenberg, Y. Zhuand, T. F. Deuel, and L. Glaser. 1983. Platelet-derived growth factor stimulates Na^+/H^+ exchange and induces cytoplasmic alkalinization in NR6 cells. *Proc. Natl. Acad. Sci. USA* **80**:6224–6228.
30. Chaillet, J. R., K. Amsler, and W. F. Boron. 1985. Optical measurements of intercellular pH in single PK_1 cells: Evidence for Cl-HCO_3 exchange. *Kidney Int.* in press.
31. Chaillet, J. R., and W. F. Boron. 1985. Intracellular calibration of a pH-sensitive dye in isolated perfused salamander proximal tubules. *J. Gen Physiol.* in press.
32. Chaillet, J. R., A. G. Lopes, and W. F. Boron. 1985. Basolateral Na–H exchange in the rabbit cortical collecting tubule. *J. Gen. Physiol.* in press.
33. Clancy, R. L., N. C. Gonzales, and R. A. Fenton. 1976. Effect of beta-adrenoreceptor blockade on rat cardiac and skeletal muscle pH. *Am. J. Physiol.* **230**:959–964.
34. Cogan, M. G. 1984. Stimulation of proximal bicarbonate reabsorption by chronic hypercapnia. *Kidney Int.* **25**:273a.
35. Cohn, D. E., S. Klahr, and M. R. Hammerman. 1983. Metabolic acidosis and parathyroidectomy increase Na^+–H^+ exchange in brush border vesicles. *Am. J. Physiol.* **345**:F217–F222.
36. DeHemptine, A. 1980. Intracellular pH and surface pH in skeletal and cardiac muscle measured with a double-barrelled pH microelectrode. *Pfluegers Arch.* **386**:121–126.
37. DeHemptine, A., R. Marranes, and B. Vanheel. 1983. Influence of organic acids on intracellular pH. *Am. J. Physiol.* **245**:C178–C183.
38. Deitmer, J. W., and D. Ellis. 1980. Interactions between the regulation of the intracellular pH and sodium activity of sheep cardiac Purkinje fibres. *J. Physiol. (London)* **304**:471–488.
39. Dennis, V. W. 1976. Influence of bicarbonate on parathyroid-induced changes in fluid absorption by the proximal tubule. *Kidney Int.* **10**:373–380.
40. Deutsch, C., J. S. Taylor, and D. F. Wilson. 1982. Regulation of intracellular pH by human peripheral blood lymphocytes as measured by ^{19}F NMR. *Proc. Natl. Acad. Sci. USA* **79**:7944–7948.
41. Ellis, D., and R. C. Thomas. 1976. Direct measurement of the intracellular pH of mammalian cardiac muscle. *J. Physiol. (London)* **262**:755–761.
42. Evans, M. G., and R. C. Thomas. 1984. Acid influx into snail neurones caused by reversal of the normal pH_i-regulating system. *J. Physiol. (London)* **346**:143–154.
43. Evans, M. G., and R. C. Thomas. 1984. The effects of acid solutions on intracellular pH and Na in snail neurones. *J. Physiol. (London)* **341**:68P.
44. Fenton, R. A., N. C. Gonazalez, and R. L. Clancy. 1978. The effect of dibutyrl cyclic AMP and glucagon on the myocardial cell pH. *Respir. Physiol.* **32**:213–223.
45. Folbergrova, J., V. MacMillan, and B. K. Siesjo. 1972. The effect of hypercapnic acidosis upon some glycolytic and Krebs cycle-associated intermediates in the rat brain. *J. Neurochem.* **19**:2507–2517.
46. Frelin, C., P. Vigne, and M. Lazdunski. 1983. The amiloride-sensitive Na^+/H^+ antiport in 3T3 fibroblasts. *J. Biol. Chem.* **258**:6272–6276.
47. Fromter, E., K. Sato, and K. Gessner. 1975. Acetazolamide inhibits passive buffer exit from rat kidney proximal tubular cells. *Pfluegers Arch.* **359**:R118.
48. Gadian, D. G., G. K. Radda, M. J. Dawson, and D. R. Wilkie. 1982. pH_i measurements of cardiac and skeletal muscle using ^{31}-NMR. In: *Intracellular pH: Its Measurement, Regulation, and Utilization in Cellular Functions*. R. Nuccitelli and D. W. Deamer, eds. Liss, New York. Pp. 61–77.

49. Gilies, R. J., J. R. Alger, J. A. den Hollander, and R. G. Shulman. 1982. Intracellular pH measured by NMR: Methods and results. In: *Intracellular pH: Its Measurement, Regulation, and Utilization in Cellular Functions*. R. Nuccitelli and D. W. Deamer, eds. Liss, New York. Pp. 79–104.
50. Heiple, J. M., and D. L. Taylor. 1982. An optical technique of measurement of intracellular pH in single living cells. In: *Intracellular pH: Its Measurement, Regulation and Utilization in Cellular Functions*. R. Nuccitelli and D. W. Deamer, eds. Liss, New York. Pp. 21–54.
51. Hinke, J. A. M. 1967. Cation-selective microelectrodes for intracellular use. In: *Glass Electrodes for Hydrogen and Other Cations: Principles and Practice*. G. Eisenman, ed. Dekker, New York. pp. 464–477.
52. Jacobs, M. H. 1920. The production of intracellular acidity by neutral and alkaline solutions containing carbon dioxide. *Am. J. Physiol.* **53**:457–463.
53. Jacobs, M. H. 1922. The influence of ammonium salts on reaction. *J. Gen. Physiol.* **5**:181–188.
54. Kahn, A. M., G. M. Dolson, S. C. Bennett, and E. J. Weinman. 1984. cAMP and PTH inhibits Na^+/H^+ exchange in brush border membrane vesicles (BBM) derived from a suspension of rabbit proximal tubules. *Kidney Int.* **25**:289a.
55. Keifer, D. W., and A. Roos. 1981. Membrane permeability to the molecular and ionic forms of DMO in barnacle muscle. *Am. J. Physiol.* **240**:C73–C79.
56. Kinsella, J. L., and P. S. Aronson. 1980. Properties of the Na^+–H^+ exchanger in renal microvillus membrane vesicles. *Am. J. Physiol.* **238**:F461–F469.
57. Kinsella, J. L., and P. S. Aronson. 1981. Amiloride inhibition of the Na^+–H^+ exchanger in renal microvillus membrane vesicles. *Am. J. Physiol.* **241**:F374–F379.
58. Kinsella, J. L., and P. S. Aronson. 1981. Interaction of NH_4^+ and Li^+ with the renal microvillus membrane Na^+–H^+ exchanger. *Am. J. Physiol.* **241**:C220–C226.
59. Kinsella, J. L., and P. S. Aronson. 1982. Determination of the coupling ratio for Na^+–H^+ exchange in renal microvillus membranes vesicles. *Biochim. Biophys. Acta* **689**:161–164.
60. Koppel, M., and K. Spiro. 1914. Uber die wirking von moderatoren (Puffern) bei der verschiebung des saure-basengleichgewichtes in biologischen flussigkeiten. *Biochem. Z.* **65**:409–439.
61. MacMillan, V., and B. K. Siesjo. 1973. The influence of hypocapnea upon intracellular pH and upon some carbohydrate substrates, amino acids and organic phosphates in the brain. *J. Neurochem.* **21**:1283–1299.
62. Meech, R. W., and R. C. Thomas. 1977. The effect of calcium injection on the intracellular sodium and pH of snail neurones. *J. Physiol. (London)* **265**:867–879.
63. Meech, R. W., and R. C. Thomas. 1980. Effect of measured calcium chloride injections on the membrane potential and internal pH of snail neurones. *J. Physiol. (London)* **298**:111–129.
64. Michaelis, L. 1922. *Die Wasserstoffionenkonzentration.* Springer-Verlag, Berlin. pp. 89–93.
65. Moody, W. J., Jr. 1981. The ionic mechanism of intracellular pH regulation in crayfish neurones. *J. Physiol. (London)* **316**:293–308.
66. Moolenaar, W. H., J. Boonstra, P. T. van-der Saag, and S. W. de Laat. 1981. Sodium/proton exchange in mouse neuroblastoma cells. *J. Biol. Chem.* **256**:12883–12887.
67. Moolenaar, W. H., R. W. Tsien, P. T. van der Saag, and S. W. Laat. 1983. Na^+/H^+ exchange and cytoplasmic pH in the action of growth factors in human fibroblasts. *Nature (London)* **304**:645–648.
68. Moolenaar, W. H., Y. Yarden, S. W. de Laat, and J. Schlessinger. 1982. Epidermal growth factor induces electrically silent Na^+ influx in human fibroblasts. *J. Biol. Chem.* **257**:8502–8506.
69. Moore, R. D. 1979. Elevation of intracellular pH by insulin in frog skeletal muscle. *Biochem. Biophys. Res. Commun.* **91**:900–904.
70. Moore, R. D., M. L. Fidelman, and S. H. Seeholzer. 1979. Correlation between insulin action upon glycolysis and change in intracellular pH. *Biochem. Biophys. Res. Commun.* **91**:905–910.
71. Murer, H., U. Hopfer, and R. Kinne. 1976. Sodium/proton antiport in brush border membrane vesicles isolated from rat small intestine and kidney. *Biochem. J.* **154**:597–604.
72. Piwnica-Worms, D., and M. Lieberman. 1983. Microfluorometric monitoring of pH_i in cultured heart cells: Na^+–H^+ exchange. *Am. J. Physiol.* **244**:C442–C448.
73. Riegle, K. M., and R. L. Clancy. 1975. Effect of norepinephrine on myocardial intracellular hydrogen ion concentration. *Am. J. Physiol.* **229**:344–349.
74. Rindler, M. J., and M. H. Saier, Jr. 1981. Evidence for Na^+/H^+ antiport in cultured dog kidney cells (MDCK). *J. Biol. Chem.* **256**:10820–10825.
75. Rindler, M. J., M. Taub, and M. H. Saier, Jr. 1979. Uptake of $^{22}Na^+$ by cultured dog kidney cells (MDCK). *J. Biol. Chem.* **254**:11431–11439.
76. Rink, T. J., R. W. Tsien, and T. Pozzan. 1982. Cytoplasmic pH and free Mg^{2+} in lymphocytes. *J. Cell Biol.* **95**:189–196.
77. Roos, A. 1975. Intracellular pH and distribution of weak acids across cell membranes: A study of D- and L-lactate and of DMO in rat diaphragm. *J. Physiol. (London)* **249**:1–25.
78. Roos, A., and W. F. Boron. 1980. The buffer value of weak acids and bases: Origin of the concept, and first mathematical derivation and application to physico-chemical systems. The work of M. Koppel and K. Spiro. *Respir. Physiol.* **40**:1–32.
79. Roos, A., and W. F. Boron. 1981. Intracellular pH. Physiol. Rev. **61**:296–434.
80. Rothenberg, P., L. Glaser, P. Schlesinger, and D. Cassel. 1983. Activation of Na^+/H^+ exchange by epidermal growth factor elevates intracellular pH in A431 cells. *J. Biol Chem.* **258**:12644–12653.
81. Russell, J. M., and W. F. Boron. 1976. Role of chloride transport in regulation of intracellular pH. *Nature (London)* **264**:73–74.
82. Russell, J. M., W. F. Boron, and M. S. Brodwick. 1983. Intracellular pH and Na fluxes in barnacle muscle with evidence for reversal of the ionic mechanism of intracellular pH regulation. *J. Gen. Physiol.* **82**:47–78.
83. Sachs, G., J. G. Spenney, and M. Lewin. 1978. H^+ transport: Regulation and mechanism in gastric mucosa and membrane vesicles. *Physiol. Rev.* **58**:106–173.
84. Seifter, J. L., and R. C. Harris. 1984. Chronic K depletion Na–H exchange in rat renal cortical brush border membrane vesicles. *Kidney Int.* **25**:282a.
85. Shuldiner, S., and E. Rozengurt. 1982. Na^+/H^+ antiport in Swiss 3T3 cells: Mitogenic stimulation leads to cytoplasmic alkalinization. *Proc. Natl. Acad. Sci. USA* **79**:7778–7782.
86. Siesjo, B. K., and K. Messeter. 1971. Factors determining intracellular pH. In: *Ion Homeostasis of the Brain.* B. K. Siesjo and S. C. Sorensen, eds. Munksgaard, Copenhagen. pp. 244–262.
87. Steinmetz, P. A., and O. S. Andersen. 1982. Electogenic proton transport in epithelial membranes. *J. Membr. Biol.* **65**:155–174.
88. Thomas, J. A., R. N. Buchsbaum, A. Zimniak, and E. Racker. 1979. Intracellular pH measements in Ehrlich ascites tumor cells utilizing spectroscopic probes generated *in situ*. *Biochemistry* **18**:2210–2218.
89. Thomas, R. C. 1974. Intracellular pH of snail neurones measured with a new pH-sensitive glass microelectrode. *J. Physiol. (London)* **238**:159–180.
90. Thomas, R. C. 1976. Ionic mechanism of the H^+ pump in a snail neurone. *Nature (London)* **262**:54–55.
91. Thomas, R. C. 1976. The effect of carbon dioxide on the intracellular pH and buffering power of snail neurones. *J. Physiol. (London)* **255**:715–735.
92. Thomas, R. C. 1977. The role of bicarbonate, chloride and sodium ions in the regulation of intracellular pH in snail neurones. *J. Physiol. (London)* **273**:317–338.
93. Trivedi, B., and H. Danforth. 1966. Effect of pH on the kinetics of frog muscle phosphofructokinase. *J. Biol. Chem.* **241**:4110–4112.
94. Tsai, C. J., H. E. Ives, R. J. Alpern, V. J. Yee, D. G. Warnock,

and F. C. Rector, Jr. 1984. The V_{max} for Na^+/H^+ antiporter activity in rabbit brush border vesicles (BBV) is increased in metabolic acidosis. *Kidney Int.* **25**:284a.

95. Van Slyke, D. 1922. On the measurement of buffer values and on the relationship of buffer value to the dissociation constant of the buffer and the concentration and reaction of the buffer solution. *J. Biol. Chem.* **52**:525–570.
96. Vaughan-Jones, R. D. 1979. Regulation of chloride in quiescent sheep heart Purkinje fibres studied using intracellular chloride and pH-sensitive microelectrodes. *J. Physiol. (London)* **295**:111–137.
97. Vaughan-Jones, R. D. 1982. Chloride–bicarbonate exchange in the sheep cardiac Purkinje fiber. In: *Intracellular pH: Its Measurement, Regulation, and Utilization in Cellular Functions*. R. Nuccitelli and D. Deamer, eds. Liss, New York. pp. 239–252.
98. Vaughan-Jones, R. D., W. J. Lederer, and D. A. Eisner. 1983. Ca^{2+} ions can affect intracellular pH in mammalian cardiac muscle. *Nature (London)* **301**:522–524.
99. Vigne, P., C. Frelin, E. J. Cragoe, Jr., and M. Lazdunski. 1983. Ethyl-isopropyl-amiloride: A new and highly potent derivative of amiloride for the inhibition of the Na^+/H^+ exchange system in various cell types. *Biochem. Biophys. Res. Commun.* **116**:86–90.
100. Vigne, P., C. Frelin, and M. Lazdunski. 1982. The amiloride sensitive Na^+/H^+ exchange system in skeletal muscle cells in culture. *J. Biol. Chem.* **257**:9394–9400.
101. Weinman, S. A., and L. Reuss. 1982. Na^+/H^+ exchange at the apical membrane of *Necturus* gallbladder. *J. Gen. Physiol.* **80**:299–321.

CHAPTER 4

Properties of Ionic Channels in Excitable Membranes

Francisco Bezanilla and Michael M. White

1. Introduction

Excitable cells respond to appropriate stimuli with changes in their transmembrane potential. We will briefly review the origin of the membrane potential and the ways that it can be modified. Two main classes of excitability will be discussed: chemical excitability, in which the stimulus is a chemical transmitter released by another cell; and electrical excitability, in which the stimulus is a change in the membrane potential itself.

The bulk of the chapter will be concerned with electrical excitability. We shall see that the generation of the action potential in nerve is due to discrete voltage-dependent permeability pathways called channels. We will focus on the processes by which these ion channels open ("gate") to form the permeability pathway. A simple two-state model for the voltage-dependent gating of ion channels will be used to illustrate the basic concepts involved. We shall then examine the gating of the Na^+ and K^+ channels of the squid giant axon. Although the gating of these channels is more complex than that of a two-state channel, all of the behavior can be accounted for by the extension of the two-state model.

2. How Do You Get a Resting Potential?

Work in the late 18th and early 19th centuries by Galvani and others led to the proposal that muscles produced "animal electricity" through the action of a resting potential difference between the inside and the outside of the cell. Bernstein[1] was the first person to satisfactorily propose a mechanism for the generation of this resting potential. He knew from work by others that the internal ionic composition of cells was quite different from the ionic composition of the surrounding extracellular fluid. The interior of nerve and muscle fibers is rich in K^+ but deficient in Na^+ and Cl^-, while the extracellular fluid is rich in Na^+ and Cl^-, but contains little K^+ (see Chapter 13). Furthermore, evidence available at the time indicated that the cell membrane was impermeable to Na^+. Bernstein proposed that the voltage across the cell membrane arose from the unequal distribution of K^+ ions according to the Nernst equation, which predicts that a voltage will be set up across a semipermeable membrane if unequal concentrations of the permeant ion exist across the membrane. For the case that Bernstein considered, that K^+ was the only permeant ion, the membrane potential is the equilibrium potential of K^+ and is given by

$$V_m = (RT/F) \ln ([K^+]_{out}/[K^+]_{in}) \quad (1)$$

where R, T, and F have their usual meanings. Equation (1) predicts a resting potential of about -70 mV (inside negative) for a nerve cell under physiological ionic conditions.

Equation (1) also predicts that the membrane potential should vary logarithmically with the ratio of K^+ concentrations. Although this is true over a wide range of $[K^+]_{out}$, it is not true for very low $[K^+]_{out}$. This deviation can be accounted for if one assumes that a small permeability for other ions exists at rest. Goldman[2] and Hodgkin and Katz[3] derived a general equation for the potential for such a situation using the Nernst–Planck electrodiffusion equation. If we reduce the ionic composition of the intracellular and extracellular fluids to just Na^+, K^+ and Cl^-, the Goldman–Hodgkin–Katz equation becomes

$$V_m = (RT/F) \ln \frac{P_{Na}[Na^+]_{out} + P_K[K^+]_{out} + P_{Cl}[Cl^-]_{in}}{P_{Na}[Na^+]_{in} + P_K[K^+]_{in} + P_{Cl}[Cl^-]_{out}} \quad (2)$$

where P_x is the permeability coefficient of ion x. Hodgkin and Katz[3] found that for the squid giant axon at rest, $P_K : P_{Na} : P_{Cl} = 1.0 : 0.04 : 0.45$. Cl^- ions are passively distributed and do not

Francisco Bezanilla and Michael M. White • Department of Physiology, Ahmanson Laboratory of Neurobiology, and the Jerry Lewis Neuromuscular Research Center, University of California Medical School, Los Angeles, California 90024.

contribute to the resting potential in nerve, so Eq. (1) is in fact a good approximation for Eq. (2). Thus, the nerve membrane can be said to be, to a first approximation, a K^+ electrode. Although the relative permeabilities for Na^+, K^+, and Cl^- may vary from cell type to cell type, most cells do seem to behave as K^+ electrodes at rest.

3. How Do You Change the Membrane Potential?

Examination of Eq. (2) shows that there are two ways to change the membrane potential: either change the concentration or the permeability of one or more ions. Since the changes in membrane potential can be quite rapid (on the order of 1 msec or less), we can safely ignore the former and concentrate on the latter as the correct mechanism. In the same paper that he proposed the basis of the resting potential, Bernstein proposed that the nerve impulse was due to the membrane permeability becoming transiently nonselective. Later work on a variety of systems showed that this is not quite true; although the ion permeability does increase, it is a fairly selective increase in the permeability of one or more (but not all) ions, rather than the nonspecific one proposed by Bernstein. Which ions become more permeable depends upon the cell type and the function of the permeability change. For example, in the action potential of nerve and muscle, the Na^+ permeability increases transiently,[3] while in the inhibitory postsynaptic potential (IPSP) of spinal motoneurons it is the Cl^- permeability that increases.[4]

The mechanism of the permeability change falls into two broad classes according to the stimulus that elicits the change: electrical and chemical. Electric, or voltage-dependent responses are found in cells such as nerve and muscle in which the signal propagates along a single cell. Chemical responses are found where the impulse is transmitted from one cell to another.

3.1. Chemical Excitability

In systems in which an electrical signal must travel from one cell to another, such as the neuromuscular junction (NMJ), a mechanism must be present to "transmit" the signal across the synapse, which is a specialized region between two excitable cells. Until the mid-19th century the ability to transmit itself across intercellular spaces was considered one of the properties of "animal electricity." In 1843, DuBois-Reymond[5] suggested that instead of an electrical signal, this transmission involved the release of a chemical from one cell that would diffuse to the other and initiate an impulse. In 1921, Loewi[6] demonstrated that repetitive stimulation of the vagus nerve (which slows the heart beat) releases a substance into the fluid bathing the tissue which, when added to a normally beating heart, mimicked the effect of vagus nerve stimulation. This was the first convincing demonstration that chemical transmission did indeed exist. This substance, which was originally termed "vagusstoff," was subsequently identified by Loewi and Navratil[7] as acetylcholine (ACh). Dale *et al.*[8] later demonstrated that ACh is also involved in transmission at the NMJ. Since then, a wide variety of neurotransmitters have been identified in many different systems. Although the mechanisms of action of the various neurotransmitters may differ on a fine scale, in general they are similar. We shall briefly discuss the mechanism of action of ACh at the frog NMJ, which is the system studied in the most detail.

Extensive electrophysiological studies of the action of ACh at the frog NMJ have shown that stimulation of the nerve elicits depolarization of the muscle, which in turn initiates an action potential in the muscle. This response can be mimicked by applying ACh to the muscle fiber. Fatt and Katz[9] proposed that ACh increases the permeability of the muscle membrane at the NMJ nonspecifically, similar to what Bernstein proposed for nerve and muscle excitation. Takeuchi and Takeuchi[10] convincingly demonstrated that the increase in permeability is not *totally* nonselective; only the permeabilities of Na^+, K^+, and Ca^{2+} increased, while that of Cl^- remained constant.

A combination of electrophysiological, pharmacological, and biochemical investigations have shown that ACh (T) binds to a specific receptor (R) in the membrane forming a complex which then becomes a permeability pathway in the membrane:

$$n\mathrm{T} + \mathrm{R} \rightleftharpoons \mathrm{T}_n\mathrm{R} \rightleftharpoons \mathrm{T}_n\mathrm{R}^* \quad (3)$$

where the asterisk indicates the activated permeability pathway. For the frog NMJ, $n = 2$ ACh/receptor.

In the case of ACh at the NMJ and other neurotransmitters, it is the binding of the transmitter to the receptor that initiates the formation of a permeability pathway in the membrane. In the next section we shall discuss electrical excitability, in which the activation of a permeability pathway is initiated not by the binding of a small molecule to a receptor, but rather a change in the transmembrane potential.

3.2. Electrical Excitability

The classical example of an electrically excitable cell is the squid giant axon. We will illustrate voltage-dependent excitable properties using data obtained from this preparation. A giant axon has a resting potential of about -60 mV (negative inside) due mainly to the asymmetric distribution of K^+. When a pulse of current is injected into the cell, the membrane potential increases or decreases depending upon the polarity of the current. A pulse of inward positive current makes the membrane potential more negative (hyperpolarizes) with a sigmoidal time course and the increase is more or less proportional to the amplitude of the pulse. A pulse of current in the opposite direction produces a similar response only when the pulse amplitude is very small. For larger pulse amplitudes the response is sudden, transient, very large, and no longer proportional to the pulse amplitude. There is a minimum pulse amplitude that produces this response, and after crossing this threshold, any increase in pulse amplitude will not change the size of the response. This response is called the action potential.

The explosive nature of the action potential made its study a very difficult task because there was practically no way to modulate its amplitude and duration by changing the stimulating current. Cole and Curtis[11] found that during the action potential there was a dramatic decrease in membrane resistance, and Hodgkin and Katz[3] found that the amplitude and rate of rise of the action potential were dependent upon the concentration of Na^+ in the external medium. However, the detailed study of the processes occurring during the action potential was only possible after Cole[12] invented the voltage clamp in 1949.

The voltage clamp is a technique that controls the membrane potential using negative feedback. Using the voltage clamp the experimenter imposes whatever time course of voltage across the membrane desired regardless of the changes in impedance occurring in the membrane. To achieve this, the negative feedback circuitry passes current through the membrane and measurement of this current gives an indication of the electrical changes occurring in the membrane. When the squid giant axon was studied under voltage clamp conditions, it was found that the currents showed no threshold. The description of the membrane currents under voltage clamp was done in detail by Hodgkin and Huxley.[13–16] Their series of classical papers is the foundation of modern electrophysiology and constitutes a physicochemical description of membrane phenomena underlying electrical excitability.

The response of the nerve membrane to a sudden depolarization to 0 mV consists of a brief capacitive current followed by a transient inward current and finally a steady outward current. By means of a series of ion-replacement experiments, Hodgkin and Huxley identified Na^+ as the carrier of the transient inward current and K^+ as the carrier of the steady outward current. Both currents are the result of an increase of the membrane permeability activated by the depolarization. When the depolarization is maintained, the Na^+ permeability decreases with time, a phenomenon known as inactivation. The increase in membrane permeability specific for K^+ develops with depolarization more slowly than the Na^+ permeability and does not show inactivation on a scale of tens of milliseconds.

The description of the Na^+ and K^+ permeabilities as a function of voltage and time allowed Hodgkin and Huxley to determine the sequence of events responsible for the action potential. These events can be summarized as follows. The stimulating current will produce enough depolarization of the membrane to activate the Na^+ permeability. Once the Na^+ permeability is increased, the membrane potential will move toward the Na^+ equilibrium potential ($\sim$ +50 mV), which in turn will activate the permeability even more. This regenerative process produces a sudden depolarization which corresponds to the rising phase of the action potential. At the same time, the depolarization turns on the inactivation process, which decreases the Na^+ permeability, and activates the K^+ permeability which initiates the repolarization of the membrane potential toward the K^+ equilibrium potential ($\sim$ −70 mV). This corresponds to the falling phase of the action potential. The K^+ permeability becomes higher than in the resting state resulting in a membrane potential very close to the K^+ equilibrium potential producing an underswing on the action potential. Finally, as the membrane repolarizes, the K^+ conductance returns to its resting level and the membrane potential approaches the resting potential. This sequence of events occurs in a patch of axon while distant patches are still in the resting state. However, the depolarization in the original patch produces a current circulation to the neighboring regions which in turn become depolarized and initiate the same sequence. The conduction of the nervous impulse is the result of this process continuing along the axon and propagating as a wave of depolarization.

In summary, the basis of electrical excitability is the presence of specific ionic permeabilities that are turned on and off by the membrane potential. Since the work of Hodgkin and Huxley, research efforts have been directed toward an understanding of the physicochemical nature of the permeability changes to explain the physical basis of ionic selectivity and the voltage dependence of the permeabilities.

4. Ionic Channels

The specific permeabilities of excitable membranes reside in discrete protein structures called channels which span the lipid bilayer. These channels are selective because they can discriminate among ionic species. The channel contains a pore or conductive pathway that allows ions to go through depending on the ion's charge and the dimensions. Chemical groups within the channel confer the *selectivity* properties. In addition, there is a *gating* apparatus consisting of charged groups and dipoles that can be oriented by the membrane electric field to change the molecular conformation from an open to a closed pore. The channel can randomly gate between open and closed states but the transition probability between open and closed states is modulated by the membrane potential. For example, in the case of a K^+ channel, as the membrane is depolarized, the probability of finding the channel in the open state increases, while as the membrane is hyperpolarized, the closed state becomes more probable. When the channel is in a closed state, no ions can go through, but when it is in the open state, several millions of ions pass per second. The open state typically lasts from a fraction to a few milliseconds and produces small bursts of current which recently have been detected in several excitable membranes for most of the known ionic channels.[17]

In this chapter we will be mostly concerned about the gating properties of channels because the selectivity, which is a property of the open channel, is discussed in detail in another chapter.

A complete description of the channel at the structural and functional level is the objective of present research. This description will be complete when all the physical states of the channel and the transitions between the states are known and correlated with the molecular structure of the channel. At present we are far from achieving this objective because the physical states have not been described uniquely, nor is the structure of the macromolecule known for any of the channels present in biological membranes. However, rapid progress has been made in the molecular characterization of the ACh receptor channel[17a] and much new information is emerging about the Na^+ channel macromolecule.[17b]

In order to describe the physical state of a channel, there are three types of electrical measurements that can be made: macroscopic currents, gating currents, and microscopic currents in the form of current fluctuations of a small population of channels or direct observation of single-channel events. Once these measurements have been made, the experimental results are compared and fitted to kinetic models to count and characterize the states of the channel and the transition among the states. Unfortunately, these measurements do not give rise to a unique model. Pharmacological and chemical modification of channel structure and function has been used to restrict the class of models that can explain the experimental data.

In the next section a very simple model will be presented that will be used to predict the minimum gating functions of the channel at the macroscopic and microscopic levels. In a later section we will present the data from some channels and comparison with this model will show that more elaborate models with many more states are required to account for the data.

5. The Two-State Model

In this section we will develop a simple model for the gating of a voltage-gated ion channel. Although few, if any, channels in excitable membranes follow such a simple model, it does provide a theoretical foundation upon which to build more complicated gating schemes. The two-state model was originally formulated by Tsien and Noble[18] to explain gating behavior in terms of a simple physical chemical model.

Let us imagine a channel molecule embedded in a membrane that can undergo a conformational change between only two states, one conducting ("OPEN") and the other nonconducting ("CLOSED"). The gating of the channel can then be described by the following scheme:

$$\text{CLOSED} \underset{\beta}{\overset{\alpha}{\rightleftharpoons}} \text{OPEN} \qquad (4)$$

where α and β are the rate constants for opening and closing, respectively. Let us make the following assumptions concerning the channels:

1. The total number of channels in the membrane, N, is constant.
2. The channels are uniformly oriented within the membrane.
3. The channels open and close independently of one another.
4. The single-channel conductance is independent of voltage.
5. Channel formation does not rely upon the aggregation of independently diffusing subunits.
6. The two conformations of the channel carry different dipole moments.

With these assumptions in mind, we can now predict all of the equilibrium and kinetic behavior of the channel at the level of both microscopic (single-channel) and macroscopic conductances, as well as the behavior of gating currents.

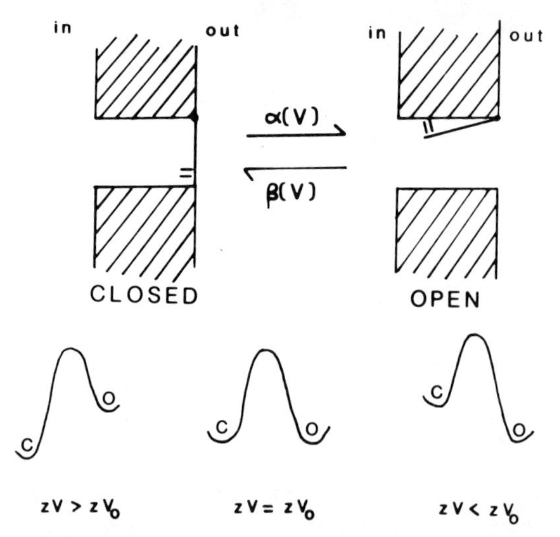

Fig. 1. Gating of a two-state channel. A schematic representation of the OPEN and CLOSED states of the channel is shown. The electrical charge on the gate is drawn such that the equilibrium will be driven toward the OPEN state as the inside of the cell is made increasingly positive with respect to the outside. The lower portion shows the gating free energy profile for three different voltages: one at which the CLOSED state is thermodynamically favored ($zV > zV_o$); one at which both states are equally likely ($V = V_o$); and one at which the OPEN state is favored ($zV < zV_o$).

5.1. Equilibrium Behavior

Since the two conformations of the channel carry different dipole moments, obviously a dipole moment change accompanies the conformational change. This means that the equilibrium between OPEN and CLOSED states will be voltage-dependent. The probability that a channel exists in the OPEN state, $p(V)$, can be expressed in terms of the total Gibbs free energy of the opening reaction, $\Delta G(V)$:

$$p(V) = (n_o/N) = \{1 + \exp[\Delta G(V)/RT]\}^{-1} \qquad (5)$$

where n_o and N are the number of open channels and the total number of channels, respectively. We can write the voltage dependence of $\Delta G(V)$ explicitly by breaking up $\Delta G(V)$ into a voltage-independent part, ΔG_i, and a voltage-dependent part given by the difference in energy of the dipole moments of the two conformations of the channel:

$$\Delta G(V) = \Delta G_i + zFV \qquad (6)$$

The parameter z is the "effective gating charge," which is the energy equivalent of the dipole moment change that accompanies channel opening. A gating charge of 1 is equivalent to a dipole moment change of 250 D in a membrane 50 Å thick. It should be emphasized that ΔG_i is merely a lumped term representing all of the interactions within the channel molecule which are not voltage-sensitive.

We can now write an expression for the probability that a channel will be in the OPEN state as a function of voltage:

$$p(V) = \{1 + \exp[(\Delta G_i + zFV)/RT]\}^{-1} \qquad (7)$$

Equation (7) shows that $p(V)$ depends sigmoidally upon voltage. The steepness of the curve depends upon z, while the midpoint of the curve is at a voltage $V_o = (-\Delta G_i/zF)$. Fig. 1 is a highly schematic drawing of the gating of a two-state channel. The lower part of Fig. 1 depicts the voltage-dependent equilibrium between OPEN and CLOSED states. When $V = V_o$, the probability of a channel being in the OPEN state is the same as it being in the CLOSED state; in other words, $p(V) = 0.50$. When $zV > zV_o$, the CLOSED state is thermodynamically favored and $p(V) < 0.5$, while if $zV < zV_o$, the OPEN state is favored and $p(V) > 0.5$.

5.2. Kinetic Behavior

The two-state scheme presented above makes several predictions concerning the kinetic behavior of the conformational transition between CLOSED and OPEN states. First, α and β should vary exponentially with voltage. This is because the free energy difference between the "gating transition state" and the OPEN and CLOSED states depends linearly upon the voltage.[19] The model makes no prediction as to the explicit (i.e., sign and value of the exponents) voltage dependence of the two rate constants, as this would depend upon the shape of the free energy barrier between OPEN and CLOSED states.

The model also predicts that the macroscopic conductance (g) after a voltage step must relax in a single-exponential fashion from the initial value determined by the number of channels open at the potential from which the voltage jump was made, V_i, to a steady-state value at the new voltage, V_f, both of which are determined by Eq. (7). The relaxation will be given by

$$1-[\Delta g(t,V_f)/\Delta g(\infty,V_f)] = \exp[-t/\tau(V_f)] \qquad (8)$$

where the time constant τ is given in terms of the individual rate constants of opening and closing;

$$\tau(V) = [\alpha(V) + \beta(V)]^{-1} \tag{9}$$

Although the relaxation time constant will in general depend upon the test voltage after the jump, V_f, it does not depend upon V_i; in other words, the channels have no "memory" of the voltage from which the jump was made. Note also that τ does not depend upon N, the number of channels. This is a consequence of the assumptions that the channels gate independently and that channel formation does not involve an aggregation mechanism.

In a relaxation experiment we can determine only the sum of the two rate constants; however, it is the individual rate constants which ultimately are of interest. In order to determine α and β, we can use Eq. (7) to obtain the equilibrium constant, (α/β), and Eq. (9) to obtain expressions for α and β in terms of measurable parameters:

$$\alpha(V) = p(V)/\tau(V) \tag{10a}$$

$$\beta(V) = 1-p(V)/\tau(V) \tag{10b}$$

The model also makes predictions concerning single-channel fluctuations at a fixed voltage. Under the assumptions of the model channel opening and closing are Poisson processes. This means that the dwell times in the OPEN (t_o) and CLOSED (t_c) states will be exponentially distributed.[20] If we define $P_o(t)$ and $P_c(t)$ as the probabilities that the dwell time in the OPEN and CLOSED states, t_o and t_c, respectively, are greater than time t, then

$$P_o(t) = \exp(-t/\tau_o) \tag{11a}$$

$$P_c(t) = \exp(-t/\tau_c) \tag{11b}$$

The relaxation time constants τ_o and τ_c are given by the reciprocals of the rate constant for leaving the OPEN (for τ_c) and CLOSED (for τ_o) states:

$$\tau_o = 1/\beta \tag{12a}$$

$$\tau_c = 1/\alpha \tag{12b}$$

Thus, from histograms of the channel open times and closed times, one can obtain the individual rate constants for opening and closing. Since α and β depend upon voltage, so will τ_o and τ_c.

5.3. Gating Currents

One of the assumptions of the two-state model (and indeed of any model for a voltage-dependent channel) is that a net dipole moment change accompanies the conformational transition as the channel opens and closes. This dipole moment change will appear as a charge movement within the membrane, i.e., a capacitive current. The rates of this charge movement should reflect the rates of the conformational transition.

We can obtain expressions for the steady-state charge–voltage (Q–V) distribution as well as the rates for the charge movement using the expressions derived for the behavior of the macroscopic conductance. If we call the net charge movement that accompanies channel opening ΔQ and the maximum charge moved ΔQ_{max}, the Q–V curve is given by

$$[\Delta Q(V)/\Delta Q_{max}] = [n_o(V)/N] = \{1 + \exp[(\Delta G_i + zFV)/RT]\}^{-1} \tag{13}$$

which is identical to the expression for the number of open channels, Eq. (7). The kinetics of this charge movement are obtained using Eq. (8):

$$1 - [\Delta Q(t,V_f)/\Delta Q(\infty,V_f)] = \exp[-t/\tau(V)] \tag{14}$$

It is not the charge movement itself that we measure, but rather the capacitive current caused by the charge movement I_g:

$$I_g(t,V_f) = [d\Delta Q(t,V_f)/dt] \propto \exp[-t/\tau(V)] \tag{15}$$

We see that $I_g(t,V_f)$ is a single exponential with a relaxation time constant identical to that obtained from macroscopic conductance measurements. The agreement between steady-state and kinetic properties of both ΔQ and n_o is a consequence of the two-state model. For more complicated kinetic schemes, this may not be true, since some of the charge movement may accompany conformational transitions that do not directly result in the opening of a channel (such as between multiple closed states), as we shall see in the next section.

Fig. 2 shows single-channel, macroscopic, and gating currents for a two-state channel in response to a voltage step which opens 50% of the channels ($\alpha = \beta$). Part A of Fig. 2 shows single-channel fluctuations for three different channels in the membrane. Note that each channel behaves differently, reflecting the fact that the channels gate independently. Part B shows the exact solution of Eq. (8), which shows the current one would obtain by summing an infinite number of fluctuations like those shown in A. Note the single-exponential rise of the currents. Part C shows the gating current for the channels.

We shall now examine the gating properties of voltage-gated channels in biological membranes. Although the gating is more complex than that of a two-state channel, we shall see that by extending this model we can understand the process underlying these more complicated systems.

6. Real Channels Have More Than Two States

Fig. 3 shows two recordings of K^+ currents when the membrane is depolarized to 0 mV from an initial potential of −60 mV for the leftmost trace and from an initial potential of −110 mV for the rightmost trace. There are two features which are not expected from the two-state model presented in the previous section. The first discrepancy is the sigmoidal shape of the current in contrast to the exponential rising phase expected from the two-state model. This characteristic shows that the gating process is of order higher than one. The second feature is that the shape of the current depends on the initial voltage showing that the gating process has some "memory." The two-state model can be extended to account for both phenomena. For example, if we assume that the channel opens only when six dipoles have flipped to the open conformation, it is easy to see that the current will turn on with a delay because six transitions will be required per channel for the channel to conduct. The effect of the initial voltage can also be explained because at very hyperpolarized potentials, all six units will be in the closed conformation, while

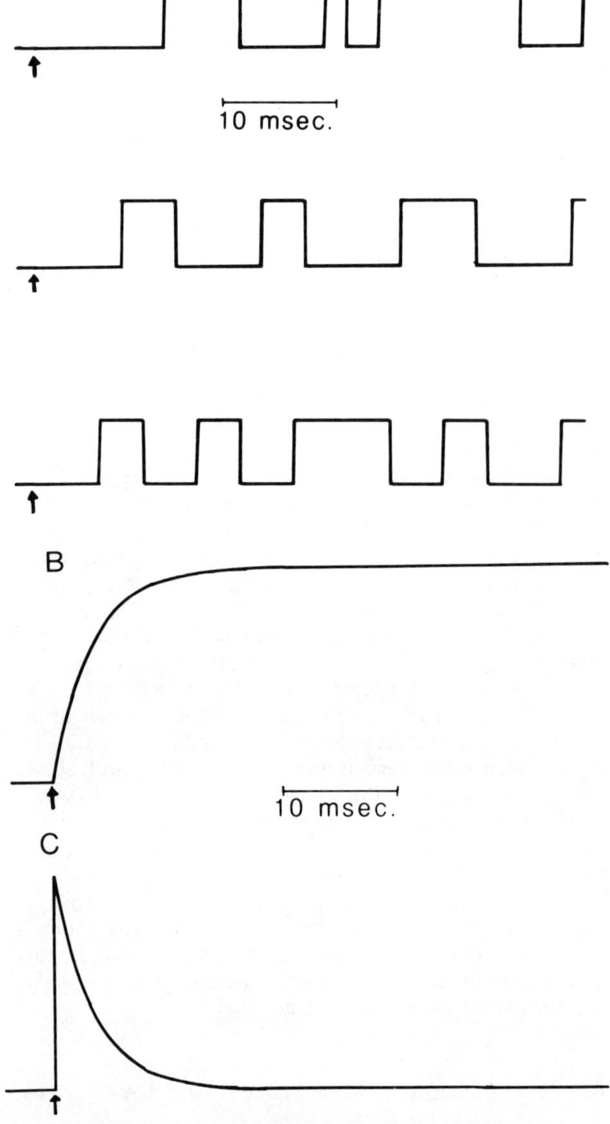

Fig. 2. Single-channel, macroscopic, and gating currents for a two-state channel. Currents were generated using the model for a voltage step from a voltage at which a channel is rarely open ($zV \gg zV_o$) to one at which both states are equally favored ($V = V_o$). In this case $\alpha = \beta = 133.3\ sec^{-1}$. Part A shows single-channel fluctuations for three different channels in the membrane. Note that each channel behaves differently. Part B shows the exact solution of eq. (8), which shows the macroscopic currents obtained by summing an infinite number of single channel records. The current rises in a single-exponential fashion with a time constant of 3.75 msec. Part C shows the gating current associated with the channel opening. Once again, it follows a single-exponential decay with a time constant of 3.75 msec. The vertical scales of each section bear no relationship to each other.

at intermediate potentials, only a few of them will be in the closed position. Therefore, a depolarization from a very negative potential will produce a current with a longer lag than the current obtained with a more positive initial condition.

This example illustrates that a full description of the kinetics and steady-state properties of the K^+ channel will require many physical states. This situation is true for most of the ionic channels in biological membranes that have been studied in detail.

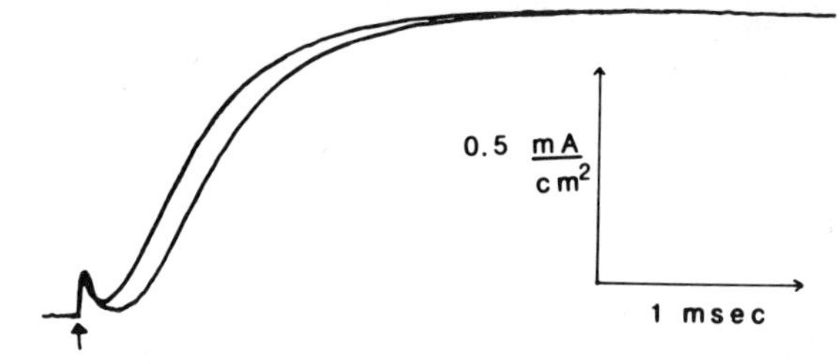

Fig. 3. Real channels are more complicated than two-state channels. Squid axon K^+ currents were recorded in response to depolarizations to 0 mV from either −60 mV (leftmost trace) or −110 mV (rightmost trace). The arrow marks the start of the voltage step. The small hump of current at the beginning of the pulse is the gating current associated with both Na^+ and K^+ channel activation. Note the effect of the initial voltage on the kinetics.

In the rest of this chapter we will analyze the general properties of Na^+ and K^+ channels based on microscopic and macroscopic eletrophysiological studies performed in several preparations.

7. Na^+ Channels

A family of Na^+ currents recorded from a squid giant axon for different depolarizations is shown in Fig. 4. The general features can be summarized as follows: the current turns on with a delay, activates, and then decays (inactivates) as the depolarization is maintained. In addition, the speed of activation and inactivation increases as the membrane potential is made more positive. In their kinetic description of the Na^+ currents, Hodgkin and Huxley assumed that the activation and inactivation processes were both voltage-dependent, but independent of each other. Their description satisfies most of the general properties of the Na^+ conductance but some fine details cannot be reproduced by this formulation. The lag in the turn-on of the conductance indicates that the channel evolves through a series of closed states before it opens. Ionic current measurements do not provide direct information about the events occurring during

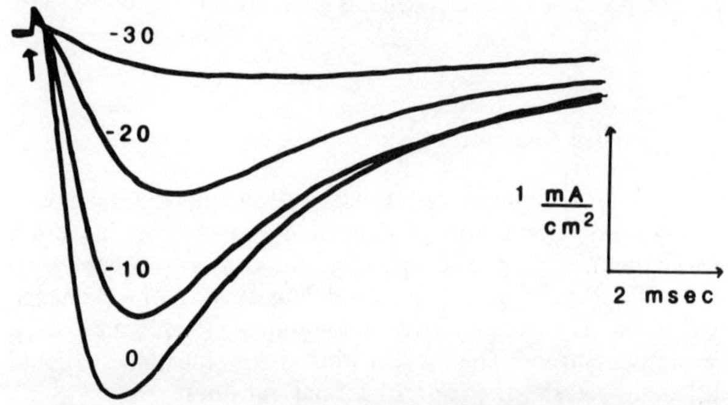

Fig. 4. Na^+ currents in response to different depolarizations. Currents were recorded at 4°C in response to voltage steps from −70 mV to the indicated voltages. The arrow marks the start of the pulse. Gating currents appear as a small hump at the beginning of the pulse. Note the pronounced inactivation as the voltage becomes more positive.

the closed state transitions because no current flows during these transitions. A more direct measurement is obtained through the use of gating currents, which are the electrical expression of conformational changes between closed states produced by intramembrane charge movement caused by membrane potential changes. We shall thus focus on results obtained through the study of gating currents.

7.1. Gating Currents

Gating currents have been briefly discussed in the section on the two-state model. We saw that in the simple two-state model the time course of the gating current coincides with the time course of the conductance. When there are several transitions between closed states, the gating current time course can be quite different from that of the conductance. Examination of the gating current time course should provide information about the transitions between closed states.

Gating currents were predicted by Hodgkin and Huxley[16] when they described the voltage dependence of the ionic conductance. In the case of the Na^+ conductance, they found that the conductance changes e-fold per 4 mV, which requires at least six electronic charges moving the whole span of the membrane field. The actual movement of these charges was detected as gating currents in 1973 by Armstrong and Bezanilla.[21]

The measurement of gating currents is by no means trivial. Since at least some of the movement of the charges or dipoles responsible for gating is concurrent with the actual opening of the channel, the gating currents will be superimposed upon the much larger ionic currents (see Fig. 4). Experimental conditions must be adjusted to remove the contaminating ionic currents by replacement of permeant ions by impermeant ones (e.g., Na^+ by $Tris^+$) and/or use of pharmacological agents to block the channels. Once the ionic currents have been eliminated, the only time-dependent current remaining is the large capacitive current, which contains both voltage-dependent and voltage-independent components. The voltage-independent part (produced by the "linear charge movement") is due to the charging of the membrane capacitance, while the voltage-dependent component (produced by the "nonlinear charge movement") is due to the movement of the gating particles. A subtraction procedure must be employed to subtract the nonlinear charge movement from the linear portion. All methods used consist of recording capacitive currents at the voltage of interest and at another voltage where no nonlinear charge movement occurs (i.e., where no channel gating takes place). Subtraction will eliminate the linear component since it will appear in both records, while the nonlinear component does not. In the original procedure,[21] the membrane voltage was held at -70 mV and the gating currents were obtained by summing capacitive currents resulting from the application of equal and opposite voltage pulses. This is known as the $\pm P$ procedure. It was later found that this procedure was not accurate because some nonlinear charge movement occurs at voltages negative to -70 mV. The subtracting pulse is now applied at very negative potentials (typically -150 mV), and to prevent membrane breakdown, the amplitude of the subtracting pulse is made ¼ the test pulse and applied four times. The currents are summed and then summed with the current from the test pulse. This is knon as the $P/4$ procedure.

7.1.1. Steady-State Properties

The total amount of charge moved in the voltage range between -150 mV and $+70$ mV is about 2000 electronic charges/μm^2 ($e^-/\mu m^2$) in the squid giant axon. The results of tetrodotoxin binding give about 300 channels/μm^2, so we can calculate that about 6 e^-/channel must move to open each channel, which is the value expected from measurements of the voltage dependence of the Na^+ conductance. The voltage dependence of the charge (Q–V curve) is less steep than the conductance (g_{Na}–V curve) and there is a sizable amount of charge that moves in the region more negative than -70 mV where the Na^+ conductance is zero. This indicates that there are a number of closed states with voltage-dependent transitions which are visible in the charge movement and responsible for the lag of the Na^+ conductance turn-on. When the Q–V and g_{Na}–V curves are normalized and plotted in the same graph, the g_{Na}–V curve saturates at a more negative potential than the Q–V curve. This can be interpreted as an indication that not all the charge is related to the gating of the Na^+ channel. This explanation is not very likely because studies of the effects of azure A on Na^+ channel gating currents show that most of the charge measured at about 8°C is related to the Na^+ conductance.[22] Furthermore, Taylor and Bezanilla[23] have shown that the charge movement in the region more negative than -70 mV is responsible for the added delay in the conductance turn-on when the voltage jump is made from increasingly more negative potentials. Stimers *et al.*[24] found that when inactivation is removed by Pronase (see below), the Q–V and g_{Na}–V curves saturate at the same potential, an indication that inactivation prevents an accurate estimation of the g_{Na}–V and Q–V relationships.

Bezanilla *et al.*[25] have shown that the Q–V curve shifts on the voltage axis depending on the initial conditions. This phenomenon has been explained by an effect of voltage on the charge movement which has both a short-term and a long-term effect. The determination of a real steady-state Q–V curve was done by Fernandez *et al.*[26] using frequency domain techniques. The voltage dependence of the steady-state Q–V differs from that determined in the time domain with different initial conditions. The apparent discrepancy of these results originates in the nonstationary nature of the charge movement which results in the inactivation processes. We have learned that in order to correlate Q–V and g_{Na}–V curves, fast inactivation must be removed and the measurement must be done on a fast time scale to avoid nonstationarity produced by slow inactivation.

7.1.2. Kinetics

When the squid axon membrane is depolarized in voltage clamp, the gating current recorded at 8°C shows an almost instantaneous rise and then decays with at least two exponential components. Since the two-state model predicts a single-exponential decay, the observed gating current reflects a more complex process. In principle, one would expect gating current components due to Na^+ activation, Na^+ inactivation, and K^+ activation. The K^+ gating current is too slow to be detected at 8°C in the squid giant axon but it is clearly visible at 20°C (see below). The expected Na^+ inactivation gating current is also slow and small and consequently would be difficult to detect; however, if the depolarization goes to very positive potentials, inactivation is fast enough to expect a gating current with the time constant of inactivation. It was found, however, that no component of gating shows the kinetics expected for inactivation even at positive potentials.[27]

An exact correlation between gating current and Na^+ conductance must be done through the use of predictive models because some features of the conductance are not apparent in the gating currents. For example, the conductance develops with a delay which is not present in the gating currents. However, inspection of both ionic and gating currents reveals a component of the gating current that has the same time course as the main component of Na^+ current activation.

Correlation between ionic and gating currents is clearer at the turn-off of the pulse. If the depolarization is given for a short duration (about 0.3 msec), the ionic and gating currents at the end of the pulse decrease with a time course very close to a single exponential. If the potential after the pulse is −60 mV, the time constants of both processes are the same, but for more negative potentials the gating currents relax more slowly than the ionic currents. The ratio of the time constants may be as high as 1.7 at −110 mV.[28] The correlation between the OFF time constants has important theoretical implications. Assume that the activation of the channel involves several *identical* subunits (three in the case of the Hodgkin and Huxley model), all of which must be in the OPEN position for the channel to activate, and that each subunit follows the two-state model. In this case the gating current turn-off will be a single exponential since it corresponds to the superposition of several identical particles all following first-order kinetics. The Na^+ current also will decay with first-order kinetics but as the channel closes when any one of the subunits flip to the CLOSED position, the time constant of decay of the current will be faster than the time constant of the gating current. The exact relationship between the time constants can be obtained analytically. Suppose that x independent subunits are needed to activate the channel, and following Hodgkin and Huxley, let us assume that the fraction of subunits in the open position is m. The conductance will be given by

$$g = k_1 m^x$$

where k_1 is a proportionality constant involving the total number of channels and the conductance of an open channel. At the end of a pulse that fully activates the m variable, the potential is returned to a hyperpolarized value, the conductance decays according to

$$g = k_2 \exp[-t/(\tau/x)]$$

where k_2 is another proportionality constant. The gating current is proportional to the decay of the m fraction (see the two-state model section):

$$I_g = k_3 \exp[-t/\tau]$$

which shows that the time constant of gating is x times slower than the time constant of conductance. As stated above, the experimental results show that the ratio varies with voltage and does not exceed 1.7, which shows that a channel with any number of identical subunits is incompatible with the available data. These results, however, do not rule out the possibility that the channel has several nonidentical subunits.

The case of x identical subunits is a special case of a sequential model of activation. In a general sequential model like the following where a_0 through a_{n-1} are closed states and a_n is an open state:

$$a_0 \underset{\beta_1}{\overset{\alpha_1}{\rightleftharpoons}} a_1 \underset{\beta_2}{\overset{\alpha_2}{\rightleftharpoons}} a_2 \ . \ . \ . \ a_{n-1} \underset{\beta_n}{\overset{\alpha_n}{\rightleftharpoons}} a_n$$

the rate constants α_i and β_i (i=1 to n) may be all different and need not be related to each other. In the case of x identical subunits the sequential model still applies but the α's and β's are related as follows: $\alpha_i = \alpha_{i+1}[(x-i+1)/(x-i)]$ and $\beta_i = \beta_{i+1}[i/(i+1)]$. This means that it is enough to specify one α and one β to determine the whole sequence. Although attractive for its simplicity, the experimental data do not support this model and we must deal with the general sequential model which is characterized by a large number of rate constants. As can be expected, this general model has been used to fit data quite successfully,[28–30] but some simplifications can be introduced which are still compatible with the experimental data. The salient features of the sequential model that fit the activation of the conductance are equal rate constants in all the transitions except the last one, which leads to opening the channel and has slower kinetics.[29,30] In fact, the slower step between the last closed state and the open state is enough in many circumstances to account for most of the data when the delays in the turn-on of the conductance are ignored.[25,26]

It is clear, however, that a unique description of all the states and transitions of the general sequential model is not possible at present and data obtained from preparations modified by chemical or drug treatment will be required to sort out the value of the rate constants. An extra complication arises in the study of the activation of the conductance and gating by the inactivation process which is not independent from activation and contributes with more uncertainty in the determination of the parameters of the model. We will now describe the characteristics of Na^+ channel inactivation using conductance and gating current measurements.

7.1.3. Na^+ Channel Inactivation

The decay of the Na^+ current during a maintained depolarization is produced by the inactivation process. This process is voltage- and time-dependent and was assumed by Hodgkin and Huxley to be independent of the activation process. Some evidence has accumulated[31] which indicates that a better assumption would be to consider the inactivation process coupled to activation. We have mentioned that no component of the gating current showed the kinetics expected from the inactivation process even at potentials where it would be visible. On the other hand, it was found that a positive prepulse decreased both Na^+ and gating currents,[32] an indication that the same charge responsible for channel activation is involved in the inactivation process. Upon further investigation it was found that inactivation of the Na^+ conductance is indeed produced by the same set of charges that after activating the channel go into a different set of physical states which are associated with the inactivated state of the conductance.[28] The basic observations that led to this conclusion were the following: (1) the charge recorded at the OFF of the pulse decreased with the duration of the pulse and the time course of the charge decrease (immobilization) followed the time course of the Na^+ current inactivation for a pulse to the same potential; (2) the voltage dependence of the charge immobilization was the same voltage dependence of the Na^+ conductance inactivation; and (3) when the membrane was stepped back to −70 mV after inactivation was complete, the charge immobilization and the Na^+ inactivation were removed with the same time course.

When one talks about charge decrease or immobilization, the real phenomenon is a change in the kinetics of the charge movement which results in an apparent decrease of charge. For example, when the depolarizing pulse is terminated after 0.3 msec, the OFF gating current carries all the charge of the ON gating current. If the pulse is 10 msec long, the charge contained in the OFF gating current integrated for 2 to 3 msec will be smaller than the ON gating charge. This is due to the finite integration period because 15 msec after the pulse all the charge is available to move again. This is because after a few milliseconds of depolarization a fraction of the charge moves back rapidly upon repolarization but the rest moves much more slowly and produces a gating current too small to detect above the baseline noise. This last slowly moving charge is the immobilized or inactivated charge which is associated with the inactivated state of the conductance. The remobilization of this inactivated charge can be observed directly by speeding up the recovery of the inactivation process. When the recovery experiment is done at -150 mV (instead of -70 mV), the OFF gating current shows two distinct components: a fast part that corresponds to the movement of the nonimmobilized charge and a slower component that represents the movement of the immobilized charge; the total charge matches that moved during the ON of the pulse because now we are able to see all the charge moving back at the end of the pulse within the integration period of 2 to 3 msec.

These results imply that the inactivation process does not have its own charge movement and that its voltage dependence is achieved by coupling it to the activation process. One can envision the inactivation step as a plug that blocks an open channel preventing the flow of ions. The blocking plug moves in and out of the mouth of the channel randomly but becomes stabilized in the blocking position when the activating charge moves in the open position. A specific model consisting of an activating particle which moves in the membrane field interacting with an inactivating particle moving outside the membrane field was presented by Bezanilla *et al.*[25] for both slow and fast inactivation processes. In its simplest version, in which all activating steps have been lumped into one, the model involves the movement of the charged activating particle and an inactivating particle, each of which has two stable positions. Consequently, the model contains four states. By introducing interaction between the particles, it is possible to reproduce many of the features of Na^+ and gating current inactivation. A model considering all of the closed states reproduced the experimental data using 16 states,[30] but it contains too many parameters which cannot be determined uniquely from the available data. It is clear, however, that the full description of the physical states of the Na^+ channel will have at least as many, if not more, states, and the real task will be to determine the rate constants between states from the experimental data by appropriate chemical or pharmacological modification and simplification of the channel operation. For example, a detailed study of Na^+ current and gating currents with inactivation removed by Pronase eliminates at least half of the physical states of the channel, making the fitting of the parameters a more feasible task. At the same time, modification induced by local anesthetic derivatives has helped in the counting of the possible states.[33–35]

7.2. Fluctuation Analysis and Single-Channel Recordings

The goal of fluctuation or noise analysis is to relate the small fluctuations ("noise") in the macroscopic currents to the properties of the single channel.[36] If the total number of channels is small enough, the random opening and closing of single channels will be visible as noise in the current records. Analysis of these fluctuations can, in principle, give information concerning both the conductance of the single channel and the gating process. In practice, however, fluctuation analysis has not proven very useful in the determination of the steps involved in the gating of voltage-dependent channels. It has proven useful in determining the single-channel conductance. Values of 3.5–5 pS[37,38] and 7 pS[39,40] were obtained for Na^+ channels in squid giant axon and myelinated nerve, respectively. More recently, channel fluctuation studies in squid giant axon with extended bandwidth up to 25 KHZ obtained a value of 7 pS.[38a]

Single-channel events have been recorded from many types of cultured cells (myotubes, neuroblastoma, chromaffin cells) which have a Na^+ conductance similar (but not identical) to the Na^+ conductance of squid giant axon. When the membrane is depolarized, the current shows discrete steps of a few picoamperes with variable duration. These events, which represent the current through a single channel, reveal a conductance of about 15 pS for the open channel and indicates that the channel does not have intermediate conducting states, i.e., they are either open or closed. When a number of identical depolarizations are given and the currents are compared, the most striking feature is the clustering of the single events at the beginning of the pulse, which is a direct expression of the inactivation process. If hundreds of these current records are averaged, the current shows activation and inactivation. Patlak and Horn[41] treated patches with *N*-bromoacetamide which is known to remove inactivation.[42] After this treatment the single-channel events seem to last longer. Furthermore, the average of individual current traces shows a current that activates but does not inactivate.

Single-channel recording has the unique advantage that the experimenter can observe the unitary event taking place. For example, it allows the calculation of conditional probability of a channel opening after a certain time given that no opening occurred before. Aldrich *et al.*[43] have used this approach to test whether activation and inactivation are completely independent events or whether the channel must open before it becomes inactivated. Their results show that neither of the two propositions is true, indicating that the channel can inactivate before it opens but that the inactivation does not proceed independently of the activation process. This result is in agreement with the evidence obtained from gating current measurements in squid axon.

8. K^+ Channels

Fig. 5 shows a family of K^+ channel currents recorded from squid giant axon in response to voltage steps from -60 mV

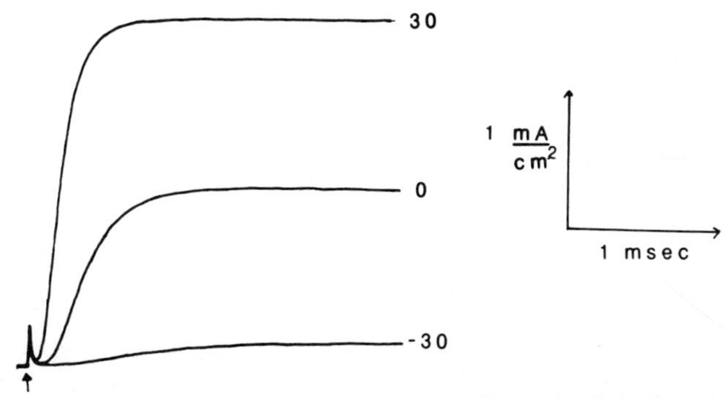

Fig. 5. K^+ channel currents in response to different depolarizations. Currents were recorded at 20°C in response to voltage steps from -70 mV to the indicated potentials. The arrow marks the start of the pulse and gating currents appear as a spike at the beginning of the pulse. These currents are approximately four times faster than they would be at 4°C.

to the indicated voltages. Several features are immediately obvious. First, the currents turn on with a delay. Second, the rate of channel opening increases as the voltage is made more positive. Both of these behaviors are seen for the Na^+ channel as well. Third, unlike the Na^+ channel, no inactivation takes place on a physiological time scale (tens of milliseconds). This lack of inactivation suggests that the K^+ channel gating process may be much simpler than that of the Na^+ channel.

Hodgkin and Huxley[16] modeled K^+ channel activation as four parallel identical first-order processes with each process (which has been termed the "n" process) representing the movement of a charged particle across the membrane, similar to that described for the two-state model. A channel is open only when all four particles have moved, so channel activation is a fourth-, rather than a first-, order process. This gives the delay seen in the ionic currents.

As mentioned before, a parallel subunit scheme can be written as a sequential model. The Hodgkin–Huxley n^4 scheme becomes

$$C_4 \underset{\beta_n}{\overset{4\alpha_n}{\rightleftharpoons}} C_3 \underset{2\beta_n}{\overset{3\alpha_n}{\rightleftharpoons}} C_2 \underset{3\beta_n}{\overset{2\alpha_n}{\rightleftharpoons}} C_1 \underset{4\beta_n}{\overset{\alpha_n}{\rightleftharpoons}} 0$$

where C_1–C_4 represent closed states, 0 the open state, and α_n and β_n are the forward and backward rate constants for the n process, respectively. Although this model can predict many of the observed properties of K^+ channel gating, it cannot reproduce everything. For example, the n^4 scheme does not predict the extra lag in activation observed when the voltage jumps are made from the increasingly more negative voltages shown in Fig. 3.

Rather than list all of the shortcomings of the n^4 model, we shall instead assemble the available data to synthesize a plausible model for K^+ channel activation. We will discuss data obtained from measurements of both ionic and gating currents.

8.1. Ionic Currents

In their investigations of the effect of divalent cations on K^+ channel activation, Gilly and Armstrong[44] found that the macroscopic currents could be adequately generated using a six-state model with four of the forward rate constants equal and the remaining one slower than the rest. The slow step could be anywhere in the sequence with no noticeable effect on the shape of the ionic currents. Observation of single K^+ channel fluctuations in squid axon by Conti and Neher[45] showed that once a K^+ channel opens, it rapidly flickers between an open state and a closed state and then closes for a longer period. This result suggests that the transition from the last closed state to the open state is much faster than the others. Gilly and Armstrong[44] and White and Bezanilla[46] arrived at a similar conclusion from macroscopic current measurements using a pulse paradigm that isolated the last step of the activation sequence. From these data we can suggest that the activation sequence contains a very fast step at the end preceded by three fast and one slow steps. Ionic current measurements do not allow us to assign a location to the slow step.

8.2. Gating Currents

Gating currents have proven to be powerful tools in the dissection of the process of Na^+ activation and inactivation.

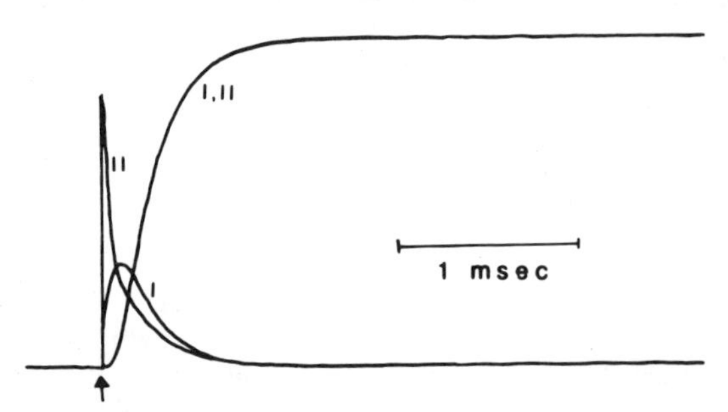

Fig. 6. Gating currents can examine transitions between closed states. Ionic and gating currents were generated using a six-state model with four fast steps ($\alpha = 34{,}000\ sec^{-1}$; $\beta = 1600\ sec^{-1}$) and one slow one ($\alpha = 4500\ sec^{-1}$; $\beta = 1600\ sec^{-1}$). In traces I, the slow step is first, while in traces II, it is the third one. Note that although the ionic currents and both sets of gating current are drawn to the same vertical scale, the scale differs between ionic and gating currents.

This is because they allow us to directly observe transitions between closed states of the channel, while with ionic current measurements we can only infer what is happening among the closed states. Fig. 6 illustrates the usefulness of gating currents in examining the transitions among closed states. The figure shows ionic and gating currents generated using a six-state model with four fast steps and one slow one with the slow step either the first (I) or the third (II) step. Note that although the ionic currents are indistinguishable, the gating currents are significantly different. When the first step is the slowest, a pronounced rising phase is seen for the gating current. Thus, the shape of the gating currents can be used to determine the location of the slow step in the model described above.

When Armstrong and Bezanilla first described Na^+ channel gating currents in 1973,[21] they detected no component of gating current that had a time course similar to that of K^+ channel activation. In retrospect, it is easy to see why this was so. The lower density and less steep voltage dependence of K^+ channels relative to Na^+ channels mean that the total amount of charge moved as K^+ channels open is only ~ 20–25% of that moved for Na^+ channels. Furthermore, at the temperatures normally employed for squid axon experiments (6–8°C), K^+ channel activation is much slower than Na^+ channel activation, so what little charge is available to move does so slowly that it would be lost in the baseline noise and drift. There is nothing one can do to increase the charge available to move (short of designing a new squid), but one can increase the rate of channel activation by performing the experiments at 20°C. When this was done, Bezanilla *et al.*[47] observed a new component of gating current that had the kinetic and equilibrium properties expected for K^+ channel gating currents. Fig. 7 shows K^+ ionic and gating currents for a voltage step to 0 mV preceded by an 8 msec prepulse to -110 mV. The fast spike in the gating current record is the residual Na^+ channel gating current. The slow phase of the gating current, the K^+ channel gating current, has a time course similar to I_k. More recent measurements of K^+ channel gating currents, made under conditions that eliminate a significant fraction of the Na^+ channel gating current, show a pronounced rising phase,[46] an indication that the first step in the activation sequence is the slowest.

With these properties in mind, we can propose a minimal model for K^+ channel activation:

$$C_5 \overset{\text{Slow}}{\rightleftharpoons} C_4 \overset{\text{Fast}}{\rightleftharpoons} C_3 \overset{\text{Fast}}{\rightleftharpoons} C_2 \overset{\text{Fast}}{\rightleftharpoons} C_1 \overset{\text{Fast}}{\rightleftharpoons} 0$$

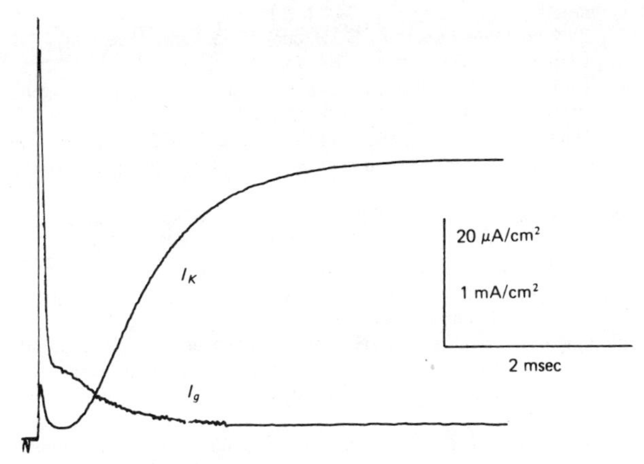

Fig. 7. K^+ channel ionic and gating currents. Currents were recorded at 20°C in response to a voltage step to 0 mV preceded by an 8 msec prepulse to −110 mV. The ionic current (I_k) was recorded in the presence of 200 mMK^+ inside the axon and the gating current (I_g) was recorded in the absence of permanent ions. The blip at the beginning of the I_k trace is gating current, while the fast spike at the beginning of the I_g trace is the residual Na^+ channel gating current.

Further investigation of the properties of the gating and ionic currents will help to refine this model and assign specific values to the rate constants. See White and Bezanilla[46] for an example of testing and refining this model.

9. Summary

The voltage-dependent conductances in electrically excitable cells are due to discrete structures embedded in the cell membrane called ionic channels. These channels contain a conducting pore that can select among different species of ions, and a gating structure that determines whether or not the pore is open. The gating structure is controlled by the transmembrane electric field via dipoles or charged groups that act as voltage sensors. We have described a simple two-state model of a voltage-gated channel that reproduces the basic properties of a voltage-gated channel. In this model, the channel has only one OPEN and one CLOSED state. When the channel is open, millions of ions pass through per second and this event can be detected as a single-channel current. Each state is associated with a different net dipole moment and it is this difference in dipole moments that gives rise to the voltage dependence of the transition between the two states. The change in dipole moment associated with the transition can be detected as a gating current.

We have shown that real channels, such as the Na^+ and K^+ channels of the squid giant axon, follow the basic principles of the two-state model and any deviations observed can be explained by extending the model to contain additional states. When a squid giant axon is depolarized, the Na^+ channel rapidly activates, and then slowly inactivates. Examination of the Na^+ channel gating currents showed that the same set of charges of dipoles responsible for the activation process are also responsible for the inactivation process.

The K^+ channel gating is a simpler process because no fast inactivation takes place. The activation process is slower than that of the Na^+ channel and experiments must be performed at higher temperatures in order to detect the K^+ channel gating currents. Examination of both ionic and gating currents shows that K^+ channel activation involves at least six steps: an initial slow step, several faster steps, and an extremely fast last step.

References

1. Bernstein, J. 1902. Untersuchungen zur Thermodynamic der bioelektrischen Ströme. *Pfluegers Arch. Gesamte Physiol.* **92**:521–562.
2. Goldman, D. E. 1943. Potential, impedance, and rectification in membranes. *J. Gen. Physiol.* **27**:37–60.
3. Hodgkin, A. L., and B. Katz. 1949. The effect of sodium ions on the electrical activity of the giant axon of the squid. *J. Physiol. (London)* **108**:37–77.
4. Coombs, J. S., J. C. Eccles, and P. Fatt. 1955. The specific ion conductances and the ionic movements across the motoneuronal membrane that produce the inhibitory post-synaptic potential. *J. Physiol. (London)* **130**:326–373.
5. DuBois-Reymond, E. 1843. *Untersuchungen uber thierische Electricitat,* Volume 1. Reimer, Berlin.
6. Loewi, O. 1921. Über humorale Übertragbarkeit der Herznervenwirkung. *Pfluegers Arch. Gesamte Physiol.* **189**:239–247.
7. Loewi, O., and E. Navratil. 1926. Über humorale Übertragarbarkeit der Herzenervewirkung. X. Mittelung. Über das Schicksal des Vagusstoffs. *Pfluegers Arch. Gesamte Physiol.* **214**:678–688.
8. Dale, H. H., W. Feldberg, and M. Vogt. 1936. Release of acetylcholine at voluntary motor nerve endings. *J. Physiol. (London)* **86**:353–380.
9. Fatt, P., and B. Katz. 1951. An analysis of the end-plate potential recorded with an intercellular electrode. *J. Physiol. (London)* **115**:320–370.
10. Takeuchi, A., and N. Takeuchi. 1960. On the permeability of the end-plate membrane during the action of transmitter. *J. Physiol. (London)* **154**:52–67.
11. Cole, K. S., and H. J. Curtis. 1939. Electric impedance of the squid axon during activity. *J. Gen. Physiol.* **22**:649–670.
12. Cole, K. S. 1949. Dynamic electrical characteristics of the squid axon membrane. *Arch. Sci. Physiol.* **3**:253–258.
13. Hodgkin, A. L., and A. F. Huxley. 1951. Currents carried by sodium and potassium ions through the membrane of the giant axon of *Loligo. J. Physiol. (London)* **116**:449–472.
14. Hodgkin, A. L., and A. F. Huxley. 1951. The components of membrane conductance in the giant axon of *Loligo. J. Physiol. (London)* **116**:473–496.
15. Hodgkin, A. L., and A. F. Huxley. 1951. The dual effect of membrane potential on the sodium conductance of *Loligo. J. Physiol. (London)* **116**:497–506.
16. Hodgkin, A. L., and A. F. Huxley. 1951. A quantitative description of membrane current and its application to conduction and excitation in nerve. *J. Physiol. (London)* **117**:500–544.
17. Hamill, O. P., A. Marty, E. Neher, B. Sakmann, and F. Sigworth. 1981. Improved patch-clamp techniques for high-resolution current recording from cells and cell-free membrane patches. *Pfluegers Arch.* **391**:85–100.

17a. Popot, J.-L., and J.-P. Changeux. 1984. Nicotinic receptor of acetylcholine: Structure of digomeric integral membrane protein. *Physiol. Rev.* **64**:1162–1239.

17b. Agnew, W. S. 1984. Voltage-regulated sodium channel molecules. *Ann. Rev. Physiol.* **46**:517–530.

18. Tsien, R. W., and D. Noble. 1969. A transition state theory approach to the kinetics of conductance changes in excitable membranes. *J. Membrane Biol.* **1**:248–273.
19. Ehrenstein, G., R. Blumenthal, R. Latorre, and H. Lecar. 1974. Kinetics of opening and closing of individual EIM channels in lipid bilayers. *J. Gen. Physiol.* **63**:707–721.
20. Larson, H. J. 1974. *Introduction to Probability Theory and Statistical Inference.* Wiley, New York.
21. Armstrong, C. M., and F. Bezanilla. 1973. Currents relating to movement of the gating particles of the sodium channel. *Nature (London)* **242**:459–461.

22. Armstrong, C. M., and R. S. Croop. 1982. Simulation of Na channel inactivation by thiazin dyes. *J. Gen. Physiol.* **80**:641–662.
23. Taylor, R. E., and F. Bezanilla. 1983. Sodium and gating current time shifts resulting from changes in initial conditions. *J. Gen. Physiol.* **81**:773–784.
24. Stimers, J. R., F. Bezanilla, and R. E. Taylor. 1985. Sodium channel activation in squid giant axon: Steady-state properties. *J. Gen. Physiol.* **85**:65–82.
25. Bezanilla, F., J. M. Fernandez, and R. E. Taylor. 1982. Distribution and kinetics of membrane dielectric polarization. I. Long-term inactivation of gating currents. *J. Gen. Physiol.* **79**:21–40.
26. Fernandez, J. M., F. Bezanilla, and R. E. Taylor. 1982. Distribution and kinetics of membrane dielectric polarization. II. Frequency domain studies of gating currents. *J. Gen. Physiol.* **79**:41–67.
27. Bezanilla, F., and C. M. Armstrong. 1977. Inactivation of the sodium channel. I. Sodium current experiments. *J. Gen. Physiol.* **70**:549–556.
28. Armstrong, C. M., and F. Bezanilla. 1977. Inactivation of the sodium channel. II. Gating current experiments. *J. Gen. Physiol.* **70**:557–590.
29. Armstrong, C. M., and W. F. Gilly. 1979. Fast and slow steps in the activation of sodium channels. *J. Gen. Physiol.* **74**:691–711.
30. Bezanilla, F., and R. E. Taylor. 1982. Voltage-dependent gating of sodium channels. In: *Abnormal Nerves and Muscles as Impulse Generators*. W. J. Culp and J. Ochoa, eds. Oxford University Press, London. pp. 62–79.
31. Goldman, L., and C. L. Schauf. 1972. Inactivation of the sodium current in *Myxicola* giant axons: Evidence of coupling to the activation process. *J. Gen. Physiol.* **59**:659–675.
32. Bezanilla, F., and C. M. Armstrong. 1974. Gating currents of the sodium channels: Three ways to block them. *Science* **183**:753–754.
33. Cahalan, M. D., and W. Almers. 1979. Block of sodium conductance and gating current in squid axons poisoned with quaternary strychnine. *Biophys. J.* **27**:57–74.
34. Yeh, J. Z., and C. M. Armstrong. 1978. Immobilization of gating charge by a substance that simulates inactivation. *Nature (London)* **273**:387–389.
35. Yeh, J. Z. 1982. A pharmacological approach to the structure of the Na channel in squid axon. In: *Proteins in the Nervous System: Structure and Function*. B. Haber, J. R. Perez-Polo, and J. D. Coulter, eds. Liss, New York. pp. 17–50.
36. DeFelice, L. J. 1981. *Introduction to Membrane Noise*. Plenum Press, New York.
37. Conti, F., L. J. DeFelice, and E. Wanke. 1975. Potassium and sodium in current noise in the membrane of the squid giant axon. *J. Physiol. (London)* **248**:45–82.
38. Llano, I., and F. Bezanilla. 1982. Analysis of sodium current fluctuations in the cut-open axon. *J. Gen. Physiol.* **83**:133–142.
38a. Levis, R. A., F. Bezanilla, and R. M. Torres. 1984. Estimate of the squid axon sodium channel conductance with improved frequency response. *Biophys. J.* **45**:11a.
39. Conti, F., B. Hille, B. Neumke, W. Nonner, and R. Stämpfli. 1976. Measurement of the conductance of the sodium channel from current fluctuations at the node of Ranvier. *J. Physiol. (London)* **262**:699–727.
40. Sigworth, F. J. 1980. The variance of sodium current fluctuations at the node of Ranvier. *J. Physiol. (London)* **307**:97–129.
41. Patlak, J., and R. Horn. 1982. Effect of N-bromoacetamide on single sodium channel currents in excised membrane patches. *J. Gen. Physiol.* **79**:333–351.
42. Oxford, G. S., C. H. Wu, and T. Narahashi. 1978. Removal of sodium channel inactivation in squid giant axons by N-bromoacetamide. *J. Gen. Physiol.* **71**:227–247.
43. Aldrich, R. W., D. P. Corey, and C. F. Stevens. 1983. A reinterpretation of mammalian sodium channel gating based on single channel recording. *Nature (London)* **306**:436–441.
44. Gilly, W. F., and C. M. Armstrong. 1982. Divalent cations and the activation kinetics of potassium channels in squid giant axons. *J. Gen. Physiol.* **79**:965–996.
45. Conti, F., and E. Neher. 1980. Single channel recordings of K currents in squid axons. *Nature (London)* **285**:140–143.
46. White, M. M., and F. Bezanilla. 1985. Activation of squid axon K^+ channels: Ionic and gating current studies. *J. Gen. Physiol.* **85**:539–554.
47. Bezanilla, F., M. M. White, and R. E. Taylor. 1982. Gating currents associated with potassium channel activation. *Nature (London)* **296**:657–659.

CHAPTER 5

Ion Movements in Skeletal Muscle in Relation to the Activation of Contraction

Hans Christoph Lüttgau and George Dimitrie Stephenson

1. Introduction

The dramatic event by which skeletal muscle, when stimulated, converts chemical energy into mechanical work has fascinated and puzzled physiologists for a long time.[1] The interest with which the process of muscular activation has been studied was probably also aroused by the possibility of measuring accurately both the electrical activity associated with the outer membranous system and the mechanical output. The whole sequence of events which bridges these two processes has been intuitively called excitation–contraction coupling.[2] Most of the current knowledge concerning this coupling is based on experiments performed in recent years, particularly on single amphibian and crustacean muscle fibers, a point to be borne in mind in any discussion on this subject.

The chain of events leading to contraction starts with the depolarization of the outer membrane beyond a critical value, which either triggers a propagated action potential, as in all fast vertebrate muscles, or spreads electrotonically along the sarcolemma as, for example, in crustacean skeletal muscles. The wave of depolarization then travels radially inward along the "transverse" tubular system.[3] In twitch frog skeletal muscle this depolarization is thought to be propagated in a regenerative manner.[4] The electrical signal is subsequently passed by some unknown mechanism to the terminal cisternae of the sarcoplasmic reticulum (SR). As a result of this sequence of events, calcium is released into the intrafibrillar space, possibly triggering the release of more calcium, which then diffuses among the myofilaments and activates both the contractile machinery and its own reaccumulation into the SR. Soon after the influence of the electrical depolarization on the SR terminates, the release is thought to stop, while the process of calcium reuptake into the SR continues, leading to the relaxation of the fiber.

The flow of information from the site of stimulation to the site at which force is developed is represented at the molecular level by the movement of different ions through the surface and inner membranes; these movements are discussed in this chapter.

Hans Christoph Lüttgau • Department of Cell Physiology, Ruhr University, Bochum, West Germany. **George Dimitrie Stephenson** • Department of Zoology, La Trobe University, Melbourne, Australia.

2. The Ultrastructure of the Tubular System in Skeletal Muscle Fibers

The way in which a muscle fiber is activated can only be understood from precise knowledge of its ultrastructure, as revealed by electron microscopy.[5–8] This section, therefore, is intended to describe those elements in frog skeletal muscle fibers which are particularly relevant to the activation process, namely the tubular systems.

The transverse tubular system (T system) consists of tubular extensions of the surface membrane which enter the fiber at the level of the Z line. The flattened tubules, 25–80 nm in diameter, form a continuous network which rings each myofibril (see Fig. 1). Their lumina are open to the extracellular fluid, and can be filled with relatively large extracellular markers such as ferritin molecules (∅ 11 nm).[9,10] Toward the periphery the tubules follow a tortuous path and have variable diameters.[11] In the region where the T tubules open to the extracellular space (I band in the frog), near-spherical vesicles, the caveolae, have been observed. They lie immediately under the plasma membrane, open to the extracellular space, and may increase the sarcolemmal surface area by 70%.[12] Since in frog muscle the caveolae are concentrated in the region of the I band, it has been suggested[13] that the T tubules open to the outside via a great number of interconnected caveolae.

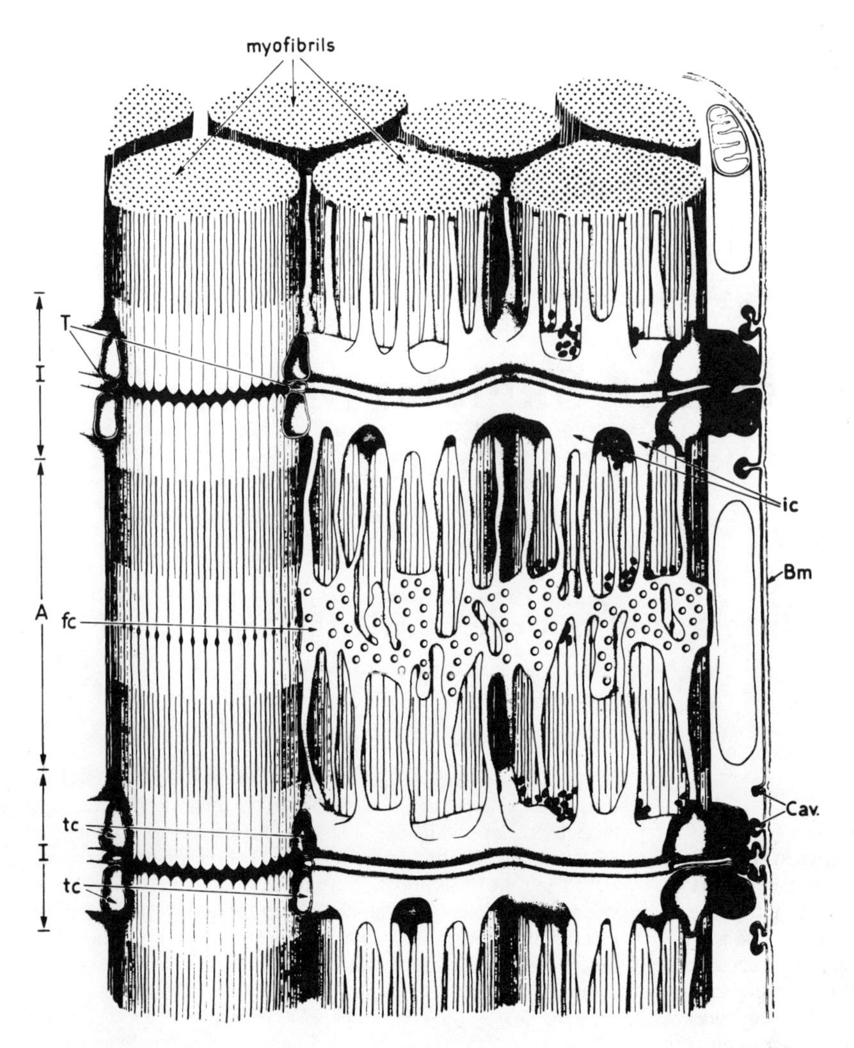

Fig. 1. Transverse tubular system and sarcoplasmic reticulum in frog twitch fibers. Bm, Basement membrane; Cav, caveolae; T, transverse tubular system; fc, fenestrated collar; ic, intermediate cisternae; tc, terminal cisternae; A, A band; I, I band. This figure is an extension of Peachey's well-known drawing of 1965.[6]

The second tubular system, namely the SR, envelops each myofibril and cannot be reached by large extracellular markers. It consists of three main elements: (1) the terminal cisternae (TC) which adjoin and cover the major part of the T system; (2) the intermediate cisternae; and (3) the fenestrated collar (see Fig. 1).

The T system comprises 0.32% of the fiber volume, the TC 4.1%, and the rest of the SR about 5%.[6,7,14] In a fiber with a diameter of 100 μm, the surface of the T system is about six to nine times larger than the smooth cylinder surface.[7] This fact and the additional increase in surface area provided by the caveolae, explain the apparently large membrane capacity of 7 $\mu F/cm^2$ (referred to the cylinder area) measured in fibers with a diameter of 100 μm.[15]

The region in which two TC of the SR and one tubule of the T system come into apposition is called a triad. About 60–80% of the surface of the T system is surrounded by the SR[6,14] whereby a gap of only 12–14 nm separates the two systems. In part of this region of close association, electron-dense "feet" cross the junctional gap and join SR and T-tubule membranes. They appear as solid structures with electron-opaque interiors and are arranged in a tetragonal disposition (forming two or multiple parallel rows) at a density of 790 per μm^2 of tubular membrane (in frog muscle).[8,16–18] With different fixation and staining methods, pillarlike structures instead of feet can be detected.[19,20] They are defined as pairs of electron-opaque lines bounding an electronlucent interior spanning the gap between the T system and the SR.[19] Under resting conditions, there are 39 pillars per μm^2 of tubular membrane.

The diversity of the tubular systems in different fiber types from various animals and the correlation between morphology and function have been well covered in two reviews.[3,8]

3. Electrical Properties of the Surface and Tubular Membrane

Table I presents the concentrations of the main ions in the extracellular fluid (Ringer's solution) and in the myoplasm of frog skeletal muscle fibers (see Refs. 21 and 22). The data, taken from different publications,[21–26] may serve as a starting point for the analysis of the origin of the membrane potential during rest and activity in these fibers. In the resting state the membrane is mainly permeable to Cl^- and K^+ ions for which, in a defined concentration range, a Donnan distribution will be displayed. At external K^+ concentrations higher than 10 mM, the membrane behaves like a K^+ or Cl^- electrode if $[K^+]_o$ and $[Cl^-{}_o]$ are

Table I. Ionic Concentrations of K+, Na+, Cl−, and Ca2+ in the Extracellular Solution (Ringer's Solution, $[X]_o$) and in the Myoplasm ($[X]_i$) of Frog Skeletal Muscle Fibers

	$[X]_o$ (mM)	$[X]_i$ (mmoles/kg fiber water)	$[X]_o/[X]_i$	E_x[a] (mV)
K^+	2.5	139	0.0180	−101.2
Na^+	120	15	8.0	+52.4
Cl^-	126	3.5	36.0	−90.3
Ca^{2+}	1.8	<0.0001	18000	>+123.4

[a]Equilibrium potential derived from the Nernst equation (T = 18°C) for the ion concerned.

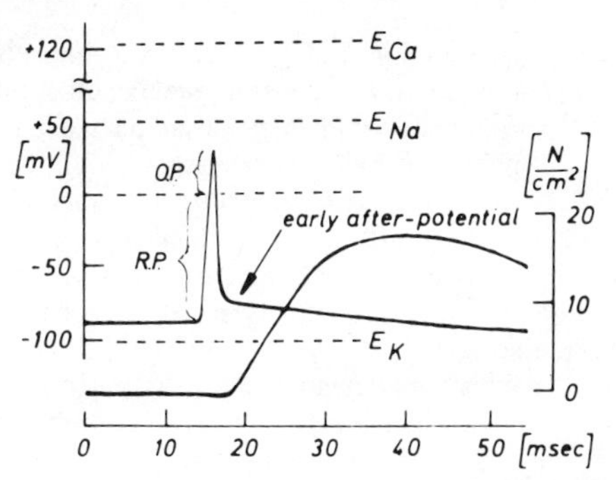

Fig. 2. Action potential and isometric force of a single skeletal muscle fiber of the frog (T = 18°C). Included are the equilibrium potentials (E) for K^+, Na^+, and Ca^{2+}. E_{Cl} is identical with the resting potential of −90.3 mV. O.P., overshoot potential; R.P., resting potential. A simultaneous record of action potential and force from an isolated fiber was first published by Hodgkin and Horowicz.(30)

changed reciprocally (to maintain a constant ion product), or simply as a K^+ electrode if $[K^+]_o$ is altered in Cl^--free solutions. Between 10 and 0.05 mM K^+, the resting potential deviates from the line for a K^+ electrode, with the slope becoming smaller than the expected 58 mV/decade (T = 18°C).(26) This deviation is explained by a slight permeability to Na^+ ions. Thus, the measured dependence of the membrane potential (E) on the $[K^+]_o$ can be fitted by a curve drawn according to a modified Goldman equation:

$$E = \frac{RT}{F} \ln \left(\frac{[K^+]_o + \alpha[Na^+]_o}{[K^+]_i + \alpha[Na^+]_i} \right) \quad (1)$$

where the subscripts o and i refer to the external and internal concentrations for K^+ and Na^+, respectively, R is the gas constant, T the absolute temperature, and F the Faraday number. α, the ratio of the permeability coefficients for Na^+ and K^+, has a value of 0.01.(26) By applying different experimental conditions (detubulation, which separates the T system from the surface membrane; low pH, which reduces the Cl^- conductance), it becomes possible to analyze the spatial distribution of the ion conductances for K^+ and Cl^-.(27) According to these measurements, the Cl^- conductance of frog twitch fibers is mainly located in the surface membrane, whereas the K^+ conductance, whose total value reaches only half of that for Cl^-, is distributed between the tubular and the surface membranes. The situation is different in mammalian muscles. Here the Cl^- conductance is mostly (at least 60%) located in the tubular system.(28,29)

The changes in membrane conductance underlying the action potential in frog skeletal muscle fibers are qualitatively similar to those found in squid axons. The upstroke of the action potential (see Fig. 2) is caused by a regenerative increase in Na^+ conductance while repolarization results from an inactivation of the Na^+, and a delayed increase of the K^+ conductance. The "overshoot" potential depends linearly on the logarithm of $[Na^+]_o$, and activation is blocked by tetrodotoxin (TTX), a specific blocker of the Na^+ channel.(31–35) Using isotopes, a net Na^+ entry of 15.6, and a net K^+ exit of 9.6 pmoles/cm² per impulse were measured.(25) In a fiber with a diameter of 100 μm, these result in an increase in internal Na^+ by 0.0077, and a decrease in internal K^+ by 0.0047 mmole/kg H_2O; i.e., the alterations in internal ion concentration can be neglected during a short tetanus. Typical of skeletal muscle fibers are the afterpotentials. The *early* afterpotential which follows a spike (see Fig. 2) is explained as being due to the lower ion selectivity of the delayed K^+ channel(32,33) as compared with that of the K^+ channel which determines the resting potential. The increase in the delayed K^+ current during activity brings the potential rapidly back to the equilibrium potential for these channels (−70 to −80 mV). The subsequent return to the resting potential with a half-time (20°C) of 15 sec depends on: (1) the rate of decline in the delayed conductance; and (2) the time constant of the membrane.(32,33) During repetitive stimulation at high frequencies, a *late* afterdepolarization develops.(36) Following a sigmoid time course, this depolarization reaches an amplitude of 5–20 mV during 5–30 stimuli at 100 Hz. The effect is mainly due to an accumulation of K^+ ions in the T system (delayed rectifier channels) and may be modified by the activation of a "slow component" of the K^+ conductance.(36,37,38) A review on active and passive transport of electrolytes in muscle has been published by Sjodin.(39)

In recent years, further details about ionic channels in muscle fibers have been investigated. They deal with kinetics, selectivity, chemical properties, etc., and have been the subject of an excellent review by Stefani and Chiarandini.(40) Here only those data will be presented which are characteristic for muscle fibers or which are in a specific way related to the activation of contraction.

1. In addition to Na^+, K^+, and Cl^- channels, skeletal muscle fibers also possess Ca^{2+} channels. These Ca^{2+} channels are mainly located in the T system and are slowly activated upon depolarization. Ca^{2+} influx is therefore negligible during a single spike but becomes large during a tetanus.(41–44). When external Ca^{2+} is reduced to values below 1–10 μM these channels loose their Ca^{2+} selectivity and become permeable for Na^+ and other small monovalent cations.(44 a–d) Almers and coworkers explained this phenomenon by assuming a single file pore with two specific Ca^{2+} binding sites.(44 c,d) If Na^+ (but not Li^+) ions enter the cell through these "Ca^{2+}"-channels in the T-tubular membrane the inflow is accompanied by a phasic force activation.(44b)

2. It has recently been demonstrated that at least the Na^+ channels are in some way "anchored" in the membranes.(45–47) The relevant observations were made with a modified patch clamp method,(45) using UV light to destroy Na^+ channels under the patch. Following the recovery of the Na^+ conductance under the patch, it was possible to set an upper limit of 10^{-12} cm²/sec for the lateral diffusion coefficient of these channels.

The very low value obtained suggests that the channels cannot freely move laterally within the membrane like other integrated proteins. This finding can explain the observed uneven distribution of Na^+ channels which run in stripes along the surface membrane of the muscle fiber, and indirectly supports the suggestion that there is a gradient of Na^+ channel density along the T system from the surface to the center of the fiber.[46,47]

3. Altogether four different types of K^+ channels have been detected so far[40]: (1) the delayed rectifier channel, responsible for the fast repolarization phase of the action potential; (2) the slow channel which becomes activated only during repetitive activity; (3) the inward rectifier responsible for the resting K^+ conductance, which decreases when the net flow of K^+ is outward and increases when it is inward (inward or anomalous rectification)[48]; and (4) the subsequently described Ca^{2+}-activated K^+ channel, which becomes activated when the $[Ca^{2+}]_i$ increases.

4. Repetitive stimulation is accompanied by two unexpected and largely unexplained phenomena:

When muscle fibers are stimulated at 100 Hz, the height of the action potential decreases rapidly, and after about 100–200 stimuli, the fibers respond only to every second stimulus (see Fig. 3). The drop in overshoot observed during the stimulation period is difficult to explain since the increase in $[Na^+]_i$ can be neglected. As the effect is absent in metabolically exhausted, i.e., mechanically inactive, fibers, it seems that repetitive contractile activity liberates some substance, or has some effect, which reduces the size of the action potential, probably by accelerating the inactivation of the Na^+ channels.[49]

During longer-lasting activity at low stimulation frequencies, a specific increase in K^+ conductance develops after ATP has fallen to about 1/10th of its resting value.[50] This increase is most certainly related to the existence of Ca^{2+}_i-activated K^+ channels observed with the patch clamp method.[51,52] The role of energy depletion (phosphorylation of membrane channels?) in this context is still unclear. It is interesting that prior to the increase in the K^+ conductance, the action potentials were no longer followed by twitches.[50,53]

Repetitive stimulation, i.e., the natural form of muscle activation, in addition leads to an increase in tubular K^+ and a decrease in tubular Na^+ and Ca^{2+}, which might influence the spread of excitation and excitation–contraction (EC) coupling.[37,54,55]

The diversity of ion channels with different characteristics, the effect of the metabolic state[50] and innervation[46] upon these characteristics, the accumulation or depletion of ion species in the T system, and, finally, the uneven channel distribution reveal a fascinating but complex system responsible for muscle activation. It is not always possible to weigh the physiological significance of the described effects. Altogether, however, the picture of a carefully optimized system emerges.[46] As an example, one could imagine why the density of the delayed K^+ rectifier channels should be lower (but not absent) in the tubular wall than in the surface membrane. These channels are needed for a fast repolarization in the course of an action potential. However, too many such K^+ channels in the T system could produce an excessive and rapid K^+ accumulation in the tubules during repetitive activity. This would result in a large afterdepolarization which could induce repetitive discharges similar to those observed in goats with congenital myotonia.[46,56] Other observations, like the dropout of action potentials during repetitive stimulation (Fig. 3), or the interruption in the coupling between excitation and contraction prior to a complete exhaustion of energy reserves (ATP), might also be interpreted as protective mechanisms against an irreversible breakdown of fibers.[50]

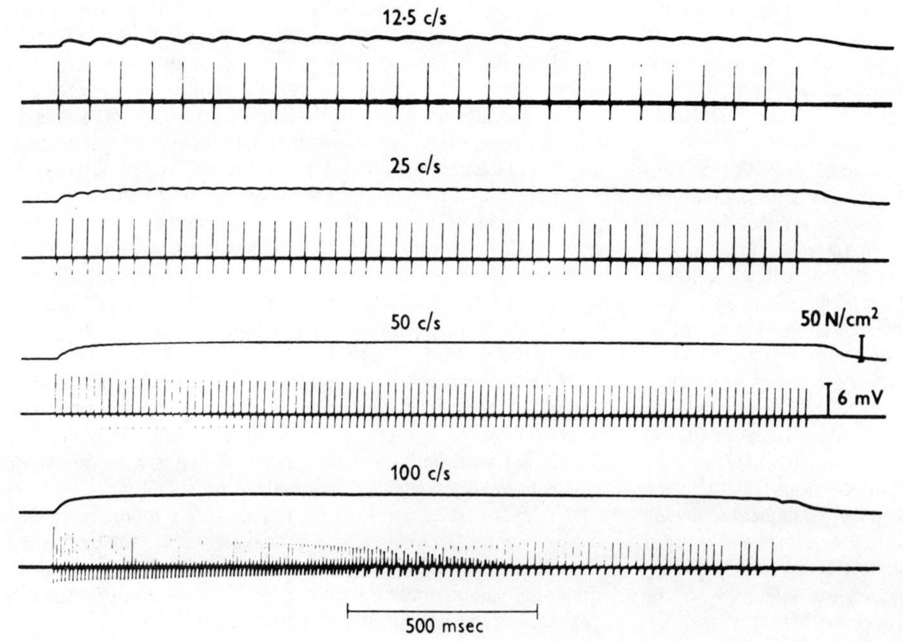

Fig. 3. Action potentials recorded with external electrodes (lower traces) and isometric force of an isolated frog skeletal muscle fiber at different frequencies (T = 21°C).[49]

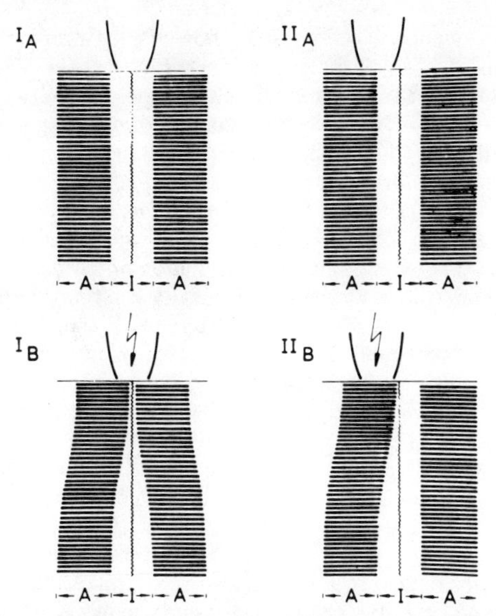

Fig. 4. Local activation of contraction in frog (I) and lizard (II) skeletal muscle fibers, before (A) and during (B) the application of a depolarizing current through a micropipette. A, A band; I, I band. For further details see the text.

4. Inward Spread of Excitation

4.1. Conductance of Excitation along the T System

In 1955, Huxley and Taylor[57] suggested that excitation spreads along some structures within the Z disk from the cell surface to the innermost fibrils. It is now well established that this structure is the T system along which excitation is conducted into the fiber in a regenerative manner (see Refs. 58–61).

In their well-known experiments, Huxley and co-workers[3,59,62] produced highly localized depolarizations by applying current pulses through extracellular micropipettes placed near the surface of isolated muscle fibers. Local contractions could be induced only when a pipette was positioned at specific sites ("sensitive spots") along the sarcomere, where elements of the T system open to the extracellular fluid. In frog twitch fibers, sensitive spots were correspondingly located at the Z disk, whereas in lizard and crustacean muscles, with a T system at the boundary between A and I bands, sensitive spots were located in that region (Fig. 4). After activation, a movement of the A band from both sides toward the Z disk was observed in frog, whereas in lizard only the A band near the activated T system at the A–I boundary moved toward the Z line.[3] The extent of contraction spread into the fiber depended on current strength, so that at all times a graded, and not an "all-or-none," reaction was found. This indicates an electrotonic, nonregenerative spread of excitation. The experiment, however, is not fully conclusive, "since stimulation at a single point on the edge of a network, which the T system forms in a single I band, might well fail to set up a propagated wave if the safety factor is low."[3]

In recent years, several arguments in favor of a regenerative spread of excitation along the T system have been obtained from experiments with twitch fibers of the frog. The radial spread of mechanical activation was investigated under voltage clamp control with a polarizing microscope.[63] TTX was used to block the regenerative Na^+ influx. With just-threshold depolarizing steps applied to the surface membrane, a contraction of the superficial fibrils only was observed. A further depolarization by 2–4 mV was needed to produce a shortening over the whole cross-section.[63] In further experiments, it was found that the voltage span between the activation of the superficial fibrils and the entire cross-section decreased when TTX was omitted from the external solution.[58] As an explanation, a regenerative and TTX-sensitive Na^+ inward current in the wall of the T system was suggested.[58] This suggestion was strengthened by the observation that, at a just-threshold depolarization, the innermost myofibrils occasionally shortened more strongly than the superficial ones. Such a finding is to be expected if one assumes that the membranes of the T system nearer the cell surface are voltage clamped at the applied potential, while the innermost tubular membranes with insufficient voltage clamping are further depolarized by an inward Na^+ current.[58] The hypothesis of a regenerative spread has recently been supported by optical investigations[64] and by the finding that about half the TTX binding sites in a muscle (i.e., Na^+ pores) reside in the T system.[65]

4.2. Transmission of Excitation at the Level of the Triad

The way in which excitation is communicated from the T system to the SR, thus causing the release of Ca^{2+} from the TC, is still a controversial matter. Three hypotheses will be discussed: (1) that Ca^{2+} acts as a transmitter[66]; (2) that Ca^{2+} release is "remotely" controlled by movable intramembranous charges in the wall of the T system[67]; and (3) that following excitation, ionic channels become activated and bridge the gap between the T system and the SR.[68]

Before going into detail it might be helpful to describe the dependence of the activation of contraction and its subsequent inactivation on membrane potential. In isolated frog muscle fibers, the appropriate experiments were first performed by Hodgkin and Horowicz.[69] These investigators changed the membrane potential by altering the $[K^+]_o$ at a constant $[K^+] \times [Cl^-]$ product. This could likewise be done by employing the voltage clamp method, for example in short toe muscle fibers.[70] The threshold for force activation is normally reached near −50 mV and the extent of force development is related to the membrane potential by a steep sigmoid curve [Fig. 5a ("activation curve")]. The fibers relax spontaneously after several seconds of maintained force. At membrane potentials less negative than about −25 mV, fibers then remain in a state of mechanical refractoriness (inactivation). However, contractility can be fully restored if fibers are repolarized to potentials more negative than −50 mV. The degree of restoration or repriming is related to membrane potential by an S-shaped curve with a half value near −40 mV [Fig. 5b ("steady-state restoration curve")].

4.2.1. Ca^{2+}-Induced Ca^{2+} Release

Fifteen years ago, two groups[73,74] published experiments in which they showed that the application of Ca^{2+} ions to skinned muscle fibers (i.e., fibers whose surface membrane has been removed, so that external solutions have free access to the contractile apparatus and the membranous systems inside the fiber) induces the release of Ca^{2+} from internal stores. This Ca^{2+}-induced Ca^{2+} release phenomenon together with the electrophysiological demonstration of a slow Ca^{2+} inward current[40] revived former proposals[2,75] that a small amount of ionic Ca^{2+}, which might either enter the cell from the T system

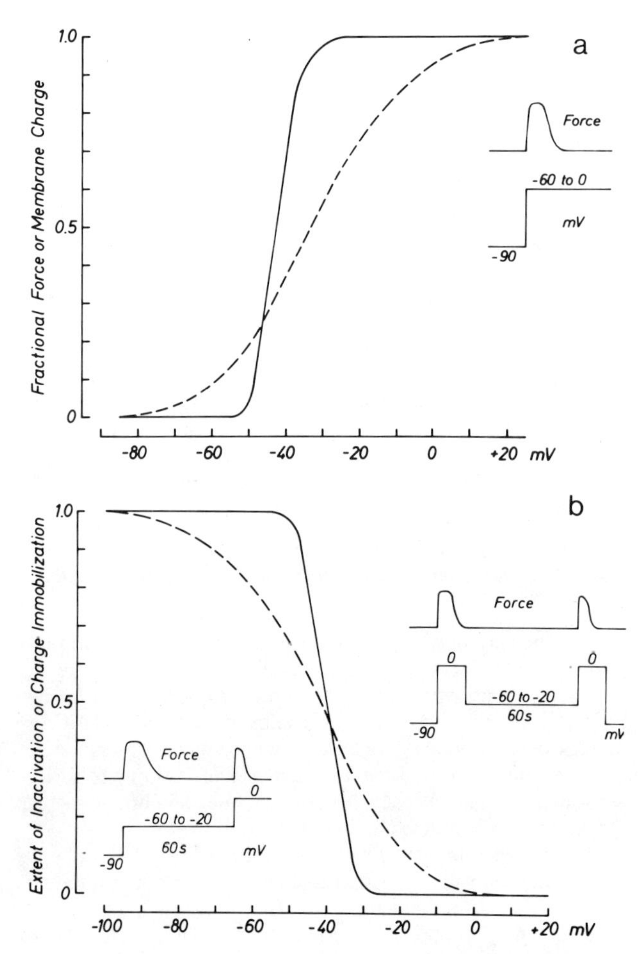

Fig. 5. The dependence of activation and restoration of force on the membrane potential.[69–71] The activation curve (a) is obtained by depolarizing the fiber from the resting potential to the potentials given on the abscissa (see inset) and measuring the peak of the developed force. The restoration curve (b) can be obtained by repolarizing a fully depolarized and mechanically refractory fiber for 60 sec to the potential values on the abscissa and measuring the extent of mechanical restoration after a second full depolarization (right inset). An identical relation can be obtained by depolarizing the fiber from the resting potential to values given on the abscissa until a steady state in inactivation is reached and subsequently activating the remaining force by a full depolarization (inactivation curve; left inset). Evidently only a quasi-steady state is reached during a period of 60 sec. If the conditioning pulse is prolonged, inactivation proceeds in a slowly developing second phase.[72]

The dashed lines show the steady-state charge distribution (a) and the steady-state charge immobilization (b) in dependence of the membrane potential (Section 4.2.2).

Force and charge movement data are probably not fully comparable since the measurements were done under different experimental conditions (composition of solutions, temperature).

or be released from membrane sites during excitation,[66] could induce the release of Ca^{2+} from the SR.

Recently, this hypothesis was subjected to rigorous tests. The influx of Ca^{2+} was either suppressed, by lowering external $[Ca^{2+}]$ below that measured intracellularly,[76,77] or inhibited, by the application of a rather specific blocker of Ca^{2+} channels.[78] Under both conditions, twitches and contractures could still be obtained. These results show that the influx of Ca^{2+} as a means of bridging the gap between T system and SR cannot be regarded as an alternative to other models of EC coupling discussed below. However, they do not exclude the possibility that the influx of Ca^{2+} under normal conditions,[40] its release from the internal side of the tubular membrane,[66] or simply its presence in the gap between T system and SR, modifies the coupling mechanism. The main effect observed when external Ca^{2+} was replaced by Mg^{2+} was an acceleration of the potential-dependent inactivation process, indicated by a shift in the steady-state inactivation curve (see Fig. 5b) by about 25 mV toward more negative potentials.[76,77] Further experiments suggest that the effect is due to a decrease in intracellular Ca^{2+} in the region of the triad as a consequence of Ca^{2+} removal from the extracellular fluid (see Refs. 76 and 79 for literature). Lüttgau and Spiecker[76] linked the described effect upon inactivation with the Ca^{2+}-induced Ca^{2+} release phenomenon described above. They suggested that an increase in internal Ca^{2+} causes a shift of the inactivation curve to more positive potentials, thus retarding or partly removing force inactivation. This process might become regenerative without necessarily inducing an all-or-none response of the whole fiber. Several recent observations might be related to regenerative processes in living muscle fibers such as the phasic alterations in $[Ca^{2+}]_i$ during tetani or contractures observed with the Ca^{2+} detector aequorin[80] or the slow contractile waves induced by caffeine which are not accompanied by potential alterations.[81] Some authors believe that the free Ca^{2+} concentration necessary to induce a regenerative release of Ca^{2+} from the SR in skinned skeletal fibers might be too high (i.e., 3×10^{-4} M) to be reached in living cells.[82,83] However, more recent observations on skinned skeletal fibers indicate that a regenerative Ca^{2+} release of physiological significance may operate in both amphibian and crustacean muscle fibers.[84,85]

Several still unexplained results obtained with so-called Ca^{2+} antagonists are of interest in this context. Single frog muscle fibers turned into a "paralyzed" state when D600, a tertiary amine, was applied at low temperatures (7°C). The failure in EC coupling, which became evident after one normal contracture in the presence of the drug, could be suspended by raising the temperature to 22°C.[86] Intramembranous charge movements were absent in paralyzed fibers.[87] Since Ca^{2+} channels were blocked by D600 at low *and* high temperature,[42,44] it appears that the drug interferes more directly in EC coupling, reaching its strategic position only in the depolarized state. The release of Ca^{2+} from the SR induced by caffeine was not affected by D600.[86]

4.2.2. Intramembranous Charge Movements

In 1973, Schneider and Chandler described voltage-dependent intramembranous charge movements in skeletal muscle fibers which they related to the activation of contraction.[67] The experiments were performed with sartorius muscle fibers under voltage clamp conditions and were evaluated by using a signal-averaging technique. NaCl and KCl were omitted from the external solutions, and voltage- and time-dependent Na^+ and K^+ currents were blocked by applying TTX (10^{-6} g/ml), RbCl (5 mM), and tetraethylammonium chloride (117.5 mM) to the external solutions. Intramembranous charge movements were detected by comparing the current measured from a voltage step toward the potential range of contraction activation (e.g., from a holding potential of -80 to -20 mV) (see Fig. 5a) with the current observed from the same voltage step at highly negative potentials (e.g., −160 to −100 mV). Differences in current between the first and the second voltage step showed a transient outward component ("on" response) associated with depolarization and a transient inward component ("off" response) on repolarization. The equality of the time integrals of the two responses, together with further arguments put forward by Schneider and Chandler, suggest that they are dealing with intramembranous charge movements rather than ionic currents through the membrane. The experiments of these workers have been confirmed and extended.[88–91]

The S-shaped dependence of charge movement (Q) on membrane potential (V) (see Fig. 5a) could be fitted by

$$Q(V) = \frac{Q_{\max}}{1 + \exp[-(V - V')/k]} \qquad (2)$$

where k is a constant ($RT/\beta zF$) and V' the potential at which an even distribution of charges between two positions within the membrane is reached.

A similar expression has been derived from a model in which charged molecules distribute themselves according to the Boltzmann relation between one position at the outer and another at the inner part of the membrane.[92] With depolarization, some of the charges move to the inner part and return after repolarization, thus producing transient currents, as observed. With $k = 8$ mV, and provided that the total membrane potential is sensed between the two positions ($\beta = 1$), the valence z would have magnitude 3, or each charged group would have a valence of magnitude 3. With 1700 charges per μm^2 of total membrane, Schneider and Chandler[67] arrived at about 600 charged groups/μm^2. If the additional assumption is made that these charges are only localized in the T system, the charge density would increase to about 700 groups/μm^2.

A further argument for associating charge movement with contraction activation is the observation that like the mechanical force, charges become inactivated (immobilized) during a prolonged depolarization and recover from immobilization after repolarization.[91,93,94] The potential dependence of steady-state immobilization is shown in Fig. 5b. Applying an equation analogous to that used to describe the charge activation process [Eq. (2)], a similar steepness factor (k) was obtained while the midpoint voltage (V') was found to be ~ 10 mV more negative than that of the activation curve.

The question must finally be discussed as to how the charge movement, presumed to be located in the wall of the T system, can activate the release of Ca^{2+} from the TC. Schneider and Chandler[67] put forward the idea that the charged groups are attached to long molecules which extend to the membrane of the adjacent terminal cisterna. This would provide a means by which the potential across the T-tubular membrane could be sensed by the SR. They further suggested that the sensor molecules might be associated with the "feet" which occur roughly at the same number of 700 per μm^2 between the two membrane systems. Probably only a single channel has to be activated by each charged group in order to release sufficient Ca^{2+} (400 Ca^{2+} ions per twitch and channel) for contractile activation.[95] Chandler *et al.*[89] published a more detailed model of excitation transmission in the triad (Fig. 6). A charged complex (valence 3) is connected by a rigid rod to a plug. The Ca^{2+} channel is closed by the plug when the charge is in the resting position (Fig. 6a). In the activating position (Fig. 6b), the plug is removed so that Ca^{2+} ions can flow from the cisterna into the myoplasm. In

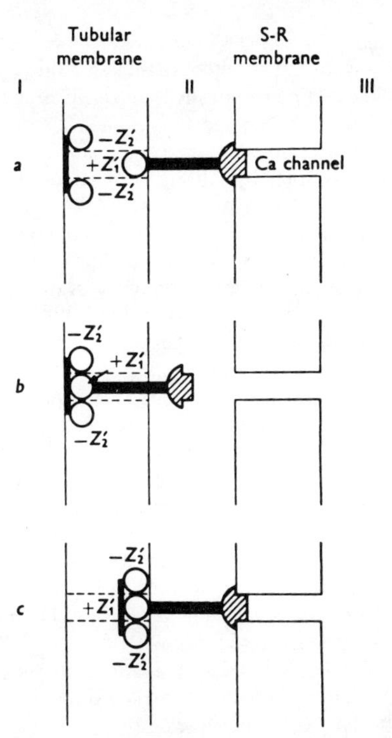

Fig. 6. Hypothetical example of how a charged complex might regulate Ca^{2+} release from the terminal cisternae. I denotes the lumen of the T tubules, the space labeled II is continuous with the sarcoplasm, and III indicates the inside of the terminal cisterna. Z_1' has a valence of 3; the other charged complex consists of two groups of valence 2 Z_2' connected by a basket which can move Z_1' to the right; $2\,Z_2' > Z_1'$. Configuration (a) corresponds to resting, (b) to activating, and (c) to refractory position.[89]

the refractory condition (Fig. 6c), the channel is again plugged. It is suggested that the movements of Z_1'(valence 3) are rapid compared to the movements of the Z_2' complex. On depolarization Z_1' moves to the left (Fig. 6b). Then, if $2\,Z_2' > Z_1'$, the Z_2' complex slowly moves to the right, taking Z_1' with it. As stressed by Chandler *et al.*, this model is only one of the many possibilities which might correspond to the experimental findings.

Without arguing about details, the model clearly illustrates the principle of the proposed coupling mechanism: Ca^{2+} release from the SR is controlled by voltage-dependent charge movement in the wall of the T system (Ref. 67; reviews: 96–99). Horowicz and Schneider[100] measured the amount of charge movement needed to induce a microscopically just-detectable contraction (Q_{th}) during depolarizing pulses of different duration (strength–duration curve). They found that Q_{th} was constant for pulses to potentials ranging from −45 mV (rheobase) to −15 mV (pulse duration 80 and 9 msec, respectively; 1–3°C). These results suggest some relation between charge movement and one or more steps in the depolarizing–contraction coupling process and may thus be regarded as a further verification of the charge movement concept.

In recent years the effect of potentiators and inhibitors of EC coupling upon charge movement kinetics has been investigated in order to test the new concept or, taking this for granted, to localize the exact site of action. Caffeine, which shifts the threshold of force activation toward more negative potentials and accelerates restoration after inactivation of force,[71,101] had no effect upon charge movement kinetics.[102] This would suggest that each releasing site in the SR sets free more Ca^{2+} after caffeine application than under normal conditions. Such an effect could be explained by the hypothesis mentioned above, that the increase in sarcoplasmic Ca^{2+}, induced by subthreshold caffeine concentrations,[103] impedes inactivation and accelerates restoration of force by shifting the inactivation curve toward more positive potentials (see Ref. 76). Tetracaine and dantrolene, both of which depress EC coupling,[71,104] alter charge movements by abolishing a shoulder or hump component which usually appears in the time course of the "on" response when the contraction threshold is reached.[105–108] From quantitative evaluations of the tetracaine effect,[105,107] in particular from the finding that the hump disappears in parallel with the Ca^{2+} transient,[108] it has been suggested that it might be the hump component of the charge movement signal that provides the voltage sensitivity for EC coupling. However, the possibility that the hump could be a secondary effect due to the massive release of Ca^{2+} into a restricted space, was not excluded.

The perchlorate anion rather specifically shifts the force activation curve toward more negative potentials without acting on the potential dependence of force inactivation or the activation of the Na^+ or K^+ channels.[109,110] The effect can be explained by a direct interference with intramembranous charge movement: the anion causes an increase in steepness of the potential dependence of charge activation and a distinct prolongation of the "off" response.[111]

The charge movement concept was in addition strengthened by recent investigations on mammalian muscles.[112] In the soleus muscle (slow-twitch fibers) the midpoint voltage (V') for charge movement was reached at −37.5 mV and in the extensor digitorum longus (edl; fast-twitch fibers) with a five- to sixfold greater maximum charge at −14 mV. The values corresponded roughly to midpoint voltages of force activation at −25 and −14 mV, respectively. In paraplegic soleus muscles the V' values for charge movement and force activation shifted by 13–14 mV to more positive values approaching the potential dependence of the edl fibers.

4.2.3. Ionic Channels between T System and SR

Recently the charge movement concept has been challenged by Eisenberg and co-workers.[68,113] They suggested that the above-described asymmetry current, believed to be intramembranous charge movement, could also arise from a membrane with linear capacitance in series with a nonlinear conductance. An electric network of this kind would exist if the pillars[20] contained ionic channels which open upon depolarization, thus allowing a direct depolarization of the terminal cisternae, i.e., an activation of the Ca^{2+} releasing sites. Evidence that such a hypothesis is plausible was presented by Eisenberg and Eisenberg[20] who showed that the number of pillars, i.e., electron microscopically detectable connections between T system and SR, increased during a K^+ contracture from 39 to 66 pillars/μm^2 tubular membrane. Interestingly, the number of pillars remained low when the contracture was caused by the application of caffeine, which induces Ca^{2+} release from the SR without a depolarization of the T-tubular membrane. The results show that pillar formation is not due to secondary reactions such as the increase in internal free Ca^{2+}, and may thus conform to the authors' hypothesis. However, they do not exclude alternative explanations, e.g., that (unknown) chemical reactions taking place in the course of muscle activation may help to preserve the fragile structures of the pillars.

The dilemma is that the experiments cited above, with substances which alter EC coupling, reveal a close relation between the kinetics of these asymmetry currents and the activation of Ca^{2+} release from the SR. Unfortunately, they do not permit a distinction to be made between the two concurrent interpretations of the observed currents. New experiments, however, have recently been performed by Huang.[114] Specifically designed pulse procedures gave results which were consistent with a charge movement model and at variance with a resistive one.

4.2.4. Electrical Properties of the Membranes of the TC

The electrical response of the TC is of great theoretical interest for the comprehension of EC coupling. Since the membrane potential of intracellular membrane systems cannot be directly measured with microelectrodes, several investigators applied indirect optical methods to detect voltage alterations (light scattering, absorption, birefringence, diffraction pattern, fluorescence; for review see Refs. 115–117). Optical studies with impermeable potentiometric dyes allow reasonable conclusions about the regenerative spread of excitation along the T system to be made.[64,116] However, it is not yet possible to unequivocally transform quantitatively optical responses into voltage alterations of SR membranes.

Recently, the problem was approached with two very promising new methods. Somlyo *et al.*[118,119] applied electron probe microanalysis for measuring elemental distribution in thin cryosections. They were able to show that the concentrations of Na^+, K^+, and Cl^- in the myoplasm and the TC are very much alike, implying that a substantial SR membrane potential cannot arise from concentration gradients of these ions. After a tetanus, a fall in the Ca^{2+} content of the TC was found, partly compensated by an increase in K^+ and Mg^{2+}. These measurements of elemental balances must, however, still be regarded as preliminary, in particular since, as stated by the authors, the balance after a tetanus of 1.2-sec duration might have been affected by recovery processes. Nevertheless, some conclusions can probably be drawn. The large differences in the content of Na^+, K^+, and Cl^- between T system and TC, together with the finding that the TC content of Na^+ and Cl^- did not alter during a tetanus, argue against the activation of continuing pathways for charges between T system and TC. Since the TC were found to be highly permeable for Cl^-,[120] the equality in Cl^- concentration between TC and myoplasm before and after a tetanus speaks against the development of a substantial TC potential induced by the opening of channels between T system and TC or an electrogenic efflux of Ca^{2+} ions from the TC.

A method which allows a more direct study of electrical parameters of the SR has been developed by Miller and Racker[121] who succeeded in fusing isolated mammalian SR vesicles with an artificial planar bilayer membrane. After the fusion process, two types of conductance pathways were observed[122]: one was voltage-independent and anion-selective, the other was voltage-gated with a large single-channel conductance of 10^{-10} S in 0.1 M K^+ and showed selectivity to small monovalent cations. Interestingly, a specific Ca^{2+} conductance was not observed with this method.

The results obtained by the described methods demonstrate the existence of channels for anions and monovalent cations in the membrane of the TC. However, no substantial ion exchange between T system and TC appears to occur during activity. Such an ion exchange would be necessary to produce a reasonable depolarization of the TC membrane as proposed in the Eisenberg model. From this point of view, the results so far favor the alternative Schneider–Chandler model in which Ca^{2+} sites in the TC are activated by mechanical or chemical reactions and not by depolarization. However, considering the complexity of the system, a definite choice of one of the two coupling models cannot be presently made. The coupling process, which is not comparable with the known events at chemical or electrical synapses, remains a problem to be solved in the future. Progress may be expected from a further improvement of the above new methods or from the detection of substances which interfere in a very specific way with the triadic structures or local chemical reactions.

5. Cellular Ca^{2+} Movements Related to the Activation of Contraction

5.1. Background

The mechanical activity of muscle is controlled ultimately by the free Ca^{2+} concentration in the sarcoplasm.[123,124] In the absence of Ca^{2+}, the interaction between the thin and the thick (actin and myosin) filaments is suppressed, and no active force is generated by the muscle. The suppression can be abolished when Ca^{2+} ions are bound to specific sites and induce various conformational changes in the filaments. The Ca^{2+}-regulating systems in striated muscle from different organisms can be linked with actin, myosin, or both.[125]

The actin-linked Ca^{2+} regulation involves the interaction between Ca^{2+} and the complex protein "troponin,"[123,126] which controls the activity of seven actin monomers via a neighboring tropomyosin molecule[123,124] (see Fig. 7).

The myosin-linked Ca^{2+} regulation can be of *two* types. One, typical for molluskan muscles, consists of Ca^{2+}-binding systems comprising myosin light chains associated with the myosin heads (S-1 subfragments).[125,127] (This myosin subfragment is considered to be the main morphological structure of cross-bridges between the thin and the thick filaments which can be observed by electron microscopy.[128]) In the other type of myosin-linked Ca^{2+} regulation, Ca^{2+} activates first a Ca^{2+}-dependent myosin light chain kinase, which in turn phosphorylates myosin light chains, resulting in the activation of the contractile proteins. Relaxation in this case is caused by dephosphorylation of the light chains by a phosphatase in the absence of Ca^{2+}. This type of regulation is typical of smooth muscle,[135] although the prerequisites for such Ca^{2+} regulation have been shown to exist also in striated vertebrate muscle fibers.[136,137]

More than one of these Ca^{2+}-controlled mechanisms can operate simultaneously in a muscle fiber. Thus, a double Ca^{2+} regulation involving both the troponin–tropomyosin and the Ca^{2+}-binding myosin light chain systems has been found to function in crustaceans and insects.[125] For some time it was considered that vertebrate skeletal muscle had a *simple* actin-linked Ca^{2+} regulation.[123,125] However, more recent studies indicate that a double Ca^{2+} regulation is not excluded.[138–141]

Direct evidence for an increase in the myoplasmic $[Ca^{2+}]$ in living skeletal muscle fibers as the result of stimulation was obtained with optical methods, based either on absorption

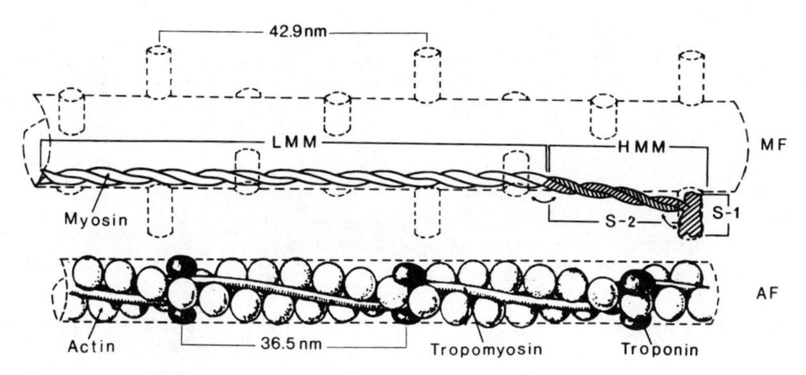

Fig. 7. Ultrastructure of the myofilaments in the relaxed vertebrate (amphibian) skeletal muscle. The myosin filament (MF) can be approximated (dashed line) to a cylinder of radius 6 nm and length 1.6 μm[128–131] made up of the light meromyosin (LMM) fragments of the myosin molecules, with projections disposed in a helical 6/2 arrangement.[129] Other arrangements for the projections have also been proposed.[130]

The projections are considered to be made of the heavy meromyosin (HMM) subfragment 1 (S-1), which contains all the ATPase activity of the myosin molecule. This subfragment (HMM-S-1) is supposed to hinge on HMM subfragment 2 (S-2), which in turn is assumed to hinge on LMM fragments,[59,128] so that the gap between the myosin and actin filaments, which varies as the inverse square root of the sarcomere length,[132] can be bridged in order to allow the direct interaction between the myosin heads (HMM-S-1) and the actin monomers. The actin filament (AF) is made up of a double helix of actin monomers and can be approximated to a cylinder of radius 3 nm and total length (on both sides of the Z line) 2.05 μm.[128,131] In the grooves of the double helix, however not centrally located, lie the tropomyosin molecules,[133] each molecule being in contact with seven actin monomers and one troponin.[123] When the Ca^{2+} ions are bound on troponin,[123] the position of the tropomyosin in the groove is supposed to change, uncovering the actin sites which can interact with the myosin heads.[134]

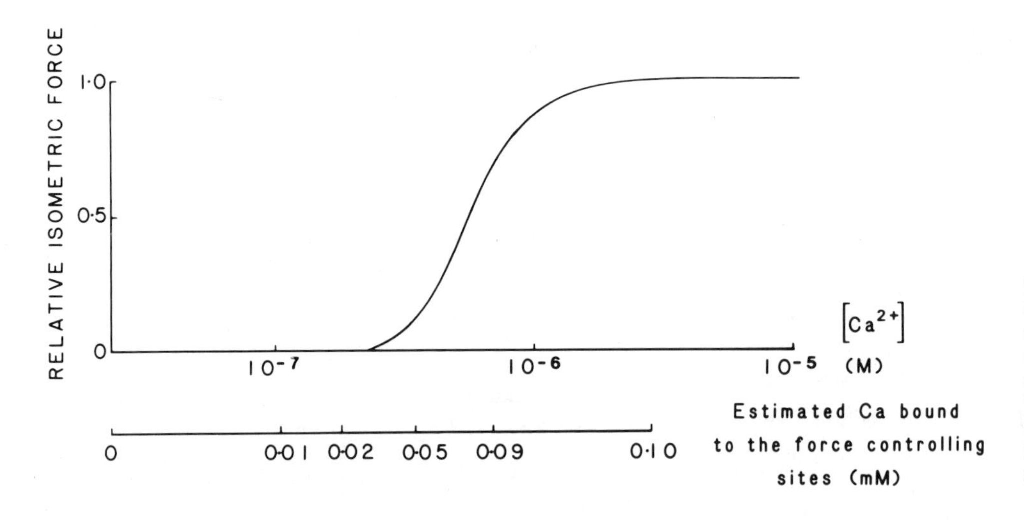

Fig. 8. The relationship between the relative isometric response and free Ca^{2+} concentration as found in frog skinned muscle fibers.[138] The lower abscissa shows the estimated amount of Ca^{2+} bound to the force-controlling sites (see text).

changes resulting from the binding of Ca^{2+} to various indicators like murexide[142] and metallochromic dyes,[143–147] or on the light emitted by the Ca^{2+}-sensitive photoproteins aequorin and obelin.[80,148–151]

In order to estimate the minimum amount of Ca^{2+} necessary to activate the contractile apparatus, one has to know the mode of muscle force dependence on both the *free* $[Ca^{2+}]$ and the amount of *Ca^{2+} bound* to the Ca^{2+} regulatory system(s). Figure 8 represents the relationship between the normalized isometric force response and the ionized Ca^{2+} level in "skinned"[152,153] frog fiber preparations activated in Ca^{2+}-buffered solutions.[138] It can be seen that the relative force increased from 10 to 90% if the free $[Ca^{2+}]$ was raised about 3-fold, from $\sim 3.5 \times 10^{-7}$ to $\sim 10^{-6}$ M. Such a steep curve relating force to ionized $[Ca^{2+}]$ was observed not only for twitch amphibian skeletal muscle[154,155] but also for mammalian fast-twitch muscle fibers.[139,156] In contrast, slow-twitch mammalian fibers[139], and under certain conditions some crustacean[157] and molluskan fibers,[158] require significantly higher changes in $[Ca^{2+}]$ ($\sim$ 10-fold) to produce a change in relative force between 10 and 90%.

It is important to note that the force–$[Ca^{2+}]$ curve for a given fiber type is shifted toward considerably higher $[Ca^{2+}]$ if the concentrations of Mg^{2+}[123,157,159,160], K^{+}[157,161], or H^{+}[157,159,162] are increased. Therefore, the $[Ca^{2+}]$ threshold for activation, defined as the smallest free $[Ca^{2+}]$ that produces a detectable force response, is not an absolute value but a relative one depending strongly both on the concentration of the other cations in the interfilament space and on the sensitivity of the force-measuring system. In addition, the $[Ca^{2+}]$ threshold for activation may also be modified by other factors including phosphorylation of various muscle proteins, sarcomere length,[163] and drugs such as caffeine.[164]

The amount of Ca^{2+} bound by the filaments, or only by troponin itself, shows a different, less steep relationship on the free $[Ca^{2+}]$ than the isometric force.[103,165,166] This indicates that the force response is not simply proportional to the amount of Ca^{2+} bound by the regulatory systems. The two quantitites can, however, be accurately related by assuming that Ca^{2+} ions are *independently* bound on two types of sites on the Ca^{2+} regulatory unit (which can involve only actin, or both actin and myosin regulatory systems), and that this regulatory unit can allow the interaction between the controlled actin and myosin heads only when both types of sites are occupied by Ca^{2+}.[124,138,139,167] Assuming, for simplicity, that each type of site involves only one Ca^{2+} ion, and this appears to be the case for skeletal muscle of barnacle,[167] it follows that the amount of Ca^{2+} bound by the Ca^{2+} regulatory sites on the myofibrils would be proportional to the *sum* of Ca^{2+} bound on each type of site, while force would be proportional to the probability that the Ca^{2+} regulatory unit had Ca^{2+} bound on both its sites, i.e., to the *product* of the relative concentrations of Ca^{2+} bound on each site.

The lower abscissa in Fig. 8 shows the estimated amount of Ca^{2+} bound by the Ca^{2+} regulatory system(s) for each relative force value on the curve, based on the foregoing type of considerations[138,139] and on a *total* amount of Ca^{2+} bound by the regulatory systems of the order of 0.1 mM.[103,124,167] It is obvious that the *free* $[Ca^{2+}]$ accounts for only a *minor* fraction of the minimal Ca^{2+} required for activation, represented by the *sum* of the Ca^{2+} bound by the regulatory system(s) and the corresponding free $[Ca^{2+}]$. In the intact fiber there are other Ca^{2+}-binding sites in addition to the Ca^{2+} sites directly associated with the Ca^{2+} regulatory system, such as those on parvalbumins,[168,172] and calmodulin,[171–173] proteins which exist freely in the sarcoplasm and which are related to the Ca^{2+}-binding subunit of troponin, troponin C. These additional sites have a strong Ca^{2+}-buffering capacity for $[Ca^{2+}]$ around 10^{-5}–10^{-4} M, and they would increase substantially the requirements of total Ca^{2+} to be released into the myoplasm if the free $[Ca^{2+}]$ were to increase above $\sim 1\ \mu M$. In addition, it has been shown that mitochondria can actively accumulate Ca^{2+}, which, however, cannot subsequently be easily released.[174–176] The mitochondria are considered to play a significant role in the regulation of Ca^{2+} movements *only* when their density is high, as in red skeletal muscles.

The free $[Ca^{2+}]$ in the sarcoplasm at any moment is determined by the extent of cellular Ca^{2+} movements. Formally, one can distinguish between Ca^{2+} movements responsible for the maintenance of the *capability* of skeletal muscle to develop force when *stimulated*, and Ca^{2+} movements *directly associated* with the activation process.

The ability of muscle to change from a relaxed to a contracted state is determined by the very low free $[Ca^{2+}]$ in the sarcoplasm, which increases following stimulation. The sarcoplasmic $[Ca^{2+}]$ in relaxed skeletal muscle is less than 0.1

μM[177–180], whereas the concentration in the external environment is of the order of several millimolar. The presence of very large electrical and chemical gradients, both favoring Ca^{2+} entry, and the finite permeability of the sarcolemma to Ca^{2+} (see Section 3), implies the existence of a continuous passive Ca^{2+} influx. If the muscle cells are not to accumulate Ca^{2+} progressively, this passive leak must be balanced in the long run by Ca^{2+}-transporting systems capable of removing cellular Ca^{2+} against a large electrochemical gradient. Such Ca^{2+}-extruding systems have been identified as operating in skeletal muscle.[22,181,183] The driving force for the extrusion of one Ca^{2+} ion seems to be supplied by the movement of several Na^+ ions entering the myoplasm down their electrochemical gradient,[22,182,184] although a direct involvement of ATP has not yet been excluded.[22,182] It is also likely that another Ca^{2+}-transporting system (Ca^{2+}, Mg^{2+}-ATPase Ca^{2+}-transporting system), recently identified in the plasma membrane of a variety of cells (see Ref. 184), also operates in skeletal muscle fibers. A theory proposed for nerve fibers, which could also apply for skeletal muscle fibers,[185] is that the latter Ca^{2+}-transporting system is relatively more important when the $[Ca^{2+}]_i$ is low, i.e., in the fiber at rest, while the Ca^{2+}/Na^+ exchange system extrudes mainly Ca^{2+} at higher $[Ca^{2+}]_i$, i.e., during rapid stimulation.

The *total* amount of Ca^{2+} released in the sarcoplasm following stimulation must be at least 0.1 mM in order to provide the necessary condition for the full activation of the contractile apparatus (see Fig. 8). This amount of Ca^{2+} must be released within several milliseconds at room temperature in vertebrate skeletal muscle. During relaxation, the same amount of Ca^{2+} must be removed from the sarcoplasm. The fluxes involved to produce this change in sarcoplasmic Ca^{2+} are far too large to be accounted for by Ca^{2+} entering from the extracellular space.[186,187] Furthermore, caffeine and K^+ contractures can be readily induced in the presence of very low Ca^{2+} concentrations in the external medium,[76,148,188] indicating that the role of extracellular Ca^{2+} in the *direct* activation of the contractile apparatus in skeletal muscle is minimal (see also Section 4.2.1).

There is additional evidence[189,190] which indicates that the Ca^{2+} directly responsible for the activation of contraction is derived from discrete intracellular sites associated with the SR.

5.2. The Mechanism of Ca^{2+} Transport Associated with the SR Membranes

In the resting state, most of the intracellular Ca^{2+}, estimated to be about 1–2 mmoles/kg fiber,[119,191–193] is contained in the SR. Since the SR volume comprises ~ 9–13%[6,14] of the fiber volume, the total $[Ca^{2+}]$ in the SR should be as high as 10–20 mM, if Ca^{2+} were uniformly distributed within the SR. However, many observations made with a variety of techniques[119,190,194–197] suggest that at rest, the total $[Ca^{2+}]$ is higher (~ 50 mM[119,197]) in the TC than in the longitudinal SR (~ 5 mM[197]). This unequal distribution of Ca^{2+} between different SR elements could be related to the two different types of Ca^{2+}-binding proteins found within the SR. One of these proteins, calsequestrin,[198] has a high Ca^{2+}-binding capacity (~ 45 moles Ca^{2+}/mole protein) but a relatively low affinity for Ca^{2+} and is located mainly in the TC.[199] The other Ca^{2+}-binding protein has two types of Ca^{2+}-binding sites: one with a high affinity for Ca^{2+} and numerous others (~ 25) with low affinity for Ca^{2+}. This Ca^{2+}-binding protein appears to be located only in the longitudinal SR.[199] A large proportion of the total Ca^{2+} identified in the SR is expected to be bound to these proteins and to other Ca^{2+}-binding sites on the luminal side of the SR membranes (phospholipids and the Ca^{2+} transport ATPase itself), implying that the *free* $[Ca^{2+}]$ in the SR is significantly lower.

The way in which depolarization of the T system may lead to the release of Ca^{2+} from the SR has already been discussed. Here, the mechanism of Ca^{2+} transport capable of maintaining the large electrochemical gradient for Ca^{2+} across the SR membranes is considered.

Most of our current knowledge about the molecular mechanism of active Ca^{2+} transport across the SR membranes is based on studies of isolated SR vesicles.[103,200–205] The vesicles are fragments of disrupted SR membranes which reseal following the homogenization of the whole muscle.

It is now established that the active translocation of Ca^{2+} across the SR membranes is stoichiometrically coupled to the hydrolysis of ATP: for each molecule of ATP hydrolyzed, two Ca^{2+} ions are transported into the SR.[201,206] It is also thought that the Ca^{2+} carrier on the side of the SR membranes facing the sarcoplasm initially binds one ATP molecule, and only after that does it bind the two Ca^{2+} ions which are going to be transported. This Ca^{2+}-binding step is followed by the splitting of ATP into ADP, which is released, and inorganic phosphate, which remains attached to the carrier. The resulting phosphorylated carrier undergoes a conformational change and the two Ca^{2+} attached to it cross the membrane. The dephosphorylation of the carrier is thought to follow the release of the two Ca^{2+} inside the SR membrane. Mg^{2+} ions are implicated in the dephosphorylation of the carrier, which is rate limiting for the whole Ca^{2+} translocation process. It is supposed that the original conformation of the carrier is regained following dephosphorylation and the carrier return to the outside (sarcoplasmic) surface of the SR membrane[201,204,205,207] to start a new cycle.

Charge displacement induced by the accumulation of Ca^{2+} ions inside the SR appears to be balanced partly by transport of protons in the opposite direction[208], and partly by passive fluxes of other ions (K^+, Cl^-, Mg^{2+}) through independent channels.[209] As a result, only a small, short-lasting membrane potential difference (~ 10 mV, inside positive) can be observed across the SR membranes during the initial phase of the Ca^{2+} translocation process.[209]

There is also evidence that the release of a certain fraction of Ca^{2+} from the SR vesicles may be coupled to the ATPase system and promote the synthesis of ATP.[204,210,211] The rate of Ca^{2+} pumping as reflected by the SR ATPase activity, and the Ca^{2+}-accumulating capacity of the SR depend strongly on the free $[Ca^{2+}]$[202,212] in the bathing medium in a way similar to that in which the isometric force response depends on $[Ca^{2+}]$ in Fig. 8. It has also been reported that the rate of formation of the phosphorylated carrier displays a very steep relationship with $[Ca^{2+}]$,[207] suggesting that more than two Ca^{2+} ions are involved in the activation of the Ca^{2+}-translocating system. This observation, taken together with the finding of two Ca^{2+} transported for each hydrolyzed ATP molecule, is in agreement with the results of aequorin experiments on living fibers,[213–215] which can be explained if the Ca^{2+} ions transported are not identical with those required for the activation of the translocating system.

A great deal has been learned about cellular Ca^{2+} movements from the study of separate muscle systems, but if one wishes to determine how the interrelation of these Ca^{2+} movements can lead to mechanical activation in living muscle fibers, the time course of the free Ca^{2+} change during an isometric force response must be explained in terms of its functional constituents.

5.3. Ca^{2+} Movements in Single Fibers

Several optical techniques(80,142–151) have been employed to demonstrate the occurrence of $[Ca^{2+}]$ changes in skeletal muscle fibers during activation. However, there are problems in accurately correlating the various optical responses observed with changes of $[Ca^{2+}]$. These difficulties stem from a number of intrinsic properties of the Ca^{2+} indicators used. Thus, some indicators like the metallochromic dyes appear to bind to certain elements in the sarcoplasm, which modify their response to Ca^{2+}.(217) In addition, some of these indicators show insufficient selectivity against competing ions, exhibit complex stoichiometric interactions with Ca^{2+} and are required in relatively high concentrations, which could modify the intracellular Ca^{2+} movements.(147,218,219) Other Ca^{2+} indicators such as photoproteins do not respond linearly to changes in $[Ca^{2+}]$ and the luminescent reaction induced by Ca^{2+} is relatively slow.(220,221) The problem is additionally compounded by the fact that the distribution of ions in the sarcoplasm may not be uniform due to the negative charges on the myofilaments which generate Donnan-type ionic distributions.(222–224) Ca^{2+} ions in particular are likely to have an even more pronounced nonhomogeneous distribution in the sarcoplasm during activation than other ions(145,222,225) and none of the techniques so far used have been able to detect local changes in $[Ca^{2+}]$. Such local $[Ca^{2+}]$ changes, like those in the immediate vicinity of the TC,(226) may be crucial in determining the magnitude of the Ca^{2+} fluxes associated with the SR.

Since one is limited in measuring simultaneously all parameters which control the various Ca^{2+} movements in the fiber, it is important to be able to link, in a quantitative way, those parameters which can be measured by devising mathematical models. Then, one can make accurate predictions for certain other parameters which could subsequently be experimentally measured, advancing our knowledge about the system in an educated way. Theoretically, one should be able to find several apparently distinct models, sharing common characteristics, which can explain certain relations between measured parameters. It is by far more important to identify those common characteristics of such groups of models than to develop one of the several possible, but distinct, models in too much detail. It is unavoidable that as more experimental facts are to be explained, the more complicated the models become. However, simple models which can adequately explain only some particular properties of the muscle fibers can be very useful if they allow certain types of generalizations to be made.

Bearing these considerations in mind, it is possible to characterize, at least in part, the relation between the Ca^{2+} transient and the force response in single skeletal muscle fibers.(167,215) Furthermore, more detailed analyses of the Ca^{2+} transients have allowed a better understanding of the interplay between Ca^{2+} movements associated with various compartments in muscle during activation.(144,171,172,215,216) Some of the conclusions from such analyses are diagrammatically summarized in Fig. 9.

Thus, all results obtained on both crustacean(146,148,149) and vertebrate muscle fibers(80,142,143,147,151) indicate that for an isometric twitch, the Ca^{2+} transient peaks well before force. Actually, force appears to have its highest rate of increase

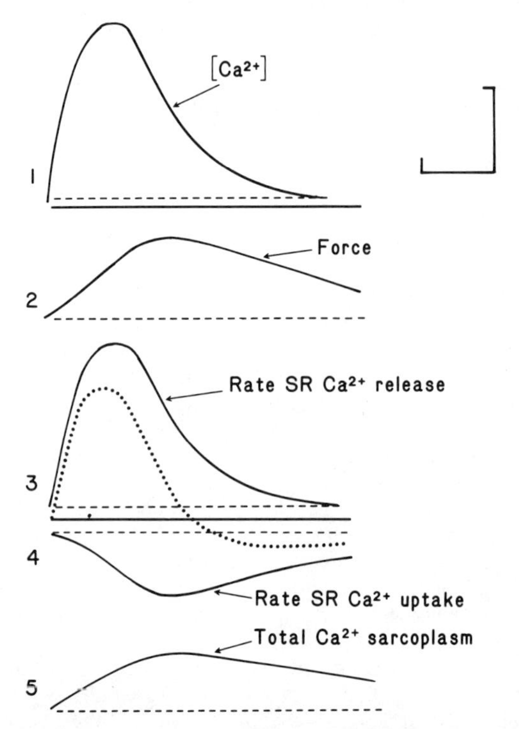

Fig. 9. Schematic diagram showing possible time relation between different Ca^{2+}-dependent processes during the activation of a skeletal muscle fiber (see also text and Ref. 216).

Trace 1 shows the Ca^{2+} transient; trace 2 represents the force response; traces 3 and 4 show the time course of SR Ca^{2+} release and SR Ca^{2+} uptake rates, respectively; trace 5 illustrates the changes in total Ca^{2+} in the sarcoplasm, including myofilaments; and the dotted trace represents the *net* SR contribution to the Ca^{2+} transient. The horizontal calibration bar could vary between about 20 msec in a vertebrate muscle fiber at ~ 15°C and 100 msec in a barnacle muscle fiber at the same temperature. The vertical calibration bar could represent 10^{-6} to 3×10^{-6} M for trace 1; 10–60% maximum isometric force for trace 2; 10^{-4} to 10^{-2} Ms^{-1} for traces 3 and 4; and between 0.1 and 1 mM for trace 5.

The dashed baseline for each trace represents the level of the respective parameter at rest and the solid lines for traces 1, 3, and 4 represent zero levels for the respective functions. Release of Ca^{2+} into the sarcoplasm is considered to be positive (trace 3) while the uptake of Ca^{2+} from the sarcoplasm is negative (trace 4).

around the peak of the Ca^{2+} transient. These and additional experiments on skinned muscle fibers(138,139,157) suggested that Ca^{2+} ions are involved in apparently slowly equilibrating steps of reaction to activate the contractile apparatus.(139,167,215) Recent observations have demonstrated that the cross-bridge turnover is very rapid compared with the rate of force development(227) so that the delay in force production cannot be explained in terms of cross-bridge kinetics (see also Refs. 138, 139, 167). Biochemical experiments have confirmed that at least two of the four Ca^{2+}-binding sites on troponin C have slow $[Ca^{2+}]$ dependent kinetics.(171,228a,b) Ca^{2+}-binding sites of similar slower kinetics have also been identified on parvalbumins.(171) Controversy exists whether these "slow" sites on troponin, called Ca^{2+}–Mg^{2+} sites, are involved in the Ca^{2+} regulation of contraction.(139,171) The main current argument against their involvement in regulation hinges on the observations from biochemical studies that the dissociation rates (k_{off}) for Ca^{2+} from these sites are very slow compared to the rates of force development and relaxation in intact muscle.[171,228] However, physiological studies on Ca^{2+} kinetics associated with the process of force development in mechanically skinned muscle fibers of twitch amphibian[138] and fast-twitch mammalian[139] muscles have suggested the involvement of different Ca^{2+}-reacting sites characterized by both fast and slow dissociation rates for Ca^{2+}. In the kinetic models derived from these physiological studies (see also Section 5.1), force was considered to develop when two independent types of Ca^{2+}-binding sites were both turned "on" by Ca^{2+}. Then, if one of the sites had rapid overall kinetics, while the other had slow kinetics, force would increase with the kinetics characteristic of the slow sites and decrease with the kinetics characteristic of the fast sites. Since the overall kinetics of the "slow" sites depends upon $[Ca^{2+}]$, becoming faster at higher $[Ca^{2+}]$, such a model would explain the time course of the force transient in intact muscle despite involving Ca^{2+} sites with slow dissociation properties in Ca^{2+} regulation.

Apparently more difficult to reconcile with the time course of the force transient in intact muscle is the observed slow dissociation rate of Mg^{2+} from the so called Ca^{2+}–Mg^{2+} sites on troponin, which at rest are likely to be occupied by Mg^{2+}.[172] Such a difficulty could arise if one assumed that the replacement of Mg^{2+} by Ca^{2+} on the Ca^{2+}–Mg^{2+} sites remained slow during a Ca^{2+} transient due to a fixed (slow) dissociation rate for Mg^{2+}. However, there is no reason to believe that the rate of dissociation of Mg^{2+} from these Ca^{2+}–Mg^{2+} sites cannot be modified by Ca^{2+}. For example, one cannot rule out without further experimentation the following simple reaction scheme for these Ca^{2+}–Mg^{2+} sites:

$$Ca^{2+} + MgTn \underset{k-}{\overset{k}{\rightleftharpoons}} CaTn + Mg^{2+} \tag{3}$$

In this reaction scheme the rate of Mg^{2+} coming off the complex MgTn is given by the product $k[MgTn]\cdot[Ca^{2+}]$ which is strongly dependent on $[Ca^{2+}]$. Thus, during a Ca^{2+} transient, which precedes contraction, the rate of dissociation of Mg^{2+} from the Ca^{2+}–Mg^{2+} sites could increase substantially, allowing for their involvement in the process of force regulation. One should point out here that a similar scheme of reaction to that described by Eq. (3), but involving protons and Ca^{2+}, is used as a model for the kinetics of the well-known Ca^{2+} buffer, EGTA, which like Ca^{2+}–Mg^{2+} sites on troponin shows slow Ca^{2+} dissociation.[229,230] Such a scheme predicts that the rate of Ca^{2+} dissociation from EGTA is not fixed, but depends on pH, a phenomenon which has recently been observed.[231] There is also unconfirmed[238a] biochemical evidence which suggests that the exchange process of bound Mg with free Ca^{2+} on troponin is a fast process.[231a]

The main message emerging from this discussion with regard to kinetics modeling is that in most situations one can use different sets of plausible assumptions which may in turn lead to quite different sets of conclusions. Therefore, caution must be exerted not to generalize one set of conclusions by removing them from the context of the assumptions originally used to derive them.

Returning to our main discussion, one could safely conclude that both biochemical and physiological observations indicate that a large component of the total amount of Ca^{2+} (released by the SR) in the sarcoplasm follows with delay the free Ca^{2+} transient due to the slower kinetics characteristics of a major group of Ca^{2+}-binding sites in the sarcoplasm (see Fig. 9). This implies that the total amount of Ca^{2+} in the sarcoplasm (including Ca^{2+} bound by parvalbumins and by the Ca^{2+}–Mg^{2+} sites on troponin) is significantly higher soon after a Ca^{2+} transient, when $[Ca^{2+}]$ has returned to the resting level, than it was before stimulation.[213–215,232] Therefore, the rate of Ca^{2+} uptake into the SR immediately following a Ca^{2+} transient must be higher than at rest in order to cope with the additional release of Ca^{2+} from the numerous Ca^{2+}-binding sites with slower kinetics, after $[Ca^{2+}]$ has returned to the resting level. Since the rate of Ca^{2+} translocation into the SR is also dependent on $[Ca^{2+}]$,[202,212] it follows that the time course for the rate of Ca^{2+} pumping into the SR should also be significantly delayed compared to the Ca^{2+} transient (see Fig. 9).

One should note that these conclusions are correct regardless of the side taken in the controversy surrounding the involvement of the Ca^{2+}–Mg^{2+} sites on troponin in the process of force regulation.

Considering that the rate of free Ca^{2+} change ($d[Ca^{2+}]/dt$) is at any moment in time the result of the rates with which the myofibrils, SR, and other sarcoplasmic binding sites release and/or take up ionized Ca^{2+} from the sarcoplasm and knowing the time course of the Ca^{2+} transient, one can make predictions regarding the Ca^{2+} movements associated with the SR.[144,213–216,232,233] However, without attempting to separate between the various components of the SR Ca^{2+} fluxes—SR Ca^{2+} release in the sarcoplasm, SR Ca^{2+} uptake from the sarcoplasm, and Ca^{2+} movements associated with the Ca^{2+}-binding sites located on the SR membrane facing the sarcoplasm—little useful information can actually be gained regarding the specific SR functions with respect to Ca^{2+} release and Ca^{2+} uptake.

The only analysis of a Ca^{2+} transient in a skeletal muscle fiber in terms of all its main components—(i) Ca^{2+} movements associated with sarcoplasmic Ca^{2+} sites characterized by fast kinetics; (ii) Ca^{2+} movements associated with sarcoplasmic Ca^{2+} sites characterized by slower kinetics than the Ca^{2+} transient; (iii) Ca^{2+} release by the SR in the sarcoplasm; (iv) Ca^{2+} uptake by the SR; and (v) Ca^{2+} movements associated with the Ca^{2+}-binding sites on the sarcoplasmic side of the SR membranes—was done on barnacle muscle fibers injected with aequorin.[213–216] In the other studies on frog twitch fibers, either the *net* Ca^{2+} flux associated with the SR was considered (i.e., only the overall response comprising components iii to v)[232] or even less satisfactory, myoplasmic Ca^{2+} movements associated with Ca^{2+} sites characterized by slow kinetics (ii) have been lumped together with the SR Ca^{2+} uptake system (iv and v), thus ignoring their contribution to the calculation of the rate of SR Ca^{2+} release in the sarcoplasm (iii) (i.e., the overall response comprising components ii, iii, and v was misleadingly interpreted to represent component iii[233]).

It is very likely that many of the general conclusions regarding the SR function in barnacle muscle will also be valid for the vertebrate muscle since the overall shape of Ca^{2+} transients are very similar for equivalent conditions. For example, the Ca^{2+} transient in cut frog fibers under voltage clamp conditions[144] has a similar time course to that seen in barnacle fibers for long depolarization pulses.[148] In these invertebrates, the electrical event associated with the outer membrane is not usually an all-or-nothing process, contrary to such events in most vertebrates. Consequently, the intensities of membrane depolarization, size of Ca^{2+} transient and force response can be uniformly graded by applying stimuli of different intensities. This situation is similar, although not identical, to that of vertebrate muscle under voltage clamp condition.

The time course of the Ca^{2+} transient in barnacle fibers can be explained if the rate of Ca^{2+} release from the SR, $(d[Ca^{2+}]/dt)_{SR\ release}$, is linearly dependent on the ionized $[Ca^{2+}]$ for the duration of a short or medium-long depolarization pulse.[213,215,216] Thus, one can write

$$(d[Ca^{2+}]/dt)_{SR\ release} = a[Ca^{2+}] + b \tag{4}$$

where $a > 0$ and $b < 0$. The positive value for a implies that the rate of Ca^{2+} release from the SR increases when the ionized $[Ca^{2+}]$ in the sarcoplasm increases and this signifies a kind of regenerative Ca^{2+} release. The negative parameter b can be interpreted as a threshold for $[Ca^{2+}]$ above which the process of regenerative release occurs. Both parameters a and b must depend on the intensity of stimulation (depolarization) in order to explain the graded Ca^{2+} transient and force responses in these fibers, and a mechanism by which changes in the electrical state of the external membranous system could influence the properties of the SR has been discussed previously. Shortly after the tubular membrane repolarizes, the rate of Ca^{2+} release from the SR is predicted to decay exponentially to a constant level representing a permanent Ca^{2+} leak from the SR.

An important feature for such interpretation of the Ca^{2+} transient in muscle is the *continuous* Ca^{2+}-accumulating activity of the SR. This overcomes the difficulty of arbitrarily assuming the moments at which the SR is "switched on" and "switched off." Furthermore, the present discussion shows that the assumption of a regenerative process for the release of Ca^{2+} [e.g., Eq. (4)] from the SR does not necessarily lead to the maximal activation of muscle, or to an all-or-none response, as has very often been supposed. Moreover, the coupling of a regenerative Ca^{2+} release with a permanent, but delayed Ca^{2+} uptake (the latter mechanism depending on a higher power of $[Ca^{2+}]$ than the first) can lead to a *transient* increase in the free $[Ca^{2+}]$, as can be seen during long depolarization pulses in barnacle fibers or in cut frog fibers under voltage clamp conditions.[144] Such coupling between Ca^{2+} release and Ca^{2+} uptake processes could also explain the induction of a spontaneous relaxation if the external membranous system is depolarized for long durations as in the case of K^+ contractures.

With this interpretation of the Ca^{2+} transient in skeletal muscle, one can also explain why optical signals thought to be associated with changes in the SR membrane properties during activation of a muscle fiber have a time course resembling that of the Ca^{2+} transient (for a review see Ref. 117). This is simply because the optical signals from the SR membranes are likely to be generated by processes which control the efflux of Ca^{2+} from the SR, which according to Eq. (4) is linearly dependent on the ionized $[Ca^{2+}]$ in sarcoplasm. Conversely, the observation that changes in fluorescence and optical retardation signals associated with the SR follow the time course of the Ca^{2+} transient could be interpreted as evidence for the existence of a Ca^{2+}-induced Ca^{2+}-release mechanism in the muscle during activity. From the representation in Fig. 9 of the various fiber parameters which depend on the Ca^{2+} transient, one can see that the *net* contribution of the SR to the Ca^{2+} transient (dotted trace) is quite different from the SR Ca^{2+} release component (trace 3). Thus, while the SR Ca^{2+} release component closely follows the Ca^{2+} transient (trace 1), the net SR contribution to the Ca^{2+} transient peaks earlier and decreases much faster compared to the Ca^{2+} transient, a conclusion which also emerged from an analysis of the Ca^{2+} transient in frog fibers.[232] Therefore, it is inadequate to regard the overall, net positive SR contribution to the Ca^{2+} transient as the SR Ca^{2+}-release component, an assumption which was made in this study on frog muscle.[232]

The Ca^{2+} transient in cut frog fibers under voltage clamp conditions[144] has been interpreted in different ways. The most favored interpretation was that the Ca^{2+} transient corresponds to a three-compartment system, one being the myoplasm and the other two being the release and uptake pools of the SR. The properties of the latter were considered to depend upon the membrane potential. However, some important experimental observations cannot be easily accommodated in this model. Thus, the myofibrillar Ca^{2+} pool (including troponin and parvalbumin) can only be accommodated in the model if it is assumed to equilibrate either *very* quickly or *very* slowly[144] when compared with the duration of depolarization (100 msec). This is difficult to accept when it is known that many myoplasmic Ca^{2+}-binding sites are characterized by slow kinetics (see before and Refs. 232 and 228a,b). In a more recent interpretation of the same results [223] the contribution of the myoplasmic Ca^{2+}-binding sites characterized by relatively slow kinetics was lumped together with that of the SR Ca^{2+}-uptake component. This was equivalent with a theoretical transfer of these myoplasmic Ca^{2+} sites from the myoplasmic compartment to the SR uptake compartment. Therefore, the so-called rate of Ca^{2+} release deduced and used in this analysis (RREL) may bear little resemblance to the actual rate of Ca^{2+} release from the SR into the myoplasm since it represents the difference between the latter and the rate of Ca^{2+} binding by the slowly equilibrating sites in the myoplasm.

Furthermore, neither of these interpretations of the Ca^{2+} transient in cut frog fibers[144,233] takes into consideration the fact that the rate of Ca^{2+} uptake by the SR is also dependent on $[Ca^{2+}]$.[202,212] Nevertheless, such models, although inadequate for a complete explanation of Ca^{2+} movements associated with the activation of the muscle, may be useful for studies of the potential-dependent steps in depolarization–contraction coupling.

Earlier results have suggested that Ca^{2+} movements are not only simply between the SR and the myofibrillar space, but also between certain elements of the SR.[190] Thus, it has been suggested that the longitudinal SR might play a major role in the process of Ca^{2+} reaccumulation and translocation to the TC.[190] However, more accurate experiments performed recently[119] did not produce much evidence in support of these suggestions.

The movements of Ca^{2+} at the level of the sarcomere are also likely to be influenced by the nonuniform electrical (Donnan) potential in various regions of the sarcoplasm.[222] Since the lowest electrical potential within the myofibrils will be found in the region of highest negative fixed charge density,[222] it follows that the largest electrical potential difference within the sarcoplasm would exist between the region around the SR and the region of filament overlap. On the one hand, this potential difference would selectively facilitate the diffusion of Ca^{2+} ions released from the SR toward the overlapping region between the two types of myofilaments, where force is generated, thus speeding up the Ca^{2+} activation process. On the other hand, the negative potential within the myofibrils (which appears to be insensitive to $[Ca^{2+}]$ up to 0.1 mM) would oppose a fast removal of Ca^{2+} from the interfilament space. However, parvalbumins could facilitate removal of Ca^{2+} from the myofibrillar space acting both as a slowly equilibrating Ca^{2+} buffer and as a Ca^{2+} carrier.[172]

6. Summary

The ionic movements which underlie EC coupling in skeletal muscle are considered in conjunction with ultrastructural and functional elements of the muscle cell.

The ultrastructure of the SR and T system is described and summarized by a three-dimensional representation in the light of recently published observations.

The electrical properties of the outer membranes, as reflected by the changes in their transverse potential following activation, are discussed in terms of the relative changes in the conductance of the physiologically occurring ions.

The main experiments which led to the current view about the inward spread of excitation along the T-tubule system are analyzed.

Three possible hypotheses are explored in the discussion concerning transmission of excitation at the level of the triad: one in which Ca^{2+} acts as a transmitter, one in which the main role is played by membranous charge movements, and another in which activation is spread to the TC through ionic channels which form and bridge the gap between T system and SR following excitation.

Possible changes in the electrical state of the SR membranes associated with the process of Ca^{2+} release are also discussed.

Current knowledge about how cellular Ca^{2+} movements are related to the process of activation of contraction is critically presented. The mechanisms by which Ca^{2+} controls force generation in muscle are summarized and the major features of the Ca^{2+} transport across SR membranes are outlined. The possible interplay at rest and following excitation between the various Ca^{2+} movements associated with different compartments in the muscle is discussed.

ACKNOWLEDGMENT. This work was supported by SFB 114 "Bionach" and ARGS.

References

1. Hill, A. V. 1965. *Trails and Trials in Physiology.* Arnold, London. p. 374.
2. Sandow, A. 1952. Excitation–contraction coupling in muscular response. *Yale J. Biol. Med.* **25**:176–201.
3. Huxley, A. F. 1971. The activation of striated muscle and its mechanical response. *Proc. R. Soc. London Ser. B* **178**:1–27.
4. Costantin, L. L. 1971. Inward spread of activation in frog skeletal muscle. In: *Contractility of Muscle Cells and Related Processes.* R. J. Podolsky, ed. Prentice–Hall, Englewood Cliffs, N.J. pp. 89–98.
5. Page, S. G. 1965. A comparison of the fine structures of frog slow and twitch muscle fibers. *J. Cell Biol.* **26**:477–497.
6. Peachey, L. D. 1965. The sarcoplasmic reticulum and transverse tubules of the frog's sartorius. *J. Cell Biol.* **25**:209–231.
7. Peachey, L. D., and R. F. Schild. 1968. The distribution of the T-system along the sarcomeres of frog and toad sartorius muscles. *J. Physiol. (London)* **194**:249–258.
8. Franzini-Armstrong, C., and L. D. Peachey. 1981. Striated muscle—Contractile and control mechanism. *J. Cell Biol.* **91**:166s–186s.
9. Page, S. G. 1964. The organization of the sarcoplasmic reticulum in frog muscle. *J. Physiol. (London)* **175**:10P–11P.
10. Huxley, H. E. 1964. Evidence for continuity between the central elements of the triads and extracellular space in frog sartorius muscle. *Nature (London)* **202**:1067–1071.
11. Franzini-Armstrong, C., L. Landmesser, and G. Pilar. 1975. Size and shape of transverse tubule openings in frog twitch muscle fibers. *J. Cell Biol.* **64**:493–497.
12. Dulhunty, A. F., and C. Franzini-Armstrong. 1975. The relative contributions of the folds and caveolae to the surface membrane of frog skeletal muscle fibres at different sarcomere lengths. *J. Physiol. (London)* **250**:513–539.
13. Zampighi, G., J. Vergara, and F. Ramón. 1975. On the connection between the transverse tubules and the plasma membrane in frog semitendinosus skeletal muscle: Are caveolae the mouths of the transverse tubule system? *J. Cell Biol.* **64**:734–740.
14. Mobley, B. A., and B. R. Eisenberg. 1975. Sizes of components in frog skeletal muscle measured by methods of stereology. *J. Gen. Physiol.* **66**:31–45.
15. Hodgkin, A. L., and S. Nakajima. 1972. Analysis of the membrane capacity in frog muscle. *J. Physiol. (London)* **221**:121–136.
16. Franzini-Armstrong, C. 1970. Studies of the triad. I. Structure of the junction in frog twitch fibers. *J. Cell Biol.* **47**:488–499.
17. Franzini-Armstrong, C. 1975. Membrane particles and transmission at the triad. *Fed. Proc.* **34**:1382–1389.
18. Franzini-Armstrong, C., and G. Nunzi. 1983. Junctional feet and particles in the triads of a fast-twitch muscle fibre. *J. Musc. Res. Cell Motil.* **4**:233–252.
19. Eisenberg, B. R., and A. Gilai. 1979. Structural changes in single muscle fibers after stimulation at a low frequency. *J. Gen. Physiol.* **74**:1–16.
20. Eisenberg, B. R., and R. S. Eisenberg. 1982. The T–SR junction in contracting single skeletal muscle fibers. *J. Gen. Physiol.* **79**:1–19.
21. Zachar, J. 1971. *Electrogenesis and Contractility in Skeletal Muscle Cells.* University Park Press, Baltimore. p. 638.
22. Baker, P. F., and H. Reuter. 1975. *Calcium Movement in Excitable Cells.* Pergamon Press, Elmsford, N.Y. p. 102.
23. Adrian, R. H. 1960. Potassium chloride movement and the membrane potential of frog muscle. *J. Physiol. (London)* **151**:154–185.
24. Adrian, R. H. 1961. Internal chloride concentration and chloride efflux of frog muscle. *J. Physiol. (London)* **156**:623–632.
25. Hodgkin, A. L., and P. Horowicz. 1959. Movements of Na and K in single muscle fibres. *J. Physiol. (London)* **145**:405–432.
26. Hodgkin, A. L., and P. Horowicz. 1959. The influence of potassium and chloride ions on the membrane potential of single muscle fibres. *J. Physiol. (London)* **148**:127–160.
27. Eisenberg, R. S., and P. W. Gage. 1969. Ionic conductances of the surface and transverse tubular membranes of frog sartorius fibers. *J. Gen. Physiol.* **53**:279–297.
28. Palade, P. T., and R. L. Barchi. 1977. On the inhibition of muscle membrane chloride conductance by aromatic carboxylic acids. *J. Gen. Physiol.* **69**:879–896.
29. Dulhunty, A. F. 1979. Distribution of potassium and chloride permeability over the surface and T-tubule membranes of mammalian skeletal muscle. *J. Membr. Biol.* **45**:293–310.
30. Hodgkin, A. L., and P. Horowicz. 1957. The differential action of hypertonic solutions on the twitch and action potential of a muscle fibre. *J. Physiol. (London)* **136**:17P.
31. Nastuk, W. L., and A. L. Hodgkin. 1950. The electrical activity of single muscle fibers. *J. Cell. Comp. Physiol.* **35**:39–74.
32. Adrian, R. H., W. K. Chandler, and A. L. Hodgkin. 1970. Voltage clamp experiments in striated muscle fibres. *J. Physiol. (London)* **208**:607–644.
33. Adrian, R. H., W. K. Chandler, and A. L. Hodgkin. 1970. Slow changes in potassium permeability in skeletal muscle. *J. Physiol. (London)* **208**:645–668.
34. Ildefonse, M., and O. Rougier. 1972. Voltage clamp analysis of the early current in frog skeletal muscle fibre using the double sucrose-gap method. *J. Physiol. (London)* **222**:373–395.
35. Ildefonse, M., and G. Roy. 1972. Kinetic properties of the sodium current in striated muscle fibres on the basis of the Hodgkin–Huxley theory. *J. Physiol. (London)* **227**:419–431.
36. Freygang, W. H., Jr., D. A. Goldstein, and D. C. Hellam. 1964.

The after-potential that follows trains of impulses in frog muscle fibers. *J. Gen. Physiol.* **47**:929–952.

37. Kirsch, G. C., R. A. Nichols, and S. Nakajima. 1978. Delayed rectification in the transverse tubule. *J. Gen. Physiol.* **70**:1–12.
38. Stanfield, P. R. 1975. The effect of zinc ions on the gating of the delayed potassium conductance of frog sartorius muscle. *J. Physiol. (London)* **251**:711–735.
39. Sjodin, R. A. 1982. Transport of electrolytes in muscle. *J. Membr. Biol.* **68**:161–178.
40. Stefani, E., and D. J. Chiarandini. 1982. Ionic channels in skeletal muscle. *Annu. Rev. Physiol.* **44**:357–372.
41. Stanfield, P. R. 1977. A calcium dependent inward current in frog skeletal muscle fibres. *Pfluegers Arch.* **368**:267–270.
42. Sánchez, J. A., and E. Stefani. 1978. Inward calcium current in twitch muscle fibres of the frog. *J. Physiol. (London)* **283**:197–209.
43. Nicola-Siri, L., J. A. Sánchez, and E. Stefani. 1980. Effect of glycerol treatment on the calcium current of frog skeletal muscle. *J. Physiol. (London)* **305**:87–96.
44. Almers, W., and P. T. Palade. 1981. Slow calcium and potassium currents across frog muscle membrane: Measurements with a Vaseline-gap technique. *J. Physiol. (London)* **312**:159–176.

44a. Takeda, K. 1977. Prolonged sarcotubular regenerative response in frog sartorius muscle. *Jap. J. Physiol.* **27**:379–389.

44b. Potreau, D., and G. Raymond. 1982. Existence of a sodium-induced calcium release mechanism in frog skeletal muscle fibres. *J. Physiol. (London)* **333**:463–480.

44c. Almers, W., E. W. McCleskey, and P. T. Palade. 1984. A non-selective cation conductance in frog muscle membrane blocked by micromolar external calcium ions. *J. Physiol. (London)* **353**:565–583.

44d. Almers, W., and E. W. McCleskey. 1984. Non-selective conductance in calcium channels of frog muscle: Calcium selectivity in a single-file pore. *J. Physiol. (London)* **353**:585–608.

45. Stühmer, W., and W. Almers. 1982. Photobleaching through glass micropipettes: Sodium channels without lateral mobility in the sarcolemma of frog skeletal muscle. *Proc. Natl. Acad. Sci. USA* **79**:946–950.
46. Almers, W., R. Fink, and N. Shepherd. 1982. Lateral distribution of ionic channels in the cell membrane of skeletal muscle. In: *Disorders of the Motor Unit.* D. L. Scotland, ed. Wiley, New York. pp. 349–366.
47. Almers, W., P. R. Stanfield, and W. Stühmer. 1983. Lateral distribution of sodium and potassium channels in frog skeletal muscle: Measurements with a patch-clamp technique. *J. Physiol. (London)* **336**:261–284.
48. Stanfield, P. R., N. B. Standen, C. A. Leech, and F. M. Ashcroft. 1981. Inward rectification in skeletal muscle fibres. *Adv. Physiol. Sci.* **5**:247–262.
49. Lüttgau, H. C. 1965. The effect of metabolic inhibitors on the fatigue of the action potential in single muscle fibres. *J. Physiol. (London)* **178**:45–67.
50. Fink, R., S. Hase, H. C. Lüttgau, and E. Wettwer. 1983. The effect of cellular energy reserves and internal Ca^{2+} on the potassium conductance in skeletal muscle of the frog. *J. Physiol. (London)* **336**:211–228.
51. Barrett, J. N., K. L. Magleby, and B. S. Pallotta. 1982. Properties of single calcium-activated potassium channels in cultured rat muscle. *J. Physiol. (London)* **331**:211–230.
52. Methfessel, C., and G. Boheim. 1982. The gating of single calcium-dependent potassium channels is described by an activation/blockade mechanism. *Biophys. Struct. Mech.* **9**:35–60.
53. Lüttgau, H. C., and E. Wettwer. 1983. Ca^{2+}-activated potassium conductance in metabolically exhausted skeletal muscle fibres. *Cell Calcium* **4**:331–341.
54. Bezanilla, F., C. Caputo, H. Gonzales-Serratos, and R. A. Venosa. 1972. Sodium dependence of the inward spread of activation in isolated twitch muscle fibres of the frog. *J. Physiol. (London)* **223**:507–523.
55. Almers, W., R. Fink, and P. T. Palade. 1981. Calcium depletion in frog muscle tubules: The decline of calcium current under maintained depolarization. *J. Physiol. (London)* **312**:177–207.
56. Adrian, R. H., and S. H. Bryant. 1974. On the repetitive discharge in myotonic muscle fibres. *J. Physiol. (London)* **240**:505–515.
57. Huxley, A. F., and R. E. Taylor. 1955. Function of Krause's membrane. *Nature (London)* **176**:1068.
58. Costantin, L. L. 1970. The role of sodium current in the radial spread of contraction in frog muscle fibers. *J. Gen. Physiol.* **55**:703–715.
59. Huxley, A. F. 1974. Review lecture: Muscular contraction. *J. Physiol. (London)* **243**:1–43.
60. Costantin, L. L. 1975. Contractile activation in skeletal muscle. *Prog. Biophys. Mol. Biol.* **29**:197–224.
61. Lüttgau, H. C., and H. G. Glitsch. 1976. Membrane physiology of nerve and muscle fibres. *Fortschr. Zool.* **24**:1–132.
62. Huxley, A. F., and R. E. Taylor. 1958. Local activation of striated muscle fibres. *J. Physiol. (London)* **144**:426–441.
63. Adrian, R. H., L. L. Costantin, and L. D. Peachey. 1969. Radial spread of contraction in frog muscle fibres. *J. Physiol. (London)* **204**:231–257.
64. Nakajima, S., and A. Gilai. 1980. Radial propagation of muscle action potential along the tubular system examined by potential-sensitive dyes. *J. Gen. Physiol.* **76**:751–762.
65. Jaimovich, E., R. A. Venosa, P. Shrager, and P. Horowicz. 1975. Tetrodotoxin (TTX) binding in normal and "detubulated" frog sartorius muscle. *Biophys. J.* **15**:255a.
66. Frank, G. B. 1982. Roles of extracellular and "trigger" calcium ions in excitation–contraction coupling in skeletal muscle. *Can. J. Physiol. Pharmacol.* **60**:427–439.
67. Schneider, M. F., and W. K. Chandler. 1973. Voltage dependent charge movement in skeletal muscle: A possible step in excitation–contraction coupling. *Nature (London)* **242**:244–246.
68. Mathias, R. T., R. A. Levis, and R. S. Eisenberg. 1980. Electrical models of excitation contraction coupling and charge movement in skeletal muscle. *J. Gen. Physiol.* **76**:1–31.
69. Hodgkin, A. L., and P. Horowicz. 1960. Potassium contractures in single muscle fibres. *J. Physiol. (London)* **153**:386–403.
70. Caputo, C., and P. Fernandez de Bolanōs. 1979. Membrane potential, contractile activation and relaxation rates in voltage clamped short muscle fibres of the frog. *J. Physiol. (London)* **289**:175–189.
71. Lüttgau, H. C., and H. Oetliker. 1968. The action of caffeine on the activation of the contractile mechanism in striated muscle fibres. *J. Physiol. (London)* **194**:51–74.
72. Nagai, T., M. Takauji, I. Kosaka, and M. Tsutsu-Ura. 1979. Biphasic time course of inactivation of potassium contractures in single twitch muscle fibers of the frog. *Jpn. J. Physiol.* **29**:539–549.
73. Ford, L. E., and R. J. Podolsky. 1970. Regenerative calcium release within muscle cells. *Science* **167**:58–59.
74. Endo, M., M. Tanaka, and Y. Ogawa. 1970. Calcium-induced release of calcium from the sarcoplasmic reticulum of skinned skeletal muscle fibres. *Nature (London)* **228**:34–36.
75. Sandow, A. 1965. Excitation–contraction coupling in skeletal muscle. *Pharmacol. Rev.* **17**:265–320.
76. Lüttgau, H. C., and W. Spiecker. 1979. The effects of calcium deprivation upon mechanical and electrophysiological parameters in skeletal muscle fibres of the frog. *J. Physiol. (London)* **296**:411–429.
77. Cota, G., and E. Stefani. 1981. Effects of external calcium reduction on the kinetics of potassium contractures in frog twitch muscle fibres. *J. Physiol. (London)* **317**:303–316.
78. Gonzalez-Serratos, H., R. Valle-Aguilera, D. A. Lathrop, and M. del Carmen Garcia. 1982. Slow inward calcium currents have no obvious role in muscle excitation–contraction coupling. *Nature (London)* **298**:292–294.
79. Lüttgau, H. C., W. Melzer, and W. Spiecker. 1981. The role of external Ca^{2+} in excitation contraction coupling. *Adv. Physiol. Sci.* **5**:375–388.
80. Blinks, J. R., R. Rüdel, and S. R. Taylor. 1978. Calcium tran-

sients in isolated amphibian skeletal muscle fibres: Detection with aequorin. *J. Physiol. (London)* **277**:291–323.
81. Kumbaraci, N. M., and W. L. Nastuk. 1982. Action of caffeine in excitation–contraction coupling of frog skeletal muscle fibres. *J. Physiol. (London)* **325**:195–211.
82. Endo, M. 1975. Conditions required for calcium-induced release of calcium from the sarcoplasmic reticulum. *Proc. Jpn. Acad.* **51**:467–472.
83. Thorens, S., and M. Endo. 1975. Calcium-induced calcium release and "depolarization"-induced calcium release: Their physiological significance. *Proc. Jpn. Acad.* **51**:473–478.
84. Moisescu, D. G., and R. Thieleczek. 1978. Calcium and strontium concentration changes within skinned muscle preparations following a change in the external bathing solution. *J. Physiol. (London)* **275**:241–262.
85. Stephenson, D. G., and D. A. Williams. 1980. Activation of skinned arthropod muscle fibres by Ca^{2+} and Sr^{2+}. *J. Musc. Res. Cell Motil.* **1**:73–87.
86. Eisenberg, R. S., R. T. McCarthy, and R. L. Milton. 1983. Paralysis of frog skeletal muscle fibres by the calcium antagonist D-600. *J. Physiol. (London)* **341**:495–505.
87. Hui, C. S., R. L. Milton, and R. S. Eisenberg. 1983. Elimination of charge movement in skeletal muscle by a calcium antagonist. *Biophys. J.* **41**:178a.
88. Chandler, W. K., R. F. Rakowski, and M. F. Schneider. 1976. A non-linear voltage dependent charge movement in frog skeletal muscle. *J. Physiol. (London)* **254**:245–283.
89. Chandler, W. K., R. F. Rakowski, and M. F. Schneider. 1976. Effects of glycerol treatment and maintained depolarization on charge movement in skeletal muscle. *J. Physiol. (London)* **254**:285–316.
90. Adrian, R. H., and W. Almers. 1976. Charge movement in the membrane of striated muscle. *J. Physiol. (London)* **254**:339–360.
91. Adrian, R. H., W. K. Chandler, and R. F. Rakowski. 1976. Charge movement and mechanical repriming in skeletal muscle. *J. Physiol. (London)* **254**:361–388.
92. Hodgkin, A. L., and A. F. Huxley. 1952. The dual effect of membrane potential on sodium conductance in the giant axon of *Loligo*. *J. Physiol. (London)* **116**:497–506.
93. Rakowski, R. F. 1981. Immobilization of membrane charge in frog skeletal muscle by prolonged depolarization. *J. Physiol. (London)* **317**:129–148.
94. Rakowski, R. F. 1981. Inactivation and recovery of membrane charge movement in skeletal muscle. In: *The Regulation of Muscle Contraction.* A. D. Grinell and M. A. B. Brazier, eds. Academic Press, New York. pp. 23–37.
95. Almers, W. 1975. Observations on intramembrane charge movements in skeletal muscle. *Philos. Trans. R. Soc. London Ser. B.* **270**:507–513.
96. Almers, W. 1978. Gating currents and charge movements in excitable membranes. *Rev. Physiol. Biochem. Pharmacol.* **82**:96–190.
97. Adrian, R. H. 1978. Charge movement in the membrane of striated muscle. *Annu. Rev. Biophys. Bioeng.* **7**:85–112.
98. Schneider, M. F. 1981. Membrane charge movement and depolarization–contraction coupling. *Annu. Rev. Physiol.* **43**:507–517.
99. Gilly, W. F. 1981. Intramembrane charge movements and excitation–contraction (E-C) coupling. In: *The Regulation of Muscle Contraction.* A. D. Grinell and M. A. B. Brazier, eds. Academic Press, New York. pp. 3–22.
100. Horowicz, P., and M. F. Schneider. 1981. Membrane charge moved at contraction thresholds in skeletal muscle fibres. *J. Physiol. (London)* **314**:595–633.
101. Caputo, C., G. Gottschalk, and H. C. Lüttgau. 1981. The control of contraction activation by the membrane potential. *Experientia* **37**:580–581.
102. Kovács,L., and G. Szücs. 1983. Effect of caffeine on intramembrane charge movement and calcium transients in cut skeletal muscle fibres of the frog. *J. Physiol. (London)* **341**:559–578.
103. Weber, A., R. Herz, and I. Reiss. 1964. The regulation of myofibrillar activity by calcium. *Proc. R. Soc. London Ser. B* **160**:489–501.
104. Desmedt, I. E., and K. Hainaut. 1977. Inhibition of the intracellular release of calcium by dantrolene in barnacle giant muscle fibres. *J. Physiol. (London)* **265**:565–585.
105. Hui, C. S. 1983. Pharmacological studies of charge movement in frog skeletal muscle. *J. Physiol. (London)* **337**:509–529.
106. Hui, C. S. 1983. Differential properties of two charge components in frog skeletal muscle. *J. Physiol. (London)* **337**:531–552.
107. Huang, C. L.-H. 1982. Pharmacological separation of charge movement components in frog skeletal muscle. *J. Physiol. (London)* **324**:375–387.
108. Vergara, J., and C. Caputo. 1983. Effects of tetracaine on charge movements and calcium signals in frog skeletal muscle fibers. *Proc. Natl. Acad. Sci. USA* **80**:1477–1481.
109. Foulks, J. G., J. A. D. Miller, and F. A. Perry. 1973. Repolarization-induced reactivation of contracture tension in frog skeletal muscle. *Can. J. Physiol. Pharmacol.* **51**:324–334.
110. Gomolla, M., G. Gottschalk, and H. C. Lüttgau. 1983. Perchlorate-induced alterations in electrical and mechanical parameters of frog skeletal muscle fibres. *J. Physiol. (London)* **343**:197–214.
111. Lüttgau, H. C., G. Gottschalk, L. Kovács, and M. Fuxreiter. 1983. How perchlorate improves excitation–contraction coupling in skeletal muscle fibers. *Biophys. J.* **43**:247–249.
112. Dulhunty, A. F., and P. W. Gage. 1983. Asymmetrical charge movement in slow- and fast-twitch mammalian muscle fibres in normal and paraplegic rats. *J. Physiol. (London)* **341**:213–231.
113. Mathias, R. T., R. A. Levis, and R. S. Eisenberg. 1981. An alternative interpretation of charge movement in muscle. In: *The Regulation of Muscle Contraction.* A. D. Grinell and M. A. B. Brazier, eds. Academic Press, New York. pp. 39–52.
114. Huang, C. L.-H. 1983. Experimental analysis of alternative models of charge movement in frog skeletal muscle. *J. Physiol. (London)* **336**:527–543.
115. Stephenson, E. W. 1981. Activation of fast skeletal muscle: Contributions of studies on skinned fibers. *Am. J. Physiol.* **240**:C1–C19.
116. Baylor, S. M., W. K. Chandler, and M. W. Marshall. 1981. Optical studies in skeletal muscle using probes of membrane potential. In: *The Regulation of Muscle Contraction.* A. D. Grinell and M. A. B. Brazier, eds. Academic Press, New York. pp. 97–130.
117. Oetliker, H. 1982. An appraisal of the evidence for a sarcoplasmic reticulum membrane potential and its relation to calcium release in skeletal muscle. *J. Musc. Res. Cell Motil.* **3**:247–272.
118. Somlyo, A. V., H. Shuman, and A. P. Somlyo. 1977. Elemental distribution in striated muscle and the effects of hypertonicity: Electron probe analysis of cryo sections. *J. Cell Biol.* **74**:828–857.
119. Somlyo, A. V., H. Gonzalez-Serratos, H. Shuman, G. McClellan, and A. P. Somlyo. 1981. Calcium release and ionic changes in the sarcoplasmic reticulum of tetanized muscle: An electron-probe study. *J. Cell Biol.* **90**:577–594.
120. Kasai, M., T. Kanemasa, and S. Fukumoto. 1979. Determination of reflection coefficients for various ions and neutral molecules in sarcoplasmic reticulum vesicles through osmotic volume change studied by stopped flow technique. *J. Membr. Biol.* **51**:311–324.
121. Miller, C., and E. Racker. 1976. Ca^{2+}-induced fusion of fragmented sarcoplasmic reticulum with artificial planar bilayers. *J. Membr. Biol.* **30**:283–300.
122. Miller, C. 1978. Voltage-gated cation conductance channel from fragmented sarcoplasmic reticulum: Steady-state electrical properties. *J. Membr. Biol.* **40**:1–23.
123. Ebashi, S., M. Endo, and I. Ohtsuki. 1969. Control of muscle contraction. *Q. Rev. Biophys.* **2**:351–384.
124. Weber, A., and J. M. Murray. 1973. Molecular control mechanisms in muscle contraction. *Physiol. Rev.* **53**:612–673.
125. Lehmann, W., and A. G. Szent-Györgyi. 1975. Regulation of muscular contraction. *J. Gen. Physiol.* **65**:1–30.

126. Potter, J. D., and J. Gergely. 1974. Troponin, tropomyosin, and actin interactions in the Ca^{2+} regulation of muscle contraction. *Biochemistry* **13**:2697–2703.
127. Chantler, P. D., and A. G. Szent-Györgyi. 1980. Regulatory light chains and scallop myosin: Full dissociation, reversibility and cooperative effects. *J. Mol. Biol.* **138**:473–492.
128. Huxley, H. E. 1969. The mechanism of muscular contraction. *Science* **164**:1356–1366.
129. Huxley, H. E., and W. Brown. 1967. The low-angle X-ray diagram of vertebrate striated muscle and its behaviour during contraction and rigor. *J. Mol. Biol.* **30**:383–434.
130. Squire, J. M. 1974. Symmetry and three-dimensional arrangement of filaments in vertebrate skeletal muscle. *J. Mol. Biol.* **90** :153–160.
131. Gordon, A. M., A. F. Huxley, and F. J. Julian. 1966. The variation in isometric tension with sarcomere length in vertebrate muscle fibres. *J. Physiol. (London)* **184**:170–192.
132. Elliott, G. F., J. Lowy, and C. R. Worthington. 1963. An X-ray and light diffraction study of the filament lattice of striated muscle in the living state and rigor. *J. Mol. Biol.* **6**:295–305.
133. Hanson, J., and J. Lowy. 1963. The structure of F-actin and of actin filaments isolated from muscle. *J. Mol. Biol.* **6**:46–60.
134. Haselgrove, J. C. 1972. X-ray evidence for a conformational change in the actin-containing filaments of vertebrate striated muscle. *Cold Spring Harbor Symp. Quant. Biol.* **37**:341–359.
135. Adelstein, R. S., M. A. Conti, D. R. Hathaway, and C. B. Klee. 1978. Phosphorylation of smooth muscle myosin light chain kinase by the catalytic subunit of adenosine 3′:5′ monophosphate-dependent protein kinase. *J. Biol. Chem.* **253**:8347–8350.
136. Morgan, M., S. V. Perry, and J. Ottaway. 1976. Myosin light-chain phosphatase. *Biochem. J.* **157**:687–697.
137. Stull, J. T., and C. W. High. 1977. Phosphorylation of skeletal muscle contractile proteins *in vivo*. *Biochem. Biophys. Res. Commun.* **77**:1078–1083.
138. Moisescu, D. G. 1976. Kinetics of reaction in Ca-activated skinned muscle fibres. *Nature (London)* **262**:610–613.
139. Stephenson, D. G., and D. A. Williams. 1981. Calcium–activation force responses in fast- and slow-twitch skinned muscle fibres of the rat at different temperatures. *J. Physiol. (London)* **317**:281–302.
140. Lehmann, W. 1978. Thick-filament-linked calcium regulation in vertebrate striated muscle. *Nature (London)* **274**:80–81.
141. Chin, T. K., and A. J. Rowe. 1982. Biochemical properties of native myosin filaments. *J. Musc. Res. Cell Motil.* **3**:118.
142. Jöbsis, F. F., and M. J. O'Connor. 1966. Calcium release and reabsorption in the sartorius muscle of the toad. *Biochem. Biophys. Res. Commun.* **25**:246–252.
143. Miledi, R., I. Parker, and G. Schalow. 1977. Measurement of calcium transients in frog muscle by the use of arsenazo III. *Proc. R. Soc. London Ser. B* **198**:201–210.
144. Kovács, L., E. Ríos, and M. F. Schneider. 1979. Calcium transients and intramembrane charge movement in skeletal muscle fibres. *Nature (London)* **279**:391–396.
145. Palade, P., and J. Vergara. 1982. Arsenazo III and antipyrylazo III calcium transients in single skeletal muscle fibers. *J. Gen. Physiol.* **79**:679–707.
146. Dubyak, G. R., and A. Scarpa. 1982. Sarcoplasmic Ca^{2+} transients during the contractile cycle of single barnacle muscle fibres: Measurements with arsenazo III-injected fibres. *J. Musc. Res. Cell Motil.* **3**:87–112.
147. Baylor, S. M., W. K. Chandler, and M. W. Marshall. 1982. Use of metallochromic dyes to measure changes in myoplasmic calcium during activity in frog skeletal muscle fibres. *J. Physiol. (London)* **331**:139–177.
148. Ashley, C. C., and E. B. Ridgway. 1970. On the relationship between membrane potential, calcium transient and tension in single barnacle muscle fibres. *J. Physiol. (London)* **209**:105–130.
149. Ashley, C. C., P. C. Caldwell, A. K. Campbell, T. J. Lea, and D. G. Moisescu. 1976. Calcium movements in muscle. *Symp. Soc. Exp. Biol.* **30**:397–422.
150. Ashley, C. C., and A. K. Campbell. 1979. Detection and measurement of free calcium ions in cells. Elsevier, Amsterdam.
151. Eusebi, F., R. Miledi, and T. Takahashi. 1980. Calcium transients in mammalian muscles. *Nature (London)* **284**:560–561.
152. Natori, R. 1954. The property and contraction process of isolated myofibrils. *Jikeikai Med. J.* **1**:119–126.
153. Hellam, D. C., and R. J. Podolsky. 1969. Force measurements in skinned muscle fibres. *J. Physiol. (London)* **200**:807–819.
154. Julian, F. J. 1971. The effect of calcium on the force–velocity relation of briefly glycerinated frog muscle fibres. *J. Physiol. (London)* **218**:117–145.
155. Godt, R. E., and B. D. Lindley. 1982. Influence of temperature upon contractile activation and isometric force production in mechanically skinned muscle fibers of the frog. *J. Gen. Physiol.* **80** :279–297.
156. Brandt, P. W., R. N. Cox, and M. Kawai. 1980. Can the binding of Ca^{2+} to two regulatory sites on troponin C determine the steep pCa/tension relationship of skeletal muscle? *Proc. Natl. Acad. Sci. USA* **77**:4717–4720.
157. Ashley, C. C., and D. G. Moisescu. 1977. The effect of changing the composition of the bathing solutions upon the isometric tension–pCa relationship in bundles of myofibrils isolated from single crustacean muscle fibres. *J. Physiol. (London)* **270**:627–652.
158. Simmons, R. M., and A. G. Szent-Györgyi. 1980. Control of tension development in scallop muscle fibers with foreign regulatory light chains. *Nature (London)* **286**:626–628.
159. Ashley, C. C., and D. G. Moisescu. 1974. The influence of $[Mg^{2+}]$ and pH upon the isometric steady state tension–Ca^{2+} relationship in isolated bundles of myofibrils. *J. Physiol. (London)* **239**:112P–114P.
160. Donaldson, S. K. B., and W. G. L. Kerrick. 1975. Characterization of the effects of Mg^{2+} on Ca^{2+}- and Sr^{2+}-activated tension generation of skinned skeletal muscle fibers. *J. Gen. Physiol.* **66** :427–444.
161. Moisescu, D. G. 1975. The effect of $[K^{+}]$ on the calcium-induced development of tension in isolated bundles of myofibrils. *Pfluegers Arch.* **355**:R62.
162. Fabiato, A., and F. Fabiato. 1978. Effects of pH on the myofilaments and the sarcoplasmic reticulum of skinned cells from cardiac and skeletal muscles. *J. Physiol. (London)* **276**:233–255.
163. Stephenson, D. G., and I. R. Wendt, 1984. Length dependence of changes in sarcoplasmic calcium concentration and myofibrillar calcium sensitivity in striated muscle fibres. *J. Musc. Res. Cell Motil.* **5**:243–272.
164. Wendt, I. R., and D. G. Stephenson. 1983. Effects of caffeine on Ca-activated force production in skinned cardiac and skeletal muscle fibres of the rat. *Pfluegers Arch.* **398**:210–216.
165. Potter, J. D., and J. Gergely. 1975. The calcium and magnesium binding sites on troponin and their role in the regulation of myofibrillar ATPase. *J. Biol. Chem.* **250**:4628–4633.
166. Fuchs, F., and C. Fox. 1982. Parallel measurements of bound calcium and force in glycerinated rabbit psoas muscle fibers. *Biochim. Biophys. Acta* **679**:110–115.
167. Ashley, C. C., and D. G. Moisescu. 1972. Model for the action of calcium in muscle. *Nature New Biol.* **237**:208–211.
168. Pechère, J. F., J. Demaille, J. P., Capony, E. Dutruge, G. Baron, and C. Pina. 1975. Muscular parvalbumins: Some explorations into their possible biological significance. In: *Calcium Transport in Contraction and Secretion*. E. Carafoli, F. Clementi, W. Drabikowski, and A. Margreth, eds. North-Holland, Amsterdam. pp. 459–468.
169. Lehky, P., H. E. Blum, E. A. Stein, and E. H. Fischer. 1974. Isolation and characterization of parvalbumins from the skeletal muscle of higher vertebrates. *J. Biol. Chem.* **249**:4332–4334.
170. Cox, J. A., D. R. Winge, and E. Stein. 1979. Calcium, magnesium and the conformation of parvalbumin during muscular activity. *Biochimie* **61**:501–605.
171. Robertson, S. P., J. D. Johnson, and J. D. Potter. 1981. The time course of Ca^{2+} exchange with calmodulin, troponin, parvalbumin

and myosin in response to transient increases in Ca^{2+}. *Biophys J.* **34**:559–569.

172. Gillis, J. M., D. Thomason, J. Lefèvre, and R. H. Kretsinger. 1982. Parvalbumins and muscle relaxation: A computer simulation study. *J. Musc. Res. Cell Motil.* **3**:377–398.
173. Cheung, W. Y. 1980. Calmodulin plays a pivotal role in cellular regulation. *Science* **207**:19–27.
174. Carafoli, E., K. Malmstrom, H. Capano, E. Sigel, and M. Crompton. 1975. Mitochondria and the regulation of cell calcium. In: *Calcium Transport in Contraction and Secretion*. E. Carafoli, F. Clementi, W. Drabikowski, and A. Margreth, eds. North-Holland, Amsterdam. pp. 53–64.
175. Scarpa, A. 1975. Kinetics and energy-coupling of Ca^{2+} transport in mitochondria. In: *Calcium Transport in Contraction and Secretion*. E. Carafoli, F. Clementi, W. Drabikowski, and A. Margreth, eds. North-Holland, Amsterdam. pp. 65–76.
176. Bygrave, F. L. 1978. Mitochondria and the control of intracellular calcium. *Biol. Rev.* **53**:43–79.
177. Portzehl, H., P. C. Caldwell, and J. C. Rüegg. 1964. The dependence of contraction and relaxation of muscle fibres from the crab *Maia squinado* on the internal concentration of free calcium ions. *Biochim. Biophys. Acta* **79**:581–591.
178. Hagiwara, S., and S. Nakajima. 1966. Effects of the intracellular [Ca^{2+}] upon the excitability of the muscle fiber membrane of a barnacle. *J. Gen. Physiol.* **49**:807–817.
179. Keynes, R. D., E. Rojas, R. E. Taylor, and J. Vergara. 1973. Calcium and potassium systems of a giant barnacle muscle fibre under membrane potential control. *J. Physiol. (London)* **229**:409–455.
180. Coray, A., C. H. Fry, P. Hess, Y. A. S. McGuigan, and R. Weingart. 1980. Resting calcium in sheep cardiac tissue and in frog skeletal muscle measured with ion-selective micro-electrodes. *J. Physiol. (London)* **305**:60P–61P.
181. Cosmos, E., and E. J. Harris. 1961. *In vitro* studies of the gain and exchange of calcium in frog skeletal muscle. *J. Gen. Physiol.* **44**:1121–1130.
182. DiPolo, R. 1973. Sodium-dependent calcium influx in dialysed barnacle muscle fibres. *Biochim. Biophys. Acta* **298**:279–283.
183. Ashley, C. C., J. C. Ellory, and K. Hainaut. 1974. Calcium movements in single crustacean muscle fibres. *J. Physiol. (London)* **242**:255–272.
184. Barritt, G. J. 1981. Calcium transport across cell membranes: Progress toward molecular mechanisms. *Trends Biochem. Sci.* **6**:322–325.
185. DiPolo, R., and L. Beauge. 1980. Mechanisms of calcium transport in the giant axon of the squid and their physiological role. *Cell Calcium* **1**:147–169.
186. Bianchi, C. P., and A. M. Shanes. 1959. Calcium influx in skeletal muscle at rest, during activity, and during potassium contracture. *J. Gen. Physiol.* **42**:803–815.
187. Curtis, B. A. 1966. Ca fluxes in single twitch muscle fibers. *J. Gen. Physiol.* **50**:255–267.
188. Ashley, C. C., P. J. Griffiths, D. G. Moisescu, and R. M. Rose. 1975. The use of aequorin and the isolated myofibrillar bundle preparation to investigate the effect of SR calcium releasing agents. *J. Physiol. (London)* **245**:12P–14P.
189. Hill, A. V. 1949. The abrupt transition from rest to activity in muscle. *Proc. R. Soc. London Ser. B* **136**:399–420.
190. Winegrad, S. 1968. Intracellular calcium movements of frog skeletal muscle during recovery from tetanus. *J. Gen. Physiol.* **51**:65–83.
191. Curtis, B. 1970. Calcium efflux from frog twitch muscle fibers. *J. Gen. Physiol.* **55**:243–253.
192. Ford, L. E., and R. J. Podolsky. 1972. Calcium uptake and force development by skinned muscle fibres in EGTA buffered solutions. *J. Physiol. (London)* **233**:1–19.
193. Ashley, C. C., P. C. Caldwell, and A. G. Lowe. 1972. The efflux of calcium from single crab and barnacle muscle fibres. *J. Physiol. (London)* **223**:735–755.
194. Winegrad, S. 1965. Autoradiographic studies of intracellular calcium in frog skeletal muscle. *J. Gen. Physiol.* **48**:455–479.
195. Costantin, L. L., C. Franzini-Armstrong, and R. J. Podolsky. 1965. Localization of calcium-accumulating structures in striated muscle fibers. *Science* **147**:158–160.
196. Pease, D. C., D. J. Jenden, and J. N. Howell. 1965. Calcium uptake in glycerol-extracted rabbit psoas muscle fibres. II. Electron microscopic localization of uptake sites. *J. Cell. Comp. Physiol.* **65**:141–154.
197. Campbell, K. P., C. Franzini-Armstrong, and A. E. Shamoo. 1980. Further characterization of light and heavy sarcoplasmic reticulum vesicles: Identification of the "sarcoplasmic feet" associated with heavy sarcoplasmic reticulum vesicles. *Biochim. Biophys. Acta* **602**:97–116.
198. MacLennan, D. H., and P. G. Wong. 1971. Isolation of a calcium-sequestering protein from sarcoplasmic reticulum. *Proc. Natl. Acad. Sci. USA* **68**:1231–1235.
199. Jorgensen, A. O., V. Kalnins, and D. H. MacLennan. 1979. Localization of sarcoplasmic reticulum proteins in rat skeletal muscle by immunofluorescence. *J. Cell Biol.* **80**:372–384.
200. Hasselbach, W. 1964. Relaxing factor and the relaxation of muscle. *Prog. Biophys. Mol. Biol.* **14**:167–222.
201. Hasselbach, W. 1979. The sarcoplasmic calcium pump: A model of energy transduction in biological membranes. *Fortschr. Chem. Forsch.* **78**:1–56.
202. Weber, A., R. Herz, and I. Reiss. 1966. Study of the kinetics of calcium transport by isolated fragmented sarcoplasmic reticulum. *Biochem. Z.* **345**:329–369.
203. Martonosi, A. 1972. Biochemical and clinical aspects of sarcoplasmic reticulum function. *Curr. Top. Membr. Transp.* **3**:83–197.
204. de Meis, L., and A. L. Vianna. 1979. Energy interconversion by the Ca^{2+}-dependent ATPase of the sarcoplasmic reticulum. *Annu. Rev. Biochem.* **48**:275–292.
205. Tada, M., T. Yamamoto, and Y. Tonomura. 1978. Molecular mechanism of active calcium transport by sarcoplasmic reticulum. *Physiol. Rev.* **58**:1–72.
206. Hasselbach, W., and M. Makinose. 1963. Über den Mechanismus des Calciumtransportes durch die Membranen des sarkoplasmatischen Retikulums. *Biochem. Z.* **339**:94–111.
207. Martonosi, A. N. 1975. The mechanism of Ca transport in sarcoplasmic reticulum. In: *Calcium Transport in Contraction and Secretion*. E. Carafoli, F. Clementi, W. Drabikowski, and A. Margreth, eds. North-Holland, Amsterdam. pp. 313–327.
208. Chiesi, M., and G. Inesi. 1980. Adenosine 5′-triphosphate dependent fluxes of manganese and hydrogen ions in sarcoplasmic reticulum vesicles. *Biochemistry* **19**:2912–2918.
209. Beeler, T. J., R. H. Farmen, and A. N. Martonosi. 1981. The mechanism of voltage-sensitive dye responses on sarcoplasmic reticulum. *J. Membr. Biol.* **62**:113–137.
210. Makinose, M., and W. Hasselbach. 1971. ATP synthesis by the reversal of the sarcoplasmic calcium pump. *FEBS Lett.* **12**:271–272.
211. Carvalho, A. P., M. G. P. Vale, and V. R. O. e Castro. 1975. Utilization of X-537A to differentiate between intravesicular and membrane bound Ca^{2+} in sarcoplasmic reticulum. In: *Calcium Transport in Contraction and Secretion*. E. Carafoli, F. Clementi, W. Drabikowski, and A. Margreth, eds. North-Holland, Amsterdam. pp. 349–358.
212. Ogawa, Y. 1970. Some properties of fragmented frog sarcoplasmic reticulum with particular reference to its response to caffeine. *J. Biochem. (Tokyo)* **67**:667–683.
213. Ashley, C. C., and D. G. Moisescu. 1973. The mechanism of the free calcium change in single muscle fibres during contraction. *J. Physiol. (London)* **231**:23P–25P.
214. Moisescu, D. G. 1973. The intracellular control and action of calcium in striated muscle and the forces responsible for the stability of the myofilament lattice. Ph.D. thesis. University of Bristol.
215. Ashley, C. C., D. G. Moisescu, and R. M. Rose. 1974. Kinetics of calcium during contraction: Myofibrillar and SR fluxes during a

single response of a skeletal muscle fibre. In: *Calcium Binding Proteins*. W. Drabikowski, H. Strzelecka-Golaszewska, and E. Carafoli, eds. North-Holland, Amsterdam. pp. 609–642.

216. Lüttgau, H. C., and D. G. Moisescu. 1978. Ion movements in skeletal muscle in relation to the activation of contraction. In: *Physiology of Membrane Disorders*. T. E. Andreoli, J. F. Hoffman, and D. D. Fanestil, eds. Plenum Press, New York. pp. 493–515.
217. Beeler, T. J., A. Schibeci, and A. Martonosi. 1980. The binding of arsenazo III to cell components. *Biochim. Biophys. Acta* **629** :317–327.
218. Rios, E., and M. F. Schneider. 1979. Stoichiometry of the reactions of calcium with the metallochromic indicator dyes antipyrylazo III and arsenazo III. *Biophys. J.* **36**:607–621.
219. Thomas, M. V. 1979. Arsenazo III forms 2:1 complexes with Ca^{2+} and 1:1 complexes with Mg under physiological conditions. *Biophys. J.* **25**:541–548.
220. Blinks, J. R., F. G. Prendergast, and D. G. Allen. 1976. Photoproteins as biological calcium indicators. *Pharmacol. Rev.* **28** :1–93.
221. Stephenson, D. G., and P. J. Sutherland. 1981. Studies on the luminescent response of the Ca-activated photoprotein obelin. *Biochim. Biophys. Acta* **678**:65–75.
222. Stephenson, D. G., I. R. Wendt, and Q. G. Forrest. 1981. Non-uniform ion distributions and electrical potentials in sarcoplasmic regions of skeletal muscle fibres. *Nature (London)* **289**:690–692.
223. Elliott, G. F., and E. M. Bartels. 1982. Donnan potential measurements in extended hexagonal polyelectrolyte gels such as muscle. *Biophys J.* **38**:195–199.
224. Naylor, G. R. S. 1982. Average electrostatic potential between the filaments in striated muscle and its relation to a simple Donnan potential. *Biophys. J.* **38**:201–204.
225. Close, R. I. 1981. Activation delays in frog twitch muscle fibres. *J. Physiol. (London)* **313**:81–100.
226. Stephenson, E. W. 1981. Ca dependence of stimulated ^{45}Ca efflux in skinned muscle fibres. *J. Gen. Physiol.* **77**:419–443.
227. Goldman, Y. E., M. G. Hibberd, J. A. McCray, and D. R. Trentham. 1982. Relaxation of muscle fibres by photolysis of caged ATP. *Nature (London)* **300**:701–705.
228. Johnson, J. D., S. C. Charlton, and J. D. Potter. 1979. A fluorescence stopped flow analysis of Ca^{2+} exchange with troponin C. *J. Biol. Chem.* **254**:3497–3502.
228a. Rosenfeld, S. S., and E. W. Taylor. 1985. Kinetic studies of calcium and magnesium binding to troponin C. *J. Biol. Chem.* **260**:242–251.
228b. Rosenfeld, S. S., and E. W. Taylor. 1985. Kinetic studies of calcium binding to regulatory complexes from skeletal muscle. *J. Biol. Chem.* **260**:251–261.
229. Hellam, D. C., and R. J. Podolsky. 1969. Force measurements in skinned muscle fibres. *J. Physiol. (London)* **200**:807–819.
230. Harafuji, H., and Y. Ogawa. 1980. Re-examination of the apparent binding constant of ethylene glycol bis (β-amino-ethyl ether)-N,N,N′,N′-tetracetic acid with calcium around neutral pH. *J. Biochem. (Tokyo)* **87**:B05–1312.
231. Stephenson, D. G. 1985. The role of calcium in contractile activation of skeletal muscle. *Prog. Biophys. Mol. Biol.* in press.
231a. Lio, T., and H. Kondo. 1981. Fluorescence titration and fluorescence stopped—flow studies of troponin C labeled with fluorescent maleimide reagent or dansylaziridine. *J. Biochem. (Tokyo)* **90**:163–173.
232. Baylor, S. M., W. K. Chandler, and M. W. Marshall. 1984. Sarcoplasmic reticulum calcium release in frog skeletal muscle fibres estimated from arsenazo III calcium transients. *J. Physiol. (London)* **344**:625–666.
233. Melzer, W., E. Rios, and M. F. Schneider. 1984. Time course of calcium release and removal in skeletal muscle fibers. *Biophys. J.* **45**:637–641.

CHAPTER 6

Excitable Tissues
The Heart

Richard W. Tsien and Peter Hess

1. Introduction

Cardiac action potentials serve many purposes. They form the cellular basis for pacemaker activity, impulse spread, and control of cardiac contraction. Despite this variety of functions, there are ample reasons for believing that impulses in cardiac cells follow the same general principles as in other excitable tissues. The preceding chapters on nerve and muscle have provided a useful foundation for understanding action potentials in the heart. We shall draw upon such similarities, but shall also focus on the unique aspects of cardiac activity.

This chapter begins with a brief description of the multicellular structure of the heart and the functional consequences of cell-to-cell coupling. We follow the spread of the impulse from one region of the heart to another, noting the diversity of action potential characteristics. The diversity arises from different expressions of ionic channels: Na^{+} channels, Ca^{2+} channels, K^{+} channels, and inward current channels underlying pacemaker activity. The properties of function of these channels are described in individual sections. We close with a brief summary of the modulatory effects of adrenergic and cholinergic transmitters.

2. Multicellular Structure of the Heart

In the normally functioning heart, individual cardiac cells do not exist in isolation, but are in continual communication with their neighbors. Thus, the functional behavior of the heart is quite unlike that in skeletal muscle, where adjoining cells may undergo very different patterns of activity, depending on the commands of the CNS. In the heart, impulses spread from one cell to the next without any help from nerves. Stimulation of one small region will lead to electrical and mechanical activity in large regions, and in some cases, the whole organ. This behavior was long interpreted as evidence that the heart was a continuous cytoplasmic structure without discrete cells. But the advent of electron microscopy in the early 1950s firmly established that each heart cell is completely bounded by plasma membrane. This raised new questions about how the individual cells might communicate.

Richard W. Tsien and Peter Hess • Department of Physiology, Yale University School of Medicine, New Haven, Connecticut 06510.

2.1. Structural Basis of Cell-to-Cell Communication

Electron microscopic studies have revealed a specialized structure for current flow from one heart cell to the next. This structure, the nexus, is a localized zone of contact between the plasma membrane of adjoining cells. It is one particular type of gap junction (see McNutt and Weinstein[1] for a review of terminology). Nexuses are only one part of the intercalated disk, which also includes other regions of specialized contact (fasciae adherentes and maculae adherentes or desmosomes) which provide mechanical connections. The nexuses are particularly interesting because of their presumed electrical role (see Fig. 1).

The relationship between nexal structure and function was first demonstrated by Barr *et al.*[2] These workers used stretch and hypertonicity to interrupt cell-to-cell current flow in strips of heart muscle. Electron microscopy showed separation of nexal membranes under these conditions, thus providing strong circumstantial evidence that the nexuses were indeed responsible for normal intercellular current flow. Other experiments by Dreifuss *et al.*[3] and Kawamura and Konishi[4] have provided arguments against the involvement of fasciae adherentes or desmosomes in electrical coupling: exposure to Ca-free media caused separation of these structures (without separation of nexuses), while electrical coupling remained intact.

Much attention has been directed toward the ultrastructure of the nexus (for reviews see Refs. 1 and 5). Electron microscopy of lanthanum-stained or freeze-fractured preparations has shown that nexuses are composed of a regular hexagonal array of "connexons" (gap junctional channels) with a 90- to 100-Å

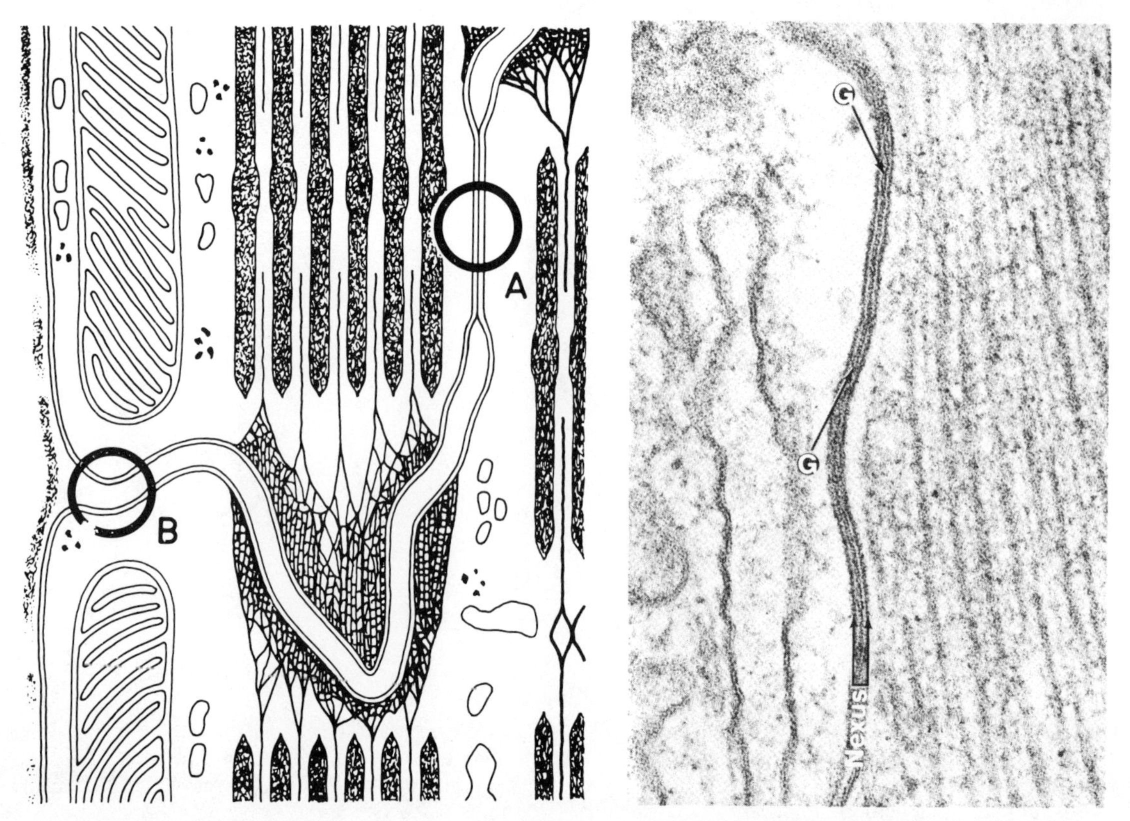

Fig. 1. (Left) A region of close contact between two heart cells (from Weidmann[181]). At B, the two surface membranes are separated by extracellular space. At A, the membrane fuse, forming a nexus. (Right) An electron micrograph of the nexus in the cat heart (from McNutt and Fawcett[182]).

center-to-center spacing. Recently, Unwin and Zampighi[6] were able to reconstruct a three-dimensional model of an individual connexon protein from electron micrographs analyzed with a powerful digital imaging technique known as Fourier averaging. Their studies were performed on liver cells but it seems likely that their general conclusions might also apply for the connexons found in cardiac gap junctions. Each connexon in one cell membrane is attached to another identical unit in the other membrane, thus forming a protein bridge across the gap junction. The connexon protein consists of six apparently identical subunits which are arranged such that they form a central channel of about 10–20 Å in diameter. This channel is the actual pathway for direct cell-to-cell communication. The size of the central channel is variable; under conditions which are known to decouple neighboring cells (see last paragraph of this section), a twisting movement of the connexon subunits can cause constriction of the central pore.[7]

The structural evidence is complemented by studies of cell-to-cell movements of various tracer substances. These began with Weidmann's measurements of $^{42}K^+$ movements within ventricular muscle bundles.[8] Tracer K^+ redistributes longitudinally over many cell lengths, with an effective diffusivity which is consistent with K^+ ions as the major carrier of intracellular current. Additional studies have characterized the diffusivity of larger particles, including tetraethylammonium,[9] procion yellow,[10] and fluorescein.[11] So far, the results show a simple inverse relationship between diffusivity and molecular weight (see Fig. 2). Procion yellow is the largest molecule studied so far, and has dimensions of 5 × 10 × 27Å (see Weingart[9]). This established 10 Å as a plausible lower limit on the diameter of a hypothetical nexal channel, in fairly good agreement with estimates from the structural data described earlier.

How much do nexuses contribute to the overall longitudinal resistance of heart muscle? Electrical and tracer K^+ experiments by Weidmann[8,12] both measured two barriers in series, namely the myoplasm and the nexus. Some further assumption is necessary to distinguish between these sources of resistance. One possibility is to assume that cardiac myoplasm resembles skeletal myoplasm, where diffusivity of tracer substances is about half that in water.[13] In the case of $^{42}K^+$, myoplasm resistance would account for 69% of the total, leaving 31% to the nexus. Another approach relies on the morphological data. Matter[14] has employed morphometric techniques to quantitate the distribution of nexuses in rat ventricular muscle. He calculated that the nexus surface occupies at least 47 μm^2, or about one-fourth the cross-sectional area of a typical 15-μm fiber. Assuming that each nexal subunit is like a doughnut, 90 Å across and with a central hole 10 Å in diameter, all the holes would combine to form an area that is 1/300th of the total cell cross-section. If the holes were filled with aqueous solution, Matter calculated that the nexus would contribute about 4 Ω-cm to the longitudinal resistivity. This value is only about 1% of the total, 470 Ω-cm.[14] The discrepancy between these different calculations could be accounted for if ionic diffusion through the nexal channel were somehow restricted. In any case, it seems clear that

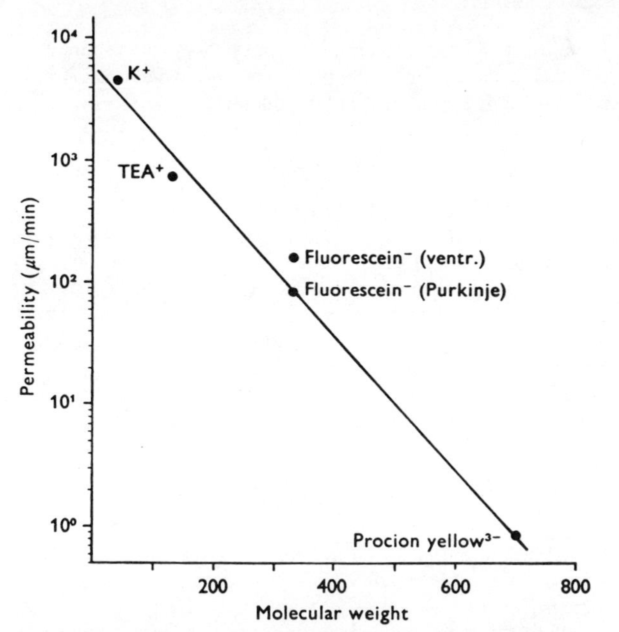

Fig. 2. Permeability of the nexal membrane to substances of varying molecular weight. Values for K^+, TEA^+, and procion yellow^{3-} were computed by Weingart[9] from data published by Weidmann,[8] Weingart,[9] and Imanaga,[10] respectively. Data for K^+ and TEA^+ obtained from ventricular muscle; procion yellow from Purkinje fibers. Fluorescein data and overall plot from Pollack.[11].

cell-to-cell resistance represents only a small fraction of the total longitudinal resistance. Apparently, nexuses can be very successful in their specialized function. They allow a multicellular structure like the myocardium to approach the behavior of the continuous cytoplasm that earlier morphologists envisioned.

A crucial property of gap junctional channels is their ability to change conductance in response to certain chemical stimuli. This is important for two reasons: (1) In the case of local cell damage it provides a mechanism for electrical and chemical decoupling of intact cells from their injured neighbors ("healing over") and thus prevents depolarization of the surviving tissue and loss of intracellular substances. (2) Increased gap junctional resistance tends to slow the rate of action potential propagation which is one of the conditions that can lead to certain types of arrhythmias (conduction block, reentrant arrhythmias). The most important regulator of gap junctional resistance is Ca^{2+}[15]; complete decoupling occurs at $[Ca^{2+}]$ greater than ~ 5–10 μM.[16] Intracellular pH also influences cell-to-cell coupling: Acidosis increases and alkalosis decreases gap junctional resistance.[17,18] Recently, the effects of Ca^{2+} and pH described above which were obtained with the use of cable theory in multicellular cardiac preparations were very elegantly confirmed in pairs of isolated cardiac cells which allow a more direct measurement of gap junctional resistance under controlled intra- and extracellular ionic conditions.[19,20]

2.2. Electrical Coupling and Impulse Spread

In the heart, as in other excitable tissues, the ease of current spread may be defined in terms of the space constant λ. In simple terms, the space constant provides a yardstick of electrical distance: It describes how far current flows along a cablelike structure before leaking passively across the surface membrane. It is usually measured by passing current at one point, and measuring the resulting voltage deflections at various distances.[21] Over the length of one space constant, the size of the electrical deflection decreases from 100% to 37%. Values of λ in the heart range from 2 mm in Purkinje fibers[21] to 0.5 mm in the sinus node.[22] In all cases in which λ is known, it is on the order of 10 times the length of an individual cell.

Most estimates of λ are obtained by constraining current to flow parallel to the long axis of the individual cells. Relatively little is known about current spread transverse to the long axis, even though this is obviously important for excitation in the intact heart. Woodbury and Crill[23] found that in rat atrial trabeculae, the decay of electrotonic potentials was twice as steep in the transverse direction as it was longitudinally. They suggested that the disparity might be related to the relative slowness of conduction velocity in the transverse direction.[24,25]

This idea has received elegant confirmation in the experiments of Clerc,[26] who found that the longitudinal conduction velocity was three times greater than the transverse conduction velocity. The ratio was fully accounted for by differences in the resistances in the longitudinal and transverse directions, as determined by independent measurements. The main factor is the disparity in intracellular resistance (R_i), and this may be explained by the structural anisotropy of heart muscle.[26,27] Current flow in the transverse direction encounters a much more tortuous pathway and more nexuses per unit distance.

Current spread from cell to cell is important for the local circuits which are set up during the passage of an impulse. As Fig. 3 indicates, the local circuit may be broken down into four individual branches:

1. Inward excitatory current in one region (largely carried by Na^+ ions in most regions of the heart)
2. Intracellular current flow along the length of the tissue (mostly K^+)
3. Escape of current across the membrane. The buildup of

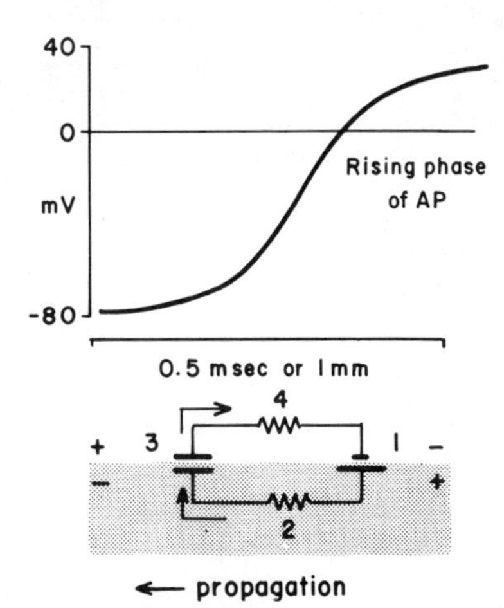

Fig. 3. Local circuit current and action potential propagation. (Top) The rising phase of a Purkinje fiber action potential.[29] The horizontal axis represents distance, using an assumed conduction velocity of 2 m/sec. (Bottom) Simplified diagram of the corresponding local circuit current. The circuit elements represent (1) excitatory Na^+ channels, (2) longitudinal intracellular resistance (including nexuses as well as myoplasm), (3) membrane capacitance of resting membrane, and (4) extracellular resistance. Note that the local circuit extends over a distance of several cell lengths.

positive charge on one side of the membrane capacitance and its removal from the other side corresponds to the flow of displacement current. The change in charge on the membrane capacitance produces a depolarization (which in turn produces local permeability changes)
4. Extracellular current flow in the longitudinal direction (carried by the ions of the surrounding medium, mostly Na^+ and Cl^-).

The scheme of local circuit currents is very similar to that proposed by Hodgkin[28] and others. The only important difference in the heart arises from its multicellular structure and, as we have discussed, this should not seriously change the mechanism of propagation. It is not surprising that the experimental evidence for this picture in cardiac tissue parallels that in nerve. Thus, it has been possible to interfere with at least two of the branches, items 1 and 4 in the foregoing list. In most regions of the heart, removal of external Na^+ slows impulse propagation and eventually abolishes the impulse.[29] Conduction may also be interrupted by raising the extracellular resistance and thus reducing local circuit current. Barr *et al.*[2] increased r_o by replacing the external solution with isotonic sucrose over a limited length of cardiac tissue. The sucrose solution contained very few ions and therefore offered a high electrical resistance. This sucrose gap blocked impulse spread beyond the gap; shunting of the gap by an appropriate electrical resistor promptly restored propagation by providing a path for local circuit current.

The importance of the intracellular longitudinal resistance was demonstrated in squid nerve by Del Castillo and Moore,[30] who speeded impulse velocity by inserting an axial wire. Shunting the myoplasmic resistance is a formidable task in cardiac tissues because of their multicellular structure. However, the influence of longitudinal resistivity is reflected in Weingart's[31] experiments. Toxic concentrations of ouabain increase the longitudinal resistance in ventricular myocardium and this increase is accompanied by slowing of conduction.

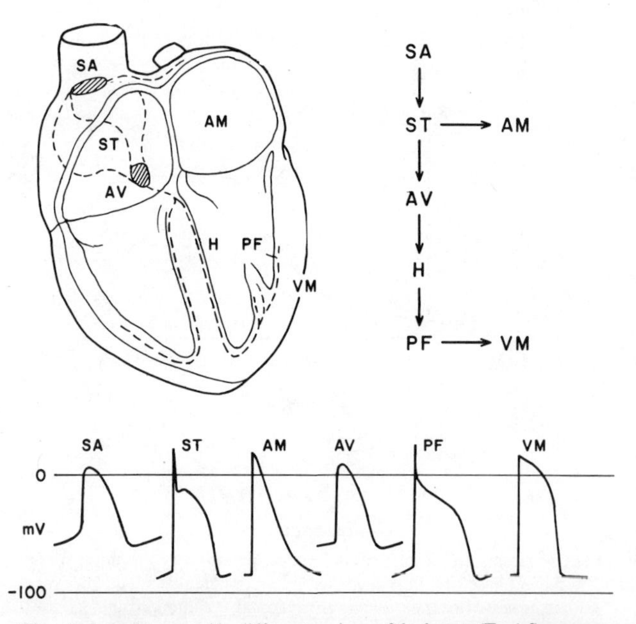

Fig. 4. Impulse spread in different regions of the heart. (Top) Sequence of impulse spread. Propagation from sinoatrial node (SA) to specialized atrial tracts (ST) and then to atrial muscle (AM) as well as atrioventricular junction to bundle of His (H), then Purkinje fibers (PF), and finally ventricular muscle (VM). Drawing of the heart after Netter.[183] (Bottom) Action potentials from various regions differ in their duration, shape, and voltage range.

2.3. Possible Chemical Coupling

Electrical coupling is the most obvious role of nexuses in the heart, but it may not be the only function. Since nexal channels are large enough to allow passage of small molecules such as amino acids or nucleotides, it seems reasonable to ask if biochemical communication also takes place. One might imagine, for example, that cell-to-cell movements of ATP or other high-energy phosphates might help equalize the metabolic state of cardiac cells. Such "metabolic cooperation" might be particularly important in cases of localized ischemia. Another interesting possibility is that cAMP diffuses from one cell to the next, promoting the even distribution of hormonal messages.

Support for such cell-to-cell movements in the heart has been provided by experiments using a cut-end method.[32] cAMP was introduced by local superfusion of the cut end of a ventricular muscle bundle. This procedure evoked a delayed increase in force generation at points many cell lengths from the region that had previously been exposed to cAMP. Application of 5′-AMP, the breakdown product of cAMP, gave no increase in force. These results strongly suggest that cAMP can move from one heart cell to the next.

3. Electrical Activity in Different Regions of the Heart

3.1. Normal Sequence of Impulse Spread

It is generally accepted that during normal beats, the impulse sweeps through the heart in an orderly and stereotyped manner. Figure 4 illustrates the sequence of events, as reflected by intracellular microelectrode recordings. This diagram differs slightly from Hoffman and Cranefield's[33] original diagram in emphasizing the role of specialized conducting tissue in the auricles. This tissue resembles ventricular Purkinje fibers in morphology, and provides avenues for rapid impulse conduction between the sinus and atrioventricular (AV) junctions (see Hogan and Davis[34]). There are three specialized tracts (anterior, middle, and posterior internodal pathways) which act as alternate routes. Under normal circumstances, conduction through the right auricular muscle itself is slower than that in the specialized pathways and does not directly lead to AV junctional excitation.

At the AV junction, the impulse comes to a spatial and temporal bottleneck. Here an important delay between atrial and ventricular excitation is introduced by slow impulse conduction. We discuss the ionic mechanism of the slow conduction later in this chapter. Once past the AV junction, the impulse is carried rapidly by the bundle of His and peripheral Purkinje fibers. The Purkinje fibers ramify extensively and undergo a gradual morphological transition into working myocardium. The overall result is more or less synchronous excitation of the ventricles.

Figure 4 conveys some idea of broad regional differences in the heart which can be seen in electrical recordings as well as in anatomical studies. Further subdivisions would be possible if the various regions were viewed on a finer scale. For example, within the AV junction itself there are several distinct types of action potential, and at least two functionally separate paths for impulse spread.[35]

3.2. Various Cardiac Action Potentials and Their Specialized Functions

Regional differences are a problem for the student or clinician who is trying to understand the cellular basis of cardiac electrical activity. It is not surprising that many textbooks explain the workings of the cardiac action potential (as though there were only one!). In the long run, it may be more sensible to face squarely the fact that the various parts of the heart have become specialized during the course of evolution.[36] Such specialization is reflected by the diversity of action potential shapes (Fig. 4), and is related to the wide variety of functions served by cardiac electrical activity. The functions include:

1. Rhythmicity: natural pacemaker activity in the sinus node, latent pacemaking in other specialized conducting tissue
2. Conduction: ranging from rapid conduction in the bundle of His and Purkinje fibers to slow conduction in the sinus and AV junctions
3. Initiating and modifying contractile activity: in the atria and ventricles

The variety of function provides some perspective for consideration of the ionic basis of action potentials in different parts of the heart.

3.3. Ionic Basis and Function of Electrical Activity in Different Regions

The diverse action potentials in various regions of the heart are generated by the activity of a number of different kinds of ionic channel. Although ionic conductance for Cl^- ions exists,[37] most of the Cl^- flux seems to be electrically silent.[38] The most important and best understood channels are cation-selective.

1. Na^+ channels underlie rapid impulse conduction and the rapidly rising upstroke of action potentials in atrial and ventricular muscle and in Purkinje tissue.

2. Ca^{2+} channels support a second type of impulse conduction, called a slow response or slow action potential, which is characteristic of the nodal regions (sinus and AV junctions). In the other regions of the heart, where rapid conduction is the normal mode of electrical activity, Ca^{2+} channels allow an influx of Ca^{2+} ions that helps support the action potential plateau. The Ca^{2+} influx also contributes to excitation–contraction coupling.

3. K^+ channels are the dominant ionic pathway in resting cardiac tissue and are largely responsible for the normal resting potential. Slowly activated outward current through K^+ channels plays an important role in terminating the action potential plateau. Rapidly activating transient outward K^+ currents are found in a number of cardiac cells; in Purkinje fibers, transient outward K^+ current underlies an early phase of rapid repolarization preceding the action potential plateau.

4. Ionic channels which select relatively poorly among cations provide inward current. Under certain conditions in a variety of cells, the inward currents can generate pacemaker depolarizations that lead to spontaneous activity.

In the following sections, we describe some of the properties of the various channels in relation to their functional roles. Owing to restrictions of space, we will not describe other important pathways for electrical current across cardiac cell membranes: the membrane capacitance, the Na^+ pump, or the Na^+/Ca^{2+} exchange.

4. Na^+ Channels and Excitability

4.1. Inferences about Na^+ Channels from Action Potential Recordings

Much of modern cardiac electrophysiology is founded on the premise that Na^+ sodium channels in the heart are very much like Na^+ channels in other excitable cells. The evidence for this view has accumulated over 30 years, beginning with Weidmann's pioneering experiments in the 1950s, and continuing up to much more recent studies of cardiac Na^+ currents under voltage clamp. Although cardiac channels have some unique pharmacological properties, the overall conclusion is that cardiac Na^+ channels are quite similar to their counterparts in nerve and skeletal muscle.

Early experiments in Purkinje fibers by Weidmann and others provide appropriate illustrations of the contribution of Na^+ current. Figure 5 shows some evidence from action potential recordings, where the effect of lowering the external $[Na^+]$[29] or exposure to tetrodotoxin (TTX)[39] was studied. Na^+ removal and TTX both decrease parameters such as overshoot, rate of rise of the upstroke, and conduction velocity. All of these parameters change together because they reflect the same underlying event: a large transient increase in Na^+ conductance.

Figure 5 also shows that the upstroke is not the only phase of the action potential which shows sensitivity to Na^+ removal or TTX. The plateau is also reduced in height and duration. Particularly prominent in Purkinje fibers, this effect indicates that a small amount of inward current may be carried by Na^+

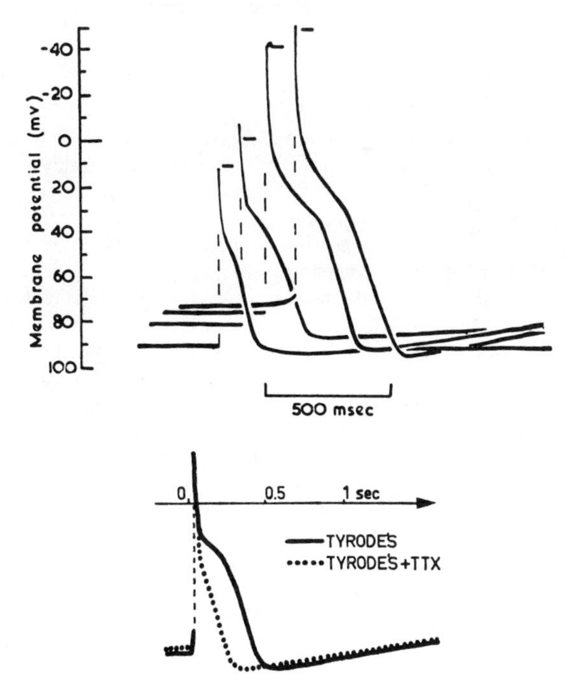

Fig. 5. Role of Na^+ current in the action potential. (Top) Effect of $[Na^+]$ on the size and shape of action potentials recorded in a kid Purkinje fiber.[184] The extracellular Na^+ content was (from left to right) 13, 22, 100, and 150% of normal. The horizontal lines indicate expected changes in the height of action potentials if membrane is assumed to be exclusively permeable to Na^+ ions. (Bottom) Influence of tetrodotoxin (10^{-5} g/ml) in a sheep Purkinje fiber. From Dudel *et al.*[185]

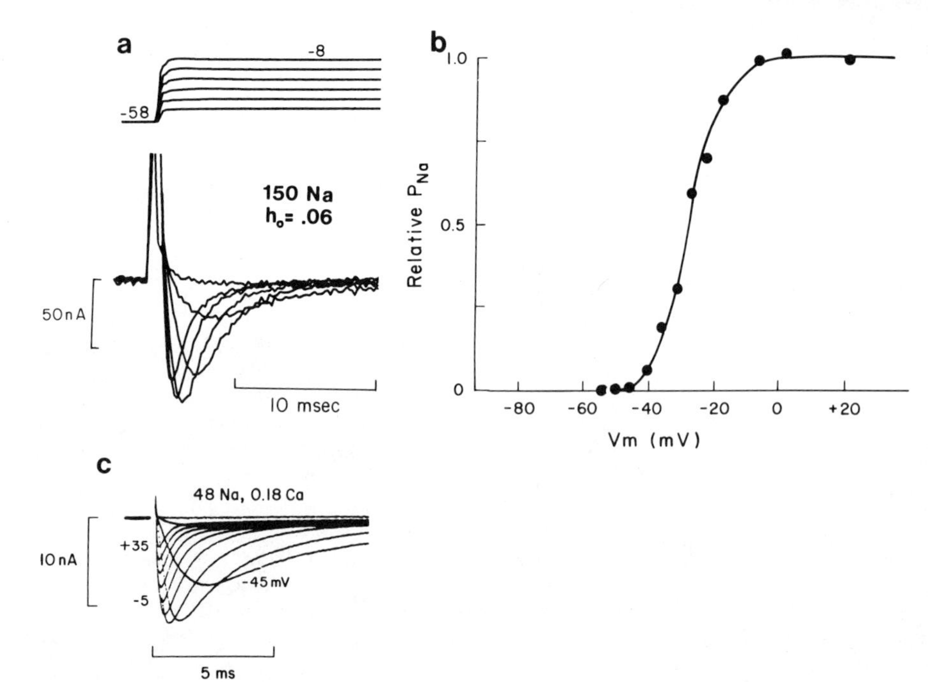

Fig. 6. Na^+ channel activation. (A) Na^+ channel currents recorded from a rabbit Purkinje fiber under two-microelectrode voltage clamp. 150 mM Na^+, 19°C. To improve voltage control during the flow of Na^+ current, the holding potential was set at −58 to inactivate all but 6% of the Na^+ channels. The family of currents accompany the step depolarizations shown above. The inward current reaches its peak value earlier as the depolarization becomes stronger. From Colatsky.[43] (B) Na^+ channel currents from a single ventricular cell from guinea pig, recorded with a suction pipette voltage clamp at 22°C. Membrane potential was held at −75 mV and stepped to potentials between −45 and +35 mV. From Lee and Tsien.[69] (C) Analysis of relative peak Na^+ permeability from the experiment in (A), using the Goldman–Hodgkin–Katz equation.

channels during the plateau, long after the large surge of inward current that generates the upstroke.[39] As will be described below, Ca^{2+} channels are an even more important factor in supporting the plateau in most regions of the heart.

4.2. Properties of Cardiac Na^+ Channels Studied with Voltage Clamp

Up until the late 1970s, direct measurements of I_{Na} in heart muscle were considered impossible because of the complex multicellular organization of cardiac muscle, which prevented any successful application of the voltage clamp technique.[40,41] Recent advances have been possible because of the development of suitable multicellular preparations with favorable electrical structure [42–44] and the development of single-cell preparations.[45–48]

Figure 6 gives some examples of voltage clamp recordings of cardiac Na^+ currents. Panel A shows the response of a rabbit Purkinje fiber under a two-microelectrode voltage clamp to increasingly strong step depolarizations. Activation of Na^+ channels becomes noticeable at about −50 mV. Peak I_{Na} reaches a maximum with stronger depolarizations before becoming progressively smaller. With even stronger depolarizations (not shown), the peak inward current continues to decrease as the membrane potential level approaches the reversal potential for current flow through the Na^+ channel. Na^+ currents from single cardiac cells show similar voltage dependence (Fig. 6B).

The voltage dependence of the peak Na^+ permeability can be determined from the magnitude of the peak current with the help of the Goldman–Hodgkin–Katz current equation (see Chapter 15). As Fig. 6C indicates, the permeability increases in a steeply voltage-dependent fashion over the potential range between −55 and 0 mV. As in other excitable tissues, this voltage dependence can lead to a strongly regenerative action potential because depolarization begets activation, activation increases inward current, and inward current causes more depolarization.

The voltage dependence of inactivation is analyzed with the protocol shown in Fig. 7A. The size of Na^+ current surge depends on the steady membrane potential before the sudden depolarizing step. This voltage dependence is also steep (Fig. 7B); contrary to activation, the more depolarized the preceding level, the smaller the subsequent surge of I_{Na}.

The development of the patch clamp technique has made it possible to carry voltage clamp studies to the level of individual Na^+ channels. Figure 7C illustrates patch clamp recordings from a patch containing at least six Na^+ channels. Depolarizations from a negative holding potential elicit overlapping activity from several channels. The horizontal lines indicate multiples of the current carried by a single open Na^+ channel. As the holding potential (V_H) is made less negative, the number of simultaneously open channels decreases (7C). The voltage dependence is steep (7D), very much like that seen in global current recordings from multicellular preparations (7B).

4.3. Consequences of Na^+ Channel Properties

4.3.1. Rapid Conduction in Certain Regions of the Heart

As already mentioned, Na^+ channels underlie rapidly conducted impulses in several regions of the heart, including the His–Purkinje system, atrial and ventricular muscle. The most rapid propagation, in Purkinje fibers, can show conduction velocities of up to 3 m/sec and a rate of rise of up to 800 V/sec.

4.3.2. Refractory Period

Variations in the excitability of heart cells take place during each cardiac cycle and are largely determined by the voltage dependence of inactivation. A stimulus that is sufficient to excite a resting myocardial preparation will fail to excite the same tissue during the action potential plateau, when the membrane is depolarized and inactivation is virtually complete (Fig. 7D).

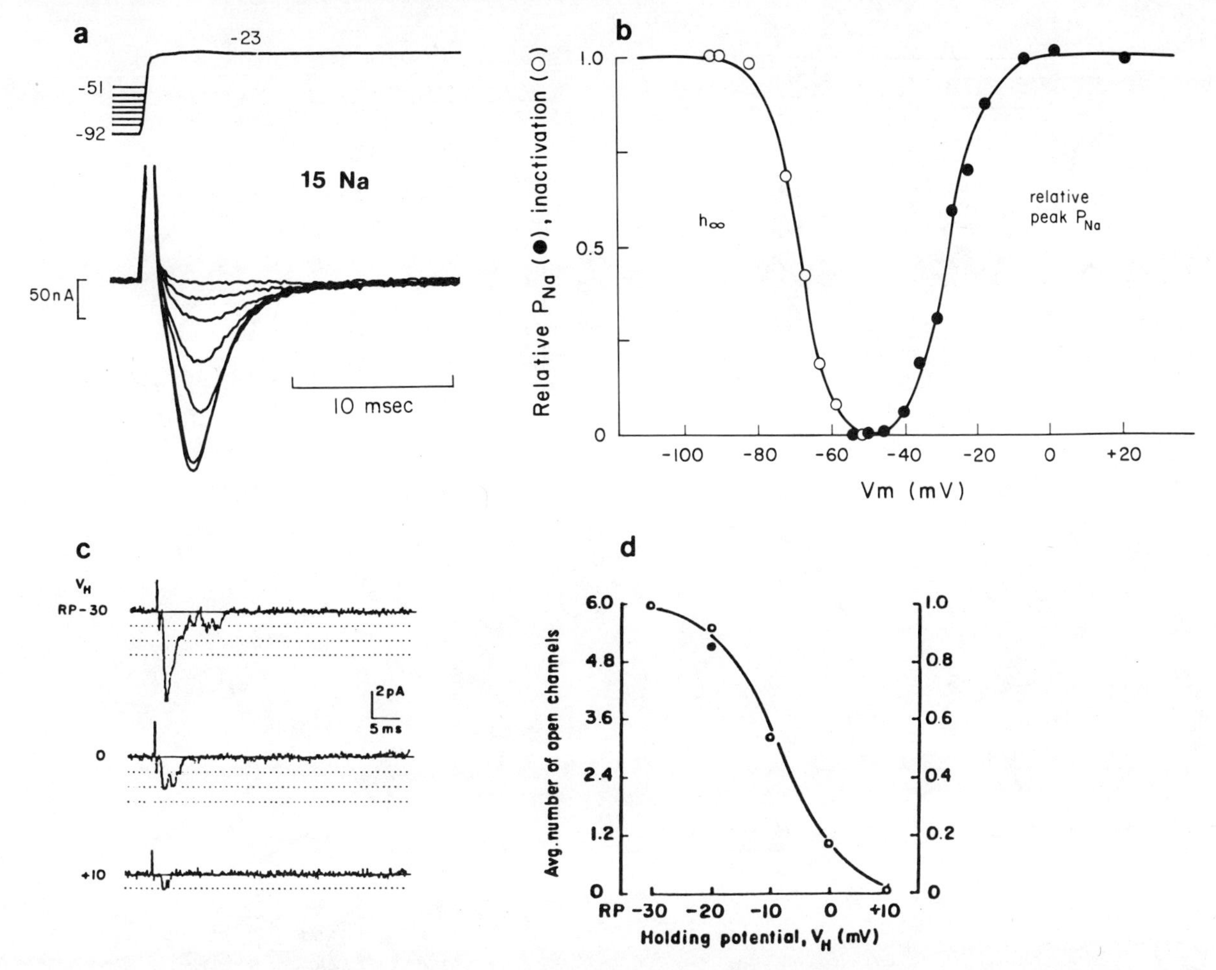

Fig. 7. Na^+ channel inactivation. (A) Global Na^+ channel currents recorded from the same rabbit Purkinje preparation as in Fig. 6A. 15 mM external Na^+. The membrane was held at various levels as indicated before a test depolarization to −23 mV. As the holding level becomes less negative, the peak inward current becomes smaller due to Na^+ inactivation. (B) Voltage dependence of inactivation. From Colatsky.[43] (C) Single-channel activity from a cell-attached patch on a neonatal rat ventricular cell. Holding potential (V_H) was varied while the test potential remained fixed. (D) Voltage dependence of inactivation, determined in the same experiment. Ordinate plots the average peak number of simultaneously open channels. From Cachelin *et al.*[49]

Thus, the length of the absolute refractory period is largely determined by the duration of the plateau. Graded removal of inactivation takes place during the action potential repolarization and accounts for the relative refractory period. Distal Purkinje fibers have a particularly long plateau, which makes their refractory period 50–100 msec longer than that of ventricular muscle further downstream. The long refractory period may allow the distal conducting tissue to protect the ventricles against premature excitation.[51] Excitation of ventricular muscle during its relative refractory period is believed hazardous because it favors impulse asynchrony.

4.3.3. Accommodation

Slow diastolic depolarization in Purkinje fibers extends over the same potential range where Na^+ inactivation is steeply voltage dependent. When the Purkinje system is excited by a higher pacemaker (e.g., in the case of normal sinus rhythm), the upstroke of the action potential takes off from different levels depending on the degree of diastolic depolarization. The greater the degree of inactivation, the slower the conduction velocity (see Singer *et al.*[52]). This helps explain why idioventricular impulses tend to propagate slowly. It also provides an indirect mechanism by which chronotropic agents such as digitalis or epinephrine can influence excitability or conduction.

4.3.4. K^+ Ions

The $[K^+]_o$ has no direct effect on Na^+ channels, but it has a powerful indirect influence that is mediated by the resting potential. Since the normal resting potential falls partway along the inactivation curve, small changes in $[K^+]_o$ produce substantial changes in the degree of availability of Na^+ channels. It is reasonable to expect that $[K^+]_o$ will also help determine the effectiveness of agents such as lidocaine and diphenylhydantoin which act in part by shifting the inactivation curve (see Singh and Vaughan-Williams[53]).

4.3.5. Na^+ Channels as Targets for Antiarrhythmic Drugs

Na^+ channels are blocked by lidocaine, quinidine, and a number of other antiarrhythmic drugs. The degree of block is enhanced by steady or repetitive membrane depolarizations

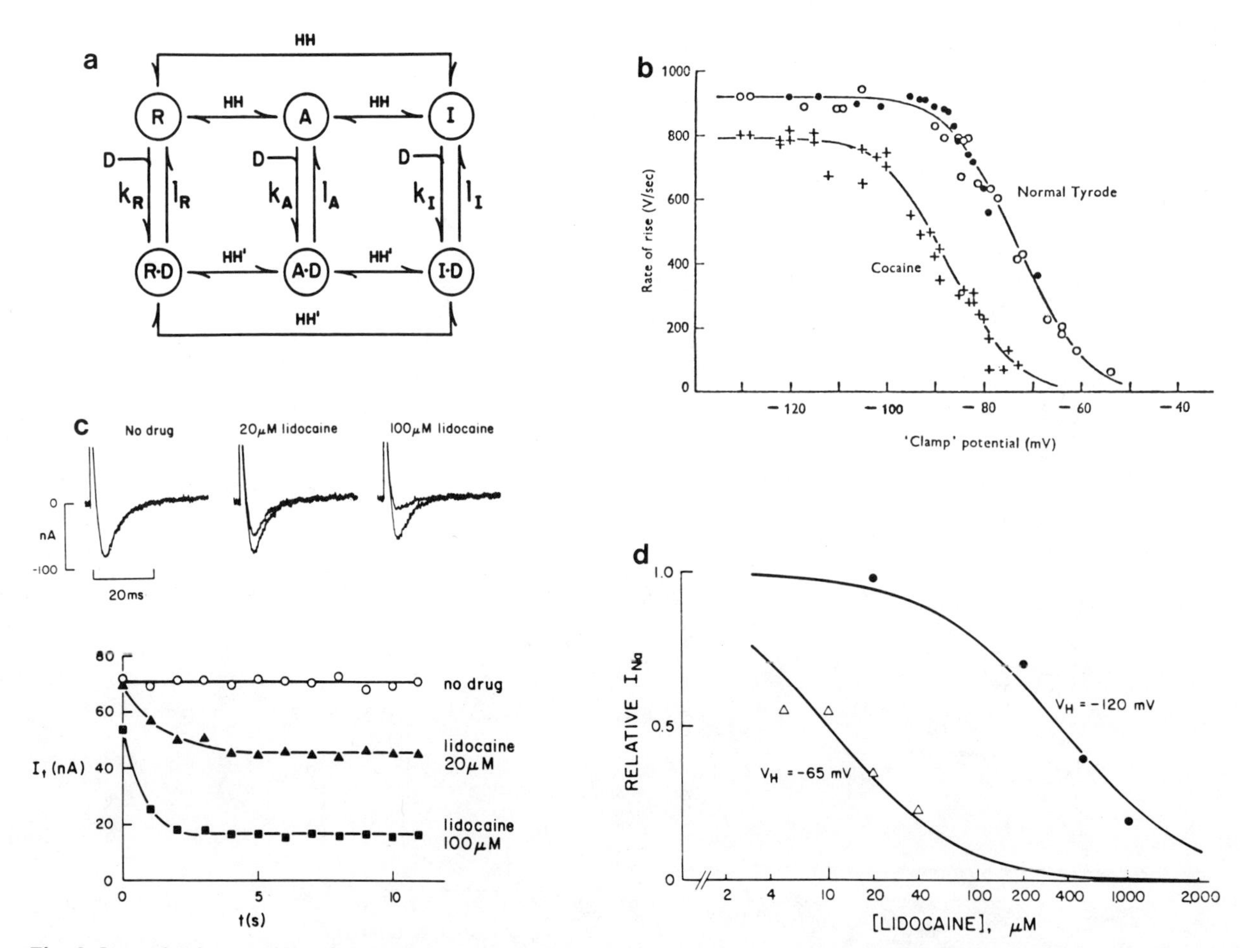

Fig. 8. Interaction between Na^+ channels and antiarrhythmic local anesthetic drugs. (A) Modulated receptor hypothesis of Hondeghem and Katzung[54] and Hille.[55] Na^+ channel gating is represented simply by transitions between a resting state (R), an open state (A), and an inactivated state (I). The rate constants for binding and unbinding of drug to the channel (k's, l's) depend on the gating state, and the presence of bound drug alters the gating transitions from their normal kinetics (HH for Hodgkin–Huxley) to modified kinetics (HH'). (B) Use-dependent block. I_{Na} was measured during trains of 500-msec pulses from −105 to −35 mV at 1 Hz following a period of rest. The traces show membrane currents associated with the 1st and 12th pulses superimposed, and the graph plots measured I_{Na} amplitudes for all the pulses. Lidocaine has relatively little effect on the first inward current signal following the rest period, but it substantially reduces peak I_{Na} following repetitive depolarization. From Bean *et al.*[56] (C) Cocaine influences the voltage-dependent equilibrium between resting and inactivated states. Na^+ channel function assessed by measurement of maximal rate of rise of stimulated action potential. The inactivation curve in the absence of drug (circles) reflects an equilibrium between resting and inactivated channels at any given potential, presumably by preferentially binding to and stabilizing channels in the inactivated condition. From Weidmann.[57] (D) Steady depolarization promotes lidocaine block of Na^+ currents. At the holding potential of −65 mV, almost all the channels are inactivated; the sensitivity to lidocaine expresses (as in C) the relative stability of the lidocaine-bound, inactivated state. Note that the ordinate plots normalized Na^+ current, so the reduction of I_{Na} by inactivation is not shown. From Bean *et al.*[56]

("voltage dependence" or "use dependence"). These phenomena probably contribute to the drugs' clinical effectiveness. Many types of arrhythmias involve regions of metabolically compromised tissue that are either partially depolarized or repetitively firing; because of their voltage-dependent or use-dependent actions, lidocaine and its congeners might act preferentially on those areas that contribute to the arrhythmia.[54]

The mechanism of voltage dependence or use dependence can be explained rather well by the "modulated receptor" hypothesis of Hille[55] and Hondeghem and Katzung.[54] The idea is that each Na^+ channel has only one receptor for drugs like lidocaine, but that the state of the receptor is altered as the channel undergoes structural changes during activation or inactivation. Figure 8 illustrates some experimental evidence that is consistent with the modulated receptor hypothesis.

5. Ca^{2+} Channels and Slow Responses

The importance of Ca^{2+} channels in the heart is now well established, although their existence was still disputed as little as 7 years ago, when the first edition of this book was published. Much of the progress can be credited to improvements in experimental methods for studying multicellular preparations, single cells, and single channels. Considerable information about the basic physiology and pharmacology of Ca^{2+} channels is now available (see Section 5.2).

5.1. Inferences about Ca^{2+} Channels from Action Potential Recordings

Some of the first insights into the importance of Ca^{2+} channels came from recordings of Ca^{2+}-dependent action po-

Table I. Properties of Fast and Slow Responses

	Fast response	Slow response
Conduction velocity (in Purkinje fibers)	2–3 m/sec	0.1 m/sec
Safety factor	High	Low
Ionic basis	Na^+	Mainly Ca^{2+}
Pharmacological blockers	Tetrodotoxin, local anesthetics	Verapamil, D600, Mn^{2+}
Sensitive to sympathetic hormone?	No (apart from changes in membrane potential)	Strong enhancement
Inactivated by mild depolarization?	Completely inactivated at −55 mV	Not inactivated at −55 mV
Sensitive to elevated serum K^+?	Yes, blocked	Not blocked even at 16 mM K_o^+

tentials, often known as "slow responses." These impulses take place without direct involvement of Na^+ channels. The word *slow* refers to several aspects of this activity: its sluggish conduction velocity and rate of depolarization, and its underlying mechanism, an inward Ca^{2+} current with turn-on and turn-off kinetics somewhat slower than Na^+ current. The term *slow response* has been used in a general sense when referring to such slowly rising action potentials, whether they be spontaneous or stimulated, propagating or nonpropagating. Slow responses are the normal form of activity in the sinus and AV junctions. They may also be evoked in working myocardium and Purkinje tissue by conditions which prevent the participation of the rapid Na^+ current.

A number of qualitative characteristics distinguish the slow response from rapid Na^+-dependent activity. Table I lists some of these features. Ionic concentration changes provide the most basic distinction. $[Na^+]$ controls the rate of rise of the conventional rapid upstroke, as mentioned earlier, but it has little influence on the rise of the slow response. Slow responses can be obtained in sodium-free solutions. Conversely, the rising phase of the slow response is directly dependent on $[Ca^{2+}]_o$, whereas changes in $[Ca^{2+}]_o$ have no great influence on the rapid upstroke, apart from their known effect on the voltage dependence of Na^+ channel gating.[57]

Rapid and slow responses can also be distinguished by their different sensitivities to pharmacological agents such as TTX and Mn^{2+} ions, an approach first applied in heart by Hagiwara and Nakajima.[58] Figure 9 illustrates the effect of TTX and a Ca^{2+} channel antagonist, D600 (a methoxy derivative of verapamil). The upper left-hand panel shows transmembrane potential in a Purkinje fiber bathed in 7.2 mM Ca^{2+} Tyrode solution. Electrical activity was evoked by a train of strong external shocks. The record begins with four rapidly rising action potentials, initiated from the normal resting potential (about −80 mV). During the later part of the record, a steady current was applied through a second intracellular microelectrode in order to partially depolarize the preparation. From the depolarized level (−50 mV), the stimuli produced action potentials with a characteristically slow rate of rise (evident when viewed on an expanded time scale). These slow responses differ from normal action potentials in their sensitivity to Ca^{2+} channel antagonists. For example, exposure to D600 (Fig. 7, lower left panel) promptly abolished the slow responses. The normal action potentials were not blocked, but showed a lower and briefer plateau. The right panels show a closely related experiment, in which TTX is used to block the rapid Na^+ channels. In this particular preparation, the steady depolarizing current evoked repetitive firing even without external stimuli. Exposure to TTX eliminated the normal action potentials, but scarcely affected the spontaneous activity in the partially depolarized state. Evidently, TTX and D600 have complementary actions on the rapid upstroke and slow response. Such pharmacological evidence reinforces the classification of electrical activity into two functional types.

5.2. Properties of Cardiac Ca^{2+} Channels Studied with Voltage Clamp

Differences in the nature of slow and fast responses are rooted in different characteristics of Ca^{2+} channels and Na^+ channels. Some basic properties of Ca^{2+} channels are illustrated in Fig. 10. Panel A shows a Ca^{2+} current recorded from a single ventricular cell. Following a sudden depolarization from a negative potential, Ca^{2+} channel current increases to a peak and then decreases. The rising phase reflects the opening or "activation" of Ca^{2+} channels; the falling phase reflects the closing of Ca^{2+} channels during maintained depolarization, a process termed *inactivation*. These processes give the Ca^{2+} channel current a phasic time course, qualitatively similar to Na^+ channel current. However, comparisons between activation and inactivation of Ca^{2+} and Na^+ channels reveal significant differences in speed, voltage dependence, and mechanism.

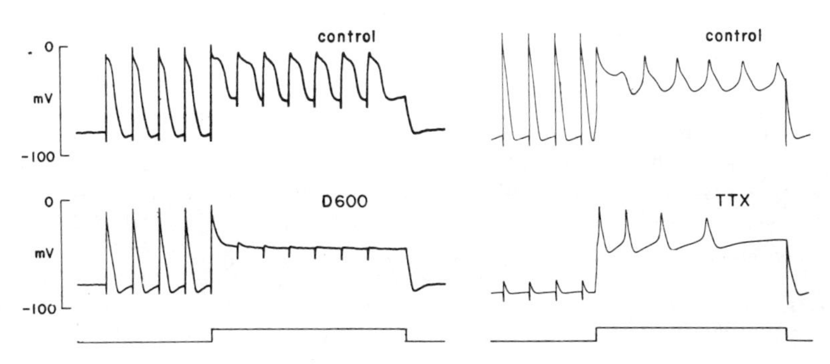

Fig. 9. Effects of D600 and tetrodotoxin on two types of electrical activity in Purkinje fibers. Each panel shows responses to external stimuli, initiated from the normal diastolic level (near −80) or from a partially depolarized voltage (near −50). The partially depolarized state was achieved by application of a rectangular current pulse. (Left) From Kass and Tsien[186]; (right) Siegelbaum and Tsien (unpublished).

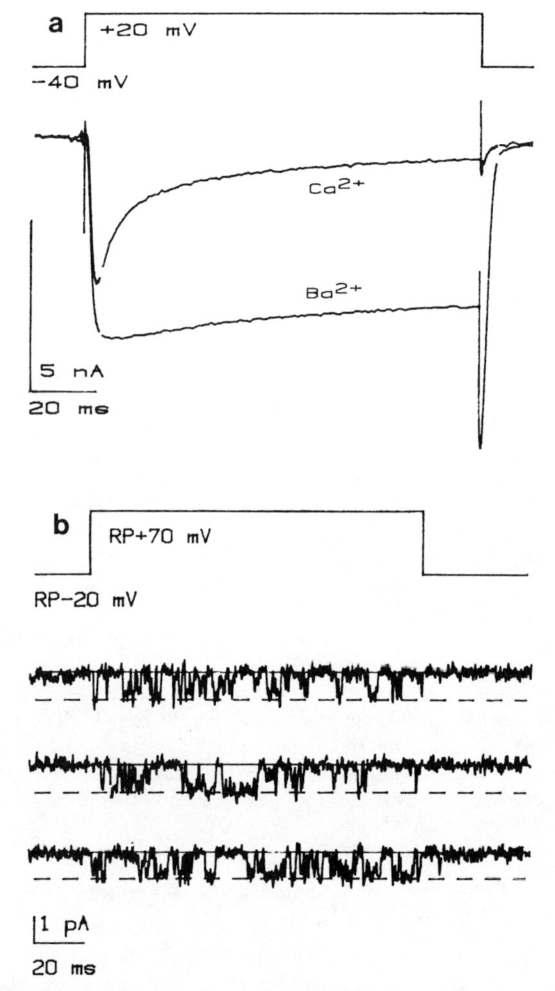

Fig. 10. Ca^{2+} channel activity at the level of single cells and single channels. (A) Whole-cell current recordings using a suction pipette for voltage clamp and internal dialysis following the approach of Lee *et al.*[45] Na^+ and K^+ channel currents were minimized by Na^+-free and K^+-free internal and external solutions containing Cs^+ as the major monovalent cation; the largely linear leak current that remained was subtracted. Under these conditions, the Ca^{2+} current declines rapidly from its peak value to a residual level, typically 10–20% of the peak amplitude.[69] The inactivation slows dramatically when Ba^{2+} replaces Ca^{2+} as the current carrier, consistent with the idea that internal Ca^{2+} promotes the onset of inactivation (see Refs. 62, 63, and 66 for review). Unpublished records of Hess and Tsien. (B) Single-channel recording from a cell-attached patch on a guinea pig ventricular cell with a pipette containing 110 mM Ba^{2+}. Resting potential (RP) was near −60 mV with 10 mM K^+_o. Single-channel activity appears as downward pulses of current throughout the patch depolarization. Unpublished records of P. Hess, J. B. Lansman, and R. W. Tsien.

5.2.1. Activation

Following a sudden depolarization to 0 mV, Ca^{2+} channel current rises to half its peak within about 2 msec at room temperature and within 0.5 msec at 37°C. For comparison, Na^+ current turns on roughly an order of magnitude faster under comparable conditions.*

*With this comparison with Na^+ channels in mind, investigators have often referred to Ca^{2+} channel current as "slow inward current," or "second inward current" or "I_{si}." In this review, we will simply speak of Ca^{2+} channel current or "I_{Ca}."

Activation of Ca^{2+} channels is not a first-order process, as once believed; like Na^+ channels, Ca^{2+} channels activate with a sigmoid time course, as if there were more than one step between the normal resting state and the open state of the channel.[59,60]

The degree of activation at the peak of the Ca^{2+} current can be expressed as a probability, p_{max}. This probability is strongly dependent on membrane potential. Under most ionic conditions, p_{max} first becomes appreciable near −40 mV, and increases steeply with membrane depolarization until it reaches a saturating value on the order of 0.5–0.8 near +20 mV or so. The midpoint of the voltage dependence lies near −10 or 0 mV, 20–30 mV more depolarized than the midpoint of the corresponding "activation curve" for Na^+ channels. This difference in voltage dependence accounts for the relatively strong depolarizations needed to trigger slow responses.

5.2.2. Inactivation

The speed of inactivation of Ca^{2+} channels is strongly dependent on membrane potential and the species of permeant ion. Figure 10A illustrates the decay in the presence of external Ca^{2+}. The half-time ($t_{1/2} \sim$ 10–20 msec under the conditions of this particular experiment) is typically an order of magnitude longer than that seen during inactivation of Na^+ channels (see Fig. 6). The decay of Ca^{2+} channel current becomes even slower when Ba^{2+} replaces Ca^{2+} as the external divalent cation (Fig. 10A). This dependence of inactivation rate on the external ion fits with the idea that the Ca^{2+} entry and accumulation of intracellular Ca^{2+} help promote inactivation (see Refs. 61–64). Intracellular Ca^{2+} seems to exert an action over and above a more conventional type of voltage-dependent inactivation, which can be seen as a slower decay with external Ba^{2+} or without any permeant divalent cation in the bathing solution.[65]

The Ca^{2+} dependence complicates the description of the steady-state voltage dependence of Ca^{2+} channel inactivation. It is clear, however, that relative to Na^+ channels, much stronger depolarizations are necessary to inactivate Ca^{2+} channels. This accounts for the finding that slow responses can be seen in the presence of up to 25 mM K^+_o.

5.2.3. Properties of Single Ca^{2+} Channels

Rises and falls in the probability of Ca^{2+} channel opening (p) control the time course of Ca^{2+} channel current (I) following a sudden depolarization. The magnitude of the current is dependent on two other factors: i, the current through an open Ca^{2+} channel, and N, the number of channels:

$$I = N \cdot p \cdot i$$

Distinctions between N, p, and i have become possible with direct recordings of the activity of single channels (Fig. 10B), and with analysis of the fluctuations in membrane current arising from the collective behavior of individual channels.

Pulses of current associated with the opening and closing of individual channels can be clearly resolved when the medium outside the cell is enriched with permeant divalent cations. Typically, near-isotonic (90–110 mM) $BaCl_2$ solutions are used. Under this condition, the amplitude of the pulses (i) near 0 mV can be of the order of 1–2 pA (with a slope conductance of 15–30 pS), corresponding to an ion flux of 3–6 million Ba^{2+} ions/sec. Since the open channel flux is much too large to be accounted for by a carrier mechanism, there is little doubt that Ca^{2+} channels are pores.

Under more physiological ionic conditions of millimolar Ca_o^{2+}, the current through an open Ca^{2+} channel is much smaller and cannot yet be resolved as a unitary current amplitude. Fortunately, the opening and closing of a group of Ca^{2+} channels produces "noise" which can be analyzed in terms of N, p, and i, with the help of a technique called ensemble fluctuation analysis.(66) At 3 mM Ca_o^{2+}, and a membrane potential of 0 mV, i is close to 0.05 pA (B. P. Bean, M. C. Nowycky, and R. W. Tsien, unpublished results).

The number of Ca^{2+} channels can be estimated from analysis of unitary currents or current fluctuations. Typical estimates are of the order of 1–10 functional channels/μm^2 surface area, a value which agrees reasonably well with the density of Ca^{2+} channels inferred from measurements of specific [^{3}H]nitrendipine binding to cells or cell fragments.(67)

5.2.4. Selectivity

Ca^{2+} channels are much more selective for Ca^{2+} than Na^+ channels are for Na^+.(65,68,69) This degree of discrimination seems appropriate in view of the fact that Ca^{2+} ions are vastly outnumbered by other ions in both blood and cytoplasm. The mechanism of ion permeation in the Ca^{2+} channel can be described by a model in which selectivity results from the presence of binding sites for which permeating ions must compete.(70,71) Because the affinity of those sites for Ca^{2+} is more than 10,000 times higher than for other physiological ions, the channel is almost continuously occupied by at least one Ca^{2+} ion even at low (millimolar) $[Ca^{2+}]_o$.(72) Therefore, under physiological ionic conditions, more than 90% of the inward current through the Ca^{2+} channel is carried by Ca^{2+} ions.

Competition of divalent ions for the binding sites in the Ca^{2+} channel also explains block of Ca^{2+} channel current by ions such as Ni^{2+}, Mn^{2+}, Cd^{2+}, and Co^{2+}. Block occurs because of extremely high affinity of the binding sites for these ions. Current carried by relatively weakly binding permeant ions is blocked more completely than that carried by more strongly binding ions (e.g., Cd^{2+} is a more potent blocker of Ba^{2+} current than of Ca^{2+} current because Ca^{2+} binds more strongly than Ba^{2+}).(73)

5.3. Ca^{2+} Channels as Targets for Organic Antagonists and Agonists

Three chemically distinct groups of drugs have been used *in vitro* and *in vivo* to reduce Ca^{2+} channel current: (1) verapamil and its methoxy derivative D600, (2) diltiazem, and (3) dihydropyridines like nifedipine and nitrendipine. D600 has been shown to reduce Ca^{2+} channel current by reducing open channel probability.(74)

Dihydropyridine compounds also change the open channel probability. Interestingly, however, while most dihydropyridines studied so far block Ca^{2+} channel current ("antagonists"), some (e.g., Bay K 8644) have the opposite effect, i.e., they can greatly enhance the current.(75) The underlying mechanism appears to be a stabilization of the channel in a gating mode characterized by very-long-lasting openings when the dihydropyridine receptor is occupied by an agonist. On the other hand, a bound antagonist seems to stabilize the channel in a mode in which the channel is not available for opening (P. Hess, J. B. Lansman, and R. W. Tsien, unpublished results). Depending on whether a dihydropyridine preferentially stabilizes one or the other gating mode, it can therefore have predominantly agonist, antagonist, or mixed (partial antagonist or agonist) effects.

No information about the mechanism of channel inhibition by diltiazem is yet available.

5.4. Possible Involvement of Slow Responses in Arrhythmias

Figure 9 illustrates the importance of partial depolarization to slow responses. Partial depolarization causes inactivation of Na^+ channels as well as bringing the potential closer to the point where slow inward current turns on. In Fig. 9, applied current was used to produce the depolarization, but it could also have been evoked by exposure to K^+-rich solution (usually ranging from 16 to 25 mM K_o^+). Both of these experimental procedures share relevance to occurrence of slow responses in the intact heart. Local circuit current flow from a damaged and depolarized region could cause partial depolarization of a neighboring area. Elevated $[K^+]_o$ can also occur in cases of damage or ischemia; high $[K^+]_o$ could act directly, via a shift in E_K, as well as by inducing local circuit current.(76) Other conditions such as overstretch or cardiac glycoside intoxication may also produce partial depolarization.

Conditions which favor slow responses in Purkinje fibers and working heart muscle preparations have attracted special attention because of the possibility that such activity may occur in cardiac arrhythmias. According to current thinking, reentry and abnormal automaticity are two important mechanisms for the genesis of arrhythmias. Although these mechanisms are conceptually very different, each of them can involve slow responses. Reentry is caused by a defect in impulse conduction, combined with a geometrical arrangement of tissue which allows the impulse to double back on itself. The result is an impulse traveling in a loop, or "circus movement." Slow responses may be important in fostering reentry because a localized region of slow conduction could provide delay in impulse spread. Such delay is necessary in allowing recovery of excitability in advance of the reentering impulse. The other major mechanism, abnormal automaticity, depends on spontaneous impulse formation rather than recycling of old impulses. Here again, slow responses come into the picture because they are one major type of sustained rhythmic activity.

5.5. Slow Responses in Nodal Tissue

Action potentials in the sinoatrial (SA) or AV nodes have many earmarks of slow response activity (see Cranefield(36) for a detailed list of references). In the nodal cells the membrane potential normally remains positive to −65 or −70 mV (Fig. 4). Conduction velocity is slow—on the order of 0.05 m/sec over a limited region of the AV node. Such slow conduction contributes to the lag between atrial and ventricular excitation and allows proper ventricular filling. Nodal action potentials are relatively insensitive to elevated $[K^+]_o$ (ranging up to 10 mM or more) and are not blocked by TTX. On the other hand, Ca^{2+} antagonists such as Mn^{2+}, La^{3+}, verapamil, or D600 markedly inhibit the ability of nodal cells to generate or conduct impulses.

In nonnodal regions, slow responses are produced when fast Na^+ channels are blocked or inactivated. This raises questions about the basis of naturally occurring slow responses, as found in nodal cells. Do nodal membrane simply lack Na^+ channels, or are their Na^+ channels inactivated by the relatively depolarized range of potentials over which their activity takes place? Recent experiments favor the second explanation for both types of nodal tissue. Rapid upstroke activity has been revealed in rabbit AV nodal cells by using a suction electrode to hyper-

Table II. Roles of Na^+ and Ca^{2+} Currents

		Na^+ current	Ca^{2+} current
Working myocardium	Ventricular atrial	Na^+ current supports conduction	Ca^{2+} current underlies plateau and activates contraction
Specialized conducting tissue	SA node, AV node	Na^+ current not operative (potentials are not normally negative enough to remove inactivation)	Ca^{2+} current generates upstroke and allows propagation
	Purkinje, His bundle	Na^+ current supports rapid conduction	Ca^{2+} current underlies plateau, and also may allow slow responses when the Na^+ current is reduced. This may lead to ectopic impulses, reentry

polarize the membrane. The rate of rise increased in a sigmoid fashion as the "resting potential" was hyperpolarized, in a manner consistent with removal of Na^+ channel inactivation. Related experiments in SA node suggest that these cells also have rapid Na^+ channels. Some of the first evidence came in experiments where rabbit SA node was hyperpolarized by exposure to carbamylcholine.[69] The hyperpolarization was accompanied by an increase in rate of rise; the elevation of $(dV/dt)_{max}$ was TTX sensitive, unlike the rising phase in the absence of drug-induced hyperpolarization. More recent voltage clamp experiments by Noma, Irisawa, and colleagues have directly demonstrated the existence of fast, TTX-sensitive Na^+ current. The conclusion is that Na^+ channels exist but remain functionally dormant because of the limited range of nodal membrane potential.

Slow responses provide a particularly dramatic manifestation of the participation of Ca^{2+} channels in the heart. But Ca^{2+} channels also participate in action potentials with rapid upstrokes dominated by the rapid Na^+ system. Table II summarizes the functions of both types of ionic current. Although contributing very little to the rapid depolarizing phase in Purkinje fibers or working heart muscle, the Ca^{2+} current plays a large part in generating the plateau. This phase is determined by a delicate balance between relatively small currents.[70]

5.6. Ca^{2+} Current and Contractile Activation

Ca^{2+} channels are crucial to activation of contraction. Ca^{2+} entry not only provides at least some activator for the contractile proteins, but also helps control the discharge and replenishment of internal Ca^{2+} stores. The relation between Ca^{2+} currents and contractile activation is complex enough to deserve a chapter by itself. For a thorough discussion of this important subject, the reader is referred to recent reviews by Chapman[77] and Fabiato.[78]

6. K^+ Channels Support the Resting Potential and Action Potential Repolarization

Several types of K^+ current coexist in cardiac membranes. This often comes as a surprise to the majority of people who have studied squid giant axon or frog node of Ranvier as an introduction to excitability: these excitable membranes make do with only one kind of K^+ channel (see Chapter 27). As it turns out, many excitable cells seem to rely on more than one type of K^+ channel for functions such as maintaining the resting potential, generating the action potential repolarization, or controlling rhythmic firing. This is true not only for heart cells, but for skeletal muscle, smooth muscle, and many nerve cell bodies as well.

The subject of K^+ currents in heart cells has been rather extensively reviewed.[37,79,80] Owing to limitations of space, we will only briefly touch upon the main components of K^+ current and their functional roles.

6.1. The Inwardly Rectifying K^+ Channel I_{K_1} Supports the Resting Potential

I_{K_1} is a K^+ current with a large conductance at the normal resting potential of cardiac cells, and the dominant factor in making the resting potential follow E_K over a wide range of $[K^+]_o$.[81] It shows the interesting property of "inward-going rectification": although the K^+ conductance is large near E_K, it decreases sharply as the driving force $(E_m - E_K)$ grows with depolarization. The mechanism of the nonlinearity is not known for certain,[82,83] but its function in heart cells is clear.[81] The restricted flow of outward current through I_{K_1} makes it possible for relatively small inward currents carried by Ca^{2+} and Na^+ ions to support the cardiac action potential plateau. Were it not for inward rectification, the plateau would be extremely wasteful of ionic gradients and metabolic energy.

Patch clamp recordings from I_{K_1} channels suggest that the rectifier properties are achieved within a period shorter than the resolution of the voltage clamp technique (1.5 msec), as though the rectification was due to the ion transfer process through open channels rather than channel gating.[84]

6.2. The Delayed Rectifier Current I_x Helps Control Action Potential Duration

The I_{K_1} channels provide substantial repolarizing current in the last phase of action potential repolarization, as the membrane potential approaches the resting potential. However, inward rectification reduces their contribution at potentials in the plateau range. Time-dependent changes in other ionic currents are rate-limiting in the termination of the plateau. In some cells, inactivation of Ca^{2+} channels is sufficient to trigger repolarization; in other cells, repolarization depends upon the additional involvement of a slowly time-dependent K^+ current, which activates over hundreds of milliseconds when the membrane is depolarized over the plateau range of potentials (−40 mV up to +20 mV). This pathway has been found in a wide variety of cardiac cells[c.g., 85,86] although its magnitude seems to be somewhat variable. It is most often labeled I_x,[85] although some investigators have chosen to call it I_K.[86]

Activation of I_x is one of the main factors in governing action potential duration. The relationship between action potential duration and diastolic interval reflects in large part the deactivation of I_x during the diastolic interval. Deactivation of I_x also plays a major role in controlling the type of pacemaker activity found in SA node, AV node, and partially depolarized Purkinje fibers (see Ref. 81 for review; see Section 7).

Despite proposals to the contrary,[e.g., 87] there is strong evidence that I_x is not a Ca^{2+}-activated K^+ current. Elevation of $[Ca^{2+}]_o$ increases Ca^{2+} channel current, but decreases activation of I_x[88]; inhibition of Ca^{2+} channels with nisoldipine blocks Ca^{2+} entry but leaves I_x unaffected.[89]

Patch clamp recordings from cultured chick heart cells have revealed a voltage-gated K^+ channel with a steady-state gating curve and rectifier properties expected for I_x.[90] A K^+-specific channel was found when calf ventricular sarcolemmal membranes were incorporated into planar lipid bilayers,[91] but its identification with I_x remains uncertain (R. Coronado, personal communication).

6.3. Transient Outward K^+ Currents Mediate Rapid Early Repolarization

In many if not all cardiac cells, a step depolarization from the resting potential to levels beyond −20 mV elicits a transient outward current which reaches a peak within 5–10 msec before decaying over tens or hundreds of milliseconds. Such transient outward current is particularly prominent in Purkinje fibers where it underlies the characteristic phase 1, a phase of rapid repolarization preceding the plateau.[e.g., 92–96] In other preparations which lack a distinct phase 1 repolarization, such as ventricular muscle of rat or cat, transient outward current is less prominent, but still significant.[97,98]

Because the outward current largely overlaps the influx of Ca^{2+} through Ca^{2+} channels, it influences the configuration of the action potential plateau and corresponding currents under voltage clamp. The outward current helps preserve a large driving force for Ca^{2+} entry during the plateau, but it also obscures the real amount of Ca^{2+} entry in undissected voltage clamp current records.[99]

In sheep Purkinje fibers, where the analysis has been most extensive, voltage clamp experiments have revealed two separate components of transient outward current[96]:

1. A large, slowly decaying outward current that is abolished by the K^+ channel blocker 4-aminopyridine (4-AP) (1 mM)
2. A more rapidly decaying outward current that is abolished by 10 mM caffeine, and other procedures that interfere with the intracellular Ca^{2+} transient

Both of these currents seem to be carried by K^+ ions. They are inhibited by replacement of intracellular K^+ with Cs^+,[99] by exposure to tetraethylammonium (TEA),[93] or by injection of TEA or tetrabutylammonium.[100] The 4-AP-sensitive component is associated with significant $^{42}K^+$ fluxes.[101]

The slowly decaying, 4-AP-sensitive current shows voltage-dependent activation and inactivation properties[94] and pharmacological responses[93,96] similar to those of the K^+ channel labeled I_A in molluskan neurons. On the other hand, the more rapidly decaying, Ca_i^{2+}-dependent current seems analogous to Ca^{2+}-activated K^+ currents in other tissues.[95] The brevity of this outward current component may be a reflection of the rapidly decaying Ca^{2+} transient in cardiac cells.[102]

Both components of transient outward current await analysis with patch clamp techniques. In neither case has it been possible to establish the involvement of K^+ ions by showing a clear-cut reversal potential that is sensitive to $[K^+]$.[93,95,96,97]

6.4. K^+ Channels Inhibited by Intracellular ATP

Noma[103] has discovered a specific K^+ channel that may account for the shortening of cardiac action potentials under hypoxic conditions. The channel carries outward currents much more readily than I_{K_1} channels. It is fully inhibited by intracellular concentrations of ATP in the physiological range (3–4 mM), but becomes active when ATP falls below ~ 0.5 mM.

7. Inward Currents and Pacemaker Activity

Under physiological conditions, only the nodal regions (sinus node and AV node) and parts of the specialized impulse conducting system (Purkinje fibers) exhibit spontaneous electrical and mechanical activity. Because the natural pacemaker frequency is highest in the sinus node, the slower spontaneous rhythms of the AV node and the Purkinje fibers are normally overridden and the heart beats at the frequency of the sinus node. However, in the event of pathological slowing of the sinus pacemaker frequency or delayed impulse conduction, the AV node or the Purkinje fibers can take over the role of the primary pacemaker, thus preventing cardiac arrest.

Pacemaker activity involves a slow diastolic depolarization which links the repolarization of one action potential to the upstroke of the next action potential. Since a spontaneous action potential occurs whenever the voltage threshold for the activation of the excitatory (inward) current is reached, the interval between successive action potentials and therefore the pacemaker frequency varies with the slope of the diastolic depolarization.

7.1. Pacemaker Current i_h

Over the last few years much progress has been made in understanding the ionic mechanism underlying the generation of pacemaker potentials in both the nodal tissues and the Purkinje fibers. All the cardiac pacemaker cells have been found to have a voltage-dependent channel which is activated by hyperpolarizations to potentials negative to −50 mV.[104–108] The channel is relatively nonselective for monovalent cations as judged from its reversal potential of around −20 mV in the presence of physiological concentrations of Na^+ and K^+.[104,109] The inward current carried by the channel in the range of pacemaker potentials becomes activated at −50 to −60 mV and saturates at −100 to −120 mV. Its rate of activation increases with hyperpolarization; at −70 mV the time constant ranges from 2 to 4 sec.[104] Reduction of $[Na^+]_o$ decreases the inward current because of the diminished driving force but does not change its slope conductance, whereas rising $[K^+]_o$ increases the current at a given potential by increasing both the driving force and the slope conductance.[109] This pacemaker current has been called i_h[104] or i_f.[87] In this chapter we will use the term i_h (h for hyperpolarization). Because of the voltage range of its activation and its relatively slow kinetics, it seems very plausible that i_h underlies the slow diastolic pacemaker depolarization in Purkinje fibers which occurs between −90 and −60 mV.

This interpretation of the pacemaker current in Purkinje

fibers has replaced earlier interpretations which attributed the pacemaker current change to deactivation of an outward K^+ current ("I_{K_2}"). For a detailed discussion of the reasons for this revision, see Refs. 105 and 109.

Although i_h is also found in the SA and AV node, its contribution to the generation of pacemaker potentials is less obvious in these cells since they show pacemaker depolarization over a voltage range (−70 to −40 mV) where i_h is only slightly activated and its kinetics are rather slow. Rather it appears that in the nodal cells the pacemaker potential is determined by a balance between three currents: decaying outward K^+ current, activation of inward Ca^{2+} current, and activation of inward pacemaker current i_h. The voltage threshold for the activation of Ca^{2+} current in the SA node was found to be about −60 mV in Tyrode solution (1.8 mM Ca^{2+}).[110] In addition to these time-dependent ("gated") currents, the nodal cells also have a significant time-independent ("background") Na^+ conductance while their "background" K^+ conductance is lower than that of other cardiac cells. These background conductances are important because they explain the relative insensitivity of the diastolic potentials of the nodal cells to changes in $[K^+]_o$ and of course the fact that the maximal diastolic potential in these cells lies about 20 mV positive to E_K as measured with K^+-selective microelectrodes.

The effects of blockers on the pacemaker potentials in nodal preparations offer a clue to the relative importance of the three time-dependent currents. While the pacemaker potentials are abolished by Ca^{2+} channel blockers (D600), they are only little affected by blockade of the K^+ current (by Ba^{2+}) or the pacemaker current i_h (by Cs^+) (but see also Ref. 111 for discussion of the validity of these results). From such experiments it can be concluded[112,113] that while the three time-dependent current systems all contribute to the diastolic depolarization in nodal cells, the most important current change is activation of slow inward Ca^{2+} current. This is an important conclusion for three reasons: (1) The chronotropic effects of drugs, hormones, and neurotransmitters on nodal cells can be studied by investigating their effects on the cardiac Ca^{2+} channel (see Section 7). (2) Since the pacemaker mechanisms in nodal cells and Purkinje fibers differ significantly, drugs can be expected to have quite different chronotropic effects on these two pacemaker systems. (3) Since all cardiac cells have Ca^{2+} currents, it can be expected that all cardiac cells including the normally quiescent ones will exhibit spontaneous pacemaker activity when depolarized to diastolic potentials comparable to those found in nodal cells. Indeed, the occurrence of such Ca^{2+} current-dependent pacemaker activity in depolarized cardiac cells has long been known. An example is given in the right panels of Fig. 9. Conditions under which such a depolarization of the diastolic potential is likely to occur in the intact heart are described in Section 4.4. The development of such unnatural pacemaker centers and the spread of impulses generated by such "ectopic foci" form the basis of one major class of cardiac arrhythmias.

A completely different mechanism by which normally quiescent cells can become spontaneously active involves the Ca^{2+}-activated inward current reviewed in the following section.

7.2. Ca^{2+}-Activated Transient Inward Current

In the previous sections, we described the contribution of voltage-dependent membrane currents to the normal electrical activity in the heart. Pacemaker activity may also be generated by another type of current whose activation depends on intracellular Ca^{2+} under conditions of intracellular Ca^{2+} overload (toxic concentrations of cardiotonic steroids, high extracellular Ca^{2+}, extreme sympathetic stimulation). This is illustrated in Fig. 11 for a canine Purkinje fiber that has been injected with the bioluminescent Ca^{2+} indicator protein aequorin to monitor the intracellular Ca^{2+} transient. The left panel shows the Ca^{2+} transient and twitch contraction signals associated with the action potential in the absence of drug. The right panel shows the effect of a toxic concentration of the cardiac glycoside ouabain. The Ca^{2+} transient and twitch are both enhanced (positive inotropic effect); the stimulated activity is followed by spontaneous variations in each of the traces, i.e., an afterdepolarization follows the action potential, an "afterglimmer" follows the main luminescence peak, and an aftercontraction follows the twitch. If the Purkinje fiber were *in situ* in an intact ventricle, propagation of the spontaneous impulse would give rise to a premature ventricular contraction, one of the first signs of digitalis arrhythmia.

The mechanism and role of the afterdepolarizations have been extensively studied (see Refs. 114–119). The after-

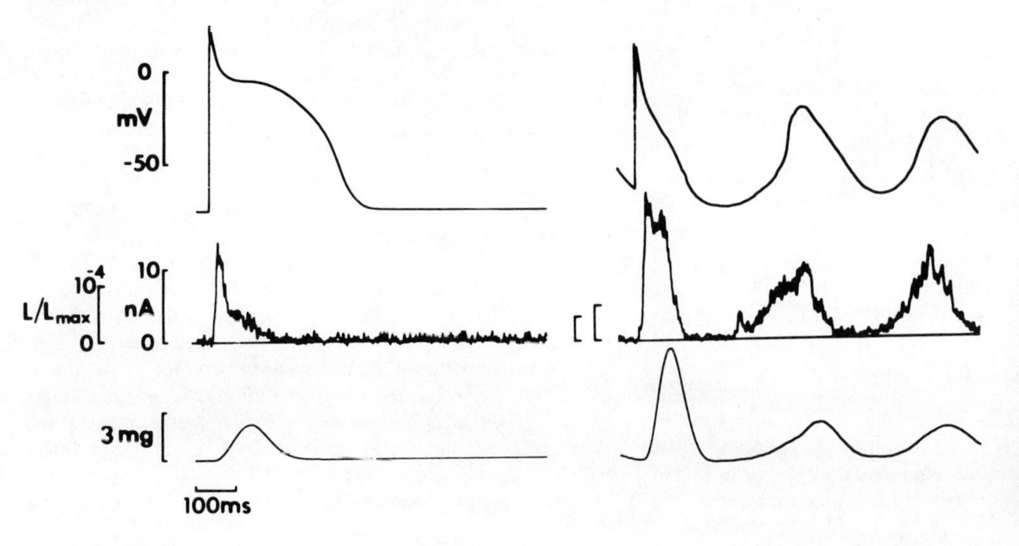

Fig. 11. Ca^{2+}-dependent arrhythmogenic activity in an ouabain-intoxicated canine Purkinje fiber bundle. Recordings of transmembrane potential (top), aequorin light signal (middle), and contractile force (bottom). Unpublished record of W. G. Wier and P. Hess (see Ref. 122 for experimental details). Similar results are presented in Ref. 123.

depolarization [or "transient depolarization" (TD)] is generated by a "transient inward current" (TI) that is small or absent under normal physiological conditions.[115] The TI has been seen in a wide variety of cardiac preparations besides Purkinje fibers, including multicellular ventricular preparations[116] and single ventricular cells.[117] Like the aftercontraction, the TI is driven by the rise in $[Ca^{2+}]_i$.[118] Ion replacement experiments demonstrate that the inward current is carried by an influx of Na^+ ions.[119] The pathway for the Na^+ movement is not known for certain, but the two main possibilities are a Ca^{2+}-activated nonselective cation channel,[119,120] and an electrogenic Na^+/Ca^{2+} exchange (see Ref. 119).

No specific pharmacological blockers for the Ca^{2+}-activated inward current have yet been found, but of course the afterdepolarizations can be inhibited by all interventions that abolish the diastolic Ca^{2+} transient, whether by reduction of the Ca^{2+} overload (Ca^{2+} channel blockers, β-receptor blockers) or by inhibition of the Ca^{2+} release from the sarcoplasmic reticulum (caffeine, ryanodine). Lidocaine and other local anesthetics are used clinically in the management of digitalis-induced arrhythmias. Lidocaine has a number of beneficial effects with regard to TI and afterdepolarizations.[56,121] One of the more important effects involves changes in ionic balance: like TTX, lidocaine reduces the Na^+ current, decreases $[Na^+]_i$,[121] and thereby relieves the Ca^{2+} overload, which is linked to $[Na^+]_i$ via the Na^+/Ca^{2+} exchange.

8. Adrenergic and Cholinergic Modulation of Cardiac Activity

Although the heart is capable of generating rhythmic activity on its own, independent of any neuronal influence, the nervous system can profoundly modify both the pace and the strength of cardiac contractions through the release of neurotransmitters and hormones. Physiologists have long been fascinated by modulatory effects on cardiac cells, not only for their importance to the heart, but also for their value as models for understanding the modulation of neuronal function.[e.g., 124,125] Sympathetic and parasympathetic agents are among the most important of the neurotransmitters, hormones, and drugs that modulate cardiac function.[126,127] In this section, we restrict ourselves to a brief overview of the regulatory effects of β-adrenergic and cholinergic stimulation (Fig. 12).

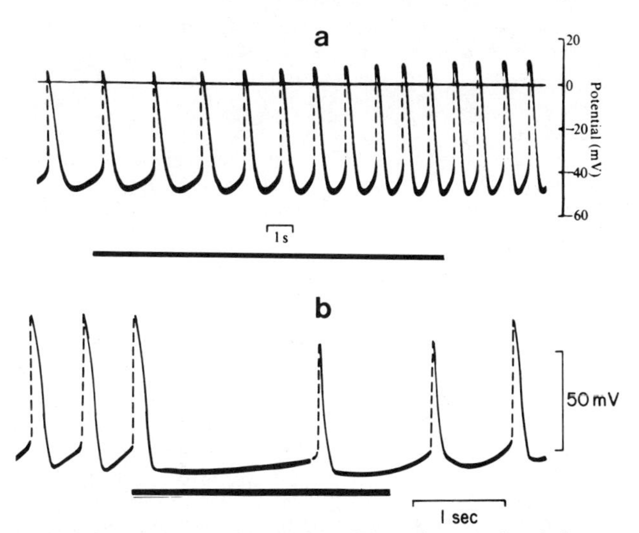

Fig. 12. Action of sympathetic and vagal transmitters on electrical activity in sinus tissue. (A) Acceleratory effect of sympathetic nerve stimulation on frog sinus venosus. Nerve stimulation indicated by the horizontal line. From Hutter and Trautwein.[187] (B) Rabbit SA node. Vagal nerve stimulation, indicated by the horizontal line, slowed firing and shortened the action potential. Redrawn from Toda and West.[188]

8.1. Overview of Adrenergic Effects

The sympathetic agents norepinephrine and epinephrine are delivered to the heart by the bloodstream (mainly epinephrine), and by local release of neurotransmitter from sympathetic nerve endings (mainly norepinephrine). They produce their most important cardiac effects by interacting with β-adrenergic receptors although significant α-adrenergic effects also exist. Specific drugs are available to stimulate β-receptor (e.g., isoproterenol) or to inhibit them (e.g., propranolol, timolol).

The β-adrenergic system serves as an excellent example of how multipronged effects of a hormone can combine to give a coordinated response overall. Aspects of the β-adrenergic response include (1) increased availability of ATP through changes in metabolic state, (2) changes in the rate of sequestration of intracellular Ca^{2+}, (3) changes in the responsiveness of the contractile filaments to myoplasmic Ca^{2+}. These effects all contribute to strengthening and abbreviating the cardiac contraction and have been extensively reviewed.[e.g., 128,129]

Our main purpose here is to summarize effects on cardiac ionic currents. Several of the current systems described earlier in this chapter show significant modulation with β-adrenergic stimulation. A notable exception is the excitatory Na^+ current: even though biochemical studies show that the α subunit of Na^+ channels from the brain is phosphorylated by cAMP-dependent protein kinase,[130] no modulatory effect has been found for the Na^+ channel in the heart or nerve.

8.2. β-Adrenergic Modulation of Ca^{2+} Channels

Interaction between catecholamines and β-adrenergic receptors leads to an increase in Ca^{2+} influx in all cardiac cells. This is one of the oldest and best-studied examples of neuromodulation of a voltage-gated channel (see Ref. 126 for review). The enhancement of Ca^{2+} current is a major factor in the response to sympathetic stimulation in different regions of the heart, which takes the form of accelerated automaticity in the SA node and Purkinje fibers, more rapid impulse conduction in the AV node, and increased contractility in all regions, particularly ventricular and atrial muscle.

Figures 13A and B show some early evidence that epinephrine modifies the action potential and I–V relation through an increase in Ca^{2+} current.[131] The recordings are from frog atrial trabeculae treated with TTX to block the rapid Na^+ current. Epinephrine increases the height of the plateau (A) and enhances the peak inward current (B). The enhancement has often been described as a simple scaling up of current, with no change in the voltage dependence or time dependence of activation or inactivation.[132] More recent patch clamp experiments suggest that β-adrenergic stimulation may slow the time course of Ca^{2+} channel activation and inactivation under at least some circumstances.[133]

There is strong evidence that the enhancement of Ca^{2+} current is mediated through the action of cAMP as an internal

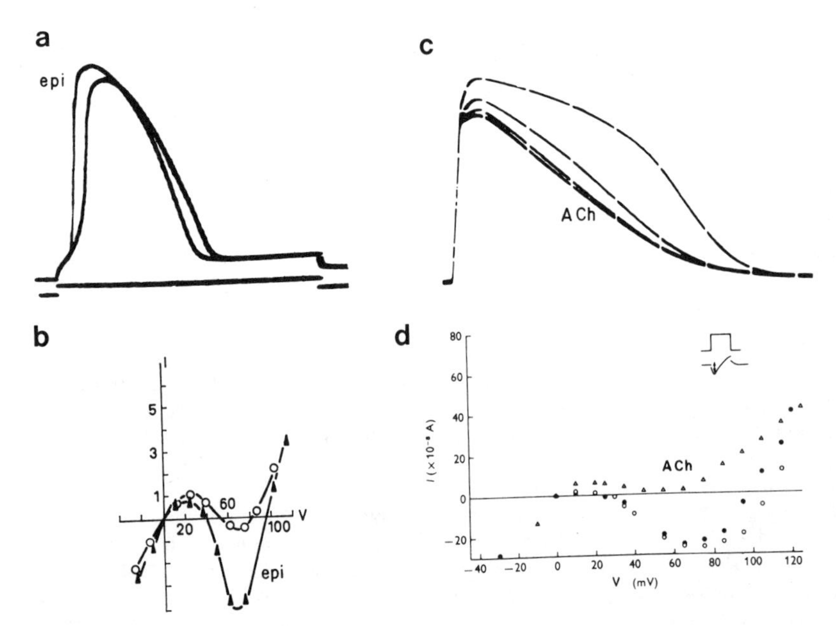

Fig. 13. Action of ACh and epinephrine on action potentials and slow inward current in frog atrial muscle. (A) Epinephrine (10^{-5} M) increases the plateau level of the action potential. Vertical scale: 20 mV; horizontal scale: 200 msec.[131] (B) Current–voltage relation in Ringer (○) and in Ringer containing 10^{-5} M epinephrine (▲). V is voltage displacement from the resting potential in millivolts.[131] (C) Action potentials recorded before and during exposure to ACh (8×10^{-7} M). ACh reduces the amplitude and duration of the plateau. Vertical scale: 20 mV; horizontal scale: 100 msec.[164] (D) Current–voltage relations showing the effect of ACh on peak slow inward current (inset). ●, control; ▲, 8×10^{-7} M ACh; ○, ACh + 10^{-5} M atropine.[164] Throughout these experiments TTX was present to block the fast Na^+ current.

messenger and cAMP-dependent protein phosphorylation as an enzymatic mechanism. β-Adrenergic effects on membrane potential activity or inward Ca^{2+} current have been mimicked by direct injection of cAMP,[134–136] or by application of exogenous derivatives of cAMP[137,138] and other agents that would be expected to elevate cAMP, such as phosphodiesterase inhibitors[137] or cholera toxin.[139] Similar effects were obtained with injection of the catalytic subunit of cAMP-dependent protein kinase.[140]

The enhancement of Ca^{2+} influx is associated with an increase in the overall Ca^{2+} conductance. The nature of the conductance increase has been the subject of considerable investigation. Selectivity of Ca^{2+} channels remains unchanged, judging by measurements of the Ca^{2+} channel reversal potential.[132,133] Likewise, the inward current through individual open channels seems unaffected.[133,141,142] What does seem to change is the fraction of time individual Ca^{2+} channels spend open.[59] This seems to involve modulatory effects on two very different time scales.

1. On a rapid time scale (milliseconds), β-adrenergic stimulation or cAMP biases channels toward the open state. There is an increase in the rate constants leading to the channel's open state, and a decrease in the rate constant of channel closing. These effects are expressed as an increased probability of channel openness in single-channel recordings[59,141,142] and in fluctuation analysis of whole-cell recordings.[133]

2. On a much slower time scale (seconds), β-adrenergic stimulation increases Ca^{2+} channel availability. This is seen in single-channel recordings as a decrease in the number of empty sweeps (sweeps where the channel is not available to undergo rapid opening and closing transitions during a depolarizing pulse). In fluctuation analysis of recordings from whole cells, the increased Ca^{2+} channel availability is expressed as an increase in the number of functional channels during individual sweeps.[133]

The relative importance of these effects seems to vary from one cardiac preparation to the next. In cultured ventricular cells from neonatal rat, the rapid time scale effect (increase in open probability) is sufficient to account for most of the ~ 1.5- to 2.0-fold increase in peak inward current.[133,141] In frog ventricular cells, on the other hand, the overall enhancement of Ca^{2+} channel current averages about 6-fold, and is dominated by an increase in the number of functional channels.[133]

In all preparations described so far, there is no convincing evidence that β-adrenergic stimulation increases the *total* number of Ca^{2+} channels (see Refs. 133, 142).

8.3. β-Adrenergic Modulation of Pacemaker Current

The spontaneous activity of cardiac Purkinje fibers is dramatically accelerated by epinephrine through a steepening of the diastolic depolarization.[143] The mechanism of the effect is a modulation of the voltage-dependent gating of the pacemaker current,[144,145] now identified as an inward current activated by hyperpolarization (see Section 7.1). The midpoint of the voltage-dependent gating curve is shifted by up to 20 mV in the depolarizing direction. This has the effect of increasing both the degree and the rate of activation of the pacemaker current over the range of potential where pacemaker activity occurs. Although modulation of i_h has been observed in sinus tissue,[146] its importance in the physiological response to sympathetic transmitter is much less certain in the natural pacemaker[113] than in Purkinje fibers.

The pacemaker current provides an unusual example of how the voltage dependence of gating may be modulated; in most systems, the predominant effect is an increase or decrease in the magnitude of an ionic current, with little or no change in the voltage dependence of activation (see Ref. 127). The mechanism of the modulation is unknown, although it has been suggested that cAMP-dependent phosphorylation might alter the electrical or chemical energy difference between open and closed states.[147]

8.4. β-Adrenergic Modulation of Outward K^+ Currents

β-Adrenergic stimulation has additional effects on outward K^+ currents. (1) β-Agonists increase I_x, the time- and voltage-dependent K^+ current that helps trigger repolarization.[137,148–

150) The enhancement counteracts the enhancement of I_{Ca}, and helps keep action potential duration from increasing. Changes in I_x or associated changes in action potential configuration can be mimicked by phophodiesterase inhibitors, exogenous cAMP, or intracellular cAMP injection,[134,137,151] suggesting once again that cAMP is the second messenger. (2) β-Agonists increase a background K^+ conductance that contributes to maintenance of the resting potential.[152]

8.5. Overview of Cholinergic Effects

The parasympathetic agent acetylcholine (ACh) is released from postganglionic nerve terminals as a result of vagal activity, and reaches muscarinic receptors on cardiac cells through diffusion. The end effects are essentially the opposite of those induced by sympathetic agents (see Fig. 12B). The two main ionic mechanisms are enhancement of K^+ efflux through ACh-sensitive K^+ channels, and inhibition of Ca^{2+} influx through voltage-dependent Ca^{2+} channels. These mechanisms act in synergy to decrease the duration of the action potential and strength of contraction in working myocardium, and to reduce automaticity and conduction velocity in nodal tissue.

8.6. ACh-Sensitive K^+ Channels

In the late 1950s, Hutter[153] showed that ACh greatly enhanced $^{42}K^+$ efflux from the sinus pacemaker region; Trautwein and Dudel[154] demonstrated that ACh markedly increased the conductance of heart muscle. Woodbury and Crill[155] showed that such increases in conductance were unchanged when Cl^- ions were replaced by SO_4^{2-}. These pioneering studies established that ACh can increase the K^+ conductance of cardiac membranes. Subsequent voltage clamp experiments showed that the ACh-induced K^+ current shows inward-going rectification[156,157] and that it is blocked by Cs^+ ions in much the same way as the normal inward-going rectifier current I_{K_1}.[158] The obvious question is whether ACh-induced K^+ conductance simply reflects an enhancement of i_{K_1}. Patch clamp experiments in nodal pacemaker cells suggest that this is not the case. Openings of single ACh-sensitive channels last ~ 1–2 msec,[159] much shorter than values of ~ 100 msec found for I_{K_1} channels in atrial or ventricular cells.[84]

ACh is known to stimulate cGMP production in the heart and other tissues. However, cGMP is almost certainly not involved as a mediator of the increase in K^+ conductance.[160–162] Indeed, participation of any diffusible second messenger seems unlikely: ACh-sensitive K^+ channels in cell-attached patches fail to respond when ACh is applied to the cell membrane outside the patch; they respond only when ACh is delivered to the extracellular side of the patch.[163] The most likely explanation is that the muscarinic receptor and K^+ channel are intimately related, possibly like the nicotinic receptor and the Na^+- and K^+-permeable endplate channel.

8.7. Cholinergic Modulation of Ca^{2+} Channels

Modulation of Ca^{2+} channels by ACh was discovered much later than regulation of K^+ current: convincing demonstration required the use of voltage clamp techniques and only came in the mid-1970s.[164,165] As briefly mentioned already, cholinergic stimulation of muscarinic receptors produces a reduction of Ca^{2+} influx, in more or less diametrical opposition to the effect of β-adrenergic stimulation. This joint regulation makes the Ca^{2+} channel a particularly interesting system for studying neuromodulation. (It should be noted that in addition to their postsynaptic interaction, parasympathetic and sympathetic systems also interact at the presynaptic level: vagal stimulation inhibits the release of adrenergic transmitter from sympathetic nerve endings.[166])

Figures 13C and D illustrate the effect of ACh on the action potential and I–V relation of TTX-treated frog atrial muscle.[164] The decrease in the height and duration of the plateau (C), and the voltage dependence of the ACh-sensitive current (D) are as expected for inhibition of Ca^{2+} channel current. Similar results have been reported for a number of cardiac preparations.[e.g., 167–171] The mechanism of the inhibition at the level of single channels has not yet been described. Ca^{2+} channel currents seem to be scaled rather uniformly with no change in the apparent reversal potential.[167,169]

Cholinergic inhibition of Ca^{2+} current can often be seen in isolation from enhancement of background K^+ current. In frog atrial tissue, I_{Ca} reduction occurs at much lower ACh concentrations than enhancement of K^+ current (see Refs. 167, 172). In chick ventricle, inhibition of I_{Ca} occurs in the absence of the K^+ current effect.[173] Unlike the enhancement of K^+ permeability, the muscarinic effect on Ca^{2+} currents is expressed most clearly against a background of β-adrenergic stimulation[e.g., 171,174] or phosphodiesterase inhibition,[e.g., 174,175,176,177] a phenomenon known as "accentuated antagonism."[166]

Two hypotheses have been proposed for the mechanism of the cholinergic suppression of Ca^{2+} channel current. One possibility is that cGMP acts as a second messenger, in opposition to the stimulatory effect of cAMP. This fits with experiments in which cholinergic effects on electrical or mechanical activity were mimicked by applying cGMP as the 8-bromo derivative[177,178] or by direct intracellular injection.[179] Another possibility is that muscarinic stimulation inhibits the activity of adenylate cyclase and the production of cAMP.[e.g., 180] These mechanisms are not mutually exclusive; it may yet turn out that the cardiac accelerator and braking systems interact at almost every possible level.

9. Summary

The heart has a multicellular structure but it behaves like a syncytium because individual cells communicate with their neighbors. There is evidence that nexuses (gap junctions) provide the pathway for electrical coupling between cells, and to some extent, chemical coupling. Electrical coupling is very effective and enables local circuit currents to support impulse propagation.

In various regions of the heart, different types of action potential serve a variety of functions, including pacemaker rhythm generation, rapid or slow impulse conduction, and control of contraction. The diversity of electrical activity arises from a number of ionic channels, working together in different combinations.

Na^+ channels underlie rapid impulse conduction in working myocardium and Purkinje tissue. During the action potential plateau, inactivation of Na^+ channels is brought about by the maintained depolarization and underlies the refractory period; removal of inactivation with repolarization allows recovery of excitability. Na^+ channels are an important target for the antiarrhythmic action of local anesthetics like lidocaine.

Ca^{2+} channels support a Ca^{2+} influx that helps support the action potential plateau, and that can generate a slow response—a slowly rising, slowly propagating impulse. Slow responses are the normal form of activity in SA node and AV node, and may also be important in certain arrhythmias. Ca^{2+} channels and slow responses are readily distinguished from Na^{+} channels and rapidly propagating impulses in their differing sensitivity to steady depolarizations and pharmacological agents. Ca^{2+} influx plays an important role in excitation–contraction coupling.

K^{+} channels show several subtypes. One set of K^{+} channels displays increasing conductance with hyperpolarization (inward rectification). These K^{+} channels support the negative resting potentials in working myocardium and Purkinje tissue but are absent in nodal regions. Another type of K^{+} channel activates slowly with depolarization beyond −30 mV and helps terminate the plateau. Transient outward K^{+} currents are activated by depolarization or intracellular Ca^{2+}, and tend to shunt the depolarizing effect of Ca^{2+} entry early during the plateau.

Inward currents largely carried by Na^{+} ions play an important part in different forms of pacemaker activity. One inward current, activated by hyperpolarization negative to about −50 mV, generates diastolic depolarization in Purkinje fibers. Another inward current is activated by intracellular Ca^{2+}, and produces transient depolarizations and arrhythmic action potentials during the Ca^{2+} overload brought on by digitalis intoxication.

The heartbeat is profoundly modulated by sympathetic and parasympathetic systems. Norepinephrine and epinephrine act through stimulation of β receptors to increase the activity of Ca^{2+} channels and delayed outward K^{+} channels. Increased Ca^{2+} entry helps accelerate sinus pacemaker activity and promotes enhanced myocardial contractility. ACh interacts with muscarinic receptors. It increases the resting K^{+} conductance through activation of a special set of K^{+} channels, while also decreasing Ca^{2+} channel activity. Both effects contribute to an inhibition of pacemaker activity and contraction.

References

1. McNutt, N. S., and R. S. Weinstein. 1973. Membrane ultrastructure at mammalian intercellular junctions. *Prog. Biophys. Mol. Biol.* **26**:45–101.
2. Barr, L., M. M. Dewey, and W. Berger. 1965. Propagation of action potentials and the structure of the nexus in cardiac muscle. *J. Gen. Physiol.* **48**:797–823.
3. Dreifuss, J. J., L. Girardier, and W. G. Forssmann. 1966. Étude de la propagation de l'éxcitation dans le ventricle de rat au moyen de solutions hypertoniques. *Pfluegers Arch.* **292**:13–33.
4. Kawamura, K., and T. Konishi. 1967. Ultrastructure of the cell junction of heart muscle with special reference to its functional significance in excitation conduction and to the concept of "disease of intercalated disc." *Jpn. Circ. J.* **31**:1533–1543.
5. Gilula, N. B. 1974. Junctions between cells. In: *Cell Communication.* R. P. Cox, ed. Wiley, New York.
6. Unwin, P. N. T., and G. Zampighi. 1980. Structure of the junction between communicating cells. *Nature (London)* **283**:545–549.
7. Unwin, P. N. T., and P. D. Ennis. 1984. Two configurations of a channel-forming protein. *Nature (London)* **307**:609–613.
8. Weidmann, S. 1966. The diffusion of radiopotassium across intercalated disks of mammalian cardiac muscle. *J. Physiol. (London)* **187**:323–342.
9. Weingart, R. 1974. The permeability to tetraethylammonium ions of the surface membrane and the intercalated disks of sheep and calf myocardium. *J. Physiol. (London)* **240**:741–762.
10. Imanaga, I. 1974. Cell to cell diffusion of procion yellow in sheep and calf Purkinje fibres. *J. Membr. Biol.* **16**:381–388.
11. Pollack, G. H. 1976. Intercellular coupling in the atrioventricular node and other tissues of the rabbit heart. *J. Physiol. (London)* **255**:275–298.
12. Weidmann, S. 1970. Electrical constants of trabecular muscle from mammalian heart. *J. Physiol. (London)* **210**:1041–1054.
13. Kushmerick, M. J., and R. J. Podolsky. 1969. Ionic mobility in muscle cells. *Science* **166**:1297–1298.
14. Matter, A. 1973. A morphometric study on the nexus of rat cardiac muscle. *J. Cell Biol.* **56**:690–696.
15. Rose, B., and W. R. Loewenstein. 1975. Permeability of cell junction depends on local cytoplasmic calcium activity. *Nature (London)* **254**:250–252.
16. Dahl, G., and G. Isenberg. 1980. Decoupling of heart muscle cells: Correlation with increased cytoplasmic calcium activity and with changes of nexus ultrastructure. *J. Membr. Biol.* **53**:63–75.
17. Reber, W. R., and R. Weingart. 1982. Ungulate cardiac Purkinje fibers: The influence of intracellular pH on the electrical cell-to-cell coupling. *J. Physiol. (London)* **328**:87–104.
18. De Mello, W. C. 1980. Influence of intracellular injection of H^{+} on the electrical coupling in cardiac Purkinje fibers. *Cell. Biol. Int. Rep.* **4**:51–58.
19. Spray, D. C., J. H. Stern, A. L. Harris, and M. V. Bennett. 1982. Gap junctional conductance: Comparison of sensitivities to H and Ca ions. *Proc. Natl. Acad. Sci. USA* **79**:441–445.
20. White, R. L., A. C. Carvalho, D. C. Spray, B. A. Wittenberg, and M. V. L. Bennett. 1983. Gap junctional conductance between isolated pairs of ventricular myocytes from rat. *Biophys. J.* **41**:217a.
21. Weidmann, S. 1952. The electrical constants of Purkinje fibres. *J. Physiol. (London)* **115**:227–236.
22. Bonke, F. I. M. 1973. Electrotonic spread in the sinoatrial node of the rabbit heart. *Pfluegers Arch.* **339**:17–23.
23. Woodbury, J. W., and W. E. Crill. 1961. On the problem of impulse conduction in the atrium. In: *Nervous Inhibition.* E. Florey, ed. Pergamon Press, Elmsford, N.Y. pp. 124–135.
24. Draper, M. H., and M. Mya-Tu. 1959. A comparison of the conduction velocity in cardiac tissues of various mammals. *Q. J. Exp. Physiol.* **44**:91–109.
25. Sano, T., N. Takayama, and T. Shimamoto. 1959. Directional difference of conduction velocity in the cardiac ventricular syncytium studied by microelectrodes. *Circ. Res.* **7**:262–267.
26. Clerc, L. 1976. Directional differences of impulse spread in trabecular muscle from mammalian heart. *J. Physiol. (London)* **255**:335–346.
27. Woodbury, J. W. 1962. Cellular electrophysiology of the heart. In: *Handbook of Physiology,* Section 2, Volume 1. W. F. Hamilton and P. Dow, eds. American Physiological Society, Washington, D.C. pp. 237–238.
28. Hodgkin, A. L. 1951. The ionic basis of electrical activity in nerve and muscle. *Biol. Rev.* **26**:339–409.
29. Draper, M. H., and S. Weidmann. 1951. Cardiac resting and action potentials recorded with an intracellular electrode. *J. Physiol. (London)* **115**:74–94.
30. Del Castillo, J., and J. W. Moore. 1959. On increasing the velocity of a nerve impulse. *J. Physiol. (London)* **148**:665–670.
31. Weingart, R. 1977. The actions of ouabain on intercellular coupling and conduction velocity in mammalian ventricular muscle. *J. Physiol. (London)* **264**:341–365.
32. Tsien, R. W., and R. Weingart. 1976. Inotropic effect of cyclic AMP in calf ventricular muscle studied by a cut-end method. *J. Physiol. (London)* **260**:117–141.
33. Hoffman, B. F., and P. F. Cranefield. 1960. *Electrophysiology of the Heart.* McGraw–Hill, New York.
34. Hogan, P. M., and L. D. Davis. 1968. Evidence for specialized fibers in the canine right atrium. *Circ. Res.* **23**:387–396.
35. Mendez, C., and G. K. Moe. 1972. Atrioventricular transmission. In: *Electrical Phenomena in the Heart.* W. C. De Mello, ed. Academic Press, New York.
36. Cranefield, P. F. 1975. *The Conduction of the Cardiac Impulse.* Futura, Mount Kisco, N.Y.
37. Carmeliet, E., and J. Vereecke. 1980. Electrogenesis of the action potential and automaticity. In: *Handbook of Physiology,* Volume I. R. M. Berne, ed. American Physiological Society, Washington, D.C. pp. 269–334.

38. Piwnica-Worms, D., R. Jacob, C. R. Horres, and M. Lieberman. 1983. Transmembrane chloride flux in tissue-cultured chick heart cells. *J. Gen. Physiol.* **81**:731–748.
39. Dudel, J., K. Peper, R. Rüdel, and W. Trautwein. 1967. The effect of tetrodotoxin on the membrane current in cardiac muscle (Purkinje fibers). *Pfluegers Arch.* **295**:213–226.
40. Johnson, E. A., and M. Lieberman. 1971. Heart: Excitation and contraction. *Annu. Rev. Physiol.* **33**:479–532.
41. Reuter, H. 1979. Properties of two inward membrane currents in the heart. *Annu. Rev. Physiol.* **41**:413–424.
42. Colatsky, T. J., and R. W. Tsien. 1979. Sodium channels in rabbit cardiac Purkinje fibers. *Nature (London)* **278**:265–268.
43. Colatsky, T. J. 1980. Voltage clamp measurements of sodium channel properties in rabbit cardiac Purkinje fibers. *J. Physiol. (London)* **305**:215–234.
44. Ebihara, L., N. Shigeto, M. Lieberman, and E. A. Johnson. 1980. The initial inward current in spherical clusters of chick embryonic heart cells. *J. Gen. Physiol.* **75**:437–456.
45. Lee, K. S., T. A. Weeks, R. L. Kao, N. Akaike, and A. M. Brown. 1979. Sodium current in single heart muscle cells. *Nature (London)* **278:**269–271.
46. Brown, A. M., K. S. Lee, and T. Powell. 1981. Sodium current in single rat heart muscle cells. *J. Physiol. (London)* **318:**479–500.
47. Bodewei, R., S. Hering, B. Lemke, L. V. Rosenshtraukh, A. I. Undrovinas, and A. Wollenberger. 1982. Characterization of the fast sodium current in isolated rat myocardial cells: Simulation of the clamped membrane potential. *J. Physiol. (London)* **325**:301–315.
48. Bustamante, J. O., and T. F. McDonald. 1983. Sodium currents in segments of human heart cells. *Science* **220**:320–321.
49. Cachelin, A. B., J. E. DePeyer, S. Kokubun, and H. Reuter. 1983. Sodium channels in cultured cardiac cells. *J. Physiol. (London)* **340**:389–401.
50. Hamill, O. P., A. Marty, E. Neher, B. Sakmann, and F. J. Sigworth. 1981. Improved patch clamp techniques for high resolution patch clamp recording from cells and cell-free membrane patches. *Pfluegers Arch.* **391**:85–100.
51. Myerburg, R. J., H. Gelband, and B. F. Hoffman. 1971. Functional characteristics of the gating mechanism in the canine A-V conducting system. *Circ. Res.* **28**:136–147.
52. Singer, D. H., R. Lazzara, and B. F. Hoffman. 1967. Interrelationships between automaticity and conduction in Purkinje fibers. *Circ. Res.* **21**:537–558.
53. Singh, B. N., and E. M. Vaughan-Williams. 1971. Effect of altering potassium concentration on the action of lidocaine and diphenylhydantoin on rabbit atrial and ventricular muscle. *Circ. Res.* **29**:286–295.
54. Hondeghem, L. M., and B. G. Katzung. 1977. Time- and voltage-dependent interactions of antiarrhythmic drugs with cardiac sodium channels. *Biochim. Biophys. Acta* **472**:373–398.
55. Hille, B. 1977. Local anesthetics: Hydrophilic and hydrophobic pathways for the drug–receptor reaction. *J. Gen. Physiol.* **69** :497–515.
56. Bean, B. P., C. J. Cohen, and R. W. Tsien. 1983. Lidocaine block of cardiac sodium channels. *J. Gen. Physiol.* **81**:613–642.
57. Weidmann, S. 1955. Effects of calcium ions and local anesthetics on the electrical properties of Purkinje fibres. *J. Physiol. (London)* **129**:568–582.
58. Hagiwara, S., and S. Nakajima. 1966. Differences in Na and Ca spikes as examined by application of tetrodotoxin, procaine and manganese ions. *J. Gen. Physiol.* **49**:793–806.
59. Reuter, H., C. F. Stevens, R. W. Tsien, and G. Yellen. 1982. Properties of single calcium channels in cardiac cell culture. *Nature (London)* **297**:501–504.
60. Cavalie, A., R. Ochi, D. Pelzer, and W. Trautwein. 1983. Elementary currents through Ca channels in guinea pig myocytes. *Pfluegers Arch.* **398**:284–297.
61. Eckert, R., D. L. Tillotson, and P. Brehm. 1981. Calcium mediated control of Ca and K currents. *Fed. Proc.* **40**:2226–2232.
62. Marban, E., and R. W. Tsien. 1982. Enhancement of calcium current during digitalis inotropy in mammalian heart: Positive feedback regulation by intracellular calcium. *J. Physiol. (London)* **329**:589–614.
63. Tsien, R. W. 1983. Calcium channels in excitable cell membranes. *Annu. Rev. Physiol.* **45**:341–358.
64. Mentrard, D., G. Vassort, and R. Fischmeister. 1984. Calcium mediated inactivation of the calcium conductance in cesium loaded frog heart cells. *J. Gen. Physiol.* **83**:105–131.
65. Lee, K. S., and R. W. Tsien. 1982. Reversal of current through calcium channels in dialysed single heart cells. *Nature (London)* **297**:498–501.
66. Sigworth, F. J. 1980. The variance of sodium current fluctuations at the node of Ranvier. *J. Physiol. (London)* **307**:97–129.
67. Janis, R. A., and D. J. Triggle. 1983. New developments in Ca channel antagonists. *J. Med. Chem.* **26**:775–785.
68. Reuter, H., and H. Scholz. 1977. A study of the ion selectivity and the kinetic properties of the calcium dependent slow inward current in mammalian cardiac muscle. *J. Physiol. (London)* **264**:17–47.
69. Lee, K. S., and R. W. Tsien. 1984. High selectivity of calcium channels in single dialyzed heart cells of the guinea pig. *J. Physiol (London)* **354**:253–272.
70. Hagiwara, S., J. Fukuda, and D. C. Eaton. 1974. Membrane currents carried by Ca, Sr, and Ba in barnacle muscle fiber during voltage clamp. *J. Gen. Physiol.* **63**:564–578.
71. Vereecke, J., and E. E. Carmeliet. 1971. Sr action potentials in cardiac Purkinje fibers. II. Dependence of the Sr conductance on the external Sr concentration and Sr–Ca antagonism. *Pfluegers Arch.* **322**:73–82.
72. Hess, P., and R. W. Tsien. 1984. Mechanism of ion permeation through calcium channels. *Nature (London)* **309**:453–456.
73. Lee, K. S., and R. W. Tsien. 1983. Mechanism of calcium channel blockade by verapamil, D600, diltiazem and nitrendipine in single dialyzed heart cells. *Nature (London)* **302**:790–794.
74. Cavalie, A., Pelzer, D., and W. Trautwein. 1985. Modulation of the gating properties of single calcium channels by D600 in guinea pig ventricular myocytes. *J. Physiol. (London)* **358**:59.
75. Schramm, M., G. Thomas, R. Towart, and G. Franckowiak. 1983. Novel dihydropyridines with positive inotropic action through activation of Ca channels. *Nature (London)* **303**:535–537.
76. Katzung, B. G., L. M. Hondeghem, and A. O. Grant. 1975. Cardiac ventricular automaticity induced by current of injury. *Pfluegers Arch.* **360**:193–197.
77. Chapman, R. 1979. Excitation–contraction coupling in cardiac muscle. *Prog. Biophys. Mol. Biol.* **35**:1–52.
78. Fabiato, A. 1983. Calcium-induced release of calcium from the cardiac sarcoplasmic reticulum. *Am. J. Physiol.* **245**:C1–C14.
79. Eisner, D. A., and R. D. Vaughan-Jones. 1983. Do calcium-activated potassium channels exist in the heart? *Cell Calcium* **4**:371–386.
80. Reuter, H. 1984. Ion channels in cardiac cell membranes. *Annu. Rev. Physiol.* **46**:473–484.
81. Noble, D. 1979. *The Initiation of the Heartbeat.* Oxford University Press, London.
82. Adrian, R. H. 1969. Rectification in muscle membrane. *Prog. Biophys. Mol. Biol.* **19**:339–369.
83. Hille, B., and W. Schwarz. 1978. Potassium channels as multi-ion single-file pores. *J. Gen. Physiol.* **72**:409–441.
84. Sakmann, B., and G. Trube. 1984. Conductance properties of single inwardly rectifying potassium channels in ventricular cells from guinea-pig heart. *J. Physiol. (London)* **347**:641–657.
85. Noble, D., and R. W. Tsien. 1969. Outward membrane currents activated in the plateau range of potentials in cardiac Purkinje fibres. *J. Physiol. (London)* **200**:205–231.
86. McDonald, T. F., and W. Trautwein. 1978. The potassium current underlying delayed rectification in cat ventricular muscle. *J. Physiol. (London)* **274**:193–216.
87. Brown, H., and DiFrancesco, D. 1980. Voltage-clamp investigations of membrane currents underlying pace-maker activity in rabbit sino-atrial node. *J. Physiol. (London)* **308**:331–351.
88. Kass, R. S., and R. W. Tsien. 1975. Multiple effects of calcium

antagonists on plateau currents in cardiac Purkinje fibers. *J. Gen. Physiol.* **66**:169–192.

89. Kass, R. S. 1982. Delayed rectification is not a calcium activated current in cardiac Purkinje fibers. *Biophys. J.* **37**:342a.
90. Clapham, D. E., and L. J. DeFelice. 1984. Voltage-activated K channels in embryonic heart. *Biophys. J.* **45**:40–42.
91. Coronado, R., and R. Latorre. 1982. Detection of K^+ and Cl^- channels from calf cardiac sarcolemma in planar lipid bilayer membranes. *Nature (London)* **298**:849–852.
92. Dudel, J., K. Peper, R. Rudel, and W. Trautwein. 1967. The dynamic chloride component of membrane current in Purkinje fibres. *Pfluegers Arch.* **295**:197–212.
93. Kenyon, J. L., and W. R. Gibbons. 1979. 4-Aminopyridine and the early outward current of sheep cardiac Purkinje fibers. *J. Gen. Physiol.* **73**:139–157.
94. DiFrancesco, D., and P. A. McNaughton. 1979. The effects of calcium on outward membrane currents in the cardiac Purkinje fibre. *J. Physiol. (London)* **289**:347–373.
95. Siegelbaum, S. A., and R. W. Tsien. 1980. Calcium-activated transient outward current in calf cardiac Purkinje fibres. *J. Physiol. (London)* **299**:485–506.
96. Coraboeuf, E., and E. Carmeliet. 1982. Existence of two transient outward currents in sheep cardiac Purkinje fibers. *Pfluegers Arch.* **392**:352–359.
97. Josephson, I., and J. Sanchez-Chapula. 1982. Plateau membrane currents in single heart cells. *Biophys. J.* **37**:238a.
98. Ito, K., J. L. Kenyon, G. Isenberg, and J. L. Sutko. 1984. The existence of two components of transient outward current in isolated cardiac ventricular myocytes. *Biophys. J.* **45**:54a.
99. Marban, E., and R. W. Tsien. 1982. Effects of nystatin-mediated intracellular ion substitution on membrane currents in calf Purkinje fibres. *J. Physiol. (London)* **329**:569–587.
100. Kass, R. S., T. Scheuer, and K. J. Malloy. 1982. Block of outward current in cardiac Purkinje fibers by injection of quaternary ammonium ions. *J. Gen. Physiol.* **79**:1041–1063.
101. Vereecke, J., G. Isenberg, and E. Carmeliet. 1980. K efflux through inward rectifying K channels in voltage clamped Purkinje fibers. *Pfluegers Arch.* **384**:207–217.
102. Wier, W. G. 1980. Calcium transients during excitation–contraction coupling in mammalian heart: Aequorin signals of canine Purkinje fibers. *Science* **207**:1085–1087.
103. Noma, A. 1983. ATP-regulated K^+ channels in cardiac muscle. *Nature (London)* **305**:147–148.
104. Yanagihira, K., and H. Irisawa. 1980. Inward current activated during hyperpolarization in the rabbit sinoatrial node cell. *Pfluegers Arch.* **385**:11–19.
105. DiFrancesco, D. 1981. A new interpretation of the pace-maker current in calf Purkinje fibres. *J. Physiol. (London)* **314**:359–376.
106. Maylie, J., M. Morad, and J. Weiss. 1981. A study of pace-maker potential in rabbit sino-atrial node: Measurement of potassium activity under voltage clamp condition. *J. Physiol. (London)* **311**:161–178.
107. Kokubun, S., M. Nishimura, A. Noma, and H. Irisawa. 1982. Membrane currents in the rabbit atrioventricular node cell. *Pfluegers Arch.* **393**:15–22.
108. Cohen, I. S., R. T. Falk, and N. K. Mulrine. 1983. Actions of barium and rubidium on membrane currents in canine Purkinje fibers. *J. Physiol. (London)* **338**:589–612.
109a. DiFrancesco, D. 1981. A study of the ionic nature of the pacemaker current in calf Purkinje fibres. *J. Physiol. (London)* **314**:377–393.
109b. Noble, D. 1984. The surprising heart. A review of recent progress in cardiac electrophysiology. *J. Physiol. (London)* **353**:1–50.
110. Noma, A., H. Kotake, and H. Irisawa. 1980. Slow inward current and its role mediating the chronotropic effect of epinephrine in the rabbit sinoatrial node. *Pfluegers Arch.* **388**:1–9.
111. Brown, H. F. 1982. Electrophysiology of the sinoatrial node. *Physiol. Rev.* **62**:505–530.
112. Yanagihira, K., and H. Irisawa. 1980. Potassium current during the pacemaker depolarization in rabbit sino-atrial node cells. *Pfluegers Arch.* **388**:255–260.
113. Noma, A., M. Morad, and H. Irisawa. 1983. Does the "pacemaker current" generate the diastolic depolarization in the rabbit SA node cells? *Pfluegers Arch.* **397**:190–194.
114. Ferrier, G. R. 1977. Digitalis arrhythmias: Role of oscillatory after-potentials. *Prog. Cardiovasc. Dis.* **19**:459–474.
115. Lederer, W. J., and R. W. Tsien. 1976. Transient inward current underlying arrhythmogenic effects of cardiotonic steroids in Purkinje fibres. *J. Physiol. (London)* **263**:73–100.
116. Karagueuzian, H. S., and B. G. Katzung. 1982. Voltage-clamp studies of transient inward current and mechanical oscillations induced by ouabain in ferret papillary muscle. *J. Physiol. (London)* **327**:255–571.
117. Matsuda, H., A. Noma, Y. Kurachi, and H. Irisawa. 1982. Transient depolarization and spontaneous voltage fluctuations in isolated single cells from guinea pig ventricles: Calcium-mediated membrane potential fluctuations. *Circ. Res.* **51**:142–151.
118. Kass, R. S., W. J. Lederer, R. W. Tsien, and R. Weingart. 1978. Role of calcium ions in transient inward currents and aftercontractions induced by strophanthidin in cardiac Purkinje fibres. *J. Physiol. (London)* **281**:187–208.
119. Kass, R. S., R. W. Tsien, and R. Weingart. 1978. Ionic basis of transient inward current induced by strophanthidin in cardiac Purkinje fibres. *J. Physiol. (London)* **281**:209–226.
120. Colquhoun, D., E. Neher, H. Reuter, and C. F. Stevens, 1981. Inward current channels activated by intracellular Ca in cultured cardiac cells. *Nature (London)* **294**:752–754.
121. Eisner, D., W. J. Lederer, and S.-S. Sheu. 1983. The role of intracellular sodium activity in the anti-arrhythmic action of local anaesthetics in sheep Purkinje fibres. *J. Physiol. (London)* **340**:239–257.
122. Wier, W. G., and P. Hess. 1984. Excitation–contraction coupling in cardiac Purkinje fibers: Effects of cardiotonic steroids on the intracellular $[Ca^{2+}]$ transient, membrane potential, and contraction. *J. Gen. Physiol.* **83**:395–415.
123. Orchard, C. H., D. A. Eisner, and D. G. Allen. 1983. Oscillations of intracellular Ca^{2+} in mammalian cardiac muscle. *Nature (London)* **304**:735–738.
124. Kehoe, J., and A. Marty. 1980. Certain slow post-synaptic responses: Their properties and possible underlying mechanisms. *Annu. Rev. Biophys. Bioeng.* **9**:437–465.
125. Hartzell, H. C. 1981. Mechanisms of slow post-synaptic potentials. *Nature (London)* **291**:539–544.
126. Reuter, H. 1983. Calcium channel modulation by neurotransmitters, enzymes and drugs. *Nature (London)* **301**:569–574.
127. Siegelbaum, S. A., and R. W. Tsien. 1984. Modulation of gated ion channels as a mode of transmitter action. *Trends Neurosci.* **6**:307–313.
128. Tsien, R. W. 1977. Cyclic AMP and contractile activity in the heart. *Adv. Cyclic Nucleotide Res.* **8**:363–420.
129. Katz, A. M. 1983. Cyclic adenosine monophosphate effects on the myocardium: A man who blows hot and cold with one breath. *J. Am. Coll. Cardiol.* **2**:143–149.
130. Costa, M. R. C., J. E. Casnellie, and W. A. Catterall. 1982. Selective phosphorylation of the α subunit of the sodium channel by cAMP-dependent protein kinase. *J. Biol. Chem.* **257**:7918–7921.
131. Vassort, G., O. Rougier, D. Garnier, M. P. Sauviat, E. Coraboeuf, and Y. M. Gargouil. 1969. Effects of adrenaline on membrane inward currents during the cardiac action potential. *Pfluegers Arch.* **309**:70–81.
132. Reuter, H., and H. Scholz. 1977. The regulation of the Ca conductance of cardiac muscle by adrenaline. *J. Physiol. (London)* **264**:49–62.
133. Bean, B. P., M. C. Nowycky, and R. W. Tsien. 1984. β-Adrenergic modulation of calcium channels in frog ventricular heart cells. *Nature (London)* **307**:371–375.
134. Tsien, R. W. 1973. Adrenaline-like effects of intracellular iontophoresis of cyclic AMP in cardiac Purkinje fibers. *Nature New Biol.* **245**:120–122.

135. Yamasaki, Y., M. Fujiwara, and N. Toda. 1974. Effects of intracellularly applied cyclic 3′,5′-adenosine monophosphate and dibutyryl cyclic 3′,5′-adenosine monophosphate on the electrical activity of sinoatrial nodal cells of the rabbit. *J. Pharmacol. Exp. Ther.* **190**:15–20.
136. Vogel, S., and N. Sperelakis. 1981. Induction of slow action potentials by microiontophoresis of cyclic AMP into heart cells. *J. Mol. Cell. Cardiol.* **13**:51–64.
137. Tsien, R. W., W. R. Giles, and P. Greengard. 1972. Cyclic AMP mediates the action of adrenaline on the action potential plateau of cardiac Purkinje fibres. *Nature New Biol.* **240**:181–183.
138. Reuter, H. 1974. Localization of beta adrenergic receptors, and effects of noradrenaline and cyclic nucleotides on action potentials, ionic currents and tension in mammalian cardiac muscle. *J. Physiol. (London)* **242**:429–451.
139. Li, T., and N. Sperelakis. 1983. Stimulation of slow action potentials in guinea pig papillary muscle cells by intracellular injection of cAMP, Gpp(NH)p, and cholera toxin. *Circ. Res.* **52**:111–117.
140. Osterrieder, W., G. Brum, J. Hescheler, W. Trautwein, V. Flockerzi, and F. Hofmann. 1982. Injection of subunits of cyclic AMP-dependent protein kinase into cardiac myocytes modulates Ca^{2+} current. *Nature (London)* **298**:576–578.
141. Cachelin, A. B., J. E. dePeyer, S. Kokubun, and H. Reuter. 1983. Calcium channel modulation by 8-bromo-cyclic AMP in cultured heart cells. *Nature (London)* **304**:462–464.
142. Brum, G., W. Osterrieder, and W. Trautwein. 1984. β-Adrenergic increase in the calcium conductance of cardiac myocytes studied with the patch clamp. *Pfluegers Arch.* **401**:111–118.
143. Otsuka, M. 1958. Die Wirkung von Adrenalin auf Purkinje-Fasern von Säugetieren. *Pfluegers Arch. Gesamte Physiol.* **266**:512–517.
144. Hauswirth, O., D. Noble, and R. W. Tsien. 1968. Adrenaline: Mechanism of action of the pacemaker potential in cardiac Purkinje fibres. *Science* **162**:916–917.
145. Tsien, R. W. 1974. Effects of epinephrine on the pacemaker potassium current of cardiac Purkinje fibers. *J. Gen. Physiol.* **64**:293–319.
146. Brown, H., D. DiFrancesco, and S. J. Noble. 1979. How does adrenaline accelerate the heart? *Nature (London)* **280**:235–236.
147. Tsien, R. W. 1974. Mode of action of chronotropic agents in cardiac Purkinje fibers: Does epinephrine act by directly modifying the external surface charge? *J. Gen. Physiol.* **64**:320–342.
148. Brown, H. F., P. A. McNaughton, D. Noble, and S. J. Noble. 1975. Adrenergic control of cardiac pacemaker currents. *Philos. Trans. R. Soc. London Ser. B* **270**:527–537.
149. Pappano, A. J., and E. Carmeliet. 1979. Epinephrine and the pacemaker mechanism at plateau potentials in sheep cardiac Purkinje fibers. *Pfluegers Arch.* **382**:17–26.
150. Kass, R. S., and S. E. Wiegers. 1982. The ionic basis of concentration-related effects of noradrenaline on the action potential of calf cardiac Purkinje fibres. *J. Physiol. (London)* **322**:541–558.
151. Scholz, H., and H. Reuter. 1976. Effect of theophylline on membrane currents in mammalian cardiac muscle. *Naunyn-Schmiedebergs Arch. Exp. Pathol. Pharmakol.* **293**:R19.
152. Gadsby, D. 1984. β-adrenoceptor agonists increase membrane K^+ conductance in cardiac Purkinje fibres. *Nature (London)* **306**:691–693.
153. Hutter, O. F. 1957. Mode of action of autonomic transmitters on the heart. *Br. Med. Bull.* **13**:176–180.
154. Trautwein, W., and J. Dudel. 1958. Zum Mechanismus der Membranwirkung des Acetylcholins an der Herzmuskelfaser. *Pfluegers Arch.* **266**:324–334.
155. Woodbury, W. W., and W. E. Crill. 1961. On the problems of impulse conduction in the atrium. In: *Nervous Inhibition*. E. Florey, ed. Pergamon Press, Elmsford, N.Y. pp. 124–134.
156. Giles, W., and S. J. Noble. 1976. Changes in membrane current in bullfrog atrium produced by acetylcholine. *J. Physiol. (London)* **261**:103–123.
157. Garnier, D., J. Nargeot, C. Ojeda, and O. Rougier. 1978. The action of acetylcholine on background conductance in frog atrial trabeculae. *J. Physiol. (London)* **274**:381–396.
158. Ojeda, C., O. Rougier, and Y. Tourneur. 1981. Effects of Cs on acetylcholine induced current: Is i_{K_1} increased by acetylcholine in frog atrium? *Pfluegers Arch.* **391**:57–59.
159. Sakmann, B., A. Noma, and W. Trautwein. 1983. Acetylcholine activation of single muscarinic K^+ channels in isolated pacemaker cells of the mammalian heart. *Nature (London)* **303**:250–253.
160. Mirro, M. J., J. C. Bailey, and A. M. Watanabe. 1979. Dissociation between the electrophysiological properties and total tissue cyclic guanosine monophosphate content of guinea pig atria. *Circ. Res.* **45**:225–233.
161. Fleming, B. P., W. Giles, and W. J. Lederer. 1981. Are acetylcholine-induced increases in ^{42}K efflux mediated by intracellular cyclic GMP in turtle cardiac pacemaker tissue? *J. Physiol. (London)* **314**:47–64.
162. Trautwein, W., J. Taniguchi, and A. Noma. 1982. The effect of intracellular cyclic nucleotides and calcium and the action potential and acetylcholine response of isolated cardiac cells. *Pfluegers Arch.* **392**:307–314.
163. Soejima, M., and A. Noma. 1983. The K channel coupled with the muscarinic ACh receptor in the heart muscle. *Proc. Int. Union Physiol. Sci.* **15**:51.
164. Giles, W., and R. W. Tsien. 1975. Effects of acetylcholine on membrane currents in frog atrial muscle. *J. Physiol. (London)* **246**:64P–66P.
165. Ikemoto, Y., and M. Goto. 1975. Nature of the negative inotropic effect of acetylcholine on the myocardium. *Proc. Jpn. Acad.* **51**:501–505.
166. Levy, M. N. 1977. Parasympathetic control of the heart. In: *Neural Regulation of the heart*. W. C. Randall, ed. Oxford University Press, London. pp. 95–129.
167. Giles, W., and S. J. Noble. 1976. Changes in membrane currents in bullfrog atrium produced by acetylcholine. *J. Physiol. (London)* **261**:103–123.
168. Ten Eick, R., H. Nawrath, T. F. McDonald, and W. Trautwein. 1976. On the mechanism of the negative inotropic effect of acetylcholine. *Pfluegers Arch.* **361**:207–213.
169. Hino, N., and R. Ochi. 1980. Effect of acetylcholine on membrane currents in guinea-pig papillary muscle. *J. Physiol. (London)* **307**:183–197.
170. Biegon, R. L., and A. J. Pappano. 1980. Dual mechanism for inhibition of calcium-dependent action potentials by acetylcholine in avian ventricular muscle. *Circ. Res.* **46**:353–362.
171. Josephson, I., and N. Sperelakis. 1982. On the ionic mechanism underlying adrenergic–cholinergic antagonism in ventricular muscle. *J. Gen. Physiol.* **79**:69–86.
172. Garnier, D., J. Nargeot, C. Ojeda, and O. Rougier. 1978. Action of carbachol on atrial fibres: Induced extra current and slow inward current inhibition. *J. Physiol. (London)* **276**:27P–28P.
173. Inoue, D., M. Hachisu, and A. J. Pappano. 1983. Acetylcholine increases resting membrane potassium conductance in atrial but not in ventricular muscle during muscarinic inhibition of Ca^{++}-dependent action potentials in chick heart. *Circ. Res.* **53**:158–167.
174. Inui, J., and H. Imamura. 1977. Effects of acetylcholine on calcium-dependent electrical and mechanical responses in the guinea-pig papillary muscle partially depolarized by potassium. *Naunyn-Schmiedeberg's Arch. Pharmacol.* **299**:1–7.
175. Biegon, R. L., P. M. Epstein, and A. J. Pappano. 1980. Muscarinic antagonism of the effects of a phosphodiesterase inhibitor (methylisobutylxanthine) in embryonic chick ventricle. *J. Pharmacol. Exp. Ther.* **215**:348–356.
176. Linden, J., S. Vogel, and N. Sperelakis. 1982. Sensitivity of Ca-dependent slow action potentials to methacholine is induced by phosphodiesterase inhibitors in embryonic chick ventricles. *J. Pharmacol. Exp. Ther.* **222**:383–387.
177. Nawrath, H. 1977. Does cyclic GMP mediate the negative inotropic effect of acetylcholine in the heart? *Nature (London)* **267**:72–74.
178. Kohlhardt, M., and K. Haap. 1978. 8-Bromo-guanosine-3′,5′-monophosphate mimics the effect of acetylcholine on slow response action potential and contractile force in mammalian atrial myocardium. *J. Mol. Cell. Cardiol.* **10**:573–586.

179. Ikemoto, Y., and M. Goto. 1976. Effects of acetylcholine and cyclic nucleotides on the bullfrog atrial muscle. *Recent Adv. Stud. Card. Struct. Metab.* **11**:57–61.
180. Watanabe, A. M., M. M. McConnaughey, R. A. Strawbridge, J. W. Fleming, and H. R. Besch. 1978. Muscarinic cholinergic receptor antagonism of β-adrenergic receptor affinity for catecholamines. *J. Biol. Chem.* **253**:4833–4836.
181. Weidmann, S. 1966. Cardiac electrophysiology in the light of recent morphological findings. *Harvey Lect.* **61**:1–15.
182. McNutt, N. S., and D. W. Fawcett. 1974. Myocardial ultrastructure. In: *The Mammalian Myocardium.* G. A. Langer and A. J. Brady, eds. Wiley, New York.
183. Netter, F. H. 1969. *The CIBA Collection of Medical Illustrations,* Volume 5. Elsevier, Amsterdam.
184. Weidmann, S. 1957. Resting and action potentials of cardiac muscle. *Ann. N.Y. Acad. Sci.* **65**:663–678.
185. Dudel, J., K. Peper, R. Rudel, and W. Trautwein. 1967. The effect of tetrodotoxin on the membrane current in cardiac muscle (Purkinje fibers). *Pfluegers Arch.* **295**:213–226.
186. Kass, R. S., and R. W. Tsien. 1975. Multiple effects of calcium antagonists in cardiac Purkinje fibers. *J. Gen. Physiol.* **66**:169–192.
187. Hutter, O. F., and W. Trautwein. 1956. Vagal and sympathetic effects on the pacemaker fibres in the sinus venosus of the heart. *J. Gen. Physiol.* **39**:715–733.
188. Toda, N., and T. C. West. 1967. Interactions of K, Na, and vagal stimulation in the S-A node of the rabbit. *Am. J. Physiol.* **212**:416–423.

CHAPTER 7

Ion Transport through Ligand-Gated Channels

Richard W. Aldrich, Vincent E. Dionne, Edward Hawrot, and Charles F. Stevens

1. Introduction and Overview

1.1. What Are Channels?

Channels are integral membrane proteins that permit charged species to traverse the lipid bilayers of cells by providing a water-filled pathway through which ions may pass. All cell types face the problem of how to regulate the transport of ions across the inhospitable hydrophobic environment of the cell membrane, and they all solve this problem in part by employing channels of various sorts. Channels generally have several characteristics:

1. Ion flow through channels is a passive—though possibly complex—process that derives the energy for ion flux solely from concentration gradients of the ion species moving through the channel. Ion motion is much like that in free diffusion but may be complicated by ion–ion and ion–protein interactions.
2. Channels are selective in that only certain ion species are permitted to pass through. Different types of channels exist, and each channel type generally has its characteristic selectivity. For example, one channel species might allow a Na^+ flux and exclude K^+ ions whereas another sort of channel would accept K^+ ions and not Na^+ ions.
3. Ion flux is regulated by the channel. A usual form of regulation is termed *gating,* in which case a pore through which ions can pass is opened and closed by conformational changes in the channel protein. Another form of regulation occurs when a cell controls the number and distribution of channels by metabolic or other means.

Richard W. Aldrich and Charles F. Stevens • Section of Molecular Neurobiology, Yale University School of Medicine, New Haven, Connecticut 06510. **Vincent E. Dionne** • Division of Pharmacology, Department of Medicine, University of California at San Diego, La Jolla, California 92093. **Edward Hawrot** • Department of Pharmacology, Yale University School of Medicine, New Haven, Connecticut 06510.

All channels that have been investigated so far are gated; that is, they have the capacity to regulate transmembrane ion flux by means of conformational changes that, in effect, open and close a gate that guards the pathway ions must use in crossing. Two modes of gating are known.

In *voltage-gated* channels the channel protein senses the transmembrane electric field, and the magnitude and direction of the field determine the conformation the protein will tend to adopt. That is, changes in the cell's membrane potential open or close the channel. The Na^+ channel that is responsible for nerve impulses is voltage gated.

Ligand-gated channels have their conformation regulated by the binding of some ligand. Binding a particular molecule at a special site will, for example, cause the channel to open. The acetylcholine receptor (AChR) channel that is responsible for one step in neuromuscular transmission is an example of a ligand-gated channel.

Each species of channel has its own particular mode of gating. Within the class of voltage-gated channels, then, every channel species has its distinctive gating characteristics like dwell time in the open state, degree of voltage sensitivity, and characteristic voltage at which half of the channels in a population would be open. For the ligand-gated channels, each species has its own ligand type as well as concentration sensitivity and dwell time in the open state. Altogether, a large number of different types of channels—perhaps four dozen—can be distinguished by their functional characteristics.

The categories of ligand- and voltage-gated channels are not mutually exclusive. These two means of achieving gating conformational changes can and do exist in the same channel protein. A good example of the coexistence is the *Ca^{2+}-activated K^+ channel* found in many different cell types. This channel can be caused to open either by binding Ca^{2+} at sites on the intracellular surface of the protein or by changes in the membrane potential. Although several examples of dual modes of gating are known, the general rule for most channel types is that one or the other mode is used, and the other mode is absent or physiologically insignificant.

1.2. Chapter Organization

This chapter deals with the ligand-gated channel whose gating conformational state is regulated by binding neurotransmitter molecules at the extracellular surface of the channel protein. The major part of our treatment will focus on the AChR channel. We concentrate on this channel type because it is the best understood integral membrane protein involved in transport. In the sections that follow the introductory overview we consider the structure, biosynthesis and regulation, ligand binding and gating, and selectivity transport characteristics of this interesting protein. In the final section we shall survey briefly what is known about other types of ligand-gated channels.

1.3. Definition of the AChR

The AChR channel was initially defined by the binding of ACh and subsequent transport events that lead finally, at the neuromuscular junction, to muscle contraction. Biochemical and other studies have depended, however, on an alternative identification of the protein by a class of elapid snake neurotoxins. These are polypeptide toxins with molecular weights near 7,000 that bind to the AChR with nanomolar dissociation constants and thus can be used for histological and biochemical identification of the receptor. The most widely used toxins in this class are α-bungarotoxin and naja toxin.

1.4. The AChR Is a Pentameric Integral Membrane Protein

Using snake neurotoxins as tags, biochemists have purified a multisubunit integral membrane protein that has been firmly identified as the AChR. This receptor is, as will be described more fully in Section 2, a pentameric protein composed of four distinct polypeptide chains, designated α, β, γ, and δ, that occur in the pentamer with the stoichiometry $\alpha_2\beta\gamma\delta$; the apparent molecular weights, as determined by gel electrophoresis, of the subunits are 40, 50, 60, and 65Kd. The primary amino acid sequence for the first 60 or so residues of each subunit has been determined directly from the peptide chains. This sequence information has been important for two reasons: First, strong homologies between the sequences of all four subunits indicate a close evolutionary relationship between them and may well provide a clue to the general scheme about how the complex proteins have evolved. Second, knowing a part of the amino acid sequence has been the key for applying molecular biological techniques to be described.

Two of the subunits have been assigned functions (see Section 2.3): α chains carry the ACh-binding sites and are thus responsible for the ligand-binding specificity that characterizes this receptor. The δ chain serves in some species to link together pairs of receptors through disulfide bonds. This dimer formation is not required for normal receptor operation and presumably serves as a means of increasing receptor concentration in membrane regions of the cell where an especially high density of AChRs is needed. Specific functions have not yet been assigned to other subunits but each has distinct characteristics. For example, both the γ and the δ subunit have specific sites for phosphorylation by a cAMP-dependent protein kinase, although the role of phosphorylation in receptor function is not yet understood.

Antibodies are available that bind specifically to each of the AChR subunits, and even to specific regions of the various chains (see Section 3). Studies using these antibodies have shown that all subunits traverse the membrane and thus have three classes of domains: cytoplasmic, intramembranous, and extracellular. The current view, then, is that the pentamer is formed by five subunits arranged like staves of a barrel, with each stave traversing the membrane; looking onto the face of the membrane, one would thus see a five-petal rosette—each petal is a different subunit. This model is a pseudosymmetric one based on the similarity of the individual subunits—that is, on the strong homologies revealed by amino acid sequence and other common structural features described below.

1.5. AChR Size and Shape

Negative-stained preparations of membrane especially rich in AChRs reveal proteins with characteristics much like those just described and various types of experiments have proved these to be the ACh receptors. Viewed face on, the receptor appears, as described in Section 2.2, as a rather ill-defined rosette with a diameter of about 0.85 nm and a central pit that is believed to represent a water-filled pore. Using the center as a point of reference, microscopists have attempted to locate the position of the various subunits by measuring the angle between labels for specific binding sites. The only definitive evidence so far, from three laboratories, is that the α subunits are not adjacent but rather lie on either side of another, as yet unidentified, subunit. Ambiguities in the various data may result from technical difficulties in applying rather tricky methods for electron microscopic localization at the limit of their resolution; alternatively, lack of clear results may indicate that the arrangement of subunits is not strictly fixed, and that some or all of them may be able to occupy different locations within the pentameric complex.

A combination of evidence—from electron microscopic images, X-ray diffraction of two-dimensional arrays, and neutron diffraction—has led to a proposed three-dimensional structure for the receptor. In this model, the bulk of the receptor is located on the extracellular face of the membrane and protrudes about 0.5 nm from the membrane surface. A 0.2-nm-wide funnel-like vestibule in the center of this extramembranous region leads to a narrow pore that passes through the membrane; this pore is formed by hydrophilic surfaces of helices contributed by each of the subunits, and additional transmembrane helices surround the five chains that form the pore. A small cytoplasmic region of the protein protrudes about 0.15 nm from the inner face of the membrane and contains the channel orifice.

1.6. Amino Acid Sequence Suggests a Three-Dimensional Structure

cDNA that codes for each of the four subunit types has been cloned and sequenced so that the complete primary structure of the AChR is known; this information is detailed in Section 2. Each subunit type is coded for by a separate gene, and each has a 5′ hydrophobic signal sequence that is characteristic of integral membrane proteins. Although every subunit type has a distinct sequence, the sequences are about 40% homologous overall, and some regions have 80% homology. Furthermore, all of the subunit types share the same general design: About a half of the protein, starting from the NH_2-terminus, is quite hydrophilic and presumably constitutes a major portion of the extracellular domain of the receptor. Four very hydrophobic regions of sufficient length to span the membrane are readily identified and are believed to form four transmembrane helices. Between the third and fourth membrane-spanning regions is a hydrophilic stretch in which can be found an amino acid sequence where residues with hydrophobic side chains are interspersed with hydrophilic amino acids in such a way that a membrane-spanning helix formed in this region would be amphipathic, i.e., one side of the helix would have charged and hydrophilic residues and the other sides would be strongly hydrophobic. The amphipathic helix contributed by each subunit are visualized as forming a charge-lined pore through which ions may pass. This pore would, because of the charged side chains present, be water filled. Furthermore, the channel formed would be stable because a great deal of energy would be required to remove one of the wall-side helices from its position in contact with water. A pore formed in this way would have a diameter of about 0.6–0.7 nm.

1.7. ACh Binding Is Responsible for Gating

ACh binds to sites on the α subunits with a dissociation constant of about 20 μM (see Section 5 on the dose–response relationship). When the sites on both α subunits are occupied, a conformational change is induced that opens the channel for ion flux (see Section 6 on gating kinetics). The channel remains in its open conformation for random lengths of time with a mean of about 1 msec. While the channel is open, a pore with a conductance of about 25 pS is available for small cations to pass through.

1.8. Biosynthesis and Regulation

Like other membrane proteins, the AChR subunits are, as discussed in Section 4, synthesized in rough endoplasmic reticulum, pass through the Golgi apparatus, and are inserted into surface membrane. Metabolism of the AChR is regulated during development and by receptor use and other factors. Because the AChR is multimeric, its regulated synthesis, assembly, and insertion into surface membrane, and degradation are all of considerable cell biological interest as a model for other complex integral membrane proteins.

1.9. Only Cations Pass through the AChR Pore

The AChR channel is highly selective for cations over anions (selectivity and ion permeation are treated in Section 7), but discriminates only moderately among cations whose diameter is less than about 0.65 nm. Because ions of 0.65 nm will pass through the channel, its minimum dimension must be at least that large; because the size cutoff for ion permeation is quite sharp, the channel probably is not much larger than that. This pore size inferred from ion permeation studies is consistent with that proposed from the structural model discussed above. Although all cations with diameters up to about 0.65 nm can pass through the channel, it does show some moderate discrimination between the various ionic species. Among the monovalent alkali metal cations, the larger ones are more permeant than smaller ones: Cs^+ (crystal radius 0.17 nm) is the most permeant, Na^+ (0.099 nm) is intermediate, and Li^+ (0.065 nm) is least permeant. For the divalent ions in the same size range, the alkaline earths, the reverse selectivity sequence is found; larger ions are less permeant than smaller ones. For example, Mg^{2+} (crystal radius 0.067 nm) is more permeant than Ca^{2+} (0.98 nm) which is in turn more permeant than Ba^{2+} (1.44 nm).

2. Structure of the Nicotinic AChR

Within the last several years the convergence of a variety of biochemical, biophysical, and morphometric techniques has combined to give a remarkably detailed representation of the three-dimensional structure of the AChR. This information, together with that being obtained from ongoing efforts to dissect further the structural features of the individual subunits, provides a strong basis for attempts to correlate the functional aspects of the AChR with its structural domains. Added to this already extensive amount of data on the tertiary structure of the AChR, recombinant DNA techniques have recently provided us with the amino acid sequence for each of the four subunits. It is clear then that the AChR holds the greatest promise for a complete description of structure–function relationships at the molecular level.

The high abundance of AChR in the electric organ tissue of various species of fish such as *Torpedo* has contributed greatly to the advances in structural information. The receptor from *Torpedo* and *Electrophorus* electric organ and from *Electrophorus* muscle has been purified and extensively characterized (for recent reviews see Refs. 1–3).

2.1. Subunit Characteristics

The four different polypeptide chains (α, β, γ, δ) that comprise the *Torpedo* AChR are believed to be typical of the nicotinic AChR found in higher vertebrate muscle. Although the apparent molecular weights from SDS gel electrophoresis of the subunits are 40, 50, 60, and 65K, respectively, these polypeptides appear to migrate anomalously under denaturing conditions since the molecular weights calculated from the primary amino acid sequences differ significantly from the apparent values. The true protein molecular weights are 50K for α, 54K for β, 56K for γ, and 58K for δ.[4–9] Biochemical analyses indicate that each subunit is glycosylated,[10,11] suggesting that the lack of correlation between the calculated and the observed molecular weights may be in part due to the presence of the carbohydrate additions. The $\alpha_2\beta\gamma\delta$ stoichiometry is consistent with the observed molecular weight of the monomeric AChR solubilized with nondenaturing detergents. Sedimentation equilibrium analyses indicate a molecular weight of about 250K for the AChR complex.[12] The monomeric form of the AChR contains two sites at which ACh competes for binding with the neurotoxins α-bungarotoxin and α-cobratoxin. The monomeric form of the AChR also contains the ion channel that is opened upon agonist binding. The evidence for this comes from the ability to reconstitute the purified AChR monomeric complex into artificial membranes and to elicit functional agonist-induced cation flux in a system where the only proteins present are the four subunits of the AChR.[13,14]

Although the four polypeptide chains, upon partial proteolytic cleavage, produce entirely distinct peptide maps,[10,15] comparison of the primary amino acid sequences reveals, as mentioned in the Introduction, a high degree of homology between the subunits.[6] There is approximately 40–50% homology in the amino acid sequence between any two of the AChR subunits, suggesting that the four subunits have been derived from a common ancestral gene.[16] Antisera raised against denatured, isolated subunits can be subunit-specific, indicating that the subunits are immunologically distinct. Some antisera, however, are capable of recognizing the analogous polypeptide chains across species lines,[17] supporting the view that the structural features of the AChR have been highly conserved during evolution. Analysis of the subunit specificity of monoclonal antibodies raised against AChR has identified some monoclonal antibodies that recognize determinants present on more than one subunit.[18] Presumably, these shared determinants represent regions of high sequence homology.

2.2. Subunit Organization

Information on the structural configuration of the AChR within the membrane has been obtained from X-ray diffraction analysis and from electron micrographs of receptor-rich membranes.[19,20] The rosette-like particles abundant in negatively stained *Torpedo* membranes correspond to the monomeric, 250K AChR and have a diameter of 8.5 nm.[21–24] The central pit is 2–3 nm in diameter and is believed to correspond to the pore through which ions may pass. Kistler and colleagues have suggested that the uranyl ions of the negative stain are filling a central channel 7.2 Å in diameter but which extends the entire length of the AChR, 11 nm,[24] thus giving rise to the observed image in electron micrographs. An analysis of the membrane density profile in receptor-rich membranes demonstrates that the AChR traverses the membrane, extending 5.5 nm above the bilayer on the synaptic side and 1.5 nm below the lipid bilayer on the cytoplasmic side. Adding in the 4-nm thickness of the bilayer region produces the overall length of the AChR as 11 nm.[23,25] By averaging images from electron micrographs of two-dimensional arrays of AChR in receptor-rich membranes, Kistler *et al.* [24] have further suggested that the AChR is funnel-shaped with a wide vestibule on the extracellular face narrowing down to the 7.2-Å pore through the membrane. Thus, the biophysical studies indicate that a large part of the mass of the receptor resides on the extracellular side of the membrane. In addition, X-ray diffraction studies have suggested that per monomeric AChR, 12–25 α helices are oriented perpendicular to the membrane.[20,25] Since these α-helical regions extend up to 0.8 nm in length, the subunits themselves seem to be situated perpendicular to the membrane bilayer.

In large part, the available biochemical evidence concerning the structure of the AChR is consistent with the biophysical data. Labeling studies with *Torpedo* membrane vesicles reveal that all four subunits are exposed on the extracellular face of the membrane.[26] Furthermore, studies using proteolytic digestion of accessible regions of the AChR have suggested that all four subunits are also exposed on the cytoplasmic face of the membrane.[27–29] In addition, newly synthesized subunits are oriented in a similar transmembrane fashion prior to assembly.[30,31] The amino acid sequence derived from cDNA analysis is also consistent with a transmembrane orientation for each of the four subunits. Each of the latter contains a hydrophobic signal peptide sequence that is not found in the mature, processed subunit. The four highly hydrophobic regions in each of the four subunits are good candidates for transmembrane domains of the protein.[6–9,32]

The initial models based on the deduced amino acid sequences place the NH_2-terminus of each subunit on the extracellular side of the membrane. This assignment is consistent with the known orientation of a number of other membrane proteins whose signal peptides are cleaved by signal peptidase localized on the luminal face of the endoplasmic reticulum. The next 251 amino acids are hypothesized to be on the extracellular surface and within this region potential glycosylation sites have been identified. Asparagine-linked glycosylation sites contain the sequence Asn-X-Ser/Thr although not all such sites are necessarily glycosylated *in vivo*. The NH_2-terminal extracellular region of the α subunit also contains the two cysteine residues believed to form a disulfide cross-bridge in the mature AChR in the region of the ACh-binding site. Amino acids 253 to 279 form the first highly hydrophobic domain followed closely by two other hydrophobic domains of similar size. The fourth hydrophobic domain extends from amino acid 505 to 524. The region between the third and fourth hydrophobic domains comprises about one-fourth of the mass of the AChR and if, as hypothesized, the hydrophobic sequences alternate in traversing the membrane as α helices, this region would be on the cytoplasmic face of the membrane. If there are only four transmembrane runs, then this model predicts that the COOH-terminus of each subunit resides on the extracellular surface.

Finer-Moore and Stroud have suggested that the proposed cytoplasmic region of each subunit (residues 345 to 504) contains at least one amphipathic region containing alternate hydrophilic and hydrophobic residues which could form yet another transmembrane helical domain.[20] They have further suggested that if each subunit contributed one such domain, then a central ion channel could be formed in which the hydrophilic residues from each of the five encircling domains would be facing the channel core. The outward-facing hydrophobic residues would thus serve to stabilize the transmembrane channel. This model therefore predicts that all five subunits contribute to the formation of a centrally located hydrophilic and thus water-containing pore. Because of the additional transmembrane traverse, this model places the COOH-terminus on the cytoplasmic face of the membrane bilayer. The localization of the COOH-termini of the subunits should be experimentally verifiable and would thus distinguish between models with even or odd numbers of transmembrane regions.

Information is now also becoming available as to the arrangement of the subunits around the central pore. Data obtained so far have been interpreted with the assumption that the arrangement is invariant from one AChR copy to the next. The α subunits have been tagged for electron microscopic examination of receptor complexes by the use of biotinylated derivatives of α-bungarotoxin.[33,34] The biotinylated toxin is nearly as active as native toxin and when bound to receptor will bind avidin. Due to its large size, avidin is clearly visible in the negatively stained complex of receptor, toxin, and avidin. The average angle between two avidins bound to the same monomer of receptor is 110°, suggesting that another subunit lies between the two αs. *Torpedo* AChR monomers are linked to form dimers via a δ–δ disulfide linkage[35] (for review see Ref. 1). In addition, the β subunits of two receptor monomers can be chemically joined *in vitro* via a disulfide bridge.[36] The localization of avidin–biotinylated toxin on such cross-linked dimers has led to the conclusion that the γ subunit is most likely to be localized between the two αs.[34] Stroud and co-workers measured the angle between αs using Fab fragments obtained from monoclonal antibodies with anti-α specificity. They obtained an angle of 144°, thus also concluding that there is one subunit between the two αs.[37] An alternative interpretation of the chemical cross-linking data has tentatively suggested that the β subunit lies between the two αs.[24] Because the different subunits have such similar overall structures, some variability of subunit arrangement may occur. This possibility has not yet been investigated.

2.3. Structure–Function Correlations

An important question currently under active investigation is the relation between AChR structure and function. In the resting state of the receptor, the channel is closed and opens only upon agonist binding. With the increased resolution of patch-clamping, it is now evident that the receptor can sometimes rapidly fluctuate on a microsecond time scale between an open configuration of the channel and one or more nonconducting states (see discussion in Section 6). At higher concentrations of agonist or upon prolonged exposure, the receptor undergoes a transition to a "desensitized" form that is functionally inactive. The role of each of the subunits in these various functional states of the AChR is presently unknown.

Affinity labeling agents have, however, identified a high-affinity cholinergic binding site on the α subunit. Karlin and co-workers have shown a readily reducible disulfide resides in close proximity to the ACh-binding site.[38] Upon exposure to mild reducing conditions, one of the sulfhydryls produced from this disulfide can be specifically alkylated with various agents resembling cholinergic ligands.[39] Two such affinity alkylating agents are bromoacetylcholine and 4-(*N*-maleimido)-α-benzyltrimethylammonium iodide (MBTA). By using radiolabeled alkylating agents, Karlin and co-workers were able to show that of the four subunits, only the α chain of *Electrophorus* and *Torpedo* AChR became labeled.[40] Labeling could be blocked by other cholinergic ligands including the snake neurotoxins α-bungarotoxin and α-cobratoxin, indicating that the α subunit contains a high-affinity ACh-binding site. Other studies have shown that there are two ACh-binding sites per monomeric AChR (containing two α subunits per monomer). In addition, the finding that channel opening is greatly enhanced by the binding of two ACh molecules per monomer[e.g., 41] provides strong indirect evidence that the sites identified by affinity alkylation are the physiologically significant ones in terms of inducing channel opening. Cysteine residues 156 and 170 have been proposed to contribute the sensitive disulfide bridge near the ACh-binding site.[6,7] A direct experimental verification of this site should aid in assigning functional activities to the primary amino acid sequences now available. Karlin and co-workers have also shown that the affinity alkylating agents appear to label one of the two sites (αs) more readily than the other.[42] Binding studies with the antagonist *d*-tubocurarine have also suggested a difference in the binding properties of the two ACh-binding sites.[43,44] These differences in binding and reactivity may be a consequence of the asymmetrical orientation of the two α subunits within the receptor complex. The two α chains will each be in contact with two different subunits and this fact may underlie the observed asymmetry in chemical reactivity.[19] Whether this asymmetry has an important functional consequence is presently unclear.

Since the concentration of ACh in the synaptic cleft may transiently approach near millimolar levels, a high-affinity binding site present on the AChR probably does not represent the physiological activating site but rather the site involved in desensitization. Lower-affinity ACh-binding sites on the AChR that may be located on the other subunits[45] could bind ACh in the physiologically effective dose range. It is possible, however, that the binding site responsible for activation and that involved in desensitization may be one and the same and that a difference in affinities may be due to a secondarily induced change in receptor conformation.

Because the curarimimetic neurotoxins compete with the affinity alkylating agents and with other cholinergic ligands for binding to AChR, the toxin-binding site is believed to overlap the ACh-binding site. Furthermore, the two toxin-binding sites per monomeric AChR correspond with the pair of sites revealed with ACh binding. It is presently unclear whether toxin binding to either α subunit or toxin binding to one particular α of the pair (e.g., the one more readily labeled with MBTA) is responsible for functional blockade. This question should be approachable with patch-clamp analysis of single AChR channels. Nevertheless, a localization of the toxin-binding sites may provide further information on the location of the ACh-binding site. Until recently, it was not known whether toxin bound to a site resident on one of the subunits alone or whether the binding site was produced by a combination of domains from several of the subunits. Haggerty and Froehner were able to test the individual subunits isolated by SDS gel electrophoresis for binding of labeled α-bungarotoxin.[46] Only in the case of the α subunit did removal of the denaturing detergent by dialysis lead to partial restoration of bungarotoxin-binding ability, although of reduced affinity (10^{-7} M vs. 10^{-11} M for native receptor complex). It appears, however, that low levels of SDS (0.02%) can improve the recovery of higher-affinity (10^{-9} M) toxin binding.[47] The addition of nondenaturing detergents did not enhance the recovery of high-affinity toxin binding.

Another approach to this question has been the use of electrophoretic transfer techniques analogous to the "Western" blotting approach used in identifying polypeptide chains recognized by specific antibodies. The four subunits of the AChR are readily resolved upon SDS or lithium dodecyl sulfate (LDS) gel electrophoresis. One advantage to LDS is that due to its enhanced solubility, gels can be electrophoresed at low temperatures (e.g., 4°C), perhaps increasing the possibility of subsequent protein renaturation. After being resolved on gels, the polypeptide chains can be electrophoretically transferred, in the orthogonal direction, from the gel matrix onto a physical support such as nitrocellulose or a positively charged membrane (e.g., Zetabind). Various labeled ligands can then be added to detect possible binding to the dissociated polypeptides.

This methodology has been successfully applied to the toxin-binding site as well as to a definition of monoclonal antibody subunit specificity (see below). When *Torpedo* membrane proteins are resolved by denaturing gel electrophoresis, transferred to a support, and probed with [^{125}I]α-bungarotoxin, only the α subunit binds the labeled toxin.[48,49] The binding of toxin can be effectively competed by *d*-tubocurarine but the affinity for toxin is reduced as compared to native receptor ($K_D \sim 10^{-7}$ M). As in the case of native receptor, labeling the cholinergic ligand-binding site with MBTA blocks the binding of labeled toxin to the isolated α subunit on transfers.[48] In addition, by treating the AChR with endoglycosidase H to remove the *N*-linked oligosaccharide of the α chain, it was possible to show that this carbohydrate moiety is not required for toxin binding.

The bungarotoxin blotting approach is now being extended to an analysis of the proteolytic fragments generated from the α subunit after treatment with various proteolytic enzymes.[50] Fragments as small as 8K retain the ability to bind toxin with an affinity comparable to that of the intact α subunit. MBTA treatment blocked toxin binding to all the fragments, indicating that the MBTA site and the toxin-binding site have not yet been separated into two polypeptide fragments. Partial NH_2-terminal sequence analysis of the polypeptide fragments of interest should identify them with regard to the known primary amino acid sequence. A further analysis of toxin-binding fragments, together with a comparison with fragments identified with specific monoclonal antibodies (see below), or with concanavalin A (glycopeptide fragments) thus should generate a physical map of the structural features of the α subunit. Similar analyses could also be performed on the other subunits using subunit-specific probes such as monoclonal antibodies. Some progress along these lines has already been made with regard to the δ and β subunits.[31]

Although the competitive blockers of ACh binding have proved extremely useful in the study of the role of the α subunits, few pharmacological agents are highly specific for any of the other subunits of the AChR. A large number of noncompetitive blockers, many of which exhibit local anesthetic properties, have been studied. In general, most of these agents are amphipathic and their pharmacological actions are not specific for the AChR. Their interaction with the AChR is often very sensitive to the presence of cholinergic agonists or antagonists, suggesting that they may be useful probes of receptor conformation. Some of these inhibitors may bind within the pore itself and others may bind to or induce the conversion of the receptor to the closed or desensitized state. Among the agents examined in this regard are: meproadifen, trimethisoquin, phencyclidine, procainamide, and histrionicotoxin.[2,19]

Various attempts to cross-link the noncompetitive inhibitors to the AChR have been pursued with the objective of identifying the subunits and the molecular domains contributing to the inhibitor-binding site. For example, radioactive, photoactivatable trimethisoquin analogs label the α chain, except in the presence of cholinergic agonists and competitive antagonists where the δ chain becomes predominantly labeled.[51] The labeling with phencyclidine and histrionicotoxin is similar to that with trimethisoquin. The labeling of the α subunit by procainamide derivatives is reduced in the presence of agonist.[52] In contrast, chlorpromazine labels all four subunits and the labeling is increased by the agonist carbachol.[53]

Kaldany and Karlin have pursued a similar approach using a reactive derivative (quinacrine mustard) of the potent local anesthetic quinacrine. In addition to pursuing labeling studies, they also examined the consequences of quinacrine labeling on receptor function. Under mild labeling conditions, channel activity as measured by ion flux could be inhibited with little effect on ACh binding.[54] The functional inhibition was enhanced by agonists or competitive antagonists such as *d*-tubocurarine. The irreversible inhibition produced by the mustard could be retarded by occupation of the local anesthetic site by quinacrine or proadifen. Under these conditions all four subunits are labeled by the radioactive mustard but the labeling of α and β is increased in the presence of cholinergic ligands. These findings are interpreted to indicate that the α and β chains are involved in a functionally detectable local anesthetic-binding site and that the binding of cholinergic ligands to their binding site somehow induces a conformational change that increases the reactivity of α and β toward quinacrine mustard. Since γ and δ chains are also labeled, it is not possible to conclude from this study that these two chains are not involved in the local anesthetic-binding site.

The AChR is phosphorylated *in vivo*.[11] Huganir and Greengard have shown that *Torpedo* membranes contain a cAMP-dependent protein kinase that phosphorylates the γ and δ chains of the AChR *in vitro*.[55] *Torpedo* membranes also contain a calmodulin-dependent protein kinase, but this kinase phosphorylates endogenous substrates distinct from the AChR subunits. Purified AChR was also phosphorylated on the γ and δ chains when incubated with the catalytic subunit of purified bovine heart protein kinase. One phosphate is added to each γ and δ subunit *in vitro*. The physiological role of a cAMP-regulated phosphorylation of the AChR is presently unclear. Phosphorylation may modulate aspects of receptor channel activity such as developmental changes in mean channel-open time or the process of receptor desensitization. These hypotheses should be testable using patch-clamp techniques to study the suitably modified AChRs reconstituted into liposomes.[e.g., 56]

3. Immunological Approaches to the Study of the Nicotinic AChR

3.1. Myasthenic Patients Produce Antibodies against Their AChRs

Since large amounts of purified AChR could be prepared from *Torpedo* electric organ tissue, the nicotinic AChR became the first neurotransmitter receptor for which specific antibodies were produced. The initial interest in using antibodies to study the receptor revolved around the potential ability of antibodies to provide additional structural information about the AChR and its subunits. It soon became clear, however, that antibodies against the AChR would also be of clinical interest. Patrick and Lindstrom[57] demonstrated that animals injected with purified eel AChR developed muscular weakness and fatigue, symptoms typically observed in the human disorder of myasthenia gravis. This animal model for myasthenia gravis, called experimental autoimmune myasthenia gravis (EAMG), has led to the recognition that myasthenia gravis is an autoimmune disorder in which antibodies are produced against the muscle AChR.

Nearly all patients with myasthenia gravis produce antibodies reactive with human muscle AChR.[58] The nature of the events that lead to a breakdown in tolerance and the production of anti-AChR antibodies is, however, presently unknown. In addition, the exact mechanism by which the antibodies produce muscle weakness is still a matter of dispute. Anti-AChR antibodies cause an increase in the rate of degradation of cell-surface AChR apparently by a process similar to antigenic modulation; bivalent antibodies capable of cross-linking AChR lead to an enhanced rate of disappearance of AChR from the surface of cultured muscle cells.[59–62] This net loss of surface receptor could lead to a decreased safety factor in synaptic transmission *in vivo*. Although some anti-AChR antibodies are capable of blocking ligand-activated channel opening, such a direct blockade seems not to be a major factor in most patients with myasthenia gravis. A third possible mechanism for which there is considerable evidence involves complement fixation. Immune complexes on the surface of muscle cells at the neuromuscular junction lead to a focal, complement-dependent lysis of the postsynaptic membrane resulting in extensive morphological alterations of the synapse. Such marked changes have been readily observed in acute EAMG as well as in myasthenia gravis,[63,64] and may represent the most important pathological effect of anti-AChR antibodies.

Although anti-AChR antibodies are clearly involved in myasthenia gravis, only a poor correlation is found between the levels of anti-AChR activity in sera and the severity of the disease in different individuals. This observation suggests either that other physiological factors may be important in setting the safety factor governing the number of postsynaptic receptors required for successful neuromuscular transmission, or that subclasses of anti-AChR antibodies of particular specificities may be more pathologically significant than simply the aggregate level of antibodies of all specificities. Clearly, some of the antibodies may be produced secondarily, as a result of postsynaptic lysis, and may have little to do with the primary events underlying the disease.

With the availability of monoclonal antibodies (mAbs) against the nicotinic AChR (see below), it has been possible to examine and compare the subunit specificities of the antibodies produced in EAMG and in myasthenia gravis. Using, as immunogen, intact AChR that had been extracted from membranes with a mild, nonionic detergent, such as Triton, Lindstrom and co-workers have determined that there is a region on the α subunit that is highly immunogenic,[65] and that specifies the binding site for more than half the mAbs generated against this so-called "native" form of the receptor. These workers have designated this region the "main immunogenic region" (MIR). The MIR is not a single antigenic determinant but rather appears to consist of a number of determinants in close apposition since mAbs to the MIR mutually exclude each other's binding. Furthermore, the antigenicity of the MIR is at least partially conformationally dependent since many of the anti-MIR mAbs do not bind to the denatured α subunit; others do bind but with reduced affinity. In antibody competition binding experiments, most of the antibodies in polyclonal EAMG sera produced with "native" receptor as antigen, also bind to the MIR. Furthermore, most of the antibodies in myasthenia gravis antisera also bind to the MIR.[66] The spectrum of antisubunit-specific antibodies in myasthenia gravis antisera did not appear to vary in any systematic manner with respect to clinical severity of the disease. Thus, no correlation could be made between clinical manifestations and the distribution of the subunit specificity of the antibodies within the antisera. Furthermore, the passive transfer of mAbs with anti-MIR specificity, and which can fix complement, is capable of reproducing the symptoms of the acute phase of EAMG. At present, then, no direct evidence is available for a specific subclass of antibody with an antigenic specificity that is particularly and uniquely pathogenic.

3.2. mAbs Reveal Aspects of AChR Organization

In the attempt to define additional substructural features of the AChR and in order to obtain additional specific probes for the AChR, a number of laboratories have obtained mAbs directed against the intact AChR purified from electric organs of *Torpedo*,[65,67–72] *Narcine*,[73] and *Electrophorus*,[74] as well as from bovine[75] and human muscle.[76,77] In addition, AChR and purified receptor subunits denatured in SDS have also been used as an immunogen for mAb production.[65,76,78] Only a very small proportion of these mAbs are directed against the ligand- or α-bungarotoxin-binding sites.[68,69,72] Some of these mAbs may, however, be capable of differentiating antigenic differences between the two α-bungarotoxin-binding sites within each AChR.[72,79] A similar phenomenon has recently been reported using a myasthenic antiserum.[80] Some of these anti-ligand-binding-site mAbs may also be useful in defining subsites within the binding site or in detecting different conformational changes produced by different ligands (e.g., carbachol vs. decamethonium vs. tubocurarine).[72]

As was indicated earlier, most mAbs prepared against intact *Torpedo* AChR bind to a single region on the receptor designated the MIR. The MIR is on the extracellular face of the membrane, is located on the α subunit, but is not the ligand-binding site or the α-bungarotoxin-binding site. In addition, it does not appear to contain carbohydrate. Based on cross-reacting mAbs, the MIR is found in the AChR of *Torpedo, Electrophorus,* human, bovine, and the amphibian *Rana pipiens*.[81] It appears to be lacking, however, in the amphibian *Xenopus laevis*. Thus, this region appears to be highly conserved across species and the MIR might play an important functional role such as in the interaction of the AChR with the extracellular matrix.[74] Exactly why such a highly conserved structure would also be so highly immunogenic is unclear.

As a first step in characterizing the binding properties of mAbs, the subunit specificity is usually determined either by immunoprecipitation of the individual, radiolabeled subunits, or by immunoblotting to electrophoretic transfers of the unlabeled subunits resolved by SDS gel electrophoresis (i.e., "Western blots").[81,82] This type of characterization is feasible only for those mAbs that are capable of recognizing the denatured subunits. About half of the mAbs raised against intact AChR will recognize the denatured proteins.[18] If denatured AChR or denatured, purified subunits are used as immunogens, then the resultant mAbs (and antisera) are directed predominantly to cytoplasmic regions of the AChR.[29,74,78] This is in marked contrast to the preferential production of mAbs to extracellular determinants when intact AChR is used as immunogen. The transmembrane distribution of mAb-binding sites can be readily demonstrated with immunohistochemical techniques. Sargent *et al.*[83] have shown, using electron microscopic localization, that some mAbs bind to the extracellular face of intact frog neuromuscular junctions, whereas others bind to the inner face of the postsynaptic membrane, but only when the tissue has been saponin treated to enhance accessibility to the antibody.

Using a competitive binding assay to map the possible overlap of mAb-binding sites on intact AChR, Tzartos *et al.* were initially able to identify nine immunogenic regions on the receptor molecule.[74] Within the subunits themselves, the mAb-binding sites could be further defined by examining the immunoprecipitation of peptide fragments prepared by subjecting the individual subunits to protease treatment.[81] For example, treating radiolabeled α subunit with *Staphylococcus aureus* V8 protease results in a characteristic set of peptide fragments, one of which (M_r 17K) contains carbohydrate, and another can be labeled with MBTA (19K). Although anti-MIR mAbs bind to both of these two fragments, they also recognize other fragments that do not contain carbohydrate and that are not labeled with MBTA. Thus, such an analysis demonstrates that the MIR is close to but distinct from these two other structural features of the α subunit. Furthermore, by comparing the profile of peptide fragments immunoprecipitated by the various mAbs, and grouping together those mAbs with a similar spectrum of specificities, the number of antigenic determinants per subunit can be estimated. At least seven determinants are present on α, nine on β, seven on γ, and five on δ.[18] These are minimal estimates since all those mAbs incapable of binding denatured subunits are not included in the groupings. Additional antigenic determinants are most probably made up of noncontiguous arrays of amino acids the recognition of which would be dependent on retention of native protein conformation. Along these lines, note that a number of laboratories have reported differential binding of mAbs to intact AChR in membranes compared to the detergent-solubilized AChR.[e.g., 84] Some antigenic determinants expressed on membrane-bound AChR may thus be obliterated by detergent solubilization, and other determinants not available in the membrane-associated form of the AChR may become accessible upon solubilization. This may be an important consideration when devising screening procedures for mAbs against membrane proteins.

By examining the peptides immunoprecipitated by the various mAbs and determining their amino acid sequences, one should, after a comparison with the deduced amino acid sequence of the subunit, arrive at an ordering of the various structural features of the subunit along the primary amino acid sequence of the protein. Additional information on the physical location of various domains within the intact AChR can be obtained by using the membrane-inserted, *in vitro* translation products produced by mRNA for *Torpedo* AChR. Anderson *et al.* have used partial proteolysis and immunoprecipitation with domain-specific mAbs to identify the topological distribution of peptide domains in transmembrane-oriented AChR subunits.[31] When intact microsomal vesicles containing *in vitro*-translated δ subunit are treated with low levels of trypsin, a discrete 12K fragment of the subunit is released. Since in intact vesicles the subunits are oriented with their NH_2-terminal, "extracellular" ends within the lumen of the vesicle and their "cytoplasmic" ends outside the vesicle,[85] only the cytoplasmic domain of the subunit is susceptible to proteolysis. Some mAbs against the δ subunit were capable of immunoprecipitating the 12K fragment and thus were specific for the cytoplasmic domain of this subunit. A similar analysis was carried out with *in vitro*-translated and membrane-inserted β subunit. In this case, one mAb was identified that bound to a cytoplasmic-derived fragment, and another mAb bound to a domain in the extracellular portion of the subunit. Thus, topological localization of domains coupled to sequence information should permit a rather detailed description of the structural distribution of functionally important parts of the receptor molecule. In addition, the visualization of mAb binding with electron microscopic techniques should facilitate mapping the structural domains of the AChR in three dimensions and should allow the verification of the orientation of the subunits within the AChR complex.

In terms of functional effects of the mAbs, very few of the mAbs have been examined in great detail. As previously mentioned, one common indirect effect seen with the mAbs that can cross-link AChR monomers, is their ability to increase the rate of receptor internalization leading to a lower surface density of receptor.[86] Also, mAbs that bind to the ligand-binding site block receptor function by preventing ligand-induced activation in a classic competitive manner. This has been ably demonstrated in the characterization of mAb 5.5 which binds to the AChR of intact chick muscle cells in culture. This mAb was raised against *Torpedo* AChR and its binding to solubilized receptor is competed by bungarotoxin and cholinergic ligands. Similarly, incubating cultured chick muscle cells with mAb 5.5 inhibits over 70% of the bungarotoxin binding to chick muscle cells, indicating a considerable level of similarity between cholinergic binding sites from these two sources.[84] When AChR-mediated ion flux was measured in these cells, mAb 5.5 is found to block up to 80% of the $^{22}Na^+$ uptake induced by the addition of carbamylcholine. Presumably, this effect is due to mAb interference with agonist binding and is not a direct action on the ion channel. In contrast to the striking inhibition by mAb 5.5, a mouse polyclonal antiserum against AChR produced only a 10% inhibition of the initial rate of $^{22}Na^+$ uptake. Great variability, however, can be found in the ability of antisera to affect receptor function. Antireceptor antisera have been reported to affect both the amplitude and the mean open time of receptor channels.[61] The ability of antisera to block bungarotoxin binding is usually low but also highly variable.[58]

Lindstrom *et al.*[3,87] have begun to test the effect of mAbs on receptor function using purified *Torpedo* AChR reconstituted into lipid vesicles. They have examined the ability of mAbs to block carbamylcholine-induced uptake of $^{22}Na^+$ into receptor-containing vesicles under conditions in which the rate of $^{22}Na^+$ flux is proportional to the amount of functional receptor in the vesicle. Of 25 mAbs tested in this way, only 2 appeared to have authentic inhibitory effects. In both cases, the inhibition was noncompetitive with respect to agonist. One of these two mAbs is directed against the β subunit but cross-reacts somewhat with the α subunit. Because this particular mAb is of low affinity, its specific peptide binding region has not been determined. Nevertheless, this observation indicates that it should be possible to identify additional mAbs with functional effects on receptor activity that are clearly independent of an interaction with the ligand-binding site. The physical mapping of the binding sites for these mAbs would then produce important information relating receptor structure with function. The pursuit of this particular question should be greatly facilitated by investigating the effect of mAbs on receptor function in single-channel "patch-clamp" recordings from isolated or reconstituted membranes.[56,88]

Another potentially very exciting application of mAbs against the nicotinic AChR is in the investigation of the neuronal nicotinic AChR. Although some central and peripheral neurons do appear to contain nicotinic AChRs, in at least some cases, the neuronal AChR does not seem to bind α-bungarotoxin.[89] Antisera against eel AChR will bind to some extent with the neuronal receptor on rat PC12 pheochromocytoma cells.[89] Other workers have described the binding of mAbs against *Torpedo* AChR to neurons in chick brain[90] and to neurons in rat sympathetic and sensory ganglia.[91] In addition, one mAb to the α subunit of *Torpedo* AChR cross-reacts with frozen section of guinea pig ileum smooth muscle, a tissue highly enriched for muscarinic receptors.[91] The further characterization of these mAbs and their peptide binding sites will be very valuable in exploring possible structural homologies between various ACh-binding proteins.

4. Biogenesis, Membrane Localization, and Regulation

4.1. The AChR Is a Useful Molecule for Cell Biologists

The nicotinic AChR, by virtue of being a multimeric, transmembrane complex of glycoproteins, provides an attractive system for the study of the mechanisms underlying the synthesis, assembly, and insertion into appropriate membrane sites of functional membrane proteins. Much information concerning the synthesis of membrane proteins in eukaryotic cells has come from the analysis of the biosynthesis of simple viral coat proteins.[e.g., 92] In addition, in recent years the study of some endogenous membrane proteins has become possible.[93–99] In most cases, such studies have become technically feasible due to the availability of: (1) specific and high-affinity antisera or mAbs that can recognize the biosynthetic intermediate forms of the membrane proteins, and (2) cultured cell lines that produce significant levels of the membrane proteins and that can be metabolically labeled. In some cases, the use of somatic variants with altered biosynthetic expression of the particular membrane protein in question has greatly facilitated the description of the biosynthetic process.[93,95] In none of these cases has a membrane protein complex consisting of more than two subunits been investigated. Furthermore, most of the membrane proteins that have been investigated are primarily extracellular in localization and have only small regions at the carboxyl end of the protein chain within the cytoplasm. The AChR, because of its important transmembrane function as a neurotransmitter receptor and ligand-activated ion channel, provides an excellent opportunity to explore the biogenesis of a more complex category of membrane protein.

4.2. AChRs Are Present in an Intracellular Compartment

Early studies on the kinetics of biosynthesis of AChR used cultured embryonic muscle cells and relied on the ability of radiolabeled α-bungarotoxin to label the receptor. These studies indicated that the synthesis of new AChR polypeptide chains was complete within 30 min and that the appearance of the newly synthesized AChR at the membrane surface was delayed by 2–3 hr.[100] As much as 20% of the total AChR of the cell resided in a pool of assembled receptor in transit to the membrane. Similar observations have been made in a murine nonfusing muscle cell line, BC3H-1.[101] Autoradiographic studies, again with labeled bungarotoxin, suggested that at least a portion of these intracellular receptors were located within the Golgi complex[102] and thus appeared to share with secretory proteins a segment of the same intracellular transport pathway.[103–105] Other studies argued against the possibility that these intracellular receptors represented the intracellular recycling of membrane AChR.[106] Further information on the early events of AChR biosynthesis, however, required the development of additional structural probes for newly synthesized AChR polypeptide chains.

4.3. AChR Subunits Are Synthesized in Rough Endoplasmic Reticulum

With the availability of specific antisera and mAbs against the AChR subunits, the molecular events in AChR biogenesis may be investigated. First of all, the mRNAs coding for the *Torpedo* AChR could be assayed by carrying out cell-free protein synthesis with labeled amino acids and immunoprecipitating the *in vitro*-synthesized polypeptide chains with anti-AChR antibodies.[30,107] By the addition of dog pancreas microsomes to the *in vitro* translation system, the four polypeptide chains of the AChR were shown to be independently integrated into the membrane by a cotranslational process analogous to that described for viral membrane glycoproteins.[30] NH_2-terminal, hydrophobic, "signal" peptides provide the mechanism by which nascent viral membrane glycoproteins are recognized and inserted into the microsomal membrane while being translated from a ribosome complex.[108] A similar "signal" peptide has been directly demonstrated in the *in vitro* translation product of the δ subunit[109] and the appropriate signal sequences have now been documented for each of the four subunits by analysis of cDNA clones. The cell-free translation studies together with the cDNA sequencing efforts indicate that the four subunits are coded for by four separate mRNAs.

The association of a nascent AChR polypeptide chain with the endoplasmic reticulum membrane requires the presence of a so-called signal recognition protein (SRP). The SRP complex serves to form a bridge bringing together the polysomes and an integral membrane protein receptor in the rough endoplasmic reticulum.[110] Subsequent to this association, translocation of the polypeptide chain across the membrane bilayer into the lumen of the microsomal vesicle occurs concomitantly with the actual translational synthesis of the peptide chain by the ribosome. In the case of secretary proteins, the entire peptide chain is translocated into the luminal space. Membrane proteins, however, are only partially translocated. They can, in general, span the membrane bilayer once or several times, often leaving a variably sized peptide region on the external "cytoplasmic" face of the microsomal vesicle. The exact mechanism by which membrane proteins achieve their final topological distribution across the membrane is unknown.

Anderson and Blobel have examined the transmembrane distribution of *in vitro*-synthesized and membrane-inserted AChR subunits.[30] For each of the four subunits, they observed that the polypeptide chain spanned the membrane at least once since a portion of each chain was accessible to externally added proteases. The size of the "cytoplasmic," untranslocated peptide domains varied between 5 and 20K for the four subunits. In a more complete analysis of the transmembrane distribution of the *in vitro*-synthesized δ subunit using domain-specific mAbs, a 12K fragment of the δ subunit was localized to the cytoplasm. A mAb that binds to the cytoplasmic region of intact, membrane-associated AChR was shown also to bind to this cytoplasmic 12K fragment. The results with the *in vitro*-synthesized δ subunit appear to agree with other studies examining the proteolytic sensitivity of the δ subunit within the plasma membrane-associated AChR complex.[111] Taken together, these findings suggest that the transmembrane localization of the newly synthesized δ subunit is very similar to the distribution of the δ subunit within the mature AChR complex. This would suggest that assembly of the subunits into a functional complex does not involve a massive redistribution of the subunit topology with respect to the membrane bilayer.

4.4. AChR Subunits Are Glycosylated

In the microsomal, cell-free translation system, all four of the *Torpedo* AChR subunits are glycosylated on sites exposed in the lumen of the membrane vesicle. As has been observed with viral membrane glycoproteins, during their translocation through the membrane, the AChR subunits receive "core" oligosaccharide chains linked to specific asparagine residues within the protein. One such "core" group is added to the α and β subunits, three are added to δ, and three or possibly four are added to γ. Further modifications to these oligosaccharide chains are believed to occur in the Golgi complex. In addition, *O*-linked oligosaccharide chains are also most probably added to the AChR subunits while in the Golgi. The precise details and intracellular location of these sugar modifications have not been carefully examined as yet.

4.5. Subunit Maturation and Assembly

Functional assembly of the AChR subunits synthesized in a cell-free translation system has not yet been achieved. In addition, the binding of bungarotoxin to *in vitro*-synthesized α subunit cannot be detected. The lack of assembly is most probably explained by technical considerations of the cell-free system that make it highly improbable that all four subunits will be integrated into the same microsomal membrane vesicle. Under the conditions used, the different subunits, synthesized from separate polysomes, are by chance segregated into separate membrane compartments and thus do not come into contact with the other subunits. To gain further information of the assembly process, therefore, *in vivo* studies with cultured cell lines have been used.

Consistent with the *in vitro* results of Anderson and co-workers using *Torpedo* mRNA, Merlie *et al.*(112) have shown that membrane-bound polysomes isolated from the mouse muscle cell line BC3H-1 are capable of directing the synthesis *in vitro* of the AChR α subunit. Synthesis was detected by immunoprecipitation with an antiserum raised against SDS-denatured *Torpedo* α subunit. The BC3H-1 α subunit, synthesized *in vitro*, in the absence of exogenous microsomal membranes was not glycosylated and was approximately 2K larger than the native α subunit found in the AChR complex.(113) The increased molecular weight is most probably due to the retention of the NH_2-terminal "signal" sequence which, in other systems, is cleaved upon translocation of the protein through the microsomal membrane. BC3H-1 α subunit synthesized in the presence of microsomal membranes became glycosylated and underwent cleavage of the "signal" sequence. The mRNA species responsible for the *in vitro* synthesis was further characterized and found to be poly(A) and approximately 2000 nucleotides in length. A cell-free translation assay of α-subunit mRNA levels revealed a 12-fold increase in the abundance of this mRNA in BC3H-1 cells that have differentiated and are producing large amounts of AChR. Merlie and co-workers used BC3H-1 poly(A) RNA obtained by a polysome fractionation procedure to construct a cDNA library.(114) A clone containing AChR α-specific sequences was identified by hybrid-selected translation *in vitro*. In this procedure, candidate cDNA clones are used to extract from total BC3H-1 mRNA those sequences that cross-hybridize with the cDNAs. The extracted mRNAs are then added to a cell-free translation system and the resultant *in vitro*-synthesized and labeled proteins are immunoprecipitated with anti-AChR α antibody. If a particular cDNA clone contains α sequences, then the immunoprecipitation of *in vitro*-synthesized α will be successful and will thus identify the appropriate cDNA. One such clone was found and is being further characterized. By using this cDNA, it was possible to show that BC3H-1 cells contain 100–1000 times more α mRNA than newborn or adult muscle.

Some of the early events in AChR biogenesis are now being investigated in the BC3H-1 cell line, because this cell line can produce relatively large amounts of AChR.(115) Merlie and co-workers have used pulse–chase experiments to explore the kinetics of receptor production in BC3H-1 cells.(116,117) By labeling cells for short periods of time (pulse) with highly labeled [^{35}S]methionine and then incubating cells for various periods of time with cold, unlabeled methionine, the time course of the processing and assembly of peptide chains synthesized *in vivo* during the short interval of pulse-labeling can be followed. This general type of analysis has been used with great success in a wide variety of experimental systems. The key to its success in a particular application, however, is the ability to isolate the protein in question from all the other proteins being synthesized in the cell. This can usually be accomplished with the use of specific, high-affinity antisera or mAbs to immunoprecipitate the polypeptides from a cell extract. The labeled polypeptides can then be examined by various techniques such as SDS gel electrophoresis.

Using antisera to denatured AChR polypeptides, Merlie and Sebbane were able to detect and quantitate the biosynthesis of AChR in cultured BC3H-1 cells.(117) Furthermore, by using α-bungarotoxin and antibodies to α-bungarotoxin, Merlie and Sebbane were able to determine the time course with which the newly synthesized AChR subunits acquire the ability to bind the neurotoxin. These experiments indicated that AChR subunits synthesized during a 5-min pulse of label required 15–30 min of intracellular processing to advance to a state that could be recognized by α-bungarotoxin, and thus could be immunoprecipitated from cell extracts by the sequential addition of α-bungarotoxin and anti-α-bungarotoxin. These findings suggested, therefore, that the AChR subunits go through some sort of precursor pool prior to acquiring the structural ability to bind α-bungarotoxin. A further analysis of this point with mAbs against the α subunit showed that only 30% of the α subunits synthesized at any one time go on to a state that can bind bungarotoxin. This result indicates that the α polypeptide may be produced in excess of that required for proper processing or assembly and that the excess α appears to be rapidly degraded.(118) Endoglycosidase (endo-β-*N*-acetylglucosaminidase H) treatment reveals that the native α subunit contained one *N*-asparagine-linked oligosaccharide chain of the so-called "high mannose" or "simple" type.(118) No evidence could be obtained for any "complex"-type *N*-linked oligosaccharides.

Initially, based on these early studies, it was proposed that the acquisition of the ability of newly synthesized AChR to bind bungarotoxin coincided with the actual assembly of the α subunit with the other subunits. This hypothesis agreed with certain other studies indicating that the α subunit alone did not appear to bind bungarotoxin with high affinity. One method for demonstrating that assembly had occurred at the time of bungarotoxin-binding acquisition would be to show that by immunoprecipitating with anti-α-subunit antibody, the other three subunit polypeptides are also immunoprecipitated. Since the assembled subunits would be in one complex, they should all be immunoprecipitated by an antibody to any one subunit. Unfortunately, at present the only available antibodies with sufficient affinity to accomplish such an immunoprecipitation are those against α and β. An additional technical difficulty is that the γ and δ subunits appear to be very susceptible to proteolysis and the newly synthesized forms of these two subunits are difficult to detect. Nevertheless, by using anti-α, anti-β, and toxin antitoxin immunoprecipitations, Merlie and Lindstrom found that the α subunit acquired toxin-binding activity prior to being associated with β subunit.(118) In addition, they observed that the α subunit acquired the highly immunogenic MIR site at the same time (10 min) that bungarotoxin binding was detected. Only during the next 60 min are α and β subunits found in association with each other; that is, either anti-α or anti-β antibodies immunoprecipitate both α and β subunits. Yet another 30–60 min is required for intracellular transport and insertion into the plasma membrane.

The newly synthesized subunits were analyzed by glycerol gradient velocity sedimentation to determine the approximate size of possible complexes that contained the subunits: At early times α and β subunits appeared in the low-molecular-weight region suggesting that, prior to assembly, the α and β subunits are not themselves associated in molecular complexes much larger than expected for individual monomers. Only after 30–60 min of chase are the α and β subunits found in a higher-molecular-weight region (9 S) consistent with incorporation into the complete receptor complex. At these later times the two subunits are coimmunoprecipitated by antibody to either one subunit. Together with the previous set of experiments, these results were interpreted to indicate that the α subunit acquires toxin-binding ability while in the monomeric form, and no evidence could be seen for homo-oligomeric complexes. Merlie and Lindstrom[118] suggest the possibility that appropriate intrachain disulfide bond formation must occur for the individual subunits to acquire their mature tertiary structure. They suggest that such a disulfide cross-linking is important in the acquisition by the α subunit of toxin-binding ability, and development of the conformation specifying the MIR site.

In contrast to the lack of evidence for homo-oligomers of the individual subunits *in vivo*, Anderson and Blobel, using their *in vitro* system, have reported that newly synthesized *Torpedo* AChR subunits do appear to self-associate.[119] They find that the subunits do not form heterologous interactions, as previously reported, but do sediment in a broad range between 7 and 13 S. The major fraction of the subunits are found at a sedimentation region appropriate for the native AChR complex run in parallel. These workers propose that such homologous associations may provide an energetically favorable mechanism *in vivo* for preserving the ionophore surfaces of the subunits during the period of time required for intracellular subunit assembly. Each of the four subunits is believed to contribute a structural domain to the makeup of the ion channel, presumably by having several hydrophilic amino acid side chains directed toward the channel. Since the region containing the channel is also one that traverses the lipid membrane, it would appear to be highly unlikely that, in the monomeric form of each subunit, the amino acid side chains necessary for channel formation would be exposed to the hydrophobic milieu of the membrane. The proposed model of an intermediate form of the AChR subunits made up of homo-oligomers would thus provide a means by which the ion channel could be formed early and retained throughout the additional subunit assembly process. The formation of such homo-oligomers would also appear to be consistent with the high degree of homology between subunits and the fact that the subunits appear to be derived from a common ancestor. Additional experiments are required, however, to resolve the sedimentation differences observed in these two experimental systems.

Merlie and co-workers used the *N*-glycosylation inhibitor tunicamycin to study the possible role of the *N*-asparagine-linked carbohydrate addition on AChR subunit biosynthesis in cultured BC3H-1 cells.[116] They found that whereas newly synthesized α subunit was normally converted with 30% efficiency into a form capable of binding bungarotoxin, in the presence of tunicamycin, only 5% of the α underwent this conversion. The remainder was rapidly degraded. Merlie and co-workers suggested that core-glycosylation was necessary for subunit assembly. This was their interpretation at that time since they believed that only the assembled complex could bind bungarotoxin. Prives and Bar-Sagi,[120] performing related experiments with tunicamycin but with cultured embryonic chick muscle cells, also found that tunicamycin treatment greatly decreased the production and membrane insertion of mature AChR complex. Nevertheless, the underglycosylated AChR that did become incorporated into the membrane in the presence of tunicamycin retained the ability to mediate agonist-induced ion permeability changes and also bound bungarotoxin. These latter results were consistent with the observation that the bulk of the *N*-linked oligosaccharide can be removed from α subunit with endoglycosidase H without affecting bungarotoxin binding as measured by protein-blots.[48] Exactly what role glycosylation plays in the biosynthesis of the AChR, as well as of other membrane proteins is still unclear. Furthermore, the AChR subunits may undergo additional carbohydrate additions (*O*-linked oligosaccharides) and modifications in the Golgi on their way to the plasma membrane. Precisely what function these modifications may have is also unclear.

Because of the extended time course required for assembly of α and β chains, the assembly of the AChR subunits presumably occurs in the Golgi complex. This hypothesis is based on the known transit times of other Golgi-processed proteins but at present no direct evidence on this point with respect to the AChR is available. The mechanism by which the subunits arrive at the Golgi is also unknown as is the means by which the assembled AChR complex is delivered to the plasma membrane. Although coated vesicles have generally been suggested to be involved in the process of intracellular sorting and delivery of membrane proteins, again very little direct information on this point is available. What is clear is that the ability to synthesize, process, assemble, and deliver the AChR complex to the plasma membrane is not restricted to muscle and nerve cells. *Xenopus* oocytes, when injected with mRNA from *Torpedo* electric organ, are capable of synthesizing and assembling into the plasma membrane a functional AChR complex that can be assayed electrophysiologically.[121,122] Furthermore, when the AChR subunit-specific mRNAs produced in COS monkey cells transfected with recombinant plasmids containing *Torpedo* cDNA AChR subunit sequences are injected into *Xenopus* oocytes, the production of a normal nicotinic response is observed only if all four subunit mRNAs are injected.[123] Besides providing evidence that all four subunits are indeed required for full receptor function, these experiments also demonstrate that no other *Torpedo* specific proteins or RNAs are needed for the functional synthesis and assembly of the *Torpedo* AChR since injection of these four mRNAs alone was sufficient for AChR biogenesis. Although *Xenopus* oocytes are clearly capable of the complete biosynthesis and assembly of functional AChR, these cells are particularly efficient at protein synthesis. Whether other cell types such as fibroblasts would be able to carry out the appropriate processing and assembly of AChR from exogenously introduced mRNAs or cDNAs is not yet known.

4.6. Developmental Regulation of the AChR

The interest in the AChR does not, of course, end with its assembly and insertion into the plasma membrane. There is a whole other arena of questions dealing with the specific localization and concentration of the AChR at particular sites in the membrane. Also, a wide variety of regulatory events, the mechanisms of which are still unclear, govern the appearance, stability, function, mobility, and degradation of the cell-surface AChR. Many of these phenomena are of particular interest to developmental neurobiologists who see their applicability to other synaptic interactions under developmental control. By fully understanding the molecular basis of AChR biosynthesis and function, one may approach the regulatory aspects of AChR metabolism at the molecular level.

4.7. AChRs Aggregate at Synapses

During development of chick muscle fibers, shortly after fusion of the myoblasts, AChR is found distributed all along the surface of the fiber.[124] At the mature neuromuscular junction, the AChR is localized specifically to the region of the membrane immediately opposed to the presynaptic nerve terminal. Furthermore, in many species, characteristic folds of the postsynaptic muscle membrane are formed and the AChR is found in extremely high densities (20,000–30,000/μm^2) at the top of these folds.[125,126] The sharp delimitation of the AChR localization can be demonstrated by the fact that by applying ACh only 2 μm from the nerve terminal region, a 50- to 100-fold fall in ACh sensitivity is seen as compared to application directly onto the terminal. One fundamental question of muscle cell biology and of synaptogenesis then is how the postsynaptic AChR molecules come to be localized so tenaciously to the nerve terminal region.

As model systems for the study of this question, several laboratories have been examining muscle cell development in culture while others have looked at receptor events during denervation and reinnervation of muscle. Chick muscle cells cultured in the absence of neurons will form small clusters or "hot spots" of AChR on their surface membrane. Innervating nerve growth cones apparently do not seek out preexisting hot spots but do cause the lateral movement and reorganization of receptor clusters to the region of the membrane in juxtaposition to the nerve terminal.[127] In fact, cell-free extracts of brain and spinal cord cause an increase in the total number of AChR on the surface of chick muscle cells. In addition, these extracts as well as extracellular matrix fractions from *Torpedo* electric organ induce an increase in the number of receptor clusters.[128,129] These aneural preparations should be extremely important in identifying the means by which receptor clustering occurs in response to neurotrophic factors.[130]

4.8. Junctional and Extrajunctional AChRs Are Different

A very distinct difference between the postsynaptic AChR (junctional) and the AChR diffusely dispersed over the membrane (extrajunctional) is in their respective turnover rates *in vivo* and in culture. Whereas the AChR on developing muscle cells and on muscle cells cultured in the absence of nerves is removed and degraded with a metabolic half-life on the order of 17 hr, the AChR within the postsynaptic cluster is metabolized much more slowly, with a half-life of greater than 10 days.[131] Furthermore, when muscles are denervated, the amount of extrajunctional AChR increases and the half-life of this receptor species is also on the order of 17 hr. The biochemical mechanism by which the various forms of the AChR are differentially metabolized is presently unknown. Muscle activity is known, however, to play an important role in controlling the appearance of the extrajunctional form of the AChR in denervated muscle.[130]

There are additional biochemical and immunological differences between junctional and extrajunctional AChRs. The isoelectric point of the two receptors appears to be slightly different[132] and certain myasthenic antisera contain antibody specificities that cross-react preferentially with rat extrajunctional AChR[133] as compared to rat junctional AChR. Besides reacting with extrajunctional AChR on adult, denervated rat muscle, these antisera also react with the embryonic form of the rat AChR[134] but do not bind to the AChR on mature rat endplates. Thus, at least in rat muscle, the extrajunctional AChR found after denervation of adult muscle bears antigenic similarity to the dispersed, embryonic form of the AChR.

At least one channel property of the AChR also appears to be developmentally regulated in rat muscle. The mean channel open time observed on neonatal rat muscle is about three times that observed on adult muscle.[135] In chick muscle, however, no such developmental change in mean channel open time is found and the channel open time is comparable to that of embryonic rat muscle.[136] At this point, the key molecular differences between junctional and extrajunctional forms of the AChR are still obscure. Use of appropriate recombinant cDNA probes should reveal whether there are two different structural genes, whether gene expression is developmentally regulated, or whether posttranslational modifications of the AChR are subject to neuronal influence.[137]

4.9. Possible Role of Associated Molecules in AChR Regulation and Localization

At the synapse, one finds the AChR in close association with various subsynaptic, intracellular proteins as well as with certain particular components of the extracellular matrix attached to the basal lamina surrounding the muscle cell. The possible involvement of these AChR-associated components in the regulation and localization of postsynaptic AChR is a very active area of current research. Immunocytochemical studies have localized various cytoskeletal components to the neuromuscular junction. These include a form of cytoplasmic actin,[138] vinculin, α-actinin, and filamin.[139] Microtubules have also been implicated in AChR clustering.[140] A 43K subsynaptic, peripheral protein, distinct from actin, is also localized to the postsynaptic region.[141–143] Chemical cross-linking studies have shown that this 43K protein is closely associated with the membrane-bound AChR and can be covalently cross-linked to the AChR β subunit.[144] Additional studies are required to identify the possible physiological functions that these AChR-associated proteins may mediate.

In studies on the reinnervation of the frog neuromuscular junction, components within the basal lamina at the synaptic site have been shown to be responsible for the fact that a reinnervating nerve forms synapses precisely at the former synaptic region.[145] This was shown by the fact that muscle regeneration is not required for the reinnervation of the appropriate former synaptic site: the location of the reinnervated synaptic site was completely specified by the basal lamina at that site. Furthermore, when nerve reinnervation was prevented, a previously destroyed myofiber will regenerate, presumably from myoblast satellite cells, and will accumulate AChR predominantly at the original synaptic site even though no nerve terminal is present.[146] Thus, at least in regenerating systems, some factors within the muscle basal lamina play a critical role in specifying where receptor will be localized. Whether such a process is also operating during normal development and synaptogenesis is presently unclear. Also unclear is the question of the cellular origin of the specifying factors. Are they produced by the muscle or the nerve? The recent findings of Carlson *et al.*[147] that *Torpedo* electric organ synaptic vesicles contain a proteoglycan material that undergoes active exocytosis and is incorporated into the extracellular matrix at the synaptic cleft suggest a possible answer to this question. It is interesting to speculate that the nerve produces certain synapse-specifying factors that are packaged into synaptic vesicles. The synaptic vesicles then serve to deliver these factors to the extracellular matrix only at synaptic sites. Such a mechanism would certainly be important in stabilizing and possibly maintaining a synaptic connection. Whether a similar mechanism is involved in *de novo* synaptogenesis will be an exciting hypothesis to investigate further. Nevertheless, this review of the regulatory and developmental aspects of AChR metabolism should serve to illustrate the great richness of biological phenomena that the study of the AChR has uncovered. The mechanisms by which nerves regulate AChR and its localization should be directly relevant to other receptor systems in neurobiology.

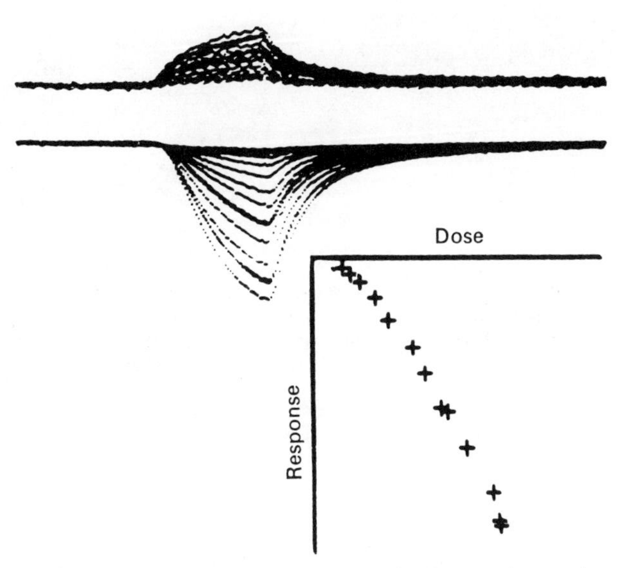

Fig. 1. Some typical raw data are shown; the photograph was taken during the course of an experiment using carbachol as agonist. The upper series of traces show digitized records of ion-selective electrode responses, while the lower traces show the corresponding endplate currents (EPCs). The bottom right quadrant shows a preliminary dose–response plot, simply plotting the EPC near the peak against the ion-selective response near the peak. This experiment showed no signs of desensitization; note that similar doses gave similar responses. The iontophoretic pulse (not shown) lasted 1 sec. The largest ion-selective response was ~ 3.9 mV (100 μM), the largest EPC was ~ 25 nA.

5. Dose–Response

5.1. What Is Measured to Establish Dose–Response Relationships

Equilibrium dose–response measurements provide a useful characterization of drug–receptor interactions because they contain information on the number of binding sites per functional receptor unit and on drug-binding affinity. Katz and Thesleff,[148] working on AChRs at the frog endplate, first reported that induced membrane depolarization varied more than linearly with agonist dosage. Interpretation of this result proved ambiguous because depolarization is a secondary effect of AChR activation. In more recent studies, agonist-induced current has been the preferred response (see Fig. 1): induced current, which may be measured rapidly and with great precision, directly monitors the open-channel receptor state. However, a comparably accurate and rapid measure of agonist concentration has proven technically difficult to obtain, so that many dose–response studies have limited value. The necessity for rapid concentration measurements arises because AChRs desensitize in the presence of agonists, reducing the activation response in a rapid, dose-dependent manner. Agonist-induced current has been measured directly by voltage-clamp and indirectly by monitoring tracer flux from vesicles using a rapid-mix quenched-flow technique. The main conclusions from several studies using different techniques are that agonist molecules bind at two sites on the AChR protein to open the channel. The apparent equilibrium dissociation constants between agonists and the undesensitized receptor range from one to several hundred micromolar; the binding steps occur independently or possibly with modest positive cooperativity.

5.2. Two ACh Molecules Generally Cooperate to Open the AChR Channel

Dose–response studies from four different laboratories have been reported using a variety of preparations and approaches which allow agonist concentration to be measured. Their results are in good agreement. Lester *et al.*[149] and Sheridan and Lester[150] bath-applied agonists to voltage-clamped *Electrophorus* electroplaques. Although the induced responses desensitized rapidly, they gave nonintegral Hill slopes between 1 and 2, and provided a lower limit for the equilibrium dissociation constants: e.g., carbamylcholine 300 μM (20–22°C); ACh 10–15 μM (15°C). Adams[151,152] bath-applied 13 different agonists to voltage-clamped frog skeletal muscle endplates. Since exposure times of approximately 1 min were used, agonist concentrations were kept low to minimize desensitization. Hill slopes of 2 rather than 1 or 3 fit the data; the apparent equilibrium dissociation constants at 18–24°C were, e.g., carbamylcholine 86 μM and methyl-TMA 240 μM. Dionne *et al.*,[41] also using voltage-clamped frog endplates, applied agonist focally with brief 1- to 2-sec current pulses from iontophoretic pipettes. Agonist concentration was measured with a sensitive ion-selective microelectrode (response time 10 msec), and dose–response curves constructed from a series of responses induced by concentration transients of different amplitudes (Fig. 1). Although exposure to agonists was brief, the Hill slopes were between 1 and 2. The apparent equilibrium dissociation constants at 15–17°C were: carbamylcholine 300 μM, HPTMA 25 μM, and suberyldicholine 25 μM. Neubig and Cohen[153] measured the initial rate of $^{22}Na^{+}$ efflux from *Torpedo* postsynaptic membrane vesicles with a quench-flow apparatus capable of millisecond time resolution. For carbamylcholine they found a Hill slope of 1.97 ± 0.06 and an apparent equilibrium dissociation constant of 600 μM at 4°C. In all these studies it was concluded that two agonist molecules bound to each channel to induce opening. The nonintegral slopes of Dionne *et al.*[41] were accounted for by the hypothesis that some channels opened after binding only one agonist, although such openings occurred with low probability. Agonist binding exhibited little or no cooperativity.

Since dose–response studies of the AChR are best performed quickly enough to avoid desensitization but slowly enough that the activation kinetics are at equilibrium, experiments have been designed to use relative rather than absolute agonist concentration. Accurate relative concentration changes have been made quickly by two-electrode iontophoresis[154] and quantitative, current-controlled iontophoresis.[155–157] These methods yield uncalibrated dose–response curves in the sense that the absolute agonist concentration is not known. From the slopes of uncalibrated curves the number of agonist-binding sites per channel can be estimated, but direct information on agonist affinity is lost. Results have been interpreted as evidence for two[154,157] and three[155,156] binding sites per AChR channel. In the latter studies the authors also estimated a nominal transfer number for their iontophoretic pipettes and used this to calculate the apparent equilibrium dissociation constants at 20–23°C for three agonists: ACh 27.8 μM, carbamylcholine 336 μM, and suberyldicholine 18 μM. These affinities are in excellent agreement with the previously sited values obtained by directly measuring agonist concentration. Thus, the disagreement on the number of binding sites per AChR channel is not easily explained. Nevertheless, quantitative iontophoretic studies are especially sensitive to the stability of the transfer characteristics of iontophoretic electrodes; this may account for the discrepant conclusions.

In a few studies the effects of extrinsic variables such as voltage and temperature upon the dose–response parameters have been reported. Sheridan and Lester[150] found that the apparent affinity constants between agonist and receptor depended on membrane voltage while the Hill slope did not. Hoffman and Dionne[157] confirmed this latter result using a rapid dose–response method which minimized desensitization; in addition, they reported that the Hill slope was independent of temperature. In recent years, since these dose–response studies were performed, additional results from antagonist binding,[44] electrophysiology,[e.g., 158] biochemistry, and electron microscopy (see Section 2) have supported the main conclusion of two agonist-binding sites per AChR channel.

6. Kinetics of Channel Gating

6.1. Introduction

The time course of AChR channel opening and closing contains information about the mechanisms of gating by this protein. Gating kinetics may be studied at equilibrium using noise and single-channel methods and under nonequilibrium conditions with selective perturbations. In the equilibrium case, one applies ACh and measures the rapidity and intensity of fluctuations in current that arise from AChR channels opening and closing; these fluctuations are due to random thermal perturbations of the gating mechanism and reflect its properties. Gating properties are similarly revealed by nonequilibrium kinetic studies in which controlled perturbations of sensitive variables such as voltage are made. Here the relaxation time course of macroscopic AChR-mediated currents induced by the perturbation is monitored as the population of AChRs returns to equilibrium.

Both equilibrium and nonequilibrium methods have been used productively by neurobiologists studying the AChR gating kinetics, and their results are summarized here. As with other physiological studies of ion-channel function, the preferred measurement is the channel-mediated passive ionic flux. The highest resolution data have been obtained with voltage-clamp methods which monitor transmembrane current. By comparison, tracer-ion flux methods provide valid information at equilibrium and for desensitization kinetics, but are too slow and insensitive to resolve AChR activation kinetics.

The AChR channel opens in response to binding suitable agonists. More accurately, the probability of channel opening increases following binding, but opening is not obligatory; when agonist is bound, the channel may reversibly make transitions between opened and closed states. In addition, the liganded AChR can desensitize, and in some cases the open channel appears to experience a brief, reversible closing process (excess closed gaps) that is distinct from the gating mode. The kinetic scheme in Fig. 2 summarizes present ideas of AChR channel activation. The actual conformational changes that open and close the channel occur with rate constants α and β. Note that this scheme does not incorporate states for desensitization or the excess closed gaps, since the relations of those processes to binding and gating are not yet clearly understood.

A mechanism for channel gating consistent with this kinetic model was first proposed by Magleby and Stevens.[159] According to their model, the physical gate was envisioned as being coupled to a charged portion of the AChR protein. In response to

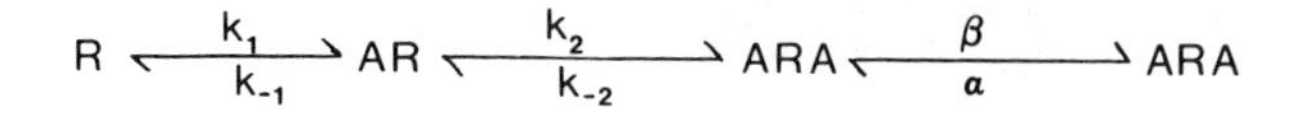

Fig. 2. Scheme 1: the basic kinetic scheme for AChR channel activation. The binding of two agonist molecules (A) to the receptor channel (R) is depicted sequentially and characterized by the macroscopic transition rates k_i, $i = \pm1, \pm2$. Here the binding occurs reversibly to the closed channel only. The gating transitions which open and close the channel are defined by the rates α, β. These are the essential steps which have been described for AChR activation. The process of desensitization and the phenomenon of excess closed gaps involve additional states not shown here. Free agonist molecules have not been depicted.

occupation of binding sites by ACh, a conformational change in the protein can open or close the ion channel. This isomerization reorients a portion of the protein within the membrane electric field, providing a source for the voltage dependence of gating kinetics. The macroscopic current responses should reflect both the binding steps and the protein conformational change, but if binding is rapid compared to isomerization, the time course of these responses will be dominated by the conformational change. High-resolution measurements made possible by single-channel recording have allowed the interdependence of binding and isomerization to be examined. To complete this model, a mechanism for desensitization and some process to account for the excess closed gaps which occur with some AChR channels should be included.

6.2. Single-Channel and Macroscopic Methods Are Complementary

Two different approaches are used now for kinetic studies: single-channel methods[160] and macroscopic (ensemble) current methods. Both rely upon the voltage-clamp, but each has its unique advantages and limitations, so that the two provide complementary information.

The macroscopic current methods give information on the average kinetic behavior of a large population of channels quickly. Both synaptic and nonsynaptic AChRs in many cell types may be examined under relatively physiological conditions. The major limitations are that resolution (both temporal and amplitude) is low, small cells are difficult to study, and regional distributions of channels on a cell membrane often may be impossible to separate.

Single-channel recording provides significantly improved performance in these areas: membrane fragments, small cells, and even cellular processes may be examined at wider bandwidth (ca. 2- to 10-fold) and with greater sensitivity (ca. 1000-fold). But to take advantage of single-channel methods, it is sometimes necessary to compromise cell and membrane integrity. Furthermore, more accurate estimates of certain rate constants can be made from a single 10-msec measurement of macroscopic currents generated by a large population of AChRs than from hours of observations of single-channel behavior.

Freshly isolated cells from animal tissues tend to be unsuited for single-channel recording, presumably because an extracellular matrix of collagen fibrils and glycoproteins interferes with seal formation.[161] Enzymes can remove the matrix, but may alter channel function. One means to avoid this problem has been the use of cultured cells which often produce less connective tissue. However, culture introduces problems of expression and cell identification; the cells in culture may not be the ones of interest or may not express channels with the desired physiology or pathophysiology.

6.2.1. Nonequilibrium Methods

Nonequilibrium kinetic studies are examples of classical perturbation methods. In general, the distribution of AChRs among the allowed kinetic states is suddenly forced to a nonequilibrium condition and the time course of relaxation back to equilibrium observed. For example, a rapid ACh concentration transient might drive a large number of channels open, and the time course of their return to the closed state might be observed. Both the amplitude and the time course of the response are determined by the transition rate constants between states, from which the rates themselves may be evaluated. Since current is the measured parameter, only the population of open AChR channels is directly monitored, and from this the presence of closed-channel states must be inferred.

Agonist concentration and membrane voltage have been the most useful variables for nonequilibrium studies. Nerve-evoked and miniature endplate current are caused by brief, intense impulses of ACh released into the synaptic cleft from the nerve terminal[(159)] and rapidly hydrolyzed. Similar impulses have been made with synthetic agonists by iontophoresis, producing responses which mimic miniature endplate currents.[(154)] Concentration jumps resembling a step change have been produced with the photoisomerizable agonists Bis Q and QBr by flash-induced interconversion between the *cis* (inactive) and the *trans* (active) isomers.[(162,163)] Each of these concentration perturbations shifts the AChR activation equilibrium, thus altering the probability that channels will be open.

Voltage provides an equally effective perturbation, since it also affects the probability that channels will be open.[(164)] Although voltage is easily manipulated, voltage-jump experiments[(165,166)] require exceptionally stable recording conditions, making these measurements especially sensitive to desensitization and other nonstationary processes.

In principle, temperature and pressure should also provide effective perturbations for kinetic studies since they affect the transition rates. However, the experimental problems of manipulating these variables on a submillisecond time scale have limited their use.

6.2.2. Equilibrium Methods

Equilibrium kinetic studies rely on spontaneous fluctuations in the number of open AChR channels. In effect, these fluctuations are random perturbations, and their formal treatment is similar to that for nonequilibrium studies. The relaxation time course from these random perturbations is given by their autocorrelation function.

Both single-channel studies and macroscopic endplate current noise provide kinetic information about channel behavior at equilibrium. However, most kinetic analyses of single-channel measurements have not been done with noise methods. The original AChR noise studies examined endplate voltage[(167,168)] and current[(169)] fluctuations produced by many simultaneously active channels; measurement of bandwidth of those macroscopic signals was less than available with modern single-channel studies, and interpretation of the results in terms of mean open-channel lifetime and conductance depended upon the assumption of a simple open–shut gating mechanism. More recently, single-channel records have verified this assumption,[(170)] and analytical techniques have been developed for single-channel kinetic studies. These techniques can be more informative than macroscopic noise, because individual transitions into and out of the open state can be resolved.

The principle used for kinetic analysis of single-channel data is identical to that for nonequilibrium macroscopic currents: specific perturbations are employed to obtain the time-dependent probability that channels will be open. However, here the perturbations are single transitions between opened and closed channel states, and the average relaxation from a perturbation is determined using a histogram.[(171,172)] Just as for the relaxations of macroscopic currents, these histograms can be described by functions of the kinetic transition rates.

Consider the histogram of open-channel duration as an example. An opening transition is the event which defines the start time of the histogram. The opening transition is a perturbation in the sense that the population of open channels is suddenly incremented by unity. The time-dependent expectation that a channel will close after this perturbation is given by the open-duration histogram which decays exponentially with a time constant equal to the mean lifetime of the open channel. This and other similarly constructed histograms are mathematically defined by probability density functions which may be derived from the kinetic models and fit to the data so portrayed.

6.3. Macroscopic Measurements Provided the First Kinetic Results

Early kinetic measurements revealed but a single rate-limiting process in the activation of AChRs. That is, EPCs, MEPCs, voltage jumps, concentration jumps, and noise autocorrelation functions all relaxed as single-exponential functions of time.[(159,165,169)] Under identical conditions at low agonist concentration, the relaxation time constant (T_d) measured from different responses was the same, indicating that the same physical processes were being revealed. At room temperature and 0 mV, T_d usually has a value of 1–2 msec; it increases *e*-fold with approximately 125-mV hyperpolarization, and has an activation enthalpy near 18 kcal/mole with ACh.[(173,174)] Since its temperature dependence is large and its voltage dependence persists in the presence of an anticholinesterase,[(159)] neither diffusion nor ACh hydrolysis seems to determine the relaxation time course. Instead, these data suggest that T_d reflects the gating process of AChR channels at the endplate.

In the limit of very low agonist concentration, Scheme 1 predicts a dominant decay time constant of

$$T_{d(\text{Scheme 1})} = 2/\{\alpha + \beta + k_{-2} - [(\alpha + \beta + k_{-2})^2 - 4\alpha k_{-2}]^{1/2}\}$$

By analogy to enzyme kinetics, it was argued that the unbinding rate k_{-2} should be much greater than the channel opening rate β. If so, then the expression for $T_{d(\text{Scheme 1})}$ reduces to

$$T_{d(\text{Scheme 1: } k_{-2} \gg \beta)} = 1/\alpha$$

That is, the relaxation process resolved by macroscopic measurements is determined by the rate of channel closure (given certain assumptions) and can be interpreted as the mean open lifetime of AChR channels.

Although these early kinetic studies showed simple relaxation characteristics, Scheme 1 actually predicts more complicated time courses for the noise autocorrelation function, voltage-jump relaxation, and evoked currents. Both the amplitudes and the rise times of these responses are expected to scale with several kinetic rates together with agonist concentration and the number of AChR channels, rather than depend on a single rate. Thus, based on Scheme 1, the simple observed behavior must reflect the limited measurement resolution. These simple relaxations have made estimates of the transition rates other than α difficult to obtain.

The macroscopic currents at tonic muscle endplates of the garter snake differ from those in twitch muscle by relaxing with two kinetic components instead of one. Although this could imply that tonic muscle AChRs differ from those in twitch muscle, that conclusion seems to be incorrect. Instead, in tonic muscles the relaxation time constants are longer and more easily resolved. Scheme 1 describes AChR kinetics in both fiber types. The differences can be explained if the lifetime of the doubly liganded, closed state of tonic muscle AChRs is long enough to detectably alter the macroscopic relaxations, while being too short to be resolved in twitch muscles. The more complicated relaxations in tonic muscles allow β and k_{-2} as well as α to be estimated by fitting Scheme 1. At 12°C and -100 mV, β = 220/sec and k_{-2} = 400/sec; β appeared to be independent of voltage while k_{-2} increased *e*-fold for 108-mV depolarization. The reciprocal of the sum of β and k_{-2} predicts the lifetime of the doubly liganded, closed state. In tonic fibers at 12°C, both doubly liganded opened and closed-channel lifetimes are estimated to have millisecond values which increase with hyperpolarization.[175,176]

The interpretation from macroscopic results that T_d reflects open-channel lifetime was based on the assumption that $k_{-2} \gg \beta$. Estimates of β and k_{-2} from snake tonic muscles and other estimates using single-channel measurements on twitch muscles (below) can be used to evaluate the accuracy of this assumption. Endplate AChRs in both twitch and tonic muscles of the garter snake show a value for the ratio of k_{-2}/β between 2 and 10. Thus, T_d only approximates the open-channel lifetime of these receptors. T_d is longer than the correct open lifetime because the channel is capable of reopening. Independent of this, both T_d and the open-channel lifetime are sensitive to different agonists; carbamylcholine gives about one-third and suberyldicholine about three times the ACh values.

6.4. Single-Channel Measurements Revealed Detail and Subtle Complexities

The first single-channel studies of AChR activation kinetics were made with bandwidth comparable to that available macroscopically; these tended to confirm the macroscopic results. Since then, more sensitive measurements have revealed a rich and unexpected kinetic complexity which is not yet fully understood. Some AChR channels have been reported with kinetics well described by Scheme 1, while other studies find kinetics that are more complicated. The major complexities relate to both channel conductance and the number of kinetic states, and may be due to different kinds of AChRs.

AChR kinetics have been carefully studied with single-channel techniques in only three preparations. Colquhoun and Sakmann used perijunctional AChRs along the margins of adult frog neuromuscular junctions in twitch muscles.[177] Sine and Steinbach used the clonal cell line BC3H-1,[178] and Leibowitz and Dionne used junctional AChRs on twitch muscle endplates in garter snakes.[179] Only the junctional AChRs in the latter study seem to be well described by kinetic Scheme 1.

6.4.1. Conductance

In several different preparations the AChR channel exhibits a single open-channel conductance value in the vicinity of 30–50 pS.[178–181] However, more than one kind of AChR channel, based upon mean conductance, has been found in some tissues, and some channels appear to have main and subconductance states. In cultured preparations of embryonic rat muscle, AChR channels are reported to have different primary conductance levels.[182,183] In addition, these channels and the AChR channels in cultured embryonic chick muscle[184,185] exhibit two nonzero conductance states, the main level and a sublevel with 10–30% of the main level conductance.

The observation of two types of AChR channels is not limited to cultured preparations of embryonic tissue. In nonjunctional muscle from mature *Xenopus* tadpole tail, two classes of AChR channels have been described.[186] These have conductances of 64 and 44 pS, similar reversal potentials, and are blocked by α-bungarotoxin. In addition to their conductances, these channels also have different mean open lifetimes, the larger conductance type being shorter-lived. Although the channels were observed in nonjunctional membrane, they closely resemble the "junctional" (larger conductance, shorter lifetime) and "nonjunctional" channel types which have been described with evoked macroscopic currents in this preparation.[187,188]

6.4.2. Open Channels

The presence of a single class of open-channel states is indicated by an exponential open-duration distribution; multiexponential distributions are evidence of more than one class. It is not necessary for different open-channel states to have different conductances. More specifically, open-duration distributions with an excess of short openings have been reported for perijunctional AChRs in frog skeletal muscle[177] and for AChRs in cultured cells.[178] These distributions are well fitted by the sum of two exponentials, with the implication that there are at least two different open-channel states of the same conductance in these preparations. In contrast, the open-duration distribution for snake endplate receptors has a single component, although two-component distributions have been recorded occasionally (M. Leibowitz and V. Dionne, unpublished). And in tadpole extrajunctional membrane where two kinds of AChR channels can be distinguished by their single-channel currents, the open-duration distribution of each seems to be exponential. Although bandwidth limitations of the recording equipment can introduce some distortion in these distributions, this does not appear to account for the discrepant observations.

The open-duration distribution expected from Scheme 1 is exponential with a time constant equal to the mean open lifetime of channels. This scheme cannot account for a multiexponential open-duration distribution. It is not clear what accounts for such distributions, but two general explanations are possible: either several channel types exist in some preparations, each with different mean open lifetimes, or the channels are kinetically complex.

6.4.3. Closed Channels

Complexities in the distributions of closed intervals between openings which are inconsistent with Scheme 1 have also been observed. As with the open durations, these complexities are seen with AChR channels from the perijunctional and nonjunctional preparations. However, there are insufficient data to conclude that only extrasynaptic AChR channels show this behavior.

The histogram of closed intervals between openings is called a first-latency histogram, since it registers the times at which the first opening transition occurs after each closure. This histogram cannot be described by a single exponential for any preparation of AChRs that has been examined, indicating the presence of more than one closed-channel state. At low agonist concentration the histogram resembles the sum of two or more exponential components.

The time course of the first-latency histogram predicted by Scheme 1 is the sum of two exponential components at low agonist concentrations. Consider a hypothetical record from a small number of Scheme-1-like AChR channels exposed to a concentration of agonist much less than its equilibrium dissociation constant. For any receptor the agonist sites will be occupied infrequently, and openings will occur even less frequently. These independent openings arising directly as the result of binding will contribute a component to the first-latency histogram whose time constant depends on agonist concentration. In addition, there will be a second component in the histogram arising from temporally correlated openings because doubly liganded receptors can flicker between opened and closed states. This component will have a concentration-independent time constant approximating the lifetime of the closed state ARA.

Studies with endplate AChRs from garter snake twitch muscles give first-latency results that agree with this prediction from Scheme 1. These studies provide estimates of the opening rate β (ca. 1000/sec at -100 mV, 22°C) and the apparent unbinding rate k_{-2} (ca. 4000/sec at -100 mV, 22°C). At -100 mV and 22°C the lifetime of the closed state ARA is nominally 200 μsec and the open-channel lifetime 2.0 msec. Both β and k_{-2} increase with hyperpolarization requiring approximately 100 mV per e-fold change. Similar qualitative agreement with Scheme 1 is found for both types of AChR channels seen in mature tadpole tail nonjunctional membranes.[186]

In contrast to the behavior predicted by Scheme 1, several single-channel studies have reported more than two components in the first-latency histogram. These indicate the presence of added closed-channel states of the AChR, some of which do not reflect the gating kinetics in which we are interested here. For example, desensitization can produce two additional components which have time constants of several hundred milliseconds and several tens of seconds.[189] At low agonist concentrations the desensitization components should have negligible amplitudes. Channel block by agonist occurs at high agonist concentration,[178,190] introducing a component in the first-latency histogram that increases in amplitude and decay rate with agonist concentration; this is not normal gating.

A most perplexing extra component in the first-latency histogram occurs in some preparations even at very low agonist concentrations. In these cases, two rapidly decaying components of the histogram can be resolved, but only one can be attributed to the mode of channel gating that has been described using Scheme 1. Typical time constants are 50 and 200–1000 μsec. The mechanism underlying the added component is not known. Furthermore, determining which if either of the components is dominated by gating is difficult. With snake endplate AChRs the faster component is not often seen, leaving the slower component to be interpreted as gating. However, with frog perijunctional AChRs both components are detected; there the faster component has been attributed to gating and the slower component suppressed.[177] This interpretive difference may account for a 10-fold difference in published estimates of the opening rate β (frog: $> 10{,}000$/sec[177]; snake: ca. 1000/sec[179]).

In AChR kinetic studies using the cell line BC3H-1, a multicomponent first-latency distribution which is qualitatively similar to that from frog perijunctional receptors has been reported.[178] Here the amplitude of the fastest component does not depend on agonist concentration in the manner expected if it were to arise from transitions between doubly liganded opened and closed states. Thus, this very brief component indicates that channels can close by a process different from the gating transition. The implication is that the slower of the two brief components in these first-latency histograms is dominated by channel gating, while the faster component reflects the occurrence of an unknown mechanism.

6.5. Do We Know as Much about the AChR Channel Now as Before?

The apparent differences between junctional, perijunctional, and nonjunctional AChR channels which have been revealed by single-channel studies relate to the number of opened and closed channel states. Although it would be simpler if all nicotinic AChR channels behaved similarly, considerable evidence here suggests otherwise.

Two extremes of kinetic models can explain these results (together with others more general), but no resolution between them is yet possible. Either a variety of AChR channel types exist with essentially different kinetic properties, or there is a single type of channel with complicated kinetics and regulatory mechanisms which modify its behavior. Distinguishing in detail between these two not mutually exclusive alternatives can be rather difficult. The results here reveal a variety of nicotinic AChR channel types differing in kinetics and conductance. Possibly the more important questions are whether different channel types can interconvert, and if so by what means and under what control. Either of these possibilities could accommodate the endplate AChR results and the spectral differences between junctional and nonjunctional AChR current noise. For example, differences between junctional and nonjunctional channels could arise if the formation of AChR clusters at a synapse affected channel gating properties. Such effects could be associated with changes in the lipid environment or with connection to a cytoskeletal structure.

7. AChR Cation Selectivity and Permeation

7.1. Selectivity

During the brief time an AChR channel remains in its open conformation, about 20,000 ions/msec normally pass through the water-filled pore. Anions are completely excluded, and some cations pass through more easily than others.[191–194] An analysis of the relative permeabilities of various ions is instructive because it explains how one thinks about ion permeation and gives some information about the environment experienced by an ion within the pore. The key concept in modern considerations of ion permeability is the channel's *energy diagram.*[195] This diagram specifies the free energy of an ion at each fraction of the distance through the channel and serves to summarize the environment a pore provides for each ionic species. The rate of ion movement through the channel is, at least under certain specific circumstances, governed by energy barriers—those locations within the channel that are energetically least favorable for ion occupancy—and one usually assumes that a particular barrier dominates and thus is rate limiting for ion transmigration. The assumption of a single rate-limiting barrier, known as the *selectivity site,* is a reasonable one because ion occupancy of a region depends exponentially on the energy at that location, so relatively small differences in energy translate into much larger differences in occupancy probability. Different ion species have different energy barriers: for the AChR channel, for example, Cs^+ has a lower barrier than Li^+ and is thus more permeant. Energies of ions within the channel are all relative to the hydration energies of the ion in question. For chemically nonreactive ions—that is, those ions with a noble gas outer electron configuration (alkali metals like Na^+ and alkaline earths like Ca^{2+})—the ion's interactions with its environment are almost entirely electrostatic. Each ion's selectivity barrier can, in principle, be calculated from a knowledge of the protein structure that forms the pore, and involves terms for ion–dipole, ion-induced dipole, and ion–quadrupole interactions, etc.

Consider the relationship between barrier height and ion size.[194] For the weakly selective AChR channel, a power series expansion of selectivity barrier height as a function of ion size (measured as the reciprocal of the ionic crystal radius) may be terminated after the linear term. This approximation is justified because "weakly selective" means that barrier height is a slowly varying function of ion size. The proportionality constant that relates ion size to barrier height is composed of two factors: the first, weighted by the ion valence, is the ion–dipole interaction, and the second, weighted by the ion valence squared, is the ion-induced dipole interaction. In both cases, the energy terms specify the difference between electrostatic interactions of the ion with its environment within the pore as compared to free solution. The fact that large monovalent ions are more permeant than smaller ones means that this proportionality constant is positive for monovalent ions; the reverse relationship for divalents implies a negative proportionality constant. This sign reversal can occur if the two factors that compose the proportionality constant have opposite signs because the weighting of the two factors is different for mono- and divalent ions: the weighting is 1 : 1 for monovalents and 2 : 4 for divalents. Data on the barrier height as a function of ion size for mono- and divalent ions permit an estimate of the ion–dipole and ion-induced dipole factors. Such estimates reveal that the environment for the ion in the channel is only a fraction of a percent different from the free aqueous environment, and suggest that the AChR pore is water-filled and that ions are, at least in the selectivity region, completely surrounded by water.

7.2. Permeation

The conductance of an AChR channel is, in normal circumstances, about 25 pS.[169] This implies that, with the resting potential as a driving force, about twenty million ions per second traverse the channel, a very considerable transport rate indeed. Ions can, however, interact within the channel in a rather complex way. For example, increasing the Ca^{2+} ion concentration in the extracellular medium from 2 to 10 mM (remember that Ca^{2+} ions can pass through the channel) causes the single-channel conductance to decrease by about a half: Ca^{2+} ions can permeate but also can block the channel.[196,197] A variety of permeation data can be quantitatively accounted for by an energy diagram for the channel that has three barriers and two wells. Such an energy diagram predicts the correct shapes for channel current–voltage relationships, and also accounts for ion blocking effects of the sort mentioned above. The challenge for future investigations will be to explain this or an alternative energy diagram for permeation in terms of a proposed channel structure of the sort described earlier.

8. Ligand-Gated Channels Other Than the AChR

Because the nicotinic AChR channel has been the most accessible to both biochemical and electrophysiological study, it has become a general model for transmitter-activated ion channels. A large number of transmitter molecules are used by the brain for synaptic transmission and it is important to ask whether details of the AChR channel can be generalized to other channel types. Recently, a number of different transmitter-activated channels have been investigated with single-channel recording and their properties are beginning to be elucidated. In this section we will summarize the results of these studies and compare their results with those of the better understood nicotinic AChR channel.

8.1. Glutamate-Activated Channels

Single glutamate-activated channel currents from locust muscle were first reported by Patlak *et al.* in 1979.[198] These channels (extrajunctional D-receptors) are activated by quisqualate, glutamate, fluoroglutamate, and cysteine-sulfinate[199,200] and were found to have a surprisingly large conductance (120 pS), about 5 times that of the AChR channel. In fact, the large currents controlled by these channels have allowed their kinetics to be studied in detail without the improvements in signal-to-noise ratio that can be obtained with gigaohm seals. The first report showed that the kinetics of the channel were complex, with transitions between mostly open and mostly closed behavior spontaneously occurring under constant conditions. It has been suggested that this nonstationary behavior is a result of changes in the affinity for the agonist in time or changes in the microenvironment of the channel.[201,202]

Desensitization of glutamate-activated channels occurs and is seen as a bursting of channel openings at higher agonist concentrations, similar to the behavior of the AChR channel. The channel gating in the absence of desensitization can be studied, however, in concanavalin A-treated muscle fibers.

Initial results suggested that, in the absence of desensitization, both open and closed time distributions could be fitted with single exponentials, indicating that the channel normally had only single open and closed states.[198,199] The rates for transitions between these states are not very dependent upon membrane potential and vary with the species of agonist in the sequence quisqualate > glutamate > fluoroglutamate. The open times are independent of agonist concentration in the 100–600 μM range,[199] as would be expected for a single binding–dissociation mechanism, but increase with higher concentrations.[201] This dependence suggests that a second agonist molecule can bind to an open channel and stabilize its conformation. The closed times decrease with agonist concentration as the 1.4th or the second power of the concentration.[199,200] This higher-order dependence suggests that, like the AChR channel, there are two agonist molecules bound for the channel to open and would predict a double-exponential closed time distribution. The fact that the closed time distribution can be fitted with a single exponential could mean that the binding of one of the agonist molecules is much faster than the binding of the other.

These results are consistent with the following state diagram for the channel:

$$\mathrm{C} \overset{A}{\rightleftharpoons} \mathrm{C} \overset{A}{\rightleftharpoons} \mathrm{O} \overset{A}{\rightleftharpoons} \mathrm{O}$$

Two agonist molecules bind to the channel, causing it to open. An additional agonist can bind to the open state, further stabilizing it. This binding site probably has a much lower affinity for agonist because the concentration dependence of the open times is not seen at lower concentrations.

With improvements in the frequency response of the recording system, a faster class of openings and closings has been found.[203] What seemed to be individual opening events before were now found to be fast bursts of openings interrupted by brief closings. Open and closed time distributions were now found to be fitted by a sum of two exponentials. Because of the similarity to AChR kinetics, the authors have interpreted the data in a similar way to the analysis of Colquhoun and Sakmann in 1981[177] on fast openings and closings of the AChR channel.

According to this scheme, agonist molecules bind to the channel which then quickly opens. Upon closing, the channel can either reopen or the agonist can unbind, terminating the burst. This translates to the following kinetic scheme:

$$\mathrm{A} + \mathrm{C} \underset{k_{-1}}{\overset{k_1}{\rightleftharpoons}} \mathrm{AC} \underset{\alpha}{\overset{\beta}{\rightleftharpoons}} \mathrm{AO}$$

where A represents the agonist molecule, C a closed channel, and O an open channel. The rate constants k_1 and k_{-1} are the rates for agonist binding and unbinding and α and β are the rates for opening and closing. According to this scheme, fast closings during a burst result from the transitions AO–AC–AO. Their mean duration is $1/(\beta + k_{-1})$. The mean number of openings per burst is $1 + \beta/k_{-1}$. Cull-Candy and Parker have found that, in terms of this model, a channel opens an average of 1.6 times during the time an agonist molecule is bound.[203] An alternative mechanism for the brief interruptions is that they result from transient blockage of the channel. This possibility has not been thoroughly tested, but it seems clear that channel block by agonist molecules is not responsible because the number of brief closings per burst did not depend on agonist concentration.

The occurrence of fast openings in the open time distributions has not been studied in detail and has not been incorporated into a model for channel kinetics. Based on analogy with work on AChR channels, they have suggested that the brief openings may be the result of channels opening when only one agonist molecule is bound. This interpretation is qualitatively consistent with an increase in open duration with agonist concentration.

Other workers have found that there are fewer shorter openings than would be expected from a single-exponential distribution.[204]

These workers have proposed a cyclic model to explain the nonexponential open duration distributions and the dependence of the open duration upon agonist concentration:

$$\begin{array}{ccccl} n\mathrm{A} + \mathrm{R} & \rightleftharpoons & \mathrm{A}_n\mathrm{R} & & \text{closed} \\ k_4 \uparrow & & \downarrow k_2 & & \\ n\mathrm{A} \quad \mathrm{R}^* & \leftarrow & \mathrm{A}_n\mathrm{R}^* & & \text{open} \end{array}$$

In this scheme, opening and closing are irreversible steps (k_4 and k_2). Because closing of channels in state AnR* must occur through state R*, there will be a deficit of short openings. The presence of irreversible steps requires that energy be supplied to the channel for its gating to occur in the steady state. This suggestion departs from the common assumption that channel gating can be explained in terms of equilibrium behavior. Because of the conflicting experimental observations made by the two groups, however, it remains to be seen whether a nonequilibrium scheme is necessary to explain the gating of the channel.

A single glutamate-activated channel type has also been studied in dissociated mouse CNS neurons.[205] This channel is activated by glutamate and *N*-methyl-D-aspartate, but not by quisqualate or kainate. Its conductance is 48 pS in magnesium-free solutions and it is strongly selective for cations over anions but does not strongly select among cations. When external magnesium is present in micromolar concentrations, a voltage dependence is seen in the single-channel current and the channel openings are interrupted with brief closing events. This behavior is likely due to a voltage-dependent block of the channel by magnesium. The voltage dependence of the block can account for the voltage dependence of the macroscopic current that is recorded when glutamate is applied to the cell.

8.2. GABA-Activated Channels

Single-channel currents from GABA-activated channels were first reported in cultured mouse spinal cord neurons.[206] These channels are activated by GABA, the mushroom toxin muscimol, and the general anesthetic (−)-pentobarbital. They have a conductance of 20–25 pS and are selective for chloride. The mean open duration of these channels depends upon the species of agonist, with (−)-pentobarbital- and muscimol-activated channels remaining open longer. For all three agonists there is evidence for a double-exponential open duration distribution, suggesting that a channel has two open states (or that two distinct channel populations are present). More recent work, with better frequency response, has shown that when muscimol is used as an agonist, both open and closed duration distributions could be fitted with the sum of two exponentials. The open duration distribution was found to be voltage dependent, with the time constants of both exponential components decreasing at more hyperpolarized potentials.[207]

The kinetics of the GABA-activated channel in mouse and rabbit spinal neurons and rat hippocampal cells have been studied in greater detail by Sakmann *et al.*[208] and Bormann *et al.*[209] Channel openings are found to have a burstlike appearance with brief closings on the order of 1 msec interrupting the opening. As in the case of the AChR channel[177] and the glutamate-activated channel,[203] these bursts of openings have been interpreted to reflect reopening of the channel during a single occupancy of the receptor by the agonist. Additional behavior which is similar to that of the AChR channel is a double-exponential open duration distribution and a marked bursting behavior at higher agonist concentrations, reflecting desensitization of the receptor. The mean duration of closed intervals within a burst decreases more than linearly with agonist concentration, indicating that the binding of more than one agonist molecule is required for the channel to open. This agrees with the results of macroscopic recordings that show that, at low agonist concentrations, the current rises more than linearly with increasing agonist concentration. The occurrence of two components in the open duration distribution and of three components in the closed time distribution requires a kinetic model with at least two open and three closed states, similar to that proposed for the AChR channel.

Glycine-activated channels have also been studied in the same cells and it has been found that, although the single-channel conductance is larger than that for GABA-activated channels (46 pS as compared to 29), the kinetic behavior and ion selectivity are quite similar. In fact, further studies have led to the suggestion that GABA and glycine receptors may be coupled to the same ion channel.[210]

Detailed study of single-channel currents from GABA- and glycine-activated channels reveal multiple conductance states. GABA activates two conductance levels of 19 and 30 pS, whereas glycine activates three levels of 45, 31, and 21 pS. The similarity of the lower two conductance levels of the glycine-activated channel to the two levels of the GABA-activated channel has led the authors to suggest that a single multistate Cl^- channel is coupled to glycine or GABA receptor subunits. The main substate of the channel is determined by the receptor subunit to which it is coupled. Further evidence for this scheme is that the ion selectivity of the shared conductance levels is identical for both agonists. The fact that the GABA- and glycine-activated channels are independently distributed among patches suggests that the channel is independently linked to either receptor type.

The selectivity sequence among anions for this channel (measured by reversal potentials) is $Br^- > I^- > Cl^- > F^-$, with sulfate and acetate ions being impermeant. On the other hand, the sequence of single-channel conductance is $Cl^- > Br^- > F^- > I^-$. This difference between permeability and conductance sequences indicates that permeant ions physically interact with the channel in some way.

8.3. Muscarinic ACh-Activated Channels

K^+ channels that are activated by muscarinic ACh receptors have been studied in the pacemaker cells of mammalian heart.[211] The kinetic behavior of this channel is similar to that of the nicotinic AChR channel. The openings occur in bursts, and there are multiexponential open duration distributions. One difference between nicotinic and muscarinic AChR channel kinetics is that the muscarinic receptor channel operates much more slowly.

8.4. Some Channels Use Second Messengers

Synaptic transmission in the nervous system can operate on many different time scales. In this chapter we have focused on the molecular properties of ion channels that underlie fast synaptic transmission, but important recent work also deals with slow synaptic transmission between various cells.[212,213] A common mechanism for some slow synaptic potentials is the agonist-induced closing of channels which are normally open. This closing is mediated through second messenger systems. Little work on the gating properties of these channels has appeared to date, but their study is rapidly progressing.

One system in which the properties of single channels underlying a slow synaptic potential have been studied with single-channel recording techniques is the sensory neurons of *Aplysia*. In these cells, serotonin elicits a long-lasting EPSP. The ionic basis for this synaptic potential is a decrease in the cell's K^+ conductance. Siegelbaum *et al.*[214] have found a class of K^+ channels that close for long durations when serotonin is applied to the cell. The channels are isolated from the serotonin in the bathing solution due to the presence of the gigaohm seal between the patch pipette and the plasma membrane. This leads to the conclusion that the serotonin effect is mediated by a second messenger. Accordingly, the results can be duplicated by the injection of cAMP into the cell. The gating of the serotonin-sensitive channels is not voltage or calcium dependent and serotonin and cAMP do not change the single-channel conductance or the lifetime of the channel in the open state. These channels seem to represent a class of channels that underlie slow synaptic transmission. Given their importance in neuronal transmission and the little that is now known about them, the study of slow synaptic channels should progress rapidly in the next few years.

8.5. Conclusion

In conclusion, the knowledge of the gating of transmitter-activated channels other than the nicotinic AChR is much more rudimentary than for that channel. It seems, however, that the mechanism of activation of fast synaptic channels is qualitatively similar for a number of receptor types. On the other hand, too few systems have been studied in enough detail to make generalizations about the properties of the channels and the gating mechanisms underlying slow synaptic transmission. It is clear, however, that the integration of molecular biology with single-channel recording methods will add enormously to our knowledge of the molecular basis of synaptic transmission. Both nicotinic AChR and GABA-activated channels have been expressed in frog oocytes which have been injected with foreign mRNA and studied with single-channel recording.[215] It is evident that there will be rapid progress in the study of the molecular properties of many transmitter-activated channels and that questions concerning the generality of mechanisms among these channels can be answered at a much more fundamental level.

9. An Emerging View of Transmitter-Activated Channels

The accumulated results of single-channel studies of a variety of transmitter-activated channels suggest that they may have a common mechanism of gating. Channels that are operated by nicotinic and muscarinic ACh receptors, GABA, and glycine all show similar patterns of channel opening and closing: Channel opening events are interrupted by brief closings, two exponential components exist in the open duration distributions, and multiple closed states exist. At high agonist concentrations, desensitization gives rise to a slow bursting behavior as channels transiently return from the desensitized state. In addition, it may be a general feature that multiple agonist bindings are required for the channel to open. Although there is some controversy concerning the experimental results, glutamate-activated channels probably fall into the same category. This type of kinetic behavior has generally been interpreted in terms of the Colquhoun and Sakmann scheme for gating of the nicotinic AChR channel.[177] The fast bursting behavior is interpreted as multiple openings of the channel during a single occupancy of the receptor by an agonist molecule. The two components of the open duration distribution are thought to reflect openings with different numbers of agonist molecules bound to the receptor.

Although the molecular details are as yet unclear for all channel types, common mechanisms seem to be operating in all cases.

References

1. Karlin, A. 1980. Molecular properties of nicotinic acetylcholine receptors. In: *The Cell Surface and Neuronal Function.* C. W. Cotman, G. Poste, and G. L. Nicholson, eds. Elsevier/North-Holland, Amsterdam. pp. 191–260.
2. Conti-Tronconi, B. M., M. W. Hunkapiller, J. M. Lindstrom, and M. A. Raftery. 1982. Subunit structure of the acetylcholine receptor from *Electrophorus electricus. Proc. Natl. Acad. Sci. USA* **79**:6489–6493.
3. Lindstrom, J., S. Tzartos, W. Gullick, S. Hochschwender, L. Swanson, M. Jacob, P. Sargent, and M. Montal. 1983. Use of monoclonal antibodies to study acetylcholine receptors from electric organs, muscles and brain and the autoimmune response to receptor in myasthenia gravis. *Cold Spring Harbor Symp. Quant. Biol.* **48**:89–100.
4. Noda, M., H. Takahashi, T. Tanabe, M. Toyosato, Y. Furutani, T. Hirose, M. Asai, S. Inayama, T. Miyata, and S. Numa. 1982. Primary structure of alpha-subunit precursor of *Torpedo californica* acetylcholine receptor deduced from cDNA sequence. *Nature (London)* **299**:793–797.
5. Noda, M., H. Takahashi, T. Tanabe, M. Toyosato, S. Kikyotani, T. Hirose, M. Asai, H. Takashima, S. Inayama, T. Miyata, and S. Numa. 1983. Primary structures of beta- and gamma-subunit precursors of *Torpedo californica* acetylcholine receptor deduced from cDNA sequences. *Nature (London)* **301**:251–255.
6. Noda, M., H. Takahashi, T. Tanabe, M. Toyosato, S. Kikyotani, T. Miyata, and S. Numa. 1983. Structural homology of *Torpedo californica* acetylcholine receptor subunits. *Nature (London)* **302**:528–532.
7. Claudio, T., M. Ballivet, J. Patrick, and S. Heinemann. 1983. Nucleotide and deduced amino acid sequences of *Torpedo californica* acetylcholine receptor gamma subunit. *Proc. Natl. Acad. Sci. USA* **80**:1111–1115.
8. Sumikawa, K., M. Houghton, J. C. Smith, L. Bell, B. M. Richards, and E. A. Barnard. 1982. The molecular cloning and characterisation of cDNA coding for the alpha subunit of the acetylcholine receptor. *Nucleic Acid Res.* **10**:5809–5822.
9. Devillers-Thiery, A., J. Giraudat, M. Bentaboulet, and J.-P. Changeux. 1983. Complete mRNA coding sequence of the acetylcholine binding alpha-subunit of *Torpedo marmorata* acetylcholine receptor: A model for the transmembrane organization of the polypeptide chain. *Proc. Natl. Acad. Sci. USA* **80**:2067–2071.
10. Lindstrom, J., J. Merlie, and G. Yogeeswaran. 1979. Biochemical properties of acetylcholine receptor subunits from *Torpedo californica. Biochemistry* **18**:4465–4470.
11. Vandlen, R. L., W. C.-S. Wu, J. C. Eisenach, and M. A. Raftery. 1979. Studies of the composition of purified *Torpedo californica* acetylcholine and of its subunits. *Biochemistry* **18**:1845–1854.
12. Reynolds, J., and A. Karlin. 1978. Molecular weight in detergent solution of acetylcholine receptor in *Torpedo californica. Biochemistry* **17**:2035–2038.
13. Lindstrom, J., R. Anholt, B. Einarson, A. Engel, M. Osame, and M. Montal. 1980. Purification of acetylcholine receptors, reconstitution into lipid vesicles and study of agonist induced cation channel regulation. *J. Biol. Chem.* **255**:8340–8350.
14. Wu, W. C.-S., H.-P. H. Moore, and M. A. Raftery. 1981. Quantitation of cation transport by reconstituted membrane vesicles containing purified acetylcholine receptor. *Proc. Natl. Acad. Sci. USA* **78**:775–779.
15. Froehner, S. C., and S. Rafto. 1979. Comparison of the subunits of *Torpedo californica* acetylcholine receptor by peptide mapping. *Biochemistry* **18**:301–307.
16. Raftery, M. A., M. W. Hunkapiller, C. D. Strader, and L. E. Hood. 1980. Acetylcholine receptor: Complex of homologous subunits. *Science* **208**:1454–1457.
17. Lindstrom, J., B. Walter, and B. Einarson, 1979. Immuno-chemical similarities between subunits of acetylcholine receptors from *Torpedo, Electrophorus,* and mammalian muscle. *Biochemistry* **18**:4470–4480.
18. Gullick, W., and J. Lindstrom. 1983. Mapping the binding of monoclonal antibodies to the acetylcholine receptor from *Torpedo californica. Biochemistry* **22**:3312–3320.
19. Karlin, A. 1983. The anatomy of a receptor. *Neurosci. Comment.* **1**:111–123.
20. Finer-Moore, J., and R. M. Stroud. 1984. Amphipathic analysis and possible formation of the ion channel in an acetylcholine receptor. *Proc. Natl. Acad. Sci. USA* **81**:155–159.
21. Cartaud, J. E., L. Bendetti, J. B. Cohen, J. C. Meunier, and J.-P. Changeux. 1973. Presence of a lattice structure in membrane fragments rich in nicotinic receptor protein from the electric organ of *Torpedo marmorata. FEBS Lett.* **33**:109–113.
22. Nickel, E., and L. T. Potter. 1973. Ultrastructure of isolated membranes of *Torpedo* electric tissue. *Brain Res.* **57**:508–517.
23. Klymkowsky, M. W., and R. M. Stroud. 1979. Immunospecific identification and three-dimensional structure of a membrane-bound acetylcholine receptor from *Torpedo californica. J. Mol. Biol.* **128**:319–334.
24. Kistler, J., R. M. Stroud, M. W. Klymkowsky, R. A. Lalancette, and R. H. Fairclough. 1982. Structure and function of an acetylcholine receptor. *Biophys. J.* **37**:371–383.
25. Ross, M. J., M. W. Klymkowsky, D. A. Agard, and R. M. Stroud. 1977. Structural studies of a membrane bound acetylcholine receptor from *Torpedo californica. J. Mol. Biol.* **116**:635–659.
26. St. John, P. A., S. C. Froehner, D. A. Goodenough, and J. B. Cohen. 1982. Nicotinic postsynaptic membranes from *Torpedo:* Sidedness, permeability to macromolecules, and topography of major polypeptides. *J. Cell Biol.* **92**:333–342.
27. Wennogle, L. P., and J.-P. Changeux. 1980. Transmembrane orientation of proteins present in acetylcholine receptor-rich membranes from *Torpedo marmorata* studied by selective proteolysis. *Eur. J. Biochem.* **106**:381–393.
28. Strader, C. D., and M. A. Raftery. 1980. Topographic studies of *Torpedo* acetylcholine receptor subunits as a transmembrane complex. *Proc. Natl. Acad. Sci. USA* **77**:5807–5811.
29. Froehner, S. C. 1981. Identification of exposed and buried determinants of the membrane bound acetylcholine receptor from *Torpedo californica. Biochemistry* **20**:4905–4515.
30. Anderson, D., and G. Blobel. 1981. *In vitro* synthesis, glycosylation and membrane insertion of the four subunits of *Torpedo* acetylcholine receptor. *Proc. Natl. Acad. Sci. USA* **78**:5598–5602.
31. Anderson, D., G. Blobel, S. Tzartos, W. Gullick, and J. Lindstrom. 1983. Transmembrane orientation of an early biosynthetic form of acetylcholine receptor delta subunit determined by proteolytic dissection in conjunction with monoclonal antibodies. *J. Neurosci.* **3**:1773–1784.
32. Ballivet, M., J. Patrick, J. Lee, and S. Heinemann. 1982. Molecular cloning of cDNA coding for the gamma subunit of *Torpedo* acetylcholine receptor. *Proc. Natl. Acad. Sci. USA* **79**:4466–4470.
33. Holtzman, E., D. Wise, J. Wall, and A. Karlin. 1982. Electron microscopy of complexes of isolated acetylcholine receptor, biotinyl-toxin and avidin. *Proc. Natl. Acad. Sci. USA* **79**:310–314.
34. Karlin, A., E. Holtzman, N. Yodh, P. Lobel, J. Wall, and J. Hainfeld. 1983. The arrangement of the subunits of the acetylcholine receptor of *Torpedo californica. J. Biol. Chem.* **258**:6678–6681.
35. Hamilton, S. L., M. McLaughlin, and A. Karlin. 1977. Disulfide

bond cross-linked dimer in acetylcholine receptor from *Torpedo californica. Biochem. Biophys. Res. Commun.* **79**:692–699.

36. Hamilton, S. L., M. McLaughlin, and A. Karlin. 1979. Formation of disulfide-linked oligomers of acetylcholine receptor in membrane from *Torpedo* electric tissue. *Biochemistry* **18**:155–163.
37. Fairclough, R. H., J. Finer-Moore, R. A. Love, D. Kristofferson, P. J. Desmueles, and R. M. Stroud. 1983. Subunit organization and structure of an acetylcholine receptor. *Cold Spring Harbor Symp. Quant. Biol.* **48**:9–20.
38. Reiter, M. J., D. A. Cowburn, J. M. Prives, and A. Karlin. 1972. Affinity labeling of the acetylcholine receptor in the electroplax: Electrophoretic separation in sodium dodecyl sulfate. *Proc. Natl. Acad. Sci. USA* **69**:1168–1172.
39. Karlin, A. 1969. Chemical modification of the active site of the acetylcholine receptor. *J. Gen. Physiol.* **54**:245s–264s.
40. Weill, C. L., M. G. McNamee, and A. Karlin. 1974. Affinity labeling of purified acetylcholine receptor from *Torpedo californica. Biochem. Biophys. Res. Commun.* **61**:997–1003.
41. Dionne, V. E., J. H. Steinbach, and C. F. Stevens. 1978. Voltage dependence of agonist effectiveness at the frog neuromuscular junction. *J. Physiol. (London)* **281**:421–444.
42. Damle, V., and A. Karlin. 1978. Affinity labeling of one of two alpha-neurotoxin binding sites in acetylcholine receptor from *Torpedo californica. Biochemistry* **17**:2039–2045.
43. Neubig, R. R., and J. B. Cohen. 1979. Equilibrium binding of [^{3}H]acetylcholine by *Torpedo* postsynaptic membranes: Stoichiometry and ligand interactions. *Biochemistry* **18**:5464–5475.
44. Sine, S. M., and P. Taylor. 1981. Relationship between reversible antagonist occupancy and the functional capacity of the acetylcholine receptor. *J. Biol. Chem.* **256**:6692–6699.
45. Dunn, S. M. J., and M. A. Raftery. 1982. Multiple binding sites for agonists on *Torpedo californica* acetylcholine receptor. *Biochemistry* **21**:6264–6272.
46. Haggerty, J. G., and S. C. Froehner. 1981. Restoration of ^{125}I-alpha-bungarotoxin binding activity to the alpha subunit of *Torpedo* acetylcholine receptor isolated by gel electrophoresis in sodium dodecyl sulfate. *J. Biol. Chem.* **256**:8294–8297.
47. Tzartos, S. J., and J.-P. Changeux. 1983. High affinity binding of alpha-bungarotoxin to the purified alpha-subunit and to its 27,000-dalton proteolytic peptide from *Torpedo marmorata* acetylcholine receptor: Requirement for sodium dodecyl sulfate. *EMBO J.* **2**: 381–387.
48. Gershoni, J. M., E. Hawrot, and T. L. Lentz. 1983. Binding of alpha-bungarotoxin to isolated alpha subunit of the acetylcholine receptor of *Torpedo californica:* Quantitative analysis with protein blots. *Proc. Natl. Acad. Sci. USA* **80**:4973–4977.
49. Oblas, B., N. D. Boyd, and R. H. Singer. 1983. Analysis of receptor–ligand interactions using nitrocellulose gel transfer: Application to *Torpedo* acetylcholine receptor and alpha bungarotoxin. *Anal. Biochem.* **130**:1–8.
50. Wilson, P. T., J. M. Gershoni, E. Hawrot, and T. L. Lentz. 1984. Binding of alpha-bungarotoxin to proteolytic fragments of the alpha subunit of *Torpedo* acetylcholine receptor analyzed by protein transfer on positively charged membrane filters. *Proc. Natl. Acad. Sci. USA* **81**:2553–2557.
51. Oswald, R. E., and J.-P. Changeux. 1981. Selective labeling of the delta subunit of the acetylcholine receptor by a covalent local anesthetic. *Biochemistry* **20**:7166–7174.
52. Blanchard, S. G., and M. A. Raftery. 1979. Identification of the polypeptide chains in *Torpedo californica* electroplax membranes that interact with a local anesthetic analog. *Proc. Natl. Acad. Sci. USA* **76**:81–85.
53. Oswald, R. E., and J.-P. Changeux. 1981. Ultraviolet light-induced labeling by noncompetitive blockers of the acetylcholine receptor from *Torpedo marmorata. Proc. Natl. Acad. Sci. USA* **78**:3925–3929.
54. Kaldany, R.-R., and A. Karlin. 1983. Reaction of quinacrine mustard with the acetylcholine receptor from *Torpedo californica:* Functional consequences and sites of labeling. *J. Biol. Chem.* **258**: 6232–6242.
55. Huganir, R. L., and P. Greengard. 1983. cAMP-dependent protein kinase phosphorylates the nicotinic acetylcholine receptor. *Proc. Natl. Acad. Sci. USA* **80**:1130–1134.
56. Tank, D. W., R. L. Huganir, P. Greengard, and W. W. Webb. 1983. Patch-recorded single-channel currents of the purified and reconstituted *Torpedo* acetylcholine receptor. *Proc. Natl. Acad. Sci. USA* **80**:5129–5133.
57. Patrick, J., and J. Lindstrom. 1973. Autoimmune response to acetylcholine receptors. *Science* **180**:871–872.
58. Lindstrom, J. M., M. E. Seybold, V. A. Lennon, S. Whittingham, and D. Duane. 1976. Antibody to acetylcholine receptor in myasthenia gravis: Prevalence, clinical correlates and diagnostic value. *Neurology* **26**:1054–1059.
59. Appel, S. H., R. Anwyl, M. W. McAdams, and S. Elias. 1977. Accelerated degradation of acetylcholine receptor from cultured rat myotubes with myasthenia gravis sera and globulins. *Proc. Natl. Acad. Sci. USA* **74**:2130–2134.
60. Drachman, D. B., C. W. Angus, R. N. Adams, J. D. Michelson, and G. J. Hoffman. 1978. Myasthenic antibodies crosslink acetylcholine receptors to accelerate degradation. *N. Engl. J. Med.* **298**: 1116–1122.
61. Heinemann, S., S. Bevan, R. Kullberg, J. Lindstrom, and J. Rice. 1977. Modulation of the acetylcholine receptor by anti-receptor antibody. *Proc. Natl. Acad. Sci. USA* **74**:3090–3094.
62. Lindstrom, J., and B. Einarson. 1979. Antigenic modulation and receptor loss in EAMG. *Muscle Nerve* 2:173–179.
63. Engel, A., M. Tsujihata, J. Lindstrom, and V. Lennon. 1976. The motor end-plate in myasthenia gravis and in experimental autoimmune myasthenia gravis: A quantitative ultrastructural study. *Ann. N.Y. Acad. Sci.* **274**:60–79.
64. Engel, A., K. Sahashi, E. Lambert, and F. Howard. 1979. The ultrastructural localization of the acetylcholine receptor, immunoglobulin G, and the third and ninth complement components at the motor endplate and the implications for the pathogenesis of myasthenia gravis. *Excerpta Med. Int. Congr. Ser.* **455**:111–122.
65. Tzartos, S., and J. M. Lindstrom. 1980. Monoclonal antibodies used to probe acetylcholine receptor structure: Localization of the main immunogenic region and detection of similarities between subunits. *Proc. Natl. Acad. Sci. USA* **77**:755–759.
66. Tzartos, S. J., M. Seybold, and J. Lindstrom. 1982. Specificity of antibodies to acetylcholine receptors in sera from myasthenia gravis patients measured by monoclonal antibodies. *Proc. Natl. Acad. Sci. USA* **79**:188–192.
67. Lennon, V., and E. Lambert. 1980. Myasthenia gravis induced by monoclonal antibodies to acetylcholine receptors. *Nature (London)* **285**:238–240.
68. Mochly-Rosen, C., and S. Fuchs. 1981. Monoclonal anti-acetylcholine receptor antibodies directed against the cholinergic binding site. *Biochemistry* **20**:5920–5924.
69. James, R., A. Kato, M. Rey, and B. Fulpius. 1980. Monoclonal antibodies directed against the neurotransmitter binding site of nicotinic acetylcholine receptor. *FEBS Lett.* **120**:145–148.
70. Gomez, C., D. Richman, P. Berman, S. Burres, B. Arnason, and F. Fitch. 1979. Monoclonal antibodies against purified nicotinic acetylcholine receptor. *Biochem. Biophys. Res. Commun.* **88**:575–582.
71. Gomez, C., D. Richman, S. Burres, and B. Arnason. 1981. Monoclonal hybridoma anti-acetylcholine receptor antibodies: Antibody specificity and effect of passive transfer. *Ann. N.Y. Acad. Sci.* **377**:97–109.
72. Watters, D., and A. Maelicke. 1983. Organization of ligand binding sites at the acetylcholine receptor: A study with monoclonal antibodies. *Biochemistry* **22**:1811–1819.
73. Dwyer, D., J. Kearney, R. Bradley, G. Kemp, and S. Oh. 1981. Interaction of human antibody and murine monoclonal antibody with muscle acetylcholine receptor. *Ann. N.Y. Acad. Sci.* **377**:143–157.
74. Tzartos, S.J., D. E. Rand, B. E. Einarson, and J. Lindstrom. 1981. Mapping of surface structures of *Electrophorus* acetylcholine receptor using monoclonal antibodies. *J. Biol. Chem.* **256**: 8635–8645.

75. Tzartos, S., and J. Lindstrom. 1981. Production and characterization of monoclonal antibodies for use as probes of acetylcholine receptors. In: *Monoclonal Antibodies in Endocrine Research*. R. Fellows and G. Eisenbarth, eds. Raven Press, New York. pp. 69–86.
76. Tzartos, S., L. Langeberg, S. Hochschwender, and J. Lindstrom. 1983. Demonstration of a main immunogenic region on acetylcholine receptors from human muscle using monoclonal antibodies to human receptor. *FEBS Lett.* **158**:116–118.
77. Garabedian, B., and S. Morel. 1983. Monoclonal antibodies against the human acetylcholine receptor. *Biochem. Biophys. Res. Commun.* **113**:1–9.
78. Froehner, S., K. Douville, S. Klink, and W. Culp. 1983. Monoclonal antibodies to cytoplasmic domains of the acetylcholine receptor. *J. Biol. Chem.* **258**:7112–7120.
79. Mihovilovic, M., and D. Richman. 1983. Monoclonal antibody (mcab) 247G: Example of a functional probe for the acetylcholine receptor (AcChR) molecule. *Neurosci. Abstr.* **9**:158.
80. Roison, M.-P., Y. Gu, and Z. W. Hall. 1983. The specificity of a myasthenic serum for developmentally different forms of the acetylcholine receptor. *Neurosci. Abstr.* **9**:580.
81. Gullick, W. J., S. Tzartos, and J. Lindstrom. 1981. Monoclonal antibodies as probes of acetylcholine receptor structure. I. Peptide mapping. *Biochemistry* **20**:2173–2180.
82. Hawrot, E., J. M. Gershoni, T. G. Burrage, G. S. Paladino, T. L. Lentz, and L. L. Y. Chun. 1982. Monoclonal antibodies to nicotinic acetylcholine receptor characterized by electrotransfer techniques. *Neurosci. Abstr.* **8**:335.
83. Sargent, P., B. Hedges, L. Tsavaler, L. Clemmons, S. Tzartos, and J. Lindstrom. 1984. The structure and transmembrane nature of the acetylcholine receptor in amphibian skeletal muscle as revealed by crossreacting monoclonal antibodies. *J. Cell Biol.* **98**:609–618.
84. Souroujon, M., D. Mochly-Rosen, A. Gordon, and S. Fuchs. 1983. Interaction of monoclonal antibodies to *Torpedo* acetylcholine receptor with the receptor of skeletal muscle. *Muscle Nerve* **6**:303–311.
85. Anderson, D., P. Walter, and G. Blobel. 1982. Signal recognition protein is required for the integration of acetylcholine receptor delta subunit, a transmembrane glycoprotein, into the endoplasmic reticulum membrane. *J. Cell Biol.* **93**:501–506.
86. Contri-Tronconi, B., S. Tzartos, and J. Lindstrom. 1981. Monoclonal antibodies as probes of acetylcholine receptor structure. II. Binding to native receptor. *Biochemistry* **20**:2181–2191.
87. Lindstrom, J., S. Tzartos, and B. Gullick. 1981. Structure and function of acetylcholine receptors studied using monoclonal antibodies. *Ann. N.Y. Acad. Sci.* **377**:1–19.
88. Suarez-Isla, B. A., K. Wan, J. Lindstrom, and M. Montal. 1983. Single-channel recordings from purified acetylcholine receptors reconstituted in bilayers formed at the tip of patch pipets. *Biochemistry* **22**:2319–2323.
89. Patrick, J., and W. Stallcup. 1977. Immunological distinction between acetylcholine receptor and the alpha bungarotoxin binding component on sympathetic neurons. *Proc. Natl. Acad. Sci. USA* **76**:4689–4692.
90. Swanson, L., J. Lindstrom, L. Schmued, D. O'Leary, and W. Cowan. 1983. Immunohistochemical localization of monoclonal antibodies to the nicotinic acetylcholine receptor in the midbrain of the chick. *Proc. Natl. Acad. Sci. USA* **80**:4532–4536.
91. Hawrot, E., J. Holliday, B. Schweitzer, and L. L. Y. Chun. 1983. Monoclonal antibodies to *Torpedo* nicotinic acetylcholine receptor that cross-react with specific subsets of mammalian peripheral neurons and smooth muscle. *Neursci. Abstr.* **9**:577.
92. Lodish, H. F., and J. E. Rothman. 1978. The assembly of cell membranes. *Sci. Am.* **240**:48–63.
93. Sidman, C., M. J. Potash, and G. Kohler. 1981. Roles of protein and carbohydrate in glycoprotein processing and secretion: Studies using mutants expressing altered IgM mu chains. *J. Biol. Chem.* **256**:13180–13187.
94. Krangel, M. S., H. T. Orr, and J. L. Strominger. 1979. Assembly and maturation of HLA-A and HLA-B antigens *in vivo*. *Cell* **18**:979–991.
95. Krangel, M. S., D. Pious, and J. L. Strominger. 1982. Human histocompatibility antigen mutants immunoselected *in vitro:* Biochemical analysis of a mutant which synthesizes an altered HLA-A2 heavy chain. *J. Biol. Chem.* **257**:5296–5305.
96. Omary, M. B., and I. S. Trowbridge. 1981. Biosynthesis of the human transferrin receptor in cultured cells. *J. Biol. Chem.* **256**:12888–12892.
97. Owen, M. J., A.-M. Kissonerghis, and H. F. Lodish. 1980. Biosynthesis of HLA-A and HLA-B antigens *in vivo*. *J. Biol. Chem.* **255**:9678–9684.
98. Owen, M. J., A.-M. Kissonerghis, H. F. Lodish, and M. J. Crumpton. 1981. Biosynthesis and maturation of HLA-DR antigens *in vivo*. *J. Biol. Chem.* **256**:8987–8993.
99. Mains, P. E., and C. H. Sibley. 1983. The requirement of light chain for the surface deposition of the heavy chain of immunoglobulin M. *J. Biol. Chem.* **258**:5027–5033.
100. Fambrough, D. 1979. Control of acetylcholine receptors in skeletal muscle. *Physiol. Rev.* **59**:165–227.
101. Patrick, J., J. McMillan, H. Wolfson, and J. C. O'Brien. 1977. Acetylcholine receptor metabolism in a nonfusing muscle cell line. *J. Biol. Chem.* **252**:2143–2153.
102. Fambrough, D. M., and P. N. Devreotes. 1978. Newly synthesized acetylcholine receptors are located in the Golgi apparatus. *J. Cell Biol.* **76**:237–244.
103. Palade, G. 1975. Intracellular aspects of the process of protein synthesis. *Science* **189**:347–358.
104. Farquhar, M. G., and G. E. Palade. 1981. The Golgi apparatus (complex)—(1954–1981)—from artifact to center stage. *J. Cell Biol.* **91**:77s–103s.
105. Sabatini, D. D., G. Kreibich, T. Morimoto, and M. Adesnik. 1982. Mechanisms for the incorporation of proteins in membranes and organelles. *J. Cell Biol.* **92**:1–22.
106. Gardner, J. M., and D. M. Fambrough. 1979. Acetylcholine receptor degradation measured by density labeling: Effects of cholinergic ligands and evidence against recycling. *Cell* **16**:661–674.
107. Mendez, B., P. Valenzuela, J. A. Martial, and J. D. Baxter. 1980. Cell-free synthesis of acetylcholine-receptor polypeptides. *Science* **209**:695–697.
108. Blobel, G. 1980. Intracellular protein topogenesis. *Proc. Natl. Acad. Sci. USA* **77**:1496–1500.
109. Anderson, D. J., P. Walter, and G. Blobel. 1982. Signal recognition protein is required for the integration of acetylcholine receptor delta subunit, a transmembrane glycoprotein, into the endoplasmic reticulum membrane. *J. Cell Biol.* **93**:501–506.
110. Gilmore, R., P. Walter, and G. Blobel. 1982. Protein translocation across the endoplasmic reticulum. I. Detection in the microsomal membrane of a receptor for the signal recognition particle. *J. Cell Biol.* **95**:470–477.
111. Wennogle, L. P., R. Oswald, T. Saitoh, and J.-P. Changeux. 1981. Dissection of the 66,000-dalton subunit of the acetylcholine receptor. *Biochemistry* **20**:2492–2497.
112. Merlie, J., J. Hofler, and R. Sebbane. 1981. Acetylcholine receptor synthesis from membrane polysomes. *J. Biol. Chem.* **256**:6995–6999.
113. Sebbane, R., G. Clokey, J. Merlie, S. Tzartos, and J. Lindstrom. 1983. Characterization of the mRNA for mouse muscle acetylcholine receptor alpha subunit by quantitative translation *in vitro*. *J. Biol. Chem.* **258**:3294–3303.
114. Merlie, J., R. Sebbane, S. Gardner, and J. Lindstrom. 1983. A cDNA clone for the alpha subunit of the acetylcholine receptor from the mouse muscle cell line BC3H-1. *Proc. Natl. Acad. Sci. USA* **80**:3845–3849.
115. Boulter, J., and J. Patrick. 1977. Purification of an acetylcholine receptor from a nonfusing muscle cell line. *Biochemistry* **16**:4900–4908.
116. Merlie, J. P., R. Sebbane, S. Tzartos, and J. Lindstrom. 1982. Inhibition of glycosylation with tunicamycin blocks assembly of newly synthesized acetylcholine receptor subunits in muscle cells. *J. Biol. Chem.* **257**:2694–2701.

117. Merlie, J. P., and R. Sebbane. 1981. Acetylcholine receptor subunits transit a precursor pool before acquiring alpha-bungarotoxin binding activity. *J. Biol. Chem.* **256**:3605–3608.
118. Merlie, J. P., and J. Lindstrom. 1983. Assembly *in vivo* of mouse muscle acetylcholine receptor: Identification of an alpha subunit species that may be an assembly intermediate. *Cell* **34**:747–757.
119. Anderson, D. J., and G. Blobel. 1983. Identification of homo-oligomers as potential intermediates in acetylcholine receptor subunit assembly. *Proc. Natl. Acad. Sci. USA* **80**:4359–4363.
120. Prives, J., and D. Bar-Sagi. 1983. Effect of tunicamycin, an inhibitor of protein glycosylation, on the biological properties of acetylcholine receptor in cultured muscle cells. *J. Biol. Chem.* **258**: 1775–1780.
121. Barnard, E. A., R. Miledi, and K. Sumikawa. 1982. Translation of exogenous messenger RNA coding for nicotinic acetylcholine receptors produces functional receptors in *Xenopus* oocytes. *Proc. R. Soc. London. Ser. B* **215**:241–246.
122. Sumikawa, K., M. Houghton, J. S. Emtage, B. M. Richards, and E. A. Barnard. 1981. Active multi-subunit ACh receptor assembled by translation of heterologous mRNA in *Xenopus* oocytes. *Nature (London)* **292**:862–864.
123. Mishina, M., T. Kurosaki, T. Tobimatsu, Y. Morimoto, M. Noda, T. Yamamoto, M. Terao, J. Lindstrom, T. Takahashi, M. Kuno, and S. Numa. 1984. Expression of functional acetylcholine receptor from cloned cDNAs. *Nature (London)* **307**:604–608.
124. Burden, S. 1977. Development of the neuromuscular junction in the chick embryo: The number, distribution, and stability of acetylcholine receptors. *Dev. Biol.* **57**:317–329.
125. Fertuck, H. C., and M. M. Salpeter. 1976. Quantitation of junctional and extrajunctional acetylcholine receptors by electron microscope autoradiography after ^{125}I-alpha-bungarotoxin binding at mouse neuromuscular junctions. *J. Cell Biol.* **69**:144–158.
126. Salpeter, M. M., and R. Harris. 1983. Distribution and turnover rate of acetylcholine receptors throughout the junction folds at a vertebrate neuromuscular junction. *J. Cell Biol.* **96**:1781–1785.
127. Frank, E., and G. D. Fischbach. 1979. Early events in neuromuscular junction formation *in vitro*. *J. Cell Biol.* **83**:143–158.
128. Jessell, T., R. E. Siegel, and G. D. Fischbach. 1979. Induction of acetylcholine receptors on cultured skeletal muscle by a factor extracted from brain and spinal cord. *Proc. Natl. Acad. Sci. USA* **76**:5397–5401.
129. Nitkin, M., E. W. Godfrey, B. G. Wallace, and U. J. McMahan. 1983. Characterization of the AChR aggregating molecules in extracellular matrix fractions from electric organ and muscle. *Neurosci. Abstr.* **9**:1179.
130. Hasegawa, S., H. Kuromi, and Y. Hagihara. 1982. Neuronal regulation of muscle properties and the trophic substances. *Trends Pharmacol. Sci.* **August**:340–342.
131. Burden, S. 1977. Acetylcholine receptors at the neuromuscular junction: Developmental change in receptor turnover. *Dev. Biol.* **61**:79–85.
132. Brockes, J. P., and Z. W. Hall. 1975. Acetylcholine receptors in normal and denervated rat diaphragm muscle. II. Comparison of junctional and extrajunctional receptors. *Biochemistry* **14**:2100–2106.
133. Weinberg, C. B., and Z. W. Hall. 1979. Antibodies from patients with myasthenia gravis recognize determinants unique to extrajunctional acetylcholine receptors. *Proc. Natl. Acad. Sci. USA* **76**:504–508.
134. Reiness, C. G., and Z. W. Hall. 1981. The developmental change in immunological properties of the acetylcholine receptor in rat muscle. *Dev. Biol.* **81**:324–331.
135. Fishbach, G. D., and S. M. Schuetze. 1980. A post-natal decrease in acetylcholine channel open time at rat end-plates. *J. Physiol. (London)* **303**:125–137.
136. Schuetze, S. M. 1980. The acetylcholine channel open time in chick muscle is not decreased following innervation. *J. Physiol. (London)* **303**:111–124.
137. Patrick, J., and S. Heinemann. 1982. Outstanding problems in acetylcholine receptor structure and regulation. *Trends Neurosci.* **5**:300–302.
138. Hall, Z. W., B. W. Lubit, and J. H. Schwartz. 1981. Cytoplasmic actin in postsynaptic structures at the neuromuscular junction. *J. Cell Biol.* **90**:789–792.
139. Bloch, R. J., and Z. W. Hall. 1983. Cytoskeletal components of the vertebrate neuromuscular junction: Vinculin, alpha-actinin, and filamin. *J. Cell Biol.* **97**:217–223.
140. Bloch, R. J. 1983. Acetylcholine receptor clustering in rat myotubes: Requirment for Ca^{2+} and effects of drugs which depolymerize microtubules. *J. Neurosci.* **3**:2670–2680.
141. Porter, S., and S. C. Froehner. 1983. Characterization and localization of the M_r = 43,000 proteins associated with acetylcholine receptor-rich membranes. *J. Biol. Chem.* **258**:10034–10040.
142. Froehner, S. C., V. Gulbrandsen, C. Hyman, A. Y. Jeng, R. R. Neubig, and J. B. Cohen. 1981. Immunofluorescence localization at the mammalian neuromuscular junction of the M_r 43,000 protein of *Torpedo* postsynaptic membranes. *Proc. Natl. Acad. Sci. USA* **78**:5230–5234.
143. Sealock, R. 1982. Cytoplasmic surface structure in postsynaptic membranes from electric tissue visualized by tannic acid-mediated negative contrasting. *J. Cell Biol.* **92**:514–522.
144. Burden, S. J., R. L. DePalma, and G. S. Gottesman. 1983. Crosslinking of proteins in acetylcholine receptor-rich membranes: Association between the beta-subunit and the 43 kd subsynaptic protein. *Cell* **35**:687–692.
145. Sanes, J. R., L. M. Marshall, and U. J. McMahan. 1978. Reinnervation of muscle fiber basal lamina after removal of myofibers: Differentiation of regenerating axons at original synaptic sites. *J. Cell Biol.* **78**:176–198.
146. Burden, S. J., P. B. Sargent, and U. J. McMahan. 1979. Acetylcholine receptors in regenerating muscle accumulate at original synaptic sites in the absence of the nerve. *J. Cell Biol.* **82**:412–425.
147. Carlson, S. S., K. M. Buckley, P. Caroni, and R. B. Kelly. 1983. Synaptic vesicles and the synaptic cleft contain an identical proteoglycan. *Neurosci. Abstr.* **9**(part 2):1028.
148. Katz, B., and S. Thesleff. 1957. A study of the "desensitization" produced by acetylcholine at the motor end-plate. *J. Physiol. (London)* **138**:63–80.
149. Lester, H. A., J.-P. Changeux, and R. E. Sheridan. 1975. Conductance increases produced by bath application of cholinergic agonists to *Electrophorus* electroplaques. *J. Gen. Physiol.* **65**:797–816.
150. Sheridan, R. E., and H. A. Lester. 1977. Rates and equilibria at the acetylcholine receptor of *Electrophorus* electroplaques. *J. Gen. Physiol.* **70**:187–219.
151. Adams, P. R. 1975. An analysis of the dose–response curve at voltage-clamped frog-endplates. *Pfluegers Arch.* **360**:145–153.
152. Adams, P. R. 1977. Relaxation experiments using bath-applied suberyldicholine. *J. Physiol. (London)* **268**:271–289.
153. Neubig, R. R., and J. B. Cohen. 1980. Permeability control by cholinergic receptors in *Torpedo* postsynaptic membranes: Agonist dose–response relations measured at second and millisecond times. *Biochemistry* **19**:2770–2779.
154. Hartzell, H. C., S. W. Kuffler, and D. Yoshikami. 1975. Postsynaptic potentiation: Interaction between quanta of acetylcholine at the skeletal neuromuscular synapse. *J. Physiol. (London)* **251**:427–464.
155. Dreyer, F., and K. Peper. 1975. Density and dose–response curve of acetylcholine receptors in frog neuromuscular junction. *Nature (London)* **253**:641–643.
156. Dreyer, F., K. Peper, and R. Sterz. 1978. Determination of dose–response curves by quantitative iontophoresis at the frog neuromuscular junction. *J. Physiol. (London)* **281**:395–419.
157. Hoffman, H. M., and V. E. Dionne. 1980. The Hill coefficient of the acetylcholine receptor dose–response relation is independent of membrane voltage and temperature. *Soc. Neurosci. Abstr.* **6**:753.
158. Land, B. R., E. E. Salpeter, and M. M. Salpeter. 1980. Acetylcholine receptor site density affects the rising phase of miniature endplate currents. *Proc. Natl. Acad. Sci. USA* **77**:3736–3740.

159. Magleby, K. L., and C. F. Stevens. 1972. A quantitative description of endplate currents. *J. Physiol. (London)* **223**:173–197.
160. Sakmann, B., and E. Neher, eds. 1983. *Single-Channel Recording*. Plenum Press, New York.
161. Corey, D. P., and C. F. Stevens. 1983. Science and technology of patch-recording electrodes. In: *Single-Channel Recording*. B. 68.
162. Lester, H. A., and H. W. Chang. 1977. Response of acetylcholine receptors to rapid photochemically produced increases in agonist concentration. *Nature (London)* **266**:373–374.
163. Nass, M. M., H. A. Lester, and M. E. Krouse. 1978. Response of acetylcholine receptors to photoisomerizations of bound agonist molecules. *Biophys. J.* **24**:153–160.
164. Dionne, V. E., and C. F. Stevens. 1975. Voltage dependence of agonist effectiveness at the frog neuromuscular junction: Resolution of a paradox. *J. Physiol. (London)* **251**:245–270.
165. Adams, P. R. 1975. Kinetics of agonist conductance changes during hyperpolarization at frog endplates. *Br. J. Pharmacol.* **53**:308–310.
166. Sheridan, R. E., and H. A. Lester. 1975. Relaxation measurements on the acetylcholine receptor. *Proc. Natl. Acad. Sci. USA* **72**:3496–3500.
167. Katz, B., and R. Miledi. 1970. Membrane noise produced by acetylcholine. *Nature (London)* **226**:962–963.
168. Katz, B., and R. Miledi. 1972. The statistical nature of the acetylcholine potential and its molecular components. *J. Physiol. (London)* **224**:665–699.
169. Anderson, C. R., and C. F. Stevens. 1973. Voltage clamp analysis of acetylcholine produced end-plate current fluctuations at frog neuromuscular junction. *J. Physiol. (London)* **235**:655–691.
170. Neher, E., and B. Sakmann. 1976. Single-channel currents recorded from membrane of denervated frog muscle fibers. *Nature (London)* **260**:799–802.
171. Colquhoun, D., and A. G. Hawkes. 1981. On the stochastic properties of single ion channels. *Proc. R. Soc. London Ser. B* **211**:205–235.
172. Dionne, V. E., and M. D. Leibowitz. 1982. Acetylcholine receptor kinetics: A description from single channel currents at snake neuromuscular junctions. *Biophys. J.* **39**:253–261.
173. Gage, P. W., and R. N. McBurney. 1975. Effects of membrane potential, temperature and neostigmine on the conductance change caused by a quantum of acetylcholine at the toad neuromuscular junction. *J. Physiol. (London)* **244**:385–407.
174. Nelson, D. J., and F. Sachs. 1979. Single ionic channels observed in tissue-cultured muscle. *Nature (London)* **282**:861–863.
175. Dionne, V. E., and R. L. Parsons. 1981. Characteristics of the acetylcholine-operated channel at twitch and slow fiber neuromuscular junctions of the garter snake. *J. Physiol. (London)* **310**:145–158.
176. Dionne, V. E. 1981. The kinetics of slow muscle acetylcholine-operated channels in the garter snake. *J. Physiol. (London)* **310**:159–190.
177. Colquhoun, D., and B. Sakmann. 1981. Fluctuations in the microsecond time range of the current through single acetylcholine receptor ion channels. *Nature (London)* **294**:464–466.
178. Sine, S. M., and J. H. Steinbach. 1984. Activation of a nicotinic acetylcholine receptor. *Biophys. J.* **45**:175–185.
179. Leibowitz, M. D., and V. E. Dionne. 1984. Single-channel acetylcholine receptor kinetics. *Biophys. J.* **45**:153–163.
180. Horn, R., and J. Patlak. 1980. Single channel currents from excised patches of muscle membrane. *Proc. Natl. Acad. Sci. USA* **77**: 6930–6934.
181. Fenwick, E. M., A. Marty, and E. Neher. 1982. A patch-clamp study of bovine chromaffin cells and of their sensitivity to acetylcholine. *J. Physiol. (London)* **333**:577–597.
182. Hamill, O. P., and B. Sakmann. 1981. Multiple conductance states of single acetylcholine receptor channels in embryonic muscle cells. *Nature (London)* **294**:962–963.
183. Trautmann, A. 1982. Curare can open and block ionic channels associated with cholinergic receptors. *Nature (London)* **298**:272–275.
184. Auerbach, A., and F. Sachs. 1982. Flickering of a nicotinic ion channel to a subconductance state. *Biophys. J.* **42**:1–10.
185. Auerbach, A., and F. Sachs. 1984. Single-channel currents from acetylcholine receptors in embryonic chick muscle. *Biophys. J.* **45**:187–198.
186. Brehm, P., R. Kullberg, and F. Moody-Corbett. 1984. Properties of non-junctional acetylcholine receptor channels on innervated muscle of *Xenopus laevis*. *J. Physiol. (London)* **350**:631–648.
187. Kullberg, R. W., P. Brehm, and J. H. Steinbach. 1981. Nonjunctional acetylcholine receptor channel open time decreases during development of *Xenopus* muscle. *Nature (London)* **289**:411–413.
188. Brehm, P., J. H. Steinbach, and Y. Kidokoro. 1982. Channel open time of acetylcholine receptors on *Xenopus* muscle cells in dissociated cell culture. *Dev. Biol.* **91**:93–102.
189. Sakmann, B., J. Patlak, and E. Neher. 1980. Single acetylcholine-activated channels show burst-kinetics in presence of desensitizing concentrations of agonist. *Nature (London)* **286**:71–73.
190. Adams, P. R., and B. Sakmann. 1978. Decamethionium both opens and blocks end-plate channels. *Proc. Natl. Acad. Sci. USA* **75**:2994–2998.
191. Huang. L.-Y., W. A. Catterall, and G. Ehrenstein. 1978. Selectivity of cations and nonelectrolytes for acetylcholine-activated channels in cultured muscle cells. *J. Gen. Physiol.* **71**:397–410.
192. Gage, P. W., and D. Van Helden. 1979. Effects of permeant monovalent cations on end-plate channels. *J. Physiol. (London)* **288**:509–528.
193. Adams, D. J., T. M. Dwyer, and B. Hille. 1980. The permeability of endplate channels to monovalent and divalent metal cations. *J. Gen. Physiol.* **75**:493–510.
194. Lewis, C. A., and C. F. Stevens. 1983. Acetylcholine receptor channel ionic selectivity: Ions experience an aqueous environment. *Proc. Natl. Acad. Sci. USA* **80**:6110–6113.
195. Lewis, C. A., and C. F. Stevens. 1979. Mechanism of ion permeation through channels in a postsynaptic membrane. In: *Membrane Transport Processes,* Volume 3. Stevens, C. F. and R. W. Tsien, eds. Raven Press, New York. pp. 133–151.
196. Lewis, C. A. 1979. The ion concentration dependence of the reversal potential and the single channel conductance of ion channels at the frog neuromuscular junction. *J. Physiol. (London)* **286**: 417–445.
197. Patlak, J. B., K. A. F. Gration, and P. N. R. Usherwood. 1979. Single glutamate-activated channels in locust muscle. *Nature (London)* **278**:643–645.
198. Cull-Candy, S. G., R. Miledi, and I. Parker. 1980. Single glutamate-activated channels recorded from locust muscle fibres with perfused patch-clamp electrodes. *J. Physiol. (London)* **321**:195–210.
199. Gration, K. A. F., J. J. Lampert, R. L. Ramsey, R. P. Rand, and P. N. R. Usherwood. 1981. Agonist potency determination by patch clamp analysis of single glutamate receptors. *Brain Res.* **230**:400–405.
200. Gration, K. A. F., J. J. Lambert, R. Ramsey, and P. N. R. Usherwood. 1981. Non-random openings and concentration-dependent lifetimes of glutamate-gated channels in muscle membrane. *Nature (London)* **291**:423–425.
201. Gration, K. A. F., R. L. Ramsey, and P. N. R. Usherwood. 1983. Analysis of single-channel data from glutamate receptor-channel complexes on locust muscle. In: *Single-Channel Recording*. B. Sakmann and E.Neher, eds. Plenum Press, New York. pp. 377–388.
202. Cull-Candy, S. G., and I. Parker. 1982. Rapid kinetics of single glutamate-receptor channels. *Nature (London)* **295**:410–412.
203. Gration, K. A. F., J. J. Lambert, R. L. Ramsey, R. P. Rand, and P. N. R. Usherwood. 1982. Closure of membrane channels gated by glutamate receptors may be a two-step process. *Nature (London)* **295**:599–601.
204. Nowak, L., P. Bregestovski, P. Ascher, A. Herbet, and A. Prochiantz. 1984. Magnesium gates glutamate-activated channels in mouse central neurones. *Nature (London)* **307**:462–465.
205. Jackson, M. B., H. Lecar, D. A. Mathers, and J. L. Barker. 1982. Single channel currents activated by gamma-aminobutyric acid,

muscimol, and (−)-pentobarbital in cultured mouse spinal neurons. *J. Neurosci.* **2**:889–894.

206. Redmann, G. A., J. L. Barker, and H. Lecar. 1983. Single muscimol-activated ion channels show voltage-sensitive kinetics in cultured mouse spinal cord neurons. *Soc. Neurosci. Abstr.* **9**:507.
207. Sakmann, B., O. P. Hamill, and J. Bormann. 1983. Patch-clamp measurements of elementary chloride currents activated by the putative inhibitory transmitters GABA and glycine in mammalian spinal neurons. *J. Neural Trans. Suppl.* **18**:83–95.
208. Bormann, J., B. Sakmann, and W. Seifert. 1983. Isolation of GABA-activated single-channel Cl^- currents in the soma membrane of rat hippocampal neurones. *J. Physiol. (London)* **341**:9P–10P.
209. Hamill, O. P., J. Bormann, and B. Sakmann. 1983. Activation of multiple-conductance state chloride channels in spinal neurons by glycine and GABA. *Nature (London)* **305**:805–808.
210. Sakmann, B., A. Noma, and W. Trautwein. 1983. Acetylcholine activation of single muscarinic K^+ channels in isolated pacemaker cells of the mammalian heart. *Nature (London)* **303**:250–253.
211. Kehoe, J., and A. Marty. 1980. Certain slow synaptic responses: Their properties and possible underlying mechanisms. *Annu. Rev. Biophys. Bioeng.* **9**:437–465.
212. Hartzell, H. C. 1981. Mechanisms of slow postsynaptic potentials. *Nature (London)* **291**:539–544.
213. Siegelbaum, S. A., J. S. Camardo, and E. R. Kandel. 1982. Serotonin and cyclic AMP close single K^+ channels in *Aplysia* sensory neurones. *Nature (London)* **299**:413–417.
214. Miledi, R., I. Parker, and K. Sumikawa. 1983. Recording of single gamma-aminobutyrate and acetylcholine activated receptor channels translated by exogenous mRNA in *Xenopus* oocytes. *Proc. R. Soc. London Ser. B* **218**:481–484.

PART II

Transport in Epithelia

Vectorial Transport through Parallel Arrays

CHAPTER 8

Cellular Models of Epithelial Ion Transport

Stanley G. Schultz

1. Introduction

The first cellular model of epithelial ion transport was proposed in 1958 by Koefoed-Johnsen and Ussing[1] (KJU) to account for the relation between active Na^+ transport and the electrical potential difference across isolated frog skin. The essential feature of this now-classic model (Fig. 1) was that the epithelial cell could be viewed as two membranes arranged in series separated by a homogeneous cytoplasmic compartment with net transcellular or vectorial transport resulting from the asymmetric properties of the two limiting barriers. In the case of frog skin, the outer or apical membrane was presumed to be permselective to Na^+ and scarcely if at all permeable to K^+. The inner or basolateral membrane, on the other hand, was presumed to be permselective to K^+, scarcely if at all permeable to Na^+, and to possess an active pump mechanism that extrudes Na^+ from the cell in exchange for K^+. This asymmetric arrangement of pump and leaks could simultaneously account for active transcellular Na^+ transport as well as the maintenance of the low intracellular Na^+ concentration and the high intracellular K^+ concentration characteristic of virtually all cells of higher animals. Because of its simplicity—involving only transport mechanisms that had already been reasonably well established by studies on nonepithelial cells—this model was rapidly accepted and elevated to the status of a paradigm.

In the 27 years since the introduction of the KJU model, its essential features have been repeatedly confirmed and extended to a number of other epithelia such as amphibian and mammalian urinary bladder[2,3] and colon[4] as well as segments of mammalian distal nephron.[5] All of these tissues are characterized by relatively large transepithelial electrical potential differences and resistances and are capable of absorbing Na^+ against considerable concentration gradients thereby markedly reducing the Na^+ concentration in the luminal or outer solution; because of their relatively high paracellular resistances, they are often referred to as "tight" or "moderately tight" epithelia.[2,6,7]

However, within a few years after the introduction of the KJU model it became clear that this simplest of all constructs could not account for Na^+ transport by a number of other "leakier" epithelia and that additional, more complex, transport mechanisms had to be invoked. Indeed, in hindsight, it is fortunate that Ussing,[8] influenced by his mentor Krogh, chose to study the isolated frog skin rather than, say, amphibian or mammalian gallbladder. It is doubtful whether current models of NaCl absorption by gallbladder would have been received with the same enthusiasm that greeted the KJU model a quarter of a century ago.

During the past two decades, studies of numerous epithelia using a variety of *in vitro* techniques have uncovered a number of apical and basolateral membrane transport processes; the best established of these processes are illustrated in Fig. 2A (apical) and Fig. 2B (basolateral). In this chapter we will review the properties of these transport processes and illustrate how, when appropriately combined, they can yield cellular models that ac-

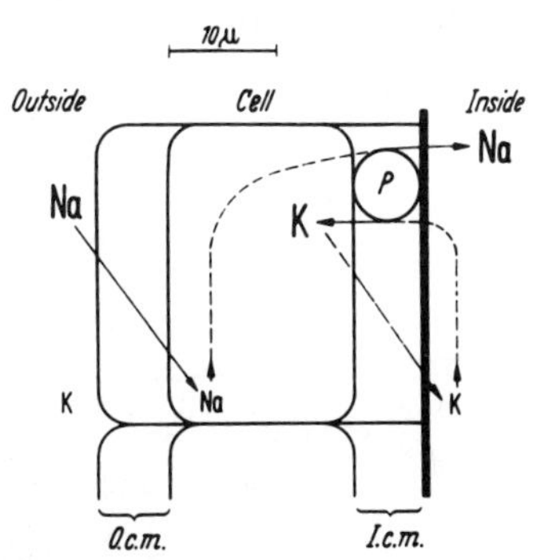

Fig. 1. Koefoed-Johnsen–Ussing model for transcellular Na^+ transport by isolated frog skin.[1]

Stanley G. Schultz • Department of Physiology and Cell Biology, University of Texas Medical School, Houston, Texas 77225.

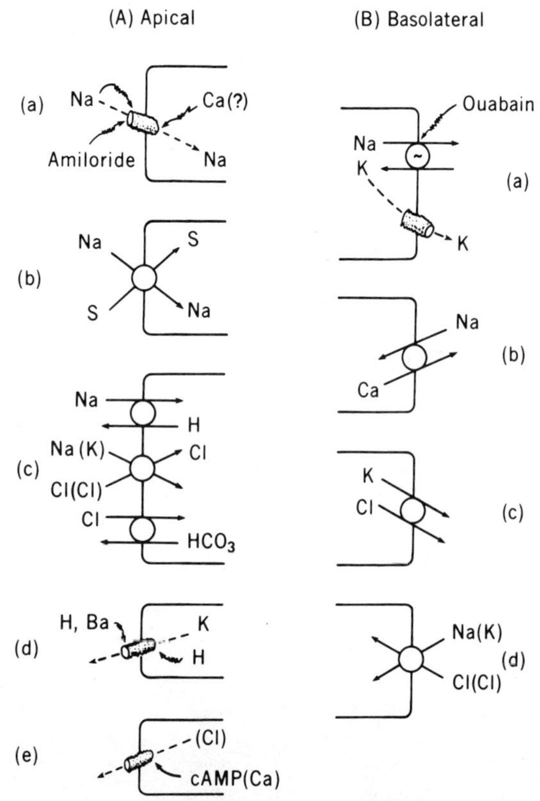

Fig. 2. Transport processes located at apical (A) or basolateral (B) membranes of epithelial cells.

count for most of the well-defined ion absorptive and secretory processes observed in epithelia from animals throughout the phylogenetic scale. In essence, Figs. 2A and B can be viewed much like a "menu in a Chinese restaurant"; that is, combining one or more mechanisms from Fig. 2A with one or more from Fig. 2B results in a cellular model of epithelial ion transport consistent with available data.

Before proceeding, several points should be emphasized. First, all of the models we will consider assume that the cell can be adequately represented by two limiting membranes separated by a homogeneous intracellular milieu as originally postulated by KJU[1,2]; there are certainly no compelling reasons to doubt the validity of this assumption.

Second, none of the processes we will consider are well understood at the "molecular level." Thus, terms like "pore" or "channel," and "carrier" are for the most part employed to describe, and distinguish between, the macroscopic kinetic properties of these transport processes. Co- and countertransport processes will be considered "carrier-mediated" whereas "simple diffusional pathways" will be referred to as "pores" or "channels." The kinetic criteria for the distinction between "channels" and "carriers" have been discussed elsewhere.[9–13]

Finally, in the limited space available it is patently impossible to review exhaustively all of the literature bearing on epithelial ion absorption and secretion. We will therefore only summarize major, common, points and refer the interested reader to more comprehensive treatments whenever possible. In addition, we will not consider the problem of active proton secretion by epithelia such as gastric mucosa and turtle urinary bladder.

2. Models of Sodium- and Chloride-Absorbing Epithelial Cells

One of the gratifying results of numerous studies on a wide variety of Na^+-, and in some instances Cl^--absorbing epithelial cells is that, with few exceptions, the properties of the basolateral membranes appear to be qualitatively similar. Thus, we will begin by considering the different ways Na^+, and in some instances Cl^-, may enter the cells across the apical membranes and conclude this section by summarizing what is known about the properties of the basolateral membranes.

2.1. Uncoupled Na^+ Entry across the Apical Membrane

The simplest, and by far the best understood, mechanism by which Na^+ gains entry into absorptive cells is that characteristic of "tight" and "moderately tight" epithelia such as amphibian skin, colon, and urinary bladder; mammalian colon and urinary bladder; and segments of the distal nephron. In most of these epithelia the apical membrane is almost exclusively permeable to Na^+ and Li^+ and the electrical resistance of that barrier is 8–9 times greater than that of the basolateral membrane; thus, the rate at which Na^+ can enter the absorptive cell across the apical membrane is often considered the rate-limiting step in overall active transcellular Na^+ transport.[2] The apical membranes of some of these epithelia also possess a significant K^+ conductance making them capable of actively secreting K^+; this matter will be discussed in Section 2.4.

Na^+ entry across the apical membranes of all of these epithelia is completely, rapidly, and reversibly blocked by the pyrazine diuretic amiloride, and this agent has become as important a tool in the study of this entry mechanism as is tetrodotoxin in the study of Na^+ channels in excitable membranes or the digitalis glycosides in the study of the Na^+,K^+ pump.[2,14] The kinetics and precise mechanism by which this positively charged quanidinium derivative blocks the Na^+ entry step have not been entirely resolved but the concentrations required for half-maximal inhibition are in the submicromolar range and it appears that the stoichiometry of this interaction is 1 : 1.[14–24] The mechanism of the inhibitory action of this agent is as yet unclear. The simplest explanation is that it simply "plugs" the openings of the Na^+ entry pathway. However, there are kinetic data consistent with the notion that the Na^+ entry site and the amiloride-binding site are separate and that the diuretic interacts with a "regulatory unit" that modulates the activity of the Na^+ entry site.[16,20,22–24]

The availability of this specific blocking agent has permitted several groups to define the current–voltage (I–V) relations of the Na^+ entry step across the apical membranes of frog skin,[25] toad urinary bladder,[26–30] *Necturus* urinary bladder,[31,32] rabbit descending colon,[33–35] and turtle colon (S. M. Thompson, personal communication); in every instance the relation between the rate of Na^+ entry across the apical membrane and the electrical potential difference across that barrier conforms closely to the Goldman–Hodgkin–Katz (GHK) "constant field flux equation"[36,37] for a single cation over a reasonably wide range of voltages and under a variety of experimental

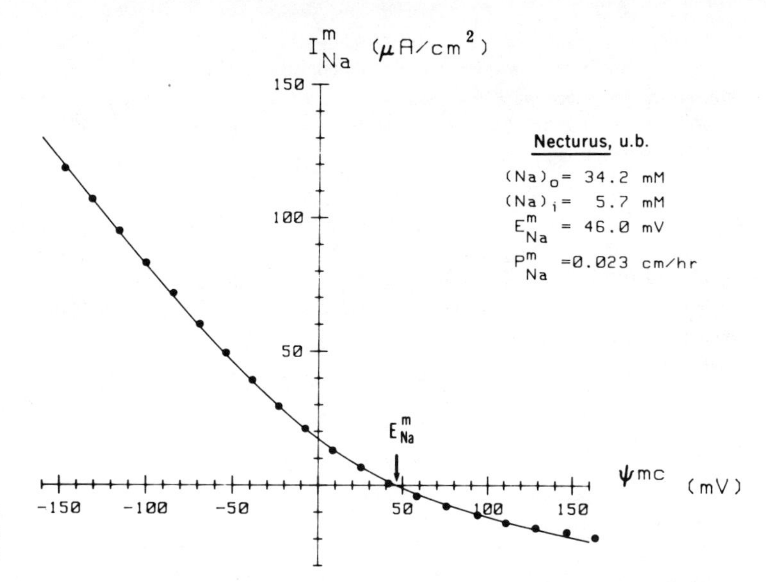

Fig. 3. Relation between the Na^+ current across the apical membrane (I^m_{Na}) and the electrical potential difference across that barrier (ψ^{mc}) for *Necturus* urinary bladder.[32]

conditions (Fig. 3). In addition, the bidirectional fluxes of Na^+ across the apical membranes of frog skin[38] and toad urinary bladder[39] appear to conform to the Ussing flux-ratio equation[40] over a wide range and thus conform to the "independence principle" that underlies the GHK flux equation. These findings are consistent with the simplest form of electrodiffusion through relatively homogeneous pores or channels uncomplicated by "single filing," "exchange diffusion," "binding to membrane components," etc.[41]

The results of noise (fluctuation) analyses of Na^+ entry across the apical membranes of frog skin[42,43] and toad urinary bladder[27,28] in the presence of submaximal (partially) blocking concentrations of amiloride are consistent with the notion that each entry "site" is capable of transferring about 10^6 Na^+ ions/sec; this transfer rate is consistent with those reported for known channel-forming ionophores in artificial lipid bilayers[9,10,44,45] but is four orders of magnitude greater than the turnover rates of known, artificial[9,10] or natural[46] carrier systems. Further, the I–V relations of *single* Na^+ channels in the outer membrane of frog skin also appear to conform to the GHK constant field flux equation.[47]

The combined results of all of these studies strongly suggest that uncoupled, amiloride-sensitive Na^+ entry across the apical membranes of these "tight" and "moderately tight" epithelia is the result of the simplest possible process—electrodiffusion—through the simplest possible channels (i.e., low charge density at the channel openings, a near-symmetrical potential energy profile within the channels, and minimal occupancy—no more than 2 Na^+ ions per channel at any time[41,48]). It follows that the rate of Na^+ entry across the apical membrane at any moment will be determined by the number of open channels, the electrochemical potential difference for Na^+ acting across each open channel, and the conductance of each open channel. The density of these channels in the outer membrane of frog skin[42] and apical membrane of toad urinary bladder[27] is approximately 1–2 × 10^8/cm^2 (or 1–2/μm^2) when the Na^+ concentration in the mucosal solution is 60 mM.

2.1.1. Regulation of Apical Membrane Na^+ Permeability

There are a number of factors that influence the permeability of the apical membrane to Na^+; some have obvious physiological significance, others are of interest largely because they may shed light on underlying mechanisms.

1. Increasing the concentration of Na^+ in the outer or mucosal solution, $[Na^+]_m$, results in a decrease in the permeability of the apical membrane to Na^+[25–28,32,35] (Fig. 4). This results in the saturating (Michaelis–Menten type) relation between Na^+ concentration and the rate of Na^+ transport that has been recognized for many years and suggested to a number of early investigators that Na^+ entry was not simply diffusional but instead involved binding to a "carrier-like" component of that barrier.[49–51] However, the results of noise analysis clearly indicate that this "self-inhibition" is the result of a decrease in

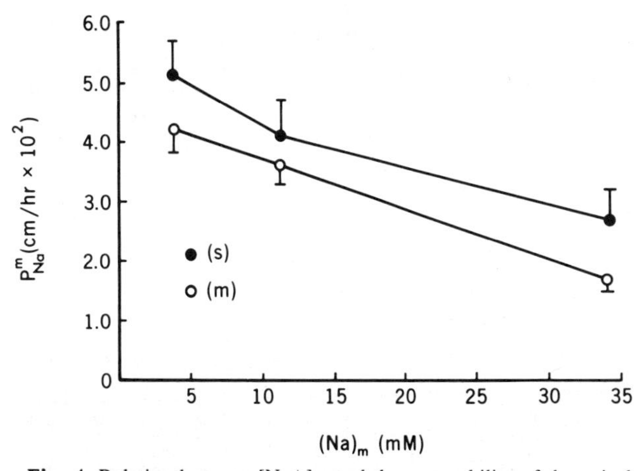

Fig. 4. Relation between $[Na^+]_m$ and the permeability of the apical membrane of *Necturus* urinary bladder to Na^+ (P^m_{Na}) at constant intracellular Na^+ activity.[32]

the number of open channels with no change in the transport capacity of individual channels.[43] The mechanism of this self-inhibition is unclear. Because it is a relatively slow process, Lindemann[52,53] has suggested that it is the result of an interaction between Na^+ and nearby regulatory sites rather than with the Na^+ channel directly. In any event, the physiological usefulness of this regulatory process is immediately apparent when one considers its inverse; i.e., the permeability of the apical membrane to Na^+ *increases* as the luminal $[Na^+]$ *decreases*. Clearly, such a regulatory process is ideally suited for epithelia whose functions are to conserve Na^+ by reducing the concentration of this cation in the luminal contents to minimal values.

2. There also appears to be an inverse relation between intracellular Na^+ activity, $[Na^+]_c$, and the permeability of the apical membrane to that cation ("negative feedback") such that an increase in $[Na^+]_c$ resulting from inhibition of the basolateral Na^+ pump brings about a decrease in apical membrane Na^+ permeability and, conversely, depletion of cell Na^+ results in an increase in apical Na^+ permeability.[54–62] Preliminary results of noise analysis suggest that these changes in permeability are also the result of changes in the number of open Na^+ channels with no change in the conductive properties of individual channels.[63]

The underlying mechanism of this "negative feedback" is not as yet completely established but there is strong, suggestive evidence that the direct mediator of this effect is cell Ca^{2+}, $[Ca^{2+}]_c$. As will be discussed below, there is evidence for the presence of a carrier-mediated Na^+–Ca^{2+} countertransport mechanism in the basolateral membrane of several epithelia[64–69] which enables the downhill movement of Na^+ into the cell to energize the uphill extrusion of Ca^{2+} from the cell. Thus, an increase in $[Na^+]_c$ would reduce the electrochemical potential difference for that ion across the basolateral membrane and, in turn, the energy available for Ca^{2+} extrusion. This *could* lead to a higher steady-state level of $[Ca^{2+}]_c$ which *might* directly reduce the apical Na^+ permeability. The finding that Ca^{2+} in the micromolar range inhibits amiloride-sensitive Na^+ uptake by apical membrane vesicles from toad urinary bladder[64] lends strong support to the plausibility of this hypothesis. However, compelling evidence that this purported mechanism exerts a significant regulatory effect on apical Na^+ permeability under physiological conditions is lacking.

The ability of cell Na^+ to regulate the Na^+ permeability of the apical membrane directly, or indirectly (via cell Ca^{2+}), also has obvious physiological utility. Clearly, such a mechanism can "gear" the rate of Na^+ entry across the apical membrane to the rate of pump-mediated active Na^+ extrusion across the basolateral membrane and protect the cell against inordinate changes in $[Na^+]_c$ in response to changes in basolateral pump activity. Thus, a slowing of pump activity (for whatever reason) would lead to an increase in cell Na^+ which, in turn, would result in a decrease in the ease with which Na^+ can enter the cell. Conversely, a primary increase in pump activity would lead to a decrease in cell Na^+ which, in turn, would result in an increase in the ease with which Na^+ can enter the cell. Finally, a low $[Na^+]_c$ due to the presence of a low mucosal (luminal) $[Na^+]$ would contribute to maximizing the ease with which Na^+ can enter the cell from a dilute mucosal solution; under these conditions (i.e., low $[Na^+]_m$ *and* low $[Na^+]_c$) the permeability of the apical membrane to Na^+ approaches its maximum.[35,62]

3. Na^+ absorption by all of these epithelia is stimulated by aldosterone[2,4,70] and, in some, by ADH[2,71] as well. Both hormones bring about a marked increase in the permeability of the apical membrane to Na^+ by increasing the number of open Na^+ channels in that barrier.[27–29] There is evidence for the case of ADH that these channels were not previously present in the apical membrane but are inserted into that barrier (from the cytoplasm) in response to the humoral stimulus.[72] For the case of aldosterone, these new channels appear to have been present in the membrane in an inactive or latent form and are somehow activated in response to the humoral stimulus.[72]

4. The replacement of Cl^- in the mucosal solution with a number of anions such as sulfate, isethionate, and several organic-sulfonic acid derivatives results in a prompt and marked stimulation of Na^+ absorption by rabbit colon due to an increase in apical Na^+ permeability.[73] Similar findings have been reported for toad urinary bladder.[74] The mechanism of action of these anions is not established but it may simply involve "screening" of positively charged groups in the vicinity of the Na^+ channels.

5. Finally, a number of organic reagents affect the permeability of the apical membrane to Na^+ when present in the outer or mucosal solution.[75,76] Of particula interest is the sulfhydryl-reactive agent PCMBS. In frog skin[75,76] and toad urinary bladder[77] this agent increases the permeability of the apical membrane apparently by increasing the number of open Na^+ channels[76]; at the same time, "self-inhibition" by $[Na^+]_m$ is abolished and the affinity for amiloride inhibition of the apical Na^+ permeability is increased.[76] Somewhat different results have been obtained in the case of rabbit descending colon where PCMBS and mersalyl stimulate active Na^+ transport when it is initially low but inhibit it when it is initially high; in addition, these agents completely block or reverse the inhibitory action of amiloride.[78,79]

Finally, it is of considerable interest that at very low concentrations, amiloride exerts a stimulatory effect on apical Na^+ permeability and also appears to "compete" with Na^+ for the mechanism responsible for "self-inhibition."[18,76]

Clearly, these observations raise a number of important questions regarding the mechanisms responsible for the regulation of the number of open Na^+ channels in the apical membrane. Does amiloride directly "plug" open Na^+ channels or does it act as a "super Na^+" on the "self-inhibition" mechanism[18–20,52,53]? Why does PCMBS increase the inhibitory effectiveness of amiloride in frog skin but abolish or reverse the inhibitory action of this agent in rabbit colon? For a more detailed description and discussion of the effects and possible mechanisms of action of a number of reagents that stimulate and/or inhibit apical Na^+ permeability, the reader is referred to Refs. 18, 76, 80.

In conclusion: It is now reasonably well established that Na^+ entry across the apical membranes of a number of "tight" and "moderately tight" epithelia is the result of simple electrodiffusion through the simplest imaginable channels that are about equally permeable to Na^+ and Li^+ but essentially exclude other cations.[2,52,59,81–85] Changes in macroscopic apical Na^+ permeability under a variety of conditions appear to be due entirely to changes in the *number* of open channels with no significant changes in the transport capacity of each channel. The basis of the permselective properties of these channels, the determinants of the fraction of channels in the apical membrane that are "open" or "closed" during any interval of time, and the mechanisms responsible for "openness" or "closedness" remains to be clarified at the microscopic levels. The resolution of some of these seminal issues is likely to await successful

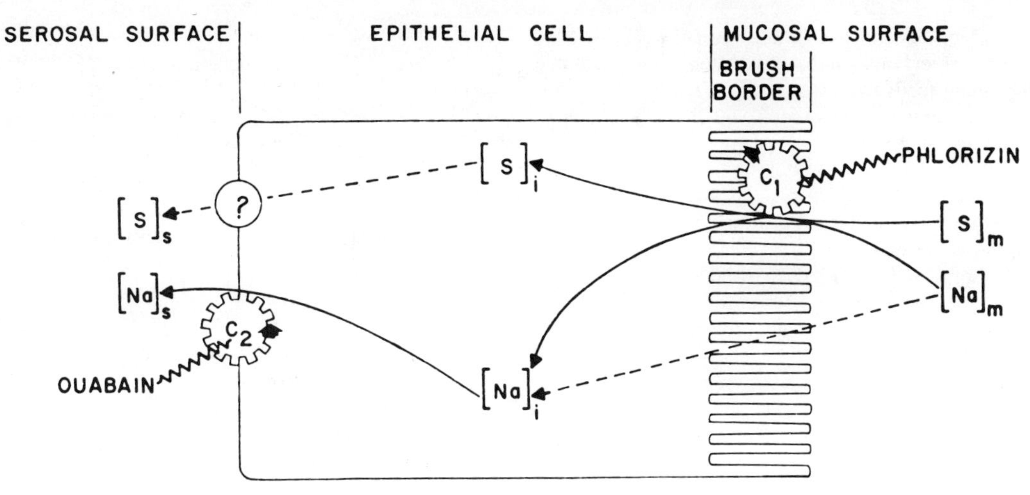

Fig. 5. Cellular model for Na$^+$-coupled sugar (and amino acid) transport by small intestine.[91]

"patch-clamping" of the apical membrane and/or reconstitution of amiloride-sensitive Na$^+$ channels into planar lipid bilayers.

2.2. Na$^+$ Entry across Apical Membrane Coupled to Organic Solutes

Our current views of Na$^+$-coupled solute transport by epithelia evolved from two sets of observations reported in the early 1960s. The first was that "active"* transepithelial transport of some sugars and amino acids[86,87] and the "active" intracellular accumulation of some sugars[88,89] by *in vitro* small intestine are dependent upon the presence of Na$^+$ in the mucosal solution and are blocked by ouabain. The second was that the addition of "actively" transported sugars or amino acids to the mucosal solution bathing *in vitro* small intestine results in an immediate increase in the rate of active transcellular Na$^+$ absorption.[90–92] On the basis of these findings, Schultz and Zalusky[91] proposed a model (Fig. 5) for intestinal sugar and amino acid absorption which featured a carrier-mediated Na$^+$-coupled entry of these solutes across the apical membrane (Fig. 2A, part b) with Na$^+$ subsequently being actively extruded from the cell by the ouabain-inhibitable basolateral pump and the sugar or amino acid exiting across that barrier "downhill."† Inasmuch as the current status of Na$^+$-coupled transport mechanisms will be reviewed in some detail by this author in another chapter, we will here simply summarize the salient features of these processes as related specifically to the focus of this chapter.

1. Na$^+$-coupled cotransport processes have been identified in the apical membranes of small intestine and renal proximal tubule in every animal studied ranging throughout the phylogenetic scale from man through the roundworm *Ascaris* and the sea cucumber, a holothurian echinoderm,[93] and have been extended to include a wide variety of water-soluble organic solutes other than sugars and amino acids[94]; indeed, this appears to be the mechanism for the "active" absorption or reabsorption of all water-soluble nutrients throughout the animal kingdom analogous to proton-coupled solute transport in the plant kingdom.

2. Studies on vesicles prepared from apical membranes of small intestine and renal proximal tubule have amply confirmed the location of these coupled transport processes and the fact that the downhill entry of Na$^+$ can energize the uphill movement of coupled solutes in the absence of coupling to metabolic energy.[95,96]

3. Microelectrophysiological studies have demonstrated that Na$^+$-coupled sugar and amino acid transport processes are rheogenic; that is, they involve the carrier-mediated translocation of charge.[97–104]* Thus, part of the driving force for these

*The word "active" is used to describe net transport of a solute in the absence of, or against, an electrochemical driving force. We now know that for the case of Na$^+$-coupled transport, the term "secondary active" is more appropriate inasmuch as these processes are not *directly* coupled to a source of metabolic energy (see Ref. 41).

†The word "downhill" is used to describe the movement of a solute in the direction of its electrochemical potential difference. For the cases of sugar and amino exit across the basolateral membranes, these "downhill" transport processes appear to be carrier-mediated, facilitated diffusion (see Ref. 95).

*Some investigators use the term "electrogenic" to describe carrier-mediated transport mechanisms that *directly* result in the translocation of charge (i.e., current generators). I prefer the term "rheogenic" for the following reasons.

First, the term "electrogenic" was originally, and is still commonly, employed to describe processes that somehow result in a change in the electrical potential difference across a membrane or a sheet of epithelial cells.[71] Indeed, in his classic review, Thomas[165] defined "electrogenic" as ". . . directly contributing to the membrane potential by generating a current across the cell membrane" (p. 564). However, whether or not a change in membrane potential is observed (detectable), the magnitude of this change is largely dependent on properties of the membrane that are *extrinsic to* and parallel the "current generator." Clearly, the "electrogenicity" of a weak current generator in parallel with a large extrinsic conductance may be undetectable. Indeed, the first compelling demonstration of the "electrogenicity" of the Na$^+$,K$^+$ pump[165] involved the measurement of an increase in "pump current" across the membrane of snail neuron following the microinjection of Na$^+$ under *voltage-clamped conditions;* thus, what was actually measured was "rheogenicity," not "electrogenicity."

Second, electroneutral carrier-mediated processes can bring about a change in the electrical potential difference across a membrane by simply changing the ionic gradients across that barrier.

In short, the "intrinsic property" of a carrier-mediated process that results in the translocation of charge is "current generation." This movements must be derived from the electrical potential difference across the apical membranes in addition to the chemical potential difference for Na^+ across those barriers. The rheogenicity of these processes has also been confirmed using apical membrane vesicles.[95,96,106]

4. Important questions that remain to be resolved include the stoichiometry of these coupled systems and their I–V relations; both questions have important bearing on the kinetics and energetics of these processes.

5. Finally, some progress has been made in the direction of isolating the Na^+–sugar cotransporter in relatively pure form and reconstituting the functional protein in artificial liposomes[108–111]; but, as yet, our understanding of the molecular basis of these ubiquitous cotransport processes is minimal to say the least.

2.3. Neutral Ion Co- and Countertransport across Apical Membranes

The first evidence that "all epithelia are not frog skins" came from the elegant studies of Diamond on fish[112] and rabbit[113] gallbladder; both of these epithelia were found to vigorously absorb NaCl without generating a significant transepithelial electrical potential difference. Further, absorption of both ions ceased when one or the other was omitted from the mucosal solution. On the basis of these findings, Diamond proposed the existence of a neutral NaCl pump but did not specify its location or its source of energy. Subsequently, Nellans *et al.*[114] provided evidence for the presence of a neutral NaCl influx mechanism in the apical membranes of rabbit ileum and shortly thereafter Frizzell *et al.*[115] reported the presence of a similar mechanism in the apical membrane of rabbit gallbladder; both processes could be inhibited by elevation of intracellular cAMP.

Thus, it appeared that NaCl cotransport across the apical membrane of the two "leaky epithelia"—small intestine and gallbladder—was analogous to the Na^+-coupled entry of organic solutes across the apical membranes of small intestine and renal proximal tubule with the downhill movement of Na^+ providing the energy for the uphill movement of Cl^- (Fig. 6). Direct support for this notion came from the findings of Duffey *et al.* on rabbit gallbladder[116] and the small intestine of the winter flounder[117] that in the presence of Na^+ the intracellular Cl^- activity, $[Cl^-]_c$, was 2–3 times greater than that predicted for a passive distribution but that $[Cl^-]_c$ fell toward the value predicted by the Nernst equation in the absence of Na^+. These findings have been repeatedly confirmed for the cases of bullfrog small intestine,[118] *Necturus* gallbladder,[119–121] and *Necturus* proximal[122,123] and distal[124] renal tubules.

But, what appeared to be a rather simple explanation for neutral NaCl absorption by "leaky epithelia" only a few years ago has evolved into a somewhat complicated and controversial issue today.

While it is generally accepted that these epithelia are capable of effecting "overall" neutral NaCl absorption, there are two schools of thought regarding underlying mechanisms.

One school maintains that neutral NaCl absorption is the result of two parallel countertransport processes at the apical

intrinsic "rheogenic" property can be measured precisely using voltage-clamp techniques. The "electrogenicity" of a "rheogenic" or a "neutral" carrier-mediated mechanism is not an "intrinsic" quality but depends on extrinsic, parallel properties of the membrane.

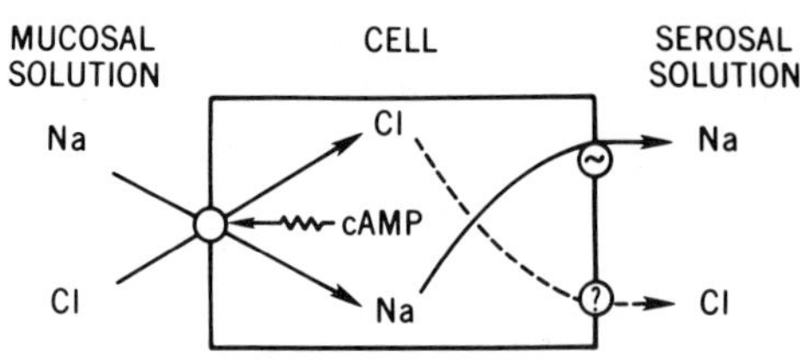

Fig. 6. Original model for neutral, coupled NaCl transport by gallbladder and small intestine.

membranes: (1) a Na^+/H^+ exchange process whereby the *uphill* extrusion of H^+ from the cell is energized by the *downhill* movement of Na^+ into the cell; and (2) a Cl^-/HCO_3^- (or OH^-) exchange whereby the *downhill* movement of HCO_3^- out of the cell energizes the uphill movement of Cl^- into the cell.* The "link" between these two parallel processes that assures obligatory NaCl absorption is presumed to be the intracellular pH. Thus, removal of Na^+ from the mucosal solution would lead to intracellular acidification, a fall in intracellular HCO_3^-, and, consequently, an inhibition of uphill Cl^- influx. Conversely, removal of Cl^- from the mucosal solution would result in an increase in cell HCO_3^- leading to a decrease in cell H^+ and inhibition of Na^+ influx.

Original support for this "dual countertransport model" came from studies on membrane vesicles[125–127] where it was also shown that high concentrations of amiloride (1 mM) inhibit the Na^+/H^+ exchange[127–131] and that Cl^-/HCO_3^- exchange could be inhibited by the disulfonic stilbene derivative SITS.[126,127,132] Recently, evidence for this model of neutral NaCl absorption has been presented for intact preparations of *Necturus*[133] and guinea pig gallbladder[134] and mouse renal cortical thick ascending limb of Henle.[135]

The other school maintains that neutral NaCl transport across the apical membranes is the direct result of carrier-mediated NaCl cotransport or, in some instances, the cotransport of a neutral complex involving Na^+, K^+, and Cl^-[136–142] 1 : 1 : 2. This cotransport mechanism is not affected by amiloride or SITS but is exquisitely sensitive to the "loop diuretics" furosemide and bumetanide. Indeed, sensitivity to bumetanide as opposed to high concentrations (i.e., 1 mM) of amiloride or SITS has become a "pharmacological criterion" for distinguishing between (K)NaCl cotransport and parallel Na^+–H^+/Cl^-–HCO_3^- countertransport.[133,136,137] However, some caution should be exercised in applying these criteria since furosemide is also known to inhibit anion exchange in erythrocytes.[143] In this regard, it is of interest to note that in studies of Liedtke and Hopfer[126,127] on apical membrane vesicles from rat small intestine, 36% of Cl^- uptake was inhibited by furosemide and 63% by SITS.

In the opinion of this author, the mechanism(s) responsible for neutral NaCl absorption by small intestine, gallbladder, and

*The electrical potential difference across the apical membranes of these epithelia is approximately −40 to −60 mV, cell interior negative. The intracellular pH does not differ markedly from that in the luminal solution.[249,250] Thus, the intracellular H^+ activity is *lower* than that predicted for an equilibrium distribution so that H^+ exit is an "uphill" process and the chemical potential difference for Na^+ across the apical membrane is sufficient to energize this movement.

Conversely, the intracellular HCO_3^- (or OH^-) activity is considerably greater than the value predicted for an equilibrium distribution so that the "downhill" movement of that ion could energize the "uphill" movement of Cl^-.[240]

renal epithelia is not a matter of *either/or*. There is ample evidence for both of the models described above and suggestive evidence that in some tissues these two modes of Na^+ and Cl^- absorption coexist.[136–138] For example, Ericson and Spring[136,137] have presented evidence that a neutral coupled NaCl cotransport process is responsible for normal fluid absorption by *Necturus* gallbladder but that parallel Na^+–H^+/Cl^-–HCO_3^- mechanisms are responsible for the volume regulatory increase by these cells in response to a hypertonic medium. Indeed, at present one cannot rule out the possibility that these two models are pleiotropic expressions of the same molecular mechanism(s).

Finally, it should be noted that there are examples of apical Na^+–H^+ countertransport mechanisms that do not appear to be associated with parallel Cl^-–HCO_3^- countertransport mechanisms and therefore result in the acidification of the mucosal (luminal) solution (see Ref. 129). Likewise, there are examples of apical Cl^-–HCO_3^- (or OH^-) countertransport mechanisms that do not appear to be associated with parallel H^+ secretory processes and thereby bring about an alkalinization of the luminal solution (see Ref. 144).

2.3.1. Regulation by Cyclic Nucleotides

In their description of neutral, coupled NaCl influx across the apical membranes of rabbit ileum, Nellans *et al.*[114] also reported that both Na^+ and Cl^- influxes were equally inhibited by elevated levels of intracellular cAMP resulting from treatment of the tissue with theophylline. Shortly thereafter, Frizzell *et al.*[115] reported that elevated intracellular cAMP inhibits NaCl absorption by rabbit gallbladder. These findings have been confirmed for NaCl absorption by *Necturus* gallbladder[145] and (K)NaCl absorption by flounder small intestine.[139,140] Rao *et al.*[146] have recently demonstrated that cGMP is far more effective in inhibiting coupled (K)NaCl influx across the apical membrane of flounder small intestine than is cAMP.

The mechanism of action of these cyclic nucleotides is far from clear but it is of interest that they have been shown to stimulate[147] as well as inhibit[148] cotransport processes in erythrocytes and it may very well be that sensitivity to these agents can also be used to distinguish between (K)NaCl cotransport and parallel Na^+–H^+/Cl^-–HCO_3^- countertransport.

Finally, Peterson and Reuss[149] have recently presented evidence that cAMP induces a large Cl^- conductance in the apical membrane of *Necturus* gallbladder and have raised the possibility that this nucleotide does not directly inhibit the NaCl entry process but that its effect can be attributed to opening a new pathway for recycling of Cl^- across the apical membrane. While this notion could explain the inhibition of fluid absorption by this epithelium, it is difficult to see how it can account for the inhibition of Na^+ *and* Cl^- *influxes* and the *transepithelial* flux of Na^+ observed in rabbit ileum[114] and gallbladder.[115]

In conclusion: The mechanism(s) involved in "overall" neutral NaCl uptake across the apical membranes of a variety of epithelia is not as yet fully understood even at the phenomenological level. There is indisputable evidence for NaCl and/or $KNaCl_2$ cotransport as well as parallel Na^+–H^+/Cl^-–HCO_3^- (or OH^-) countertransport processes. The extent to which one or both of these processes may be present in the same epithelium or indeed whether they are the results of *different* molecular mechanisms or *different* expression of the same underlying mechanism remain to be fully resolved. The differing sensitivities to pharmacological inhibitors suggest different molecular mechanisms, but this is not conclusive. In both cases, the energy for uphill movement of Cl^- into the cell is ultimately derived from the Na^+ gradient across the apical membrane.*

Finally, the regulation of these transport processes by cyclic nucleotides certainly has physiological as well as pathophysiological implications[150] but the mechanism of these regulatory influences remains to be resolved.

2.4. K^+ Channels in Apical Membranes

In recent years the apical membranes of several "leaky"[151–154] as well as "tight" and "moderately tight"[155–160] epithelia have been found to possess K^+ conductances which appear to be channels similar to those found in excitable membranes.[161] Thus, in every instance studied, these conductive pathways are blocked by Ba^{2+} and, in some instances, by other divalent cations in the mucosal solution, and recent elegant microelectrophysiological studies on isolated segments of rabbit cortical collecting duct[160] indicate that this effect is voltage-dependent in a manner consistent with the "plugging" of channels. Further, noise analysis of K^+ conductive pathways across the apical membranes of frog skin,[156] *Necturus* gallbladder,[152] and rabbit colon[157] are consistent with single-site conductances expected for channels but orders of magnitude greater than those expected for carriers. Finally, in some instances, these channels are blocked by H^+.[160]

Clearly, inasmuch as there is a significant electrochemical driving force for K^+ exit across the apical membrane, the presence of K^+ channels in that barrier permits active K^+ secretion driven by the Na^+,K^+ pump. In some leaky epithelia such as gallbladder and small intestine, where the apical membranes possess mimimal, if any, conductances to Na^+ and Cl^-, the K^+ conductance naturally is predominant but the rates of K^+ secretion are very low and are probably of little physiological significance. In skin from "mature" frogs the density of amiloride-insensitive K^+ channels is approximately 3 orders of magnitude lower than the density of amiloride-sensitive Na^+ channels[155,156] so that K^+ secretion by this epithelium is probably also of minimal physiological significance.

However, in epithelia such as renal collecting duct[160] and, perhaps, mammalian colon,[162,163] *active* K^+ secretion via these channels certainly plays an important physiological role in body K^+ homeostasis. Since K^+ secretion by these epithelia is increased by mineralocorticoids,[70,71] it will be of interest to determine whether this is due to an increase in the driving force for K^+ exit across the apical membrane, to an increase in the number (and/or conductance) of channels in response to the hormonal stimulus, or perhaps some combination of each.

2.5. Properties of Basolateral Membranes

As noted above, the basolateral membranes of every Na(Cl)-absorbing epithelium studied to date share at least two common features: (1) an ouabain-inhibitable Na^+,K^+-ATPase that is capable of actively extruding Na^+ from the cell and actively pumping K^+ into the cell energized by the hydrolyses of ATP[164] and (2) a K^+ conductance that permits the "recycling" of K^+ across that barrier. In addition, there is a growing body of evidence for the presence of a Na^+–Ca^{2+}

*Refer back to the footnote on p. 524.

countertransport mechanism at the basolateral membranes of several "tight" and "leaky" epithelia. We will briefly consider each of these processes in turn.

2.5.1. The Na^+,K^+ Pump

Although the Na^+/K^+ exchange pump at the basolateral membranes of epithelia as well as the plasma membranes of nonepithelial cells was originally considered to be electroneutral, mediating the exchange of 1 Na^+ for 1 K^+ for every ATP consumed,[1,71] it is now generally accepted that these pumps are rheogenic (generate a flow of current)* and extrude approximately 3 Na^+ in exchange for 2 K^+ for every ATP consumed.[165–168] Evidence for this stoichiometry in epithelia is derived both from studies of transepithelial transport[169–171] as well as from studies dealing with the relation between $[Na^+]_c$ and pump rate. As illustrated in Fig. 7, an increase in $[Na^+]_c$ in rabbit colon is associated with a sigmoidal increase in pump rate (I_p) and the curve conforms with the predictions of the Hill equation for a strongly cooperative interaction between 2–3 Na^+ and the pump mechanism.[172] Similar sigmoidal relations have been observed for frog skin[173] and rabbit urinary bladder.[174–176] Further, Lewis and Wills[174] have demonstrated that the relation between the K^+ concentration in the serosal solution, $[K^+]_s$, and the pump rate is sigmoidal and consistent with the notion that 2 K^+ interact with the pump mechanism per cycle. Similar results have recently been reported for turtle colon by Halm and Dawson.[176a] The value of $[Na^+]_c$ at which pump rate is half-maximal has been estimated to be between 2 and 24 mM with the higher values in reasonable agreement with those reported for human erythrocytes[177] and the activation of purified renal Na^+,K^+-ATPase.[178]

Unfortunately, little more is known about the kinetics of this pump mechanism under physiological conditions. Although the Na^+,K^+-ATPase must rank among the most extensively investigated enzymes derived from higher animals, most studies have been carried out on suspensions where the essential asymmetric properties of this "pump enzyme" are lost or in cell systems where precise control of important parameters is difficult if not impossible. Thus, it is still not clear whether the stoichiometry of the Na^+/K^+ exchange is fixed or whether it may vary under physiological conditions. And, the I–V relations of this rheogenic process are still conjectural.[167,168] Clearly, the reconstitution of the purified, functional Na^+,K^+-ATPase in planar lipid bilayers which would permit electrophysiological studies under carefully defined conditions would represent a major breakthrough toward the resolution of these seminal issues.

Finally, it is well established that mineralocorticoids, Na^+ depletion, or K^+ loading bring about an increase in ATPase activity in basolateral membranes of mammalian colon[179–181] and distal nephron.[182–189] Recently, intriguing evidence has been presented that this increase is *secondary* to the increased rate of Na^+ entry across the apical membrane induced by mineralocorticoids inasmuch as it can be blocked by amiloride[185–187] or a reduction in $[Na^+]_m$.[188] Thus, it seems as if an increase in the rate of Na^+ entry into the cell can result in a "recruitment" of new, active pump units into the basolateral membrane. Whether this is specific for mineralocorticoid-treated tissues or is a more general response of epithelia to an increase in Na^+ entry, and the time course of this response, are important issues that remain to be resolved. In short, the possibility that an increase in *overall* pump rate is not due entirely to an increase in the *turnover rates* of a *fixed* number of pumps but may also involve an *increase* in the *number* of pump units must be entertained.

*Refer back to the footnote on p. 523.

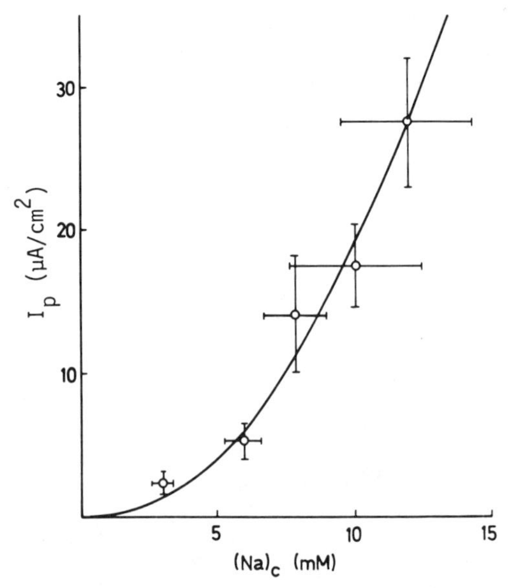

Fig. 7. Relation between $[Na^+]_c$ and the activity of the Na^+,K^+ pump in rabbit descending colon.[172]

2.5.2. The K^+ "Leak"

As noted above, a general feature of all Na(Cl)-absorbing epithelia is the presence of a major conductive pathway for K^+ in the basolateral membrane that permits passive ("downhill") recycling of most, if not all, of the K^+ pumped into the cell across that barrier.* The properties of this conductive pathway are poorly understood except for the fact that it, like the K^+ conductive pathways across the apical membrane of some epithelia (Section 2.4), can be blocked by Ba^{2+}[169,171,190–193] suggesting the presence of "channels" rather than "carriers."

Finally, in recent years, evidence has been presented that the K^+ conductance of the basolateral membranes of *Necturus*,[32] toad, and frog[194,195] urinary bladders, *Necturus* small intestine,[101,193,196] and cannine trachea[192,197–200] parallels the Na^+,K^+ pump rate. In addition, the K^+ conductance of the basolateral membranes of frog and toad urinary bladders[195] and *Necturus* gallbladder[201] and small intestine[193] are somehow influenced by changes in cell volume.

There are a number of seminal issues regarding this ubiquitous basolateral K^+ conductance that await resolution. For example: (1) Is there only one "type" of basolateral membrane K^+ channel or are there several "types" as found in some

*Clearly, if the cell possesses a K^+ conductance in the apical membrane (see Section 2.4), some of the K^+ pumped into the cell across the basolateral membrane will exit across that barrier, resulting in active K^+ secretion. But in every instance studied the rate of K^+ secretion is much lower than the rate of Na^+ absorption so that most of the K^+ pumped into the cell across the basolateral membrane is recycled across that barrier.

other membranes?[202–204] (2) What, if any, is the relation between the Na^+,K^+ pump and the K^+ leak? Are they part of the same molecular mechanism or are they distinct entities? (3) If they are distinct molecular entities, what intracellular "signal(s)" is responsible for the close parallelism between pump rate and K^+ conductance found in some epithelia? (4) What "signals" change in K^+ conductance in response to changes in cell volume? (5) Last but by no means final, what are the kinetic characteristics of these channels?

Clearly, the list of "unknowns" considerably exceeds the list of "knowns."

2.5.3. Na^+–Ca^{2+} Countertransport

There is both direct and reasonably compelling indirect evidence for the presence of Na^+–Ca^{2+} countertransport mechanisms in the basolateral membranes of frog skin,[65,205] toad[58,64,66,67,69] and frog[69] urinary bladder, *Necturus* renal proximal tubule,[206,207] and rat small intestine.[68] Assuming that these mechanisms are similar to those described for a number of nonepithelial cells,[208] they are rheogenic and couple the downhill movement of at least 3 Na^+ into the cell across the basolateral membrane to the uphill extrusion of 1 Ca^{2+} from the cell across that barrier. Thus, both the chemical potential difference for Na^+ across the basolateral membrane and the electrical potential difference across that barrier energize the counterflow of Ca^{2+} against a considerable electrochemical potential difference. It follows that an increase in $[Na^+]_c$, a decrease in $[Na^+]_s$, and/or a depolarization of the basolateral membrane (cell interior less negative) would reduce the energy available for Ca^{2+} extrusion and *could* lead to an increase in $[Ca^{2+}]_c$. Indeed, an increase in $[Ca^{2+}]_c$ has been observed in *Necturus* renal proximal tubule following reduction in $[Na^+]_s$[206] and treatment with ouabain which brings about an increase in $[Na^+]_c$.[207]

As discussed in Section 2.1, there is reasonable evidence that the "negative feedback" between $[Na^+]_c$ and the permeability of the apical membrane to Na^+ is directly mediated by changes in $[Ca^{2+}]_c$ which parallel changes in $[Na^+]_c$. But, again, it is not as yet clear whether this "feedback loop" plays a meaningful role in "fine-tuning" the Na^+ permeability of the apical membrane under physiological conditions or whether it is a "fail-safe" mechanism that comes into play only when $[Na^+]_c$ is grossly elevated beyond physiological levels.

2.5.4. KCl Cotransport

As discussed in Section 2.3, Cl^- entry across the apical membranes of a number of epithelia is the result either of cotransport coupled to the flows of Na^+ (and K^+) or of countertransport coupled to the flow of HCO_3^- (or OH^-). Either of these mechanisms can result in levels of $[Cl^-]_c$ that considerably exceed those predicted by the Nernst equation for an equilibrium distribution across the basolateral membrane so that, in principle, Cl^- exit from the cell could be the result of simple diffusion across that barrier. However, estimates of the conductances of the basolateral membranes of rabbit[209–212] and *Necturus*[213–215] gallbladder and *Necturus* proximal tubule[216] suggest that they are far too low to permit Cl^- exit by simple electrodiffusion. Forced to seek other possible exit mechanisms, several groups suggested that downhill Cl^- exit may be a neutral process coupled to the downhill flow of K^+[212,215,216] and, recently, evidence in support of this notion has been presented for *Necturus* gallbladder.[217] It should be emphasized that this purported mechanism is an "escape route" forced by estimates of basolateral membrane conductances that are open to question.[218] More refined measurements of the Cl^- conductances of these epithelia will be required to determine the extent to which Cl^- exit can be attributed to simple electrodiffusion and the extent to which nonconductive exit processes must be invoked.

3. A Model for Active Chloride Secretion by Epithelial Cells

In 1952, Koefoed-Johnsen *et al.*[219] reported that treatment of isolated frog skin with adrenaline elicits active Cl^- secretion presumably arising from the epidermal glandular cells, and 3 years later Hogben[220] reported that the electrical potential difference and short-circuit current across isolated frog stomach were entirely attributable to active Cl^- secretion. However, these remained somewhat isolated and poorly understood observations until a general interest in ion secretion by epithelia was kindled by the milestone contributions of Field and his collaborators commencing in 1968. These investigators demonstrated that active Cl^- secretion by *in vitro* rabbit ileum could be elicited by elevation of intracellular cAMP levels and by a host of physiological and pathophysiological agents that activate the membrane-bound adenylate cyclase in that epithelium including cholera toxin, exotoxins from other enteropathic bacteria, and several gastrointestinal hormones and neurotransmitters.[221–227]

cAMP-induced active Cl^- secretion is now a widely recognized phenomenon involving a variety of epithelia including rabbit colon,[228–230] frog cornea,[231–233] killifish operculum,[234] dogfish rectal gland,[235] gastric mucosa,[236] and mammalian tracheal epithelium.[237–239]

The common features of all of these secretory processes include[240]:

1. Stimulation by agents that elevate intracellular levels of cAMP and/or PGE_2
2. An absolute dependence on the presence of Na^+ in the serosal solution
3. Inhibition by ouabain in the serosal solution
4. Inhibition by furosemide or bumetanide in the serosal solution
5. In some instances, stimulation by treatment of the tissue with the Ca^{2+} ionophore A23187, which presumably increases $[Ca^{2+}]_c$

These "black box" findings strongly suggested the model for Cl^--secreting epithelial cells illustrated in Fig. 8.[240] According to this model, Cl^- entry into the cell across the basolateral membrane is mediated by a furosemide-sensitive, neutral NaCl cotransport mechanism similar to that found in some NaCl-absorbing epithelia; by virtue of this coupling, the downhill movement of Na^+ energizes the uphill movement of Cl^-. As a result, the intracellular electrochemical potential of Cl^- is considerably greater than that in the mucosal solution and the cell is thus "poised" for active Cl^- secretion which can be "triggered" by any agent (e.g., cAMP, PGE, and/or Ca^{2+}) that induces a conductive pathway for Cl^- (i.e., "leak") in the apical membrane. The Na^+ that enters the cell coupled to the movement of Cl^- is subsequently extruded by the basolateral Na^+,K^+ pump.

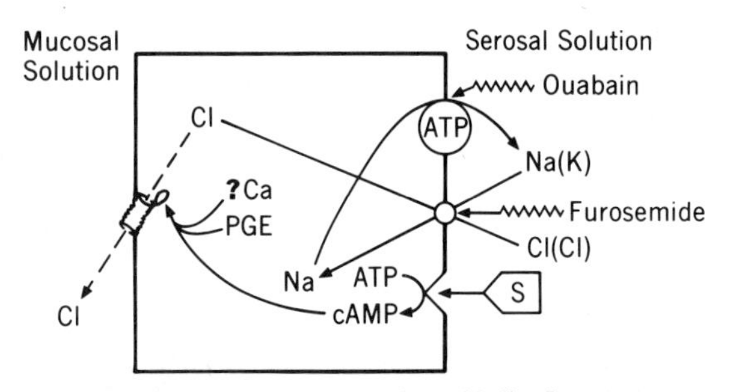

Fig. 8. Cellular model for Cl^--secreting epithelia. S represents a natural secretagogue which interacts with specific receptors on the basolateral membranes to initiate the series of intracellular processes that lead to an increase in the Cl^- conductance of the apical membrane. Bacterial enterotoxins interact with receptors on the apical membrane and somehow appear to initiate the same cascade of events. Modified from Ref. 240.

Direct microelectrophysiological support for this cellular model was initially hampered by the facts that the cells responsible for active Cl^- secretion by small intestine and colon appear to reside in the crypts[241] and, in other tissues, the cells are either dispersed among other nonsecreting cells[234] or embedded in glandular structures[235]; in both instances, successful impalement of the Cl^--secreting cells with microelectrodes is difficult and uncertain.

In recent years the mammalian tracheal mucosa, which is comprised of pseudostratified columnar cells, has proved to be an excellent tissue for the microelectrophysiological study of Cl^--secreting cells and the results of these studies have amply confirmed the model illustrated in Fig. 8.

Specifically:

1. In the nonsecreting cell (indomethacin-treated to reduce endogenous levels of PGE), the resistance of the apical membrane accounts for 75–85% of the transcellular resistance[197–199] and $[Cl^-]_c$ is considerably greater than the value predicted for a passive distribution of that anion across the apical and basolateral membranes[242,243]; in the absence of Na^+, $[Cl^-]_c$ approaches the value expected for a passive distribution.[242] Addition of furosemide to the serosal solution results in a decrease in $[Cl^-]_c$ with little or no change in the electrical properties of the basolateral membrane.[242] These findings are consistent with the notion that the "uphill" entry of Cl^- across the basolateral membrane is the result of a neutral Na^+-coupled cotransport process.

2. Stimulation of secretion with cAMP, PGE, or epinephrine is accompanied by a dramatic fall in the resistance of the apical membrane due to the appearance of a highly specific Cl^- conductance.[197–199,242,243] At the same time, $[Cl^-]_c$ either remains unchanged[242] or declines[243] but, in any event, remains above the equilibrium value.

Similar, but less extensive, findings have been reported for cornea[244,245] and dogfish rectal gland.[246,247]

Thus, the model proposed by Frizzell *et al.*[240] appears to be established beyond reasonable doubt. One major question that remains to be resolved is whether the basolateral entry process is a neutral NaCl cotransport or a $KNaCl_2$ cotransport process as appears to be the case for several NaCl-absorbing epithelia (Section 2.3). This point is difficult to establish in the intact tissue because inhibition of Cl^- secretion by removal of serosal K^+ could be attributable to inhibition of the Na^+,K^+ pump responsible for the maintenance of the Na^+ gradient across the basolateral membrane. Recent studies on basolateral membrane vesicles from dogfish rectal gland support the notion of a $KNaCl_2$ influx process.[248] Other major questions that remain to be elucidated concern the nature of the Cl^- conductance at the apical membrane and the mechanism(s) of its regulation by the ubiquitous intracellular "signals," cAMP, PG, and/or Ca^{2+}.

4. Summary

As indicated in the Introduction, the purpose of this chapter was not to exhaustively review the subject of ion transport by epithelia; such a review would easily occupy a volume of its own. Instead, the aim was to summarize, briefly, some of the major transport properties of the apical and basolateral membranes that have been uncovered during the past quarter of a century and illustrate how cellular models that account for most Na^+ and Cl^- absorptive as well as secretory processes can be constructed by combining these elements in parallel (same membrane) and series (both limiting membranes) arrays. It should be emphasized that these properties are not simply the products of "flights of imagination" in an attempt to explain "black box" observations (i.e., "suppose we put this here, that there"). Instead, these transport properties (processes) have been established using microelectrophysiological approaches on intact tissues and by the study of isolated membrane vesicles from disrupted cells. These two approaches—aimed at "peering into" and/or "breaking open" the "black box"—each have their own, quite different but complementary, sets of difficulties and uncertainties. It is thus comforting that the membrane properties deduced using one approach have been, for the most part, confirmed using the other. In short, by all accepted scientific criteria, the processes we have considered are "real"; that is, they have molecular counterparts.

Thus, we are now in a position where transcellular Na^+ and Cl^- transport can be explained at the cellular and single-membrane levels using transport processes (elements) that appear to be highly conserved throughout the animal kingdom. The major questions that remain to be resolved at the cellular and single-membrane levels deal with the regulation of these processes. When these questions are resolved, we will have "narrowed" the "black box" from the level of the "cell" to that of the "membranes" and we will then be faced with the formidable problems of understanding these transport processes at the microscopic, molecular levels.

References

1. Koefoed-Johnsen, V., and H. H. Ussing. 1958. The nature of the frog skin potential. *Acta Physiol. Scand.* **42**:298–308.
2. Macknight, A. D. C., D. R. DiBona, and A. Leaf. 1980. Sodium transport across toad urinary bladder: A model tight epithelium. *Physiol. Rev.* **60**:615–715.
3. Lewis, S. A. 1977. A reinvestigation of the function of the mammalian urinary bladder. *Am. J. Physiol.* **232**:F187–F195.
4. Schultz, S. G. 1984. A cellular model for active sodium absorption by mammalian colon. *Annu. Rev. Physiol.* **46**:435–451.
5. Giebisch, G. 1978. Amiloride effects on distal nephron function. In: *Cell Membrane Receptors for Drugs and Hormones: A Multidisciplinary Approach.* R. W. Straub and L. Bohs, eds. Raven Press, New York. pp. 337–342.
6. Frömter, E., and J. Diamond. 1972. Route of passive ion permeation in epithelia. *Nature New Biol.* **235**:9–13.

7. Schultz, S. G. 1977. The role of paracellular pathways in isotonic fluid transport. *Yale J. Biol. Med.* **50**:99–113.
8. Ussing, H. H. 1980. Life with tracers. *Annu. Rev. Physiol.* **42**:1–16.
9. Hladky, S. B. 1974. Pore or carrier? Gramicidin A as a simple pore. In: *Drugs and Transport Processes*. B. A. Callingham, ed. University Park Press, Baltimore. pp. 193–210.
10. Laüger, P. 1972. Carrier-mediated ion transport. *Science* **178**:24–30.
11. Laüger, P. 1980. Kinetic properties of ion carriers and channels. *J. Membr. Biol.* **57**:163–178.
12. Haydon, D. A., and S. B. Hladky. 1972. Ion transport across thin lipid membranes: A critical discussion of mechanisms in selected systems. *Q. Rev. Biophys.* **5**:187–282.
13. Kolb, H.-A., and P. Laüger. 1978. Spectral analysis of current noise generated by carrier-mediated ion transport. *J. Membr. Biol.* **41**:167–187.
14. Bentley, P. J. 1968. Amiloride: A potent inhibitor of sodium transport across toad bladder. *J. Physiol. (London)* **195**:317–330.
15. Benos, D. J., L. J. Mandel, and R. S. Balaban. 1979. On the mechanism of the amiloride–sodium entry site interaction in anuran skin epithelia. *J. Gen. Physiol.* **73**:307–326.
16. Turnheim, K., A. Luger, and M. Grasl. 1981. Kinetic analysis of the amiloride–sodium entry site interaction in rabbit colon. *Mol. Pharmacol.* **20**:543–550.
17. O'Neil, R. G., and E. L. Boulpaep. 1979. Effect of amiloride on the apical cell membrane cation channels of a sodium-absorbing potassium-secreting renal epithelium. *J. Membr. Biol.* **50**:365–387.
18. Li, J. H.-Y., and B. Lindemann. 1983. Competitive blocking of epithelial sodium channels by organic cations: The relationship between macroscopic inhibition constants. *J. Membr. Biol.* **76**:235–251.
19. Cuthbert, A. W. 1974. Interactions of sodium channels in transporting epithelia: A two state model. *Mol. Pharmacol.* **10**:892–903.
20. Moran, M., R. Hudson, and S. G. Schultz. 1984. Kinetic analysis of the effect of amiloride on sodium entry across the apical membrane of rabbit colon. *J. Membr. Biol.* in press.
21. Cuthbert, A. W., and W. K. Shum. 1974. Amiloride and the sodium channel. *Naunyn-Schmiedebergs Arch. Pharmacol.* **281**:261–269.
22. Cuthbert, A. W. 1981. Sodium entry step in transporting epithelia: Results of ligand-binding studies. In: *Ion Transport by Epithelia*. S. G. Schultz, ed. Raven Press, New York. pp. 181–195.
23. Benos, D. 1982. Amiloride: A molecular probe of sodium transport in tissues and cells. *Am. J. Physiol.* **242**:C131–C145.
24. Lindemann, B., and U. Gebhardt. 1973. Delayed changes of Na permeability in response to steps of [Na] at the outer surface of frog skin and frog bladder. In: *Transport Mechanisms in Epithelia*. H. H. Ussing and N. A. Thorn, eds. Munksgaard, Copenhagen. pp. 115–130.
25. Fuchs, W., E. H. Larsen, and B. Lindemann. 1977. Current–voltage curve of sodium channels and concentration dependence of sodium permeability in frog skin. *J. Physiol. (London)* **267**:137–166.
26. Palmer, L. G., I. S. Edelman, and B. Lindemann. 1980. Current–voltage analysis of apical sodium transport in toad urinary bladder: Effects of inhibitors of transport and metabolism. *J. Membr. Biol.* **57**:59–71.
27. Palmer, L. G., J. H.-Y. Li, B. Lindemann, and I. S. Edelman. 1982. Aldosterone control of the density of sodium channels in the toad urinary bladder. *J. Membr. Biol.* **64**:91–102.
28. Li, J. H.-Y., L. G. Palmer, I. S. Edelman, and B. Lindemann. 1982. The role of sodium-channel density in the natriferic response of the toad urinary bladder to an antidiuretic hormone. *J. Membr. Biol.* **64**:77–89.
29. Helman, S. I., T. C. Cox, and W. van Driessche. 1983. Hormonal control of apical membrane Na transport in epithelia: Studies with fluctuation analysis. *J. Gen. Physiol.* **82**:201–220.
30. Garty, H., I. S. Edelman, and B. Lindemann. 1983. Metabolic regulation of apical sodium permeability in toad urinary bladder in the presence and absence of aldosterone. *J. Membr. Biol.* **74**:15–24.
31. Frömter, E., J. T. Higgins, and B. Gebler. 1981. Electrical properties of amphibian urinary bladder epithelia. IV. The current–voltage relationship of the sodium channels in the apical cell membrane. In: *Ion Transport by Epithelia*. S. G. Schultz, ed. Raven Press, New York. pp. 31–45.
32. Thomas, S. R., Y. Suzuki, S. M. Thompson, and S. G. Schultz. 1983. The electrophysiology of *Necturus* urinary bladder. I. "Instantaneous" current–voltage relations in the presence of varying mucosal sodium concentrations. *J. Membr. Biol.* **73**:157–175.
33. Thompson, S. M., Y. Suzuki, and S. G. Schultz. 1982. The electrophysiology of rabbit descending colon. I. Instantaneous transepithelial current–voltage relations and the current–voltage relation of the Na-entry mechanism. *J. Membr. Biol.* **66**:41–54.
34. Thompson, S. M., Y. Suzuki, and S. G. Schultz. 1982. The electrophysiology of rabbit descending colon. II. Current–voltage relations of the apical and baso-lateral membranes and the paracellular pathway. *J. Membr. Biol.* **66**:55–61.
35. Turnheim, K., S. M. Thompson, and S. G. Schultz. 1983. Relation between intracellular sodium and active sodium transport in rabbit colon. *J. Membr. Biol.* **76**:299–309.
36. Goldman, D. E. 1943. Potential, impedance and rectification in membranes. *J. Gen. Physiol.* **27**:37–60.
37. Hodgkin, A. L., and B. Katz. 1949. The effect of sodium ions on the electrical activity of the giant axon of the squid. *J. Physiol. (London)* **108**:37–77.
38. Benos, D. J., B. A. Hyde, and R. Latorre. 1983. Sodium flux ratio through the amiloride-sensitive entry pathway in frog skin. *J. Gen. Physiol.* **81**:667–685.
39. Palmer, L. G. 1982. Na transport and flux ratio through apical channels in toad bladder. *Nature (London)* **297**:688–690.
40. Ussing, H. H. 1949. The distinction by means of tracers between active transport and diffusion. *Acta Physiol. Scand.* **19**:43–56.
41. Schultz, S. G. 1980. *Basic Principles of Membrane Transport*. Cambridge University Press, London.
42. Lindemann, B., and W. van Driessche. 1977. Sodium-specific membrane channels of frog skin are pores: Current fluctuations reveal high turnover. *Science* **195**:292–294.
43. van Driessche, W., and B. Lindemann. 1979. Concentration dependence of currents through single sodium-selective pores in frog skin. *Nature (London)* **282**:519–520.
44. Finkelstein, A., and O. S. Anderson. 1981. The gramicidin A channel: A review of its permeability characteristics with special reference to the single-file aspect of transport. *J. Membr. Biol.* **59**:155–171.
45. Anderson, O. S., and R. U. Muller. 1982. Monazomycin-induced single channels. I. Characterization of the elementary conductance events. *J. Gen. Physiol.* **80**:403–426.
46. Hoffman, J. F., B. G. Kennedy, and G. Lunn. 1981. Modulation of red cell Na/K pump rates. In: *Erythrocyte Membranes 2: Recent Clinical and Experimental Advances*. W. C. Kruckeberg, J. W. Eaton, and G. T. Brewer, eds. Liss, New York. pp. 5–9.
47. Henrich, M., and B. Lindemann. 1983. Fluctuation analysis of apical Na channels: Voltage dependence of channel currents and channel densities. In Intestinal Absorption and Secretion. E. Skadhauge and K. Heintze, eds. M.T.P. Press, Lancaster. pp. 209–220.
48. Lindemann, B. 1982. Dependence of ion flow through channels on the density of fixed charges at the channel opening. *Biophys. J.* **39**:15–22.
49. Kirschner, L. B. 1955. On the mechanism of active sodium transport across the frog skin. *J. Cell. Comp. Physiol.* **45**:65–87.
50. Frazier, H. S., E. F. Dempsey, and A. Leaf. 1962. Movement of sodium across the mucosal surface of the isolated toad bladder and its modification by vasopressin. *J. Gen. Physiol.* **45**:529–543.
51. Cereijido, M., F. C. Herrera, W. J. Flanigan, and P. F. Curran.

1964. The influence of Na concentration on Na transport across frog skin. *J. Gen. Physiol.* **47**:879–893.
52. Lindemann, B., and C. Voute. 1976. Structure and function of the epidermis. In: *Frog Neurobiology*. R. Llinas and W. Precht, eds. Springer-Verlag, Berlin. pp. 169–210.
53. Lindemann, B. 1977. Steady-state kinetics of a floating receptor model for the inhibition of sodium uptake by sodium in frog skin. In: *Renal Function*. G. H. Giebisch and E. F. Purcell, eds. J. C. Macy, Jr. Foundation, New York. pp. 110–131.
54. Larsen, E. H. 1973. Effect of amiloride, cyanide and ouabain on the active transport pathway in toad skin. In: *Transport Mechanisms in Epithelia*. H. H. Ussing and N. A. Thorn, eds. Munksgaard, Copenhagen. pp. 131–143.
55. Lewis, S. A., D. C. Eaton, and J. M. Diamond. 1976. The mechanism of Na transport by rabbit urinary bladder. *J. Membr. Biol.* **28**:41–70.
56. Turnheim, K., R. A. Frizzell, and S. G. Schultz. 1978. Interaction between cell sodium and the amiloride-sensitive sodium entry step in rabbit colon. *J. Membr. Biol.* **39**:233–256.
57. Helman, S. I., W. Nagel, and R. S. Fisher. 1979. Ouabain on active transepithelial Na transport in frog skin: Studies with microelectrodes. *J. Gen. Physiol.* **74**:105–127.
58. Chase, H. S., Jr., and Q. Al-Awqati. 1981. Regulation of the sodium permeability of the luminal border of toad bladder by intracellular sodium and calcium. *J. Gen. Physiol.* **77**:693–712.
59. Leblanc, G., and F. Morel. 1975. Na and K movements across the membranes of frog skin associated with transient current changes. *Pfluegers Arch.* **358**:159–177.
60. Erlij, D., and M. W. Smith. 1973. Sodium uptake by frog skin and its modification by inhibitors of transepithelial sodium transport. *J. Physiol. (London)* **228**:221–239.
61. Essig, A., and A. Leaf. 1963. The role of potassium in active transport of sodium by the toad bladder. *J. Gen. Physiol.* **46**:505–515.
62. Schultz, S. G. 1981. Homocellular regulatory mechanisms in sodium-transporting epithelia: Avoidance of extinction by "flush-through." *Am. J. Physiol.* **241**:F579–F590.
63. Erlij, D., and W. van Driessche. 1983. Noise analyses of inward and outward current in ouabain treated frogs. *Fed. Proc.* **42**:1101.
64. Chase, H. S., Jr., and Q. Al-Awqati. 1983. Calcium reduces the sodium permeability of luminal membrane vesicles from toad bladder: Studies using a fast reaction apparatus. *J. Gen. Physiol.* **81**:643–665.
65. Grinstein, S., and D. Erlij. 1978. Intracellular calcium and the regulation of sodium transport in the frog skin. *Proc. R. Soc. London Ser. B* **202**:353–360.
66. Taylor, A., and E. E. Windhager. 1979. Possible role of cytosolic calcium and Na–Ca exchange in regulation of transepithelial sodium transport. *Am. J. Physiol.* **236**:F505–F512.
67. Windhager, E. E., and A. Taylor. 1983. Regulatory role of intracellular calcium ions in epithelial Na transport. *Annu. Rev. Physiol.* **45**:519–532.
68. Hildemann, B., A. Schmidt, and H. Murer. 1982. Ca transport across basal-lateral plasma membranes from rat small intestine epithelial cells. *J. Membr. Biol.* **65**:55–62.
69. Arruda, J. A. L., S. Sabatini, and C. Westenfelder. 1982. Serosal Na/Ca exchange and H and Na transport by the turtle and toad bladders. *J. Membr. Biol.* **70**:135–146.
70. Feldman, D., J. W. Funder, and I. S. Edelman. 1972. Subcellular mechanisms in the action of adrenal steroids. *Am. J. Med.* **53**:545–560.
71. Ussing, H. H. 1960. *The Alkali Metal Ions in Biology*. Springer-Verlag, Berlin.
72. Garty, H., and I. S. Edelman. 1983. Amiloride-sensitive trypsinization of apical sodium channels: Analysis of hormonal regulation of sodium transport in toad bladder. *J. Gen. Physiol.* **81**:785–803.
73. Turnheim, K., R. A. Frizzell, and S. G. Schultz. 1977. Effect of anions on amiloride-sensitive, active sodium transport across rabbit colon, *in vitro*. *J. Membr. Biol.* **37**:63–84.
74. Singer, I., and M. M. Civan. 1971. Effects of anions on sodium transport in toad urinary bladder. *Am. J. Physiol.* **221**:1019–1026.
75. Lindemann, B., and W. van Driessche. 1978. The mechanism of Na uptake through Na-selective channels in the epithelium of frog skin. In: *Membrane Transport Processes*, Volume 1. J. Hoffman, ed. Raven Press, New York. pp. 155–178.
76. Li, J. H.-Y., and B. Lindemann. 1983. Chemical stimulation of Na-transport through amiloride blockade channels of frog skin. *J. Membr. Biol.* **75**:179–192.
77. Spooner, P. M., and I. S. Edelman. 1976. Stimulation of Na transport across the toad urinary bladder by *p*-chloromercuribenzene sulfonate. *Biochim. Biophys. Acta* **455**:272–276.
78. Gottleib, G. P., K. Turnheim, R. A. Frizzell, and S. G. Schultz. 1978. *p*-Chloromercuribenzene sulfonate blocks and reverses the effect of amiloride on sodium transport across rabbit colon *in vitro*. *Biophys. J.* **22**:125–129.
79. Luger, A., and K. Turnheim. 1981. Modification of cation permeability of rabbit descending colon by sulphydryl reagents. *J. Physiol. (London)* **317**:49–66.
80. Lindemann, B. 1984. Fluctuation analysis of sodium channels in epithelia. *Annu. Rev. Physiol.* **46**:497–515.
81. Thompson, S. M., and D. C. Dawson. 1978. Cations selectivity of the apical membrane of the turtle colon: Sodium entry in the presence of lithium. *J. Gen. Physiol.* **72**:269–282.
82. Palmer, L. G. 1982. Ion selectivity of the apical membrane Na channel in the toad urinary bladder. *J. Membr. Biol.* **67**:91–98.
83. Nagel, W. 1977. Influence of lithium upon the intracellular potential of frog skin epithelium. *J. Membr. Biol.* **37**:347–359.
84. Herrera, F. C. 1972. Inhibition of lithium transport across toad bladder by amiloride. *Am. J. Physiol.* **222**:499–502.
85. Benos, D. J., L. J. Mandel, and S. A. Simon. 1980. Cation selectivity and competition at the sodium entry site in frog skin. *J. Gen. Physiol.* **76**:233–247.
86. Csaky, T. Z., and L. Zollicoffer. 1960. Ionic effect on intestinal transport of glucose in the rat. *Am. J. Physiol.* **198**:1056–1058.
87. Csaky, T. Z., and M. Thale. 1960. Effect of ionic environment on intestinal sugar transport. *J. Physiol. (London)* **151**:59–65.
88. Bihler, I., K. A. Hawkins, and R. K. Crane. 1962. Studies on the mechanism of intestinal absorption of sugars. VI. The specificity and other properties of Na^+ dependent entrance of sugars into intestinal tissue under anaerobic conditions *in vitro*. *Biochim. Biophys. Acta* **59**:94–102.
89. Crane, R. K. 1962. Hypothesis for mechanism of intestinal active transport of sugars. *Fed. Proc.* **21**:891–895.
90. Schultz, S. G., and R. Zalusky. 1964. Ion transport in isolated rabbit ileum. II. The interaction between active sodium and active sugar transport. *J. Gen. Physiol.* **47**:1043–1059.
91. Schultz, S. G., and R. Zalusky. 1965. Interactions between active sodium transport and active amino acid transport in isolated rabbit ileum. *Nature (London)* **204**:292–294.
92. Schultz, S. G., and P. F. Curran. 1970. Coupled transport of sodium and organic solutes. *Physiol. Rev.* **50**:637–718.
93. Schultz, S. G. 1977. Sodium-coupled solute transport by small intestine: A status report. *Am. J. Physiol.* **233**:E249–E254.
94. Schultz, S. G. 1978. Ion-coupled transport across biological membranes. In: *Physiology of Membrane Disorders*. T. E. Andreoli, J. F. Hoffman, and D. D. Fanestil, eds. Plenum Press, New York. pp. 273–286.
95. Kinne, R., and E. Kinne-Saffran. 1978. Differentiation of cell faces in epithelia. In: *Molecular Specialization and Symmetry in Membrane Function*. A. K. Solomon and M. Karnovsky, eds. Harvard University Press, Cambridge, Mass. pp. 272–293.
96. Sacktor, B. 1982. Na gradient-dependent transport systems in renal proximal tubule brush border membrane vesicles. In: *Membranes and Transport*, Volume 2. A. N. Martonosi, ed. Plenum Press, New York. pp. 197–206.
97. Rose, R. C., and S. G. Schultz. 1971. Studies on the electrical potential profile across rabbit ileum: Effects of sugars and amino acids on transmural and transmucosal electrical potential differences. *J. Gen. Physiol.* **57**:639–663.

98. White, J. F., and W. M. Armstrong. 1971. Effect of transported solutes on membrane potentials in bullfrog small intestine. *Am. J. Physiol.* **221**:194–201.
99. Maruyama, T., and T. Hoshi. 1972. The effect of D-glucose on the electrical potential profile across the proximal tubule of newt kidney. *Biochim. Biophys. Acta* **282**:214–225.
100. Okada, Y., W. Tsuchiya, A. Irimajiri, and A. Inouye. 1977. Electrical properties and active solute transport in rat small intestine. I. Potential profile changes associated with sugar and amino acid transports. *J. Membr. Biol.* **31**:205–219.
101. Gunther-Smith, P., E. Grasset, and S. G. Schultz. 1982. Sodium-coupled amino acid sugar transport by *Necturus* small intestine: An equivalent electrical circuit analysis of a rheogenic co-transport system. *J. Membr. Biol.* **66**:25–39.
102. Frömter, E. 1982. Electrophysiological analysis of rat renal sugar and amino acid transport. I. Basic principles. *Pfluegers Arch.* **393**:179–189.
103. Samarzija, I., and E. Frömter. 1982. Electrophysiological analysis of rat renal sugar and amino acid transport. III. Neutral amino acids. *Pfluegers Arch.* **393**:199–209.
104. Samarzija, I., and E. Frömter. 1982. Electrophysiologic analysis of rat renal sugar and amino acid transport. IV. Basic amino acids. *Pfluegers Arch.* **393**:210–214.
105. Samarzija, I., and E. Frömter. 1982. Electrophysiological analysis of rat renal sugar andamino acid transport. V. Acidic amino acids. *Pfluegers Arch.* **393**:215–221.
106. Murer, H., and U. Hopfer. 1974. Demonstration of an electrogenic Na-dependent D-glucose transport in intestinal brush border membranes. *Proc. Natl. Acad. Sci. USA* **71**:484–488.
107. Beck, J. C., and B. Sacktor. 1975. Energetics of the Na^+-dependent transport of D-glucose in renal brush border membrane vesicles. *J. Biol. Chem.* **250**:8674–8680.
108. Fairclough, P., P. Malathi, H. Preiser, and R. K. Crane. 1979. Reconstitution into liposomes of glucose active transport from the rabbit renal proximal tubule. Characteristics of the system. *Biochim. Biophys. Acta* **553**:295–306.
109. Koepsell, H., H. Menuhr, I. Ducis, and T. F. Wissmuller. 1983. Partial purification and reconstitution of the Na–D-glucose cotransport protein from pig renal proximal tubule. *J. Biol. Chem.* **258**:1888–1894.
110. Ducis, I., and H. Koepsell. 1983. A simple liposomal system to reconstitute and assay highly efficient Na/D-glucose cotransport from kidney brush-border membranes. *Biochim. Biophys. Acta* **730**:119–129.
111. Schmidt, O. M., B. Eddy, C. M. Fraser, J. C. Venter, and G. Semenza. 1983. Isolation of (a subunit of) the Na^+/D-glucose cotransporter(s) of rabbit intestinal brush border membranes using monoclonal antibodies. *FEBS Lett.* **61**:279–293.
112. Diamond, J. M. 1962. The mechanism of solute transport by the gallbladder. *J. Physiol. (London)* **161**:474–502.
113. Diamond, J. M. 1964. Transport of salt and water in rabbit and guinea pig gallbladder. *J. Gen. Physiol.* **48**:1–14.
114. Nellans, H. N., R. A. Frizzell, and S. G. Schultz. 1973. Coupled sodium–chloride influx across the brush border of rabbit ileum. *Am. J. Physiol.* **225**:467–475.
115. Frizzell, R. A., M. C. Dugas, and S. G. Schultz. 1975. Sodium chloride transport by rabbit gallbladder: Direct evidence for a coupled NaCl influx process. *J. Gen. Physiol.* **65**:769–795.
116. Duffey, M. E., K. Turnheim, R. A. Frizzell, and S. G. Schultz. 1978. Intracellular chloride activities in rabbit gallbladder: Direct evidence for the role of the sodium-gradient in energizing "uphill" chloride transport. *J. Membr. Biol.* **42**:229–245.
117. Duffey, M. E., S. M. Thompson, R. A. Frizzell, and S. G. Schultz. 1979. Intracellular chloride activities and active chloride absorption in the intestinal epithelium of the winter flounder. *J. Membr. Biol.* **50**:331–341.
118. Armstrong, W. McD., W. R. Bixenman, K. F. Frey, J. F. Garcia-Diaz, M. G. O'Regan, and J. L. Owens. 1979. Energetics of coupled Na and Cl entry into epithelial cells of bullfrog small intestine. *Biochim. Biophys. Acta* **551**:207–219.
119. Reuss, L., and S. A. Weinman. 1979. Intracellular ionic activities and transmembrane electrochemical potential differences in gallbladder epithelium. *J. Membr. Biol.* **49**:345–362.
120. Reuss, L., and T. P. Grady. 1979. Effects of external sodium and cell membrane potential on intracellular Cl activity in gallbladder epithelium. *J. Membr. Biol.* **51**:15–31.
121. Garcia-Diaz, J. F., and W. M. Armstrong. 1980. The steady-state relationship between sodium and chloride transmembrane electrochemical potential differences in *Necturus* gallbladder. *J. Membr. Biol.* **55**:213–222.
122. Spring, K. R., and G. Kimura. 1978. Chloride reabsorption by renal proximal tubules of *Necturus*. *J. Membr. Biol.* **38**:233–254.
123. Kimura, G., and K. R. Spring. 1979. Luminal Na entry into *Necturus* proximal tubule cells. *Am. J. Physiol.* **236**:F295–F301.
124. Oberleithner, H., W. Guggino, and G. Giebisch. 1982. Mechanism of distal tubular chloride transport in *Amphiuma* kidney. *Am. J. Physiol.* **242**:F331–F339.
125. Murer, H., U. Hopfer, and R. Kinne. 1976. Sodium/proton antiport in brush border membrane vesicles isolated from rat small intestine and kidney. *Biochem. J.* **154**:597–604.
126. Liedtke, C. M., and U. Hopfer. 1982. Mechanism of Cl translocation across small intestinal brush-border membrane. I. Absence of Na–Cl cotransport. *Am. J. Physiol.* **242**:G263–G271.
127. Liedtke, C. M., and U. Hopfer. 1982. Mechanism of Cl translocation across small intestinal brush border membrane. II. Demonstration of Cl–OH exchange and Cl conductance. *Am. J. Physiol.* **242**:G272–G280.
128. Kinsella, J. L., and P. S. Aronson. 1980. Properties of the Na–H exchanger in renal microvillus membrane vesicles. *Am. J. Physiol.* **238**:F461–F469.
129. Aronson, P. S. 1981. Identifying secondary active solute transport in epithelia. *Am. J. Physiol.* **240**:F1–F11.
130. Knickerbein, R., P. S. Aronson, W. Atherton, and J. W. Dobbins. 1983. Sodium and chloride transport across rabbit ileal brush border. I. Evidence for Na–H exchange. *Am. J. Physiol.* **245**:G504–G510.
131. Dubinsky, W. B., and R. A. Frizzell. 1983. A novel effect of amiloride on H^+-dependent Na^+ transport. *Am. J. Physiol.* **245**:C157–C159.
132. Cabantchik, Z. I., and A. Rothstein. 1972. The nature of the membrane sites controlling anion permeability of human red blood cells as determined by studies with disulfonic stilbene derivatives. *J. Membr. Biol.* **10**:311–330.
133. Weinman, S. A., and L. Reuss. 1982. Na^+–H^+ exchange at the apical membrane of *Necturus* gallbladder. *J. Gen. Physiol.* **80**:299–321.
134. Heintze, K., K.-U. Peterson, P. Olles, S. H. Saverymuttu, and J. R. Wood. 1979. Effects of bicarbonate on fluid and electrolyte transport by the guinea pig gallbladder: A bicarbonate–chloride exchange. *J. Membr. Biol.* **45**:43–59.
135. Friedman, P. A., and T. E. Andreoli. 1982. CO_2-stimulated NaCl absorption in the mouse renal cortical thick ascending limb of Henle. *J. Gen. Physiol.* **80**:683–711.
136. Ericson, A.-C., and K. R. Spring. 1982. Coupled NaCl entry into *Necturus* gallbladder epithelial cells. *Am. J. Physiol.* **243**:C140–C145.
137. Ericson, A.-C., and K. R. Spring. 1982. Volume regulation by *Necturus* gallbladder: Apical Na–H and Cl–HCO_3 exchange. *Am. J. Physiol.* **243**:C146–C150.
138. Cremaschi, D., G. Meyer, S. Bermano, and M. Marcati. 1983. Different sodium chloride cotransport systems in the apical membrane of rabbit gallbladder epithelial cells. *J. Membr. Biol.* **73**:227–235.
139. Frizzell, R. A., P. L. Smith, E. Vosburgh, and M. Field. 1979. Coupled sodium–chloride influx across brush border of flounder intestine. *J. Membr. Biol.* **46**:27–40.
140. Musch, M. W., S. A. Orellana, L. S. Kimberg, M. Field, D. R. Halm, E. J. Krasny, Jr., and R. A. Frizzell. 1982. Na–K–Cl cotransport in the intestine of a marine teleost. *Nature (London)* **300**:351–353.

141. Greger, R., E. Schlatter, and F. Lang. 1983. Evidence for electroneutral sodium chloride cotransport in the cortical thick ascending limb of Henle's loop of rabbit kidney. *Pfluegers Arch.* **396**:308–314.
142. Oberleithner, H., G. Giebisch, F. Lang, and W. Wang. 1982. Cellular mechanism of the furosemide sensitive transport system in the kidney. *Klin. Wochenschr.* **60**:1173–1179.
143. Brazy, P. C., and R. B. Gunn. 1976. Furosemide inhibition of chloride transport in human red blood cells. *J. Gen. Physiol.* **68**:583–599.
144. Duffey, M. E., and C. Bebernitz. 1983. Intracellular chloride and hydrogen activities in rabbit colon. *Fed. Proc.* **42**:1353.
145. Diez de los Rios, A., N. E. DeRose, and W. McD. Armstrong. 1981. Cyclic AMP and intracellular ionic activities in *Necturus* gallbladder. *J. Membr. Biol.* **63**:25–30.
146. Rao, M. C., N. T. Nash, and M. Field. 1984. Differing effects of cGMP and cAMP on ion transport across flounder intestine. *Am. J. Physiol.* **246**:C167–C171.
147. Palfrey, H. C., and P. Greengard. 1981. Hormone-sensitive ion transport systems in erythrocytes as models for epithelial ion pathways. *Ann. N.Y. Acad. Sci.* **373**:291–308.
148. Garay, R. P. 1982. Inhibition of the Na/K cotransport system by cyclic AMP and intracellular Ca in human red cells. *Biochim. Biophys. Acta* **688**:786–792.
149. Peterson, K.-U., and L. Reuss. 1983. Cyclic AMP-induced chloride permeability in the apical membrane of *Necturus* gallbladder epithelium. *J. Gen. Physiol.* **81**:705–729.
150. Field, M. 1979. Intracellular mediators of secretion in the small intestine. In: *Mechanisms of Intestinal Secretion*. H. J. Binder, ed. Liss, New York. pp. 83–91.
151. Reuss, L., L. Y. Cheung, and T. P. Grady. 1981. Mechanisms of cation permeation across apical cell membrane of *Necturus* gallbladder: Effects of luminal pH and divalent cations on K and Na permeability. *J. Membr. Biol.* **59**:211–224.
152. Gogelein, H., and W. van Driessche. 1981. Noise analysis of the K current through the apical membrane of *Necturus* gallbladder. *J. Membr. Biol.* **60**:187–198.
153. Stewart, C. P., P. L. Smith, M. J. Welsh, R. A. Frizzell, M. W. Musch, and M. Field. 1980. Potassium transport by the intestine of the winter flounder, *Pseudopleuronectes americanus*. Mount Desert Island Biological Laboratory Bulletin **20**:92–95.
154. Musch, M. W., M. Field and R. A. Frizzell. 1981. Active K transport by the intestine of the flounder, *Pseudopleuronectes americanus:* Evidence for cotransport with Na and Cl. *Bull. Mt. Desert Is. Biol. Lab.* **21**:95–99.
155. Nagel, W., and W. Hirschmann. 1980. K-permeability of the outer border of the frog skin (*R. temporaria*). *J. Membr. Biol.* **52**:107–113.
156. van Driessche, W., and W. Zeiske. 1980. Ba-induced conductance fluctuations of spontaneously fluctuating K channels in the apical membrane of frog skin (*Rana temporaria*). *J. Membr. Biol.* **56**:31–42.
157. Wills, N. K., W. Zeiske, and W. van Driessche. 1982. Noise analysis reveals K channel conductance fluctuations in the apical membrane of rabbit colon. *J. Membr. Biol.* **69**:187–197.
158. O'Neil, R. 1983. Voltage-dependent interaction of barium and cesium with the potassium conductance of the cortical collecting duct apical cell membrane. *J. Membr. Biol.* **74**:165–173.
159. Greger, R., and E. Schlatter. 1983. Properties of the lumen membrane of the cortical thick ascending limb of Henle's loop of rabbit kidney. *Pfluegers Arch.* **396**:315–324.
160. O'Neil, R. C., and S. C. Sansom. 1984. Characterization of apical cell membrane Na and K channels of cortical collecting duct using microelectrode techniques. *Am. J. Physiol.* **247**: F14–F24.
161. Armstrong, C. M., and S. R. Taylor. 1980. Interaction of barium ions with potassium channels in squid giant axons. *Biophys. J.* **30**:473–488.
162. Wills, N. K., and B. Biagi. 1982. Active potassium transport by rabbit descending colon epithelium. *J. Membr. Biol.* **64**:195–203.
163. McCabe, R., H. J. Cook, and L. P. Sullivan. 1982. Potassium transport by rabbit descending colon. *Am. J. Physiol. 242*:C81–C86.
164. Dibona, D. R., and J. W. Mills, 1979. Distribution of Na^+-pump sites in transporting epithelia. *Fed. Proc.* **38**:134–143.
165. Thomas, R. C. 1972. Electrogenic sodium pump in nerve and muscle cells. *Physiol. Rev.* **52**:563–594.
166. Hoffman, J. F., H. Kaplan, and T. J. Callahan. 1979. The Na:K pump in red cells is electrogenic. *Fed. Proc.* **38**:2440–2441.
167. Thomas, R. C. 1982. Electrophysiology of the sodium pump in a snail neuron. *Curr. Top. Membr. Transp.* **16**:3–16.
168. Nelson, M. T., and W. J. Lederer. 1983. Stoichiometry of the electrogenic Na pump in barnacle muscle: Simultaneous measurement of Na efflux and membrane current. *Curr. Top. Membr. Transp.* **19**:707–711.
169. Nielsen, R. 1979. A 3 to 2 coupling of the Na-K pump responsible for the transepithelial Na transport in frog skin as disclosed by the effect of Ba. *Acta Physiol. Scand.* **107**:189–191.
170. Nielsen, R. 1979. Coupled transepithelial sodium and potassium transport across isolated frog skin: Effect of ouabain, amiloride and the polyene antibiotic filipin. *J. Membr. Biol.* **51**:161–184.
171. Kirk, K. L., D. R. Halm, and D. C. Dawson. 1980. Active sodium transport by turtle colon via an electrogenic Na–K exchange pump. *Nature (London)* **287**:237–239.
172. Turnheim, K., S. M. Thompson, and S. G. Schultz. 1983. Relation between intracellular sodium and active sodium transport in rabbit colon. *J. Membr. Biol.* **76**:299–309.
173. Nielsen, R. 1982. Effect of ouabain, amiloride, and antidiuretic hormone on the sodium-transport pool in isolated epithelia from frog skin (*Rana temporaria*). *J. Membr. Biol.* **65**:221–226.
174. Lewis, S. A., and N. K. Wills. 1981. Interaction between apical and baso-lateral membranes during sodium transport across tight epithelia. In: *Ion Transport by Epithelia*. S. G. Schultz, ed. Raven Press, New York. pp. 93–107.
175. Eaton, D. C. 1981. Intracellular sodium ion activity and sodium transport in rabbit urinary bladder. *J. Physiol. (London)* **316**:527–544.
176. Eaton, D. C., A. M. Frace, and S. U. Silverthorn. 1982. Active and passive Na fluxes across the basolateral membrane of rabbit urinary bladder. *J. Membr. Biol.* **67**:219–229.
176a. Halm, D. R. and D. C. Dawson. 1983. Cation activation of the basolateral sodium-potassium pump in turtle colon. *J. Gen. Physiol.* **82**:315–329.
177. Hoffman, J. F., B. G. Kennedy, and G. Lunn. 1981. Modulation of red cell Na/K pump rates. In: *Erythrocyte Membranes 2: Recent Clinical and Experimental Advances*. Liss, New York. pp. 5–9.
178. Jorgensen, P. L. 1980. Sodium and potassium ion pump in kidney tubules. *Physiol. Rev.* **60**:864–917.
179. Charney, A. N., M. D. Kinsey, L. Myers, R. A. Giannella, and R. E. Gots. 1975. Na-K-activated adenosine triphosphatase and intestinal electrolyte transport. *J. Clin. Invest.* **56**:653–660.
180. Silva, P., A. N. Charney, and F. H. Epstein. 1975. Potassium adaptation and Na-K-ATPase activity in mucosa of colon. *Am. J. Physiol.* **229**:1576–1579.
181. Will, P. C., R. C. DeLisle, R. N. Cortright, and U. Hopfer. 1981. Induction of amiloride sensitive sodium transport in the intestines by adrenal steroids. *Ann. N.Y. Acad. Sci.* **372**:64–78.
182. Katz, A. I., and F. H. Epstein. 1967. The role of sodium-potassium-activated adenosine triphosphatase in the reabsorption of sodium by the kidney. *J. Clin. Invest.* **46**:1999–2011.
183. Jorgensen, P. L. 1972. The role of aldosterone in the regulation of (Na-K)-ATPase in rat kidney. *J. Steroid Biochem.* **3**:181–191.
184. Charney, A. N., P. Silva, A. Beserab, and F. H. Epstein. 1974. Separate effects of aldosterone, DOCA and methylprednisolone on renal Na-K-ATPase. *Am. J. Physiol.* **227**:345–350.
185. Petty, K. J., J. P. Kokko, and D. Marver. 1981. Secondary effect of aldosterone on Na-K-ATPase activity in the rabbit cortical collecting tubule. *J. Clin. Invest.* **68**:1514–1521.
186. Petty, K. J. 1982. The role of sodium-potassium-activated ade-

nosine triphosphatase in the mechanism of mineralocorticoid action in the mammalian nephron. Ph.D. thesis. The University of Texas Health Science Center, Dallas, Texas.

187. Handler, J. S., A. S. Preston, F. M. Perkins, M. Matsumura, J. P. Johnson, and C. O. Watlington. 1981. The effect of adrenal steroid hormones on epithelia formed in culture by A6 cells. *Ann. N.Y. Acad. Sci.* **372**:442–454.
188. O'Neil, R. G., and W. P. Dubinsky. 1983. Na-dependent mineralocorticoid regulation of cortical collecting duct (CCD) Na-K-ATPase. *Fed. Proc.* **42**:475.
189. Katz, A. 1982. Renal Na-K-ATPase: Its role in tubular sodium and potassium transport. *Am. J. Physiol.* **242**:F207–F219.
190. Nagel, W. 1979. Inhibition of potassium conductance by barium in frog skin epithelium. *Biochim. Biophys. Acta* **552**:346–357.
191. Bello-Reuss, E. 1982. Electrical properties of the basolateral membrane of the straight portion of the rabbit proximal renal tubule. *J. Physiol. (London)* **326**:49–63.
192. Welsh, M. J. 1983. Barium inhibition of basolateral membrane potassium conductance in tracheal epithelium. *Am. J. Physiol.* **244**:F639–F645.
193. Lau, K., R. L. Hudson, and S. G. Schultz. 1984. Cell swelling increases a barium-inhibitable energy dependent potassium conductance in the basolateral membrane of *Necturus* small intestine. *Proc. Natl. Acad. Sci. USA* **81**:3591–3954.
194. Davis, C. W., and A. L. Finn. 1982. Sodium transport effects on the basolateral membrane in toad urinary bladder. *J. Gen. Physiol.* **80**:733–751.
195. Davis, C. W., and A. L. Finn. 1982. Sodium transport inhibition by amiloride reduces basolateral membrane potassium conductance in tight epithelia. *Science* **216**:525–527.
196. Grasset, E., P. Gunter-Smith, and S. G. Schultz. 1983. Effects of Na-coupled alanine transport on intracellular K activities and the K conductance of the basolateral membranes of *Necturus* small intestine. *J. Membr. Biol.* **71**:89–94.
197. Welsh, M. J., P. L. Smith, and R. A. Frizzell. 1982. Chloride secretion by canine tracheal epithelium. II. The cellular electrical potential profile. *J. Membr. Biol.* **70**:227–238.
198. Welsh, M. J., P. L. Smith, and R. A. Frizzell. 1983. Chloride secretion by canine tracheal mucosa. III. Membrane resistances and electromotive forces. *J. Membr. Biol.* **71**:209–218.
199. Shorofsky, S. R., M. Field, and H. A. Fozzard. 1983. Electrophysiology of Cl secretion in canine trachea. *J. Membr. Biol.* **72**:105–115.
200. Smith, P. L., and R. A. Frizzell. 1984. Chloride secretion by canine tracheal epithelium. IV. Basolateral membrane K permeability parallels secretion rate. **77**:187–199.
201. Foskett, J. K., and K. R. Spring. 1983. Control of epithelial cell volume regulation. *J. Gen. Physiol.* **82**:21a.
202. Stevens, C. F. 1980. Ionic channels in neuromembranes: Methods for studying their properties. In: *Molluscan Nerve Cells: From Biophysics to Behavior.* J. Koester and J. H. Byrne, eds. Cold Spring Harbor Laboratory, Cold Spring Harbor, N.Y. pp. 11–31.
203. Miller, C. 1983. Integral membrane channels: Studies in model membranes. *Physiol. Rev.* **63**:1209–1242.
204. Kandel, E. 1980. The multichannel model of the nerve cell membrane: A perspective. In: *Molluscan Nerve Cells: From Biophysics to Behavior.* J. Koester and J. H. Byrne, eds. Cold Spring Harbor Laboratory, Cold Spring Harbor, N.Y. pp. 1–10.
205. Grinstein, S., O. Candia, and D. Erlij. 1978. Nonhormonal mechanisms for the regulation of transepithelial sodium transport: The role of surface potential cell calcium. *J. Membr. Biol.* **40**:261–280.
206. Lee, C. O., A. Taylor, and E. E. Windhager. 1980. Cytosolic calcium ion activity in epithelial cells of *Necturus* kidney. *Nature (London)* **287**:859–861.
207. Lorenzen, M., C. O. Lee, and E. E. Windhager. 1981. Effect of quinidine and ouabain on intracellular calcium and sodium activities in isolated perfused proximal tubules of *Necturus* kidney. *Kidney Int.* **21**:281a.
208. Blaustein, M. P. 1974. The interrelationship between sodium and calcium fluxes across cell membranes. *Rev. Physiol. Biochem. Pharmacol.* **70**:33–82.
209. Cremaschi, D., and S. Henin. 1975. Na and Cl transepithelial routes in rabbit gallbladder: Tracer analysis of the transports. *Pfluegers Arch.* **361**:33–41.
210. Henin, S., and D. Cremaschi. 1975. Transcellular ion route in rabbit gallbladder: Electric properties of the epithelial cells. *Pfluegers Arch.* **355**:125–139.
211. Van Os, C. H., and J. F. G. Slegers. 1975. The electrical potential profile of gallbladder epithelium. *J. Membr. Biol.* **24**:341–363.
212. Gunter-Smith, P., and S. G. Schultz. 1982. Intracellular potassium activities and potassium transport by rabbit gallbladder. *J. Membr. Biol.* **65**:41–47.
213. Reuss, L. 1979. Electrical properties of the cellular transepithelial pathway in *Necturus* gallbladder. III. Ionic permeability of the basolateral cell membrane. *J. Membr. Biol.* **47**:239–259.
214. Reuss, L., and S. A. Weinman. 1979. Intracellular ionic activities and transmembrane electrochemical potential differences in gallbladder epithelium. *J. Membr. Biol.* **49**:345–362.
215. Reuss, L., S. A. Weinman, and T. P. Grady. 1980. Intracellular K activity and its relation to basolateral membrane ion transport in *Necturus* gallbladder. *J. Gen. Physiol.* **76**:33–52.
216. Shindo, T., and K. R. Spring. 1981. Chloride movement across the basolateral membrane of proximal tubule cells. *J. Membr. Biol.* **58**:35–42.
217. Corcia, A., and W. McD. Armstrong. 1983. KCl cotransport: A mechanism for basolateral chloride exit in *Necturus* gallbladder. *J. Membr. Biol.* **76**:173–182.
218. Suzuki, K., G. Kottra, L. Kampmann, and E. Frömter. 1982. Square wave pulse analysis of cellular and paracellular conductance pathways in *Necturus* gallbladder epithelium. *Pfluegers Arch.* **394**:302–312.
219. Koefoed-Johnsen, V., H. H. Ussing, and K. Zerahn. 1952. The origin of the short-circuit current in the adrenaline stimulated frog skin. *Acta Physiol. Scand.* **27**:38–48.
220. Hogben, C. A. M. 1955. Active transport of chloride by isolated frog gastric mucosa. *Am. J. Physiol.* **180**:641–649.
221. Field, M., G. R. Plotkin, and W. Silen. 1968. Effects of vasopressin, theophylline and cyclic adenosine monophosphate on short-circuit current across isolated rabbit ileal mucosa. *Nature (London)* **217**:469–471.
222. Kimberg, D. V., M. Field, J. Johnson, A. Henderson, and E. Gershon. 1971. Stimulation of intestinal mucosal adenyl cyclase by cholera enterotoxin and prostaglandins. *J. Clin. Invest.* **50**:1218–1230.
223. Field, M., D. Fromm, Q. Al-Awqati, and W. B. Greenough, III. 1972. Effect of cholera enterotoxin on ion transport across isolated ileal mucosa. *J. Clin. Invest.* **51**:796–804.
224. Field, M. 1971. Intestinal secretion: Effect of cyclic AMP and its role in cholera. *N. Engl. J. Med.* **284**:1137–1144.
225. Field, M. 1974. Intestinal secretion. *Gastroenterology* **66**:1063–1084.
226. Field, M. 1979. Intracellular mediators of secretion in the small intestine. In: *Mechanisms of Intestinal Secretion.* H. J. Binder, ed. Liss, New York. pp. 83–91.
227. Field, M. 1980. Regulation of small intestinal ion transport by cyclic nucleotides and calcium. In: *Secretory Diarrhea.* M. Field, J. S. Fordtran, and S. G. Schultz, eds. American Physiological Society, Washington, D.C. pp. 21–30.
228. Frizzell, R. A., M. J. Koch, and S. G. Schultz. 1976. Ion transport by rabbit colon. I. Active and passive components. *J. Membr. Biol.* **27**:297–316.
229. Frizzell, R. A. 1977. Active chloride secretion by rabbit colon: Calcium-dependent stimulation by ionophore A23187. *J. Membr. Biol.* **35**:175–187.
230. Frizzell, R. A., K. Heintze, and C. P. Stewart. 1980. Mechanism of intestinal chloride secretion. In: *Secretory Diarrhea.* M. Field,

J. S. Fordtran, and S. G. Schultz, eds. American Physiological Society, Washington, D.C. pp. 11–19.

231. Zadunaisky, J. A. 1966. Active transport of chloride in frog cornea. *Am. J. Physiol.* **211**:506–512.
232. Zadunaisky, J. A. 1972. Sodium activation of chloride transport in the frog cornea. *Biochim. Biophys. Acta* **282**:255–257.
233. Zadunaisky, J. A., M. A. Lande, M. Chalfie, and A. H. Neufeld. 1973. Ion pumps in the cornea and their stimulation by epinephrine and cyclic AMP. *Exp. Eye Res.* **15**:577–584.
234. Degnan, K. J., K. J. Karnaky, and J. A. Zadunaisky. 1977. Active chloride transport in the *in vitro* opercular skin of a teleost (*Fundulus heteroclitus*), a gill-like epithelium rich in chloride cells. *J. Physiol. (London)* **271**:155–191.
235. Silva, P., J. Stoff, M. Field, L. Fine, J. N. Forrest, and F. H. Epstein. 1977. Mechanism of active chloride secretion by shark rectal gland: Role of Na-K-ATPase in chloride transport. *Am. J. Physiol.* **233**:F298–F306.
236. Sachs, G., J. G. Spenney, and M. Lewin. 1978. H^+ transport: Regulation and mechanism in gastric mucosa and membrane vesicles. *Physiol. Rev.* **58**:106–173.
237. Al-Bazzaz, F., and Q. Al-Awqati. 1979. Interaction between sodium and chloride transport in canine tracheal mucosa. *Am. J. Physiol.* **46**:111–119.
238. Smith, P. L., M. J. Welsh, J. S. Stoff, and R. A. Frizzell. 1982. Chloride secretion by canine tracheal epithelium. I. Role of intracellular cAMP levels. *J. Membr. Biol.* **70**:217–226.
239. Welsh, M. J. 1983. Inhibition of chloride secretion by furosemide in canine tracheal epithelium. *J. Membr. Biol.* **71**:219–226.
240. Frizzell, R. A., M. Field, and S. G. Schultz. 1979. Sodium-coupled chloride transport by epithelial tissues. *Am. J. Physiol.* **236**:F1–F8.
241. Welsh, M. J., P. L. Smith, M. Fromm, and R. A. Frizzell. 1982. Crypts are the site of intestinal fluid and electrolyte secretion. *Science* **218**:1219–1221.
242. Welsh, M. J. 1983. Intracellular chloride activities in canine tracheal epithelium: Direct evidence for sodium-coupled intracellular chloride accumulation in a chloride-secreting epithelium. *J. Clin. Invest.* **71**:1391–1401.
243. Shorofsky, S. R., M. Field, and H. A. Fozzard. 1984. Mechanism of Cl secretion in canine trachea: Changes in intracellular chloride activity with secretion. *J. Membr. Biol.* **81**:1–8.
244. Klyce, S. D., and R. K. S. Wong. 1977. Site and mode of adrenaline action on chloride transport across the rabbit corneal epithelium. *J. Physiol. (London)* **266**:777–799.
245. Zadunaisky, J. A., K. R. Spring, and T. Shindo. 1979. Intracellular chloride activity in the corneal epithelium. *Fed. Proc.* **38**:1059.
246. Eveloff, J. R., R. Kinne, E. Kinne-Saffran, H. Murer, P. Silva, H. Epstein, J. Stoff, and W. B. Kinter. 1978. Coupled sodium and chloride transport into plasma membrane vesicles prepared from dogfish rectal gland. *Pfluegers Arch.* **378**:87–92.
247. Welsh, M. J., P. L. Smith, and R. A. Frizzell. 1981. Intracellular chloride activities in the isolated perfused shark rectal gland. *Clin. Res.* **29**:480a.
248. Hannafin, J., E. Kinne-Saffran, D. Friedman, and R. Kinne. 1983. Presence of a sodium–potassium chloride cotransport system in the rectal gland of *Squalus acanthias*. *J. Membr. Biol.* **75**:73–83.
249. Roos, A., and W. F. Boron. 1981. Intracellular pH. *Physiol. Rev.* **61**:296–434.
250. Boron, W. F., and E. L. Boulpaep. 1983. Intracellular pH regulation in the renal proximal tubule of the salamander: Na–H exchange. *J. Gen. Physiol.* **81**:29–52.

CHAPTER 9

Ion Transport by Gastric Mucosa

John G. Forte and Terry E. Machen

1. Introduction

The stomach serves important alimentary functions, as an organ for food storage and in providing the secretory juices that serve to initiate the digestive process. These functions are carefully regulated by neural and hormonal mechanisms to liquify, sterilize, macerate, and partially degrade the components of a meal. The heavy muscular wall of the stomach provides the peristaltic mixing waves, while the epithelial cells of the gastric mucosa supply the secretory products of the juice. Gastric secretory products can conveniently be subdivided into a mucus component, principally serving a lubricating function, an enzyme component, in the form of the enzyme precursor pepsinogen, and an aqueous component, which largely consists of hydrochloric acid. The purpose of this chapter is to review the mechanisms responsible for the secretion of the ions and water that constitute the so-called aqueous component of gastric juice. We will focus our discussion on those ion transport processes that are directly or indirectly related to HCl secretion.

Our present appreciation of gastric ion transport comes from studies utilizing a wide variety of methodological approaches and tissue preparations. Since our objective was to identify, localize, and specify the mechanics of the particular transport system, we have restricted our discussion to various *in vitro* gastric preparations. These include (1) the isolated gastric mucosa, where the electrophysiological and ion flux properties have been characterized in terms of an epithelial composite, (2) isolated gastric gland and cell preparations, where functional activities can be ascribed to identifiable cell types, and (3) isolated membrane fractions, in which the characteristics of active and passive ion flux can be evaluated in simple membrane vesicles.

Ion transport by the gastric epithelium is exquisitely controlled by neural and humoral regulators, so that widely different states of physiological secretory activity can occur. Major morphological changes accompany the functional changes, especially in the ultrastructure of the cell responsible for HCl secretion, the oxyntic cell. Section 2 describes the structural organization of the gastric epithelium, with particular emphasis on oxyntic cell morphology, and Section 3 reviews the action of secretagogues in stimulating secretory function. These sections serve as background for the discussion of specific ion transport processes that follow. We have taken the approach of presenting the biochemical studies of gastric secretion with an emphasis on the characterization of the enzymatic and ion transport properties of isolated membrane fractions (Sections 4 and 5). While these studies have provided unequivocal evidence for specific transport processes in simple membrane systems, and offer provocative insights into molecular mechanisms at that level, they must be examined in the context of the intact epithelium in order to appreciate tissue transport function. Thus, Section 6 is devoted to studies of ion transport in more composite systems. We have reviewed the electrophysiology of gastric epithelium, relating these studies to the structural complexities of the tissue, and, where possible, to the localization and nature of the ion transport events. Later in the section, we have surveyed the specific characteristics for the transepithelial transport of H^+, Cl^-, Na^+, K^+, and HCO_3^-. Finally in our summary, Section 7, we present a comprehensive representation of the ion transport process in the gastric epithelium, drawing on the studies of morphology, isolated membranes, and intact tissue, to specify transport sites and possible mechanisms.

2. Organization of Gastric Epithelial Cells

2.1. Histology

The gastric mucosa is a complex epithelium made up of several cell types and organized in a highly folded pattern of glandular invaginations. The relative abundance of particular cell types and their specific organization varies somewhat in different regions of the stomach. The most luminal aspect of the entire stomach is covered by a layer of columnar surface epithelial cells that are responsible for secreting much of the lubricating mucus formed by the tissue. Low-power magnification of the mucosal surface reveals the presence of numerous pits; these gastric pits lead to the gastric glands which extend 100–500 μm

John G. Forte and Terry E. Machen • Department of Physiology–Anatomy, University of California, Berkeley, California 94720.

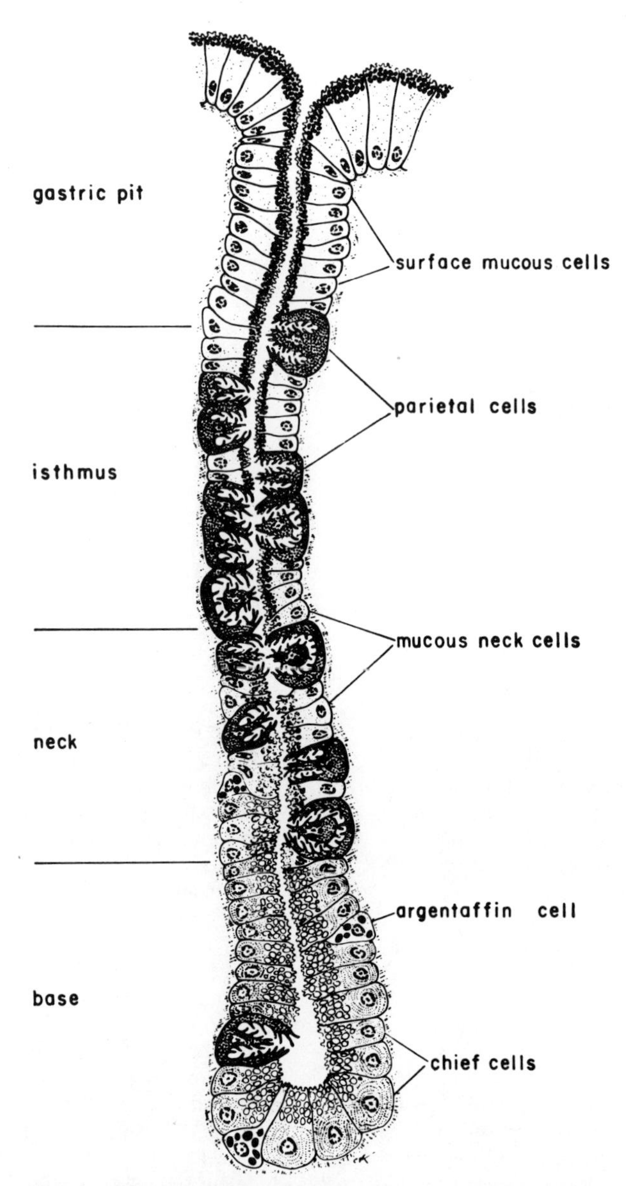

Fig. 1. Schematic diagram of tubular oxyntic gland from corpus or fundus of the stomach. Structural arrangement is similar in all mammalian species with some variations in the relative proportion of cell types. However, amphibian and reptilian gastric mucosas do not have separate parietal (oxyntic) and chief peptic cells. A single oxyntopeptic cell predominates in the gastric glands of these species. From Ito[76] with permission.

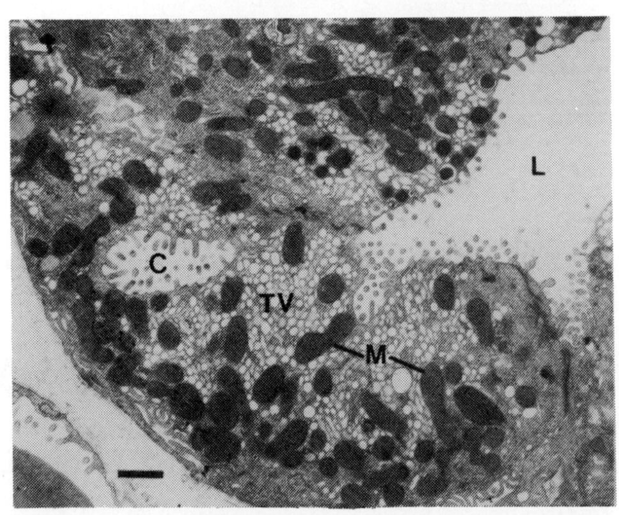

Fig. 2. Electron micrograph of oxyntic cells from nonsecreting piglet gastric mucosa. Tubulovesicles (TV) and large mitochondria (M) are prominent within the cells. Another specialized feature of mammalian oxyntic cells are the canaliculi (C) extending from the gland lumen (L) to deep within the cell. The apical membrane of the cell, including the canaliculi, possesses short microvilli. Bar = 1 μm.

below the surface and are responsible for other secretory products of the stomach. Although there are many variations among animal species (e.g., see Ref. 66), HCl is generally secreted by the gastric glands in the fundus and main body (corpus) of the stomach. We will refer to these as oxyntic glands, the word coming from the Greek *oxyntos,* meaning to make acid. About four to five oxyntic glands lead into a single gastric pit. A schematic representation of an oxyntic gland, showing the composite cell types, is shown in Fig. 1. In the mammal the oxyntic gland is composed of three major cell types: (1) mucous neck cells, which abound in the neck region between the gastric pit and the gland, and produce a mucoid secretion; (2) chief cells, which secrete the gastric zymogen, pepsinogen; and (3) parietal or oxyntic cells, which secrete the HCl. In most other vertebrates (e.g., fish, amphibians, reptiles, and birds) the oxyntic glands contain a single cell type, the oxyntopeptic cell, that is capable of secreting both pepsinogen and HCl. Other smaller cell types (e.g., argentaffin, endocrinelike, mast cells, etc.) can also be found in oxyntic glands. Their occurrence and abundance vary according to species and they are considered to play an important role in the regulation of gastric secretory processes.

2.2. Ultrastructure of the Oxyntic Cell

The oxyntic cell is in itself a highly complex and unique cell type which undergoes dramatic ultrastructural transformations related to the secretory state of the tissue. The nonsecreting, or resting, oxyntic cell has an interesting morphology (Fig. 2). A network of narrow canals, or cellular canaliculi, project into the cell from the apical surface. Short clubby microvilli are found on the entire apical surface including the canalicular lumina. The cytoplasm of the resting oxyntic cell abounds with tubular and vesicular membrane profiles, constituting the unique tubulovesicular system. Numerous, large mitochondria are also characteristic of oxyntic cells, consistent with the high oxidative capacity of this tissue.

Two principal cytological changes are immediately obvious for the stimulated, or maximally secreting, oxyntic cell: the apical and canalicular surfaces are greatly expanded with long elaborate microvilli; and there is a large reduction in tubulovesicles within the cytoplasm (Fig. 3). Quantitative morphometry has shown that the 6- to 10-fold increase in apical plasma membrane surface area is very nearly matched by the reduction in tubulovesicular membrane area.[8,68,77] Withdrawal of the secretory stimulus initiates a series of events beginning with collapse of canalicular surfaces and leading to the eventual uptake of plasma membrane and return to the resting morphology. On the basis of ultrastructural transformations dur-

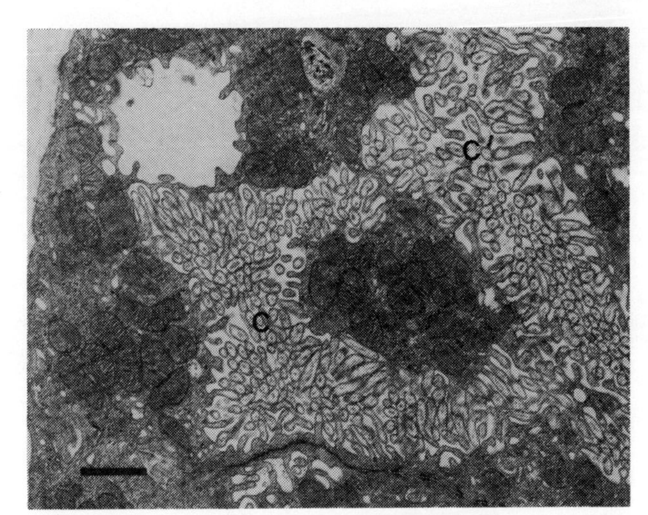

Fig. 3. Electron micrograph of oxyntic cells from piglet gastric mucosa that had been stimulated to maximal HCl secretion by histamine. Note the elaborate extension of microvilli at the apical canalicular (C) surface. The large increase in apical membrane surface area, compared to the nonsecreting tissue (see Fig. 2), can be accounted for by the decrease in surface area of the tubulovesicular membrane. Bar = 1 μm.

ing various phases of the secretory cycle, that is, in going from rest to maximal secretion and return to rest after stimulus removal, a membrane recycling hypothesis of HCl secretion has been proposed.[51] A schematic representation of the secretory cycle is shown in Fig. 4. The tenet of the hypothesis is conservation of membrane, and the associated transport systems, during the massive fluctuations in secretory activity. The membrane recycling hypothesis contends that fusion of tubulovesicles with the apical plasma membrane is the cytological event that leads to the great surface expansion while providing immediate extracellular access for the incumbent transport systems. While this idea was originally based on morphological data, it conforms very well with other observations using widely varied approaches of electrophysiology, transport activity, and isolated membrane studies. These will be discussed below.

An alternative hypothesis, not requiring membrane fusion, has also been proposed. According to this view, the tubulovesicular compartment is actually confluent with the apical surface, but remains in a supercollapsed state in the resting cell.[2] Hydrostatic pressure associated with secretion and volume flow would expand the collapsed tubules and give the appearance of an increased apical cell surface. There is no doubt that considerable volume flow accompanies HCl secretion, and observed measurements of the canalicular and glandular luminal spaces in the stimulated state[8,77] are in keeping with some expansion force associated with bulk flow. In fact, Gibert and Hersey[56] used an osmotic force to expand the luminal spaces in a resting preparation of isolated oxyntic glands. Although the oxyntic canaliculi were swollen, the observed volume change was not due to surface expansion but rather to a flattening of apical microvilli; the tubulovesicles remained unaffected and distinct of apical microvilli; the tubulovesicles remained unaffected and distinct from the apical surface. In addition, a large body of evidence supports the observation that the cytoplasmic tubulovesicular membrane system is isolated from the apical surface in the resting cell. The evidence comes from direct microscopic observation, membrane structure from freeze–fracture microscopy, cytochemical staining patterns of extracellular surfaces, and direct biochemical analysis of isolated membrane fractions (e.g., see review of Forte *et al.*[44]). Thus, some process of membrane fusion must lead to the surface expansion and general conservation of membrane that is observed when appropriate secretagogues are used to stimulate the oxyntic cell. Unfortunately, actual fusion events have been detected relatively rarely, and we have no clear idea as to the biochemical mechanisms underlying the fusion process. Recent studies suggest that microfilaments and microtubules may play a role in the membrane rearrangements associated with the secretory cycle of the oxyntic cell.[9,81] Work in the near future may provide insight into how these cytoskeletal elements may serve to organize and direct the membrane traffic during the secretory cycle.

3. Stimulus–Secretion Coupling in Oxyntic Cells

3.1. Background

We have already discussed the elaborate ultrastructural changes associated with stimulation of the oxyntic cell, which appear to involve the fusion of cytoplasmic tubulovesicles with the apical membrane. In addition to the fusion events, it is likely that the apical membranes of oxyntic cells undergo some activa-

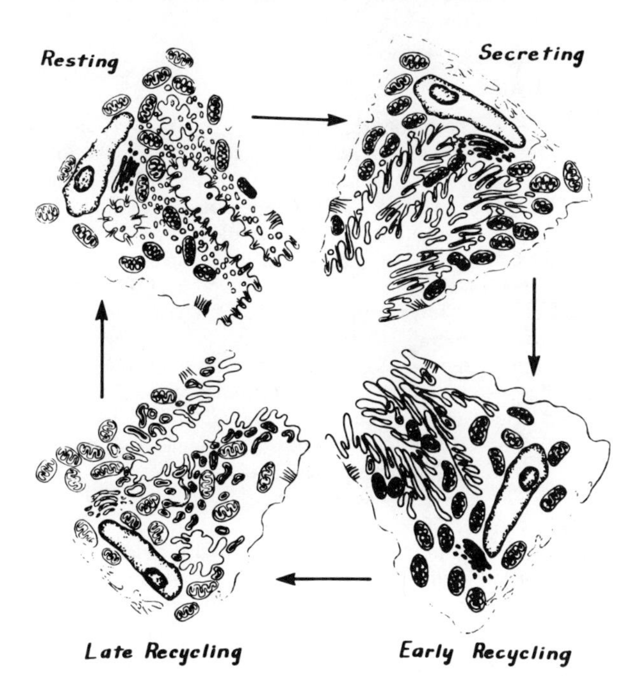

Fig. 4. A schematic summary of salient features of oxyntic cell membrane cycling during various stages of the secretory process. The resting cell is characterized by short microvilli on the secretory surface and numerous intracellular tubulovesicles. The actively secreting cell is characterized by an enlarged, patent canalicular lumen that contains elongated microvilli. Tubulovesicles are depleted, since they have contributed to elaboration of the secretory membrane by fusing with the secretory surface. In stages of early recycling, elongated microvilli collapse upon each other, and the canaliculus is occluded. This step is followed by endocytosis of collapsed surface membranes, which often assume closely apposed whorls of membranes, appearing pentalaminar in thin-section microscopy. Pentalaminar membranes can be traced deep within the cell during late stages of membrane recycling. In this latter stage, some tubulovesicles are regenerated, and the secretory surface has assumed an almost "resting" configuration. From Forte *et al.*[44]

tion of the transport sites for H^+ and/or K^+. For example, recent experiments in which oxyntic cell ultrastructure and H^+ secretion have been simultaneously monitored suggest that membrane fusion occurs at a time significantly before the cells have begun to transport H^+.[56] The nature of the pump activation is not completely known, but it seems to involve at a minimum the "turn on" of a system of rapid KCl transport (see Sections 5.5 and 5.6). The KCl transport system has been observed in vesicles isolated from stimulated cells but not in vesicles from resting cells.[162] *In vivo,* these stimulation events are controlled by the variable interactions of neural (e.g., vagal release of acetylcholine), hormonal (e.g., gastrin), and paracrine (e.g., histamine) secretagogues. The nature of these interactions has been the subject of continued debate, and we will review some recent experiments on isolated cells and glands which have markedly clarified the picture.

Gastric secretagogues all appear to act, at least in part, by binding to specific receptors on the basolateral membranes of oxyntic cells. The question naturally arises as to the cellular factors (also called second messengers) that couple the receptor-mediated events at the serosal membrane to the fusion and transport-related events at the mucosal membrane. The answer is partially known for both acetylcholine and histamine. Since the documentation for the action of histamine is the most extensive, we will begin there.

3.2. Histamine*

cAMP has been considered to be of major importance as a second messenger in hormone action, and over the years there has accumulated a large literature on its possible role in oxyntic cells. Many earlier reports were conflicting, at least partially because of problems inherent to the use of *in vivo* preparations and to the cellular heterogeneity of the gastric mucosa. Here we will review the information from *in vitro* experiments in which a second messenger role for cAMP has been established according to four of Sutherland's criteria.[124]

1. Experiments on a variety of gastric preparations have shown that histamine causes large and rapid increases in oxyntic cell cAMP levels, which precede the H^+ secretory response by several minutes.[e.g., 18,126,146] The implication is that cAMP is involved in the early events of activation.

2. Exogenous administration of the dibutyryl derivative of cAMP (Bt_2cAMP) mimics the effects of histamine in isolated preparations of amphibian or mammalian gastric mucosa[63,126] and in isolated gastric glands.[18]

3. Inhibitors of cAMP phosphodiesterase enhance the effects of histamine. In both piglet gastric mucosa and isolated dog parietal cells, a low-dose background concentration of isobutylmethylxanthine (IBMX, a potent phosphodiesterase inhibitor) causes an increase in histamine-induced cAMP production as well as increases in apparent H^+ secretion (see Refs. 126, 145, 146).

4. A histamine-stimulated adenylate cyclase has been identified in gastric mucosa from many different species.[18,108,111,126,151]

We conclude that cAMP serves as a second messenger for the action of histamine on oxyntic cells. We should stress, though, that cAMP is most likely not the only cellular second messenger for histamine's action. For example, in the steady state, the H^+ secretory rate and [cAMP] are not directly related (see Ref. 100). This could mean that only small changes of [cAMP] are required to initiate large changes in secretory activity, but it may also mean that other second messengers are involved. We will discuss below possible roles for Ca^{2+} and calmodulin, protein kinases, and phosphorylation of specific membrane proteins. The role of cGMP in regulation of oxyntic cells has been largely ignored, mainly because exogenous addition has no stimulatory effect on any preparation in which it has been tested.

3.3. Potentiation with Gastrin and Acetylcholine*

There has been a long-standing question of whether the oxyntic cell is regulated *in vivo* exclusively by histamine or whether gastrin and acetylcholine can also act directly on the oxyntic cell—as opposed to acting via the release of histamine from glandular cells near the oxyntic cells (e.g., see Refs. 10, 17, 58, 145). However, recent experiments have shed some interesting insights into this controversy.

A good part (80%) of the stimulatory action of gastrin, both *in vivo* and *in vitro,* appears to occur by virtue of the ability of the hormone to release histamine from within the gland itself.[6,17] However, both Chew and Hersey[17] and Soll[142,143,145] have provided evidence that gastrin itself has a small (10–20% of the histamine response), yet reproducible direct effect to stimulate isolated oxyntic cells and glands. This gastrin response seems not to be mediated via increases in [cAMP] since the glands contain no gastrin-sensitive adenylate cyclase, and [cAMP] does not increase during gastrin treatment.[17,146] As discussed below, part of this response seems to be mediated through changes in cellular [Ca^{2+}]. Treatment of isolated cells[142,143,145] or glands[4] with the cholinergic agonist carbachol also induces only small and transient stimulatory effects. This stimulation is inhibited by atropine and largely blocked by removal of Ca^{2+} from the bathing solutions.

At present, the data indicate that the oxyntic cell must have three separate receptors—one each for histamine, acetylcholine, and gastrin—and there appear to be potentiating interactions between histamine and the other secretagogues. For example, while gastrin and carbachol do not show mutual potentiation, they have each been shown to potentiate the action of histamine.[17,145] Finally, Soll[145] has provided convincing evidence for three-way potentiation with histamine, gastrin, and carbachol.

Such three-way potentiation implies that there must be a

*The role of histamine in the control of a variety of cellular activities has been furthered to an enormous extent by the synthesis of specific histamine agonists and antagonists[e.g., 10] and the classification of histamine receptors and responses as being of one of two different types, H_1 or H_2. The histamine receptor in the gastric mucosa is clearly of the H_2 type, being inhibited by cimetidine and metiamide and stimulated by dimaprit. However, this classification and all the relevant experimental findings are really of more pharmacological than physiological significance, so we have largely excluded these data from our discussion. We refer the reader to the above-mentioned references as well as to more recent papers by Chew *et al.*[18] and Hersey[70] for further information about the H_1 vs. H_2 classification of the oxyntic cells' response to histamine.

*It is likely that a variety of peptides and neurotransmitters can affect the secretory activity of the oxyntic cells. Because little work has been conducted *in vitro* on these (e.g., cholecystokinin, somatostatin, vasoactive intestinal peptide, gastric inhibitory peptide, norepinephrine, epinephrine, etc.), we have limited our discussion to the work on the actions of histamine, gastrin, and acetylcholine.

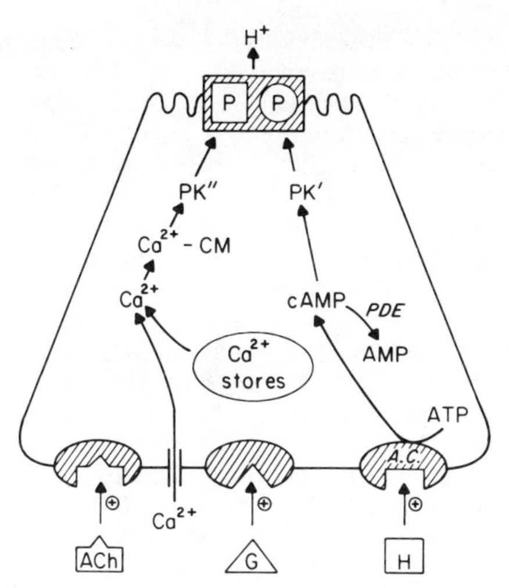

Fig. 5. Model to explain stimulus–secretion coupling to the oxyntic cell. The initial steps in the activation by histamine (H), gastrin (G), or acetylcholine (ACh) involve their binding to three different receptors at the serosal membrane of oxyntic cells. The H receptor is most likely coupled to adenylate cyclase (A.C.), which converts ATP to cAMP (whose action is terminated by an active phosphodiesterase, PDE). cAMP in turn activates a protein kinase (PK′) which catalyzes the phosphorylation of one or more membrane proteins that are intimately involved in the H^+ secretory process. Gastrin and ACh cause increases in cytoplasmic [Ca^{2+}]. This Ca^{2+}, which may enter from the serosal solution and/or from cellular bound "stores," promotes the activity of calmodulin (CM). The Ca^{2+}–CM system in turn activates another kinase (PK″), ultimately leading to additional phosphorylation events. Full activation of H^+ secretion is generated only when both PK′ and PK″ are activated and phosphorylation is complete. Partial stimulation (by gastrin or ACh) can be achieved through activation of the Ca^{2+}–CM pathway alone. Potentiation (i.e., H + G, H + ACh, or H + G + ACh, but *not* G + ACh) is achieved through interactions between the cAMP and Ca^{2+}–CM pathways. It should be stressed that many specific details of this hypothetical scheme are not known: (1) the presence of Ca^{2+} stores and of Ca^{2+}–CM-dependent phosphorylation events; (2) the cellular localization of CM (cytoplasm?, membranes?); (3) the effects of gastrin; and (4) the participation of other second messengers besides cAMP and Ca^{2+}–CM.

site of convergence and amplification in the cellular responses of these secretagogues. Because cAMP is the intracellular mediator for histamine, this cyclic nucleotide is one obvious possibility. However, neither carbachol nor gastrin potentiates a histamine-induced increase in cellular [cAMP].[146] Since carbachol and gastrin both show some Ca^{2+} dependence, and because Ca^{2+} and cAMP often interact during cellular activation processes, another possibility is that gastrin and carbachol alter cellular [Ca^{2+}] (see Fig. 5).

3.4. Is Calcium a Second Messenger?

Ca^{2+} plays numerous roles in cellular physiology. Ca^{2+} often acts as an intracellular messenger in tissues in which secretion involves exocytosis (e.g., exocrine pancreas[160]), a membrane fusion event,[149] as well as in those where Ca^{2+} regulates the permeability of membranes and the movement of ions across their surfaces (e.g., insect salivary gland[7]). Such Ca^{2+}-mediated events are often associated with the actions of the Ca^{2+}-binding protein, calmodulin.[106]

In the stomach, both membrane fusion as well as alterations of membrane pumps and carriers occur on a grand scale, so it has been assumed that intracellular Ca^{2+} must play a role in the ultimate activation events in oxyntic cells. However, experimental approaches to define specific mechanisms have met with only moderate success.

The evidence at present is most convincing for a direct role for Ca^{2+} in cholinergic stimulation. Thus, in isolated parietal cells[144] or gastric glands,[5] acetylcholine or carbachol induces increases in apparent H^+ secretion and $^{45}Ca^{2+}$ uptake; these effects are abolished by treating the cells with Ca^{2+}-free external solutions, atropine, or La^{3+} (a blocker of Ca^{2+} channels). Results with gastrin indicate that its small stimulatory effect is largely (70%), but not completely, reduced in Ca^{2+}-free solutions or during treatment with La^{3+}. Histamine and cAMP appear to be even less dependent on extracellular Ca^{2+} than the other secretagogues: extracellular Ca^{2+} may be partially required for histamine to exert its full stimulatory effect, while Bt_2cAMP can stimulate in the complete absence of extracellular Ca^{2+}.[5,94,144]

We summarize these experiments dealing with the importance of extracellular Ca^{2+} by listing the relative secretory response (compared to control) of various isolated preparations to different secretagogues in Ca^{2+}-free bathing media: Bt_2cAMP (100%) ≥ histamine (60–100%) > gastrin (70%) > acetylcholine or carbachol (8–30%).

We must point out that the bulk of the experimental evidence has been focused on the role for extracellular Ca^{2+}. However, it is well known for some tissues that secretagogues can exert their actions by releasing intracellular stores of Ca^{2+}. Although gastrin and histamine show only moderate sensitivity to external Ca^{2+}, the question still remains whether these secretagogues might operate in part by releasing intracellular Ca^{2+} from some bound stores. There have been very few experiments which have approached this problem directly (e.g., see Refs. 78, 79, 144), and this area of research remains cloudy. Future progress may result from actual measurements of cellular Ca^{2+} activity.

3.5. Regulation via Protein Kinases

Early work showed that oxyntic cells contain both soluble and membrane-bound cAMP-dependent protein kinases.[112,123] More recent work[134] has shown that the apical membranes of oxyntic cells contain proteins which are phosphorylated specifically by the action of a variety of membrane-bound and soluble protein kinases (PK): an intrinsic PK; a soluble cAMP-dependent PK; and a membrane-bound Ca^{2+}–calmodulin-dependent PK. Thus, phosphorylation of membrane proteins (through the actions of cAMP and Ca^{2+}–calmodulin-dependent PKs) is likely to be a means by which the extracellular secretagogues exert their cellular effects. Further studies in this interesting area will be directed toward isolating the membrane proteins which exhibit specific phosphorylation as well as toward trying to understand the interactions which occur between the cyclic nucleotides and Ca^{2+}–calmodulin. We have summarized our present ideas about stimulus–secretion coupling in the oxyntic cell in Fig. 5.

4. Metabolism and Energetics Associated with Gastric HCl Secretion

Owing to the large gradients and relatively voluminous solute and solvent flows, the metabolic requirements for gastric secretion are relatively exacting. The stomach is capable of transporting H^+ and Cl^- against extraordinary electrochemical gradients. Mammalian gastric juice can achieve an acidity of approximately isotonic HCl (150 mM, pH 0.8). An estimate can be made of the energetic demands in terms of the work required for H^+ and Cl^- ($W_{H,Cl}$) transport using the expression

$$W_{H,Cl} = n_H\left(RT \ln \frac{[H^+]_m}{[H^+]_s} + zF\psi_{ms}\right)$$

$$+ n_{Cl}\left(RT \ln \frac{[Cl^-]_m}{[Cl^-]_s} + zF\psi_{ms}\right)$$

where the number of equivalents of H^+ (n_H) and Cl^- (n_{Cl}) transported are multiplied by their respective chemical and electrical gradients. Ψ_{ms} is the transmucosal potential difference the subscripts m and s refer to the mucosal and serosal sides) and z, R, T, and F have their usual meanings. As a minimum estimate for the total process, we assume the secretory product is ultimately derived from the blood, with pH = 7.4 and $[Cl^-] = 120$ mM. Using a typical value for the transmucosal potential difference = −40 mV, mucosal side negative with respect to serosal side, we can calculate from the above equation that at 37°C, 9577 calories are required per mole HCl secreted (8417 cal/equiv H^+; 1060 cal/equiv Cl^-).

This high energetic demand is served by the metabolic activity of the tissue, and primarily the oxidative metabolism. In fact, the close correlation between H^+ output and oxidative metabolism of gastric mucosa led to a great deal of experimentation and debate concerning the immediate energy source for gastric H^+ transport. One view, the redox hypothesis, proposed that H^+ in the gastric juice was derived directly by oxidation of substrate hydrogen atoms, the electrons being passed to molecular oxygen via a suitable membrane redox system, e.g., cytochrome chain. The alternative view held that ATP, derived from oxidative phosphorylation, was the direct energy substrate for the H^+ pump. Rather than belabor the older data and arguments (see Ref. 96), we shall refer to more recent experiments that clearly show ATP is the principal energy source for gastric HCl secretion. The evidence comes from work with isolated oxyntic glands for which the accumulation of weak bases, such as aminopyrine, into acidic glandular and canalicular lumina can be used to assess the relative acid secretory response. Oxyntic glands were treated with mild detergent[103] or electric shock[3] in order to ''permeabilize'' their basolateral membranes to small molecules; thus, agents added to the bulk solution had immediate access to the cytoplasm. The permeabilized glands were shown to accumulate aminopyrine (index of H^+ secretion) when ATP was added, even in the presence of cyanide or in the absence of oxygen.

Before reviewing the findings concerning the identification of ATP-utilizing transport pumps, or ATPases, we will first consider some theoretical limitations on ATP as the protonmotive substrate. The free energy change (ΔG) for the hydrolysis of ATP into ADP and inorganic phosphate (P_i) can be calculated from the expression

$$\Delta G = \Delta G_0 + RT \ln \frac{[ADP][P_i]}{[ATP]}$$

where ΔG_0 is the standard free energy change for the reaction, which at 37°C is equal to −7400 cal/mole. Using the available measurements for nucleotide and P_i concentrations in gastric mucosa,[43] ΔG under physiological conditions is estimated to be about −12 to −13 kcal/mole ATP. This is a typical estimate for ATP in many cells, often quoted in textbooks. We can compare this assumed maximum available energy with the minimum energy requirement for physiological H^+ secretion given above (ca. 8 to 9 kcal/equiv. H^+ transported), and make the estimate that the transport of 1 equivalent of H^+ would require the utilization of about 1 mole of ATP. There are some precautions that must be considered before this relationship can be unequivocally accepted, e.g., the efficiency of converting chemical energy into ion transport work and the determination of precise chemical and electrical gradients at the membrane transport site. However, within reasonable limits, if ATP were the sole energy substrate, we would conclude that a gastric H^+-transport ATPase would have a stoichiometry of 1.0 H^+ transported/ATP utilized. This is clearly different from the Na^+ pump enzyme for which there are 3 Na^+/ATP transported by the Na^+,K^+-ATPase.

5. Studies with Isolated Cell Fractions and Membranes

5.1. Background

As has been the case for many tissues, our understanding of gastric transport processes has been greatly advanced through the study of cell and membrane fractions isolated from the epithelial cells. Although the disruption process destroys the asymmetry that is characteristic of the epithelium, the information gained from isolated organelles, separation of apical and basolateral membranes, and the identification and localization of specific transport enzymes is not available using any other technique. Furthermore, the membrane fragments can often be re-formed into vesicles and the characteristics of intrinsic transport processes can be studied in these model systems. The aim, of course, is to ''reconstitute'' the information from the cell-free systems and to incorporate the observations into the physiological activity of the tissue.

Kasbekar and Durbin[80] pioneered the study of membrane fractionation in gastric mucosa. Unfortunately, the enzyme they suggested might be the source of gastric acid, a HCO_3^--stimulated ATPase, turned out to be a mitochondrial contaminant, and is probably not directly involved in the H^+ secretory process.[48] To date, there is no convincing evidence for the existence of an anion pump ATPase (Cl^- or HCO_3^-), and this will obviously influence our interpretation of the cellular mechanisms responsible for net Cl^- transport by the gastric epithelium. However, Na^+,K^+-ATPase is present in gastric cells, and another enzyme occurs in great abundance in the oxyntic cells, the H^+,K^+-ATPase which is now considered as the primary protonmotive pump of the stomach. We will review the functions of these two cationic ATPases, paying particular attention to properties of the unique gastric H^+,K^+-ATPase as well as studies of transport in isolated vesicular systems that can be interpreted in terms of transepithelial transport.

5.2. Na^+,K^+-ATPase

The gastric mucosa contains an ample quantity of Na^+,K^+-ATPase, and like most of the gastrointestinal epithelium, it is localized to the basolateral membrane.[22,67] All gastric epithelial cells contain Na^+,K^+-ATPase, but it is of significance that the oxyntic cells are the richest in the Na^+ pump enzyme.[23] The principal function of Na^+,K^+-ATPase is to maintain low intracellular $[Na^+]$ and high $[K^+]$. The high cellular $[K^+]$ is important for normal functioning of many cell reactions, and in particular, the H^+-transporting machinery of the oxyntic cell is absolutely dependent upon K^+, as we shall elaborate below. The Na^+ gradient, from ECF to cell, maintained by Na^+K^+-ATPase serves as a driving force for Na^+-dependent transport functions at the basolateral membrane (see Section 6.2.4). In the mammalian stomach the Na^+ pump enzyme is also responsible for a small net movement of Na^+ from lumen to blood.[149,101] This net Na^+ flux probably occurs only through the surface epithelial cells, and its physiological significance is uncertain.

Inhibition of the Na^+ pump activity by ouabain leads to marked reduction in H^+ secretion and net Cl^- transport,[21,49] an elevation of cellular $[Na^+]$, and a reduction of cellular $[K^+]$.[24] The reduction of H^+ secretion is most likely related to decreased cellular $[K^+]$. After treatment with ouabain, if $[K^+]$ of the serosal solution is elevated, H^+ secretory rates and cellular ionic composition can be maintained at essentially normal values.[24] Moreover, in isolated gastric glands that have been permeabilized by detergent, the restoration of aminopyrine uptake (index of H^+ secretion) requires not only the addition of ATP as an energy substrate, but also the elevation of K^+ in the bathing medium.[3] The decrease in Cl^- transport associated with inhibition of the Na^+ pump may be related to the altered Na^+ gradient at the basolateral surface, and hence the driving force associated with Na^+-dependent Cl^- entry. However, the concomitant reduction in cellular $[K^+]$ may also play a role in the reduction of Cl^- flux that is directly associated with H^+ transport, i.e., acidic Cl^-.[96]

5.3. H^+,K^+-ATPase

Tubulovesicles that abound in the cytoplasm of oxyntic cells can be isolated from gastric mucosal homogenates as the microsomal fraction (i.e., membrane fraction that sediments between 12,000*g* × 10 min and 100,000*g* × 60 min). These membranes can be further purified of contaminants, such as mitochondria and RER, by sucrose density gradient centrifugation, yielding a highly homogeneous, low-density (1.08–1.11), fraction of vesicles. The purified gastric microsomal vesicles contain a unique enzyme, a K^+-stimulated ATPase.[45,46,55]

Evidence that the K^+-stimulated ATPase activity is the primary H^+ transport enzyme in gastric oxyntic cells comes from several sources. The enzyme occurs almost exclusively in oxyntic cells, and is present in all species examined that secrete HCl.[47] Ontogenetic studies in the frog have shown that the enzyme appears at the metamorphic stage when the oxyntic cell differentiates and the capacity for HCl secretion first develops.[92] Finally, in the isolated vesicles the enzyme catalyzes an ATP-dependent K^+/H^+ exchange reaction, and is thus called the H^+,K^+-ATPase.[48,86,128] We will describe some of the characteristics of this enzyme and proceed to discuss those studies of H^+ transport by isolated vesicles that provide insight into how the system operates in intact tissue.

Partial reactions of the gastric H^+,K^+-ATPase have been studied by several laboratories, and all agree that there is a complex sequence of reactions leading to ATP hydrolysis, but there is still some uncertainty about the specific details of the catalytic cycle.[12,113,157] For simplicity, the scheme here is presented as the following two-step reaction sequence:

$$\mathrm{ATP} + \mathrm{E} \xrightarrow{Mg^{2+}} \mathrm{E} \sim \mathrm{P} + \mathrm{ADP}$$

$$\mathrm{E} \sim \mathrm{P} \xrightarrow{K^{+}} \mathrm{E} + \mathrm{P_i}$$

The first step includes the binding of ATP to the enzyme (E) and the formation of a phosphoenzyme intermediate (E~P) via a rapid Mg^{2+}-dependent phosphokinase reaction. This step will also catalyze an ATP/ADP exchange reaction. The phosphoenzyme is in the form of an acyl phosphate linkage onto a peptide of 95,000 daltons.[47] There is some suggestion that within the membrane the enzyme may occur as a dimer, or even a trimer of 95K peptides.[127] The second enzymatic step shown above is the hydrolysis of the phosphoenzyme via a K^+-catalyzed phosphatase reaction. Without K^+ this is a rate-limiting step and E~P accumulates; acceleration of phosphatase by K^+ leads to disappearance of E~P.[113] Other monovalent cations substitute for K^+ in the overall ATPase or phosphatase reaction, with the sequence of selectivity being $Tl^+ > K^+ > Rb^+ > Cs^+ > NH_4^+ >> Na^+$.

It is evident that several characteristics of the H^+,K^+-ATPase described above are similar to what is known of the ubiquitous Na^+,K^+-ATPase, with the major exceptions that for the gastric enzyme Na^+ is not involved and cardiac glycosides (e.g., ouabain) do not inhibit the K^+-dependent step. Although these and other differences are clear, one cannot help but be impressed by many of the analogies between these two cation-transporting enzymes, and even extending to the Ca^{2+}-ATPase of the sarcoplasmic reticulum. Similarities include (1) molecular weight of the catalytic subunit, (2) phosphoenzyme linkage via an aspartyl residue, (3) homologies in hydrolyzed peptide subunits,[156] and (4) sensitivity of phosphoenzyme to K^+, all of which suggest some commonalities of the molecular organization and function of these ion-translocating ATPases. This class of what we might liberally call plasma membrane cation-transport ATPases is clearly distinct from the H^+-transport ATPases of mitochondria, chloroplasts, and bacterial membranes.

5.4. Ion Transport by Gastric Microsomal Vesicles

Using dog gastric microsomes supplemented with ATP, Mg^{2+}, and K^+, Lee *et al.*[86] were the first to demonstrate that the H^+,K^+-ATPase was capable of transporting protons. Vigorous subsequent experimentation by several groups has provided the following picture of ion transport by gastric microsomes from a variety of species.[48,87–89,128,129,133]

The purified gastric microsomes occur as small (1000–2000 Å in diameter), well-sealed vesicles (e.g., see Fig. 8a), and for the most part the ATP catalytic site faces the vesicular exterior. The vesicles are relatively impermeable to K^+ and somewhat more permeable to Cl^-. ATP-driven H^+ transport into the vesicular interior occurs as a H^+/K^+ exchange reaction, and hence requires that K^+ be present within the vesicles. This can be accomplished only by long-term preincubation of the vesicles in KCl (on the order of hours) or by facilitating K^+ entry by the K^+ ionophore, valinomycin. Given these conditions, a relatively vigorous ATP-dependent H^+ uptake occurs and pH gradients of up to 4 to 4.5 pH units have been measured.[88] Sachs *et al.*,[128] concluded that the H^+/K^+ exchange

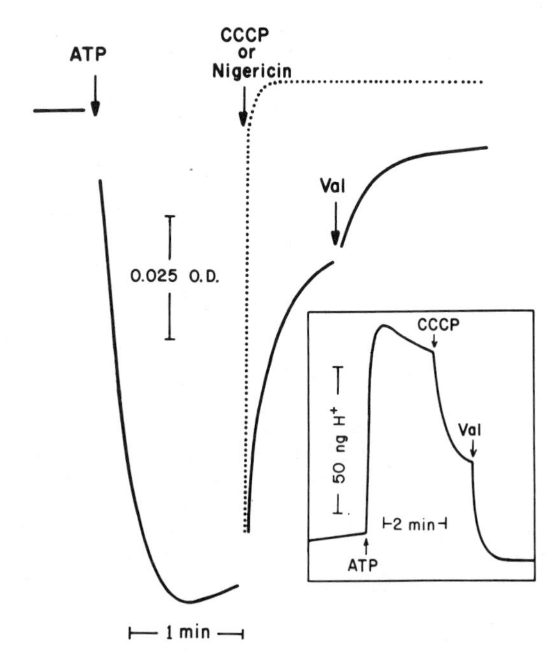

Fig. 6. H^+ uptake by isolated gastric microsomes. The microsomes were preequilibrated with 150 mM KCl for 2 days at 0°C, and then transferred to a cuvette containing KCl and 5 μM acridine orange. Addition of MgATP (1 mM; arrow) caused an immediate decrease in absorption as the weak base acridine orange was trapped within the acidic vesicular interiors. The magnitude of the absorption decrease, or the degree of fluorescence quenching, is related to the magnitude of the pH gradient. Addition of the H^+/K^+ exchange ionophore, nigericin (dotted line), brought about immediate dissipation of the pH gradient and restored O.D. to its initial value. The addition of a H^+ ionophore (CCCP) produced a partial dissipation of the pH gradient (solid line); subsequent addition of the K^+ ionophore, valinomycin, completely dissipated the gradient. (Inset) An experiment similar to that given above except that the pH of the external medium was monitored by a glass electrode. ATP caused an alkalinization of the medium as an equivalent amount of protons was being pumped into the vesicular interiors. Again the ATP-generated pH gradient was dissipated by ionophores. From Lee *et al.*[87] with permission.

was electroneutral (i.e., 1 : 1 exchange), based on their studies showing a parallelism between $^{86}Rb^+$ (as K^+ substitute) efflux and H^+ uptake, and through the deployment of membrane potential probes. Experiments showing the rate of ATP-dependent H^+ transport by gastric microsomal vesicles, as measured by accumulation of fluorescent amine probe and by pH electrode, are shown in Fig. 6.

Given the characteristics of the microsomal H^+-transport process, a pump-leak model was developed to account for pH gradient formation in terms of the H^+,K^+ pump, diffusive K^+ entry, H^+ leak pathways, and the role of anions (Fig. 7).[87] Assuming an adequate supply of ATP, pump activity depends on intravesicular $[K^+]$ which, in turn, is dependent upon the rate of K^+ entry. Charge considerations require that K^+ must enter either in exchange for H^+ (which is nonproductive in terms of pH gradient formation) or in conjunction with an anion. For the latter case when K^+ entry is not rate limited (i.e., valinomycin present), the ATPase activity and microsomal pH gradient formation are dependent upon the rate of anion penetration by passive diffusion with the observed sequence for permeability being $SCN^- > NO_3^- > Br^- > Cl^- > I^- >>$ isethionate, acetate.[87]

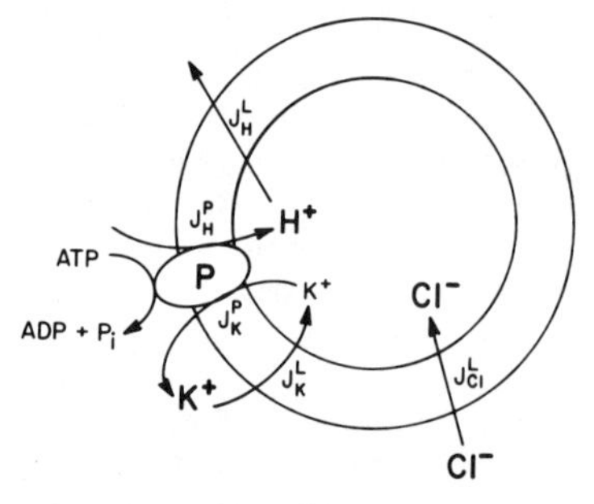

Fig. 7. Schematic representation of ion movements across gastric microsomal vesicles in a pump-leak model. The *J*'s are ion fluxes with the superscripts designating the pump flux (P) or leak pathway (L). The model consists of an ATP-driven H^+/K^+ exchange pump and the passive leak pathways for the principal ions, H^+, K^+, and Cl^-. From Lee *et al.*[87] with permission.

SCN^- is a special case in that the initial H^+ uptake rates can be very fast in KSCN supplemented with valinomycin, but the accumulated product, HSCN, leaks out very rapidly, and thus SCN^- supports very poor pH gradients.[114] This action of SCN^- as an "uncoupler" of the gastric H^+ pump may be the mechanism by which SCN^-, and several other pseudohalogens, operate as effective inhibitors of gastric H^+ secretion.[59,132]

The general pump-leak model of gastric microsomal ATPase and ion transport has served a useful purpose in that it demonstrates (1) ATP-dependent vectorial H^+ transport, (2) dependence on K^+, a well-known feature of gastric H^+ secretion, and (3) there is sufficient quantity of H^+,K^+-ATPase in gastric homogenates to account for *in vivo* H^+ secretory rates.[46] However, there are several questions that must be addressed before the system can be accepted as the gastric proton pump. These include the placement of the system within the oxyntic cell and the obvious kinetic limitation that K^+ permeability is extremely low, and rate limiting, in the absence of exogenous ionophores. Recent studies that have examined membrane fractions from oxyntic mucosa under distinct phases of physiological secretory activity have answered these questions.

5.5. ATPase Activity and H^+ Transport in Apical Membrane Vesicles

In an effort to determine whether the H^+,K^+-ATPase showed functional changes associated with gastric secretory activity, Wolosin and Forte[162,163] monitored the enzyme in gastric homogenates from "resting" stomachs (not secreting HCl) and those that were maximally secreting HCl. In the resting stomachs, the enzyme was distributed exactly as described above, i.e., the microsomal fraction, and the transport properties coincided with the microsomal model given in Fig. 6, including the absolute requirement of K^+–valinomycin for maximal H^+-transport rates. It was of great interest to find that in the secreting stomach the enzyme had largely disappeared (60–80% decrease) from the light microsomal fraction, and had stoichiometrically redistributed to the much larger and heavier membrane fractions. Density gradient centrifugation of these latter fractions purified the H^+,K^+-ATPase from mitochondria and other contaminants yielding a preparation of large, complex structures that were clearly distinct from the classical microsomes (Fig. 8). It was most encouraging to find that the larger vesicles, which were called stimulation-associated (SA) vesicles, showed ATP-dependent H^+ uptake, subject to $[K^+]$ and

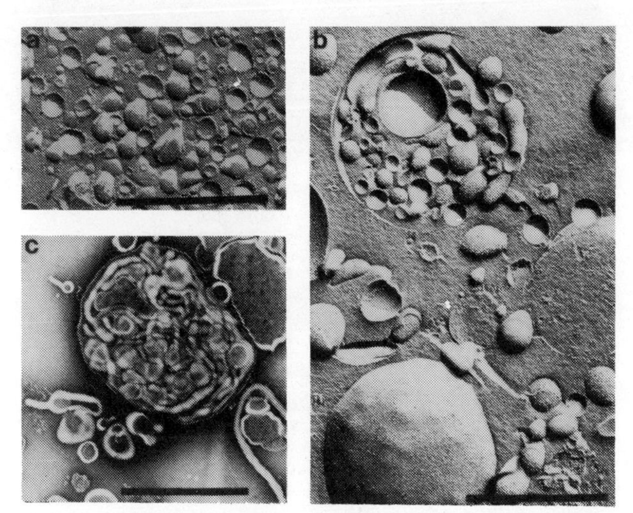

Fig. 8. Electron micrographs of gastric microsomes and stimulation-associated (SA) vesicles isolated respectively from resting and secreting gastric mucosa. (a) Freeze–fracture replica of microsomal vesicles prepared from hog stomach. Note the homogeneity and relatively small size of this population of microsomal vesicles. (b) Freeze–fracture replica of SA membrane vesicles prepared from histamine-stimulated rabbit gastric mucosal homogenates by the method of Wolosin and Forte.[163] Note the large size and complex structure of many of the membranes. It often appears as though vesicular and microvillar structures are incorporated into larger membrane sacs. (c) Preparation of SA vesicles was as described in (b), but 2% sodium phosphotungstate was used as a negative stain. Structural analogies between (b) and (c) are apparent, and these can be distinguished from the microsomal vesicles in (a). Bars = 1 μm. From Forte *et al.*[44]

[Cl −] in the medium, but did not require valinomycin. Figure 9 shows the patterns of H^+ uptake in SA vesicles as compared to gastric microsomes. The initial rate of vesicular H^+ uptake is presented in Fig. 10A as a function of various anionic concentrations, at constant [K^+], and Fig. 10B as a function of various [K^+], at constant [Cl^-]. From these experiments, we conclude that the proton pump in SA vesicles (1) does not require exogenous ionophores, (2) the system is saturated at relatively low concentrations of K^+ and Cl^-, and (3) various anions can substitute for Cl^- to a greater or lesser extent. These data and other isotopic flux experiments[166] are consistent with the existence of a system for the rapid transport of K^+ and Cl^- (or other anion) in SA membranes that was not present in the gastric microsomal vesicles.

A body of evidence supports the conclusion that SA vesicles are derived directly from the apical plasma membrane of the secreting oxyntic cell, while the gastric microsomal vesicles are from the tubulovesicles that predominate in the cytoplasm of the resting cell.[162–164] Gently treated SA vesicles are large, morphologically complex structures which often contain microvilluslike processes, reminiscent of the apical surface. The SA membranes are dense, probably because of abundant adhering proteins, one of which has been identified as actin, in the microfilamentous form, also characteristic of the apical cell surface. Wolosin and Forte[164] proposed that the H^+,K^+-ATPase-rich tubulovesicles fuse with the apical surface when the oxyntic cell is stimulated, greatly expanding that surface, and placing the H^+/K^+ exchange pump in parallel with the system of K^+ and Cl^- transport (Fig. 11). Thus, the resulting HCl secretory process would require activation of the fusion process, and possibly some activation or modulation of the system of K^+ and Cl^-

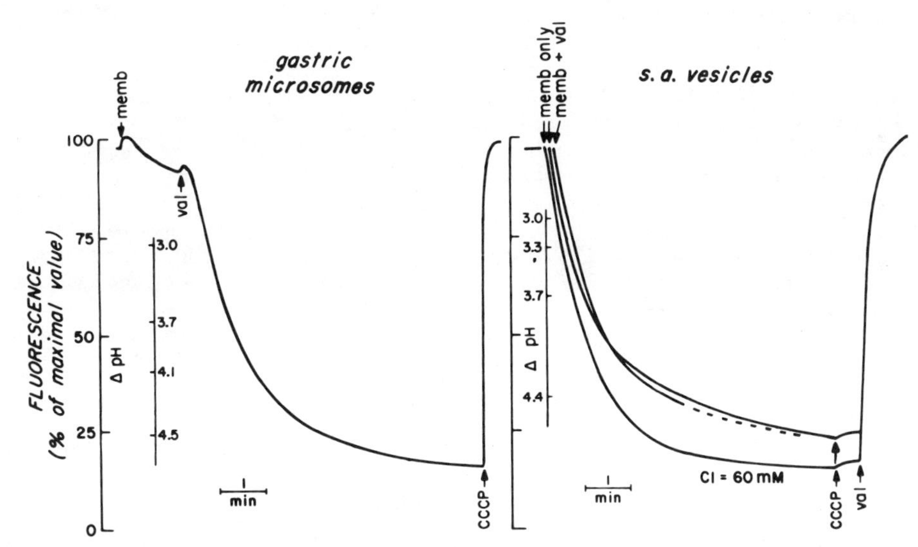

Fig. 9. Comparison of H^+ uptake in gastric microsomes and SA vesicles. ATP-generated H^+ uptake was monitored by the uptake and fluorescence quenching of the weak base probe, acridine orange. The pH gradients were calibrated and an indication is given for intravesicular pH (pH_i). (Left) Gastric microsomes were added to uptake medium containing 150 mM KCl, 0.5 mM MgATP, and 5 μM acridine orange. At first, very little H^+ uptake occurred, in contrast to conditions where the vesicles had been preequilibrated with KCl (see Fig. 6). Addition of the K^+ ionophore, valinomycin (val), produced a prompt increase in H^+ uptake, since this promoted KCl entry and, therefore, ATP-generated K^+/H^+ exchange. The gradient was immediately dissipated by the protonophore, CCCP. (Right) SA vesicles were added to uptake medium as for gastric microsomes. In this case, there was a rapid uptake of H^+ that was independent of valinomycin, suggesting that the endogenous pathway for KCl entry was sufficient to provide K^+ for the K^+/H^+ exchange pump. In the absence of valinomycin, 5 μM CCCP was very weak in dissipating the pH gradient, whereas this protonophore was effective when valinomycin was present. Adapted from Wolosin and Forte.[162]

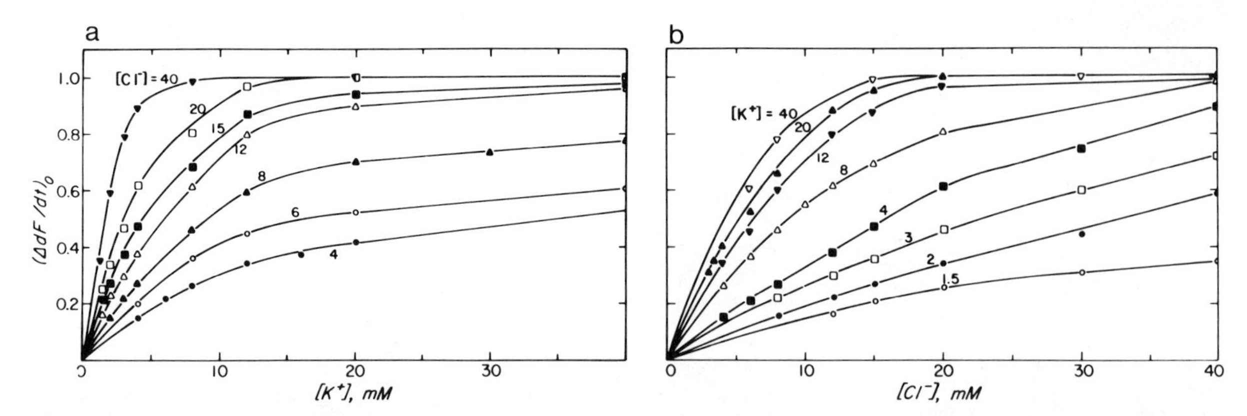

Fig. 10. Relative rates of H^+ uptake in SA vesicles as a function of $[K^+]$ and $[Cl^-]$. H^+ uptake was monitored by the acridine orange fluorescence quench method and the initial rates of uptake are expressed in the parameter $(dF/dt)_0$. (A) The dependence on $[K^+]$ is shown at various fixed $[Cl^-]$. (B) The dependence on $[Cl^-]$ is shown at various fixed $[K^+]$. The data can be replotted in double reciprocal form to give apparent Michaelis constants for $K^+ = 15$ mM and $Cl^- = 46$ mM. Adapted from Wolosin and Forte.[165]

transport. Support for this general hypothesis of redistribution of ATPase through membrane fusion has also come from the subcellular localization of monclonal antibodies to the H^+,K^+-ATPase.[141] In resting oxyntic cells of rabbit and hog stomach, the enzyme was heavily localized to tubulovesicles, with only weak staining of the unelaborated apical surface; however, the expanded apical surface of stimulated cells was heavily laden with the marker.

5.6. Ion Transport by Apical Membrane Vesicles

The most unique functional feature of the apical membrane fragments of the stimulated oxyntic cell (SA vesicles) is the novel system for the transport of K^+ and Cl^-. It will be an important future development to isolate and ascertain the properties of the apical membrane from the resting cell. However, the characteristics of K^+ and Cl^- transport by SA vesicles have already provided some useful insights.

It was first thought that KCl was transported via a totally electroneutral cotransport system, i.e., that K^+ and Cl^- moved via a nonconductive mechanism. This was based on the lack of apparent effects of the classical protonophore, CCCP, on dissipating ATP-generated H^+ gradients (e.g., Fig. 9B), and on the demonstrated cooperativity between K^+ and Cl^- in the transport system.[165] The experiments with CCCP have now been called into question since we now know that more effective protonophores (e.g., tetrachlorosalicylanilide), or the use of more massive doses of CCCP, will lead to dissipation of proton gradients under appropriate conditions of $[K^+]$ and $[Cl^-]$ on either side of the vesicles. Either an outward-directed Cl^- gradient ($[Cl^-]_{in} > [Cl^-]_{out}$) or an inward-directed K^+ gradient ($[K^+]_{out} > [K^+]_{in}$) will accelerate the dissipation of accumulated intravesicular H^+ in the presence of appropriate protonophore; thus, it appears that intrinsic conductive pathways exist for both K^+ and Cl^- in SA vesicles.[167]

The interdependence between K^+ and Cl^- was established in a study of KCl uptake by SA vesicles, using the H^+/K^+ exchange pump and vesicular accumulation of HCl as an index.[165] It was found that increased levels of K^+ enhanced Cl^- uptake, and increased levels of Cl^- enhanced K^+ uptake. The kinetics of uptake closely conformed to a 1 : 1 random sequential binding model of cotransport, with defined affinities for K^+ and Cl^-. The apparent Michaelis constants determined in this study were about 15 mM for K^+ and 46 mM for Cl^-. The assumption of electroneutral cotransport must now be reexamined, and conceivably fit with other possibilities such as those shown in Fig.

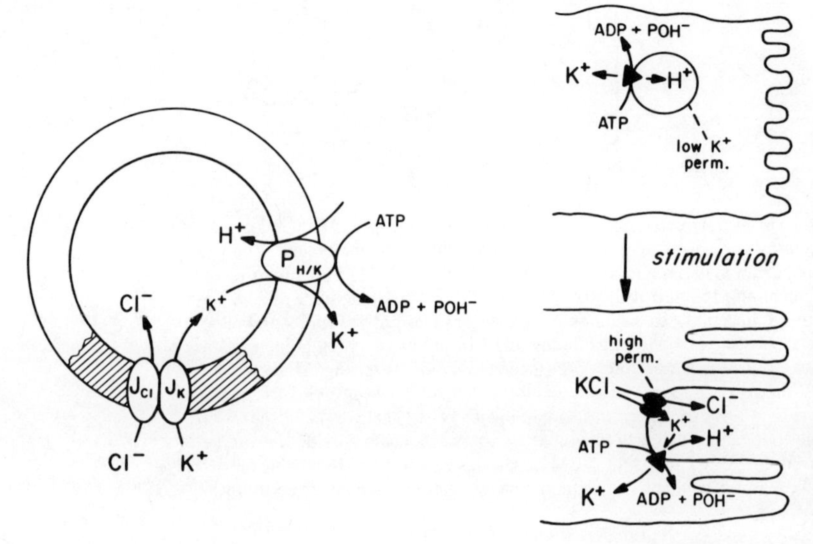

Fig. 11. Schematic representation of the mechanisms proposed for HCl accumulation in SA vesicles and secretion of HCl at the apical membrane of the intact oxyntic cell. (Left) The combined action of a system for rapid transport of K^+ and Cl^- and the H^+,K^+-ATPase pump results in net accumulation of HCl in the SA vesicle. Uptake of KCl may occur through a single carrier system or through separate pathways, each of which would have to be electrically compensated. (Right) The integration of the fusion process with stimulation of the KCl pathway in the oxyntic cell. Fusion of tubulovesicles (i.e., derivative microsomes) with the apical surface puts the H^+,K^+ pump in parallel with the KCl cotransport. It is assumed that, *in vivo*, the H^+/K^+ exchange pump closely matches the KCl efflux so that K^+ is recycled with little net K^+ loss. Adapted from Wolosin and Forte.[164]

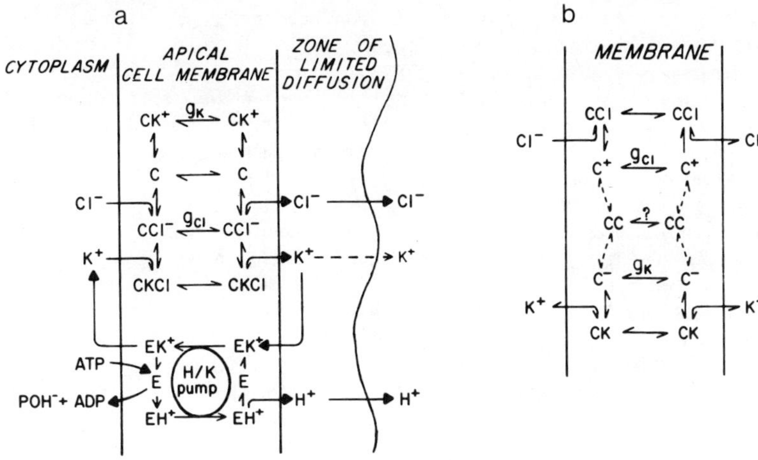

Fig. 12. Schematic representation of possible KCl transport systems operating in parallel with the H^+/K^+ exchange pump in the apical membrane of the oxyntic cell. (A) Single carrier model. A proposed membrane carrier (C) combines with K^+ and/or Cl^-. As the ternary CKCl complex, the flux of KCl from cytoplasm is rapid and effectively electroneutral. H^+/K^+ exchange by the pump provides the force for net HCl secretion. The proposed existence of the charged binary complexes, CK^+ and CCl^-, permits the system to operate in electrogenic modes and would be consistent with the carriage of current through these pathways (g_K and g_{Cl}) as observed in intact tissue studies. (B) Separate carrier model. Here it is proposed that K^+ and Cl^- each combine and flow through separate electrogenic pathways. The possibility of cooperativity between the separate systems (e.g., formation of CC) would permit a high KCl flux without the need for an external return circuit. Recycling of K^+ through the H^+,K^+-ATPase would be as shown in (A).

12. In the first case (Fig. 12A), it is assumed that a single membrane carrier (E) randomly combines with both K^+ and Cl^- to form the EKCl complex. Thus, KCl transport could be electroneutral, but depending on the relative mobilities and concentrations (i.e., conductance) of the charged EK^+ and ECl^- forms, the system could carry current. The second scheme (Fig. 12B) proposes that K^+ and Cl^- bind with separate and oppositely charged carriers, E^- and E^+. The transmembrane movements of the ions would be a function of the "return" rate of the respective carriers. This latter pathway could account for the observed ionic conductivities; moreover, an interaction (either molecular association or charge/charge interaction) between the carrier forms could result in a neutralized complex that rapidly translocates the membrane.

It is clear that these are just working models to account for a highly complex system of transport, and future work will improve our view of the detailed mechanisms. In any event, we should now summarize the findings of the current studies on SA vesicles. These vesicles are obtained only from stimulated oxyntic cells and are most likely derived from the apical plasma membrane. They possess a system for rapid transport of K^+ and Cl^- which operates in parallel with the H^+,K^+-ATPase to produce a net intravesicular accumulation of HCl. Conductance of SA membranes to K^+ and Cl^- is relatively high. These properties generally conform to the requirements of the principal mechanism for the transport of HCl at the apical surface of the oxyntic cell. They may also provide a basis for interpreting some of the electrophysiological and ion flux data obtained from intact tissue.

6. Electrophysiological and Tracer Flux Studies of Gastric Ion Transport

6.1. Background

In the previous section we have discussed biochemical and biophysical studies of ion transport mechanisms in isolated gastric membranes. In this section we will review electrophysiological and ion flux studies which have been conducted *in vitro* on intact gastric mucosa or the recently developed preparation of isolated gastric glands. Ultimately, one must integrate the observations from isolated membranes with those obtained from intact tissue so as to have a comprehensive view of the mechanisms and control of gastric ion transport. As a first step and as background we will present some basic information about the electrophysiology of this tissue.

The gastric mucosa of most vertebrate species maintains a transepithelial electrical potential difference (Ψ_{ms}) with the mucosal (luminal) side negative to the serosal (blood, nutrient) side by −10 to −60 mV (see Ref. 32). The gastric Ψ_{ms} is generated by the active transport of monovalent ions traversing a transepithelial resistance of about 100 to 600 Ω-cm². The transepithelial resistance decreases quite markedly and characteristically during the transition from the resting to the fully stimulated state. There are also changes in transepithelial potential associated with stimulation; however, the direction and magnitude of the change vary among the species studied. We will have more to say about these changes below, most specifically about the drop in resistance.

From the previous described studies on isolated apical membrane vesicles, the net secretion of HCl at the mucosal aspect of oxyntic cells would normally appear as an overall

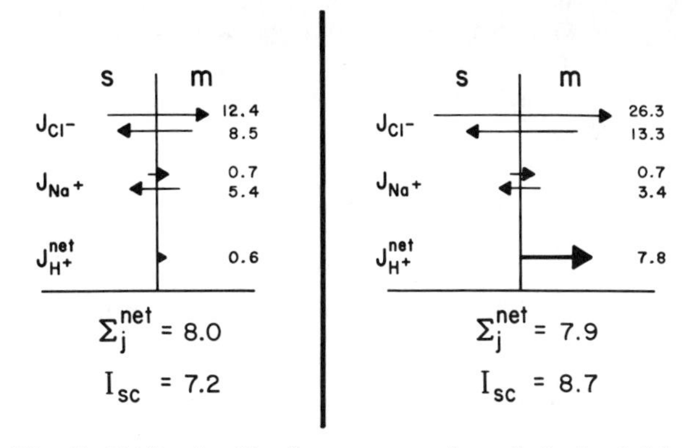

Fig. 13. Unidirectional ion fluxes across resting and stimulated piglet gastric mucosa in the short-circuited state (i.e., ψ_{ms} clamped to zero). Tissues were bathed with buffered Ringer's solution on the serosal (s) side and unbuffered solution on the mucosal (m) side; Na^+ and Cl^- fluxes were measured with radioisotopes; H^+ secretion was measured by titration. Note that in both resting and stimulated tissues there is a net secretion of Cl^- and absorption of Na^+. Histamine stimulation causes approximately equivalent increases in net H^+ and Cl^- secretion. Note that $I_{sc} = J^{Cl}_{net} + J^{Na}_{net} - J^H$ (i.e., Σ^{net}_j), so these ions represent the majority of active transport. Data from Forte and Machen.[49]

electroneutral process (i.e., KCl transport combined with neutral H^+ for K^+ exchange). In the absence of Cl^-, the frog gastric mucosa exhibits a mucosa-positive Ψ_{ms}, and the magnitude of this voltage is proportional to the rate of H^+ secretion. This type of finding led to the idea that the H^+ secretory mechanism was electrogenic.[122] We will discuss below this conflict between the apparent electrogenicity observed in the whole tissue and the neutral H^+/K^+ exchange ATPase of isolated membranes. In addition to the active HCl secretion, most gastric mucosas also actively secrete Cl^-. The active Cl^- secretion is a major determinant of the lumen-negative Ψ_{ms} in most species. Gastric mucosas from reptiles and mammals also actively absorb Na^+, and this transport process contributes to the Ψ_{ms} in these species. We have summarized these active transport processes by presenting data on the unidirectional and net fluxes of H^+, Cl^-, and Na^+ across piglet gastric mucosa in Fig. 13; the specific transport mechanisms will be discussed later in this section. In the following subsection we will abstract the present state of our knowledge in a model for ion transport and membrane conductance in surface cells and oxyntic cells to the gastric mucosa.

6.2. Permeability and Conductances of the Tight Junctions, Apical Membrane, and Basolateral Membrane

6.2.1. Conductance of Surface Cells vs. Glandular Cells

Definition of the conductive pathways is important for a complete characterization of gastric mucosal function, but the anatomical complexity of this tissue, both histologically (i.e., presence of multiple cell types) and ultrastructurally (i.e., enormously folded apical membrane), has made comprehensive analysis difficult.

Rehm and his colleagues[109] made an interesting microelectrode analysis of the K^+ conductances of the two major epithelial cell types of frog gastric mucosa. Their quantitative analysis showed that the surface cell population contributes less than 25% to the total epithelial conductance and the oxyntic cells must therefore contribute at least 75% of the conductive pathway. This is perhaps not too surprising, given the large surface area the gastric glands provide.[11] Recent measurements of the transepithelial impedance of frog gastric mucosa indicate that the tubules may provide as much as 90% of the total tissue conductance.[19]

6.2.2. Tight or Leaky Junctions?

It has been generally assumed that the conductance properties of the gastric mucosa are dominated by those of the cell membranes rather than the paracellular pathways.* And with respect to the ability of the stomach to maintain large gradients of $[H^+]$, the tissue does indeed appear to be a "tight" epithelium according to the designation of Fromter and Diamond.[54] However, like the small intestine, proximal tubule, and gallbladder, the stomach transports large volumes of nearly isotonic fluid (see Refs. 34, 96), and in the stimulated state the transepithelial resistance of the stimulated frog gastric mucosa is as low as those of several "leaky" epithelia in which ion conductance is known to be largely paracellular, i.e., via the junctions. Thus, an evaluation of the magnitude of the paracellular shunt in gastric mucosa is of some interest.

Three different techniques have been used to quantitate the junctional tightness in stomach. Microelectrode studies of *Necturus* gastric mucosa[147] have yielded the estimate that the paracellular pathway contributes less than 25% of the total transepithelial conductance. Measurements of transepithelial fluxes of ions (e.g., Na^+) and nonelectrolytes (e.g., mannitol) that are assumed to move from serosa to mucosa via the junctional pathway have indicated that this paracellular conductance contributes somewhere between 15 and 30% of total tissue conductance in resting mammalian gastric mucosa.[101,152]

Despite the consistency of the data obtained from microelectrode and tracer flux experiments, each technique rests on uncertain assumptions. The microelectrode measurements assume good electrical coupling between surface cells and oxyntic cells. Coupling has been demonstrated among surface cells[147] and among gland cells,[11] but the critical coupling between surface cells and gland cells has never been shown. The tracer experiments suffer from the possibilities that a finite fraction of the flux may penetrate through the cells and/or that the junctions might be Cl^- specific. Thus, confirmation from yet a third technique, impedance analysis, was of some interest.

The basic idea of impedance analysis is to examine and model the tissue's potential response during passage of a constant sinusoidal current at a number of different frequencies.[30] An advantage of AC impedance analysis is that one can obtain measurements of membrane areas in a living preparation. This is obviously of importance in a tissue like the gastric mucosa which exhibits extensive membrane folding. The reasoning here is that capacitances of diverse biological membranes cluster around 1 $\mu F/cm^2$.[20] Hence, a membrane capacitance of 1 μF indicates a true membrane area of 1 cm^2. Application of AC impedance analysis to the gastric mucosa has provided data that are consistent with the morphological complexity; the reason for the low apparent transepithelial resistance is that there is an enormous area of cell membrane packed into every square centimeter of tissue mounted in the *in vitro* chamber.[19] For example, when the transepithelial resistance is normalized to 1 cm^2 of chamber area, the stimulated frog gastric mucosa often attains values as low as 50–90 Ω-cm^2. If, however, the transepithelial resistance is normalized to the area of the serosal membrane (equal to ~ 100 cm^2 serosal membrane/cm^2 chamber area), the values become 5000–9000 Ω-cm^2, values similar to those of frog skin, toad bladder, and rabbit urinary bladder. The conclusion is that although the gastric mucosa has a nominal resistance in the range for leaky epithelia, a correction for the surface area of this highly invaginated glandular epithelium indicates that transtissue resistance values are in the range of other tight epithelia (see Ref. 30).

6.2.3. Conductance of the Cell Membranes to Ions

Accepting the conclusion that the conductance of gastric mucosa is determined by the properties of the cell membranes, one can obtain qualitative information about the permeability properties of the apical and basolateral membranes. The method

*For clarity, let us explicitly distinguish the two types of junctional pathways. The paracellular pathway permits (or fails to permit) ions and perhaps other very small molecules to pass between the luminal and the serosal solution, via the tight junctions and lateral spaces. This pathway has a high conductance in leaky epithelia and a low conductance in tight epithella.[54] The cell-to cell pathway, alias electrical coupling, permits (or fails to permit) ions and other molecules of up to medium molecular weight to pass between adjacent cells, via specialized channels in the gap junctions.

is to manipulate the bathing solution composition while monitoring transepithelial potential difference and resistance. A few workers have used microelectrode measurements which have the advantage that one can measure the electrical properties of the individual membranes directly.

In normal solutions, containing Cl^- and buffered with HCO_3^-/CO_2, the mucosal membrane of amphibian gastric mucosa appears to be predominantly conductive to Cl^-. This conclusion has been reached because changes of $[Na^+]$, $[K^+]$, $[HCO_3^-]$, or pH in the mucosal solution typically cause only small changes of transepithelial potential. In contrast, changing $[Cl^-]$ in the mucosal solution has much larger effects on Ψ_{ms}. However, there is a puzzling aspect to this apparent Cl^- conductance. For example, changes of Ψ_{ms} in response to changes of $[Cl^-]$ in the mucosal solution are much smaller than one would predict for a perfectly Cl^--selective membrane (e.g., averaging 11 mV per decade change in $[Cl^-]$[62,147]). It is possible that changes in bulk mucosal solution $[Cl^-]$ do not produce equivalent changes in $[Cl^-]$ at the apical membrane surface within glandular lumina (e.g., due to Cl^- secretion into this unstirred zone). Thus, the change in Ψ_{ms} would be less than predicted. Microelectrode studies of surface cells have shown that the movement of Cl^- across the apical membranes of surface cells has all the characteristics of a completely passive (i.e., conductive) process.[102] Isolated apical membrane vesicles show conductance properties to both Cl^- and K^+. Future work should be directed at characterizing the oxyntic apical membrane *in situ* by microelectrode analysis.

In addition to this mucosal Cl^- permeability, which seems to be present in the gastric mucosas of all species, the apical membrane of lizard[61] and mammalian[e.g., 101] gastric mucosa also contains a Na^+ conductance. We will discuss this more extensively in the section concerned with active Na^+ transport (Section 6.5). In Cl^--free solutions the mucosal membrane of gastric mucosa exhibits a K^+ conductance which may contribute to the apparent electrogenicity of H^+ secretion (Section 6.3.4).

At the serosal membrane, gastric cells appear to have conductance channels for K^+ and perhaps for Cl^-. In the first experiments designed to quantitate these conductances, Harris and Edelman[62] measured changes in Ψ_{ms} while serosal $[K^+]$ and/or $[Cl^-]$ was altered. They presented data which indicated that the serosal membrane was almost exclusively conductive to K^+ and Cl^- and that the ratio of K^+ to Cl^- conductance at this surface was 2.6. However, in direct measurements of serosal membrane potential with microelectrodes, Spenney *et al.*[147] found that a 10-fold change of serosal $[K^+]$ exerted approximately the same effect on Ψ_{cs} in both Cl^- and Cl^--free solutions. Though these authors did not stress this particular point, these microelectrode data indicate that changes in $[Cl^-]$ have a very small direct effect on the serosal membrane potential, i.e., the Cl^- conductance of this membrane is relatively small, certainly not equal to one-third of the serosal K^+ conductance.

What, then, is the explanation for the findings of many investigators who have shown that changes in serosal $[Cl^-]$ have relatively large effects on Ψ_{ms}[e.g., 42,62]? The answer may be that interactions between membrane conductances and rates of active ion transport can yield data which appear to indicate a Cl^- conductance. For example, as we will discuss at greater length below (Section 6.4.5), it has been proposed for resting tissues that there is some sort of feedback between rates of neutral Cl^- accumulation across the serosal membrane and the magnitude of K^+ conductance at this membrane. If there were such feedback, then changes of serosal $[Cl^-]$ might elicit, through indirect cellular effects, changes of K^+ conductance and concomitant changes in Ψ_{ms} which would be indistinguishable from those expected for a serosal Cl^- conductance. Such feedback may also include simultaneous effects on the mucosal membrane conductance to Cl^-.

6.2.4. Electrical Potential Profile of Surface Cells and Oxyntic Cells

Measurement of the electrical potential across the membranes of surface cells with microelectrodes has consistently shown that the intracellular voltage is negative with respect to both mucosal (Ψ_{mc}) and serosal (Ψ_{sc}) solutions. The magnitudes of the potentials measured have depended on the species used, the secretory state of the tissue (i.e., resting or stimulated), and the tip resistance of the microelectrodes used by the particular investigators (e.g., see Refs. 14, 102, 135, 147). Taking all these factors into account, it appears that surface cells have $\Psi_{mc} = -35$ to -55 mV and $\Psi_{sc} = -55$ to -75 mV. Although there are often large changes of transepithelial potential during gastric stimulation (up to 30 mV), there have been no systematic studies to investigate the locus of these changes.

Measurement of the potentials across the membranes of oxyntic cells in the intact tissue has proven difficult due to the geometrical configuration of the gastric tubules. There has been only one published report[155] on this subject, and the measurements have not been systematically reported. Because of the technical difficulties in making the measurements, we must conclude that the potential profile of the oxyntic cells is still uncertain (see also Ref. 109). Resolution of this problem will have important consequences for such issues as possible electrical coupling between surface cells and oxyntic cells and the electrical driving forces on K^+, Na^+, H^+, and Cl^- across the membranes of resting and stimulated oxyntic cells.

6.2.5. Changes in Gastric Membranes and Tissue Resistance during Stimulation

Stimulation of the gastric mucosa induces large changes in the transport and ultrastructural characteristics of oxyntic cells (Section 2.2). The most characteristic and consistent change in the electrophysiology is a large decrease in the transepithelial resistance (R_t). For example, in the well-studied frog gastric mucosa, R_t often decreases from about 600 Ω-cm^2 to 100 Ω-cm^2 or less. The cause of this large drop in tissue resistance has been the subject of some debate. On the one hand, such a decrease in R_t is consistent with the activation of electrogenic H^+ and Cl^- pumps which can contribute to the tissue's conductance properties (e.g., see Ref. 117). On the other hand, it has been proposed that the large increase in apical membrane area of oxyntic cells which also occurs during stimulation, could give rise to a large increase in tissue conductance even though the specific resistance properties of the cell membranes had not changed.[51]

Recent experiments utilizing transepithelial impedance analysis of frog gastric mucosa have yielded information which favors the latter possibility.[19] It was demonstrated that the drop in tissue resistance was indeed caused by an increase in conductance of the apical membrane, but this increase in conductance could be accounted for by an increase in area of the apical membrane. The resistance of the serosal membrane was relatively independent of the H^+ secretory state. Put another way, when the membrane conductances were normalized to actual membrane area (by the measurements of membrane capacitances equivalent to membrane areas), it turned out that H^+ secretion was not associated with a change in specific ionic conductance (change in conductance per unit area) at the apical membrane, as would be predicted by the so-called electrogenic hypothesis. Thus, these data indicate that the change in R_t during gastric stimulation can be fully explained by the accompanying proliferation of apical membrane area observed in electron micrographs. It should be added, though, that the demonstrated close correlation between apical membrane conductance and area holds only for the resting/stimulated transition of tissues bathed in normal Ringer's solutions. There are other situations for which there are changes in tissue conductance with no apparent change in membrane proliferation, e.g., when H^+ secretion was inhibited by SCN^- solutions.[8] Rehm *et al.*[119] have recently suggested that changes in volume flow with secretory state might alter glandular luminal diameter and hence the series resistance of solution with the luminal spaces. Perhaps impedence studies of those variant states may serve to distinguish whether the conductance changes can be ascribed to quantitative (membrane area) or qualitative (membrane properties) ones.

6.3. H^+ Transport

6.3.1. Comparison of Isolated Membranes with Intact Tissue

The mechanism of acid secretion has now been fairly well established to involve H^+,K^+-ATPase which operates in parallel with a system for K^+ and Cl^- transport (Section 5). K^+ and Cl^- move from cell cytoplasm to gland lumen and the secreted K^+ is then recycled directly back into the cell in a neutral exchange for H^+ as ATP is hydrolyzed. Thus, there are obvious direct roles for K^+ and Cl^- in H^+ secretion. In addition, because K^+ is accumulated by the operation of the Na^+,K^+-ATPase of gastric cells, Na^+ also plays an important role. Finally, during the secretion of H^+ into the stomach, there is a concomitant cellular production of OH^- and neutralization by combination with CO_2 to form HCO_3^-, and thus CO_2 and HCO_3^- also assume central roles in the H^+ secretory process. The main purpose of this section is to review the involvement of these cations and anions in the H^+ secretory process of oxyntic cells. Since much of the early work on gastric mucosa indicated that H^+ secretion was generated by an electrogenic process, we will also discuss this possibility.

6.3.2. Cation Dependence of H^+ Secretion

Although the gastric mucosa can secrete HCl in the absence of K^+, Na^+, and Ca^{2+} in the mucosal solution, the tissue requires all three of these cations in the serosal solution for normal function. In keeping with the ideas gleaned from isolated membrane studies, K^+ seems to play the most direct role in maintenance of H^+ secretion. K^+ also seems to be important for ultrastructural transformations in the oxyntic cell during secretagogue-induced stimulation.[95]

A number of studies could be cited that bear on the question of the Na^+ and K^+ requirements for H^+ secretory activity; however, the recent work of Koelz *et al.*[85] has provided the most comprehensive picture of the interdependencies. These investigators studied the ability of the isolated gastric gland preparation to produce acid, as measured by the uptake of the weak base aminopyrine (AP), as a function of variations of the intra- and extracellular $[K^+]$ and $[Na^+]$. They found that K^+ was essential at some membrane site for H^+ production (most likely the H^+,K^+-ATPase) and that the intracellular $[K^+]$ required for half-maximal stimulation of AP uptake was about 10 mM for histamine-stimulated glands. Cellular Na^+ was frankly inhibitory to AP uptake. Both ouabain and Na^+-free solutions inhibit acid secretion because the Na^+,K^+-ATPase can no longer operate to maintain cellular $[K^+]$ at sufficiently high levels. Thus, the role of the basolateral Na^+ pump enzyme can be seen to play an important, albeit indirect, role in gastric H^+ secretion.

The possible role of Ca^{2+} as a cellular second messenger in stimulus–secretion coupling has been discussed above. Ca^{2+} is also important for the maintenance of the integrity of the tight junctions which bind the epithelial cells togther.[139] The threshold for this effect is somewhere between 10^{-5} and 10^{-4} M Ca^{2+}. At $[Ca^{2+}] < 10^{-5}$ M, the tight junctions are disrupted and tissue resistance falls to zero. An interesting aspect of the role for Ca^{2+} is that Ca^{2+}-free solutions cause nearly immediate disruption of the tight junctions in resting tissues while the junctions of stimulated tissues are much more resistant to such Ca^{2+}-free treatment.[94] It would be of obvious teleological advantage for the tight junctions of the stomach to become more stable during stimulation of H^+ secretion.

6.3.3. Anion and HCO_3^-/CO_2 Dependence

As noted for the actions above, major changes in the anionic composition of the mucosal solution do not affect rates of H^+ secretion. However, Cl^--free serosal solution causes large (> 75%) decreases in H^+ secretion in amphibian gastric mucosa[29,31,42] and complete cessation of H^+ secretion in that of the mammal.[1,49] This dependence on Cl^- is not absolute—a variety of other anions can substitute for Cl^-. The relative abilities of different anions to support H^+ secretion by frog gastric mucosa are: Br^- (1.05) > Cl^- (1.00) > I^- (0.87) > ClO_4^- (0.66) > NO_3^- (0.12) ≈ isethionate (0.12) > BrO_3^- (0.04).[31,97] This anion requirement might relate to the operation of the KCl transport system at the apical membrane and/or to a Cl^-/HCO_3^- exchanger at the basolateral membrane.

Concomitant with the secretion of H^+ at the apical membrane, an equivalent amount of base is necessarily formed within the cell. Recent experiments indicate that during gastric stimulation, intracellular pH rises by 0.5–1.0 pH unit.[35,69] Davies[26] has shown that the OH^- is neutralized by the CO_2/HCO_3^- system, most likely via a carbonic anhydrase-catalyzed reaction (see also Ref. 50). The HCO_3^- generated in the reaction of CO_2 with OH^- is liberated at the serosal side of the cell; one HCO_3^- reaches the blood (the so-called "alkaline tide") for each H^+ secreted into the stomach lumen.[150] Under conditions in which rates of H^+ secretion exceed the rate of CO_2 production from metabolism, the mammalian gastric mucosa takes up CO_2 from the blood. In the absence of exogenous CO_2, normal rates of H^+ secretion can be maintained by reducing the pH of the serosal solution to approximately pH 4.[131] A CO_2-recycling (between cell and serosal unstirred layer) scheme has been proposed to account for these findings.

Acetazolamide, an inhibitor of carbonic anhydrase, causes severe reduction of HCl secretion in mammalian preparations, but variable effects have been observed with amphibian gastric mucosa. These findings have been reviewed by Forte and Solberg,[50] and fit with the conclusions that carbonic anhydrase activity is absolutely required to catalyze the hydration of CO_2 at high secretory rates. Low or residual secretory rates can be accounted for by uncatalyzed CO_2 utilization, or even by residual carbonic anhydrase activity. For frog gastric mucosa, both Hogben[73] and Durbin and Heinz[33] have observed that acetazolamide causes a much more pronounced reduction in I_{sc} than in H^+ secretion. This raises the possibility that the effects of carbonic anhydrase inhibitors on active Cl^- transport may be mediated by their action on a specific Cl^- receptor,[73] but an indirect effect via changes in intracellular pH should not be ignored (see below). Acetazolamide has very little effect on I_{sc} in the resting frog or mammalian gastric mucosa (Ref. 33; Machen and Forte, unpublished), or in actively secreting tissues in which the gassing phase is 100% O_2 instead of the usual 95% O_2–5% CO_2.[140]

Once HCO_3^- is formed in the oxyntic cell, it most likely moves via a neutral nonconductive exchange for Cl^- across the serosal membrane, thereby serving to maintain cellular acid–base balance at the mucosal membrane as well as provide the cell with Cl^- for operation of the required apical KCl transport. Rehm[118] was the first to offer evidence for neutral HCO_3^-/Cl^- exchange across the serosal surface of the oxyntic cell (see also Ref. 121), which is now generally acknowledged as an important exchange mechanism for other cell types (e.g., red cells).

6.3.4. Is H^+ Secretion Electrogenic?

The electrogenic hypothesis for H^+ (and Cl^-) secretion was originated and developed by Rehm (Refs. 115, 116; see also Ref. 122). According to this formulation, these transport systems operate in parallel in the mucosal membranes of gastric cells, each with their own EMFs and internal resistances. It has been found recently that the active step for Cl^- secretion, at least for surface cells, occurs at the serosal membrane,[102] and the electrogenicity of active Cl^- transport may relate to the passive, conductive movement of Cl^- through the mucosal membrane and of K^+ through the serosal membrane (e.g., see Ref. 105). With regard to the electrogenicity of the H^+ secretory system, Rehm and his co-workers accumulated a large body of data which appeared to support the model (reviewed previously by Rehm and Sanders[122] and Machen and Forte[96]). For example, in perhaps the most convincing experiments, Cl^--free solutions caused the transepithelial potential difference of frog gastric mucosa to invert its normal orientation, becoming mucosa positive, and changes in H^+ secretory rates (induced by a variety of inhibitors) were linearly related to Ψ_{ms}.[e.g., 120,121] Thus, the case for electrogenic H^+ transport has appeared to be fairly solid. How, then, do we rationalize such data with the studies on isolated membranes which show clearly that the H^+,K^+-ATPase is neutral?

One possible explanation may lie in the fact that the mucosal membranes of oxyntic cells contain conductance channels for K^+ which become particularly evident in Cl^--free solutions (e.g., see Refs. 28, 167, 169). It is therefore possible that H^+ secretion is always a neutral (H^+/K^+ exchange) process and that the electrogenicity of H^+ secretion observed in the intact tissue is only apparent.[130,169] In Cl^--free solutions the apical K^+ conductive pathway becomes manifest. The apparent electrogenic H^+ secretion may then be provided by the conductive cell-to-lumen movement of K^+, which is recycled back into the cell via the neutral ATPase. This recycling of K^+ must be extremely efficient, for net K^+ secretion by the intact tissue is very small.[72]

6.4. Cl^- Transport

6.4.1. Active Transepithelial Transport

In most of the early theories of the mechanism of gastric acid secretion, it was assumed that the Cl^- ions followed the secreted H^+ ions passively so as to maintain electroneutrality.[27,65] It has now been established that Cl^- movement through the gastric epithelium occurs by three distinct processes: active transport, exchange diffusion, and simple ionic diffusion. (For a discussion of these pathways and some of the influencing conditions in frog mucosa, see Forte.[41]) Hogben[72] demonstrated that with identical bathing solutions on each side of frog gastric mucosa and no electrical potential difference (short-circuit), there was a net secretion of Cl^- from the serosal to the mucosal solution. He recorded unidirectional fluxes of Cl^- and showed that the short-circuit current (I_{sc}) was equal to net transport of Cl^- (J^{Cl}_{net}) in excess of that secreted as HCl, i.e.,

$$I_{sc} = J^{Cl}_{net} - \text{HCl production} = J^{Cl}_{net} - J^{H}_{net}$$

A more recent study of Forte and Machen[49] has shown analogous relationships for both resting and stimulated mammalian mucosa (piglet) with the added feature of a component of net Na^+ absorption in gastric preparations from the mammal. As the flux data of Fig. 13 show for the short-circuited piglet mucosa, there is a net secretion of Cl^- (J^{Cl}_{net}) from serosal to mucosal solution and a net absorption of Na^+ from mucosal to serosal solution (J^{Na}_{net}). J^{Cl}_{net} was increased about threefold in going from "resting" to histamine-stimulated states, while J^{Na}_{net} was somewhat reduced. Of course, J^{H}_{net} was greatly stimulated by histamine. The algebraic summation of these ionic flows was essentially equivalent to the measured I_{sc}. Thus, for the mammalian preparation:

$$I_{sc} = J^{Cl}_{net} - J^{H}_{net} + J^{Na}_{net}$$

In resting tissues, i.e., when $J^{H}_{net} = 0$, the gastric mucosa still secretes Cl^-, at least in the short-circuited state, such that these above equations still hold.[49,98,102,136]

This Cl^- secretion manifests itself as a negative I_{sc} directed toward the lumen and is the primary reason that the stomach lumen is electrically negative with respect to the blood in both resting and stimulated states. This negative potential tends to counter active Cl^- transport such that there is little, if any, net Cl^- secretion in the open-circuited, resting tissue.[49,102] And even in stimulated tissues, net Cl^- secretion is approximately equal to H^+ secretion such that the excess Cl^- transport is not expressed.[49,72] The role of this excess Cl^- transport which can really only be observed in the (unnatural) short-circuited state is not presently clear.

This current-generating transport of Cl^- which the stomach secretes in excess of neutral HCl and even in the absence of any H^+ secretion has been termed "nonacidic" Cl^-. This is in contrast to the "acidic" Cl^- which appears to accompany the secreted protons quite specifically (see Ref. 97). Acidic Cl^- transport is contributed by the oxyntic cells,[e.g., 165] while nonacidic Cl^- transport appears to be generated both by oxyntic cells and by surface cells.[102,105] The mechanisms of Cl^- transport at the serosal membrane may be quite similar in the two cell types (e.g., Cl^-/HCO_3^- exchange and/or NaCl cotransport). At the mucosal membrane, Cl^- appears to move purely passively and conductively through both surface cells[102] and oxyntic cells.[165]

6.4.2. Exchange Diffusion

A significant component of the isotopically measured flux of Cl^- across gastric mucosa has been shown to occur by the phenomenon of exchange diffusion, namely a one-for-one neutral exchange of Cl^- on one side of the tissue for Cl^- on the other side. This was first pointed out by Hogben[72] through his analysis of flux data. The partial ionic conductance due to Cl^- (G_i) can be calculated from the equation developed by Ussing[154] and Hodgkin[71]:

$$G_i = \frac{z^2F^2}{RT} \cdot J_i$$

where J_i is the unidirectional flux of the ion (in this case Cl^-) and z, F, R, and T have their usual meanings. Hogben found that the calculated partial ionic conductance of Cl^-, using this equation and the passive mucosal-to-serosal Cl^- flux, was greater than the measured total transepithelial conductance. Thus, he concluded that a significant fraction of Cl^- movement across the gastric epithelium must occur via an uncharged, nonconductive combination of Cl^- with some membrane carrier.

Independent demonstrations of this proposed exchange diffusion for Cl^- were provided by Heinz and Durbin[64] and Forte,[41] who showed that the unidirectional flux of isotopic Cl^- (from serosa to mucosa) was dependent on the presence of Cl^- in the *trans* solution. Thus, during short-circuit conditions, replacement of Cl^- in the mucosal solution with a variety of anions and nonelectrolytes caused serosal-to-mucosal flux of Cl^- to decrease. Heinz and Durbin modeled and characterized the system as an ion exchange operating through some mobile membrane carrier.

The components of Cl^- flux across bullfrog gastric mucosa were further studied by Forte,[41] who characterized the fractional flux occurring by active transport, passive diffusion, and exchange diffusion. Since these and other studies[49] showed some correlation between the magnitude of acidic Cl^- transport and Cl^- exchange diffusion, it was suggested that these processes may be occurring in the oxyntic cell, even possibly using the same membrane carrier. Cl^-/Cl^- exchange diffusion is very small[49] or absent[102] in resting tissues.

Recent experiments indicate that exchange diffusion may be operating at both apical and basolateral membranes of oxyntic cells. Thus, apical membranes isolated from stimulated oxyntic cells exhibit rapid Cl^-/Cl^- exchange.[166] And electrophysiological experiments have indicated that Cl^-/HCO_3^- (and, most likely, Cl^-/Cl^-) exchange also occurs at the serosal membrane of oxyntic cells.[121]

6.4.3. Nonacidic Cl^- Secretion by Surface Cells: Active Accumulation and Passive Efflux

Until recently, it was assumed that active Cl^- secretion by gastric mucosa was localized to the mucosal membrane of gastric epithelial cells.[e.g., 122] However, indirect measurements showing that cellular $[Cl^-]$ = 30–60 mM,[25,62] coupled with microelectrode measurements of serosal membrane potentials of −50 to −70 mV,[135,147] gave indications that the "active" step for Cl^- transport was more likely located at the serosal membrane. This idea has been confirmed by direct measurements of $[Cl^-]$ and membrane potentials of surface cells in resting *Necturus* gastric mucosa.[102] Thus, during both open- and short-circuit current conditions, cellular $[Cl^-]$ was found to be about 29 mM, and Cl^- was out of equilibrium (too high) across the serosal membrane by about 20 mV (Fig. 14). During open-circuit conditions, the chemical potential for Cl^- and the electrical potential across the mucosal membrane were approximately equal (i.e., Cl^- was in approximate equilibrium), and, therefore, there was little, if any, net movement of Cl^- across this membrane. This result explains the findings mentioned above that during open-circuit conditions when transepithelial potential difference = 20–40 mV, net Cl^- secretion in excess of H^+ secretion is zero.[49,102] During short-circuit current conditions (i.e., transepithelial voltage clamped to zero), Cl^- was found to be distributed out of equilibrium across both membranes: an active accumulation of Cl^- must be present at the serosal membrane; the negative intracellular potential could then tend to drive Cl^- out the cell into the mucosal solution.

It was also possible from the experiments of Machen and Zeuthen[102] to measure the permeability of the mucosal membrane to Cl^- (P_m^{Cl}). Thus, it was estimated that P_m^{Cl} was sufficiently large so that the negative intracellular potential could drive Cl^- out of the cell passively at normal rates of transport. Neither facilitated nor active transport mechanisms need be invoked to explain the movement of Cl^- across the mucosal membrane. Because the mucosal membrane of gastric mucosa is predominantly conductive to Cl^-, it may be that the passive movement occurs through Cl^- conductance channels similar to those observed in other Cl^--secreting epithelia (e.g., corneal epithelium[83]).

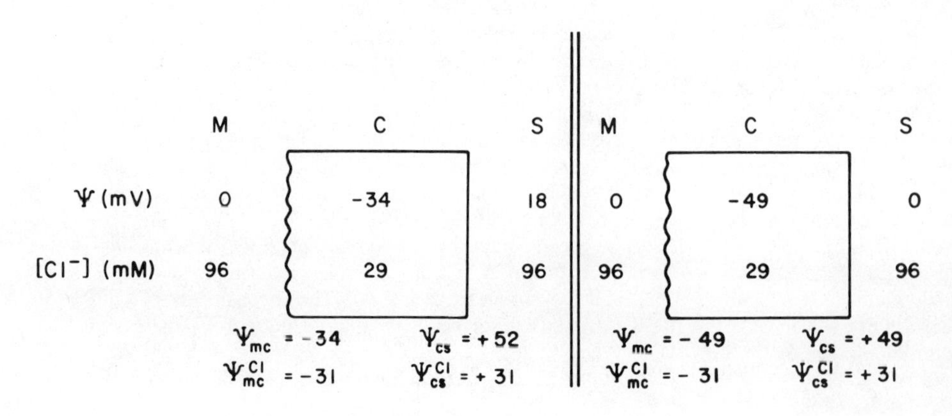

Fig. 14. Membrane potentials, Cl^- concentrations, and chemical potentials for Cl^- in surface cells of *Necturus* gastric mucosa during open-circuit and short-circuit conditions. Membrane potentials (ψ) were measured with microelectrodes with reference to the mucosal (M) solution; $[Cl^-]$ was measured with Cl^--sensitive microelectrodes. The chemical potential for Cl^- (ψ^{Cl}) was calculated as the Nernst equilibrium potential. Data from Machen and Zeuthen.[102]

6.4.4. Na^+–Cl^- Cotransport or Parallel Na^+/H^+ and Cl^-/HCO_3^- Exchangers?

It has been proposed for many epithelia,[53] including the gastric mucosa,[105] that net Cl^- secretion is generated by first being "actively" accumulated across the serosal membrane and then passively "leaking" across the mucosal membrane. According to this model, Cl^- is accumulated "uphill" to a level above its equilibrium by being coupled to the "downhill" movement of Na^+ from the serosal solution to the cell. Although this entry step is purportedly electrically neutral, direct evidence of electroneutral transfer has only been provided for the shark rectal gland.[36] Once accumulated, the Cl^- is then driven across the mucosal membrane by the negative intracellular potential. Na^+, on the other hand, is recycled out of the serosal membrane by the Na^+,K^+-ATPase, which thereby maintains the Na^+ gradient across this membrane. An interesting aspect of this model is that the transepithelial short-circuit current is generated initially by a neutral accumulation; current is carried across the mucosal membrane by Cl^-, but across the serosal membrane by K^+ moving from cell to serosal solution.

We have just discussed the hypothesis that Cl^- is indeed "actively" accumulated across the serosal membrane of surface cells and that Cl^- appears to move passively and conductively across the mucosal membrane. In addition, several other steps in this scheme have been tested for a variety of gastric mucosal preparations: (1) Na^+-free serosal solutions (replaced by choline or tetramethylammonium) cause rapid and reversible abolition of net Cl^- secretion and I_{sc}.[98,152] (2) Treatment of the mucosal membrane of resting frog gastric mucosa with the cation ionophore, nystatin, opens this membrane to Na^+ and allows Na^+ to "flood" the cell. Under these conditions, the Na^+ gradient across the serosal membrane is most likely reduced, and net Cl^- secretion is abolished.[99] (3) Ouabain, the well-known inhibitor of the Na^+,K^+-ATPase, causes I_{sc} and, presumably, net Cl^- secretion to be reduced to zero.[49,101,152] (4) K^+-free solutions (which also inhibit the Na^+,K^+ pump) also appear to reduce Cl^- secretion.[101] (5) Depolarization of the negative internal potential[147] by high serosal $[K^+]$ also causes reductions in net Cl^- secretion and I_{sc}.[74,98,105] The simplest interpretation here is that the driving force to move Cl^- from the cell to the mucosal solution is reduced by the high serosal $[K^+]$. (6) Blockade of serosal K^+ channels by Ba^{2+} reduced Cl^- secretion and I_{sc} with no effect on H^+ secretion.[105] The implication is that Ba^{2+} acts by blocking K^+ channels and depolarizing the cell membrane potential. Both of these effects would tend to reduce net Cl^- secretion across the mucosal membrane of gastric cells.

Thus, there is a great deal of evidence which fits with the general concept of NaCl cotransport. However, it is difficult to distinguish between this specific model and one in which Na^+/H^+ and Cl^-/HCO_3^- exchangers operate in parallel. In the former case, Cl^- entry is coupled directly to the "downhill" movement of Na^+ into the cells. In the parallel exchanger model (Fig. 15), the Na^+ gradient is used to drive H^+ out of the cell, leading to an elevation of both cellular pH and $[HCO_3^-]$ above their passive, electrochemical equilibrium levels. The HCO_3^- gradient then becomes the proximal driving force for accumulation of Cl^-.

The argument over cotransport vs. parallel exchangers has been discussed and reviewed at some length for tissues like the proximal tubule, small intestine and gallbladder which absorb Na^+ and Cl^- (e.g., see Refs. 82, 90, 91, 158, 159). The issue has not been raised for Cl^--secreting tissues (with the exception of the shark rectal gland; see Ref. 36), at least partially because the isolation of basolateral membranes has proven more difficult than isolation of apical membrane vesicles from the intestine, renal tubules, or gastric mucosa. At the present time, we favor the parallel exchanger model for the gastric mucosa, primarily because both $[HCO_3^-]$ and pH of the serosal solution have large effects on nonacidic Cl^- transport (see Ref. 49).

Recent experiments from two separate laboratories have shown for resting tissue that I_{sc} and net Cl^- secretion are both reduced nearly to zero when the serosal solution is changed, at constant pH, from one containing 18 mM HCO_3^- (and 5% CO_2) to a HCO_3^-- and CO_2-free solution.[104,136] In addition, this Cl^- secretion is strongly dependent on the pH in both CO_2-containing and CO_2-free bathing solutions. These effects were only observed by changing the $[HCO_3^-]$ or pH of the serosal solution; Cl^- transport was relatively insensitive to changes or pH or $[HCO_3^-]$ in the mucosal solution.[104]

It has been argued that the bulk of the physiological data are most compatible with the parallel exchanger model.[49] However, the data at present do not allow us to exclude the possibility that neutral NaCl cotransport and the parallel exchange of Na^+/H^+ and Cl^-/HCO_3^- are all operating in the serosal membranes of gastric cells. The possibility even exists that neutral Na^+–K^+–$2Cl^-$ cotransport like the observed in duck red cells[60] and flounder intestine[107] is present. Suffice it to say that the active accumulation of Cl^- across the serosal membrane of gastric cells appears to be dependent on the cell-to-serosal concentration gradients of Na^+, HCO_3^-, H^+, and CO_2 as well as on the activity of the Na^+,K^+-ATPase at the serosal membrane. Movement of Cl^- across the mucosal membrane appears to depend on the permeability/conductance of this membrane and the transmembrane voltage. As we will now discuss, the magnitude of the K^+ conductance at the serosal membrane is also a critical determinant of rates of nonacidic Cl^- secretion.

6.4.5. Possible Interactions between Neutral Exchangers and Membrane Conductances

Despite the fact that we have discussed only neutral accumulation of NaCl across the serosal membrane, we have no direct evidence that the system is electrically neutral. In fact, it has been variously proposed that perhaps the NaCl "carrier" exhibits a net charge, because of unequal numbers of Na^+ and Cl^- ions moving together (see Refs. 15, 98). Such a charged system might even be expected to be conductive, which would help to explain the fact that I_{sc} and conductance are closely related in this tissue.[97] However, the neutral mechanisms are so well established for a variety of other systems that it seems unlikely that the stomach would be an exception.

Another possibility, as mentioned above, may be that the activity of the neutral exchangers somehow affects the magnitude of the predominant membrane conductances: Cl^- conductance at the mucosal side and K^+ conductance at the serosal border. Accordingly, one might postulate that increases of neutral NaCl uptake may be reflected in increases in these specific ionic conductances, e.g., as a result of regulation by a cytoplasmic component that "senses" the rate of NaCl entry (Fig. 15). Possible "sensors" might be pH_c or $[Ca^{2+}]_c$.

A similar circumstance arises in Na^+-transporting tight epithelia, where basolateral Na^+ transport somehow controls apical Na^+ conductance and basolateral K^+ conductance (see Ref. 138). The former control involves negative feedback: decreased basolateral Na^+ extrusion tends to raise intracellular

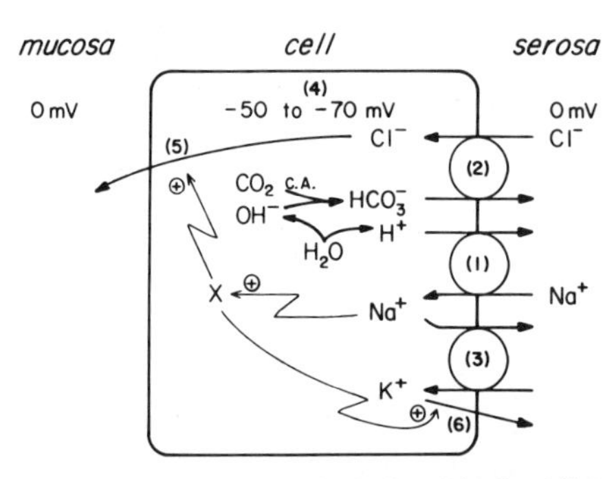

Fig. 15. Parallel exchanger model, including "feedback" for active, nonacidic Cl^- transport in gastric mucosa. The model accounts for energy-dependent accumulation of Cl^- across the serosal membrane and passive efflux across the mucosal membrane into the mucosal solution. In the short-circuited state depicted here, secretion of Cl^- ($I_{sc} = J_{net}^{Cl}$) is dependent on the activity of the parallel (1) Na^+/H^+ and (2) Cl^-/HCO_3^- exchangers, (3) the Na^+,K^+-ATPase to maintain the Na^+ gradient across the serosal membrane, and (4) the intracellular potential to drive Cl^- across the mucosal membrane through (5) the Cl^- conductance channel. C.A., carbonic anhydrase. Current across the serosal membrane is carried by (6) K^+ movement through conductance channels. The membrane conductances (5 and 6) may be controlled by some cytoplasmic factor (X) which senses the rate of NaCl entry. With increasing rates of NaCl entry, X increases, thereby eliciting increases in the conductance pathways. The nature of X is unknown, but a likely candidate is cell [Ca^{2+}].

[Na^+], suppressing apical Na^+ conductance, perhaps through secondary changes in intracellular Ca^{2+} produced by Na^+/Ca^{2+} exchange dependent on cell Na^+.[16,57] In gastric mucosa (and other Cl^--secreting tissues), the effector might be the same, but the direction of the postulated feedback would be opposite to that in the Na^+-absorbing epithelia: increases in Na^+ (and Cl^-) entry would increase cellular [Ca^{2+}] but this increase in [Ca^{2+}] would then increase Cl^- conductance of the mucosal membrane (as observed in colon[52]) and also K^+ conductance at the serosal membrane (e.g., through a Gardos-like effect). A similar scheme utilizing changes in cellular pH could easily be devised. Such a scheme would allow gastric cells to secrete variable amounts of Cl^- (e.g., in response to nervous or hormonal stimuli) without overloading themselves with K^+

6.5. Na^+ and K^+ Transport

Although the principal ionic secreting product of the stomach is HCl, the operation of active transport processes for both Na^+ and K^+ is of enormous functional importance in this (and most every) tissue. We have already discussed the active K^+ transport associated with the K^+,H^+-ATPase of the apical membranes of oxyntic cells. In normal, intact tissue, cellular [K^+] must be maintained high (~ 100–120 mM; see Refs. 85, 135). to provide a substrate for the apical KCl transport system and subsequent recycling of K^+ on the ATPase. Conversely, cellular [Na^+] must be maintained at a low level. Direct measurements have shown that [K^+] is maintained in surface cells at a level above its equilibrium voltage,[135] and because there are K^+ conductance channels in the serosal membranes of these cells,[147] there must be a constant leak of K^+ out of the cells, most likely in exchange for Na^+. The ubiquitous Na^+,K^+-ATPase of the serosal membrane appears to serve the function of recycling leaked K^+ into the cell and Na^+ out of the cell.

In addition to its role in recycling Na^+ and K^+ at the serosal membrane of gastric cells, the Na^+,K^+-ATPase also appears to take a direct part in the transepithelial transport of Na^+ in reptiles[61] and mammals.[e.g., 49,101,152] Flemstrom[37] suggested that the Na^+ transport observed for *in vitro* mammalian preparations was the artifactual result of tissue hypoxia; however, active Na^+ absorption by gastric mucosa is also observed *in situ*.[13,49] The exact percentage that Na^+ contributes to the I_{sc} depends on the particular species and whether the tissue is in the resting or actively secreting state (e.g., see Fig. 13).

The gastric epithelial cell type that effects net Na^+ transport is most likely the surface cell. Fetal rabbit stomach, which has not yet developed oxyntic cells or chief cells (these develop relatively late during gestation), can still actively absorb Na^+.[168] The physiological role of this Na^+ transport is unknown, though it might be speculated that it has something to do with the ability of the surface cells to withstand an acid load, i.e., to help maintain the barrier function of the stomach.

The gastric Na^+ transport system has much in common with those in other "tight" epithelia such as frog skin and urinary bladder. For example, Na^+ enters the cells through specific Na^+ conductance channels that are reversibly blocked by amiloride, the K^+-sparing diuretic.[101,152] Moreover, the Na^+ entry step in gastric mucosa and other tight epithelia exhibits very similar dependencies on the pH and Cl^- content of the bathing solution (e.g., cf. Refs. 101, 137, 153). In view of these similarities, it is interesting to note that Palmer[110] has found that H^+ ions are very permeable through the amiloride-sensitive Na^+ channels of toad bladder. If, as seems likely, the same is true for Na^+ channels of mammalian gastric mucosa, the surface cells may, under some circumstances, be particularly susceptible to luminal acidity.

It is likely that the modes of Na^+ transport in gastric mucosa and in other tight (like the urinary bladders and frog skin) and moderately tight (like the colon) epithelia are identical. It has now been generally accepted that the basic aspects of the model of Koefoed-Johnsen and Ussing[84] for Na^+ transport are correct. According to this model, Na^+ leaks into the cell through Na^+ conductance channels[93] down a concentration and electrical gradient; cellular Na^+ is pumped out across the serosal membrane on the Na^+,K^+-ATPase. Recent modifications of the original model have included discussions of feedback mechanisms which we mentioned in Section 6.4.5. In all other Na^+-transporting epithelia, aldosterone is an effective stimulant of net Na^+ transport (see Ref. 138). The hormonal control of net Na^+ transport by gastric mucosa has not been investigated.

6.6. Alkaline Secretion

In addition to H^+, Cl^-, Na^+, and K^+ transport described above, it had been inferred by early workers that other gastric ion transport mechanisms also existed. Hollander[75] suggested that the stomach was capable of secreting a Na^+-containing nonacidic fluid, though the cellular (or, even, paracellular) mechanism was unknown. If such an alkaline secretion did exist, this could be "trapped" at the mucosal surface in the unstirred layer provided by the mucus, the [H^+] at the surface would be reduced, and a protective mechanism for the surface cells would be provided. And, indeed, it has recently been found using pH microelectrodes that even when the bulk solution is highly acidic, (pH 2.3), the surface of the mucus-secreting cells is about pH 7.3.[161] Metabolic inhibitors and acetylsalicylate abolish this surface pH gradient.[125] Thus, alkaline secretion is a clinically relevant process which has recently attracted great interest. A review on this subject has appeared.[39]

Both the fundus and the antrum secrete small amounts of base (J^{OH}). When fundic H^+ secretion is completely abolished by treatment with either histamine antagonists or thiocyanate, J^{OH} occurs at about 1/10th the rates of maximal H^+ secretion. Careful experiments have shown that J^{OH} is the result of some active transport process and that this transport is affected by a wide variety of drugs, inhibitors, and stimulants of transport.[38,39,40,148] Because of the similarities between the properties of alkaline secretion by the antrum (which has surface cells but no oxyntic cells) and by the fundus, it has been assumed that the surface epithelial cells are the sources of active HCO_3^- secretion in both antrum and fundus.

The ionic mechanisms involved in this alkaline secretion have not been well established, and experimental results are sometimes conflicting. For example, Cl^-/HCO_3^- exchange at the apical or basolateral membrane has been variously proposed as the source of the alkaline flow (cf. Refs. 39, 148). Whatever the final explanation for the experimental differences, it appears that there may be some close relationship between the current-generating Cl^- transport system and the alkaline secretion. Thus, when $[HCO_3^-]$ in the serosal solution is increased, both J^{OH} and I_{sc} increase in parallel.[148] Also, Na^+-free serosal solutions reduce both J^{OH} and net Cl^- secretion. This latter effect is more pronounced when the mucosal solution is Cl^--free. To account for such findings, Takeuchi *et al.*[148] have proposed that alkaline secretion is dependent on the presence of HCO_3^- and Na^+ in the serosal solution, and also on an adequate cellular $[Cl^-]$.

We can speculate (based on the absence of a plasma membrane HCO_3^--stimulated ATPase in this tissue) that the active alkaline secretion is most likely the result of secondary active transport. Thus, active HCO_3^- secretion would be coupled to the movements of other ions, similar to the situation for Cl^-. One possibility involves the operation at the serosal membrane of Na^+/H^+ (and Cl^-/HCO_3^-) exchange and at the mucosal membrane an anion conductance channel. According to such a scheme, Na^+/H^+ exchange at the serosal membrane alkalinizes the cell interior, and HCO_3^- could then move across the mucosal membrane through channels which are predominantly conductive to Cl^- but which may also be permeable to HCO_3^-. Understanding these relationships among Na^+, Cl^-, and HCO_3^- transport will likely be very important for our ultimate understanding of the stomach's barrier function in normal circumstances and also its breakdown and ulceration in diseased conditions.

7. Summary

In this chapter, we have shown the enormous functional and structural responsiveness of the gastric mucosa in its capacity to transport H^+ and Cl^-. While HCl is the principal ionic secretory component, its interdependence with the specific pathways and mechanisms for the transport of Na^+, K^+, and HCO_3^- is now well established. A schematic overview of the various ion transport properties in the gastric epithelium is given in Fig. 16, where we have attempted to summarize the nature and location of the ionic pathways in the oxyntic cell and surface cell.

There are only two direct ion-pumping systems that have been clearly identified in the gastric epithelial cells. The Na^+,K^+-ATPase is present in the basolateral membrane of all the cells, and in especially high density in oxyntic cells. This is the classical Na^+ pump enzyme that maintains low intracellular $[Na^+]$ and high $[K^+]$. As in many other tissues, these ion gradients are employed by the gastric cells for additional, secondary, transport functions. The other primary pump enzyme is the H^+,K^+-ATPase, relatively unique to the gastric oxyntic cell, and responsible for the primary protonmotive force of gastric secretion.

The H^+,K^+-ATPase is localized on a system of intracellular membranes during resting conditions; oxyntic cell stim-

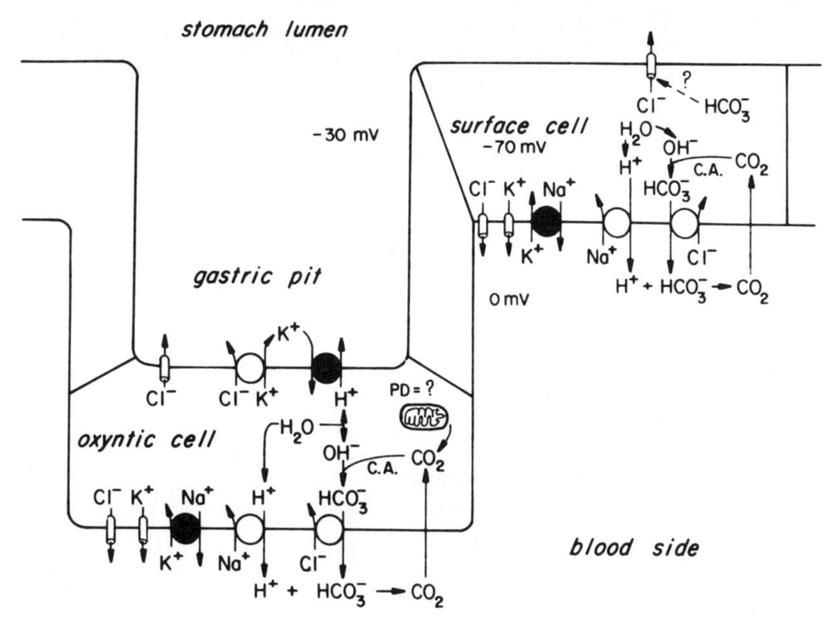

Fig. 16. Summary of ion transport and conductance pathways in oxyntic cells and surface epithelial cells of gastric mucosa. These are the necessary pathways for secretion of H^+, acidic and nonacidic Cl^-, and HCO_3^-, as well as the absorption of Na^+. Open circles are cotransport or exchangers that do not consume energy; filled circles are ATP-dependent processes. Cylinders are conductance channels, with the predominant direction of net flux shown by an arrow. C.A., carbonic anhydrase. A possible mode for regulation of the membrane conductances via feedback interaction with NaCl cotransport is shown in Fig. 15. See text for details. Adapted from Diamond and Machen.[30]

ulation initiates a redistribution of these membranes, probably by fusion, to apical plasma membrane. In addition to ATP, H^+-pumping function requires K^+ at some external site, and the system has been characterized as a H^+/K^+ exchange pump. An additional system for the rapid transport of K^+ and Cl^- has been identified in apical plasma membranes from secreting oxyntic cells. Thus, in the stimulated cell, the flux of K^+ and Cl^- from cytoplasm to the lumen of the gastric gland operates in parallel with the H^+/K^+ exchange pump to produce a recycling of the K^+ and net flow of KCl into the lumen. The resulting net OH^- production in the cytoplasm combines with CO_2 (both endogenous and plasma CO_2 are necessary for optimal secretory rates), and HCO_3^- egress can be accommodated by a Cl^-/HCO_3^- exchange system at the basolateral membrane. These systems in the oxyntic cell can account for transepithelial transport of H^+ and the so-called acidic Cl^- transport that accompanies H^+.

It is also well established that there is a component of gastric Cl^- transport that is independent of H^+ secretion, the so-called nonacidic, electrogenic, Cl^- transport. This appears to be generated by both surface cells and oxyntic cells, and the mechanism of transport is most likely the same for both cell types. Nonacidic Cl^- transport is dependent upon serosal $[Na^+]$ and the activity of the serosally directed Na^+ pump. We suggest in Fig. 16 that this component of Cl^- transport is generated by the operation of (1) neutral Na^+/H^+ and Cl^-/HCO_3^- exchangers at the basolateral membrane, and (2) conductance pathways for Cl^- and K^+ at the apical membrane and basolateral membrane, respectively. The Na^+,K^+-ATPase is the driving force for the system through the resulting Na^+ and K^+ gradients. Na^+ enters the cell down a steep chemical gradient in exchange for H^+, thereby alkalinizing the cell interior. In the presence of CO_2 and carbonic anhydrase, this alkalinity occurs in the form of increased intracellular $[HCO_3^-]$, which then drives an "active" accumulation of Cl^- through the Cl^-/HCO_3^- exchanger. In the open-circuit condition, the system achieves a steady state with the transmucosal potential difference primarily being the result of a diffusion potential for Cl^- at the apical membrane and an even larger, and oppositely oriented, diffusion potential for K^+ at the basolateral membrane. On open circuit, net transport of Cl^- by the system would depend on whether the trans-apical membrane Cl^- gradient were in equilibrium with the membrane potential. The dominant apical Cl^- conductance of amphibian surface cells is consistent with little or no net Cl^- flux. The additional Na^+ conductance in the apical membrane of mammalian surface cells can actually lead to net Cl^- absorption on open circuit. In the short-circuited state, Cl^- exits across the apical plasma membrane, driven by the negative cellular potential. The electrical circuit for Cl^- secretion is completed by the cell-to-serosa movement of K^+ across the basolateral membrane. Na^+ and K^+ must recycle through the ATPase to keep the system going. There may also be an intrinsic regulatory system operating; e.g., the rate of NaCl entry may regulate the K^+ and Cl^- conductive pathways, thereby preventing the cell from overfilling itself with KCl (and water).

The small secretion of HCO_3^- by surface cells at first appears to be a contradiction in the major functional activity of HCl secretion, but may serve some cytoprotective function for these cells. This apparent HCO_3^- transport may be generated by a system rather similar to that for nonacidic Cl^- transport. Thus, Na^+/H^+ exchange of the basolateral membrane alkalinizes the cell interior. Movement of HCO_3^- across the apical membrane may be through some specific HCO_3^- pathway or through the Cl^- conductance channels which may also be somewhat permeable to HCO_3^-. This scheme, in combination with the feedback proposed above for nonacidic Cl^- transport, may explain the Na^+ and Cl^- dependence of HCO_3^- transport.

Mammalian and some reptilian gastric mucosas are also capable of net Na^+ absorption, although the actual flux values are small relative to the stimulated rates of H^+ and Cl^- secretion. Na^+ conductance channels at the apical membrane of the surface cells provide the route for Na^+ entry from the mucosal solution; the Na^+,K^+-ATPase provides the pump at the basolateral surface.

References

1. Berglindh, T. 1977. Absolute dependence on chloride for acid secretion in isolated gastric glands. *Gastroenterology* **73**:874–880.
2. Berglindh, T., D. R. DiBona, S. Ito, and G. Sachs. 1980. Probes of parietal cell function. *Am. J. Physiol.* **238**:G115–G176.
3. Berglindh, T., D. R. DiBona, C. S. Pace, and G. Sachs. 1980. ATP dependence of H^+ secretion. *J. Cell Biol.* **85**:392–401.
4. Berglindh, T., H. F. Helander, and K. J. Obrink. 1976. Effects of secretagogues on oxygen consumption, aminopyrine accumulation and morphology in isolated gastric glands. *Acta Physiol. Scand.* **97**:401–414.
5. Berglindh, T., G. Sachs, and N. Takeguchi. 1980. Ca^{2+}-dependent secretagogue stimulation in isolated rabbit gastric glands. *Am. J. Physiol.* **239**:G90–G94.
6. Bergqvist, E., and K. J. Obrink. 1979. Gastrin–histamine as a normal sequence in gastric acid stimulation. *Upsala J. Med. Sci.* **84**:145–154.
7. Berridge, M. J. 1975. Interaction of cyclic nucleotides and calcium in the control of cellular activity. *Adv. Cyclic Nucleotide Res.* **6**:1–98.
8. Black, J. A., T. M. Forte, and J. G. Forte. 1981. Inhibition of HCl secretion and the effects on ultrastructure and electrical resistance in isolated piglet gastric mucosa. *Gastroenterology* **81**:509–519.
9. Black, J. A., T. M. Forte, and J. G. Forte. 1982. The effects of microfilament disrupting agents on HCl secretion and ultrastructure of piglet gastric oxyntic cells. *Gastroenterology* **83**:595–604.
10. Black, J. W., W. A. M. Duncan, C. J. Durant, C. R. Ganellin, and E. M. Parsons. 1972. Definition and antagonism of histamine H_2 receptors. *Nature (London)* **236**:385–390.
11. Blum, A. L., B. I. Hirschowitz, H. F. Helander, and G. Sachs. 1971. Electrical properties of isolated cells of *Necturus* gastric mucosa. *Biochim. Biophys. Acta* **241**:261–272.
12. Bonting, S. L., J. J. Schrijen, and J. J. H. H. M. DePont. 1980. Mg^{2+}-induced conformational state of $(K^+$-$H^+)$-ATPase deduced from chemical modification and substrate binding studies. In: *Hydrogen Ion Transport in Epithelia*. I. Schultz, G. Sachs, J. Forte, and K. Ullrich, eds. Elsevier, Amsterdam. pp. 185–192.
13. Bornstein, A. M., W. H. Dennis, and W. S. Rehm. 1959. Movement of water, sodium, chloride and hydrogen ions across the resting stomach. *Am. J. Physiol.* **197**:332–336.
14. Canosa, C. A., and W. S. Rehm. 1968. Microelectrode studies of dog's gastric mucosa. *Biophys. J.* **8**:415–430.
15. Carrasquer, G., T. C. Chu, M. Schwartz, and W. S. Rehm. 1982. Evidence for electrogenic Na–Cl symport in *in vitro* frog stomach. *Am. J. Physiol.* **242**:G620–G627.
16. Chase, H. S., and Q. Al-Awqati. 1981. Regulation of the sodium permeability of the luminal border of toad bladder by intracellular sodium and calcium. *J. Gen. Physiol.* **77**:693–712.
17. Chew, C. S., and S. J. Hersey. 1982. Gastrin stimulation of isolated gastric glands. *Am. J. Physiol.* **242**:G504–G512.
18. Chew, C. S., S. J. Hersey, G. Sachs, and T. Berglindh. 1980. Histamine responsiveness of isolated gastric glands. *Am. J. Physiol.* **238**:G312–G320.

19. Clausen, C., T. E. Machen, and J. M. Diamond. 1983. Use of AC impedance analysis to study membrane changes related to acid secretion in amphibian gastric mucosa. *Biophys. J.* **41**:167–178.
20. Cole, K. S. 1972. *Membranes, Ions, and Impulses*. University of California Press, Berkeley.
21. Cooperstein, I. L. 1959. The inhibitory effect of strophanthidin on secretion by isolated gastric mucosa. *J. Gen. Physiol.* **42**:1233–1239.
22. Culp, D. J., and J. G. Forte. 1981. An enriched preparation of basolateral plasma membranes from gastric glandular cells. *J. Membr. Biol.* **59**:135–142.
23. Culp, D. J., J. M. Wolosin, A. H. Soll, and J. G. Forte. 1983. Muscarinic receptors and guanylate cyclase in mammalian gastric glandular cells. *Am. J. Physiol.* **245**:G760–G768.
24. Davenport, H. W. 1962. Effect of ouabain on acid secretion and electrolyte content of frog gastric mucosa. *Proc. Soc. Exp. Biol. Med.* **110**:613–615.
25. Davenport, H. W., and F. Alzamora. 1962. Sodium, potassium, chloride and water in frog gastric mucosa. *Am. J. Physiol.* **202**:711–715.
26. Davies, R. E. 1948. Hydrochloric acid production by isolated gastric mucosa. *Biochem. J.* **42**:609–621.
27. Davies, R. E. 1951. The mechanism of hydrochloric acid production by the stomach. *Biol. Rev.* **26**:87–120.
28. Davis, T. L., J. R. Rutledge, and W. S. Rehm. 1963. Effect of potassium on secretion and potential of frog's gastric mucosa in Cl^--free solutions. *Am. J. Physiol.* **205**:873–877.
29. Davis, T. L., J. R. Rutledge, D. C. Keesee, F. J. Bajandes, and W. S. Rehm. 1965. Acid secretion, potential difference and resistance of frog stomach in K^+-free solutions. *Am. J. Physiol.* **209**:146–152.
30. Diamond, J. M., and T. E. Machen. 1983. Impedance analysis in epithelia and the problem of gastric acid secretion. *J. Membr. Biol.* **72**:17–41.
31. Durbin, R. P. 1964. Anion requirements for gastric acid secretion. *J. Gen. Physiol.* **47**:735–748.
32. Durbin, R. P. 1967. Electrical potential difference of the gastric mucosa. In: *Handbook of Physiology,* Section 6, Volume 2. C. F. Code, ed. American Physiological Society, Washington, D.C. pp. 879–888.
33. Durbin, R. P., and E. Heinz. 1958. Electromotive chloride transport and gastric acid secretion in the frog. *J. Gen. Physiol.* **41**:1035–1047.
34. Durbin, R. P., and H. F. Helander. 1978. Distribution of osmotic flow in stomach and gall-bladder. *Biochim. Biophys. Acta* **513**:179–181.
35. Ekblad, E. B. M. 1980. Increase of intracellular pH in secreting frog gastric mucosa. *Biochim. Biophys. Acta* **632**:375–385.
36. Eveloff, J., R. Kinne, E. Kinne-Safran, H. Murer, P. Silver, F. H. Epstein, J. Stoff and W. B. Kinter. 1980. Coupled sodium and chloride transport into plasma membrane vesicles prepared from dogfish rectal gland. *Pfluegers Arch.* **378**:87–92.
37. Flemstrom, G. 1971. Na^+ transport and impedance properties of the isolated frog gastric mucosa at different O_2 tensions. *Biochim. Biophys. Acta* **225**:35–45.
38. Flemstrom, G. 1977. Active alkalinization by amphibian gastric fundic mucosa *in vitro*. *Am. J. Physiol.* **233**:E1–E12.
39. Flemstrom, G., and A. Garner. 1982. Gastroduodenal HCO_3^- transport: Characteristics and proposed role in acidity regulation and mucosal protection. *Am. J. Physiol.* **242**:G183–G193.
40. Flemstrom, G., J. R. Heylings, and A. Garner. 1982. Gastric and duodenal HCO_3^- transport *in vitro*: Effects of hormones and local transmitters. *Am. J. Physiol.* **242**:G100–G110.
41. Forte, J. G. 1969. Three components of Cl^- flux across isolated bullfrog gastric mucosa. *Am. J. Physiol.* **216**:167–174.
42. Forte, J. G., P. H. Adams, and R. E. Davies. 1963. The source of the gastric mucosal potential difference. *Nature (London)* **197**:874–876.
43. Forte, J. G., P. H. Adams, and R. E. Davies. 1965. Acid secretion and phosphate metabolism in bullfrog gastric mucosa. *Biochim. Biophys. Acta* **104**:25–38.
44. Forte, J. G., J. A. Black, T. M. Forte, T. E. Machen, and J. M. Wolosin. 1981. Ultrastructural changes related to functional activity in gastric oxyntic cells. *Am. J. Physiol.* **241**:G349–G358.
45. Forte, J. G., T. M. Forte, and P. Saltman. 1967. K^+-stimulated phosphatase in microsomes isolated from gastric mucosa. *J. Cell. Physiol.* **69**:293–304.
46. Forte, J. G., A. L. Ganser, R. Beesley, and T. M. Forte. 1975. Unique enzymes of purified microsomes from pig fundic mucosa. *Gastroenterology* **69**:175–189.
47. Forte, J. G., A. L. Ganser, and T. K. Ray. 1976. The K^+-stimulated ATPase from oxyntic glands of gastric mucosa. In: *Gastric Hydrogen Ion Secretion*. D. K. Kasbekar, G. Sachs, and W. Rehm, eds. Dekker, New York. pp. 302–330.
48. Forte, J. G., and H. C. Lee. 1977. Gastric adenosine triphosphatases: A review of their possible role in HCl secretion. *Gastroenterology* **73**:921–926.
49. Forte, J. G., and T. E. Machen. 1975. Transport and electrical phenomena in resting and secreting piglet gastric mucosa. *J. Physiol. (London)* **244**:33–51.
50. Forte, J. G., and L. Solberg. 1973. Pharmacology of isolated amphibian gastric mucosa. In: *International Encyclopedia of Pharmacology and Experimental Therapeutics,* Section 29A, Volume 1, Pergamon Press, Elmsford, N.Y. pp. 195–260.
51. Forte, T. M., T. E. Machen, and J. G. Forte. 1977. Ultrastructural changes in oxyntic cells associated with secretory function: A membrane recycling hypothesis. *Gastroenterology* **73**:941–955.
52. Frizzell, R. A. 1976. Active chloride transport by rabbit colon: Calcium-dependent stimulation by ionophore A23187. *J. Membr. Biol.* **35**:175–187.
53. Frizzell, R. A., M. Field, and S. G. Schultz. 1979. Sodium-coupled chloride transport in epithelial tissues. *Am. J. Physiol.* **236**:F1–F8.
54. Fromter, E., and J. M. Diamond. 1972. Route of passive ion permeation in epithelia. *Nature New Biol.* **235**:9–13.
55. Ganser, A. L., and J. G. Forte. 1973. K^+-stimulated ATPase in purified microsomes of bullfrog oxyntic cells. *Biochim. Biophys. Acta* **307**:169–180.
56. Gibert, A. J., and S. J. Hersey. 1982. Morphometric analysis of parietal cell membrane transformations in isolated gastric glands. *J. Membr. Biol.* **67**:113–124.
57. Grinstein, S., and D. Erlij. 1978. Intracellular calcium and the regulation of sodium transport in frog skin. *Proc. R. Soc. London Ser. B* **202**:353–360.
58. Grossman, M. I., and S. J. Konturek. 1974. Inhibition of acid secretion in dog by metiamide, a histamine antagonist acting on H_2 receptors. *Gastroenterology* **66**:517–521.
59. Gutknecht, J., and A. Walter. 1982. SCN^- and HSCN transport through lipid bilayer membranes: A model for SCN^- inhibition of gastric acid secretion. *Biochim. Biophys. Acta* **685**:233–240.
60. Haas, M., W. F. Schmidt, and T. J. McManus. 1982. Catecholamine-stimulated transport in duck red cells: Gradient effects in electrically neutral [Na + K + 2Cl] co-transport. *J. Gen. Physiol.* **80**:125–147.
61. Hansen, T., J. F. G. Slegers, and S. L. Bonting. 1975. Gastric acid secretion in the lizard: Ionic requirements and effects of inhibitors. *Biochim. Biophys. Acta* **382**:590–608.
62. Harris, J. B., and I. S. Edelman. 1964. Chemical concentration gradients and electrical properties of gastric mucosa. *Am. J. Physiol.* **206**:769–782.
63. Harris, J. B., K. Nigon, and D. Alonso. 1969. Adenosine 3′,5′-monophosphate: Intracellular mediator for methylxantine stimulation of gastric secretion. *Gastroenterology* **57**:377–382.
64. Heinz, E., and R. P. Durbin. 1957. Studies of chloride transport in the gastric mucosa of the frog. *J. Gen. Physiol.* **41**:101–117.
65. Heinz, E., and K. J. Obrink. 1954. Acid formation and acidity control in the stomach. *Physiol. Rev.* **34**:643–673.
66. Helander, H. F. 1981. The cells of the gastric mucosa. *Int. Rev. Cytol.* **70**:279–351.

67. Helander, H. F., and R. P. Durbin. 1982. Localization of ouabain binding sites in frog gastric mucosa. *Am. J. Physiol.* **243**:G297–G303.
68. Helander, H. F., and B. I. Hirschowitz. 1972. Quantitative ultrastructural studies on gastric parietal cells. *Gastroenterology* **63**:951–961.
69. Hersey, S. J. 1979. Intracellular pH measurements in gastric mucosa. *Am. J. Physiol.* **237**:E82–E89.
70. Hersey, S. J. 1981. Histamine receptor in bullfrog gastric mucosa. *Am. J. Physiol.* **241**:G93–G97.
71. Hodgkin, A. L. 1951. The ionic basis of electrical activity in nerve and muscle. *Biol. Rev.* **26**:339–409.
72. Hogben, C. A. M. 1955. Active transport of chloride by isolated frog gastric epithelium: Origin of the gastric mucosa potential. *Am. J. Physiol.* **180**:641–649.
73. Hogben, C. A. M. 1967. The chloride effect of carbonic anhydrase inhibitors. *Mol. Pharmacol.* **3**:318–326.
74. Hogben, C. A. M. 1968. Observations on ionic movement through the gastric mucosa. *J. Gen. Physiol.* **51**:2485–2495.
75. Hollander, F. 1954. The two-component mucous barrier. *Arch. Intern. Med.* **94**:107–120.
76. Ito, S. 1967. Anatomic structure of the gastric mucosa. In: *Handbook of Physiology,* Section 6, Volume 2. C. F. Code, ed. American Psychological Society, Washington, D.C. p. 607.
77. Ito, S., and G. C. Schofield, 1974. Studies on the depletion and accumulation of microvilli and changes in the tubulovesicular compartment of mouse parietal cells in relation to gastric acid secretion. *J. Cell Biol.* **63**:364–382.
78. Jiron, C., M. C. Ruiz, and F. Michelangeli. 1981. Role of Ca^{2+} in stimulus–secretion coupling in the gastric oxyntic cell: Effect of A23187. *Cell Calcium* **2**:573–585.
79. Kasbekar, D. K., and H. Chugani. 1976. Role of calcium ion in *in vitro* gastric acid secretion. In: *Gastric Hydrogen Ion Secretion.* D. K. Kasbekar, G. Sachs, and W. S. Rehm, eds. Dekker, New York. pp. 187–211.
80. Kasbekar, D. K., and R. P. Durbin. 1965. An adenosine triphosphatase from frog gastric mucosa. *Biochim. Biophys. Acta* **105**:472–482.
81. Kasbekar, D. K., and G. S. Gordon. 1979. The effects of colchicine and vinblastine on *in vitro* gastric secretion. *Am. J. Physiol.* **236**:E550–E555.
82. Kinsella, J. L., and P. S. Aronson. 1981. Properties of the Na^+–H^+ exchanger in renal microvillus membrane vesicles. *Am. J. Physiol.* **241**:F374–F379.
83. Klyce, S. D., and R. K. S. Wong. 1977. Site and mode of adrenaline action on chloride transport across the rabbit corneal epithelium. *J. Physiol. (London)* **266**:777–799.
84. Koefoed-Johnsen, V., and H. H. Ussing. 1958. The nature of the frog skin potential. *Acta Physiol. Scand.* **42**:298–308.
85. Koelz, H. R., G. Sachs, and T. Berglindh. 1981. Cation effects on acid secretion in rabbit gastric glands. *Am. J. Physiol.* **241**:G431–G442.
86. Lee, J., G. Simpson, and P. Scholes. 1974. An ATPase from dog gastric mucosa; changes of outer pH in suspensions of membrane vesicles accompanying ATP hydrolysis. *Biochem. Biophys. Res. Commun.* **60**:825–832.
87. Lee, H. C., H. Breitbart, M. Berman, and J. G. Forte. 1979. Potassium-stimulated ATPase activity and H^+ transport in gastric microsomal vesicles. *Biochim. Biophys. Acta* **553**:107–131.
88. Lee, H. C., and J. G. Forte. 1978. A study of H^+ transport in gastric microsomal vesicles using fluorescent probes. *Biochim. Biophys. Acta* **508**:339–356.
89. Lee, H. C., A. Quintanilha, and J. G. Forte. 1976. Energized gastric microsomal membrane vesicles—An index using metachromatic dyes. *Biochem. Biophys. Res. Commun.* **72:**1179–1186.
90. Liedtke, C. M., and U. Hopfer. 1982. Mechanism of Cl^- translocation across small intestinal brush-border. I. Absence of Na^+–Cl^- contransport. *Am. J. Physiol.* **242**:G263–G271.
91. Liedtke, C. M., and U. Hopfer. 1982. Mechanism of Cl^- translocation across small intestinal brush-border. II. Demonstration of Cl^-–OH^- exchange and Cl^- conductance. *Am. J. Physiol.* **242**:G272–G280.
92. Limlomwongse, L., and J. G. Forte. 1970. Developmental changes of ATPase and K^+-stimulated phosphatase in microsome of tadpole gastric mucosa. *Am. J. Physiol.* **219**:1717–1722.
93. Lindemann, B., and W. van Driessche. 1977. Sodium-specific channels of frog skin are pores: Current fluctuations reveal high turnover. *Science* **195**:292–294.
94. Logsdon, C. D., and T. E. Machen. 1981. Involvement of extracellular Ca^{2+} in gastric stimulation. *Am. J. Physiol.* **241**:G365–G375.
95. Logsdon, C. D., and T. E. Machen. 1982. Ionic requirements for H^+ secretion and membrane elaboration in frog oxyntic cells. *Am. J. Physiol.* **242**:G388–G399.
96. Machen, T. E., and J. G. Forte. 1978. Gastric secretion. In: *Membrane Transport in Biology,* Volume IVB. G. Giebisch, D. C. Tosteson, and H. H. Ussing, eds. Springer-Verlag, Berlin. pp. 693–747.
97. Machen, T. E., and J. G. Forge. 1983. Anion secretion by gastric mucosa. In: *Chloride Transport Coupling in Biological Membranes and Epithelia.* G. Gerencser, ed. Elsevier, Amsterdam. pp. 415–446.
98. Machen, T. E., and W. L. McLennan. 1980. Na^+-dependent H^+ and Cl^- secretion in *in vitro* frog gastric mucosa. *Am. J. Physiol.* **238**:G403–G413.
99. Machen, T. E., W. McLennan, and T. Zeuthen. 1980. Electrogenic Cl^- secretion by resting gastric mucosa: Na–Cl co-transport model. In: *Hydrogen Ion Transport in Epithelia.* I. Schultz, K. J. Ullrich, G. Sachs, and J. Forte, eds. Elsevier, Amsterdam. pp. 379–390.
100. Machen, T. E., M. J. Rutten, and E. B. M. Ekblad. 1982. Histamine, cAMP, and activation of piglet gastric mucosa. *Am. J. Physiol.* **242**:G79–G84.
101. Machen, T. E., W. Silen, and J. G. Forte. 1978. Na^+ transport by mammalian stomach. *Am. J. Physiol.* **234**:E228–E235.
102. Machen, T. E., and T. Zeuthen. 1982. Cl^- transport by gastric mucosa: Cellular Cl^- activity and membrane permeability. *Proc. R. Soc. London Ser. B* **299**:559–573.
103. Malinowska, D. H., H. R. Koelz, S. J. Hersey, and G. Sachs. 1981. Properties of the gastric proton pump in unstimulated permeable gastric glands. *Proc. Natl. Acad. Sci. USA* **78**:5908–5912.
104. Manning, E. C., and T. E. Machen. 1983. Effects of bicarbonate and pH on chloride transport by gastric mucosa. *Am. J. Physiol.* **243**:G60–G68.
105. McLennan, W. L., T. E. Machen, and T. Zeuthen. 1980. Ba^{2+} inhibition of electrogenic Cl^- secretion in *in vitro* frog and piglet gastric mucosa. *Am. J. Physiol.* **239**:G151–G160.
106. Means, A. R., J. S. Tash, and J. G. Chafouleas. 1982. Physiological implications of the presence, distribution and regulation of calmodulin in eukaryotic cells. *Physiol. Rev.* **62**:1–39.
107. Musch, M. W., S. A. Orellana, L. S. Kimberg, M. Field, D. R. Halm, E. J. Krasny, and R. A. Frizzell. 1982. Na^+-, K^+-, Cl^- and CO-transport in the intestine of a marine teleost. *Nature (London)* **300**:351–353.
108. Nakajima, S., B. I. Hirschowitz, and G. Sachs. 1971. Studies of adenyl cyclase in *Necturus* gastric mucosa. *Arch. Biochem. Biophys.* **143**:123–126.
109. O'Callaghan, J., S. S. Sanders, R. L. Shoemaker, and W. S. Rehm. 1974. Barium and K^+ on surface and tubular cell resistances of frog stomach with microelectrodes. *Am. J. Physiol.* **227**:273–288.
110. Palmer, L. G. 1982. Ion selectivity of the apical membrane Na channel in the toad urinary bladder. *J. Membr. Biol.* **67**:91–98.
111. Ray, T. K., and J. G. Forte. 1974. Adenyl cyclase of oxyntic cells. *Biochim. Biophys. Acta* **363**:320–339.
112. Ray, T. K., and J. G. Forte. 1974. Soluble and bound protein kinases of rabbit gastric secretory cells. *Biochem. Biophys. Res. Comm.* **61**:1199–1206.
113. Ray, T. K., and J. G. Forte. 1976. Studies on the phosphorylated

intermediate of a K^+-stimulated ATPase from rabbit gastric mucosa. *Biochim. Biophys. Acta* **443**:451–467.
114. Reenstra, W. W., and J. G. Forte. 1983. Action of thiocyanate on pH gradient formation by gastric microsomal vesicles. *Am. J. Physiol.* **244**:G308–G313.
115. Rehm, W. S. 1945. The effect of electric current on gastric secretion and potential. *Am. J. Physiol.* **144**:115–125.
116. Rehm, W. S. 1950. A theory of the formation of HCl by the stomach. *Gastroenterology* **14**:401–417.
117. Rehm, W. S. 1965. Electrophysiology of the gastric mucosa in chloride-free solutions. *Fed. Proc.* **24**:1387–1395.
118. Rehm, W. S. 1967. Ion permeability and electrical resistance of the frog's gastric mucosa. *Fed. Proc.* **26**:1303–1313.
119. Rehm, W. S., T. C. Chu, M. Schwartz, and G. Carrasquer. 1983. Mechanism responsible for thiocyanate increase in resistance of *in vitro* frog gastric mucosa. *Am. J. Physiol.* **245**:G143–G156.
120. Rehm, W. S., and M. E. LeFevre. 1965. Effect of dinitrophenol on potential, resistance, and H^+ rate of frog stomach. *Am. J. Physiol.* **208**:922–930.
121. Rehm, W. S., and S. S. Sanders. 1975. Implications of the neutral Cl^-–HCO_3^- exchange mechanism in gastric mucosa. *Ann. N.Y. Acad. Sci.* **264**:442–455.
122. Rehm, W. S., and S. S. Sanders. 1977. Electrical event during activation and inhibition of gastric HCl secretion. *Gastroenterology* **73**:959–969.
123. Reimann, E. M., and N. G. Rapino. 1974. Partial purification and characterization of an adenosine 3′,5′-monophosphate-dependent protein kinase from rabbit mucosa. *Biochim. Biophys. Acta* **350**:201–214.
124. Robinson, G. A., R. W. Butcher, and E. W. Sutherland. 1971. *Cyclic AMP*. Academic Press, New York.
125. Ross, I. N., H. M. M. Baharai, and L. A. Turnberg. 1981. The pH gradient across mucus adherent to rat fundic mucosa *in vivo* and the effects of potential damaging agents. *Gastroenterology* **81**:713–718.
126. Rutten, M. J., and T. E. Machen. 1981. Histamine, cyclic AMP and activation events in piglet gastric mucosa. *Gastroenterology* **80**:928–937.
127. Saccomani, G., H. F. Helander, S. Crago, H. H. Chang, and G. Sachs. 1979. Characterization of gastric mucosal membranes. X. Immunological studies of gastric (H^+ + K^+)-ATPase. *J. Cell Biol.* **83**:271–283.
128. Sachs, G., H. Chang, E. Rabon, R. Schackmann, M. Lewin, and G. Saccomani. 1976. A nonelectrogenic H^+ pump in plasma membranes of hog stomach. *J. Biol. Chem.* **251**:7690–7698.
129. Sachs, G., H. Chang, E. Rabon, R. Schackmann, H. M. Saran, and G. Saccomani. 1977. Metabolic and membrane aspects of gastric H^+ transport. *Gastroenterology* **73**:931–940.
130. Sachs, G., L. D. Faller, and E. Rabon. 1982. Proton, hydroxyl transport in gastric and intestinal epithelia. *J. Membr. Biol.* **64**:123–135.
131. Sanders, S. S., V. B. Haynes, and W. S. Rehm. 1973. Normal H^+ rate in frog stomach in absence of exogenous CO_2 and a note on pH stat method. *Am. J. Physiol.* **225**:1311–1321.
132. Sanders, S. S., J. A. Pirkle, R. L. Shoemaker, and W. S. Rehm. 1978. Effects of inhibitors and weak bases on electrophysiology and secretion in frog stomach. *Am. J. Physiol.* **234**:E120–E128.
133. Schackmann, R., A. Schwartz, G. Saccomani, and G. Sachs. 1977. Cation transport by gastric H^+ − K^+ ATPase. *J. Membr. Biol.* **32**:361–381.
134. Schaltz, L. J., C. Bool, and E. M. Reimann. 1981. Phosphorylation of membranes from the rat gastric mucosa. *Biochim. Biophys. Acta* **673**:539–551.
135. Schettino, T., and S. Curci. 1980. Intracellular potassium activity in epithelial cells of frog fundic gastric mucosa. *Pfluegers Arch.* **383**:99–103.
136. Schiessel, R., A. Merhav, J. B. Matthews, L. A. Fleisher, A. Barzilai, and W. Silen. 1980. Role of nutrient HCO_3^- in protection of amphibian gastric mucosa. *Am. J. Physiol.* **239**:G536–G542.
137. Schoffeniels, E. 1955. Enfluence du pH sur le transport actif de sodium a travers la peau de grenouille. *Arch. Int. Physiol. Biochem.* **63**:513–530.
138. Schultz, S. G. 1981. Homocellular regulation mechanism in sodium-transporting epithelia: Avoidance of extinction by "flush-through." *Am. J. Physiol.* **241**:F579–F590.
139. Sedar, A. W., and J. G. Forte. 1964. Effects of calcium depletion on the junctional complex between oxyntic cells of gastric glands. *J. Cell Biol.* **22**:173–188.
140. Silen, W., T. E. Machen, and J. G. Forte. 1975. Acid–base balance in amphibian gastric mucosa. *Am. J. Physiol.* **229**:721–730.
141. Smolka, A., H. F. Helander, and G. Sachs. 1984. Monoclonal antibodies against the gastric (H^+ − K^+)-ATPase. *Am. J. Physiol.* **245**:G589–G596.
142. Soll, A. H. 1978. The actions of secretagogues on oxygen uptake by isolated mammalian parietal cells. *J. Clin. Invest.* **61**:370–380.
143. Soll, A. H. 1978. The interactions of histamine with gastrin and carbamylcholine on oxygen uptake by isolated mammalian parietal cells. *J. Clin. Invest.* **61**:381–389.
144. Soll, A. H. 1981. Extracellular calcium and cholinergic stimulation of isolated canine parietal cells. *J. Clin. Invest.* **68**:270–278.
145. Soll, A. H. 1982. Potentiating interactions of gastric stimulants on [^{14}C]aminopyrine accumulation by isolated canine parietal cells. *Gastroenterology* **83**:216–223.
146. Soll, A. H., and A. Wollin. 1979. Histamine and cyclic AMP in isolated canine parietal cells. *Am. J. Physiol.* **237**:E444–E450.
147. Spenney, J. G., R. L. Shoemaker, and G. Sachs. 1974. Microelectrode studies of fundic gastric mucosa: Cellular coupling and shunt conductance. *J. Membr. Biol.* **15**:105–128.
148. Takeuchi, K., A. Merhav, and W. Silen. 1982. Mechanism of luminal alkalinization by frog fundic mucosa. *Am. J. Physiol.* **243**:G377–G388.
149. Tanaka, Y., P. De Camilli, and J. Meldolesi. 1980. Membrane interactions between secretion granules and plasmalemma in three exocrine glands. *J. Cell Biol.* **84**:438–453.
150. Teorell, T. 1951. The acid–base balance of the secreting isolated gastric mucosa. *J. Physiol. (London)* **114**:267–276.
151. Thompson, W. J., L. K. Chang, and G. C. Rosenfeld. 1980. Histamine regulation of adenylyl cyclase of enriched rat gastric parietal cells. *Am. J. Physiol.* **240**:G76–G84.
152. Tripathi, S., and P. K. Rangachari. 1980. *In vitro* primate gastric mucosa: Electrical characteristics. *Am. J. Physiol.* **239**:G77–G82.
153. Turnheim, K., R. A. Frizzell, and S. G. Schultz. 1978. Interaction between cell sodium and the amiloride-sensitive sodium entry step in the rabbit colon. *J. Membr. Biol.* **39**:233–256.
154. Ussing, H. H. 1949. The distinction by means of tracers between active transport and diffusion: The transfer of iodide across the isolated frog skin. *Acta Physiol. Scand.* **19**:43–56.
155. Villegas, L. 1962. Cellular location of the electrical potential difference in frog gastric mucosa. *Biochim. Biophys. Acta* **64**:359–367.
156. Walderhaug, M. O., G. Saccomani, G. Sachs, and R. L. Post. 1982. The active site of phosphorylation of hog gastric adenosine triphosphatase. *Fed. Proc.* **41**:1369a.
157. Wallmark, B., H. B. Stewart, E. Rabon, G. Saccomani, and G. Sachs. 1980. The catalytic cycle of gastric (H^+ − K^+)-ATPase. *J. Biol. Chem.* **255**:5313–5319.
158. Warnock, D. G., and J. Eveloff. 1982. NaCl entry mechanisms in the luminal membrane of the renal tubule. *Am. J. Physiol.* **242**:F561–F574.
159. Weinman, S. A., and L. Reuss. 1982. Na^+–H^+ exchange at the apical membrane of *Necturus* gallbladder: Extracellular and intracellular pH studies. *J. Gen. Physiol.* **80**:299–321.
160. Williams, J. A. 1980. Regulation of pancreatic acinar cell function by intracellular calcium. *Am. J. Physiol.* **238**:G269–G279.
161. Williams, S. E., and L. A. Turnberg. 1981. Demonstration of a pH gradient across mucus adherent to rabbit gastric mucosa: Evidence for a mucus–bicarbonate barrier. *Gut* **22**:94–96.
162. Wolosin, J. M., and J. G. Forte. 1981. Functional differences

between K^+-ATPase membranes isolated from resting or stimulated rabbit fundic mucosa. *FEBS Lett.* **125**:208–212.
163. Wolosin, J. M., and J. G. Forte. 1981. Changes in the membrane environment of the ($K^+ + H^+$)-ATPase following stimulation of the gastric oxyntic cell. *J. Biol. Chem.* **256**:3149–3152.
164. Wolosin, J. M., and J. G. Forte. 1981. Isolation of the secreting oxyntic cell apical membrane: Identification of an electroneutral KCl symport. In *Membrane Biophysics: Structures and Function in Epithelia*. M. Dinno and A. Callahan, eds. Liss, New York. pp. 189–204.
165. Wolosin, J. M., and J. G. Forte. 1983. Kinetic properties of KCl transport at the secreting apical membrane of the oxyntic cell. *J. Membr. Biol.* **71**:195–207.
166. Wolosin, J. M., and J. G. Forte. 1983. Anion exchange in oxyntic cell apical membrane; relationship to thiocyanate inhibition of acid secretion. *J. Membr. Biol.* in press.
167. Wolosin, J. M., and J. G. Forte. 1984. Stimulation of oxyntic cell triggers K^+ and Cl^- conductances in apical ($H^+ + K^+$)-ATPase membrane. *Am. J. Physiol.* **246**:C537–C545.
168. Wright, G. H. 1962. Net transfers of water, sodium, chloride and hydrogen ions across the gastric mucosa of the rabbit foetus. *J. Physiol. (London)* **163**:281–293.
169. Zeiske, W., T. E. Machen, and W. van Driessche. 1983. Cl^-- and K^+-related fluctuations of the ionic current through oxyntic cells in frog gastric mucosa. *Am. J. Physiol.* **245**:G797–G807.

CHAPTER 10

Ion and Water Transport in the Intestine

Don W. Powell

1. Introduction

Disorders of membrane function in the gastrointestinal tract are frequent and cause significant morbidity and mortality. In the developed nations, diarrheal disease is a cause of loss of productivity at work, and death of neonatal livestock; thus, it is primarily of economic significance. However, in the Third World, diarrheal diseases account for the deaths of over 5 million infants and young children each year. Whereas such economic or social issues might direct the investigator to the gastrointestinal tract, its easy accessibility is perhaps the main reason why such research has flourished. As a result, studies of gastrointestinal tissue have added much to our understanding of mechanisms of transport in all epithelia.

Up-to-date reviews of intestinal transport are available in several excellent monographs[1,2] and texts.[3,4] The reader is referred to these for details and historic perspective that space here does not allow. The present review assumes that the reader is familiar with the general concepts of epithelial transport; e.g., the double-membrane model proposed by Koefoed-Johnsen and Ussing[5]; the Na^+ gradient hypothesis of nonelectrolyte transport[6,7]; the concept of cellular and paracellular routes across epithelia with the ionic permeabilities of the paracellular junctions dividing epithelia into "tight" and "leaky" tissues[8]; and the "double-membrane"[9] or "standing osmotic gradient" models[10] of isotonic water movement. Indeed, these paradigms are the very foundation of our current understanding of epithelial physiology.

Over the past decade, investigators have refined these basic concepts, and one of the major goals of the present review is to bring the reader up to date. A particularly fruitful area of research in epithelia has been directed toward understanding how Na^+ enters the cell across the apical membrane and how anions, particularly Cl^-, are secreted. Interesting and novel ideas about K^+ and HCO_3^- transport and water movement have emerged and are considered here. The other major area reviewed in this chapter is the control of transport. We discuss briefly the current concepts of the intracellular messengers—calcium and cyclic nucleotides—which regulate ion and water movement, and discuss in more detail the systemic factors and neurohumoral mechanisms that regulate intestinal transport. The clinically oriented investigator will recognize that these latter two areas are likely the ones in which aberrations result in diarrhea. Conversely, these are also the areas that are points of attack for developing antisecretory therapy. In this regard, it should be obvious that there is a natural overlap between this chapter and the chapter by Field *et al.* on secretory diarrheas. While it is necessary that our chapters stand alone, we have endeavored where possible to make them complimentary.

2. Models of Intestinal Na^+, Cl^-, and H_2O Transport

In the past decade, three general models of intestinal electrolyte transport have emerged. A consideration of these models provides the conceptual framework for rigorously analyzing recent research in this field. One model of transport comes from the studies of Field, Frizzell, and Schultz[11–14] (Fig. 1). This model anatomically separates the absorptive from the secretory functions of the intestine: the villi of the small intestine and the surface epithelium of the colon are assigned an absorptive capacity, while the small intestinal and colonic crypts are thought to be the site of electrolyte and fluid secretion. This model proposes that increased cellular levels of cyclic nucleotides and/or of Ca^{2+} ions inhibit brush border coupled NaCl influx across the apical cell membrane, thus having an antiabsorptive effect in small intestine. The colonic absorptive cells of some species have channels for Na^+ entry that are unaffected by secretagogues, and in these animals, the antiabsorptive effects are not

Don W. Powell • Department of Medicine, University of North Carolina School of Medicine, Chapel Hill, North Carolina 27514.

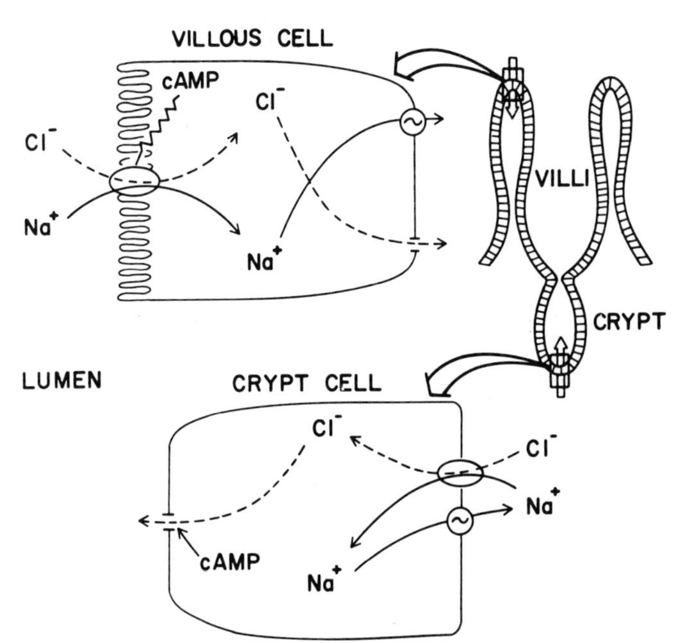

Fig. 1. This model of small intestinal electrolyte transport separates the absorptive from the secretory functions of the intestinal mucosa and assigns both antiabsorptive and secretory actions to intracellular messengers such as cAMP. The surface cells and crypt cells in the colonic mucosa would be analogous sites of absorption and secretion in the large intestine. Increased intracellular levels of Ca^{2+} are postulated to have the same effects as cAMP. Reproduced by permission from Ref. 11.

seen. In this model, intestinal secretion from either the small or the large intestine is thought to come about by cyclic nucleotide- or Ca^{2+}-mediated increases in crypt cell apical membrane permeability to Cl^-.

The second model of intestinal electrolyte transport is that articulated by Turnberg, Krejs, and Fordtran.[15,16] This hypothesis replaces the apical cell membrane coupled NaCl transport mechanism with a double, simultaneous exchange (antiport) of Na^+ for H^+ and Cl^- for HCO_3^- (or OH^-) (Fig. 2). Mechanistically, these processes yield the same result—electrically neutral NaCl transport into the cell. The two antiports are coupled by intracellular pH. If one exchange limb is inhibited, then the pH inside the cell either rises or falls and the resulting acidification or alkalinization abolishes the driving force for the other exchange limb. An attractive feature of this model is that in various intestinal segments the two antiports might be to some degree uncoupled, thus explaining the various patterns of electrolyte transport observed in different parts of the mammalian gut. If only the Na^+–H^+ antiport were present, it would explain the jejunal absorption of Na^+ and HCO_3^- by virtue of the H^+ titration of luminal HCO_3^-. In the ileum and in the colon of some animal species, both antiport systems would explain the apparent coupled NaCl absorption. In rabbit descending colon, only the Cl^-/HCO_3^- exchange is present in conjunction with an amiloride-sensitive Na^+ entry mechanism, and this combination of transporters would account for the apparent electrogenic Na^+ absorption and Cl^-/HCO_3^- exchange.

The third model of electrolyte transport, presented by Naftalin and colleagues, concerns itself primarily with mechanisms of electrolyte secretion.[17–19] Cyclic nucleotide- and Ca^{2+}-stimulated secretion in rabbit ileum has been shown by numerous investigators to be the result of decreases in the mucosal to serosal fluxes of Na^+ and Cl^- rather than increases in serosal to mucosal fluxes of these ions. This finding is one of the reasons why Field, Frizzell, and Schultz proposed an antiabsorptive effect of cAMP. Naftalin and colleagues suggested that the changes in Na^+ flux come about by regurgitation of Na^+ back across the intracellular tight junction from the intracellular space where it has been transported (Fig. 3). The cation permeability of the tight junction would allow passive Na^+ movement to be electrically coupled to the movement of Cl^- across the apical cell membrane. The most compelling proof for their model comes from studies with triaminopyrimidine (TAP), an organic polyvalent cation which blocks tight junctional conductance of Na^+ by neutralizing the negative charges that line the tight junction.[20] When TAP was added to the Ussing-chambered rabbit ileum, it blocked theophylline-induced decreases in J^{Na} and J^{Cl} but it did not alter electrogenic Cl^- secretion.[17]

The three models described above can be analyzed in terms of transport mechanisms at the apical and basolateral cell mem-

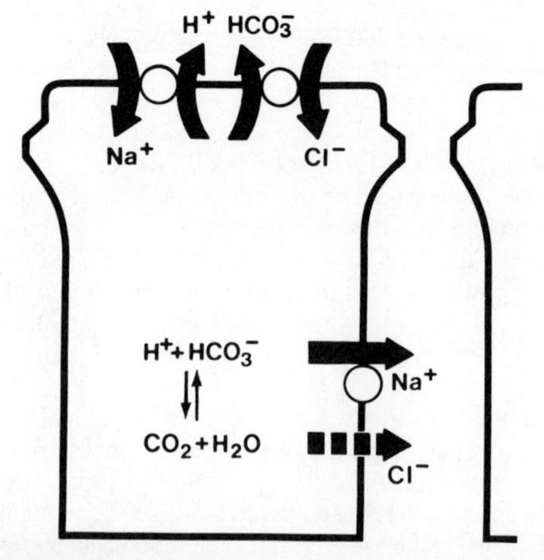

Fig. 2. This modification of the model of intestinal electrolyte transport places coupled antiport systems (Na^+–H^+/Cl^-–HCO_3^-) at the apical cell membrane instead of NaCl cotransport as the major process for Na^+ and Cl^- entry. Reproduced by permission from Ref. 16.

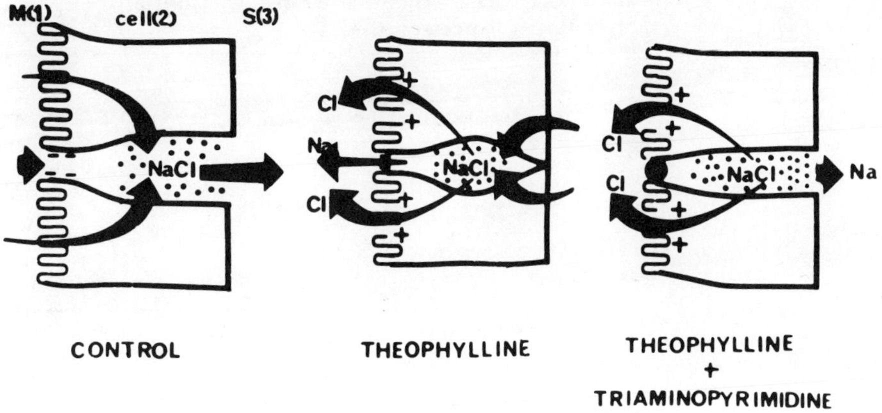

Fig. 3. This model for intestinal secretion states that the antiabsorptive effect of secretory stimuli is the result of electrical coupling of Na^+ to Cl^- secretion with subsequent regurgitation of Na^+ back across the cation-selective tight junction. Compounds such as triaminopyridine, which block tight junction cation conductance, would prevent the serosal to mucosal movement of Na^+. Reproduced by permission from Ref. 17.

branes and new information concerning the role and properties of the intercellular shunt pathway.

3. Intestinal Na^+ and Cl^- Absorption

3.1. Apical Cell Membrane Entry Processes

3.1.1. NaCl Coupled Entry

Nellans *et al.* were the first to suggest that Na^+ and Cl^- absorption across the rabbit ileal epithelium was initiated by coupled NaCl entry at the apical cell membrane.[21] They measured unidirectional influx of $^{36}Cl^-$ and $^{22}Na^+$ in the first 30–45 sec after exposure to the isotopes (thus reflecting only the entry of Na^+ or Cl^-) and corrected this entry of electrolytes for shunt path permeation. These studies indicated that approximately 20% of Cl^- entry was Na^+-dependent and a similar proportion of Na^+ entry was Cl^--dependent (Fig. 4). Furthermore, this process was noncompetitively inhibited to the same degree by elevation of cellular cyclic nucleotide concentrations, suggesting an important role for this coupled entry process in hormonal or toxin-regulated electrolyte transport. The process was also inhibited by acetazolamide, a carbonic anhydrase inhibitor, by a mechanism *not* involving carbonic anhydrase.[22] Competitive inhibition was demonstrated with Li^+[23] and with SCN^-.[24] A similar coupled NaCl process was demonstrated by the same investigators in the brush border membrane of the flounder intestine[25–27] and in the rabbit gallbladder[28] and the mechanism was shown to be inhibited by loop diuretics such as furosemide.

Because membrane vesicles allow study of transport processes in a single cell membrane and also permit rigorous control

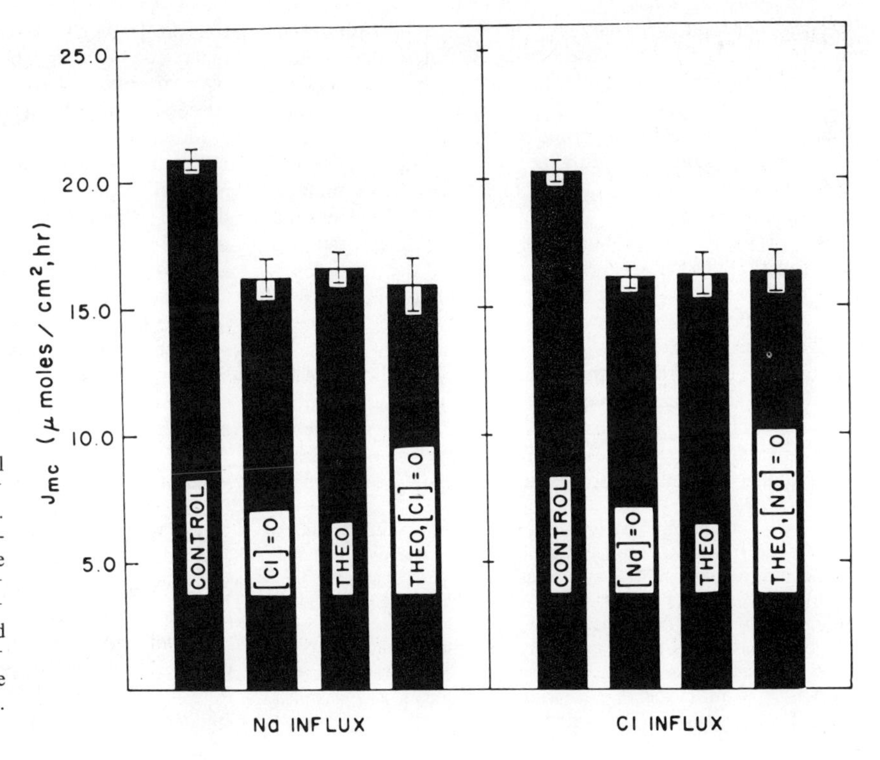

Fig. 4. Unidirectional influx from mucosal solution into the cell (J_{mc}) of Na^+ and Cl^- across the rabbit ileal apical cell membrane. Influx is measured in the presence and absence of the corresponding co-ion and in the presence and absence of theophylline. Na^+ influx from Cl^--free solutions and Cl^- influx from Na^+-free solutions are inhibited to the same degree (~ 20%) as Na^+ and Cl^- influx in tissues treated with theophylline which increases levels of cyclic nucleotides. Reproduced by permission from Ref. 21.

Table I. Apparent Half-Maximal Transport Constants (K_m's) for Na^+-Coupled Electrolyte Entry Processes, mM[a]

	Na^+–Cl^-			
	Cl^--stimulated Na^+	Na^+-stimulated Cl^-	Na^+–H^+	Cl^-–HCO_3^-
Intestine	11[21]	29[24]	16.0 (ileum)[32]	—
	4.5[44]	—	4.7 (jejunum)[31]	—
Gallbladder	27[48]	20[48]	2.8[40]	1.9[40]
			11.3[41]	
Kidney	2[52]	—	13.3[35]	—
			5.0[38,39]	

[a]These K_m's were derived from influx studies in whole tissue, studies of cell volume, and uptake studies in vesicles.

of transmembrane potentials, it is not surprising that investigators turned to these techniques to study the nature of the coupling of Na^+ to Cl^-. Using vesicles, the bulk of evidence has suggested that the coupling of Na^+ to Cl^- in intestine is indirect—the result of simultaneous Na^+–H^+/Cl^-–HCO_3^- (or OH^-) antiport mechanisms.*

Murer *et al.*[29] and Liedtke and Hopfer[30] showed Na^+/H^+ exchange by demonstrating proton ejection from rat small intestinal brush border vesicles when Na^+ was added to the outside (transport) solution. In addition, a pH gradient across the vesicle membrane ($[H^+]_{in} > [H^+]_{out}$) could be shown to induce the uphill transport and accumulation of Na^+ with an "overshoot." These phenomena were not altered by "short circuiting" the vesicle with K^+ and the K^+ ionophore valinomycin or when the proton conductance was increased with an H^+ ionophore such as carbonylcyanide-*p*-trifluoromethoxyphenylhydrazone (CFCCP), indicating that the movement of Na^+ in exchange for H^+ was an electrically neutral process not driven through a conductance channel by any electrical potential. The best characterization of intestinal Na^+/H^+ exchange has been done in rabbit jejunum by Gunther and Wright[31] and in rabbit ileum by Knickelbein *et al.*[32] These investigators demonstrated that an outwardly directed proton gradient (pH_{in} 5.5:pH_{out} 7.5) would stimulate Na^+ uptake with a two- to fourfold overshoot. The process could be inhibited by high dose ($> 10^{-6}$ M) amiloride† and harmaline, but not by diuretics such as acetazolamide, furosemide, or bumetanide or by anion-exchange inhibitors such as 4,4′-diisothiocyano-2,2′-stilbene disulfonate (DIDS) and 4-acetamido-4′-isothiocyano-2,2′-stilbene disulfonate (SITS). It has been suggested that the inhibition of Na^+/H^+ exchange by amiloride, a weak base, is due to its ability to enter the vesicle and titrate the H^+ ion.[34] This seems unlikely since Knickelbein *et al.*[32] have shown: (1) amiloride inhibition of Na^+ influx in the absence of a pH gradient or when the pH gradient was reversed (inside alkaline); (2) inhibition occurs at amiloride concentrations insufficient to overcome the intravesicular buffer; and (3) the effect of amiloride appears to be instantaneous, unlike a process whereby the base must move across the membrane to titrate H^+ and thus collapse the pH gradient. $^{22}Na^+$ uptake in gut was also inhibited by other monovalent cations with the following sequence of substitution: $Na^+ = Li^+ = NH_4^+ \gg K^+ = Rb^+ = Cs^+ >$ trimethylammonium (TMA^+),[32] a sequence also demonstrated in brush border kidney membranes.[34–36] The apparent K_m values for the Na^+/H^+ exchange process were 4.7 mM in the jejunum and 16 mM in the ileum (Table I), very similar to those demonstrated in the renal tubule[35–39] or in the gallbladder.[40,41]

The Cl^-/HCO_3^- exchange has also been demonstrated in intestinal brush border membrane vesicles,[42–44] as well as in the kidney[45] and gallbladder.[40] Kinetic values for this process are not available except for the gallbladder, where the K_m appears to be 1.9 mM.[40] This antiport system is inhibited by SITS and DIDS and by furosemide.[40–43]

A series of investigations from Spring's laboratory[40,46–49] presents evidence for both the NaCl symport and the two exchangers in the apical cell membrane of *Necturus* gallbladder. They propose that the function of the coupled NaCl influx mechanism is to accomplish transepithelial Na^+ and Cl^- (and H_20) transport, whereas the exchange processes, which could be demonstrated only after exposure of the cell to hypertonic solutions, play an important role in cell volume regulation. These two processes could be separated on the basis of solution HCO_3^- requirement, by their sensitivity to loop diuretics, SITS, and DIDS, and to amiloride (Table II), as well as by differences in Michaelis–Menten kinetics (Table I). Others have suggested that the antiport systems might be important in volume regulation,[33] pH regulation,[33,49] ovum fertilization,[33] cell proliferation,[33] or cell differentiation.[33]

These processes have been studied in the gut and kidney. Liedtke and Hopfer[42,50] could not demonstrate NaCl symport in rat small intestinal vesicles. However, they explored the effects of Na^+ on Cl^- self-exchange and Cl^- on Na^+ self-exchange at concentrations greater than 25 mM, which is probably above the K_m of the postulated NaCl symport mechanism (see Table I). Thus, it is possible that they might have missed such a system. Fan *et al.*[44] in rabbit ileum, Eveloff *et al.*[51] in flounder small intestine, and Eveloff and Kinne[52] in rabbit kidney have presented evidence for Cl^--stimulated Na^+ transport in brush border membrane vesicles. Fan *et al.* also found Na^+-stimulated Cl^- uptake.[44] These investigators studied transport at low Na^+ (2 mM) and low Cl^- (20 mM) concentrations and reported a sequence of anion stimulation of Na^+ transport of $Cl^- = SCN^- > NO_3^- >$ gluconate and cation stimulation of Cl^- transport of $Na^+ = Li^+ > K^+ >$ choline. The apparent K_m for the Cl^--stimulated Na^+ transport was 4.5 mM.

*For ease of presentation, we will refer to the anion-exchange mechanism as a Cl^-–HCO_3^- antiport. It should be understood that it is difficult, if not impossible, to differentiate such a mechanism from a Cl^-–OH^- antiport or a Cl^-–H^+ symport that is accompanied by shifts in CO_2 across the membrane.

†Benos[33] has discussed the amiloride dose responsiveness of Na^+ entry mechanisms: Na^+/H^+ exchange requires concentrations greater than 10^{-6} M for inhibition whereas the Na^+ conductance channels require concentrations of 10^{-6} M or less.

Table II. Comparison of the Effect of Various Inhibitors on Coupled Electrolyte Transport in Intestine and Gallbladder[a]

	Intestine[21,26,27,31,41–44,51]			Gallbladder[28,40,41,47–49]		
	Na^+–Cl^-	Na^+–H^+	Cl^-–HCO_3^-	Na^+–Cl^-	Na^+–H^+	Cl^-–HCO_3^-
Loop diuretic (bumetanide or furosemide)	+	–	+	+	–	–
Amiloride	+	+	–	–	+	–
Disulfonic stilbenes (SITS and/or DIDS)	+	–	+	–	–	+
Cyclic nucleotides or Ca^{2+}	+	–	?	+	?	?

[a] +, inhibition; –, no inhibition; ?, data not available.

This NaCl transport was inhibited by bumetanide and furosemide but also by inhibitors of Na^+ transport (harmaline and amiloride) and by anion-exchange inhibitors such as SITS and DIDS (Ref. 44; Fan and Powell, unpublished observation). Thus, although the pattern of inhibition (Table II) suggests that all three processes—Na^+/Cl^-, Na^+/H^+, and Cl^-/HCO_3^-—are present in the gallbladder apical cell membrane, the pattern of response to inhibitors in the intestine suggests only the simultaneous exchanges of Na^+ for H^+ and Cl^- for HCO_3^-. Similar studies in renal tubules also yield conflicting answers, with some investigators proposing coupled NaCl transport and others Na^+–H^+/Cl^-–HCO_3^- exchange.[35–39,52,53] A final answer as to the exact nature of these mechanisms must await more specific inhibitors or else purification and reconstitution of the separate processes.

Recently, studies of the cotransport process in flounder[54] and *Necturus* gallbladder[55] have shown that K^+ is involved with Na^+ as the cation translocated with two Cl^- molecules, i.e., Na^+–K^+–2Cl-cotransport mechanism. Whether K^+ is involved in mammalian epithelial NaCl transport remains to be determined. This process is discussed in more detail in the section on K^+ transport.

HCO_3^- appears to stimulate Na^+ and Cl^- absorption without itself being apparently absorbed. The mechanism of luminal HCO_3^- stimulation of NaCl transport has been studied in gallbladder.[56,57] The proposed mechanism (Fig. 5) depends on the presence of coupled antiport systems in the brush border, and therefore, this process could also apply to mammalian ileum or colon. In this scheme, un-ionized H_2CO_3 diffuses across the brush border and is converted to HCO_3^- and H^+ by carbonic anhydrase. This increases the supply of these ions for participation in the Na^+/H^+ and Cl^-/HCO_3^- exchangers. Such a process could also account for the stimulation of Na^+ (and Cl^-) absorption by luminal SO_3^{2-}[58] or by weak organic acids[59] without having to postulate involvement of a specific ion pump such as an anion-stimulated ATPase.[60] For that matter, careful studies by Van Os *et al.*[61] have not demonstrated a HCO_3^--stimulated ATPase in purified rat apical membranes. Glucose in the bathing solution also stimulates coupled NaCl transport.[62,63] Perhaps metabolic substrates increase the availability of intracellular H^+ and HCO_3^- for the exchange mechanism.

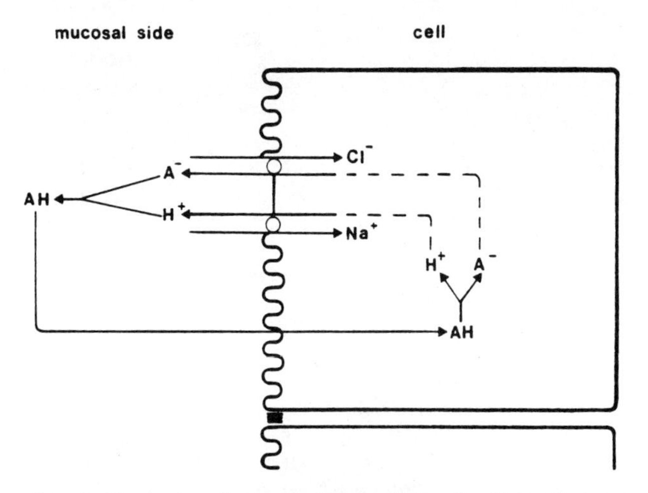

Fig. 5. Proposed mechanism for the stimulation of epithelial Na^+ and Cl^- absorption by luminal HCO_3^-. The un-ionized weak acid H_2CO_3 diffuses across the apical cell membrane, dissociates, and the H^+ and HCO_3^- drive the coupled Na^+ and Cl^- antiports. Reproduced by permission from Ref. 56.

3.1.2. Na^+–Glucose Coupled Entry

In the past decade the important basic advances regarding Na^+–glucose coupling have been the emerging concepts that it is the electrochemical gradient, not just the chemical gradient, that drives Na^+-coupled glucose transport and that the stoichiometry of the Na^+–glucose coupling is probably 2 : 1 and not 1 : 1 as originally thought. Studies which showed the inadequacy of the Na^+ chemical gradient for the energetics of Na^+-coupled glucose transport threatened the entire concept of the Na^+ gradient hypothesis.[64–66] The general concept was revived, however, by the demonstration with microelectrodes that Na^+-coupled glucose movement across the apical cell membrane depolarized the mucosal membrane; thus, it is an electrogenic (rheogenic) process.[67–70] Studies with brush border membrane vesicles confirmed the electrogenicity of Na^+–glucose entry; alterations in vesicle transmembrane electrical potential difference (PD) with ion gradients and ionophores were shown to significantly affect glucose uptake.[71,72] A similar effect of transmembrane PD was demonstrated in isolated intestinal cells that were ATP depleted and then loaded with K^+ and short-circuited with valinomycin.[73]

Thus, glucose entry across the brush border membrane of the cell appears to take place on a carrier that is positively charged, and the movement of charge is not compensated for by the simultaneous movement of anion or the countermovement of cation out of the cell on the carrier. If one takes into account the total electrochemical potential, there is sufficient energy to achieve 30-fold chemical gradients for sugar across the cell membrane.[74] However, if Na^+-independent sugar efflux routes are blocked by a variety of inhibitors such as phloretin or cytochalasin B,[75] or if one studies uptake of a sugar such as α-methylglucoside, which does not exit on the serosal carrier,[76] gradients as large as 70- to 100-fold can be demonstrated. The electrochemical gradient required for sugar accumulation of this degree demands a coupling ratio of Na:glucose that is greater than 1 (see Refs. 64–66, 74 for a discussion of the thermodynamics and energetic adequacy of the Na^+ gradient).

Several investigators have now presented clear evidence for a coupling ratio of 2 or greater. Kimmich and Randles used $^{22}Na^+$ and 3-*O*[^{14}C]methylglucose influx into valinomycin-treated, ATP-depleted intestinal cells to show a ratio of 1.93.[77] In similar studies of normally energized (ATP intact) cells with the usual transmembrane electrical PDs, the coupling ratio was 1. Previous studies of Na^+–glucose uptake into renal[78,79] and intestinal[80] brush border membrane vesicles also supported a 1 : 1 stoichiometry. However, if Na^+ and glucose uptake into such a leaky system is maximized by short-circuiting the vesicle with valinomycin and KCl, by adding CCFP (an H^+ ionophore) to prevent the establishment of H^+ gradients across the membrane, and by adding amiloride to block Na^+ entry via the Na^+/H^+ exchange mechanism, a coupling coefficient of 1.9 was found using a Hill analysis and a coupling ratio of 3.2 was demonstrated by direct measurements of Na^+ and glucose uptake.[81] Using α-methylglucoside as the substrate, a stoichiometry of 2 has also been demonstrated in brush border membrane vesicles from cultured kidney cells.[82]

The next question, given the new findings above, is: what is the nature of the carrier—how does it interact with Na^+ and glucose? Studies with phlorizin, a glycoside that competitively inhibits glucose transport but is not moved across the apical cell membrane by the glucose carrier, suggest that the carrier in renal microvillous membranes is negatively charged. Na^+ appears to increase the affinity of the carrier for glucose, and Na^+ binding to the carrier without glucose prevents its translocation.[83] While some studies suggest that Na^+ and phlorizin bind to the carrier in random fashion,[84] equilibrium uptake studies in renal and intestinal brush border membrane vesicles indicate that the binding of glucose and Na^+ to the carrier is by an order reaction with either ligand binding to the carrier first, and that the translocation of ligands and carriers has glide symmetry (i.e., the first in, the first out).[80] In kinetic terminology, this would fit an isoordered bi-bi model. This model is consistent with a gated pore concept where transport is achieved by the transport protein undergoing a rocker-type conformational change which moves the permeability barrier past the solute. It is hoped that purification studies of the Na^+–glucose carrier will clarify this process.[85–92]

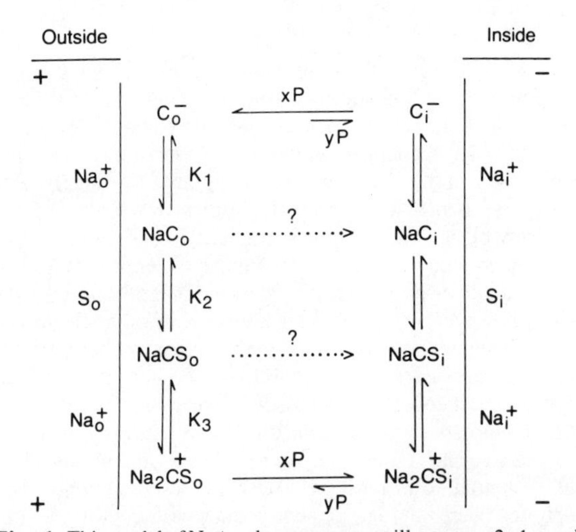

Fig. 6. This model of Na^+–glucose transport illustrates a 2 : 1 coupling ratio and a role for transmembrane electrical potential difference. The electrical orientation would drive the negatively charged carrier to the luminal (outside) phase boundary where it would bind to Na^+, then to glucose, and then to a second Na^+ ion. The positively charged quaternary complex would then be pulled to the cellular (inside) phase boundary where the low Na^+ concentration would cause the carrier and solutes to dissociate. It is not clear whether the intermediate form of the carrier (NaC_o and $NaCS_o$) might cross the membrane, but probably does not. Reproduced by permission from Ref. 65.

Based on the considerations above, a new model for Na^+-coupled glucose transport can be described (Fig. 6).

3.1.3. Other Na^+-Coupled Entry Processes

There are relatively few new additions to the list published by Schultz and Curran in 1970 of solutes whose transport is Na^+-dependent.[93] Some recent findings concerning Na^+-dependent organic solute transport are: (1) neutral amino acids enter across the apical cell membrane on a carrier that is also sensitive to electrical potential[94,95]; (2) the Na^+ gradient model may explain the transport of many vitamins such as biotin, choline, inositol, nicotinic acid, thiamine, folic acid, ascorbic acid, and perhaps riboflavin[96]; (3) the transport of monocarboxylic acids,[97,98] dicarboxylic amino acids,[99] and tricarboxylic acids[100] across both renal and intestinal brush border membrane vesicles is Na^+-dependent; (4) xylose transport is Na^+-dependent[101]; and (5) the active transport of bile acids across the terminal ileum is a Na^+-dependent process.[102–104]

Two multivalent inorganic anions, sulfate and phosphate, have been extensively investigated. Studies in small intestinal vesicles[105] and intact ileum[106–108] indicate that SO_4^{2-} uptake across the brush border is a Na^+-dependent, electrically neutral process; i.e., two Na^+ (or one Na^+ and another cation) move across the membrane with one SO_4^{2-}. Changes in $Na^+ : SO_4^{2-}$ stoichiometry at different SO_4^{2-} concentrations and pHs suggest cellular entry by way of Na^+ coupling to ion pairs ($NaSO_4^-$) or by Na^+–H^+–SO_4^{2-} cotransport depending on the concentration of the ion pair available.[108] Sulfate exit across the basolateral membrane is by an anion-exchange process that is inhibited by SITS and DIDS. Interestingly, the exchange of SO_4^{2-} does not appear to involve HCO_3^-,[107] an anion involved in the basolateral Cl^- exchange.

The transport of phosphate is interesting because phosphate exists at physiological pHs in both the monovalent ($H_2PO_4^-$) and the divalent (HPO_4^{2-}) form. Phosphate transport in brush border membrane vesicles is a Na^+-dependent, electrically neutral process that is competitively inhibited by arsenate. Thus, the monovalent phosphate is accompanied by one Na^+ ion and the divalent phosphate by two Na ions.[109–111] There is the suggestion that the intestine preferentially transports monovalent phosphate while the renal brush borders transport divalent phosphate.[110,111] In the intestine, the Na^+-dependent phosphate influx mechanism appears to be stimulated by vitamin D_3 (1,25-dihydroxycholecalciferol).[112,113]

3.1.4. Noncoupled and Amiloride-Sensitive Na^+ Entry

Removal of either Na^+ or Cl^- from or addition of furosemide or theophylline to glucose-free, amino acid-free mucosal solutions reduces Na^+ and/or Cl^- entry across the rabbit ileal,[22–24] flounder intestinal,[25–27] and rabbit gallbladder[28] apical cell membrane by only 20–30%. There remains a significant Na^+-dependent short-circuit current (I_{sc}) in rabbit ileum bathed in glucose-, amino acid-, Cl^--, and HCO_3^--free solutions that is not inhibited by 10^{-4} M amiloride,[113] a concentration that should have some effect on Na^+/H^+ exchange.[33]* Thus, it appears from studies *in vivo* and with whole tissue *in vitro* preparations that a significant component (> 50%) of Na^+ entry in the small intestine is not identified and presumably takes place by simple diffusion. Na^+-uptake studies in vesicles are contradictory. Vesicles made from flounder small intestine[51] and rabbit ileum[32] show no influence of imposed PD, indicating that virtually all Na^+ uptake was predominantly by electrically neutral, carrier-mediated pathways (Na^+/H^+ exchange or Na^+–Cl^- symport). In contrast, simple diffusion appears to be the predominant entry pathway in rabbit jejunal vesicles.[31] Similar studies of Cl^- uptake in vesicles provide clear evidence of a significant Cl^- conductance.[42,43] Therefore, noncoupled Na^+ entry in small intesting remains poorly defined.

The colon is less perplexing because there the noncoupled Na^+ entry mechanism is inhibited by low dose (< 10^{-6} M) amiloride* and thus is easily identified. However, the percentage of Na^+ entry that is amiloride-sensitive varies in different species,[115] varies in different colonic segments,[116] and varies in the same segment under different physiological conditions.[63] Tissues in which the amiloride-sensitive component is the major Na^+ entry process manifest characteristic *in vitro* electrolyte fluxes in which $J^{Na}_{net} = I_{sc}$. In these tissues there is either no net Cl^- transport or else net Cl^- absorption is exactly equal to the residual flux (HCO_3^- secretion).[115] Thus, in amphibian colon,[117–119] rabbit descending colon,[120,121] and bird cloaca,[122] appropriately 90% of Na^+ transport across the cell originates from amiloride-sensitive apical membrane channels. In contrast, NaCl coupling accounts for 50–100% of entry in rat colon.[62,123] The rat ascending colon appears to have exclusively electrically neutral entry while the rectum transports Na^+ electrogenically.[62] In man, both types of Na^+ transport are found.[115,124,125] In the monkey, both types of transport are seen in the same segments of different animals when studied with the same techniques by the same investigators.[63] Thus, the proportions of electrogenic and neutral Na^+ transport are, to a certain extent, species- and colonic segment-specific, but are also controlled by other factors (e.g., mineralocorticoids and cell [Na^+]; see below).

The process of Na^+ entry across the membrane by the amiloride-sensitive channel is a saturable function of mucosal solution [Na^+].[117,119,126] The process is also conductive: microelectrode studies in rabbit colon indicate that addition of amiloride to the mucosal solution hyperpolarizes the apical membrane by 53 mV as it abolishes the net Na^+ transport.[127,128] Li^+ will also be transported across the amiloride-sensitive mechanism, but when both Li^+ and Na^+ are in the mucosal solution, Li^+ does not inhibit Na^+ entry.[129,130] These latter results are *not* compatible with a carrier-mediated process and are more explicable by postulating a cation-sensitive "pore" which allows noninteractive movement of both Na^+ and Li^+ at the samg time. "Noise" analysis,[131] binding studies with ^{14}C-labeled amiloride analogs,[132] and evaluation of Ussing's flux-ratio equation[133] all reach similar conclusions. The possible mechanisms of amiloride inhibition of the Na^+ channel, questions such as competitive vs. noncompetitive inhibition, the requirements for divalent cations such as Ca^{2+}, and the chemical nature of the amiloride receptor have recently been reviewed.[33]

An interesting teleological explanation for amiloride-sensitive channels has been suggested by Legris *et al.*[134] Noting the structural similarity between amiloride and naturally occurring indoleamines such as serotonin and melatonin, they suggest that release of these hormones by mucosal endocrine cells into the solution bathing the apical membrane would provide a means of instantaneous regulation of fluid and electrolyte transport. The practicality of this suggestion hinges on whether serotonin and melatonin concentrations in the intestinal mucosal unstirred layer can reach the 6–8 mM necessary for effective inhibition of the amiloride-sensitive channels.

3.2. Interactions between Apical and Basolateral Cell Membranes

There is a close interaction between the pumps and leaks of the membranes at the opposite poles of the epithelial cell which may prevent explosion of the cell through loss of volume regulation. This concept has been reviewed succinctly by Schultz.[135] This interaction has two main components: first, the rate-limiting step to Na^+ transport by epithelia is the entry process across the apical cell membrane. Second, intracellular [Na^+] regulates Na^+ entry through a negative feedback mechanism and also regulates K^+ leak (conductance) across the basolateral membrane. The increase in K^+ conductance would be useful to dissipate the K^+ entry across the basolateral membrane as the result of the 3:2 Na^+/K^+ exchange via the Na^+,K^+-ATPase located there.

In tight epithelia, maneuvers that increase [Na^+] in the cell, such as increased Na^+ entry by amphotericin B,[126] organic anions,[136] and aldosterone,[137] as well as inhibited Na^+ exit by ouabain,[138] will decrease amiloride-sensitive Na^+ conductance across the apical membrane. Conversely, decreases in cell [Na^+] by exposure to Na^+-free solutions increase amiloride-sensitive Na^+ channels.[138] Thus, the number of open Na^+ channels in the apical membrane is synchronized to the activity of the basolateral Na^+ pump with the number of open channels decreasing as cell [Na^+] rises. Aldosterone appears to circumvent this regulatory step, maximizing the number of open Na^+ channels.

Microelectrode studies in *Necturus* intestine have shown interactions between the two membranes of leaky epithelia as well.[70] The addition of amino acids (alanine) and actively transported monosaccharides (galactose) to the mucosal solution depolarizes the apical membrane due to the electrogenic movement of Na^+ on the coupled carrier. However, within a few minutes the apical membrane is repolarized as a result of the changes in the basolateral membrane PD (due to an increase in

*See daggered footnote on p. 562.

K^+ conductance) with translation of the electrical signal via the low resistance of the shunt pathway.

Schultz has suggested that a change in intracellular free [Ca^{2+}] is the intracellular signal which regulates the apical Na^+ and the basolateral K^+ conductances, and he quotes ample circumstantial evidence in other epithelia to substantiate this speculation.[135]

3.3. Basolateral Cell Membrane Exit Processes

3.3.1. Na^+ Exit

There has been indisputable localization of intestinal Na^+,K^+-ATPase to the basolateral membrane, with both autoradiographic techniques using [^{3}H]ouabain binding[139] and cell fractionation and membrane localization technique[140,141] demonstrating that 95% of the cellular content of the enzyme is localized there. Furthermore, the identity of Na^+,K^+-ATPase the Na^+ pump seems certain based on recent reconstitution experiments in which relatively purified enzyme placed in liposomes demonstrates characteristic Na^+ pumping.[142–147] The stoichiometry of cation exchange by this enzyme in the intestine[148] and in other tissues[142–147] has been substantiated as 3 Na^+ : 2 K^+. Thus, the Na^+ pump is electrogenic and it accomplishes net charge transfer across the basolateral membrane contributing to the transepithelial PD.[148,149] In spite of studies to the contrary,[150,151] most investigations confirm its role as the primary driving force for transepithelial Na^+ transport.[152]

In the rat, the rank order of ATPase activity among different intestinal segments is jejunum > colon ≫ ileum.[153] The enzyme in this species is relatively insensitive to cardiac glycosides; the K_m for ouabain binding to purified rat basolateral membrane enzyme is 1.5×10^{-5} M and for inhibition of transport is 3×10^{-5} M,[141] which is two orders of magnitude higher than those in red cell and rabbit kidney. Harms and Wright[141] have purified rat brush border membranes and have found $1.5–15 \times 10^6$ pump site per enterocyte with a turnover number approximating 3500 sec^{-1} at 22°C. These studies confirmed that there is sufficient Na^+,K^+-ATPase to accomplish measured rates of intestinal Na^+ transport.

ATP utilization and O_2 consumption have been studied in several epithelia in relation to Na^+ transport.[154] The stoichiometry of Na^+ transport to O_2 consumption in rat intestine is 11,[155] similar to the 10–30 moles of Na^+ per mole of O_2 reported in other epithelia.[154] One Na^+ transported per ATP molecule broken down was found in rat intestine[155] as compared to two or four in other epithelia.[154] Cl^- and HCO_3^- transport also require energy as demonstrated in experiments where O_2 consumption is measured as these ions are sequentially removed from the solution.[156]

Studies of O_2 consumption and CO_2 production in the basal state and after addition of various oxidative fuels give insight into the oxidative metabolism of the intestine. The basal RQ of rabbit[157] and canine[158] small intestine is 0.70–0.75, a value consistent with lipid being the major source of energy in the interdigestive period. Furthermore, this rate of metabolism in the small intestine can be sustained for several hours *in vitro*. Presumably, fatty acid metabolism takes place via the tricarboxylic acid pathway or more likely via utilization of ketone bodies such as β-hydroxybutyrate. After ingestion of food, the small intestine and colon can metabolize glucose anaerobically to lactate via the glycolytic pathway.[159,160] The ileum, but not the jejunum, exhibits the Pasteur effect, i.e., a decrease in lactate production on exposure to O_2, signifying a switch from anaerobic to aerobic metabolism. In fact, there appears to be a gradient of glycolytic enzymes proceeding from proximal to distal small bowel.[161] The addition of glucose, fructose, and short-chain fatty acids (SCFAs) to rabbit small intestine has little effect on O_2 consumption[157] but does increase consumption in the dog and chick.[158,162] Glutamine and glutamate increase O_2 consumption, increasing the RQ toward 1 in several species.[157,158,162,163] Presumably, these substrates enter the tricarboxylic acid cycle after conversion to α-ketoglutarate.

Studies of colonic aerobic fuel utilization in the rabbit,[164] rat,[165–167] and human[165,166] indicate that the proximal colon utilizes glucose, butyrate, and glutamine for respiratory fuels; the distal colon depends almost entirely upon SCFAs, especially on butyrate, via ketone formation. Thus, SCFAs, which are generated by bacterial action on nonabsorbed carbohydrates, represent a major source of metabolic fuel for the colon. Metabolic studies in ulcerative colitis indicate an unexplained depression of butyrate utilization and an increase in glucose and glutamate utilization in the diseased colon.[167]

The practical message of these studies for investigators is that optimal energy for *in vitro* studies can be obtained by the addition of glucose, glutamine, and ketone bodies (β-hydroxybutyrate) to the serosal bathing solution.

3.3.2. Cl^- Exit

Apical cell membrane entry of Cl^- in the small intestine via the coupled NaCl entry or exchange processes results in an intracellular Cl^- activity that is 2–3 times above electrochemical equilibrium as predicted by the Nernst equation for passive distribution of the ion (Table III). Therefore, there is no energetic need for a special process for Cl^- exit across the basolateral cell membrane; it can simply move passively down its electrochemical gradient. Even so, evidence suggests that epithelial Cl^- exit is by a carrier-mediated process. When measured in gallbladder, the Cl^- permeability of the basolateral membrane is quite low, and less than 6% of the net Cl^- transport can be attributed to passive diffusion across the cell membrane.[168] Some carrier-mediated process seems necessary. HCO_3^- has been known to stimulate Na^+, Cl^-, and water ab-

Table III. Epithelial Intracellular Cl^- Activities[a,b]

			aCl/aCl^{eq}	
	mc (mV)	aCl (mM)	Na^+–Ringer	Na^+-free Ringer
Gallbladder				
Rabbit	−49	35	2.4	1.2
Necturus	−53 to −65	17 to 35	1.8 to 4.3	1.0
Renal tubule				
Necturus	−52	25	2.3	1.4
Small intestine				
Amphiuma	−33	28	1.4	—
Bullfrog	−24	71	2.3	—
Flounder	−69	24	3.4	1.3

[a]Modified from Ref. 14.
[b]mc is the electrical potential difference across the apical cell membrane. aCl is the intracellular Cl^- activity. aCl/aCl^{eq} is the ratio of the measured to predicted Cl^- activity.

sorption in the gallbladder[169–171] and is effective when present in the serosal as well as the mucosal solution.[172] Removal of HCO_3^- from the bathing solution reduces Na^+ and Cl^- transport in bullfrog,[173] *Amphiuma*,[174] and flounder[26] small intestine, as does the addition of SITS to the serosal bathing solution.[173,174] These studies would suggest that basolateral Cl^-/HCO_3^- exchange is the process of Cl^- exit in these species.

3.3.3. Glucose and Amino Acid Transport across the Basolateral Membrane

Sugars and amino acids entering across the brush border can probably be used by the intestine for metabolic fuel and for protein synthesis, but most exit across the basolateral membrane to supply the nutrient needs of the animal. In the interdigestive period or during prolonged fast, it is necessary to have mechanisms for reentry of these important metabolites back across the basolateral membrane.

Glucose uptake across the basolateral membrane is a carrier-mediated, but not a Na^+-dependent process.[175–179] It is 25 times more rapid for D-glucose than L-glucose and is inhibited by all sugars with the D-glucose pyranose ring chair conformation.[177] The K_m for basolateral uptake (30–50 mM) is much higher than for brush border Na^+-dependent processes (< 10 mM). It is inhibited by phloretin, cytochalasin B, theophylline, and many other substrates that do not inhibit the brush border process, but it is not well inhibited by phlorizin.[177] Thus, the basolateral membrane transport system for glucose resembles the facilitated glucose transport system seen in red cell membranes.

In contrast, amino acid uptake across the basolateral membrane takes place by both Na^+-independent and -dependent mechanisms.[175,180] The Na^+-independent mechanism is stereospecific for L-amino acids and has a K_m (alanine) of 0.7 mM. The affinity for this carrier is that classic for the "L" system as described by Christensen.[181] The Na^+-dependent system has an affinity for alanine an order of magnitude higher (K_m 0.04 mM). Competition experiments with various amino acids indicate the presence of other classic amino acid systems such as the "A" and "ASC" mechanisms[181] and also a novel transport system.[175,180] Since the V_{max} for the intestine for the Na^+-independent system is an order of magnitude higher (100 nmoles/mg protein) than for the Na^+-dependent system (5 nmoles/mg protein), the Na^+-independent system is probably the major pathway for amino acid exit across the basolateral membrane after absorbed amino acids have accumulated to high concentrations in the cell. Conversely, the low K_m of the Na^+-dependent process would be useful for basolateral membrane transport of amino acids into the cell from the blood during the interdigestive period.

4. Intestinal Na^+ and Cl^- Secretion

All models of intestinal electrolyte transport (see above) suggest that modulation of net transport comes about by the stimulation of secretion, i.e., the direction of net water and electrolyte transport is dependent on the degree to which Cl^- secretion is stimulated. In the basal state, a finite rate of stimulated secretion can be demonstrated in the human jejunum perfused with glucose-, amino acid-, and HCO_3^--free solutions which lower active absorption of Na^+ (and passive absorption of Cl^-) such that active Cl^- secretion can be seen.[182] Because intestinal secretion plays such a significant role in diarrhea, the normal physiology and more explicit details of this process are described more completely in the chapter in this volume by Field *et al.* Nevertheless, for completeness and a broad understanding of intestinal ion transport, the bare elements of intestinal secretion of water and electrolytes are considered here.

4.1. Intracellular Mediators of Secretion

Intestinal secretion, like most other secretory systems, is controlled by intracellular levels of cyclic nucleotides and Ca^{2+}. The stimuli which elevate intracellular levels of these agents; the transducing mechanisms which couple hormone- or toxin-receptors to the respective cyclase; the processes of Ca^{2+} gating or release from internal reservoirs; concepts and information on cyclic nucleotide-, Ca^{2+}–calmodulin-, and Ca^{2+}–phospholipid-dependent protein kinases; and the various possible interactions between the cyclic nucleotide- and Ca^{2+}-dependent systems are summarized in Fig. 7. Additional information on these subjects is available in various reviews.[1,2,183–187]

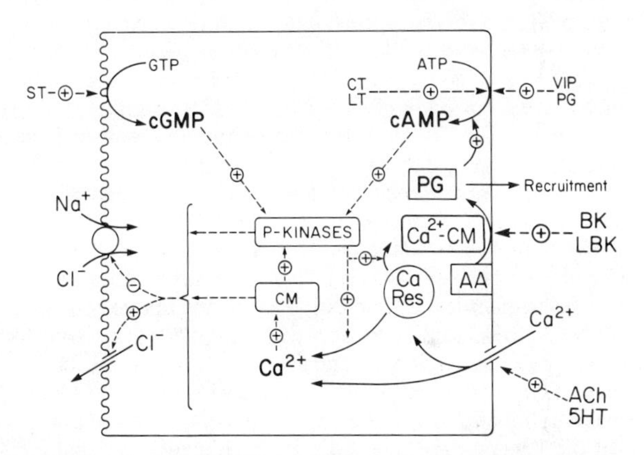

Fig. 7. The three intracellular mediators of intestinal secretion are cGMP, cAMP, and Ca^{2+}. Intracellular levels of these agents are increased by bacterial toxins such as *E. coli* heat-stable toxin (ST) and *E. coli* heat-labile toxin (LT), cholera toxin (CT), and by hormones such as vasoactive intestinal polypeptide (VIP), prostaglandins and leukotrienes (PG), acetyl choline (ACh), serotonin (5HT), bradykinin (BK), or kallidin (LBK). The cyclic nucleotides may inhibit coupled NaCl influx or stimulate Cl^- conductance across the apical cell membrane through protein kinase-mediated (P-KINASE) phosphorylation of membrane proteins. Alternatively, cyclic nucleotide-dependent kinases might release Ca^{2+} ions from intracellular reservoirs (Ca Res). Increases in intracellular Ca^{2+}, originating from reservoirs or from hormones that cause gating of Ca^{2+} across the basolateral cell membrane, might inhibit coupled NaCl uptake and promote Cl^- conductance via calmodulin (CM)-dependent protein kinases. Alternatively, Ca^{2+}–CM might activate phospholipase which releases membrane-bound arachidonic acid (AA) with the subsequent formation of prostaglandins and leukotrienes by the cyclooxygenase and lipoxygenase pathways. These PGs might stimulate the production of cAMP (or cGMP) in some specific, transport-related cyclic nucleotide pool. Evidence suggests that BK and LBK work through cellular PG production; it is possible that other stimuli might do so also. PGs thus formed might also stimulate adjacent cells to secrete (recruitment), and could modulate intestinal motility and/or blood flow. Ca^{2+}–CM is also a necessary cofactor for cyclic nucleotide synthesis and degradation (cyclase and phosphodiesterase activity).

4.2. Secretory Transport Mechanisms

The models of Cl^- secretion discussed previously all depend on Cl^- attaining an intracellular activity above electrochemical equilibrium. As noted in Table III, intracellular Cl^- activity has indeed been found to be 2–4 times greater than equilibrium. Presumably, this occurs as the result of a coupled NaCl entry mechanism at the basolateral aspect of the Cl^--secreting cell. The findings that intestinal cyclic nucleotide-stimulated Cl^- secretion is dependent on bathing solution Na^+[188–190] and is inhibited by the serosal addition of loop diuretics such as ethacrynic acid,[191,192] furosemide,[191] and ozolinone[193] are evidence in favor of this. A dependence of Cl^- secretion on serosal K^+[190] also raises the question whether this coupled process, like the one on the apical membrane, may really be a Na^+–K^+–$2Cl^-$ cotransport mechanism.

The models of Cl^- secretion also require a stimulus-mediated increase in apical cell membrane Cl^- conductance to allow Cl^- egress into the luminal solution. Such has been demonstrated in corneal[194] and tracheal epithelia,[195] but not in gut. This issue is even more complicated in the vertebrate small intestine where cAMP, cGMP, and Ca^{2+} reduce the mucosal to serosal flux of Na^+ and Cl^- and also stimulate Cl^- secretion by increasing the serosal to mucosal fluxes of this ion.[1–3] These changes in the fluxes suggest both antiabsorptive and secretory effects of these intracellular mediators.

The basic issue, therefore, is whether these intracellular messengers actually inhibit coupled NaCl influx (antiabsorptive effect in villous cells, Fig. 1) or whether the antiabsorptive effect is really only secondary to the increase in Cl^- permeability (Fig. 3). There have been relatively few studies that directly demonstrate the effects of cyclic nucleotides and Ca^{2+} on intestinal ion transport processes. Hyun and Kimmich showed cAMP inhibition of Na^+ uptake but no effect on Cl^- uptake by cholera toxin in isolated enterocytes, suggesting an antiabsorptive action.[196] Microelectrode studies in *Necturus* gallbladder, a tissue like intestine that absorbs Na^+ and Cl^- by a coupled mechanism, present conflicting results. Diez de los Rios *et al.*[197] reported that cAMP had no effect on the apical cell membrane potential but did decrease the intracellular activities of Na^+ and Cl^-, indicating an antiabsorptive effect of cyclic nucleotides on the coupled NaCl influx mechanism. In contrast, Petersen and Reuss[198] found depolarization of the apical cell membrane on exposure of *Necturus* gallbladder to theophylline or cAMP with reduction of the intracellular Cl^- activity to equilibrium levels. Thus, cAMP caused a marked increase in apical cell membrane Cl^- permeability, and the apparent inhibition of NaCl uptake was secondary to this; i.e., the reduction in mucosal to serosal fluxes of Na^+ and Cl^- seemed to be due to the rapid efflux of Cl^- and electrical coupling of Na^+ to that movement rather than to a direct inhibition of coupled NaCl uptake. This would be direct evidence in favor of Naftalin's model (Fig. 3). Obviously, these conflicting studies need to be reconciled and similar experiments performed in the intestine.

Brush border membrane vesicles are another system in which to directly test these hypotheses.[199] Fan and Powell have demonstrated a Ca^{2+}–calmodulin inhibition of NaCl uptake in rabbit brush border membrane vesicles with no effect on Cl^- permeability.[200] In those studies, calmodulin (20 μM) was found to shift the K_m for Ca^{2+} inhibition of coupled NaCl uptake from 200 to 0.2 mM free Ca^{2+}.

Taken at face value, these studies would suggest that Ca^{2+} mediates the antiabsorptive effect of stimuli, whereas cAMP is more directly responsible for Cl^- secretion. Obviously, because of the complex interactions between cyclic nucleotides and Ca^{2+} (Fig. 7), either of these agents could have both effects in the intact cell.

5. HCO_3^-, Short-Chain Fatty Acid, and K^+ Transport

5.1. HCO_3^- Transport

Of all the areas of electrolyte transport research, study of HCO_3^- movement is the most difficult. Not only does the gut both absorb and secrete HCO_3^-, but both processes may go on simultaneously in the same segments. In addition, HCO_3^- secretion occurs from both endogenous and exogenous sources. Furthermore, it is very difficult to determine if the transported species is HCO_3^- or OH^- in one direction, or H^+ in the other. Indeed, this controversy has existed for years with regards to acidification of urine by turtle bladder (see Ref. 201). The final complicating feature is that HCO_3^- transport may take place by more than one mechanism in the same tissue.

With regard to mammalian gut, it can be generally stated that the jejunum absorbs HCO_3^-, while the duodenum, ileum, and colon exhibit net HCO_3^- secretion. Teleologically, such patterns of net bicarbonate transport are useful for neutralizing gastric acid that is presented to the duodenum,[202,203] for recovering excess HCO_3^- from the jejunum,[204–206] for regulating the pH of colonic contents to promote bacterial growth,[115] and to aid in regulation of systemic acid–base balance (see below).

5.1.1. HCO_3^- Absorption

Human,[204] rat,[205] and *Amphiuma*[206] jejunum all have been shown to absorb bicarbonate. In the mammal, HCO_3^- absorption is, at least in part, Na^+-dependent and the two ions mutually stimulate the absorption of each other.[204–210] The Na^+-dependent fraction could be due either to Na^+/H^+ exchange or to cotransport of $NaHCO_3$. An acid microclimate has been demonstrated adjacent to the brush border membrane,[211,212] which is increased by metabolizable substrates such as glucose[213] and is inhibited by acetazolamide and aminophylline.[204,213,214] The process of bicarbonate absorption is accompanied by an increase in jejunal pCO_2[205,209,210] suggesting a shift in the equilibrium of $H^+ + HCO_3^- \rightarrow H_2O + CO_2$ to the right (Fig. 8). Therefore, it has been concluded that the major mechanism for jejunal HCO_3^- absorption is the Na^+/H^+ exchange. However, not all HCO_3^- absorption is Na^+-dependent[209,210,212,214] and additional mechanisms must be considered.

White and Imon[206,215–217] have investigated an electrogenic HCO_3^- absorptive process in *Amphiuma* jejunum that is totally independent of Na^+ and, therefore, cannot be due to Na^+/H^+ exchange. It is inhibited by acetazolamide, theophylline, metabolic poisons, ouabain, and, most interestingly, by the removal of luminal K^+. Microelectrode measurements of intracellular K^+ activity together with measured decreases in HCO_3^- absorption by luminal removal of K^+ suggest $KHCO_3$ symport or K^+–H^+ antiport as the mechanism of HCO_3^- absorption in this species.[215–217] It is not known if K^+-dependent HCO_3^- uptake (K^+/H^+ exchange) occurs in the mammalian gut; however, Lucus[214] has reported an ouabain- and

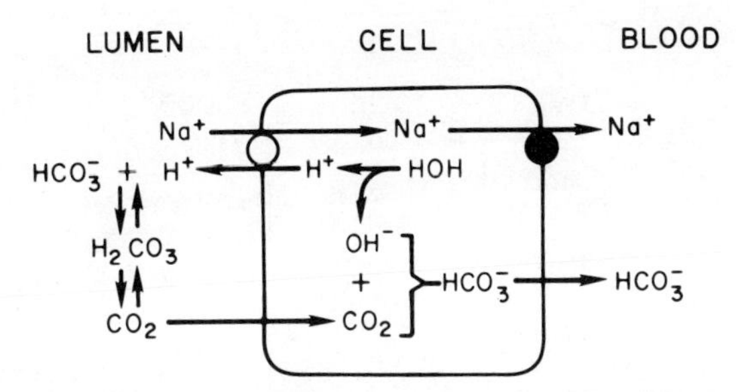

Fig. 8. Jejunal HCO_3^- absorption may be due in part to Na^+/H^+ exchange at the luminal membrane with a resulting increase in luminal CO_2. CO_2 diffusing across the apical cell membrane is converted to HCO_3^- by carbonic anhydrase. The process overall results in $NaHCO_3$ absorption. Reproduced by permission from Ref. 201.

theophylline-inhibitable acidification in rat jejunum that occurs concomitantly with mucosal K^+ uptake.

5.1.2. HCO_3^- Secretion

The study of HCO_3^- secretion by gut is complicated by the fact that there are at least three simultaneously occurring mechanisms[203]: passive paracellular diffusion, HCO_3^-/Cl^- exchange, and electrogenic, cAMP-stimulated HCO_3^- secretion (Fig. 9). In addition, some tissues appear to transport HCO_3^- from endogenous sources (rat colon,[205] rabbit duodenum,[218,219] frog colon,[220–222] rabbit colon[120]); others transport exogenously available HCO_3^- (bullfrog duodenum,[223] rat ileum[224]); and some tissues appear to transport bicarbonate from both sources (guinea pig[225] and rabbit ileum[226–228]). The cAMP-stimulated ileal and colonic bicarbonate secretion is difficult to study because even though such tissues secrete bicarbonate *in vivo*,[229,230] when studied *in vitro*

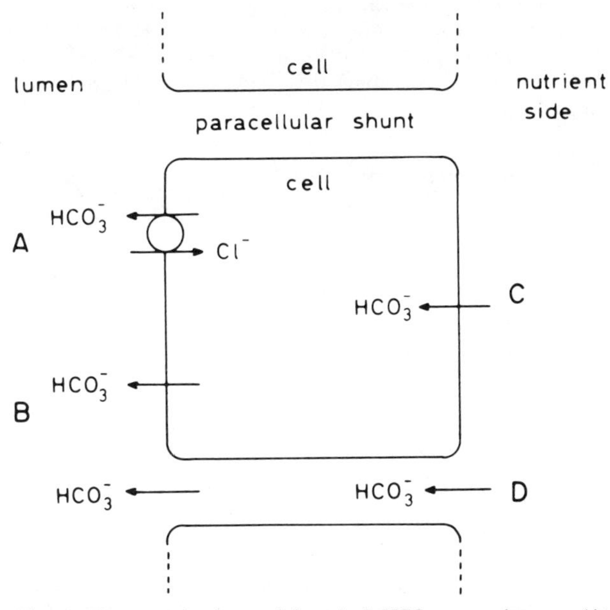

Fig. 9. Three mechanisms of intestinal HCO_3^- secretion are (A) HCO_3^-/Cl^- exchange; (B, C) electrogenic, transcellular cAMP-stimulated HCO_3^- secretion; and (D) passive paracellular diffusion. Reproduced by permission from Ref. 203.

cAMP stimulates only net Cl^- secretion.[120,227,228] In one of the few *in vivo* studies of mechanism, Hubel has suggested that it is OH^-, rather than HCO_3^-, that is transported in this segment in response to cholera toxin.[229] In addition, Field and McColl have shown that basal *in vitro* HCO_3^- secretion in mammalian ileum is inhibited by α-adrenergic agents.[231]

HCO_2^-/Cl^- exchange can be easily demonstrated in the duodenum, ileum, and colon: removal of Cl^- from the lumen *in vivo* abolishes all or part of the HCO_3^- secretion.[15,232–235] Indeed, net HCO_3^- absorption can be demonstrated under such circumstances. *In vitro* studies in rat[123] and rabbit colon[120] also demonstrate a decrease in HCO_3^- secretion when Cl^- is removed from the luminal solution. This is brought out most clearly in rabbit colon when net Na^+ absorption is inhibited by mucosal amiloride[136]—experiments which demonstrate the electrically silent nature of the HCO_3^-/Cl^- exchange process. When studied *in vitro*, HCO_3^- secretion by mammalian ileum is less clearly dependent on luminal Cl^- but probably is.[205,236]

Many studies in mammalian ileum and colon report that the HCO_3^- secretion is inhibited by acetazolamide.[62,123,234,235] However, the concentration of acetazolamide utilized (1–10 mM) is far above that that can be considered as a physiological dose (< 0.1 mM) for uncomplicated carbonic anhydrase inhibition.[237]

Amphiuma duodenum, like the mammalian ileum and colon, has HCO_3^-/Cl^- exchange, but unlike ileum and colon, also demonstrates cAMP-stimulated HCO_3^- secretion. Thus, *Amphiuma* duodenum is a good model for studying these processes in more detail.[238–245] White and his colleagues have demonstrated that HCO_3^- secretion by *Amphiuma* duodenum is dependent on active Cl^- absorption. This Cl^- absorption is blocked by serosal SITS, suggesting Cl^-/HCO_3^- exchange at the basolateral membrane as the mechanism for Cl^- exit. The cAMP-stimulated fraction of HCO_3^- secretion could be resolved into several components. One component was dependent on bathing solution Na^+, another on bathing solution Cl^-, and the third component appeared to be independent of either Na^+ or Cl^-. Therefore, the initial model proposed for these processes in *Amphiuma* duodenum was really not fundamentally different from that proposed for ileum (see Fig. 2) except that they believed that the entry step for Cl^- and exit step for HCO_3^- were electrogenic processes rather than via exchange carriers.[240]

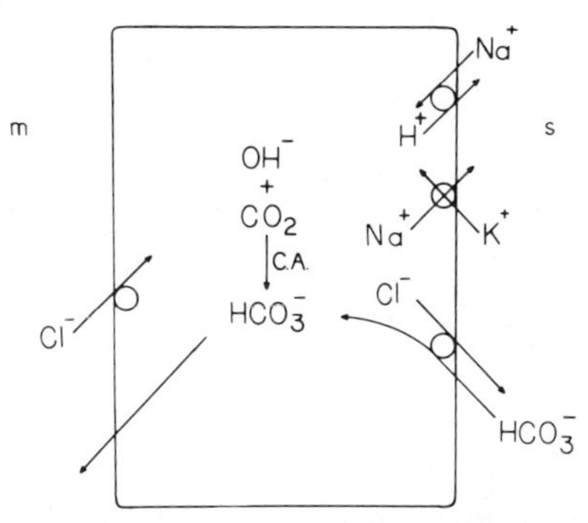

Fig. 10. HCO_3^- secretion by *Amphiuma* duodenum. Intracellular HCO_3^- is produced by carbonic anhydrase activity which is regulated by a basolateral Na^+/H^+ exchange and by a basolateral Cl^-/HCO_3^- exchange. Cl^- entry and HCO_3^- exit across the apical membrane are both electrogenic processes rather than antiport mechanisms. Reproduced by permission from Ref. 245.

More recent studies by White and colleagues[245] have led to a different conclusion (see Fig. 10). The Na^+ requirement for HCO_3^- secretion is for Na^+ in the serosal solution, whereas Cl^- is required in the mucosal solution. In addition, the Cl^- requirement for a portion of HCO_3^- secretion could be demonstrated only if serosal Na^+ were present. Thus, they believe that intracellular HCO_3^- generated by carbonic anhydrase activity and regulated by serosal Na^+/H^+ and Cl^-/HCO_3^- exchange leads to the accumulation of HCO_3^- at a concentration above electrochemical equilibrium. HCO_3^- exits across the mucosal membrane down this electrochemical gradient may be aided by cAMP-induced increases in apical cell membrane anion permeability. They further propose that intracellular H^+ regulates Cl^- uptake. The removal of Na^+ from the serosal solution leads to accumulation of intracellular H^+, reduced Cl^- absorption, less Cl^- available for HCO_3^- exchange at the serosal border, and, therefore, diminished HCO_3^- secretion. A similar model would account for HCO_3^- secretion by bullfrog duodenum, although $NaHCO_3$ coupled entry across the basolateral membrane could also.[223,246]

5.2. Short-Chain Fatty Acid Absorption

SCFAs, also known as volatile fatty acids, are those fatty acids with less than six carbon atoms. They are present in high concentrations (50–250 mM) in the colonic contents of monogastric animals where acetate, propionate, and butyrate are the predominant species formed by bacterial action on unabsorbed carbohydrate.[247] These weak acids have pK_a values of 4–5 and are, therefore, more than 90% ionized at physiological pH. However, the molecular form in which they are absorbed by mammalian colon is not known. On theoretical grounds, a three-compartment system in which the middle compartment has a pH different from the others and is separated by barriers of different permeabilities to ionized and un-ionized forms of acid is sufficient for net translocation of weak electrolytes.[115,248,249] More specifically, SCFA transport could take place by several mechanisms. In the jejunum, Na^+-dependent SCFA absorption could occur by Na^+-coupled processes.[97,98] In the ileum and in the colon of some species where Na^+/H^+ and Cl^-/HCO_3^- exchangers exist, the processes depicted in Fig. 5 might also be responsible. Indeed, in both small intestine and colon, SCFA absorption appears to stimulate Na^+ and Cl^- absorption as well.[58,250–253] However, this stimulatory effect is not seen in all colonic segments or animal species, and other mechanisms must be considered. In human rectum[254] and horse colon,[255] the process of SCFA absorption is accompanied by the appearance of HCO_3^- in the lumen and by a decrease in luminal pCO_2.[252,255] This has led Argenzio *et al.*[255] to postulate that SCFAs, like Cl^-, may participate in the brush border anion-exchange mechanism (see Fig. 11B) or else luminal H^+ formed by carbonic anhydrase activity might supply protons for the un-ionized absorption of SCFA (Fig. 11A).

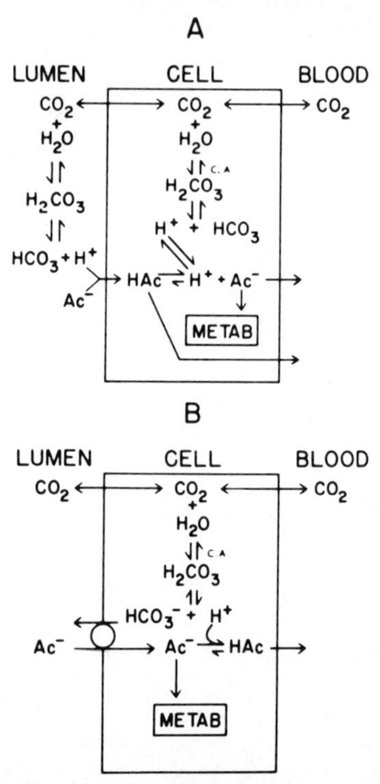

Fig. 11. Short-chain fatty acid absorption could occur through several mechanisms. (A) H^+ formed from carbonic anhydrase activity or else secreted into the unstirred layer (acid microclimate) would allow movement of lipid-soluble HAc across the cell membrane. (B) Ionized fatty acids (Ac^-) might enter the cell on an anion exchanger. Reproduced by permission from Ref. 255.

5.3. K^+ Transport

The regulation of high intracellular $[K^+]$ and low $[Na^+]$ is the role of Na^+,K^+-ATPase and, for mammals, is a fundamental tenet of biological life. Most of the metabolic energy produced by the organism is used to maintain these transmembrane concentration differences. It is not surprising that the intestine, especially the distal gut, takes part in the adaptation process to regulate intracellular K^+ just as does the kidney, particularly the distal nephron.[256] K^+ movement in the small intestine appears to be passive, moving across the mucosa in the same direction as net water flow (solvent drag) and/or in response to the electrical PD.[257] These mechanisms might also suffice for the colon. However, there is now compelling evidence that K^+ transport in the colon is by active carrier-mediated mechanisms.

5.3.1. K^+ Secretion

When reviewed in 1978,[115] there was considerable question as to whether K^+ transport in colon was active or passive. Frizzell, Schultz, and colleagues were unable to demonstrate net K^+ transport that could not be explained by PD-driven movement through a paracellular shunt path that has a very high K^+ permeability.[120,258,259] In contrast, Edmonds and colleagues had repeatedly shown that colonic K^+ secretion was not explicable on the basis of potential difference.[260–263] This failure of steady-state luminal K^+ concentrations to obey the Nernst equation was reconfirmed by more recent studies.[264] In addition, active K^+ secretion was seen under special circumstances, e.g., in response to K^+ loading or in response to endogenous or exogenous aldosterone.[265–271] All of these states were shown to increase the basolateral Na^+,K^+-ATPase activity by virtue of an absolute increase in the number of pump sites. This was due, at least in part, to an increase in the area of the basolateral cell membrane and perhaps also due to an increased density of ATPases per given area.[256,270] The other conditions that promoted net K^+ secretion were those that increased the apical cell

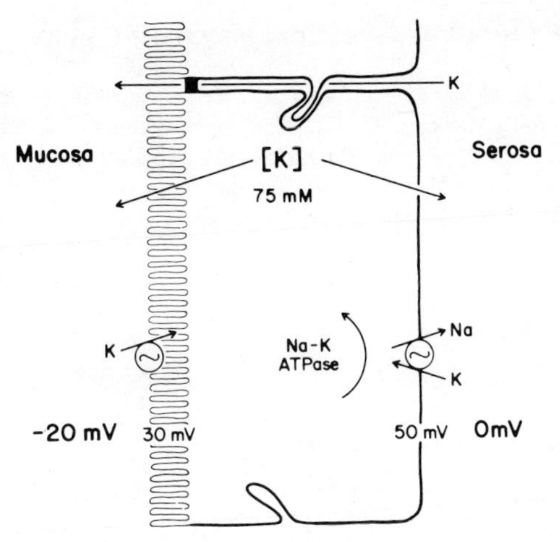

Fig. 12. Mechanisms of active K^+ absorption and secretion are depicted in the normal epithelial cell. Active absorption could be due to brush border-located ouabain-insensitive (K^+, H^+-ATPase or to a K^+–Na^+–$2Cl^-$ cotransport mechanism. Active K^+ secretion could come about by Na^+,K^+-ATPase-mediated K^+ entry across the basolateral membrane (such that K^+ attains a level in the cell above electrochemical equilibrium) together with K^+ exit across the brush border membrane by a Ba^{2+}-sensitive conductance pathway. Increased K^+ secretion results from stimuli that increase the number of K^+ pumps on the basolateral membrane or by processes that increase the K^+ conductance across the apical cell membrane. Reproduced by permission from Ref. 256.

membrane permeability, e.g., luminal application of amphotericin B, low mucosal solution pH, the apical membrane effect of aldosterone itself, hydroxy fatty acids such as ricinoleic acid, increased intracellular cAMP and Ca^{2+} levels, and possibly disease states such as inflammatory bowel disease and villous adenoma.[115,256,272–275]

More recently, it has been appreciated that there is actual net K^+ secretion in the basal state in the short-circuited colon if at least 45–50 min is allowed after $^{42}K^+$ addition to the bath to allow for the attainment of steady-state fluxes.[273–277] In such studies, the active component of serosal to mucosal K^+ movement was inhibited by cardiac glycosides and by metabolic inhibitors such as DNP and also by the luminal application of Ba^{2+}.[276–278] These flux studies, the demonstration that intracellular K^+ activity is above electrochemical equilibrium,[277,279–282] and the demonstration of a finite K^+ conductance in the luminal apical cell membrane[277,282] suggest a mechanism of net K^+ secretion as depicted in Fig. 12.

5.3.2. K^+ Absorption

The *in vitro* studies in the short-circuited rabbit colon by McCabe *et al.*[276] and by Wills and Biagi[277] demonstrated active K^+ absorption in ouabain-treated tissues. Active K^+ absorption has also been shown *in vivo* by the luminal perfusion of K^+-depleted animals with a 20 mM Na^+ perfusion solution.[278] Thus, it seems highly likely that the colon has an active mechanism for K^+ absorption and secretion. There are two possible candidates for the absorptive process. First, Gustin and Goodman[283,284] have demonstrated a novel, K^+-stimulated, ouabain-insensitive ATPase in the brush border of the rabbit colon that has the same characteristics as the gastric H^+,K^+-ATPase. Second, Musch *et al.*[54] have demonstrated a furosemide-inhibitable K^+–Na^+–$2Cl^-$ cotransport process in the brush border of flounder intestine. A similar K^+–Na^+–Cl^- transport mechanism has been demonstrated recently in erythrocytes, cultured tumor and kidney cells, squid giant axon, the kidney tubule, *Necturus* gallbladder, and *Amphiuma* jejunum. Whether this process is present in mammalian ileum and colon remains to be determined.

6. Shunt Pathway and Water Transport

The mammalian gut demonstrates a decreasing gradient of passive permeability from duodenum to rectum.[285–287] It is the tight junction–intercellular space complex (shunt pathway) that accounts for this.[7,8,288] Electrical resistances of epithelia range between 20 and 10,000 ohm-cm^2, but cell membrane resistances (100–5000 ohm-cm^2) bear no relationship to overall tissue resistance (Table IV). Thus, it is the shunt path resistance that determines whether the epithelium is "leaky" or "tight." In the intestine, the ratio of cell to shunt resistance (R_c/R_s) increases

Table IV. Electrical Resistance and Shunt Permselectivity of Intestine[a,b]

Segment (species)	Resistance hm cm^2			Resistance and conductance ratios		Permselectivity		
	R_c	R_s	R_t	R_c/R_s	$100\ G_s/G_t$	pK^+ :	pNa^+ :	pCl^-
Duodenum (rat)	2019	103	98	20	95	—	—	—
Jejunum (rat)	214	67	51	3	76	1.20	1.00	0.10
(rat)	1155	70	66	17	94	1.60	1.00	0.20
Ileum (rat)	1379	94	88	15	94	—	—	—
(rabbit)	767	115	100	6	85	1.14	1.00	0.55
(rabbit)	84	28	21	3	75	1.14	1.00	0.04
Colon (rabbit)	1670	345	286	5	83	1.00	1.00	—
(rabbit)	800	730	385	1	53	14.3[c]	1.00	1.57
(turtle)	1351	781	495	2	63	—	—	—
Free solution						1.47	1.00	1.52

[a] Modified from Ref. 288, which should be consulted for original references.
[b] R_c, R_s, R_t, G_s, and G_t = resistance (R) and conductance (G) of the cell (c), shunt (s), and total tissue (t).
[c] Hyperpolarized and depolarized tissue; the passage of electrical current may alter permselectivity as discussed in Ref. 288.

from 1 to 20 moving from colon to duodenum. Viewed another way, from 50 to 95% of the entire tissue ionic conductance is via the shunt path with the small intestine (a leaky epithelium) having a shunt conductance greater than 75% of the total conductance, while the colon (a moderately tight epithelium) has a shunt conductance 50–75% of total.

As previously reviewed,[288] shunt path resistance can be decreased by passing high electrical current, by luminal hypertonicity, by volume expansion, by Ca^{2+} removal or chelation, and by an extremely low pH in the mucosal solution. Shunt path resistance can be increased by polyvalent cations such as TAP and by Ca^{2+}. Recent studies have also suggested that the shunt path resistance in leaky epithelia is controlled by intracellular cyclic nucleotides. Agents that increase cAMP levels in both gut and gallbladder gradually increase shunt path resistance.[188,189,288–290] Whether this is due to some action of cyclic nucleotides or Ca^{2+} on the tight junction proper or whether it is due to cell swelling is not clear.

In addition to general permeability, as demonstrated by electrical resistance, the shunt path also exhibits ionic permselectivity. Thus, the intestinal shunt path is cation-selective such that the passive permeabilities of the pathway are $pK^+ > pNa^+ \gg pCl^-$ (see Table IV). A very high shunt path K^+ selectivity ($10\times > Na^+$) has been reported in rabbit colon although this was not confirmed by others. Cation permselectivity may allow the backflux of Na^+ from intercellular space across the tight junction into the intestinal lumen that may account for the serosa-negative PD found in gallbladder and flounder intestine.[13] Shunt path permselectivity may be altered by H^+ and by polyvalent cations such as TAP, Th^{4+}, La^{3+}, and Ca^{2+} which titrate the negative charges that confer the cation selectivity to the tight junction. cAMP also appears to alter cation/anion selectivity, making the shunt path less cation- or more anion-selective.[188,189,288]

Although the studies of Curran[9] and Diamond[10] demonstrated the role of the intercellular space in epithelial water transport, discrepancies between measured and predicted cell membrane water permeabilities (L_p) and compartment osmolalities have questioned both models of water transport.[291–293] Two issues were raised: how do epithelia transport fluid isosmotically, and does fluid move across the tight junctions or through the cells? These issues remained unsettled until 1982 when Persson and Spring[294] measured the hydraulic conductivity (L_p) of *Necturus* gallbladder membranes by altering cell volume in a chamber that minimized unstirred layers. They found a very high L_p for both apical and basolateral membranes ($1–2 \times 10^{-3}$ cm/S per osm) and calculated that the osmolality in the cell need be only 2.4 mOsm greater than the luminal solution to achieve a normal rate of fluid absorption. In addition, their calculations showed that the intercellular space osmolality need be only 1.1 mOsm hypertonic to the cell (or 3.5 mOsm hypertonic to the mucosal bathing solution) to account for the rates of water flow. Thus, there is no need to postulate specialized models such as standing osmotic gradients or electroosmosis to account for water flow in epithelia. The degree of hypertonicity of the transported solution, though truly hypertonic to the luminal solution, would be unmeasurable given the sensitivity of current techniques. Furthermore, the high L_p of the membranes and the large area of the cell membranes with respect to the cross-sectional area of the intercellular junctions (ratio of 10^4) make it highly unlikely that significant amounts of water move across the tight junctions during fluid transport.

7. Control of Intestinal Electrolyte Transport

Several forces exert an influence on intestinal electrolyte and water transport: (1) systemic factors such as acid–base status and intestinal mucosal blood flow; (2) hormones such as aldosterone and glucocorticoids and perhaps insulin and secretin; (3) substances difficult to classify, such as bradykinin and arachidonic acid metabolites; and (4) neurotransmitters, neuromodulators, and neurohormones which number in the score.

7.1. Systemic Factors

7.1.1. Acid–Base Status

A decrease in pH and HCO_3^- content in the serosal bath of Ussing-chambered rabbit ileum stimulates net Na^+ and Cl^- absorption and decreases HCO_3^- secretion, whereas high serosal pH and HCO_3^- content have the opposite effect.[227] Studies in Charney's laboratory have extended these findings and have suggested a role for the gut in the maintenance of systemic acid–base balance in mammals.[295–299] These investigators created acute metabolic and respiratory acidosis in separate groups of animals by gavage feeding alkaline ($NaHCO_3$) and acidic [NH_4Cl and $(NH_4)_2SO_4$] salts and by ventilating the animals with oxygen gases of varying (0, 3, and 8%) CO_2 content.[295,296] The results of their experiments are summarized in Table V. Alterations in systemic acid–base balance had no effect on electrolyte transport by the jejunum. In the ileum and colon, acidosis in general increased Na^+ absorption and Cl^- absorption, while alkalosis by and large decreased Na^+ and Cl^- absorption. The directions of net HCO_3^- transport in these segments correlated with the level of blood HCO_3^-; increased HCO_3^- secretion was seen when there was a high blood HCO_3^- content (metabolic alkalosis and respiratory acidosis), and the opposite was seen in low blood HCO_3^- states (metabolic acidosis and respiratory alkalosis). To define which aspect of systemic acid–base balance governed changes in transport, Charney and Haskell repeated the experiments but then corrected for the acid–base imbalance by acutely ventilating with 0, 3, or 8% CO_2.[297,298] They found that changes in blood HCO_3^- correlated best with changes in active HCO_3^- transport. In the ileum,

Table V. Effects of Acute Changes in Systemic Acid–Base Balance in Rat Intestinal Electrolyte Transport[a]

	Jejunum	Ileum	Colon
Metabolic alkalosis	No effect	↓ Na^+ absorption ↑ HCO_3^- secretion	↑ Cl^- absorption ↑ HCO_3^- secretion
Metabolic acidosis	No effect	↑ Na^+ absorption ↓ HCO_3^- secretion	↓ Cl^- absorption ↓ HCO_3^- secretion
Respiratory alkalosis	No effect	↓ Na^+ absorption ↓ Cl^- absorption ↓ HCO_3^- secretion	↓ Na^+ absorption ↓ Cl^- absorption ↓ HCO_3^- secretion
Respiratory acidosis	No effect	↑ Na^+ absorption ↑ Cl^- absorption ↑ HCO_3^- secretion	↑ Na^+ absorption ↑ Cl^- absorption ↑ HCO_3^- secretion

[a]Modified from Ref. 296.

blood pH correlated best with Na^+ and Cl^- absorption, whereas in the colon, blood pCO_2 levels seemed to be more important.

While these complex effects of systemic acid–base balance are difficult to interpret, they are best explained by the model in Fig. 2 where Na^+ and Cl^- transport can be governed by intracellular pH via the Na^+–H^+/Cl^-–HCO_3^- antiport systems.[15,16] To test this hypothesis, Charney and Haskell[297,298] studied the effects of acute respiratory acidosis and alkalosis on transport by rabbit descending colon and gallbladder, tissues where the coupled antiports are not thought to be present, and by rabbit ileum where they are.[299] There was no effect in colon and gallbladder, but the response in rabbit ileum was similar to rat. Thus, the effects of systemic acid–base balance may be due to the intracellular availability of H^+ for the Na^+/H^+ exchange mechanism. However, intracellular H^+ also modulates the Na^+/H^+ exchange carrier itself to enhance or inhibit its activity,[300] and intracellular H^+ can probably alter intracellular Ca^{2+} levels[301] which could, in turn, regulate NaCl coupled entry into the cell.[200]

7.1.2. Blood Flow

As described by Mailman,[302] there are several potential interactions between intestinal blood flow and fluid transport: (1) Increased absorption might increase O_2 consumption which, in turn, could stimulate more blood flow through metabolic feedback mechanisms. This could increase O_2 delivery that would then allow increased absorption. (2) Increase fluid delivery to the intestinal tract might alter Starling forces (oncotic pressure and permeability) that might promote the net movement of fluid from or into the lumen. (3) Increased blood flow might "wash out" absorbed solutes. (4) Altered blood flow or absorbed agents could stimulate neural, hormonal, or paracrine responses that increase or decrease blood flow.

The relationship between blood flow and transport is not an easy subject to investigate. Total gut flow is divided into three compartments: muscle, submucosal, and absorptive site (villus) flow.[302–304] Submucosal blood flow represents from 50 to 95% of the total gut flow, while the absorptive site blood flow may comprise up to 30%. Also, vasoactive agents may have different effects on absorptive blood flow when injected intravenously or even intraarterially than when the agents are released onto the serosal aspect of the vascular endothelium after stimulation of nerves or paracrine cells.

There seem to be two general schools of thought concerning the relationship between blood flow and intestinal transport: (1) Mailman[302] and Winne[305] suggest a positive correlation between blood flow and water absorption. (2) Granger indicates that blood flow also increases with stimulated secretion.[303,304] As shown in Table VI, with the exception of α-adrenergic agents, which increase absorption and yet decrease blood flow, most agents which stimulate *either* absorption or secretion also increase total gut blood flow. Therefore, it is likely that both stimulated absorption and stimulated secretion, each energy-dependent processes, are accompanied by increases in blood flow.[306]

The necessity for blood flow to increase with both absorption and secretion suggests an important integrative role for substances such as bradykinin, prostaglandins, serotonin, neurohormones, and neurotransmitters. The release of these substances by luminal secretagogues such as bacterial toxins or cAMP may increase blood flow which allows the stimulated enterocyte secretion to proceed. This could be the explanation for the data that suggest that the enteric nervous system takes part in toxin- and cAMP-stimulated intestinal secretion.[307–309]

Table VI. The Effect of Neurohumoral Agents on Total Gut Blood Flow and on Water and Electrolyte Transport[a]

Neurohumoral agent	Transport[c]	Blood flow[b]: Small intestine	Blood flow[b]: Colon
α-Adrenergics	abs	↓	↓
β-Adrenergics	abs	↑[d]	↑
Dopamine	abs	↑	—
Morphine	abs	↑	—
Acetylcholine	sec	variable	variable
Serotonin	sec	↑	↑
Bradykinin	sec	↑	↑
ATP	sec	↑	↑
Histamine	sec	↑	↑
Gastrin	sec	↑	↓
Secretin	sec	↑[d]	—
Cholecystokinin	sec	↑	0
Glucagon	sec	↑	↑
Vasoactive intestinal polypeptide	sec	↑[d]	↑
Gastric inhibitory peptide	sec	↑	—
Motilin	sec	↑	—
Pancreatic polypeptide	sec	↑	—
Prostaglandins	sec	variable	↓

[a]Summarized from Refs. 302–305.
[b]—, no data available; ↑, increased; ↓, decreased; 0, no effect.
[c]abs = increased absorption or decreased secretion; sec = decreased absorption or increased secretion.
[d]Total gut blood increased, absorptive site flow decreased.

7.2. Hormonal Control

Certain agents traditionally thought of as being secreted into the bloodstream and affecting distal targets can be considered as *hormones*. Finding an effect on water and electrolyte transport after *in vivo* injection or *in vitro* addition of the agent does not, however, prove a physiological role. For example, the stimulation of intestinal absorption with prolactin,[310,311] of secretion with calcitonin,[312–314] and the contradictory stimulation of absorption,[315–318] secretion,[319–321] or neither[322] with antidiuretic hormone are all of uncertain physiological significance. Four hormones—mineralocorticoids, glucocorticoids, and (less certainly) insulin and secretin—appear to be physiological regulators of intestinal water and electrolyte transport.

7.2.1. Mineralocorticoids

The physiological role of mineralocorticoids, especially aldosterone, can be demonstrated by: (1) the increase in ileal and colonic electrolyte transport following Na^+ depletion[260–263,271,323]; (2) the decrease in transport following adrenal ablation, with partial restoration upon replacement of aldosterone[269]; (3) the development of aldosterone-related Na^+ conservation by the neonatal colon[324]; and (4) the decrease in

Na^+ absorption that occurs with use of competitive inhibitors of aldosterone such as spironolactone.[269,325]

Aldosterone increases Na^+ and water absorption and stimulates K^+ secretion.[115,138,326] The dose of aldosterone that alters K^+ secretion in rats is much lower than that affecting Na^+ absorption, suggesting a dissociation of these two effects of the hormone.[269] Aldosterone is thought to act primarily on the apical cell membrane, where it increases the number[138,271,323,327,328] and perhaps the sensitivity[327,328] of amiloride-sensitive Na^+ conductance channels. The effect is variable in different gut segments, as determined by the degree of amiloride sensitivity of the I_{sc}. In rat colon, aldosterone converts the descending colon from one of essentially no amiloride sensitivity to 95% sensitivity,[271,323] and it increases the sensitivity of the ascending colon, the cecum, and the terminal ileum by 10–20%.[322,323,327] Foster *et al.*[271] have shown that these changes in rat colon occur by a switch from an electrically neutral NaCl entry mechanism to an electrogenic Na^+ transport process.

A secondary effect of mineralocorticoids is to stimulate Na^+,K^+-ATPase,[325,327,329] a response inhibited by protein synthesis inhibitors such as cycloheximide[330] and by spironolactone.[325] This stimulation of ATPase is presumed to occur in response to the increase in apical membrane Na^+ entry. Evidence for this sequence of events is the observation that the acute addition of aldosterone *in vitro* stimulates the I_{sc} of descending rabbit colon, but no more so than does amphotericin B which increases Na^+ entry across this membrane by nonspecific mechanisms.[138] Even more compelling evidence is the dissociation in the time necessary for stimulation of amiloride sensitivity ($t_{1/2}$ 14 hr) from that for increased synthesis of Na^+,K^+-ATPase ($t_{1/2}$ 7–10 days).[327]

Aldosterone-stimulated K^+ secretion appears to be the result of hormone-induced stimulation of Na^+,K^+-ATPase, which elevates intracellular K^+ activity, coupled with aldosterone-induced increases in apical membrane K^+ conductance (see Fig. 12). However, aldosterone also inhibits mucosal to serosal K^+ fluxes, suggesting an effect on the apical membrane K^+-absorbing mechanism as well.[271]

7.2.2. Glucocorticoids

These hormones have both physiological and pharmacological effects on intestinal water and electrolyte transport. The physiological role can be demonstrated in adrenalectomized animals,[269] or in those in which glucocorticoid synthesis is inhibited by administration of aminoglutethimide,[114,269] where hypoglucocorticoidemia results in decreased Na^+ and water absorption and a reversal of Cl^- transport from absorption to secretion. Such low glucocorticoid states have no effect on the maximal response to secretory stimuli such as cAMP.[114] Replacement of glucocorticoids in ''physiological amounts'' will return Na^+ and H_2O absorption to normal in the adrenalectomized rat, a response that aldosterone alone cannot do.[269] Pharmacological doses of glucocorticoids have effects quite similar to mineralocorticoids: stimulated Na^+ absorption, amiloride sensitivity, and Na^+,K^+-ATPase activity.[114,267,271,325,327,329,331]

There are several distinct differences in the effects of glucocorticoids and mineralocorticoids, suggesting separate mechanisms of action: (1) glucocorticoids, but not aldosterone, stimulate Na^+,K^+-ATPase in the jejunum and ileum as well as in the large intestine[325,329,331]; (2) glucocorticoids acutely activate guanylate cyclase in the rat small bowel, thus initially stimulating intestinal Cl secretion prior to the later stimulation of Na^+ absorption[332]; (3) although the glucocorticoid effect in small bowel is inhibited by cycloheximide, it is not reduced by spironolactone[325,330]; (4) the half-time for the effect of glucocorticoids on amiloride sensitivity is approximately the same as that for stimulated Na^+,K^+-ATPase[327]; (5) apical cell membrane entry processes for Na^+ do not seem to be increased by glucocorticoids in rabbit ileum[332,333]; and (6) there is evidence for cytosolic receptors for both mineralo- and glucocorticoids in rat colon.[334] These studies have led to the tentative conclusion that the primary effect of glucocorticoids is to increase Na^+,K^+-ATPase with the increase in amiloride sensitivity being a secondary or an additional specific event. It seems certain from the evidence quoted in these two sections that both of these hormones are necessary for normal physiological regulation of intestinal water and electrolyte transport.

7.2.3. Insulin

Mammalian insulin added to the serosal bath of Ussing-chambered rabbit ileum[335] or toad colon[336] stimulates net Na^+ absorption and I_{sc}. The dose required is quite large: 0.1–1 unit/ml bathing solution. A similar effect has been found with amphibian skin and urinary bladder.[336] Presumably, this action is mediated by insulin receptors which have been found on the basolateral membrane of the intestinal cells.[337] The mechanism of insulin effect and whether it occurs *in vivo* are not known.

7.2.4. Secretin

This hormone causes small intestinal secretion or inhibits absorption both *in vivo*[338,339] and *in vitro*,[340,341] apparently through a cAMP-mediated process.[342] The finding of specific receptors from small intestinal cells of the guinea pig suggests that secretin may well be a physiological hormone.[342] However, the K_d for inhibition of binding (and for stimulation of adenylate cyclase) is some 1000 times that for VIP, raising some question about its physiological role.

7.3. Kinins and Arachidonic Acid Metabolites

7.3.1. Bradykinin

Bradykinin (BK) might be considered as a hormone inasmuch as the precursor is synthesized at a site (liver) distal from the target (intestine). Several recent studies suggest that there may be an intestinal kallikrein–kinin–prostaglandin system concerned with diarrhea, just as such a system in the kidney relates to hypertension (see Chapter 48 and Smith and Dunn[343]). As shown in Fig. 13, activation of tissue-bound serine proteases (glandular kallikrein) in the colon[344,345] cleaves kinins from acidic glycoproteins synthesized in the liver and circulated in blood. Aminopeptidases and kininases such as angiotensin II converting enzyme (ACE or kininase II)[346] cleave off other amino acid fragments to convert kallidin (LBK) to BK and then to inactive peptides.[347]

In 1975, Crocker and Willavoys demonstrated an effect of BK on rat jejunal water transport: when basal absorption was low, the kinin stimulated absorption; when basal absorption was high, it inhibited transport.[348] Hardcastle *et al.*[349] showed that BK increased the PD in rat jejunum and colon and that this

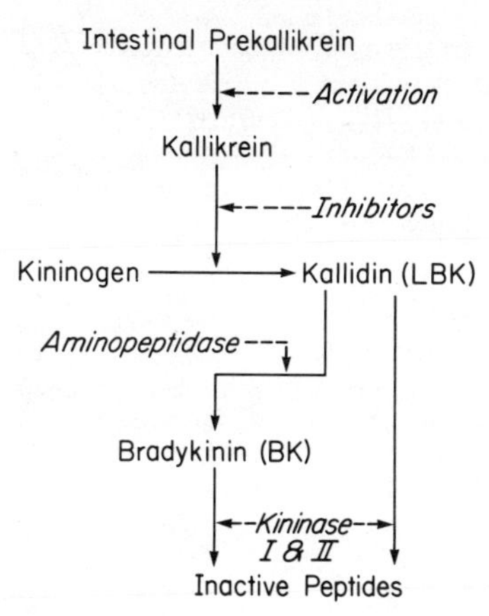

Fig. 13. The intestinal kallikrein system might be similar to that envisioned for the kidney. Tissue kallikreins form kallidin from serum kininogens; kallidin is further metabolized by serum enzymes to bradykinin or to inactive peptides. Modified from Ref. 343.

effect could be inhibited in part by indomethacin. These observations were extended by Cuthbert and Margolius[350] who showed that kallikrein from rat kidney and mellitin (a bee venom polypeptide activator of tissue-bound kallikrein) produced an indomethacin- and mepacrine-inhibitable stimulation of Cl^- secretion. Musch *et al.* convincingly demonstrated receptor-mediated release of arachidonic acid metabolites as the intermediates in BK-stimulated Cl^- secretion.[351–353] Because of differences in receptor sensitivity,[351] high doses of angiotensin II are required for the stimulation of intestinal Cl^- secretion.[354] Low doses of angiotensin II stimulate intestinal Na^+ and water absorption via the release of norepinephrine (NE), a phenomenon discussed further in Section 7.4.3 (Neuromodulation).

7.3.2. Prostaglandins, Prostanoids, and Leukotrienes

Arachidonic acid metabolites of cyclooxygenase activity such as the various prostaglandins, prostacyclin, and thromboxane (designated here as PGs) and of lipoxygenase activity such as hydroperoxyeicosatetraenoic acids (HPETE), hydroxyeicosatetraenoic acids (HETE), and leukotrienes (designated here all inclusively as LT) are unique compounds because they can act as hormones, as paracrine agents, or as intracellular messengers. Studies over the past decade indicate that PGs stimulate intestinal secretion with a rank-order activity of $PGE_2 > F_{2\alpha} > A_2 > D_2 >$ prostacyclin (PGI_2).[355–363] The relative activity of thromboxane is unknown. In fact, it has been found that PGI_2[363] and its stable analog[364] will inhibit intestinal secretion stimulated by other agents. There have been few studies of lipoxygenase metabolites, but in the colon, 5-HPETE and 5-HETE are stimulants of secretion whereas 9-, 11-, and 12-HPETE and the leukotrienes C_4 and B_4 are not.[353]

Prostaglandins have their effect by activating adenylate cyclase.[355,357,365,366] However, it seems likely that prostaglandins also stimulate pools of cAMP that have nothing to do with secretion; e.g., PGI_2 stimulates adenylate cyclase in the colon[365,366] with little effect on electrolyte transport.

These studies of prostaglandin and electrolyte transport are complicated by the fact that PGs also affect intestinal mucosal blood flow, transcapillary fluid exchange, lymph flow, and interstitial fluid dynamics.[367–370] Furthermore, prior stimulation with one prostaglandin appears to desensitize the intestine to the subsequent stimulation of that or other prostaglandins.[371]

Whereas little is known about enterocyte production of lipoxygenase products, studies of the intestinal cyclooxygenase pathway indicate differing rates of synthesis of the individual PGs depending on species, preparation, and the starting substrates. If the cyclooxygenase step is bypassed in canine mucosal scrapings by supplying the intermediate PGH_2, the metabolites isolated from the small and large intestine are the degradation products of PGI_2 and thromboxane A_2 at a greater rate than PGE_2, $PGF_{2\alpha}$, and PGD_2.[372,373] In contrast, using microsomes of purifed rabbit small intestinal enterocytes, cyclooxygenase products formed from ^{14}C-labeled arachidonic acid are $PGF_{2\alpha} > PGE_2 > PGD_2 > TxB_2 \gg PGI_2$.[374]

7.4. Neural and Neuroendocrine Control

7.4.1. Classic Concepts of Neural Control

Prior to 1960, gut innervation was thought of as a simple, excitatory–inhibitory system (Fig. 14a).[375] Postganglionic cholinergic nerves were presumed to release acetylcholine (ACh) onto muscarinic receptors located on epithelial cells, causing water and electrolyte secretion. In this scheme,

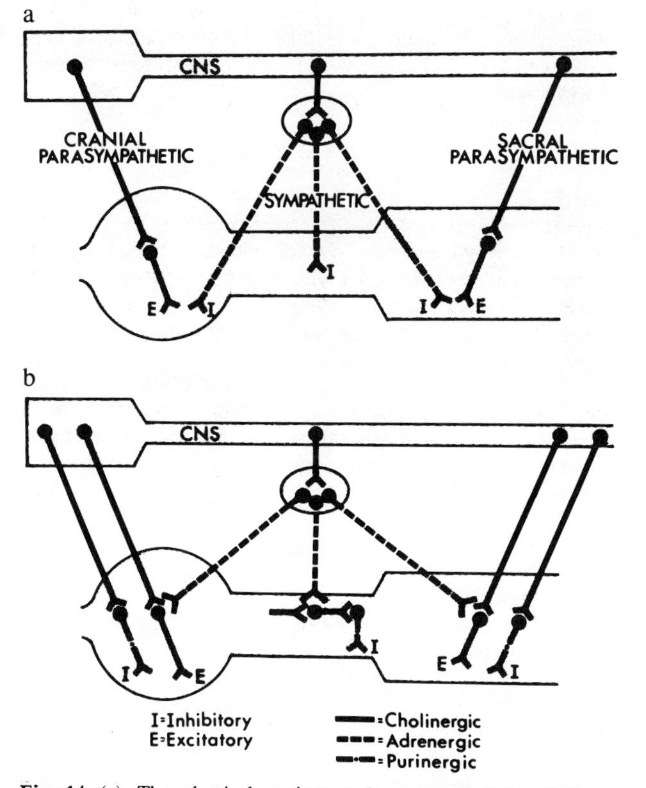

Fig. 14. (a) The classical excitatory (parasympathetic)–inhibitory (sympathetic) scheme of gut innervation and (b) the revised cholinergic, adrenergic, purinergic scheme which shows both inhibitory and excitatory responses after stimulation of cranial and sacral preganglionic parasympathetic processes and modulation of those processes in the eneteric plexuses by the sympathetic system. Reproduced by permission from Ref. 375.

postganglionic adrenergic nerves were thought to release catecholamines onto inhibitory receptors on the enterocyte or muscle cells proper. However, this simplistic view of control was questioned in the 1960s by a number of discoveries. Histochemical methods suggested that many, if not most, of the adrenergic fibers ended on the intestinal plexuses and not on the muscle and mucosal cells. Also, Burnstock suggested the presence of nonadrenergic, noncholinergic inhibitory nerves in the gut.[375] This led to a proposal for a more complex form of gut innervation (Fig. 14b) that eventually developed into the emerging concepts of the "enteric nervous system" (ENS) and of neuromodulation. The complexities of these modern concepts as they relate to the neurohumoral control of intestinal water and electrolyte transport have been reviewed.[376–381]

7.4.2. The Enteric Nervous System

There are several reasons why the ENS should be considered as a system separate from the classic sympathetic and parasympathetic systems[382]: (1) There is a great degree of independence of intestinal function from CNS control; sectioning cranial or sacral parasympathetic and sympathetic nerves does little to gut function. (2) The complexity and diversity of gut neurons and plexuses are evident histochemically and by electron microscopy. (3) There are clear-cut reflex arcs confined to the gut (e.g., the peristaltic reflex). (4) The number of neurons found in the ENS (10^8) approaches that found in the spinal cord.

Immunocytochemistry and electron microscopy establish three major components of the ENS[383–390]: (1) extrinsic neurons containing cholinergic and adrenergic agents but with some fibers containing various peptide transmitters; (2) intrinsic enteric neurons that contain a variety of other amines, chemicals, and peptides in addition to ACh and NE; and (3) neuroendocrine cells in the epithelium that release chemicals or peptide messengers.

The *extrinsic autonomic nervous supply* of the intestine arises primarily from cranial (vagal) and sacral spinal cord nuclei and from the prevertebral sympathetic ganglia. ACh and NE are the principle transmitters present in these nerves; however, substance P, enkephalin, vasoactive intestinal polypeptide (VIP), gastrin/cholecystokinin (gastrin/CCK), and somatostatin are also found.[382,391] Current investigation ascribes an important role to substance P in afferent, nociceptive (pain) transmission,[392] although this role in the gut remains to be proven.[393]

Table VII. Putative Neurotransmitters, Neuromodulators, and Hormones of the Enteric Nervous System[a]

Acetylcholine	Pancreatic polypeptide
Vasoactive intestinal polypeptide	Motilin
Substance P	Bombesin
Neurotensin	Norepinephrine
Serotonin	Dopamine
Secretin	Methionine enkephalin
Enteroglucagon	Leucine enkephalin
Gastrin	Dynorphin
Cholecystokinin	Somatostatin
Prostaglandin	Insulin
Angiotensin II	γ-Aminobutyric acid
Bradykinin	Thyrotropin-releasing hormone
Gastric inhibitory peptide	Corticotropin-releasing factor
Adenosine or ATP	

[a]Compiled from Refs. 377–403.

The ganglia of the submucosal (Meissner's) and myenteric (Auerbach's) plexuses form the "minibrain" of the gut which is capable of autologous regulation of gut function (Fig. 15). These ganglia and fibers make up the *intrinsic neurons* of the ENS.[390,394] Various histochemical methods, including indirect immunofluorescence techniques using antisera raised against the various peptides, have localized over 20 putative neurotransmitters in the nerve bodies, fibers, and the endocrine cells within the mucosa (Table VII). The myenteric plexus, located between the longitudinal and the circular muscle layers, is composed of neurons and glial cells that extend throughout the intestinal tract from the esophagus to the anus.[390] The submucosal plexus lies in the connective tissue between the inner circular muscle and the lamina propria. Both plexuses receive neuronal input from the extrinsic nerves and from the other plexus and both send processes to muscle, to blood vessels, and to epithelial cells proper. Most of the nerves in a plexus originate from the same or from the other plexus, in keeping with the autonomous nature of the ENS. It is tempting to ascribe more importance to the submucosal plexus for regulating water and electrolyte transport since many of the nerve fibers found in the core of the villus originate there. Such an assumption may be unwarranted, however, because of the complexity of the interconnecting neurons and because, in fact, the majority of nerves in the ENS originate in the ganglia of the myenteric plexus.[390]

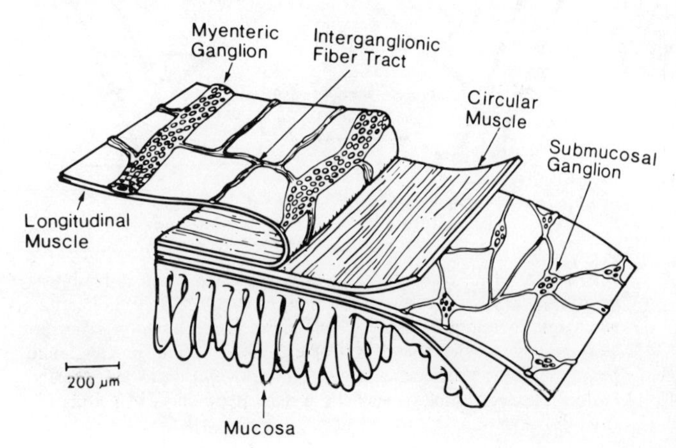

Fig. 15. The intrinsic autonomic nervous system with its two principal components: the myenteric (Auerbach's) plexus and the submucosal (Meissner's) plexus. Reproduced by permission from Ref. 394.

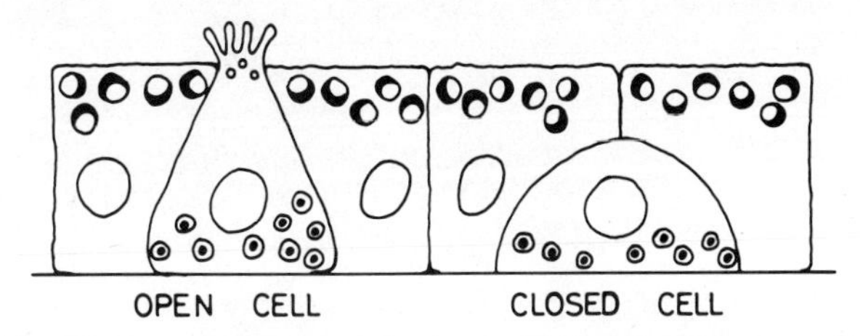

Fig. 16. Two types of gut endocrine–paracrine cells: the open cells (with luminal microvilli) and the closed cells. Both cells may be innervated but presumably only the open cells respond to luminal stimuli. Reproduced by permission from Ref. 402.

One way to determine if a putative neurotransmitter has a primary function in regulating fluid and electrolyte transport is to assess whether fibers containing that neurotransmitter end on or near the enterocyte. Electron microscopy suggests but does not prove cholinergic innervation of enterocytes and of neuroendocrine cells.[395,397] Histochemical stains for catecholamines indicate that the majority of NE-containing fibers end in the two plexuses where they modulate the release of other neurotransmitters.[398–401] However, some adrenergic fibers follow the blood vessels into the villus and pass near the basolateral membrane of the intestinal epithelial cells, especially around the crypts.[400,401] Immunofluorescence histology indicates neuronal cell bodies or fibers containing substance P, VIP, serotonin (5HT), enkephalin, gastrin/CCK, neurotensin, bombesin, and somatostatin in the plexuses of the large and small intestine of several species.[382–387,395,401] Peripheral projections of nerve fibers containing many of these peptides are found in the lamina propria, often around the base of the crypts.[382–387,395,401] In particular, VIP fibers form an impressive network extending up to the villus core to the tip.[401]

The third part of the ENS is the *endocrine–paracrine system*. Over 20 different endocrine cells have been tentatively identified in the pancreatic and gastrointestinal mucosa (Table VII). Anatomically, the cells are of two types[386,389,402,403] (Fig. 16): (1) The "open cell" has contact with both the lumen and the basement membrane of the epithelium and has luminal microvilli which are thought to respond to stimuli with a receptor–secretory function triggering the release of chemical transmitters from the basal aspect of the cells. Gastrin cells in the gastric antrum and secretin or CCK cells in the duodenum are the classic example of open cells.[402] Some gut endocrine cells (closed cells) have no luminal contact and are sandwiched between the basal aspect of columnar epithelial cells. Cholinergic and adrenergic terminal nerve fibers have been identified next to both types of endocrine cells, suggesting that they may either withhold or expel their hormones in response to inhibitory or excitatory neural stimuli.[396] Forsberg and Miller[404] have shown that it is a muscarinic cholinergic receptor that controls 5HT release from enterochromaffin cells; catecholamines and peptides such as bombesin, substance P, neurotensin, and VIP will not. Furthermore, the 5HT released by cholinergic agents in rabbit duodenum is directed into the submucosa where it might have hormonal or paracrine effects. Once the chemical messengers are expelled from these endocrine cells, it can be transmitted via the bloodstream to affect distant epithelia (hormonal or endocrine peptide secretion), or it may affect local adjacent cells (paracrine peptide secretion) (Fig. 17). Some paracrine cells (e.g., somatostatin cells in the stomach, gastrin/CCK cells in the ileum, and 5HT cells in the rectum) have been found to give off long cytoplasmic extensions which terminate on the basal aspect of adjacent epithelial cells (modified paracrine secretion).[402,403] Locally released messengers can also stimulate other nerves or blood vessels and thus regulate blood flow.

7.4.3. Neuromodulation

In the strictest sense, several criteria must be met for an agent to be classified as a neurotransmitter[381,385,395]: (1) syn-

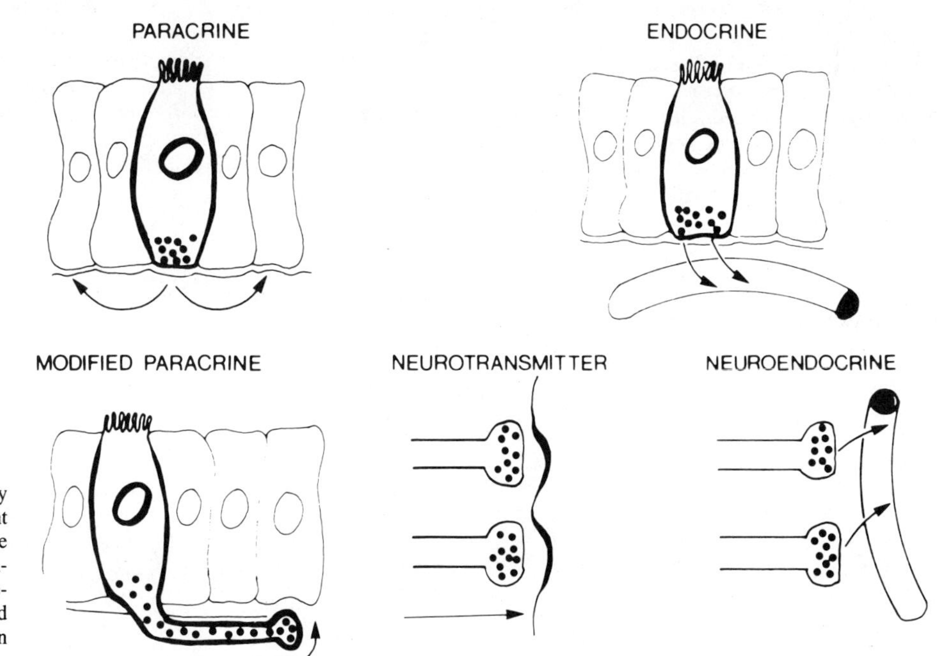

Fig. 17. Various mechanisms by which chemical messengers might modify gut function. All of these mechanisms might operate in the intestine, although neuroendocrine secretion has not been documented there. Reproduced by permission from Ref. 389.

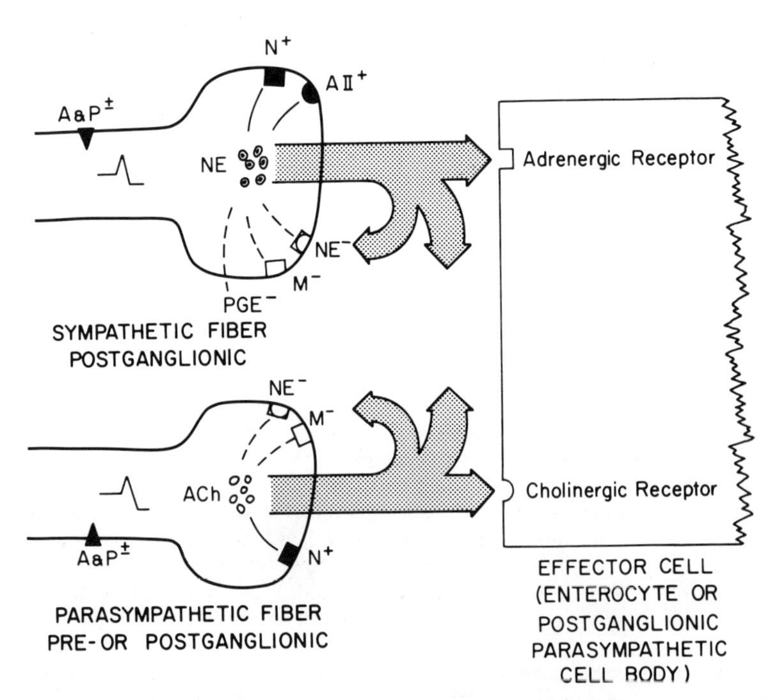

Fig. 18. Examples of neuromodulation in the gut. Both sympathetic and parasympathetic fibers contain receptors which stimulate (+) or inhibit (−) release of the neurotransmitter from that fiber. Postganglionic sympathetic fibers release norepinephrine (NE) in response to nicotinic (N^+) and angiotensin II (AII^+) stimulation, whereas NE release is reduced by NE itself (NE^-), by muscarinic agonists (M^-), and by prostaglandins (PGE^-). Postganglionic parasympathetic nerves release acetylcholine (ACh) upon nicotinic stimulation whereas NE and muscarinic agonists inhibit release. Various other amines and peptides (A & $P^\pm$) may either stimulate or inhibit transmitter release from either fiber. See text. Modified from Refs. 377, 378, 381.

thesis and storage of the chemical must be demonstrated in the nerve terminals; (2) neurotransmitter release must be demonstrated during nerve stimulation; (3) the response to nerve stimulation should be mimicked by exogenous application of the agent; (4) there must be mechanisms for the uptake and inactivation of the neurotransmitter; and (5) there should be nerves that either block or potentiate the responses to both the exogenous transmitter and nerve stimulation. The definition of a neuromodulator is somewhat less strict[381]: Neuromodulators are substances released synaptically or nonsynaptically, either alone or with other neuroactive chemicals, to act only on a *presynaptic site* of the neuron from which it originates or on another neuron to decrease or enhance the release of a primary neurotransmitter. Thus, neuromodulators may act in (1) a negative feedback mode (where the neurochemical released inhibits its own further release), (2) in an interneuronal mode where the modulator inhibits or stimulates release of the same or a different neurotransmitter from an adjacent neuron, or (3) in a transsynaptic mode where a true neurotransmitter causes receptor-related release of a neuromodulator, which diffuses back across the synaptic cleft to alter (usually to decrease) release of the primary neurotransmitter. Neuromodulation has been excellently reviewed in detail by Tapper.[381]

Figure 18 illustrates some important neuromodulators that have been demonstrated in the classic cholinergic and adrenergic systems of the gut and other tissues. Postganglionic parasympathetic cell bodies in gut plexuses and nerve fibers in the lamina propria may contain receptors for adrenergic agents which inhibit the release of ACh from the terminal fiber.[381,405,406] Thus, adrenergic agents would decrease cholinergic activity, and this could be at both the ganglion level and the periphery near the site of the release of ACh onto the enterocyte. Conversely, various agents will modulate adrenergic activity in postganglionic sympathetic fibers. For example, NE itself and prostaglandins will inhibit NE release from the colon mucosa.[407,408] Neuromodulation also explains certain dichotomous dose-related responses of the intestine to various agents. Tapper *et al.*[409] found that low-dose carbachol ($< 10^{-5}$ M) caused an atropine-inhibitable intestinal secretion, while high doses ($> 10^{-5}$ M) induced absorption that was inhibited by nicotinic blockers (hexamethonium) and by α-adrenergic blockers (phentolamine). Such a response would be explained by nicotinic receptors on adrenergic nerves which release catecholamines onto the enterocyte.

Another example is the biphasic response of rat jejunum to angiotensin II (AII).[354,410–417] At low doses, AII stimulates water and electrolyte absorption by a mechanism that is blocked by protein synthesis inhibitors such as cycloheximide, by peripheral sympathectomy, and by α-receptor blockade. At high doses, AII inhibits absorption by a process that is inhibited by prostaglandin synthesis inhibitors. The low-dose effect is likely due to AII-stimulated release of NE from adrenergic neurons in the gut, while the high-dose effect is likely due to AII stimulation of prostaglandin synthesis by intestinal cells (see Section

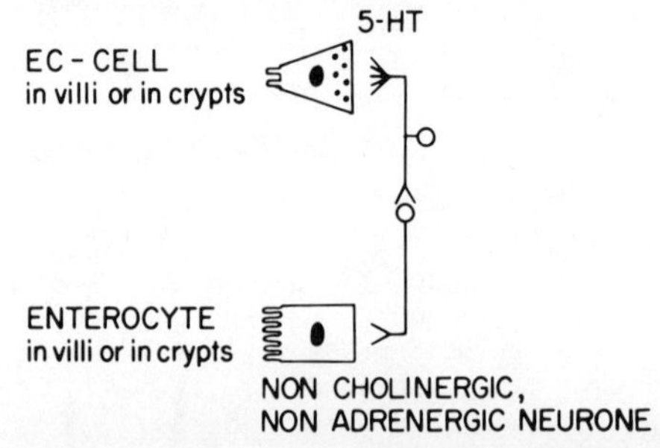

Fig. 19. Proposed example of neuromodulation involving mucosal endocrine cells. Luminal stimuli such as cholera toxin or bile salts might release 5HT from mucosal enterochromaffin cells (EC-cell) which would, in turn, cause release of a secretory neurotransmitter (e.g., VIP) onto the enterocyte. Reproduced by permission from Ref. 309.

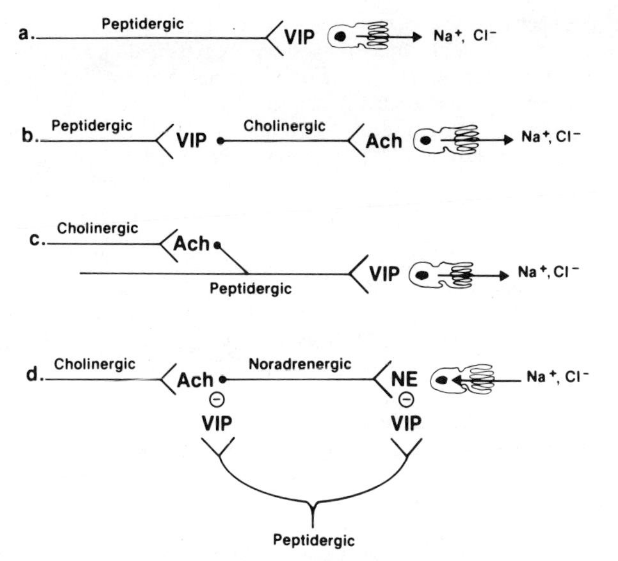

Fig. 20. Mechanisms whereby classic and peptide neurotransmitters might alter intestinal water and electrolyte transport. (a) A peptidergic neuron releases transmitter (VIP) onto the enterocyte. (b) A peptide (VIP) causes release of a classic neurotransmitter (ACh) from postganglionic cholinergic nerves whose cell bodies reside in the gut plexuses but whose fibers end on the enterocyte. (c) ACh modulates release of peptide transmitters. (d) A peptide modulates ACh or NE release at a pre- or postganglionic level. Reproduced by permission from Ref. 421.

7.3). These prostaglandins would directly activate the secretory mechanisms and, in addition, would decrease NE release from postganglionic sympathetic fibers through the prostaglandin feedback mechanism.

Cassuto *et al.* have suggested a possible example of neuromodulation involving mucosal endocrine cells (Fig. 19).[308,309,418–420] They present evidence that cholera toxin or bile salts release 5HT from endocrine cells in the mucosa which, in turn, excites noncholinergic, nonadrenergic nerves to release secretory neurotransmitters (? VIP) onto the enterocyte. Such a phenomenon also could be important in the vasodilation and increased mucosal blood flow that accompany secretion. While these remain speculative proposals, it is clear that neuromodulator release could have significant and diverse effects on electrolyte transport and represent a mechanism for reflex-type stimulated intestinal secretion in response to luminal stimuli.

7.4.4. Mechanisms of Action of Neuroactive Agents

As indicated above, neurohumoral agents might affect water and electrolyte transport by three mechanisms[421]: (1) by altering blood flow or motility; (2) by releasing other neurotransmitters or neuromodulators which act on other neurons through classic neuromodulation of a presynaptic receptor or as an interneuron to stimulate release of a second neurotransmitter (Fig. 20); and (3) by acting directly on the enterocyte.

The task of the physiologist is to differentiate between these various mechanisms. This can be done in several ways: (1) Effects due to alterations in blood flow and motility can be alleviated by studying the response in an *in vitro* system, e.g., in the Ussing chamber. Such studies do not, however, indicate how the agent is working under normal conditions in the live animal; neuroeffectors may have different effects *in vivo* and *in vitro*. Therefore, both *in vivo* and *in vitro* systems should be studied to truly understand their mechanism of action. (2) Neuromodulation or interneuron function can be differentiated from a direct effect on the enterocyte by determining whether classic antagonists of either the cholinergic muscarinic system (atropine) or the adrenergic system (α and β blockers) alter the response to the putative neurotransmitter. One can also determine if desensitization to one agent abolishes the response to a second. (3) Abolition of an agent's effect with tetrodotoxin (TTX), a toxin that blocks Na^+ channels in neural tissue thus abolishing transmission without changing receptor sensitivity, is good evidence for participation of a neural intermediate in the mechanism of action of that agent. (4) The demonstration of specific receptors on the enterocyte to the neuroeffector agent in question gives strong evidence in favor of a direct neurotransmitter (or hormone) action. However, failure to demonstrate receptors is less meaningful because the receptors might be altered or destroyed during the process of isolating and purifying the intestinal cells or their membranes. With these strategies in mind, we can draw some tentative conclusions regarding the putative neurotransmitters or physiological hormones and their mechanisms of action (see Tables VIII and IX).

7.4.5. Agents Stimulating Intestinal Secretion

7.4.5a. Cholinergic Muscarinic Agents. Classic studies prior to 1940 demonstrated an atropine-inhibitable intestinal secretion in response to vagal or pelvic nerve stimulation and following section of splanchnic sympathetic fibers.[376] The response to parasympathetic stimulation documents the role of the cholinergic system in the secretory processes of the intestine, and the sympathetic ablation experiments suggest a constant state of cholinergic secretory "tone," a concept that gains further support from increases in absorption observed after atro-

Table VIII. Neuroactive Secretory Agents of the Enteric Nervous System[a,b]

	Effects abolished by			
Secretory agent	Cholinergic antagonists	TTX	Enterocyte receptors	Classification
Muscarinic cholinergic	Yes	No	Yes	NT[c]
VIP	No[d]	—	Yes	Probable NT
Bombesin	No	No	—	Probable NT
Substance P	No	No	—	Probable NT
Bradykinin	—	—	Yes	Probable NT
Angiotensin II (high dose)	—	—	Yes	Possible NT
Secretin	—	—	Yes	Possible NT
Serotonin	No	No	No	?
Neurotensin	No[e]	Yes	—	NM[f]

[a]Modified from Ref. 422.
[b]No information is available for gastrin, CCK, glucagon, GIP, motilin, histamine, or ATP.
[c]NT = primary neurotransmitter or physiological hormone.
[d]One report of inhibition by atropine *in vivo*.[453]
[e]Effect of neurotensin is blocked by desensitization to substance P.
[f]NM = neuromodulator or interneuron.

Table IX. Neuroactive Absorptive Agents of the Enteric Nervous System[a]

Absorptive agent	Effects abolished by: Adrenergic antagonists	Effects abolished by: TTX	Enterocyte receptors	Classification
α-Adrenergics	Yes	No	Yes	NT[b]
β-Adrenergics	Yes	—	—	Possible NT
Dopamine	Partially[c]	—	—	Possible NT
Insulin	—	—	Yes	Possible NT
Nicotinic cholinergics	Yes	Yes	No	NM[d]
Angiotensin II (low dose)	Yes	—	No[e]	NM
Somatostatin	No	Yes	No	Probable NM
Enkephalin	No	Yes	No	Probable NM

[a]Modified from Ref. 422.
[b]NT = primary neurotransmitter or physiological hormone.
[c]Dopamine's effect is partially blocked by α-adrenergic antagonists and completely blocked by adrenergic plus dopamineric antagonists.
[d]NM = neuromodulator or interneuron.
[e]The enterocyte receptor has low affinity for AII.

pine.[423–425] These investigators also demonstrated that cholinergic fibers participate in secretion that originates from tactile stimuli to the mucosa of the gut and is mediated by neuroconnections confined to the ENS. This general scheme of both extrinsic and intrinsic cholinergic-mediated secretion has been confirmed by a host of modern studies in which ACh or other parasympathetomimetic drugs have been shown to stimulate ion and water secretion in both small and large intestine of several species including man.[424–427] In general, the classic criteria for neurotransmitter status for ACh in gut have been met including [^{3}H]choline uptake with subsequent release of [^{3}H]-ACh.[428]

The secretory response to cholinergic agonists *in vitro* indicates that a major part of cholinergic-mediated secretion occurs independent of changes in either motility or blood flow.[409,429–432] Atropine has a small effect on basal Cl^- transport *in vitro*, increasing absorption, further proof for "cholinergic tone."[409] The response to muscarinic and low-dose nicotinic agents *in vitro* is an increased I_{sc} that is due to electrogenic Cl^- secretion. This response is inhibited by muscarinic blockers such as atropine, but not by TTX. When nicotinic agents or high-dose cholinergics are added to the Ussing chamber, a subsequent late inhibition of I_{sc} is accompanied by increased Na^+ and Cl^- absorption which is blocked by hexamethonium, by α-adrenergic agents, and by TTX.[409,432] Therefore, it appears that stimulation of nicotinic receptors in the gut elicits release of adrenergic neuroeffector agents which then act on the intestinal cells to stimulate absorption (see Section 7.4.3).

The crypt location of the muscarinic secretion has been demonstrated by the inhibition of secretion that comes from selectively damaging the crypts with protein synthesis inhibitors such as cycloheximide, but with no effect after damaging the gut surface cells or villi with hypertonic salts.[433]

[^{3}H]-quinuclidinyl benzylate (QNB) is a potent muscarinic antagonist that binds to intestinal cells of the small bowel and colon in a specific manner that meets the common criteria for receptor identification (temperature dependence, saturation kinetics, high affinity, pharmacological specificity, and stereoselectivity).[434–437] In the gastrointestinal tract, the stomach and colon have the highest number of QNB-binding sites, followed by the ileum, jejunum, and duodenum.[434–438] The number of binding sites in the duodenum was decreased by vagotomy,[439] but was increased in parotid gland by long-term atropine treatment,[440] leaving open the question whether receptors are involved in the phenomenon of postdenervation hypersensitivity. There appear to be as many binding sites on the basolateral cell membranes of surface cells or villus cells as in the crypts.[441] Furthermore, the enterocyte receptor seems to be a different subtype (M1) from those on intestinal muscle (M2).[442]

The stimulation of Cl^- secretion by muscarinic cholinergic agents involves an increase in cytosolic Ca^{2+}.[186,443–445] Cholinergic agents also stimulate guanylate cyclase in rabbit ileum, increasing cGMP content.[446] However, there is no evidence that cGMP is an important intermediate in muscarinic cholinergic secretion. α-Adrenergic agents also increase cGMP content in rabbit ileum,[446] and the increase in cGMP with carbachol occurs at a dose that stimulates absorption through the nicotinic receptor-mediated catecholamine release.[409]

Based on the studies summarized above, there is concrete evidence that ACh is a primary neurotransmitter in the gut that effects receptor-mediated Cl^- secretion. However, the response of the *in vitro* rabbit and human ileum to electrical field stimulation (EFS) suggests that there are other secretory neurotransmitters released by the ENS.[447–450] EFS-induced Cl^- secretion is inhibited by agents that block neurotransmission (TTX) or that polarize nerves (veratrine, K^+, and scorpion venom), but is only partially inhibited by atropine. Some of the agents discussed below are candidate neurotransmitters for this atropine-insensitive secretion.

7.4.5b. VIP. The significant number of VIP fibers identified in the gut mucosa by immunohistological techniques,[401] the demonstrated secretory effect *in vivo* and *in vitro* in both small bowel and colon,[451–458] and the finding of VIP receptors on intestinal cell membranes[342,459] suggest a neurotransmitter role for this peptide. The intracellular mediator of stimulated secretion is cAMP.[451,452,458] The ED_{50} for adenylate cyclase stimulation of basolateral VIP receptors in rat small intestine is exceedingly low, 4×10^{-9} M, and is some 3–4 logs less than the stimulation of secretion noted by secretin or parasympathetic agents.[459,460]

7.4.5c. Bombesin. The tetradecapeptide bombesin, first demonstrated in frog skin, has been found in both nerves and mucosal endocrine cells of the gut. This peptide may be working there as a neuromodulator since it has been shown to cause release of a number of other peptides such as neurotensin, CCK, and substance P.[461–463] In addition, its effect on canine colon motility is blocked by TTX.[461] However, Kachur *et al.* have presented compelling evidence that bombesin may be a neurotransmitter for small intestinal secretion.[464] It causes a short-lived, anion-dependent secretory electrical potential (PD) that is not blocked by atropine or TTX (or for that matter by indomethacin, diphenylhydramine, etorphine, somatostatin, epinephrine, or verapamil). Cyclic nucleotides have been proposed as the intermediate of secretory action, although this has not been proven. Since the response to bombesin is short-lived, the physiological significance of any neurotransmitter-related secretion remains to be determined.

7.4.5d. Substance P. Substance P is distributed in the PNS in a manner that suggests an afferent role in pain transmission. However, it also stimulates a short-lived Cl^- secretory potential in rat[314] and guinea pig ileum.[465] Its effect is not blocked by atropine or TTX, nor is it altered by indomethacin, diphenylhydramine, somatostatin, or etorphine. Prior desensitization of gut with neurotensin or 5HT partially blocks the effect of substance P, suggesting that these peptides might release substance P. Alternatively, this may mean that substance P, neurotensin, and 5HT stimulate secretion through a similar cellular mechanism such as Ca^{2+} gating. Indeed, substance P's secretory effect is inhibited by Ca^{2+}-free solutions and by verapamil, a response demonstrated also for 5HT and neurotensin (see below). Thus, current evidence would suggest that substance P is a neurotransmitter for intestinal secretion.

7.4.5e. 5HT. 5HT inhibits NaCl absorption or stimulates Cl^- secretion in rabbit small intestine by a Ca^{2+}-dependent process.[466–471] The hormone does not stimulate adenylate cyclase,[468] but its action *in vitro* is blocked by use of Ca^{2+}-free bathing solutions and by verapamil.[470] Donowitz *et al.* have suggested that one of the effects of 5HT is to inhibit coupled NaCl influx across the epithelial cell membrane via increases in free cytosolic $[Ca^{2+}]$, perhaps in concert with calmodulin.[469,470] 5HT also inhibits NaCl transport in rabbit gallbladder, but not in rabbit descending colon.[469] It remains to be clarified whether this is because rabbit descending colon does not have a Ca^{2+}-dependent NaCl influx mechanism, or whether it is because 5HT is acting as a neuromodulator or as a transmitter in an interneuron and the colon does not contain the necessary elements for an effect.

It is not clear whether 5HT is a neurotransmitter or neuromodulator.[472] Ca^{2+} and calmodulin are important for neurotransmitter release and synaptic function,[473] and investigators have suggested that 5HT may act as a Ca^{2+} ionophore, causing neuronal release of ACh or other peptides.[309,379,474] The enterochromaffin cells of the gut are innervated and these cells can be shown to liberate 5HT both to the lumen and to the blood (or serosal solution) after transmural EFS,[475] after vagal stimulation,[476,477] or after application of carbachol to the serosal surface of the gut.[404] In addition, 5HT release can be demonstrated in response to luminal stimuli such as a decrease in pH.[478] The most compelling evidence against a neurotransmitter role for 5HT comes from failure to find 5HT receptors on small intestinal and colonic epithelial cells by a technique that did demonstrate such receptors in brain tissue.[479] In favor of a neurotransmitter role is the absence of inhibition of 5HT secretory effects by cholinergic blockers such as atropine or hexamethonium or by TTX (Ref. 472; Donowitz, personal communication). Therefore, at this point, the role of 5HT as a primary neurotransmitter is uncertain.

7.4.5f. Neurotensin. In guinea pig ileum and canine jejunum, neurotensin induces Cl^- secretion or inhibits its absorption.[465,480] Desensitization of guinea pig ileum with substance P completely blocks the effect of neurotensin, but desensitization with neurotensin has little effect on substance P-induced secretion. The effects of neurotensin are also blocked by TTX. The action of neurotensin is inhibited in Ca^{2+}-free solutions and by verapamil, suggesting that its action (presumably via substance P) is Ca^{2+}-dependent. Its action is also attenuated by prior desensitization with 5HT. Therefore, it is probable that neurotensin releases 5HT as well as substance P from endocrine cells.[465] Others have shown neurotensin-mediated release of NE and histamine (presumably from mast cells) in the gut and a reduction of release of ACh following neurotensin.[481,482] Thus, neurotensin, concentrated primarily in the ileum,[483] appears to be a fairly important and ubiquitous neuromodulator of both secretory and motility functions of gut.

7.4.5g. Gastrin, CCK, Gastric Inhibitory Peptide (GIP), and Motilin. With one exception,[484] both gastrin and CCK have been shown either to inhibit water and electrolyte absorption or to cause frank secretion in the small intestine both *in vivo* and *in vitro*.[339,340,485,486] There are no studies to date allowing conclusions as to their status as either neurotransmitters or neuromodulators. Little is known about GIP or motilin other than isolated reports that they cause intestinal secretion.[487–491]

7.4.5h. Glucagon and Histamine. Glucagon usually induces intestinal secretion,[487,488,492–495] although the opposite effect has also been reported.[341,496,497] Histamine more clearly causes secretion both *in vivo* and *in vitro* by a process that involves H_1 receptors.[498–500] These two hormones are listed together because they have significant effects on intestinal vasculature and, therefore, could well be affecting intestinal secretion by changing blood flow. For instance, Turnberg's group has demonstrated that glucagon stimulates intestinal secretion in the human jejunum *in vivo*,[493] but not in Ussing-chambered gut.[497] This suggests that glucagon works through changes in Starling forces, i.e., the secretion may be due to secretory filtration.[495] However, since glucagon actually stimulates Cl^- secretion in mouse intestine *in vitro*,[494] it is still possible that it (as well as histamine)[499,500] may be working directly on the enterocyte.

7.4.5i. ATP. Although ATP was one of the first noncholinergic, nonadrenergic neurotransmitters to be postulated,[375] its role as a stimulant of intestinal secretion is not clear. In high concentrations ($>$ 1 mM), ATP inhibits absorption or stimulates Cl^- secretion[501–503] possibly via a cAMP-dependent process.[502] In isolated intestinal cells, ATP also permits Ca^{2+} gating and stimulates Na^+ uptake.[504–506] In other systems (kidney, muscle, etc.), ATP mediates stimulus–secretion coupling through prostaglandin intermediates,[507,508] presumably via Ca^{2+} gating which activates phospholipase activity, and this may be the mechanism of stimulation of adenylate cyclase in gut as well (see Fig. 7).

7.4.6. Agents Stimulating Intestinal Absorption

7.4.6a. Adrenergic Agents. Catecholamines stimulate intestinal water and electrolyte absorption *in vivo*.[424,509,510] The response also occurs *in vitro*[231] and therefore cannot be due to changes in blood flow. Furthermore, sympathetic nerve stimulation[511] and α-adrenergic agents (Table VI) also diminish intestinal blood flow, a response not likely to but does promote fluid absorption. NE has met the criteria necessary to be termed a neurotransmitter in gut. Synthesis and catabolism of NE has been shown in gut mucosa[512] and specific uptake mechanisms demonstrated.[407] Sympathetic nerve stimulation[511,513] and α-adrenergic agents[231,514–517] have the same effects on transport as does release of NE by reserpine[510] or tyramine.[408,518] The absorptive effects of adrenergic agents are blocked by adrenergic blockers[231] and specific α_2 receptors have been demonstrated on the enterocyte[519–521] as well as on nerves in Auerbach's plexus.[522] After depletion of NE in the rabbit by chemical sympathectomy with 6-hydroxydopamine, tyramine no longer has an effect on transport.[408,518] Finally, the transport effects of epinephrine are not blocked by TTX.[523]

The absorptive effect in rabbit ileum[515,516,521] and rat jejunum[517] appears to be through a specific α_2 mechanism. Dose–response curves with various adrenergic agents show the ED_{50} for clonidine, a specific α_2 agonist, to be less than 10^{-5} M, whereas the dose for α_1 agonists such as methoxamine or phenylephrine is greater. In addition, prazosin (a specific α_1 antagonist) does not alter the epinephrine dose–response curve, whereas yohimbine (α_2 antagonist) completely inhibits the epinephrine effect. The stimulatory effect of catecholamines on Na^+ and water absorption in the rabbit colon is also an α-agonist effect, although the studies do not clearly define whether this is an α_1 or α_2 response.[524] In rat colon and human small intestine, however, β agonists may also stimulate electrolyte and water absorption. In rat colon, both isoproterenol and NE affect transport, and the response to epinephrine is partially blocked by either α-antagonists or β-antagonists and is totally blocked by their combination.[523] Similarly, triple lumen tube perfusion studies in human jejunum and ileum indicate that isoproterenol stimulates fluid and electrolyte absorption.[525] In addition, β-blockers such as propranolol inhibit human jejunal absorption and reverse ileal transport from absorption to secretion.[525] These studies suggest that both α and β mechanisms are involved in the small and large intestine of many species. Furthermore, the decrease in absorption with adrenergic blockers *in vivo*[525] suggests the presence of "adrenergic tone" in the basal state.

The mechanism of enhanced Na^+ and Cl^- transport by adrenergic agents is not known. In the Ussing chamber there is a characteristic decrease in I_{sc} and stimulation of net Na^+ and Cl^- absorption.[231] The decrease in I_{sc} in the rabbit ileum appears to be due to inhibition of HCO_3^- secretion,[231] while in the rat colon it may represent stimulation of electrogenic Cl^- absorption.[523]

Current models of electrolyte transport would require the stimulation of Na^+ and Cl^- absorption to occur at the brush border of the intestinal cell via an increase in coupled NaCl influx, a process presumably controlled by intracellular cyclic nucleotide or Ca^{2+} levels. There is evidence that α-adrenergics prevent stimulation of adenylate cyclase by cholera toxin, VIP, and prostaglandins.[526,527] However, adrenergics do not alter the increase in cyclic nucleotide levels induced by theophylline,[526] and adrenergics have no effect on basal cyclic nucleotide levels in rat colon or ileum.[524,526] In some tissues, e.g., lacimal glands,[528] catecholamines alter Ca^{2+} gating and stimulate electrolyte and enzyme secretion through a Ca^{2+}-mediated process. There is also evidence that α-adrenergics act as Ca^{2+} channel blockers in brain.[529] However, Ca^{2+}-free solutions have no effect on α-adrenergic-stimulated colonic Na^+ absorption in rabbit colon.[524] These experiments do not rule out the possibility that catecholamines might decrease Ca^{2+} stores within the cell. Sequestration of Ca^{2+} would, under the currently proposed models, increase NaCl influx across the brush border membrane.[186] Thus, additional studies are necessary to understand the basic mechanism of action of adrenergic agents.

7.4.6b. Dopamine. Dopamine, another catecholamine found in significant quantities in the gut mucosa, also stimulates electrolyte and water absorption *in vitro*[530] and *in vivo*.[531] Like epinephrine, it causes a dose-dependent decrease in I_{sc} and stimulates Na^+ and Cl^- transport. This response has also been noted with bromocryptine, a long-acting dopamine agonist that can be given orally.[531] The effect of dopamine is partially inhibited by α_2 antagonists (yohimbine) and by dopamine receptor antagonists (haloperidol and domperidone), but not by α_1 or β antagonists. It does not appear that dopamine releases NE from the gut mucosa, because peripheral sympathectomy did not subsequently alter the response to dopamine. Dopamine does not alter basal cAMP levels in the gut, but lowers total Ca^{2+} content and may inhibit Ca^{2+} influx across the basolateral cell membrane. While these studies suggest that dopamine works through both α_2 and dopamine receptors, studies with TTX have not been done nor have enterocyte dopaminergic receptors been sought.

7.4.6c. Somatostatin. Somatostatin decreases the I_{sc} in rabbit ileum and rat colon in a fashion similar to adrenergic agents and stimulates Na^+ and Cl^- absorption.[532–535] Animals that are maximally absorbing, i.e., those with no apparent spontaneous secretion, do not respond to this agent.[535] In the rabbit, the effect appears to be localized to the ileum.[535] In spite of its similarity to the α-adrenergic effect, somatostatin is not inhibited by α-adrenergic blockers, chemical sympathectomy, or opiate antagonists.[532] From these studies, one can surmise that it has no effects on basal water and electrolyte transport, but that it will inhibit spontaneous or stimulated secretion and return transport to the basal state. Somatostatin stimulates coupled NaCl influx across rabbit ileal brush border membranes, but it does not appear to do this by altering cyclic nucleotide levels.[536] The mechanism of somatostatin's effects on coupled NaCl transport is unknown, but alterations in Ca^{2+} levels in the cell are proposed. Somatostatin receptors have not been found in rabbit ileal cells, and its effects on transport are partially inhibited by TTX.[337] Thus, it appears that somatostatin is a neuromodulator. Whether it works by release of a yet unknown absorptive neurotransmitter, or whether it inhibits release of secretory transmitters, remains to be determined.

7.4.6d. Enkephalin. In 1980, Dobbins *et al.* reported that various enkephalins including leucine enkephalin, methionine enkephalin, and the more poorly metabolized [D-Ala²]-methionine enkephalin amide decreased the I_{sc} and stimulated Na^+ and Cl^- absorption in the Ussing-chambered rabbit ileum.[537] These effects were reversed by naloxone. McKay *et al.* reported similar effects with a synthetic enkephalin analog and also with morphine, although these investigators did not find an increase in Na^+ absorption.[538,539] In McKay's studies, the opiates appeared to stimulate Cl^-/HCO_3^- exchange. The reasons for these discrepancies in the same species are not clear, although, as is the case for somatostatin, it may have something to do with the level of spontaneous secretion in the experimental animals. Dobbins *et al.* reported spontaneous net Na^+ secretion, and their animals responded to enkephalin with a decrease in I_{sc} of greater than 50%. Conversely, McKay *et al.* reported high levels of Na^+ absorption in their animals, and the I_{sc} in their studies, which was low to begin with, responded to enkephalin with less than a 25% decrease.

The response to enkephalin is not blocked by α-blockers or dopamine blockers or by peripheral chemical sympathectomy, indicating that the absorptive effect is not via catecholamine release.[534,537] The response was decreased in Ca^{2+}-free solution, suggesting that it is a Ca^{2+}-dependent process. However, this Ca^{2+} dependence could be one of nerve transmission rather than enterocyte function. Most importantly, TTX appeared to block the effect of the enkephalin. Enkephalins have certainly been shown to inhibit ACh release from gut myenteric plexuses.[540–542] However, in the studies of Dobbins *et al.*, the opiate effect was seen in the presence of atropine, making it unlikely that opiates work by inhibition of ACh release. These observations suggest that the opiates either release a non-adrenergic stimulant of absorption or else they inhibit the release of some noncholinergic stimulant of secretion.

The opiate effect on intestinal electrolyte transport by guinea pig appears to be of the δ receptor type.[543,544] This contrasts with the inhibition of smooth muscle function in the same species which is of the μ receptor type. However, there may be species differences,[545] and transport studies need to be done with the gut opiate dynorphin, which is an extremely potent κ-receptor agonist.[546] These differences in opiate receptors do raise the possibility that the receptors controlling intestinal electrolyte transport and intestinal motility might be quite different.

Studies of isolated intestinal cells have failed to demonstrate opiate receptors.[480] This and the inhibition of enkephalin effect with TTX make it more likely that the effect of enkephalins on transport is mediated through neuromodulation.

7.4.6e. Nicotinic Agents and Low-Dose Angiotensin. The absorptive effect of nicotinic agents and low-dose angiotensin has been described above (Section 7.4.3) as classic examples of neuromodulation.

8. Summary and Conclusions

In a dynamic field such as electrolyte transport research, drawing dogmatic conclusions is hazardous. However, based on the data presented here, I believe the following statements are warranted:

1. The model of intestinal electrolyte transport shown in Fig. 1 is the best general framework for understanding current data.
2. Most evidence favors Na^+–H^+/Cl^-–HCO_3^- antiport (Fig. 2) as the mechanism of coupling of Na^+ to Cl^- entry across absorptive cell apical cell membranes. In some tissues, K^+ entry is also coupled to NaCl entry.
3. The coupling of Na^+ to glucose entry is electrogenic with 2 : 1 stoichiometry.
4. Amiloride-sensitive conductance channels are the primary mechanism of Na^+ entry in distal segments of mammalian colon, but in most species (especially in proximal colonic segments) coupled NaCl entry mechanisms are present as well.
5. Amiloride-sensitive conductance channels, NaCl coupled entry, Na^+-coupled nonelectrolyte entry, and passive diffusion are the rate-limiting steps for Na^+ transport in gut, not the Na^+,K^+-ATPase-regulated exit step. Entry is tightly coupled to exit through intracellular [Na^+].
6. The tight junctions of gut cells are cation-selective and some regurgitation of Na^+ does occur from intracellular space to lumen, especially in the flounder. The junctions govern the leakiness of epithelia.
7. Cyclic nucleotides and Ca^{2+}, via protein kinases and calmodulin, modulate stimulus–secretion coupling in gut. Whether both agents have the same effects on transport, or whether cyclic nucleotides alter apical membrane Cl^- permeability while Ca^{2+} only inhibits coupled NaCl entry, remains to be determined.
8. The gut both absorbs and secretes HCO_3^-. Absorption occurs as the result of H^+ secretion. Net secretion is the result of Cl^-/HCO_3^- exchange and/or electrogenic cAMP-mediated HCO_3^- transport.
9. There is clear evidence in colon for both active absorptive and secretory processes for K^+.
10. Both mineralo- and glucocorticoids are important hormones physiologically; insulin and secretin are possible physiological regulators of transport. Systemic acid–base balance and gut blood flow may be important permissive factors regulating transport. The kinins and prostaglandins are probably important pathophysiological stimulants of secretion.
11. ACh is the primary neurotransmitter for secretion, while α-adrenergics are absorptive neurotransmitters. There may be other neurotransmitters acting directly on the enterocyte, i.e., VIP and dopamine, but many of the peptides, amines, and chemicals newly located in the gut are neuromodulators acting on the nerves of the gut plexuses.

In the coming decade, we should see a clarification of many of these issues. Hopefully, this will lead to effective treatments of diarrhea through the development of useful antisecretory drugs.

ACKNOWLEDGMENTS. The author is indebted to Drs. H. J. Binder, A. N. Charney, J. W. Dobbins, M. Donowitz, M. Field, T. S. Gaginella, G. A. Kimmich, and E. J. Tapper for reprints of manuscripts in press and to Jane McLelland and Leslie McDonough for excellent editorial assistance.

References

1. Binder, H. J., ed. 1979. *Mechanisms of Intestinal Secretion*. Liss, New York.
2. Field, M., J. S. Fordtran, and S. G. Schultz, eds. 1980. *Secretory Diarrhea*. American Physiological Society, Bethesda, Md.
3. Johnson, L. R., ed.-in-chief. 1981. *Physiology of the Gastrointestinal Tract*. Raven Press, New York.
4. Giebisch, G., D. C. Tosteson, and H. H. Ussing, eds. 1979. *Membrane Transport in Biology*, Volumes I–IVB. Springer-Verlag, Berlin.
5. Koefoed-Johnsen, V., and H. H. Ussing. 1958. The nature of the frog skin potential. *Acta Physiol. Scand.* **42**:298–308.
6. Crane, R. K. 1962. Hypothesis for mechanism of intestinal active transport of sugars. *Fed. Proc.* **21**:891–895.
7. Curran, P. F. 1968. Twelfth Bowditch Lecture. Coupling between transport processes in intestine. *Physiologist* **1**:3–23.
8. Diamond, J. M. 1977. Twenty-first Bowditch Lecture. The epithelial junction: Bridge, gate, and fence. *Physiologist* **20**:10–18.
9. Curran, P. F., and J. R. Macintosh. 1962. A model system for biological water transport. *Nature (London)* **193**:347–348.
10. Diamond, J. M., and W. H. Bossert. 1967. Standing-gradient osmotic flow: A mechanism for coupling of water and solute transport in epithelia. *J. Gen. Physiol.* **50**:2061–2083.
11. Field, M. 1979. Intracellular mediators of secretion in the small

intestine. In: *Mechanisms of Intestinal Secretion*. H. J. Binder, ed. Liss, New York. pp. 83–91.
12. Frizzell, R. A., and K. Heintze. 1979. Electrogenic chloride secretion by mammalian colon. In: *Mechanisms of Intestinal Secretion*. H. J. Binder, ed. Liss, New York. pp. 101–110.
13. Frizzell, R. A., M. Field, and S. G. Schultz. 1979. Sodium-coupled chloride transport by epithelial tissues. *Am. J. Physiol.* **236**:F1–F8.
14. Frizzell, R. A., and M. E. Duffey. 1980. Chloride activities in epithelia. *Fed. Proc.* **39**:2860–2864.
15. Turnberg, L. A., F. A. Bieberdorf, S. G. Morawski, and J. S. Fordtran. 1970. Interrelationships of chloride, bicarbonate, sodium, and hydrogen transport in the human ileum. *J. Clin. Invest.* **49**:557–567.
16. Krejs, G. J., and J. S. Fordtran. 1978. Physiology and pathophysiology of ion and water movement in the human intestine. In: *Gastrointestinal Disease: Pathophysiology, Diagnosis, and Management,* Volume II, 2nd ed. M. J. Sleisenger and J. S. Fordtran, eds. Saunders, Philadelphia. pp. 297–312.
17. Naftalin, R. J., and N. L. Simmons. 1979. The effects of theophylline and choleragen on sodium and chloride ion movements within isolated rabbit ileum. *J. Physiol. (London)* **290**:331–350.
18. Holman, G. D., and R. J. Naftalin. 1979. Fluid movements across rabbit ileum coupled to passive paracellular ion movements. *J. Physiol. (London)* **290**:351–366.
19. Holman, G. D., R. J. Naftalin, N. L. Simmons, and M. Walker. 1979. Electrophysiological and electron-microscopical correlations with fluid and electrolyte secretion in rabbit ileum. *J. Physiol. (London)* **290**:367–386.
20. Moreno, J. H. 1975. Blockage of gallbladder tight-junction cation selective channels by 2,4,6-triaminopyrimidine (TAP). *J. Gen. Physiol.* **66**:97–115.
21. Nellans, H. N., R. A. Frizzell, and S. G. Schultz. 1973. Coupled sodium–chloride influx across the brush border of rabbit ileum. *Am. J. Physiol.* **225**:467–475.
22. Nellans, H. N., R. A. Frizzell, and S. G. Schultz. 1975. Effect of acetazolamide on sodium and chloride transport by *in vitro* rabbit ileum. *Am. J. Physiol.* **228**:1808–1814.
23. Schultz, S. G., P. F. Curran, R. A. Chez, and R. E. Fuisz. 1967. Alanine and sodium fluxes across mucosal border of rabbit ileum. *J. Gen. Physiol.* **50**:1241–1260.
24. Frizzell, R. A., H. N. Nellans, R. C. Rose, L. Markscheid-Kaspi, and S. G. Schultz. 1973. Intracellular Cl concentrations and influxes across the brush border of rabbit ileum. *Am. J. Physiol.* **224**: 328–337.
25. Frizzell, R. A., P. L. Smith, E. Vosburgh, and M. Field. 1979. Coupled sodium–chloride influx across brush border of flounder intestine. *J. Membr. Biol.* **46**:27–39.
26. Field, M., K. J. Karnaky, P. L. Smith, J. E. Bolton, and W. B. Kinter. 1978. Ion transport across the isolated intestinal mucosa of the winter flounder, *Pseudopleuronectes americanus*. I. Functional and structural properties of cellular and paracellular pathways for Na and Cl. *J. Membr. Biol.* **41**:265–293.
27. Field, M., P. L. Smith, and J. E. Bolton. 1980. Ion transport across the isolated intestinal mucosa of the winter flounder, *Pseudopleuronectes americanus*. II. Effects of cyclic AMP. *J. Membr. Biol.* **55**:157–163.
28. Frizzell, R. A., M. C. Dugas, and S. G. Schultz. 1975. Sodium chloride transport by rabbit gallbladder: Direct evidence for a coupled NaCl influx process. *J. Gen. Physiol.* **65**:769–795.
29. Murer, H., U. Hopfer, and R. Kinne. 1976. Sodium/proton antiport in brush-border-membrane vesicles isolated from rat small intestine and kidney. *Biochem. J.* **154**:597–604.
30. Liedtke, C. M., and U. Hopfer. 1977. Anion transport in brush border membranes isolated from rat small intestine. *Biochem. Biophys. Res. Commun.* **76**:579–585.
31. Gunther, R. D., and E. M. Wright. 1983. Na^+, Li^+, and Cl^- transport by brush border membranes from rabbit jejunum. *J. Membr. Biol.* **74**:85–94.
32. Knickelbein, R., P. S. Aronson, W. Atherton, and J. W. Dobbins. 1983. Na and Cl transport across rabbit ileal brush border. I. Evidence for Na:H exchange. *Am. J. Physiol.* **245**:G504–G510.
33. Benos, D. J. 1982. Amiloride: A molecular probe of sodium transport in tissues and cells. *Am. J. Physiol.* **242**:C131–C145.
34. Dubinsky, W., and R. A. Frizzell. 1983. A novel effect of amiloride on H^+-dependent Na^+ transport. *Am. J. Physiol.* **245**:C157–C159.
35. Warnock, D. G., and J. Eveloff. 1982. NaCl entry mechanisms in the luminal membrane of the renal tubule. *Am. J. Physiol.* **242**:F561–F574.
36. Aronson, P. S., and S. E. Bounds. 1980. Harmaline inhibition of Na-dependent transport in renal microvillus membrane vesicles. *Am. J. Physiol.* **238**:F210–F217.
37. Kinsella, J. L., and P. S. Aronson. 1980. Properties of the Na^+–H^+ exchanger in renal microvillus membrane vesicles. *Am. J. Physiol.* **238**:F461–F469.
38. Reenstra, W. W., D. G. Warnock, V. J. Yee, and J. G. Forte. 1981. Proton gradients in renal cortex brush-border membrane vesicles: Demonstration of a rheogenic proton flux with acridine orange. *J. Biol. Chem.* **256**:11663–11666.
39. Warnock, D. G., W. W. Reenstra, and V. J. Yee. 1982. Na^+/H^+ antiporter of brush border vesicles: Studies with acridine orange uptake. *Am. J. Physiol.* **242**:F733–F739.
40. Ericson, A.-C., and K. R. Spring. 1982. Volume regulation by *Necturus* gallbladder: Apical Na^+–H^+ and Cl^-–HCO_3^- exchange. *Am. J. Physiol.* **243**:C146–C150.
41. Weinman, S. A., and L. Reuss. 1982. Na^+–H^+ exchange at the apical membrane of *Necturus* gallbladder: Extracellular and intracellular pH studies. *J. Gen. Physiol.* **80**:299–321.
42. Liedtke, C. M., and U. Hopfer. 1982. Mechanism of Cl^- translocation across small intestinal brush-border membrane. II. Demonstration of Cl^-–OH^- exchange and Cl^- conductance. *Am. J. Physiol.* **242**:G272–G280.
43. Knickelbein, R. G., P. S. Aronson, and J. W. Dobbins. 1982. Na–H and Cl–HCO_3 exchange in rabbit ileal brush border membrane vesicles. *Fed. Proc.* **41**:1494.
44. Fan, C.-C., R. G. Faust, and D. W. Powell. 1983. Coupled Na–Cl transport by rabbit ileal brush border membrane vesicles. *Am. J. Physiol.* **244**:G375–G385.
45. Warnock, D. G., and V. J. Yee. 1981. Chloride uptake by brush border membrane vesicles isolated from rabbit renal cortex: Coupling to proton gradients and K^+ diffusion potentials. *J. Clin. Invest.* **67**:103–115.
46. Spring, K. R., and A. Hope. 1978. Size and shape of the lateral intercellular spaces in a living epithelium. *Science* **200**:54–58.
47. Fisher, R. S., B. E. Persson, and K. R. Spring. 1981. Epithelial cell volume regulation: Bicarbonate dependence. *Science* **214**:1357–1360.
48. Ericson, A.-C., and K. R. Spring. 1982. Coupled NaCl entry into *Necturus* gallbladder epithelial cells. *Am. J. Physiol.* **243**:C140–C145.
49. Spring, K. R., and A.-C. Ericson. 1982. Epithelial cell volume modulation and regulation. *J. Membr. Biol.* **69**:167–176.
50. Liedtke, C. M., and U. Hopfer. 1982. Mechanism of Cl^- translocation across small intestinal brush-border membrane. I. Absence of Na^+Cl^- cotransport. *Am. J. Physiol.* **242**:G263–G271.
51. Eveloff, J., M. Field, R. Kinne, and H. Murer. 1980. Sodium-cotransport systems in intestine and kidney of the winter flounder. *J. Comp. Physiol.* **135**:175–182.
52. Eveloff, J., and R. Kinne. 1983. Sodium–chloride transport in the medullary thick ascending limb of Henle's loop: Evidence for a sodium–cbloride cotransport system in plasma membrane vesicles. *J. Membr. Biol.* **72**:173–181.
53. Friedman, P. A., and T. E. Andreoli. 1982. CO_2-stimulated NaCl absorption in the mouse renal cortical thick ascending limb of Henle: Evidence for synchronous Na^+/H^+ and Cl^-/HCO_3^- exchange in apical plasma membranes. *J. Gen. Physiol.* **80**:683–711.
54. Musch, M. W., S. A. Orellana, L. S. Kimberg, M. Field, D. R. Halm, E. J. Krasny, Jr., and R. A. Frizzell. 1982. Na^+–K^+–

Cl^- co-transport in the intestine of a marine teleost. *Nature (London)* **300**:351–353.
55. Davis, C. W., and A. L. Finn. 1983. Coupled cellular salt entry in *Necturus* gallbladder requires mucosal Na, Cl, and K. *Biophys. J.* **41**:81a.
56. Heintze, K., K.-U. Petersen, and J. R. Wood. 1981. Effects of bicarbonate on fluid and electrolyte transport by guinea pig and rabbit gallbladder: Stimulation of absorption. *J. Membr. Biol.* **62**:175–181.
57. Petersen, K.-U., J. R. Wood, G. Schulze, and K. Heintze. 1981. Stimulation of gallbladder fluid and electrolyte absorption by butyrate. *J. Membr. Biol.* **62**:183–193.
58. Humphreys, M. H., and L. Y.-N. Chou. 1983. Anion effects on fluid absorption from rat jejunum perfused *in vivo*. *Am. J. Physiol.* **244**:G33–G39.
59. Forsyth, G. W., R. A. Kapitany, and D. L. Hamilton. 1981. Organic acid proton donors decrease intestinal secretion caused by enterotoxins. *Am. J. Physiol.* **241**:G227–G234.
60. Humphreys, M. H., and L. Y.-N. Chou. 1979. Anion-stimulated ATPase activity of brush border from rat small intestine. *Am. J. Physiol.* **236**:E70–E76.
61. Van Os, C. H., A. K. Mircheff, and E. M. Wright. 1977. Distribution of bicarbonate-stimulated ATPase in rat intestinal epithelium. *J. Cell Biol.* **73**:257–260.
62. Gazitua, S., and J. W. L. Robinson. 1982. Ion fluxes and electrical characteristics of the short-circuited rat colon *in vitro*. *Pflügers Arch.* **394**:32–37.
63. Powell, D. W., P. T. Johnson, J. C. Bryson, R. C. Orlando, and C.-C. Fan. 1982. Effect of phenolphthalein on monkey intestinal water and electrolyte transport. *Am. J. Physiol.* **243**:G268–G275.
64. Kimmich, G. A. 1981. Intestinal absorption of sugar. In: *Physiology of the Gastrointestinal Tract*. L. R. Johnson, ed. Raven Press, New York. pp. 1035–1061.
65. Kimmich, G. 1983. Coupling stoichiometry and the energetic adequacy question. In: *Intestinal Transport: Fundamental Comparative Aspects*. M. Gilles-Baillien and R. Gilles, eds. Springer-Verlag, Berlin. pp. 87–102.
66. Freel, R. W., and A. M. Goldner. 1981. Sodium-coupled nonelectrolyte transport across epithelia: Emerging concepts and directions. *Am. J. Physiol.* **241**:G451–G460.
67. White, J. F., and W. M. Armstrong. 1971. Effect of transported solutes on membrane potentials in bullfrog small intestine. *Am. J. Physiol.* **221**:194–201.
68. Armstrong. W. M., W. R. Bixenman, K. F. Frey, J. F. Garcia-Diaz, M. G. O'Regan, and J. L. Owens. 1979. Energetics of coupled Na^+ and Cl^- entry into epithelial cells of bullfrog small intestine. *Biochim. Biophys. Acta* **551**:207–219.
69. Rose, R. C., and S. G. Schultz. 1971. Studies on the electrical potential profile across rabbit ileum: Effects of sugars and amino acids on transmural electrical potential difference. *J. Gen. Physiol.* **57**:639–663.
70. Gunter-Smith, P. J., E. Grasset, and S. G. Schultz. 1982. Sodium-coupled amino acid and sugar transport by *Necturus* small intestine: An equivalent electrical circuit analysis of a rheogenic co-transport system. *J. Membr. Biol.* **66**:25–39.
71. Murer, H., and U. Hopfer. 1974. Demonstration of electrogenic Na^+-dependent D-glucose transport in intestinal brush border membranes. *Proc. Natl. Acad. Sci. USA* **71**:484–488.
72. Beck, J. C., and B. Sacktor. 1975. Energetics of the Na^+-dependent transport of D-glucose in renal brush border membrane vesicles. *J. Biol. Chem.* **250**:8674–8680.
73. Carter-Su, C., and G. A. Kimmich. 1980. Effect of membrane potential on Na^+-dependent sugar transport by ATP-depleted intestinal cells. *Am. J. Physiol.* **238**:C73–C80.
74. Schultz, S. G. 1977. Sodium-coupled solute transport by small intestine: A status report. *Am. J. Physiol.* **233**:E249–E254.
75. Kimmich, G. A., and J. Randles. 1979. Energetics of sugar transport by isolated intestinal epithelial cells: Effects of cytochalasin B. *Am. J. Physiol.* **237**:C56–C63.
76. Kimmich, G. A., and J. Randles. 1981. α-Methylglucoside satisfies only Na^+-dependent transport system of intestinal epithelium. *Am. J. Physiol.* **241**:C227–C232.
77. Kimmich, G. A., and J. Randles. 1980. Evidence for an intestinal Na^+:sugar transport coupling stoichiometry of 2.0. *Biochim. Biophys. Acta* **596**:439–444.
78. Hilden, S., and B. Sacktor. 1982. Potential-dependent D-glucose uptake by renal brush border membrane vesicles in the absence of sodium. *Am. J. Physiol.* **242**:F340–F345.
79. Beck, J. C., and B. Sacktor. 1978. The sodium electrochemical potential-mediated uphill transport of D-glucose in renal brush border membrane vesicles. *J. Biol. Chem.* **253**:5531–5535.
80. Hopfer, U., and R. Groseclose. 1980. The mechanism of Na^+-dependent D-glucose transport. *J. Biol. Chem.* **255**:4453–4462.
81. Kaunitz, J. D., R. Gunther, and E. M. Wright. 1982. Involvement of multiple sodium ions in intestinal D-glucose transport. *Proc. Natl. Acad. Sci. USA* **79**:2315–2318.
82. Moran, A., J. S. Handler, and R. J. Turner. 1982. Na^+-dependent hexose transport in vesicles from cultured renal epithelial cell line. *Am. J. Physiol.* **243**:C293–C298.
83. Aronson, P. S. 1978. Energy-dependence of phlorizin binding to isolated renal microvillus membranes: Evidence concerning the mechanism of coupling between the electrochemical Na^+ gradient and sugar transport. *J. Membr. Biol.* **42**:81–98.
84. Turner, R. J., and M. Silverman. 1981. Interaction of phlorizin and sodium with the renal brush-border membrane D-glucose transporter: Stoichiometry and order of binding. *J. Membr. Biol.* **58**:43–55.
85. Crane, R. K., P. Malathi, and H. Preiser. 1976. Reconstitution of specific Na^+-dependent D-glucose transport in liposomes by Triton X-100 extracted proteins from purified brush border membranes of hamster small intestine. *Biochem. Biophys. Res. Commun.* **71**:1010–1016.
86. Crane, R. K., P. Malathi, H. Preiser, and P. Fairclough. 1978. Some characteristics of kidney Na^+-dependent glucose carrier reconstituted into sonicated liposomes. *Am. J. Physiol.* **234**:E1–E5.
87. Fairclough, P., P. Malathi, H. Preiser, and R. K. Crane. 1979. Reconstitution into liposomes of glucose active transport from the rabbit renal proximal tubule: Characteristics of the system. *Biochim. Biophys. Acta* **553**:295–306.
88. Kinne, R., and R. G. Faust. 1977. Incorporation of D-glucose-, L-alanine- and phosphate-transport systems from rat renal brush-border membranes into liposomes. *Biochem. J.* **168**:311–314.
89. Lin, J. T., M. E. M. Da Cruz, S. Riedel, and R. Kinne. 1981. Partial purification of hog kidney sodium–D-glucose cotransport system by affinity chromatography on a phlorizin polymer. *Biochim. Biophys. Acta* **640**:43–54.
90. Klip, A., S. Grinstein, and G. Semenza. 1979. Partial purification of the sugar carrier of intestinal brush border membranes: Enrichment of the phlorizin-binding component by selective extractions. *J. Membr. Biol.* **51**:47–73.
91. Im, W. B., K. Y. Ling, and R. G. Faust. 1982. Partial purification of the Na^+-dependent D-glucose transport system from renal brush border membranes. *J. Membr. Biol.* **65**:131–137.
92. Bissonnette, J. M., J. A. Black, K. L. Thornburg, K. M. Acott, and P. L. Koch. 1982. Reconstitution of D-glucose transporter from human placental microvillous plasma membranes. *Am. J. Physiol.* **242**:C166–C171.
93. Schultz, S. G., and P. F. Curran. 1970. Coupled transport of sodium and organic solutes. *Physiol. Rev.* **50**:637–718.
94. Fass, S. J., M. R. Hammerman, and B. Sacktor. 1977. Transport of amino acids in renal brush border membrane vesicles: Uptake of the neutral amino acid L-alanine. *J. Biol. Chem.* **252**:583–590.
95. Sigrist-Nelson, K., H. Murer, and U. Hopfer. 1975. Active alanine transport in isolated brush border membranes. *J. Biol. Chem.* **250**:5674–5680.
96. Rose, R. C. 1981. Transport and metabolism of water-soluble vitamins in intestine. *Am. J. Physiol.* **240**:G97–G101.
97. Hildmann, B., C. Storelli, W. Haase, M. Barac-Nieto, and H. Murer. 1980. Sodium ion/L-lactate co-transport in rabbit small-

intestinal brush-border-membrane vesicles. *Biochem. J.* **186**:169–176.

98. Nord, E., S. H. Wright, I. Kippen, and E. M. Wright. 1982. Pathways for carboxylic acid transport by rabbit renal brush border membrane vesicles. *Am. J. Physiol.* **243**:F456–F462.
99. Corcelli, A., G. Prezioso, F. Palmieri, and C. Storelli. 1982. Electroneutral Na^+/dicarboxylic amino acid cotransport in rat intestinal brush border membrane vesicles. *Biochim. Biophys. Acta* **689**:97–105.
100. Wright, S. H., S. Krasne, I. Kippen, and E. M. Wright. 1981. Na^+-dependent transport of tricarboxylic acid cycle intermediates by renal brush border membranes: Effect on fluorescence of a potential-sensitive cyanine dye. *Biochim. Biophys. Acta* **640**:767–778.
101. Heyman, M., A.-M. Dumontier, and J. F. Desjeux. 1980. Xylose transport pathways in rabbit ileum. *Am. J. Physiol.* **238**:G326–G331.
102. Wilson, F. A. 1981. Intestinal transport of bile acids. *Am. J. Physiol.* **241**:G83–G92.
103. Lücke, H., G. Stange, R. Kinne, and H. Murer. 1978. Taurocholate–sodium co-transport by brush-border membrane vesicles isolated from rat ileum. *Biochem. J.* **174**:951–958.
104. Beesley, R. C., and R. G. Faust. 1979. Sodium ion-coupled uptake of taurocholate by intestinal brush-border membrane vesicles. *Biochem. J.* **178**:299–303.
105. Lücke, H., G. Stange, and H. Murer. 1981. Sulfate–sodium cotransport by brush-border membrane vesicles isolated from rat ileum. *Gastroenterology* **80**:22–30.
106. Smith, P. L., S. A. Orellana, and M. Field. 1981. Active sulfate absorption in rabbit ileum: Dependence on sodium and chloride and effects of agents that alter chloride transport. *J. Membr. Biol.* **63**:199–206.
107. Langridge-Smith, J. E., and M. Field. 1981. Sulfate transport in rabbit ileum: Characterization of the serosal border anion exchange process. *J. Membr. Biol.* **63**:207–214.
108. Langridge-Smith, J. E., J. H. Sellin, and M. Field. 1983. Sulfate influx across the rabbit ileal brush border membrane: Sodium and proton dependence, and substrate specificities. *J. Membr. Biol.* **72**:131–139.
109. Cheng, L., and B. Sacktor. 1981. Sodium gradient-dependent phosphate transport in renal brush border membrane vesicles. *J. Biol. Chem.* **256**:1556–1564.
110. Berner, W., R. Kinne, and H. Murer. 1976. Phosphate transport into brush-border membrane vesicles isolated from rat small intestine. *Biochem. J.* **160**:467–474.
111. Hoffmann, N., M. Thees, and R. Kinne. 1976. Phosphate transport by isolated renal brush border vesicles. *Pflügers Arch.* **362**:147–156.
112. Danisi, G., J.-P. Bonjour, and R. W. Straub. 1980. Regulation of Na-dependent phosphate influx across the mucosal border of duodenum by 1,25-dihydroxycholecalciferol. *Pflügers Arch.* **388**:227–232.
113. Murer, H., and B. Hildmann. 1981. Transcellular transport of calcium and inorganic phosphate in the small intestinal epithelium. *Am. J. Physiol.* **240**:G409–G416.
114. Sellin, J. H., and M. Field. 1981. Physiologic and pharmacologic effects of glucocorticoids on ion transport across rabbit ileal mucosa *in vitro*. *J. Clin. Invest.* **67**:770–778.
115. Powell, D. W. 1978. Transport in large intestine. In: *Membrane Transport in Biology*. G. Giebisch, D. C. Tosteson, and H. H. Ussing, eds. Springer-Verlag, Berlin. pp. 781–809.
116. Fromm, M., U. Hegel, with S. Lüderitz. 1978. Segmental heterogeneity of epithelial transport in rat large intestine. *Pflügers Arch.* **378**:71–83.
117. Dawson, D. C., and P. F. Curran. 1976. Sodium transport by the colon of *Bufo marinus:* Na uptake across the mucosal border. *J. Membr. Biol.* **28**:295–307.
118. Dawson, D. C. 1977. Na and Cl transport across the isolated turtle colon: Parallel pathways for transmural ion movement. *J. Membr. Biol.* **37**:213–233.
119. Thompson, S. M., and D. C. Dawson. 1978. Sodium uptake across the apical border of the isolated turtle colon: Confirmation of the two-barrier model. *J. Membr. Biol.* **42**:357–374.
120. Frizzell, R. A., M. J. Koch, and S. G. Schultz. 1976. Ion transport by rabbit colon. I. Active and passive components. *J. Membr. Biol.* **27**:297–316.
121. Schultz, S. G., R. A. Frizzell, and H. N. Nellans. 1977. Active sodium transport and the electrophysiology of rabbit colon. *J. Membr. Biol.* **33**:351–384.
122. Bindslev, N., A. W. Cuthbert, J. M. Edwardson, and E. Skadhauge. 1982. Kinetics of amiloride action in the hen coprodaeum *in vitro*. *Pflügers Arch.* **392**:340–346.
123. Binder, H. J., and C. L. Rawlins. 1973. Electrolyte transport across isolated large intestinal mucosa. *Am. J. Physiol.* **225**:1232–1239.
124. Hawker, P. C., K. E. Mashiter, and L. A. Turnberg. 1978. Mechanisms of transport of Na, Cl, and K in the human colon. *Gastroenterology* **74**:1241–1247.
125. Rask-Madsen, J., and K. Hjelt. 1977. Effect of amiloride on electrical activity and electrolyte transport in human colon. *Scand. J. Gastroenterol.* **12**:1–6.
126. Frizzell, R. A., and K. Turnheim. 1978. Ion transport by rabbit colon. II. Unidirectional sodium influx and the effects of amphotericin B and amiloride. *J. Membr. Biol.* **40**:193–211.
127. Thompson, S. M., Y. Suzuki, and S. G. Schultz. 1982. The electrophysiology of rabbit descending colon. I. Instantaneous transepithelial current–voltage relations and the current–voltage relations of the Na-entry mechanism. *J. Membr. Biol.* **66**:41–54.
128. Thompson, S. M., Y. Suzuki, and S. G. Schultz. 1982. The electrophysiology of rabbit descending colon. II. Current–voltage relations of the apical membrane, the basolateral membrane, and the parallel pathways. *J. Membr. Biol.* **66**:55–61.
129. Thompson, S. M., and D. C. Dawson. 1978. Cation selectivity of the apical membrane of the turtle colon: Sodium entry in the presence of lithium. *J. Gen. Physiol.* **72**:269–282.
130. Sarracino, S. M., and D. C. Dawson. 1979. Cation selectivity in active transport: Properties of the turtle colon in the presence of mucosal lithium. *J. Membr. Biol.* **46**:295–313.
131. Lindemann, B., and W. van Driessche. 1976. Sodium-specific membrane channels of frog skin are pores: Current fluctuations reveal high turnover. *Science* **195**:292–294.
132. Aceves, J., and A. W. Cuthbert. 1979. Uptake of [^{3}H] benzamil at different sodium concentrations: Inferences regarding the regulation of sodium permeability. *J. Physiol. (London)* **295**:491–504.
133. Palmer, L. G. 1982. Na^+ transport and flux ratio through apical Na^+ channels in toad bladder. *Nature (London)* **297**:688–690.
134. Legris, G. J., P. C. Will, and U. Hopfer. 1982. Inhibition of amiloride-sensitive sodium conductance by indoleamines. *Proc. Natl. Acad. Sci. USA* **79**:2046–2050.
135. Schultz, S. G. 1981. Homocellular regulatory mechanisms in sodium-transporting epithelia: Avoidance of extinction by "flush-through." *Am. J. Physiol.* **241**:F579–F590.
136. Turnheim, K., R. A. Frizzell, and S. G. Schultz. 1977. Effect of anions on amiloride-sensitive, active sodium transport across rabbit colon, *in vitro:* Evidence for "trans-inhibition" of the Na entry mechanism. *J. Membr. Biol.* **37**:63–84.
137. Frizzell, R. A., and S. G. Schultz. 1978. Effect of aldosterone on ion transport by rabbit colon *in vitro*. *J. Membr. Biol.* **39**:1–26.
138. Turnheim, K., R. A. Frizzell, and S. G. Schultz. 1978. Interaction between cell sodium and the amiloride-sensitive sodium entry step in rabbit colon. *J. Membr. Biol.* **39**:233–256.
139. Stirling, C. E. 1972. Radioautographic localization of sodium pump sites in rabbit intestine. *J. Cell Biol.* **53**:704–714.
140. Quigley, J. P., and G. S. Gotterer. 1969. Distribution of (Na^+-K^+)-stimulated ATPase activity in rat intestinal mucosa. *Biochim. Biophys. Acta* **173**:456–468.
141. Harms, V., and E. M. Wright. 1980. Some characteristics of Na/K-ATPase from rat intestinal basal lateral membranes. *J. Membr. Biol.* **53**:119–128.

142. Hokin, L. E. 1981. Topical review: Reconstitution of "carriers" in artificial membranes. *J. Membr. Biol.* **60**:77–93.
143. Hilden, S., H. M. Rhee, and L. E. Hokin. 1974. Sodium transport by phospholipid vesicles containing purified sodium and potassium ion-activated adenosine triphosphatase. *J. Biol. Chem.* **249**:7432–7440.
144. Hilden, S., and L. E. Hokin. 1975. Active potassium transport coupled to active sodium transport in vesicles reconstituted from purified sodium and potassium ion-activated adenosine triphosphatase from the rectal gland of *Squalus acanthias*. *J. Biol. Chem.* **250**:6296–6303.
145. Anner, B. M., L. K. Lane, A. Schwartz, and B. J. R. Pitts. 1977. A reconstituted $Na^+ + K^+$ pump in liposomes containing purified $(Na^+ + K^+)$-ATPase from kidney medulla. *Biochim. Biophys. Acta* **467**:340–345.
146. Anner, B. M. 1980. Reconstitution of the Na^+, K^+-transport system in artificial membranes. *Acta. Physiol. Scand. Suppl.* **481**:15–19.
147. Dixon, J. F., and L. E. Hokin. 1980. The reconstituted (Na, K)-ATPase is electrogenic. *J. Biol. Chem.* **255**:10681–10686.
148. Kirk, K. L., D. R. Halm, and D. C. Dawson. 1980. Active sodium transport by turtle colon via an electrogenic Na–K exchange pump. *Nature (London)* **287**:237–239.
149. Thomas, R. C. 1972. Electrogenic sodium pump in nerve and muscle cells. *Physiol. Rev.* **52**:563–593.
150. Nellans, H. N., and S. G. Schultz. 1976. Relations among transepithelial sodium transport, potassium exchange, and cell volume in rabbit ileum. *J. Gen. Physiol.* **68**:441–463.
151. Schultz, S. G. 1978. Is a coupled Na–K exchange "pump" involved in active transepithelial Na transport? A status report. *Membr. Transp. Process* **1**:213–227.
152. Charney, A. N., and M. Donowitz. 1978. Functional significance of intestinal Na^+-K^+-ATPase: *In vivo* ouabain inhibition. *Am. J. Physiol.* **234**:E629–E636.
153. Hafkenscheid, J. C. M. 1973. Occurrence and properties of a $(Na^+$-$K^+)$ activated ATPase in the mucosa of the rat intestine. *Pflügers Arch.* **338**:289–294.
154. Mandel, L. J., and R. S. Balaban. 1981. Stoichiometry and coupling of active transport to oxidative metabolism in epithelial tissues. *Am. J. Physiol.* **240**:F357–F371.
155. Esposito, G., A. Faelli, and V. Capraro. 1966. Metabolism and sodium transport in the isolated rat intestine. *Nature (London)* **210**:307–308.
156. Jackson, M. J., and L. M. Kutcher. 1977. Influence of ionic environment on intestinal oxygen consumption. *Experientia* **33**:1061.
157. Frizzell, R. A., L. Markscheid-Kaspi, and S. G. Schultz. 1974. Oxidative metabolism of rabbit ileal mucosa. *Am. J. Physiol.* **226**:1142–1148.
158. Lester, R. G., and E. Grim. 1975. Substrate utilization and oxygen consumption by canine jejunal mucosa *in vitro*. *Am. J. Physiol.* **229**:139–143.
159. Gilman, A., and E. S. Koelle. 1960. Substrate requirements for ion transport by rat intestine studied *in vitro*. *Am. J. Physiol.* **199**:1025–1029.
160. Srivastava, L. M., and G. Hübscher. 1966. Glucose metabolism in the mucosa of the small intestine: Glycolysis in subcellular preparations from the cat and rat. *Biochem. J.* **100**:458–466.
161. Nakayama, H., and E. Weser. 1972. Adaptation of small bowel after intestinal resection: Increase in the pentose phosphate pathway. *Biochim. Biophys. Acta* **279**:416–423.
162. Watford, M., P. Lund, and H. A. Krebs. 1979. Isolation and metabolic characteristics of rat and chicken enterocytes. *Biochem. J.* **178**:589–596.
163. Windmueller, H. G., and A. E. Spaeth. 1978. Identification of ketone bodies and glutamine as the major respiratory fuels *in vivo* for postabsorptive rat small intestine. *J. Biol. Chem.* **253**:69–76.
164. Henning, S. J., and F. J. R. Hird. 1972. Ketogenesis from butyrate and acetate by the caecum and the colon of rabbits. *Biochem. J.* **130**:785–790.
165. Roediger, W. E., and S. C. Truelove. 1979. Method of preparing isolated colonic epithelial cells (colonocytes) for metabolic studies. *Gut* **20**:484–488.
166. Roediger, W. E. W. 1980. Role of anaerobic bacteria in the metabolic welfare of the colonic mucosa in man. *Gut* **21**:793–798.
167. Roediger, W. E. W. 1980. The colonic epithelium in ulcerative colitis: An energy-deficiency disease? *Lancet* **2**:712–718.
168. Reuss, L. E. 1979. Electrical properties of the cellular transepithelial pathway in *Necturus* gallbladder. III. Ionic permeability of the basolateral cell membrane. *J. Membr. Biol.* **47**:239–259.
169. Diamond, J. M. 1968. Transport mechanisms in the gallbladder. In: *Handbook of Physiology*, Section 6, Volume V. W. Heidel, ed. American Physiological Society, Washington, D.C. pp. 2451–2482.
170. Cremaschi, D., S. Henin, and G. Meyer. 1979. Stimulation by HCO_3^- of Na^+ transport in rabbit gallbladder. *J. Membr. Biol.* **47**:145–170.
171. Heintze, K., K.-U. Petersen, P. Olles, S. H. Saverymuttu, and J. R. Wood. 1979. Effects of bicarbonate on fluid and electrolyte transport by the guinea pig gallbladder: A bicarbonate–chloride exchange. *J. Membr. Biol.* **45**:43–59.
172. Duffey, M. E., K. Turnheim, R. A. Frizzell, and S. G. Schultz. 1978. Intracellular chloride activities in rabbit gallbladder: Direct evidence for the role of the sodium-gradient in energizing "uphill" chloride transport. *J. Membr. Biol.* **42**:229–245.
173. Armstrong, W. M., and S. J. Youmans. 1980. The role of bicarbonate ions and of adenosine 3′,5′-monophosphate (cAMP) in chloride transport by epithelial cells of bullfrog small intestine. *Ann. N.Y. Acad. Sci.* **341**:139–154.
174. White, J. F. 1980. Bicarbonate-dependent chloride absorption in small intestine: Ion fluxes and intracellular chloride activities. *J. Membr. Biol.* **53**:95–107.
175. Wright, E. M., V. Harms, A. K. Mircheff, and C. H. Van Os. 1981. Transport properties of intestinal basolateral membranes. *Ann. N.Y. Acad. Sci.* **372**:626–636.
176. Hopfer, U., K. Sigrist-Nelson, E. Ammann, and H. Murer. 1976. Differences in neutral amino acid and glucose transport between brush border and basolateral plasma membrane of intestinal epithelial cells. *J. Cell. Physiol.* **89**:805–810.
177. Wright, E. M., C. H. Van Os, and A. K. Mircheff. 1980. Sugar uptake by intestinal basolateral membrane vesicles. *Biochim. Biophys. Acta* **597**:112–124.
178. Ling, K. Y., W. B. Im, and R. G. Faust. 1981. Na^+-independent sugar uptake by rat intestinal and renal brush border and basolateral membrane vesicles. *Int. J. Biochem.* **13**:693–700.
179. Ling, K. Y., and R. G. Faust. 1983. Reconstitution of a partially purified Na^+-independent D-glucose transport system from rat jejunal basolateral membranes. *Int. J. Biochem.* **15**:27–34.
180. Mircheff, A. K., C. H. Van Os, and E. M. Wright. 1980. Pathways for alanine transport in intestinal basal lateral membrane vesicles. *J. Membr. Biol.* **52**:83–92.
181. Christensen, H. N. 1979. Exploiting amino acid structure to learn about membrane transport. *Adv. Enzymol.* **49**:41–101.
182. Davis, G. R., C. A. Santa Ana, S. Morawski, and J. S. Fordtran. 1980. Active chloride secretion in the normal human jejunum. *J. Clin. Invest.* **66**:1326–1333.
183. Kuo, J. F. 1980. Cyclic nucleotide-dependent protein kinases: An overview. In: *Progress in Pharmacology*, Volume 4/1. H. Vapaatalo, ed. Fischer Verlag, Stuttgart. pp. 21–30.
184. De Jonge, H. R. 1981. Cyclic GMP-dependent protein kinase in intestinal brushborders. *Adv. Cyclic Nucleotide Res.* **14**:315–333.
185. Berridge, M. J. 1981. Phosphatidylinositol hydrolysis: A multifunctional transducing mechanism. *Mol. Cell. Endocrinol.* **24**:115–140.
186. Donowitz, M. 1983. Ca^{++} in the control of active intestinal Na and Cl transport: Involvement in neurohumoral action. *Am. J. Physiol.* **245**:G165–G177.
187. Cheung, W. Y. 1980. Calmodulin plays a pivotal role in cellular regulation. *Science* **207**:19–27.
188. Powell, D. W., R. K. Farris, and S. T. Carbonetto. 1974. The-

ophylline, cyclic AMP, choleragen, and electrolyte transport by rabbit ileum. *Am. J. Physiol.* **227**:1428–1435.
189. Powell, D. W. 1974. Intestinal conductance and permselectivity changes with theophylline and choleragen. *Am. J. Physiol.* **227**:1436–1443.
190. Heintze, K., C. P. Stewart, and R. A. Frizzell. 1983. Sodium-dependent chloride secretion across rabbit descending colon. *Am. J. Physiol.* **244**:G357–G365.
191. Al-Awqati, Q., M. Field, and W. B. Greenough, III. 1974. Reversal of cyclic AMP-mediated intestinal secretion by ethacrynic acid. *J. Clin. Invest.* **53**:687–692.
192. Heintze, K., M. Lies, H. Dohnen, and K.-H. Sehring. 1982. Inhibition of chloride secretion and sodium absorption of rabbit colonic mucosa by ethacrynic acid. In: *Electrolyte and Water Transport across Gastrointestinal Epithelia.* R. M. Case, A. Garner, L. A. Turnberg, and J. A. Young, eds. Raven Press, New York. pp. 77–84.
193. Heintze, K., K.-U. Petersen, and O. Heidenreich. 1982. Stereospecific inhibition by ozolinone of stimulated chloride secretion in rabbit colon descendens. *Naunyn-Schmiedeberg's Arch. Pharmacol.* **318**:363–367.
194. Klyce, S. D., and R. K. S. Wong. 1977. Site and mode of adrenaline action on chloride transport across the rabbit corneal epithelium. *J. Physiol. (London)* **266**:777–799.
195. Welsh, M. J., P. L. Smith, and R. A. Frizzell. 1982. Chloride secretion by canine tracheal epithelium. II. The cellular electrical potential profile. *J. Membr. Biol.* **70**:227–238.
196. Hyun, C. S., and G. A. Kimmich. 1982. Effect of cholera toxin on cAMP levels and Na^+ influx in isolated intestinal epithelial cells. *Am. J. Physiol.* **243**:C107–C115.
197. Diez de los Rios, A., N. E. DeRose, and W. M. Armstrong. 1981. Cyclic AMP and intracellular ionic activities in *Necturus* gallbladder. *J. Membr. Biol.* **63**:25–30.
198. Petersen, K.-U., and L. Reuss. 1983. Cyclic AMP-induced chloride permeability in the apical membrane of *Necturus* gallbladder epithelium. *J. Gen. Physiol.* **81**:705–729.
199. De Jonge, H. R., and F. S. Van Dommelen. 1981. Cyclic GMP-dependent phosphorylation and ion transport in intestinal microvilli. *Cold Spring Harbor Conference on Cell Proliferation—Protein Phosphorylation* **8**:1313–1332.
200. Fan, C.-C., and D. W. Powell. 1983. Calcium–calmodulin inhibition of coupled NaCl transport in brush border membrane vesicles from rabbit ileum. *Proc. Natl. Acad. Sci. USA* **80**:5248–5252.
201. Rector, F. C., Jr. 1983. Sodium, bicarbonate, and chloride absorption by the proximal tubule. *Am. J. Physiol.* **244**:F461–F471.
202. Flemström, G., A. Garner, O. Nylander, B. C. Hurst, and J. R. Heylings. 1982. Surface epithelial HCO_3^- transport by mammalian duodenum *in vivo*. *Am. J. Physiol.* **243**:G348–G358.
203. Flemström, G., and A. Garner. 1982. Gastroduodenal HCO_3^- transport: Characteristics and proposed role in acidity regulation and mucosal protection. *Am. J. Physiol.* **242**:G183–G193.
204. Turnberg, L. A., J. S. Fordtran, N. W. Carter, and F. C. Rector, Jr. 1970. Mechanism of bicarbonate absorption and its relationship to sodium transport in the human jejunum. *J. Clin. Invest.* **49**:548–556.
205. Powell, D. W., L. I. Solberg, G. R. Plotkin, D. H. Catlin, R. M. Maenza, and S. B. Formal. 1971. Experimental diarrhea. III. Bicarbonate transport in rat salmonella enterocolitis. *Gastroenterology* **60**:1076–1086.
206. White, J. F., and M. A. Imon. 1981. Bicarbonate absorption by *in vitro* amphibian small intestine. *Am. J. Physiol.* **241**:G389–G396.
207. Fordtran, J. S., F. C. Rector, Jr., and N. W. Carter. 1968. The mechanisms of sodium absorption in the human small intestine. *J. Clin. Invest.* **47**:884–900.
208. Sladen, G. E., and A. M. Dawson. 1968. Effect of bicarbonate on sodium absorption by the human jejunum. *Nature (London)* **218**:267–268.
209. Hubel, K. A. 1973. Effect of luminal sodium concentration on bicarbonate absorption in rat jejunum. *J. Clin. Invest.* **52**:3172–3179.
210. Podesta, R. B., and D. F. Mettrick. 1977. HCO_3 transport in rat jejunum: Relationship to NaCl and H_2O transport *in vivo*. *Am. J. Physiol.* **232**:E62–E68.
211. Lucas, M. L., W. Schneider, F. J. Haberich, and J. A. Blair. 1975. Direct measurement by pH-microelectrode of the pH microclimate in rat proximal jejunum. *Proc. R. Soc. London Ser. B* **192**:39–48.
212. Jackson, M. J., and B. N. Morgan. 1975. Relations of weak-electrolyte transport and acid–base metabolism in rat small intestine *in vitro*. *Am. J. Physiol.* **228**:482–487.
213. Blair, J. A., M. L. Lucas, and A. J. Matty. 1975. Acidification in the rat proximal jejunum. *J. Physiol. (London)* **245**:333–350.
214. Lucas, M. L. 1976. The association between acidification and electrogenic events in the rat proximal jejunum. *J. Physiol. (London)* **257**:645–662.
215. White, J. F. 1982. Intestinal electrogenic HCO_3^- absorption localized to villus epithelium. *Biochim. Biophys. Acta* **687**:343–345.
216. White, J. F., and M. A. Imon. 1983. A role for basolateral anion exchange in active jejunal absorption of HCO_3^-. *Am. J. Physiol.* **244**:G397–G405.
217. Imon, M. A., and J. F. White. 1984. Association between HCO_3^- absorption and K^+ uptake by *Amphiuma* jejunum: Relation among HCO_3^- absorption, luminal K^+ and intracellular K^+ activity. *Am. J. Physiol.* **246**:G732–G744.
218. Fiddian-Green, R. G., and W. Silen. 1975. Mechanisms of disposal of acid and alkali in rabbit duodenum. *Am. J. Physiol.* **229**:1641–1648.
219. Fromm, D. 1973. Na and Cl transport across isolated proximal small intestine of the rabbit. *Am. J. Physiol.* **224**:110–116.
220. Lew, V. L., and N. J. Carlisky. 1967. Evidence for a special type of bicarbonate transport in the isolated colonic mucosa of *Bufo arenarum*. *Biochim. Biophys. Acta* **135**:793–796.
221. Carlisky, N. J., and V. L. Lew. 1970. Bicarbonate secretion and non-Na component of the short-circuit current in the isolated colonic mucosa of *Bufo arenarum*. *J. Physiol. (London)* **206**:529–541.
222. Lew, V. L. 1970. Short-circuit current and ionic fluxes in the isolated colonic mucosa of *Bufo arenarum*. *J. Physiol. (London)* **206**:509–528.
223. Simson, J. N. L., A. Merhav, and W. Silen. 1981. Alkaline secretion by amphibian duodenum. I. General characteristics. *Am. J. Physiol.* **240**:G401–G408.
224. Tai, Y., and R. A. Decker. 1980. Mechanisms of electrolyte transport in rat ileum. *Am. J. Physiol.* **238**:G208–G212.
225. Powell, D. W., H. J. Binder, and P. F. Curran. 1972. Electrolyte secretion by the guinea pig ileum *in vitro*. *Am. J. Physiol.* **223**:531–537.
226. Field, M., D. Fromm, and I. McColl. 1971. Ion transport in rabbit ileal mucosa. I. Na and Cl fluxes and short-circuit current. *Am. J. Physiol.* **220**:1388–1396.
227. Sheerin, H. E., and M. Field. 1975. Ileal HCO_3 secretion: Relationship to Na and Cl transport and effect of theophylline. *Am. J. Physiol.* **228**:1065–1074.
228. Dietz, J., and M. Field. 1973. Ion transport in rabbit ileal mucosa. IV. Bicarbonate secretion. *Am. J. Physiol.* **225**:858–861.
229. Hubel, K. A. 1974. The mechanism of bicarbonate secretion in rabbit ileum exposed to choleragen. *J. Clin. Invest.* **53**:964–970.
230. Donowitz, M., and H. J. Binder. 1976. Effect of enterotoxins of *Vibrio cholerae, Escherichia coli,* and *Shigella dysenterize* type 1 on fluid and electrolyte transport in the colon. *J. Infect. Dis.* **134**:135–143.
231. Field, M., and I. McColl. 1973. Ion transport in rabbit ileal mucosa. III. Effects of catecholamines. *Am. J. Physiol.* **225**:852–857.
232. Davis, G. R., S. G. Morawski, C. A. Santa Ana, and J. S. Fordtran. 1983. Evaluation of chloride/bicarbonate exchange in the human colon *in vivo*. *J. Clin. Invest.* **71**:201–207.

233. Hubel, K. A. 1967. Bicarbonate secretion in rat ileum and its dependence on intraluminal chloride. *Am. J. Physiol.* **213**:1409–1413.
234. Hubel, K. A. 1969. Effect of luminal chloride concentration on bicarbonate secretion in rat ileum. *Am. J. Physiol.* **217**:40–45.
235. Phillips, S. F., and P. F. Schmalz. 1970. Bicarbonate secretion by the rat colon: Effect of intraluminal chloride and acetazolamide. *Proc. Soc. Exp. Biol. Med.* **135**:116–121.
236. Binder, H. J., D. W. Powell, Y.-H. Tai, and P. F. Curran. 1973. Electrolyte transport in rabbit ileum. *Am. J. Physiol.* **225**:776–780.
237. Maren, T. H. 1977. Use of inhibitors in physiological studies of carbonic anhydrase. *Am. J. Physiol.* **232**:F291–F297.
238. Gunter, P. J., and J. F. White. 1978. Evidence for electrogenic bicarbonate transport in *Amphiuma* small intestine. *Biochim. Biophys. Acta* **507**:549–551.
239. Gunter-Smith, P. J., and J. F. White. 1979. Contribution of villus and intervillus epithelium to intestinal transmural potential difference and response to theophylline and sugar. *Biochim. Biophys. Acta* **557**:425–435.
240. White, J. F. 1980. Bicarbonate-dependent chloride absorption in small intestine: Ion fluxes and intracellular chloride activities. *J. Membr. Biol.* **53**:95–107.
241. Imon, M. A., and J. F. White. 1981. Intestinal bicarbonate secretion in *Amphiuma* measured by pH stat *in vitro:* Relationship with metabolism and transport of sodium and chloride ions. *J. Physiol. (London)* **314**:429–443.
242. White, J. F. 1981. Chloride transport and intracellular chloride activity in the presence of theophylline in amphiuma small intestine. *J. Physiol. (London)* **321**:331–341.
243. Imon, M. A., and J. F. White. 1981. The effect of theophylline on intestinal bicarbonate transport measured by pH stat in amphiuma. *J. Physiol. (London)* **321**:343–354.
244. Imon, M. A., and J. F. White. 1981. Intestinal bicarbonate secretion in amphiuma measured by pH *in vitro:* Relationship with metabolism and transport of sodium and chloride ions. *J. Physiol. (London)* **314**:429–443.
245. White, J. F., and M. A. Imon. 1982. Intestinal HCO_3^- secretion in *Amphiuma:* Stimulation by mucosal Cl^- and serosal Na^+. *J. Membr. Biol.* **68**:207–214.
246. Simson, J. N. L., A. Merhav, and W. Silen. 1981. Alkaline secretion by amphibian duodenum. II. Short-circuit current and Na^+ and Cl^- fluxes. *Am. J. Physiol.* **240**:G472–G479.
247. Cummings, J. H. 1981. Short chain fatty acids in the human colon. *Gut* **22**:763–779.
248. Jackson, M. J. 1973–1974. Transport of short chain fatty acids. *Biomembranes* **4B**:673–709.
249. Jackson, M. J., C.-Y. Tai, and J. E. Steane. 1981. Weak electrolyte permeation in alimentary epithelia. *Am. J. Physiol.* **240**:G191–G198.
250. Umesaki, Y., T. Yajima, T. Yokokura, and M. Mutai. 1979. Effect of organic acid absorption on bicarbonate transport in rat colon. *Pflügers Arch.* **379**:43–47.
251. Schmitt, M. G., Jr., K. H. Soergel, C. M. Wood, and J. J. Steff. 1977. Absorption of short-chain fatty acids from the human ileum. *Dig. Dis.* **22**:340–347.
252. Ruppin, H., S. Bar-Meir, K. H. Soergel, C. M. Wood, and M. G. Schmitt, Jr. 1980. Absorption of short-chain fatty acids by the colon. *Gastroenterology* **78**:1500–1507.
253. McNeil, N. I., J. H. Cummings, and W. P. T. James. 1978. Short chain fatty acid absorption by the human large intestine. *Gut* **19**:819–822.
254. McNeil, N. I., J. H. Cummings, and W. P. T. James. 1979. Rectal absorption of short chain fatty acids in the absence of chloride. *Gut* **20**:400–403.
255. Argenzio, R. A., M. Southworth, J. E. Lowe, and C. E. Stevens. 1977. Interrelationship of Na, HCO_3, and volatile fatty acid transport by equine large intestine. *Am. J. Physiol.* **233**:E469–E478.
256. Hayslett, J. P., and H. J. Binder. 1982. Mechanism of potassium adaptation. *Am. J. Physiol.* **243**:F103–F112.
257. Turnberg, L. A. 1971. Potassium transport in the human small bowel. *Gut* **12**:811–818.
258. Fromm, M., and S. G. Schultz. 1981. Potassium transport across rabbit descending colon *in vitro:* Evidence for single-file diffusion through a paracellular pathway. *J. Membr. Biol.* **63**:93–98.
259. Schultz, S. G. 1981. Potassium transport by rabbit descending colon, *in vitro. Fed. Proc.* **40**:2408–2411.
260. Edmonds, C. J. 1967. The gradient of electrical potential difference and of sodium and potassium of the gut contents along the caecum and colon of normal and sodium-depleted rats. *J. Physiol. (London)* **193**:571–588.
261. Edmonds, C. J. 1967. Transport of sodium and secretion of potassium and bicarbonate by the colon of normal and sodium-depleted rats. *J. Physiol. (London)* **193**:589–602.
262. Edmonds, C. J. 1967. Transport of potassium by the colon of normal and sodium depleted rats. *J. Physiol. (London)* **193**:603–617.
263. Edmonds, C. J., and T. Smith. 1979. Epithelial transport pathways of rat colon determined *in vivo* by impulse response analysis. *J. Physiol. (London)* **269**:471–485.
264. Kliger, A. S., H. J. Binder, C. Bastl, and J. P. Hayslett. 1981. Demonstration of active potassium transport in the mammalian colon. *J. Clin. Invest.* **67**:1189–1196.
265. Silva, P., A. N. Charney, and F. H. Epstein. 1975. Potassium adaptation and Na-K-ATPase activity in mucosa of colon. *Am. J. Physiol.* **229**:1576–1579.
266. Fisher, K. A., H. J. Binder, and J. P. Hayslett. 1976. Potassium secretion by colonic mucosal cells after potassium adaptation. *Am. J. Physiol.* **231**:987–994.
267. Binder, H. J. 1978. Effect of dexamethasone on electrolyte transport in the large intestine of the rat. *Gastroenterology* **75**:212–217.
268. Bastl, C. P., A. S. Kliger, H. J. Binder, and J. P. Hayslett. 1978. Characteristics of potassium secretion in the mammalian colon. *Am. J. Physiol.* **234**:F48–F53.
269. Bastl, C. P., H. J. Binder, and J. P. Hayslett. 1980. Role of glucocorticoids and aldosterone in maintenance of colonic cation transport. *Am. J. Physiol.* **238**:F181–F186.
270. Hayslett, J. P., N. Myketey, H. J. Binder, and P. S. Aronson. 1980. Mechanism of increased potassium secretion in potassium loading and sodium deprivation. *Am. J. Physiol.* **239**:F378–F382.
271. Foster, E. S., T. W. Zimmerman, J. P. Hayslett, and H. J. Binder. 1983. Corticosteroid alteration of active electrolyte transport in rat distal colon. *Am. J. Physiol.* **245**:G668–G675.
272. Foster, E. S., G. I. Sandle, J. P. Hayslett, and H. J. Binder. 1983. Cyclic adenosine monophosphate stimulates active potassium secretion in the rat colon. *Gastroenterology* **84**:324–330.
273. Archampong, E. Q., J. Harris, and C. G. Clark. 1972. The absorption and secretion of water and electrolytes across the healthy and the diseased human colonic mucosa measured *in vitro. Gut* **13**:880–886.
274. Bentley, P. J., and M. W. Smith. 1975. Transport of electrolytes across the helicoidal colon of the new-born pig. *J. Physiol. (London)* **249**:103–117.
275. Yorio, T., and P. J. Bentley. 1977. Permeability of the rabbit colon *in vitro. Am. J. Physiol.* **232**:F5–F9.
276. McCabe, R., H. J. Cooke, and L. P. Sullivan. 1982. Potassium transport by rabbit descending colon. *Am. J. Physiol.* **242**:C81–C86.
277. Wills, N. K., and B. Biagi. 1982. Active potassium transport by rabbit descending colon epithelium. *J. Membr. Biol.* **64**:195–203.
278. Hayslett, J. P., J. Halevy, P. E. Pace, and H. J. Binder. 1982. Demonstration of net potassium absorption in mammalian colon. *Am. J. Physiol.* **242**:G209–G214.
279. Civan, M. M. 1980. Potassium activities in epithelia. *Fed. Proc.* **39**:2865–2870.

280. White, J. F. 1976. Intracellular potassium activities in *Amphiuma* small intestine. *Am. J. Physiol.* **231**:1214–1219.
281. Wills, N. K., S. A. Lewis, and D. C. Eaton. 1979. Active and passive properties of rabbit descending colon: A microelectrode and mystatin study. *J. Membr. Biol.* **45**:81–108.
282. Gunter-Smith, P. J., and S. G. Schultz. 1982. Potassium transport and intracellular potassium activities in rabbit gallbladder. *J. Membr. Biol.* **65**:41–47.
283. Gustin, M. C., and D. B. P. Goodman. 1981. Isolation of brush-border membrane from the rabbit descending colon epithelium. *J. Biol. Chem.* **256**:10651–10656.
284. Gustin, M. C., and D. B. P. Goodman. 1982. Characterization of the phosphorylated intermediate of the K^+–ouabain-insensitive ATPase of the rabbit colon brush-border membrane. *J. Biol. Chem.* **257**:9629–9633.
285. Fordtran, J. S., F. C. Rector, Jr., M. F. Ewton, N. Soter, and J. Kinney. 1965. Permeability characteristics of the human small intestine. *J. Clin. Invest.* **44**:1935–1944.
286. Soergel, K. H., G. E. Whalen, and J. A. Harris. 1968. Passive movement of water and sodium across the human small intestinal mucosa. *J. Appl. Physiol.* **24**:40–48.
287. Davis, G. R., C. A. Santa Ana, S. G. Morawski, and J. S. Fordtran. 1982. Permeability characteristics of human jejunum, ileum, proximal colon and distal colon: Results of potential difference measurements and unidirectional fluxes. *Gastroenterology* **83**:844–850.
288. Powell, D. W. 1981. Barrier function of epithelia. *Am. J. Physiol.* **241**:G275–G288.
289. Bentzel, C. J., B. Hainau, S. Ho, S. W. Hui, A.Edelman, T. Anagnostopolus, and E. L. Benedetti. 1980. Cytoplasmic regulation of tight-junction permeability: Effect of plant cytokinins. *Am. J. Physiol.* **239**:C75–C89.
290. Goerg, K. J., M. Gross, G. Nell, W. Rummel, and L. Schulz. 1980. Comparative study of the effect of cholera toxin and sodium deoxycholate on the paracellular permeability and on net fluid and electrolyte transfer in rat colon. *Naunyn-Schmiedebergs Arch. Pharmacol.* **312**:91–97.
291. Boulpaep, E. L., and H. Sackin. 1977. Role of the paracellular pathway in isotonic fluid movement across the renal tubule. *Yale J. Biol. Med.* **50**:115–131.
292. Diamond, J. M. 1978. Solute-linked water transport in epithelia. In: *Membrane Transport Processes*. J. F. Hoffman, ed. Raven Press, New York. pp. 257–276.
293. Hill, A. 1980. Salt–water coupling in leaky epithelia. *J. Membr. Biol.* **56**:177–182.
294. Persson, B. E., and K. R. Spring. 1982. Gallbladder epithelial cell hydraulic water permeability and volume regulation. *J. Gen. Physiol.* **79**:481–505.
295. Feldman, G. M., and A. N. Charney. 1980. Effect of acute metabolic alkalosis and acidosis on intestinal electrolyte transport *in vivo*. *Am. J. Physiol.* **239**:G427–G436.
296. Feldman, G. M., and A. N. Charney. 1982. Effect of acute respiratory alkalosis and acidosis on intestinal ion transport *in vivo*. *Am. J. Physiol.* **242**:G486–G492.
297. Charney, A. N., and L. P. Haskell. 1983. Relative effects of systemic pH, PCO_2 and HCO_3 concentration on ileal ion transport. *Am. J. Physiol.* **245**:G230–G235.
298. Kurtin, P., and A. N. Charney. 1984. Intestinal ion transport and intercellular pH during acute respiratory alkalosis and acidosis. *Am. J. Physiol.* **247**:G24–G31.
299. Charney, A. N., M. Arnold, and N. Johnstone. 1983. Acute respiratory alkalosis and acidosis and rabbit intestinal ion transport *in vivo*. *Am. J. Physiol.* **244**:G145–G150.
300. Aronson, P. S., J. Nee, and M. A. Suhm. 1982. Modifier role of internal H^+ in activating the Na^+–H^+ exchanger in renal microvillus membrane vesicles. *Nature (London)* **299**:161–163.
301. Haynes, D. H. 1983. Mechanism of Ca^{2+} transport by Ca^{2+}-Mg^{2+}-ATPase pump: Analysis of major states and pathways. *Am. J. Physiol.* **244**:G3–G12.
302. Mailman, D. 1982. Blood flow and intestinal absorption. *Fed. Proc.* **41**:2096–2100.
303. Granger, D. N., P. D. I. Richardson, P. R. Kvietys, and N. A. Mortillaro. 1980. Intestinal blood flow. *Gastroenterology* **78**:837–863.
304. Kvietys, P. R., and D. N. Granger. 1982. Regulation of colonic blood flow. *Fed. Proc.* **41**:2106–2110.
305. Winne, D. 1979. Influence of blood flow on intestinal absorption of drugs and nutrients. *J. Pharmacol. Exp. Ther.* **6**:333–393.
306. Jacobson, E. D. 1982. Physiology of the mesenteric circulation. *Physiologist* **25**:439–443.
307. Cedgard, S., D.-A. Hallback, M. Jodal, O. Lundgren, and S. Redfors. 1978. The effects of cholera toxin on intramural blood flow distribution and capillary hydraulic conductivity in the cat small intestine. *Acta Physiol. Scand.* **102**:148–158.
308. Cassuto, J., M. Jodal, R. Tuttle, and O. Lundgren. 1981. On the role of intramural nerves in the pathogenesis of cholera toxin-induced intestinal secretion. *Scand. J. Gastroenterol.* **16**:377–384.
309. Cassuto, J., M. Jodal, H. Sjövall, and O. Lundgren. 1981. Nervous control of intestinal secretion. *Clin. Res. Rev.* **1**(Suppl. 1):11–21.
310. Mainoya, J. R. 1975. Further studies on the action of prolactin on fluid and ion absorption by the rat jejunum. *Endocrinology* **96**:1158–1164.
311. Mainoya, J. R., H. A. Bern, and J. W. Regan. 1974. Influence of ovine prolactin on transport of fluid and sodium chloride by the mammalian intestine and gall bladder. *J. Endocrinol.* **63**:311–317.
312. Gray, T. K., P. Brannan, D. Juan, S. G. Morawski, and J. S. Fordtran. 1976. Ion transport changes during calcitonin-induced intestinal secretion in man. *Gastroenterology* **71**:392–398.
313. Kisloff, B., and E. W. Moore. 1977. Effects of intravenous calcitonin on water, electrolyte, and calcium movement across *in vivo* rabbit jejunum and ileum. *Gastroenterology* **72**:462–468.
314. Walling, M. W., T. A. Brasitus, and D. V. Kimberg. 1977. Effects of calcitonin and substance P on the transport of Ca, Na, and Cl across rat ileum *in vitro*. *Gastroenterology* **73**:89–94.
315. Blickenstaff, D. D. 1954. Increase in intestinal absorption of water from isosmotic saline following pitressin administration. *Am. J. Physiol.* **179**:471–472.
316. Green, K., and A. J. Matty. 1966. Effects of vasopressin on ion transport across intestinal epithelia. *Life Sci.* **5**:205–209.
317. Cofré, G., and J. Crabbé. 1967. Active sodium transport by the colon of *Bufo marinus:* Stimulation by aldosterone and antidiuretic hormone. *J. Physiol. (London)* **188**:177–190.
318. Bridges, R. J., G. Nell, and W. Rummel. 1983. Influence of vasopressin and calcium on electrolyte transport across isolated colonic mucosa of the rat. *J. Physiol. (London)* **338**:463–475.
319. Field, M., G. R. Plotkin, and W. Silen. 1968. Effects of vasopressin, theophylline and cyclic adenosine monophosphate on short-circuit current across isolated rabbit ileal mucosa. *Nature (London)* **217**:469–471.
320. Soergel, K. H., G. E. Whalen, J. A. Harris, and J. E. Geenen. 1968. Effect of antidiuretic hormone on human small intestinal water and solute transport. *J. Clin. Invest.* **47**:1071–1082.
321. Levitan, R., and I. Mauer. 1968. Effect of intravenous antidiuretic hormone administration on salt and water absorption from the human colon. *J. Lab. Clin. Med.* **72**:739–746.
322. Bentley, P. J. 1962. Studies on the permeability of the large intestine and urinary bladder of the tortoise (*Testudo graeca*) with special reference to the effects of neuro-hypophysial and adrenocortical hormones. *Gen. Comp. Endocrinol.* **2**:323–328.
323. Will, P. C., J. L. Lebowitz, and U. Hopfer. 1980. Induction of amiloride-sensitive sodium transport in the rat colon by mineralocorticoids. *Am. J. Physiol.* **238**:F261–F268.
324. Ferguson, D.R., P. S. James, J. Y. F. Paterson, J. C. Saunders, and M. W. Smith. 1979. Aldosterone induced changes in colonic sodium transport occurring naturally during development in the neonatal pig. *J. Physiol. (London)* **292**:495–504.

325. Charney, A. N., J. Wallach, S. Ceccarelli, M. Donowitz, and C. L. Costenbader. 1981. Effects of spironolactone and amiloride on corticosteroid-induced changes in colonic function. *Am. J. Physiol.* **241**:G300–G305.
326. Edmonds, C. J. 1972. Effect of aldosterone on mammalian intestine. *J. Steroid Biochem.* **3**:143–149.
327. Will, P. C., R. C. DeLisle, R. N. Cortright, and U. Hopfer. 1981. Induction of amiloride-sensitive sodium transport in the intestines by adrenal steroids. *Ann. N.Y. Acad. Sci.* **372**:64–78.
328. Lewis, S. A., and N. K. Wills. 1981. Localization of the aldosterone response in rabbit urinary bladder by-electrophysiological techniques. *Ann. N.Y. Acad. Sci.* **372**:56–63.
329. Charney, A. N., M. D. Kinsey, L. Myers, R. A. Giannella, and R. E. Gots. 1975. Na^+-K^+-activated adenosine triphosphatase and intestinal electrolyte transport: Effect of adrenal steroids. *J. Clin. Invest.* **56**:653–660.
330. Charney, A. N., J. D. Wallach, M. Donowitz, and N. Johnstone. 1982. Effect of cycloheximide on corticosteroid-induced changes in colonic function. *Am. J. Physiol.* **243**:G112–G116.
331. Charney, A. N., and M. Donowitz. 1976. Prevention and reversal of cholera enterotoxin-induced intestinal secretion by methylprednisolone induction of Na^+-K^+-ATPase. *J. Clin. Invest.* **57**:1590–1599.
332. Marnane, W. G., Y.-H. Tai, R. A. Decker, E. C. Boedeker, A. N. Charney, and M. Donowitz. 1981. Methylprednisolone stimulation of guanylate cyclase activity in rat small intestinal mucosa: Possible role in electrolyte transport. *Gastroenterology* **81**:90–100.
333. Tai, Y.-H., R. A. Decker, W. G. Marnane, A. N. Charney, and M. Donowitz. 1981. Effects of methylprednisolone on electrolyte transport *in vitro* rat ileum. *Am. J. Physiol.* **240**:G365–G370.
334. Marusic, E. T., J. P. Hayslett, and H. J. Binder. 1981. Corticosteroid-binding studies in cytosol of colonic mucosa of the rat. *Am. J. Physiol.* **240**:G417–G423.
335. Fromm, D., M. Field, and W. Silen. 1969. Effects of insulin on sugar, amino acid, and ion transport across isolated small intestine. *Surgery* **66**:145–151.
336. Crabbé, J. 1981. Stimulation by insulin of transepithelial sodium transport. *Ann. N.Y. Acad. Sci.* **372**:220–234.
337. Binder, H. J., J. Reinprecht, K. Dharmsathaphorn, and J. W. Dobbins. 1980. Intestinal peptide receptors. *Regul. Pept. Suppl.* **1**:S10.
338. Hicks, T., and L. A. Turnberg. 1973. The influence of secretin on ion transport in the human jejunum. *Gut* **14**:485–490.
339. Moritz, M., G. Finkelstein, H. Meshkinpour, J. Fingerut, and S. H. Lorber. 1973. Effect of secretin and cholecystokinin on the transport of electrolyte and water in human jejunum. *Gastroenterology* **64**:76–80.
340. Gardner, J. D., G. W. Peskin, J. J. Cerda, and F. P. Brooks. 1966. Alterations of *in vitro* fluid and electrolyte absorption by gastrointestinal hormones. *Am. J. Surg.* **113**:57–64.
341. Hubel, K. A. 1972. Effects of secretin and glucagon on intestinal transport of ions and water in the rat. *Proc. Soc. Exp. Biol. Med.* **139**:656–658.
342. Binder, H. J., G. F. Lemp, and J. D. Gardner. 1980. Receptors for vasoactive intestinal peptide and secretin on small intestinal epithelial cells. *Am. J. Physiol.* **238**:G190–G196.
343. Smith, M. C., and M. J. Dunn. 1981. Renal kallikrein, kinins, and prostaglandins in hypertension. In: *Hypertension.* B. M. Brenner and J. H. Stein, eds. Churchill Livingstone, London. pp. 168–202.
344. Seki, T., T. Nakajima, and E. G. Erdös. 1972. Colon kallikrein, its relation to the plasma enzyme. *Biochem. Pharmacol.* **21**:1227–1235.
345. Al-Dhahir, H. A. R., and I. J. Zeitlin. 1982. Bile-induced colonic motility increase may be mediated by activation of a kallikrein-like enzyme. *Br. J. Pharmacol.* **76**:188p.
346. Ward, P. E., M. A. Sheridan, K. J. Hammon, and E. G. Erdos. 1980. Angiotensin I converting enzyme (kininase II) of the brush border of human and swine intestine. *Biochem. Pharmacol.* **29**:1525–1529.
347. Regoli, D., and J. Barabé. 1980. Pharmacology of bradykinin and related kinins. *Pharmacol. Rev.* **32**:1–46.
348. Crocker, A. D., and S. P. Willavoys. 1975. Effect of bradykinin on transepithelial transfer of sodium and water *in vitro*. *J. Physiol. (London)* **253**:401–410.
349. Hardcastle, J., P. T. Hardcastle, R. J. Flower, and P. A. Sanford. 1978. The effect of bradykinin on the electrical activity of rat jejunum. *Experientia* **34**:617–618.
350. Cuthbert, A. W., and H. S. Margolius. 1982. Kinins stimulate net chloride secretion by the rat colon. *Br. J. Pharmacol.* **75**:587–598.
351. Manning, D. C., S. H. Snyder, J. F. Kachur, R. J. Miller, and M. Field. 1982. Bradykinin receptor-mediated chloride secretion in intestinal function. *Nature (London)* **299**:256–259.
352. Musch, M. W., J. F. Kachur, R. J. Miller, M. Field, and J. S.Stoff. 1983. Bradykinin-stimulated electrolyte secretion in rabbit and guinea pig intestine: Involvement of arachidonic acid metabolites. *J. Clin. Invest.* **71**:1073–1083.
353. Musch, M. W., R. J. Miller, M. Field, and M. I. Siegel. 1982. Stimulation of colonic secretion by lipoxygenase metabolites of arachidonic acid. *Science* **217**:1255–1256.
354. Levens, N. R., M. J. Peach, and R. M. Carey. 1981. Interactions between angiotensin peptides and the sympathetic nervous system mediating intestinal sodium and water absorption in the rat. *J. Clin. Invest.* **67**:1197–1207.
355. Kimberg, D. V., M. Field, J. Johnson, A. Henderson, and E. Gershon. 1971. Stimulation of intestinal mucosal adenyl cyclase by cholera enterotoxin and prostaglandins. *J. Clin. Invest.* **50**:1218–1230.
356. Al-Awqati, Q., and W. B. Greenough, III. 1972. Prostaglandins inhibit intestinal sodium transport. *Nature New Biol.* **238**:26–27.
357. Racusen, L. C., and H. J. Binder. 1980. Effect of prostaglandin on ion transport across isolated colonic mucosa. *Dig. Dis. Sci.* **25**:900–904.
358. Pierce, N. F., C. C. J. Carpenter, Jr., H. L. Elliott, and W. B. Greenough, III. 1971. Effects of prostaglandins, theophylline, and cholera exotoxin upon transmucosal water and electrolyte movement in the canine jejunum. *Gastroenterology* **60**:22–32.
359. Coupar, I. M., and I. McColl. 1975. Stimulation of water and sodium secretion and inhibition of glucose absorption from the rat jejunum during intraarterial infusions of prostaglandins. *Gut* **16**:759–765.
360. Bukhave, K., and J. Rask-Madsen. 1980. Saturation kinetics applied to *in vitro* effects of low prostaglandin E_2 and $F_{2\alpha}$ concentrations on ion transport across human jejunal mucosa. *Gastroenterology* **78**:32–42.
361. Cummings, J. H., A. Newman, J. J. Misiewicz, G. J. Milton-Thompson, and J. A. Billings. 1973. Effect of intravenous prostaglandin $F_{2\alpha}$ on small intestinal function in man. *Nature (London)* **243**:169–171.
362. Milton-Thompson, G. J., J. H. Cummings. A. Newman. J. A. Billings, and J. J. Misiewicz. 1975. Colonic and small intestinal response to intravenous prostaglandin $F_{2\alpha}$ and E_2 in man. *Gut* **16**: 42–46.
363. Robert, A., A. J. Hanchar, C. Lancaster, and J. E. Nezamis. 1979. Prostacyclin inhibits enteropooling and diarrhea. In: *Prostacyclin.* J. R. Vane and S. Bergström, eds. Raven Press, New York. pp. 147–158.
364. Vischer, P., and J. Casals-Stenzel. 1982. Pharmacological properties of ciloprost, a stable prostacyclin analogue, on the gastrointestinal tract. *Prostaglandins Leukotrienes Med.* **9**:517–529.
365. Simon, B., and H. Kather. 1980. Human colonic adenylate cyclase: Stimulation of enzyme activity by vasoactive intestinal peptide and various prostaglandins via distinct receptor sites. *Digestion* **20**:62–67.
366. Simon, B., H. Kather, and B. Kommerell. 1978. Prostacyclin: A potent activator of human colonic adenylate cyclase activity. *Z. Gastroenterol.* **12**:748–751.

367. Robert, A., and M. J. Ruwart. 1982. Effects of prostaglandins on the digestive system. In: *Prostaglandins*. J. B. Lee, ed. Elsevier, Amsterdam. pp. 113–176.
368. Beubler, E., and H. Juan. 1977. The function of prostaglandins in transmucosal water movement and blood flow in the rat jejunum. *Naunyn-Schmiedebergs Arch. Pharmacol.* **299**:89–94.
369. Granger, D. N., P. R. Kvietys, M. A. Perry, and A. E. Taylor. 1980. Relationship between intestinal volume secretion and oxygen uptake. *Dig. Dis. Sci.* **27**:42–48.
370. Granger, D. N., J. S. Shackleford, and A. E. Taylor. 1979. PGE_1-induced intestinal secretion: Mechanism of enhanced transmucosal protein efflux. *Am. J. Physiol.* **236**:E788–E796.
371. Field, M., M. W. Musch, and J. S. Stoff. 1981. Role of prostaglandins in the regulation of intestinal electrolyte transport. In: *Prostaglandins* **21**(Suppl.):73–79.
372. LeDuc, L. E., and P. Needleman. 1979. Regional localization of prostacyclin and thromboxane synthesis in dog stomach and intestinal tract. *J. Pharmacol. Exp. Ther.* **211**:181–188.
373. LeDuc, L. E., and P. Needleman. 1980. Prostaglandin synthesis by dog gastrointestinal tract. In: *Advances in Prostaglandin and Thromboxane Research,* Volume 8. B. Samuelsson, P. W. Ramwell, and R. Paoletti, eds. Raven Press, New York. pp. 1515–1517.
374. Balaa, M. A., and D. W. Powell. 1983. Prostaglandin synthesis by isolated rabbit small intestinal enterocytes. *Gastroenterology* **84**:1096a.
375. Burnstock, G. 1979. Non-adrenergic, non-cholinergic nerves in the intestine and their possible involvement in secretion. In: *Mechanisms of Intestinal Secretion*. H. J. Binder, ed. Liss, New York. pp. 147–174.
376. Florey, H. W., R. D. Wright, and M. A. Jennings. 1941. The secretions of the intestine. *Physiol. Rev.* **21**:36–69.
377. Powell, D. W., and E. J. Tapper. 1979. Intestinal ion transport: Cholinergic–adrenergic interactions. In: *Mechanisms of Intestinal Secretion*. H. J. Binder, ed. Liss, New York. pp. 175–192.
378. Powell, D. W., and E. J. Tapper. 1979. Autonomic control of intestinal electrolyte transport. In: *Frontiers of Knowledge in the Diarrheal Diseases*. H. D. Janowitz and D. B. Sachar, eds. Projects in Health, Inc., Upper Montclair, N.J. pp. 37–52.
379. Gershon, M. D. 1981. The enteric nervous system: An apparatus for intrinsic control of gastrointestinal motility. *Viewpoints Dig. Dis.* **13**:13–16.
380. Miller, R. J., J. F. Kachur, M. Field, and J. Rivier. 1981. Neurohumoral control of ileal electrolyte transport. *Ann. N.Y. Acad. Sci.* **372**:571–593.
381. Tapper, E. J. 1983. Local modulation of intestinal ion transport by enteric neurons. *Am. J. Physiol.* **244**:G457–G468.
382. Furness, J. B., and M. Costa. 1980. Types of nerves in the enteric nervous system. *Neurosciences* **5**:1–28.
383. Polak, J. M., and S. R. Bloom. 1978. Peptidergic nerves of the gastrointestinal tract. *Invest. Cell Pathol.* **1**:301–326.
384. Bishop, A. E., G.-I. Ferri, L. Probert, S. R. Bloom, and J. M. Polak. 1981. Peptidergic nerves. *Scand. J. Gastroenterol. Suppl.* **71**:43–59.
385. Furness, J. B., M. Costa, R. Murphy, A. M. Beardsley, J. R. Oliver, I. J. Llewellyn-Smith, R. L. Eskay, A. A. Shulkes, T. W. Moody, and D. K. Meyer. 1981. Detection and characterisation of neurotransmitters, particularly peptides, in the gastrointestinal tract. *Scand. J. Gastroenterol. Suppl.* **71**:61–70.
386. Polak, J. M., A. M. J. Buchan, L. Probert, F. Tapia, J. DeMey, and S. R. Bloom. 1981. Regulatory peptides in endocrine cells and autonomic nerves. *Scand. J. Gastroenterol. Suppl.* **70**:11–23.
387. Gershon, M. D. 1981. Serotonergic neurotransmission in the gut. *Scand. J. Gastroenterol. Suppl.* **71**:27–42.
388. Mutt, V. 1982. Gastrointestinal hormones: A field of increasing complexity. *Scand. J. Gastroenterol. Suppl.* **77**:133–152.
389. Dockray, G. J. 1981. Brain–gut peptides. *Viewpoints Dig. Dis.* **13**:5–8.
390. Gabella, G. 1981. On the ultrastructure of the enteric nerve ganglia. *Scand. J. Gastroenterol. Suppl.* **71**:15–25.
391. Lundberg, J. M., T. Hökfelt, G. Nilsson, L. Terenius, J. Rehfeld, R. Elde, and S. Said. 1978. Peptide neurons in the vagus, splanchnic and sciatic nerves. *Acta Physiol. Scand.* **104**:499–501.
392. Leeman, S. E., and R. Gamse. 1981. Substance P in sensory neurons. *Trends Pharm. S.* **2**:119–121.
393. Jiang, Z.-G., N. J. Dun, and A.G. Karczmar. 1982. Substance P: A putative sensory transmitter in mammalian autonomic ganglia. *Science* **217**:739–741.
394. Wood, J. D. 1981. Physiology of the enteric nervous system. In: *Physiology of the Gastrointestinal Tract*. L. R. Johnson, ed. Raven Press, New York. pp. 1–37.
395. Burnstock, G. 1981. Ultrastructural identification of neurotransmitters. *Scand. J. Gastroenterol. Suppl.* **70**:1–9.
396. Lundberg, J. M., A. Dahlström, A. Bylock, H. Ahlman, G. Pettersson, I. Larsson, H.-A. Hansson, and J. Kewenter. 1978. Ultrastructural evidence for an innervation of epithelial enterochromaffine cells in the guinea pig duodenum. *Acta Physiol. Scand.* **104**:3–12.
397. Newson, B., H. Ahlman, A. Dahlström, T. K. Das Gupta, and L. M. Nyhus. 1979. On the innervation of the ileal mucosa in the rat—A synapse. *Acta Physiol. Scand.* **105**:387–389.
398. Jacobowitz, D. 1965. Histochemical studies of the autonomic innervation of the gut. *J. Pharmacol. Exp. Ther.* **149**:358–364.
399. Norberg, K.-A. 1964. Adrenergic innervation of the intestinal wall studied by fluorescence microscopy. *Int. J. Neuropharmacol.* **3**:379–382.
400. Thomas, E. M., and D. Templeton. 1981. Noradrenergic innervation of the villi of rat jejunum. *J. Auton. Nerv. Syst.* **3**:25–29.
401. Schultzberg, M., T. Hökfelt, G. Nilsson, L. Terenius, J. F. Rehfeld, M. Brown, R. Elde, M. Goldstein, and S. Said. 1980. Distribution of peptide- and catecholamine-containing neurons in the gastro-intestinal tract of rat and guinea-pig: Immunohistochemical studies with antisera to substance P, vasoactive intestinal polypeptide, enkephalins, somatostatin, gastrin/cholecystokinin, neurotensin and dopamine β-hydroxylase. *Neuroscience* **5**:689–744.
402. Larsson, L.-I. 1980. Peptide secretory pathways in GI tract: Cytochemical contributions to regulatory physiology of the gut. *Am. J. Physiol.* **239**:G237–G246.
403. Solcia, E., C. Capella, R. Buffa, L. Usellini, R. Fiocca, B. Frigerio, P. Tenti, and F. Sessa. 1981. The diffuse endocrine–paracrine system of the gut in health and disease: Ultrastructural features. *Scand. J. Gastroenterol. Suppl.* **70**:25–36.
404. Forsberg, E. J., and R. J. Miller. 1982. Cholinergic agonists induce vectorial release of serotonin from duodenal enterochromaffin cells. *Science* **217**:355–356.
405. Ennis, C., P. A. J. Janssen, H. Schnieden, and B. Cox. 1979. Characterization of receptors on postganglionic cholinergic neurons in the guinea-pig isolated ileum. *J. Pharm. Pharmacol.* **31**:217–221.
406. Manber, L., and M. D. Gershon. 1979. A reciprocal adrenergic–cholinergic axoaxonic synapse in the mammalian gut. *Am. J. Physiol.* **236**:E738–E745.
407. Wu, Z.-A. C., and T. S. Gaginella. 1981. Functional properties of noradrenergic nervous system in rat colonic mucosa: Uptake of [^{3}H]norepinephrine. *Am. J. Physiol.* **241**:G137–G142.
408. Wu, Z.-A. C., and T. S. Gaginella. 1981. Release of [^{3}H]norepinephrine from nerves in rat colonic mucosa: Effects of norepinephrine and prostaglandin E2. *Am. J. Physiol.* **241**:G416–G421.
409. Tapper, E. J., D. W. Powell, and S. M. Morris. 1978. Cholinergic–adrenergic interactions on intestinal ion transport. *Am. J. Physiol.* **235**:E402–E409.
410. Crocker, A. D., and K. A. Munday. 1970. The effect of the renin–angiotensin system on mucosal water and sodium transfer in everted sacs of rat jejunum. *J. Physiol. (London)* **206**:323–333.
411. Davies, N. T., K. A. Munday, and B. J. Parsons. 1970. The effect of angiotensin on rat intestinal fluid transfer. *J. Endocrinol.* **48**:39 46.
412. Davies, N. T., K. A. Munday, and B. J. Parsons. 1972. Studies on

the mechanism of action of angiotensin on fluid transport by the mucosa of rat distal colon. *J. Endocrinol.* **54**:483–492.

413. Hornych, A., P. Meyer, and P. Milliez. 1973. Angiotensin, vasopressin, and cyclic AMP: Effects on sodium and water fluxes in rat colon. *Am. J. Physiol.* **224**:1223–1229.
414. Dolman, D., and C. J. Edmonds. 1975. The effect of aldosterone and the renin–angiotensin system on sodium, potassium and chloride transport by proximal and distal rat colon *in vivo*. *J. Physiol. (London)* **250**:597–611.
415. Levens, N. R., K. A. Munday, B. J. Parsons, J. A. Poat, and C. P. Stewart. 1979. Noradrenaline as a possible mediator of the actions of angiotensin on fluid transport by rat jejunum *in vivo*. *J. Physiol. (London)* **286**:351–360.
416. Levens, N. R., M. J. Peach, R. M. Carey, J. A. Poat, and K. A. Munday. 1981. Response of rat jejunum to angiotensin II: Role of norepinephrine and prostaglandins. *Am. J. Physiol.* **240**:G17–G24.
417. Dorey, P. G., K. A. Munday, B. J. Parsons, J. A. Poat, and M. E. Upsher. 1981. Effect of chemical sympathectomy and ganglion blockade on angiotensin-stimulated fluid absorption in the rat jejunum. *J. Endocrinol.* **91**:205–211.
418. Cassuto, J., M. Jodal, R. Tuttle, and O. Lundgren. 1979. The effect of lidocaine on the secretion induced by cholera toxin in the cat small intestine. *Experientia* **35**:1467–1468.
419. Cassuto, J., J. Fahrenkrug, M. Jodal, R. Tuttle, and O. Lundgren. 1981. Release of vasoactive intestinal polypeptide from the cat small intestine exposed to cholera toxin. *Gut* **22**:958–963.
420. Karlström, L., J. Cassuto, M. Jodal, and O. Lundgren. 1980. The effect of hexamethonium on the secretion induced by sodium deoxycholate in the rat jejunum. *Experientia* **37**:991–992.
421. Gaginella, T. S., and T. M. O'Dorisio. 1979. Vasoactive intestinal polypeptide: Neuromodulator of intestinal secretion? In: *Mechanisms of Intestinal Secretion*. H. J. Binder, ed. Liss, New York. pp. 231–247.
422. Powell, D. W. 1983. Neurological control mechanisms: Neurohumoral control of intestinal secretion. In: *Intestinal Secretion*. L. A. Turnberg, ed. Smith, Kline and French Laboratories Limited Oxford, pp. 42–45.
423. Blickenstaff, D. D., and L. J. Lewis. 1952. Effect of atropine on intestinal absorption of water and chloride. *Am. J. Physiol.* **170**:17–23.
424. Hubel, K. A. 1976. Intestinal ion transport: Effect of norepinephrine, pilocarpine, and atropine. *Am. J. Physiol.* **231**:252–257.
425. Morris, A. I., and L. A. Turnberg. 1980. The influence of a parasympathetic agonist and antagonist on human intestinal transport *in vivo*. *Gastroenterology* **79**:861–866.
426. Hubel, K. A. 1977. Effects of bethanechol on intestinal ion transport in the rat. *Proc. Soc. Exp. Biol. Med.* **154**:41–44.
427. Browning, J. G., J. Hardcastle, P. T. Hardcastle, and P. A. Sanford. 1977. The role of acetylcholine in the regulation of ion transport by rat colon mucosa. *J. Physiol. (London)* **272**:737–754.
428. Wu, Z.-A. C., S. D. Kisslinger, and T. S. Gaginella. 1982. Functional evidence for the presence of cholinergic nerve endings in the colonic mucosa of the rat. *J. Pharmacol. Exp. Ther.* **221**:664–669.
429. Isaacs, P. E. T., C. L. Corbett, A. K. Riley, P. C. Hawker, and L. A. Turnberg. 1976. *In vitro* behavior of human intestinal mucosa: The influence of acetyl choline on ion transport. *J. Clin. Invest.* **58**:535–542.
430. Browning, J. G., J. Hardcastle, P. T. Hardcastle, and J. S. Redfern. 1977. Site of action of acetylcholine in regulating intestinal epithelial ion transport in the rat. *J. Physiol. (London)* **270**:78–79.
431. Zimmerman, T. W., J. W. Dobbins, and H. J. Binder. 1982. Mechanism of cholinergic regulation of electrolyte transport in rat colon *in vitro*. Am. J. Physiol. **242**:G116–G123.
432. Zimmerman, T. W., and H. J. Binder. 1983. Effect of tetrodotoxin on cholinergic-mediated colonic electrolyte transport. *Am. J. Physiol.* **244**:G386–G391.
433. Browning, J. G., J. Hardcastle, P. T. Hardcastle, and J. S. Redfern. 1978. Localization of the effect of acetylcholine in regulating intestinal ion transport. *J. Physiol. (London)* **281**:15–27.
434. Rimele, T. J., T. M. O'Dorisio, and T. S. Gaginella. 1981. Evidence for muscarinic receptors on rat colonic epithelial cells: Binding of [^{3}H] quinuclidinyl benzilate. *J. Pharmacol. Exp. Ther.* **218**:426–434.
435. Rimele, T. J., and T. S. Gaginella. 1982. Binding of [^{3}H] quinuclidinyl benzilate to intestinal mucus. *Biochem. Pharmacol.* **31**:515–520.
436. Rimele, T. J., and T. S. Gaginella. 1982. *In vivo* identification of muscarinic receptors on rat colonic epithelial cells: Binding of [^{3}H]quinuclidinyl benzilate. *Naunyn-Schmiedebergs Arch. Pharmacol.* **319**:18–21.
437. Zimmerman, T. W., and H. J. Binder. 1982. Muscarinic receptors on rat isolated colonic epithelial cells: A correlation between inhibition of [^{3}H]-quinuclidinyl benzilate binding and alteration in ion transport. *Gastroenterology* **83**:1244–1251.
438. Morisset, J., L. Geoffrion, L. Larose, J. Lanöe, and G. G. Poirier. 1981. Distribution of muscarinic receptors in the digestive tract organs. *Pharmacology* **22**:189–195.
439. Isaacs, P. E. T., J. S. Whitehead, and Y. S. Kim. 1982. Muscarinic acetylcholine receptors of the small intestine and pancreas of the rat: Distribution and the effect of vagotomy. *Clin. Sci.* **62**:203–207.
440. Hedlund, B., J. Abens, and T. Bartfai. 1983. Vasoactive intestinal polypeptide and muscarinic receptors: Supersensitivity induced by long-term atropine treatment. *Science* **220**:519–521.
441. Wahawisan, R., L. J. Wallace, and T. S. Gaginella. 1983. Muscarinic receptors exist on ileal crypt and villus cells of the rat. *Fed. Proc.* **42**:761.
442. Tien, X. Y., R. Wahawisan, L. J. Wallace, and T. S. Gaginella. 1985. Intestinal epithelial cells and musclulature contain different muscarinic binding sites. *Life Sci.* in press.
443. Bolton, J. E., and M. Field. 1977. Ca ionophore-stimulated ion secretion in rabbit ileal mucosa: Relation to actions of cyclic 3′, 5′-AMP and carbamylcholine. *J. Membr. Biol.* **35**:159–173.
444. Zimmerman, T. W., J. W. Dobbins, and H. J. Binder. 1983. Role of calcium in the regulation of colonic secretion in the rat. *Am. J. Physiol.* **244**:G552–G560.
445. Donowitz, M., R. Fogel, L. Battisti, and N. Asarkof. 1982. The neurohumoral secretagogues carbachol, substance P and neurotensin increase Ca^{++} influx and calcium content in rabbit ileum. *Life Sci.* **31**:1929–1937.
446. Brasitus, T. A., M. Field, and D. V. Kimberg. 1976. Intestinal mucosal cyclic GMP: Regulation and relation to ion transport. *Am. J. Physiol.* **231**:275–282.
447. Hubel, K. A. 1978. The effects of electrical field stimulation and tetrodotoxin on ion transport by the isolated rabbit ileum. *J. Clin. Invest.* **62**:1039–1047.
448. Hubel, K. A. 1981. Effect of veratrine and 50 mM K on ileal transport and electrically induced secretion. *Am. J. Physiol.* **240**:G211–G216.
449. Hubel, K. A., and S. Shirazi. 1982. Human ileal ion transport *in vitro:* Changes with electrical field stimulation and tetrodotoxin. *Gastroenterology* **83**:63–68.
450. Hubel, K. A. 1983. The effects of scorpion venom on electrolyte transport by rabbit ileum. *Am. J. Physiol.* **244**:G501–G506.
451. Schwartz, C. J., D. V. Kimberg, H. E. Sheerin, M. Field, and S. I. Said. 1974. Vasoactive intestinal peptide stimulation of adenylate cyclase and active electrolyte secretion in intestinal mucosa. *J. Clin. Invest.* **54**:536–544.
452. Racusen, L. C., and H. J. Binder. 1977. Alteration of large intestinal electrolyte transport by vasoactive intestinal polypeptide in the rat. *Gastroenterology* **73**:790–796.
453. Mailman, D. 1978. Effects of vasoactive intestinal polypeptide on intestinal absorption and blood flow. *J. Physiol. (London)* **279**:121–132.
454. Krejs, G. J., R. M. Barkley, N. W. Read, and J. S. Fordtran. 1977. Intestinal secretion induced by vasoactive intestinal poly-

peptide: A comparison with cholera toxin in the canine jejunum *in vivo*. *J. Clin. Invest.* **78**:1337–1345.

455. Wu, Z.-A. C., T. M. O'Dorisio, S. Cataland, H. S. Mekhjian, and T. S. Gaginella. 1979. Effects of pancreatic polypeptide and vasoactive intestinal polypeptide on rat ileal and colonic water and electrolyte transport *in vivo*. *Dig. Dis. Sci.* **24**:625–630.
456. Krejs, G. J., and J. S. Fordtran. 1980. Effect of VIP infusion on water and ion transport in the human jejunum. *Gastroenterology* **78**:722–727.
457. Davis, G. R., C. A. Santa Ana, S. G. Morawski, and J. S. Fordtran. 1981. Effect of vasoactive intestinal polypeptide on active and passive transport in the human jejunum. *J. Clin. Invest.* **67**:1687–1694.
458. Beubler, E. 1980. Influence of vasoactive intestinal polypeptide on net water flux and cyclic adenosine 3', 5'-monophosphate formation in the rat jejunum. *Naunyn-Schmiedebergs Arch. Pharmacol.* **313**:243–247.
459. Dharmsathaphorn, K., V. Harms, D. J. Yamashiro, R. J. Hughes, H. J. Binder, and E. M. Wright. 1983. Preferential binding of vasoactive intestinal polypeptide to basolateral membrane of rat and rabbit enterocytes. *J. Clin. Invest.* **71**:27–35.
460. Amiranoff, B., M. Laburthe, C. Dupont, and G. Rosselin. 1978. Characterization of a vasoactive intestinal peptide-sensitive adenylate cyclase in rat intestinal epithelial cell membranes. *Biochim. Biophys. Acta* **544**:474–481.
461. Angel, F., V. L. W. Go, and J. H. Szurszewski. 1982. Evidence for interaction between substance P and bombesin containing neurons in the muscularis mucosa of the canine colon. *Gastroenterology* **82**:1008a.
462. Jansen, J. B. M. J., and C. B. H. W. Lamers. 1982. Bombesin releases cholecystokinin in man. *Gastroenterology* **82**:1093a.
463. Jaffee, B. M., B. Akande, I. M. Modlin, P. Reilly, and D. Albert. 1982. Cholinergic modulation of substance P release. *Dig. Dis. Sci.* **27**:28–32.
464. Kachur, J. F., R. J. Miller, M. Field, and J. Rivier. 1982. Neurohumoral control of ileal electrolyte transport. I. Bombesin and related peptides. *J. Pharmacol. Exp. Ther.* **220**:449–455.
465. Kachur, J. F., R. J. Miller, M. Field, and J. Rivier. 1982. Neurohumoral control of ileal electrolyte transport. II. Neurotensin and substance P. *J. Pharmacol. Exp. Ther.* **220**:456–463.
466. Kisloff, B., and E. W. Moore. 1976. Effect of serotonin on water and electrolyte transport in the *in vivo* rabbit small intestine. *Gastroenterology* **71**:1033–1038.
467. Sheerin, H. E. 1979. Serotonin action on short-circuit current and ion transport across isolated rabbit ileal mucosa. *Life Sci.* **24**:1069–1616.
468. Donowitz, M., A. N. Charney, and M. Heffernan. 1977. Effect of serotonin treatment on intestinal transport in the rabbit. *Am. J. Physiol.* **232**:E85–E94.
469. Donowitz, M., Y. H. Tai, and N. Asarkof. 1980. Effect of serotonin on active electrolyte transport in rabbit ileum, gallbladder, and colon. *Am. J. Physiol.* **239**:G463–G472.
470. Donowitz, M., N. Asarkof, and G. Pike. 1980. Calcium dependence of serotonin-induced changes in rabbit ileal electrolyte transport. *J. Clin. Invest.* **66**:341–352.
471. Hardcastle, J., P. T. Hardcastle, and J. S. Redfern. 1981. Action of 5-hydroxytryptamine on intestinal ion transport in the rat. *J. Physiol. (London)* **320**:41–55.
472. Costa, M., and J. B. Furness. 1979. Commentary on the possibility that an indoleamine is a neurotransmitter in the gastrointestinal tract. *Biochem. Pharmacol.* **28**:565–571.
473. DeLorenzo, R. J. 1982. Calmodulin in neurotransmitter release and synaptic function. *Fed. Proc.* **41**:2265–2272.
474. Johnson, S. M., Y. Katayama, and R. A. North. 1980. Multiple actions of 5-hydroxytryptamine on myenteric neurones of the guinea-pig ileum. *J. Physiol. (London)* **304**:459–470.
475. Pettersson, G., H. Ahlman, A. Dahlström, J. Kewenter, I. Larsson, and P. A. Larsson. 1979. The effect of transmural field stimulation on the serotonin content in rat duodenal enterochromaffin cells—*in vitro*. *Acta Physiol. Scand.* **107**:83–87.
476. Zinner, M. J., B. M. Jaffe, L. DeMagistris, A. Dahlström, and H. Ahlman. 1982. Effect of cervical and thoracic vagal stimulation on luminal serotonin release and regional blood flow in cats. *Gastroenterology* **82**:1403–1408.
477. Larsson, I. 1981. Studies on the extrinsic neural control of serotonin release from the small intestine. *Acta Physiol. Scand. Suppl.* **499**:1–43.
478. Kellum, J. M., M. McCabe, J. Schneier, and M. Donowitz. 1982. Neural mediation of acid-stimulated serotonin release from rabbit duodenum. *Gastroenterology* **82**:1098.
479. Gaginella, T. S., T. J. Rimele, and M. Wietecha. 1983. Studies on rat intestinal epithelial cell receptors for serotonin and opiates. *J. Physiol. (London)* **335**:101–111.
480. Reasbeck, P., G. Barbezat, A. Shulkes, and D. Fletcher. 1982. Effect of neurotensin at physiological blood levels on the canine small bowel. *Gastroenterology* **82**:1156.
481. Fox, J. E. T., J. McLean, E. E. Daniel, Y. Sakai, and J. Jury. 1982. Neurotensin, evidence for multiple receptors for gastrointestinal motility action in dogs. *Gastroenterology* **82**:1060.
482. Teitelbaum, D. H., T. M. O'Dorisio, and T. S. Gaginella. 1982. Caerulein, somatostatin, and neurotensin modulate the release of acetylcholine from the guinea-pig myenteric plexus. *Gastroenterology* **82**:1194.
483. Sundler, F., R. Hakanson, S. Leander, and R. Uddman. 1982. Light and electron microscopic localization of neurotensin in the gastrointestinal tract. *Ann. N.Y. Acad. Sci.* **400**:94–104.
484. Hubel, K. A. 1972. Effects of pentagastrin and cholecystokinin on intestinal transport of ions and water in the rat. *Proc. Soc. Exp. Biol. Med.* **140**:670–672.
485. Modigliani, R., J.-Y. Mary, and J. J. Bernier. 1976. Effects of synthetic human gastrin I on movements of water, electrolytes, and glucose across the human small intestine. *Gastroenterology* **71**:978–984.
486. El Masri, S. H., M. R. Lewin, and C. G. Clark. 1977. *In vitro* effects of gastrin on the movement of electrolytes across the human colon. *Scand. J. Gastroenterol.* **12**:999–1002.
487. Barbezat, G. O., and M. I. Grossman. 1971. Intestinal secretion: Stimulation by peptides. *Science* **174**:422–424.
488. Barbezat, G. O. 1973. Stimulation of intestinal secretion by polypeptide hormones. *Scand. J. Gastroenterol.* **8**:1–21.
489. Helman, C. A., and G. O. Barbezat. 1977. The effect of gastric inhibitory polypeptide on human jejunal water and electrolyte transport. *Gastroenterology* **72**:376–379.
490. Mitchenere, P., T. E. Adrian, R. M. Kirk, and S. R. Bloom. 1981. Effect of gut regulatory peptides on intestinal luminal fluid in the rat. *Life Sci.* **29**:1563–1570.
491. Kachel, G. W., L. L. Frase, W. Domschke, and G. J. Krejs. 1982. Effect of motilin infusion on water and ion transport in the human jejunum. *Gastroenterology* **82**:1094.
492. Gottesbüren, H., H. Leising, H. Menge, H. Lorenz-Meyer, and E. O. Riecken. 1974. Einfluss von glucagon auf die glucose-, wasser- und elektrolytresorption des menschlichen jejunums. *Klin. Wochenschr.* **52**:926–929.
493. Hicks, T., and L. A. Turnberg. 1974. Influence of glucagon on the human jejunum. *Gastroenterology* **67**:1114–1118.
494. Kaufman, M. E., M. A. Dinno, and K. C. Huang. 1980. Effect of glucagon on ion transport in mouse intestine. *Am. J. Physiol.* **238**:G491–G494.
495. Granger, D. N., P. R. Kvietys, W. H. Wilborn, N. A. Mortillaro, and A. E. Taylor. 1980. Mechanism of glucagon-induced intestinal secretion. *Am. J. Physiol.* **239**:G30–G38.
496. MacFerran, S. N., and D. Mailman. 1977. Effects of glucagon on canine intestinal sodium and water fluxes and regional blood flow. *J. Physiol. (London)* **266**:1–12.
497. Isaacs, P. E. T., and L. A. Turnberg. 1977. Failure of glucagon to influence ion transport across human jejunal and ileal mucosa *in vitro*. *Gut* **18**:1059–1061.
498. Lee, J. S., and J. W. Silverberg. 1976. Effect of histamine on intestinal fluid secretion in the dog. *Am. J. Physiol.* **231**:793–798.
499. Fromm, D., and N. Halpern. 1979. Effects of histamine receptor

antagonists on ion transport by isolated ileum of the rabbit. *Gastroenterology* **77**:1034–1038.

500. Linaker, B. D., J. S. McKay, N. B. Higgs, and L. A. Turnberg. Mechanisms of histamine stimulated secretion in rabbit ileal mucosa. *Gut* **22**:964–970.
501. Kohn, P. G., H. Newey, and D. H. Smyth. 1970. The effect of adenosine triphosphate on the transmural potential in rat small intestine. *J. Physiol. (London)* **208**:203–220.
502. Korman, L. Y., G. F. Lemp, M. J. Jackson, and J. D. Gardner. 1982. Mechanism of action of ATP on intestinal epithelial cells: Cyclic AMP-mediated stimulation of active ion transport. *Biochim. Biophys. Acta* **721**:47–54.
503. Dobbins, J. W., J. P. Laurenson, and J. N. Forrest, Jr. 1983. Effects of adenosine and adenosine analogs on ion transport in rabbit ileum. *Gastroenterology* **84**:1138.
504. Kimmich, G. A., and J. Randles. 1982. An ATP- and Ca^{2+}-regulated Na^+ channel in isolated intestinal epithelial cells. *Am. J. Physiol.* **243**:C116–C123.
505. Kimmich, G., and J. Randles. 1980. Regulation of Na^+-dependent sugar transport in intestinal epithelial cells by exogenous ATP. *Am. J. Physiol.* **238**:C177–C183.
506. Richards, N. W., L. J. Wallace, and T. S. Gaginella. 1983. Effect of ATP on ion transport in isolated rat intestinal epithelial cells. *Pharmacologist* **25**:182.
507. Frew, R., and H. J. Baer. 1979. Adenosine-α, β-methylene diphosphate effects in intestinal smooth muscle: Sites of action and possible prostaglandin involvement. *J. Pharmacol. Exp. Ther.* **211**:525–530.
508. Anderson, G. F. 1982. Evidence for a prostaglandin link in the purinergic activation of rabbit bladder smooth muscle. *J. Pharmacol. Exp. Ther.* **220**:347–352.
509. Aulsebrook, K. A. 1965. Intestinal absorption of glucose and sodium: Effects of epinephrine and norepinephrine. *Biochem. Biophys. Res. Commun.* **18**:165–169.
510. Aulsebrook, K. A. 1965. Intestinal transport of glucose and sodium: Stimulation by reserpine and the humoral mechanism involved. *Proc. Soc. Exp. Biol. Med.* **119**:387–389.
511. Brunsson, I., S. Eklund, M. Jodal, O. Lundgren, and H. Sjövall. 1979. The effect of vasodilatation and sympathetic nerve activation on net water absorption in the cat's small intestine. *Acta Physiol. Scand.* **106**:61–68.
512. Landsberg, L., M. B. Berardino, J. Stoff, and J. B. Young. 1978. Further studies on catechol uptake and metabolism in rat small bowel *in vivo:* (1) a quantitatively significant process with distinctive structural specifications; and (2) the formation of a dopamine glucuronide reservoir after chronic L-dopa feeding. *Biochem. Pharmacol.* **27**:1365–1371.
513. Cassuto, J., H. Sjövall, M. Jodal, J. Svanvik, and O. Lundgren. 1982. The adrenergic influence on intestinal secretion in cholera. *Acta Physiol. Scand.* **115**:157–158.
514. Newsome, P. M., M. N. Burgess, G. D. Holman, N. A. Mullan, D. H. Richards, and M. R. Smith. 1981. α_2-Adrenoceptors controlling intestinal secretion. *Biochem. Soc. Trans.* **9**:413–414.
515. Chang, E. B., M. Field, and R. J. Miller. 1982. α_2-Adrenergic receptor regulation of ion transport in rabbit ileum. *Am. J. Physiol.* **242**:G237–G242.
516. Durbin, T., L. Rosenthal, K. McArthur, D. Anderson, and K. Dharmsathaphorn. 1982. Clonidine and lidamidine (WHR-1142) stimulate sodium and chloride absorption in the rabbit intestine. *Gastroenterology* **82**:1352–1358.
517. Nakaki, T., T. Nakadate, S. Yamamoto, and R. Kato. 1982. α_2-Adrenoceptors inhibit the cholera-toxin-induced intestinal fluid accumulation. *Naunyn-Schmiedebergs Arch. Pharmacol.* **318**:181–184.
518. Tapper, E. J., A. S. Bloom, and D. L. Lewand. 1981. Endogenous norepinephrine release induced by tyramine modulates intestinal ion transport. *Am. J. Physiol.* **241**:G264–G269.
519. Tanaka, T., and K. Starke. 1979. Binding of ^3H-clonidine to an α-adrenoceptor in membranes of guinea-pig ileum. *Naunyn-Schmiedebergs Arch. Pharmacol.* **309**:207–215.
520. Cotterell, D. J., K. A. Munday, and J. A. Poat. 1982. The binding of (^3H)-prazosin and (^3H)-clonidine to crude basolateral membranes from rat jejunum. *Br. J. Pharmacol. Proc. Suppl.* **76**: 277P.
521. Chang, E. B., M. Field, and R. J. Miller. 1983. Enterocyte α_2-adrenergic receptors: Yohimbine and ρ-aminoclonidine binding relative to ion transport. *Am. J. Physiol.* **244**:G76–G82.
522. Wikberg, J. E. S., and R. J. Lefkowitz. 1982. $Alpha_2$ adrenergic receptors are located prejunctionally in the Auerbach's plexus of the guinea pig small intestine: Direct demonstration by radioligand binding. *Life Sci.* **31**:2899–2905.
523. Racusen, L. C., and H. J. Binder. 1979. Adrenergic interaction with ion transport across colonic mucosa: Role of both α and β adrenergic agonists. In: *Mechanisms of Intestinal Secretion*. H. J. Binder, ed. Liss, New York. pp. 201–215.
524. Albin, D., and Y. Gutman. 1980. The effect of adrenergic agents and theophylline on sodium fluxes cross the rabbit colon *in vitro*. *Biochem. Pharmacol.* **29**:1271–1273.
525. Morris, A. I., and L. A. Turnberg. 1981. Influence of isoproterenol and propranolol on human intestinal transport *in vivo*. *Gastroenterology* **81**:1076–1079.
526. Field, M., H. E. Sheerin, A. Henderson, and P. L. Smith. 1975. Catecholamine effects on cyclic AMP levels and ion secretion in rabbit ileal mucosa. *Am. J. Physiol.* **229**:86–92.
527. Laburthe, M., B. Amiranoff, and C. Boissard. 1982. α-Adrenergic inhibition of cyclic AMP accumulation in epithelial cells isolated from rat small intestine. *Biochim. Biophys. Acta* **721**:101–108.
528. Parod, R. J., B. A. Leslie, and J. W. Putney. 1980. Muscarinic and α-adrenergic stimulation of Na and Ca uptake by dispersed lacrimal cells. *Am. J. Physiol.* **239**:G99–G105.
529. Atlas, D., and M. Adler. 1981. α-Adrenergic antagonists as possible calcium channel inhibitors. *Proc. Natl. Acad. Sci. USA* **78**:1237–1241.
530. Donowitz, M., S. Cusolito, L. Battisti, R. Fogel, and G. W. G. Sharp. 1982. Dopamine stimulation of active Na and Cl absorption in rabbit ileum. *J. Clin. Invest.* **69**:1008–1016.
531. Donowitz, M., G. Elta, L. Battisti, R. Fogel, and E. Label-Schwartz. 1983. Effect of dopamine and bromocriptine on rat ileal and colonic transport. *Gastroenterology* **84**:516–523.
532. Dharmsathaphorn, K., H. J. Binder, and J. W. Dobbins. 1980. Somatostatin stimulates sodium and chloride absorption in the rabbit ileum. *Gastroenterology* **78**:1559–1565.
533. Dharmsathaphorn, K., L. Racusen, and J. W. Dobbins. 1980. The effect of somatostatin on ion transport in the rat colon. *J. Clin. Invest.* **66**:813–820.
534. Dobbins, J. W., K. Dharmsathaphorn, L. Racusen, and H. J. Binder. 1981. The effect of somatostatin and enkephalin on ion transport in the intestine. *Ann. N.Y. Acad. Sci.* **372**:594–612.
535. Guandalini, S., J.F. Kachur, P. L. Smith, R. J. Miller, and M. Field. 1980. *In vitro* effects of somatostatin on ion transport in rabbit intestine. *Am. J. Physiol.* **238**:G67–G74.
536. Freedman, J., H. Rasmussen, and J. W. Dobbins. 1980. Somatostatin stimulates coupled sodium chloride influx across the brush border of the rabbit ileum. *Biochem. Biophys. Res. Commun.* **97**:243–247.
537. Dobbins, J., L. Racusen, and H. J. Binder. 1980. Effect of D-alanine methionine enkephalin amide on ion transport in rabbit ileum. *J. Clin. Invest.* **66**:19–28.
538. McKay, J. S., B. D. Linaker, and L. A. Turnberg. 1981. Influence of opiates on ion transport across rabbit ileal mucosa. *Gastroenterology* **80**:279–284.
539. McKay, J. S., B. D. Linaker, N. B. Higgs, and L. A. Turnberg. 1982. Studies of the antisecretory activity of morphine in rabbit ileum *in vitro*. *Gastroenterology* **82**:243–247.
540. Waterfield, A. A., R. W. J. Smokcum, J. Hughes, H. W. Kosterlitz, and G. Henderson. 1977. *In vitro* pharmacology of the opioid peptides, enkephalins and endorphins. *Eur. J. Pharmacol.* **43**:107–116.
541. Kromer, W., and H. Schmidt. 1982. Opioids modulate intestinal

peristalsis at a site of action additional to that modulating acetylcholine release. *J. Pharmacol. Exp. Ther.* **223**:271–274.

542. Gaginella, T. S., and Z.-A. C. Wu. 1983. [D-Ala^2,D-Met^{5}NH$_2$]-enkephalin inhibits acetylcholine release from the submucosal plexus of rat colon. *J. Pharm. Pharmacol.* **35**:823–825.
543. Kachur, J. F., R. J. Miller, and M. Field. 1980. Control of guinea pig intestinal electrolyte secretion by a δ-opiate receptor. *Proc. Natl. Acad. Sci. USA* **77**:2753–2756.
544. Kachur, J. F., and R. J. Miller. 1982. Characterization of the opiate receptor in the guinea-pig ileal mucosa. *Eur. J. Pharmacol.* **81**:177–183.
545. Oka, T. 1980. Enkephalin receptor in the rabbit ileum. *Br. J. Pharmacol.* **68**:193–195.
546. Corbett, A. D., S. J. Paterson, A. T. McKnight, J. Magnan, and H. W. Kosterlitz. 1982. Dynorphin$_{1-8}$ and dynorphin$_{1-9}$ are ligands for the κ-subtype of opiate receptor. *Nature (London)* **299**:79–81.

CHAPTER 11

The Uptake of Lipids into the Intestinal Mucosa

Henrik Westergaard and John M. Dietschy

1. Introduction

Fats, proteins, and complex carbohydrates represent the major sources of calories in the typical diet found in the Western world. During digestion within the proximal small intestine, proteins and carbohydrates are broken down into simpler peptides and saccharides that are very polar and so are soluble in the aqueous environment of the intestinal contents. Because of the high degree of interaction between these polar products and the water phase, carrier-mediated and energy-linked transport processes are required to bring about net transfer of these molecules from the intestinal lumen into the cytosolic compartment of the columnar absorptive cell of the jejunum and ileum.[1,2] In contrast, the digestion of complex dietary lipids releases products that are still very unpolar and, therefore, have very low aqueous solubilities. Such molecules are, however, readily absorbed across the microvillus membrane of the intestinal cell by passive mechanisms. This chapter reviews the unique biochemical and physiological mechanisms that interact to bring about the absorption of dietary lipids from the external environment, i.e., the contents of the intestinal lumen, into the body, i.e., the cytosolic compartment of the columnar epithelial cell of the small intestine.[3]

2. Chemical Species of Lipids That Are Involved during Fat Absorption

Although the diet contains many types of complex lipids of both plant and animal origin, only the two quantitatively important fractions are considered here. As shown in column A of Fig. 1, the typical daily Western diet contains 50–100 g of triglyceride which consists of various long-chain fatty acids esterified to the three carbon atoms of glycerol. In addition, 0.5–2.0 g of cholesterol is present either as the free sterol or as cholesterol esterified to a long-chain fatty acid at the 3β position. Within the proximal jejunum, pancreatic enzymes having specific esterase activity attack both of these types of ester bonds.[4] One type of pancreatic enzyme specifically cleaves the fatty acids from the α and α′ positions (1 and 3 positions) of the triglyceride molecule, releasing free fatty acids and β-monoglycerides, whereas a specific cholesterol ester esterase cleaves the esterified cholesterol, releasing free sterol and fatty acid. Thus, as shown in column B of Fig. 1, the major products of digestion of dietary lipids in the intestine are β-monoglyceride, free fatty acid, and free cholesterol. These products differ from the esterified dietary lipids from which they are derived in that they are amphipathic, i.e., they have different regions on the molecule that are either polar, and so can interact with water, or are nonpolar, and so can interact with lipids. These molecules are, nevertheless, still relatively insoluble under the conditions that exist in the aqueous environment of the proximal intestinal contents.[5,6]

During digestion, bile acids also are secreted into the intestinal lumen. Above a certain critical concentration, these detergentlike molecules spontaneously aggregate to form macromolecular structures known as micelles.[7] As shown diagrammatically in column C of Fig. 1, the amphipathic products of fat digestion interact with the bile acid micelles to form mixed micelles which principally contain free fatty acid, β-monoglyceride, free cholesterol, and bile acids. The nonpolar portion of each of these amphipathic molecules presumably interdigitates into the hydrophobic interior of the micelle while the more polar region sticks outwardly to interact, principally by hydrogen bonding, with the surrounding aqueous phase. It should be emphasized, however, that this mixed micelle is not a rigid structure of fixed composition; rather, each component of the complex is in equilibrium with a monomer phase of that component present in the surrounding aqueous environment. Thus, the relationship of the amount of bile acid, fatty acid, β-monoglyceride, and cholesterol present in the mixed micelle and

Henrik Westergaard and John M. Dietschy • Department of Internal Medicine, Southwestern Medical School, University of Texas Health Science Center, Dallas, Texas 75235.

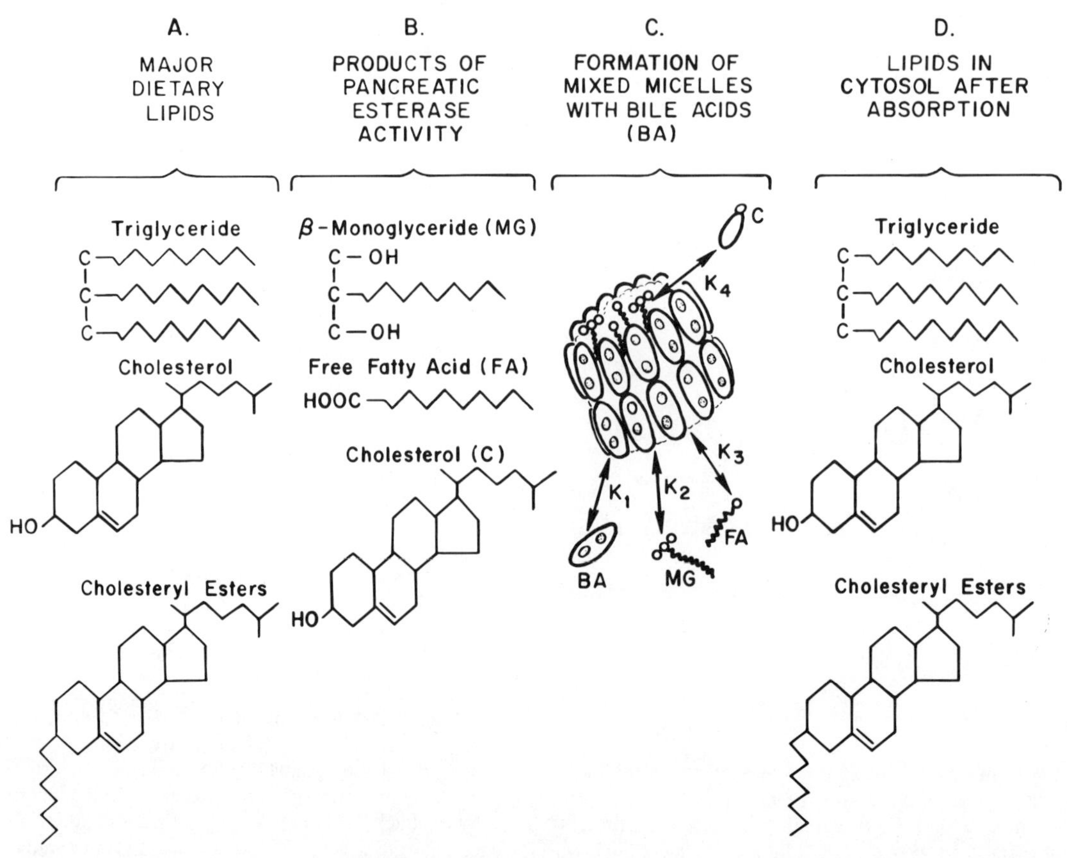

Fig. 1. Diagrammatic representation of the major types of lipids present in the intestinal contents and in the cytosol of the intestinal absorptive cell during the process of fat absorption. In column A is represented the major types of dietary lipids including triglyceride and free and esterified cholesterol, and column B shows the major products of pancreatic esterase activity that are found in the intestinal lumen. Column C shows in schematic form how these major products of fat digestion interact with bile acids in the aqueous phase of the intestinal contents to form mixed micelles. Column D shows the major lipid species found in the cytosolic compartment of the intestinal epithelial cell after lipid absorption has taken place.

present as a monomer phase in the aqueous solution can be described by a series of partitioning ratios, designated K_1 through K_4 in Fig. 1, that are analogous to classical lipid–water partitioning ratios. Thus, any process that, for example, decreases the concentration of fatty acid monomers in the aqueous phase results in the rapid movement of fatty acid out of the mixed micelle until a new equilibrium state, dictated by K_3, is achieved.

The actual process of lipid absorption across the microvillus membrane of the intestinal cell takes place from this micellar solution of bile acid and lipids, and involves the membrane translocation of fatty acids, β-monoglycerides, cholesterol, and, to some extent, bile acids. Once in the cytosolic compartment of the intestinal epithelial cell, further biochemical transformations rapidly occur. Essentially all the free fatty acid and β-monoglyceride is reesterified to triglyceride which accumulates as droplets visible under the microscope within the cytosol.[8] In addition, a large proportion of the free cholesterol also is esterified to long-chain fatty acids. The cholesterol esters partition into the lipid phase of the triglyceride droplet and the particle then becomes coated with small amounts of free cholesterol, phospholipid, and specific proteins to form chylomicrons.[9] It is in this specific lipoprotein fraction that both the triglyceride and the cholesterol absorbed from the diet leave the intestinal absorptive cell, travel through the intestinal lymph, and, ultimately, reach the peripheral circulation for disposition to various target organs.[10]

In summary, the process of fat absorption involves the membrane translocation principally of free fatty acids, β-monoglycerides, and free cholesterol. A detailed description of this uptake process necessarily involves an understanding of (1) the nature of the barriers that exist in the intestine that influence the rate of absorption; (2) the mechanisms responsible for the actual membrane translocation of the lipids; and (3) the role of bile acid micelles in facilitating this lipid uptake process.

3. The Barriers to Lipid Absorption in the Intestine

As a molecule moves from the bulk phase of the intestinal contents into the absorptive cell, it must, in effect, cross two barriers or "membranes," an unstirred layer of water and the lipid–protein matrix of the microvillus membrane on the luminal surface of the columnar cell. The unstirred water layer consists of a series of water lamellae interpositioned between the cell surface and the bulk phase of the intestinal contents. As a molecule moves from the bulk phase, which is subjected to thorough

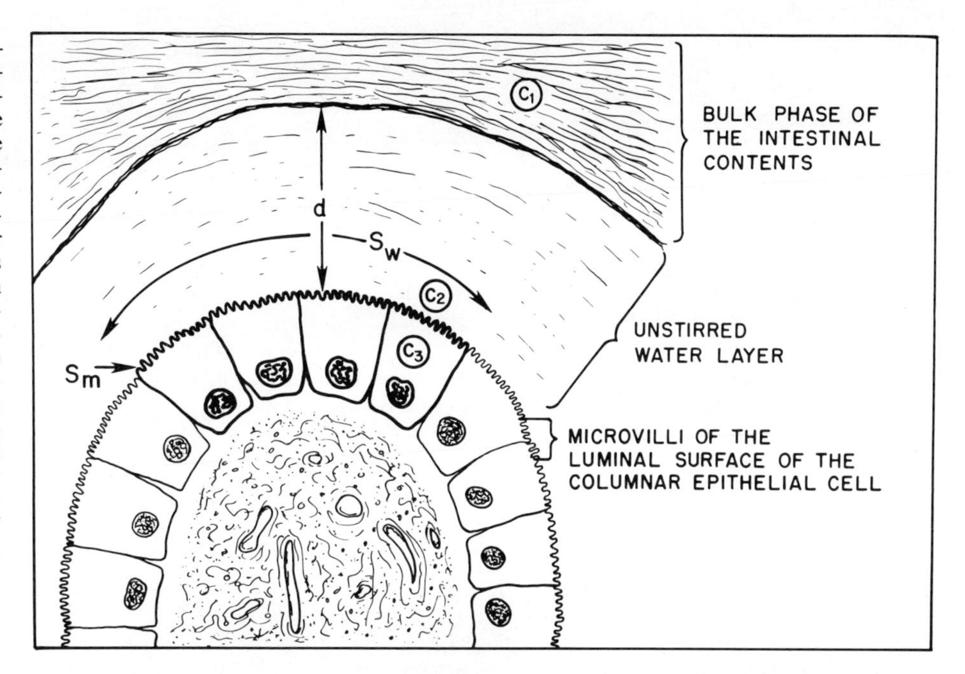

Fig. 2. Diagrammatic representation of the two major barriers to absorption in the intestine. The concentration of a particular solute molecule in the bulk phase of the perfusing medium or intestinal contents is designated C_1 and its concentration at the aqueous–membrane interface is given as C_2. After absorption across the microvillus membrane of the epithelial cell, a concentration of C_3 is attained within the cytosolic compartment. During movement from the bulk phase of the perfusing medium or intestinal contents into the cytosol of the absorptive cell, a solute molecule must diffuse across an unstirred water layer having a functional thickness d and a functional surface area S_w and must cross a portion of the microvillus membrane having a finite surface area S_m. For the purpose of this diagram, all absorption is considered to occur across the four accentuated epithelial cells.

mixing by peristaltic activity, toward the cell membrane, it enters layers of water that are subjected to progressively less bulk mixing and, therefore, regions in which diffusion becomes essentially the sole means for molecular movement. Although there is no sharp demarcation between such water layers, experimental means are now available to measure the functional thickness, designated d in Fig. 2, and functional surface area, designated S_w, of such unstirred water layers in various tissue preparations.[11,12] The rate of net movement of a molecule across the unstirred water layer (J_d) is directly proportional to the concentration gradient that exists for that solute molecule between the bulk phase (C_1) and the aqueous–membrane interface (C_2), its diffusion coefficient (D), and the surface area of the unstirred water layer, and is inversely related to the thickness of the diffusion barrier:

$$J_d = (C_1 - C_2)\left(\frac{DS_w}{d}\right) \tag{1}$$

The rate of net flux of the molecule across the microvillus membrane depends on the mechanism of membrane translocation and the concentration gradient that exists for the solute molecule between the aqueous–membrane interface (C_2) and the cytosolic compartment of the epithelial cell (C_3). However, since essentially all of the free fatty acid and the great majority of the cholesterol that is absorbed are rapidly esterified, in the case of lipid absorption it is reasonable to assume that C_3 is, for practical purposes, essentially equal to zero. Under these circumstances, the net flux of a passively absorbed molecule across the microvillus membrane will be

$$J_d = (P)(C_2) \tag{2}$$

where P represents the passive permeability coefficient for the molecule. In the case of a carrier-mediated transport process, the flux is

$$J_d = \frac{J^m C_2}{K_m + C_2} \tag{3}$$

where J^m and K_m, respectively, equal the maximal transport rate and the true Michaelis constant for the carrier-mediated process. Both Eqs. (2) and (3), however, are written in terms of the concentration of the solute molecule at the aqueous–membrane interface, C_2, not, as is conventionally the case, the bulk phase concentration. However, by rearranging Eq. (1), C_2 can be expressed in terms of C_1 so that

$$C_2 = C_1 - \frac{dJ_d}{S_w D} \tag{4}$$

Substituting this term for C_2 into Eqs. (2) and (3) yields Eqs. (5) and (6), respectively, which are appropriate for describing the uptake rate of a molecule into the intestine either by passive or by carrier-mediated mechanisms in terms of C_1:

$$J_d = \frac{PC_1}{1 + Pd/S_w D} \tag{5}$$

$$J_d = (0.5)\left\{\frac{DS_w}{d}C_1 + K_m + \frac{dJ^m}{S_w D} \pm \left[\left(C_1 + K_m + \frac{dJ^m}{S_w D}\right)^2 - 4C_1\left(\frac{dJ^m}{S_w D}\right)\right]^{1/2}\right\} \tag{6}$$

It is apparent from these equations that two extreme situations may be encountered by biological systems, depending on whether, in a given circumstance, the overwhelming resistance to molecular absorption is the unstirred water layer or the cell membrane. Under conditions in which unstirred water layer resistance is very low, the rate of uptake of a particular molecule is determined essentially completely by the permeability characteristics of the cell membrane so that in the case of a solute absorbed by passive or carrier-mediated means, the rate of uptake is given by Eqs. (7) and (8), respectively:

$$J_d = (P)(C_1) \tag{7}$$

$$J_d = \frac{J^m C_1}{K_m + C_1} \qquad (8)$$

Alternatively, the resistance of the unstirred water layer may be very high relative to the resistance encountered during membrane translocation. C_2 for a passively absorbed molecule in this situation essentially equals zero, so that the rate of uptake is given by

$$J_d = \frac{C_1 D S_w}{d} \qquad (9)$$

The rate of carrier-mediated absorption in this condition is still given by Eq. (6), but the very high unstirred layer resistance term will drastically alter the kinetic characteristics of the transport process. For example, the relationship between J_d and C_1 will tend to become linear rather than being hyperbolic, and there will be a marked increase in the apparent K_m value for the transport process. Furthermore, demonstration of competition for transport sites by different solute molecules, or of changes in the K_m value for a given transport process, will be obscured in the presence of such a high unstirred layer resistance.

Obviously, then, the detailed understanding of a specific absorption process such as the uptake of various lipids by the intestinal mucosa requires knowledge of the dimensions and permeability characteristics of these two barriers. Such information is currently available only in the case of the rabbit jejunum so that data from this species are utilized for illustrative purposes in the following discussion. There are a variety of indirect lines of evidence, however, to suggest that the principles derived from the experimental data obtained in the rabbit also apply to other species including man.

4. Characteristics of the Intestinal Microvillus Membrane Barrier to Lipid Absorption

As is evident in Eqs. (5) and (6), the characteristics of lipid movement across the microvillus membrane will critically depend on whether the process takes place by diffusion of the solute through the bulk lipid phase of the membrane or involves interaction with a finite number of recognition sites prior to the translocation step. A variety of experimental data support the concept that mucosal uptake of lipids such as fatty acids and sterols occurs by passive diffusion.[13–15] For example, the rate of uptake of such compounds is a linear function of their concentration in the bulk perfusate. In well-performed studies, no saturable uptake process has been demonstrated. Furthermore, competition for mucosal uptake has not been shown for structurally related molecules such as cholesterol and β-sitosterol or stearate and palmitate; rather, absorption of each of these molecules seems to occur at a rate that is independent of the presence or absence of other solutes in the perfusate. Finally, inhibition of metabolic sources of high-energy phosphate bonds within the mucosal cell, and the presence or absence of Na^+ ions or various sugars in the mucosal perfusate do not alter the rate of lipid absorption.

Since these various lines of evidence indicate that lipid uptake occurs by passive diffusion through the microvillus membrane, the rate at which this process occurs is properly defined in terms of the passive permeability coefficient (P) appropriate for each lipid species. The passive permeability coefficient for any molecule, in turn, is determined by three separate variables:

$$P = (K)\left(\frac{D_m}{d_m}\right) \qquad (10)$$

where K is the partitioning ratio for the molecule between the lipid phase of the membrane and the aqueous phase of the perfusate, D_m the diffusion coefficient for the molecule through the cell membrane, and d_m the functional thickness of the membrane.[16] Of these three variables, K is the overwhelmingly important term since, for any group of solute molecules of biological importance, this term may vary over a range of 10^{10}, whereas D_m/d_m varies over a very much smaller interval. The K term, in turn, is proportional to the free energy change associated with the transfer of 1 mole of a particular solute from the aqueous phase to the lipid phase of the cell membrane:

$$K = \exp\left(\frac{-\Delta F_{w\rightarrow l}}{RT}\right) \qquad (11)$$

Ideally, one would like to characterize the microvillus membrane barrier in terms of K, D_m, and d_m but this has not been possible experimentally. It is feasible, however, to characterize a given membrane by determining the manner in which the addition of a particular substituent group, e.g., $\text{-}CH_2\text{-}$ or -OH, to a solute molecule alters the rate of movement of that molecule across the cell membrane. This incremental change in the rate of passive penetration can be related to the incremental change in K for the two solute molecules and, ultimately, to the incremental change in $\Delta F_{w\rightarrow l}$ brought about by the addition of the substituent group. Thus, as shown by

$$\delta\Delta F_{w\rightarrow l} = -RT \ln \frac{P^+}{P^0} \qquad (12)$$

the incremental free energy change, $\delta\Delta F_{w\rightarrow l}$, associated with the movement of 1 mole of the substituent group from the aqueous to the membrane phase, can be calculated from the passive permeability coefficient determined for solute molecules with (P^+) and without (P^0) a particular substituent group.[16,17]

Incremental free energy values associated with the movement of the $\text{-}CH_2\text{-}$ and -OH groups across several biological membranes and into several bulk organic solvents have now been measured, and representative values are shown in Fig. 3. The addition of the $\text{-}CH_2\text{-}$ group to a solute enhances membrane penetration or partitioning into an organic solvent and so is associated with a negative $\delta\Delta F_{w\rightarrow l}$ value. As seen in the left panel, the incremental free energy change found for the partitioning of the $\text{-}CH_2\text{-}$ group into very nonpolar solvents such as olive oil, triglyceride, and *n*-heptane is more negative than 600 cal/mole. However, the addition of this substituent group has significantly less effect on enhancing passive penetration of the intestinal mucosa so that $\delta\Delta F^{CH_2}_{w\rightarrow l}$ in this organ equals only approximately -258 to -356 cal/mole.[18,19] These values correspond to increases in the passive permeability coefficient of only 1.52- to 1.78-fold. As shown in the right panel of Fig. 3, the addition of the -OH group reduces the rate of passive membrane penetration or partitioning into an organic solvent of a solute molecule. Again, however, it has been found that the addition of this substituent group has significantly less effect in lowering the passive permeability coefficient for solute movement across the microvillus border of the intestine ($\delta\Delta F^{OH}_{w\rightarrow l}$ of $+564$ cal/mole) than in reducing partitioning into very nonpolar solvents such as triglyceride and olive oil.

From such observations it has been concluded that epithelial membranes in general, and the microvillus border of the

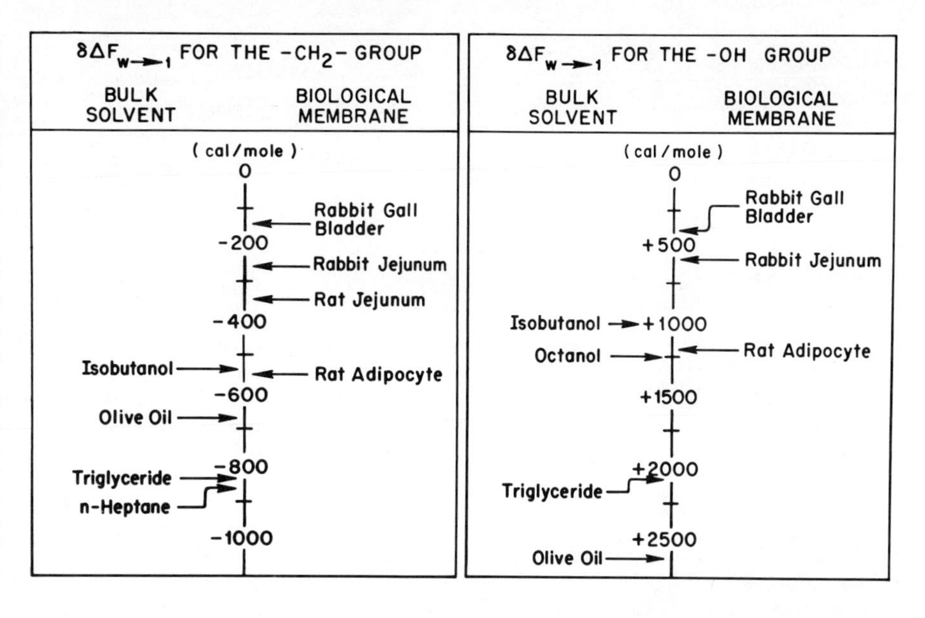

Fig. 3. Comparison of the relative polarity of several different cell membrane and bulk solvents. This diagram shows the effect of the addition of either (A) the $-CH_2-$ group or (B) the -OH group on the passive penetration of a probe molecule across a cell membrane or on the partitioning of that molecule between water and various bulk solvents. These effects are given in terms of the incremental free energy changes, $\delta\Delta F_{w\rightarrow 1}$, associated with the addition of these two substituent groups to the probe molecule and are calculated from the relationship $\delta\Delta F_{w\rightarrow 1} = -RT \ln (P^+/P^0)$ or $\delta\Delta F_{w\rightarrow 1} = -RT \ln (K^+/K^0)$. In this formulation, P^+ and K^+, respectively, represent the passive permeability coefficient and the lipid–aqueous partitioning coefficient for the probe molecule with the particular substituent group present, and P^0 and K^0 represent the same parameters for a probe molecule not containing the particular substituent group. The addition of the $-CH_2-$ group is associated with an increased rate of passive uptake of the probe molecule across the cell membrane (or increased partitioning into the bulk solvents), whereas the addition of the -OH group reduces transmembrane movement of the probe molecule (and partitioning into the bulk solvent).

intestinal absorptive cell in particular, behave as if passively absorbed solute molecules are partitioned into a lipid phase in the membrane that is much more polar than substances such as olive oil, triglyceride, or *n*-heptane. In fact, data such as those shown in Fig. 3 indicate that the membrane in the intestine that is rate limiting to passive uptake is considerably more polar than bulk isobutanol. This finding is of considerable importance with respect to the quantitative aspects of lipid absorption in the gastrointestinal tract as will be apparent later in this discussion.

5. Characteristics of the Intestinal Unstirred Water Layer Barrier to Lipid Absorption

Whereas the resistance encountered as a solvent molecule passively penetrates the microvillus membrane is related to a single term, the reciprocal of the passive permeability coefficient, the resistance encountered by the same molecule in crossing the unstirred water layer outside the microvillus membrane is dependent on the four terms given in the expression dJ_d/S_wD. Of these four terms, one is determined by the properties of the solute molecule itself(D) and another is determined ultimately by the rate at which the solute molecule permeates the microvillus membrane (J_d). The resistance of the unstirred water layer itself is characterized by the two terms d and S_w. In a tissue with such complex anatomy as the intestine, the values of these latter two terms critically depend on the characteristics of the mixing process taking place in the mucosal perfusate under either *in vivo* or *in vitro* conditions. For example, as seen in columns B and C

Table I. Representative Values for the Functional Dimensions of the Unstirred Water Layer and Microvillus Membrane of the Jejunum of the Rabbit under Different Conditions of Stirring of the Bulk Solution[a]

A. Stirring rate of bulk solution (rpm)	B. Functional thickness of unstirred water layer, d (cm)	C. Functional surface area of unstirred water layer, S_w (cm^2/100 mg)	D. Calculated surface area of rate-limiting membrane, S_m (cm^2/100 mg)	E. Limiting value of J_d/D when C_1 equals 1.0 mM (nmoles/cm^2 per 100 mg)	F. Maximum rate of uptake of FA 18:0 assuming C_1 equals 0.0044 mM (nmoles/min per 100 mg)
0	0.033	2.4	58	72,700	0.116
200	0.026	4.4	106	169,000	0.269
400	0.016	7.7	185	481,000	0.766
600	0.014	11.7	281	836,000	1.33
800	0.013	11.3	271	869,000	1.39

[a]This table shows representative values for the dimensions of the two "membranes" that determine the rate of fat absorption in the jejunum of the rabbit. These values are taken from studies performed *in vitro* where the jejunal mucosa was exposed to bulk perfusates that were stirred at different rates, as shown in column A, by means of a magnetic stirring device.[19] The functional thickness (column B) and functional surface area (column C) of the unstirred water layer were measured experimentally at each of these rates of mixing. The surface area of the microvillus membrane underlying the functional surface area of the unstirred water layer is shown in column D. These values were calculated by multiplying the data in column C by a factor of 24 to correct for the increase in the surface area caused by the presence of microvilli on the mucosal surface of the columnar cells. The surface area terms shown in columns C and D are given as the square centimeters of area present in any amount of jejunum having a dry weight of 100 mg. Column E gives the highest value of J_d/D that can be attained for any solute when C_2 essentially equals zero and the unstirred water layer becomes totally rate limiting to absorption. Column F shows illustrative data on how changes in unstirred layer resistance brought about by altering the rate of stirring alter the rate of absorption of FA 18:0, a long-chain saturated fatty acid the uptake of which is limited by its rate of diffusion through the unstirred water layer.

Table II. Representative Values for the Anatomical Dimensions of the Mucosal Surface of the Jejunum of the Rabbit[a]

Anatomical parameter	Anatomical surface area (cm^2/100 mg)
Minimum cylindrical surface area at tips of villi	3.2
Mucosal cell surface area	78.9
Microvillus membrane surface area	1890.0

[a]This table shows representative values for three different anatomical surface area measurements obtained from histological preparations of the jejunum of the rabbit.[19] The minimum cylindrical surface area is that area obtained by disregarding the intestinal villi and treating the jejunum as a simple cylindrical tube. The mucosal cell surface area represents the area of the columnar epithelial membrane covering the villi. The third surface area parameter was calculated by assuming that the microvilli at the tip of each columnar cell increase the suface area by a factor of 24 and, therefore, this value presumably represents the actual anatomical area of the cell membrane across which movement of solute molecules into the cytosolic compartment can take place. In all cases the values represent the square centimeters of surface area present in an amount of jejunum having a dry weight of 100 mg.

of Table I, in rabbit jejunum studied *in vitro* where the rate of mixing of the bulk perfusate is controlled by varying the rate of rotation of a magnetic stirring device, the functional thickness of the unstirred water layer decreases from 0.033 to 0.013 cm (330 to 130 μm) and the functional surface area of this diffusion barrier increases from 2.4 to 11.3 cm^2/100 mg as the bulk solution perfusing the intestinal mucosa is subjected to a progressively higher rate of stirring.[19] Under these circumstances, then, the actual resistance of the unstirred water layer, denoted by the ratio d/S_w, decreases 11.5-fold from 0.0138 to 0.0012 100 mg/cm.

These functional dimensions of the unstirred water layer can be compared to the anatomical dimensions of the underlying intestinal mucosa as given in Table II. It is apparent that when the stirring rate of the bulk solution equals zero, the functional surface area of the unstirred water layer corresponds very closely to the anatomical minimal cylindrical surface area of the gut. Even in the more highly stirred solution, S_w equals only approximately one-seventh of the anatomical surface area of the mucosal cells covering the intestinal villi (11.3 versus 78.9 cm^2/100 mg). The functional thickness of this diffusion barrier is very great relative to the height of the microvilli on the mucosal cells (130–330 μm versus approximately 1 μm), but is only a fraction of the height of the villus in this species (1000 μm). Thus, under most *in vivo* and in *in vitro* conditions, the unstirred water layer can be visualized as a blanket of water several hundred micrometers thick that overlies the upper portions of the villi and that acts as a barrier to the diffusion of solute molecules from the bulk contents of the intestinal lumen to the sites of absorption on the microvilli.

Assuming that the surface area of the villi is increased by a factor of 24 because of the microvillus folds, the surface area of the cell membranes beneath the unstirred water layer through which absorption takes place at each of these stirring rates can be calculated and is designated S_m in Fig. 2 and in column D of Table I. Measurements of S_m are necessary in order to calculate passive permeability coefficients in the intestine having the units centimeters per second, but the exact value of this term in most experimental circumstances is very uncertain.

Table III. Experimentally Determined Passive Permeability Coefficients for Various Saturated Fatty Acids and Alcohols and for Several Bile Acids in the Jejunum of the Rabbit[a]

Probe molecule	A. Passive permeability coefficient P (nmoles/min per 100 mg per mM)	B. Passive permeability coefficient $P \times 10^6$ (cm/sec)
Saturated fatty acids (FA)		
FA 4:0	22.4	1.33
FA 5:0	16.7	0.99
FA 6:0	26.4	1.57
FA 7:0	24.4	1.45
FA 8:0	28.8	1.71
FA 9:0	49.9	2.96
FA 10:0	74.4	4.41
FA 12:0	185.1	10.98
Saturated alcohols (Alc)		
Alc 6:0	119.9	7.11
Alc 8:0	344.2	20.41
Bile acids		
Taurocholate	1.3	0.08
Taurodeoxycholate	4.2	0.25
Cholate	18.1	1.07
Deoxycholate	36.7	2.18

[a]This table shows two types of passive permeability coefficients, both of which have been corrected for unstirred layer resistance, for a variety of solute molecules. In column A the coefficients have been normalized to the amount of jejunal mucosa having a dry weight of 100 mg and to a concentration of the probe molecule at the aqueous–membrane interface (C_2) of 1.0 mM. Since these coefficients are derived from flux measurements made when the bulk mucosal solution was stirred at 600 rpm, S_w equaled 11.7 cm^2/100 mg and S_m presumably equaled 281 cm^2/100 mg (Table I). Using this assumed value of S_m, the passive permeability coefficients in column A have been recalculated to yield the values in column B that are normalized to the square centimeters of microvillus membrane and so reduced to the conventional units used for passive permeability coefficients, i.e., centimeters per second.

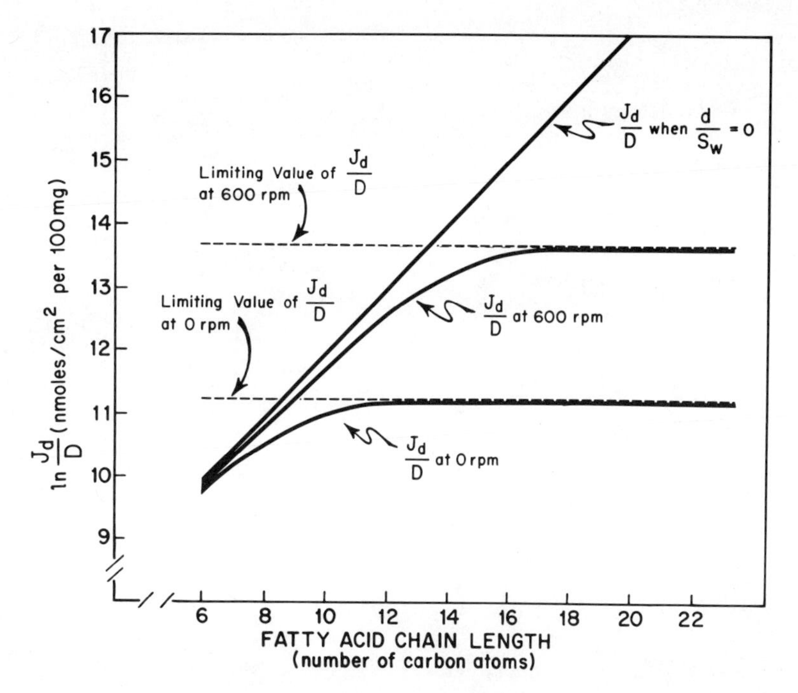

Fig. 4. The effect of different unstirred layer resistances on the intestinal uptake of saturated fatty acids of different chain lengths. The natural logarithm of the quantity J_d/D is plotted on the vertical axis. In these circumstances, J_d represents the flux rate appropriate for each probe molecule when C_1 for the fatty acid is normalized to 1.0 mM. The upper solid line represents the expected results if the intestinal villi were not overlaid with an unstirred water layer, while the two lower solid curves are derived from experimental data obtained at stirring rates of 600 and 0 rpm. The two dashed lines represent the theoretical limiting values of J_d/D at the two stirring rates using the appropriate values of d and S_w given in Table I and assuming that C_2 essentially equals zero.

6. Characteristics of Fatty Acid and Cholesterol Absorption in the Intestine

The overall characteristics of lipid absorption in the gut are determined by the ability of fatty acids, β-monoglycerides, and cholesterol to penetrate each of these two barriers to molecular uptake. The passive permeability coefficients, expressed in two different units, for fatty acids of different chain lengths and for several other molecules that are also absorbed passively across the jejunal mucosa are given in Table III. In general, the less polar a molecule, the greater is its passive permeability coefficient. In more specific terms, it is evident that in the case of the two alcohols and the fatty acids of longer chain lengths, P increases by a constant factor of 1.52 for each -CH_2- group added to the hydrocarbon chain. As discussed earlier, this value derives from the fact that the microvillus membrane in the rabbit is a relatively polar structure with a $\delta\Delta F^{CH_2}_{w \rightarrow l}$ value of only −258 cal/mole (Fig. 3).[19] The logarithm of the passive permeability coefficients experimentally determined for these relatively soluble medium-chain-length fatty acids plotted as a function of chain length yields a linear regression curve that can be extrapolated to yield P values for the longer-chain-length dietary fatty acids. Thus, for example, the passive permeability coefficient for FA 18:0 equals approximately 3000 nmoles/min per 100 mg per mM.

Such large passive permeability coefficients for more hydrophobic molecules as the long-chain fatty acids and sterols imply that these solutes would be absorbed at very high rates by the intestine since the rate of passive uptake (J_d) of a solute equals the product of its passive permeability coefficient and its concentration in the bulk perfusate. However, as J_d increases, the resistance of the unstirred water layer also increases, so that this diffusion barrier becomes progressively more rate limiting, and the concentration of the solute molecule at the aqueous–membrane interface (C_2) becomes significantly lower than the concentration of the molecule in the bulk perfusate (C_1). This is best seen in Eq. (4) where it is evident that the magnitude of the correction term for unstirred layer resistance varies directly with J_d. Furthermore, it is also apparent that at sufficiently high values of J_d, the value of the correction term equals C_1 so that C_2 essentially equals zero. This represents the situation in which diffusion across the unstirred water later becomes totally rate limiting to solute uptake by the gut. In this special circumstance, the rate of intestinal absorption is given by

$$J_d = \frac{C_1 D S_w}{d} \tag{13}$$

In this diffusion-limited situation, it is evident that the maximum rate of absorption of any solute is directly related to its free diffusion coefficient (D) and, furthermore, at any value of C_1 the quantity J_d/D is a constant that is determined by the characteristics of the unstirred water layer as given by the term S_w/d, i.e.:

$$\frac{J_d}{D} = (C_1)\left(\frac{S_w}{d}\right) \tag{14}$$

The actual limiting values of J_d/D at five different stirring rates and, consequently, five different values of S_w and d, are shown in column E of Table I.

The combined effect of the unstirred water layer and the microvillus membrane resistances on fatty acid uptake is illustrated in Fig. 4. In this formulation, C_1 is assumed to be constant at 1.0 mM for all fatty acids. The upper solid line illustrates the anticipated experimental findings if the unstirred water layer could be totally eliminated from the intestine, i.e., if d/S_w equals zero. In this case, $\ln J_d/D$ would essentially be a linear function of fatty acid chain length since P increases by a constant fraction for each -CH_2- group added to the fatty acid chain. The lower

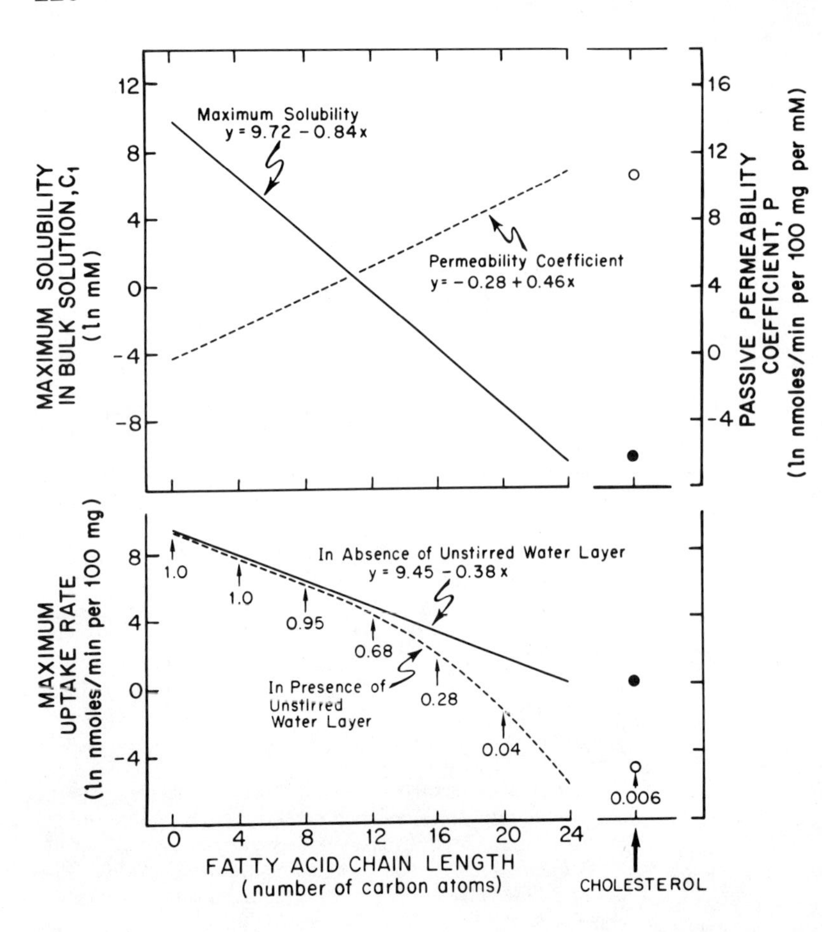

Fig. 5. The maximum rate of intestinal uptake of fatty acids of different chain lengths and of cholesterol in the presence and absence of an unstirred water layer. In the upper panel the natural logarithm of the maximum solubility of various lipids in Krebs bicarbonate buffer at 37°C is plotted as the solid line,[13] and the natural logarithm of the passive permeability coefficients for these same molecules is shown as the dashed line.[19] In the absence of an unstirred water layer, the maximum rate of uptake of any particular lipid molecule equals its maximum solubility times its passive permeability coefficient. This product for each of the lipids is shown as the solid line in the lower panel. In the presence of an unstirred water layer, the maximum rate of mucosal uptake is reduced, particularly for the less polar members of the series, as shown by the dashed line in the lower panel. These latter data were calculated assuming values of 0.014 cm and 11.7 cm²/100 mg for d and S_w, respectively. The numbers shown below the dashed curve represent the factors by which the presence of an unstirred water layer of these dimensions reduced the maximum rates of uptake of the various lipids.

solid curve is derived from experimental measurements made *in vitro* where the mucosal perfusate was unstirred: In this situation, J_d becomes proportional to D so that J_d/D becomes essentially constant for all fatty acids with 10 or more carbon atoms in the chain. Furthermore, for the longer-chain-length compounds the experimentally derived curve is essentially superimposable upon the limiting value of J_d/D (72,700 nmoles/cm² per 100 mg) derived from the values of S_w (2.4 cm²/100 mg) and d (0.033 cm) appropriate for this experimental circumstance as given in Table I. The middle solid curve in Fig. 4 is derived from experimental measurements made under highly stirred conditions where the limiting value of J_d/D is increased to 836,000 nmoles/cm² per 100 mg. Again, it is seen that the microvillus membrane primarily is rate limiting to the uptake of the shorter-chain-length members of the series. However, with the longer-chain-length fatty acids, even under these highly stirred conditions, significant unstirred layer resistance becomes evident with deviation of the curve downward until the diffusion barrier becomes totally rate limiting to uptake and a constant value of J_d/D is achieved.

The important conclusion derived from this type of study is that the absorption of longer-chain-length fatty acids, sterols such as cholesterol, and other molecules with very high passive permeability coefficients is diffusion limited in the intestine. Thus, the rate of uptake of these types of molecules is very sensitive to alterations in unstirred layer resistance. This is illustrated by the data in column F of Table I which shows that the maximum uptake of FA 18:0 is increased 12-fold simply by increasing the degree of stirring of the bulk solution and so reducing the effective resistance of the diffusion barrier.

In the illustrative data presented in Fig. 4, the concentration of each fatty acid is fixed at 1.0 mM. Under physiological conditions, however, it is likely that the concentration of each fatty acid in the bulk solution, as well as the passive permeability coefficient, varies for each fatty acid. Thus, the overall rate at which different lipids are absorbed under *in vivo* conditions is a complex function of three sets of variables: the maximum solubility, passive permeability coefficient, and unstirred layer resistance appropriate for each compound. Figure 5 illustrates the interaction of these three variables. In this example it is assumed that the fatty acids of all chain lengths are present in the bulk phase of the intestinal contents in an amount that exceeds their respective aqueous solubilities. Under these assumed conditions, the maximum aqueous solubility for each fatty acid decreases by a factor of 2.32 for each -CH_2- group in the hydrocarbon chain. This decrease in solubility as a function of fatty acid chain length is shown by the solid line in the upper panel. Thus, for example, the maximum solubility of FA 8:0 equals 19.8 mM whereas that of the FA 16:0 equals 0.024 mM. In contrast, the passive permeability coefficient for each fatty acid increases by a factor of 1.58 for each -CH_2- group in the hydrocarbon chain as shown by the dashed line.

In the absence of an unstirred water layer, the maximum rate of uptake of any fatty acid equals its maximum solubility in the bulk solution times its passive permeability coefficient. This product, calculated for each fatty acid, is shown as the solid line in the lower panel of Fig. 5. It is apparent that since solubility decreases by a relatively greater amount than the values of P increase with increasing chain length, the maximum uptake rate decreases by a factor of 1.47 for each -CH_2- group added to the hydrocarbon chain. Thus, the less polar a molecule, the lower is its rate of maximum absorption. For example, the maximum uptake values derived from the lower panel of Fig. 5 for FA 16:0 and FA 18:0 equal 27.5 and 12.8 nmoles/min per 100 mg, respectively, whereas the corresponding value for cholesterol absorption is 1.6 nmoles/min per 100 mg. It should be emphasized that this relationship exists because of the relatively polar nature of the microvillus membrane. If this membrane behaved as a more hydrophobic structure, so that $\delta\Delta F_{w\rightarrow 1}^{CH_2}$ was more negative than 517 cal/mole, then the opposite situation would occur, and the more hydrophobic lipids would have higher rates of maximum absorption.

The introduction of an unstirred water layer further decreases the maximum rate of absorption of the less polar lipids as illustrated by the dashed line in the lower panel of Fig. 5. The magnitude of this decrease for each lipid is given by the ratio below the dashed line. Thus, for fatty acids with less than eight carbon atoms, the presence of the unstirred water layer has essentially no effect on the maximum rate of uptake. There is a progressively greater effect on the longer-chain-length fatty acids, and maximum cholesterol uptake is suppressed 167-fold by the introduction of this diffusion barrier.

Two additional points concerning these curves deserve emphasis. First, the data used in Fig. 5 are calculated assuming a relatively well-stirred situation in which d equals 0.014 cm and S_w equals 11.7 cm^2/100 mg. *In vivo* where unstirred water layer resistance is greater, it would be anticipated that the dashed line in Fig. 5 would deviate even further downward from the solid linear curve. Second, the relative effect of the unstirred water layer on the uptake of passively absorbed molecules such as fatty acids and cholesterol is independent of C_1. This derives from the fact that the unstirred layer resistance term in Eq. (5) increases in proportion to J_d which, in turn, increases as a linear function of C_1. This is not true for transport processes involving a finite number of transport sites where the unstirred layer resistance term approaches a maximum value dictated by J^m. In this situation, the magnitude of the unstirred layer effect is a more complex function of C_1.

7. Role of Bile Acid Micelles in Facilitating Lipid Absorption in the Intestine

Several lines of evidence now support the concept that the major function of bile acid micelles in facilitating lipid absorption is to overcome the resistance to diffusion engendered by the presence of the unstirred water layer between the bulk phase of the intestinal contents and the microvillus membrane. Stated in the quantitative terms shown in Fig. 5, the presence of bile acid micelle eliminates the deviation of the dashed curve from the solid curve shown in the lower panel and so enhances the uptake of fatty acids and sterols to the extent that uptake of particular compounds is limited by the unstirred water layer. Thus, the presence of bile acids in the intestinal lumen profoundly enhances the rate of cholesterol absorption, significantly increases the rate of long-chain fatty acid uptake, but has relatively little effect on the maximum mucosal absorption of short- and medium-chain-length fatty acids.[13]

The ability of the micelle to enhance uptake of the relatively nonpolar compounds presumably derives from the capacity of bile acids to solubilize large concentrations of fatty acids, β-monoglycerides, and sterols in the structure of the mixed micelle and to carry these substances across the unstirred water layer. The addition of the various lipids to the bile acid micelle expands its size and so slows diffusion of the particle across the aqueous diffusion barrier. However, since the rate of diffusion varies only as the cube root of the micellar molecular weight, the net effect of micellar solubilization is that a much greater mass of fatty acids, β-monoglycerides, and sterols can be delivered per unit time at the aqueous–microvillus interface.

Once the micelle reaches the vicinity of the aqueous–membrane interface, the constituent lipid molecules carried in the micelle theoretically could be absorbed into the mucosal cell by at least three different mechanisms: (1) the micelle could be taken up into the cell intact; (2) the constituent lipids might partition into the cell membrane during direct interaction or "collision" between the micelle and the cell membrane; or (3) absorption might occur only through the monomer phase of lipid molecules present in the aqueous environment in equilibrium with the lipids in the micelle.[13] There are adequate data in the literature to exclude the first possibility since the various constituent molecules in the mixed micelle are absorbed at essentially independent rates.[14,17,20] In the second possible mechanisms, it is assumed that the micelle and the membrane interact in such a way that water is excluded from the interface between the two structures following which there is direct transfer of fatty acids and cholesterol from the micelle to the microvillus membrane. If the number of such interactions is proportional to the concentration of micelles at the aqueous–membrane interface, then the observed rate of uptake of a particular lipid molecule, J_d is given by the expression

$$J_d = (P_a)\left(\frac{M_L}{V_L}\right)(V_L) \tag{15}$$

where P_a is the passive permeability coefficient for the movement of the lipid molecule from the micellar phase into the cytosol of the mucosal cell, V_L the volume of the micellar phase, and M_L the mass of lipid carried in the micelle. For a given bile acid and lipid, P_a is constant and the V_L term cancels, so that the observed velocity of uptake would be proportional to the mass of lipid in the micellar phase; i.e., if this model is correct, then the rate of lipid uptake should be proportional to M_L. In the third possible model, it is postulated that lipid uptake occurs only by way of the molecules present in aqueous solution at the interface which are in equilibrium with lipid molecules in the micelle. In this situation, the rate of uptake would be given by the equation

$$J_d = (P_b)\left(\frac{M_w}{V_w}\right) \tag{16}$$

where P_b is the passive permeability coefficient for the uptake of the lipid from the aqueous phase of the perfusate into the cytosol (Table III), M_w the mass of lipid in the aqueous phase, and V_w the volume of the aqueous phase. For a given lipid, P_b will be constant and, for practical purposes, V_w will not change, so that if this model is correct, the observed rate of uptake of a given lipid should be proportional to M_w.

Thus, the second and third possible models to explain the interaction between the micelle and the microvillus membrane can be discriminated by determining whether the rate of mucosal uptake of a particular lipid molecule is proportional to M_L or

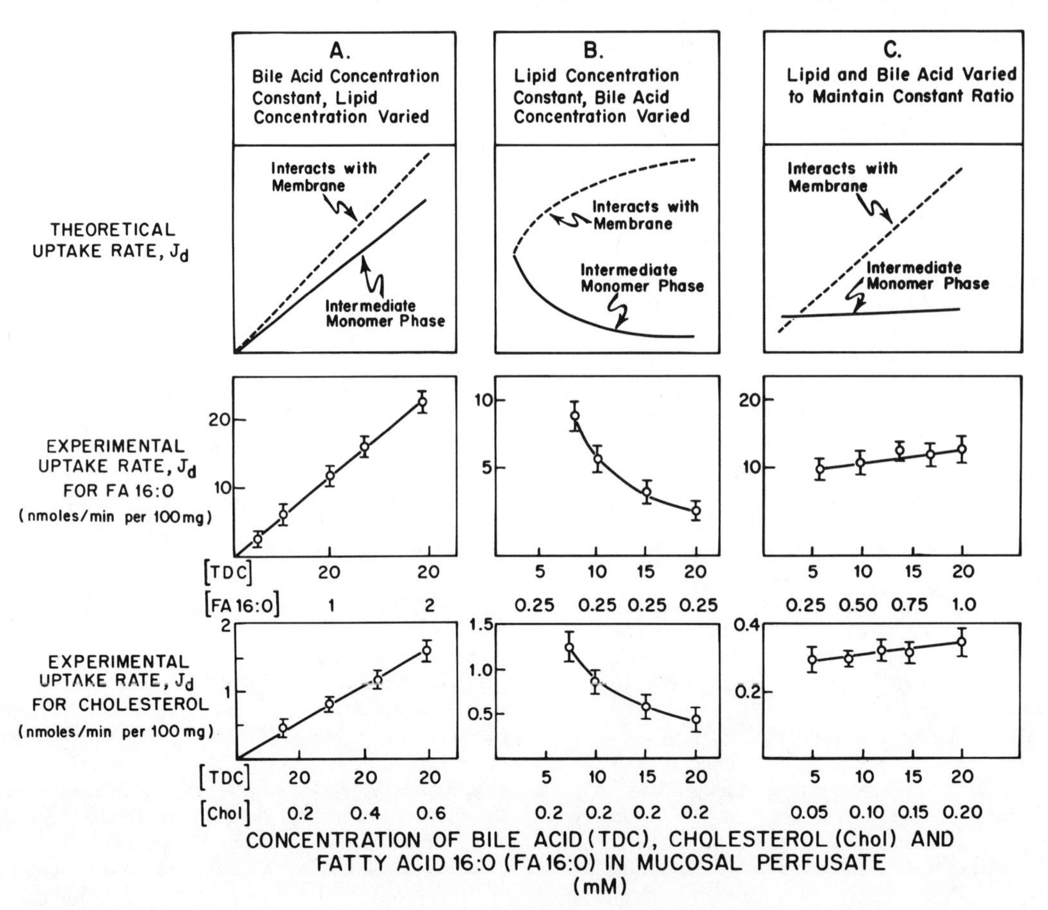

Fig. 6. The effect on lipid uptake into the intestinal mucosa of altering the relationship between the concentration of fatty acid or cholesterol and the concentration of bile acid in the bulk perfusate. The three upper panels show the theoretical rates of uptake of a lipid that would be anticipated if the micelle interacted directly with the cell membrane, as shown by the dashed line, or if uptake occurred only through a monomer phase of the lipid molecules in equilibrium with the mixed micelle, as shown by the solid lines. The middle and lower panels show actual experimental results depicting the uptake of FA 16:0 and cholesterol from micellar solutions containing the bile acid taurodeoxycholate (TDC). In column A the concentration of the bile acid was kept constant while that of the lipid molecule was progressively increased. In column B the concentration of the lipid molecule was kept constant while that of the bile acid was progressively increased. In column C the concentrations of both the bile acid and the lipid molecule were increased in parallel to maintain a constant ratio between these two molecules.

M_w. In any experimental setting, the value of M_L can be calculated from

$$M_L = \frac{KM_T}{V_w/V_L + K} \tag{17}$$

Similarly, the value of M_w is given by

$$M_w = \frac{V_w M_T}{KV_L + V_w} \tag{18}$$

In these two expressions, K represents the partitioning ratio for a particular lipid between the bile acid micelle and the aqueous bulk phase, and M_T equals the total amount of lipid dissolved in both aqueous and micellar phases of the solution.[13] The theoretical results that would be anticipated from Eqs. (17) and (18) in three different experimental situations are shown in the upper three panels of Fig. 6, and the lower panels show actual experimental results obtained for the mucosal uptake of FA 16:0 and cholesterol.

The panels in column A show the theoretical and experimental results obtained when the concentration of bile acid is kept constant but the concentration of solubilized lipid is increased, whereas the panels in columns B and C, respectively, show the effects of increasing bile acid concentration while maintaining a constant lipid concentration or increasing the concentration of both bile acid and lipid to maintain a constant ratio between the two. It is apparent that the experimental results shown in the lower panels parallel very closely the results that would be anticipated if the mechanism of uptake is described by the third model where membrane translocation takes place from the monomer phase of lipids in the aqueous solution rather than by uptake directly from the micelle.

8. Nonpolar Lipids

A variety of nonpolar lipids are ingested in small to trace quantities. These include the fat-soluble vitamins (A, D, E, and K), halogenated hydrocarbons from industrial pollutants, hydro-

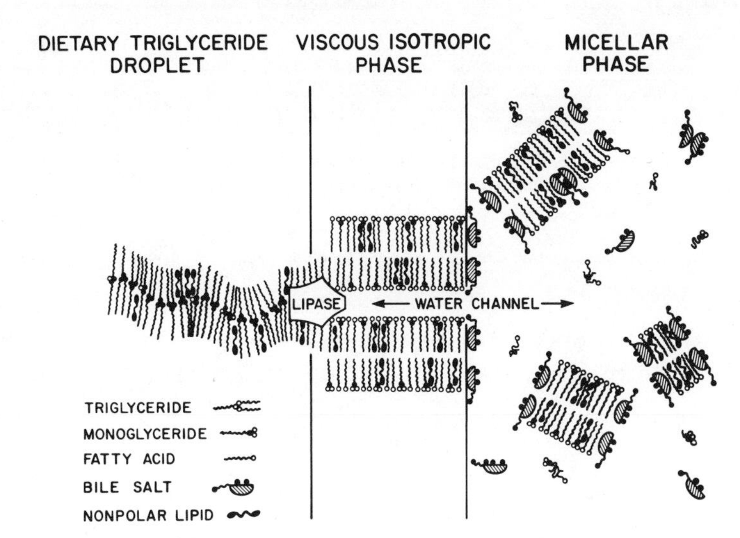

Fig. 7. Diagrammatic illustration of triglyceride lipolysis by pancreatic lipase which depicts the generation of lipolytic products (monoglycerides, fatty acids, and nonpolar lipids) from the triglyceride oil phase and the movements of these products into first a viscous isotropic phase and then into a micellar phase. Modified from Ref. 22.

carbons from combustion of coal and petroleum, very-long-chain fatty acids (> 20 carbon atoms) in surface lipids, and toxic compounds such as insecticides and fungicides. The common feature of these diverse, nonpolar lipids is minimal aqueous solubility. For instance, the water solubility of the hydrocarbon octadecane has been estimated to be about 1×10^{-12}M. Conversely, the nonpolar lipids have high solubility in triglyceride oil. The absorption of nonpolar lipids is not well understood. Recent studies of the phase changes occurring during lipid digestion *in vitro* have demonstrated that nonpolar lipids move quantitatively from the oil phase to a viscous isotropic phase under conditions simulating normal lipid digestion.[21] Presumably, these nonpolar molecules are then incorporated into the lipophilic interior of unsaturated bile acid micelles. This movement of nonpolar lipids from the oil phase to the micellar phase is illustrated in Fig. 7. This model allows nonpolar lipids to flow within a "hydrocarbon continuum" and to avoid contact with the aqueous environment.[22] The movement of nonpolar lipids from the micellar phase into the absorptive cells has not been characterized. It is difficult to envision that nonpolar lipids diffuse through the water phase at the microvillus membrane since the driving force for diffusion would be exceedingly low. Thus, it is conceivable that the transfer of trace amounts of these very nonpolar lipids occurs during "collision" of the mixed micelle with the microvillus membrane. In any event, the molecular mechanisms operative in the absorption of nonpolar lipids require further elucidation.

9. Summary Description of the Process of Lipid Uptake

In general terms, then, the process of lipid uptake into the intestinal mucosal cell takes place as follows. Fatty acids and free cholesterol are generated by enzymatic activity in the bulk intestinal contents and partition into bile acid micelles. These mixed micelles then diffuse down the existing concentration gradient through the unstirred water layer surrounding the villi and reach the immediate vicinity of the aqueous–microvillus interface. Uptake of the lipid molecules occurs at a rate determined by the product of the passive permeability coefficient appropriate for each species and the aqueous concentration of that molecule present at the microvillus surface. The concentration of the various lipids just inside the mucosal cell is presumably maintained at very low levels because of rapid esterification,[4] whereas the concentration of these molecules outside the cell membrane is maintained at a relatively high value because of constant partitioning of fatty acids, β-monoglycerides, and cholesterol out of the mixed micelle as dictated by the partitioning coefficient appropriate for each lipid species. The net effect of these processes is the rapid movement of various lipids and sterols into the mucosal cell. Trace quantities of nonpolar lipids also are absorbed, but the mechanism of this process is not well understood. There is also a finite concentration of bile acid monomers in solution in equilibrium with the bile acid molecules present in the micelle so that there is also passive uptake of these substances.[17] The rate of such uptake is low, however, because of the very low passive permeability coefficients for the various bile acids. Since the critical micelle concentration should increase at the aqueous–membrane interface as lipids are absorbed from the mixed micelle,[6,7] the concentration of bile acid monomers is presumably higher at the aqueous–membrane interface than in the bulk phase of the intestinal contents. Consequently, there is a gradient down which bile acid molecules can back-diffuse into the bulk phase of the intestinal contents to be reutilized in the creation of new mixed micelles as additional amounts of free cholesterol, fatty acids, and β-monoglycerides are generated from dietary lipids. In a very real sense, the bile acid micelle acts as a shuttle between the bulk intestinal contents and the aqueous–microvillus interface.

ACKNOWLEDGMENTS. Support for this work came from U.S. Public Health Service Research Grants HL-09610 and AM-16386, the American Heart Association, Texas Affiliate, and a grant from the Moss Heart Fund.

References

1. Crane, R. K. 1968. Absorption of sugars. In: *Handbook of Physiology*, Section 6. C. F. Code, section ed. American Physiological Society, Washington, D.C. pp. 1323–1351.
2. Wiseman, G. 1968. Absorption of amino acids. In: *Handbook of*

Physiology, Section 6. C. F. Code, section ed. American Physiological Society, Washington, D.C. pp. 1277–1307.
3. Wilson, F. A., and J. M. Dietschy. 1971. Differential diagnostic approach to clinical problems of malabsorption. *Gastroenterology* **61**:911–931.
4. Johnston, J. M. 1977. Lipid metabolism in gastrointestinal tissue. In: *Lipid Metabolism in Mammals,* Volume 1. F. Snyder, ed. Plenum Press, New York. pp. 151–187.
5. Hofmann, A. F., and D. M. Small. 1967. Detergent properties of bile salts: Correlation with physiological function. *Annu. Rev. Med.* **18**:333–376.
6. Carey, M. C., and D. M. Small. 1969. Micellar properties of dihydroxy and trihydroxy bile salts: Effects of counterion and temperature. *J. Colloid Interface Sci.* **31**:382–396.
7. Small, D. M. 1968. Size and structure of bile salt micelles: Influence of structure, concentration, couterion concentration, pH and temperature. *Adv. Chem. Ser.* **84**:31–52.
8. Polheim, D., J. S. K. David, F. M. Schultz, M. B. Wylie, and J. M. Johnston. 1973. Regulation of triglyceride biosynthesis in adipose and intestinal tissue. *J. Lipid Res.* **14**:415–421.
9. Zilversmit, D. B. 1969. Chylomicrons. In: *Structural and Functional Aspects of Lipoproteins in Living Systems.* E. Tria and A. M. Scanu, eds. Academic Press, New York. pp. 329–368.
10. Nervi, F. O., and J. M. Dietschy. 1975. Ability of six different lipoprotein fractions to regulate the rate of hepatic cholesterogenesis *in vivo. J. Biol. Chem.* **250**:8704–8711.
11. Diamond, J. M. 1966. A rapid method for determining voltage–concentration relations across membranes. *J. Physiol. (London)* **183**:83–100.
12. Wilson, F. A., and J. M. Dietschy. 1974. The intestinal unstirred layer: Its surface area and effect on active transport kinetics. *Biochim. Biophys. Acta* **363**:112–126.
13. Westergaard, H., and J. M. Dietschy. 1976. The mechanism whereby bile acid micelles increase the rate of fatty acid and cholesterol uptake into the intestinal mucosal cell. *J. Clin. Invest.* **58**:97–108.
14. Simmonds,W. J. 1974. Absorption of lipids. In: *Gastrointestinal Physiology.* E. D. Jacobson and L. L. Shanbour, eds. University Park Press, Baltimore. pp. 343–376.
15. Watt, S. M., and W. J. Simmonds. 1971. Uptake and efflux by everted intestinal sacs of micellar cholesterol in bile salts and in nonionic detergent. *Biochim. Biophys. Acta* **225**:347–355.
16. Diamond, J. M., and E. M. Wright. 1969. Biological membranes: The physical basis of ion and nonelectrolyte selectivity. *Annu. Rev. Physiol.* **31**:581–646.
17. Schiff, E. R., N. C. Small, and J. M. Dietschy. 1972. Characterization of the kinetics of the passive and active transport mechanisms for bile acid absorption in the small intestine and colon of the rat. *J. Clin. Invest.* **51**:1351–1362.
18. Sallee, V. L., and J. M. Dietschy. 1973. Determinants of intestinal mucosal uptake of short- and medium-chain fatty acids and alcohols. *J. Lipid Res.* **14**:475–484.
19. Westergaard, H., and J. M. Dietschy. 1974. Delineation of the dimensions and permeability characteristics of the two major diffusion barriers to passive mucosal uptake in the rabbit intestine. *J. Clin. Invest.* **54**:718–732.
20. Hoffman, N. E. 1970. The relationship between uptake *in vitro* of oleic acid and micellar solubilization. *Biochim. Biophys. Acta* **196**:193–203.
21. Patton, J. S., and M. C. Carey. 1979. Watching lipid digestion. *Science* **204**:145–148.
22. Patton, J. S. 1981. Gastrointestinal lipid digestion. In: *Physiology of the Gastrointestinal Tract.* L. R. Johnson, ed. Raven Press, New York. pp. 1123–1146.

CHAPTER 12

Mechanisms of Bile Secretion and Hepatic Transport

James L. Boyer

1. Introduction

Bile is a complex aqueous secretion that is elaborated by the liver of all vertebrate species, stored in the gallbladder, and discharged into the common hepatic duct and intestine. Its primary source is formed from transport processes in the liver parenchymal cell. This hepatic bile originates at the bile canaliculus (Fig. 1) but can be modified by absorption or secretion at more distal sites along the bile ductules and ducts (Fig. 2). Not all of these transport phenomenon are clearly understood but together they generate an isosmotic electrolyte solution into which a variety of organic and inorganic solutes are also excreted. (Table I) These solutes comprise 5% of the weight of bile by volume in man, and include mixed lipid micelles composed of bile acids, cholesterol, and phospholipid, as well as amino acids, hormones, enzymes, metals, vitamins, prophyrins, and other miscellaneous endogenous and exogenous drugs, xenobiotics, and toxins.

Brauer first grouped biliary substances into three classes according to their bile/plasma ratios.[1] Class A included solutes with a bile/plasma ratio of approximately 1 such as sodium, potassium, chloride, and bicarbonate; class B substances were concentrated in bile and included bile salts, bilirubin, and a number of drugs, dyes, and xenobiotics. Class C contained compounds whose bile/plasma ratios were less than 1 and thus were partially excluded from bile such as proteins, biliary lipids, and solutes like inulin, sucrose, and ferricyanide. Although useful in focusing attention on compounds that might be concentrated in bile or restricted by diffusion, this classification has not advanced our understanding of how these substances enter bile, and thus has largely been abandoned.

Bile formation has been difficult to study not only because it drains into the intestine but also because of the minute size and anatomical arrangement of hepatocyte canaliculi that have made it impossible to directly sample canalicular bile or to use traditional electrophysiological approaches. Nonetheless, knowledge has increased rapidly in recent years by utilizing a number of indirect techniques in a variety of animal species including man. Several recent reviews have summarized this progress.[2–10]

Bile secretion serves many functions (Table II); a major one is to provide an excretory pathway for organic lipid-soluble compounds of molecular weight above 300–500, since these substances cannot be easily filtered at the glomerulus or secreted by the renal tubule.[11] Many lipophilic compounds are excreted in bile after metabolic conversion to more polar water-soluble substances by the hepatic mixed function oxidase system.[12] Still other organic solutes are transported as part of a continuing enterohepatic circulation.[13] Bile is the sole route of excretion for cholesterol and the predominant means for excretion of end products of metabolism such as bilirubin. Bile also serves an important digestive function by supplying bile salts for the intestinal phase of fat digestion and an immunological role in the intestine by transporting IgA.[14]

Many species differ with respect to bile volume, the biliary composition of lipids and bile acids, as well as the relative contribution of the bile duct epithelium to the final secretion.[15] The 24-hr volumes of bile that are excreted by man following cholecystectomy have been estimated to be between 500 and 600 ml.[16,17]

2. Structural Determinants of Bile Secretory Function

2.1. Membrane Polarity and Intracellular Organelles

The primary secretory unit consists of two adjacent hepatic parenchymal cells. Each hepatocyte maintains a well-

James L. Boyer • Liver Study Unit, Department of Medicine, Yale University School of Medicine, New Haven, Connecticut 06510.

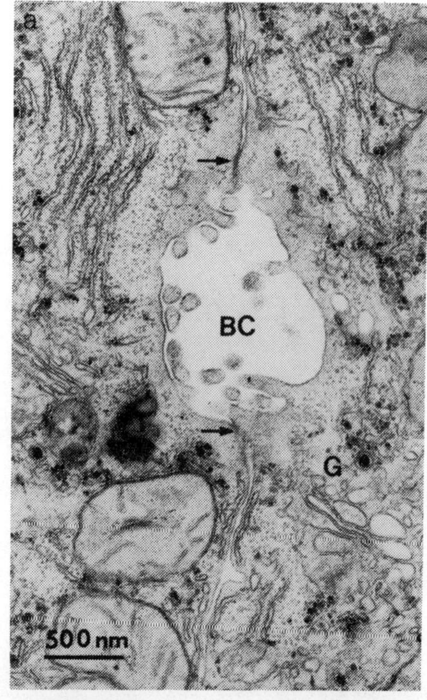

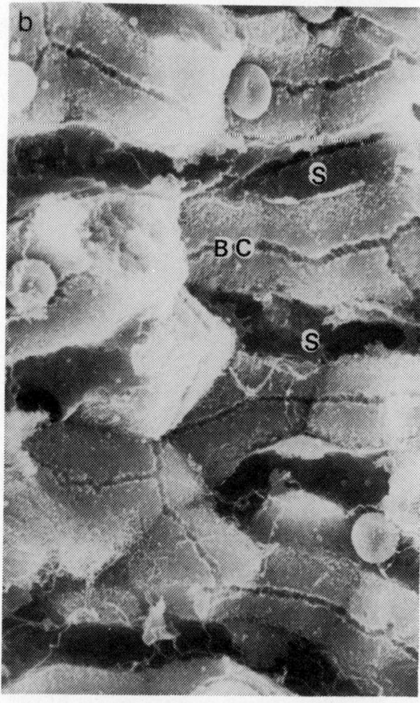

Fig. 1. (A) The bile canaliculus (BC) is formed between two adjacent hepatocytes and is sealed by tight junctions (arrows). The Golgi apparatus (G) is often seen within the vicinity of the canalicular membrane. Magnification × 27,383. (B) Scanning electron micrograph of an *en face* fracture of rat liver revealing sinusoids (S) and hepatic plates with the exposed lumen of the bile canaliculus. Original magnification × 1000.

Table I. Hepatic Bile Composition[a]

		B/P ratios
Solids (% wt/vol) ~5%		
Electrolytes (meq/liter)		
Na^+ (141–165)		~ 1
K^+ (2.7–6.7)		~ 1
Cl^- (77–117)		~ 1
HCO_3^- (12–55)		~ 1
Ca^{2+} (2.5–6.4)		~ 1
Mg^{2+} (1.5–3.0)		~ 1
SO_4^{2-} (4–5)		
PO_4^{3-} (1–2)		
Organic anions		
Total bilirubin (1–2 mM)		>1
Bile acids (3–45 mM)		>1
Lipids		
Cholesterol (free)	97–320 mg/dl	<1
Lecithin	140–810 mg/dl	<1
Heavy metals		
Cu		>1
Mn		>1
Fe		>1
Zn		>1
Total proteins (2–20 mg/ml)		
Albumin		<1
IgA		>1
Haptoglobin		>1
Secretory component		>1
Transferrin		>1
Enzymes		
Alkaline phosphatases		<1
Acid glycosidases		<1
β-Glucuronidase		<1
β-Galactosidase		<1
N-Acetyl-β-glucosaminidase		<1
Esterase		<1
5′-Nucleotidase		<1
Acid phosphatase		<1
Peptides		
GSH (3–5 mM)		>1
GSSG (0–5 mM)		>1
Cystyl glycine		>1
Amino acids ($>$ 1)		
Glutamic acid (0.8–2.5 mM)		>1
Aspartic acid (0.4–1.1 mM)		>1
Glycine (0.6–2.6 mM)		>1
Vitamins		
25-Hydroxyvitamin D, folic acid		
1,25-Dihydroxyvitamin D, pyridoxine		
Transcobalamin		
Steroid and hormones		
Estrogens, thyroxine, prolactin, insulin, and others		
Miscellaneous		
Protoporphyrin I and III		
Coproporphyrin		
CEA		

[a]Values were obtained from measurements of human, rat and rabbit hepatic bile.

Table II. Functions of Bile

Major excretory mechanism for lipid-soluble organic solutes, e.g., bilirubin, bile acids, and drugs with $MW > 300$–500
Major excretory mechanism for cholesterol
Digestion of dietary fats
Part of the enterohepatic circulation
Excretion of intestinal IgA

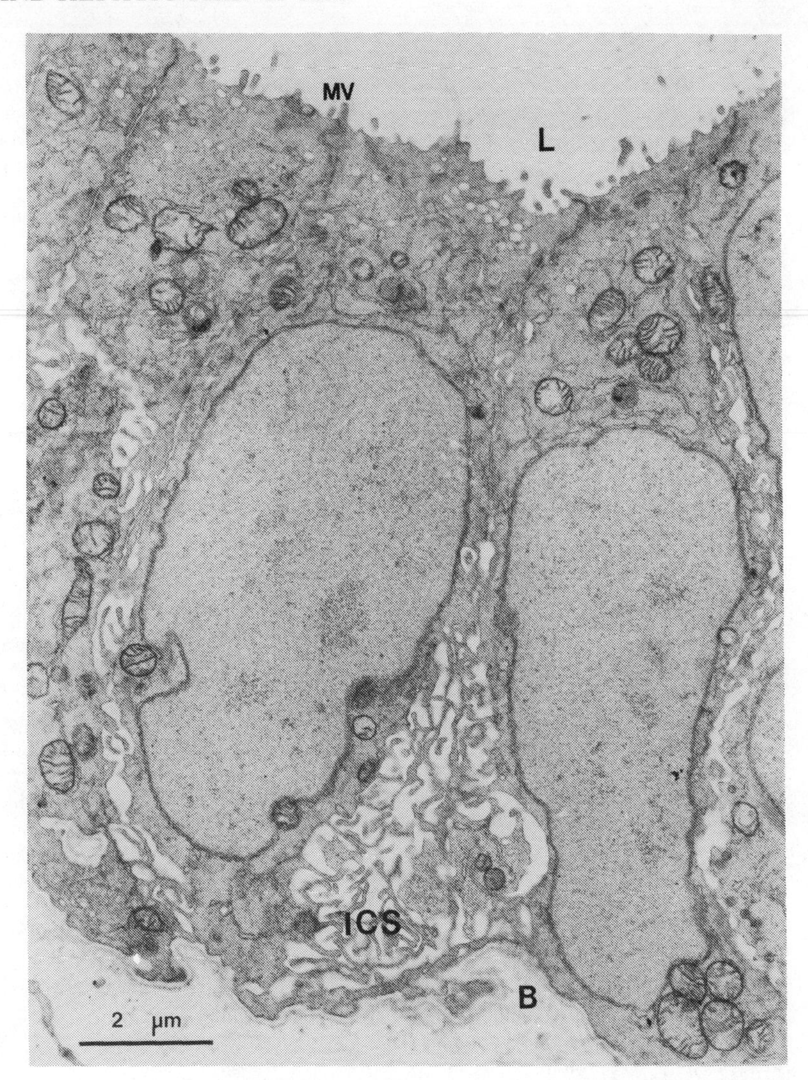

Fig. 2. Bile duct columnar epithelium (rat) illustrating the duct lumen (L) lined by microvilli (MV), and an expanded intercellular space (ICS). Reproduced with permission of the *American Journal of Physiology* **242:**G52 (1982).

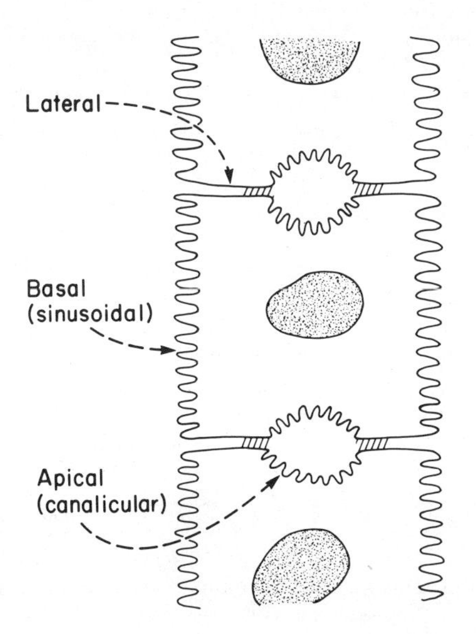

Fig. 3. Surface membrane domains of the liver.

defined structural and functional polarity along its sinusoidal, lateral, and canalicular surface membranes (Fig. 3). Portal blood perfuses the hepatic syncytium through sinusoids lined by highly fenestrated endothelial cells that permit plasma proteins to bathe the sinusoidal membranes of hepatocytes without diffusion across a capillary membrane. The sinusoidal or "basal" surface of the hepatocyte is covered with microvilli that increase the surface membrane area thereby facilitating hepatic uptake transport processes. Each hepatocyte is joined to adjacent liver cells at the "apical" or canalicular pole of the cell. The canalicular membrane is demarcated by two beltlike tight junctions that encircle each hepatocyte and define the lumen of the bile canaliculus. The canalicular membrane consists of only 13% of the plasma membrane surface of the hepatocyte[18] but this is the only portion of the cell surface across which bile excretory and secretory processes are directed. The canalicular and sinusoidal membrane are separated by a smooth surface lateral cell membrane that lines the intercellular space. Each of these plasma membrane domains is specialized for different transport functions and can be easily identified in scanning electron micrographs of liver tissue (Fig. 4).

Cell fractionation and cytochemical, ultrastructural, and

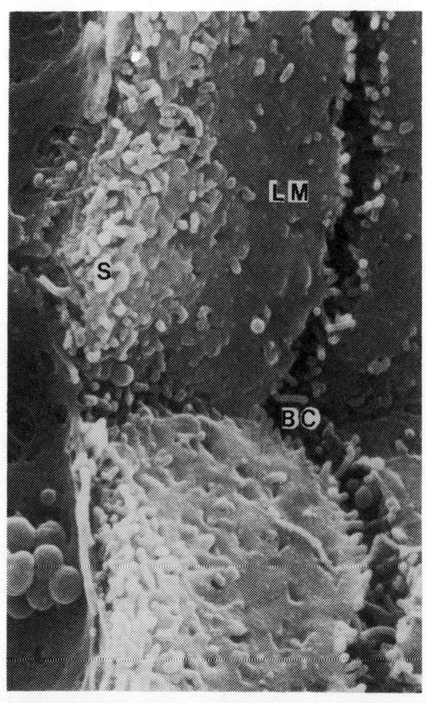

Fig. 4. Scanning electron micrograph of the surface domains of two adjacent hepatocytes. A blood sinusoid is to the left enclosed by a fenestrated endothelial lining. A sinusoidal membrane (S) containing many microvilli lies next to a smooth lateral cell membrane (LM) that merges with the membrane of the bile canaliculus (BC). Original magnification × 5000.

biochemical studies have localized distinct transport functions to the different plasma membrane domains.[19] Sinusoidal membranes are enriched in receptors for the uptake of proteins such as asialoglycoproteins, low-density lipoproteins, transferrin, and IgA. Hormone receptors also exist for insulin, glucagon, and epinephrine, and there are enzymes for transport functions such as hormone-sensitive adenylate cyclase and Na^+,K^+-ATPase. Carriers are also present within the sinusoidal membranes that facilitate the uptake of many organic anions including bile acids, bilirubin, and many exogenous drugs and xenobiotics. The lateral cell membranes are specialized for cell–cell communication and adhesion properties and contain the gap junctions, desmosomes, and tight junctions. The latter are well-defined structures, comprised of four to five parallel elements linked by cross strands. These junctions form the only structural barrier between the sinusoidal blood and bile and thus play an important role in determining the permeability properties of the canalicular system. Canalicular membranes are specialized for the excretion of bile and are associated with high specific activities of many transport enzymes including Mg^{2+}-ATPase, alkaline phosphodiesterase, leucylnaphthylamidase, and γ-glutamyl transpeptidase.[20–22] However, the functional relationship of these enzymes to bile formation is still not clear. The lipid composition of canalicular membranes is different from lateral and sinusoidal membranes in that the concentrations of sphingomyelin and ratios of cholesterol to phospholipids are both increased.[22] These modifications in lipid composition result in increased resistance to detergents and decreased membrane fluidity that could prevent the high concentrations of bile acids in bile from solubilizing the canalicular membrane. Sialic acid concentrations are also higher in this pole of the hepatocyte.[19,22]

Many intracellular structures are also specialized for bile excretory functions; endocytotic vesicles move rapidly across the cell to deliver proteins such as IgA to the canalicular lumen[23] and lysosomes fuse with the canalicular membrane and discharge enzymes and lysosomal debris.[24] The Golgi apparatus is also located at the canalicular pole of the hepatocyte. Although direct evidence for involvement of Golgi function in bile production is still lacking, the organelle increases in volume density and surface area following sustained bile acid stimulation of secretion.[25]

Cytoskeletal elements are other important structures of the hepatocytes that can also influence bile production.[26] Actin-containing microfilaments (60–70 Å) are particularly prominent within the pericanalicular area where they insert into the microvilli of the bile canaliculus and into the cytoplasmic surface at regions of the tight junctions and desmosomes. Intermediate filaments (100 Å) and thick filaments or microtubules (> 200 Å) can also be identified although they are more difficult to discern in routine electron micrographs. Microfilament poisons (phalloidin, cytochalasins) result in the loss of canalicular microvilli, an increase in permeability of the tight junctions, and a diminution of bile secretion.[27] Microtubule poisons (colchicine, vinblastin) result in a randomly positioned Golgi apparatus,[28] and impairment of bile acid-stimulated bile flow, although basal bile secretion remains unchanged.[29] The canalicular membrane appears to undergo periodic segmental contractions that are dependent on the functions of these microfilaments and other contractile proteins in the cell, a process that could facilitate the flow of bile through the irregular meshwork of canaliculi.[30,31]

2.2. Tight Junctions

The tight junction or zonulae occludens between adjacent hepatocytes are strategically located structures that not only define the lumen of the bile canaliculus but represent the only barrier to diffusion between the sinusoidal and intercellular spaces, and the bile. Thus, they appear to be major determinants of the permeability of the bile canaliculus.[32] The tight junction can be best viewed by freeze–fracture replicas (Fig. 5) and consists of four to five parallel strands that line the lateral borders of the bile canaliculi. Considerable heterogeneity exists along the junction and strand number may vary from 1 to 8, suggesting that this permeability barrier may also be highly variable.[33] Gaps in the junctional strands appear to be artifacts of the fixation and fracture process.[34] The chemistry of these structures is still unknown but there is evidence that they may be negatively charged.[35] The junction is a dynamic structure, and is quite sensitive to cell injury particularly during cholestatic liver injury where the permeability of the paracellular pathway between the intercellular space and the bile may be increased.[32] However, under physiological conditions, the junction is impermeable to solutes as large as proteins although water and small ions may diffuse across this barrier. Whether molecules such as bile acids diffuse back across this barrier during the normal process of bile production is not known.

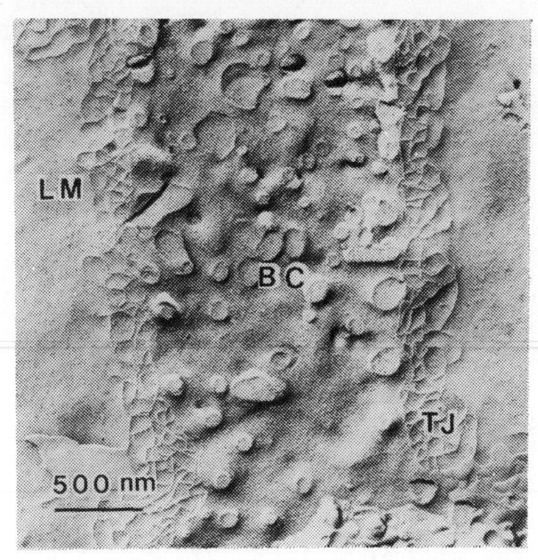

Fig. 5. Freeze–fracture replica of the tight junction of a rat hepatocyte. Magnification × 23,325. BC, bile canaliculus (TJ, tight junction); LM, lateral cell membrane. Reproduced with permission of *Hepatology* **3:** 614 (1983).

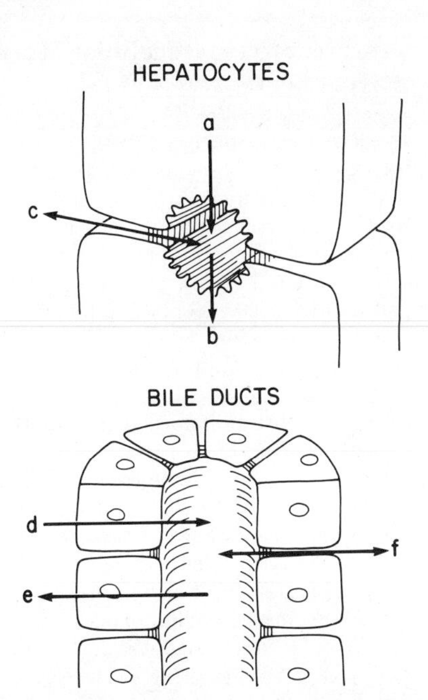

Fig. 7. Diagram of sites of water and solute movement across membrane domains of the hepatocyte and the bile ducts. Arrows represent potential sites of absorptive or secretory transport mechanisms.

2.3. Bile Ducts

Bile flows from central to portal zones of the lobule counter to the flow of blood through the liver; canaliculi from several adjacent cells appear to join together and empty into a common cistern called the canal of Hering (Fig. 6). Beginning with small interlobular ducts, bile then traverses a series of epithelial cell lined structures and ultimately empties into a common extrahepatic bile duct where bile flows to the gallbladder and intestine. Smaller ducts are lined by a cuboidal epithelium but this changes to columnar cells as the size of the ducts enlarge. Bile duct epithelium may modify canalicular bile further through both secretory and absorptive transport processes (Fig. 7).[15]

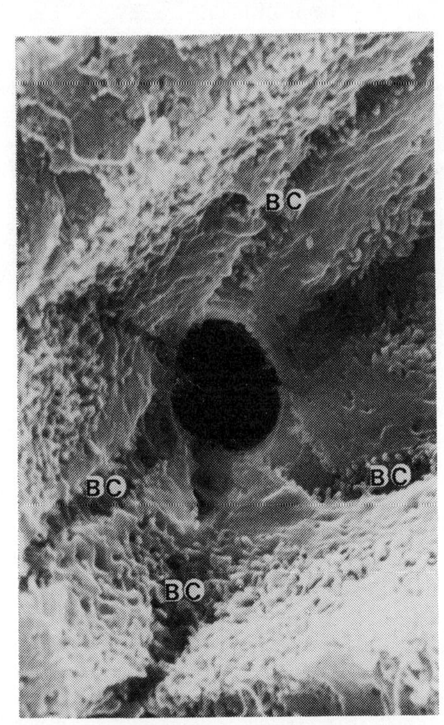

Fig. 6. Bile canaliculi (BC) from several hepatocytes merge together to drain into a conduit called the canal of Hering near the portal triad of the liver. From there bile flows into the bile ducts. Original magnification × 5000. Reproduced with permission of *Gastroenterology* **69:**724 (1975).

2.4. Lobular Gradient

Each hepatocyte exists within an environment defined by a microcirculatory unit or acinus divided into zones I, II, and III (the lobular gradient).[36] Periportal hepatocytes are confined to zone I in the periphery of the acinus whereas pericentral cells are localized to zone III around the terminal central vein. The midzone, II, lies in between. Portal blood perfusing the liver first contacts cells within zone I where surface volume relationships optimize interactions with the sinusoidal membranes.[37] Solutes that undergo a first-pass clearance by the liver will be extracted by the time blood reaches the central zone of the lobule, creating a gradient within the lobule for these solutes. Diffusible substances such as oxygen also are affected by this lobular gradient. These anatomically determined lobule gradients have important implications for bile secretory function since the transport of bile acids occurs predominantly at periportal cells resulting in secretory processes that are relatively bile acid independent in pericentral portions of the lobule.[38] There is considerable lobular heterogeneity for a variety of other metabolic events in the liver as well[39,40] (see reviews, refs. 39 and 40).

During feeding and fasting, the enterohepatic circulation of bile acids and other solutes waxes and wanes so that there is a rise and fall in the concentration of these substances in portal blood and a corresponding recruitment and derecruitment of hepatocytes for bile acid transport up and down the lobule.

3. Mechanisms of Hepatocellular Water and Electrolyte Secretion

Since the classic experiments of Brauer and co-workers in the isolated perfused rat liver established that the secretory pressure of bile exceeds the hydrostatic perfusion pressure of portal blood, it has generally been accepted that bile is formed by energy-dependent active secretory processes and not by filtration as with urine.[41] This has been confirmed by the demonstration of temperature dependence for bile secretion and the cessation of secretion following administration of metabolic inhibitors such as KCN, or dinitrophenol.[42] Net movement of water from liver to canaliculus could theoretically occur through active transport, exocytosis of intracellular membrane vesicles, or passive movement of water in response to the generation of osmotic gradients either within the bile canaliculus or smaller intracellular vesicles that fuse with the canalicular membrane. However, active movement of water has never been demonstrated in any transport system, and exocytosis, although it does occur at the canaliculus, is probably not a major source of water movement, so that the major stimulus to bile formation must be related to the establishment of osmotic gradients within a small confined space such as the canaliculus or vesicular precursor. The final driving force for the generation of fluid secretion is the sum total of the effective osmotic solutes in bile.

As pointed out by Moore and Dietschy,[43] inorganic electrolytes account for the predominant osmotic activity in bile. Although the concentration of total solutes in bile is in excess of the plasma, bile is actually an isosmotic solution because certain constituents are associated either in mixed lipid micelles or in self aggregates, thereby diminishing their osmotic activity.[44] To generate secretion, solutes must be transported into the bile canaliculus and produce an effective osmotic force so that water and other solutes can move passively into the canaliculus to maintain osmotic equilibration between the bile and blood.[45,46] Thus, the osmolarity of canalicular bile always equals the osmolarity of the portal blood.

By confining biliary solutes within a small closed space (the canaliculus) through the sealing properties of the tight junctions, a hydrostatic secretory pressure is generated which then promotes the forward movement of bile. Delineation of the physical and chemical forces that move solutes into bile is an ongoing area of current research interest that is attempting to define the solute pumps, carriers, and channels that are responsible for bile formation.

3.1. Sites for Bile Formation (Hepatocyte and Bile Ductules)

Estimates of the quantity of bile produced by the hepatocytes (canalicular bile) rely on indirect technqiues that are dependent on the biliary clearance of inert nontransportable solutes such as mannitol or erythritol. Ideally, these solutes should enter bile at the hepatocyte either across or between the cells but be excluded from movement into bile at more distal sites in the ducts of the biliary tree. Although the precise site of entry of these solutes has not been accurately determined, a number of studies indicate that these solutes and others such as inulin, sucrose, and ferricyanide enter bile by a combination of diffusion and solvent drag (bulk water flow) through pore- and charge-restricted channels and that their clearance varies directly with changes in canalicular secretion as modified by choleretics, such as the bile acids.[47,48] In contrast, clearance of these solutes is usually not altered when ductular secretion is increased following administration of hormones such as secretin that act primarily at these distal sites (ductular secretion) (see Section 11).

In general, these markers of canalicular flow have been validated experimentally in studies from a variety of animal species including dog,[47,48] guinea pig,[49], monkey,[50] baboon,[51] and man.[17] In species such as the rat, ductular secretion cannot be stimulated by secretin and the clearance of erythritol or mannitol is similar to the rates of total bile flow, suggesting that bile production is either entirely canalicular in origin or that there is a constant fractional addition or subtraction of erythritol-containing fluid at the level of the bile ducts.[2] Based on these assumptions, estimates of canalicular and ductular components of bile secretion have been made in man (and several animal species) usually after an overnight fast when meal-related stimuli to bile secretion are minimized. The components of bile flow are easily and variably modifed after eating as a result of gallbladder contractions and subsequent augmentation of portal vein bile acid concentrations as well as the release of gastrointestinal hormones such as secretin. Estimates of the components of bile flow under basal conditions are diagrammed in Fig. 8, and summarized for several species in Table III.

Despite the common use of these putative markers of canalicular flow, they are undoubtedly subject to a variable degree of inaccuracy, depending on the species. In the dog, secretin stimulates erythritol or mannitol clearance at unidentified

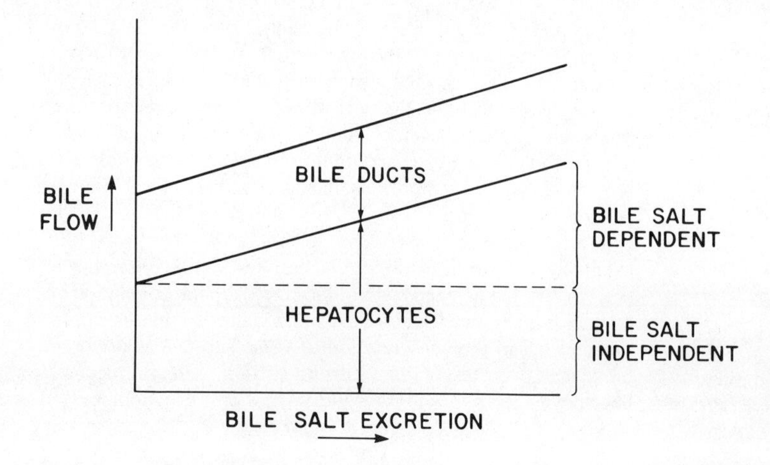

Fig. 8. Schematic representation of components of bile flow. Extrapolation of the regression line of bile salt excretion results in the estimated value for bile acid-independent secretion from the hepatocyte.

Table III. Components of Bile Flow[a]

	Man	Dog	Rhesus monkey	Rat	Rabbit
Total flow (μl/min per kg)	5	10	15	70	90
Bile salt-stimulated flow (μl/μmole bile salt)	7	8	16	15	70
Bile salt-independent flow (μl/min per kg b.w.)	2	5	7	40	60
Duct flow (μl/min per kg b.w.)	2	—[b]	—[b]	NS[c]	NS

[a]Representative values from the literature.
[b]Ducts reabsorb canalicular flow following cholecystectomy and cannula insertion.
[c]NS, not significant.

sites distal to the bile canaliculus.[52–54] In the rat, erythritol, sucrose, and inulin diffuse from plasma to bile across isolated perfused segments of hepatic ducts in proportion to their molecular size.[55] Furthermore, these markers of water movement may traverse the hepatocyte, by way of either transcellular or paracellular pathways or some combination of both. To date, no marker has been identified that convincingly distinguishes between these three possibilities. The observation that inulin and sucrose equilibrate between bile and blood before reaching diffusion equilibrium between blood and liver water has been cited as evidence of a paracellular route[56] for these solutes. However, other studies demonstrate rapid equilibration of inulin and sucrose in intracellular intravesicular compartments suggesting that they may mark a rapidly communicating but small vesicular transcellular shunt pathway.[10,57] Polyethylene glycol-900 and orthophosphate[58,59] which seems to be excluded from hepatocyte water might provide more accurate measurements in rats of paracellular fluid movement at the canalicular level but this possibility remains to be confirmed.

The isolated perfused rat liver has provided a convenient model to help separate canalicular and bile duct transport properties since the liver is perfused only by way of the portal vein. Thus, the bile ducts, which are perfused by the hepatic artery, should be in a relatively functionless state. Graf and Peterlik have confirmed this impression by showing that $^{25}Na^+$, $^{36}Cl^-$, and an organic dye (sulfobromophthalein) are excreted into bile and appear simultaneously in the biliary cannula after the estimated volume of bile in the biliary tree is washed out.[60] Figure 7 represents a schematic of the possible sites for addition and subtraction of water and inorganic solutes to bile.

3.2. Bile Salt-Dependent and Independent Secretion

Almost 100 years ago, Schiff demonstrated that bile was choleretic when refed to dogs with biliary fistulae.[61] Since that time, bile acids, the most plentiful organic solutes in bile, have been viewed as the principal determinant of bile secretion. Sperber observed that other organic solutes were also choleretic when concentrated in bile presumably because of the creation of favorable osmotic gradients.[45,46]

The relationship between the rate of bile acid excretion and canalicular bile production is conventionally defined as "bile acid-dependent flow" (BADF). By extrapolating a regression line relating bile flow (y) and bile acid excretion rates (x), to the ordinate, a positive intercept is obtained that is defined as "bile acid-independent bile flow" (BAIF). These relationships are diagrammed in Fig. 8. However, wide species differences exist with respect to the volume of water obligated by a given rate of bile acid excretion, (e.g., the slope of the regression line). For example, 7 ml of bile is excreted per mmole bile acid in man compared to 70 ml in the rabbit. In addition, the magnitude of the bile acid-independent fraction also varies greatly between species[62] (Table III).

Because the osmotic activity of a bile acid (the critical micellar concentration) often cannot account for its choleretic properties, bile acids may have other effects on secretion that are related to their specific structures. For example, the conjugated triketo analog of cholic acid, taurodehydrocholic acid, is twice as potent a choleretic in dogs, rats, and man compared with the trihydroxy bile acid, taurocholic acid.[63–65] In contrast, the unconjugated cholic acid stimulates more flow than its taurine conjugate, taurocholic acid, in both dogs and cats.[65,66] Klaassen found that the choleretic potential of bile acids in dogs was deoxycholate, > cholate > chenodeoxycholate and that conjugation of these bile acids diminished this potential.[63] In a preliminary study, O'Maille *et al.* found that side chain charge was a significant determinant of transport and that only negatively charged bile acids or analogs were excreted into bile.[67] Monohydroxybile acids are usually cholestatic.[68,69] In another study in the taurine-depleted dog, cholic acid infusions actually stimulated more flow than taurodehydrocholate despite equal bile salt secretion rates and the inability of the latter to form micelles.[65] Finally, the bile acid ursodeoxycholic acid is also a potent choleretic and stimulates biliary bicarbonate secretion.[70,71] Although the osmotic effects of bile acids (theoretically small compared to the contribution of inorganic electrolytes in bile) are probably predominant at lower rates of bile acid excretion, factors other than osmotic activity must influence the secretory potential of many bile acids.

Bile acids can also alter the specific activity of important membrane transport enzymes such as Na^+,K^+-ATPase, both *in vitro* and *in vivo*.[72–74a] In addition, bile acids may modify membrane function and alter biliary permeability. Acute infusions of taurocholate in the rhesus monkey result in augmentation of the biliary permeability to inulin and a rise in volume of secretion per mole of excreted bile acid.[73] Chronic infusion of taurocholate in rats with partially obstructed bile ducts also leads to an increase in the volume of water secreted per mole of bile acid,[75] a phenomenon that has also been observed in postcholeretic bile following release of complete bile duct obstruction,[76] and in the chronic bile duct–vena cava shunt rat.[77] Biliary clearances of [^{3}H] inulin are also elevated following relief of bile duct obstruction in the rat reflecting an increase in biliary permeability at an unidentified site.[76] Thus, a clear separation between effects on bile acid-dependent and -independent secretion may often be difficult to make. Therefore the choleretic effect of a given bile acid (or other organic solute) would seem to depend on at least four factors[66,67]: (1) the osmotic effect of the secreted solute; (2) its potential for stimulating or inhibiting other canalicular secretory pumps; (3) the permeability characteristics of the canalicular membrane and/or the tight junction sealing the lumen of the canaliculus; and (4) the effect of the solute on secretory and absorptive processes in bile duct epithelium.

Since the primary driving forces for bile formation relate to other transport processes as well, including the active secretion

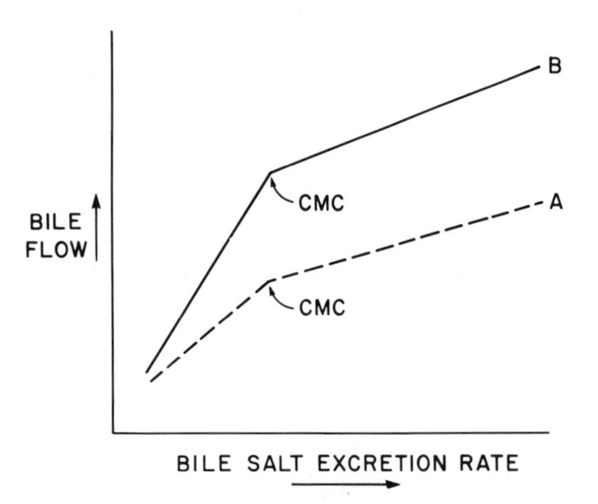

Fig. 9. Relationship between bile salt excretion and bile flow. The traditional regression line is illustrated in Fig. 8. In this figure, line A and B illustrate that bile salts exert different osmotic activities above and below their critical micellar concentrations (CMC). In addition, some bile salts stimulate more flow (line B) than others (line A) even though excreted at the same rate.

of certain inorganic solutes, BADF and BAIF, rather than independent components, are probably closely interrelated by the ability of bile acids to modify BAIF. BADF might therefore be divided into two components: (1) that component of secretion generated solely by the osmotic activity of the transported bile acid anion—*bile acid-dependent osmotic flow,* and (2) that component of secretion indirectly stimulated or inhibited by the bile acid's effects on other membrane pumps and functions—*bile acid-regulated flow.*

Although it is conventional to define bile acid-dependent and -independent bile secretory components in animals by linear regression analysis of flow vs. bile acid excretion, it should be clear from the preceding considerations that this formulation represents an oversimplification that is not entirely justified. Not only must the regression line relating bile flow and bile acid excretion be extrapolated to the ordinate to represent the proportion of flow that would be present were bile acid secretion to cease altogether (a theoretical consideration that is impossible to reproduce experimentally because some endogenous bile acid synthesis and excretion always exists), but as emphasized from studies in rat and rhesus monkey,[73,78] bile flow varies in a curvilinear fashion at low rates of bile acid excretion. This phenomenon is observed probably because the concentration of canalicular bile acids falls below critical micellar concentrations and these bile acids are no longer aggregated in micelles. Such theoretical considerations predict that a given increment of bile acids will stimulate larger volumes of secretion at concentrations, below, compared to above, their critical micellar concentration, as illustrated in Fig. 9. Although linear regression lines will not accurately define the magnitude of BAIF, such representations may still be useful to compare the effects of stimulators or inhibitors of bile secretion since they distinguish between primary effects on bile acid transport and excretion and other factors that determine the volume of fluid accompanying a given unit of excreted bile acid.[79–81]

3.3. Bile Acid Transport

Bile acids undergo an enterohepatic circulation.[13,13a] After absorption from the intestines, all bile acids are transported to the liver in portal blood and enter the hepatic sinusoids. The hepatic transport of these major biliary organic solutes can be considered in three steps: (1) hepatic uptake, (2) intrahepatic binding and transcellular transport, and (3) canalicular excretion. Hepatic uptake is dependent on hepatic blood flow which increases during meals and reduces the fractional clearance of bile acid by the liver. Yet the v_{max} is so large that the net extraction is minimally reduced and results in only a small increment in postprandial peripheral blood bile acid levels.[82,83] Portal vein concentrations of bile acids range from 12.9 ± 1.5 to 22.2 ± 5.1 μM in fasting blood from anesthetized patients[82–85] and from 50 to 170 μM in the rat.[86,87] Most portal vein bile acids are in conjugated form. Based on a portal flow of 1000 ml/min, in the adult human 1200 μmoles of bile acid will be transported by the liver each hour during fasting, a rate slightly higher than the calculated rate of bile acid excretion during an overnight fast (300–900 μmoles/hr).[88,89]

Bile acids are bound to both albumin[90] and high-density lipoproteins; smaller amounts bind to low-density lipoproteins (< 15%). Unconjugated bile acids bind more tightly and binding is greatest for monohydroxybile acids (lithocholate), less for dihydroxybile acids (chenodeoxycholic acid and deoxycholic acid), and least for trihydroxybile acids (cholic acid).[91] These properties may account for a first-pass extraction efficiency for cholates that is more than 90% compared to a lower extraction efficiency for chenodeoxycholate and deoxycholate (75–80%).[13,88] In general, the conjugated bile acid is extracted with higher efficiency than its unconjugated form and trihydroxybile acids are extracted more efficiently than dihydroxybile acids.[91,91a,92] The role of high-density lipoproteins as bile acid transporters in blood is not clearly defined but is greater for the more hydrophobic bile acids which tend to associate more with the lipids than the lipoprotein.[93] The binding of bile acids to serum constituents not only facilitates their transport to the liver but diminishes their free concentrations at the hepatocyte plasma membrane. Recent evidence suggests that albumin-bile acid complexes may interact directly with the membrane carrier and facilitate the uptake of taurocholate.[94] Thus, uptake rates may not be driven solely by the free concentration of the bile acid.

The hepatic uptake of bile acids is also greatly influenced by the lobular gradient.[95] Autoradiographic studies with [^{3}H]taurocholate[96] and bile acid analogs (cholylhistylglycine[97] and tyrosine-labeled cholic acid[98]) indicate that the majority of these anions are removed by zone I hepatocytes in the periportal regions of a lobule. Pericentral cells also have the capacity to transport bile acids as demonstrated by retrograde injections[96,97] as well as cell elutriation techniques and by the maintenance of bile acid excretion during selective injury to periportal cells with the hepatic toxin, allyl alcohol.[38] Morphometric studies also demonstrate that although the canalicular diameters in zone III are normally smaller than those in zone I, the former selectively increase in size during bile acid infusions,[95,97] suggesting that pericentral hepatocytes can be recruited for bile acid transport as the concentrations of sinusoidal bile acids increase. These studies emphasize that the hepatic lobule is heterogeneous with respect to bile secretory processes, being relatively bile acid dependent in periportal hepatocytes and independent of bile acids in central zone regions. A lobular gradient for bile acid transport and bile secretion may have important implications for both normal physiology as well as the pathogenesis of bile secretory failure (cholestasis). Cholestasis is often accompanied by pericentral canalicular bile plugs, a light microscopic manifestation of this lobular heterogeneity in cell function.[99]

3.4. Hepatic Uptake of Bile Acids

The hepatic uptake of bile acids has been extensively studied and is characterized by saturation and competition with other bile acids and certain organic anions, suggesting a carrier-mediated transport process. Whether examined in the isolated perfused liver or isolated hepatocytes or membrane vesicles in the rat or by indicator dilution techniques in the dog, the maximal rate of bile acid uptake (V_{max}) always greatly exceeds the secretory maxima for bile acid excretion in bile and thus is not the rate-limiting step in the transport of bile acids from blood to bile.[100–106] Furthermore, the K_m for taurocholate uptake as estimated in isolated rat hepatocytes and in a membrane vesicle system (20–50 μM) usually exceeds the physiological concentration of this bile acid in portal blood.

Based on fasting concentrations of bile acids in portal blood, sinusoidal carriers must transport bile acids against 10- to 20-fold higher intracellular concentrations.

Photoaffinity derivatives of taurocholate bind to membrane proteins with molecular weights of 54,000 and should lead to isolation, purification, and structural determination of this transport protein.[107,107a]

Numerous studies in isolated perfused rat liver, skate liver, hepatocyte suspensions and monolayers and membrane vesicles indicate that the uptake of taurocholate is dependent on the presence of Na^+ at physiological concentrations[102–107] Na^+-independent uptake (diffusion) can also be demonstrated, particularly at concentrations of the bile acid that exceed the K_m[104–110] Uptake of taurocholate is inhibited by ouabain, or substitution of Na^+ by Li^+ and other cations, and is thus dependent on the Na^+ gradient created by the activity of hepatocyte plasma membrane Na^+,K^+-ATPase. This process is characteristic of a Na^+-coupled secondary active cotransport system where the energy for "uphill" transport is provided by the enzymatic hydrolysis of ATP. Taurocholate uptake is also inhibited by bumetanide and furosemide, known inhibitors of several Na^+-coupled ion transport processes in epithelia and red cells.[105] Taurocholate uptake in cultured rat hepatocytes is accompanied by equimolar amounts of $^{23}Na^+$ suggesting that one Na^+ ion is transported with each taurocholate anion.[106] Na^+ is not required for binding of taurocholate to the membrane carriers but maximal rates of uptake occur at about 50 μM Na^+.[108,109] Whether or not the cotransport system is facilitated by the negative intracellular potential is controversial.[108,109,111] The sinusoidal membrane transport carrier appears to be shared by other bile acids, including cholate and glycocholate although only taurocholate demonstrates strict Na^+ dependency.[110] Other bile acids, cholic acid, chenodeoxycholic acid, and deoxycholic acid, and their conjugates are concentrated in hepatocytes primarily by non-Na^+-dependent carriers.[110]

Intracellular concentrations of bile acids are estimated at 100–300 μM,[112] but if bile acids are compartmentalized, local intracellular concentrations could be considerably higher. Binding to proteins would substantially reduce their effective concentration but accurate measurements of chemical activity must await development of a bile acid electrode.

3.5. Intracellular Bile Acid Transport

It is still unclear how bile acids and other solutes that are destined for excretion in bile move across the cell. Autoradiographic studies initially localize bile acid analogs to the smooth endoplasmic reticulum, followed by translocation to Golgi vesicles and ultimately the bile canaliculus,[98] but the specificity of these morphological studies has been questioned. Other morphometric ultrastructural studies of rat hepatocytes following bile acid-induced choleresis demonstrate an increase in number and volume of 100-nm or larger vesicles and vacuoles in pericanalicular regions suggesting that bile acids may stimulate a transcellular vesicle pathway.[25,113] Both colchicine and phalloidin, inhibitors of microtubules and microfilaments, respectively, diminish bile acid transport in the rat.[114] The finding that taurocholate stimulates phospholipid synthesis in smooth endoplasmic reticulum and can produce micelles of similar size to biliary mixed lipid micelles when exposed *in vitro* or *in vivo* to both smooth endoplasmic reticulum and Golgi membrane provides further indirect support suggesting that bile acids participate in some form of membrane flow that is vectorially directed to the bile canaliculus.[115] Bile acids also bind avidly to soluble cytoplasmic proteins in 100,000*g* supernatants derived from rat liver homogenates and two have been characterized as glutathione S-transferases.[116,117] Glutathione S-transferase B constitutes 3–5% of cell supernatant protein and catalyzes glutathione–ligand conjugates.[118] The protein also has important binding properties for other organic anions such as bilirubin and sulfobromophthalein. Two other major cytosolic bile acid-binding proteins have also recently been identified.[119] Bile acids may bind to these intracellular proteins prior to transfer to specific carriers within other intracellular or canalicular membranes. Bile acid "receptors" have been identified in rat liver plasma membrane subfractions[120] and a functional role is suggested by finding parallel changes between the number of membrane "receptors" and the transport secretory maximum for taurocholate in rats induced by bile acid feeding or diminished by inhibitors of protein synthesis.[121] These receptors are found only in tissues that transport bile acids (liver, ileum, and kidney) and are specific for bile acids as opposed to organic anions such as sulfobromophthalein. The membrane domain of these bile acid-binding proteins has not been defined.

3.6. Canalicular Bile Acid Excretion

Excretion of bile acids across the canalicular membrane is the rate-limiting step in the transport of bile acids from blood to bile. The driving forces for excretion are not yet clearly defined but the hepatocyte must transport bile acids from plasma to bile against concentration gradients that are at least 1 : 100 and perhaps as much as 1 : 1000. Information is needed regarding intracellular localization and the extent of protein binding of bile acids before a clearer picture can emerge. However, bile acids are anions and therefore may be driven out of the cell by the resting membrane potential that measures −30 to −40 mV across the sinusoidal membrane of the hepatocyte.[122,123] The negative intracellular potential is maintained by Na^+,K^+-ATPase through the extrusion of 3 Na^+ from the cell in exchange for 2 K^+ ions and also by the diffusion of K^+ back out of the cell. Direct measurements of canalicular potentials in isolated rat hepatocyte couplets give values of approximately −5 mV, so that the potential across the canalicular membrane may range from −25 to −35 mV.[124] Thus, according to the Nernst equation, monovalent anions could accumulate within the canaliculus by potential-driven forces at concentrations three- to fourfold higher than within the cell.[4,7] Recent studies in purified canalicular membrane vesicles confirm the existence of a conductive pathway for bile acid excretion.[111,125] Further accumulation of bile acids in bile could occur passively by solubilization within the biliary mixed lipid micelle. According to this "micellar sink" hypothesis, the free concentration of bile acids should be maintained at a lower intermicellar concentration of 3–5 mM as the bile acids diffuse into the lipid core of the micelle.[126] However, the intracellular chemical activity of bile acids should also be reduced by protein binding, thereby increasing their cell-bile concentration gradients so that it seems unlikely that the transmembrane potential and the micellar sink are the only driving forces for bile acid excretion. Rather, additional mechanisms must be postulated to account for their accumulation in bile such as canalicular membrane carrier-mediated anion transport or vesicle-mediated fusion and exocytosis.

4. Other Primary Driving Forces for Canalicular Bile Secretion (Bile Acid-Independent Secretion)

Although bile acids are important regulators of bile secretion, their contribution to the overall osmotic activity in bile is small compared to the ion species, particularly Na^+. Thus, other primary driving forces for biliary secretion have been postulated such as hydrostatic filtration, bile salt-independent determinants of osmotic filtration and fluid exocytosis.

4.1. Hydrostatic Filtration

Once a critical vascular opening pressure is exceeded, and oxygen delivery and regional perfusion of the liver is optimized, hydrostatic perfusion pressure has no further effect on total bile production[41] or on bile acid-independent components of secretion in the isolated perfused rat liver.[127] However, at low perfusion flow rates, bile volume diminishes if peripheral regions of the hepatic lobes are underperfused.[127] Hydrostatic pressure is a determinant of bile formation only in marine elasmobranches where the studies in the isolated perfused skate liver demonstrate a small component that is sensitive to perfusion pressure.[128,129] Bile flow rates are 100 times slower than in the rat in this species so that small effects of perfusion pressure may be more apparent. In mammals, hydrostatic filtration does not appear to be a significant driving force for the production of bile.

4.2. Na^+,K^+-ATPase-Generated Driving Forces for Osmotic Filtration

Bile acid-independent bile flow was originally thought to be mediated by the activity of a canalicular Na^+,K^+-ATPase and the pumping of Na^+ ions into the canalicular lumen.[74,75,79–81,130] However, parallel changes in liver plasma membrane Na^+,K^+-ATPase and BAIF are not always observed.[9,131–133] More importantly, Na^+,K^+-ATPase was subsequently localized by cytochemical techniques to the sinusoidal and lateral surfaces of the hepatocyte, rather than the bile canaliculus,[134,135] findings that have been confirmed biochemically with more highly purified fractions of canalicular and sinusoidal membrane.[20,22,136,137] These observations led to a reinterpretation of the role of Na^+,K^+-ATPase in bile formation that is similar to most secreting and absorbing epithelia.[4] The primary role for Na^+,K^+-ATPase in bile formation is now believed to be related to the maintenance of Na^+ and K^+ gradients across the basolateral cell membrane, and to the generation of a negative intracellular electrical potential.[138] These ion gradients and electrical potentials can then be utilized as driving forces for the transmembrane movement of other ions as illustrated for Na^+-coupled bile acid uptake (secondary active transport) and for the potential-driven extrusion of bile acids across the canalicular membrane. Other ion pumps (Na^+/H^+ antiport and H^+-ATPases) might also contribute to these chemical and electrical gradients but present evidence suggests a major role for Na^+,K^+-ATPase. Substitution of Li^+ for Na^+ in perfused rat liver preparations and inhibition of Na^+,K^+-ATPase by ouabain or by hypothermia substantially but not completely diminishes the electrical potential[139–141] and the ability of the liver cell to maintain its cell volume.[142]

Thus, the liver cell membrane potential appears to be primarily related to the transmembrane distribution of Na^+ and K^+. Readmission of K^+ into guinea pig or mouse liver after equilibration in a K^+-free Na^+ medium results in hyperpolarization of the membrane with an increase in sensitivity to cardiac glycosides, and an increased efflux of Cl^-[141–144] consistent with a potential-driven passive movement of this anion. To date, it is not known whether the Na^+ pump-generated ion gradients and electrical potential can drive other ions or solutes important to the generation of bile secretion, although there is clear evidence that neutral amino acids such as alanine can be transported into hepatocytes by an electrogenic Na^+-coupled carrier.[145]

Na^+-coupled Cl^- transport is a prominent secretory mechanism in tissues such as the avian salt gland and rectal gland of the dogfish.[146–148] In these tissues, the Na^+ gradient is utilized to transport Cl^- into the cell and the membrane potential provides the driving force for extrusion of Cl^- across the selectively permeable apical membrane thereby generating secretion.[149] Na^+ follows by predominantly paracellular routes to maintain electrical neutrality. However, in the isolated perfused rat liver, substitution for Cl^- with other anions does not consistently diminish bile production,[60,150–152] nor do similar substitutions reduce the rate of $^{24}Na^+$ entry into hepatocytes.[106] Furthermore, intracellular hepatocyte Cl^- concentrations are close to their predicted electrochemical equilibrium, suggesting passive distribution of Cl^- under steady-state conditions.[140] Although an increase in bile flow and biliary Cl^- efflux occurs when the liver cell membrane is hyperpolarized in K^+ readmission experiments,[144] a finding consistent with passive distribution according to the transmembrane potential, present evidence suggests that Na^+-coupled Cl^- transport is not a major driving force in the formation of hepatic bile.

The Na^+ gradient may also drive Na^+/H^+ exchange (antiport) across the basolateral membrane.[152a,b,c] The concentration of H^+ in hepatocytes is below predicted equilibrium[153] and its extrusion would result in intracellular alkalinization. Hydroxyl anions and CO_2 would yield bicarbonate anions for subsequent excretion possibly by a chloride-bicarbonate anion-exchange mechanism[154] as demonstrated in rat liver canalicular membranes.[154a] Substitution for bicarbonate with Tris in isolated perfused liver systems results in reduction of bile acid-independent secretion,[155–157,161] but this effect might also be due to a reduced permeation from the substituted anion. A bicarbonate ATPase has not been definitively identified in liver plasma membrane fractions.

A Na^+/H^+ exchange mechanism could also drive OH^- anion self-exchange, but at present these speculative models await further experimental validation.

In summary, Na^+,K^+-ATPase has a primary regulatory role in the formation of bile through generation of Na^+ and K^+ transmembrane gradients that establish a high intracellular K^+ and low intracellular Na^+; the exchange of 3 Na^+ for 2 K^+ together with outward K^+ diffusion accounts for the generation of a negative intracellular potential that creates an additional important driving force, facilitating intracellular accumulation of positively charged solutes and the extrusion of negatively charged anions. Future studies in canalicular and sinusoidal membrane vesicles should define these secretory events more precisely.

4.3. Biliary Permeability to Cations and Anions

According to ion substitution experiments for Na^+ and Cl^-, the biliary system in the isolated perfused rat liver is more permeable to Li^+ and NO_3^- and less permeable to ions such as choline and isethionate.[60,150] Thus, a transient choleresis can often be demonstrated when perfusates are changed to the more permeant cations or anions whereas a transient reduction in flow is usually observed when the perfusate is changed to the less permeant ion. Similar observations have been made following bile acid infusions with perfusates of differing ion composition[151,152] and indicate that the permeability properties of the biliary membranes must influence the secretory response to osmotic choleretics. However, neither Na^+ nor Cl^- is specifically required to sustain bile formation and NO_3^- can substitute effectively for bicarbonate. Therefore, the specificity of ion pumps in the generation of bile secretion remains in doubt.[154]

4.4. Unidentified Anions

A significant cation–anion gap of 15–20 meq/liter is present in rat bile. Klos *et al.* have described a choleretic anionic substance in rat bile that is heat stable and less than 1000 daltons in mass, but the identify of this anion or anions is yet to be determined.[156] High biliary plasma ratios of aspartic acid and glutamic acid have been described as well as increased concentrations of unidentified ninhydrin-positive compounds in bile.[10] Glutathione concentrations of 3–4 mM have also been described in acidified bile.[157] Graf has proposed that these impermeant anions may account for much of the unidentified biliary anions and that their transport across the canalicular membrane may be driven by the negative intracellular potential.[10] Studies in liver canalicular membrane vesicles provide evidence for an electrogenic carrier-mediated transport system for glutathione.[158] These preliminary observations suggest that the canalicular excretion of impermeant unidentified anions, probably peptides related to glutamyl conjugates, might play an important role in the generation of BAIF.

4.5. Exocytosis

Although exocytotic vesicles mediate protein secretion into bile, this process does not contribute a quantitatively important volume of secretion to net flow, as judged from studies where colchicine has inhibited endocytotic vesicle movement to the canalicular membrane.[159,160]

4.6. Paracellular Shunt Pathway

Following formation of osmotic gradients produced by the active biliary transport of bile acids, ions, or other impermeant solutes, water and electrolytes enter the bile canaliculus to maintain isosmolarity and electrical neutrality. Osmotic equilibration most likely occurs at the level of the hepatocyte since bile remains isotonic with perfusates in the isolated perfused rat liver and bile ducts are not functional in this system[4] Furthermore, diffusion equilibrium should occur maximally across the bile canalicular lumen since surface/volume ratios are optimal there but diminish abruptly upon entry of bile into the biliary duct system. Therefore, the site of entry of water and electrolytes should occur across the cell at the canalicular luminal membrane and between the cells (paracellularly) by way of the tight junction. The tight junctions are normally impermeable to large molecules such as hemoglobin, horseradish peroxidase, and colloidal lanthanum, but may be leaky to smaller molecules and to ions and water.[32] Electrical measurements in hepatocyte couplets suggest that these junctions are low-resistance pathways.[124] During bile acid choleresis, ionic lanthanum, with a molecular radius similar to Na^+, readily penetrates the tight junction of the canaliculus from blood to bile.[161,162] Furthermore, "blistering" occurs in the intercellular space adjacent to the tight junction as observed by both scanning and transmission electron microscopy. The latter findings are analogous to morphological changes in the tight junctions of amphibian epithelia during osmotic induction of paracellular fluid movement, imposed by hypertonic solutions placed on the apical side of these epithelia.[163,164] Blistering in amphibian tissues is associated with an increase in ionic conductance and a decrease in electrical resistance and is thought to represent focal accumulation of fluid within the junction during paracellular water and ion flux.

Abrupt increases in plasma or perfusate osmolarity by additions of mannitol, xylose, glucose, or sucrose always result in immediate reductions in bile volume consistent with osmotic equilibration across the junctional complex rather than the more slowly equilibrating hepatocytes.[166,167] Solutes such as inulin and sucrose equilibrate more rapidly between blood and bile than between blood and liver cell water, although these findings could be explained equally well by a small intracellular vesicular shunt pathway between sinusoidal membranes and the bile canaliculus.[32] According to Bradley and Herz, the biliary entry of sucrose and ferricyanide is also regulated by a charge barrier.[35] These molecules are similar in molecular size but the clearance of negatively charged ferricyanide is reduced compared to sucrose. If the junctional barrier were negatively charged, anionic solutes such as bile acid monomers would be selectively retained within the canalicular lumen by the electrical properties of the junction in addition to its pore size. Finally, paracellular entry of $^{24}Na^+$, $^{36}Cl^-$, and $^{42}K^+$ has been demonstrated in isolated perfused rat livers by Graf and Peterlik.[60] According to the patterns of perfusate and bile washout curves from prelabeled livers, the fraction of ions entering bile from the hepatocyte amounted to only 4.9, 25.6, and 27.1% for Na^+, K^+, and Cl^-, respectively. If the assumption is made that little of the label that entered bile from the liver cell exchanged with unlabeled perfusate ions, the conclusion would be valid that the majority of these ions enter bile by paracellular routes.

Taken together, all of these observations suggest that biliary permeability and secretion may be regulated by the properties of the tight junctions between hepatocytes.

5. Model for Hepatocyte Water and Electrolyte Secretion

A model for the generation of hepatocyte water and electrolyte secretion has been proposed, based on experimental data and the anatomical arrangement of the hepatocytes.[4,7,10] As illustrated in Fig. 3, the hepatocyte is a polarized cell, where uptake of solutes occurs across sinusoidal membranes while

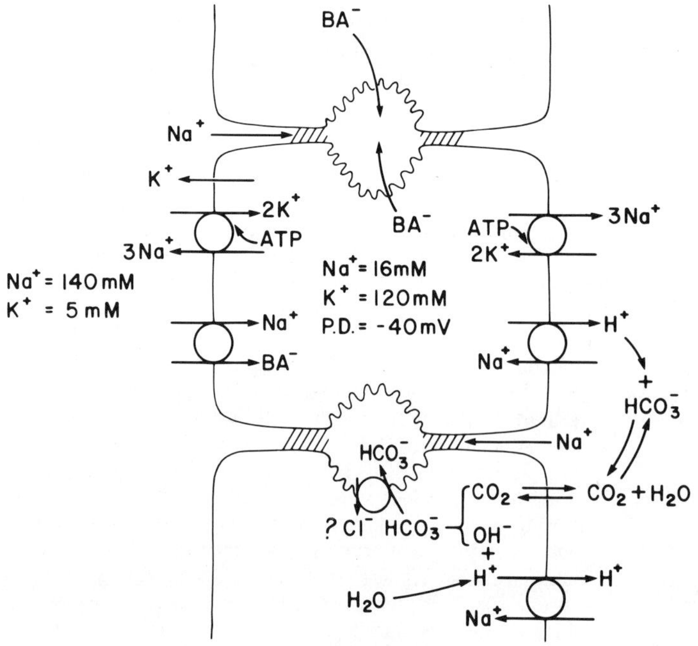

Fig. 10. A model for the generation of hepatic bile, illustrating pumps and carriers identified on the plasma membrane of the liver cell.

their excretion is directed into a confined canalicular space. The principal driving forces are the cation gradients and the intracellular potential which are maintained by the activity of the plasma membrane enzyme, Na^+,K^+-ATPase (Fig. 10). Energy for certain hepatic uptake processes is provided by the chemical gradient for Na^+ which couples to taurocholate anions and drives bile acids "uphill" against concentration gradients into the hepatocyte. A Na^+/H^+ antiport (equivalent to active transport of OH^- or HCO_3^-) also is present at the sinusoidal membrane. Na^+-independent carriers for bile acid transport and other organic anions are also present. Once these organic solutes are taken up into hepatocytes, their transcellular transport to the canalicular membrane may occur via the endoplasmic reticulum or membrane vesicles with specificity confered by specific binding proteins or receptors. The driving force for canalicular water flow is provided by osmotically active solutes that accumulate within the canaliculus, as a result of electrical potential-driven "conductive" pathways, canalicular membrane carriers and fluid exocytosis. Bile acids stimulate flow osmotically but may also regulate secretion, possibly by modifying other ion pumps or membrane permeability. Bicarbonate may accumulate in bile, facilitated by a chloride-bicarbonate exchanger in the canalicular membrane. The permeability and charge characteristics of the canalicular membrane and tight junction determine both the osmotic effectiveness of the secreted biliary solutes as well as the back diffusion of water resulting from the generation of a hydrostatic secretory pressure. Cytoskeletal elements participate in the regulation of junctional permeability, as well as generating contractile forces that help propel bile from the canaliculi to the biliary duct system.

6. Physiological Modifiers of Hepatocyte Bile Formation

6.1. Ca^{2+}

This cation is the only component of plasma that is essential for canalicular bile production. Addition of citrate to isolated perfused rat liver perfusates completely abolishes bile flow which returns with the readmission of calcium chloride.[168] Secretory pressure falls to the level of perfusion pressure.[165] Although Ca^{2+} is required for many intracellular functions, the primary effect of Ca^{2+} omission is to dissociate hepatocytes at the tight junction.

6.2. cAMP and Dibutyryl cAMP

The role of these second messengers in bile flow regulation has not yet been clearly established despite their importance in epithelial transport and the dependency of bile flow on Ca^{2+}. cAMP administration to isolated perfused rat livers stimulates secretion,[165] hyperpolarizes the cell, and is associated with an efflux in perfusate of intracellular Na^+, K^+, and Ca^{2+}.[139] Dibutyryl cAMP and hormones that stimulate adenylate cyclase (glucagon and β-sympathomimetics) produce transient increases in bile flow in similar preparations in association with an increase in biliary K^+ and the vesicular transport of macromolecules.[10,165] However, the site of action of these hormones is not clear. Others have found no choleretic effect of glucagon, or the phosphodiesterase inhibitors aminophylline or 1-methyl-3-isobutylxanthine (MIX) in the rat despite increases in cAMP after glucagon and MIX.[169,170] In the dog, neither the timing nor the magnitude of changes in liver cAMP correlated with dibutyryl cAMP-, aminophylline-, and glucagon-induced bile acid-independent secretion.[170] Still others have found that theophylline, a phosphodiesterase inhibitor, increased canalicular BAIF in the dog.[171,172] Thus, no clear picture emerges. A more detailed study of the role of the adenylate cyclase system in bile formation is needed in view of the general importance of this enzyme in the regulation of membrane permeability, metabolism, and secretion in other cell systems.

6.3. Thyroid Hormone

Thyroxine administration to rats stimulates BAIF after induction of a hyperthyroid state.[80] Conversely, thyroidectomy diminishes secretion.[80] These changes are paralleled by increases in liver plasma membrane Na^+,K^+-ATPase and membrane fluidity.[80,131]

6.4. Cortisol and Hydrocortisone

Hydrocortisone infusions stimulate bile flow but not bile salt secretion in the dog in association with an increase in Cl^- concentration. The effect cannot be attributed to osmotic effects of the steroid.[174] Cortisol acetate administration for 3 days to rats modestly enhances bile acid-independent secretion.[175] The mechanism of the choleretic effects is not known.

6.5. Microsomal Enzyme Inducers

Pregnenolone-16α-carbonitrile results in proliferation of smooth endoplasmic reticulum, mixed function oxygenases, and UDP-glucuronyl transferases and doubles BAIF when administered to rats.[175] Spironolactone has similar but less dramatic effects.[175] Other inducers of microsome function including phenobarbital, pentobarbital and thiopental result in endoplasmic reticulum hypertrophy. Although pentobarbital does not, the other four sedatives stimulate bile acid-independent secretion.[176,177] The mechanism by which these potent microsomal enzyme inducers stimulate bile formation remains unclear and paradoxical, particularly because inducers like methylcholanthrene have no effect on bile formation.[178]

6.6. Estrogens

Pharmacological doses of estrogens, in particular ethinyl estradiol, inhibit both bile acid-dependent and -independent secretion in the rat, in association with an increased viscosity of liver plasma membranes.[179–181] Na^+,K^+-ATPase activity is diminished in these membranes which contain an increased content of cholesterol esters.[180,181] Organic anion transport is also inhibited.[182,182a] Retrograde infusions of sucrose and lanthanum in the biliary tree result in diminished recovery of these solutes, supporting the view that the diminished secretion may be related in part to altered membrane permeability.[183]

6.7. Antidiuretic Hormone

Little information exists except that the hormone stimulates both hepatocellular and biliary duct secretion in the dog.[184]

6.8. Prostaglandins

PGA_1 stimulates bile flow, presumably from the hepatocytes in the rat,[185] but the effect is on ductular secretion in the dog.[186] PGE_2 apparently does not effect bile flow in the dog[187] but does stimulate this secretion in the cat.[188]

6.9. Insulin

In several species, insulin stimulates bile flow presumably at the level of the hepatocytes without an effect on bile acid.[189–191] In the dog, this is associated with a significant increase in Cl^- secretion.[192] A direct hepatic effect of insulin has been shown in the isolated perfused rat liver when normal hematocrit levels are maintained in the perfusate.[193] Insulin appears to have its greatest choleretic effect in the rat at higher bile acid infusion rates and increases bile to plasma ratios of [^{14}C]sucrose and inulin, suggesting an effect on biliary permeability.[194] Thus insulin appears to stimulate bile acid independent canalicular bile formation.[194a]

6.10. Glucagon

Glucagon stimulates bile flow[191] in man and several animals.[172,194a,195–197] It stimulates BAIF in the intact rat and in well-perfused isolated rat liver systems. Others have found glucagon without effect in perfused bovine and dog liver.[198,199] Secretin, cholecystokinin, and histamine also stimulate bile flow but their effect is predominantly at the level of the bile duct.

6.11. Neurogenic Factors

The liver appears to be well innervated with cholinergic and adrenergic fibers according to histochemical studies, particularly in portal triads, around vascular structures and the bile ducts.[200,201] Only adrenergic nerve endings have been identified within the hepatic lobule.[200] Because of the complexity of neuroendocrine metabolism, it has not been possible to outline a primary role for neurogenic stimuli in bile production. Administration of large concentrations (0.5 mM) of acetylcholine to the isolated perfused rat liver results in an initial reduction of flow followed by a sustained choleresis that can be abolished by atropine.[202] Truncal vagotomy in dogs diminishes bicarbonate secretion and the choleretic response to insulin and 2-deoxy-D-glucose which stimulate vagal-mediated gastric secretion; but these effects may be mediated through the bile ducts.[48,192,203] Atropine also blocks the effect of insulin in man.[204]

Adrenergic effects can be demonstrated in the perfused rat liver where adrenaline and isoprenaline increase the production of bile and the excretion of K^+ and Ca^{2+}[165] Dopamine infusions stimulate BAIF in dogs characterized by an augmentation of HCO_3^- and Cl^- concentrations. Bile flow and Cl^- output but not HCO_3^- could be blocked by propranolol.[205]

In these and other studies previously reviewed,[2,5] it is often impossible to be certain whether the changes in bile flow are primary responses to the neurogenic stimuli or whether the effects are occurring through hormonal mediation of a cAMP effect or from the release of other messengers, or from changes in hepatic blood flow.

6.12. Temperature

Bile flow is closely dependent on body temperature, rising in the rat to maximal values at 40–41°C, and then diminishing at higher temperatures.[41] This temperature activation curve is paralleled by the specific activity of liver plasma membrane Na^+,K^+-ATPase in the rat, whereas the temperature maximum for Mg^{2+}-ATPase is approximately 35°C.[74] A Q_{10} of 3.0 to 3.2 has been observed between 28 and 38°C in the isolated perfused rat liver and 2.8 to 3.5 between 30 and 40°C.[41]

6.13. Diurnal Rhythm

A diurnal rhythm for bile formation has been best described in the rat where biliary flow reaches a peak at 12 pm.[206] Rats are nocturnal feeders and the regulation of this pattern of secretion is linked to the dark–light cycle. Cyclical variations in bile flow are paralleled by variations in bile acid excretion and their regulatory enzymes, hepatic HMG-CoA reductases and cholesterol-7α-hydroxylase.[207,208] Fasting and feeding also clearly modify the enterohepatic circulation of bile acids, leading to an increase in bile acid output following meals.[13,13a] Attention to these rhythms is critical in experimental situations and animals must have similar feeding patterns and be sacrificed at the same time during the day for valid comparisons of flow rates and bile acid outputs.

7. Organic Anion Solute Transport

The clearance of organic anions from blood to bile is a major function of the hepatic excretory system and can be resolved into several discrete transport steps: (1) the binding of organic anions to plasma proteins, particularly albumin in the extracellular space; (2) transfer of the ligand from plasma proteins to the hepatocyte membrane carrier and translocation across the sinusoidal membrane; (3) binding of the ligand to intracellular proteins; (4) conjugation of the ligand to more water-soluble metabolites (some organic anions do not require conjugation); and (5) translocation across the canalicular membrane into bile.

Bile acids, bilirubin, and fatty acids are three naturally occurring organic anions of major physiological importance. Many drugs, dyes, and xenobiotics including organic cations and neutral compounds are also transported into bile by similar but not necessarily identical pathways.[11,12] In general, lipophilic compunds that range in molecular weight from 300 to 1000 are excreted in bile. The process is concentrative and requires the input of cellular sources of energy. The classic experiments of Hanzon provided the most direct evidence that organic solutes are progressively transported against concentration gradients.[209] Hanzon injected the fluorescent organic anion fluorescein intravenously and illuminated the peripheral lobe of a mouse liver. Fluorescein was observed initially in the plasma space and immediately was taken into the hepatocytes which became diffusely fluorescent. Within a few minutes the dye accumulated within the bile canaliculi which gradually became intensely fluorescent as the dye was concentrated in bile and eliminated from the cell. Similar visual demonstrations of organic anion transport have been reported utilizing rat hepatocyte cultures.[210] Bile/plasma ratios of organic anions generally range on the order of 1 : 100–1 : 1000.

7.1. Albumin–Ligand–Membrane Interactions

Many organic solutes transported by the liver are insoluble in aqueous solutions and are therefore, usually bound to albumin (e.g., bilirubin, fatty acids, bromsulfophthalein (BSP), and bile acids). Bilirubin also binds minimally to red blood cells[211] and bile acids are also bound by lipoproteins. Binding to albumin occurs at several sites with variable degrees of affinity. This association can be described by the following equation for each binding site on the albumin molecule:

$$\text{Ligand–albumin complex} \underset{K_a}{\overset{K_d}{\rightleftarrows}} \text{ligand} + \text{albumin}$$

where K_a is the association constant and K_d the dissociation constant. For ligands that are tightly bound to albumin such as bilirubin,[212] fatty acids,[213] or BSP,[214] free concentrations of the ligand represent a very small fraction ($< 0.1\%$) of the total circulating ligand. In contrast, for less tightly associated ligands such as taurocholate, free concentrations of ligand may represent a significant percentage of their total concentrations in portal blood.[94]

The hepatic uptake of these solutes can be examined following intravascular injection of labeled ligand with albumin and red cells by rapidly collecting multiple samples from the hepatic vein effluent, as originally described by Goresky in single-pass clearance studies.[215] Goresky observed that the red cell preceded the appearance of albumin and ligand since the latter were distributed into the space of Dissé. Albumin and the ligand appeared simultaneously but the ligand concentrations were less than albumin because a fraction was removed by the liver. Similar studies have been performed in isolated perfused rat livers, and in dogs and humans utilizing BSP and bilirubin, and demonstrate conclusively that essentially 100% of the albumin is recovered in the hepatic effluent at the same time that extensive amounts of BSP or bilirubin have been transported into the liver cells.[216] From these and other observations, it is postulated that a membrane–ligand–albumin interaction dissociates the tightly bound organic anions from albumin while simultaneously transferring them to a hepatic membrane carrier that facilitates their uptake into the cell. The nature of this membrane interaction remains to be established. However, for oleate and BSP, their fractional hepatic uptake rates exceed the values predicted from determination of the free concentrations of the ligand. Thus, some as yet unidentified process presumably augments the dissociation rates of ligand from the albumin–ligand complex when it makes contact with the hepatocyte sinusoidal membrane.[214] More recently, an albumin receptor on the plasma membrane has been suggested on the basis of kinetic studies utilizing varying concentrations of albumin, oleate, and BSP in isolated perfused rat liver studies.[217,218] In these experiments, increasing concentrations of oleate or BSP resulted in a linear increase in oleate uptake when albumin concentrations were fixed at 1%. In contrast, when 1 : 1 molar ratios of oleate and albumin were utilized, saturable uptake was observed at ligand concentrations below those obtained in the previous experiment, implying saturation of an ''albumin receptor.'' The specificity of this receptor has been questioned, however, since a specific binding protein with affinity for albumin has not been demonstrated in liver plasma membranes,[216] and similar saturation phenomenon can be demonstrated with albumin–BSP complexes in isolated perfused elasmobranch livers, a marine species that does not normally contain albumin in the plasma.[219] It seems likely therefore that a nonspecific interaction of the ligand–albumin complex takes place at the plasma membrane that increases the ligand dissociation rate and that this phenomenon is significant for tightly bound ligands where spontaneous albumin dissociation rates are slow compared to rates of hepatic uptake. This ligand–albumin facilitated dissociation is of less importance when free ligand concentrations can be maintained by rapid spontaneous dissociation from albumin.

7.2. Carrier-Mediated Uptake of Organic Anions

Many kinetic studies suggest that the translocation of organic anions across hepatic sinusoidal membranes is carrier mediated.[220–224] These studies demonstrate that the initial uptake rates are saturable as the concentration of ligand is increased; that uptake rates can be competitively inhibited (e.g., BSP and bilirubin inhibit each other's rate of uptake); that countertransport or efflux of ligand from the cell occurs against the concentration gradient or refluxes when an excess of cold ligand chases the label; and that accelerated exchange diffusion occurs[222,223] (e.g., when the ligand is preloaded in the hepatocyte, it stimulates the turnover rate of the carrier, thereby increasing the rate of transport of the external labeled ligand). Most studies indicate that bilirubin, BSP, indocyanin green, and many radiographic dyes share a common carrier mechanism for hepatic uptake since they demonstrate competitive inhibition kinetics for the uptake process.[216] In contrast, bile acids seem to be transported into hepatocytes by a separate mechanism,[100–109] although a noncompetitive-type inhibition with bilirubin and BSP has been demonstrated in some systems.[108,109,225–228] Separate transport systems for bile acids like taurocholate are suggested by their requirement for coupled transport with Na^+ in contrast to most other organic anions where Na^+ dependence has not been demonstrated.[100–109] NAP-taurine is one exception since it is a competitive inhibitor of Na^+-dependent taurocholate uptake in isolated hepatocytes. NAP-taurine also binds covalently to a 54,000-dalton bile acid-binding protein in rat liver sinusoidal membranes.[229]

A diffusional component can also be demonstrated for most organic anions as the plasma concentrations are increased. However, at physiological concentrations in portal blood, most studies suggest that hepatic organic anion uptake takes place against a concentration gradient and requires energy.

7.3. Membrane Binding Proteins

Specific and saturable binding of bilirubin, fatty acids, BSP, and bile acids to isolated rat liver plasma membranes have prompted a search for membrane proteins that may represent their carriers. Using both photoaffinity labeling and a combination of affinity column chromatography with BSP or bilirubin and Triton X-100 extracts of liver plasma membranes, a 55,000-dalton protein has been identified by several groups with affinity for BSP that seems to exceed albumin[230–232] (BSP $K_a = 5.5 \times 10^7\ M^{-1}$ compared to a K_a of $8 \times 10^6\ M^{-1}$ for albumin). The same protein also appears to bind bilirubin and bile acid but not fatty acids and is found only in hepatocytes and not in other liver cells or tissues. Others have described another bilirubin–BSP-binding protein, called bilitranslocase, that consists of two nonidentical subunits of 37,000 and 35,500 daltons, but the relationship of these binding proteins to the previously mentioned 55,000-dalton protein is uncertain.[233] A fatty acid-binding protein of 40,000 daltons has also been described in rat liver plasma membranes that has a greater affinity for oleate than albumin,[234] and several bile acid-binding proteins have been demonstrated in rat liver plasma membranes by photoaffinity labeling.[107,235] A 54,000-dalton polypeptide, has been reconstituted in membrane vesicles and facilitates Na^+-dependent taurocholate uptake.[236] To date, this is the only membrane binding protein that has been definitively shown to play a role in organic anion transport. Additional studies are needed before most of these binding proteins can qualify as functional membrane carriers for organic anions in hepatocytes.

In the hereditary disorder of hepatic bilirubin uptake (Gilbert's syndrome), the clearance of other organic anions such as BSP and indocyanin green may also be impaired, suggesting that the defect in organic anion transport may be more generalized.[237–240] However, as yet there is no evidence that these abnormalities in transport relate to a deficiency of plasma membrane binding proteins or carriers.

7.4. Intracellular Binding Proteins

Uptake studies demonstrating saturation kinetics and competition between ligands in the intact animal or isolated perfused liver do not clearly distinguish between binding sites on the plasma membrane, submembranous proteins, or proteins in the cytosol. Indeed, these conventional kinetic criteria for carrier-mediated systems must be questioned since they can be reproduced with artificial lipid membranes or liposomes in the complete absence of proteins.[241,242] A cytosolic protein was first postulated as a determinant of hepatic organic anion uptake when a major 46,000-dalton protein in hepatic cytosol was identified that specifically bound bilirubin, BSP, iodinated radiocontrast media, and many other anions.[243] This protein (originally called Y protein) was distinguishable from a fatty acid-binding protein (Z protein) but later was named ligandin when it was found to be identical to cortisol-binding protein and azo-dye carcinogen metabolite binders.[244] Subsequent work proved that ligandin was identical to glutathione transferase B and that other proteins in the glutathione transferase family also bound organic anions.[118] The binding site for anions that are not substrates for these enzymes (e.g., bilirubin) is distinct from the substrate and glutathione sites and has an affinity for anions like bilirubin that is greater than albumin only when the purified transferase B is associated with other cytosolic proteins.[245] A role for these enzymes in hepatic transport was suggested by their high concentrations in liver (5–10% of cytosolic proteins) and other organs that transport organic anions (kidney, intestine) and studies that demonstrated parallel changes in transferase concentration and rates of hepatic organic anion clearance.[118] For example, newborn animals have low concentrations of these transferases and reduced rates of hepatic organic anion uptake and both reach normal levels as the animals mature.[246,247] In adult animals, glutathione transferase concentrations and storage capacity for BSP and indocyanin green increase after phenobarbital treatment at the same time that initial hepatic uptake rates rise for bilirubin, BSP, and indocyanin green.[248,249] Yet discrepancies existed in this relationship that challenged the concept that these proteins were determinants of uptake. The affinity of purified rat glutathione S-transferase B for bilirubin was 10 times less than for rat albumin.[250] Evans blue dye binds avidly to the transferases but its transport into the liver is limited[243] whereas marine elasmobranches which have very low concentrations of these cytosolic binding proteins selectively transport bilirubin, BSP, and other organic anions from their plasma into liver and bile.[251] It was later realized that these cytosolic binding proteins influence hepatic clearance of organic anions by sequestering the ligands in the hepatocyte and preventing countertransport and efflux from the liver back into the sinusoid.[252] More recently, a new family of binding proteins have been isolated that bind other organic anions and bile acids[119,253] and are distinct from glutathione transferase B.

7.5. Hepatic Conjugation

Many organic anions like bilirubin and BSP undergo conjugation to more hydrophilic metabolites after uptake into the liver, a process which facilitates their excretion into bile by increasing their solubility in aqueous solutions rather than lipid membranes.[2,11,12] This biotransformation may in turn protect the cell from toxic injury. The broad substrate specificity and low-affinity binding for many of these biotransformations undoubtedly offer a selective advantage to these cells and the organism. However, some organic anions like indocyanin green or DBSP do not require conjugation for excretion to readily occur. Other organic anion conjugation reactions demonstrate cellular compartmentalization and lobular heterogeneity; conjugation with glutathione occurs preferentially in the pericentral zone cytosol, but can occur throughout the lobule.[254a]

Bile acids are normally conjugated with amino acids, primarily taurine and glycine. Bile acid analogues (Norcholate) that are not substrates for the bile acid conjugating enzymes are poorly excreted across the canalicular membrane into bile.[254] Bilirubin is conjugated with alkali-labile sugar esters; predominantly with glucuronic acid, less than 10% is conjugated with glucose.[255] BSP is conjugated with glutathione in mammalian liver although it is excreted largely in unconjugated form in lower vertebrates.[251,256]

The more lipid soluble the organic anion, the greater is the requirement for conjugation to more hydrophilic substances in order to facilitate biliary excretion. Bilirubin diglucuronide is the predominant form of bilirubin in bile and very little bilirubin can be excreted when glucuronyltransferase deficiency is complete. This occurs in Crigler–Najjar syndrome Type I and in the homozygous Gunn rat. The former is a cause of jaundice in the newborn and leads to progressive brain damage from retention of bilirubin in the serum. Ordinarily, hepatic conjugating reactions take place at rates that exceed the maximal excretory capacity of the liver. However, in the newborn, and in disorders where conjugating systems are partially deficient or impaired by drugs or disease, hepatic conjugation may become the rate-limiting step in the hepatic clearance of organic anions into bile.[257]

7.6. Canalicular Excretion of Organic Anions

The final excretory step in the clearance of organic anions from blood to bile is their transport across the canalicular membrane. Excretion occurs against a steep chemical gradient usually exceeding hepatic concentrations by 1–2 orders of magnitude. Canalicular excretion is an energy-requiring process, and is usually rate limiting in the overall transport of organic anions from blood to bile. Secretory or transport maxima can usually be defined experimentally for any given organic solute that is concentrated in bile and provides the most direct evidence for a saturable, presumably carrier-mediated transport mechanism. At least two transport systems exist, one for bile acids, and another for organic anions such as bilirubin, BSP, indocyanin green, iodinated organic compounds, prophyrins, fluorescein, and certain antibiotics. Evidence for two distinct transport systems is based in part on studies in patients with Dubin–Johnson syndrome, who inherit a defect in excretion of bilirubin, porphyrins, and other organic anions but appear to retain a normal excretory mechanism for bile acids.[258] Corriedale sheep are mutants that share a similar defect and have reduced secretory maxima for many organic anions and yet maintain a normal excretory maximum for taurocholate.[259,260] Other organic anions compete with one another for the same pathway of excretion.[261] Thus, if bilirubin transport is diminished, the biliary excretion of other organic anions that share this pathway will also be reduced.[262–264]

The mechanisms for excretion of organic anions are poorly understood, and have been reviewed previously in the discussion of bile acid transport. Possibilities include carrier-mediated transport, exocytosis, and electrical potential-driven conductive transport.

Bile acids may stimulate the transport maxima for certain organic anions, particularly bilirubin, BSP, DBSP, and iopanoic acid.[262–264] Several theories exist to explain this phenomenon. Bile acid-induced bile flow may facilitate organic anion transport by reducing the biliary concentration of the solute and thereby increasing its rate of dissociation from the membrane carrier, or reducing the rate of back diffusion into the cell.[262,263] Alternatively, bile acids may stimulate organic anion T_m by increasing micelle formation and sequestering the organic anion in a "micellar sink."[126] This hypothesis has been difficult to accept as a result of experiments where organic anion excretion is stimulated with non-micelle-forming bile acids such as dehydrocholate or glycodihydrofusidate and other choleretics.[265,266] Bile acids might alter the membrane carrier directly by an allosteric or detergent interaction or some other membrane effect, or by increasing the number of membrane carriers facilitating transport in the cell or by increasing the number of cells participating in organic anion transport. Taurocholate infusions can increase the binding sites for bile acids to rat liver plasma membranes but do not affect the binding sites for BSP.[121] A recruitment of additional cells within the lobule cannot be excluded in many of the studies. Difficulties in interpretation arise when high levels of organic anions and bile acids are studied since nonspecific interactions, and competition for hepatic uptake, transcellular transport, and canalicular excretion may coexist.

7.7. Excretion of Organic Cations and Neutral Compounds

Even less is known about the hepatic transport of organic cations such as procainamide, ethobromide, and other quaternary compounds.[267–269] They also demonstrate a saturable excretory maximum and competition with other cationic solutes. This pathway is largely independent of the organic anion transport mechanisms.[270,271] Taurocholate does not stimulate their transport maxima. Compounds that are uncharged (e.g., ouabain) are also concentrated in bile. Many are drugs in the clinical armamentarium.

The classical view states that there are three completely separate transport pathways for organic anions, cations, and neutral substances. However, these concepts must be questioned in light of accumulating experimental data.[107,109] For example, BSP and NAP-taurine can compete with pathways for bile acid transport, and dehydrocholate can block the hepatic clearance of a nonionic compound, e.g., ouabain.[270] Other organic anions like bilirubin and BSP inhibit ouabain transport by the liver.[271] Differences in dose, routes and rates of administration, polarity, and charge as well as experimental models and species make it difficult to generalize about these observations, but they suggest that some of these chemically diverse anions, cations, and neutral compounds may share portions of the pathways for hepatic organic solute transport between blood and bile.

8. Lipid Excretion in Bile

8.1. Composition and Source of Lipids in Bile

Sterols and phospholipids represent the major lipids in bile and consist almost entirely of unesterified cholesterol (monohydrate and anhydrous forms) and the phospholipid lecithin (phosphatidylcholine). Sphingomyelin, phosphatidylethanolamine, phosphatidylserine, and lysolecithin comprise less than 5% of the phospholipids. Bile acids and the lipopigments (bilirubin and its conjugates) comprise the other lipid compounds. Not only is lecithin the major biliary phospholipid, but its fatty acid composition is distinctive when compared to other cellular sources of lecithin and is composed predominantly of palmitoyl-linoleyl and -oleyl lecithins.[272] In general, minor species differences exist. Guinea pig is the only species whose bile does not contain lecithin.[273] The origin of these distinctive species of lecithins has not been determined. Lecithin is only slightly soluble in aqueous solutions whereas cholesterol is essentially insoluble, exceeding its solubility in bile water by a factor of more than 2000. However, normal bile is a true solution and the lipids in bile are maintained in soluble form by aggregation into mixed lipid micelles with bile acids. Based on techniques of laser light scattering and column chromatography in model solutions and in native unsaturated rat bile, hydrodynamic radii of the mixed lipid micelles have been estimated between 13 and 21 Å.[274,275] These studies also suggest that bilirubin metabolites also aggregate in larger particles (500–600 Å) that diminish in size upon dilution of bile. In contrast to cholesterol which partitions into bile acid micelles, bilipigments and organic anions such as BSP appear to be only loosely adherent to the mixed lipid micelle surface.[275] Thus, these micelles do not seem to be important "transporters" of other organic solutes as initially believed.[126] According to current models, the biliary lecithins do not form micelles by themselves but exist in water as insoluble liquid crystals in the form of bilayers. Bile acids interact with these lipid bilayers in a fixed stoichiometry, depending on the bile acid species, in lecithin/bile salt ratios of 1.3–2.0 forming a mixed bilayer disk with the periphery of the disk containing the bile acids.[13a] In bile, these mixed lipid micelles coexist with both aggregates of bile acids and the monomeric forms of bile acids which are determined by their respective critical micellar concentrations, which for taurocholate ranges between 3 and 5 mM. When bile is diluted by a bile acid free electrolyte solution (e.g., ductular fluid), the size of the micelle enlarges as bile acids equilibrate in the aqueous solution in monomeric form. The mixed lipid micelle disks swell and transform into large lipid vesicles.

The site of origin of biliary cholesterol and lecithin is unknown. Cholesterol excretion in bile is largely independent of cholesterol synthesis in man and rats and less than 30% of the lipids in bile are newly synthesized, the remainder being derived from circulating lipids in the plasma.[276,277,279] The source of biliary cholesterol depends in part on the rate of endogenous cholesterol synthesis, particularly high-density lipoproteins, diminishing to only a few percent of the total excreted in bile when cholesterol synthesis is suppressed. In the rat, newly synthesized biliary cholesterol may be derived from a microsomal subpool.[278] In the rat, only small amounts of lecithin are also derived from newly synthesized sources, presumably the microsomes.[279] The intracellular origin of lipids in bile and the site of formation of the biliary micelle are major unknowns that present important challenges for future research.

8.2. Mechanism of Biliary Lipid Secretion

The mechanism of biliary lipid excretion is also poorly understood although numerous studies in a variety of species including man indicate that the rate of excretion of phospholipid and cholesterol is tightly coupled to the excretion of micelle-forming bile acids.[13a]

Small proposed that bile acids regulate biliary lipid excretion by dissecting portions of the canalicular membrane as they efflux from the cell into bile. However, studies in the isolated perfused rat liver[280] indicate that labeled precursors of cholesterol and lecithin appear rapidly in microsomal lipids prior to their appearance in liver plasma membranes and in bile,[278] and the plasma membrane of hepatocytes does not contain enzymes that are necessary for *de novo* phospholipid synthesis.[281] In addition, rat canalicular membranes are relatively high in cholesterol and sphingomyelin compared to sinusoidal plasma membranes, indicating that they should be relatively resistant to the detergent action of bile acids.[22] This possibility is supported by *in vitro* studies demonstrating that plasma membranes do not form bile-size micelles when exposed to taurocholate, in contrast to microsomal or Golgi membranes and content.[115] Moreover, the phospholipid composition of all of these membranes is dissimilar to biliary phospholipids. Thus, the site of the hepatic origin of biliary phospholipids remains unclear. Presumably, it represents a small intrahepatic pool not presently separated from other subcellular membranes yet transportable specifically to the canalicular lumen.

The role of biliary proteins in lipid excretion in bile is controversial. Some investigators claim that a biliary protein is derived from canalicular membranes[282] and is specifically associated with the biliary micelle but this has been disputed. Apolipoproteins A-I, A-II, C-II, C-III, and B have all been detected in human bile in low concentrations.[283] Whether these proteins are transported nonspecifically into bile via vesicle transport from the sinusoid or play a specific role in directing lipid secretion into bile is not known.

Numerous studies in rats, dogs, rhesus monkey, and man demonstrate that bile acids are the major determinants of biliary lipid secretion and that both biliary lecithin and cholesterol are tightly coupled to the rate at which bile acids are excreted into bile.[283] Although species differences are prevalent, a linear relationship exists between bile acid output and biliary lipids at lower bile acid excretion rates that reaches a maximum biliary lipid excretion rate at higher rates of bile acid excretion.[284] However, in man and several primate species, bile is highly enriched in cholesterol compared to other species. The tight coupling of biliary lipids to bile acid is usually maintained when the enterohepatic circulation is intact but may be dissociated if bile acid excretion reaches low levels as occurs in man during an overnight fast. Then the molar ratio of bile acids + lecithin/cholesterol decreases and the capacity of the micelle to solubilize cholesterol is exceeded. These findings suggest that a small amount of cholesterol may be excreted by bile acid-independent mechanisms.

The detergent properties of bile acids, which are attributed to their amphipathic structure, are responsible for the coupling of bile acids and biliary lipids. Thus, infusions of sodium fusidate, a fungal steroid antibiotic with detergent properties, stimulates biliary lipid excretion[285] whereas dehydrocholate, a non-micelle-forming triketo analog of cholic acid, has no effect in animals or in man.[286–289]

Biliary lipid excretion can also be uncoupled from bile acids during the infusion of other organic anions such as bilirubin, BSP, or iodipamide (a biliary contrast agent).[288,289] This effect seems directed to inhibition of the lipid excretory mechanism since bile acid excretion and hepatic lipid synthesis remain unaffected and the phenomenon can also be demonstrated in the isolated perfused liver in the absence of serum lipids. In addition, bilirubin glucuronide but not bilirubin reproduces this phenomenon in the Gunn rat, a species incapable of excreting unconjugated bilirubin.[290] Uncoupling of biliary lipids from bile acids can also be produced by inhibitors of microtubules[159] as well as microsomal enzyme inducers.[291] Finally, during elevation of biliary tract pressures in the rhesus monkey and following recovery from transient biliary obstruction, biliary cholesterol concentrations fall and then rise disproportionately to changes in biliary bile acids.[292]

Table IV. Bile Proteins[a]

Proteins from plasma
- Albumin, α_1-acid glycoprotein, α_1-antitrypsin, α_2-macroglobulin, apolipoproteins A-I, A-II, B, C-II, C-III
- Carcinoembryonic antigen (CEA), ceruloplasmin
- Epidermal growth factor (EGF)*
- Hemoglobin–haptoglobin complex*
- Hemopexin
- Immunoglobulins (monomeric and polymeric IgA*; IgG, IgM)
- Insulin

Proteins from the hepatocyte
- Canalicular membranes: alkaline phosphatase, alkaline phosphodiesterase I, leucine β-naphthylamidase
- Lysosomes: acid phosphatase, β-galactosidase, β-glucuronidase, N-acetyl-β-glucosaminidase
- Other: alcohol dehydrogenase, acyl esterase, acyl sulfatase, aspartate aminotransferase, lactate dehydrogense, malate dehydrogenase

Major proteins in bile or bile related
- Albumin, secretory IgA*, secretory component, hemoglobin–haptoglobin*, transferrin*, bile-specific antigen, bililipoprotein-X, mucin glycoproteins

[a]Asterisks indicate receptor-mediated proteins.

9. Proteins in Bile

Small amounts of proteins are detectable in bile, ranging from 2 to 20 mg/ml, and 14 to 16 distinct peptides can be demonstrated by SDS electrophoresis.[115,293–297] The source of most of these proteins is the circulation, but they are usually detectable in bile at less than 1% of their respective plasma concentrations. There are several major exceptions, including albumin, the haptoglobin–hemoglobin complex, and IgA and its receptor, secretory component. Except for albumin, these proteins are present in much higher concentrations in bile than in serum and are transported from the blood across the hepatocyte in endocytotic vesicles.[298–300] Bile contains other "bile-specific" proteins in small amounts consisting of lysosomal and plasma membrane enzymes.[24,294,296] A bile-specific protein associated with biliary lipid has also been described but not confirmed.[282] Table IV lists some of the proteins found in bile.

In contrast to the secretion of biliary lipids, net biliary protein secretion is essentially unaffected by the rate of bile acid excretion and remains at control levels in the bile acid-depleted fistula animal, rising in concentration as the volume of secretion diminishes.[295] There are considerable species differences in biliary protein concentrations. Low concentrations of protein have been reported in guinea pig bile (0.64 mg/ml) and high concentrations in the rabbit and rat (14.2 and 20 mg, respectively). A wide range of protein concentrations have been reported within single species and values between 2 and 20 mg/ml have been noted for rat bile.[275,295,297] Diurnal variations, sex and age differences, and feeding patterns as well as technical problems associated with measurements of protein in bile may account for some of these discrepancies.

Cytosolic proteins are essentially excluded from bile. Lactic acid dehydrogenase and aspartate aminotransferase are found in bile in less than 0.007% of their total liver enzyme content.[24]

9.1. Mechanisms of Biliary Protein Secretion

The majority of biliary proteins enter bile from plasma nonspecifically. A rapid decline in biliary proteins occurs when isolated rat livers are perfused with protein and bile acid-free perfusate.[301] When single proteins are added to the perfusate or to the circulation in the intact animal, they appear in bile after 15–20 min with appearance times that are directly related to their molecular size and amounts proportionate to the administered doses of protein.[293,302] These observations suggest that most proteins enter bile by diffusion through a semipermeable barrier, perhaps through leaky portions of the tight junctional network between hepatocytes or by bulk fluid pinocytosis across the hepatocytes or bile ducts, or through a leak in the peribiliary capillary endothelium.[293,302–304] Because fluid pinocytosis might not discriminate between proteins of different size, a sieving mechanism may be operative. The size of such theoretical pores has been estimated at 127 Å in man and 133 Å in the dog.[293,305]

Several proteins are transported across the hepatocyte in association with coated endocytotic vesicles and are mediated by both receptor-specific and -nonspecific mechanisms.[300] The best example of receptor-mediated transfer is IgA which binds to its receptor, secretory component, which is synthesized in the liver cell and transported into the plasma membrane by way of the Golgi apparatus. Secretory component consists of a dimer in the sinusoidal plasma membrane and a cleavage product that is present in bile.[22] In the rat, IgA binds to secretory component at the blood sinusoidal pole of the cell at the site of clathrin coated vesicles that move directly into the cell through the endosomes.[23] IgA and the cleavage product of secretory component are probably released into vesicles prior to fusion at the canalicular membrane since the secretory component is not found in isolated canalicular membrane fractions.[22]

In man, IgA may enter bile predominantly by transfer across the bile duct epithelium where it binds to secretory component.[306] Colchicine pretreatment in rats inhibits the biliary excretion of IgA, supporting the notion that microtubules are involved in the transfer of the endosomes.[160] Endocytotic coated vesicles may also trap other proteins nonspecifically since autoradiographic techniques have demonstrated such a mechanism for the movement of horseradish peroxidase from blood to bile.[307] Whether this process can account for the sieving effect observed with proteins of differing size has not been established.

Hemoglobin–haptoglobin complexes are also transported by receptor-specific endocytosis.[299] Exocytotic mechanisms also account for the excretion of lysosomal enzymes, and approximately 3–10% of their content in rat liver is discharged into bile each day.[24,295] Glycosidases, β-glucuronidases, β-galactosidases, and N-acetyl-β-glucosaminidase are excreted at variable but comparable rates, a process refered to as lysosomal defecation. The physiological significance of this process is not understood but there are several diseases where either copper (Wilson's disease and primary biliary cirrhosis) or iron (hemochromatosis) may accumulate in hepatic lysosomes in part because of a defect in the elimination of these heavy metals into bile.

Several liver plasma membrane enzymes are also found in bile, including 5′-nucleotidase, alkaline phosphodiesterase I, and γ-glutamyl transpeptidase, and 2–3% of their content in rat liver is discharged each day.[24,295] All of these proteins are ectoenzymes that are located at the external face of the canalicular membrane. Their biliary output is stimulated by infusions of taurocholate but not by secretin or nondetergent choleretics such as dehydrocholate, suggesting that they are solubilized in micellar solutions by the detergent effects of the bile acid.[301]

Some of these enzymes can be recovered in sedimentable material in bile, particularly if the enterohepatic circulation of bile acids is interrupted for a number of hours, leading to the suggestion that portions of canalicular or intracellular membranes may be extruded into bile, perhaps during the process of exocytosis.[24]

Bile also contains mucin glycoproteins which are thought to be produced by the bile duct and gallbladder epithelium. These high-molecular-weight glycoproteins are found in human bile in concentrations between 0.15 and 2.0 g/liter and may play a role in gallstone formation.[308,309]

9.2. Peptides and Amino Acids

In addition to intact proteins, a variety of amino acids and peptides can also be identified in bile.[10,310,311] The process is selective and may involve specific transport processes, since the acidic amino acids (aspartic acid and glutamic acid), sulfur-containing amino acids (cystine and methionine), and glycine are all present in bile in millimolar concentrations that are significantly higher than obtained in plasma. In contrast, neither basic amino acids nor neutral amino acids such as alanine accumulate in bile despite active concentration within the hepatocyte.

The most biologically important polypeptides in bile are GSH and its oxidized form GSSG. GSH is transported from the hepatocyte into both blood and bile[312–315] and oxidizes within minutes to GSSG, probably accounting for the variable reports of their relative distribution in bile.[314] When rat bile is collected under reducing conditions, 80% of the total GSH content (4–5 mM) is in the GSH form. Hepatic bile ratios of GSSG (1 : 10) suggest that GSSG may be transported into bile by a concentrative mechanism.[314] Although GSH may diffuse into bile, carrier-mediated transport has been demonstrated in canalicular membrane vesicles.[158]

Other unidentified ninhydrin-positive polypeptides have been demonstrated in bile and many contain γ-glutamyl residues.[10] Whether these peptides represent proteolytic products of the tripeptide GSH or other polypeptides is unknown. There is speculation that canalicular membrane γ-glutamyl transpeptidase may play some functional role in their transport and that their osmotic activity contributes to the formation of "bile acid-independent" secretory bile flow.[10]

10. Miscellaneous Substances Found in Bile

10.1. Monosaccharides

Compounds such as inulin, sucrose, erythritol, and mannitol enter bile by a process of diffusion and solvent drag, and have been used experimentally as markers of water flow, as discussed in Section 3. The precise sites of entry of these solutes has not been clearly determined and may be species dependent. Both paracellular and transcellular pathways may be implicated.

10.2. Metals

Bile is a major route of excretion for the heavy metals, including copper, mercury, lead, magnesium, silver, cadmium, and zinc.[316–318] The process is important because of their potential toxicity and relationship to environmental pollutants. In Wilson's disease, copper accumulates within the liver because of the failure of a fundamental although poorly understood mechanism for Cu^{2+} excretion into bile.[319] Chelation of heavy metals to sulfhydryl-containing peptides such as GSH is one mechanism that has been postulated to account for their transport into bile but little definitive work has been performed.[320]

10.3. Vitamins

Cobalamin is transported from blood to bile as part of an efficient enterohepatic circulation in animals and man.[321,322] The process is apparently concentrative and mediated by a carrier. Vitamin D metabolites are also found in bile and undergo an enterohepatic circulation.[323]

10.4. Steroids and Hormones

Estrogens (estradiol, estriol, and estrone), thyroxine, insulin, prolactin, and pheromones are all examples of endogenous steroids and hormones that are transported into bile.[317,324–326] Many undergo degradation within the liver lysosomal system, but variable portions as well as the intact compounds or their metabolites are excreted in bile. Most are conjugated with glucuronide or sulfate.

10.5. Porphyrins

Bile is the major route of excretion for the more lipid-soluble porphyrins, particularly protoporphyrin. Experimental work in the rat suggests that its excretion can be stimulated by taurocholate, probably by incorporation of protoporphyrin into mixed lipid micelles.[327]

10.6. Drugs and Xenobiotics

Over 200 drugs and xenobiotics and their metabolites are excreted in bile to variable degrees depending on both the substance and other modulating factors, such as age, sex, and metabolism. Some drugs and metabolites directly alter biliary secretion either by stimulating its formation through their osmotic effects or by inducing hepatic enzyme functions. Others may be cholestatic because they interfere with various steps in the bile secretory process.[98,328] Comprehensive reviews of drugs and xenobiotics that are excreted in bile have been published.[11,12,317]

11. Bile Duct Function

Primary hepatic bile may be modified by secretory and reabsorptive transport functions in the biliary ductules and ducts that are species specific.[2,5,8] Duct modifications are minimal in animals that feed continuously like rodents but are more significant in intermittent feeders like the dog and man.[2] The biliary epithelium is extensively supplied with blood from the hepatic artery rather than the portal vein and begins in the canal of Hering which is formed by the confluence of several hepatic bile canaliculi immediately prior to entry into the portal triads. From there bile traverses conduits of increasing diameter eventually entering the common hepatic duct and the gallbladder or intestine. Although little is known about the function of bile duct epithelial cells, these structures are highly permeable to small solutes such as $^{24}Na^{+}$ and $^{42}K^{+}$ which rapidly equilibrate between blood and bile[15,329] and are washed out prior to the appearance of simultaneously administered organic anions. In isolated portal vein perfused rat liver preparation, the appearance times of small solutes and organic anions are the same, supporting the notion that hepatic arterial perfusion is necessary for transport to occur across the bile ducts.[60] Urea, sucrose, mannitol, and inulin can diffuse across this epithelia into bile from the blood in inverse proportion to their molecular weight,[55] emphasizing that solutes such as erythritol or mannitol are not ideal markers of canalicular water flow as previously believed. Permeability coefficients for the rat bile duct are similar to those of the rabbit gallbladder.[55] Stop-flow experiments have also demonstrated selective reabsorption of solutes such as glucose. Glucose which is normally absent from bile may appear if the glucose load exceeds the apparent maximum reabsorptive capacity of the biliary duct system.[167] These studies suggest that the biliary epithelia may function in an analogous fashion to the renal tubules.

It has long been known that the bile ducts are responsive to stimulation with several gastrointestinal hormones, particularly secretin, glucagon, histamine, cholecystokinin, gastrin, VIP, and caerulin.[2,5,15]

The prototype is secretin, which stimulates bile flow and HCO_3^- excretion in dogs, primates, cats, guinea pigs, pigs, and man without augmentation of organic solutes such as bile salts or bilirubin.[9] This observation together with minimal effects on erythritol or mannitol clearance has supported the view that secretin stimulates secretion only at the bile ducts. However, other studies indicate that secretin also stimulates mannitol and erythritol clearance but whether this results from an effect on hepatocytes or on transductular movement of erythritol has not been clarified.[52–54]

Secretin has minimal if any effects on bile flow in several species including the rat and the rabbit. These animals have high spontaneous bile flow rates and large "bile acid-independent" fractions. They are also continuous feeders and the suggestion has been made that secretin responsiveness may be correlated with eating habits so that animals such as dogs and primates which eat intermittently may have developed means to modify biliary secretion during periods of increased digestive activity.[2] HCO_3^- concentration increases and Cl^- concentration may diminish during hormonal stimulation of ductular flow, suggesting that an anion-exchange system may be operative in this epithelium. An adenylate cyclase-mediated transport system has also been proposed.[330]

Bile ducts are also capable of reabsorbing water. This phenomenon is most marked in the cholecystectomized dog where the bile ducts undergo an adaptive change after removal of the gallbladder and concentrate hepatic bile constituents.[15]

Somatostatin inhibits bile acid-independent secretion in the bile fistula dog, apparently by antagonizing the effects of secretin.[331,332] Stimulation of reabsorptive process cannot be entirely excluded. Somatostatin does not effect biliary bile acid or lipid excretion.[331] In the isolated perfused rat liver, somatostatin inhibited both canalicular flow and bile acid excretion.[333] It is not known if somatostatin is a physiological regulator of bile formation or if it is present in bile duct epithelium.

12. Summary

This chapter has attempted to cover a broad range of subjects related to transport mechanisms determining hepatobiliary secretion. Although an attempt has been made to be comprehensive, the literature in this field is expanding so rapidly that certain areas were either not covered in as great a depth as they deserved or other important contributions were inadvertently overlooked. Despite these problems, it is encouraging to see how much progress has been made in this field in the last few years and to glimpse ahead at what should prove to be an exciting and enlightening period when many of the more fundamental processes of bile secretion and hepatic transport will be better understood.

References

1. Brauer, R. W. 1959. Mechanisms of bile secretion. *J. Am. Med. Assoc.* **169**:1462–1466.
2. Forker, E. L. 1977. Mechanisms of hepatic bile formation. *Annu. Rev. Physiol.* **39**:323–347.
3. Javitt, N. B. 1976. Hepatic bile formation. *N. Engl. J. Med.* **295**:1464–1469, 1511–1516.
4. Boyer, J. L. 1980. New concepts of mechanisms of hepatocyte bile formation. *Physiol. Rev.* **60**:303–326.
5. Reichen, J., and G. Paumgartner. 1980. Excretory function of the liver. In: *Liver and Biliary Tract Physiology*. N. B. Javitt, ed. University Park Press, Baltimore. pp. 103–150.
6. Paumgartner, G., and D. Paumgartner. 1982. Current concepts of bile formation. *Prog. Liver Dis.* **7**:207–220.
7. Blitzer, B. L., and J. L. Boyer. 1982. Cellular mechanisms of bile formation. *Gastroenterology* **82**:346–357.
8. Erlinger, S. 1981. Hepatocyte bile secretion: Current views and controversies. *Hepatology* **1**:352–359.
9. Scharschmidt, B. F. 1982. Bile formation and cholestasis, metabolism and enterohepatic circulation of bile acids, and gallstone formation. In: *Hepatology: A Textbook of Liver Disease*. D. Zakem and T. D. Boyer, eds. Saunders, Philadelphia. pp. 297–351.
10. Graf, J. 1983. Canalicular bile salt independent bile formation: Concepts and clues from electrolyte transport in rat liver. *Am. J. Physiol.* **7**:233–246.
11. Smith, R. L. 1973. *The Excretory Function of Bile: The Elimination of Drugs and Toxic Substances in Bile*. Chapman & Hall, London.
12. Levine, W. G. 1981. Biliary excretion of drugs and other xenobiotics. *Prog. Drug Res.* **25**:362–420.
13. Hofmann, A. F. 1976. The enterohepatic circulation of bile acids in man. *Adv. Intern. Med.* **21**:501–534.

13a. Carey, M. C. 1982. The enterohepatic circulation. In: *The Liver: Biology and Pathobiology*. I. Arias, H. Popper, D. Schacter, and D. A. Schafritz, eds. Raven Press, New York. pp. 429–465.

14. Lemaitre-Coelho, I., G. D. F. Jackson, and J. P. Vaerman. 1977. Rat bile as a convenient source of secretory IgA and free secretory component. *Eur. J. Immunol.* **7**:588–590.
15. Wheeler, H. O. 1968. Water and electrolytes in bile. In: *Handbook of Physiology*, Section 6. C. F. Code, ed. American Physiological Society, Washington, D.C. pp. 2409–2431.
16. Rundle, F. F., B. Robson, and M. Middleton. 1955. Bile drainage after cholecystectomy in man, with some observations on biliary fistula. *Surgery* **37**:903–910.
17. Boyer, J. L., and J. R. Bloomer. 1974. Canalicular bile secretion in man: Studies utilizing the biliary clearance of ^{14}C-mannitol. *J. Clin. Invest.* **54**:773–781.
18. Blouin, A., R. P. Bolender, and E. R. Weibel. 1977. Distribution of organelles and membranes between hepatocytes and nonhepatocytes in the rat liver parenchyma: A sterological study. *J. Cell Biol.* **72**:441–455.
19. Evans, W. H. 1980. A biochemical dissection of the functional polarity of the plasma membrane of the hepatocyte. *Biochim. Biophys. Acta* **604**:27–64.
20. Boyer, J. L., R. M. Allen, and O.-C. Ng. 1983. Biochemical separation of Na^+, K^+-ATPase from a "purified" light density, "canalicular" enriched plasma membrane fraction from rat liver. *Hepatology* **3**:18–28.
21. Inoue, M., R. Kinne, T. Tran, L. Biempica, and I. M. Arias. 1983. Rat liver canalicular membrane vesicles. *J. Biol. Chem*, **258**:5183–5188.
22. Meier, P. J., E. S. Sztul, A. Reuben, and J. L. Boyer. 1984.

Structural and functional polarity of canalicular and basolateral plasma membrane vesicles isolated in high yield from rat liver. *J. Cell Biol.* **98**:991–1000.
23. Renston, R. H., A. L. Jones, W. D. Christiansen, and G. T. Hradek. 1980. Evidence for a vesicular transport mechanism in hepatocytes for biliary secretion of immunoglobulin A. *Science* **208**:1276–1278.
24. LaRusso, N. F., and S. Fowler. 1979. Coordinate secretion of acid hydrolases in rat bile—Hepatocyte exocytosis of lysosomal protein? *J. Clin. Invest.* **64**:948–954.
25. Jones, A. L., D. L. Schmucker, J. S. Mooney, R. K. Ockner, and R. D. Adler. 1979. Alterations in hepatic pericanalicular cytoplasm during enhanced bile secretory activity. *Lab. Invest.* **40**:512–517.
26. Phillips, M. J., M. Oda, E. Mak, M. M. Fisher, and K. N. Jeejeebhoy. 1975. Microfilament dysfunction as a possible cause of intrahepatic cholestasis. *Gastroenterology* **69**:48–58.
27. Elias, E., Z. Hruban, J. B. Wade, and J. L. Boyer. 1980. Phalloidin-induced cholestasis, a microfilament-mediated change in junctional complex permeability. *Proc. Natl. Acad. Sci. USA* **77**: 2229–2233.
28. DeBrabander, M., J. C. Wanson, R. Mosselmans, G. Geuns, and P. Drochmans. 1978. Effects of antimicrotubular compounds on monolayer cultures of adult rat hepatocytes. *Biol. Cell.* **31**:127–140.
29. Dubin, M., M. Maurice, G. Feldmann, and S. Erlinger. 1980. Influence of colchicine and phalloidin on bile secretion and hepatic ultrastructure in the rat. *Gastroenterology* **79**:646–654.
30. Oshio, C., and M. J. Phillips. 1981. Contractility of bile canaliculi: Implications for liver function. *Science* **212**:1041–1042.
31. Phillips, M. J., C. Oshio, M. Miyairi, and C. R. Smith. 1983. Intrahepatic cholestasis as a canalicular motility disorder—Evidence using cytochalasin. *Lab. Invest.* **48**:205–211.
32. Boyer, J. L. 1983. Tight junctions in normal and cholestatic liver: Does the paracellular pathway have functional significance? *Hepatology* **3**:614–617.
33. Lagarde, S., E. Elias, J. B. Wade, and J. L. Boyer. 1981. Structural heterogeneity of hepatocyte "tight junctions": A quantitative analysis. *Hepatology* **1**:193–203.
34. Easter, D. W., J. B. Wade, and J. L. Boyer. 1983. Structural integrity of hepatocyte tight junctions. *J. Cell Biol.* **96**:745–749.
35. Bradley, S. E., and R. Herz. 1978. Permselectivity of biliary canalicular membrane in rats: Clearance probe analysis. *Am. J. Physiol.* **235**:E570–E576.
36. Rappaport, A. M. 1973. The microcirculatory hepatic unit. *Microvasc. Res.* **6**:212–228.
37. Miller, D. L., C. S. Zanolli, and J. J. Gumucio. 1979. Quantitative morphology of sinusoids of the hepatic acinus-quantimet analysis of rat liver. *Gastroenterology* **76**:965–969.
38. Gumucio, J. J., C. Balabaud, D. L. Miller, L. J. Mason, H. D. Appelman, T. J. Stoecker, and D. R. Franzblau. 1978. Bile secretion and liver cell heterogeneity in the rat. *J. Lab. Clin. Med.* **91**:350–362.
39. Gumucio, J. J., and D. L. Miller. 1981. Functional implications of liver cell heterogeneity. *Gastroenterology* **80**:393–403.
40. Jungerman, K., and N. Katz. 1982. Functional hepatocellular heterogeneity. *Hepatology* **2**:385–395.
41. Brauer, R. W., G. F. Leong, R. F. McElroy, and R. J. Holloway. 1954. Mechanisms of bile secretion: Effect of perfusion pressure and temperature on bile flow and secretion pressure. *Am. J. Physiol.* **177**:103–112.
42. Bizard, G. 1965. Enzyme inhibitors and biliary secretion. In: *The Biliary System*. W. Taylor, ed. Blackwell, Oxford. pp. 315–324.
43. Moore, E. W., and J. M. Dietschy. 1964. Na and K activity coefficients in bile and bile salts determined by glass electrodes. *Am. J. Physiol.* **206**:1111–1117.
44. Wheeler, H. O., and O. L. Ramos. 1960. Determinants of the flow and composition of bile in the unanesthetized dog during constant infusions of sodium taurocholate. *J. Clin. Invest.* **39**:161–170.
45. Sperber, I. 1963. Biliary secretion and choleresis. In: *Proc. 1st Int. Pharmacol. Meet.* Volume 4. Pergamon Press, Elmsford, N.Y. pp. 137–143.
46. Sperber, I. 1965. Biliary secretion of organic anions and its influence on bile flow. In: *The Biliary System*. W. Taylor, ed. Blackwell, Oxford. pp. 457–467.
47. Wheeler, H. O., E. D. Ross, and S. E. Bradley. 1968. Canalicular bile production in dogs. *Am. J. Physiol.* **214**:866–874.
48. Preisig, R., H. L. Cooper, and H. O. Wheeler. 1962. The relationship between taurocholate secretion rate and bile production in the unanesthetized dog during cholinergic blockade and during secretin administration. *J. Clin. Invest.* **41**:1152–1162.
49. Forker, E. L. 1968. Bile formation in guinea pigs: Analysis with inert solutes of graded molecular radius. *Am. J. Physiol.* **215**:56–62.
50. Strasberg, S. M., R. G. Ilson, K. H. Siminovitch, D. Brenner, and J. E. Palaheimo. 1975. Analysis of the components of bile flow in the rhesus monkey. *Am. J. Physiol.* **228**:115–121.
51. Strasberg, S. M., R. G. Ilson, and C. N. Petrunka. 1982. ^{14}C-erythritol clearance and canalicular bile acid independent flow in the baboon. *Am. J. Physiol.* **242**:G475–G480.
52. Barnhart, J. L., and B. Combes. 1978. Erythritol and mannitol clearances with taurocholate and secretin induced choleresis. *Am. J. Physiol.* **234**:E146–E156.
53. Nicholls, R. J. 1979. Biliary mannitol clearance and bile salt output before and during secretin choleresis in the dog. *Gastroenterology* **76**:983–987.
54. Lewis, M. H., A. L. Baker, J. M. Dhorajiwala, and A. R. Moossa. 1981. Secretin enhances ^{14}C-erythritol clearance in unanesthetized dogs. *Dig. Dis. Sci.* **27**:57–64.
55. Smith, N. D., and J. L. Boyer. 1982. Permeability characteristics of bile duct in the rat. *Am. J. Physiol.* **242**:G52–G57.
56. Javitt, N. B. 1977. Bile formation. In: *Chemistry and Physiology of Bile Pigments*. P. D. Berk and N. Berlin, eds. Fogarty Int. Cent. Proc. No. 35. U.S. Government Printing Office. Bethesda. pp. 377–382.
57. Lorenzini, I., T. Ilter, P. Meier, and J. L. Boyer. 1982. Taurochenodeoxycholic acid (TCDA) stimulates hepatic uptake of ^{3}H-methoxyinulin (^{3}HMI) into membrane bound compartments. *Hepatology* **2**:737a.
58. Anwer, M. S., and J. L. Barnhart. 1982. Polyethylene glycol-900 (PEG-900): A possible marker for paracellular water movement. *Hepatology* **2**:688a.
59. Krell, H., H. Hoke, and E. Pfaff. 1982. Development of intrahepatic cholestasis by α-naphthylisothiocyanate in rats. *Gastroenterology* **82**:507–514.
60. Graf, J., and M. Peterlik. 1975. Mechanism of transport of inorganic ions into bile. In: *The Hepatobiliary System—Fundamental and Pathological Mechanisms*. W. Taylor, ed. Plenum Press, New York. pp. 43–58.
61. Schiff, M. 1890. Gallenbildung, abhangig non der Aufsaugung der Gallenstoffe. *Pfluegers Arch. Gesamte Physiol.* **3**:598–613.
62. Wheeler, H. O. 1972. Secretion of bile acids by the liver and their role in the formation of hepatic bile. *Arch. Intern. Med.* **130** :533–541.
63. Klaassen, C. D. 1972. Species differences to the choleretic response to bile salts. *J. Physiol. (London)* **224**:259–269.
64. Vonk, R. J., P. Jekel, and D. K. F. Meijer. 1975. Choleresis and hepatic transport mechanism. *Naunyn-Schmiedebergs Arch. Pharmacol.* **290**:375–387.
65. O'Maille, E. R. L. 1980. The influence of micelle formation on bile secretion. *J. Physiol. (London)* **302**:107–120.
66. Sewell, R. B., N. E. Hoffman, R. A. Smallwood, and S. Cockbain. 1980. Bile acid structure and bile formation: A comparison of hydroxy and keto bile acids. *Am. J. Physiol.* **238**:G10–G17.
67. O'Maille, E. R. L., M. S. Anwer, A. F. Hofmann, E. B. Ljunge, and R. G. Danzinger. 1982. Side chain charge: A key determinant of hepatic bile acid transport. *Gastroenterology* **82**:1140a.
68. Javitt, N. B., and S. Emerman. 1968. Effect of sodium taurolithocholate on bile flow and bile acid excretion. *J. Clin. Invest.* **47**:1002–1014.

69. Layden, T. J., and J. L. Boyer. 1977. Taurolithocholate induced cholestasis: Taurocholate, but not dehydrocholate, reverses cholestasis and bile canalicular membrane injury. *Gastroenterology* **73**:120–128.
70. Kitani, K., and S. Kanai. 1981. Biliary transport maximum of tauroursodeoxycholate is twice as high as that of taurocholate in the rat. *Life Sci.* **29**:260–275.
71. Dumont, M., S. Uchman, S. Erlinger, and N. Dumont. 1980. Hypercholeresis induced by ursodeoxycholic acid and 7-ketolithocholic acid in the rat: Possible role of bicarbonate transport. *Gastroenterology* **79**:82–89.
72. Scharschmidt, B. F., E. B. Keefe, D. Vessey, N. M. Blankenship, and R. K. Ochner. 1981. *In vitro* effect of bile salts on rat liver plasma membrane lipid fluidity and ATPase activity. *Hepatology* **1**:137–145.
73. Baker, A. L., R. A. B. Wood, A. R. Moossa, and J. L. Boyer. 1978. Sodium taurocholate modifies the bile acid-independent fraction of canalicular bile flow in the rhesus monkey. *J. Clin. Invest.* **64**:312–320.
74. Boyer, J. L., and D. Reno. 1975. Properties of (Na^+ -K^+) activated ATPase in rat liver plasma membranes enriched with bile canaliculi. *Biochim. Biophys. Acta* **401**:59–72.
74a. Nemchausky, B., D. Reno, and J. L. Boyer. 1975. Synthetic and naturally occurring bile salts—Modifiers of ATPase activity in canalicular enriched liver plasma membrane. *Clin. Res.* **23**: 254a.
75. Wannagat, F. J., R. D. Alder, and R. K. Ochner. 1978. Bile acid-induced increase in bile acid independent flow and plasma membrane Na^+, K^+-ATPase in the rat liver. *J. Clin. Invest.* **61**: 297–307.
76. Accatino, L., A. Contreras, E. Berdichevsky, and C. Qunitana. 1981. The effect of complete biliary obstruction on bile secretion: Studies on the mechanisms of post cholestatic choleresis in the rat. *J. Lab. Clin. Med.* **97**:525–534.
77. Miyai, K., and W. G. Hardison. 1979. Bile duct ligation vs. retention of bile: Pericanalicular microfilaments form bundles only with bile duct ligation. *Gastroenterology* **76**:1292a.
78. Balabaud, C., K. A. Korn, and J. J. Gumucio. 1977. The assessment of the bile salt nondependent fraction of canalicular bile water in the rat. *J. Lab. Clin. Med.* **89**:393–399.
79. Reichen, J., and G. Paumgartner. 1977. Relationship between bile flow and Na^+,K^+-adenosine triphosphatase in liver plasma membranes enriched in bile canaliculi. *J. Clin. Invest.* **60**:429–434.
80. Layden, T. J., and J. L. Boyer. 1976. The effect of thyroid hormone on bile salt-independent bile flow and Na^+,K^+-ATPase activity in liver plasma membranes enriched in bile canaliculi. *J. Clin. Invest.* **57**:1009–1018.
81. Simon, F. R., E. Sutherland, and L. A. Accatino. 1977. Stimulation of hepatic sodium and potassium-activated adenosine triphosphatase activity by phenobarbital—Its possible role in regulation of bile flow. *J. Clin. Invest.* **59**:849–861.
82. LaRusso, N. F., M. G. Korman, N. E. Hoffman, and A. F. Hofmann. 1974. Dynamics of the enterohepatic circulation of bile acids: Postprandial serum concentrations of conjugates of cholic acid in health, cholecystectomized patients, and patients with bile acid malabsorption. *N. Engl. J. Med.* **291**:689–692.
83. Angelin, B. O., I. Bjorkhem, K. Einarsson, and S. Ewerth. 1982. Hepatic uptake of bile acids in man—Fasting and postprandial concentrations of individual bile acids in portal venous and systemic blood serum. *J. Clin. Invest.* **70**:724–731.
84. Ahlberg, J., B. Angelin, I. Bjorkhem, and K. Einarsson. 1979. Individual bile acids in portal venous and systemic blood of fasting man. *Gastroenterology* **73**:1377–1382.
85. Lindblad, L., K. Lundholm, and T. Schersten. 1977. Bile acid concentrations in systemic and portal serum in presumably normal man and in cholestatic and cirrhotic conditions. *Scand. J. Gastroenterol.* **12**:395–400.
86. Olivecrona, T., and J. Sjovall. 1959. Bile acids in rat portal blood. *Acta Physiol. Scand.* **46**:284–290.
87. Cronholm, T., and J. Sjovall. 1967. Bile acids in portal blood of rats fed different diets and cholestyramine. *Eur. J. Biochem.* **2**:375–383.
88. Matern, S., and W. Gerok. 1979. Pathophysiology of the enterohepatic circulation. *Rev. Physiol. Biochem. Pharmacol.* **85**:126–204.
89. Mok, H. Y. I., K. von Bergman, and S. M. Grundy. 1980. Kinetics of the enterohepatic circulation during fasting: Biliary lipid secretion and gallbladder storage. *Gastroenterology* **78**:1023–1033.
90. Rudman, D., and F. E. Kendall. 1957. Bile acid content of human serum. II. The binding of cholanic acids by human plasma proteins. *J. Clin. Invest.* **36**:538–542.
91. Hoffman, N. E., J. H. Iser, R. A. Smallwood. 1975. Hepatic bile acid transport: Effect of conjugation and position of hydroxyl groups. *Am. J. Physiol.* **229**:298–302.
91a. Aldini, R., A. Roda, A. M. Morselli, G. Cappelleri, E. Roda, and L. Barbara. 1982. Hepatic bile acid uptake: Effect of conjugation, hydroxyl and keto groups, and albumin binding. *J. Lipid Res.* **23**:1167–1173.
92. Iga, T., and C. D. Klaassen. 1982. Hepatic extraction of bile acids in rats. *Biochem. Pharmacol.* **31**:205–209.
93. Kramer, W., H.-P. Buscherg, W. Gerok, and G. Kurz. 1979. Bile salt binding to serum components: Taurocholate incorporation into high-density lipoproteins revealed by photoaffinity labelling. *Eur. J. Biochem.* **102**:1–9.
94. Forker, E. L., and B. A. Luxon. 1981. Albumin helps mediate removal of taurocholate by rat liver. *J. Clin. Invest.* **67**:1517–1522
95. Layden, T. J., and J. L. Boyer. 1978. Influence of bile acids on bile canalicular size. *Lab. Invest.* **39**:110–119.
96. Groothuis, G. M. M., M. Hardonk, K. P. T. Keulemans, P. Nieuwenhuis, and D. K. F. Meijer. 1982. Autoradiographic and kinetic demonstration of acinar heterogeneity of taurocholate transport. *Am. J. Physiol.* **243**:G455–G462.
97. Jones, A. L., G. T. Hradek, R. H. Renston, K. W. Wong, G. Karlaganis, and G. Paumgartner. 1980. Autoradiographic evidence for hepatic lobular concentration gradient of bile acid derivative. *Am. J. Physiol.* **238**:G233–G237.
98. Suchy, F. J., W. F. Balistreri, J. Hung, P. Miller, and S. A. Garfield. 1983. Intracellular bile acid transport in rat liver as visualized by electron microscope autoradiography using a bile acid analogue. *Am. J. Physiol.* **245**:G681–G689.
99. Elias, E., and J. L. Boyer. 1979. Mechanisms of intrahepatic cholestasis. *Prog. Liver Dis.* **6**:457–470.
100. Reichen, J., and G. Paumgartner. 1975. Kinetics of taurocholate uptake by the perfused rat liver. *Gastroenterology* **68**:132–136.
101. Glasinovic, J.-C., M. Dumont, M. Duval, and S. Erlinger. 1975. Hepatocellular uptake of taurocholate in the dog. *J. Clin. Invest.* **55**:419–426.
102. Reichen, J., and G. Paumgartner. 1976. Uptake of bile acids by perfused rat liver. *Am. J. Physiol.* **231**:734–742.
103. Schwartz, L. R., R. Burr, M. Schwenk, E. Pfaff, and H. Greim. 1975. Uptake of taurocholic acid into isolated rat liver cells. *Eur. J. Biochem.* **55**:617–623.
104. Anwer, M. S., and D. Hegner. 1978. Effect of Na^+ on bile acid uptake by isolated rat hepatocytes. *Hoppe-Seylers Z. Physiol. Chem.* **359**:181–192.
105. Blitzer, B. L., S. L. Ratoosh, C. B. Donovan, and J. L. Boyer. 1982. Effects of inhibitors of Na^+-coupled ion transport on bile acid uptake by isolated rat hepatocytes. *Am. J. Physiol.* **243**: G48–G53.
106. Scharschmidt, B. F., and J. E. Stephens. 1981. Transport of sodium, chloride, and taurocholate by cultured rat hepatocytes. *Proc. Natl. Acad. Sci. USA* **78**:986–990.
107. von Dippe, P., and D. Levy. 1983. Characterization of the bile acid transport system in normal and transformed hepatocytes—Photoaffinity labelling of the taurocholate carrier proteins. *J. Biol. Chem.* **258**:8896–8901.
107a. Abberger, H., U. Bickel, H. P. Buscher, K. Fuchte, W. Gerok,

W. Krammer, and G. Kurz. 1981. Transport of bile acids: Lipoproteins, membrane polypeptides and cytosolic proteins as carriers. In: *Bile Acids and Lipids*. G. Paumgartner, A. Stiehl, and W. Gerok, eds. MTP Press, Lancaster. pp. 233–246.
108. Inoue, M., R. Kinne, T. Tran, and I. M. Arias. 1982. Taurocholate transport by rat liver sinusoidal membrane vesicles: Evidence for sodium cotransport. *Hepatology* **2**:572–579.
109. Duffy, M. C., B. L. Blitzer, and J. L. Boyer. 1983. Direct determination of the driving forces for taurocholate uptake into rat liver plasma membrane vesicles. *J. Clin. Invest.* **72**:1470–1481.
110. Van Dyke, R. W., J. E. Stephens, and B. F. Scharschmidt. 1982. Bile acid transport in cultured rat hepatocytes. *Am. J. Physiol.* **243**:G484–G492.
111. Meier, P. J., A. S. Meier-Abt, C. Barrett, and J. L. Boyer. 1984. Mechanisms of taurocholate transport in canalicular and basolateral rat liver plasma membrane vesicles. *J. Biol. Chem.* **259**: 10614–10622.
112. Okishio, T., and P. P. Nair. 1966. Studies on bile acids: Some observations on the intracellular localization of major bile acids in rat liver. *Biochemistry* **5**:3662–3668.
113. Boyer, J. L., M. Itabashi, and Z. Hruban. 1979. Formation of pericanalicular vacuoles during sodium dehydrocholate choleresis—A mechanism for bile acid transport: In: *The Liver: Quantitative Aspects of Structure and Function*. R. Preisig and J. Bircher, eds. Editio Cantor Aulendorf, Berne. pp. 163–178.
114. Reichen, J., M. D. Berman, and P. D. Berk. 1981. The role of microfilaments and of microtubules in taurocholate uptake by isolated rat liver cells. *Biochim. Biophys. Acta* **643**:126–133.
115. Reuben, A., R. M. Allen, and J. L. Boyer. 1983. Intrahepatic source of "biliary-like" bile acid–phospholipid–cholesterol micelles. In: *Bile Acids and Cholesterol in Health and Disease*. G. Paumgartner, A. Stiehl, and W. Gerok, eds. MTP Press, Lancaster. pp. 61–66.
116. Strange, R. C., R. Cramb, J. D. Hayes, and I. W. Percy-Robb. 1977. Partial purification of two lithocholic acid-binding proteins from rat liver 100,000g supernatants. *Biochem. J.* **165**:425–429.
117. Strange, R. C., I. A. Nimmo, and I. W. Percy-Robb. 1977. Binding of bile acids by 100,000g supernatants from rat liver. *Biochem. J.* **162**:659–664.
118. Kaplowitz, N. 1980. Physiological significance of glutathione S-transferases. *Am. J. Physiol.* **239**:G439–G444.
119. Sugiyama, Y., T. Yamada, and N. Kaplowitz. 1983. Newly identified bile acid binders in rat liver cytosol—Purification and comparison with glutathione transferases. *J. Biol. Chem.* **258**:3602–3607.
120. Accatino, L., and F. R. Simon. 1976. Identification and characterization of a bile acid receptor in isolated liver surface membranes. *J. Clin. Invest.* **57**:496–508.
121. Gonzalez, M., E. Sutherland, and F. R. Simon. 1979. Regulation of hepatic transport of bile salts: Effects of protein synthesis inhibition on excretion of bile salts and their binding to liver surface membrane fractions. *J. Clin. Invest.* **63**:684–694.
122. Graf, J., and O. H. Peterson. 1978. Cell membrane potential and resistance in liver. *J. Physiol. (London)* **284**:105–126.
123. Rollins, D. E., J. W. Freston, and D. M. Woodbury. 1980. Transport of organic anions into liver cells and bile. *Biochem. Pharmacol.* **29**:1023–1028.
124. Graf, J., A. Gautam, and J. L. Boyer. 1984. Isolated rat hepatocyte couplets: A primary secretory unit for electrophysiologic studies of bile secretory function. *Proc. Natl. Acad. Sci.* **81**:6516–6520.
125. Inoue, M., R. Kinne, J. Tran, and I. M. Arias. 1984. Taurocholate transport by rat liver canalicular membrane vesicles-evidence for the presence of an Na^+-independent transport system. *J. Clin. Invest.* **73**:659–663.
126. Scharschmidt, B. F., and R. Schmid. 1978. The micellar sink: A quantitative assessment of the association of organic anions with mixed micelles and other macromolecular aggregates in rat bile. *J. Clin. Invest.* **62**:1122–1131.
127. Tavoloni, N., J. S. Reed, and J. L. Boyer. 1978. Hemodynamic effects on determinants of bile secretion in isolated rat liver. *Am. J. Physiol.* **234**:E584–E592.
128. Reed, J. S., N. D. Smith, and J. L. Boyer. 1982. Hemodynamic effects on oxygen consumption and bile flow in isolated skate liver. *Am. J. Physiol.* **242**:G313–G318.
129. Reed, J. S., N. D. Smith, and J. L. Boyer. 1982. Determinants of biliary secretion in isolated perfused skate liver. *Am. J. Physiol.* **242:**G319–G325.
130. Erlinger, S., and D. Dhumeaux. 1974. Mechanism and control of secretion of bile water and electrolytes. *Gastroenterology* **66**:281–304.
131. Keeffe, E. B., B. F. Scharschmidt, N. M. Blankenship, and R. K. Ockner. 1979. Studies of relationships among bile flow, liver plasma membrane Na^+,K^+-ATPase, and membrane microviscosity in the rat. *J. Clin. Invest.* **64**:1590–1598.
132. Shaw, H. M., and T. J. Heath. 1974. Regulation of bile formation in rabbits and guinea pigs. *Q. J. Exp. Physiol.* **53**:93–102.
133. Graf, J., and M. Peterlik. 1976. Quabain-mediated sodium uptake and bile formation by isolated perfused liver. *Am. J. Physiol.* **230**:876–885.
134. Blitzer, B. L., and J. L. Boyer. 1978. Cytochemical localization of Na^+K^+-ATPase in the rat hepatocyte. *J. Clin. Invest.* **62**:1104–1108.
135. Latham, P. S., and M. Kashgarian. 1979. The ultrastructural localization of transport ATPase in the rat liver at non-bile canalicular plasma membrane. *Gastroenterology* **76**:988–996.
136. Poupon, R. E., and W. H. Evans. 1979. Biochemical evidence that Na^+,K^+-ATPase is located at the lateral region of the hepatocyte surface membrane. *FEBS Lett.* **108**:374–378.
137. Scharschmidt, B. F., and E. B. Keeffe. 1981. Isolation of a rat liver plasma membrane fraction of probable canalicular origin—Preparative technique, enzymatic profile, composition, and solute transport. *Biochim. Biophys. Acta* **646**:369–381.
138. Claret, M. 1979. Transport of ions in liver cells. In: *Membrane Transport in Biology,* Volume IVB. G. Giebisch, D.C. Tosteson, and H. H. Ussing, eds. Springer-Verlag, Berlin. pp. 899–920.
139. Dambach, G., and N. Friedmann. 1974. The effects of varying ionic composition of the perfusate on liver membrane potential, gluconeogenesis and cyclic AMP responses. *Biochim. Biophys. Acta* **332**:374–386.
140. Claret, B., M. Claret, and J. L. Mazet. 1973. Ionic transport and membrane potential of rat liver cells in normal and low chloride solutions. *J. Physiol. (London)* **230**:87–101.
141. Graf, J., and O. H. Petersen. 1974. Electrogenic sodium transport in mouse liver parenchymal cells. *Proc. R. Soc. London Ser. B* **187**:363–367.
142. van Rossum, G. D. V., and M. A. Russo, 1981. Ouabain-resistant mechanism of volume control and the ultrastructural organization of liver slices recovering from swelling *in vivo*. *J. Membr. Biol.* **59**:191–209.
143. Haylett, D. G., and D. H. Jenkinson. 1972. Effects of noradrenaline on potassium efflux, membrane potential and electrolyte levels in tissue slices prepared from guinea pig liver. *J. Physiol. (London)* **225**:721–750.
144. Graf, J. 1976. Sodium pumping and bile secretion. In: *The Liver: Quantitative Aspects of Structure and Function*. R. Preisig, J. Bircher, and G. Paumgartner, eds. Edito Cantor, Aulendorf. pp. 370–385.
145. Sips, H. J., M. M. van Amelsvoort, and F. van Dam. 1980. Amino acid transport in plasma-membrane vesicles from rat liver—Characterization of L-alanine transport. *Eur. J. Biochem.* **105**:217–224.
146. Ernst, S. A., and J. W. Mills. 1977. Basolateral plasma membrane localization of ouabain-sensitive sodium transport sites in the secretory epithelium of the avian salt gland. *J. Cell Biol.* **75**:74–94.
147. Epstein, F. H. 1979. The shark rectal gland: A model for the active transport of chloride. *Yale J. Biol. Med.* **52**:517–523.
148. Eveloff, J., R. Kinne, E. Kinne-Saffran, H. Murer, P. Silva, F. Epstein, J. Stoff, and W. B. Kinter. 1978. Coupled sodium and

chloride transport into plasma membrane vesicles prepared from dogfish rectal gland. *Pfluegers Arch.* **378**:87–92.
149. Frizzell, R. A., M. Field, and S. G. Schultz. 1979. Sodium-coupled chloride transport by epithelial tissues. *Am. J. Physiol.* **236**:F1–F8.
150. Anwer, M. S., and D. Hegner. 1982. Importance of solvent drag and diffusion in bile acid-dependent bile formation: Ion sbustitution studies in isolated perfused rat liver. *Hepatology* **2**:580–586.
151. Anwer, M. S., and D. Hegner. 1983. Role of inorganic electrolytes in bile acid-independent canalicular bile formation. *Am. J. Physiol.* **244**:116–124.
152. Van Dyke, R. W., J. E. Stephens, and B. F. Scharschmidt. 1982. Effect of ion substitution on bile formation by the isolated perfused rat liver. *J. Clin. Invest.* **70**:505–517.
152a. Arias, I. M., and M. Forgac. 1984. The sinusoidal domain of the plasma membrane of rat hepatocytes contains an amiloride-sensitive Na^+/H^+ antiport. *J. Biol. Chem.* **259**:5406–5408.
152b. Fuchs, R., J. Graf, M. Peterlick, and T. Thalhammer. 1984. Sodium-proton antiport in sinusoidal liver cell membrane. *Hepatology* **4**:761a.
152c. Mosley, R. H., P. J. Meier, R. Knickelbein, P. S. Aronson, and J. L. Boyer. 1984. Evidence for Na^+ -H^+ exchange in rat liver basolateral but not canalicular membrane vesicles. *Hepatology* **4**: 1040a.
153. Williams, J. A., C. D.Withrow, and D. M. Woodbury. 1971. Effects of ouabain and diphenylhydantoin on transmembrane potentials, intracellular electrolytes, and cell pH of rat muscle and liver *in vivo*. *J. Physiol. (London)* **212**:101–115.
154. Scharschmidt, B. F., and R. W. Van Dyke. 1983. Mechanisms of hepatic electrolyte transport. *Gastroenterology* **85**:1199–1214.
154a. Meier, P. J., R. Knickelbein, R. H. Mosley, J. W. Dobbins, and J. L. Boyer. 1985. Evidence for carrier modiated C1:HCO_3 exchange in canalicular rat liver plasma membrane vesicles. *J. Clin Invest.* **75**:1256–1263.
155. Hardison, W. G. M., and C. A. Wood. 1978. Importance of bicarbonate in bile salt-independent fraction of bile flow. *Am. J. Physiol.* **235**:E158–E164.
156. Klos, C., G. Paumgartner, and J. Reichen. 1979. Cation–anion gap and choleretic properties of rat bile. *Am. J. Physiol.* **236**: E434–E440.
157. Eberle, D., R. Clarke, and N. Kaplowitz. 1981. Rapid oxidation *in vitro* of endogenous and exogenous glutathione in bile of rats. *J. Biol. Chem.* **256**:2115–2117.
158. Inoue, M., R. Kinne, T. Tran, and I. M. Arias. 1983. The mechanism of biliary secretion of reduced glutathione—Analysis of transport process in isolated rat-liver canalicular membrane vesicles. *Eur. J. Biochem.* **134**:467–471.
159. Gregory, D. H., Z. R. Vlahcevic, M. F. Prugh, and T. Swell. 1978. Mechanism of secretion of biliary lipids: Role of a microtubular system in hepatocellular transport of biliary lipids in the rat. *Gastroenterology* **74**:93–100.
160. Godfrey, P. P., L. Lembra, and R. Coleman. 1982. Effects of colchicine and vinblastine on output of proteins into bile. *Biochem. J.* **208**:153–157.
161. Layden, T. J., E. Elias, and J. L. Boyer. 1978. Bile formation in the rat: The role of the paracellular shunt pathway. *J. Clin. Invest.* **62**:1375–1385.
162. Boyer, J. L., E. Elias, and T. J. Layden. 1979. The paracellular pathway and bile formation. *Yale J. Biol. Med.* **52**:61–67.
163. Wade, J. B., J. P. Revel, and V. A. DiScala. 1973. Effect of osmotic gradients on intercellular junctions of the toad bladder. *Am. J. Physiol.* **224**:407–415.
164. DiBona, D. R., and M. M. Civian. 1973. Pathways for movement of ions and water across toad urinary bladder. I. Anatomic site of transepithelial shunt pathways. *J. Membr. Biol.* **12**:101–128.
165. Graf, J. 1976. Some aspects of the role of cyclic AMP and calcium in bile formation: Studies in the isolated perfused rat liver. In: *Stimulus Secretion Coupling in the Gastrointestinal Tract.* M. Case and H. Goebell, eds. MTP, Lancaster. pp. 305–328.
166. Chenderovitch, J., E. Phocas, and M. Matureau. 1963. Effects of hypertonic solutions on bile formation. *Am. J. Physiol.* **205**:863–867.
167. Guzelian, P., and J. L. Boyer. 1974. Glucose reabsorption from bile: Evidence for a biliohepatic circulation. *J. Clin. Invest.* **53**:526–535.
168. Owen, C. A. 1977. Isolated rat liver needs calcium to make bile. *Proc. Soc. Exp. Biol. Med.* **155**:314–317.
169. Baker, A. L., and M. M. Kaplan. 1976. Effects of cholera enterotoxin, glucagon, and dibutyryl cyclic AMP on rat liver alkaline phosphatase, bile flow, and bile composition. *Gastroenterology* **70**:577–581.
170. Poupon, R. E., M. L. Dol, M. Dumont, and S. Erlinger. 1978. Evidence against a physiological role of cAMP in choleresis in dogs and rats. *Biochem. Pharmacol.* **27**:2413–2416.
171. Morris, T. Q. 1972. Choleretic responses to cyclic AMP and theophylline in the dog. *Gastroenterology* **62**:187a.
172. Barnhart, J. L., and B. Combes. 1975. Characteristics common to choleretic increments of bile induced by theophylline, glucagon, and SQ-2009 in the dog. *Proc. Soc. Exp. Biol. Med.* **150**:591–596.
173. Klaassen, C. D. 1971. Does bile acid secretion determine bile production in rats? *Am. J. Physiol.* **220**:667–673.
174. Macarol, V., T. Q. Morris, K. J. Baker, and S. E. Bradley. 1970. Hydrocortisone choleresis in the dog. *J. Clin. Invest.* **49**:1714–1723.
175. Zsigmond, G., and B. Solymoss. 1974. Increased canalicular bile production induced by pregnenolone-16α-carbonitrile, spironolactone and cortisol in rats. *Proc. Soc. Exp. Biol. Med.* **145**:631–635.
176. Capron, J. P., M. Dumont, G. Feldmann, and S. Erlinger. 1977. Barbiturate-induced choleresis: Possible independence from microsomal enzyme induction. *Digestion* **15**:556–565.
177. Chivrac, D., M. Dumont, and S. Erlinger. 1978. Lack of parallelism between microsomal enzyme induction and phenobarbital-induced hypercholeresis in the rat. *Digestion* **17**:516–525.
178. Klaassen, C. D. 1969. Biliary flow after microsomal enzyme induction. *J. Pharmacol. Exp. Ther.* **168**:218–223.
179. Gumucio, J. J., and V. C. Valdivieso. 1971. Studies on the mechanism of ethinylestradiol impairment of bile flow and bile salt excretion in the rat. *Gastroenterology* **61**:339–344.
180. Simon, F. R., M. Gonzalez, E. Sutherland, L. Accatino, and R. A. Davis. 1980. Reversal of ethinyl estradiol-induced bile secretory failure with Triton WR-1339. *J. Clin. Invest.* **65**:851–860.
181. Davis, R. A., F. Kern, R. Showalter, E. Sutherland, M. Sinensky, and F. R. Simon. 1978. Alterations of hepatic Na^+,K^+-ATPase and bile flow by estrogen-effects on liver surface membrane lipid structure and function. *Proc. Natl. Acad. Sci. USA* **75**:4130–4134.
182. Kern, F. 1978. Effect of estrogens on the liver. *Gastroenterology* **75**:512–522.
182a. Berr, F., F. R. Simon, and J. Reichen. 1984. Ethinylestradiol impairs Bile salt uptake and Na-K pump function of rat hepatocytes. *Amer. J. Physiol.* **247**:6437–6443.
183. Boyer, J. L., S. Lagarde, O. C. Ng, and R. Groszmann. 1981. Enhanced biliary regurgitation of ^{14}C-sucrose (^{14}C-S) and lanthanum (La^{++}) in ethinyl estradiol (EE) treated rats following retrograde bile duct infusions—A possible mechanism for intrahepatic cholestasis. *Hepatology* **1**:498a.
184. Preisig, R., H. Strebel, G. Egger, and V. Macarol. 1972. Effect of vasopressin on hepatocyte and ductual bile formation in the dog. *Experientia* **28**:1436–1437.
185. Lauterburg, B., G. Paumgartner, and R. Preisig. 1975. Prostaglandin-induced choleresis in the rat. *Experientia* **31**:1191–1193.
186. Kaminski, D. L., M. Ruwart, and L. L. Willman. 1974. The effect of prostaglandin A_1 and E_1 on canine hepatic bile flow. *Surg. Res.* **18**:391–397.
187. Sokoloff, J., and R. N. Berk. 1973. The effect of prostaglandin E_2 on bile flow and the biliary excretion of iopanoic acid. *Invest. Radiol.* **8**:9–12.

188. Karup, N., J. A. Larsen, and A. Munck. 1976. Secretin like choleretic effects of prostaglandin E_1 and E_2 in cats. *J. Physiol. (London)* **254**:813–820.
189. Larsen, J. A., and K. D. Christensen. 1978. Insulin-stimulated bile formation in cats. *Acta Physiol. Scand.* **102**:301–309.
190. Snow, J. R., and R. S. Jones. 1978. The effect of insulin on bile salt-independent canalicular secretion. *Surgery* **83**:458–463.
191. Thomsen, O. Ø, and J. A. Larsen, 1981. The effect of glucagon, dibutyrlic cyclic AMP and insulin on bile production in the intact rat and the perfused rat liver. *Acta Physiol. Scand.* **111**:23–30.
192. Geist, R. E., and R. S. Jones. 1971. Effect of selective and truncal vagotomy on insulin-stimulated bile secretion in dogs. *Gastroenterology* **60**:566–571.
193. Thomsen, O. Ø, and J. A. Larsen, 1983. Importance of perfusate hematocrit for insulin- and glucagon-induced choleresis in the perfused rat liver. *Am. J. Physiol.* **245**:G59–G63.
194. Thomsen, O. Ø. 1983. Stimulatory effect of bile acids on insulin-induced choleresis in the rat. *Am. J. Physiol.* **244**:G301–G307.
194a. Thomsen, O. Ø. 1984. Mechanism and regulation of hepatic bile production. *Scand. J. Gastroenterol.* **19**:(Suppl 97) 1–52.
195. Dyck, W. P., and H. D. Janowitz. 1971. Effect of glucagon on hepatic bile secretion in man. *Gastroenterology* **60**:400–404.
196. Jones, R. S., R. E. Geist, and A. D. Hall. 1971. The choleretic effects of glucagon and secretin in the dog. *Gastroenterology* **60**:64–68.
197. Morris, T. Q., G. F. Sardi, and S. E. Bradley. 1967. Character of glucagon-induced choleresis. *Fed. Proc.* **26**:774a.
198. Beaugie, J. M. 1972. Gastrointestinal hormones and bile flow. *Ann. R. Coll. Surg. Engl.* **50**:164–181.
199. Pissidis, A. G., C. T. Bombeck, F. Merchant, and L. M. Nyhus. 1969. Hormonal regulation of bile secretion: A study in the isolated perfused liver. *Surgery* **66**:1075–1084.
200. Forsmann, W. G., and S. Ito. 1977. Hepatocyte innervation in primates. *J. Cell Biol.* **74**:299–313.
201. Sutherland, S. D. 1965. The intrinsic innervation of the liver. *Rev. Int. Hepatol.* **15**:569–578.
202. Nevasaari, K., and N. T. Kaerki. 1976. The effect of acetylcholine on bile flow. *Arch. Int. Pharmacodyn. Ther.* **221**:283–293.
203. Fritz, M. E., and F. B. Brooks. 1963. Control of bile flow in the cholecystectomized dog. *Am. J. Physiol.* **204**:825–828.
204. Baldwin, J., F. W. Heer, R. Albo, O. Paloso, L. Ruby, and W. Silen. 1966. Effect of vagus nerve stimulation on hepatic secretion of bile in human subjects. *Am. J. Surg.* **111**:66–69.
205. Harty, R. F., R. C. Rose, and D. L. Nahrwald. 1974. Stimulation of hepatic bile secretion by dopamine. *J. Surg. Res.* **17**:359–363.
206. Ho, K. J., and J. L. Drummond. 1975. Circadian rhythm of biliary excretion and its control mechanisms in rats with chronic biliary drainage. *Am. J. Physiol.* **229**:1427–1437.
207. Ho, K. J. 1976. Circadian distribution of bile acids in the enterohepatic circulatory system in rats. *Am. J. Physiol.* **230**:1331–1335.
208. Mitropoulos, K. A. 1975. Diurnal variations in bile acid metabolism. In: *The Hepatobiliary System.* W. Taylor, ed. Plenum Press, New York. pp. 409–427.
209. Hanzon, V. 1952. Liver cell secretion under normal and pathologic conditions studied by fluorescence microscopy on living rats. *Acta Physiol. Scand. Suppl.* 101 28:1–268.
210. Gebhardt, R., and W. Jung. 1982. Primary cultures of rat hepatocytes as a model system of canalicular development, biliary secretion, and intrahepatic cholestasis. I. Distribution of filipin–cholesterol complexes during de novo formation of bile canaliculi. *Eur. J. Cell Biol.* **29**:68–76.
211. Goresky, C. A. 1977. Hepatic membrane carrier transport processes: Their involvement in bilirubin uptake. In: *The Chemistry and Physiology of Bile Pigments.* P. D. Berk and N. I. Berlin, eds. U.S. Government Printing Office, Washington, D.C. pp. 265–281.
212. Lee, K., and L. M. Gartner. 1978. Bilirubin binding by plasma proteins: A critical evaluation of methods and clinical implications. In: *Reviews in Perinatal Medicine,* Volume 2. E. M. Scarpelli and E. V. Cosmi, eds. Raven Press, New York. pp. 318–343.
213. Goodman, D. S. 1958. The interaction of human serum albumin with long-chain fatty acid anions. *J. Am. Chem. Sco.* **80**:3892–3898.
214. Baker, K. J., and S. E. Bradley. 1966. Binding of sulfobromophthalein (BSP) sodium by plasma albumin: Its role in hepatic BSP extraction. *J. Clin. Invest.* **45**:281–287.
215. Goresky, C. A. 1964. Initial distribution and rate of uptake of sulfobromophthalein in the liver. *Am. J. Physiol.* **207**:13–26.
216. Stremmel, W., N. Tavoloni, and P. D. Berk. 1983. Uptake of bilirubin by the liver. *Semin. Liver Dis.* **3**:1–10.
217. Weisiger, R., J. Gollan, and R. Ockner. 1981. Receptor for albumin on the liver cell surface may mediate uptake of fatty acids and other albumin-bound substances. *Science* **211**:1048–1051.
218. Weisiger, R. A., J. L. Gollan, and R. K. Ockner. 1982. The role of albumin in hepatic uptake processes. *Prog. Liver Dis.* **7**:71–85.
219. Weisiger, R. A., C. Zacks, N. Smith, and J. L. Boyer. 1984. Effect of albumin on extraction of sulfobromophthalein by perfused elasmobranch liver: Evidence for dissociation-limited uptake. *Hepatology* **4**:492–501.
220. Goresky, C. A. 1965. The hepatic uptake and excretion of sulfobromophthalein and bilirubin. *Can. Med. Assoc. J.* **92**:851–857.
221. Hunton, D. B., J. L. Bollman, and H. N. Hoffman. 1961. II. The plasma removal of indocyanine green and sulfobromophthalein: Effect of dosage and blocking agents. *J. Clin. Invest.* **40**:1648–1655.
222. Bloomer, J. R., and J. Zaccaria. 1976. Effect of graded loads on bilirubin transport by perfused rat liver. *Am. J. Physiol.* **230**:736–742.
223. Scharschmidt, B. F., J. G. Waggoner, and P. D. Berk. 1975. Hepatic organic anion uptake in the rat. *J. Clin. Invest.* **56**:1280–1292.
224. Paumgartner, G., and J. Reichen. 1976. Kinetics of hepatic uptake of unconjugated bilirubin. *Clin. Sci. Mol. Med.* **51**:169–176.
225. Laperche, Y., A.-M. Preux, G. Feldmann, J.-L. Maha, and P. Berthelot. 1981. Effect of fasting on organic anion uptake by isolated rat liver cells. *Hepatology* **1**:617–621.
226. Anwer, M. D., and D. Hegner. 1978. Effect of organic anions on bile acid uptake by isolated rat hepatocytes. *Hoppe-Seylers Z. Physiol. Chem.* **359**:1027–1030.
227. Maha, J.-L., P. Duvaldestin, D. Dhumeaux, and P. Berthelot. 1977. Biliary transport of cholephilic dyes: Evidence for two different pathways. *Am. J. Physiol.* **232**:E445–E450.
228. Schwarz, L. R., R. Gozt, and C. D. Klaassen. 1979. Uptake of sulfobromophthalein–glutathione conjugate by isolated hepatocytes. *Am. J. Physiol.* **239**:G118–G123.
229. von Dippe, P., P. Drain, and D. Levy. 1983. Synthesis and transport characteristics of photo affinity probes for the hepatocyte bile acid transport system. *J. Biol. Chem.* **258**:8890–8895.
230. Reichen, J., and P. D. Berk. 1979. Isolation of an organic anion binding protein from rat liver plasma membrane fractions by affinity chromatography. *Biochem. Biophys. Res. Commun.* **91**:484–489.
231. Wolkoff, A. W., and C. T. Chung. 1980. Identification, purification and partial characterization of an organic anion binding protein from rat liver cell plasma membrane. *J. Clin. Invest.* **65**:1152–1161.
232. Stremmel, W., M. Gerber, V. Glezerov, S. N. Thung, S. Kochwa, and P. D. Berk. 1982. Physicochemical and immunohistological studies of a sulfobromophthalein- and bilirubin-binding protein from rat liver plasma membranes. *Hepatology* **2**: 717a.
233. Tiribelli, C., G. Lunazzi, G. L. Luciana, E. Panfivi, B. Gazzin, G. Liut, G. Sandri, and G. Sottocasa. 1978. Isolation of a sulfobromophthalein-binding protein from hepatocyte plasma membrane. *Biochim. Biophys. Acta* **532**:105–112.
234. Stremmel, W., G. Strohmeyer, F. Borchard, S. Kochwa, and P.

D. Berk. 1983. Isolation and partial characterization of a fatty acid binding protein from rat liver plasma membranes. *Hepatology* **3**:823a.

235. Kramer, W., U. Bickel, H.-P. Buscher, W. Gerok, and G. Kurz. 1980. Binding proteins for bile acids in membranes of hepatocytes revealed by photo affinity labelling. *Hoppe-Seylers Z. Physiol. Chem.* **361**:1307a.
236. Levy, D., and P. von Dippe. 1983. Reconstitution of the bile acid transport system derived from hepatocyte sinusoidal membranes. *Hepatology* **3**:837a.
237. Berk, P. D., T. F. Blaschke, and J. G. Waggoner. 1972. Defective BSP clearance in patients with constitutional hepatic dysfunction (Gilbert's syndrome). *Gastroenterology* **63**:472–481.
238. Martin, J. F., J. M. Vierling, A. W. Wolkoff, B. F. Scharschmidt, J. Vergalla, J. G. Waggoner, and P. D. Berk. 1976. Abnormal hepatic transport of indocyanine green in Gilbert's syndrome. *Gastroenterology* **70**:385–391.
239. Nambu, M., T. Namihisa, T. Yamashiro, H. Ohama, M. Maeda, and H. Ueda. 1980. Plasma disappearance of serum bile acids in patients with constitutional hyperbilirubinemia and constitutional ICG excretory defect. *Jpn. J. Gastroenterol.* **77**:1369–1377.
240. Ohkubo, H., K. Okuda, and S. Iida. 1981. A constitutional unconjugated hyperbilirubinemia combined with indocyanine green intolerance: A new functional disorder? *Hepatology* **1**:319–324.
241. Ketterer, B., B. Neumcke, and P. Lauger. 1971. Transport mechanism of hydrophobic ions through lipid bilayer membranes. *J. Membr. Biol.* **5**:225–245.
242. Benz, R., P. Lauger, and K. Janko. 1976. Transport kinetics of hydrophobic ions in lipid bilayer membranes: Charge–pulse relaxation studies. *Biochim. Biophys. Acta* **455**:701–720.
243. Levi, A. J., Z. Gatmaitan, and I. M. Arias. 1969. Two hepatic cytoplasmic protein fractions, Y and Z and their possible role in hepatic uptake of bilirubin, sulphobromphthalein, and other anions. *J. Clin. Invest.* **48**:2156–2167.
244. Littwack, G., B. Ketterer, and I. M. Arias. 1971. Ligandin, a hepatic protein which binds steroids, bilirubin, carcinogens and a number of exogenous organic anions. *Nature (London)* **234**:466–467.
245. Listowsky, I., Z. Gaitmaitan, and I. M. Arias. 1978. Ligandin retains and albumin loses bilirubin binding capacity in liver cytosol. *Proc. Natl. Acad. Sci. USA* **75**:1213–1216.
246. Hales, B. F., and A. H. Neims. 1976. Developmental aspects of glutathione S-transferase B (ligandin) in rat liver. *Biochem. J.* **160**:231–236.
247. Levi, A. J., Z. Gaitmaitan, and I. M. Arias. 1970. Deficiency of hepatic organic anion binding protein, impaired organic anion uptake by liver and "physiologic" jaundice in newborn monkeys. *N. Engl. J. Med.* **283**:1136–1139.
248. Fleischner, G., J. Robbins, and I. M. Arias. 1972. Immunological studies of Y protein: A major cytoplasmic organic anion-binding protein in rat liver. *J. Clin. Invest.* **51**:677–684.
249. Reyes, A., A. J. Levi, Z. Gaitmaitan, and I. M. Arias. 1971. Studies of Y & Z, two hepatic cytoplasmic organic anion-binding proteins: Effect of drugs, chemicals, hormones, and cholestasis. *J. Clin. Invest.* **50**:2242–2252.
250. Kamisaka, K., I. Kistowsky, Z. Gaitmaitan, and I. M. Arias. 1975. Interactions of bilirubin and other ligands with ligandin. *Biochemistry* **14**:2175–2180.
251. Boyer, J. L., J. Schwarz, and N. Smith. 1976. Biliary secretion in elasmobranchs. II. Hepatic uptake and biliary excretion of organic anions. *Am. J. Physiol.* **230**:974–981.
252. Wolkoff, A. W., C. A. Goresky, J. Sellin, Z. Gaitmaitan, and I. M. Arias. 1979. Role of ligandin in transfer of bilirubin from plasma into liver. *Am. J. Physiol.* **236**:E638–E648.
253. Sugiyama, Y., T. Yamada, and N. Kaplowitz. 1982. Newly identified organic anion-binding proteins in rat cytosol. *Biochim. Biophys. Acta* **709**:342–352.
254. Redick, J. A., W. B. Jakoby, and J. Baron. 1982. Immunohistochemical localization of glutathione S-transferases in livers of untreated rats. *J. Biol. Chem.* **257**:15200–15203.

254a. Vessey, D. A., J. Whitney, and J. L. Gollan. 1983. The role of conjugation reaction in enhancing biliary excretion of bile acids. *Biochem. J.* **214**:923–927.

255. Gollan, J., and R. Schmid. 1982. Bilirubin update: Formation, transport and metabolism. *Prog. Liver Dis.* **8**:261–283.
256. Whelan, G., J. Hoch, and B. Combes. 1970. A direct assessment of the importance of conjugation for biliary transport of sulfobromphthalein sodium. *J. Lab. Clin. Med.* **75**:542–557.
257. Wolkoff, A. W. 1983. Bilirubin metabolism and hyperbilirubinemia. *Semin. Liver Dis.* **3**:1–83.
258. Gutstein, S., S. Alpert, and I. M. Arias. 1968. Studies of hepatic excretory function. IV. Biliary excretion of sulfobromophthalein sodium in a patient with the Dubin–Johnson syndrome and a biliary fistula. *Isr. J. Med. Sci.* **4**:36–40.
259. Alpert, S., M. Mosher, A. Shanske, and I. M. Arias. 1969. Multiplicity of hepatic excretory mechanisms for organic anions. *J. Gen. Physiol.* **53**:238–247.
260. Cornelius, C. E. 1969. Organic anion transport in mutant sheep with congenital hyperbilirubinemia. *Arch. Environ. Health* **19**:852–856.
261. Clarenberg, R., and C.-C. Kao. 1973. Shared and separate pathways for biliary excretion of bilirubin and BSP in rats. *Am. J. Physiol.* **225**:192–200.
262. Boyer, J. L., R. L. Scheig, and G. Klatskin. 1970. The effect of sodium taurocholate on the hepatic metabolism of sulfobromophthalein sodium (BSP): The role of bile flow. *J. Clin. Invest.* **49**:206–215.
263. Vonk, R. J., M. Danhof, T. Coenraads, A. B. D. van Doorn, K. Keulemans, A. H. J. Scaf, and D. K. F. Meijer. 1979. Influence of bile salts on hepatic transport of dibromosulphthalein. *Am. J. Physiol.* **237**:E524–E534.
264. Loeb, P. M., J. L. Barnhart, and R. N. Berk. 1978. The dependence of biliary excretion of iopanoic acid on bile salts. *Gastroenterology* **74**:174–181.
265. Vonk, R. J., A. B. D. van Doorn, A. H. J. Scaf, and D. K. F. Meijer. 1977. Choleresis and hepatic transport mechanisms. III. Binding of ouabain and K-strophanthoiside to biliary micelles and influence of choleresis on their biliary excretion. *Naunyn-Schmiedebergs Arch. Pharmacol.* **300**:173–177.
266. Delage, Y., M. Dumont, and S. Erlinger. 1976. Effect of glycodihydrofusidate on sulfobromophthalein transport maximum in the hamster. *Am. J. Physiol.* **231**:1875–1878.
267. Schanker, L. S., and H. M. Solomon. 1963. Active transport of quaternary ammonium compounds in bile. *Am. J. Physiol.* **204**:829–832.
268. Schanker, L. S. 1968. Secretion of organic compounds in bile. In: *Handbook of Physiology,* Section 6. C. F. Code, ed. American Physiological Society, Washington, D. C. pp. 2433–2449.
269. Russell, J. Q., and C. D. Klaassen. 1972. Species variation in the biliary excretion of ouabain. *J. Pharmacol. Exp. Ther.* **103**:513–519.
270. Meijer, D. K. F., R. J. Vonk, E. J. Scholtens, and W. G. Levine. 1976. The influence of dehydrocholate on hepatic uptake and excretion of ^{3}H-taurocholate and ^{3}H-ouabain. *Drug Metab. Dispos.* **4**:1–7.
271. Erttmann, R. R., and K. H. Damm. 1975. Influence of bile flow, theophylline and some organic anions on the biliary excretion of ^{3}H-ouabain in rats. *Arch. Int. Pharmacodyn. Ther.* **218**:290–298.
272. Balint, J. A., E. C. Kyriakides, H. L. Spitzer, and E. S. Morrison. 1965. Lecithin fatty acid composition in bile and plasma of man, dogs, rats, and oxen. *J. Lipid Res.* **6**:96–99.
273. Coleman, R., S. Iqbal, P. P. Godfrey, and D. Billington. 1979. Membranes and bile formation. *Biochem. J.* **178**:201–208.
274. Mazer, N. A., and M. C. Carey. 1980. Quasielastic light scattering studies of aqueous biliary lipid systems: Cholesterol solubilization and precipitation in model bile solutions. *Biochemistry* **19**:601–615.
275. Reuben, A., K. E. Howell, and J. L. Boyer. 1982. Effects of taurocholate on the size of mixed lipid micelles and their associa-

tions with pigment and proteins in rat bile. *J. Lipid Res.* **23**:1039–1052.
276. Turley, S. D., and J. M. Dietschy. 1979. Regulation of biliary cholesterol output in the rat: Dissociation from the rate of hepatic cholesterol synthesis, the size of the hepatic cholesteryl ester pool, and the hepatic uptake of chylomicron cholesterol. *J. Lipid Res.* **20**:923–934.
277. Schwartz, C. C., M. Berman, Z. R. Vlahcevic, L. G. Halloran, D. H. Gregory, and L. Swell. 1978. Multicompartmental analysis of cholesterol metabolism in man: Characterization of the hepatic bile acid and biliary cholesterol precursor sites. *J. Clin. Invest.* **61**:408–423.
278. Gregory, D. H., Z. R. Vlahcevic, P. Schatzki, and L. Swell. 1975. Mechanism of secretion of biliary lipids. I. Role of bile canalicular and microsomal membranes in the synthesis and transport of biliary lecithin and cholesterol. *J. Clin. Invest.* **55**:105–114.
279. Robins, S. J., and H. Brunengraber. 1982. Origin of biliary cholesterol and lecithin in the rat: Contribution of new synthesis and preformed hepatic stores. *J. Lipid Res.* **23**:604–608.
280. Small, D. M. 1970. The formation of gallstones. *Adv. Inter. Med.* **16**:243–264.
281. Gregory, D. H., Z. R. Vlahcevic, P. Schatzki, and L. Swell. 1975. Mechanism of secretion of biliary lipid. I. Role of bile canalicular and microsomal membranes in the synthesis and transport of biliary lecithin and cholesterol. *J. Clin. Invest.* **55**:105–114.
282. Lafont, H., D. Lairon, N. Domingo, G. Nalbone, and J. C. Hauton. 1974. Does a lecithin–polypeptide association in bile originate from membrane structural subunits? *Biochimie* **56**:465–468.
283. Sewell, R. B., S. J. T. Mao, T. Kawamoto, and N. F. LaRusso. 1983. Apolipoproteins of high, low, and very low density lipoproteins in human bile. *J. Lipid Res.* **24**:391–401.
284. Turley, S. D., and J. M. Dietschy. 1982. Cholesterol metabolism and excretion. In: *The Liver: Biology and Pathobiology*. I. Arias, H. Popper, D. Schacter, and D. A. Shafritz, eds. Raven Press, New York. pp. 467–492.
285. Montet, J. C., A. M. Montet, A. Gerolami, and J. C. Hauton. 1975. Effect of 3-acetoxy fusidate on the biliary secretion of lipids in the rat. *Biol. Gastroenterol.* **8**:53–62.
286. Hardison, W. G. M., and J. T. Apter. 1972. Micellar theory of biliary cholesterol excretion. *Am. J. Physiol.* **222**:61–67.
287. Soloway, R. D., A. F. Hofmann, P. J. Thomas, L. J. Schoenfield, and P. D. Klein. 1973. Triketocholanic (dehydrocholic) acid: Hepatic metabolism and effect on bile flow and biliary lipid secretion in man. *J. Clin. Invest.* **52**:715–724.
288. Apstein, M. D., and S. J. Robbins. 1982. Effect of organic anions on biliary lipids in the rat. *Gastroenterology* **83**:1120–1126.
289. Schaffer, E. A., and R. M. Preshaw. 1981. Effects of sulfobromophthalein excretion on biliary lipid secretion in humans and dogs. *Am. J. Physiol.* **240**:G85–G89.
290. Apstein, M. D., and A. R. Russo. 1982. Where does bilirubin inhibit biliary phospholipid and cholesterol secretion? *Hepatology* **2**:143a.
291. Redinger, R. N., and D. M. Small. 1973. Primate biliary physiology. VIII. The effect of phenobarbital upon bile salt synthesis and pool size, biliary lipid secretion and bile composition. *J. Clin. Invest.* **52**:161–172.
292. Strasberg, S. M., R. N. Redinger, D. M. Small, and R. H. Egdahl. 1982. The effect of elevated biliary tract pressure on biliary lipid metabolism and bile flow in nonhuman primates. *J. Lab. Clin. Med.* **99**:342–353.
293. Dive, C. H., and J. F. Heremans. 1974. Nature and origin of the proteins in bile. I. *Eur. J. Clin. Invest.* **4**:235–239.
294. Mullock, B. M., M. Dobrata, and R. H. Hinton. 1978. Sources of the proteins in rat liver. *Biochim. Biophys. Acta* **543**:497–507.
295. Godfrey, P. P., M. J. Warner, and R. Coleman. 1981. Enzymes and proteins in bile. *Biochem. J.* **196**:11–16.
296. Evans, W. H., T. Kremmer, and J. G. Culvenor. 1976. Role of membranes in bile formation: Comparison of the composition of bile and a liver bile canalicular plasma membrane fraction. *Biochem. J.* **154**:589–595.
297. Kakis, G., and I. M. Yousef. 1978. Protein composition of rat bile. *Can. J. Biochem.* **56**:287–290.
298. Mullock, B. M., R. H. Hinton, M. Dobrota, J. Peppard, and E. Orlans. 1980. Distribution of secretory component in hepatocytes and its mode of transfer into bile. *Biochem. J.* **190**:819–826.
299. Hinton, R. H., M. Dobrota, and B. M. Mullock. 1980. Haptoglobin-mediated transfer of haemoglobin from serum into bile. *FEBS Lett.* **112**:247–250.
300. Jones, A. L., R. H. Renston, and S. J. Burwen. 1982. Uptake and intracellular disposition of plasma-derived proteins and apoproteins by hepatocytes. *Prog. Liver Dis.* **8**:51–69.
301. Barnwell, S. G., P. P. Godfrey, P. J. Lowe, and R. Coleman. 1983. Biliary protein output by isolated perfused rat livers. *Biochem. J.* **210**:549–557.
302. Thomas, P., C. A. Toth, and N. Zamcheck. 1982. The mechanism of biliary excretion of α_1-acid glycoprotein in the rat: Evidence for a molecular weight-dependent, nonreceptor-mediated pathway. *Hepatology* **2**:800–803.
303. Sternlieb, I. 1972. Functional implications of human portal and bile ductular ultrastructure. *Gastroenterology* **63**:321–327.
304. Hardwicke, J., J. G. Rankin, K. J. Baker, and R. Preisig. 1964. The loss of protein in human and canine hepatic bile. *Clin. Sci.* **26**:509–517.
305. Dive, C. H., R. A. Nadalini, J. P. Vaerman, and J. F. Heremans. 1974. Origin and nature of the proteins in bile. II. A comparative analysis of serum, hepatic lymph and bile proteins in the dog. *Eur. J. Clin. Invest.* **4**:241–246.
306. Nagura, H., P. D. Smith, P. K. Nakane, and W. R. Brown. 1981. IgA in human bile and liver. *J. Immunol.* **126**:587–595.
307. Renston, R. H., D. G. Maloney, A. L. Jones, G. T. Hradek, K. Y. Wong, and I. D. Goldfine. 1980. Bile secretory apparatus: Evidence for a vesicular transport mechanism for proteins in the rat, using horseradish peroxidase and ^{125}I-insulin. *Gastroenterology* **78**:1373–1388.
308. Lee, S. P., T. H. Lim, and A. J. Scott. 1979. Carbohydrate moieties of glycoproteins in human hepatic and gallbladder bile, gallbladder mucosa and gallstones. *Clin. Sci. Mol. Med.* **56**:533–538.
309. LaMont, J. T., A. S. Ventola, B. W. Trotman, and R. D. Soloway. 1983. Mucin content of human pigment gallstones. *Hepatology* **3**:377–382.
310. Folsch, U. R., and K. G. Wormsley. 1977. The amino acid composition of rat bile. *Experientia* **33**:1055–1056.
311. Fisher, M. M., and M. Kerly. 1964. Amino acid metabolism in the perfused rat liver. *J. Physiol. (London)* **174**:273–294.
312. Bartoli, G., and H. Sies. 1978. Reduced and oxidized glutathione efflux from liver. *FEBS Lett.* **86**:89–91.
313. Bartoli, G. M., D. Haeberle, and H. Sies. 1978. Glutathione efflux from perfused rat liver and the relation to glutatione uptake by the kidney. In: *Functions of Glutathione in Liver and Kidney*. H. Sies and A. Wendel, eds. Springer-Verlag, Berlin. pp. 27–31.
314. Eberle, D., R. Clarke, and N. Kaplowitz. 1981. Rapid oxidation *in vitro* of endogenous and exogenous glutathione in bile of rats. *J. Biol. Chem.* **256**:2115–2117.
315. Kaplowitz, N., D. E. Eberle, J. Petrini, J. Touloukian, M. C. Corvasce, and J. Kuhlenkamp. 1983. Factors influencing the efflux of hepatic glutathione into bile in rats. *J. Pharmacol. Exp. Ther.* **244**:141–147.
316. Prasad, A. S. 1976. *Trace Elements in Human Health and Disease*. eds. Academic Press, New York.
317. Klaassen, C. D. 1975. Biliary excretion of xenobiotics. *CRC Crit. Rev. Toxicol.* **4**:1–30.
318. Klaassen, C. D. 1976. Biliary excretion of metals. *Drug Metab. Rev.* **5**:165–196.
319. Frommer, D. 1974. Defective biliary excretion of copper in Wilson's disease. *Gut* **15**:125.
320. Ballatori, N., and T. W. Clarkson. 1983. Biliary transport of

glutathione and methylmercury. *Am. J. Physiol.* **144**:G435–G441.

321. Grasbeck, R., W. Nyberg, and P. Reizenstein. 1958. Biliary and fecal vit B_{12} excretion in man: An isotope study. *Proc. Soc. Exp. Biol. Med.* **97**:780–784.
322. Green, R., D. W. Jacobsen, S. V. Van Tonder, M. C. Kew, and J. Metz. 1981. Enterohepatic circulation of cobalamin in the non-human primate. *Gastroenterology* **81**:773–776.
323. Kumar, R., S. Nagubandi, V. R. Mattox, and J. M. Londowski. 1980. Enterohepatic physiology of 1,25-dihydroxyvitamin D_3. *J. Clin. Invest.* **65**:277–284.
324. Aldercreutz, H., and T. Luukainen. 1967. Biochemical and clinical aspects of the enterohepatic circulation of estrogens. *Acta Endocrinol. Copenhagen Suppl.* **124**:101–140.
325. Laatikainen, T. 1970. Excretion of neutral steroid hormones in human bile. *Ann. Clin. Res.* **2**(Suppl. 5):1–28.
326. Taylor, W. 1971. The excretion of steroid hormone metabolites in bile and feces. *Vitam. Horm. (N.Y.)* **29**:201–285.
327. Avner, D. L., and M. M. Berenson. 1982. Effect of choleretics on canalicular transport of protoporphyrin in the rat liver. *Am. J. Physiol.* **242**:G347–G353.
328. Plaa, G. L., and B. G. Priestly. 1976. Intrahepatic cholestasis induced by drugs and chemicals. *Pharmacol. Rev.* **28**:207–273.
329. Chenderovitch, J., S. Troupel, H. Renault, and J. Caroli. 1961. Le transfert du Na^{24} et du K^{52} du sang dans la bile chez le cobaye au cours de la cholérèse a débit bloqué (''stop-flow analysis''). *Rev. Fr. Etud. Clin. Biol.* **6**:584–589.
330. Levine, R. A., and R. C. Hall. 1976. cAMP in secretin choleresis: Evidence for regulatory role in man and baboons but not in dogs. *Gastroenterology* **70**:537–544.
331. Rene, E., R. G. Danzinger, A. F. Hofmann, and M. Nakagaki. 1983. Pharmacologic effect of somatostatin on bile formation in the dog: Enhanced ductular reabsorption on the major mechanism of anticholeresis. *Gastroenterology* **84**:120–129.
332. Lewis, M. H., A. L. Baker, and A. R. Moossa. 1982. Effect of somatostatin on determinants of bile flow in unanesthetized dogs. *Ann. Surg.* **195**:97–103.
333. Ricci, G. L., and J. Fevery. 1981. Cholestatic action of somatostatin in the rat: Effect on the different fractions of bile secretion. *Gastroenterology* **81**:555–562.

CHAPTER 13

The Regulation of Glomerular Filtration Rate in Mammalian Kidneys

L. Gabriel Navar, P. Darwin Bell, and Andrew P. Evan

1. Introduction

During organogenesis of the kidney, small S-shaped metanephric vesicles form from the undifferentiated metanephric mesoderm. The cephalic part of each primitive renal tubule invaginates to form a cleft within which mesenchymal cells, destined to become the mesangium, accumulate. The formation of a continuous basement membrane serves to separate the intracleftal mesenchymal cells from the tubular cells. As the mesenchymal cells increase in number, vascular tissue enters the cleft to form the capillaries of the glomerular tuft. Eventually, the spherical glomerular capsule is formed and encloses the capillaries, leaving the afferent and efferent arterioles to maintain continuity with the vascular system.

The ontogeny of the glomerular corpuscle reflects its functional uniqueness as an apparatus for ultrafiltering vast quantities of fluid. The capability of this structure is enhanced further by the development of vascular smooth muscle on both the preglomerular and the postglomerular arteriolar connections to the glomerular capillary tuft, the appearance of granules in the myoepithelial cells of the afferent arteriole near the glomerular tuft, and the early and continued association of the macula densa segment of the distal tubule with the vascular pole of its own glomerular tuft. The glomerular capillaries develop numerous anastomoses within the glomerular tuft, forming a true network which serves as a highly efficient filtration apparatus.[1,2]

Kidneys from various mammals differ in their degree of complexity. The compound multirenculated kidney of man weighs approximately 150 g and contains 1 million nephrons, or approximately 7000 nephrons/g kidney. Of the mammalian experimental animals used routinely in studies of renal function, the dog has approximately 580,000 nephrons per kidney (average weight of 46 g), or approximately 12,000–13,000 nephrons/g kidney, while the rat possesses a much higher glomerular density in excess of 30,000 nephrons/g kidney.[3–5] These differences in glomerular density reflect differences in size and filtration capability of the individual glomeruli. Representative values for man and several laboratory animals commonly used in renal function studies are shown in Table I. In general, the size of the glomeruli corresponds to the calculated single-nephron GFR (SNGFR). The diameters of glomeruli range from 120 to 160 μm for the rat, 140 to 160 μm for the rabbit, 180 to 250 μm for the dog, to as large as 300 μm for man.[3–6] Estimates of glomerular capillary surface area require knowledge of the number and length of the individual glomerular capillaries, as well as the relative area available for volume flow from the capillaries into Bowman's capsule. These uncertainties have led to a wide range of values for the effective surface area, being about 0.19–0.23 mm^2 for the rat, 0.27–0.54 mm^2 for the dog, and 0.29–0.99 mm^2 for man.[4] More recently, reconstructions of glomeruli have yielded more rigorously derived values for Wistar[7] and Sprague–Dawley[8] rats. Calculated surface area in these studies ranged from 0.176 to 0.236 mm^2 for glomeruli taken from the outer cortex. Total capillary length varied from 6.65 to 8.7 mm. The highly anastomotic nature of the glomerular capillaries was also demonstrated; indeed, even the major adjacent lobes were shown to be interconnected by several capillaries. These morphological studies have prompted a greater focus on the specific interactions among the intraglomerular elements contributing to the effective filtering surface area, the permeability characteristics of the glomerular capillary wall, and the distribution of hydraulic and oncotic forces within the glomerular capillaries.

In this revised chapter, the basic organization and content of the original chapter in the first edition[9] have been maintained. However, some sections have been condensed in order to allow consideration of recent anatomical findings, the contributions of local hormonal interactions to the control of GFR, and

L. Gabriel Navar and P. Darwin Bell • Nephrology Research and Training Center and Department of Physiology and Biophysics, University of Alabama School of Medicine, Birmingham, Alabama 35294. **Andrew P. Evan** • Department of Anatomy, Indiana University Medical Center, Indianapolis, Indiana 46223.

Table I. Representative Values for Glomerular Function in Man and Selected Laboratory Mammals[a]

	Kidney weight (g)	GFR (ml/min)	GFR/g (ml/min)	Glomeruli/g	Average GFR per glomerulus (nl/min)
Man (70 kg)	290.0	130.0	0.45	7,000	64
Dog (20 kg)	84.0	60.0	0.71	12,900	55
Rat (250 g)	2.0	2.0	1.00	30,000	30
Rabbit (2.5 kg)	13.0	8.4	0.65	32,500	20

[a]Data taken from Refs. 3–6, 9, 25, 165.

recent developments which have enhanced our understanding of the tubuloglomerular feedback mechanism.

2. Ultrastructural Considerations

2.1. The Afferent and Efferent Arterioles

The afferent arterioles that provide the blood supply to the glomerular capillaries possess two distinct cell types. Most of the wall of the afferent arteriole is lined by smooth muscle cells of uniform shape and size (Figs. 1b and d); the juxtaglomerular (JG) cells constitute the second cell type and are located near the vascular pole of the glomerulus (Fig. 2b). Proximal to the JG cells, the smooth muscle cells of the afferent arteriole are typical ring-shaped structures which wrap around the endothelium. These smooth muscle cells possess abundant myofilaments and dense bodies. The JG cells are modified myoepithelial cells with cytoplasmic granules that contain renin.[10] These cells are characterized by a prominent cell body from which extend several processes containing contractile filaments and dense bodies.

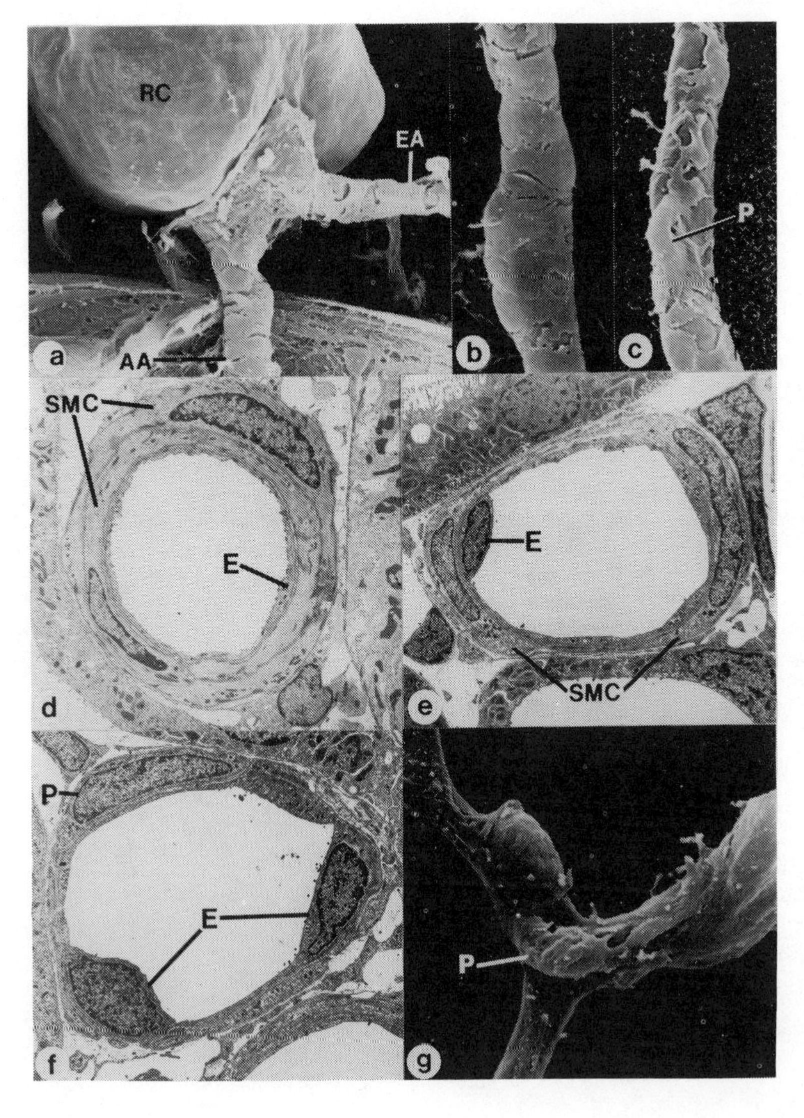

Fig. 1. Microvasculature of the glomerulus. (a) Scanning electron micrograph of renal corpuscle (RC) and associated arterioles: the afferent arteriole (AA) which delivers blood to the glomerulus and the efferent arteriole (EA) which transports blood from the glomerulus to either the peritubular capillaries or the vasa rectae. The afferent arteriole has one or two layers of ring-shaped smooth muscle cells. (b) Scanning micrograph of proximal efferent arteriole with smooth muscle cells which wrap around the vessel. The intercellular borders are irregular. (c) Scanning micrograph of distal efferent arteriole. The wall has pericyte (P) processes which extend around and along the vessel. (d) Transmission electron micrograph of a cross-section of an afferent arteriole. Note the typical vascular smooth muscle cells (SMC) which encircle the endothelium (E). The SMC have abundant myofilaments and dense bodies. (e) Transmission micrograph of a cross-section of the proximal efferent arteriole. The proximal efferent has an endothelium (E) and smooth muscle cells (SMC). The efferent SMC differ from the afferent SMC, with reduced thickness and fewer myofilaments and dense bodies. (f) Transmission micrograph of distal efferent illustrating an incomplete medium consisting of pericytes (P) and the endothelium (E). (g) Scanning micrograph of peritubular capillary that has a pericyte (P) located at a branching point.

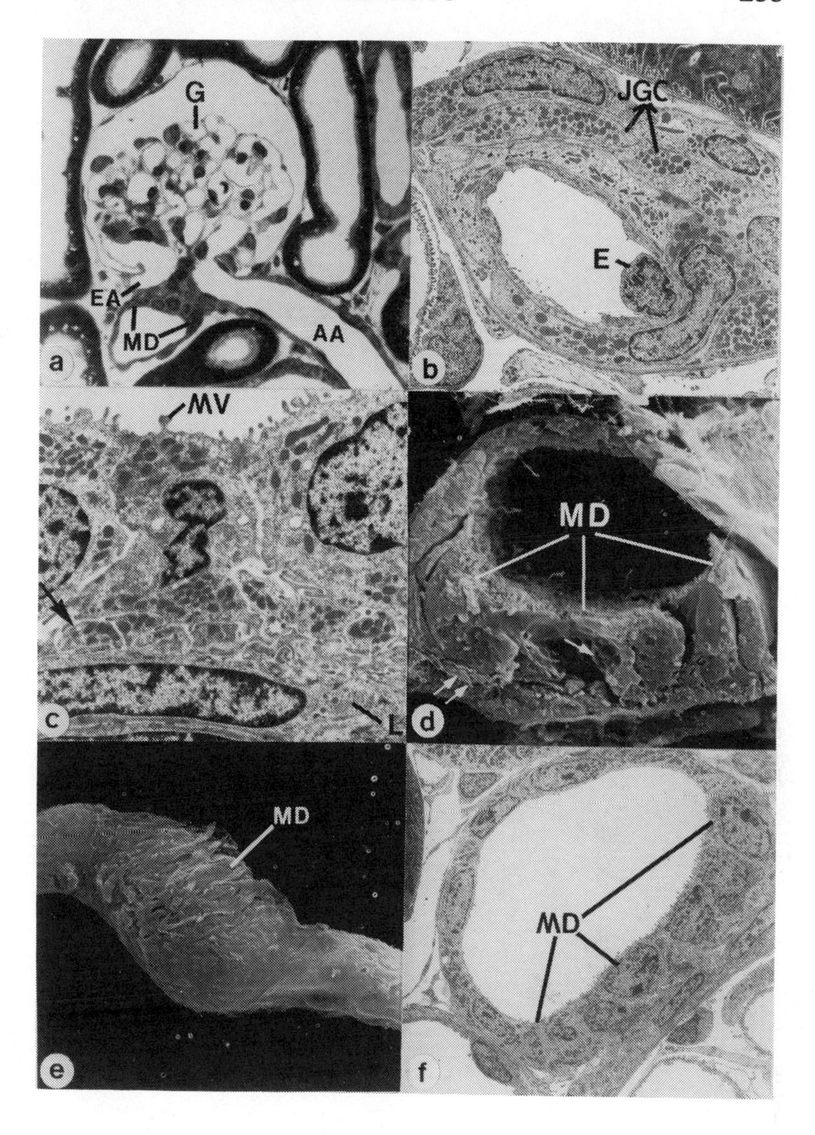

Fig. 2. The juxtaglomerular apparatus–macula densa complex. (a) Light micrograph of glomerulus (G) showing the juxtaglomerular apparatus: the afferent arteriole (AA), efferent arteriole (EA), and macula densa (MD). (b) Transmission electron micrograph of a distal afferent arteriole with myoepithelial or juxtaglomerular cells (JGC) surrounding the endothelium (E). The JGC have numerous granules that contain renin. (c) Transmission electron micrograph of the macula densa and underlying lacis cells (L). The apical surface of the macula densa cells has numerous, short microvilli (MV) while the basal region splits into many irregular processes (arrow). The macula densa cells contain mitochondria that are distributed throughout the cytoplasm. The lacis cells are closely positioned beneath the macula densa. (d) Scanning electron micrograph of a cross-section of the distal tubule in the area of the macula densa (MD). The MD cells are typically cuboidal to low columnar and lie adjacent to the vascular pole of the glomerulus. The luminal surface of the MD cells is characterized by numerous microvilli and a single cilium. The lateral surface has a more limited number of microvilli (arrow) while the basal surface presents numerous irregular processes (double arrow). (e) Scanning electron micrograph of a microdissected distal tubule treated with collagenase to remove the basal lamina and reveal the basal surface of the macula densa (MD) cells. The base of these cells is characterized by numerous large processes. (f) Transmission electron micrograph of a cross-section of the distal tubule in the area of the macula densa (MD). The cuboidal shape and luminal microvilli of these cells are evident.

The afferent arterioles are lined by endothelial cells, each of which is characterized by a fusiform-shaped nucleus (Fig. 1d).

At the vascular pole of each renal corpuscle, the glomerular capillaries coalesce to form an efferent arteriole. Occasionally, (< 10%), the juxtamedullary glomeruli possess some type of vascular shunt that allows a portion of the blood flow to bypass the glomerular capillaries.[11] The smooth muscle cells of the efferent arterioles differ from those of the afferent arteriole in that they are usually smaller, not as thick, and do not possess as many myofilaments or dense bodies. In addition, the smooth muscle cells located at the proximal end of the efferent arteriole do not have the same uniform shape as those of the afferent arteriole; instead they are irregularly shaped and not closely apposed to one another (Figs. 1b and c). Occasionally, a JG cell is found in close approximation to the glomerulus. More distally, there may be a gradual or abrupt change in the morphology of the smooth muscle cells.[12] In both conditions, the tunica media is eventually composed of pericytes characterized by numerous branching processes (Figs. 1c and f). The pericytes continue for varying distances along the efferent arteriole and sometimes extend onto the peritubular capillaries (Fig. 1g). The efferent arterioles exhibit structural variations, those of the cortical glomeruli being primarily short straight trunks breaking into the peritubular capillary network, while those of the juxtamedullary glomeruli are longer and more muscular prior to their transformation to capillaries or vasa rectae.[13–16]

2.2. The Juxtaglomerular Complex

An invariant feature of the renal corpuscle is the JG apparatus–macula densa complex which has three principal components (Fig. 2): (1) the macula densa cells of the ascending loop of Henle segment of the distal nephron, (2) the renin-producing, myoepithelioid cells of the afferent and occasionally efferent arterioles, and (3) the lacis, Goormaghtigh cells, or extraglomerular mesangial cells.[17–18] During embryogenesis, the portion of the tubule destined to become the distal tubule remains in contact with the glomerulus. This segment of the distal tubule never loses its continuity with the renal corpuscle; instead, the intermediate segments elongate to form the proximal tubule and the loop of Henle.[2] Only the cells of the terminal portion of the loop of Henle[19] that are juxtapositioned to the

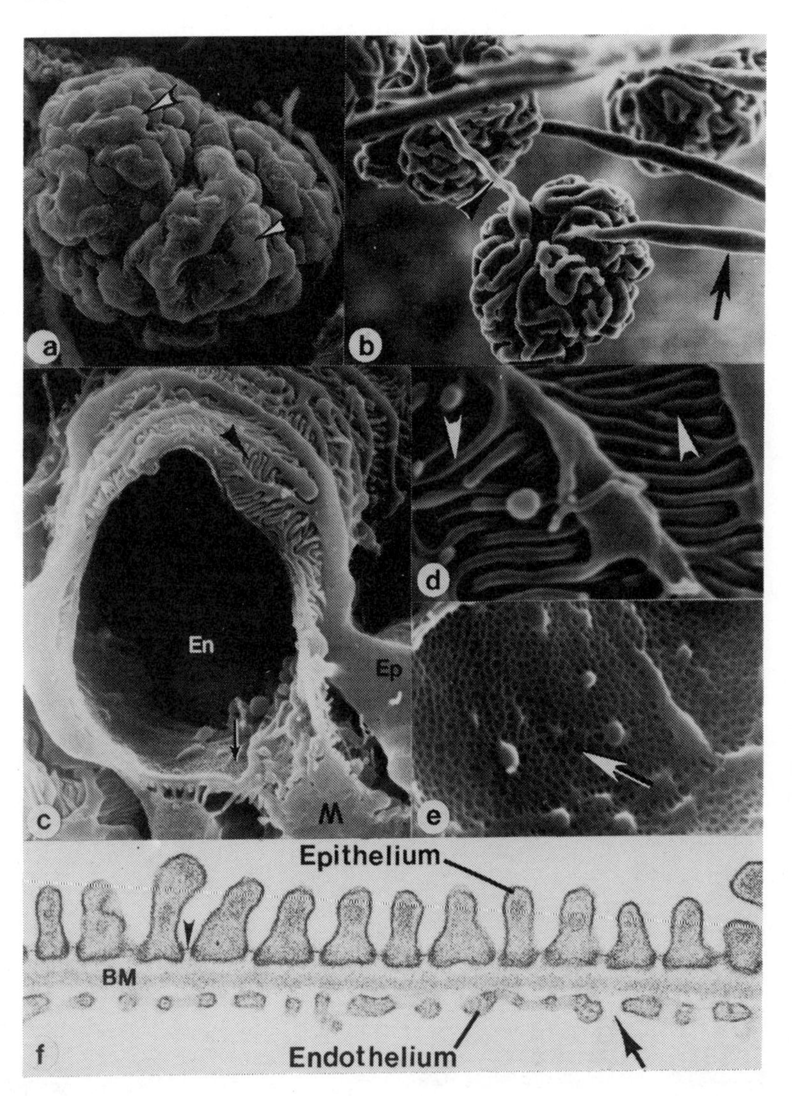

Fig. 3. Structural characteristics of glomerulus. (a) Scanning electron micrograph of a glomerulus showing the capillary loops covered by visceral epithelium [podocytes (arrowheads)]. (b) Scanning micrograph of a plastic cast of the glomerular microvasculature illustrating the relationship between the afferent arteriole (arrow) and the glomerulus with its numerous capillary loops and efferent arteriole (arrowhead). (c) Scanning micrograph of glomerular capillary loop with fenestrated (arrow) endothelium (En), visceral epithelium (Ep) with foot processes (arrowhead), and the mesangium (M). (d) High-magnification scanning micrograph of visceral epithelium with interdigitating foot processes (arrowheads). (e) High-magnification scanning micrograph of glomerular capillary endothelium. Note the numerous rounded fenestrae (arrow). (f) Transmission electron micrograph of glomerular capillary wall consisting of: the visceral epithelial foot processes which are interconnected by filtration slit membranes (arrowhead), the basement membrane (BM), and the glomerular endothelium with its numerous fenestrae (arrow).

vascular pole of the renal corpuscle form the macula densa (Fig. 2a). The cells of the JG apparatus are embedded in a delicate fibrillar network which partially or completely surrounds the afferent arteriole and also occupies the region between the angle of the afferent and efferent arteriole. The granular cells are associated primarily with the wall of the arteriole while the nongranular extraglomerular mesangial cells are in continuity with the intraglomerular mesangial cells. All lacis, myoepithelioid, extraglomerular mesangial and ordinary smooth muscle cells of the afferent and efferent arterioles are extensively coupled to each other by gap junctions.[20,21] However, such gap junctions do not extend to the macula densa cells. The nerves to the cellular elements in this region go not only to the smooth muscle cells, but also to the granular cells, thus providing the anatomical basis for the suggestion that these nerves participate in the secretion of renin[18] (Fig. 2b).

The unique relationship of the macula densa segment of the distal tubule to the vascular pole of its own glomerulus has generated a great degree of interest and enthusiasm. The macula densa segment consists of an elongated plaque of epithelial cells within the distal tubule (Figs. 2a, d–f).[19,22] These cells lie only on the side of the distal tubule which is closest to the glomerulus (Figs. 2a and f). The cell bodies of the macula densa are closely packed, and the Golgi apparatus is quite prominent and closer to the basal side, a characteristic distinct from that in the other cells of the distal tubule (Fig. 2c). The mitochondria are short and irregularly arranged. These cells appear compacted primarily because of the scantiness of the cytoplasm. The basement membrane between the macula densa cells and the afferent arterioles is not highly ordered and appears to have spaces.[22] The luminal surface of the macula densa cells is characterized by numerous microvilli and a single cilium per cell (Fig. 2d). As shown in Fig. 2e, the base of the macula densa segment is distinct from the surrounding cells and is characterized by numerous coarse processes.

There exists an extensive and relatively continuous contact between the distal tubule, the extraglomerular mesangium, and the arterioles.[23] In particular, there is a very intimate association between the macula densa segment of the distal tubule and the afferent arteriole.[24,25] This is characterized by cytoplasmic extensions from macula densa cells which actually project into the space between the vascular and the tubular components. Because of this striking relationship between the distal tubule and the vascular elements of the glomerular tuft, many suggestions have been presented that one or more aspects of glomerular function or the JG apparatus are controlled, at least in part, by feedback signals originating from the macula densa segment. At least two major hypotheses associating macula densa cell function with the function of the vascular elements have been under intensive investigation. These are that renin release is in part regulated by feedback signals from the macula densa;[26,27] and that arteriolar resistance is regulated by feedback signals from the macula densa cells.[28–30] It has also been suggested that they interact in such a way that the renin–angiotensin system participates in the intrarenal control of glomerular filtration dynamics.[31] For this chapter, emphasis is placed on the control of glomerular dynamics; the renin–angiotensin system is discussed only within this context.

2.3. The Glomerular Capillary Wall

The filtration apparatus of the renal corpuscle is unique in that it filters large volumes of fluid while maintaining a highly efficient capability to restrict macromolecules. These properties are due to its rather complex ultrastructure (see Fig. 3). After the afferent arteriole enters the renal corpuscle, it expands into a reservoir from which three to five lobular units are formed (Fig. 3b).[7,9] From these initial large capillaries, numerous smaller intercapillary channels branch and serve to connect all lobules.[29] The larger capillaries return to the vascular pole and coalesce into the efferent arteriole (Fig. 3b).

The wall of the glomerular capillaries is composed of several layers: the endothelial cells, a thick basement membrane, and the visceral epithelium or podocytes[24] (Fig. 3f). In addition, the mesangial cells are found in intercapillary locations. The outer visceral coat of the glomerular capillary wall consists of a nonsyncytial lining of discontinuous cells, the podocytes (Figs. 3d and f). These have large oval nuclei with abundant large cytoplasmic projections, or foot processes, which interdigitate in an alternating manner to cover the capillary loops (Fig. 3c). The adjacent foot processes are separated by narrow spaces, termed filtration slits, of about 25 nm in width (Fig. 3d). A thin membrane bridging adjacent foot processes is often seen.[33] Foot processes are generally characterized by large terminal portions. It has also been shown that foot processes contain contractile filaments which may be functional and may serve to regulate the width of the filtration slit.[34] Approximately 10% of the surface area on the epithelial side of the basement membrane constitutes an open slit area.[35] The free surfaces of the epithelial cells are covered with a thick glycoprotein anionic coat[36] approximately 12 nm thick.

The basement membrane has a distinct electron-dense central layer (lamina densa) about 50 nm thick and inconsistently defined outer and inner layers of 20–50 nm (lamina rara externa and lamina rara interna). The thickness of the basement membrane varies greatly among species as well as with age and with degree of glomerular disease. Structurally, the basement membrane appears finely granular; it is a complex hydrated gel structure composed of collagenous and noncollagenous glycoproteins and proteoglycans.[24,33,37] The collagenous components include several type IV and type V (AB2) collagens and are uniformly distributed over the entire width of the glomerular basement membrane.[38] Laminin, a large noncollagenous glycoprotein, has also been localized in the basement membrane, predominantly in the lamina rara interna.[39] The glycoproteins are of particular importance in that they contain sialic acid and other anionic residues, which provides these structures with a strong negative charge.[36,40] Several glycosaminoglycans are strong anionic polymers which are distributed in a regular latticelike network in both the lamina rara interna and externa.[40] These anionic residues present in the basement membrane and the endothelium may contribute significant electrostatic resistance to the passage of negatively charged macromolecules.[41]

The endothelium of the glomerular capillary loops possesses large visible fenestrations which constitute about 9% of the capillary surface area and range in size from approximately 50 to 100 nm (Figs. 3c, e, f).[35] The individual glomerular endothelial cells are characteristically attenuated except at the cell body which usually lies next to the mesangial areas. Cytoplasmic prolongations vary in thickness from 30 to 100 nm and form the wall of the capillary. The fenestrations provide the major avenue through which the filtrate traverses. While the fenestrations are sufficiently large that they probably offer little geometrical restraint to the plasma macromolecules, collectively they may control the surface area available for flow across the entire glomerular capillary wall. The size of the fenestrations may be controlled by factors that regulate volume of the endothelial cells which may be sensitive to a variety of humoral and physical stimuli.[42,43]

The mesangial cells are situated in the central, centrilobular, and intercapillary regions of the glomerulus and are not separated from the endothelial cells by a basement membrane.[24,33] The actual function of the mesangial cells has been the subject of considerable interest, but these cells do not participate as a barrier to the filtration process. Since they are topographically intimately related to the capillary loops, it has long been thought that one of their functions is to provide structural support. Mesangial cells also possess phagocytic properties and may phagocytize macromolecules which permeate the basement membrane and migrate to the area of the mesangial cells.[44,45] Other studies have raised the concept regarding the possible role of the mesangial cells in the regulation of glomerular hemodynamics.[33,45,46] Bernik originally suggested that cells of mesangial origin exhibited contractile properties.[47] It was also demonstrated that mesangial cells have receptors to angiotensin and other vasoconstrictor agents.[48] In studies on homogeneous cell cultures, Ausiello *et al.*[49] demonstrated that cells of apparent mesangial origin exhibited a contractile response upon addition of angiotensin II. These studies have implicated the mesangial cells as intraglomerular elements which may have the potential of regulating flow along the glomerular capillaries. Possible mechanisms will be discussed in a subsequent section.

2.4. Innervation of the Glomerular Structures

Both nonmyelinated and myelinated fibers are found coursing with the large renal vessels to reach the renal corpuscle. Most of these fibers appear to be associated with the sympathetic celiac plexus. A rich plexus consisting of predominantly α-adrenergic fibers is associated with the renal artery and its branches, including the afferent arteriole, the efferent arteriole, vasa recta, and large veins.[50,51] As the adrenergic nerves reach the afferent arteriole, they innervate the JG apparatus and efferent arteriole. A few fibers are then thought to innervate portions of the proximal and distal tubular segments.[18] It must be noted that a controversy exists regarding the presence or absence of cholinergic nerves. Recent evidence indicates that the previously observed acetylcholinesterase is contained within adrenergic nerves and that the innervation of the kidney is exclusively adrenergic.[51] Of particular interest is the finding of dopamine-containing neuronal elements in the dog kidney associated with the glomerular vascular pole, whereas the rest of the arterial tree possessed norepinephrine histofluorescence.[52]

Myelinated fibers have been found running in the paravasal tissue of the interlobar and arcuate arteries.[53] These afferent fibers are few in number and have not been found associated with arteriolar vessels.

3. Characteristics of the Filtration Process

3.1. Forces Governing GFR

The movement of fluid across the glomerular capillary membrane is generally described by the Starling filtration–reabsorption principle[4,54] which is based on the fundamental premise that water and solute flow through fluid-filled extracellular channels in the filtering membrane. These pathways may be direct or tortuous, but are considered to be relatively large with respect to water molecules, hydrated ions, and smaller solutes that readily permeate across the membrane. From this basic principle, it also follows that the larger solutes that either approach or exceed the average size of the channels are restricted or constrained from passage.[55]

From these basic considerations, GFR can be described as resulting from the interaction of the forces across the glomerular membrane. These include glomerular capillary pressure (P_g), Bowman's space pressure (P_B), glomerular plasma colloid osmotic pressure (π_g), and colloid osmotic pressure of filtrate (π_f). The membrane parameters are the hydraulic conductivity of the glomerular membrane (L_p), and the total filtering surface area (S_f). The complete equation describing GFR is

$$\mathrm{GFR} = L_p S_f \int_0^1 (P_{g(x)} - P_B - \sigma(\pi_{g(x)} - \pi_f)dx \quad (1)$$

In Eq. (1), x represents the normalized length of the glomerular capillaries with 0 designating the afferent arteriolar end and 1 the efferent arteriolar end. σ is the reflection coefficient which has a range of 0 to 1.[56] When σ is 1, the colloid osmotic pressure has maximum effectiveness. Normal glomerular capillaries are extremely efficient in restricting the passage of macromolecules, and the amount of protein present in the filtrate is less than 0.1% of the plasma protein.[57–59] For practical considerations therefore, σ can be considered as having a value of unity and the colloid osmotic pressure in Bowman's space (π_f) can be disregarded. Thus, the effective colloid osmotic pressure ($\Delta\pi$) can be considered to be equivalent to that of the plasma in the glomerular capillaries (π_g).

The coefficient that relates effective filtration pressure to GFR has been expressed in many different ways. Essentially, it is a product of two variables: the hydraulic conductivity L_p, which should be considered as volume flow per unit time per unit of driving force per unit of available surface area, and S_f, the total surface area available for filtration.[4,60] The product of the two is the filtration coefficient (K_f). K_f values can be expressed for individual glomeruli (nl/min per mm Hg or nl/sec per mm Hg), per gram of cortex or per gram of total kidney weight (ml/min per mm Hg). These considerations plus the use of average values for glomerular pressure and glomerular colloid osmotic pressure allow the formulation of the more commonly used relationship:

$$\mathrm{GFR} = K_f (P_g - P_B - \pi_g) \quad (2)$$

3.2. Colloid Osmotic Pressure

As fluid enters the glomerular capillary area, ultrafiltration of a protein-free fluid increases the plasma protein concentration, leading to a progressive increase in plasma colloid osmotic pressure; thus, filtration is greatest in the initial segments of the glomerular capillaries. This characteristic of the plasma protein imposes important constraints on the filtration process especially when relatively large quantities of plasma are filtered. The increase in plasma protein concentration is a direct function of the relative, and not the absolute, amount of plasma traversing across the glomerular membrane; this proportionality factor is the well-known filtration fraction term (FF). The increase in colloid osmotic pressure that actually occurs is more complex; however, it depends not only on the filtration fraction but also on the actual magnitude of the initial plasma colloid osmotic pressure. For the same filtration fraction, the relative increase in colloid osmotic pressure from the afferent to the efferent arteriole increases progressively with higher initial plasma colloid osmotic pressure values. This is due to the fact that the relationship between colloid osmotic pressure and plasma protein concentration is not linear, but shows a slightly greater slope as the absolute protein concentration is increased.[54] Several different empirical relationships have been developed to allow prediction of colloid osmotic pressure, but they can be used accurately only for known albumin to globulin (A/G) ratios. For any given plasma protein concentration, plasma colloid osmotic pressure will be higher with progressively greater A/G ratios.[54] The commonly used Landis–Pappenheimer relationship

$$\pi = 2.1C + 0.16C^2 + 0.009C^3 \quad (3)$$

applies to an A/G ratio of about 1.2 and is considered normal for man and rat. In the dog, the A/G ratio is highly variable but generally averages about 0.5.[61]

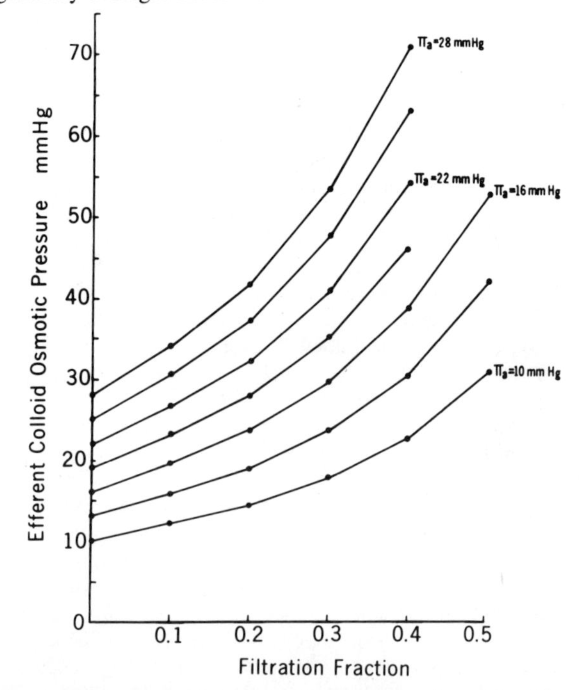

Fig. 4. Representative nomogram allowing estimation of the efferent arteriolar colloid osmotic pressure (π_e) from specific values of filtration fraction and afferent colloid osmotic pressure.[62]

It is possible to directly predict the efferent arteriolar colloid osmotic pressure for any given level of plasma colloid osmotic pressure and filtration fraction.[62] Figure 4 depicts a nomogram that shows the relationship between efferent arteriolar colloid osmotic pressure and the filtration fraction for a range of initial plasma colloid osmotic pressures. Although the A/G ratio affects the colloid osmotic pressures for any given protein concentration, it has only a minor influence on the efferent colloid osmotic pressure for any given filtration fraction and plasma colloid osmotic pressure. Thus, efferent colloid osmotic pressure can be predicted from just FF and afferent colloid osmotic pressure. One mathematical expression that defines this relationship is

$$\pi_e = \frac{R \cdot \pi_a}{R \cdot (1 - FF) - \pi_a \cdot FF} \quad (4)$$

The coefficient "R" is affected only slightly by the A/G ratio and varies from 43 to 48, with the higher values applying to higher A/G ratios.[62] As can be seen from Fig. 4, the efferent colloid osmotic pressures rise quite rapidly with progressive increases in filtration fraction, especially when the initial colloid osmotic pressure values are over 20 mm Hg. The actual plasma colloid osmotic pressure can vary substantially within a given species and among various species, being about 25–27 mm Hg for man, with dog and rat values ranging from 15 to 20 mm Hg.[61]

Although direct means to measure plasma colloid osmotic pressure are readily available, the techniques have not been miniaturized to a degree sufficient to allow direct measurements of colloid osmotic pressure in small nanoliter samples such as that collected from efferent arterioles. Efferent colloid osmotic pressure is thus estimated from the protein concentration in micropuncture experiments where efferent blood samples are collected.[63] Alternatively, the efferent arteriolar protein concentration can be calculated from plasma protein concentration and the filtration fraction, measured either at the whole kidney level or at the single nephron level.[62]

3.3. Hydrostatic Pressure Gradients

The magnitude of the pressure drop within the arterial tree itself is variable, but is generally considered to be small up to the terminal segments of the afferent arterioles. Studies in the rat have suggested that the arteries and larger arterioles contribute significantly to the arterial pressure drop within blood vessels leading to the superficial nephrons of the rat.[64,65] This pressure drop is much less in juxtamedullary nephrons or in superficial nephrons of the dog.[66,67] Recent studies in dogs indicate that the drop proximal to the afferent arterioles of outermost cortical glomeruli is only 10–15 mm Hg,[66] so that perhaps more than 70% of the preglomerular pressure drop occurs in the immediate vicinity proximal to the glomerulus.[67] Nevertheless, the entire pressure drop up to the glomerular capillary is normally much less than occurs in other microcirculatory beds such that glomerular hydrostatic pressure is relatively high. This high level of hydrostatic pressure predominates over the plasma colloid osmotic pressure and is thus responsible for the ultrafiltration of fluid into Bowman's space. The hydrostatic pressure gradient (ΔHP) itself is dependent not only on P_g, but also on pressure in Bowman's space (P_B).

In mammals, there is very little hindrance of flow from Bowman's space into the proximal tubule and these pressures are essentially equivalent.[68] Tubular pressures have been measured directly using micropuncture techniques in many different studies and in a variety of experimental models. These are highly responsive to different experimental manipulations, but normal values are in the range of 11–15 mm Hg for the rat[69,70] and 18–22 mm Hg for the dog.[71,72] Most studies have indicated that these proximal tubule pressures are maintained relatively constant throughout the length of the proximal tubule and are generally slightly higher than the pressures in the surrounding peritubular capillaries. The finding that there is a rather consistent relationship between proximal tubule pressure and peritubular capillary pressure during various experimental manipulations[70,71] coupled with the general equivalence between peritubular capillary pressures and intrarenal venous pressures[73,74] allows a means of assessing peritubular capillary pressures and estimating proximal tubule pressures in man. Measurements of intrarenal venous pressure can be made without resorting to micropuncture methodology and have been conducted in human subjects. Normal values have been in the range of 20–25 mm Hg[75,76] and would presumably be representative of peritubular capillary pressures in man; these values suggest that proximal tubule pressure is approximately 22 to 27 mm Hg. These data are helpful in making basic assessments of glomerular pressure in man.

Many different approaches have been used to estimate or measure glomerular capillary pressure;[4,9] however, current emphasis is predominantly on micropuncture measurements of glomerular pressure based on either direct puncture of single glomerular capillaries or proximal tubule stop-flow pressures. Direct measurements of glomerular capillary pressure have been reported for rats, dogs, and monkeys. Independent studies by Brenner *et al.*[60,68] and Blantz *et al.*[77,78] pioneered the application of the servo-null micropressure measurement technique for the determination of glomerular capillary pressure. Most of these studies have been conducted in Munich–Wistar rats, a mutant strain having superficial glomeruli accessible to micropuncture. Glomerular capillary pressures in these rats during hydropenia have generally been in the range of 45 to 50 mm Hg. They are somewhat higher when modest volumes of plasma are infused to compensate for blood loss due to surgery and experimental procedures.[79] Glomerular pressures in superficial glomeruli of some colonies of Munich–Wistar rats[80] and in Sprague–Dawley rats have been higher, averaging about 55 mm Hg.[81] However, since superficial glomeruli are generally scarce, special techniques have usually been necessary to expose glomeruli or to obtain access to glomerular capillaries. Aukland *et al.* developed a corticotomy technique to expose glomeruli and measured values of 58 to 63 mm Hg in the Sprague–Dawley rat for superficial and deeper glomeruli.[82] Using a similar technique, glomerular pressure measured in dogs averaged 58 to 60 mm Hg,[62,83] values similar to those obtained by Marchand[84] using a random puncture technique that did not require corticotomy.

As detailed in the first edition of this chapter,[9] the proximal tubule stop-flow pressure has been used extensively to estimate glomerular pressure. With this technique, an oil or wax block is introduced into an early segment of the proximal tubule and the maximal pressure that develops in the obstructed tubule is measured.[72,85,86] As maximal stop-flow pressure is approached, the process of filtration–reabsorption should be minimized and the effective filtration pressure should become negligible. Under these conditions, the average glomerular capillary hydrostatic pressure of that nephron is equivalent to the sum of the stop-flow pressure and the plasma colloid osmotic pressure. Most studies have yielded stop-flow pressures in the range of 36 to 42 mm Hg; these coupled with plasma colloid osmotic pressures of 14 to 16 mm Hg in anesthetized rats prepared for micropuncture provide a range of glomerular pressure of 48 to 58 mm Hg.[77,81,86–90]

The validity of the stop-flow pressure technique as a means of estimating normal glomerular capillary hydrostatic pressure in undisturbed filtering glomeruli remains uncertain due to at least two theoretical complications. First, the stopped-flow condition diverts the volume flow that is normally filtered to the efferent arteriole and may lead to an increase in the pressure in the glomerular capillaries if the efferent resistance remains unchanged.[87,91,92] It is not clear how much of an increase in glomerular pressure will actually occur since the diversion of filtrate to the efferent arteriolar blood would presumably lower the blood viscosity; however, the relationship between viscosity and blood hematocrit in small vessels remains uncertain.[91] It is also not clear if efferent arteriolar resistance remains unchanged during stop-flow conditions and the possibility exists that a higher pressure induces a dilatation.[91,93] Due to these uncertainties, the magnitude of this potential source of error remains unknown. The second complication is that this technique requires the interposition of a block in the proximal tubule which interrupts normal volume flow to the more distal parts of the nephron. As discussed in detail in a later section, to the extent that a distal tubular feedback mechanism is active and responsive to decreases in flow, the stop-flow pressure technique may result in an overestimate of normal glomerular pressure due to afferent arteriolar vasodilation mediated by the tubuloglomerular feedback mechanism.[72,74] This potential problem should be taken into consideration when evaluating glomerular pressure or glomerular pressure responses on the basis of stop-flow pressure measurements. Studies performed in our laboratory have indicated that the magnitude of the feedback-mediated overestimation is greatest at normal spontaneous arterial pressure and is minimized when the arterial pressure is close to the lower limit of the autoregulatory range.[72,94,95] It has often been argued that the possible errors involved in the stop-flow pressure method for estimating P_gFR may contribute only insignificantly since most studies comparing direct and stop-flow measurements of glomerular pressure in rats have shown very close agreement. However, recent studies by Blantz[87] and by Ichikawa[92] have indicated that the stop-flow pressure method may significantly overestimate true glomerular capillary pressure in the rat.

With regard to measurements based on direct puncture of glomerular capillaries, these also have inherent uncertainties. There is no assurance that the pressures in the outer capillaries punctured are representative of those in undisturbed capillaries, or that the pressures in these relatively rare superficially located glomeruli are indeed representative of those existing in the majority of the glomeruli in the kidney. Therefore, appropriate caution is necessary in the evaluation of glomerular pressure data, regardless of the method of measurement.

3.4. The Effective Filtration Pressure

Regardless of the exact magnitude of the hydrostatic pressure in the glomerular capillaries, it is of critical importance that the high pressure be coupled with a relatively high blood flow in order to achieve optimum conditions for the filtration of fluid into Bowman's space. Only the glomerular capillary pressure serves as a positive driving force because both proximal tubule pressure and plasma colloid osmotic pressure counter this force. When blood flow is low relative to the filtration pressure, the loss of protein-free filtrate rapidly increases the protein concentration in the remaining plasma. The progressive increase in plasma proteins during filtration leads to proportionally greater increases in colloid osmotic pressure such that the effective filtration pressure decreases continuously. As illustrated in Fig. 5, the filtration process can be operating under two basic conditions. The first is the relatively simple case where colloid osmotic pressure progressively increases such that filtration continues throughout the entire length of the glomerular capillaries and a finite positive filtration pressure remains at the efferent end of the glomerular capillaries.[80,88,94] The second condition is

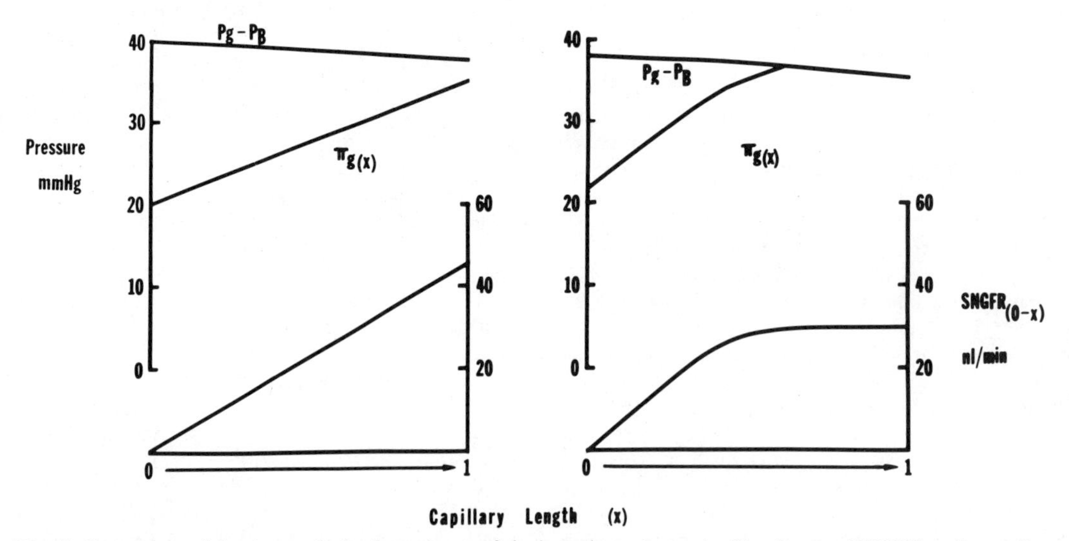

Fig. 5. Comparison of the changes in the determinants of single-nephron glomerular filtration rate (SNGFR) under conditions where filtration occurs along the entire capillary length (left panel) and during filtration equilibrium (right panel). For each panel, the capillary length is designated from 0 to 1 and the cumulative SNGFR is plotted in the lower graph. The upper graphs show the rise in π_g as filtration occurs. P_g, glomerular capillary hydrostatic pressure; P_B, hydrostatic pressure in Bowman's space; $\pi_{g(x)}$, plasma colloid osmotic pressure.

based on studies in the Munich–Wistar rat which have suggested that the magnitude of the increase in colloid osmotic pressure occurs sufficiently rapid such that an equilibrium of the Starling forces occurs at some point within the filtering capillary; when this condition exists, the filtration process is said to be in "filtration equilibrium."[6,60,78] Under these conditions, not all of the total available filtering surface area is utilized. By simply increasing the plasma flow to a system in filtration equilibrium, the rate of increase of colloid osmotic pressure along the length of the glomerular capillaries is diminished and the actual site of equilibration of hydrostatic and colloid osmotic forces is delayed.[91,94,96,97] In effect, with increases in plasma flow, there is a recruitment of additional filtering surface area (S_f) and thus the functional K_f is increased. The basic patterns of the filtration forces occurring under conditions of filtration equilibrium and disequilibrium are depicted in Fig. 5.

As pointed out, studies in the dog have suggested that the glomerular capillary hydrostatic pressure even in superficial nephrons is sufficiently high to prevent the Starling forces from reaching equilibrium by the end of the glomerular capillaries.[62,88,94,98] The condition for man remains somewhat uncertain since glomerular capillary pressure can only be estimated by indirect techniques. However, the very low filtration fraction and the relative lack of plasma flow dependence of GFR in man[6,99] would suggest that the filtration process in man continues throughout the entire length of the glomerular capillaries.

One additional factor regards the magnitude of the glomerular capillary hydrostatic pressure drop along the length of the glomerular capillary. Most measurement techniques only evaluate one value, representative of average or overall glomerular pressure. Even when measured directly, it is not possible to distinguish between an early or a late segment of the glomerular capillary loop, or whether the pressure in that loop is representative of pressures in other capillary loops. Thus, experimental assessment of the magnitude of the pressure drop is not possible. Nevertheless, there are abundant parallel glomerular capillaries that collectively have a large cross-sectional area relative to the cross-sectional area of the afferent and efferent arterioles. Accordingly, the overall pressure drop along the glomerular capillaries would be expected to be very small compared to the pressure drop across the afferent arteriole or the efferent arteriole.[60,100,101] Computations based on the number and dimensions of the glomerular capillaries have resulted in values ranging from a low of 1 mm Hg[102] to about 3 to 4 mm Hg.[8]

3.5. Hydraulic Conductivity and the Filtration Coefficient

As indicated earlier, the filtration coefficient (K_f) is a function of the hydraulic conductivity (L_p) expressed per unit of surface area and the total filtering surface area (S_f). The total surface area can be further fractionated in terms of the number of functioning capillaries and the average surface area per capillary. Alterations in any of these factors could contribute to changes in transcapillary fluid flux. The hydraulic conductivity is a specific property of the glomerular capillary membrane which is a rather complex structure potentially capable of altering its characteristics in several ways. These include the size and number of endothelial fenestrations, the thickness or permeability of the basement membrane, and the number or structural configuration of the slit pores between the foot processes.[24,32–35,37–39,43,103] At present, the quantitative contribution of each of these factors has not been analyzed rigorously and the precise means by which fluid flows across the glomerular capillary wall remains uncertain. While it is readily accepted that inflow is through the endothelial fenestrations and outflow is through the slits of the podocytes,[4] it remains undetermined if the entire area of the basement membrane is utilized for filtration. Since 90 to 95% of the basement membrane is covered by the endothelial cells,[35] it is possible that fluid flows through preferential pathways within the basement membrane as suggested in Fig. 6. To the extent that this occurs, it can be recognized that the mechanisms controlling the size of the slits

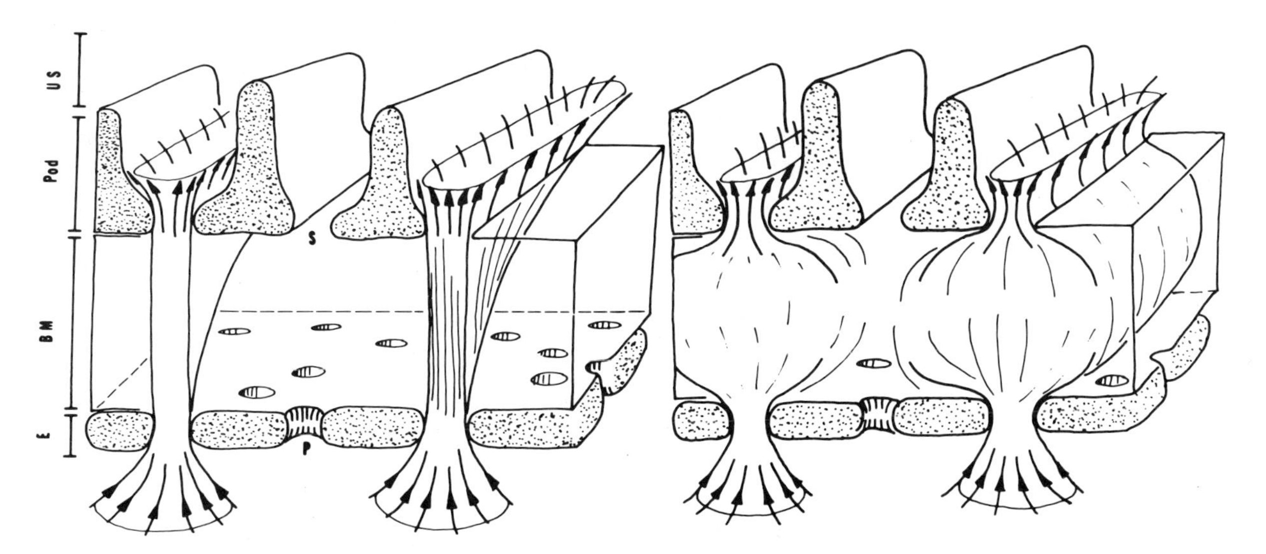

Fig. 6. Schematic of the glomerular capillary wall depicting two alternative pathways for fluid flow across the glomerular basement membrane (BM). In the preferential pathway model, flow may be restricted to the basement membrane which is directly between the endothelial fenestrations (E) and the slits of the epithelial processes (Pod). With such a system the size of the endothelial fenestrations could influence the effective area of the basement membrane and thus directly influence K_f. With the other model, flow may occur along the entire basement membrane so that small changes in the size of the fenestrations or slits would not influence K_f to any significant extent.

Table II. Representative Values of the Various Indices of Glomerular Function

	Dog[a]	Rat[b]
Glomerular pressure (mm Hg)	57.00	50.0
Proximal tubular pressure (mm Hg)	20.00	13.0
ΔP (mm Hg)	37.00	37.0
Mean colloid osmotic pressure (mm Hg)	22.00	25.0
Effective filtration pressure (mm Hg)	15.00	12.0
GFR		
Per nephron (nl/min)	55.00	30.0
Per g kidney (ml/min)	0.71	1.0
K_f		
nl/min per mm Hg per glomerulus	3.70	2.5
ml/min per mm Hg per 100-g kidney	4.70	8.3
S_f		
Per glomerulus (nm^2)	0.40	0.2
Per 100-g kidney (cm^2)	5160.00	6600.0
L_p (nl/sec per mm Hg per cm^2)	14–18	18–25

[a]Data for the dog are compiled from Refs. 4, 62, 83, 84, 88, 98.
[b]Data for the rat are representative for the Munich–Wistar strain (euvolemic conditions) and are compiled from Refs. 60, 63, 69. Other studies[80–82] suggest that glomerular pressure in other rat strains may be higher, indicating that K_f and L_p may be lower than that shown.

between the podocytes[104] or the endothelial fenestrations[43] could also influence the area of basement membrane utilized for bulk flow.

For the most part, requisite measurements allowing a meaningful quantitative analysis of the filtration parameters have been done primarily in the rat and the dog. Although controversies continue, representative values can be obtained for these two species, and are shown in Table II. The data for GFR, K_f, S_f, and L_p are derived for both single nephrons as well as per 100 g of kidney weight. Data are shown for both Munich–Wistar and Sprague–Dawley rats. The greater glomerular density in the rat compensates for the smaller S_f per glomerulus, yielding a greater total S_f per 100 g of kidney, and a substantially greater K_f for 100 g of kidney.

The data concerning glomerular dynamics in man are meager.[99] Overall GFR per gram of kidney weight is about 0.45 ml/min, which is lower than GFR in either rats or dogs. Based on glomerular density analysis of about 7000 g^{-1}, an average SNGFR of about 64 nl/min can be estimated (Table I). With regard to the filtration forces, the plasma colloid osmotic pressures are somewhat higher than those in either rats or dogs. This is counteracted by the generally lower values for filtration fraction (about 0.2) yielding an average glomerular colloid osmotic pressure of about 30 mm Hg. Using predicted values of 36 mm Hg for efferent arteriolar colloid osmotic pressure and 24 mm Hg for proximal tubule pressure, a value based on intrarenal venous pressure measurements,[75,76] minimal values of 60 mm Hg for average glomerular pressure and 6 mm Hg for effective filtration pressure can be derived. It is not known how much higher glomerular pressure actually is in man. Furthermore, in the absence of knowledge concerning the net hydrostatic pressure, it is not possible to estimate true EFP or K_f for man from such hydrodynamic considerations. Nevertheless, some insight concerning the filtration characteristics of the glomerular membrane in man has been achieved by means of the molecular sieving approach.

3.6. Permeability of the Glomerular Capillaries to Macromolecules

The molecular sieving approach to explore the permeability characteristics of the glomerular membranes is conceptually intriguing in that it evaluates directly the filtering membrane with respect to the size of the channels through which the water and solute must flow and to the relative area available for the flow. Implicit in the "molecular sieving" technique is the assumption that there exist a finite number of aqueous channels with a characteristic configuration which restrict the passage of large molecules, such as the plasma proteins, but freely allow the passage of water and solute molecules up to a specific size. The larger molecules that approach the effective size of the channels will be "restricted" or "sieved." Passage of these molecules will be dependent on a number of factors including the charge and configuration of the solute molecule but primarily on the size of the molecule with respect to the size of the channels in the membrane. By utilizing certain assumptions, the quantitative magnitude of restriction for various pore sizes and configurations can be predicted, and when coupled with experimental sieving data utilizing macromolecules of different sizes, certain conclusions regarding the hydraulic conductivity and permeability properties of the filtering membrane are possible.[4,41,54–56,105]

In essence, the application of molecular sieving data to the estimation of the filtration coefficient rests on the assumption that flow through the membrane channels follows Poiseuille's law.[4] Morphological studies have failed to establish specific configurations of the channels; accordingly, there is no justification for the use of anything other than the simplest approach assuming cylindrical pores having an average radius, r, and specific length equivalent to the wall thickness, x. Thus, volume flow through one pore (f) can be expressed as

$$f = \pi r^4 \, \Delta P / 8\eta\Delta x \tag{5}$$

where ΔP is the effective pressure gradient and η is the viscosity of the fluid. In Eqs. (5)–(8), π designates its standard mathematical meaning of 3.14. For a number of pores, N, the total volume flow (F) is described by

$$F = (N\pi r^4/8\eta)\,(\Delta P/\Delta x) \tag{6}$$

The total area available for water flow (A_w) is $N\pi r^2$; incorporating this into Eq. (6), one obtains

$$F = (r^2\Delta P/8\eta)\,(A_w/\Delta x) \tag{7}$$

Molecular sieving data yield values for average pore radius and $A_w/\Delta x$, leaving only the terms F and ΔP. Thus, when the total flow is known, ΔP can be approximated for specific values of pore radius and $A_w/\Delta x$. Furthermore, since the actual K_f term is equal to $F/\Delta P$ (ΔP in this case is the effective filtration pressure gradient and includes the contribution of colloid osmotic pressure), K_f values can be determined from molecular sieving data from the relation

$$K_f = (r^2/8\eta)\,(A_w/\Delta x) \tag{8}$$

Although this technique would initially appear to be very useful, there have always been major problems with this approach because of the substantial uncertainties inherent in the determination of the necessary parameters for channel radius and for the

$A_w/\Delta x$ term.(55,59,105–107) In particular, the techniques for determination of appropriate steric hindrance coefficients have remained empirical or based on macroscopic phenomena.

In recent years, there has been more emphasis on the use of thermodynamic principles to evaluate molecular sieving data. Although a variety of approaches are used, all such derivations for solute flux (J_s)(106–110) across a constraining membrane include a convection term, which is the solute flux that occurs as a consequence of the bulk volume flow, and a diffusion flux, which is a consequence of the concentration gradient. Obviously, if the molecules used to assess the "effective size" of the pores are large enough so that they do not readily enter the convection channels, then they are restricted from entry. σ refers to this restriction factor which varies from 0 to 1 with 1 designating complete exclusion. In its most elementary form, solute flux due to convection is

$$J_s = J_v C_p (1 - \sigma) \qquad (9)$$

and solute flux due to diffusion is

$$J_s = PS (C_p - C_B) \qquad (10)$$

where J_v is the volume flow (in this case the GFR) and C_p is the concentration in the plasma. C_B is the concentration in Bowman's space and PS is the diffusional permeability surface area product coefficient.(109,110) With small uncharged molecules, such as glucose, σ approaches zero and thus glucose flux is simply defined by the product of GFR and the plasma glucose concentration. Furthermore, no concentration gradient across the glomerular capillary is established and therefore the diffusion term is zero. For very large molecules that are restricted with almost complete efficiency, σ approaches values of 1 and thus $J_v C_p$ is factored by near-zero values. The most relevant example is for plasma albumin. Using a value of 1 mg/dl for albumin concentration in early tubular fluid(57–59) and an approximate initial plasma albumin concentration of 3600 mg/dl, σ is estimated to be 0.9997. Under these conditions, the PS coefficient is so low (0.001 ml/min) that solute flux due to diffusion also approaches zero.

The utility of the molecular sieving approach is best exemplified with the use of uncharged probe molecules that, while restricted, do permeate the glomerular membrane to a sufficient extent to have σ values in the range of 0.3 to 0.9.(105–113) In actual experimental studies, the sieving coefficient (ϕ), which is $1 - \sigma$, is obtained from the ratio of the clearance of the solute being evaluated (which must not be reabsorbed, secreted, or metabolized by the tubules) and the clearance of a filtration marker such as inulin. Most experimental studies have used neutral uncharged macromolecules such as dextran, Ficoll, or polyvinylpyrrolidone (PVP) of varying, but well-defined sizes. The experimental data can be plotted as a function of effective molecular radius and examples of such relationships obtained in dogs are shown in Fig. 7.(111–113) The data plotted were obtained from studies in dogs and also demonstrate the changes in the sieving relationships that occur when molecules having an anionic charge are used. For any given effective radius, molecules having a negative charge are restricted in their passage across the glomerular membrane whereas polycations are transported with greater efficiency as compared to neutral molecules of equivalent size.(113–116) These differences in transport appear to be due to the characteristic polyanionic constituents of the glomerular basement membrane. The glycoproteins are rich in sialic acid and dicarboxylic amino acids which set up a negative electrostatic field that repels polyanions and accelerates the passage of polycations.(36,37,40) Theoretical considerations indicate that the most important of the barriers is the lamina rara interna and its associated glycocalyceal complex.(59)

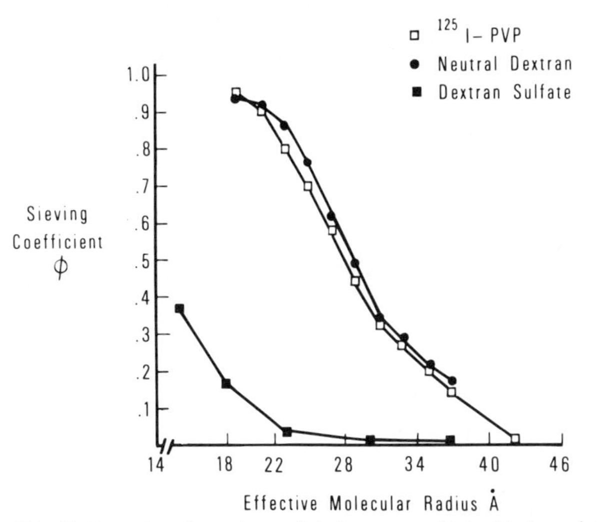

Fig. 7. Examples of experimental sieving curves obtained in dogs plotting effective molecular radius against the sieving coefficient. Data from Refs. 111–113. It is clear that anionic molecules of equivalent size are restricted to a much greater extent than neutral molecules.

Another factor affecting solute flux is the actual molecular configuration.(117,118) Studies have demonstrated that rigid molecules such as horseradish peroxidase or Ficoll have lower sieving coefficients for any given molecular size than neutral dextran.(117,118) Thus, shape, flexibility, and deformability may play important roles in determining the quantitative relationship between molecular size and transglomerular solute flux.

In order to arrive at values for average pore radius and the $A_w/\Delta x$ term, the solute flux equation including both the convection and diffusion term is transformed to include steric hindrance terms and wall correction factors.(106,107,112–114,119) These can be used to arrive at actual dimensions of the channels. However, even the determination of actual molecular radius is a problem since gel permeabilities are somewhat variable and are influenced by factors other than size.(120) Thus, the true dimensions of the extracellular channels can only be estimated. In spite of these reservations, data currently available based on sieving coefficients of uncharged molecules indicate that the effective radius of the channels in the glomerular membrane is in the range of 45–60 Å. Further analysis incorporating steric wall hindrance terms and assumed values for transcapillary glomerular pressure has yielded approximate values for K_f in man from as low as 2.5 to as high as 6.6 ml/min per mm Hg per 100 g kidney.(121,122) Using a glomerular density of 7000 nephrons/g kidney, one can estimate the average K_f per glomerulus in man to be in the range of 3.6 to 9.4 nl/min per mm Hg. In a recent mathematical analysis of glomerular dynamics in rat, dog, and man, Oken(99) calculated a K_f value of 6 nl/min per mm Hg for man. If one uses a reasonable estimate of 14 mm Hg for average effective filtration pressure in man,(122) this results in a K_f value of 4.6 nl/min per mm Hg.

4. Quantitative Description of Glomerular Dynamics

In order to appreciate the quantitative interactions among the many factors that contribute to the final level of GFR, it is helpful to use appropriate mathematical models. This type of analysis has provided a greater understanding of fluid flow dynamics as they exist within the glomerular capillary structures. Most studies of glomerular dynamics have used a single channel model,(87,91,96,99,107,123–126) but multiple parallel pathway models have also been evaluated.(8,100–102)

4.1. Single Capillary Model

The single cylinder model is quite useful in that it is simple, but still incorporates many of the features that are required to evaluate a substantial number of interactions among the various components of the system. However, such a model does not take into consideration the true structural dimensions of the system; therefore, parameters which depend on these, such as hydraulic resistances and filtration coefficients (K_f), cannot be related directly to their anatomical counterparts.

For the description of fluid flow along the length of the capillary (defined as x and varying from 0 to 1), the following relationships apply. As filtration occurs along the glomerular capillary, the plasma flow (F_p) decreases according to the Starling filtration principle described earlier:

$$-\frac{dF_{p(x)}}{dx} = K_f[P_{g(x)} - P_B - \pi_{g(x)}] \quad (11)$$

where $P_{g(x)} = P_{g(0)}$ at $x = 0$ and $P_{g(0)} - P_{g(\Delta)}$ at $x = 1$. $P_{g(\Delta)}$ is the pressure drop along the glomerular capillary and is usually only 1 to 3 mm Hg. Also, $[P_{g(x)} - P_B - \pi_{g(x)}]$ defines the effective filtration pressure ($EFP_{(x)}$). As already discussed, the colloid osmotic pressure ($\pi_{g(x)}$) is a nonlinear function of the protein concentration ($C_{(x)}$). In the case of the glomerulus, it is valid to assume that protein is not lost from the system, and thus

$$C_{(x)} = C_a \cdot F_{p(a)}/F_{p(x)} \quad (12)$$

where C_a and $F_{p(a)}$ represent protein concentration and plasma flow at the afferent arteriole. The cumulative filtration rate at any point x is equal to

$$GFR_{(x)} = F_{p(a)} - F_{p(x)} \quad (13)$$

yielding the total GFR at the end of the filtering area with $F_{p(1)}$, the plasma flow at the efferent arteriole ($F_{p(e)}$). From hemodynamic considerations, one obtains the following relationship for blood flow at the afferent arteriole ($F_{b(a)}$) and the efferent arterioles ($F_{b(e)}$):

$$F_{b(a)} = (P_a - P_{g(0)})/R_a \quad (14)$$

$$F_{b(e)} = (P_{g(0)} - P_{g(\Delta)} - P_e)/R_e \quad (15)$$

For plasma flow the hematocrit must be considered such that

$$F_{p(a)} = \frac{(1 - H_a)(P_a - P_{g(0)})}{R_a} \quad (16)$$

In these equations, P_a represents renal arterial pressure, P_e the end efferent arteriolar pressure, R_a the afferent arteriolar resistance, and R_e the efferent arteriolar resistance, and H_a the arteriolar blood hematocrit.

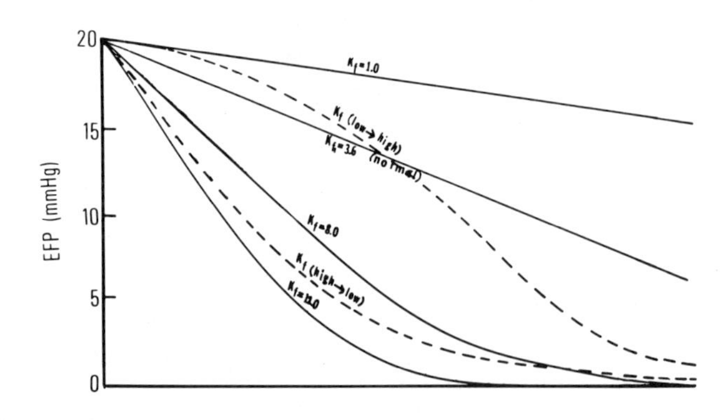

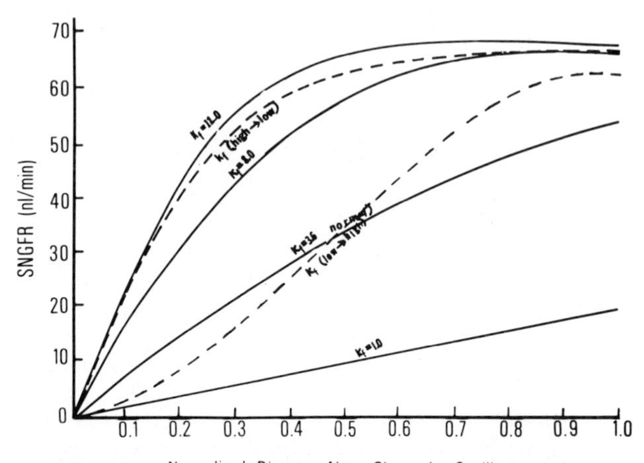

Fig. 8. Effects of alterations in K_f on the SNGFR and effective filtration pressure profiles. Base data were taken from a recent study in dogs(191) where single-nephron plasma flow was 157 nl/min. Control relationships represent K_f values calculated to provide the most consistent agreement with micropuncture data. Also shown are the model predictions at very low K_f values and at high K_f values where filtration equilibrium is achieved. The dashed lines represent model predictions when K_f values are distributed along the length of the glomerular capillaries from low to high and high to low. (Help in modeling was provided by Dr. George Yuscavage.)

Using computer simulation techniques, these equations can be solved to compute the GFR for any given set of conditions or, in turn, to compute parameters, such as K_f, when the GFR is known. They have been used extensively to predict changes in glomerular dynamics in response to a variety of specific alterations in one or more of the parameters or inputs.

Since many different studies have already provided an exhaustive evaluation of glomerular dynamics as it applies to the rat,(60,87,91,96,100–102,124–126) the analysis presented in this chapter focuses on glomerular dynamics using representative data obtained in experiments on dogs as expressed in Table II. Applying these data to the model described, one can obtain a description of the profile of effective filtration pressure (EFP) and cumulative SNGFR as a function of the normalized length of the glomerular capillary. Such profiles are plotted in Fig. 8 with the dashed lines showing the responses considered to be most representative for the dog.(88,94,98) In addition, the predicted profiles as affected by different K_f values ranging from 4 to 15 nl/min per mm Hg are shown. As can be seen, specific increases in K_f lead to increases in SNGFR and filtration fraction as long as the EFP at the end of the capillary is positive. Once the condition of "filtration equilibrium" is reached, further increases in K_f will not cause any additional increases in SNGFR or filtration fraction. The very high K_f values allow a very rapid rate of filtration in the early segments of the glomerular capillary and thus the protein concentration increases rapidly, leading to an equilibration of the hydrostatic and colloid osmotic pressures before the end of the glomerular capillary. Figure 8 also demonstrates the differences in the EFP and SNGFR profiles that occur if K_f is not uniform throughout the length of the glomerular capillaries.

As pointed out, data obtained in the dog indicate that filtration occurs throughout the length of the glomerular capillary, and that the normal K_f in the dog is approximately 3.5 to 4 nl/min mm Hg. In contrast, data obtained in the Munich–Wistar rat indicate the existence of "filtration equilibrium" in this strain. Because of apparent misconceptions that have arisen related to the implications of "filtration equilibirum," it would seem appropriate to compare the behavioral characteristics of systems that exhibit "filtration equilibirum" to those that filter throughout the length of the capillary. This has been done in the present model by comparing the responses of the model using values of K_f and ΔP that yield the two situations. For the model exhibiting "filtration equilibrium," K_f was increased and the glomerular pressure was decreased to obtain the same control SNGFR.[123]

4.2. Effects of Singular Perturbations on GFR

In Fig. 9, the predicted effects of singular changes in transglomerular pressure (Fig. 9a), plasma protein concentration (Fig. 9b), and glomerular plasma flow (Fig. 9c) on GFR are shown. The solid lines represent the responses of the model not in filtration equilibrium and the dashed lines represent the responses of the model in filtration equilibrium. As shown in

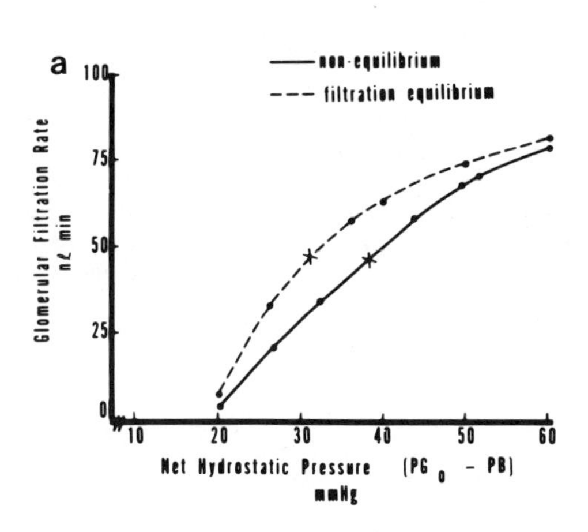

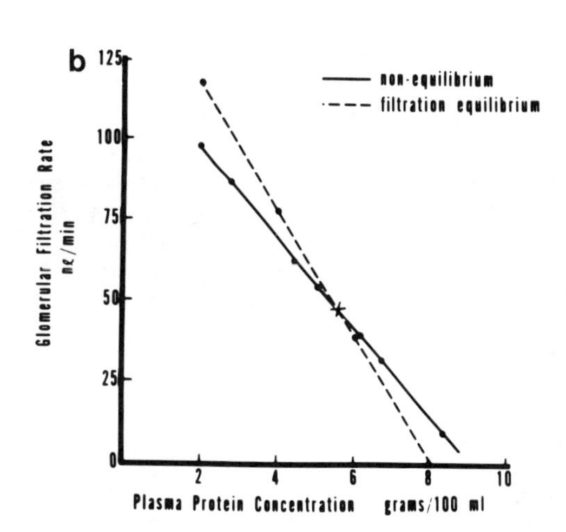

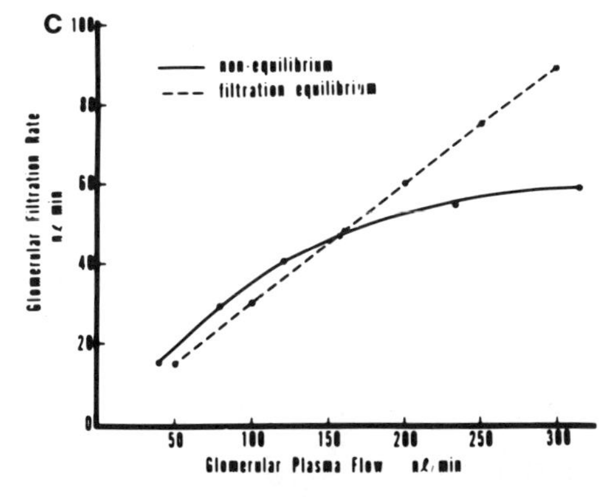

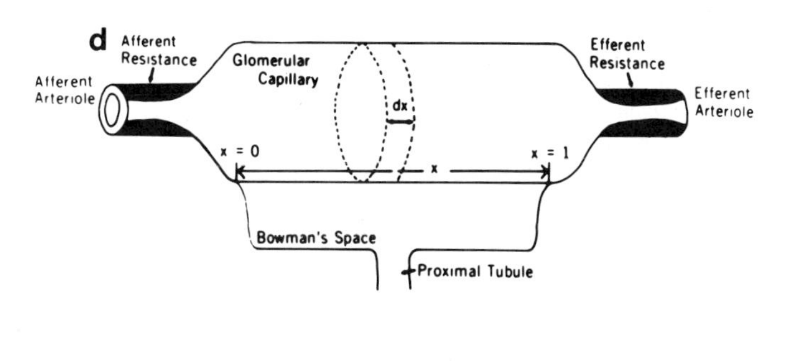

Fig. 9. Responses of the glomerular model to specific changes in (a) transglomerular pressure, (b) plasma protein concentration, and (c) glomerular plasma flow. The solid lines represent the response of the model when it does not exhibit filtration equilibrium, and the dashed lines represent the responses when the K_f was increased and the glomerular pressure decreased so as to obtain filtration equilibrium. A schematic drawing of a single-channel glomerular capillary model is also shown. From White and Navar.[123]

Fig. 9a, differences in responses to changes in the net hydrostatic pressure difference are not marked, with the nonequilibrium system showing slightly greater responsiveness to changes in pressure in the normal operating range. Under both conditions, there is a decreased sensitivity to increases in pressure as the net hydrostatic pressure increases, with the equilibrium model becoming progressively less sensitive. The effects of changes in plasma protein concentration in the absence of changes in any of the other determinants are shown in Fig. 9b. Similar qualitative responses in GFR are indicated for both conditions; however, there is a greater sensitivity of the system in filtration equilibrium. For example, a 10% decrease in protein concentration would lead to a 16% increase in GFR for the system not in equilibrium, versus a 23% increase for the system exhibiting filtration equilibrium. The GFR responses to changes in plasma flow are shown in Fig. 9c and demonstrate the major difference between the two conditions. It is important to emphasize that in both conditions, there are similar responses to decreases in plasma flow, but the responses to increases in plasma flow are quantitatively distinct. Increases in plasma flow increase GFR under both conditions; however, the response is essentially a linear and proportional one for the system in filtration equilibrium. This fundamental response has provided the basis for the concept of plasma flow dependence of GFR.[97] Increases in plasma flow will also increase GFR in the model not in filtration equilibrium, but the magnitude of the response is substantially less and is due exclusively to the decrease in the average glomerular colloid osmotic pressure.

Therefore, there are two major differences between the filtration equilibrium system and the nonequilibrium system. The first is the virtual insensitivity of the equilibrium system to increases in K_f; as long as filtration equilibrium obtains, increases in K_f will not alter the total GFR further. However, decreases in K_f could take the system out of equilibrium and eventually lead to reductions in GFR unless compensatory changes in other determinants occurred. The second major difference is the high degree of plasma flow dependence of the filtration equilibrium model as compared to the nonequilibrium condition. This characteristic of a near-proportional degree of plasma flow dependency has stimulated a great deal of interest in the potential role of changes in renal plasma flow, *per se*, in the control of GFR. A major point that is often neglected, however, is that the net hydrostatic pressure difference remains the single most important determinant of GFR regardless of the status of the system.[123]

There continues to exist a considerable degree of uncertainty regarding the general applicability of the concept of "filtration equilibrium," and its potential role in the regulation of filtration dynamics as it pertains to overall nephron function. The direct measurements of glomerular pressure from superficial glomeruli coupled with the relatively high filtration fractions found in the Munich–Wistar rat have indicated the existence of filtration equilibrium.[60,68,78,96,97,100] Whether this condition is unique to these superficial nephrons or is representative of the total nephron populations has not been established. The "Chapel Hill" substrain of Munich–Wistar rats is apparently not in filtration equilibrium when studied in similar situations.[80] With respect to other species, a rather thorough analysis conducted recently by Oken[99] indicates that the patterns of responses in dog and man are quite similar and distinctly different from those in the rat. Also, it has been demonstrated in humans and in dogs that GFR exhibits a relative independence from renal plasma flow changes under a variety of circumstances. For example, during acute extracellular fluid volume or plasma volume expansion and during the administration of vasodilators, the changes in renal plasma flow are substantially greater than the changes in GFR.[6,74,98,127,128] The fact that GFR decreases with decreases in plasma flow cannot be used as evidence for the existence of the equilibrium state since this would be expected in either equilibrium or nonequilibrium conditions. However, some of the whole kidney responses could involve complex alterations in more than one of the determinants of GFR, such as K_f, glomerular pressure, and/or proximal tubule pressure. In spite of these uncertainties, it would seem that filtration equilibrium exists only under special circumstances and is not a generally applicable concept.

4.3. Analysis of Glomerular Dynamics Using Multiple Capillary Models

One of the reasons for using more complex models of the glomerulus is to have the model parameters more closely approximate actual morphological parameters such as total surface area available for filtration. In addition, such approaches allow for the actual calculation of the hydrostatic pressure drop along the glomerular capillaries. For example, Huss *et al.*[100] calculated that the actual hydrostatic pressure drop for glomerular capillaries was less than 1 mm Hg and studied the effects of variations in the total number of capillary loops. In general, the model was shown to be rather insensitive to the actual number of glomerular capillary loops. Based on the analysis of a reconstructed glomerular capillary system, Shea[8] suggested a slightly greater pressure drop of about 3 mm Hg within the glomerular capillaries.

A rather sophisticated microrheological network model was developed by Lambert *et al.*[102] based on the reconstructed vascular network from glomeruli of Wistar rats.[7] Attempts were made to mathematically describe the actual intraglomerular pathways for blood flow. Their average glomerulus consisted of several lobules having four categories of blood vessels from large capillaries (radius between 5 and 6 μm) to very narrow capillaries (radius less than 3 μm); there were a total of 178 branches with 120 nodes. Various aspects were evaluated including the hydrostatic pressure drop along both large and small capillaries, cell velocity along the capillaries, viscosity of blood in large and small capillaries, actual capillary lengths, capillary volumes, and surface areas. For a representative glomerulus having a capillary volume of 1.75×10^{-7} cm^3 and average glomerulus capillary pressure of 49.2 mm Hg, a total hydrostatic pressure drop of 1.5 mm Hg (3.1%) was derived. Also, an average K_f of 2.7 nl/min per mm Hg was determined when the input data were taken from a study by Blantz *et al.*[77] The total surface area per glomerulus was calculated to be somewhat lower than shown in Table II and thus L_p values were rather substantial, varying in the range of 38 to 81 nl/sec mm Hg per cm. A number of interesting simulations were conducted. The consequences of having short-circuiting pathways and segmental flow along preferential pathways were evaluated. This study also demonstrated that pressure equilibrium at the local level had to be distinguished from pressure equilibrium at the distal end of the capillary bed. It is possible to achieve equilibrium in individual capillaries without attaining filtration equilibrium for the total glomerulus. Although this approach is quite complicated, further analysis of such network models may provide much greater understanding of intraglomerular dynamics than the traditional single capillary model.

5. Physiological Regulation of Glomerular Filtration Rate

In recent years, intensive investigation in this area has substantially broadened the scope of our understanding regarding mechanisms that exert controlling influences on GFR. A variety of extrinsic neural and humoral influences as well as intrinsic mechanisms have been found to alter one or more of the determinants of GFR. Much of the current investigation is focused on the delineation of the precise roles of such mechanisms in controlling GFR.

5.1. The Controlling Elements

While all stimuli that influence GFR must alter one or more of the determinants of GFR, the specific determinants which are subject to physiological control are limited. Almost all such adjustments involve changes in vascular resistance or changes in K_f. Other factors, such as arterial pressure, plasma colloid osmotic pressure, or proximal tubular pressure, are either independent variables not subject to direct acute control or, as in the case of proximal tubular pressure, are partially a consequence of changes in GFR.

5.1.1. Factors That Influence the Glomerular Filtration Coefficient

It has been demonstrated in both dogs and rats that several conditions or agents can alter K_f. The nature of the mechanisms for altering K_f remains largely unknown; however, the actual change in K_f must be due to an alteration in the filtration barrier and/or a change in the surface area available for filtration. It has been proposed that intraglomerular cellular elements, in particular the mesangial cells, have contractile capability that allows them to regulate glomerular capillary surface area available for filtration and/or the number of open capillary loops.[33,46,49] The existence of hormone receptors including receptors for angiotensin II and vasopressin[48,121] on mesangial cell membranes has supported the notion that mesangial cells may participate in the regulation of K_f. However, the centrilobular position of the mesangial cells makes it difficult to conceptualize an efficient mechanism by which changes in the contractile level of mesangial cells could regulate K_f. One possibility is that mesangial cells which are near the origin or termination of the capillary loops could serve as functional capillary sphincters and regulate the number of flowing capillary loops at any given time. It is also possible that the average luminal diameter of the capillaries is altered by these and other humoral agents.[130]

Other intraglomerular elements such as the endothelial cells, epithelial cells, or the basement membrane could influence or alter K_f by directly altering hydraulic conductivity. This may be particularly true in certain disease processes where evidence for anatomical changes in number or size of endothelial fenestrae, shape and size of the podocyte foot processes, or thickness of the basement membrane has been found.[33,34,42,43,103,104] For example, in an ischemic acute renal failure model in the dog, Williams *et al.*[103] found a decrease in the number and size of endothelial fenestrae and in the shape and size of the podocyte foot processes in association with the marked decreases in K_f.

Regardless of the precise mechanism(s) for altering K_f, changes in the ultrafiltration coefficient have been found to occur in response to several different conditions or agents. Studies in the Munich–Wistar rat have shown that decreases in K_f occur in response to infusions of several vasoactive agents.[78,129–133] Also, alterations in the physiological state of the animal result in large changes in K_f. It has been postulated that most K_f alterations are mediated by direct effects of angiotensin II or vasopressin.[130] In contrast to results obtained in the rat, studies in the dog have indicated that K_f may be more stable in this species in that it remains unchanged or is only slightly altered under a variety of conditions.[62,98,134] As indicated in Fig. 10, the intrarenal administration of vasodilators does not reduce K_f in the dog, in contrast to the reductions seen in the rat. In addition, when changes in K_f in the dog have been found, the magnitude of these changes has been much smaller than that reported for the rat. Studies in both dogs and rats, however, support the concept that angiotensin infusions can reduce K_f.[131] It is also agreed that K_f increases in response to increases in colloid osmotic pressure although the magnitude of the changes is less in dogs.[62,136] Also marked sodium depletion leads to a reduction in K_f as shown in Fig. 10.[137] In dogs, moderate sodium restriction did not significantly lower K_f.[138]

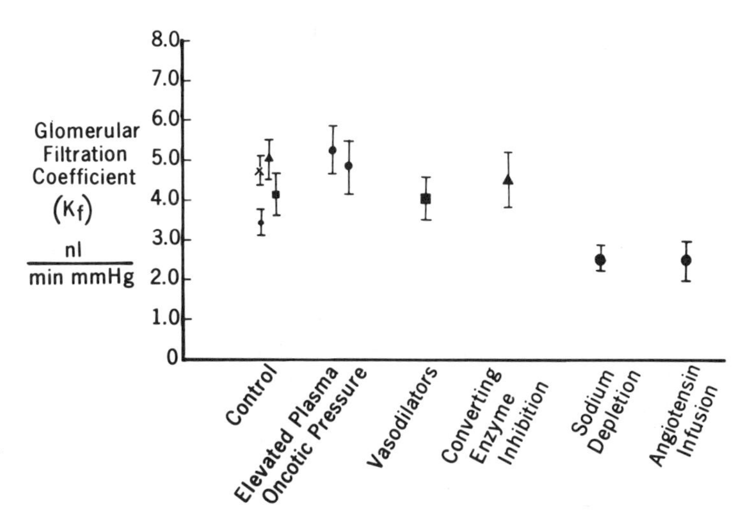

Fig. 10. Experimental data of glomerular filtration coefficients (K_f) obtained in dogs during control conditions and during various experimental situations as designated. Data from Refs. 62, 88, 94, 98, 134, 135. In dogs, vasodilators have not been shown to decrease K_f; however, average K_f values were lower during angiotensin infusions and in dogs maintained on low-sodium diets.

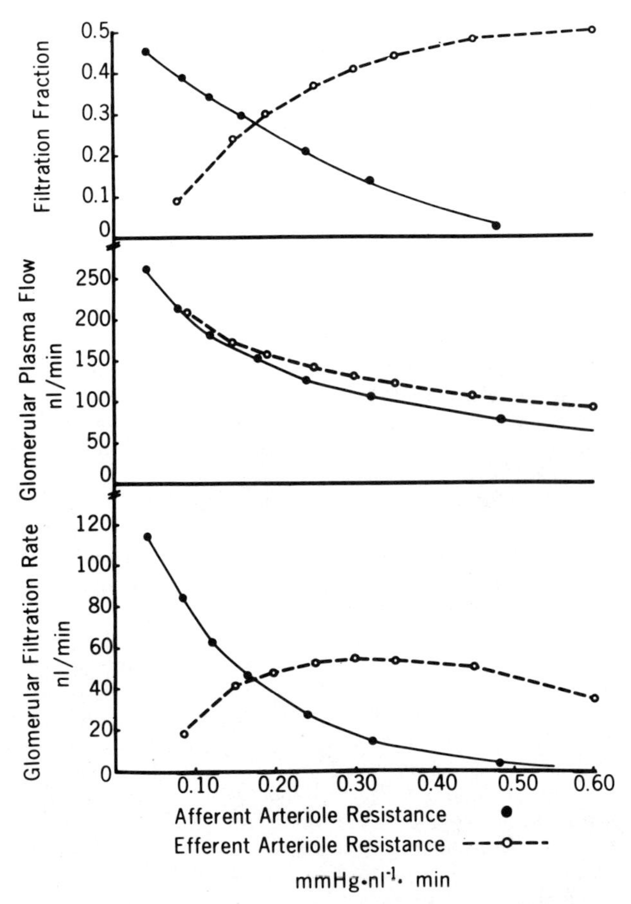

Fig. 11. Predicted effects of singular changes in either afferent arteriolar resistance or efferent arteriolar resistance on single-nephron glomerular filtration rate, glomerular plasma flow, and filtration fraction. Control values were 0.16 mm Hg/nl per min for afferent resistance and 0.19 mm Hg/nl per min for efferent resistance. Other input data are the same as for the normal dog glomerular model not in equilibrium.

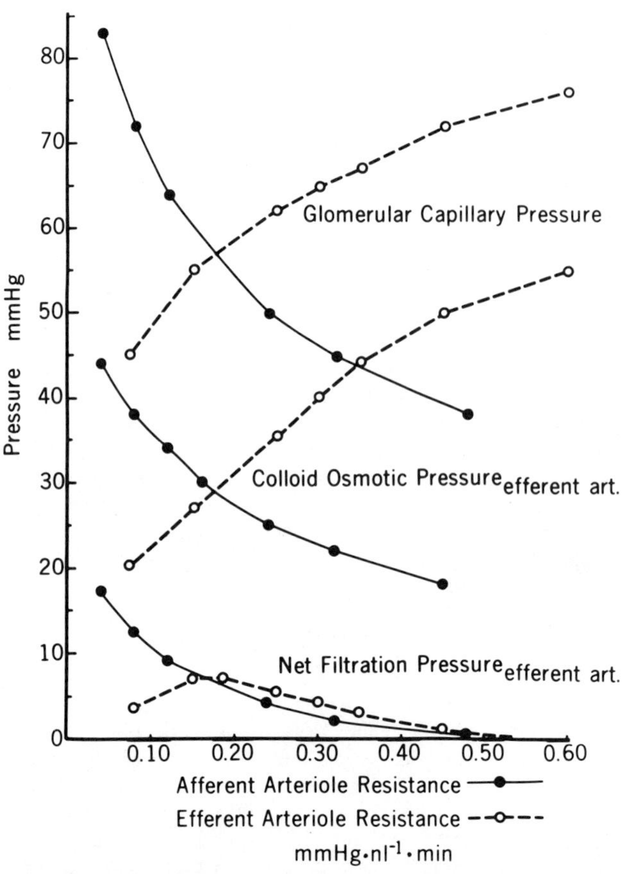

Fig. 12. Predicted effects of specific variations in afferent arteriolar resistance or efferent arteriolar resistance on glomerular capillary pressure, colloid osmotic pressure at terminal end of the glomerular capillaries (π_e), and net (effective) filtration pressure at end of the capillaries (EFP_e). The control values are the same as for Fig. 10.

5.1.2. Effects of Changes in Afferent and Efferent Vascular Resistances

It is generally appreciated that the afferent and efferent arteriolar resistance elements have a major and, most likely, the predominant regulatory influence on GFR. Since these arteriolar segments are also responsible for regulating hemodynamic resistance to blood flow, it is useful to consider the theoretical effects of selective changes in pre- and postglomerular resistances on factors such as glomerular blood flow, glomerular pressure, and SNGFR. A model similar to that described in the preceding section and similar to that used by Oken *et al.*[91,99] was used for this analysis.[123] In order to evaluate the singular effects of resistance changes, all other input factors remained unchanged from control, a case that is clearly not valid physiologically, but is helpful in understanding the effects of specific perturbations. As shown in Figs. 11 and 12, selective increases in afferent arteriolar resistance lead to concomitant decreases in plasma flow and glomerular pressure along with large changes in GFR. The filtration fraction does not remain constant, but decreases in response to increases in afferent resistance. Also of interest is the model prediction that, at high levels of afferent resistance, the effective filtration pressure at the terminal portion of the glomerular capillary approaches zero and an equilibrium state can be reached, even though filtration fraction is relatively low.

Selective increases in efferent arteriolar resistance lead to complex responses due to the fact that plasma flow decreases while the glomerular pressure increases. These relationships are also shown in Figs. 11 and 12. GFR increases slightly due to increases in glomerular pressure and then decreases because of the marked increase in the protein concentration profile within the glomerulus. Filtration fraction increases progressively, leading to a marked increase in glomerular colloid osmotic pressure and, at high levels of vascular resistance, a complete dissipation of the effective filtration pressure. Thus, under normal conditions, GFR is very close to the maximum GFR that can be achieved with only efferent arteriolar constriction. However, agents which lead to preferential vasodilation of efferent arterioles could certainly lead to decreases in GFR due to reductions in glomerular capillary pressure.

In simulations where the afferent and efferent arteriolar resistances were changed at the same time, glomerular pressure was not altered; therefore, GFR was influenced primarily by the altered colloid osmotic pressure profile within the glomerular capillaries. For example, in response to 20% changes in both afferent and efferent arteriolar resistances, SNGFR was altered only slightly (10–12%) and to a lesser extent than renal plasma flow. Therefore, filtration fraction increased with increases in resistance and decreased with decreases in resistance due to proportionately greater changes in renal plasma flow than in GFR. This analysis demonstrates that it is incorrect to conclude that increases in filtration fraction are necessarily due to preferential increases in efferent arteriolar resistance.

On the basis of these general considerations, it can be recognized that the resistance element having the greatest capability to actually "regulate GFR" is the afferent arteriolar segment. Alterations in the status of afferent arteriolar resistance have the potential for causing large changes in GFR along with concomitant effects on glomerular pressure and plasma flow. The extent to which selective changes in efferent resistance control GFR remains somewhat controversial at present. It has been suggested that the efferent resistance element, in particular that of the superficial nephrons, may be relatively passive, responding to, rather than being responsible for, changes in glomerular pressure.[93] In other studies, however, there are clear indications that efferent resistance does increase substantially with infusion of vasoconstrictors and decreases upon administration of vasodilators.[131,132,134,135,139]

5.2. Intrinsic Control Mechanisms

The intrinsic capability of the kidney to adjust renal vascular resistance was first demonstrated in experiments where arterial pressure to the kidney was altered and it was determined that renal blood flow and GFR remained nearly constant ovcr a relatively wide range.[140,141] This phenomenon of renal autoregulation was intriguing in that it demonstrated the presence of a very powerful and rapidly acting local control mechanism.[142,143] It was subsequently shown that autoregulatory responses are not specific to changes in arterial pressure, but that several other maneuvers such as increases in renal venous pressure, increases in ureteral pressure, and increases in the plasma colloid osmotic pressure also elicit autonomous adjustments in renal vascular resistance.[144] The GFR responses to these maneuvers, as well as the fact that autoregulation of renal plasma flow and GFR are normally closely coupled in response to changes in arterial pressure, have led to the conclusion that the observed changes in renal vascular resistance are localized primarily to the preglomerular segments.[30,143,144] Micropuncture studies have substantiated this concept where it has been demonstrated that proximal tubular pressure, peritubular capillary pressure, and SNGFR,[73] as well as directly measured glomerular pressure[145] all exhibit appropriate autoregulatory responses to changes in arterial pressure. Specifically defined, autoregulatory behavior is evident when there is a change in renal vascular resistance that is mediated by an intrarenal vasoactive mechanism in response to an extrinsic stimulus which itself does not have the direct capability to alter vascular resistance.[146] Importantly, hemodynamic responses to exogenous agents that can directly vasodilate or vasoconstrict *should not* be viewed as autoregulatory responses. Since autoregulatory behavior has been demonstrated to occur at the single nephron level, it can be considered that the intrinsic autoregulatory behavior of whole kidney function is but the reflection of the collective behavior of the control mechanisms responsible for regulation of afferent arteriolar resistance at the individual nephron level. Substantial evidence now exists in support of the hypothesis that such autoregulatory capability existing at the single nephron level is mediated primarily by the tubuloglomerular feedback mechanism.

5.2.1. Tubuloglomerular Feedback Mechanism

It is widely recognized that the maintenance of an optimum balance between the metabolically dependent reabsorptive functions of the tubules and the hemodynamically determined filtered load to the tubules requires effective mechanisms regulating GFR. This notion, coupled to the fact that there is an intimate morphological association between the distal tubule and the vascular pole of its own nephron unit (see Section 2), has led to the hypothesis that the macula densa region may serve as a "feedback" regulator of glomerular function.[28–30,142,143] The fundamental hypothesis states that alterations in some physicochemical component of the tubular fluid at the level of the macula densa occur in response to alterations in the flow of fluid along the nephron segments prior to the macula densa. These changes in luminal composition are thought to be capable of mediating a sequence of cellular events culminating in an adjustment of one or more of the determinants of GFR.[28,29,90,147] Many of the autoregulatory responses observed at the whole kidney level are associated with diminished distal tubular flow and could be mediated through this feedback mechanism.[66,144] In addition, autoregulatory capability has been demonstrated to be markedly diminished in an ischemic acute renal failure model in which GFR and presumably distal tubular flow were markedly reduced.[103,148] These general findings are consistent with the hypothesis that inappropriate decreases in distal flow lead to preglomerular vasodilation whereas increases in flow result in vasoconstriction. More recent studies have also shown that conditions associated with an impairment in proximal and loop reabsorptive function, may be associated with depressed levels of GFR due to an enhanced activity of the tubuloglomerular feedback mechanism.[149,150]

Evidence at the single nephron level for the existence of an interaction between some component of the distal tubule and glomerular function has accumulated rapidly over the last two decades. In rat micropuncture experiments, Schnermann *et al.* demonstrated that increased distal volume delivery elicited by perfusing into the late portions of the proximal tubule led to feedback-mediated decreases in SNGFR and in stop-flow pressure, generally regarded as an index of glomerular pressure.[151,152] These basic findings have been substantiated by various investigators in other species including dogs[153,154] and *Amphiuma*.[155] It has also been reported that feedback-mediated decreases in SNGFR and stop-flow pressure can be elicited in juxtamedullary nephrons of the rat. Using papillary micropuncture techniques, Muller-Suur *et al.*[156] observed decreases in both indices of glomerular function in response to increases in the rate of perfusion into the thin ascending limb. In fact, the feedback-mediated decreases in stop-flow pressure and SNGFR were larger than those observed in superficial nephrons. These results substantiate the concept that the feedback mechanism is a general characteristic of the entire nephron population.

Representative data from both rats and dogs showing the magnitude and characteristic shape of the feedback-mediated changes in SNGFR and stop-flow pressure during increases in distal nephron perfusion are shown in Fig. 13. For these experiments, blocked proximal tubules were perfused at a point distal to the wax block. Tubular fluid collections or pressure measurements were made from a segment proximal to the block and the changes in SNGFR or stop-flow pressure observed in response

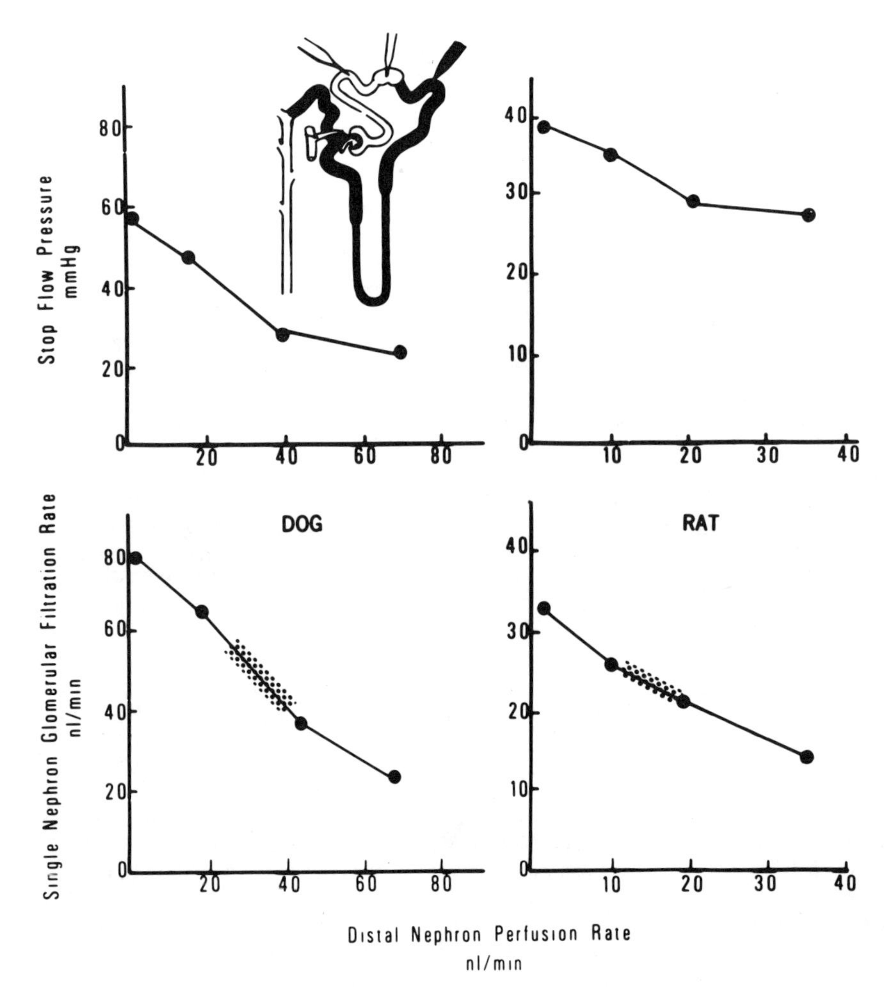

Fig. 13. The effects of changes in perfusion rate from a late proximal tubule on stop-flow pressure (upper panels) and single-nephron glomerular filtration rate (lower panels). The single-nephron orthograde microperfusion technique is illustrated. Left panels show results of studies in the dog and right panels, those in the rat. Note that the scale for SNGFR and stop-flow pressure in the rat is one-half of that for the studies in the dog. All tubules were perfused with isotonic electrolyte solutions. The hatched areas indicate the normal range for late proximal flow rate and SNGFR in both dog and rat.[151,152,154,157,169]

to increases in perfusion rate were assessed. In both dogs and rats there are substantial decreases in stop-flow pressure and SNGFR in response to microperfusion.[154,157] In the rat, SNGFR can be reduced by 40–50%, while in the dog as much as a 70 to 80% decrease in SNGFR can be achieved. The decreases in stop-flow pressure have generally been interpreted to indicate that glomerular pressure changes are responsible for the SNGFR responses.[90,144] A recent report that directly measured glomerular capillary pressure is not normally responsive to changes in distal nephron perfusion[158] has not been confirmed and other studies have demonstrated that directly measured glomerular capillary pressure does decrease in response to increases in distal nephron flow rate.[159,160] Therefore, the bulk of the available data support the contention that feedback-mediated changes in vascular tone are localized primarily in the afferent arteriole. It is also possible that the mesangial cells may participate in the contractile response.[158]

Coincident with the microperfusion experiments that demonstrated the existence of the tubuloglomerular feedback mechanism, other studies were conducted to evaluate the role of the feedback mechanism in the maintenance of SNGFR during autoregulatory-induced changes in renal vascular resistance. One prediction of the feedback hypothesis is that if afferent arteriolar tone is above the minimal physiological level, and is not unduly exposed to extrinsic vasoconstricting stimuli, then cessation of distal delivery should elicit reductions in afferent arteriolar tone.[66] Consequently, SNGFR based on proximal tubular fluid collections in the presence of an oil block would be expected to overestimate true SNGFR of the undisturbed nephron. Presumably, this interference would not occur during collections of distal tubular fluid so that these could be utilized to provide valid estimates of SNGFR. Experiments designed to test this hypothesis have demonstrated that SNGFR measured from proximal tubular fluid collections is greater than SNGFR based on distal fluid collections.[161,162] Representative data from these studies are shown in Fig. 14. In the rat, the difference between SNGFR measured from proximal collections and SNGFR measured from distal collections has averaged around 5–8 nl/min when evaluated on the basis of paired collections from the same tubule.[162–164] In the dog, the difference between SNGFR measured from complete proximal tubular fluid collections and SNGFR measured from distal tubular fluid collections has been shown to be substantially greater, averaging between 25 to 35 nl/min.[66,161] Also, values for SNGFR based on complete proximal tubular fluid collections were found to be greater than those obtained with an indicator-dilution technique in which SNGFR was measured from partial proximal tubular fluid collections during maintained distal volume delivery.[165] It has also been shown that the degree to which SNGFR measured from complete proximal tubular fluid collections overestimates true SNGFR is de-

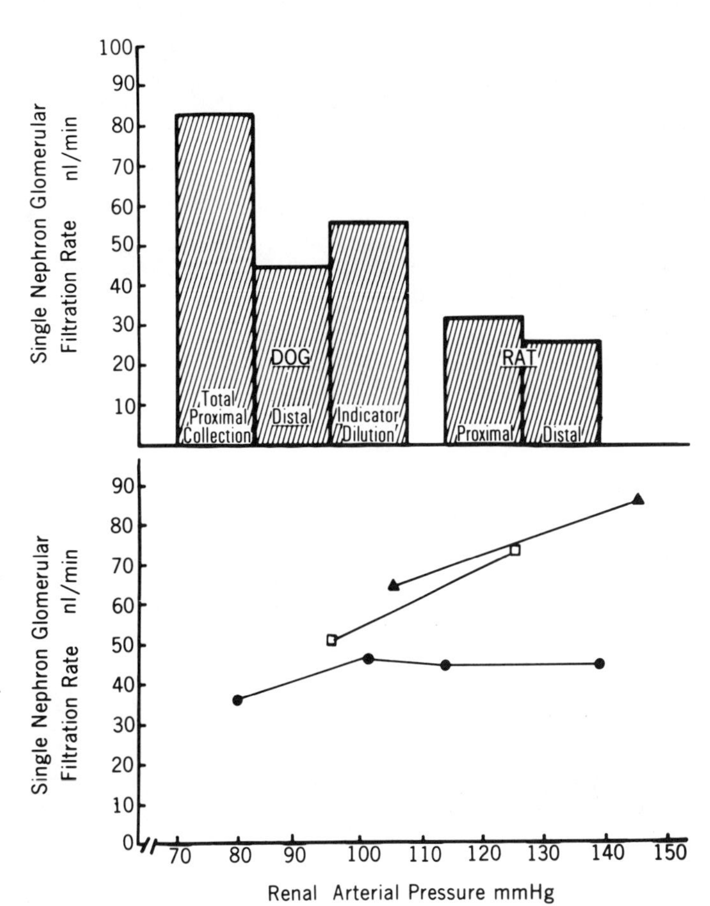

Fig. 14. Comparison of SNGFR measurements (top panel) and SNGFR autoregulation (lower panel) during intact and interrupted distal delivery. Data from dog studies indicate that SNGFR based on total proximal tubular fluid collections is higher than SNGFR measured from distal tubular fluid collections or SNGFR obtained by using an indicator-dilution technique. SNGFR values based on complete proximal tubular fluid collections are significantly higher than those based on complete distal tubular fluid collections from the same nephrons.

The lower panel demonstrates the effects of cessation of distal volume delivery on SNGFR autoregulation in the dog. The solid circles represent measurements of SNGFR based on complete distal tubular fluid collections with maintained distal volume delivery; the solid triangles and open squares depict SNGFR values obtained by total proximal tubular fluid collections. At the lower limit of the autoregulatory range (75–85 mm Hg), SNGFR values based on the two collection techniques would be similar.(161–166)

pendent in part on the length of the tubular fluid collection.(66) SNGFR reached maximum values only with collection times in excess of 1 min and approached values obtained with distal collections with collection times of 20 sec or less. These results provide evidence that feedback-mediated vasodilation occurs in response to decreases in distal volume delivery resulting from the placement of a block in the proximal tubule. These effects appear to occur primarily at the higher arterial pressures above the lower limit of the autoregulatory range. As is also shown in Fig. 14, SNGFR based on distal collections exhibits appropriate autoregulatory capability in response to decreases in renal perfusion pressure.(73,161,162) In contrast, SNGFR based on complete proximal tubular fluid collections has usually been found to have impaired autoregulation.(161,162) In the study by Williams *et al.*,(165) SNGFR autoregulation was also demonstrated with the indicator-dilution technique. Collectively, these studies support the hypothesis that the tubuloglomerular feedback mechanism is necessary in order to achieve normal efficiency of GFR autoregulatory capability. It should be acknowledge, however, that other studies have indicated that the tubuloglomerular feedback mechanism may not be solely responsible for autoregulatory behavior and that the myogenic mechanism may also contribute to renal autoregulatory capability.(146,166)

Since the existence and basic characteristics of the tubuloglomerular feedback mechanism are generally accepted, current attention is being focused on the process for the initiation and transmission of feedback signals from the receptor cells to the effector cells. Based on morphological and functional considerations, this process can be considered to involve four basic steps: first, a flow-related change in the concentration of the tubular fluid; second, detection of this change by the receptor cells, presumably the macula densa cells; third, transmission of signals to the vascular smooth muscle and/or mesangial cells; fourth, contraction or relaxation of the vascular contractile elements leading to a change in glomerular function.

Because of the relative water impermeability and the powerful sodium chloride reabsorptive capability of the thick ascending limb of the loop of Henle, the emerging tubular fluid sodium chloride concentration and osmolality are about one-third of those normally found in plasma. These tubular fluid concentrations are directly dependent upon volume flow rate through the loop such that sodium chloride concentration and osmolality approach isotonic levels at high loop flow rates.(157,167) This relationship between flow rate and composition allows factors that influence flow rate into the thick ascending limb (such as changes in GFR or tubular reabsorption rate) to be expressed as changes in distal tubular fluid composition. Under physiological conditions, changes in sodium chloride concentration and osmolality occur concurrently. One current issue of controversy is related to the precise mechanism by which the macula densa cells detect the change in concentration.

Studies by Schnermann *et al.*(168) suggested that a specific Cl^--dependent transport event mediates feedback responses. However, in a comprehensive series of orthograde microperfu-

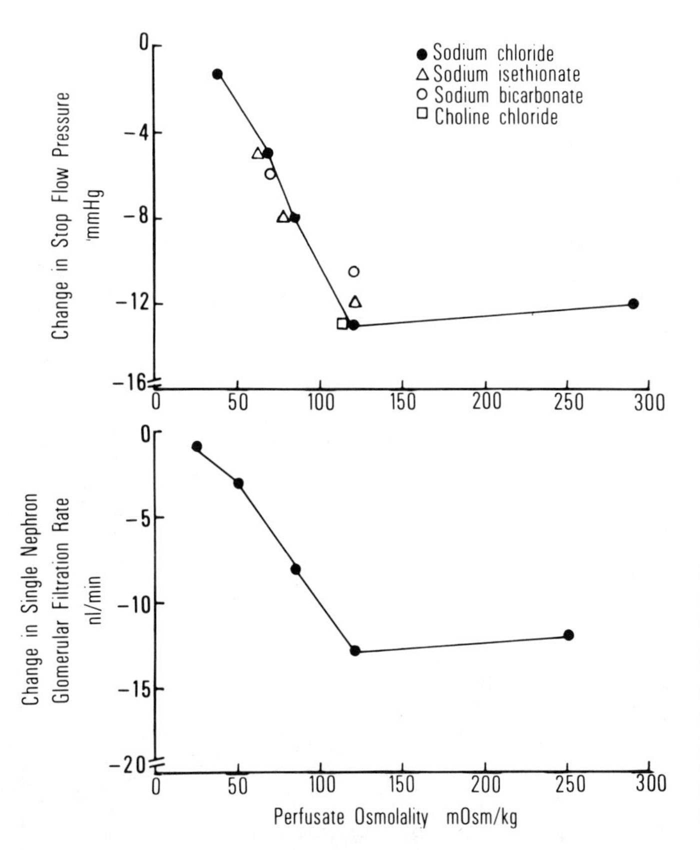

Fig. 15. The relationship between the change in stop-flow pressure (upper panel) and single-nephron glomerular filtration rate (lower panel) and perfusate osmolality. The change in these indices of glomerular function was obtained from the difference between the values obtained at no perfusion and during retrograde microperfusion at 15–20 nl/min. The SNGFR responses were adapted from studies by Schnermann *et al.*[168] The stop-flow pressure feedback responses are from Bell and Navar.[170] In these studies, the composition of the basic solution was altered by substituting most of the sodium chloride for either sodium isethionate, sodium bicarbonate, or choline chloride.

sion experiments,[154,157,69] it was demonstrated that Na^+ or Cl^- could be substituted by other electrolytes such as K^+, choline, acetate, isethionate, without altering the magnitude of the feedback responses. Also, feedback responses were obtained during perfusion with isotonic solutions containing predominantly nonelectrolytes such as mannitol or urea.[157] With these solutions the normal direct relationship between loop perfusion rate and distal tubular fluid NaCl concentration was disrupted and feedback-mediated changes in stop-flow pressure were obtained in the absence of changes in distal tubular fluid Cl^- concentration. These results have suggested that feedback-mediated changes in stop-flow pressure can be dissociated from changes in distal tubular fluid Na^+ or Cl^- concentrations or NaCl reabsorptive rates.

This issue has been approached more directly with the retrograde microperfusion technique which has the advantage that the site of perfusion in the early distal tubule is very close to the macula densa segment. With retrograde perfusion, the magnitude of the feedback responses is dependent primarily upon the concentration of the perfusate and is dissociated from the perfusion rate.[168–170] As shown in Fig. 15, changes in stop-flow pressure and SNGFR occur over a relatively narrow concentration range of perfusate osmolalities. Maximum responses occur at osmolalities greater than 120 mOsm/kg or NaCl concentrations greater than 60 mM. Schnermann *et al.*[168,171] reported that feedback responses during retrograde microperfusion were obtained most consistently with solutions of NaCl and other Cl^--containing electrolytes. Substitution of Cl^- with other anions reduced the consistency of the feedback responses. Also, mannitol or urea added to hypotonic NaCl solutions did not significantly enhance the feedback responses.[172] It was suggested that NaCl concentration or absorption is the specific signal responsible for the initiation of the feedback response.

Instead of using single salt solutions, Bell and Navar[170] used more complete solutions containing most of the electrolytes and nonelectrolytes normally found in tubular fluid. With this approach, solutions consisting predominantly of NaCl, sodium isethionate, sodium bicarbonate, or choline chloride were equally effective in eliciting feedback responses of similar magnitude at the same total osmolality. Specifically, the magnitude of the feedback response was dependent on the total concentration of the perfusate solution irrespective of the specific salt used to alter the concentration.[173] These studies indicate that the detection process may be related to a general property of the tubular fluid osmolality. In addition, it is possible that the functional integrity of the receptor system is dependent on the maintenance of an intraluminal environment that allows normal cell transport and metabolic function.

The cellular mechanism responsible for detecting changes in luminal fluid composition presents an intriguing problem since it requires a system that is highly sensitive to small changes in tubular fluid composition, has a fast response time, and is capable of transmitting signals to the contractile cells. Recent experimental findings have suggested an important role for intracellular Ca^{2+} in the transmission of feedback signals.[173,174] In retrograde microperfusion experiments, it was found that the Ca^{2+} ionophore, A23187, which increases membrane permeability to Ca^{2+}, markedly augmented the magnitude of feedback responses obtained during perfusion with a 70 mOsm/kg hypotonic solution. Since the exaggerated responses did not

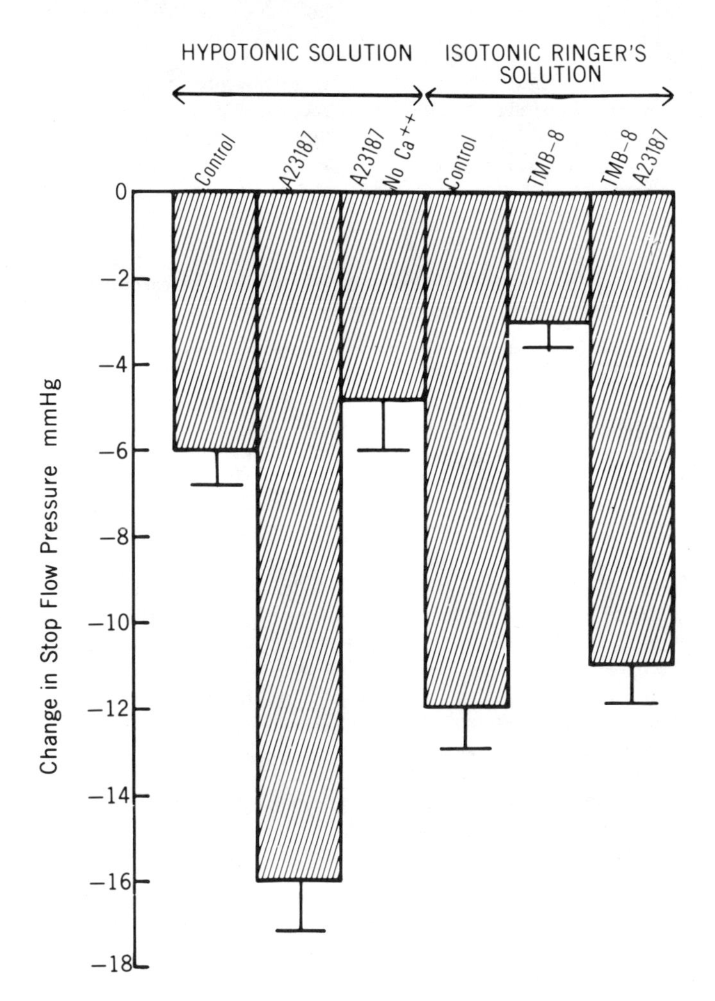

Fig. 16. The effects of retrograde microperfusion at 15 nl/min on stop-flow pressure feedback responses (SFP). The change in SFP is calculated as the difference in SFP obtained before microperfusion from the value obtained during microperfusion. The hypotonic solution had an osmolality of 70 mOsm/kg. The Ca^{2+} ionophore A23187 was added at concentrations of either 500 or 5 μM in the presence of 1 mEq/liter of calcium.

In the series using isotonic Ringer's solution, TMB-8 (500 μM), which inhibits the mobilization of Ca^{2+} from intracellular storage sites, produces a significant inhibition of SFP feedback responses after perfusion for 5 min. Addition of 5 μM A23187 reversed the inhibition produced by TMB-8.[174,175,178]

occur when Ca^{2+} was not present in the solution, the results suggested that cytosolic Ca^{2+} within the macula densa cells may serve in an important intermediary capacity.

In further studies,[175] 8-(*N,N*-diethylamino)-octyl-3,4,5-trimethoxybenzoate (TMB-8), which blocks the release of Ca^{2+} from intracellular stores,[176] was shown to be effective in attenuating feedback responses obtained during microperfusion with an isotonic Ringer's solution. The results from these studies are summarized in Fig. 16. The inhibitory effects of TMB-8 could be overcome by the addition of A23187 to a solution that contained Ca^{2+} (1 meq/liter). Presumably, TMB-8 entered the receptor cells and prevented increases in cytosolic Ca^{2+} which would occur in the response to an increased luminal solution concentration. The restoration of the response by addition of A23187 indicates that elevations in cytosolic Ca^{2+}, *per se,* are sufficient to fully activate the feedback mechanism. Furthermore, these results suggest that the source for the increase in cytosolic Ca^{2+} concentration during the transmission of feedback signals is from intracellular storage sites such as mitochondria or endoplasmic reticulum. Presumably, the cells can sense a change in tubular fluid osmolality which in some manner elicits the mobilization of Ca^{2+} from intracellular storage sites which then propagates the stimulus.

Microperfusion experiments have also demonstrated that methylxanthines such as theophylline or 3-isobutyl-1-methylxanthine (IBMX) can inhibit feedback responses in a dose-dependent fashion when added to an isotonic Ringer's perfusion solution.[177,178] In further studies,[178] forskolin, a diterpene agent that markedly stimulates adenylate cyclase activity and thereby elevates endogenous cAMP generation,[179] caused marked inhibition of stop-flow pressure feedback responses obtained during retrograde microperfusion. Since the effects of IBMX to inhibit the magnitude of feedback responses were counteracted by the addition of A23187, it would appear that cAMP may inhibit feedback responses by promoting Ca^{2+} entry into Ca^{2+}-sequestering organelles. This is consistent with studies in platelets[180] which have shown that elevations in cAMP can directly decrease cytosolic Ca^{2+} concentration, presumably through an increase in the movement of Ca^{2+} into intracellular storage sites. Thus, cAMP may serve to regulate feedback responsiveness such that elevations in intracellular cAMP may be an important factor that reduces the sensitivity of the tubuloglomerular feedback mechanism.

The specific mechanism responsible for the transmission of feedback signals from the macula densa cells to the vascular elements remains unknown. Low-resistance gap junctions have been found to connect extraglomerular and intraglomerular mesangial cells, granular cells, and vascular smooth muscle cells.[21] Thus, once the signal has been transmitted to the extraglomerular mesangial cells, it may be propagated to the smooth muscle cells through the low-resistance pathways. However, no gap junctions have been found between the macula densa cells and extraglomerular mesangial cells and it would appear that a chemical mediator is necessary for this communication step. Preliminary evidence suggests that the macula densa cells may release an arachidonic acid metabolite that elicits vasoconstriction.(181)

5.3. Influence of the Renin–Angiotensin System and Renal Prostaglandins on GFR

Despite intensive investigation on the role of the renin–angiotensin system in the control of GFR, many of the basic issues remain unsettled. There is as yet no clear understanding of the relative contributions of circulating angiotensin II versus intrarenally formed angiotensin II to the regulation of hemodynamics. Also, it is not clear if intrarenally formed angiotensin II occurs as the result of local release of renin with subsequent formation of angiotensins I and II or as a consequence of delivery of circulating angiotensin I to intrarenal conversion sites.(182) While it is generally acknowledged that the renal vasculature is highly responsive to angiotensin II, debate has continued regarding the site(s) of action of angiotensin II, and in particular whether angiotensin II has predominant effects on either afferent or efferent arterioles.(182) As mentioned in previous sections, angiotensin infusions have been shown to lower K_f which may be the result of mesangial cell contraction or other intraglomerular actions.(129,133) Interest has continued in determining the role of the renin–angiotensin system in the tubuloglomerular feedback mechanism. Obviously, all of these putative sites and mechanisms of action have substantive consequences on the potential role of angiotensin in the regulation of GFR.

Recent studies have indicated that all of the enzymatic processes necessary to form angiotensins I and II are available within the kidney. Studies by Celio and Inagami(184) and Taugner *et al.*(10,185) have localized converting enzyme in vascular structures and glomerular elements. In addition, receptors for angiotensin II have been localized in the arterioles and glomerular capillaries.(45,183) The glomerular angiotensin receptors have been localized predominantly to the mesangial cells,(86) and mesangial cells in culture appear to contract in response to angiotensin II addition.(46,49) Since the kidney can convert systemically delivered AI with about 20% efficiency,(182) the intrarenal formation of angiotensin II can occur either as the result of increased systemic delivery of angiotensin I or as a consequence of enhanced local angiotensin I formation due to local release of renin.(183)

There have been numerous investigations assessing the possible sites of action of angiotensin II. Based on experimental studies, two viewpoints have emerged. First, it has been argued that angiotensin II can influence both afferent and efferent arteriolar resistances. Various studies in both normal and hypertensive preparations(138,183,187–190) have shown that administration of converting enzyme inhibitors that block formation of angiotensin II can increase renal blood flow and GFR due to proportional decreases in afferent and efferent arteriolar resistances. Likewise, infusion of angiotensin I or II can decrease GFR due to a combination of effects including decreases in K_f and increases in afferent arteriolar resistance.(131,135,191) The second viewpoint is that the major site of direct action of angiotensin II is on the efferent arteriole.(192) It has also been proposed that angiotensin II formed locally helps control GFR by specifically controlling efferent arteriolar resistance. Hall *et al.*(193) reported the loss of GFR autoregulation in response to decreases in blood pressure during converting enzyme inhibition in sodium-restricted dogs. They proposed that angiotensin II participates in normal GFR autoregulation by counteracting autoregulatory-mediated decreases in efferent arteriolar resistance occurring during decreases in blood pressure. However, this issue remains perplexing since other studies have reported maintained GFR autoregulation in the dog during converting enzyme inhibition in either sodium-depleted or sodium-replete dogs.(138,183) In other studies, renin-depleted kidneys were shown to exhibit normal GFR autoregulatory capabilities,(194) even though it was earlier reported that GFR autoregulatory capability was compromised in renin-depleted dogs.(195) The reasons for the differences in results are not apparent. It is possible that differences in the status of other humoral systems such as the prostaglandin system or the kallikrein–kinin mechanism which can be activated by converting enzyme inhibition could help explain some of the discrepancies.

In this regard, there appears to be a complex interaction between the prostaglandin system and the renin–angiotensin system. Both the stimulation of prostaglandin formation by angiotensin, as well as the release of renin by prostaglandin infusions have been reported.(196,197) The concept has emerged that prostaglandins may "protect" the kidney from the vasoconstrictor effects of angiotensin II and other vasoconstrictors.(198) This may be particularly true during conditions where the level of the renin–angiotensin system is enhanced, such as during the administration of anesthetics, during surgery, during hemorrhage, or in conditions of salt depletion.(196–199) Under these conditions, blockade of prostaglandin formation with such agents as indomethacin leads to decreases in renal blood flow and GFR, whereas under normal awake conditions, indomethacin does not greatly affect blood flow. In sodium-depleted dogs, the indomethacin-induced decreases in renal blood flow are dependent upon elevated levels of angiotensin II.(200,201)

The reason why angiotensin blockade sometimes increases GFR and other times decreases GFR remains unknown. It is tempting to speculate that a prostaglandin–angiotensin interaction may be involved in the loss of GFR autoregulation sometimes observed. Elevations in renal prostaglandins may be induced under some experimental conditions and may decrease the responsiveness of the afferent arteriole and intraglomerular elements to angiotensin II. If the efferent arteriole failed to exhibit this reduced sensitivity, then administration of converting enzyme inhibitors would still vasodilate the efferent arteriole and thus could decrease GFR under a unique set of circumstances.

With regard to the role of the renin–angiotensin system in the mediation of the tubuloglomerular feedback mechanism, no complete understanding has emerged. Initially, Thurau *et al.*(31,202) suggested that transmission of tubuloglomerular feedback signals might directly involve an angiotensin II-dependent step such that increases in distal tubular fluid NaCl concentration would lead to an increased local angiotensin II generation which, in turn, would directly constrict the afferent arterioles. This concept has been challenged by experimental observations from both whole kidney and single nephron studies. To the extent that whole kidney autoregulatory responses reflect feedback-mediated changes in vascular tone, it has generally been observed that vascular resistance adjustments are directionally opposite to coincident changes in renin release. For example, decreases in renal arterial pressure lead to decreases in resistance of the afferent arteriole, yet renin release is increased.[26,30,182,183] In addition, increases in ureteral pressure and increases in renal venous pressure both lead to increases in renin release, but are associated with decreases in renal vascular resistance presumed to be located at the level of the afferent arteriole.[27,144,203] These basic whole kidney observations argue against a role of the renin–angiotensin system as the effector component responsible for mediating autoregulatory-induced alterations in afferent arteriolar resistance.

Other studies have attempted to define the relationship between direct tubuloglomerular feedback-mediated changes in glomerular function and the release of renin. In general, events which occur during transmission of feedback signals usually have effects on renin release which are not consistent with the coincident changes in vascular resistance. For example, increases in distal tubular fluid NaCl concentration lead to increased vascular resistance, but are usually associated with decreases in renin release.[204] Also, conditions or maneuvers that attenuate feedback responses, such as loop diuretics, systemic infusions of cAMP, or agents that elevate cAMP, result in increases in renin release.[26,205] Thus, the feedback-mediated changes in vascular tone do not appear to be a direct consequence of enhanced local renin release with subsequent local and immediate angiotensin II formation. Since there is evidence to indicate that at least one mechanism for renin release operates by means of a distal tubular fluid mechanism,[26,27] two hypotheses are tenable regarding the association between the tubuloglomerular feedback mechanism and macula densa control of renin release. First, it is quite possible that there are two completely independent pathways, one controlling renin release and the other mediating adjustments in vascular resistance through a mechanism that does not involve angiotensin II. Second, there may be only one pathway which ultimately exerts opposite effects on renin release and vascular resistance.

Recent studies have suggested that changes in cytosolic Ca^{2+} levels within the renin-containing granular cells may serve as a common pathway by which various humoral agents, neural influences, and physical factors influence renin release.[206–208] In the JGA granular cells, the stimulus–secretion mechanism is apparently reversed such that increases in Ca^{2+} entry or influx, which presumably lead to increases in cytosolic Ca^{2+} concentration, lead to an inhibition of renin release. Ca^{2+} channel blockers, such as verapamil, can block smooth muscle contraction and can also prevent the inhibition of renin release induced by various agents that promote cellular entry of Ca^{2+}.[206] Also, systemic administration of verapamil can block feedback-mediated decreases in stop-flow pressure[209] during increases in orthograde microperfusion flow rate indicating a Ca^{2+} entry requirement for vasoconstriction of the cells responsive to feedback signals. Thus, the transmission of tubuloglomerular feedback signals may increase cytosolic Ca^{2+} concentrations in the vascular and granular cells which would result in constriction of the vascular elements and inhibition of renin release.

While it is unlikely that the tubuloglomerular feedback mechanism is directly mediated by the renin–angiotensin system, there is evidence that it can influence feedback responsiveness. In particular, converting enzyme inhibitors or angiotensin II antagonists can reduce feedback responsiveness by 30–50%[164,210] However, studies have indicated that major alterations in the status of the renin–angiotensin system through various physiological or pharmacological maneuvers do not lead to predictable changes in the sensitivity of feedback responsiveness.[211,212] Since systemic infusions of angiotensin II during the administration of converting enzyme inhibitors result in a partial restoration of feedback responses,[213] it is possible that the complete blockade of the effects of angiotensin II may reduce vascular reactivity or contractility leading to a partial inhibition of feedback responses.

5.4. Control of GFR by Renal Nerves and Catecholamines

As indicated in Section 2.4, the renal vasculature possesses a dense innervation of α-adrenergic nerves to the afferent and efferent arterioles.[50,51] In addition, the kidney is very responsive to infusions of epinephrine and norepinephrine.[93,139,214] Therefore, it is generally acknowledged that increased activity of neural stimuli to the kidney or infusions of catecholamines can reduce renal blood flow and GFR markedly. Nevertheless, the precise physiological role exerted by the renal nerves in the control of glomerular function remains uncertain. There is a growing awareness, however, that more subtle changes in renal nerve activity may contribute, at least under certain conditions, to the regulation of glomerular dynamics both directly and as a consequence of increasing the activity of the renin–angiotensin system. Studies utilizing direct renal nerve stimulation have shown that renal blood flow decreases in a consistent pattern but GFR may decrease slightly or remain unchanged.[215,216] When the renal nerves were activated by carotid occlusion, renal blood flow also decreased, but in this series GFR and sodium excretion decreased more consistently. It was suggested that carotid occlusion exerted influences attributable to other factors in addition to enhanced renal nerve activity. With modest renal stimulation, the renal blood flow and GFR autoregulatory responses to decreases in renal perfusion pressure can be maintained.[216] Thus, the autoregulation mechanism can elicit changes in renal vascular resistance even when renal nerve activity is exerting a vasoconstrictive influence.

Hermansson *et al.*[89] evaluated the influence of renal nerve stimulation on glomerular filtration dynamics. Renal nerve stimulation elicited approximately equivalent effects on both afferent and efferent arteriolar conductances. Single-nephron plasma flow and GFR were reduced to about the same extent at both low levels (2 Hz) and higher levels (5 Hz) of renal nerve stimulation. At the higher level of stimulation, plasma flow and GFR were reduced to approximately 50% of control values and both glomerular capillary pressure and peritubular capillary pressure were reduced. Recent studies have demonstrated that renal nerve stimulation can also reduce K_f markedly[217] and that this reduction contributes to the reduced SNGFR. At low levels of stimulation, the increases in afferent and efferent resistance were similar; with higher levels of stimulation, it appeared that the responsiveness of the preglomerular resistance vessels was somewhat greater than that of the efferent arterioles. With very marked stimulation, glomerular ischemia was observed.[89]

The effects of renal denervation on glomerular filtration dynamics depend primarily on the preexisting level of renal nerve tone which may be highly variable and dependent on the experimental preparation used. It has been reported that acute renal denervation in anesthetized dogs can result in denervation natriuresis without any evidence of significant changes in renal blood flow or GFR.[218] Also, Pelayo *et al.*[219] were unable to demonstrate an effect of renal denervation on SNGFR or on any of the determinants of GFR in ''euvolemic'' Munich–Wistar rats. Takabatake[220] failed to observe an effect of denervation on the tubuloglomerular feedback mechanism. Neither SNGFR nor late distal tubular flow increased significantly and there was no change in the feedback-mediated increase in SNGFR determined during complete interruption of distal flow rate. In addition, there was no apparent difference in the sensitivity of tubuloglomerular feedback mechanism assessed from microperfusion studies. Thus, under conditions where basal renal nerve activity is not excessively elevated, denervation appears to exert relatively modest effects on GFR because the slight reductions in afferent and efferent resistance are proportional.

In general, the effects of norepinephrine and epinephrine infusions on renal hemodynamics resemble those exerted by renal nerve stimulation.[221] However, the vasoconstrictor effects may not be directly comparable with the effects of endogenous release of these vasoconstrictors because their intrarenal distribution could be different. With moderate doses of epinephrine, the effects to decrease renal blood flow are somewhat greater than those observed for GFR and, thus, filtration fraction increases.[6,222] Smith[6] concluded that these types of effects were best explained on the basis of approximately equivalent changes in afferent and efferent resistance. Recent studies have evaluated several effects of catecholamines on the renal microvessels. Click *et al.*[223] directly demonstrated the effects of norepinephrine and angiotensin on the afferent and efferent arterioles of kidney tissue grafted into the cheek pouch of hamsters. Both afferent and efferent arterioles responded to direct administration of norepinephrine, although the response of the efferent vessels was apparently less than that of the afferent arterioles. Edwards[224] observed dose-dependent decreases in the diameters of interlobular arteries, afferent arterioles, and efferent arterioles.

Micropuncture studies measuring glomerular pressure in Munich–Wistar rats have provided additional information regarding the effects of catecholamines on glomerular function. Myers *et al.*[139] administered norepinephrine systemically (2 to 4 μg/min per kg) and elicited decreases in glomerular blood flow without changes in GFR. Both glomerular pressure and peritubular capillary pressure increased. When arterial pressure was allowed to increase, afferent and efferent arteriolar resistance increased significantly. The increase in afferent resistance was markedly reduced when arterial pressure was maintained at control levels. Andreucci *et al.*[93] observed that norepinephrine decreased glomerular pressure and efferent arteriolar pressure indicating that this agent could increase preglomerular resistance to an extent greater than could be explained by an autoregulatory response to the elevated arterial pressure. The infusions of epinephrine did not decrease the intrarenal pressures even though the increase in systemic arterial pressure was comparable indicating that norepinephrine and epinephrine may have somewhat different quantitative effects on the pre- and postglomerular resistances.

In summary, while it is clear that renal nerve stimulation or administration of catecholamines can reduce renal blood flow and GFR, the tonic influence of the sympathetic nervous system on glomerular filtration dynamics under normal unstressed conditions remains uncertain, but may be rather subtle.

6. Intrarenal Distribution of Glomerular Filtration Rate

Much of the preceding discussion treated aspects of GFR as if the entire nephron population were homogeneous. Indeed, most studies have been directed toward this approach; in whole kidney experiments, the data obtained are for the total nephron population, and the distribution of quantitative variables such as GFR, filtration fraction, and glomerular pressure among the various subgroups of nephrons is difficult to assess. Micropuncture experiments primarily involve analysis of function of the nephrons of the superficial portions of the kidney. Therefore, most of the detailed knowledge regarding glomerular dynamics at the single nephron level is applicable specifically to superficial nephrons, and only inferentially to the remaining portion of the nephron population. However, several approaches have been utilized to evaluate differences in function of the various populations of glomeruli.

For example, it has been demonstrated by means of the ferrocyanide technique[225–228] that a gradient exists from outer cortex to inner cortex, with the SNGFR of the outer cortical nephrons being lower than the SNGFR of the middle and deep cortical nephrons. There is also evidence that the SNGFR responses to various different manipulations may be different among the various subgroups of nephrons. As an extreme case, Horster and Thurau[229] suggested a marked and true redistribution of SNGFR in response to high levels of sodium intake, such that the superficial SNGFR increased greatly whereas the SNGFR of deep nephrons exhibited a marked fall. However, this true redistribution in response to changes in sodium intake has not been observed in other studies.[230,231] The redistribution of SNGFR with increased sodium intake has been primarily due to a proportionately greater increase in the SNGFR of superficial nephrons than of deeper or juxtamedullary nephrons.[226,228,232]

It is still quite uncertain whether or not redistribution of SNGFR or renal blood flow occurs in response to moderate changes in chronic sodium intake or even moderate acute expansion of the extracellular fluid volume. In some experiments, increases in sodium intake are associated with moderate increases in whole kidney GFR, and proportionately greater increases in the superficial cortical SNGFR.[232–234] Conversely, sodium restriction appears to lead to a smaller decrease in the whole kidney GFR as compared to proportionately greater reductions in superficial nephron GFR.[232,235] The simplest explanation for these shifts in SNGFR distribution is that the deep cortical nephrons are somewhat less responsive to variations in sodium intake and extracellular fluid volume. Even this type of redistribution of SNGFR has not always been observed. Other experiments have shown that rats given a chronic salt load simply have higher SNGFR values in both superficial and juxtamedullary nephrons[230,236] and that SNGFR redistribution is not a necessary mechanism for maintaining sodium homeostasis.[230,236]

Using the ferrocyanide technique, Bonvalet *et al.*[227] demonstrated that GFR of both superficial and deep nephrons exhibited autoregulatory behavior to approximately the same extent in response to reductions in renal arterial pressure. It was concluded that no substantive redistribution of GFR occurs during autoregulatory adjustments, indicating that the deep nephrons autoregulate GFR in a manner similar to the superficial nephrons. These data are consistent with the results from micropuncture experiments showing that the ratio between SNGFR (based on distal tubular fluid collections) and whole kidney GFR remained unchanged during autoregulatory adjustments to reduced renal arterial pressure.[161] More direct recent studies have also demonstrated that the deep nephrons exhibit a high degree of autoregulatory efficiency;[237] furthermore, juxtamedullary nephrons have a very efficient tubuloglomerular feedback mechanism that may even have a greater sensitivity than the feedback mechanism in superficial nephrons.[156]

No clear-cut conclusions seem to emerge from the various studies related to SNGFR redistribution. It can certainly be appreciated that the deeper, generally larger, glomeruli may have greater SNGFR values, but the magnitude of the difference remains uncertain. The micropuncture studies would seem to suggest that juxtamedullary SNGFR in the rat may be as much as twice the superficial SNGFR. It should be realized, however, that such studies require complete tubular fluid collections in nephron segments proximal to the macula densa. Based on the recent studies showing that the feedback-mediated increases in SNGFR due to cessation of distal flow may be very substantial in these deep nephrons,[156] it seems possible that the very large juxtamedullary SNGFR values reported from micropuncture studies are spuriously elevated. Thus, SNGFR determinations based on the ferrocyanide technique may be more valid; these measurements show relatively small differences between SNGFR in the various nephron populations. With regard to the presence or absence of SNGFR redistribution in physiological circumstances, the data are conflicting and there have not been convincing mechanisms postulated to explain such redistribution processes.

7. Summary

The glomerulus has a unique capillary structure that provides the capability for filtering large quantities of plasma with very little protein passage. Tracer and histochemical studies suggest that the primary filtration barrier is located in the basement membrane. The high concentration of anionic charges in the basement membrane and on the epithelial cells retards filtration of negatively charged molecules. The macula densa is an area of specialized cells of the distal tubule that is closely associated with the glomerular structure of its own nephron unit and provides the morphological basis for the concept of feedback control of glomerular filtration.

During the process of filtration, plasma proteins are concentrated and colloid osmotic pressure increases; thus, net filtration pressure progressively decreases as blood flows toward the end of the capillary network. The ultrafiltration coefficient, K_f, describes the rate of hydraulic flow across the capillary and is equal to the product of the hydraulic permeability, L_p, and surface area available for filtration. L_p and K_f values for glomerular capillaries are 20–50 times greater than those for other capillaries. Several studies have suggested that K_f is dynamic and can change in response to a variety of factors including infusion of vasoactive agents and changes in plasma protein concentration.

Direct measurements of the glomerular hydrostatic pressure in a mutant strain of Wistar rats have led to the suggestion that the net filtration pressure at the efferent end of the glomerular capillary is zero and that filtration equilibrium normally obtains. However, studies in the dog and in other rat strains have not provided support for this concept and evidence is provided indicating that filtration normally occurs throughout the length of the glomerular capillaries. The quantitative analysis of glomerular dynamics in both single capillary and multiple capillary loop models has provided further insight about the complex interactions among the various direct and indirect determinants of glomerular function.

A substantial body of evidence exists to suggest that GFR is regulated, in part, by an intrinsic mechanism that links distal filtrate delivery to afferent arteriolar resistance in an inverse manner. This tubuloglomerular feedback mechanism provides a means for achieving a balance between the physically determined input at the glomerulus and the metabolically determined reabsorptive capabilities of the tubules. The tubuloglomerular feedback mechanism has been assessed from filtration rate responses to cessation of distal volume delivery and by microperfusion studies. Data are presented showing that a normal feedback mechanism is requisite for the manifestation of genuine autoregulatory responses. Direct microperfusion studies have shown that increasing perfusion rate to distal sites from a late proximal tubule can decrease both SNGFR and stop-flow pressure. Feedback responses obtained during retrograde perfusion of the macula densa segment indicate that the mechanism responds to changes in total solute concentration of the perfusate and not to flow directly. The magnitude of feedback responses is associated with the luminal fluid osmolality independent of the specific Na^+ or Cl^- concentrations. The intracellular mechanisms responsible for the tubuloglomerular feedback mechanism includes a step that involves activation of cytosolic Ca^{2+}.

The regulation of hemodynamics by the renin–angiotensin system remains controversial. Angiotensin II may increase both afferent and efferent arteriolar resistances although some workers contend that angiotensin II exerts a predominant efferent effect. Finally, angiotensin does not appear to be a direct mediator of tubuloglomerular feedback signals, but may be necessary for maintenance of vascular sensitivity.

The role of the sympathetic nervous system in the control of renal hemodynamics remains an active area of investigation. Current views are that under normal conditions, sympathetic control of renal function is minimal. However, increased sympathetic stimulation such as in stressful conditions may decrease GFR and blood flow.

Some investigators have suggested that juxtamedullary nephrons have a low capacity to excrete sodium whereas superficial cortical nephrons have a large excretory capacity, and that redistribution of GFR to outer cortical nephrons occurs in response to high salt intakes. Studies designed to test this hypothesis have failed to yield experimental evidence consistently supporting it; in addition, the basic premise underlying the redistribution concept remains unproven.

ACKNOWLEDGMENTS. The authors gratefully acknowledge the secretarial assistance of Mrs. Carol Grimes and Ms. Tracy Morrison and the graphical assistance of Mrs. Ellen Bernstein. The authors' experimental work has been supported by research grants from the National Heart, Lung and Blood Institute and the American Heart Association. A special note of thanks to all our colleagues with whom we have collaborated in our previous efforts.

References

1. DuBois, A. M., C. Rouiller, and A. F. Muller. 1969. The embryonic kidney. In: *The Kidney: Morphology, Biochemistry, Physiology,* Volume I. Academic Press, New York. pp. 1–50.
2. McCrory, W. W. 1972. *Developmental Nephrology.* Harvard University Press, Cambridge, Mass. pp. 1–50.
3. Oliver, J. 1968. *Nephrons and Kidneys.* Harper & Row, New York. pp. 47–50.
4. Renkin, E. M., and J. P. Gilmore. 1973. Glomerular filtration. In: *Handbook of Physiology,* Section 8. J. Orloff and R. W. Berliner, eds. American Physiological Society, Washington, D.C. pp. 185–248.

5. Kunkel, P. A. 1930. The number and size of the glomeruli in the kidney of several mammals. *Bull. Johns Hopkins Hosp.* **47**:285–291.
6. Smith, H. W. 1951. *The Kidney*. Oxford University Press, London. pp. 562–574.
7. Aeikens, B., A. Eenboom, and A. Bohle. 1979. Untersuchungen zur Struktur des Glomerulum Rekonstruktion eines Ratten glomerulum am .5u dicken serien schnitten. *Virchows Arch. A* **381**:283–293.
8. Shea, S. M. 1979. Glomerular hemodynamics and vascular structure: The pattern and dimensions of a single rat glomerular capillary network reconstructed from ultrathin sections. *Microvasc. Res.* **18**:129–143.
9. Navar, L. G. 1978. The regulation of glomerular filtration rate in mammalian kidneys. In: *Physiology of Membrane Disorders*. T. E. Andreoli, J. Hoffman, and D. Fanestil, eds. Plenum Press, New York. pp. 593–626.
10. Taugner, C., K. Poulsen, E. Hackenthal, and R. Taugner. 1979. Immunocytochemical localization of renin in mouse kidney. *Histochemistry* **62**:19–27.
11. Casellas, D., and A. Mimran. 1981. Shunting in renal microvasculature of the rat: A scanning electron microscopic study of corrosion casts. *Anat. Rec.* **201**:237–248.
12. Gattone, V. H., F. C. Luft, and A. P. Evan. 1984. The renal afferent and efferent arterioles of the rabbit. *Am. J. Physiol.* **247**: F219–F228.
13. Barger, A. C., and J. A. Herd. 1973. Renal vascular anatomy and distribution of blood flow. In: *Handbook of Physiology*, Section 8. J. Orloff and R. W. Berliner, eds. American Physiological Society, Washington, D.C. pp. 249–313.
14. Beeuwkes, R., and J. V. Bonventre. 1975. Tubular organization and vascular tubular relations in the dog kidney. *Am. J. Physiol.* **229**:695–713.
15. Evan, A. P., and W. G. Dail. 1977. Efferent arterioles in the cortex of the rat kidney. *Anat. Rec.* **187**:135–145.
16. Weinstein, S. W., and J. Szyjewicz. 1978. Superficial nephron tubular–vascular relationships in the rat kidney. *Am. J. Physiol.* **234**:F207–F214.
17. Rouiller, C., and L. Orci. 1971. The structure of the juxtaglomerular complex. In: *The Kidney: Morphology, Biochemistry, Physiology*, Volume IV. C. Rouiller and A. F. Muller, eds. Academic Press, New York. pp. 1–80.
18. Barajas, L., and J. Muller. 1973. The innervation of the juxtaglomerular apparatus and surrounding tubules: A quantitative analysis by serial sections electron microscopy. *J. Ultrastruct. Res.* **43**:107–132.
19. Kaissling, B., S. Peter, and W. Kriz. 1977. The transition of the thick ascending limb of Henle's loop into the distal convoluted tubule in the nephron of the rat kidney. *Cell Tissue Res.* **182**:111–118.
20. Pricam, C., F. Humbert, A. Perrelet, and L. Orci. 1974. Gap junctions in mesangial and lacis cells. *J. Cell Biol.* **63**:349–354.
21. Taugner, R., A. Schiller, B. Kaissling, and W. Kriz. 1978. Gap junctional coupling between the J.G.A. and the glomerular tuft. *Cell Tissue Res.* **186**:279–285.
22. Kaissling, B., and W. Kriz. 1982. Variability of intercellular spaces between macula densa cells: A transmission electron microscopic study in rabbits and rats. *Kidney Int.* **22**:S9–S17.
23. Spanidis, A., H. Wunsch, B. Kaissling, and W. Kriz. 1982. Three-dimensional shape of a Goormaghtigh cell and its contact with a granular cell in the rabbit kidney. *Anat. Embryol.* **165**:239–252.
24. Latta, H. 1973. Ultrastructure of the glomerulus and juxtaglomerular apparatus. In: *Handbook of Physiology*, Section 8. J. Orloff and R. W. Berliner, eds. American Physiological Society, Washington, D.C. pp. 1–29.
25. Navar, L. G., A. P. Evan, and L. Rosivall. 1983. Microcirculation of the kidneys. In: *The Physiology and Pharmacology of the Microcirculation*, Volume I. N. Mortillaro, ed. Academic Press, New York. pp. 397–488.
26. Davis, J. O., and R. H. Freeman. 1976. Mechanisms regulating renin release. *Physiol. Rev.* **56**:1–56.
27. Vander, A. J. 1967. Control of renin release. *Physiol. Rev.* **47**:359–382.
28. Wright, F. S., and J. B. Briggs. 1979. Feedback control of glomerular blood flow, pressure, and filtration rate. *Physiol. Rev.* **59**:958–1006.
29. Schnermann, J. 1981. Localization, mediation and function of the glomerular vascular response to alterations of distal fluid delivery. *Fed. Proc.* **40**:109–115.
30. Navar, L. G., D. W. Ploth, and P. D. Bell. 1980. Distal tubular feedback control of renal hemodynamics and autoregulation. *Annu. Rev. Physiol.* **42**:557–571.
31. Thurau, K., and J. Mason. 1974. The intrarenal function of the juxtaglomerular apparatus. In: *Kidney and Urinary Tract Physiology*. K. Thurau, ed. University Park Press, Baltimore. pp. 357–389.
32. Rovenska, E. 1983. Two types of capillaries in the rat renal glomerulus. *Acta Anat.* **115**:31–39.
33. Karnovsky, M. J. 1979. The ultrastructure of glomerular filtration. *Annu. Rev. Med.* **30**:213–224.
34. Andrews, P. M. 1981. Investigations of cytoplasmic contractile and cytoskeletal elements in the kidney glomerulus. *Kidney Int.* **20**:549–562.
35. Larsson, L., and A. B. Maunsbach, 1980. The ultrastructural development of the glomerular filtration barrier in the rat kidney: A morphometric analysis. *J. Ultrastruct. Res.* **72**:392–406.
36. Mohos, S. C., and L. Skoza. 1970. Histochemical demonstration and localization of sialoproteins in the glomerulus. *Exp. Mol. Pathol.* **12**:316–323.
37. Kanwar, Y. S., and M. G. Farquhar. 1979. Presence of heparan sulfate in the glomerular basement membrane. *Proc. Natl. Acad. Sci. USA* **76**:1303–1307.
38. Roll, F. J., J. A. Madri, J. Albert, and H. Furthmayr. 1980. Codistribution of collagen types IV and AB2 in basement membranes and mesangium of the kidney: An immunoferritin study of ultrathin frozen sections. *J. Cell Biol.* **85**:597–616.
39. Madri, J. A., F. J. Roll, H. Furthmayr, and J. M. Foidart. 1980. Ultrastructural localization of fibronectin and laminin in the basement membranes of the murine kidney. *J. Cell Biol.* **86**:682–687.
40. Kanwar, Y. S., and M. G. Farquhar. 1979. Anionic sites in the glomerular basement membrane: *In vivo* and *in vitro* localization to the laminae rarae by cationic probes. *J. Cell Biol.* **81**:137–153.
41. Rennke, H. G., and M. A. Venkatachalam. 1977. Glomerular permeability: *In vivo* tracer studies with polyanionic and polycationic ferritins. *Kidney Int.* **11**:44–53.
42. Gattone, V. H., A. P. Evan, S. A. Mong, B. A. Connors, G. R. Aronoff, and F. C. Luft. 1983. The morphology of the renal microvasculature in glycerol- and gentamicin-induced acute renal failure. *J. Lab. Clin. Med.* **101**:183–195.
43. Avasthi, P. S., A. P. Evan, and D. Hay. 1980. Glomerular endothelial cells in uranyl nitrate induced acute renal failure in rats. *J. Clin. Invest.* **65**:121–127.
44. Michael, A. F., W. F. Keane, L. Ray, R. L. Vernier, and S. M. Mauer. 1980. The glomerular mesangium. *Kidney Int.* **17**:141–154.
45. Sterzel, R. B., D. H. Lovett, H. D. Stein, and M. Kashgarian. 1982. The mesangium and glomerulonephritis. *Klin. Wochenschr.* **60**:1077–1094.
46. Kreisberg, J. I., and M. J. Karnovsky. 1983. Glomerular cells in culture. *Kidney Int.* **23**:439–447.
47. Bernik, M. B. 1969. Contractile activity of human glomeruli in culture. *Nephron* **6**:1–10.
48. Brown, G. P., J. G. Douglas, and J. Krontiris-Litowitz. 1980. Properties of angiotensin II receptors of isolated rat glomeruli: Factors influencing binding affinity and comparative binding of angiotensin analogs. *Endocrinology* **106**:1923–1929.
49. Ausiello, D. A., J. I. Kreisberg, C. Roy, and M. J. Karnovsky. 1980. Contraction of cultured rat glomerular cells of apparent

mesangial origin after stimulation with angiotensin II and arginine vasopressin. *J. Clin. Invest.* **65**:754–760.

50. Barajas, L. 1978. Innervation of the renal cortex. *Fed. Proc.* **37**:1192–1201.
51. DiBona, G. F. 1982. The functions of the renal nerves. *Rev. Physiol. Biochem. Pharmacol.* **94**:76–181.
52. Dinerstein, R. J., J. Vannici, R. C. Henderson, L. J. Roth, L. I. Goldberg, and P. C. Hoffman. 1979. Histofluorescence techniques provide evidence for dopamine-containing neural elements in canine kidney. *Science* **205**:497–499.
53. Dieterich, H. J. 1974. Electron microscopic studies of the innervation of the rat kidney. *Z. Anat. Entwicklungsgesch.* **145**:169–186.
54. Landis, E. M., and J. R. Pappenheimer. 1963. Exchange of substances through the capillary walls. In: *Handbook of Physiology*, Volume II. American Physiological Society, Washington, D.C. pp. 961–1034.
55. Pappenheimer, J. R., E. M. Renkin, and C. M. Borrero. 1951. Filtration, diffusion and molecular sieving through peripheral capillary membranes. *Am. J. Physiol.* **167**:13–46.
56. Solomon, A. K. 1968. Characterization of biological membranes by equivalent pores. *J. Gen. Physiol.* **S52**:335–364.
57. Oken, D. E., and W. Flamenbaum. 1971. Micropuncture studies of proximal tubule albumin concentrations in normal and nephrotic rats. *J. Clin. Invest.* **50**:1498–1505.
58. Von Baeyer, H., J. B. VanLiew, J. Klassen, and J. W. Boylan. 1976. Filtration of protein in the anti-glomerular basement membrane nephritic rat: A micropuncture study. *Kidney Int.* **10**:425–437.
59. Oken, D. E., D. C. Mikulecky, G. K. Smith, and D. M. Landwehr. 1984. On the sieving of macromolecules by the glomerular basement membrane—Theoretical considerations of steric and charge properties. In: *Proceedings of the International Symposium on Pathogenetic Role of Cationic Proteins with Biological Membranes*. P. Bergmann, R. Beauwers, and P. P. Lambert, eds. Presses Universitaires de Bruxelles,Brussels.
60. Brenner, B. M., W. M. Deen, and C. R. Robertson. 1974. The physiological basis of glomerular ultrafiltration. In: *Kidney and Urinary Tract Physiology*. K. Thurau, ed. University Park Press, Baltimore. pp. 335–356.
61. Navar, P. D., and Navar, L. G. 1977. Relationship between colloid osmotic pressure and plasma protein concentration in the dog. *Am. J. Physiol.* **233**:H295–H298.
62. Thomas, C. E., P. D. Bell, and L. G. Navar. 1979. Glomerular filtration dynamics in the dog during elevated plasma colloid osmotic pressure. *Kidney Int.* **15**:502–512.
63. Brenner, B. M., I. F. Ueki, and T. M. Daugharty. 1972. On estimating colloid osmotic pressure in pre- and post-glomerular plasma in the rat. *Kidney Int.* **2**:51–53.
64. Kallskog, O., L. O. Lindbom, H. R. Ulfendahl, and M. Wolgast. 1976. Hydrostatic pressures within the vascular structures of the rat kidney. *Pfluegers Arch.* **363**:205–210.
65. Tonder, K. J. H., and K. Aukland. 1979–1980. Interlobular arterial pressure in the rat kidney. *Renal Physiol.* **2**:214–221.
66. Navar, L. G., P. D. Bell, and T. J. Burke. 1982. Role of macula densa feedback mechanism as a mediator of renal autoregulation. *Kidney Int.* **22**:S157–S164.
67. Casellas, D., and L. G. Navar. 1984. In vitro perfusion of juxtamedullary nephrons in rats. *Am. J. Physiol.* **246**:F349–F358.
68. Brenner, B. M., J. L. Troy, and T. M. Daugharty. 1971. The dynamics of glomerular ultrafiltration in the rat. *J. Clin. Invest.* **50**:1776–1780.
69. Falchuk, K. H., and R. W. Berliner. 1971. Hydrostatic pressures in peritubular capillaries and tubules in the rat kidney. *Am. J. Physiol.* **220**:1422–1426.
70. Allison, M. E., E. M. Lipham, and C. W. Gottschalk. 1972. Hydrostatic pressure in the rat kidney. *Am. J. Physiol.* **223**:975–983.
71. Knox, F. G., L. R. Willis, J. W. Strandhoy, and E. G. Schneider. 1972. Hydrostatic pressures in proximal tubules and peritubule capillaries in the dog. *Kidney Int.* **2**:11–16.
72. Navar, L. G., B. Chomdej, and P. D. Bell. 1975. Absence of estimated glomerular capillary pressure autoregulation during interrupted distal delivery. *Am. J. Physiol.* **229**:1596–1603.
73. Navar, L. G., P. D. Bell, and T. J. Burke. 1977. Autoregulatory responses of superficial nephrons, and their association with sodium excretion during arterial pressure alterations in the dog. *Circ. Res.* **41**:487–496.
74. Baer, P. G., and L. G. Navar. 1973. Renal vasodilation and uncoupling of blood flow and filtration rate autoregulation. *Kidney Int.* **4**:12–21.
75. Lowenstein, J., E. R. Beranbaum, H. Chasis, and D. S. Baldwin. 1970. Intrarenal pressure and exaggerated natriuresis in essential hypertension. *Clin. Sci.* **38**:359–374.
76. Willassen, Y., and J. Ofstad. 1980. Renal sodium excretion and the peritubular capillary physical factors in essential hypertension. *Hypertension* **2**:771–779.
77. Blantz, R. C., F. C. Rector, and D. W. Seldin. 1974. Effect of hyperoncotic albumin expansion upon glomerular ultrafiltration in the rat. *Kidney Int.* **6**:209–221.
78. Blantz, R. C. 1977. Dynamics of glomerular ultrafiltration in the rat. *Fed. Proc.* **36**:2602–2608.
79. Ichikawa, I., D. A. Maddox, M. G. Cogan, and B. M. Brenner. 1978. Dynamics of glomerular ultrafiltration in euvolemic Munich–Wistar rats. *Renal Physiol.* **1**:121–131.
80. Arendshorst, W. J., and C. W. Gottschalk. 1980. Glomerular ultrafiltration dynamics: Euvolemic and plasma volume expanded rats. *Am. J. Physiol.* **239**:F171–F186.
81. Kallskog, O., L. O. Lindbom, H. R. Ulfendahl, and M. Wolgast. 1975. Kinetics of the glomerular ultrafiltration in the rat kidney: An experimental study. *Acta Physiol. Scand.* **95**:293–300.
82. Aukland, K., K. Heyeraas-Tonder, and G. Naess. 1977. Capillary pressure in deep and superficial glomeruli of the rat kidney. *Acta Physiol. Scand.* **101**:418–427.
83. Heller, J., and V. Horacek. 1980. Comparison of directly measured and calculated glomerular capillary pressure in the dog kidney at varying perfusion pressure. *Pfluegers Arch.* **385**:253–258.
84. Marchand, G. R. 1981. Direct measurement of glomerular capillary pressure in dogs. *Proc. Soc. Exp. Biol. Med.* **167**:428–433.
85. Gertz, K. H., M. Brandis, G. Braun-Schubert, and J. W. Boylan. 1969. The effect of saline infusion and hemorrhage on glomerular filtration pressure and single nephron filtration rate. *Pfluegers Arch.* **310**:193–205.
86. Andreucci, V. E., J. Herrera-Acosta, F. C. Rector, Jr., and D. W. Seldin. 1971. Effective glomerular filtration pressure and single nephron filtration rate during hydropenia, elevated ureteral pressure, and acute volume expansion with isotonic saline. *J. Clin. Invest.* **80**:2230–2234.
87. Blantz, R. 1974. Effect of mannitol on glomerular ultrafiltration in the hydropenic rat. *J. Clin. Invest.* **54**:1135–1143.
88. Ott, C. E., G. R. Marchand, J. A. Diaz-Buxo, and F. G. Knox. 1976. Determinants of glomerular filtration rate in the dog. *Am. J. Physiol.* **231**:235–239.
89. Hermansson, K., M. Larson, O. Kallskog, and M. Wolgast. 1981. Influence of renal nerve activity on arteriolar resistance, ultrafiltration dynamics and fluid reabsorption. *Pfluegers Arch.* **389**:85–90.
90. Briggs, J. P., and F. S. Wright. 1979. Feedback control of glomerular filtration rate: Site of the effector mechanism. *Am. J. Physiol.* **236**:F40–F47.
91. Oken, D. E., S. R. Thomas, and D. Mikulecky. 1981. A network thermodynamic model of glomerular dynamics: Application in the rat. *Kidney Int.* **19**:359–373.
92. Ichikawa, I. 1982. Evidence for altered glomerular hemodynamics during acute nephron obstruction. *Am. J. Physiol.* **242**:F580–F585.
93. Andreucci, V. E., A. Dal Canton, A. Corradi, R. Stanziale, and L. Migone. 1976. Role of the efferent arteriole in glomerular hemodynamics of superficial nephrons. *Kidney Int.* **9**:475–480.
94. Navar, L. G., P. D. Bell, R. W. White, R. L. Watts, and R. H.

Williams. 1977. Evaluation of the single nephron glomerular filtration coefficient in the dog. *Kidney Int.* **12**:137–149.
95. Bell, P. D., and L. G. Navar. 1979. Stop flow pressure feedback responses during reduced renal vascular resistance in the dog. *Am. J. Physiol.* **237**:F204–F209.
96. Deen, W. M., C. R. Robertson, and B. M. Brenner. 1972. A model of glomerular ultrafiltration in the rat. *Am. J. Physiol.* **223**:1178–1183.
97. Brenner, B. M., J. L. Troy, T. M. Daugharty, W. M. Deen, and C. R. Robertson. 1972. Dynamics of glomerular ultrafiltration in the rat. II. Plasma-flow dependence of GFR. *Am. J. Physiol.* **223**:1184–1190.
98. Thomas, C. E., C. E. Ott, P. D. Bell, F. G. Knox, and L. G. Navar. 1983. Glomerular filtration dynamics during vasodilation with acetylcholine in the dog. *Am. J. Physiol.* **244**:F606–F611.
99. Oken, D. E. 1982. An analysis of glomerular dynamics in rat, dog and man. *Kidney Int.* **22**:136–145.
100. Huss, R. E., D. J. Marsh, and R. E. Kalaba. 1975. Two models of glomerular filtration rate and renal blood flow in the rat. *Ann. Biomed. Eng.* **3**:72–99.
101. Papenfuss, H. D., and J. F. Gross. 1978. Analytic study of the influence of capillary pressure drop and permeability on glomerular ultrafiltration. *Microvasc. Res.* **16**:59–72.
102. Lambert, P. P., B. Aeikens, A. Bohle, F. Hanus, S. Pegoff, and M. Vandamme. 1982. A network model of glomerular function. *Microvasc. Res.* **23**:99–128.
103. Williams, R. H., C. E. Thomas, L. G. Navar, and A. P. Evan. 1981. Hemodynamic and single nephron function during the maintenance phase of ischemic acute renal failure in the dog. *Kidney Int.* **19**:503–515.
104. Andrews, P. M., and A. K. Coffey. 1983. Cytoplasmic contractile elements in glomerular cells. *Fed. Proc.* **42**:3046–3052.
105. Lambert, P. P., A. Verniory, J. P. Gassee, and P. Ficheroulle. 1972. Sieving equations and effective glomerular filtration pressure. *Kidney Int.* **2**:131–146.
106. Chang, R. L. S., C. R. Robertson, W. M. Deen, and B. M. Brenner. 1975. Permselectivity of the glomerular capillary wall to macromolecules. I. Theoretical considerations. *Biophys. J.* **15** :861–886.
107. DuBois, R., P. Decoodt, J. P. Gassee, A. Verniory, and P. P. Lambert. 1975. Determination of glomerular intracapillary and transcapillary pressure gradients from sieving data. I. A mathematical model. *Pfluegers Arch.* **356**:299–316.
108. Kedem, O., and A. Katchalsky. 1958. Thermodynamic analysis of the permeability of biological membranes to non-electrolytes. *Biochem. Biophys. Acta* **27**:229–246.
109. Bresler, E. H., and L. S. Groome. 1982. On equations for combined convective and diffusive transport of neutral solute across porous membranes. *Am. J. Physiol.* **241**:F469–F476.
110. Taylor, A. E., and D. N. Granger. 1984. Exchange of macromolecules across the microcirculation. In: *Handbook of Physiology: Cardiovascular system microcirculation,* Voume 4. I. E. Renkin and C. C. Michelf, eds. American Physiological Society, Bethesda, Md. pp. 467–520.
111. Whiteside, C., and M. Silverman. 1983. Determination of glomerular permselectivity to neutral dextrans in the dog. *Am. J. Physiol.* **245**:F485–F495.
112. Lambert, P. P., R. DuBois, P. Decoodt, J. P. Gassee, and A. Verniory. 1975. Determination of glomerular intracapillary and transcapillary pressure gradients from sieving data. II. A physiological study in the normal dog. *Pfluegers Arch.* **359**:1–22.
113. Vanrenterghem, Y., R. Vanholder, M. Lammens-Verslijpe, and P. P. Lambert. 1980. Sieving studies in urea-induced nephropathy in the dog. *Clin. Sci.* **58**:65–75.
114. Chang, R. L. S., W. M. Deen, C. R. Robertson, and B. M. Brenner. 1975. Permselectivity of the glomerular capillary wall. III. Restricted transport of polyanions. *Kidney Int.* **8**:212–218.
115. Bohrer, M. P., C. Baylis, H. D. Humes, R. J. Glassock, C. R. Robertson, and B. M. Brenner. 1978. Permselectivity of the glomerular capillary wall: Facilitated transport of circulating polycations. *J. Clin. Invest.* **61**:72–78.
116. Rennke, H. G., Y. Patel, and M. Venkatachalam. 1978. Glomerular filtration of proteins: Clearance of anionic, neutral, and cationic horseradish peroxidase in the rat. *Kidney Int.* **13**:324–328.
117. Bohrer, M. P., W. M. Deen, C. R. Robertson, J. L. Troy, and B. M. Brenner. 1979. Influence of molecular configuration on the passage of macromolecules across the glomerular capillary wall. *J. Gen. Physiol.* **74**:583–593.
118. Rennke, H. G., and M. A. Venkatachalam. 1979. Glomerular permeability of macromolecules: Effect of molecular configuration on the fractional clearance of uncharged dextran and neutral horseradish peroxidase in the rat. *J. Clin. Invest.* **63**:713–717.
119. Verniory, A., R. DuBois, P. Decoodt, J. P. Gassee, and P. P. Lambert. 1973. Measurement of the permeability of biological membranes: Application to the glomerular wall. *J. Gen. Physiol.* **62**:489–507.
120. Jorgensen, K. E., and J. V. Moller. 1979. Use of flexible polymers as probes of glomerular pore size. *Am. J. Physiol.* **236**:F103–F111.
121. Lambert, P. P., J. P. Gassee, A. Verniory, and P. Ficheroulle. 1971. Measurement of the glomerular filtration pressure from sieving data for macromolecules. *Pfluegers Arch.* **329**:34–58.
122. Bridges, C. R., B. D. Myers, B. M. Brenner, and W. M. Deen. 1982. Glomerular charge alterations in human minimal change nephropathy. *Kidney Int.* **22**:677–684.
123. White, R. J., and L. G. Navar. 1975. Modeling of renal hemodynamics in the dog: Glomerular and peritubular capillary dynamics. *Summer Computer Simulation Conf.* **2**:878–888.
124. Marshall, E. A., and E. A. Trowbridge. 1974. A mathematical model of the ultrafiltration process in a single glomerular capillary. *Theor. Biol.* **48**:389–412.
125. Steven, K., and S. Strobaek. 1974. Renal corpuscular hydrodynamic: digital computer simulation. *Pfluegers Arch.* **348**:317–331.
126. Kallskog, O., L. O. Lindbom, H. R. Ulfendahl, and M. Wolgast. 1975. Kinetics of the glomerular ultrafiltration in the rat kidney: A theoretical study. *Acta Physiol. Scand.* **95**:191–200.
127. Knox, F. G., J. L. Cuche, C. E. Ott, J. A. Diaz-Buxo, and G. Marchand. 1975. Regulation of glomerular filtration and proximal tubule reabsorption. *Circ. Res.* **36**:107–118.
128. Earley, L. E., and R. W. Schrier. 1973. Intrarenal control of sodium excretion by hemodynamic and physical factors. In: *Handbook of Physiology,* Section 8. J. Orloff and R. W. Berliner, eds. American Physiological Society, Bethesda, Md. pp. 721–762.
129. Dworkin, L. D., I. Ichikawa, and B. M. Brenner. 1983. Hormonal modulation of glomerular function. *Am. J. Physiol.* **244**:F95–F104.
130. Scharschmidt, L. A., E. Lianos, and M. J. Dunn. 1983. Arachidonate metabolites and the control of glomerular function. *Fed. Proc.* **42**:3058–3063.
131. Blantz, R. C., K. S. Konner, and B. J. Tucker. 1976. Angiotensin II effects upon the glomerular microcirculation and ultrafiltration coefficient of the rat. *J. Clin. Invest.* **57**:419–434.
132. Baylis, C., W. M. Deen, B. D. Myers, and B. M. Brenner. 1976. Effects of some vasodilator drugs on transcapillary fluid exchange in renal cortex. *Am. J. Physiol.* **230**:1148–1158.
133. Schor, N., I. Ichikawa, and B. M. Brenner. 1981. Mechanisms of action of various hormones and vasoactive substances on glomerular ultrafiltration in the rat. *Kidney Int.* **20**:442–451.
134. Thomas, C. E., P. D. Bell, and L. G. Navar. 1982. Influence of bradykinin and papaverine on renal and glomerular hemodynamics in dogs. *Renal Physiol.* **5**:197–205.
135. Rosivall, L., P. K. Carmines, and L. G. Navar. 1984. Effects of renal arterial angiotensin I infusions on glomerular dynamics in sodium replete dogs. *Kidney Int.* **26**:263–268.
136. Baylis, C., I. Ichikawa, W. T. Willis, C. B. Wilson, and B. M. Brenner. 1977. Dynamics of glomerular ultrafiltration. IX. Effects of plasma protein concentration. *Am. J. Physiol.* **232**:F58–F71.
137. Schor, N., I. Ichikawa, and B. M. Brenner. 1980. Glomerular

adaptations to chronic dietary salt restriction or excess. *Am. J. Physiol.* **238**:F428–F436.

138. Navar, L. G., D. Jirakulsomchok, P. D. Bell, C. E. Thomas, and W. C. Huang. 1982. Influence of converting enzyme inhibition on renal hemodynamics and glomerular dynamics in sodium restricted dogs. *Hypertension* **4**:58–68.
139. Myers, B. D., W. M. Deen, and B. M. Brenner. 1975. Effects of norepinephrine and angiotensin II on the determinants of glomerular ultrafiltration and proximal tubule fluid reabsorption in the rat. *Circ. Res.* **37**:101–110.
140. Selkurt, E. E., P. W. Hall, and M. P. Spencer. 1949. Influence of graded arterial pressure decrement on renal clearance of creatinine, *p*-amino hippurate and sodium. *Am. J. Physiol.* **159**:369–378.
141. Shipley, R. E., and R. S. Study. 1951. Changes in renal blood flow, extraction of inulin, GFR, tissue pressure and urine flow with acute alterations of renal artery blood pressure. *Am. J. Physiol.* **167**:676–688.
142. Guyton, A. C., J. B. Langston, and G. Navar. 1964. Theory for renal autoregulation by feedback at the juxtaglomerular apparatus. *Circ. Res.* **15**:187–196.
143. Thurau, K. 1966. Nature of autoregulation of renal blood flow. *Proc. 3rd Int. Congr. Nephrol.* **1**:162–173.
144. Navar, L. G. 1978. Renal autoregulation: Perspectives from whole kidney and single nephron studies. *Am. J. Physiol.* **234**:F357–F370.
145. Robertson, C. R., W. M. Deen, J. L. Troy, and B. M. Brenner. 1971. Dynamics of glomerular ultrafiltration in the rat. III. Hemodynamics and autoregulation. *Am. J. Physiol.* **223**:1191–1200.
146. Navar, L. G., D. J. Marsh, R. C. Blantz, J. Hall, D. W. Ploth, and A. Nasjletti. 1982. Intrinsic control of renal hemodynamics. (Symposium Report). *Fed. Proc.* **41**:3022–3030.
147. Bell, P. D., and L. G. Navar. 1982. Macula densa feedback control of glomerular filtration: Role of cytosolic calcium. *Miner. Electrolyte Metab.* **8**:61–77.
148. Adams, P. L., F. F. Adams, P. D. Bell, and L. G. Navar. 1980. Impaired renal blood flow autoregulation in ischemic acute renal failure. *Kidney Int.* **18**:68–76.
149. Tucker, B. J., R. W. Steiner, L. C. Gushwa, and R. C. Blantz. 1978. Studies on the tubulo-glomerular feedback system in the rat: The mechanism of reduction in filtration rate with benzolamide. *J. Clin. Invest.* **62**:993–1004.
150. Luke, R. G., and J. H. Galla. 1983. Chloride-depletion alkalosis with a normal extracellular fluid volume. *Am. J. Physiol.* **245**:F419–F424.
151. Schnermann, J., F. S. Wright, J. M. Davis, W. V. Stackelberg, and G. Grill. 1970. Regulation of superficial nephron filtration rate by tubulo-glomerular feedback. *Pfluegers Arch.* **318**:147–175.
152. Schnermann, J., A. E. G. Persson, and B. Agerup. 1973. Tubuloglomerular feedback: Nonlinear relation between glomerular hydrostatic pressure and loop of Henle perfusion. *J. Clin. Invest.* **52**:862–869.
153. Burke, T. J., L. G. Navar, J. R. Clapp, and R. R. Robinson. 1974. Response of single nephron glomerular filtration rate to distal nephron microperfusion. *Kidney Int.* **6**:230–240.
154. Bell, P. D., C. Thomas, R. H. Williams, and L. G. Navar. 1978. Filtration rate and stop-flow pressure feedback responses to nephron perfusion in the dog. *Am. J. Physiol.* **234**:F154–F165.
155. Persson, B. E., and A. E. G. Persson. 1981. The existence of a tubulo-glomerular feedback mechanism in the amphiuma nephron. *Pfluegers Arch.* **391**:129–134.
156. Muller-Suur, R., H. R. Ulfendahl, and A. E. G. Persson. 1983. Evidence for tubuloglomerular feedback in juxtamedullary nephrons of young rats. *Am. J. Physiol.* **244**:F425–F431.
157. Bell, P. D., L. G. Navar, D. W. Ploth, and C. B. McLean. 1980. Tubulo-glomerular feedback responses during perfusion with nonelectrolyte solutions in the rat. *Kidney Int.* **18**:460–471.
158. Ichikawa, I. 1982. Direct analysis of the effector mechanism of the tubuloglomerular feedback system. *Am. J. Physiol.* **243**:F447–F455.
159. Blantz, R. C., L. C. Gushwa, and A. E. G. Persson. 1983. Tubuloglomerular feedback (TGF) released response in glomerular capillary hydrostatic pressure (P_G) and nephron filtration rate (SNGFR) in hydropenic (H) and in angiotensin (AII) and prostaglandin (PG) blocked rats. *Am. Soc. Nephrol.* **16**:149A.
160. Bell, P. D., M. Reddington, D. Ploth, and L. G. Navar. 1984. Tubuloglomerular feedback mediated decreases in glomerular pressure in Munich–Wistar rats. *Am. J. Physiol.* **247**:F877–F880.
161. Navar, L. G., T. J. Burke, R. R. Robinson, and J. R. Clapp. 1974. Distal tubular feedback in the autoregulation of single nephron glomerular filtration rate. *J. Clin. Invest.* **53**:516–525.
162. Ploth, D. W., J. Schnermann, H. Dahlheim, M. Hermle, and E. Schmidmeier. 1977. Autoregulation and tubuloglomerular feedback in normotensive and hypertensive rats. *Kidney Int.* **12**:253–267.
163. Schnermann, J., J. M. Davis, P. Wunderlich, D. Z. Levine, and M. Horster. 1971. Technical problems in the micropuncture determination of nephron filtration rate and their functional implications. *Pfluegers Arch.* **329**:307–320.
164. Ploth, D. W., J. Rudolph, R. LaGrange, and L. G. Navar. 1979. Tubulo-glomerular feedback and single nephron function after converting enzyme inhibition in the rat. *J. Clin. Invest.* **64**:1325–1335.
165. Williams, R. H., C. Thomas, P. D. Bell, and L. G. Navar. 1977. Autoregulation of nephron filtration rate in the dog assessed by indicator-dilution technique. *Am. J. Physiol.* **233**:F282–F289.
166. Moore, L. C., J. Schnermann, and S, Yarimizu. 1979. Feedback mediation of SNGFR autoregulation in hydropenic and DOCA-salt loaded rats. *Am. J. Physiol.* **237**:F63–F74.
167. Morgan, T., and R. W. Berliner. 1969. A study by continuous microperfusion of water and electrolyte movements in the loop of Henle and distal tubule of the rat. *Nephron* **6**:388–405.
168. Schnermann, J., D. W. Ploth, and M. Hermle. 1976. Activation of tubuloglomerular feedback by chloride transport. *Pfluegers Arch.* **362**:229–240.
169. Bell, P. D., C. B. McLean, and L. G. Navar. 1981. Dissociation of tubuloglomerular feedback responses from distal tubule chloride concentration in the rat. *Am. J. Physiol.* **240**:F111–F119.
170. Bell, P. D., and L. G. Navar. 1982. Relationship between tubuloglomerular feedback responses and perfusate hypotonicity. *Kidney Int.* **22**:234–239.
171. Briggs, J., G. Schubert, and J. Schnermann. 1982. Further evidence for an inverse relationship between macula densa NaCl concentration and filtration rate. *Pfluegers Arch.* **392**:372–378.
172. Briggs, J. P., J. Schnermann, and F. S, Wright. 1980. Failure of tubule fluid osmolality to affect feedback regulation of glomerular filtration. *Am. J. Physiol.* **239**:F427–F432.
173. Bell, P. D. 1982. Luminal and cellular mechanisms for the mediation of tubular-glomerular feedback responses. *Kidney Int.* **22**:S97–S103.
174. Bell, P. D., and L. G. Navar. 1982. Cytoplasmic calcium in the mediation of macula densa tubulo-glomerular feedback responses. *Science* **215**:670–673.
175. Bell, P. D., and M. Reddington. 1983. Intracellular calcium in the transmission of tubuloglomerular feedback signals. *Am. J. Physiol.* **245**:F295–F302.
176. Shaw, J. O. 1981. Effects of extracellular Ca^{++} and the intracellular Ca^{++} antagonist 8-(N, N, diethyl amino)-octyl 3,4,5-trimethoxybenzoate on rabbit platelet conversion of arachidonic acid to thromboxane. *Prostaglandins* **21**:571–579.
177. Schnermann, J., H. Osswald, and M. Hermle. 1977. Inhibitory effect of methylxanthines on feedback control of glomerular filtration in the rat kidney. *Pfluegers Arch.* **369**:39–48.
178. Bell, P. D., and M. Reddington. 1983. Cyclic nucleotide–calcium interaction in the mediation of macula densa tubuloglomerular feedback signals. *Fed. Proc.* **42**:298.
179. Seamon, K. B., and J. W. Daley. 1981. Forskolin: A unique diterpene activator of cyclic AMP-generating systems. *J. Cyclic Nucleotide Res.* **7**:201–224.
180. Kaser-Glanzmann, R., M. Jakabova, J. W. George, and E. F. Luscher. 1977. Stimulation of calcium uptake in platelet mem-

brane vesicles by adenosine 3′,5′-cyclic monophosphate and protein kinase. *Biochim. Biophys. Acta* **466**:429–440.
181. Bell, P. D. 1984. Participation of prostaglandins in the mediation of tubuloglomerular feedback responses. *Fed. Proc.* **43**:409.
182. Rosivall, L., D. F. Rinder, J. Champion, M. C. Khosla, L. G. Navar, and S. Oparil. 1983. Intrarenal angiotensin I conversion at normal and reduced renal blood flow in the dog. *Am. J. Physiol.* **245**:F408–F415.
183. Navar, L. G., and L. Rosivall. 1984. Contribution of the renin–angiotensin system to the control of intrarenal hemodynamics. *Kidney Int.* **25**:857–868.
184. Celio, M. R., and T. Inagami. 1981. Angiotensin II immunoreactivity coexists with renin in the juxtaglomerular granular cells of the kidney. *Proc. Natl. Acad. Sci. USA* **78**:3897–3900.
185. Taugner, R., E. Hackenthal, E. Rix, R. Nobling, and K. Poulsen. 1982. Immunocytochemistry of the renin–angiotensin system: Renin, angiotensinogen, angiotensin I, angiotensin II, and converting enzyme in the kidneys of mice, rats, and tree shrews. *Kidney Int.* **22**:S533–S543.
186. Osborne, M. J., B. Droz, P. Meyer, and F. Morel. 1975. Angiotensin II: Renal localization in glomerular mesangial cells by autoradiography. *Kidney Int.* **8**:245–254.
187. Ploth, D. W., and L. G. Navar. 1979. Intrarenal effects of the renin–angiotensin system. *Fed. Proc.* **38**:2280–2285.
188. Huang, W. C., D. W. Ploth, P. D. Bell, J. Work, and L. G. Navar. 1981. Bilateral renal function responses to converting enzyme inhibitor (SQ20,881) in two-kidney, one clip Goldblatt hypertensive rats. *Hypertension* **3**:285–293.
189. Steiner, R. W., B. J. Tucker, and R. C. Blantz. 1979. Glomerular hemodynamics in rats with chronic sodium depletion. *J. Clin. Invest.* **64**:503–512.
190. Steiner, R. W., and R. C. Blantz. 1979. Acute reversal by saralasin of multiple intrarenal effects of angiotensin II. *Am. J. Physiol.* **237**:F386–F391.
191. Rosivall, L., and L. G. Navar. 1983, Effects on renal hemodynamics of intra-arterial infusions of angiotensin I and II. *Am. J. Physiol.* **245**:F181–F187.
192. Hall, J. E. 1982. Regulation of renal hemodynamics. *Int. Rev. Physiol.* **26**:243–321.
193. Hall, J. E., A. C. Guyton, T. E. Jackson, T. G. Coleman, T. E. Lohmeier, and N. C. Trippodo. 1977. Control of glomerular filtration rate by renin–angiotensin system. *Am. J. Physiol.* **233**:F366–F372.
194. Murray, R. B., and R. L. Malvin. 1979. Intrarenal renin and autoregulation of renal plasma flow and glomerular filtration rate. *Am. J. Physiol.* **236**:F559–F566.
195. Hall, J. E., A. C. Guyton, and A. W. Cowley, Jr. 1977. Dissociation of renal blood flow and filtration rate autoregulation by renin depletion, *Am. J. Physiol.* **232**:F215–F221.
196. Nasjletti, A., and K. U. Malik. 1981. Interrelationships among prostaglandins and vasoactive substances. *Med, Clin. North Am.* **65**:881–889.
197. Gerber, J. G., R. D. Olson, and A. S, Nies, 1981. Interrelationship between prostaglandins and renin release. *Kidney Int.* **19**:816–821.
198. Baer, P. G., and J. C. McGiff. 1980. Hormonal systems and renal hemodynamics. *Annu. Rev. Physiol.* **42**:582–601.
199. Baer, P. G. 1981. Contribution of prostaglandins to renal blood flow maintenance is determined by the level of activity of the renin–angiotensin system. *Life Sci.* **28**:587–593.
200. Blasingham, M. C., and A. Nasjletti. 1980. Differential renal effects of cyclooxygenase inhibition in sodium-replete and sodium-deprived dog. *Am. J. Physiol.* **239**:F360–F365.
201. Carmines, P. K., L. Rosivall, M. F. Till, and L. G. Navar. 1983. Renal hemodynamic effects of captopril in anesthetized sodium-restricted dogs: Relative contributions of prostaglandin stimulation and suppressed angiotensin activity. *Renal Physiol.* **6**:281–287.
202. Thurau, K., J. Schnermann, W. Nagel, M, Horster, and M. Wahl. 1967. Composition of tubular fluid in the macula densa segment as a factor regulating the function of the juxtaglomerular apparatus. *Circ. Res.* **20–21**(II):79–81.
203. Kishimoti, T., M. Maekawa, M. Miyazaki, K. Yamamoto, and J. Ueda. 1972. Effects of renal venous pressure elevations on renal hemodynamics, urine formation and renin release. *Jpn. Circ. J.* **36**:439–449.
204. Churchill, P. C., M. C. Churchill, and F. D. McDonald. 1978. Renin secretion and distal tubule Na^+ in rats. *Am. J. Physiol.* **235**:F611–F616.
205. Hofbauer, K. G., A. Konrads, K. Schwarz, and U. Werner. 1978. Role of cyclic AMP in the regulation of renin release from the isolated perfused kidney. *Klin. Wochenschr.* **56**:51–59.
206. Park, C. S., D. S. Han, and J. C. S. Fray. 1981. Calcium in the control of renin secretion: Ca^{2+} influx as an inhibitory signal. *Am. J. Physiol.* **240**:F70–F74.
207. Fray, J. C. S. 1980. Stimulus–secretion coupling of renin: Role of hemodynamic and other factors. *Circ. Res.* **47**:485–492.
208. Baumbach, L., and P. P. Leyssac. 1977. Studies on the mechanism of renin release from isolated superfused rat glomeruli: Effects of calcium, calcium ionophore and lanthanum. *J. Physiol. (London)* **273**:745–764.
209. Muller-Suur, R., H.-U. Gutsche, and H. J. Schurek. 1976. Acute and reversible inhibition of tubulo-glomerular feedback mediated afferent vasoconstriction by the calcium-antagonist verapamil. *Curr. Probl. Clin. Biochem.* **6**:291–298.
210. Stowe, N., J. Schnermann, and M. Hermle. 1979. Feedback regulation of nephron filtration rate during pharmacologic interference with the renin angiotensin and adrenergic systems in rats. *Kidney Int.* **15**:473–486.
211. Muller-Suur, R., H.-U. Gutsche, K. F. Samwer, W. Oelkers, and K. Hierholzer. 1975. Tubuloglomerular feedback in rat kidneys of different renin contents. *Pfluegers Arch.* **359**:33–56.
212. Moore, L. C., S. Yarimizu, G. Schubert, P. C. Weber, and J. Schnermann. 1980. Dynamics of tubuloglomerular feedback adaptation to acute and chronic changes in body fluid volume. *Pfluegers Arch.* **387**:39–45.
213. Ploth, D. W., and R. N. Roy. 1982. Renin angiotensin influence on tubuloglomerular activity in the rat. *Kidney Int.* **22**:S114–S121.
214. Kiil, F., J. Kjekshus, and E. Loyning. 1969. Renal autoregulation during infusion of noradrenaline, angiotensin, and acetylcholine. *Acta Physiol. Scand.* **76**:10–23.
215. Katz, M. A., and L. Shear. 1975. Effects of renal nerves on renal hemodynamics. I. Direct stimulation and carotid occlusion. *Nephron* **14**:246–256.
216. Johns, E. J. 1980. A comparison of the ability of two angiotensin II receptor blocking drugs, 1-Sar-8-Ala angiotensin II and 1-Sar-8-Ile angiotensin II, to modify the regulation of glomerular filtration rate in the cat. *Br. J. Pharmacol.* **71**:499–506.
217. Kon, V., and I. Ichikawa. 1983. Effector loci for renal nerve control of cortical microcirculation. *Am. J. Physiol.* **245**:F545–553.
218. Nomura, G., Y. Kibe, S. Arai, D. Uno, and J. Takeuchi. 1976. Distribution of intrarenal blood flow after renal denervation in the dog. *Nephron* **16**:126–133.
219. Pelayo, J. C., M. G. Ziegler, A. J. Pedro, and R. C. Blantz. 1983. Renal denervation in the rat: Analysis of glomerular and tubular function. *Am. J. Physiol.* **244**:F70–F77.
220. Takabatake, T. 1982. Feedback regulation of glomerular filtration rate in the denervated kidney. *Kidney Int.* **22**(Suppl. 12):S129–S135.
221. Aukland, K. 1968. Effect of adrenaline, nonadrenaline, angiotensin and renal nerve stimulation on intrarenal distribution of blood flow in dogs. *Acta Physiol. Scand.* **72**:498–509.
222, Barclay, J., W. T. Cooke, and R. A. Kenney. 1947. Observations on the effects of adrenalin on renal function and circulation in man. *Am. J. Physiol.* **151**:621–625.
223. Click, R. L., W. L. Joyner, and J. P. Gilmore. 1979. Reactivity of glomerular afferent and efferent arterioles in renal hypertension. *Kidney Int.* **15**:109–115.

224. Edwards, R. M. 1982. Response of isolated microvessels to intraluminal pressure, nonrepinephrine, and angiotensin II. *Am. Soc. Nephrol.* **15**:150A.
225. de Rouffignac, C., S. Deiss, and J. P. Bonvalet. 1970. Détermination du taux individual de filtration glomiralaire des néphrons accessibles et inaccessibles á la microponction. *Pfluegers Arch.* **315**:273–290.
226. Baines, A. D. 1973. Redistribution of nephron function in response to chronic and acute solute loads. *Am. J. Physiol.* **224**:237–244.
227. Bonvalet, J. P., P. Bencsáth, and C. de Rouffignac. 1972. Glomerular filtration rate of superficial and deep nephrons during aortic constriction. *Am. J. Physiol.* **222**:599–606.
228. Bruno, F. J., E. A. Alexander, A. L. Riley, and N. G. Levinsky. 1974. Superficial and juxtamedullary nephron function during saline loading in the dog. *J. Clin. Invest.* **53**:971–979.
229. Horster, M., and K. Thurau. 1968. Micropuncutre studies on the filtration rate of single superficial and juxtamedullary glomeruli in the rat kidney. *Pfluegers Arch.* **301**:162–181.
230. Coelho, J. B. 1974. Sodium metabolism and intrarenal distribution of nephron glomerular filtration rates in the unanesthetized rat. *Proc. Soc. Exp. Biol. Med.* **146**:225–231.
231. Chalbardes, D., P. Doujiol, S. Diess, J. P. Bonvalet, and C. de Rouffignac. 1974. Intrarenal glomerular filtration rate distribution in salt-loaded rats. *Pfluegers Arch.* **349**:191–202.
232. Davis, J. M., H. Brechtelsbauer, P. Prucksunand, J. Weigl, J. Schnermann, and K. Kramer. 1974. Relationship between salt loading and distribution of nephron filtration rates in the dog. *Pfluegers Arch.* **350**:259–272.
233. Barratt, J. L., J. D. Wallin, F. C. Rector, and D. W. Seldin. 1973. Influence of volume expansion on single nephron filtration rate and plasma flow in the rat. *Am. J. Physiol.* **224**:643–650.
234. Mandin, H., A. H. Israelit, F. C. Rector, Jr., and D. W. Seldin. 1971. Effect of saline infusions on intrarenal distribution of glomerular filtrate and proximal reabsorption in the dog. *J. Clin. Invest.* **50**:514–522.
235. Poujeol, P., D. Chabardés, J. P. Bonvalet, and C. de Rouffignac. 1975. Glomerular filtration rate and microsphere distributions in single nephron of rat kidney. *Pfluegers Arch.* **357**:291–301.
236. Bonvalet, J. P., and C. de Rouffignac. 1981. Distribution of ferrocyanide along the proximal tubular lumen of the rat kidney: Its implications upon hydrodynamics. *J. Physiol. (London)* **318**:85–98.
237. Cohen, H. J., D. J. Marsh, and B. Kayser. 1983. Autoregulation in vasa recta of the rat kidney. *Am. J. Physiol.* **245**:F32–F40.

CHAPTER 14

The Proximal Nephron

Gerhard Giebisch and Peter S. Aronson

1. General Properties of the Proximal Nephron

This chapter deals with an analysis of proximal tubular fluid and electrolyte transport. The proximal tubule of the mammalian kidney is the nephron site where the major portion, some two-thirds to three-fourths, of the filtered sodium salts is reabsorbed. The unique importance of this transport operation derives from the fact that this active, energy-consuming transport process provides the main driving force for transepithelial water movement.[1,2] Thus, the maintenance of an adequate and constant extracellular fluid and plasma volume depends crucially on the integrity of the proximal tubular Na^+ transport system. This is not to ignore the fact that distal portions of the nephron may enhance their rate of Na^+ and fluid transport when Na^+ and fluid escape proximal reabsorption, and that the fine adjustment of salt and water balance is regulated within the distal nephron. Nevertheless, it is at the proximal tubular level where the bulk of water, salts, and nonelectrolytes are normally retrieved from the filtrate.

In addition to the reabsorption of a very large fraction of the filtered Na^+ and fluid, proximal tubular Na^+ transport also plays a key role in several other tubular transport processes. First, by virtue of the fact that a substantial fraction of proximal Na^+ reabsorption is closely associated with that of HCO_3^-, and the fact that HCO_3^- reabsorption is related to the secretion of H^+ ions, an important fraction of proximal tubular Na^+ transfer is firmly linked to the maintenance of a normal acid–base balance.[3–5]

Second, recent evidence has shown clearly that Na^+ transport across the proximal tubular epithelium also plays an essential role in the reabsorption of some electrolytes and nonelectrolytes, among them glucose, amino acids, phosphate, sulfate, as well as mono- and dicarboxylic acids.[6,7] It appears that downhill Na^+ movement from lumen to the cell compartments across the luminal brush border membranes of proximal tubule cells provides the driving force for "uphill" transport of these solutes into the cell by what has been termed "secondary active transport." This mode of transport is defined as uphill movement that results form the interaction with the flow of Na^+ from lumen to cell. In this case, the term "primary active transport" has been reserved for Na^+ since it is the movement of this ion that is directly coupled to cellular energy input. The latter establishes the transmembrane downhill Na^+ gradient across the luminal brush border which in turn is coupled to the uphill movement of nonelectrolytes such as glucose, amino acids, and other solutes from lumen to cell.[8–10]

Third, in a more nonspecific manner, the primary active transport of Na^+ across the proximal tubular epithelium, by virtue of moving water across this nephron site, also provides the driving force for the transport of such nonelectrolytes as urea and other similar uncharged substances.[11,12] Since the permeability of the tubular wall to urea is less than that to water, removal of water from the proximal tubular lumen by active Na^+ transport effects a significant increase of the urea concentration in the proximal tubular lumen to levels that exceed those in plasma water. This event in turn provides a favorable diffusion gradient for passive movement of urea from the tubular lumen to the peritubular space. Ultimately, however, the energy input occurs at the site of active Na^+ transport which is responsible for water reabsorption from the tubular fluid.

From an electrophysiological point of view, the active transport of Na^+, either directly or indirectly, is also responsible for the generation of the electrical potential difference that normally exists along the proximal tubule between lumen and peritubular fluid compartment.[13–15] When Na^+ transport is blocked by metabolic inhibitors, or when Na^+ ions are replaced by non- or poorly transported cations, the lumen negative electrical potential difference across the early proximal tubule is promptly abolished.

From a general point of view of functional organization, the proximal tubular epithelium represents a typical "leaky" epithelium.[1,2,13–17] Its electrical conductance, a measure of transepithelial ionic mobility, is the highest measured across any

Gerhard Giebisch • Department of Physiology, Yale University School of Medicine, New Haven, Connecticut 06510. **Peter S. Aronson** • Departments of Physiology and Internal Medicine, Yale University School of Medicine, New Haven, Connecticut 06510.

transporting epithelium, Only small transepithelial electrical potential differences and ionic gradients can be maintained. The fact that ionic conductances are high is also demonstrated by the finding that bidirectional tracer fluxes of most ion species are high. Hence, *net* ion fluxes constitute in general only a small fraction of large transepithelial bidirectional ion movements.

The proximal tubular epithelium also shares with other leaky epithelia the presence of a very prominent intercellular shunt pathway which functions as a fairly nonselective route of transepithelial exchange of electrolytes and nonelectrolytes in parallel with the transcellular transport pathway. Thus, the high ionic conductance is not caused by an unusually high permeability of proximal tubular cell membranes but rather by the presence of an additional extracellular compartment constituting an intercellular transport path. Despite the inherent inefficiency that such a low-resistance transport system with a high back-flux rate from peritubular to luminal fluid compartment imposes on Na^+ transfer, the presence of such a specialized extracellular compartment between proximal epithelial cells is not only an integral part of the mechanism by which salt and water movements are coupled, but it is also the site at which the tonicity of the proximal reabsorbate is kept close to that of the luminal fluid from which it originates. It will also become apparent that proximal tubular cell models including intercellular shunt pathways have been useful to interpret the complex interrelationship between peritubular hemodynamic events and proximal tubular Na^+ transport.

The most important changes in composition of tubular fluid along the proximal tubule are summarized in Fig. 1.[18] From the observation that TF/P inulin ratios increase with tubular length,

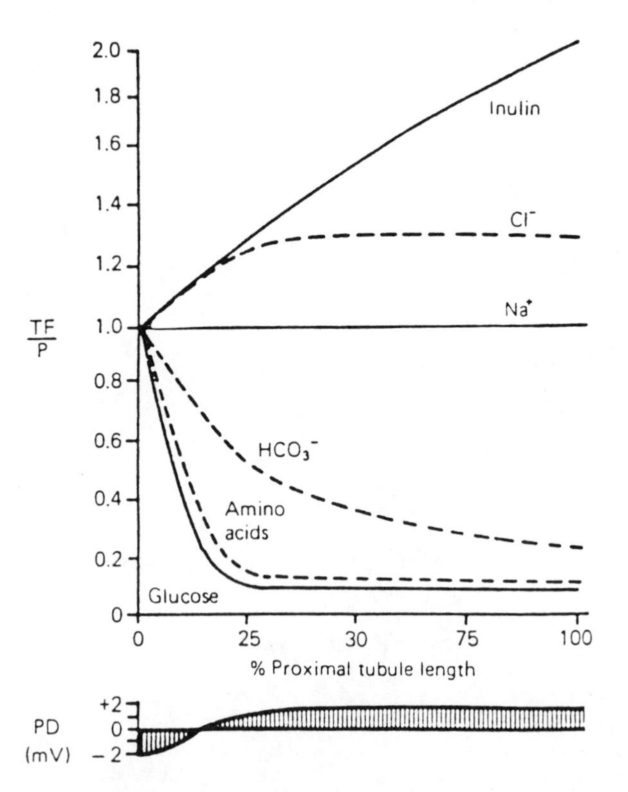

Fig. 1. Changes in composition along the superficial proximal mammalian tubule. From Rector, ref. 18.

it is clear that substantial fractions of the glomerular filtrate are reabsorbed. Glucose and amino acids, representative examples of Na^+-linked cotransport, disappear rapidly along the very first segment of the tubule; at about 25–30% proximal tubular length, their concentrations fall to a few percent of that in the filtrate. Similarly, preferential HCO_3^- reabsorption lowers its concentration along the early proximal tubule so that its concentration falls to some 5 to 8 mM as fluid leaves the proximal tubule. HCO_3^- reabsorption along the initial segment of the proximal tubule exceeds that of Cl^-, as there is only insignificant net Cl^- reabsorption. Accordingly, the Cl^- concentration rises sharply and reaches a plateau in the later part of the proximal tubule. This tubular site is characterized by reabsorption of a fluid richer in Cl^- than in HCO_3^-.

We note also from inspection of Fig. 1 that in the initial segment of the proximal tubule, the transepithelial voltage is lumen-negative.[19] However, with the rise of the Cl^- concentration in the lumen, the potential reverses its polarity and becomes lumen-positive. This voltage difference is generated by the Cl^- diffusion potential across the epithelium of the proximal tubule, as evidenced by the high Cl^- permeability of the tubule and the sensitivity of the lumen-positive potential to factors that affect the Cl^- concentration gradient.[19]

2. Distribution of Transport Functions along the Proximal Tubule

Recent evidence indicates clearly that there are significant inherent differences of transport properties along the proximal tubule. These concern differences between not only the proximal convoluted and proximal straight tubules but also the early and late proximal convoluted tubules. Since only a very small part of the proximal straight tubule is present at the kidney surface, our information of the transport properties of the *straight* segment of the proximal tubule is based largely on perfusion, *in vitro,* of straight segments of the proximal tubule, isolated by microdissection of rabbit renal cortex. Such studies and appropriate comparison with the function of the proximal *convoluted* tubule have shown that these nephron segments differ in several transport functions.

Figure 2 summarizes some morphological features of tubule cells along the proximal tubule. Morphological changes along the proximal tubule have led to a subdivision into three segments.[20] The early part of the proximal tubule, the S_1 segment, is characterized by a well-developed brush border system, an extensive basal labyrinth, and many mitochondria. The transition to the S_2 segment is gradual, and this segment constitutes the late proximal convoluted tubule and early proximal straight tubule; it is also the late proximal convoluted tubule of juxtamedullary nephrons. This tubular segment is characterized by fewer microvilli, less densely packed mitochondria, and less surface amplification of the basolateral cell membrane. The last segment of the proximal tubule, the S_3 segment, constitutes the last part of the superficial proximal straight tubule and the proximal straight tubule of the juxtamedullary nephrons. Its cells are flatter, have even less mitochondria and less basolateral membrane amplification than cells of the S_2 segment. Careful morphological studies have shown that the area of the apical and basolateral cell membranes are of similar magnitude in the S_1 and S_3 segments of the rabbit proximal tubule but that, in absolute terms, the S_1 segment has more than twice the reabsorptive area of the S_3 segment.[21] The absorptive area in tubule cells

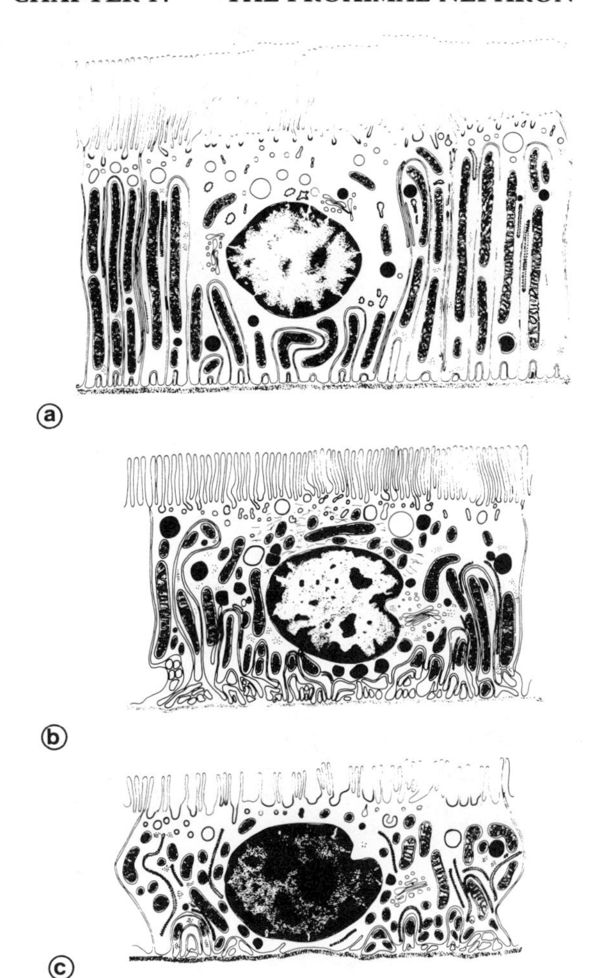

Fig. 2. Diagrams showing morphological features which distinguish three segments of the proximal tubule, emphasizing differences in cell shape, distribution of mitochondria, and brush border organization. Shown are proximal tubule cells of the (a) S_1, (b) S_2, and (c) S_3 segment of the rabbit nephron. From Kaissling and Kriz, ref. 20.

lining the proximal tubule of amphibian nephrons is significantly less than in mammalian species.[22]

2.1. Transport along the Proximal Convoluted Tubule

Several constituents of the glomerular filtrate are avidly reabsorbed along the very first portion of the mammalian proximal tubule. As stated, HCO_3^- is reabsorbed preferentially to Cl^- in the very first portion of the proximal convoluted tubule.[23–25] Organic compounds of the filtrate are also rapidly reabsorbed along the initial segment of the proximal convoluted tubule. Thus, the concentrations of glucose,[23,26,27] amino acids,[28–30] and lactate[31] have been sharply reduced to low levels by the time the tubular fluid has traversed the first one-third of the proximal convolution. Several significant differences in absolute reabsorption rates have been observed along the proximal *convoluted* tubule (S_1 and S_2 segments). Amino acid reabsorption is higher in the early than the late proximal tubule when the tubule is perfused with solutions of similar concentrations of amino acids.[29] This observation is indicative of different inherent transport rates for amino acids along the proximal convoluted tubule. In addition, inorganic phosphate and glucose are more avidly reabsorbed along the very first part of the proximal convoluted tubule than along the remainder of this nephron segment.[32,33] There is also evidence, albeit indirect, that the early proximal tubule is more effective in lowering tubular pH,[34] as measured by the steeper glycodiazine gradients which this nephron sgement can establish.

In the rabbit proximal tubule, perfused *in vitro* with solutions mimicking "early" and "late" proximal tubular fluid, Jacobson and Kokko have observed a significantly higher fluid absorption in early than in late segments.[35] Similar findings have been reported by McKeown *et al.*[33] However, in the rat no significant differences of fluid transport have been noted between early and late segments.[36] Similarly, no differences could be detected between early and late proximal nephron sites of the rat nephron with respect to the ability to establish steady-state transepithelial Na^+ gradients.[36]

2.2. Transport along the Proximal Straight Tubule

Important differences exist also with respect to transport properties between convoluted and straight proximal tubule. Net fluid reabsorption in *straight* tubules is only about one-half that measured in isolated *convoluted* tubules.[37,38] Active glucose transport is also less but proceeds against steeper concentration gradients across straight proximal tubules.[39]

Glycine[40] and HCO_3^- transport[41,42] rates are also considerably smaller in the proximal straight than the proximal convoluted tubule. Fluid absorption along the proximal straight tubule is relatively insensitive to the deletion of HCO_3^- and organic solutes in the lumen; an apparently "simple" NaCl reabsorptive mechanism can drive about one-half of the normally observed fluid transport.[43] Additional features of the proximal straight tubule are its lower lumen-negative potential,[35,44] smaller glucose-induced cell potential changes,[45] smaller surface to volume ratio of the cells lining this segment,[21] and lower Na^+,K^+-ATPase activity[46,47] (see also Refs. 37, 38).

An additional intrinsic difference between convoluted and straight proximal tubules is the presence of a more powerful secretory system for organic acids at the level of the straight proximal segment. *p*-Aminohippurate and other organic acids are secreted four times faster in the straight than in the convoluted tubule.[48] At high peritubular *p*-aminohippurate concentrations, organic anion secretion may even induce net fluid secretion into the lumen of straight proximal tubules.[49] Net fluid secretion into straight proximal tubule has also been observed *in vitro* when such nephron segments are exposed to uremic plasma.[50] Such reversal of the normal direction of net fluid movement is thought to be due to the high level of organic acids that may accumulate in the blood during the progression of uremia.

2.3. Differences between Superficial and Juxtamedullary Proximal Tubules: Internephron Heterogeneity

Experiments carried out on single, perfused proximal tubules *in vitro* have led to the recognition that there are differences in transport functions depending where within the kidney the tubules originate. This has led to renewed interest in the distinction between superficial and juxtamedullary nephrons and to the definition of internephron heterogeneity, defined as the functional characteristics of similar tubular structures in superficial and juxtamedullary nephrons. It should be noticed, as pointed out recently by Berry,[37] that in the rabbit, the species in which most of the tubule perfusions were carried out, superficial and juxtamedullary nephrons constitute only some 28 and 9% of the total nephron population, with the remainder being of *mid-cortical* origin.

The observed functional differences between superficial and juxtamedullary proximal tubules have first to do with the formation of a significantly larger rate of glomerular filtrate in juxtamedullary nephrons.[51–53] Accordingly, it is not surprising that the rate of fluid absorption in juxtamedullary proximal tubules exceeds that of their superficial counterparts.[42,54] The reabsorption of NaCl and of HCO_3^- are also higher in early juxtamedullary than in early superficial convoluted proximal tubules.[54] On the other hand, nonionic solutes are reabsorbed at similar rates in the two proximal tubule populations.[42] It is of interest that in the absence of organic solutes, a lumen-negative electrical potential difference persists in juxtamedullary but not in superficial proximal convoluted tubules.[55,56] The rate of fluid reabsorption in proximal convoluted tubules of juxtamedullary nephrons is also quite insensitive to the deletion of organic solutes, of HCO_3^-, and the abolition of a transepithelial Cl^- concentration gradient, all factors that have been shown to stimulate fluid transport across the proximal convoluted tubule of superficial nephrons.[56] This behavior has led to the proposal that juxtamedullary proximal convoluted tubules have a "simple," electrogenic Na^+ transport system similar to that found in superficial straight proximal tubules.[43]

3. Transepithelial Potentials and Passive Permeabilities

In accordance with the low resistance of the proximal tubule, the transepithelial potential difference across the proximal convoluted tubule is also low. It ranges from −6 to −15 mV in the amphibian kidney, and from −6 to +2 mV in the mammalian proximal convoluted tubule[13–15,19,57,58] (the sign indicates the luminal polarity). Problems concerning some technical aspects of these potential measurements have recently been reviewed.[59] Again, the technique of perfusing isolated mammalian proximal tubules has been particularly useful in elucidating the mechanisms generating these electrical potential differences across both proximal convoluted and straight tubules.

3.1. Proximal Convoluted Tubule

3.1.1. Transepithelial Potential Differences

There is general agreement, based on studies *in vivo* in tubules of a mutant strain of rats with surface glomeruli, that the electrical potential difference in the first tubular loop is negative, ranging from −4.5 to −0.8 mV.[19,60,61] Similarly, isolated rabbit proximal tubules exposed to physiological perfusion and bathing solutions of identical composition show consistently negative potential differences.[62–65] In contrast, later segments of the proximal convoluted tubule *in vivo* have either a smaller luminal negativity than earlier loops, or exhibit a lumen-positive potential difference ranging from +1.0 to +4.5 mV. Figure 3

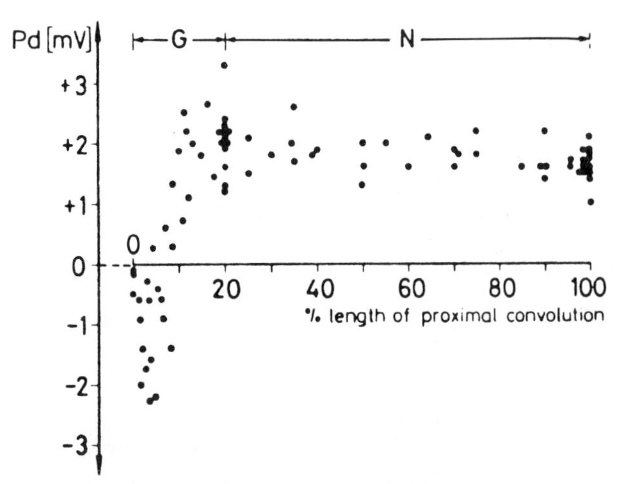

Fig. 3. Free-flow transepithelial potential difference as a function of tubular length. Potential difference in millivolts (sign indicating polarity of tubule lumen) plotted as a function of puncture site. Experiments N were done on normal Wistar rats, experiments G on Wistar rats with superficially located glomeruli. From Frömter and Gessner, ref. 19.

shows the progression of transepithelial differences along the proximal convoluted tubule.[19] Several aspects of this proximal tubular potential profile deserve further comment.

Experiments carried out in several laboratories indicate that changes in the luminal glucose, amino acid, HCO_3^- and Cl^- concentration have a dramatic effect on the transepithelial potential difference.[15,61,64,66] Using artificial solutions, Kokko first demonstrated by selective removal that glucose, amino acids, and the anionic species Cl^- and HCO_3^- contribute significantly to the generation of the lumen-negative potential difference.[64,66] Absence of glucose and alanine in the lumen, and replacement of HCO_3^- by Cl^- led to the reversal of the negative potential difference and established a lumen-positive potential of about +3 mV. Whereas glucose and alanine deletion shifted the lumen-negative potential toward zero, the lumen-positive potential difference was generated by and was proportional to the transepithelial Cl^- concentration difference ($Cl^-_{lumen} > Cl^-_{peritubular}$).

Burg *et al.* extended these studies by perfusing isolated rabbit proximal tubules with "early" proximal tubular solutions, simulating conditions in the very first proximal nephron segment *in vivo*.[15,65] When these tubules were perfused with isosmotic ultrafiltrate of serum, the transepithelial voltage was negative with reference to the peritubular bath. Studying the effects on the transepithelial potential difference and on fluid transport, these authors showed that removal of glucose, lactate, alanine, and citrate from the bath solution produced no change in either fluid reabsorption or electrical polarization (see Fig. 4). In sharp contrast, removal of these organic compounds from the *luminal* perfusion fluid led to a significant reduction of the electrical potential difference. Addition of glucose and alanine in millimolar concentrations to the perfusion fluid led to an increase in the lumen-negative electrical potential difference, and the magnitude of the potential increase caused by glucose and alanine could account for the transepithelial potential difference that is normally maintained across the isolated proximal convoluted tubule. An extensive analysis of the electrical events associated with proximal tubular sugar and amino acid transports has been presented by Frömter and his associates.[67–71]

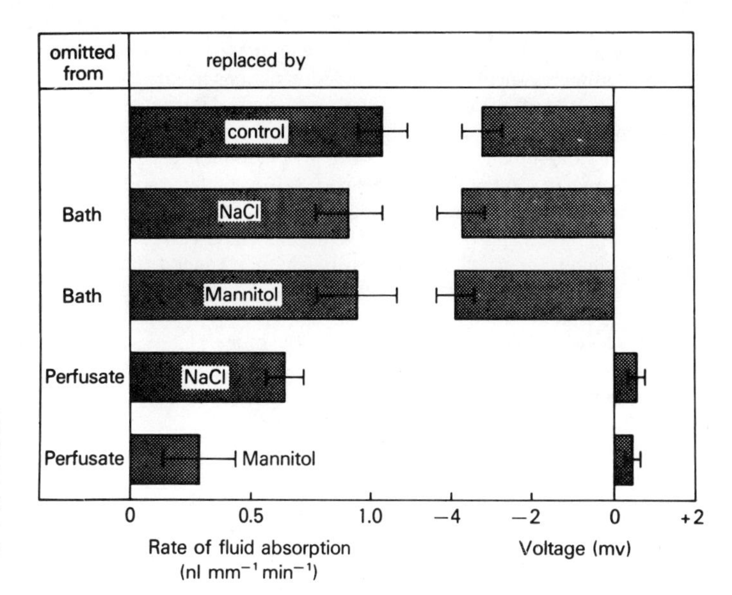

Fig. 4. Effect of removing organic solutes (glucose, lactate, alanine, and citrate) from the solutions bathing isolated perfused rabbit proximal tubules. During "control" the perfusate and bath contain organic material. It can be seen that removing all organic solutes from the bath had little effect, whereas removing them from the perfusate and replacing with either NaCl or mannitol resulted in a large decrease in fluid absorption and the magnitude of the transepithelial voltage. From Burg, ref. 15.

These studies have underscored the electrogenic effects of D-glucose and of several amino acids, and have defined the kinetic aspects of these Na^+-coupled transport events.

3.1.2. Passive Permeabilities

Both the convoluted proximal tubule and the straight segment have a high NaCl and water permeability.[1,2,13–17] However, there are differences in the relative permeability to these ions. The pars recta of superificial proximal tubules has a significantly greater Cl^- than Na^+ permeability,[44,72] whereas the reverse permeability relationship, such that the Na^+ permeability exceeds that of Cl^-, is found in the proximal convoluted tubule.[35] Based on work carried out on isolated convoluted and straight tubules, evidence began to emerge that there are also differences in the relative permeabilities to Na^+ and Cl^- between superficial cortical and deep, juxtamedullary proximal tubules.[35,44] Thus, it was observed that the high Cl^-/Na^+ permeability ratio ($P_{cl} > P_{Na}$) in *superficial* straight tubules is reversed in *juxtamedullary* straight tubules such that their Cl^- permeability, compared to that of Na^+, is low. Similar intrinsic differences in Cl^- permeability were also observed between proximal convoluted tubules, such that the Cl^- permeability was much lower in juxtamedullary tubules.[35] It has also been universally observed that in all segments of the proximal tubule, the Cl^- permeability is greater than that to HCO_3^-.[73,74] Consistent with the observation of relatively low Cl^- permeability in proximal tubules of *deep* nephrons is the observation that alterations of the Cl^- and HCO_3^- concentration of the fluid perfusing these tubule segments were relatively ineffective in altering the transepithelial electrical potential difference. In contrast, such changes in anion content do produce significant alterations in the transmembrane electrical potential differences in the corresponding superficial tubular segments.

These results on isolated tubules *in vitro* bear importantly on the *in vivo* situation since we have noted that both organic solutes (glucose, amino acids) and HCO_3^- fall rapidly along the early part of the proximal tubule. With the decline in luminal HCO_3^- concentration, Cl^- rises early along the proximal tubule and remains elevated throughout the remainder of this nephron segment, The changes in electrical potential difference along the intact proximal tubule *in vivo* are consistent with the known permeability properties and concentration differences. Thus, in *superficial* proximal convoluted tubules, a lumen-negative potential is generated early and reflects avid glucose and amino acid–Na^+ cotransport. With the fall in organic solutes and of HCO_3^- along the proximal convoluted tubule and the rise in Cl^-, the potential reverses and becomes lumen positive (see Figs. 3 and 5). In sharp contrast, in proximal tubules of *juxtamedullary* nephrons, the low Cl^- permeability renders the high Cl^- in the lumen ineffective; as a consequence, the lumen remains electrically negative along the whole proximal tubule.[56]

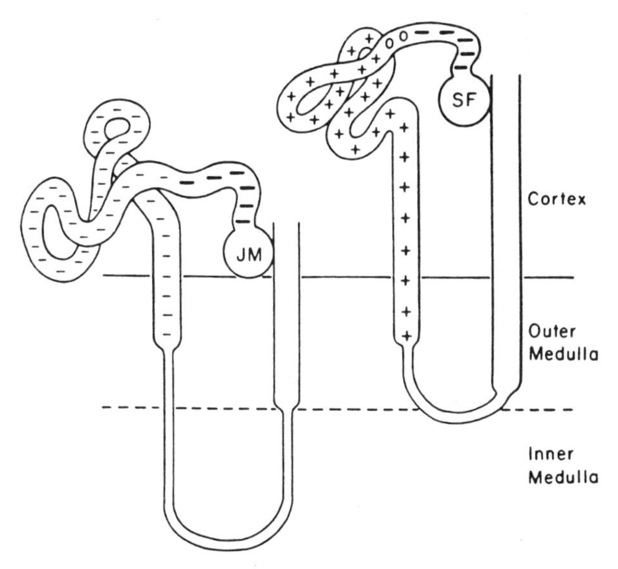

Fig. 5. Progression of transepithelial potential difference along the proximal tubule in superficial (cortical) and deep (juxtamedullary) nephrons. From Jacobson, ref. 38.

Berry and Berry and Rector have drawn attention to two additional aspects concerning the observed variations in ion selectivity.[37,75] They point out that the P_{Na}/P_{Cl} ratio determines the relationship between the electrical driving force and reabsorptive Na^+ movement. The lumen-negative potential, generated by an organic solute–Na^+ cotransport, acts on both passive Na^+ and passive Cl^- movement such that Na^+ ions diffuse back into the lumen and Cl^- ions out of the lumen. Clearly, in tubules in which P_{Na}/P_{Cl} is low, the rate of Cl^- movement from lumen to peritubular fluid exceeds the Na^+ movement in the opposite direction. On the other hand, a high P_{Na}/P_{Cl} ratio favors Na^+ recycling. Thus, the effectiveness of the electrical potential as a driving force for passive Cl^- transport depends on the selectivity pattern along the tubule and between nephron populations.

A second point concerns the relationship between P_{Na}/P_{Cl} and P_{HCO_3}/P_{Cl} and the magnitude of the lumen-positive potential that is generated in relatively Cl^--selective tubules (see above) by the diffusion potential generated by the transepithelial Cl^- gradient. It is clear that the magnitude of the lumen-positive potential will be affected by P_{Na}/P_{Cl} in such a way that with higher P_{Na}/P_{Cl} a lower lumen-positive potential is obtained. From the observation, reported by Jacobson,[56] that fluid absorption ceases in juxtamedullary proximal convoluted tubules in the absence of active transport despite the presence of a favorable Cl^- gradient, it can be concluded that the total conductance to passive ion movement in these tubules must be lower than in superficial proximal segments.

The observation that the transepithelial potential depends on the Cl^- concentration gradient is relevant to those instances in which the rise in Cl^- is small or absent. Most notably, in the *Necturus* kidney, in which HCO_3^- is reabsorbed with no change in concentration,[76,77] the Cl^- concentration fails to increase along the proximal tubule. In accord with the prediction that the lumen-positive potential difference is caused by a transepithelial Cl^- diffusion potential, the potential in *Necturus* remains negative along the proximal tubule.

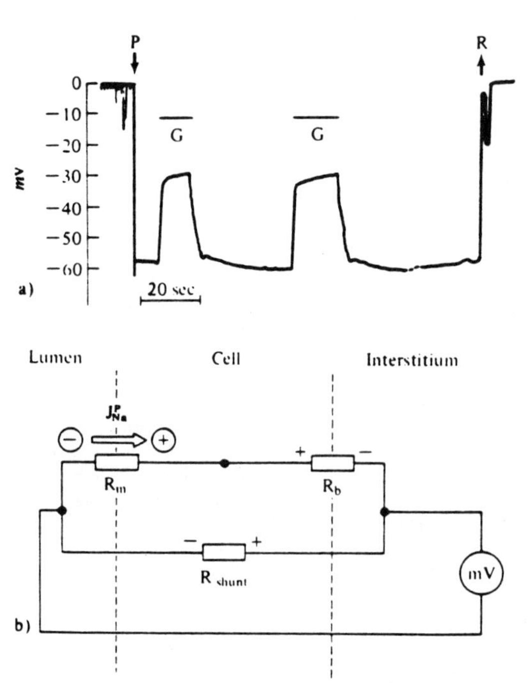

Fig. 6. (a) Electrical potential differences across the peritubular cell membrane of single proximal tubule cell of the rat kidney during luminal perfusion with a solution containing D-glucose. At P: impalement of tubule cell with microelectrode. During the time marked G: perfusion with a bicarbonate–Ringer's solution containing 5 mM of D-glucose. (b) Electrical equivalent circuit R_m, R_b, and R_{shunt} are electrical resistances of the luminal, basal (peritubular), and intercellular pathway. Upon luminal perfusion with glucose, the passive entrance of Na^+ into the cell (J^P_{Na}) increases sharply. Note that both cell membranes in connection with the shunt pathway form a closed circuit, and that current, after having passed through R_m, may also pass through R_b and R_{shunt}, thereby rendering the lumen electrically negative with respect to the interstitial fluid. From Ullrich.[6]

3.2. Cotransport of Na^+ with Sugars and Amino Acids

A large body of evidence supports the view that influx of Na^+ across the luminal brush border membrane via cotransport with other solutes, particularly sugars and amino acids, depolarizes the luminal cell membrane and thereby enhances the transepithelial negative potential difference.[6,67–71,78,79] For example, Fig. 6 presents results of a representative experiment and a circuit diagram indicating the sequence of electrical events. The glucose-coupled Na^+ translocation across the brush border occurs along a favorable electrical and chemical activity gradient for Na^+. This event depolarizes the luminal cell membrane, as shown by direct measurements of transmembrane potential differences of single amphibian[78,79] and mammalian[67,68] proximal tubule cells. The movement of Na^+ is downhill inasmuch as the activity of the Na^+ pump in the peritubular cell membrane maintains a low intracellular Na^+ concentration.

The process of Na^+-coupled entry of sugars across the luminal membrane shows considerable specificity with respect to both Na^+ and the steric requirement of individual sugars.[6] Replacement of Na^+ by other cations (Li^+, tetramethylammonium) abolishes the electrogenic response to sugars, and a large number of sugars (those lacking an OH group in the *d*-glucose configuration on C-2) are ineffective in inducing the depolarizing effect across the brush border membrane. The fact that the process of coupled Na^+–sugar transport and its electrical manifestations shows saturation kinetics, and can be inhibited by phlorizin, is further evidence in support of a carrier-mediated process in the brush border membrane.

Important aspects of the interaction between *d*-glucose and Na^+ movement have also been demonstrated in studies on closed vesicles of brush border membranes.[80,81] Figure 7 shows results from a relevant experiment.[80] Incubation in a sodium-free solution leads to a progressive influx of glucose until equilibrium is reached. At equilibrium, the internal and external concentrations of glucose are the same. If, however, Na^+ is added to the incubation medium, the influx of *d*-glucose is increased dramatically and the D-glucose concentration within the vesicles transiently reaches a concentration that is strikingly higher than the equilibrium concentration. The "overshoot" phenomenon of the intravesicular glucose is of particular interest, as it represents uphill glucose accumulation driven by the inward gradient of Na^+. It is caused by Na^+ moving from incubation medium into the vesicles and transporting D-glucose by a common carrier. Measurements utilizing voltage-sensitive fluorescent dyes in the vesicle preparation confirm that the Na^+–glucose movement renders the intravesicular space electrically

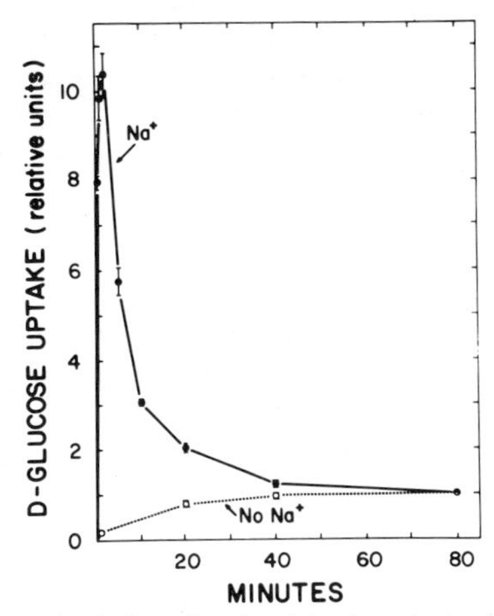

Fig. 7. Glucose uptake into closed vesicles from the brush border of the proximal renal tubule. Upper curve is from experiments in which glucose was added at a time when a large initial Na^+ gradient was present. Lower curve is representative of a situation in which the vesicles were studied in the absence of Na^+.

positive to the external medium.[82] Clearly, these data obtained *in vitro* on vesicular membranes support fully the Na^+ gradient hypothesis of glucose movement across the brush border of proximal tubule cells.

It is of interest that the transfer of D-glucose through the basolateral membrane of proximal tubule cells differs in several aspects from that across the luminal cell membrane.[6,81,83] D-Glucose uptake by basolateral membrane vesicles is almost independent of Na^+ and is inhibited by phloretin rather than by phlorizin. Although carrier-mediated, the glucose transfer system in the basolateral membranes also has sugar specificity that is different from its luminal counterpart.

There are significant differences in glucose transport between early and late segments of the proximal tubule. Studies of isolated perfused rabbit tubules demonstrate that active glucose reabsorption occurs by a low-affinity, high-capacity process in early proximal segments, and by a higher-affinity, lower-capacity process in later segments.[39] Studies of isolated brush border membranes reveal a low-affinity, high-capacity glucose transport system in vesicles prepared from outer cortex (early proximal tubule), and a high-affinity, low-capacity system in vesicles from outer medulla (late proximal tubule).[84] Interestingly, the Na^+–glucose cotransport stoichiometry is 1 : 1 for the early proximal tubular system and 2 : 1 for the late proximal tubular system.[85,86] The higher the Na^+–solute stoichiometry of a cotransport process, the steeper is the concentration gradient against which the Na^+-coupled absorption of the solute can proceed.[8] Thus, the high-affinity glucose transport system of the late proximal tubule is well suited for reclaiming the last traces of filtered glucose remaining after bulk reabsorption of glucose has occurred via the low-affinity, high-capacity system of the early proximal tubule.

Several proximal tubular amino acid transport systems share important properties with the Na^+-dependent D-glucose transfer operation. The transport path of several amino acids has been shown to be Na^+-dependent[6,7] and amino acid movement also occurs as electrogenic cotransport with Na^+.[69–71] These transport systems include one for neutral amino acids (phenylalanine, histidine), one for basic amino acids (lysine, ornithine, arginine), one for acidic amino acids (aspartic acid, glutamic acid), one for β-amino acids, and the iminoglycine system (proline, glycine).

In addition, electrophysiological studies of the type first utilized in studies on glucose transport have shown that Na^+-dependent amino acid transport is electrogenic.[69–71] In the rat kidney, the transport of all amino acids across the luminal cell membrane takes place largely via carrier systems associated with a flow of positive charge, and depends on cotransport with Na^+. In isolated brush border vesicles, the influx of amino acids into vesicles can be stimulated dramatically by imposing either a Na^+ or an appropriate electrical potential gradient across the vesicular membrane[87,88] Moreover, Na^+-coupled amino acid transport renders the intravesicular space electrically positive.[89] These data provide further evidence that secondary active transport of amino acids through the brush border operates through electrogenic cotransport with Na^+. Similar to the situation with respect to glucose, peritubular amino acid transport is largely Na^+-independent. However, in contrast to the situation with respect to glucose, there are Na^+ cotransport mechanisms for active uptake of certain amino acids across the basolateral membrane. For example, Na^+-coupled transport of the acidic amino acids L-glutamate and L-aspartate has been described in studies using basolateral membrane vesicles,[90] and in electrophysiological studies in which peritubular application of these amino acids resulted in cell depolarization.[71] Although neutral amino acids including glycine do not depolarize rat proximal convoluted tubules when applied peritubularly,[69] there is uphill, Na^+-coupled influx of glycine across the basolateral surface of rabbit proximal straight tubules.[91] There is also evidence for glutamine uptake across the basolateral membrane of dog proximal tubule cells.[92]

The Na^+-coupled transport of several substances other than sugars and amino acids is also electrogenic and associated with a flow of positive charge across the brush border and/or basolateral membrane of the proximal tubule cell. A system for Na^+ cotransport of monocarboxylic acid anions such as lactate is electrogenic[93] and is localized predominantly in the luminal membrane.[94] A separate system for Na^+ cotransport of dicarboxylic and tricarboxylic acid anions (e.g., succinate, citrate) is also electrogenic[95,96] but localized to both the luminal and the basolateral membranes.[97,98] Although Na^+-coupled phosphate transport across the luminal membrane is largely electroneutral, a small component is electrogenic[99] and associated with cell depolarization.[13]

The transepithelial electrical potential difference which renders the lumen negative is thus importantly dependent on the presence of active Na^+ transport and on cotransported solutes, particularly sugars and amino acids. The dependence of the luminal negativity upon active Na^+ transport across the proximal convoluted tubule is supported by observations that the potential can be abolished upon the substitution of Na^+ by Li^+, tetramethylammonium, or choline.[100] Also, the addition of ouabain to the bath of isolated perfused proximal convolutions abolishes the electrical potential difference.[44,66,100,101] The transepithelial potential difference is also sharply suppressed when K^+ in the bath is replaced by Na^+.[15] This observation places strong emphasis on active peritubular Na^+/K^+ exchange in the generation of the transepithelial potential difference. This view is further supported by the high temperature sensitivity of the lumen-negative potential difference in the presence of D-glucose or alanine.[102]

It is of interest that, in the mammalian proximal tubule,

only that fraction of Na^+ movement associated with D-glucose or amino acid transport generates a negative transepithelial electrical potential.[64,66,102] Although a very substantial fraction of active Na^+ transport continues after D-glucose and amino acid deletion, this remaining moiety of Na^+ transport is ineffective in rendering the lumen negative. In the straight part of the proximal tubule, cotransport of Na^+ with glucose and amino acids is not necessary to render the lumen negative.[44,101] The situation is similar in the amphibian proximal tubule. Here, despite the early disappearance of glucose from the lumen, the tubule remains electrically negative throughout its whole length.[57]

In sharp contrast, the lumen-positive potential, which is obtained in the absence of D-glucose and amino acids and in the presence of a high luminal Cl^- concentration, has only a small temperature coefficient, as expected for an electrical potential difference generated by passive ionic diffusion across a permselective membrane.[102]

3.3. The Intercellular Shunt Path

The ionic permeabilities of the proximal convoluted tubule have been explored extensively by means of measuring transepithelial potential changes after single ion substitutions and after biionic salt dilutions in the lumen.[35,72,103,104] It is clear from such studies that the overall transepithelial ionic selectivity of the proximal convoluted tubule resides mainly in its paracellular shunt pathway. In mammalian proximal convoluted tubules, the permeability to cations exceeds that to anions, and the ranking of ion permeabilities within the cation or anion series is quite similar to that in free solutions.[13–15] In sharp contrast, the cell membranes of proximal tubule cells are highly permselective, as demonstrated, for instance, by the much smaller permeability of proximal tubule cells to Na^+ than to K^+. It has already been pointed out that the transepithelial specific resistance of the proximal convoluted tubule is quite low. Calculations show that it cannot be accounted for by resistance values assigned to proximal tubule cells.[13–15] To reduce the resistance values to the low values observed, it is necessary to have a low-resistance path shunt the tubular cell layer.

Further supportive evidence for the presence of an important intercellular shunt pathway is the observation of extensive electrical coupling between luminal and peritubular cell membrane potentials.[14] This can best be accounted for by a low-resistance path between lumen and peritubular fluid and the generation of intraepithelial current loops. Finally, anatomical evidence is consistent with the shallow tight junctions and the interspaces between cells being the site of the extracellular shunt path.

An interesting expression of tubular heterogeneity is the observation of a progressive fall of the specific electrical resistance with tubular length along the proximal convoluted tubule.[105] It is virtually certain that this is due to altered resistance properties of the intercellular shunt path.

Two aspects of electrical current flow between proximal tubule cells are of particular functional significance. Both have to do with effective transmission of electrical potential changes between the luminal and the peritubular cell membranes.

Attention has been drawn recently to a feedback mechanism that is activated during stimulation of Na^+ transport and which involves electrical coupling between the luminal and the antiluminal cell membrane by current flow between epithelial cells both in the small intestine[106,107] and in the proximal tubule.[108,109] The addition of sugars or of amino acids leads to increased Na^+ entry from lumen into the cell. The current depolarizes the luminal cell membrane and renders the lumen more negative. If the paracellular shunt pathway were electrically tight, no net current could cross the luminal membrane, since no significant current can leave the tubular lumen without the same current reentering via some route. The potential difference across the luminal cell membrane would approach the new electromotive force, i.e., a value at which all currents directed into the cell across this membrane (e.g., substrate-coupled Na^+ entry) would be completely matched by currents leaving the cell into the lumen (e.g., K^+ exit). The peritubular membrane would at least temporarily maintain its potential.

Due to the leakiness of the tight junctions, however, net current may enter the cell, and leave the cell via the basolateral cell membrane (the cell, like the lumen, cannot accumulate significant charge). Current of the same polarity and magnitude must then reenter the lumen via the paracellular shunt. At each of the barriers, the luminal and basolateral cell membranes and at the shunt, the current must produce a voltage deflection, the magnitude of which depends on the resistance of that barrier. Due to the high conductance of the shunt in proximal convoluted tubules, the voltage deflection across the shunt (i.e., the transepithelial potential) is small (some −4 mV), whereas the voltage deflection across the cell membranes is large (some 40 mV).

In a second phase following glucose-induced Na^+ entry into proximal tubule cells, a sequence of events takes place that involves stimulation of basolateral Na^+ extrusion via the ATPase driven Na^+,K^+ exchange pump and, importantly, a significant increase of the K^+ conductance that reverses the *initial* potential change and results in repolarization of the antiluminal cell membrane. This initiates the generation of an intraepithelial current that tends to repolarize the apical cell membrane. Thus, the electrical driving force for Na^+-coupled transport is restored and maintained at optimal levels as long as apical Na^+ entry is stimulated.

A second aspect of intraepithelial current flow has to do with its role in coupling the electrical effects of transepithelial ionic concentration gradients between the individual barriers of the proximal tubular epithelium. For instance, it has been shown that ionic concentration gradients across the proximal tubule do not only generate *transepithelial* electrical potential differences by current flow across the shunt pathway but also may alter the polarization of the apical and basolateral cell membranes.[14] This is due to direct entry or exit of ions and generation of diffusion potentials across the membrane that is directly exposed to the altered ion concentration. Importantly, however, even the cell membranes not exposed to altered ion concentrations (e.g., the *luminal* cell membrane when ionic changes are limited to the *peritubular* cell membrane) may alter its potential difference. This effect is due to current carried by ions through the intercellular shunt path, and the closing of the current loop by flow of ions across the apical and basolateral cell membranes. The magnitude of these effects of transepithelial concentration changes upon potential changes across individual cell membranes depends not only on the ionic selectivity of the shunt pathway and exposed cell membranes but also on the relationship between the resistances of the shunt pathway and that of the individual cell barriers (see above).[14]

For instance, a transepithelial K^+ gradient not only affects the electrical potential difference across the basolateral cell membrane *directly* in view of the significant permselectivity of this membrane to K^+, but also *indirectly* by induction of current flow across both cell membranes and the paracellular shunt path.

This flow of current (positive charge transfer) tends to depolarize the luminal and to hyperpolarize the peritubular cell membrane. Accordingly, the depolarizing effect on the peritubular cell membrane of the step increase of K^+ is *opposed* by the voltage drop across the peritubular membrane that is induced by the transepithelial current flow through the shunt pathway. Two strategies have been used to correct for these secondary, shunt-mediated voltage changes that tend to alter the potential difference across the membrane where the change in ion concentration takes place. First, it is possible from circuit analysis (see Fig. 6) to correct for the voltage displacement, provided one has first measured the requisite cellular and paracellular resistance values.(14) Second, one may reduce the circular current flow by blocking ion-conductive pathways in the apical cell membrane by perfusing the lumen with appropriate channel-blocking agents. This approach has recently been used by Burckhardt and Frömter in a study on the conductive properties of the basolateral membrane of the rat proximal tubule.(110,111) By using luminal Ba^{2+}, current flow through luminal K^+ channels (see below) was minimized. The deletion of sugars and amino acids from the lumen is another maneuver that effectively blocks the currents that will flow through the electrogenic Na^+–glucose or Na^+–amino acid system.

The precise origin of the transepithelial potential difference has been discussed in several reviews.(13–15) When transepithelial ionic concentration differences exist, such as for Cl^- and HCO_3^- in the middle and late parts of the mammalian proximal tubule, diffusion potentials render the lumen positive as a result of a higher Cl^- than HCO_3^- permeability.

Essential to an analysis of the origin of the transepithelial potential difference is the point that ionic currents either originate from passive ion movement down an electrochemical potential gradient or, alternatively, are generated directly by active ion movement (electrogenic or rheogenic ion movement).* Boulpaep has analyzed various alternatives and suggested an equivalent electrical circuit model as shown in Fig. 8.(14) With respect to the *peritubular* electromotive force, parallel diffusion potentials for K^+, Cl^-, Na^+, and HCO_3^- have been demonstrated and their magnitude may account for a transmembrane potential difference of some −70 mV (cell negative). This value is close to the normal peritubular potential difference.

Similar considerations with respect to the luminal cell membrane have shown the presence of finite K^+ and Na^+ conductances.(13) However, there is presently not enough information to account fully for the electrical potential across the luminal cell membrane on the basis of these two opposing diffusion potentials alone. Hence, it is possible that in addition, rheogenic active ion movement of either cations from cell to lumen or of anions from lumen to cell takes place. An additional possibility is that an external current source may generate a potential drop. The existence of a low-resistance paracellular path makes such current flow likely. It could originate in the peritubular membrane but the details of such a mechanism have not yet been elucidated.

*Ion pumps may generate electrical potential differences across renal cell membranes in at least two ways. A neutral ion-exchange pump, for instance, in which Na^+ and K^+ movements are rigidly linked to each other, does not directly generate a transmembrane potential difference. Rather, it does so by the establishment of a high transmembrane difference, i.e., of K^+, across the peritubular membrane, which has a high K^+ permeability. Alternatively, if Na^+ were selectively extruded across the peritubular cell membranes, a charge separation and an electrical potential difference would be directly generated by such "electrogenic" or "rheogenic" pump activity. Under this condition, K^+ uptake into kidney cells would be passive and there would be no fixed stoichiometric coupling between Na^+ and K^+ movement since, depending on other ion species and their respective mobilities, the potential difference would be partly shunted.

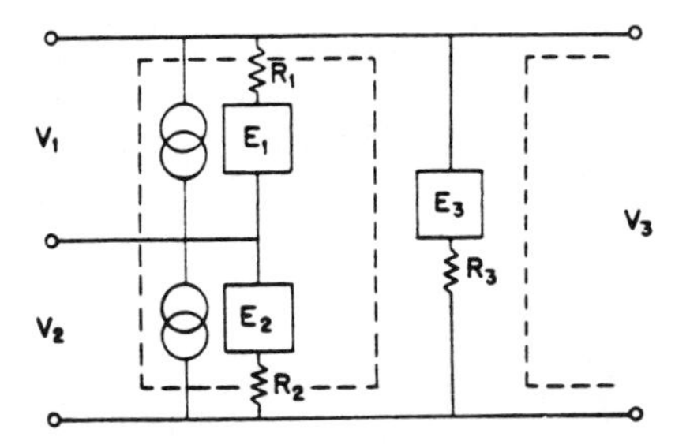

Fig. 8. Electrical equivalent circuit for proximal tubule cell. Dashed line indicates cell border. V_1, potential difference across peritubular cell membrane; E_1, electromotive force of the peritubular membrane which is equivalent to the combined electromotive force of all diffusional pathways; R_1, resistance of the peritubular (and basolateral) membrane of the diffusional pathway. Same elements for the luminal membrane with subscript 2. V_3, transepithelial potential difference; E_3, paracellular electromotive force resulting from dissipative ion leaks; R_3, paracellular resistance. The upper open circles represent a constant-current source designating an ion current driven by one or several rheogenic ion pumps. The lower open circles represent a similar, but not necessarily identical, constant-current source driven by one or more rheogenic pumps. From Boulpaep, ref. 14.

3.4. Proximal Straight Tubule

The transepithelial electrical potential difference across isolated superficial straight tubules is slightly negative by −1.3 to −2.1 mV when the lumen is perfused with symmetrical bicarbonate–Ringer's containing organic solutes.(44,101,102,112,113) As mentioned earlier, the presence of glucose and amino acids is not necessary to generate the lumen-negative potential difference as long as Na^+ is present. Cooling or ouabain treatment abolishes the potential difference. Active Na^+ transport can be shown to be responsible both for the electrical potential difference as well as for fluid reabsorption. When HCO_3^- is replaced by Cl^- in the perfusion solution, the lumen becomes slightly positive.(44,101,102,112) This lumen-positive voltage is the result of the Cl^- and HCO_3^- concentration differences across the straight proximal tubular segment. It can be shown that the Cl^- permeability exceeds that of HCO_3^-, so that its transepithelial potential difference ($Cl^-_{lumen} > Cl^-_{peritubular}$) generates a lumen-positive potential difference.(44,101)

4. NaCl and $NaHCO_3$ Transport

4.1. Evidence for the Active Nature of Transepithelial Na^+ Transport

Available evidence strongly supports the view that both the convoluted and the straight segment of the proximal tubule transport Na^+ actively. This view was first expressed by Wesson and Anslow(114) on the basis of indirect evidence. Minimizing the role of the distal tubule in modifying the composition of tubular fluid during maximal osmotic diuresis with mannitol,

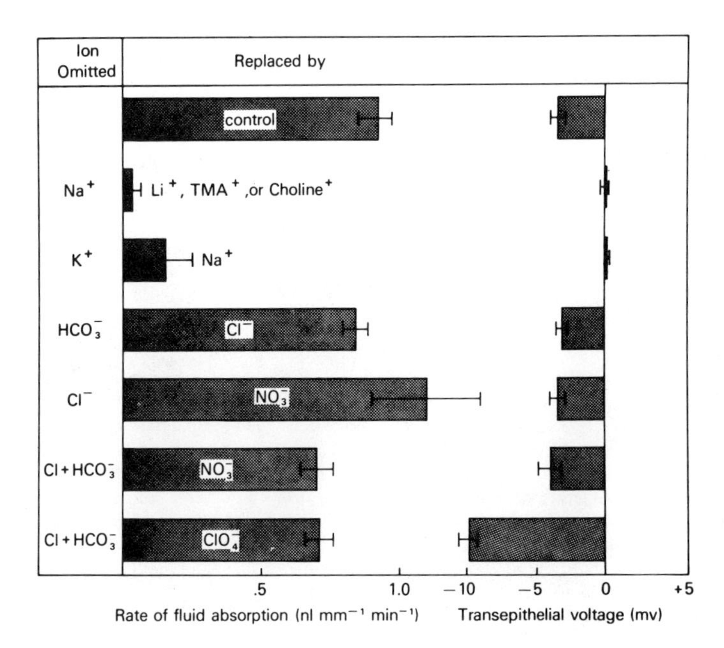

Fig. 9. Effect of ion substitutions on rate of fluid movement and the transepithelial voltage in isolated rabbit proximal convoluted tubules which were perfused with an ultrafiltratelike solution and bathed in a serumlike solution. The ion species to be investigated was removed entirely from the perfusate and bath and replaced isosmotically with other ions as indicated. (Only K^+ was removed from the bath: its removal from the lumen was without effect.) It is apparent that omission of the cations Na^+ and K^+ had a great inhibitory effect. Removal of HCO_3^- caused a smaller inhibition; removal of Cl^- had virtually no effect. TMA^+, tetramethylammonium. From Burg, ref. 15.

these authors correctly interpreted the relatively low urinary Na^+ concentrations to originate in the proximal tubule, and concluded that this nephron segment has the ability to pump Na^+ actively against a sizable concentration gradient. Several lines of direct evidence may be cited which corroborate this thesis.

First, isosmotic reabsorption of fluid in the proximal tubule depends predominantly on Na^+. Replacement of Na^+ by cations such as Li^+, tetraethylammonium, or choline stops fluid absorption completely.[15,100,115] Figure 9 summarizes the effects of various ionic substitution upon the rate of fluid reabsorption and the transepithelial voltage difference in isolated proximal convoluted tubules of rabbit kidneys *in vitro.*[15] It is apparent that both net fluid absorption and the transepithelial potential difference are highly sensitive to removal of Na^+. Also, the substitution of K^+ in the peritubular bath by Na^+ abolishes net fluid transfer, a finding that is consistent with a key role of K^+ in the operation of active Na^+ transport across the antiluminal cell membrane. Other epithelia share this property.[116,117]

In contrast to the importance of cations such as Na^+ and K^+ in the maintenance of normal rates of proximal tubular fluid transport, the effect of various anionic substitutions is far less dramatic. With the notable exception of HCO_3^-, substitution of which by Cl^- effects a moderate reduction of net fluid transport in the isolated tubule preparation, the omission of Cl^- and its replacement by several other anions has no significant inhibitory effect[15,100] on transepithelial fluid movement. Essentially similar observations have been made in the rat kidney in which tubules and peritubular capillaries were separately perfused with solutions of varying ionic composition.[115] The only difference between the results of *in vivo* and *in vitro* perfusions is that different fractions of Na^+ and fluid transport are subject to control by peritubular HCO_3^- omission: In the rat kidney, as much as 50–65% of Na^+ transport could be suppressed by peritubular HCO_3^- removal,[115,118] whereas only some 30% of fluid transport was suppressed by HCO_3^- omission in the *in vitro* perfusions.[15,100] We may conclude that Na^+, K^+, and HCO_3^- are the main ions needed for the optimal operation of fluid transport across the convoluted proximal tubule. In contrast to the situation in the proximal *convoluted* tubule, Na^+ and fluid transport across the ***straight*** segment of the proximal tubule can be shown to be insensitive to the deletion of HCO_3^-.[101] These ionic requirements for Na^+ and fluid transport are discussed in detail later.

A second line of evidence supporting the active nature of proximal tubular Na^+ transport is based on the ability of the proximal tubular epithelium to reabsorb Na^+ against a sizable electrochemical potential gradient.[119–122] Figure 10 shows the results of a series of stationary perfusion experiments in which net fluid (and Na^+) movement was studied at different luminal Na^+ concentrations. Isotonicity was achieved in the different perfusion solutions by the addition of the poorly reabsorbable nonelectrolyte raffinose. It is apparent that despite the restriction of fluid movement from the tubular lumen by raffinose and the ensuing dilution of Na^+ concentration, fluid reabsorption continues until a Na^+ concentration some 40 meq/liter below that of plasma was reached. Such Na^+ transport clearly proceeds against an electrochemical potential gradient, and is consistent with the usual definition of active transport.

A final argument in favor of the active nature of proximal Na^+ transport is the finding of complete inhibition of Na^+ and fluid transport after cooling or exposure of the tubular epithelium to a variety of active transport inhibitors.[15] A representative example is shown in Fig. 11 which summarizes the effects of ouabain on fluid absorption and the lumen-negative electrical potential difference in isolated proximal tubules of the rabbit kidney.[15] Ouabain, a cardiac steroid known to inhibit Na^+, K^+-ATPase and active K^+/Na^+ exchange in a large number of tissues, including epithelia,[116,117] decreased fluid absorption and electrical polarization to almost zero. Similar results have been made in the rat kidney, although this preparation is inherently less sensitive to the inhibitory effect of cardiac steroids.[123] Complete transport inhibition also takes place when proximal tubules and peritubular capillaries are both perfused with artifical, cyanide-containing solutions.[124] These

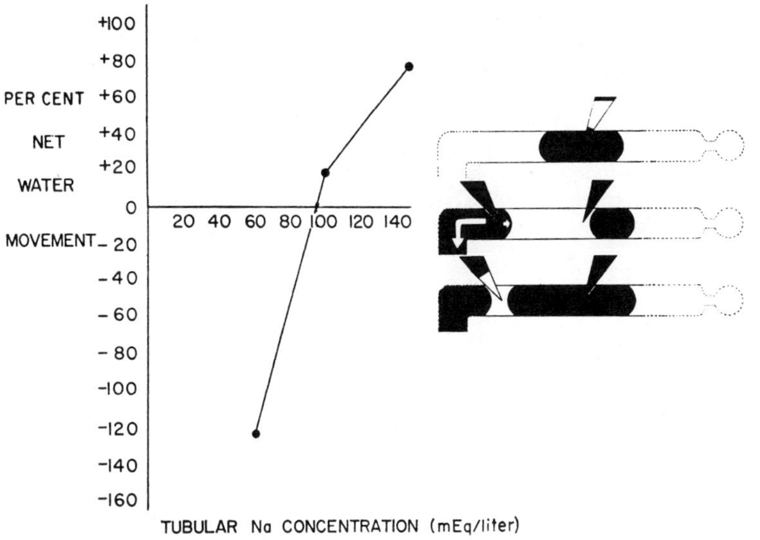

Fig. 10. Data from stationary microperfusion experiments on rat proximal tubules, showing fractional net water movement as a function of intratubular Na^+ concentration. Schematic drawing on right shows method of isolating fluid column and collection of perfusate. A column of mineral oil is injected into a suitable proximal tubule and split by the injection of perfusate fluid. Collection of fluid is accomplished by injection of oil through the proximally located pipette and simultaneous aspiration of the perfusate through the distally located micropipette. Perfusion fluids of varying Na^+ content were kept isotonic by the addition of raffinose. Plus sign indicates fluid reabsorption; minus sign indicates fluid entry into the tubule. From Giebisch and Windhager, ref. 120.

observations also support the active nature of proximal tubular Na^+ transport.

4.2. Cell Model of Proximal Na^+ Transport

The proximal tubule cell shares with other transporting epithelia the asymmetrical distribution of active and passive transport components. Thus, it is based on the original transport model prepared by Koefoed-Johnsen and Ussing.[125]

Several rate-limiting steps are involved during the translocation of Na^+ across the proximal tubular epithelium. Taking into account the direction of net Na^+ transport and of the electrical and chemical potential profile across the epithelium, a simple cell model, shown in Fig. 12, has evolved,[126] The cell compartment of proximal tubule cells has both a lower Na^+ concentration and an electrical potential more negative than either the luminal or the peritubular fluid compartment. Hence, the entry of Na^+ into the cell from the lumen takes place along an electrical and chemical potential gradient. In sharp contrast, transport of Na^+ out of the cell into the peritubular fluid is uphill with respect to both chemical and electrical potential gradients, and thus imparts an active, energy-consuming nature to overall Na^+ transport.

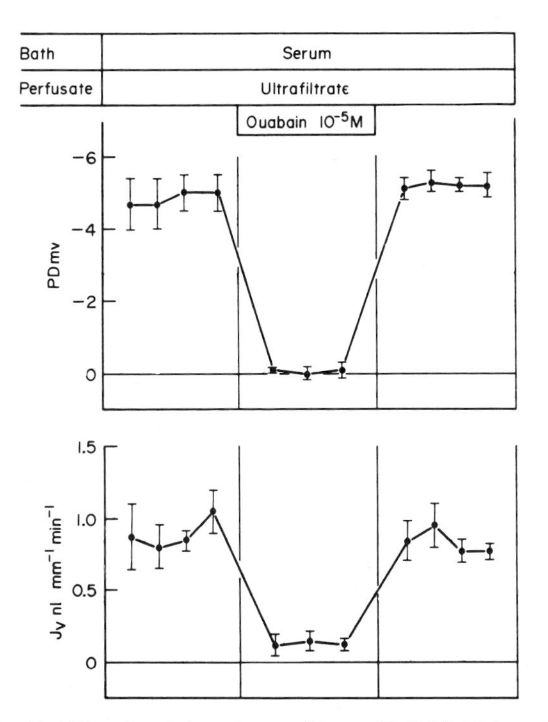

Fig. 11. Effect of ouabain on the rate of transepithelial fluid absorption (J_v) and transepithelial voltage (PD) across isolated perfused rabbit proximal convoluted tubules. Note complete inhibition of PD and J_v and reversibility of the ouabain effect. From Burg, ref. 15.

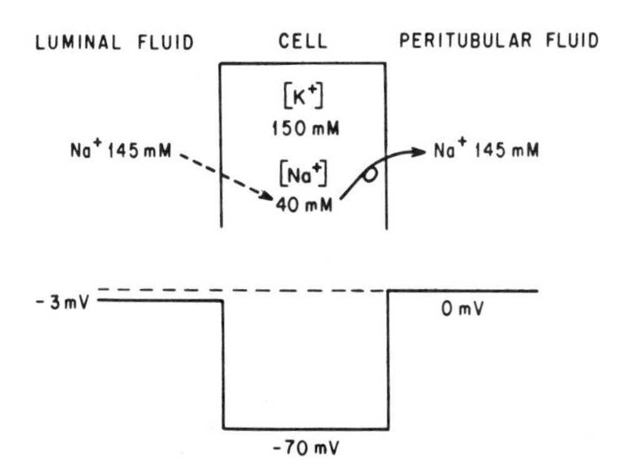

Fig. 12. Chemical concentration and electrical-potential profile across the early mammalian proximal convoluted tubule. Na^+ enters the cell along a combined electrochemical potential gradient and is pumped out against an electrochemical potential gradient across the peritubular cell boundary. From Giebisch and Windhager, ref. 120.

4.3. Na^+ Entry across the Luminal Cell Membrane

It is well known that several mechanisms involving co- and countertransport of Na^+ with organic and inorganic solutes participate in the initial translocation step from the lumen into the cell. All of these mechanisms are energetically dependent upon the electrochemical Na^+ gradient across the apical cell membrane. In the following we shall first consider the overall characteristics of transepithelial Na^+ reabsorption across the luminal cell membrane.

4.3.1. Luminal Membrane as Rate-Limiting Transport Step

Several lines of evidence based on experiments in nonrenal epithelia indicate important interactions between transport events in the apical and basolateral cell membranes.[127,128] An important feature of these mechanisms is that they protect epithelial cells from being exposed to an overload of Na^+ and that they coordinate Na^+ entry into the cell to active Na^+ extrusion across the basolateral cell membrane. It follows that *apical* Na^+ entry be subject to regulation and respond to stimuli that are somehow related to the Na^+ activity of the transporting cell. One would thus expect that the luminal cell membrane also be a site that controls Na^+ transport.

In both the amphibian and the mammalian nephron, available evidence is compatible with the view that some rate-limiting interaction takes place between Na^+ and the luminal cell membrane during the transfer of Na^+ across this membrane barrier. Several lines of evidence support this view. First, it can be shown in the proximal tubule of *Necturus* that exposure of the lumen of the proximal tubular epithelium to amphotericin B increases sharply net Na^+ reabsorption.[129,130] Amphotericin B is a polyene antibiotic which increases the cation permeability of several epithelia,[131–134] including that of the renal tubule. The most reasonable explanation is that exposure of the luminal surface of proximal tubule cells to amphotericin B increases its permeability to Na^+, and thus accelerates Na^+ entry into the cell. This in turn stimulates active extrusion of Na^+ across the peritubular membrane and results in enhancement of overall transepithelial Na^+ transport. These results indicate that the luminal cell membrane normally presents a rate-limiting barrier to overall transepithelial Na^+ movement.

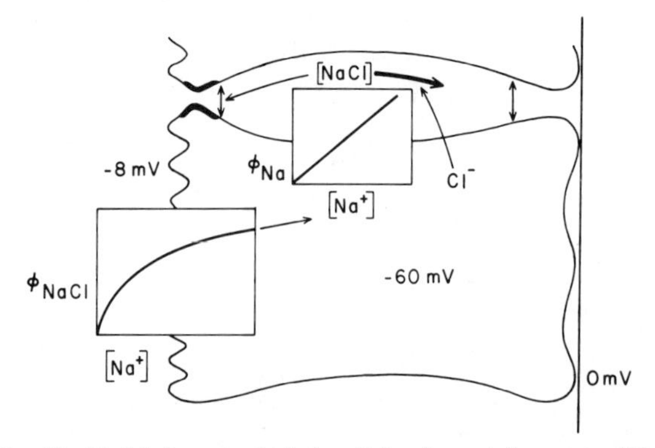

Fig. 13. Model of proximal tubule cell showing certain aspects of Na^+ movement. Note rate-limiting and saturable nature of the translocation process across the luminal cell membrane but the unsaturated nature of the peritubular basolateral Na^+ pump. Based on data from Spring and Giebisch, ref. 130.

Some kinetic aspects of proximal tubular Na^+ transport in the doubly perfused *Necturus* kidney also support the notion of a rate-limiting role of the luminal cell membrane.[130] This preparation allows extensive changes to be made in the extracellular fluid composition. When the Na^+ concentration in the *extracellular* fluid is raised over a wide range from a few milliequivalents to normal plasma levels, net Na^+ transport does not increase linearly but in a curvilinear fashion, approaching maximal transport rates at extracellular Na^+ concentrations slightly above normal plasma levels. In sharp contrast, the rate of net Na^+ transport is linearly related to the *intracellular* Na^+ concentration and does not saturate over the same range of extracellular Na^+ concentrations. Accordingly, the rate limitation of transepithelial Na^+ movement is imposed not by the peritubular active Na^+ extrusion step but by the entry of Na^+ from the tubular lumen into the cells of the proximal tubule.

The effect of amphotericin B is again of interest. When added to the aortal perfusion fluid of the perfused kidney preparation, it increases both intracellular Na^+ concentrations and net tubular Na^+ transport rates. Again, this finding is consistent with a major site of action of amphotericin B at the luminal cell membrane. In contrast to the effects of amphotericin B, elevation of the luminal Ca^{2+} concentration inhibits Na^+ entry from the lumen into tubule cells.[130]

Figure 13 depicts a cell model in which the different behavior of luminal and peritubular cell membranes is shown with respect to their response to Na^+ concentration changes. Clearly, the luminal membrane of tubule cells emerges as a potentially important control site of proximal tubular Na^+ reabsorption. It is of interest that the rate-limiting properties of the entry step of Na^+ have also been demonstrated in a variety of other Na^+-transporting epithelia.[116,117]

4.3.2. Na^+-Coupled Transport Operations in the Luminal Cell Membrane

It has already been pointed out that Na^+ plays a key role in the movement of several organic and inorganic solutes from lumen into the tubule cell. Of these, the cotransports of Na^+ with sugars and amino acids, with mono-, di-, and tricarboxylic acids, and with PO_4^{3-} and SO_4^{2-} have been best defined.[7] There is also a large body of evidence demonstrating that a significant fraction of filtered Na^+ exchanges for H^+, a process that partakes in tubular acidification and HCO_3^- reabsorption.[5]

In addition to these well-defined Na^+ transport pathways, two further Na^+ entry mechanisms deserve mention. These concern the possibility of coupled NaCl translocation along a favorable gradient across the luminal cell membrane, and finally, the movement of Na^+ by electrodiffusion. We should state at the outset that there is no consensus on the functional role of these two modes of Na^+ movement in overall reabsorption across the proximal tubule.

Concerning evidence favoring electroneutral entry of NaCl, Spring and Kimura have shown a significant fall of intracellular Cl^- activity in proximal tubule cells of *Necturus* following the drastic reduction of either Na^+ or Cl^- in the lumen of perfused proximal tubules.[135] Together with the observation that electrical polarization had little effect on cell Cl^-, this set of results was taken to support an electrically neutral NaCl cotransport mechanism. Energetically, such a mode of transport is feasible since the concentrations of both Na^+ and Cl^- exceed those in the cytoplasm of proximal tubule cells.[135] Studies utilizing isolated luminal membrane vesicles from

Necturus kidney failed to demonstrate directly coupled cotransport of Na^+ and Cl^-.[136] However, Na^+/H^+ and Cl^-/HCO_3^- exchange processes were identified in *Necturus* renal microvillus membranes.[136] It was suggested that the electroneutral, Na^+-coupled Cl^- transport observed by Spring and Kimura may actually have represented the parallel operation of Na^+/H^+ and Cl^-/HCO_3^- exchangers. According to the parallel exchanger model of Cl^- transport, it is the cell-to-lumen electrochemical gradient for HCO_3^- that is the immediate driving force for Cl^- uptake across the luminal membrane. Because this cell-to-lumen electrochemical gradient for HCO_3^- results from uphill H^+ extrusion mediated by Na^+/H^+ exchange (see below), Cl^- uptake via this mechanism is indirectly Na^+-coupled.

In mammalian proximal tubules, evidence for participation of some electrically neutral NaCl transport is also available. It is based on several observations. First, furosemide, a diuretic agent known to inhibit several types of NaCl cotransport processes, inhibits a significant fraction of proximal tubular fluid transport.[137] Second, SITS, an inhibitor of anion exchange, also has been shown to block proximal tubular Cl^- transport, particularly when the lumen is perfused with high-Cl^-, low-HCO_3^- solutions, mimicking late proximal tubular conditions.[137–139] Both of these inhibitor studies have to be interpreted with some caution, since they do not allow a precise definition of the transport operations involved. Na^+–Cl^- cotransport,[140,141] anion-exchange processes (Cl^-/HCO_3^-, Cl^-/OH^-),[136] and anion conductances[142] all can be inhibited by furosemide; SITS has been shown to inhibit several modes of anion exchange[136,143] as well as anion conductances.[110,111,142]

A third line of evidence supporting the view of electroneutral NaCl transport is based on a careful comparison of net fluid and salt transport and electrochemical driving forces.[144] When proximal convoluted tubules of the rabbit are perfused with high-Cl^- solutions, and HCO_3^- in the bath is also replaced by Cl^-—thus eliminating transepithelial Cl^- gradients—fluid and NaCl transport continue and yet *no* electrical potential difference across the epithelium is measured. These experiments confirm previous perfusion experiments of proximal tubules and peritubular capillaries in the rat in which it was shown that NaCl transport continues when the epithelium is exposed to symmetrical NaCl solutions.[145] Since it has now been shown in the rabbit that such salt transport takes place in the absence of a lumen-negative potential, it is concluded that Na^+ and Cl^- are not electrically coupled.[144]

This situation is also similar to that during extensive proximal HCO_3^- transport inhibition by carbonic anhydrase inhibitors. With HCO_3^- reabsorption suppressed, Cl^- fails to rise along the proximal convoluted tubule and the free-flow lumen-positive potential is abolished.[147] Under such conditions, Cl^- reabsorption continues at a highly significant rate,[146] again suggesting the possibility of transepithelial NaCl movement by an electrically neutral coupled mechanism.

As already discussed for the case of the *Necturus* proximal tubule, it has been suggested that net NaCl transport in the mammalian proximal tubule could also be the result of the parallel activity of two countertransport systems, one exchanging Na^+ for H^+ and the other exchanging Cl^- for HCO_3^- or OH^-.[148] However, studies on isolated brush border vesicles from mammalian tubules have not uniformly supported such a dual exchange mechanism,[142,149,150] and neither have perfusion studies on rabbit proximal convoluted tubules.[151,152] It is possible that different modes of electrically neutral NaCl transport are present in the amphibian and mammalian proximal convoluted tubule.

Although studies on isolated microvillus membrane vesicles from mammalian kidneys have failed to demonstrate direct Na^+–Cl^- cotransport or Cl^-/HCO_3^- exchange,[142] Cl^- transport in exchange for certain anions other than HCO_3^- has recently been observed in renal brush border membranes.[153] In particular, appreciable rates of exchange of Cl^- with formate were found. For exchange of intratubular Cl^- with intracellular formate to be a major mechanism of Cl^- absorption, there must exist mechanisms for recycling formate across the luminal membrane back into the cell. Otherwise, the rate of Cl^- absorption by anion exchange could not exceed the negligibly small or nonexistent net rates of secretion of anions such as formate. Formate transport across biological membranes probably occurs principally via nonionic diffusion.[154] Because luminal acidification in the mammalian proximal tubule largely results from Na^+/H^+ exchange, as will be discussed in more detail, formate reabsorption by nonionic diffusion may be regarded as an indirectly Na^+-coupled process. Thus, it is possible that formate/Cl^- exchange in parallel with Na^+-coupled formate absorption is a mechanism for Na^+-coupled Cl^- transport across the luminal membrane of the mammalian proximal tubule cell.[153] Whether significant Cl^- reabsorption occurs via this mechanism under physiological conditions is not known.

The studies cited above are of interest since they also provide some clue to the *transport route* of Cl^- across the proximal convoluted tubule. The cited observation that reabsorptive Cl^- flux continues in the absence of significant electrical or chemical driving forces (see above) suggests the presence of electroneutral Cl^- transport, be it by directly or indirectly Na^+-coupled processes. This fraction of NaCl movement must traverse the epithelium via a *transcellular* route since there are no driving forces, either chemical or electrical, that could account for passive Cl^- transport via the intercellular shunt path. Thus, at least under conditions that minimize the establishment of adequate transtubular electrochemical driving forces, Cl^- must enter the cell. Moreover, Cl^- must enter the cell by an active (presumably secondary active) process inasmuch as the intracellular Cl^- activity is significantly higher than the value at which Cl^- would be in electrochemical equilibrium across the luminal membrane.[155] It is presently not known to what extent such a mechanism of cellular NaCl transport is operative under physiological free-flow conditions in which the early proximal tubule negativity and the rising luminal Cl^- concentration provide additional and/or alternative driving forces for proximal NaCl transport. For example, in one study of the rat proximal convoluted tubule, it was estimated that less than 10% of Cl^- reabsorption occurs via the transcellular route.[155]

The precise "mix" of different NaCl transport operations along the proximal *straight* tubule differs somewhat from that in the proximal *convoluted* tubule. Under physiological conditions, at least in *cortical* nephron populations, the low-HCO_3^-–high-Cl^- concentration profile provides a favorable passive driving force for NaCl movement.[156] If exposed to symmetrical NaCl solutions, to which only small amounts of phosphate buffers, Ca^{2+}, and K^+ have been added, NaCl reabsorption continues.[43] In sharp contrast though, such salt movement is associated with a lumen-negative potential,[43] and this could provide electrical coupling between active Na^+ and passive Cl^- transport. Whether electrically neutral NaCl is also present in the straight portion of the proximal tubule of *cortical* nephrons is presently unknown. Such a possibility has to be seriously considered for juxtamedullary proximal tubules in which the anion permeability is low and in which passively driven salt movement will hence be less significant.[56]

Finally, one also has to consider the possibility that Na^+ could diffuse passively from the lumen into the proximal tubule cells. Despite the fact that both the chemical and the electrical potential gradient are favorable for such a translocation mode, it is doubtful whether the diffusion permeability for Na^+ across the luminal membrane is large enough to account for a sizable fraction of Na^+ to enter the cell by electrodiffusion.

Evidence favoring a small but finite Na^+ conductance in the luminal cell membrane of proximal tubule cells is derived from electrophysiological measurements which have demonstrated a small hyperpolarization of the luminal membrane following deletion of Na^+ from the lumen.[108] Studies on mammalian brush border membranes also indicate a Na^+ conductive pathway.[157]

At least three arguments can be raised against Na^+ diffusion being an important mechanism for translocation from lumen to cell. First, it was observed on the basis of electrophysiological measurements on rat proximal convoluted tubules that intraepithelial current flow could be effectively blocked by luminal deletion of cotransported solutes and of HCO_3^- (Na^+/H^+ exchange inhibition) and by luminal Ba^{2+}.[110,111] Accordingly, no evidence for an *additional* conductive pathway was obtained. Second, the insensitivity of proximal Na^+ transport to amiloride in doses that inhibit Na^+ channels in tight epithelia supports the view that "distal-type" diffusive Na^+ movement is not important.[158] A third argument has to do with the dependence of ouabain effects upon luminal Na^+ transport.[159] When Na^+ reabsorption is supported by appropriate luminal substrates (glucose, amino acids, HCO_3^-), administration of ouabain to the perfusion fluid of perfused frog kidneys and inhibition of active Na^+/K^+ exchange lead to rapid electrical depolarization as K^+ leaves and Na^+ enters the cell. If co- and countertransport operations involving Na^+ are inhibited, ouabain can be shown to be ineffective in depolarizing proximal tubule cells. The logical interpretation of this protective effect is the effective curtailment of Na^+ entry by inhibition of luminal carrier-mediated translocation processes. Had a significant diffusion pathway existed, Na^+ could have entered across the luminal cell membrane, K^+ loss would have occurred, and ouabain would have continued to exert its depolarizing effect.

In conclusion, several distinct transport pathways for Na^+ have been identified in proximal tubule cells. Whereas co- and countertransport processes involving organic solutes, PO_4^{3-} and SO_4^{2-}, as well as H^+ have been well defined, there remains a potentially significant fraction of NaCl that may be reabsorbed by an electrically neutral mode of operation that involves a transcellular route. Its nature and physiological significance need further investigation.

4.4. Peritubular Na^+ Extrusion

4.4.1. Na^+,K^+-ATPase-Driven Na^+ Extrusion

Na^+ extrusion across the peritubular and basolateral cell membranes of proximal tubule cells is active in nature, and involves exchange with extracellular K^+.

The properties of the active ATPase-driven cation exchange have been extensively investigated.[160] It is now firmly established that decomposition of 1 molecule of ATP is associated with outward pumping of 3 Na^+ for the simultaneous inward movement of 2 K^+. This relationship of cation exchange directly generates current flow across the basolateral membrane, and a component of the electrical potential difference has been attributed to the directly "electrogenic" nature of Na^+,K^+-ATPase activity.

Several lines of evidence support the view of the electrogenic nature of basolateral Na^+ extrusion. All of these are based on the observation that the cell potential can exceed the potential that can maximally be generated by the known ionic conductances and electromotive forces. Particularly useful have been several preparations in which Na^+ has been allowed to rise by pump inhibition. Monitoring both cell potential and cell K^+ levels during the period following sudden pump stimulation, one can unequivocally demonstrate the potential difference to exceed the level of the K^+ equilibrium potential. Figure 14 provides a summary of relevant data for the perfused rabbit proximal tubule.[161] Similar results have also been obtained in the guinea pig kidney slice[162] and the amphibian tubule preparation of *Ambystoma*.[163] The observation that the potential "overshoot" following pump stimulation is ouabain-sensitive is consistent with the dependence of a fraction of the basolateral potential upon ATPase-driven Na^+/K^+ exchange. Although the 3 Na^+ : 2 K^+ stoichiometry of the ATPase in the kidney is based on several lines of evidence,[160] some observations have suggested a variable ratio of Na^+/K^+ exchange.[164] This topic needs further studies.

Three factors are of importance for the rate of peritubular Na^+ extrusion by ATPase activity[160]: the level of ATP, intracellular Na^+, and peritubular K^+. The effects of these variables upon Na^+ transport or, alternatively, upon the electrical correlates of Na^+ transport have been examined.[164,174,175] Information on the precise relationship between cell electrolyte activity and Na^+ transport is available for the amphibian proximal tubule[130,164] but not yet available for its mammalian counterpart.

The key features of the ATPase-driven Na^+/K^+ exchange are that an increase in cytosolic Na^+ stimulates both net trans-

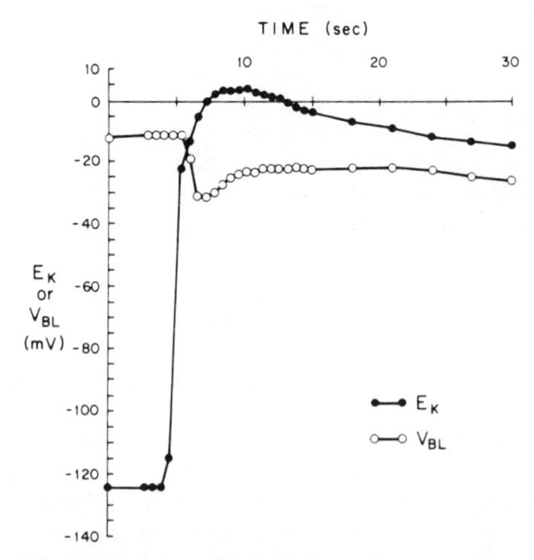

Fig. 14. The course of changes in K^+ equilibrium potential and peritubular membrane potential following stimulation of peritubular Na^+, K^+ pump activity in the perfused proximal straight tubule of rabbit. From Biagi *et al.* ref. 161.

port of Na^+ and the electrical potential.[164] In the amphibian tubule, it has already been shown (see Fig. 13) that isotonic fluid transport is stimulated following an increase in cell Na^+.[130] No saturation of Na^+ transport was noted up to a cell Na^+ concentration reaching about twice that of normal. The limiting role of the luminal membrane upon transepithelial Na^+ transport has already been mentioned. Peritubular K^+ has also been shown to stimulate Na^+/K^+ exchange. Particularly relevant are studies demonstrating suppression of net Na^+ absorption in perfused rabbit tubules by low K^+ concentration in the bath,[15] and saturation of fluid transport at a level of 2.5 mM,[165] a level of K^+ to be compared with a K_m of 1.0 mM for the *in vitro* ATPase system.[160] Na^+,K^+-ATPase microassays have disclosed significant amounts of this enzyme system in proximal tubules with lower concentrations in the proximal straight tubule than in the proximal convoluted tubule.[46] Finally, it should be noted that under certain conditions, the intracellular ATP level may be rate-limiting for active Na^+ transport.[174,175]

4.4.2. Additional Modes of Peritubular NaCl Transport

Under special experimental conditions, the possibility has been considered that mechanisms other than the classical Na^+,K^+-ATPase system may be responsible for Na^+ movement across the peritubular membrane of proximal tubule cells.

We have already referred to experiments in proximal convoluted tubules of the rabbit perfused symmetrically with high-Cl^- solutions from which the electrically neutral mode of Cl^- transport was deduced.[144] Considering the nature of Cl^- *exit* across the basolateral cell membrane, observations were made that also suggested a mode of Cl^- transport in an electrically neutral manner. First, the ineffectiveness of an anthracene analog, known to inhibit Cl^- permeability in the thick ascending limb of Henle[166] and the diluting segment of *Amphiuma*,[167] to affect NaCl transport implied a very low basolateral Cl^- conductance. Similar conclusions of a low Cl^- conductance have also been made in the rat proximal tubule[110] and the proximal tubule of *Necturus*.[168] A second observation made by Baum and Berry focused on the apparent potential-insensitive nature of NaCl transport. Addition of Ba^{2+} to the bath in concentrations that had previously been shown to reduce sharply the basolateral electrical potential[169,170] did not affect transepithelial salt and fluid transport.[144] This observation was taken to support the thesis of electroneutral Cl^- exit since the presence of an electrodiffusion pathway of Cl^- transport would have led to a curtailment of Cl^- extrusion with cell depolarization. The nature of electroneutral Cl^- exit across the basolateral membrane has so far been elusive. A Cl^-/HCO_3^- exchange system exists,[171,172] but, given the relative gradients of Cl^- and HCO_3^- across the basolateral membrane, probably mediates the net exchange of intracellular HCO_3^- with extracellular Cl^-. The presence of electrically neutral K^+–Cl^- cotransport, a mode of Cl^- transport that exists in the gallbladder epithelium,[173] is yet to be established.

There has also been some evidence that a "second" Na^+ pump may be present in renal proximal tubule cells.[176–178] Its main distinctive features are its independence of K^+, its insensitivity to the inhibitory action of ouabain, and its sensitivity to furosemide. Presence of such Na^+ transport has been deduced from experimental results in renal tissue slices in which ouabain-insensitive and K^+-independent NaCl extrusion can be shown to take place following rewarming the preparation after a period of cooling.[176] The most direct evidence derives from recent studies on basolateral plasma membrane vesicles that were prepared from rat proximal tubule cells. In this preparation, the presence of two Na^+ extrusion mechanisms could be shown, one having the properties of the classical Na^+,K^+-ATPase pump, and a second Na^+ extrusion mechanism that could be shown to be K^+-independent, ouabain-insensitive, and inhibitable by furosemide.[177,178] It should be noted though that the presence of a K^+-independent Na^+ transport mechanism could not be demonstrated in another study on renal basolateral vesicles.[179]

It is presently not clear what physiological role could be assigned to the "second" Na^+ pump. While some Na^+ reabsorption has been reported to continue in the perfused kidney of the rat in the presence of large amounts of ouabain,[180] experiments on single perfused tubules of the rabbit have repeatedly demonstrated that net NaCl and fluid transport as well as the lumen-negative potential are completely abolished after ouabain, or following the deletion of K^+ from the bathing solution.[15] There are possibly some similarities between the electrically neutral NaCl transport system referred to above[144] and the K^+-insensitive Na^+ extrusion mechanism,[176–178] but the latter is ouabain-insensitive whereas the former can be inhibited by cardiac steroids. It is also possible that the "second" Na^+ pump is normally dormant and activated only under conditions that need to be further defined.

4.5. Interaction of Transport between Apical and Basolateral Cell Membranes

Attention has been drawn to the special requirements of epithelial cells to maintain their cytosolic ionic environment within reasonably narrow limits during widely varying conditions of *transcellular* transport.[127,128] Two aspects shall be addressed. One has to do with the feedback between peritubular *Na^+* extrusion and luminal Na^+ entry and the other with the coordination between net Na^+ transport and peritubular *K^+* permeability.

4.5.1. Control of Luminal Na^+ Entry by Peritubular Na^+/Ca^{2+} Exchange

It is now well established that the basolateral membrane of proximal tubule cells has two mechanisms for Ca^{2+} extrusion[181]: a Ca^{2+}-dependent ATPase and a Na^+/Ca^{2+} exchanger. The latter is thought to operate in an electrogenic fashion with a Na^+/Ca^{2+} exchange ratio in excess of 2:1.[182,183] Its sensitivity to changes in both cytosolic and peritubular Na^+ concentration are well established and have served to evaluate its role in regulating Na^+ transport.[184]

Two aspects of Na^+ and Ca^{2+} transport constitute the basis of the control system linking basolateral with luminal transport events. First, it has been shown that the peritubular transmembrane Na^+ concentration gradient and the peritubular electrical potential difference strongly affect the cell Ca^{2+} activity. A fall in peritubular Na^+ and/or a rise in cell Na^+ lead to diminished peritubular Na^+/Ca^{2+} exchange and a rise in cell Ca^{2+}.[183] Second, it has been clearly shown that a rise in cell Ca^{2+} activity reduces luminal Na^+ entry although it is not yet clear which transport mode of luminal Na^+ entry is subject to inhibition by high cell Ca^{2+} levels.[183]

Figure 15 schematically summarizes the key elements by which increased cell Na^+ may depress Na^+ entry across the luminal cell membrane via secondary effects on the cell Ca^{2+} level. The sequence of events includes an increase in cell Na^+—

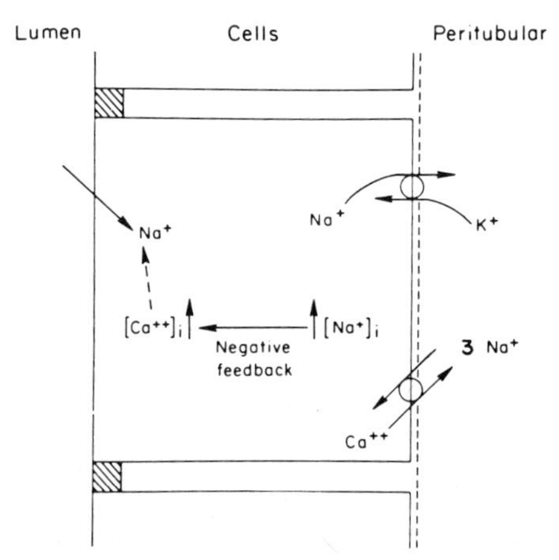

Fig. 15. Model of proximal tubule cell summarizing the essential features of feedback control system linking rate of luminal Na^+ entry and peritubular Na^+ entry and peritubular Na^+/K^+ exchange via changes in cell Ca^{2+} activity. From Taylor and Windhager, ref. 184.

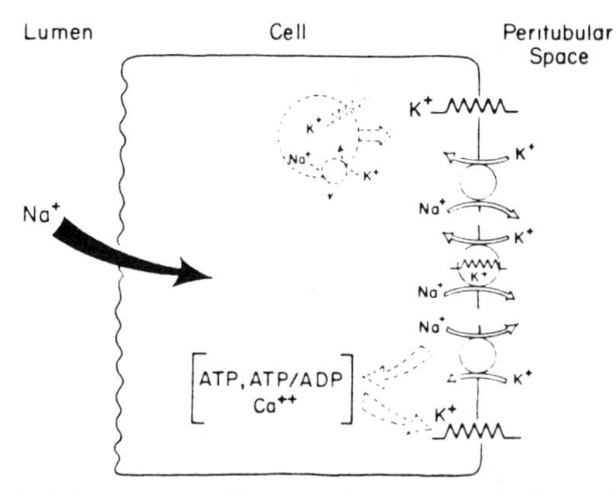

Fig. 16. Schema summarizing several possible mechanisms of coupling between Na^+,K^+ pump and K^+ conductance of the basolateral membrane.

initiated either by a phase of initially enhanced Na^+ entry across the luminal cell membrane or by basolateral Na^+,K^+-ATPase inhibition—and a reduced transmembrane Na^+ gradient across the peritubular cell membrane. The latter results then in a reduction of Na^+/Ca^{2+} exchange, increased cell Ca^{2+} levels, and a Ca^{2+}-mediated inhibition of Na^+ transport across the luminal cell membrane. On the other hand, a fall in cell Na^+ would inducc a decline in cell Ca^{2+} by steepening the Na^+ concentration gradient across the basolateral cell membrane and accelerating Ca^{2+} extrusion. As a consequence, Na^+ entry across the luminal membrane would be stimulated. It is apparent from these considerations that Ca^{2+} plays a key role in the tight coupling between apical and basolateral Na^+ transport. They prevent cell-flooding with Na^+ and potentially deleterious fluctuations in cell Na^+ and cell volume that might occur with alterations in net transport were it not for the homeostatic protection afforded by the feedback system just described.

4.5.2. Coordination between Net Na^+ Transport and Peritubular K^+ Permeability

Changes in basolateral K^+ permeability that coincide with alterations in Na^+ transport have been observed in several epithelia,[106,107] including the proximal tubule.[108,109] In general, the basolateral K^+ permeability increases with transport stimulation whereas it declines significantly following several maneuvers that block or reduce net Na^+ transport.

Since significant K^+ secretion is not a feature of the proximal tubule, the rate of basolateral K^+ uptake via the Na^+,K^+-ATPase must match its rate of passive egress across the same cell barrier.* If Na^+ extrusion is coupled to K^+ uptake at a fixed exchange ratio, stimulation of the pump rate must be coordinated with increased K^+ leakage to prevent either cell K^+ activity or cell volume from changing. Indeed, it has been observed in several epithelia[106,185] including the proximal tubule of *Necturus* that the cell K^+ activity remains constant during dramatic changes in net Na^+ transport. Since there were no significant changes in the electrochemical gradient for K^+ across the basolateral membrane, a change in basolateral K^+ permeability was strongly suggested. Evidence that the basolateral K^+ permeability of proximal tubule cells rises with Na^+ transport stimulation and falls when Na^+ transport is inhibited, is now available.[108,109]

The precise mechanism that could account for the tight coupling between Na^+ transport and K^+ permeability is not clear but several possibilities are included in Fig. 16, and the reader is referred to reports dealing with this issue.[127,128]

Three possibilities have to be considered and it should be stated that the issue is not resolved. First, it is possible that with transport changes, "pump" and "leak" units are inserted into or removed from the membrane. Second, it is also possible that alterations in pump turnover rate directly affect the functional state of already present K^+ channels.

The relationship between cell Ca^{2+} and/or cell metabolism and K^+ permeability also deserves careful attention, since both are potential candidates for a messenger system. In particular, it is tempting to speculate that Ca^{2+} is involved. In general, Ca^{2+} stimulates K^+ channels.[186,187] It is attractive to speculate that a fall in cell Na^+ (decreased Na^+ entry) steepens the Na^+ gradient across the basolateral cell membrane, a condition known to lower cell Ca^{2+}. With the decline in cell Ca^{2+}, the peritubular K^+ conductance could also decline. Conversely, stimulation of Na^+ entry could initiate an increase in cell Na^+ and cell Ca^{2+},[188] the high Ca^{2+} leading to activation of basolateral K^+ channels. A difficulty with such a simple control system is the observation that not all maneuvers that lead to an increase in cell Na^+ also increase the peritubular K^+ permeability. Thus, ouabain and a low peritubular Na^+ concentration block Na^+ transport and increase cell Ca^{2+} activity,[189] yet K^+ permeability is reduced and not activated as might have been expected if cytosolic Ca^{2+} were the prime mediator.[109] It is also of interest that peritubular membrane depolarization *per se* lowers the K^+ permeability.[190] Further studies are necessary to provide the requisite information necessary to define the relationship between Na^+ transport, electrical potential, Ca^{2+}, and peritubular K^+ permeability.

*The luminal brush border membrane is not impermeable to K^+ (see discussion of luminal transport) but it is likely that the absolute rate of K^+ loss across the basolateral cell membrane greatly exceeds that taking place across the luminal cell membrane.

4.6. Acid Extrusion across the Luminal Membrane

As stated earlier, the proximal tubule reabsorbs the bulk of the filtered HCO_3^- load. In the amphibian proximal tubule, HCO_3^- reabsorption occurs isohydrically such that a transtubular pH or HCO_3^- gradient does not develop.[76,77] Along the mammalian proximal tubule, the luminal HCO_3^- concentration falls to a limiting value of 5–8 mM, and a lumen-acid transtubular pH gradient of approximately 0.5–0.7 unit is generated.[4] This gradient is far in excess of what may be attributed to the small, lumen-negative transtubular potential difference in the early proximal tubule, and indicates the presence of active, transcellular acid secretion and/or base absorption. (In this chapter, the processes of acid secretion and base absorption will be collectively termed H^+ secretion.) Because intracellular pH in the proximal tubule cell is above the level at which H^+ and HCO_3^- would be in electrochemical equilibrium across the luminal or basolateral membrane,[191,192] it is clear that the uphill step in active, transcellular H^+ secretion must be localized to the luminal membrane.

4.6.1. Luminal Membrane Pathways for H^+ Transport

Several possible pathways for active H^+ secretion across the luminal membrane have been identified, as illustrated in Fig. 17. Directly coupled Na^+/H^+ (Fig. 17A) exchange was first demonstrated in rat renal microvillus membrane vesicles by Murer *et al.*[193] and was subsequently confirmed in microvillus membranes from rabbit,[194] dog,[195] and *Necturus*.[136] In renal membrane vesicles, imposing an outward H^+ gradient stimulates Na^+ uptake, and imposing an inward Na^+ gradient drives uphill H^+ extrusion, indicating coupling between the influx of Na^+ and the efflux of H^+.[193,194] This coupling is direct and cannot be attributed to electrical interactions between passive fluxes of Na^+ and H^+ occurring through conductive pathways.[193,194] Both H^+-coupled Na^+ influx and Na^+-coupled H^+ efflux are sensitive to inhibition by the diuretic amiloride.[196] Uphill acid extrusion across the luminal membrane via amiloride-sensitive, electroneutral Na^+/H^+ exchange has also been demonstrated in studies of isolated perfused proximal tubules of the salamander[191] and rabbit.[197] Clearly, Na^+/H^+ exchange is a secondary active mechanism of H^+ secretion for which the driving force is the electrochemical Na^+ gradient directed inward across the luminal membrane as a consequence of primary active Na^+ extrusion across the basolateral membrane.

There are additional luminal membrane transport processes by which active H^+ secretion can be coupled to active Na^+ transport. For example, in rabbit renal microvillus membrane vesicles, imposing an outward H^+ gradient can drive uphill PO_4^{3-} accumulation via the Na^+–PO_4^{3-} cotransport system in the absence of any other ion gradients.[198] This finding indicates that the process of Na^+–PO_4^{3-} cotransport is coupled to a net flux of H^+ in the opposite direction. This may arise because Na^+–PO_4^{3-} cotransport involves exchange for H^+ (as shown in Fig. 17B), cotransport with OH^-, or preferential absorption of HPO_4^{2-}.

The process of Na^+-coupled organic anion absorption across the luminal membrane can also lead to net H^+ secretion. For example, such organic anions as lactate, β-hydroxybutyrate, succinate, and α-ketoglutarate are accumulated within the proximal tubule cell by cotransport with Na^+.[93–98] Each of these anions can be transported across the luminal membrane in exchange for OH^- via a Na^+-independent, anion-exchange system identified in renal microvillus vesicles from dog[199,200] and rat.[201] Na^+-coupled organic anion uptake into the cell followed by exchange of intracellular organic anion for luminal OH^- is therefore another possible mechanism for mediating net H^+ secretion across the luminal membrane (Fig. 17C). A variation on this theme is Na^+-coupled absorption of organic anions followed by nonionic diffusion of the undissociated organic acid back into the lumen (Fig. 17D). Short straight-chain monocarboxylic acid anions such as acetate, propionate and butyrate traverse lipid membranes rapidly via nonionic diffusion[154] and are each substrates for the transport system that mediates Na^+-coupled absorption of monocarboxylates across the brush border.[202]

In addition to the Na^+-coupled mechanisms of H^+ secretion just described, at least two mechanisms for primary active H^+ secretion have been identified in preparations of renal microvillus membrane vesicles (Figs. 17E, F). One is ATP-driven, electrogenic H^+ transport via a H^+-ATPase that is stimulated by such oxyanions as sulfite and HCO_3^-.[203–205] The other is redox-driven, electrogenic H^+ transport via a NADH oxidase system.[206,207] However, as discussed in more detail elsewhere,[5] there is not yet definitive evidence that these processes are intrinsic to the luminal membrane itself. Instead, their presence in luminal membrane preparations may reflect contamination of these preparations by intracellular organelles pos-

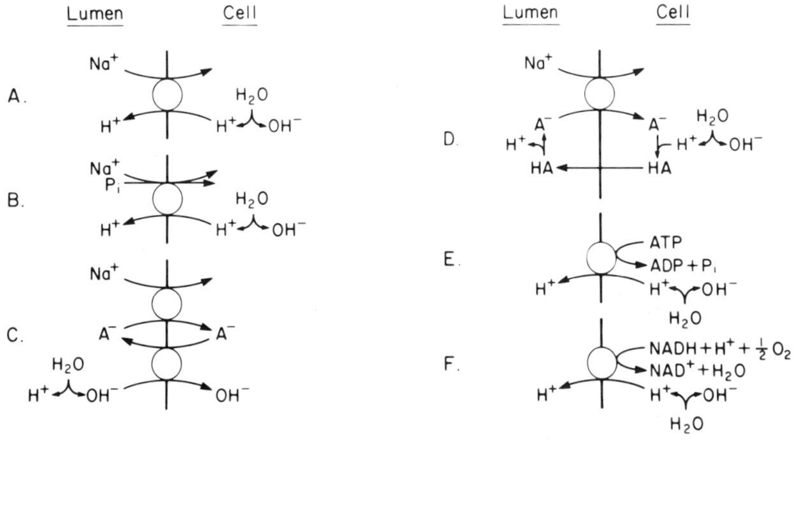

Fig. 17. Summary of luminal pathways of H^+ secretion.

sessing primary active H^+ pumps. For example, ATP-driven H^+ transport has been identified in clathrin-coated vesicles,[208,209] endosomes,[210] and lysosomes[211] from a variety of cell types. The possibility that the anion-stimulated ATPase activity[203,204] and the oligomycin-insensitive, ATP-driven H^+ transport[205] observed in renal brush border preparations might be due to contamination by these intracellular organelles has not been carefully evaluated. Of course, if coated vesicles in proximal tubule cells, like those in other tissues, have ATP-driven H^+ pumps, then it is possible that these pumps are also present on the coated areas of plasma membrane from which the vesicles are derived. Accordingly, H^+-transporting ATPase may indeed be present on the intermicrovillar areas of the luminal membrane that are extensively clathrin-coated,[212] but this remains to be directly demonstrated.

Finally, it should be noted that the process(es) responsible for active H^+ secretion across the luminal membrane are opposed by a variety of mechanisms mediating downhill H^+ absorption and/or base secretion. These include passive diffusion of HCO_3^- and H^+ or OH^- across the paracellular pathway,[213] and such luminal membrane processes as passive diffusion of H^+ or OH^-,[214] nonionic diffusion of weak acids and bases,[215] H^+–dipeptide cotransport,[216] Na^+–H^+–glutamate cotransport,[217] organic cation/H^+ exchange,[218] organic anion/OH^- exchange,[199–201] and, in amphibians[136] but not mammals,[142,149,150] $Cl^-/OH^-(HCO_3^-)$ exchange.

4.6.2. Relative Contributions of Na^+-Coupled and Primary Active Mechanisms of H^+ Secretion

Current evidence suggests that at least 80% and perhaps all of net active H^+ secretion across the luminal membrane occurs by Na^+-coupled processes, the most important of which is probably Na^+/H^+ exchange. For example, the rate of acidification in proximal tubules of rabbits and rats is inhibited 80–90% when luminal Na^+ is removed[41,42,219] or when primary active Na^+ transport is totally inhibited by addition of ouabain or removal of peritubular K^+.[41,42,219] In fact, in one study in which extreme care was taken to remove K^+ completely and to correct for errors in HCO_3^- measurement, K^+ removal inhibited acidification by 100%, raising the possibility that the small residual rate of acidification obtained after Na^+ removal, ouabain addition, or K^+ removal in other studies may have been artifactual.[220] Supporting the concept that Na^+/H^+ exchange is the principal mechanism for active H^+ secretion is that uphill H^+ extrusion across the luminal membrane of acid-loaded cells of the salamander proximal tubule is almost completely Na^+-dependent, is accompanied by an approximately 1 : 1 influx of Na^+, and has a $K_{1/2}$ for Na^+ and a sensitivity to amiloride that are very similar to those of the Na^+/H^+ exchanger as estimated in studies on membrane vesicles.[191] Moreover, the number or activity of Na^+/H^+ exchangers per unit of luminal membrane increases in response to maneuvers or conditions that are associated with enhanced rates of proximal tubular acidification, such as reduction in renal mass,[195] parathyroidectomy,[221] metabolic acidosis,[221] K^+ depletion,[222] and glucocorticoid administration.[223]

Nevertheless, several arguments have been advanced to suggest that a significant fraction of proximal tubular acid secretion occurs by primary active, electrogenic H^+ transport. However, none of these arguments is conclusive. First, net Na^+ reabsorption and net H^+ secretion in the proximal tubule can be easily dissociated in response to a variety of experimental maneuvers.[224,225]. But given the multiple pathways for Na^+ and H^+ fluxes through transcellular and paracellular routes across this leaky epithelium, it would not be unexpected that net transtubular fluxes of Na^+ and H^+ could be dissociated even if these fluxes were tightly coupled via one particular pathway, the Na^+/H^+ exchanger.

Second, replacement of luminal Na^+ inhibits acidification only partially (20–65%) in rat proximal tubules studied by stationary microperfusion[34,224]. However, in these studies luminal Na^+ concentration was only reduced to 4–5 mM, not much below the range of $K_{1/2}$ values that have been reported for the luminal membrane Na^+/H^+ exchanger.[196] Moreover, amiloride inhibits most of the residual rate of acidification persisting at 5 mM luminal Na^+, suggesting that it occurs via Na^+/H^+ exchange.[26]

Third, ouabain inhibits the rate of proximal tubular acidification only partially (30%) in the rat and not at all in the hamster.[224,227] However, ouabain inhibition of active Na^+ transport in these studies was not complete, and the gradients of H^+ and Na^+ across the luminal membrane in the presence of ouabain were not measured. Surely, 1 : 1 Na^+/H^+ exchange could not be the sole mechanism of H^+ secretion if it were observed that H^+ is secreted against a H^+ gradient that is larger than the Na^+ gradient existing across the luminal membrane in the presence of ouabain. However, in only one study has this been claimed.[228] Active H^+ transport was found to maintain an extracellular-to-intracellular H^+ gradient of 1.01-fold at a time when the corresponding Na^+ gradient was only 0.83-fold. However, a 20% overestimation of intracellular Na^+ concentration could explain this result without requiring the presence of primary active H^+ transport. Such an overestimation of intracellular Na^+ was not unlikely in this study inasmuch as in another study employing the same chemical methods the authors reported intracellular Na^+ under normal conditions to be 70 mM,[229] a value over twofold higher than that measured in proximal tubule cells by electron probe analysis.[230]

Fourth, in rat proximal tubules microperfused without organic solutes, there is a 1-mV, lumen-positive, transtubular potential difference that is inhibited by acetazolamide, suggesting electrogenic H^+ secretion.[13,147] However, as discussed below, there is an appreciable HCO_3^- conductance at the basolateral membrane of the rat proximal tubule cell.[110,111] Hence, any process, including electroneutral Na^+/H^+ exchange, that elevates intracellular HCO_3^- above equilibrium will tend to depolarize the basolateral membrane and thereby generate a lumen-positive transtubular potential. Thus, the presence of a lumen-positive acidification potential is not necessarily inconsistent with the concept that Na^+ H^+ exchange is the sole mechanism of active H^+ secretion.

Fifth, there is an increase in anion-stimulated ATPase activity in renal brush border preparations from rats with metabolic acidosis,[231] However, ATPase activity was assayed under conditions that would not have provided access of ATP to the inside of the vesicles. Inasmuch as isolated brush border vesicles are almost exclusively right-side-out,[232] it is unclear whether measuring hydrolysis of extravesicular ATP is a reliable means for assaying activity of any H^+-ATPase that is actually present on the luminal membrane.

4.7. Base Exit across the Basolateral Membrane

As mentioned previously, intracellular pH in proximal tubule cells is above the value at which H^+, OH^-, and HCO_3^-

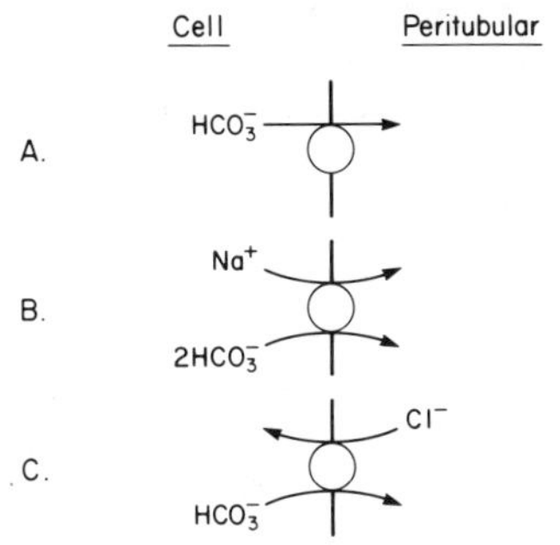

Fig. 18. Summary of basolateral pathways of base exit.

would be in electrochemical equilibrium across the basolateral membrane. Accordingly, there is a passive driving force for the downhill exit of OH^- and HCO_3^-, or downhill entry of H^+ across the basolateral membrane.

4.7.1. Basolateral Pathways for Base Transport

Several possible mechanisms for the net exit of base across the basolateral membrane have been described, as illustrated in Fig. 18. Raising the peritubular HCO_3^- concentration causes hyperpolarization, and lowering the peritubular HCO_3^- induces depolarization of the basolateral membrane potential in rat proximal convoluted tubules.[110] Varying the peritubular pH at a constant HCO_3^- concentration causes much smaller changes in basolateral membrane potential than does varying the peritubular HCO_3^- concentration at constant pH.[110] These electrical effects of peritubular HCO_3^- are reduced by acetazolamide, are abolished by SITS, are independent of Na^+, and cannot be attributed to changes in K^+ conductance.[110,111] Thus, these studies strongly suggest the presence of a SITS-sensitive, Na^+-independent, conductive pathway for HCO_3^- across the basolateral membrane of rat proximal tubule cells (Fig. 18A). The basolateral conductances of OH^- and H^+, if present at all, must be much more modest. Lowering the peritubular HCO_3^- concentration also induces depolarization of the basolateral membrane potential in rabbit proximal straight tubules.[169] However, this effect is inhibited by Ba^{2+}, an inhibitor of the basolateral membrane K^+ conductance. Thus, in rabbit proximal straight tubules, the depolarization caused by lowering the peritubular HCO_3^- concentration may result from intracellular pH-induced changes in K^+ conductance rather than from the presence of a conductive HCO_3^- pathway across the basolateral membrane. In salamander proximal tubules,[233] there is a large, SITS-sensitive HCO_3^- conductance at the basolateral membrane, but this system appears to involve cotransport with Na^+ (Fig. 18B).

Another possible mechanism of base exit from the proximal tubule cell is Cl^-/HCO_3^- exchange (Fig, 18C). Imposing an outward HCO_3^- gradient drives DIDS-sensitive, uphill Cl^- uptake into rabbit renal cortical basolateral membrane vesicles.[171] This phenomenon is independent of Na^+ and independent of anion diffusion potentials, indicating directly coupled Cl^-/HCO_3^- exchange. The rate of Cl^-/OH^- exchange is much smaller or nonexistent in basolateral membrane vesicles.[150,171] The intracellular pH of rabbit proximal tubule cells studied in tubular suspensions[234] or in isolated perfused straight tubules[172] increases when Cl^- is removed from the extracellular media, suggesting Cl^-/HCO_3^- exchange. There is also evidence for Cl^-/HCO_3^- exchange across the basolateral membrane of amphibian proximal tubule cells.[235] In rat renal cortical basolateral membrane vesicles, HCO_3^- is a substrate for a Na^+-stimulated anion exchanger shared by a variety of organic and inorganic anions.[236] The net direction of HCO_3^- transport via this system under physiological conditions is not known.

Finally, it should be noted that the processes responsible for net base exit across the basolateral membrane are opposed by several mechanisms mediating base uptake and/or H^+ extrusion. For example, Na^+/H^+ exchange occurs across the basolateral membrane of proximal tubule cells of the salamander[191] although not of the rabbit[237] or rat.[150] A SO_4^{2-}/HCO_3^- exchange process has been described in rat basolateral membrane vesicles[238] and likely exchanges intracellular SO_4^{2-} for peritubular HCO_3^- under *in vivo* conditions. There is evidence for Na^+-dependent exchange of intracellular Cl^- for extracellular HCO_3^- across the basolateral membrane of *Necturus* proximal tubule cells.[239]

4.7.2. Physiological Roles of Basolateral Pathways for Base Transport

As discussed above, the two principal mechanisms for base exit across the basolateral membrane of mammalian proximal tubule cells are via a Na^+-independent HCO_3^- conductance and via a Na^+-independent Cl^-/HCO_3^- exchange process. It is noteworthy that both mechanisms for base exit strongly prefer HCO_3^- over OH^- as the substrate for transport in the physiological pH range. This underscores the importance of the carbonic anhydrase-catalyzed conversion of intracellular OH^- to HCO_3^- in order to facilitate base exit from the proximal tubule cell. Consistent with this concept, the carbonic anhydrase inhibitor acetazolamide elevates the intracellular pH of proximal tubule cells.[192,240]

Inasmuch as the basolateral membrane HCO_3^- conductance and Cl^-/HCO_3^- exchange are both sensitive to stilbene inhibitors of anion transport, the fact that peritubular SITS inhibits the rate of acidification in rat proximal tubules by over 70%[227] does not help in assessing the relative contribution of these two mechanisms of base exist. Arguing against a significant contribution of Cl^-/HCO_3^- exchange is the finding that acidification in isolated perfused proximal tubules of rabbit is not inhibited when Cl^- is replaced by nitrate or isethionate.[241] However, the rate of HCO_3^- reabsorption in rabbit proximal convoluted tubules is inhibited 30% when Cl^- is replaced by gluconate,[241] an anion with little or no affinity for the basolateral Cl^-/HCO_3^- exchanger.[171] Thus, Cl^-/HCO_3^- exchange may indeed mediate a significant fraction of the steady-state exit of base from the proximal tubule cell. An important role for the HCO_3^- conductance pathway is suggested by the observation that depolarizing the basolateral membrane with Ba^{2+}, an inhibitor of the K^+ conductance, reduces the rate of HCO_3^- reabsorption in rabbit proximal convoluted tubules by about 40%.[241]

4.8. Role of Cl^- and HCO_3^-—Driven Fluid Movement across the Proximal Tubular Epithelium

The possible role of passive proximal tubular NaCl and fluid transport has been raised with reference to the transepithelial Cl^- and HCO_3^- gradients which develop along the proximal convoluted and proximal straight tubules. Related to this problem is the role played by HCO_3^- and the role of H^+ secretion in the process of proximal tubular reabsorption of Na^+ and fluid.

With respect to the importance of transepithelial proximal tubular *Cl*$^-$ concentration gradients in accounting for a fraction of proximal tubular fluid reabsorption, it has been argued that preferential tubular HCO_3^- reabsorption in the very first part of the proximal tubule increases the luminal Cl^- concentration, and generates a Cl^- gradient directed from the tubule to the peritubular fluid. It has already been pointed out that such a Cl^- concentration gradient ($Cl^-_{lumen} > Cl^-_{peritubular}$) renders the tubular lumen electrically positive by generating a diffusion potential across the proximal tubular epithelium. Thus, a driving force for Na^+ reabsorption might exist along that part of the proximal tubule in which the Cl^- concentration has risen significantly above peritubular plasma levels. This hypothesis puts key emphasis on early proximal tubular transport operations such as tubular H^+ secretion since this event is primarily responsible for the establishment of the Cl^- concentration difference and the lumen-positive electrical potential. It has also been argued on theoretical grounds that a significant fraction of proximal tubular Na^+ transport in rats may be driven by the small luminal electropositivity across the more distal parts of the mammalian proximal tubule.(61,242–246)

The view that the transepithelial Cl^- concentration gradient drives a significant fraction of fluid across the proximal convoluted and proximal straight tubules of superficial nephrons is supported directly by several perfusion studies carried out on rat and rabbit tubules.(138,139,145,247–249) First, it can be shown in perfused proximal convoluted tubules of the rat that the imposition of known Cl^- gradients across the epithelium leads to predictable and symmetrical changes of NaCl and fluid reabsorption. Thus, some 20 to 30% of the control rate of fluid reabsorption occurs when a favorable, "physiological" Cl^- gradient ($[Cl^-]_{lumen} > [Cl^-]_{blood}$) is imposed across the proximal tubule whereas net NaCl and fluid reabsorption are inhibited by the same fraction when the Cl^- concentration in peritubular capillaries exceeds that in the lumen.(247)

A second line of evidence was obtained in tubular perfusion experiments in which active transport processes had been abolished. Thus, during exposure of proximal convoluted and straight tubules to either cold, cyanide, or ouabain, NaCl and fluid movement were maintained, albeit at a significantly reduced rate, as long as a favorable transepithelial Cl^- gradient was maintained.(139,145,249,251,252) Such net transport completely disappeared following abolition of the anion asymmetry across the proximal tubular epithelium. It is also of interest that the fraction of NaCl and fluid movement subject to modulation by the Cl^- gradient is similar in conditions of preserved active Na^+ transport and in situations in which transport is inhibited.

4.9. Active and Passive Components of Proximal Na^+ Transport

An analysis of the participation of *active* transport processes in NaCl reabsorption of the proximal convoluted tubule has emphasized its importance even in the presence of anion gradients that are effective driving forces for passive proximal fluid transport. The situation is particularly relevant in the late proximal convoluted tubule, a segment of high Cl^- and low HCO_3^- concentration. While it had previously been assumed that such transport was largely passive, driven both by the Cl^- concentration difference and by the lumen-positive transepithelial potential, there is now general agreement that some two-thirds of NaCl and fluid transport are active in nature.(75,248) In the presence of an appropriate Cl^- concentration gradient, the removal of K^+ from the peritubular bath,(250) or the addition of cyanide to the luminal and peritubular perfusion solutions(248) leads to extensive reduction of fluid movement, but only on the order of 60 to 70%. Accordingly, only 30 to 40% of NaCl and fluid movement could be assigned to be driven by the transepithelial Cl^- gradient.

5. Solute–Solvent Coupling—Role of the Intercellular Shunt Pathway

An important element in the control of proximal Na^+ transport is the intercellular shunt pathway. It has already been pointed out that the proximal tubular epithelium is a highly permeable structure, a property quite consistent with a low-resistance transfer route connecting the luminal compartment directly with the peritubular extracellular fluid compartment. Based on both mor-

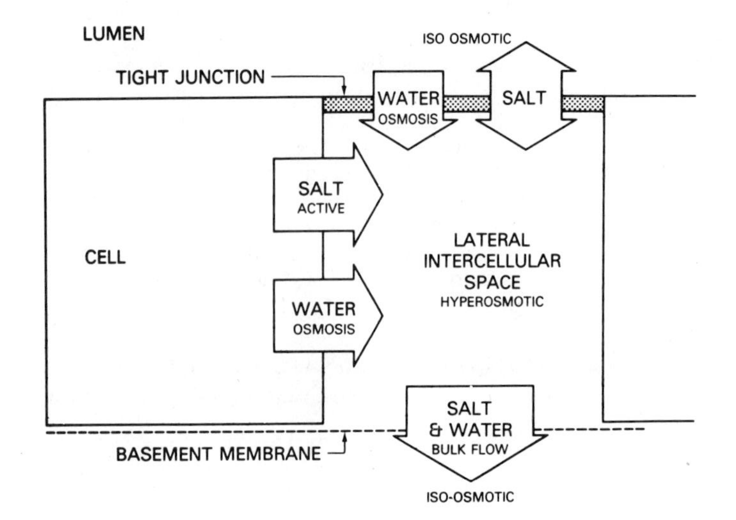

Fig. 19. Model of fluid transport in proximal convoluted tubule involving the cell compartment and the lateral intercellular space. From Burg, ref. 15.

phological and electrophysiological evidence, the intercellular shunt pathway has been incorporated into a cell model of proximal tubular reabsorption schematically shown in Fig. 19. Given the presence of a shunt pathway, consisting of shallow junctional complexes, the lateral intercellular space, and the basement membrane, it has been proposed that this fluid compartment is not only the epithelial site of high electrolyte and water permeation but also a potentially important locus where peritubular renal hemodynamic factors affect the efficiency of proximal tubular Na^+ transport.[1,2]

An essential feature of the original proximal tubule cell model depicted in Fig. 19 is that sodium salts, mainly NaCl and $NaHCO_3$, are pumped into the lateral interspaces at a fast enough rate to generate a fluid of slight hypertonicity. In response to the establishment of such an osmotic gradient, water moves either across the tight junction directly and/or from cellular to extracellular compartment into the interspaces. As water moves into the interspace region, the hydraulic pressure increases and provides a driving force for fluid movement toward the basement membrane. In a final step, the reabsorbate is transferred from the interspaces across the basement membrane and the peritubular capillary wall by the balance of hydrostatic and colloid osmotic pressure difference.

Some problems related to this model of solute–water coupling are not resolved. First, it is currently not clear to what extent water is transferred into the interspace area via the apical junction and/or via a transcellular route. Since measurements of overall transepithelial water permeability include both pathways, the problem of the relative contribution of these two transfer routes to overall water movement has not yet been fully resolved. If a large fraction of reabsorbed water were to move between cells, the salt permeability of the apical junction would probably also be high, in which case an excessive back-flux of Na^+ into the lumen would be likely and might impose an excessively high demand on the transport system. On the other hand, if a large fraction of the reabsorbate were to move through tubule cells, very high water permeabilities of tubule cells would be necessary. There has been increasing evidence lately that cell permeabilities to water in several "leaky" epithelia, including the proximal tubule,[253–259] may be high enough to allow a very significant fraction of the reabsorbed water to move via a transcellular route. However, there is presently no agreement as to the precise magnitude of such movement through cells compared to that between cells.

A second source of uncertainty has to do with the site within the proximal tubular epithelium where the osmotic gradient responsible for fluid movement is established. To account for the isosmotic or near-isosmotic character of volume absorption, the initial models of proximal fluid transport had all assumed that a region of hypertonicity exists within the proximal tubular epithelium within the intercellular space.[260–263] However, the central role of interspace hypertonicity has become less certain in view of recent experiments that have shown that the fluid within the lumen may be hypotonic with respect to the peritubular environment[264] and that the reabsorbate may become hypertonic during the reabsorption process.[265,266]

In the following, we shall briefly consider the information bearing on three key questions. First, where is the osmotic gradient across the proximal tubule established? Second, to what extent do differences in anion reflection coefficients contribute to proximal tubular fluid movement? And third, what is the precise route of fluid movement through the epithelium?

The reader is referred to several reviews for details of the mechanisms that couple solute and solvent movement in the proximal tubule.[253,258,267–269]

5.1. Site of Osmotic Gradient

Two mechanisms have always occupied a central role to account for isosomotic volume absorption across the proximal tubule. First, active solute transport provides ultimately the major driving force, and second, fluid transport is in response to osmotic pressure gradients. The second problem of the mechanism of generation of the osmotic gradient has led to the idea of a "middle" or intercellular compartment, a region of intraepithelial hypertonicity, where coupling between solute and solvent was thought to take place.[123,262,270] The main features of this model include a lateral intercellular space with appropriate diffusion constraints as to prevent dissipation of interspace hypertonicity. As pointed out by Schafer,[268] "the standing osmotic gradient is a steady-state balance of active solute pumping into the intercellular space, osmotic water flow into the space, and convective and diffusional movement out of the space."

It has been generally assumed since the classical micropuncture experiments by Walker *et al.*[23] that the reabsorption of fluid from the proximal tubule occurs without a change in osmotic pressure of the luminal fluid. The necessity of an osmotic pressure gradient to account for transepithelial fluid movement has led to the conclusion that the intercellular space is the sole site of hypertonicity. However, recent theoretical considerations as well as experimental observations have made this argument less cogent.

Schafer,[268] Schafer *et al.*,[271] and Andreoli and Schafer[272] have proposed that at least in one tubule preparation, that of the rabbit proximal tubule, the prerequisite of interspace hypertonicity is not likely to be met. They point out that the dimensions and morphometric property of the lateral interspaces are such as to make it unlikely that these spaces offer enough diffusion resistance to maintain a significant degree of hypertonicity. This is due to the small dimension of the interspace (on the order of 7 μm in length), the absence of tortuosity, and the large area that the interspace occupies between the extensively infolded and interdigitated lateral cell membranes.[268]

With regard to the organization of the intercellular space, studies on the ultrastructure of proximal tubules have led to the recognition of significant axial heterogeneity[273] and of species differences.[274] In studies focusing on the surface areas of the brush border and the lateral cell walls in the *rabbit* proximal nephron, it became apparent that the two surface areas of about *equal* magnitude are the *apical* brush border membrane and the cell surface facing the *lateral* intercellular space, i.e., the intercellular shunt path. These two membranes have been reported to be about 20 times larger than the *peritubular* cell membrane in proximal convoluted cells, and some 10 times larger in proximal straight tubule cells.[273] This is of interest since this difference of cell surface area in proximal convoluted and proximal straight tubule cells correlates with the lower transepithelial reabsorption rites of fluid. The very extensive lateral cell membrane amplification due to complex infoldings is shown in Fig. 20 and demonstrates clearly the importance of the *intercellular* compartment as a key site of transport events in this type of mammalian tubule.[273]

Quantitative ultrastructural studies in two amphibian proximal tubule preparations, that of *Necturus* and of *Ambystoma*, have recently shown significant differences in the organization

Fig. 20. Schematic three-dimensional reproduction of proximal convoluted tubule cell (top), and proximal straight tubule cell (bottom). From Welling and Welling, ref. 273.

of the intraepithelial extracellular fluid space.(274) It has been pointed out above that in the rabbit, the *lateral* intercellular space predominates as evidenced by the fact that the lateral surface area predominates.(273) In *Ambystoma,* the situation is quite different in that the *basal* intercellular labryinth predominates. The organization of the extracellular space system in *Necturus* is intermediate between that of the rabbit and that of *Ambystoma.* Thus, the geometry of the intraepithelial extracellular spaces varies greatly, a fact that has implications for transport models of solute–solvent coupling (see below). For instance, in *Ambystoma,* fluid transport takes place largely into the *basal* extracellular compartment, since the basolateral membrane only sparingly faces the lateral intercellular space. As a further consequence of this anatomical arrangement, solute transport across the basal membrane in *Ambystoma* will be ineffective in affecting fluid translocation across the *junctional* complex facing the tubule lumen. Hence, transepithelial fluid movement is most likely to follow a *transcellular* route. The situation is different in *Nectu*rus and the rabbit, where, depending on the morphological arrangement, the *lateral* intercellular spaces assume a key role in the transepithelial pathway of solute and fluid movement.(274)

Two approaches have been used recently to test whether small osmotic pressure differences could be present *outside* the epithelium, i.e., whether luminal hypotonicity or peritubular hypertonicity could provide a significant driving force for water movement. Figures 21 and 22 provide relevant data.

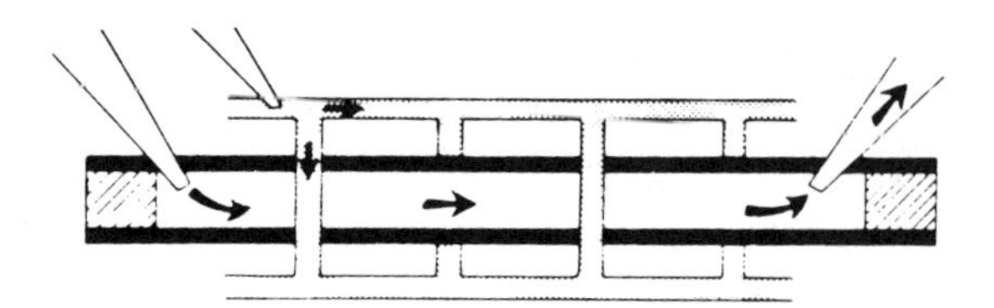

Perfusate: 154 mM NaCl (290 mOsm)

Perfusion rate	Reabsorption rate	Osmolality difference (Collectate-Perfusate)
nl min	*nl (min·mm)*	*mOsm*
10	0.41	− 1.73 ± 0.48
45	0.89	− 3.90 ± 0.64
45 + NaCN	0.06	0.47 ± 0.38

Fig. 21. Development of luminal hypotonicity in the perfused rat proximal nephron *in vivo.* Simultaneous perfusions of tubule lumen and peritubular capillary network are shown schematically. The results are for differing tubule perfusion rates in the absence or presence of NaCl. Reabsorption rates of tubular perfusate and the differences in osmolality between collected and perfused tubular fluid are given. From Schafer (ref. 268), based on data in ref. 264.

Figure 21 illustrates the experimental approach and the results of luminal and peritubular perfusion studies in the rat kidney in which simple, symmetrical NaCl solutions were used in both perfusion systems.(264) Although the rate of fluid absorption is less than that observed with more physiological solutions, we note that the proximal convoluted tubule is capable of establishing a small but significant osmotic gradient. With increased perfusion rate, absorption rate increases and, at the same time, the osmolality difference rises as well. Cyanide abolishes both the fluid movement and the transepithelial osmotic pressure difference. These experiments clearly demonstrate the possibility

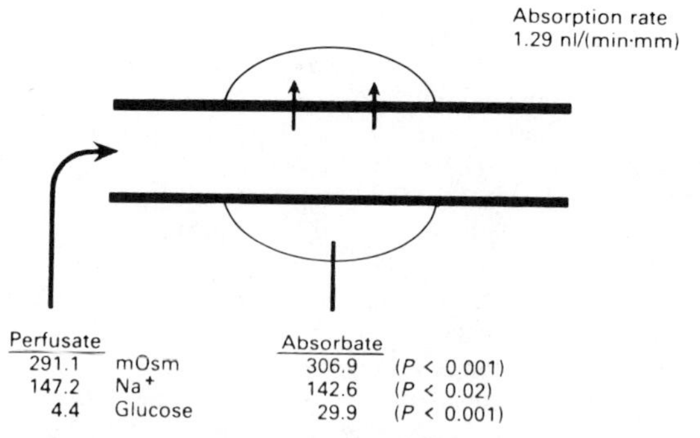

Fig. 22. Hyperosmolality of absorbate from rabbit proximal convoluted tubules. A segment of the perfused tubule under oil is shown schematically with a droplet of absorbate on the peritubular surface. The composition of the perfusate, shown on the left, is the average of measured perfusate and the collectate concentrations. Concentrations indicated for Na^+ and glucose are in millimolar. P values indicate the significance of the paired difference in average perfusate and absorbate compositions. From Barfuss and Schafer, ref. 265.

that luminal hypotonicity ("cryoscopic hypotonicity") could contribute to fluid absorption. This possibility has to be seriously considered, given the relatively high hydraulic water permeability of the proximal epithelium.

Barfuss and Schafer have recently published data that also suggest the possibility that osmotic pressure gradients can be generated across the proximal convoluted and proximal straight tubules.[265,266] Figure 22 illustrates their experimental approach. It consists of perfusing isolated segments of proximal tubules of the rabbit under oil. As fluid is reabsorbed, a peritubular fluid droplet develops. Its rate of formation and its composition can be monitored by appropriate collection and analysis. Results of perfusion experiments are also summarized in Fig. 22. The key observation is the demonstration of significant hyperosmolality of the absorbate. Figure 23 shows the relationship between observed fluid absorption and the transepithelial osmotic gradient.[266] Different perfusates were used that changed the reabsorption rate. It should be realized that in this experimental arrangement, peritubular absorbate hypertonicity can develop because solutes are transferred into a small compartment, the fluid droplet under oil. In contrast, during tubular and peritubular microperfusion, the peritubular composition is maintained "clamped" because of the high capillary perfusion. As a consequence, hypotonicity of the fluid in the lumen develops.

An important question is whether cryoscopic luminal hypotonicity is sufficiently large to account for proximal fluid absorption. This depends critically on the hydraulic permeability of the proximal tubular epithelium and the rates of fluid absorption. Table I summarizes relevant estimates from a recent review by Berry.[253] It is apparent that in the mammalian proximal tubule at least, additional driving forces must be involved since the osmotic pressure gradient that must be present for fluid transport to proceed at the observed rates exceeds the osmotic gradient that is generated by luminal hypotonicity. Nevertheless, it is very likely that in the early proximal tubule at least, preferential HCO_3^- reabsorption and reabsorption of glucose and amino acids could generate a modestly hypotonic fluid by solute reabsorption in excess of water. This could account for a significant fraction of the filtrate to be reabsorbed by this mechanism. Significant hypotonicity has been reported not only in perfusion of the proximal tubule but also in free-flow micropuncture studies along the rat proximal tubule.[275,276] What is presently unresolved is the extent to which interspace hypertonicity may contribute to fluid movement as an additional driving force.

Table I. Epithelial Water Permeability[a]

Preparation	P_f (cm/sec)	J_v (nl/mm per min)	Δ osmolality (mOsm)
Rat PCT	0.1	2–4	25.3–50.6
	0.3		8.5–17
Rabbit PCT	0.1	0.8–1.5	10–19
	0.3		3.4–6.3
Rabbit PST	0.1	0.2–0.4	2.5–5
	0.3		0.85–1.7
Necturus PCT	0.02	0.4	4.7
	0.04		2.3
Necturus gallbladder	0.04	0.1[b]	3.5

[a]From Berry, ref. 253.
[b]The rate of volume absorption in the *Necturus* gallbladder is 144 nl/cm² per min, which is equivalent to 0.1 nl/mm per min of epithelia in a cylinder with diameter of 20 μm.

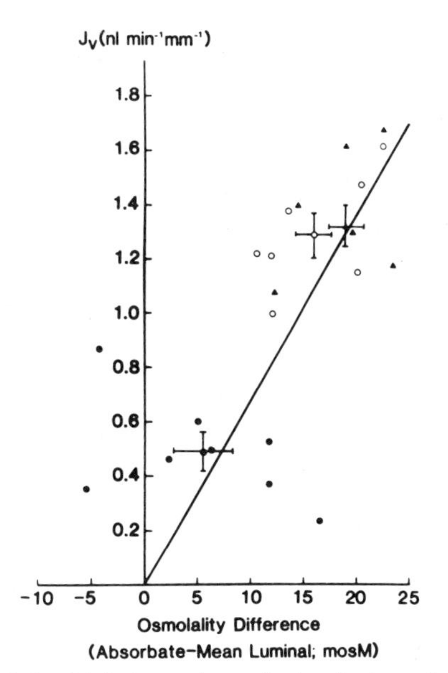

Fig. 23. Relationship between observed rates of volume absorption and transepithelial osmotic gradient. Data are from 21 proximal convoluted tubules. Experimental points indicated by closed triangles are from tubules perfused with an ultrafiltrate of rabbit serum; by open circles, artificial perfusate A; and by closed circles, artificial perfusate B. The composition of these two perfusates was different with respect to their organic solute content to achieve different fluid absorption rates. Mean for each solution is also indicated together with brackets denoting S.E. of both the volume absorption and the osmolality difference parameters. Line was drawn from least-squares regression of individual data points, forced through origin. Equation of the line J_v = 0.0678 (osmolality difference). From Barfuss and Schafer, ref. 266.

5.2. The Role of Differences in Anion Reflection Coefficients and of NaCl Diffusion in Proximal Fluid Transport

It has been pointed out since the initial suggestion by Rector *et al.*[245] that luminal hypotonicity, or interspace hypertonicity, are not the only driving forces for transepithelial fluid movement. An additional driving force can be effective without generating finite luminal hypoosmolality. The very extensive reabsorption of organic solutes and of HCO_3^- could generate an effective osmotic gradient across the proximal tubular wall, provided these solutes have a reflection coefficient exceeding that of NaCl. In addition to the selective reabsorption of organic solutes, the development of Cl^- and HCO_3^- gradients provides also a potentially significant driving force for solute and fluid translocation along the proximal convoluted tubule.[245,253,268]

We have already discussed the role of transepithelial anion gradients as a driving force for fluid movement (see Section 4.8). It is very likely that the Cl^- reflection coefficient is less than that of HCO_3^-.[244,277] Hence, in the presence of opposing Cl^- and HCO_3^- gradients, the luminal fluid would become effectively hyposmotic to the peritubular fluid. As a consequence, an additional fraction of fluid would be reabsorbed. It has been proposed that in the presence of *physiological* Cl^- and HCO_3^- gradients across the proximal straight tubule of the rabbit, an effective osmotic gradient of some 4 mOsm/kg H_2O

would be generated across the epithelium.[268] This estimate is based on the assumption of quite small differences in the reflection coefficient; a difference of 0.2 between the reflection coefficients of HCO_3^- and Cl^- would suffice to lower the effective osmolality to the extent described above. Similar calculations for organic solutes, again assuming that the reflection coefficients of organic solutes such as glucose and amino acids are 1.0 whereas that of NaCl is 0.8, would generate an additional effective osmotic difference of about 2.4 mOsm/kg H_2O. Clearly, such "effective" osmotic gradients are important additional driving forces for fluid absorption across the proximal tubular epithelium.*

An alternative, or additional, mechanism of solute coupling would be diffusion of NaCl across the proximal tubular epithelium. In this mode of transport, the lumen-positive potential and the Cl^- concentration difference would be the relevant driving forces. Schafer has calculated that the amount of fluid that could traverse the epithelium of the proximal straight tubule of the rabbit by this mechanism, is relatively small.[268] Nevertheless, this is an additional transport mechanism that has to be considered. Its precise role in solute–solvent coupling across the proximal convoluted tubule is not clear. However, this transport mode has to be considered to contribute to volume absorption in the late proximal convoluted and the proximal straight tubule.

5.3. The Route of Fluid Movement across the Proximal Tubule

Two pathways of fluid movement across the proximal tubular epithelium are available. One envisages that fluid transport in response to osmotic pressure differences occurs largely via the junctional complex between cells and along the intercellular shunt pathway. An additional, or alternative, pathway across the epithelium is through tubule cells. Such fluid movement would occur sequentially across the luminal and the basolateral cell membrane and into the basolateral or basal interspace region. Clearly, apportionment of water movement to either an intercellular or a transcellular pathway depends critically on knowledge of the driving forces and the hydraulic permeabilities of each of the rate-limiting barriers. Such information is presently incomplete.

Several arguments have been made that suggest a significant role of the transjunctional fluid pathway.[253,258] First, measurements of the relationship between osmotic and diffusive water permeability have yielded ratios far in excess of unity, implying relatively large pores that are difficult to be located within the apical cell membrane.[280] Hence, a junctional site was assumed. A second argument has to do with the very high leakiness of the junctional pathways to ions, i.e., its low electrical resistance and the assumption that such high ionic conductance implies also a high water permeability.[253] Third, if the differences between the reflection coefficients of Cl^- and HCO_3^- play a significant role as a driving force of transepithelial fluid movement, it is more likely that the requisite differences of reflection coefficients exert their effects on a single barrier—the junctional complex.[268,272,280,280a] Finally, evidence in favor of a significant paracellular pathway has to do with the presence of a significant streaming potential that, at least in the perfused proximal tubule of *Ambystoma,* was not due to solute polarization, and clearly of paracellular origin.[281]

The view that a transcellular route of water transport is functionally important in the proximal tubule, is presently gaining support. The main argument favoring a cellular path of water movement is the recognition of relatively high water permeabilities of both the apical[257] as well as the basolateral[254–256] cell membranes of proximal tubule cells. Given a high osmotic water permeability, even small osmotic gradients can be shown to drive significant amounts of water across the proximal tubule via a cellular route.

One of the important consequences of the intercellular shunt pathway is its role in regulating the efficiency of the proximal tubular Na^+ transport system. The interspace region has emerged as a control site of tubular Na^+ transport which becomes apparent when salt and water reabsorption across the proximal convoluted tubule are modulated by "physical forces," such as changes in peritubular oncotic or hydrostatic pressure.[124,282–286] It is thought that such Starling forces influence the uptake of proximal tubular reabsorbate across the basement membrane and the peritubular capillary wall. For instance, as the balance of peritubular oncotic and hydrostatic pressure is poised against fluid translocation into the capillaries, the pressure in the interspaces may rise and this increase of interstitial pressure would result in an increased back-leak of Na^+ into the tubular lumen. Conversely, such back-diffusion would become smaller as fluid is more efficiently transferred into peritubular capillaries.

Particularly relevant are studies in which changing the rate of translocation of tubular reabsorbate out of the interspace region by manipulation of such forces as the peritubular oncotic or hydrostatic pressure could be shown to lead to the expected alterations of the resistance of the intercellular shunt path. As an example, the resistance of the intercellular shunt path decreases with an increase in hydrostatic or a decrease in peritubular plasma oncotic pressure.[287] Such resistance changes would be expected to enhance passive back-flux of Na^+ from interspaces into the lumen and reduce the overall efficiency of net Na^+ transport. A relevant example is the proximal diuresis, i.e., the inhibition of proximal tubular Na^+ and fluid transport, following extracellular volume expansion. This is associated solely with an increase in passive Na^+ permeability but apparently without a change in active transport of Na^+.[289] It is most likely that changes in peritubular physical factors, such as a fall in colloid oncotic pressure and an increase in hydrostatic pressure, enhance the back-leak of Na^+ across the junctional complex. These effects of changes in peritubular oncotic pressure upon net Na^+ transfer are accentuated whenever the extracellular volume has been expanded and the interstitial pressure is elevated prior to variations in the peritubular oncotic pressure.[290] According to this view, active transport and capillary uptake of Na^+ are occurring in series, and it is the latter step that is subject to control by modulation of peritubular physical factors.

The role of "physical factors" in controlling proximal tubular fluid reabsorption has also been studied in isolated perfused proximal tubules *in vitro*.[291–294] The conclusions reached by such studies agree in general with the *in vivo* results

*The existence of differences in Cl^- and HCO_3^- reflection coefficients has been directly examined by testing whether solvent drag of anions occurs in response to osmotically induced fluid movement across the proximal tubular epithelium in the absence of active transport.[278,279] In another study, fluid transport was stimulated by addition of HCO_3^- to the peritubular perfusion fluid.[264] In such studies, no enhancement of net solute transport was observed. This could be taken to exclude solvent drag as a significant factor in anion transport and to indicate that the reflection coefficient for Cl^- is not different from 1.0. However, it has been pointed out that small differences of reflection coefficients between Cl^- and HCO_3^-, of the order that could generate significant osmotic pressure differences, could escape detection by the approach used.[264,268]

in that proximal tubular fluid reabsorption is proportional to the concentration of proteins in the bath surrounding the isolated proximal tubules.

Several aspects of the peritubular control of proximal Na^+ transport have been carefully reviewed by Knox *et al.*[295] Studies in dogs confirmed a significant role of physical factors on fluid reabsorption but the effect was strongly volume-dependent.[296] When extracellular volume had been increased—a situation characterized by a high rate of passive Na^+ back-flux into the lumen—changes in peritubular oncotic pressure were found to be quite effective in regulating net reabsorption. However, no effect of peritubular protein was seen during hydropenia.

The role of the colloid osmotic and hydrostatic pressure of postglomerular capillaries acting as a *direct* driving force has also been recently reinvestigated by measuring the effects of intraluminal and peritubular oncotic pressure changes on transepithelial water movement, and by assessing the hydraulic conductivity of the proximal tubular epithelium by imposing known solute gradients across the proximal tubular wall.[124,292,293,297] With some notable exceptions,[297,298] most investigators agree that the role of the colloid oncotic or hydrostatic pressure difference as a direct driving force for transepithelial fluid movement in the regulation of proximal tubular Na^+ transfer cannot be a major one. This is due to the large body of evidence that a wide variety of maneuvers which block the operation of the active Na^+ transport system, also completely block net Na^+ reabsorption. Treatment of the proximal tubular epithelium with metabolic inhibitors such as cyanide,[122,124] ouabain,[15,35,72,279] and a combination of metabolic blocking agents,[300] as well as deletion of K^+ from the extracellular fluid,[299] the replacement of Na^+ by a variety of nontransported cations,[15,115,299] or the imposition of an unfavorable anion gradient in the presence of a carbonic anhydrase inhibitor,[225] all completely abolish proximal tubular net Na^+ transport at a time when the full transepithelial oncotic driving force is operative. Clearly, if the colloid oncotic pressure difference were of significance as an important driving force, a significant fraction of proximal tubular Na^+ transport should have remained intact. Hence, the available body of evidence supports the view that the colloid oncotic pressure operates mainly on that moiety of NaCl and fluid that has already been transferred into the interspace secondary to active Na^+ transport. The observation that the colloid oncotic effects, produced by changing the protein concentration in peritubular capillaries, are markedly attenuated in the absence of active Na^+ transport across the proximal tubule, is consistent with such an interpretation.

The proposal that oncotic and hydrostatic pressure changes in the peritubular environment can alter proximal tubular fluid reabsorption is of further interest since these forces may play a role in setting the rate of fluid transport in accordance with changes in the glomerular filtration rate.[1,284,301] For a given rate of blood flow, the postglomerular protein concentration is determined by the rate of glomerular filtrate formation. It has been pointed out that the phenomenon of glomerulotubular balance of proximal tubules with respect to sodium salts and fluid may be related to such postglomerular alterations in oncotic pressure. According to this view, an increase in glomerular filtration rate would have a concentrating effect on peritubular proteins and lead to stimulation of net Na^+ transport by curtailing Na^+ back-flux from the interspaces into the lumen.[284,285] It should be noted, however, that not all experimental observations support a major role of peritubular proteins in the regulation of proximal tubular Na^+ and fluid transport.[301–304] Some experimental results, for instance, indicate that tubular flow rate *per se* may play an important role in glomerulotubular balance,[249,305] such that an increased flow rate *per se* stimulates tubular Na^+ reabsorption. The mechanism by which flow rate could directly alter proximal tubular Na^+ transport has not been elucidated but it has been suggested that a flow-dependent change in unstirred layers in the brush border might be involved.[306] Supporting such a view is the observation of a strong flow-dependence of proximal tubular HCO_3^- transport at subsaturating levels of luminal HCO_3^-.[307] Changes of the delivery of organic solutes (glucose, amino acids) have also been considered to be involved. An increase of glomerular filtration rate would effect a change in the solute profile along the proximal tubule such that a longer part of the tubule would be exposed to elevated glucose or amino acid concentrations.[308] Finally, it has been suggested that a humoral substance, of glomerular origin, may play a major role in stimulating tubular reabsorption and in effecting glomerulotubular balance.[309]

References

1. Giebisch, G., and E. E. Windhager. 1973. Electrolyte transport across renal tubular membranes. In: *Handbook of Physiology,* Section 8. J. Orloff and R. W. Berliner, eds. American Physiological Society, Washington, D.C. pp. 315–376.
2. Giebisch, G., E. L. Boulpaep, and G. Whittembury. 1971. Electrolyte transport in kidney tubule cells. *Proc. R. Soc. London Ser. B.* **262**:175–196.
3. Rector, F. C. 1973. Acidification of the urine. In: *Handbook of Physiology,* Section 8. J. Orloff and R. W. Berliner, eds. American Physiological Society, Washington, D.C. pp. 431–454.
4. Warnock, D. G., and F. C. Rector, Jr. 1981. Renal acidification mechanisms. In: *The Kidney.* B. M. Brenner and F. C. Rector, Jr., eds. Saunders, Philadelphia. pp. 440–494.
5. Aronson, P. S. 1983. Mechanisms of active H^+ secretion in the proximal tubule. *Am. J. Physiol.* **245**:F647–F659.
6. Ullrich, K. J. 1976. Renal tubular mechanisms of organic solute transport. *Kidney Int.* **9**:172–188.
7. Sacktor, B. 1982. Na^+ gradient-dependent transport systems in renal proximal tubule brush border membrane vesicles. In: *Membranes and Transport,* Volume II. A. N. Martonosi, ed. Plenum Press, New York. pp. 197–206.
8. Aronson, P. S. 1981. Identifying secondary active solute transport in epithelia. *Am. J. Physiol.* **240**:F1–F11.
9. Schultz, S. G. 1985. Cellular models of epithelial ion transport. This volume.
10. Schultz, S. G., and P. F. Curran. 1970. Coupled transport of sodium and organic solutes. *Physiol. Rev.* **50**:637–718.
11. Shannon, J. A. 1938. The renal reabsorption and excretion of urea under conditions of extreme diuresis. *Am. J. Physiol.* **123**:182–190.
12. Lassiter, W. E., M. Mylle, and C. W. Gottschalk. 1961. Micropuncture study of net transtubular movement of water and urea in nondiuretic mammalian kidney. *Am. J. Physiol.* **200**:1139–1147.
13. Frömter, E. 1979. Solute transport across epithelia: What can we learn from micropuncture studies on kidney tubules? *J. Physiol. (London)* **188**:1–31.
14. Boulpaep, E. L. 1979. Electrophysiology of the kidney. In: *Membrane Transport in Biology,* Volume IV A. G. Giebisch, D. C. Tosteson, and H. H. Ussing, eds. Springer-Verlag, Berlin. pp. 97–144.
15. Burg, M. B. 1981. Renal handling of sodium, chloride, water, amino acids, and glucose. In: *The Kidney.* B. M. Brenner and F. C. Rector, Jr., eds. Saunders, Philadelphia. pp. 328–370.
16. Whittembury, G., and F. A. Rawlins. 1971. Evidence of a paracellular pathway for ion flow in the kidney proximal tubule: Electronmicroscopic demonstration of lanthanum precipitate in the tight junction. *Pfluegers Arch.* **330**:302–309.

17. Whittembury, G., F. A. Rawlins, and E. L. Boulpaep. 1973. Paracellular pathway in kidney tubules: Electrophysiological and morphological evidence. In: *Transport Mechanisms in Epithelia*. H. H. Ussing and N. A. Thorn, eds. Academic Press, New York. pp. 577–588.
18. Rector, F. C., Jr. 1983. Sodium, bicarbonate and chloride absorption by the proximal tubule. *Am. J. Physiol.* **244**:F461–F471.
19. Frömter, E., and K. Gessner. 1974. Free-flow profile along rat kidney proximal tubule. *Pfluegers Arch.* **351**:69–83.
20. Kaissling, B., and W. Kriz. 1979. *Structural Analysis of the Rabbit Kidney*. Springer-Verlag, Berlin.
21. Welling, L. W., and D. J. Welling. 1976. Surface area of brush border and lateral cell walls in the rabbit proximal nephron. *Kidney Int.* **9**:385–394.
22. Maunsbach, A. B., and E. L. Boulpaep. 1984. Quantitative ultrastructure and functional correlates in proximal tubule of *Ambystoma* and *Necturus*. *Am. J. Physiol.* **246**:F710–F724.
23. Walker, A. M., P. A. Bott, J. Oliver, and M. C. MacDowell. 1941. The collection and analysis of fluid from single nephrons of the mammalian kidney. *Am. J. Physiol.* **134**:580–595.
24. Gottschalk, C. W. 1963. Renal tubular function: Lessons from micropuncture. *Harvey Lect.* **58**:99–123.
25. Le Grimellec, C., P. Poujeol, and C. de Ruffignac. 1975. Micropuncture study along proximal convoluted tubule: Electrolyte reabsorption in first convolutions. *Pfluegers Arch.* **354**:133–150.
26. Walker, A. M., and C. L. Hudson. 1937. Reabsorption of glucose from the renal tubule in Amphibia and the action of phlorizin on it. *Am. J. Physiol.* **118**:130–141.
27. Frohnert, P., B. Hohmann, R. Zwiebel, and K. Baumann. 1970. Free flow micropuncture studies of glucose transport in the rat nephron. *Pfluegers Arch.* **315**:66–85.
28. Bergeron, M., and F. Morel. 1969. Amino acid transport in rat renal tubules. *Am. J. Physiol.* **216**:1139–1149.
29. Lingard, J., G. Rumrich, and J. S. Young. 1973. Reabsorption of L-glutamine and L-histidine from various regions of the rat proximal convolution studied by stationary microperfusion: Evidence that the proximal tubule is not homogeneous. *Pfluegers Arch.* **342**:1–12.
30. Silbernagl, S. 1975. Renal transport of amino acids. *Rev. Physiol. Biochem. Pharmacol.* **74**:105–167.
31. Hohmann, B., P. P. Frohnert, R. Kinne, and K. Baumann. 1974. Proximal tubular lactate transport in rat kidney: A micropuncture study. *Kidney Int.* **5**:261–270.
32. Baumann, K., C. de Rouffignac, N. Roinel, G. Rumrich, and K. J. Ullrich. 1975. Renal phosphate transport: Inhomogeneity of local proximal transport rates and sodium dependence. *Pfluegers Arch.* **356**:287–297.
33. McKeown, J., P. C. Brazy, and V. W. Dennis. 1979. Intrarenal heterogeneity for fluid, phosphate and glucose absorption in the rabbit. *Am. J. Physiol.* **234**:F312–F318.
34. Ullrich, K. J., G. Rumrich, and K. Baumann. 1975. Renal proximal tubular buffer (glycodiazine) transport: Inhomogeneity of local transport dependence on sodium, effect of inhibitors and chronic adaptation. *Pfluegers Arch.* **357**:140–163.
35. Jacobson, H. R., and J. P. Kokko. 1976. Intrinsic differences in various segments of the proximal convoluted tubule. *J. Clin. Invest.* **57**:818–825.
36. Gjöri, A. Z., J. M. Lindgard, and J. A. Young. 1974. Relation between active sodium transport and distance along the proximal convolutions of rat nephrons: Evidence for homogeneity of sodium transport. *Pfluegers Arch.* **348**:205–210.
37. Berry, C. A. 1982. Heterogeneity of tubular transport processes in the nephron. *Annu. Rev. Physiol.* **44**:181–201.
38. Jacobson, H. R. 1981. Functional segmentation of the mammalian nephron. *Am. J. Physiol.* **241**:F203–F218.
39. Barfuss, D. W., and J. A. Schafer. 1981. Differences in active and passive glucose transport along the proximal nephron. *Am. J. Physiol.* **240**:F322–F332.
40. Barfuss, D. W., and J. A. Schafer. 1979. Active amino acid absorption by proximal convoluted and proximal straight tubules. *Am. J. Physiol.* **236**:F149–162.
41. Burg, M., and N. Green. 1977. Bicarbonate transport by isolated perfused rabbit proximal convoluted tubules. *Am. J. Physiol.* **233**:F307–F314.
42. McKinney, T. D., and M. B. Burg. 1977. Bicarbonate and fluid absorption by renal proximal straight tubules. *Kidney Int.* **12**:1–8.
43. Schafer, J. A., S. L. Troutman, M. L. Watkins, and T. E. Andreoli. 1978. Volume absorption in the pars recta. I. "Simple" active Na transport. *Am. J. Physiol.* **234**:F332–F329.
44. Kawamura, S., M. Imai, D. W. Seldin, and J. P. Kokko. 1975. Characteristics of salt and water transport in superficial and juxtamedullary straight segments of proximal tubules. *J. Clin. Invest.* **55**:1269–1277.
45. Biagi, B., T. Kubota, M. Sohtell, and G. Giebisch. 1981. Intracellular potentials in rabbit proximal tubules perfused *in vitro*. *Am. J. Physiol.* **240**:F200–F210.
46. Katz, A. I., A. Doucet, and F. Morel. 1979. Na-K-ATPase activity along the rabbit, rat and mouse nephron. *Am. J. Physiol.* **237**:F114–F120.
47. Garg, L., M. Knepper, and M. Burg. 1981. Mineralocorticoid effects on Na-K-ATPase in individual nephron segments. *Am. J. Physiol.* **240**:F536–F544.
48. Tune, B. M., M. B. Burg, and C. S. Patlak. 1969. Characteristics of *p*-amino hippurate transport in proximal renal tubules. *Am. J. Physiol.* **217**:1057–1063.
49. Grantham, J., P. Qualizza, and R. Irwin. 1974. Net fluid secretion in proximal straight renal tubules *in vitro:* Role of PAH. *Am. J. Physiol.* **226**:191–197.
50. Grantham, J., R. Irwin, P. Qualizza, D. Tucker, and F. Whittier. 1973. Fluid secretion in isolated proximal straight renal tubules: Effect of human uremic serum. *J. Clin. Invest.* **52**:2441–2450.
51. Jacobson, H. R., and J. P. Kokko. 1985. Intrarenal heterogeneity: Vascular and tubular. In: *The Kidney: Normal and Abnormal Function*. D. Seldin and G. Giebisch, eds. Raven Press, New York, in press.
52. Horster, M., and K. Thurau. 1968. Micropuncture studies on the filtration of single superficial and juxtamedullary glomeruli in the rat kidney. *Pfluegers Arch* **301**:162–181.
53. Jamison, R. L., and J. B. Lacy. 1971. Effect of saline infusion on superficial and juxtamedullary nephrons in the rat. *Am. J. Physiol.* **221**:690–697.
54. Jacobson, H. R. 1981. Effects of CO_2 and acetazolamide on bicarbonate and fluid transport in rabbit proximal tubules. *Am. J. Physiol.* **240**:F54–F62.
55. Berry, C. A. 1981. Electrical effects of acidification in the rabbit proximal convoluted tubule. *Am. J. Physiol,* **240**:F459–F470.
56. Jacobson, H. R. 1979. Characteristics of volume reabsorption in rabbit superficial and juxtamedullary proximal convoluted tubules. *J. Clin. Invest.* **63**:410–418.
57. Giebisch, G. 1961. Measurements of electrical potential difference in perfused single proximal tubules in *Necturus* kidney. *J. Gen. Physiol.* **44**:659–678.
58. Koeppen, B. M., G. Giebisch, and B. A. Biagi. 1983. Electrophysiology of mammalian renal tubules: Inferences from intracellular microelectrode studies. *Annu. Rev. Physiol.* **45**:497–517.
59. Laprade, R., and J. Cardinal. 1983. Liquid junctions and isolated proximal tubule transepithelial potentials. *Am. J. Physiol.* **244**:F304–F319.
60. Seeley, J. F., and E. Chirito. 1975. Studies of the electrical potential difference in rat proximal tubules. *Am. J. Physiol.* **229**:72–80.
61. Barratt, L. J., F. C. Rector, J. P. Kokko, and D. W. Seldin. 1974. Factors governing the transepithelial potential difference across the proximal tubule of the rat kidney. *J. Clin. Invest.* **53**:454–460.
62. Burg, M. B., L. Isaacson, J. Grantham, and J. Orloff. 1968. Electrical properties of isolated perfused rabbit renal tubules. *Am. J. Physiol.* **215**:788–794.
63. Burg, M. B., and J. Orloff. 1970. Electrical potential difference across proximal convoluted tubules. *Am. J. Physiol.* **219**:1714–1716.
64. Kokko, J. P. 1973. Proximal tubular potential difference: Dependence on glucose, HCO_3 and amino acids. *J. Clin. Invest.* **52**: 1362–1367.

65. Burg, M., C. Patlak, N. Green, and D. Villey. 1976. The role of organic solutes in fluid absorption by renal proximal convoluted tubules. *Am. J. Physiol.* **231**:627–637.
66. Kokko, J., and F. C. Rector. 1971. Flow dependence of transtubular potential difference in isolated perfused segments of rabbit proximal convoluted tubule. *J. Clin. Invest.* **50**:2745–2750.
67. Frömter, E. 1982. Electrophysiological analysis of rat renal sugar and amino acid transport. I. Basic phenomena. *Pfluegers Arch.* **393**:179–189.
68. Samarzija, I,, B. T. Hinton, and E. Frömter. 1982. Electrophysiological analysis of rat renal sugar and amino acid transport. II. Dependence on various transport parameters and inhibitors. *Pfluegers Arch.* **393**:190–197.
69. Samarzija, I., and E. Frömter. 1982. Electrophysiological analysis of rat renal sugar and amino acid transport. III: Neutral amino acids. *Pfluegers Arch.* **393**:199–200.
70. Samarzija, I., and E. Frömter. 1982. Electrophysiological analysis of rat renal sugar and amino acid transport. IV. Basic amino acids. *Pfluegers Arch.* **393**:210–214.
71. Samarzija, I., and E. Frömter. 1982. Electrophysiological analysis of rat renal sugar and amino acid transport. V. Acidic amino acids. *Pfluegers Arch.* **393**:215–221.
72. Schafer, J. A., S. L. Troutman, and T. E. Andreoli. 1974. Isotonic volume reabsorption, transepithelial potential differences and ionic permeability properties in mammalian proximal straight tubules. *J. Gen. Physiol.* **64**:582–607.
73. Berry, C. A., D. G. Warnock, and F. C. Rector, Jr. 1978. Ion selectivity and proximal salt reabsorption. *Am. J. Physiol.* **235**:F234–F245.
74. Warnock, D., and V. Yee. 1982. Anion permeabilities of the isolated perfused rabbit proximal tubule. *Am. J. Physiol.* **241**:F395–F405.
75. Berry, C. A., and F. C. Rector, Jr. 1981. Active and passive sodium transport in the proximal tubule. *Miner. Electrolyte Metab.* **4**:149–160.
76. Giebisch, G. H. 1956. Measurement of pH, chloride and inulin concentrations in proximal tubule fluid of *Necturus*. *Am. J. Physiol.* **185**:171–175.
77. Montgomery, H., and J. A. Pierce. 1937. The site of acidification of the urine within the renal tubule of Amphibia. *Am. J. Physiol.* **118**:114–152.
78. Maruyama, T., and T. Hoshi. 1972. The effect of *d*-glucose on the proximal electrical potential profile across the proximal tubule of newt kidney. *Biochim. Biophys. Acta* **282**:214–225.
79. Hoshi, T. 1976. Electrophysiological studies on amino acid transport across the luminal membrane of the proximal tubular cells of *Triturus* kidney. In: *Amino Acid Transport and Uric Acid Transport.* S. S. Silbernagl, F. Lang, and R. Greger, eds. Thieme, Stuttgart. pp. 96–103.
80. Aronson, P. S., and B. Sacktor. 1975. The Na^+ gradient-dependent transport of *d*-glucose in renal brush border membranes. *J. Biol. Chem.* **250**:6032–6039,
81. Kinne, R., H. Murer, E. Kinne-Saffran, M. Thees, and G. Sachs. 1975. Sugar transport by renal plasma membrane vesicles: Characterization of the systems in the brush-broder microvilli and the basal lateral plasma membranes. *J. Membr. Biol.* **21**:375–395.
82. Beck, J. C., and B. Sacktor. 1978. Membrane potential-sensitive fluorescence changes during Na^+-dependent D-glucose transport in renal brush border membrane vesicles. *J. Biol. Chem.* **253**:7158–7162.
83. Kleinzeller, A. 1976. Renal sugar transport systems and their specificity. In: *Proceedings of the VI International Congress on Nephrology, Florence, 1975.* Karger, Basel. pp. 130–133.
84. Turner, R. J., and A. Moran. 1982. Heterogeneity of sodium-dependent D-glucose transport sites along the proximal tubule: Evidence from vesicle studies. *Am. J. Physiol.* **242**:F406–F414.
85. Turner, R. J., and A. Moran. 1982. Further studies of proximal tubular brush border membrane D-glucose transport heterogeneity. *J. Membr. Biol.* **70**:37–45.
86. Turner, R. J., and A. Moran. 1982. Stoichiometric studies of the renal outer cortical brush border membrane D-glucose transporter. *J. Membr. Biol.* **67**:73–80.
87. Fass, S. J., M. R. Hammerman, and B. Sacktor. 1977. Transport of amino acids in renal brush border membrane vesicles: Uptake of the neutral amino acid L-alanine. *J. Biol. Chem.* **252**:583–590.
88. Evers, J., H. Murer, and R. Kinne. 1976. Phenylalanine uptake in isolated renal brush border vesicles. *Biochim. Biophys. Acta* **426** :598–615.
89. Schell, R. E., B. R. Stevens, and E. M. Wright. 1983. Kinetics of sodium-dependent solute transport by rabbit renal and jejunal brush-border vesicles using a fluorescent dye. *J. Physiol. (London)* **335**:307–318.
90. Sacktor, B., I. L. Rosenbloom, C. T. Liang, and L. Cheng. 1981. Sodium gradient- and sodium plus potassium gradient-dependent L-glutamate uptake in renal basolateral membrane vesicles. *J. Membr. Biol.* **60**:63–71.
91. Barfus, D. W., J. M. Mays, and J. A. Schafer. 1980. Peritubular uptake and transepithelial transport of glycine in isolated proximal tubules. *Am. J. Physiol.* **238**:F324–F333.
92. Silverman, M., P. Vinay, L. Shinobu, A. Gougoux, and G. Lemieux. 1981. Luminal and antiluminal transport of glutamine in dog kidney: Effect of metabolic acidosis. *Kidney Int.* **20**:359–365.
93. Barac-Nieto, M., H. Murer, and R. Kinne. 1980. Lactate–sodium cotransport in rat renal brush border membranes. *Am. J. Physiol.* **239**:F496–F506.
94. Barac-Nieto, M., H. Murer, and R. Kinne. 1982. Asymmetry in the transport of lactate by basolateral and brush border membranes of rat kidney cortex. *Pfluegers Arch.* **392**:366–371.
95. Wright, S. H., S. Krasne, I. Kippen, and E. M. Wright. 1981. Na^+-dependent transport of tricarboxylic acid cycle intermediates by renal brush border membranes: Effect on fluorescence of a potential-sensitive cyanine dye. *Biochim. Biophys. Acta* **640**:767–778.
96. Kragh-Hansen, U., K. E. Jorgensen, and M. I. Sheikh. 1982. The use of a potential-sensitive cyanine dye for studying ion-dependent electrogenic renal transport of organic solutes. *Biochem. J.* **208**: 369–376.
97. Kahn, A. M., S. Branham, and E. J. Weinman. 1984. Mechanism of L-malate transport in rat renal basolateral membrane vesicles. *Am. J. Physiol.* **246**:F779–F784.
98. Jorgensen, K. E., U. Kragh-Hansen, E. Roigaard-Petersen, and M. I. Sheikh. 1983. Citrate uptake by basolateral and luminal membrane vesicles from rabbit kidney cortex. *Am. J. Physiol.* **244**:F686–F695.
99. Hoffman, N., M. Thees, and R. Kinne. 1976. Phosphate transport by isolated renal brush border vesicles. *Pfluegers Arch.* **362**:147–156.
100. Burg, M., and N. Green. 1976. Role of monovalent ions in the reabsorption of fluid by isolated perfused proximal renal tubules of the rabbit. *Kidney Int.* **10**:221–228.
101. Schafer, J. A., S. L. Troutman, and T. E. Andreoli. 1974. Isotonic volume reabsorption, transepithelial potential differences and ionic permeability properties in mammalian proximal straight tubules. *J. Gen. Physiol.* **64**:582–607.
102. Biagi, B., and G. Giebish. 1979. Temperature dependence of transepithelial potential in isolated perfused rabbit proximal tubule. *Am. J. Physiol* **236:**F302–F310.
103. Boulpaep, E. L., and J. F. Seely. 1971. Electrophysiology of proximal and distal tubules in autoperfused dog kidney. *Am. J. Physiol.* **221**:1084–1096.
104. Frömter, E., and K. Gessner. 1974. Active transport potentials, membrane diffusion potentials and streaming potentials across rat kidney proximal tubule. *Pfluegers Arch.* **351**:85–98.
105. Seely, J. F. 1973. Variation in electrical resistance along length of rat proximal convoluted tubule. *Am. J. Physiol.* **225**:48–57.
106. Grasset, E., P. Gunter-Smith, and S. G. Schultz. 1983. Effects of Na-coupled alanine transport on intracellular K activities and the K conductance of the basolateral membranes of *Necturus* small intestine. *J. Membr. Biol.* **71**:89–94.
107. Lau, K. R., R. L. Hudson, and S. G. Schultz. 1984. Cell swelling

increases a barium-inhibitable, potassium conductance in the basolateral membrane of *Necturus* small intestine. *Proc. Natl. Acad. Sci. USA* **81**:3591–3594.

108. Cardinal, J., J.-Y. LaPointe, and R. Laprade. 1984. Luminal and peritubular ionic substitutions and intracellular potential of the rabbit proximal convoluted tubule. *Am. J. Physiol.* **247**:F352–F364.
109. Matsumura, Y., B. Cohen, W. B. Guggino, and G. Giebisch. 1984. Regulation of the basolateral potassium conductance of the *Necturus* proximal tubule. *J. Membr. Biol.* **79**:153–161.
110. Burckhardt, B. C., K. Sato, and E. Frömter. 1984. Electrophysiological analysis of bicarbonate permeation across the peritubular cell membrane of rat kidney proximal tubule. *Pfluegers Arch.* **401**:34–42.
111. Burckhardt, B. C., A. C. Cassola, and E. Frömter. 1984. Electrophysiological analysis of bicarbonate permeation across the peritubular cell membrane of rat kidney proximal tubule. II Exclusion of HCO_3^- effects on other ion permeabilities and of coupled electroneutral HCO_3^--transport. *Pfluegers Arch.* **401**:43–51.
112. Cardinal, J., M. D. Lutz, M. B. Burg, and J. Orloff. 1975. Lack of relationship of potential difference to fluid absorption in the proximal renal tubule. *Kidney Int.* **7**:94–102.
113. Lutz, M. D., J. Cardinal, and M. B. Burg. 1973. Electrical resistance of renal proximal tubule perfused *in vitro*. *Am. J. Physiol.* **225**:729–734.
114. Wesson, L. G., and W. P. Anslow, Jr. 1948. Excretion of sodium and water during osmotic diuresis in the dog. *Am. J. Physiol.* **153**:465–474.
115. Green, R., and G. Giebisch. 1975. Some ionic requirements of proximal tubular sodium transport: The role of bicarbonate and chloride. *Am. J. Physiol.* **229**:1216–1226.
116. Erlij, D. 1976. Solute transport across isolated epithelia. *Kidney Int.* **9**:76–87.
117. Ussing, H. H., D. Erlij, and U. Lassen. 1974. Transport pathways in biological membranes. *Annu. Rev. Physiol.* **36**:17–49.
118. Ullrich, K. J., H. W. Radtke, and G. Rumrich. 1971. The role of bicarbonate and other buffers on isotonic fluid absorption in the proximal convolution of the rat kidney. *Pfluegers Arch.* **330**:149–161.
119. Kokko, J., M. B. Burg, and J. Orloff. 1971. Characteristics of NaCl and water transport in the renal proximal tubule. *J. Clin. Invest.* **50**:69–75.
120. Giebisch, G., and E. E. Windhager. 1964. Renal tubular transfer of sodium chloride and potassium. *Am. J. Med.* **36**:643–669.
121. Windhager, E. E., G. Whittembury, D. E. Oken, H. J. Schatzmann, and A. K. Solomon. 1959. Single proximal tubules of the *Necturus* kidney. III. Dependence of H_2O movement on NaCl concentration. *Am. J. Physiol.* **197**:313–318.
122. Giebisch, G., R. M. Klose, G. Malnic, W. J. Sullivan, and E. E. Windhager. 1964. Sodium movement across single perfused tubules of rat kidney. *J. Gen. Physiol.* **47**:1175–1194.
123. Györy, A. Z., and R. Kinne. 1971. Energy source for transepithelial sodium transport in rat renal proximal tubules. *Pfluegers Arch.* **327**:234–260.
124. Green, R., E. E. Windhager, and G. Giebisch. 1974. Protein oncotic pressure effects on proximal tubular fluid movement in the rat. *Am. J. Physiol.* **226**:265–276.
125. Koefoed-Johnsen, V., and H. H. Ussing. 1958. The nature of the frog skin potential. *Acta Physiol. Scand.* **42**:298–308.
126. Giebisch, G., and E. E. Windhager. 1973. Electrolyte transport across renal tubular membranes. In: *Handbook of Physiology,* Section 8. J. Orloff and R. W. Berliner, eds. American Physiological Society, Washington, D.C. pp. 315–376.
127. Schultz, S. G. 1982. Homocellular regulatory mechanisms in sodium-transporting epithelia: An extension of the Koefoed-Johnsen–Ussing model. *Semin. Nephrol.* **2**:343–347.
128. Schultz, S. G. 1981. Homocellular regulatory mechanisms in sodium-transporting epithelia: Avoidance of extinction by "flush-through." *Am. J. Physiol.* **241**:F579–F590.
129. Stroup, R. F., E. Weinman, J. P. Hayslett, and M. Kashgarian. 1974. Effect of luminal permeability on net transport across the amphibian proximal tubule. *Am. J. Physiol.* **212**:1341–1349.
130. Spring, K. R., and G. Giebisch. 1977. Kinetics of Na^+ transport in *Necturus* proximal tubule. *J. Gen. Physiol.* **70**:307–328.
131. Bentley, P. J. 1968. Action of amphotericin-B on the toad bladder: Evidence for sodium transport along two pathways. *J. Physiol. (London)* **196**:703–711.
132. Finn, A. L. 1968. Separate effects of sodium and vasopressin on the sodium pump in toad bladder. *Am. J. Physiol.* **215**:849–856.
133. Lichtenstein, N. S., and A. Leaf. 1966. Evidence for a double permeability barrier at the mucosal surface of the toad bladder. *Ann. N.Y. Acad. Sci.* **137**:556–565.
134. Mendoza, S. A., J. S. Handler, and J. Orloff. 1967. Effect of amphotericin-B on permeability and short-circuit current in toad bladder. *Am. J. Phyisol.* **213**:1263–1268.
135. Spring, K. R., and G. Kimura. 1978. Chloride reabsorption by renal proximal tubules of *Necturus. J. Membr. Biol.* **38**:233–254.
136. Seifter, J. L., and P. S. Aronson. 1984. Cl^- transport via anion exchange in *Necturus* renal microvillus membranes. *Am. J. Physiol.* **247**:F888–F895.
137. Lucci, M. S,, and D. G. Warnock. 1979. Effects of anion-transport inhibitors on NaCl reabsorption in the rat superficial proximal convoluted tubule. *J. Clin. Invest.* **64**:570–579.
138. Green, R., G. Giebisch, and J. H. V. Bishop. 1979. Ionic requirements of proximal tubular sodium transport. III. Selective luminal anion substitution. *Am. J. Physiol.* **236**:F268–F277.
139. Giebisch, G., and R. Green. 1981. Anion-driven fluid movement across proximal tubular epithelium. In: *Water Transport across Epithelia.* H. H. Ussing, N. Bindsley, N. A. Lassen, and O. Sten-Knudsen, eds. Munksgaard, Copenhagen. pp. 376–385.
140. Murer, H., and R. Greger. 1982. Membrane transport in the proximal tubule and thick ascending limb of Henle's loop: Mechanisms and their alterations. *Klin. Wochenschr.* **60**:1103–1113.
141. Oberleithner, H., G. Giebisch, F. Lang, and W. Wang. 1982. Cellular mechanism of the furosemide sensitive transport system in the kidney. *Klin. Wochenschr.* **60**:1173–1179.
142. Seifter, J. L., R. Knickelbein, and P. S. Aronson. 1984. Absence of Cl–OH exchange and NaCl cotransport in rabbit renal microvillus membrane vesicles. *Am. J. Physiol.* **247**:F753–F759.
143. Kahn, A. M., and P. S. Aronson. 1983. Urate transport via anion exchange in dog renal microvillus membrane vesicles. *Am. J. Physiol.* **244**:F56–F63.
144. Baum, M., and C. A. Berry. 1984. Evidence for neutral transcellular NaCl transport and neutral basolateral Cl^- exit in rabbit proximal convoluted tubule. *J. Clin. Invest.* **74**:205–211.
145. Green, R., R. J. Moriarty, and G. Giebisch. 1981. Ionic requirements of proximal tubular fluid reabsorption. IV. Flow dependence of fluid transport. *Kidney Int.* **20**:580–587.
146. Cogan, M., and F. C. Rector, Jr. 1982. Determinants of proximal bicarbonate, chloride and water reabsorption during carbonic anhydrase inhibition. *Am. J. Physiol.* **242**:F274–F284.
147. Frömter, E., and K. Gessner. 1975. Effect of inhibitors and diuretics on electrical potential differences in rat kidney proximal tubule. *Pfluegers Arch.* **357**:209–224.
148. Warnock, D. G., and V. J. Yee. 1981. Chloride uptake by brush-border membrane vesicles isolated from rabbit renal cortex: Coupling to proton gradients and K^+ diffusion potentials. *J. Clin. Invest.* **67**:103–115.
149. Cassano, G., B. Stieger, and H. Murer. 1984. Na/H- and Cl/OH-exchange in rat jejunal and rat proximal tubular brush border membrane vesicles. *Pfluegers Arch.* **400**:309–317.
150. Sabolic, I., and G. Burckhardt. 1983. Proton pathways in rat renal brush-border and basolateral membranes. *Biochim. Biophys. Acta* **734**:210–220.
151. Schwartz, G. J. 1983. Absence of Cl^-–OH^- or Cl^-–HCO_3^- exchange in the rabbit renal proximal tubule. *Am. J. Physiol.* **245**: F462–F469.
152. Sasaki, S., Y. Iino, T. Shiigai, and J. Takeuchi. 1984. Intra-

cellular pH of isolated perfused rabbit proximal tubule: Effect of luminal and Na and Cl. *Kidney Int.* **25**:282.
153. Karniski, L. P., and P. S. Aronson. 1985. Recycling of Formic acid: a mechanism of chloride transport across renal microvillus membrane vesicles. Proc. Vth European Colloquium on Renal Physiology (in press).
154. Walter, A., and J. Gutknecht. 1984. Monocarboxylic acid permeation through lipid bilayer membranes. *J. Membr. Biol.* **77**:255–264.
155. Cassola, A. C., M. Mollenhauer, and E. Frömter. 1983. The intracellular chloride activity of rat kidney proximal tubular cells. *Pfluegers Arch.* **399**:259–265.
156. Schafer, J. A., C. S. Patlak, and T. E. Andreoli. 1975. A component of fluid absorption linked to passive ion flows in the superficial pars recta. *J. Gen. Physiol.* **66**:445–471.
157. Wright, E. M. 1984. Electrophysiology of plasma membrane vesicles. *Am. J. Physiol.* **246**:F363–F372.
158. Meng, K. 1975. Comparison of the local effects of amiloride hydrochloride on the isotonic fluid absorption in the distal and proximal convoluted tubule. *Pfluegers Arch.* **356**:91–99.
159. Messner, G., H. Oberleithner, and F. Lang. The effect of phenylalanine on the electrical properties of proximal tubule cells in the frog kidney. *Pfluegers Arch.* in press.
160. Jorgensen, P. L. 1980. Sodium and potassium ion pump in kidney tubules. *Physiol. Rev.* **60**:864–917.
161. Biagi, B. A., M. Sohtell, and G. Giebisch. 1981. Intracellular potassium activity in the rabbit proximal straight tubule. *Am. J. Physiol.* **241**:F677–F686.
162. Proverbio, F., and G. Whittembury. 1975. Cell electrical potentials during enhanced sodium extrusion in guinea-pig cortex slices. *J. Physiol. (London)* **250**:559–578.
163. Sackin, H., and E. L. Boulpaep. 1981. Isolated perfused salamander proximal tubule. II. Monovalent ion replacement and rheogenic transport. *Am. J. Physiol.* **241**:F540–F555.
164. Sackin, H., and E. L. Boulpaep. 1983. Rheogenic transport in the renal proximal tubule. *J. Gen. Physiol.* **82**:819–852.
165. Cardinal, J., and D. Duchesneau, 1978. Effect of potassium on proximal tubular function. *Am. J. Physiol.* **234**:F381–F385.
166. Warnock, D. G., R. Greger, P. B. Dunham, M. A. Benjamin, R. A. Frizzell, M. Field, K. R. Spring, H. E. Ives, P. S. Aronson, and J. Seifter. 1984. Ion transport processes in apical membranes of epithelia. *Fed. Proc.* **43**:2473–2487.
167. Oberleithner, H., M. Ritter, F. Lang, and W. Guggino. 1983. Anthracene-9-carboxylic acid inhibits renal chloride reabsorption. *Pfluegers Arch.* **398**:172–174.
168. Guggino, W., E. Boulpaep, and G. Giebisch. 1982. Electrical properties of chloride transport across the *Necturus* proximal tubule. *J. Membr. Biol.* **65**:185–196.
169. Bello-Reuss, E. 1982. Electrical properties of the basolateral membrane of the straight portion of the rabbit proximal renal tubule. *J. Physiol. (London)* **326**:49–63.
170. Biagi, B. S., T. Kubota, M. Sohtell, and G. Giebisch. 1981. Intracellular potentials in rabbit proximal tubules perfused *in vitro*. *Am. J. Physiol.* **240**:F200–F210.
171. Grassl, S. M., L. P. Karniski, and P. S. Aronson. 1985. Cl–HCO_3 exchange in rabbit renal cortical basolateral membrane vesicles. *Kidney Int.* **27**:282.
172. Nakhoul, N. L., and W. F. Boron, 1985. Intracellular pH regulation in rabbit proximal straight tubules: Basolateral HCO_3 transport, *Kidney Int.* **27**:286.
173. Reuss, L. 1983. Basolateral KCl co-transport in a NaCl-absorbing epithelium. *Nature (London)* **305**:723–726.
174. Mandel, L. J., and R. S. Balaban. 1981. Stoichiometry and coupling of active transport to oxidative metabolism in epithelial tissues. *Am. J. Physiol.* **240**:F357–F371.
175. Mandel, L. J. 1985. Bioenergetics of membrane transport processes. This volume.
176. Whittembury, G., and F. Proverbio. 1970. Two modes of Na extrusion in cells from guinea pig kidney cortex slices. *Pfluegers Arch.* **316**:1–25.
177. Proverbio, F., and J. R. DelCastillo. 1981. Na^+-stimulated ATPase activities in kidney basal-lateral plasma membranes. *Biochim. Biophys. Acta* **646**:99–108.
178. DelCastillo, J. R., R. Marin, T, Proverbio, and F. Proverbio. 1982. Partial characterization of the ouabain-insensitive, Na^+-stimulated ATPase activity of the kidney basal-lateral plasma membranes. *Biochim. Biophys. Acta* **692**:61–68.
179. Boumendil-Podevin, E. F., and R. A. Podevin. 1983. Effects of ATP on Na^+ transport and membrane potential in inside-out renal basolateral vesicles. *Biochim. Biophys. Acta* **728**:39–49.
180. Ross, B., A. Leaf, P. Silva, and F. H. Epstein. 1974. Na-K-ATPase in sodium transport by the perfused rat kidney. *Am. J. Physiol.* **226**:624–629.
181. Gmaj, P., H. Murer, and R. Kinne. 1979. Calcium ion transport across plasma membranes isolated from rat kidney cortex. *Biochem. J.* **178**:549–557.
182. Lee, C. O., A. Taylor, and E. E. Windhager. 1980. Cytosolic calicum ion activity in epithelial cells of *Necturus* kidney. *Nature (London)* **287**:859–861.
183. Lorenzen, M., C. O. Lee, and E. E. Windhager. 1984. Cytosolic Ca^{2+} and Na^+ activities in perfused proximal tubules of *Necturus* kidney. *Am. J. Physiol.* **247**:F93–F102.
184. Taylor, A., and E. E. Windhager. 1979. Possible role of cytosolic calcium and Na–Ca exchange in regulation of transepithelial sodium transport. *Am. J. Physiol.* **236**:F505–F512.
185. Kubota, T., B. A. Biagi, and G. Giebisch. 1983. Intracellular potassium activity measurements in single proximal tubules of *Necturus* kidney. *J. Membr. Biol.* **73**:51–60.
186. Schwartz, W., and H. Passow. 1983. Ca^{++}-activated K^+ channels in erythrocytes and excitable cells. *Annu. Rev. Physiol.* **45**:359–374.
187. Brown, C. D. A., and N. L. Simmons. 1982. K^+ transport in "tight" epithelial monolayers of MDCK cells: Evidence for calcium-activated K^+-channel. *Biochim. Biophys. Acta* **690**:95–102.
188. Lorenzen, M., C. O. Lee, and E. E. Windhager. 1985. Effect of gramicidin and reduction of luminal $[Na^+]$ or cytosolic $[Ca^{++}]$ on $[Na^+]$ activity in isolated perfused *Necturus* proximal tubule. *Kidney Int.* **27**:315.
189. Wang, W., G. Messner, H. Oberleithner, and F. Lang. 1984. The effect of ouabain on intracellular activities of K^+, Na^+, Cl^-, H^+ and Ca^{2+} in proximal tubules of frog kidneys. *Pfluegers Arch.* **401**:6–13.
190. Lang, F., G. Messner, W. Wang, and H. Oberleithner. 1983. Interaction of intracellular electrolytes and tubular transport. *Klin. Wochenschr.* **61**:1029–1037.
191. Boron, W. F., and E. L. Boulpaep. 1983. Intracellular pH regulation in the renal proximal tubule of the salamander: Na–H exchange. *J. Gen. Physiol.* **81**:29–52.
192. Struyvenberg, A., R. B. Morrison, and A. S. Relman. 1968. Acid–base behavior of separated canine renal tubule cells. *Am. J. Physiol.* **214**:1155–1162.
193. Murer, H., U. Hopfer, and R. Kinne. 1976. Sodium/proton antiport in brush-border-membrane vesicles isolated from rat small intestine and kidney. *Biochem. J.* **154**:597–604.
194. Kinsella, J. L., and P. S. Aronson. 1980. Properties of the Na^+–H^+ exchanger in renal microvillus membrane vesicles. *Am. J. Physiol.* **238**:F461–F469.
195. Cohen, D. E., K. A. Hruska, S. Klahr, and M. R. Hammerman. 1982. Increased Na^+–H^+ exchange in brush border vesicles from dogs with renal failure. *Am. J. Physiol.* **243**:F293–F299.
196. Kinsella, J. L., and P. S. Aronson. 1981. Amiloride inhibition of the Na^+–H^+ exchanger in renal microvillus membrane vesicles. *Am. J. Physiol.* **241**:F374–F379.
197. Chaillet, J. R., and W. F. Boron. 1984. Intracellular pH regulation in rabbit proximal tubules studied with a pH-sensitive dye. *Kidney Int.* **25**:273a.
198. Sacktor, B., and L. Cheng. 1981. Sodium gradient-dependent phosphate transport in renal brush border membrane vesicles: Effect of an intravesicular > extravesicular proton gradient. *J. Biol. Chem.* **256**:8080–8084.

199. Blomstedt, J. W., and P. S. Aronson. 1980. pH gradient-stimulated transport of urate and *p*-aminohippurate in dog renal microvillus membrane vesicles. *J. Clin. Invest.* **65**:931–934.
200. Guggino, S. E., G. J. Martin, and P. S. Aronson. 1983. Specificity and modes of the anion exchanger in dog renal microvillus membranes. *Am. J. Physiol.* **244**:F612–F621.
201. Kahn, A. M., S. Branham, and E. J. Weinman. 1983. Mechanism of urate and *p*-aminohippurate transport in rat microvillus membrane vesicles. *Am. J. Physiol.* **245**:F151–F158.
202. Nord, E. P., S. H. Wright, I. Kippen, and E. M. Wright. 1983. Specificity of the Na^+-dependent monocarboxylic acid transport pathway in rabbit renal brush border membranes. *J. Membr. Biol.* **72**:213–221.
203. Kinne-Saffran, E., and R. Kinne. 1974. Presence of bicarbonate stimulated ATPase in the brush border microvillus membranes of the proximal tubule. *Proc. Soc. Exp. Biol. Med.* **146**:751–753.
204. Kinne-Saffran, E., and R. Kinne. 1979. Further evidence for the existence of an intrinsic bicarbonate-stimulated Mg^{2+}-ATPase in brush border membranes isolated from rat kidney cortex. *J. Membr. Biol.* **49**:235–251.
205. Kinne-Saffran, E., R. Beauwens, and R. Kinne. 1982. An ATP-driven proton pump in brush-border membranes from rat renal cortex. *J. Membr. Biol.* **64**:67–76.
206. Gimenez-Gallego, G., J. Benavides, M. L. Garcia, and F. Valdivieso. 1980. Occurrence of a reduced nicotinamide adenine dinucleotide oxidase activity linked to a cytochrome system in renal brush border membranes. *Biochemistry* **19**:4834–4839.
207. Garcia, M. L., J. Benavides, G. Gimenez-Gallego, and F. Valdivieso. 1980. Coupling between reduced nicotinamide adenine dinucleotide oxidation and metabolite transport in renal brush border membrane vesicles. *Biochemistry* **19**:4840–4843.
208. Forgac, M., L. Cantley, B. Wiedenmann, L. Altstiel, and D. Branton. 1983. Clathrin-coated vesicles contain an ATP-dependent proton pump. *Proc. Natl. Acad, Sci. USA* **80**:1300–1303.
209. Stone, D. K., X,-S. Xie, and E. Racker. 1983. An ATP-driven proton pump in clathrin-coated vesicles. *J. Biol. Chem.* **258**: 4059–4062.
210. Galloway, C. J., G. E. Dean, M. Marsh, G. Rudnick, and I. Mellman. 1983. Acidification of macrophage and fibroblast endocytic vesicles *in vitro*. *Proc. Natl. Acad. Sci. USA* **80**:3334–3338.
211. Schneider, D. 1981. ATP-dependent acidification of intact and disrupted lysosomes: Evidence for an ATP-driven proton pump. *J. Biol. Chem.* **256**:3858–3864.
212. Rodman, J. S., D. Kerjaschki, E. Merisko, and M. G. Farquhar. 1984. Presence of an extensive clathrin coat on the apical plasmalemma of the rat kidney proximal tubule cell. *J. Cell Biol.* **98**:1630–1636.
213. Lang, F., P. Quehenberger, R. Greger, S. Silbernagl, and P. Stockinger. 1980. Evidence for a bicarbonate leak in the proximal tubule of the rat kidney. *Pfluegers Arch.* **386**:239–244.
214. Reenstra, W. W., D. G. Warnock, V. J. Yee, and J. G. Forte. 1981. Proton gradients in renal cortex brush-border membrane vesicles: Demonstration of a rheogenic proton flux with acridine orange. *J. Biol. Chem.* **256**:11663–11666.
215. Costa Silva, V. L., S. S. Campiglia, M. de Mello Aires, G. Malnic, and G. Giebisch. 1981. Role of luminal buffers in renal tubular acidification. *J. Membr. Biol.* **63**:13–24.
216. Ganapathy, V., and F. H. Leibach. 1983. Role of pH gradient and membrane potential in dipeptide transport in intestinal and renal brush-border membrane vesicles from the rabbit. *J. Biol. Chem.* **258**:14189–14192.
217. Nelson, P. J., G. E. Dean, P. S. Aronson, and G. Rudnick. 1983. Hydrogen ion co-transport by the renal brush border glutamate transporter. *Biochemistry* **22**:5459–5463.
218. Holohan, P. D., and C. R. Ross. 1983. Mechanisms of organic cation transport in kidney plasma membrane vesicles. 2. Δ pH studies. *J. Pharmacol. Exp. Ther.* **216**:294–298.
219. Chantrelle, B., M. G. Cogan, and F. C. Rector, Jr. 1982. Evidence for coupled sodium/hydrogen exchange in the rat superficial proximal convoluted tubule. *Pfluegers Arch.* **395**:186–189.
220. Sasaki, S., C. A. Berry, and F. C. Rector, Jr. 1983. Effect of potassium concentration on bicarbonate reabsorption in the rabbit proximal convoluted tubule. *Am. J. Physiol.* **244**:F122–F128.
221. Cohn, D. E., S. Klahr, and M. R. Hammerman. 1983. Metabolic acidosis and parathyroidectomy increase Na^+–H^+ exchange in brush border vesicles. *Am. J. Physiol.* **245**:F217–F222.
222. Seifter, J. L., and R. C. Harris. 1984. Chronic K depletion increases Na–H exchange in rat renal cortical brush-border membrane vesicles. *Kidney Int.* **25**:282.
223. Freiberg, J. M., J. Kinsella, and B. Sacktor. 1982. Glucocorticoids increase the Na^+–H^+ exchange and decrease the Na^+ gradient-dependent phosphate-uptake systems in renal brush border membrane vesicles. *Proc. Natl. Acad. Sci. USA* **79**:4932–4936.
224. de Mello Aires, M., and G. Malnic. 1979. Sodium in renal tubular acidification kinetics. *Am. J. Physiol.* **236**:F434–F441.
225. Green, R., and G. Giebisch. 1975. Some ionic requirements of proximal tubular sodium transport. II. The role of hydrogen ion secretion. *Am. J. Physiol.* **229**:1205–1215.
226. Chan, Y. L., and G. Giebisch. 1981. Relationship between sodium and bicarbonate transport in the rat proximal convoluted tubule. *Am. J. Physiol.* **240**:F222–F230.
227. Ullrich, K. J., G. Capaso, G. Rumrich, F. Papavassiliou, and S. Kloss. 1977. Coupling between proximal tubular transport processes. *Pfluegers Arch.* **368**:245–252.
228. Bichara, M., M. Paillard, F. Leviel, A. Prigent, and J.-P. Gardin. 1983. Na:H exchange and the primary H pump in the proximal tubule. *Am. J. Physiol.* **244**:F165–F171.
229. Bichara, M., M. Paillard, F. Leviel, and J.-P. Gardin. 1980. Hydrogen transport in rabbit kidney proximal tubules—Na:H exchange. *Am. J. Physiol.* **238**:F445–F451.
230. Beck, F., R. Bauer, U. Bauer, J. Mason, A. Dorge, R. Rick, and K. Thurau. 1980. Electron microprobe analysis of intracellular elements in the kidney. *Kidney Int.* **17**:756–763.
231. Misanko, B. S., and S. Solomon, 1981. Activity of the HCO_3^--stimulated ATPase in the acidotic rat kidney. *Miner. Electrolyte Metab.* **6**:217–226.
232. Haase, W., A. Schafer, H. Murer, and R. Kinne. 1978. Studies on the orientation of brush-border membrane vesicles. *Biochem. J.* **172**:57–62.
233. Boron, W. F., and E. L. Boulpaep. 1983. Intracellular pH regulation in the renal proximal tubule of the salamander: Basolateral HCO_3^- transport. *J. Gen. Physiol.* **81**:53–94.
234. Kleinman, J. G., R. A. Ware, and J. H. Schwartz. 1981. Anion transport regulates intracellular pH in renal cortical tissue. *Biochim. Biophys. Acta* **648**:87–92.
235. Edelman, A., M. Bouthier, and T. Anagnostopoulos. 1981. Chloride distribution in the proximal convoluted tubule of *Necturus* kidney. *J. Membr. Biol.* **62**:7–17.
236. Low, I., T. Friedrich, and G. Burckhardt. 1984. Properties of an anion exchanger in rat renal basolateral membrane vesicles. *Am. J. Physiol.* **246**:F334–F342.
237. Ives, H. E., V. J. Yee, and D. G. Warnock. 1983. Asymmetric distribution of the Na^+/H^+ antiporter in the renal proximal tubule epithelial cell. *J. Biol. Chem.* **258**:13513–13516.
238. Pritchard, J. B., and J. L. Renfro. 1983. Renal sulfate transport at the basolateral membrane is mediated by anion exchange. *Proc. Natl. Acad. Sci. USA* **80**:2603–2607.
239. Guggino, W. B., R. London, E. L. Boulpaep, and G. Giabisch. 1983. Chloride transport across the basolateral cell membrane of the *Necturus* proximal tubule: Dependence on bicarbonate and sodium. *J. Membr. Biol.* **71**:227–240.
240. Kleinman, J. G., W. W. Brown, R. A. Ware, and J. H. Schwartz. 1980. Cell pH and acid transport in renal cortical tissue. *Am. J. Physiol.* **239**:F440–F444.
241. Sasaki, S., and C. A. Berry. 1984. Mechanism of bicarbonate exit across basolateral membrane of the rabbit proximal convoluted tubule. *Am. J. Physiol.* **246**:F889–F896.
242. Frömter, E. 1974. Electrophysiology and isotonic fluid absorption of proximal tubules of mammalian kidney. In: *MTP International*

Review of Science, Physiology Series I, Volume 6. K. Thurau, ed. Butterworths/University Park Press, Baltimore. pp. 1–38.

243. Frömter, E., and K. Gessner. 1974. Free flow potential profile along rat kidney proximal tubule. *Pfluegers Arch.* **351**:69–84.
244. Frömter, E., G. Rumrich, and K. J. Ullrich. 1973. Phenomenologic description of Na^+, Cl^-, and HCO_3^- absorption from proximal tubules of the rat kidney. *Pfluegers Arch.* **343**:189–220.
245. Rector, F. C., M. Martinez-Maldonado, F. P. Brummer, and D. W. Seldin. 1966. Evidence for passive reabsorption of NaCl in proximal tubule of rat kidney. *J. Clin. Invest.* **45**:1060.
246. Kokko, J., F. C. Rector, Jr., and D. W. Seldin. 1970. Mechanism of salt and water reabsorption in proximal convoluted tubule (PCT). *Proc. Am. Soc. Nephrol.* **4**:42.
247. Schafer, J. A., C. S. Patlak, and T. E. Andreoli. 1975. A component of fluid absorption linked to passive ion flows in the superficial pars recta. *J. Gen. Physiol.* **66**:445–471.
248. Schafer, J. A., S. L. Troutman, M. L. Watkins, and T. E. Andreoli. 1981. Flow dependence of fluid transport in the isolated superficial pars recta: Evidence that osmotic disequilibrium between external solutions drives isotonic fluid absorption. *Kidney Int.* **20**:588–597.
249. Green, R., R. J. Moriarty, and G. Giebisch. 1981. Ionic requirements of proximal tubular fluid reabsorption: Flow dependence of fluid transport. *Kidney Int.* **20**:580–587.
250. Chantrelle, B., and F. C. Rector, Jr. 1980. Active and passive components of volume resorption in rat superficial proximal tubules. *Clin. Res.* **28**:441a.
251. Green, R., and G. Giebisch. 1975. Ionic requirements of proximal tubular sodium transport. I. Bicarbonate and chloride. *Am. J. Physiol.* **229**:1205–1215.
252. Green, R., and G. Giebisch. 1975. Some ionic requirements of proximal tubular sodium transport. II. The role of hydrogen ion secretion. *Am. J. Physiol.* **229**:1216–1226.
253. Berry, C. A. 1983. Water permeability and pathways in the proximal tubule. *Am. J. Physiol.* **245**:F275–F294.
254. Welling, L. W., D. J. Welling, and T. J. Ochs. 1983. Video measurement of basolateral membrane hydraulic conductivity in the proximal tubule. *Am. J. Physiol.* **245**:F123–F129.
255. Gonzalez, E., P. Carpi-Medina, and G. Whittembury. 1982. Cell osmotic water permeability of isolated rabbit proximal straight tubules. *Am. J. Physiol.* **242**:F321–F330.
256. Carpi-Medina, P., E. Gonzalez, and G. Whittembury. 1983. Cell osmotic water permeability of isolated rabbit proximal convoluted tubules. *Am. J. Physiol.* **244**:F554–F563.
257. Gonzalez, E., P. Carpi-Medina, H. Linares, and G. Whittembury. 1984. Water osmotic permeability of the apical membrane of proximal straight tubular cells. *Pfluegers Arch.* **402**:337–339.
258. Whittembury, G. 1985. Mechanisms of epithelial solute–solvent coupling. In: *The Kidney: Normal and Abnormal Function.* D. Seldin and G. Giebisch, eds. Raven Press, New York, pp. 199–214.
259. Spring, K. R., and A.-C. Ericson. 1982. Epithelial cell volume modulation and regulation. *J. Membr. Biol.* **69**:167–176.
260. Curran, P. F., and J. R. MacIntosh. 1962. A model system for biological water transport. *Nature (London)* **193**:347–348.
261. Curran, P. F. 1972. Solute–solvent interactions and water transport. In: *Role of Membranes in Secretory Processes.* L. Bolis, R. B. Keynes, and W. Wilbrandt, eds. American Elsevier/North-Holland, Amsterdam. pp. 408–419.
262. Diamond, J. M., and W. H. Bossert. 1967. Standing gradient osmotic flow: A mechanism for coupling of water and solute transport in epithelia. *J. Gen. Physiol.* **50**:2061–2083.
263. Sackin, H., and E. L. Boulpaep. 1975. Models for coupling of salt and water transport: Proximal tubular reabsorption in *Necturus* kidney. *J. Gen. Physiol.* **66**:671–733.
264. Green, R., and G. Giebisch. 1984. Luminal hypotonicity: A driving force for fluid absorption from the proximal tubule. *Am. J. Physiol.* **246**:F167–F174.
265. Barfuss, D. W., and J. A. Schafer. 1984. Rate of formation and composition of absorbate from proximal nephron segments. *Am. J. Physiol.* **249**:F117–F129.
266. Barfuss, D. W., and J. A. Schafer. 1984. Hyperosmolality of absorbate from isolated rabbit proximal tubules. *Am. J. Physiol.* **247**:F130–F139.
267. Weinstein, A. M., J. L. Stephenson, and K. R. Spring. 1981. The coupled transport of water. In: *Membrane Transport.* S. L. Bonting and J. J. H. H. M. dePont, eds. Elsevier/North-Holland, Amsterdam. pp. 311–351.
268. Schafer, J. 1984. Mechanisms coupling the absorption of solutes and water in the proximal nephron. *Kidney Int.* **25**:708–716.
269. Ussing, H. H., N. Bindslev, N. A. Lassen, and O. Sten-Knudsen. 1981. *Water Transport across Epithelia.* Munksgaard, Copenhagen.
270. Whitlock, R. T., and H. O. Wheeler. 1964. Coupled transport of solute and water across rabbit gallbladder epithelium. *J. Clin. Invest.* **43**:2249–2265.
271. Schafer, J. A., C. S. Patlak, and T. E. Andreoli. 1975. A component of fluid absorption linked to passive flows in the superficial pars recta. *J. Gen. Physiol.* **60**:445–471.
272. Andreoli, T. E., and J. A. Schafer. 1979. Effective luminal hypotonicity: The driving force for isotonic proximal tubular fluid reabsorption. *Am. J. Physiol.* **236**:F89–F96.
273. Welling, L. W., and D. J. Welling. 1976. Shape of epithelial cells and intercellular channels in the rabbit proximal tubule. *Kidney Int.* **9**:385–394.
274. Maunsbach, A. B., and E. L. Boulpaep. 1984. Quantitative ultrastructure and functional correlates in proximal tubule of *Ambystoma* and *Necturus*. *Am. J. Physiol.* **246**:F710–F724.
275. Bishop, J. H. V., R. Green, and S. Thomas. 1979. Free-flow reabsorption of glucose, sodium, osmoles, and water in rat proximal convoluted tubules. *J. Physiol. (London)* **288**:331–351.
276. Liu, F.-Y., M. G. Cogan, and F. C. Rector. 1984. Axial heterogeneity of anion and water transport and of osmotic water permeability along the rat superficial proximal convoluted tubule. *Kidney Int.* **25**:308a.
277. Neumann, K. H., and F. C. Rector, Jr. 1976. Mechanism of NaCl and water reabsorption in the proximal convoluted tubule of rat kidney. *J. Clin. Invest.* **58**:1110–1118.
278. Jacobson, H. R., J. P. Kokko, D. W. Seldin, and C. Holmberg. 1982. Lack of solvent drag of NaCl and $NaHCO_3$ in rabbit proximal tubules. *Am. J. Physiol.* **243**:F342–F348.
279. Corman, B., and A. DiStefano. 1983. Does water drag solute through kidney proximal tubule? *Pfluegers Arch.* **397**:35–41.
280. Persson, E., and H. R. Ulfendahl. 1970. Water permeability in rat proximal tubules. *Acta Physiol. Scand.* **78**:353–363.
280a. Whittembury, G., C. Verde-Martinez, H. Linas, and A. Paz-Aliage. 1980. Solvent drag of large solutes indicates paracellular water flow in leaky epithelia. *Proc. R. Soc. London Ser.* B **211**: 63–81.
281. Boulpaep, E. L., and S. Tripathi. 1984. Evidence for both paracellular and cellular water flow across the isolated perfused proximal tubule of the salamander *Ambystoma tigrinum*. *Proc. Physiol. Soc.* p. 83.
282. Lewy, J. E., and E. E. Windhager. 1968. Peritubular control of proximal tubular fluid reabsorption in the rat kidney. *Am. J. Physiol.* **214**:943–954.
283. Weinstein, A. M., and E. E. Windhager. 1985. Sodium transport along the proximal tubule. In: *The Kidney: Normal and Abnormal Function.* D. Seldin and G. Giebisch, eds. Raven Press, New York, pp. 1033–1062.
284. Windhager, E. E., J. E. Lewy, and A. Spitzer. 1969. Intrarenal control of proximal tubular reabsorption of sodium and water. *Nephron* **6**:247–259.
285. Windhager, E. E. 1973. Peritubular control of proximal tubular fluid reabsorption. In: *Transport Mechanisms in Epithelia.* H. H. Ussing and N. A. Thorn, eds. Academic Press, New York. pp. 596–606.
286. Chan, Y. L., G. Malnic, and G. Giebisch. 1983. Passive driving forces of proximal tubular fluid and bicarbonate transport: Gra-

dient dependence of H^+ secretion. *Am. J. Physiol.* **245**:F622–F633.
287. Grandchamp, A., and E. L. Boulpaep. 1974. Pressure control of sodium reabsorption and intercellular backflux across proximal kidney tubule. *J. Clin. Invest.* **54**:69–82.
289. Boulpaep, E. L. 1972. Permeability changes of the proximal tubule of *Necturus* during saline loading. *Am. J. Physiol.* **222**:517–531.
290. Ott, C. E., J. A. Hass, J. L. Cuche, and F. G. Knox. 1975. Effect of increased peritubular protein concentration on proximal tubule reabsorption in the presence and absence of extracellular volume expansion. *J. Clin. Invest.* **55**:612–620.
291. Burg, M., C. Patlak, N. Green, and D. Villey. 1976. The role of organic solutes in fluid absorption by renal proximal convoluted tubules. *Am. J. Physiol.* **231**:627–637.
292. Grantham, J. J., P. B. Qualizza, and L. W. Welling. 1972. Influences of serum proteins on net fluid reabsorption of isolated proximal tubular. *Kidney Int.* **2**:66–75.
293. Imai, M., and J. P. Kokko. 1972. Effect of peritubular protein concentration on reabsorption of sodium and water in isolated perfused proximal tubules. *J. Clin. Invest.* **51**:314–325.
294. Berry, C., and M. Cogan. 1981. Influence of peritubular protein on solute absorption in the rabbit proximal tubule: A specific effect on NaCl transport. *J. Clin. Invest.* **68**:506–516.
295. Knox, F. G., J. I. Mertz, C. Burnett, Jr., and A. Horamati. 1983. Role of hydrostatic and oncotic pressures in renal sodium reabsorption. *Circ. Res.* **52**:491–500.
296. Ott, C. E. 1981. Effect of saline expansion on peritubular capillary pressures and reabsorption. *Am. J. Physiol.* **240**:F106–F110.
297. Persson, A. E. G., J. Schnermann, B. Agerup, and N. E. Eriksson. 1975. The hydraulic conductivity of the rat proximal tubular wall determined with colloidal solution. *Pfluegers Arch.* **360**:25–44.
298. Schnermann, J. 1974. Physical forces and transtubular movement of solutes and water. In: *MTP International Review of Science,* Physiology Series I, Volume 6. K. Thurau, ed. Butterworths/University Park Press, Baltimore. pp. 157–198.
299. Burg, M., and N. Green. 1976. Role of monovalent ions in the reabsorption of fluid by isolated perfused proximal renal tubules of the rabbit. *Kidney Int.* **10**:221–228.
300. Györy, A. Z., and R. Kinne. 1971. Energy source for transepithelial sodium transport in rat renal proximal tubules. *Pfluegers Arch.* **327**:234–260.
301. Windhager, E. E., and G. Giebisch. 1976. Proximal sodium and fluid transport. *Kidney Int.* **9**:121–133.
302. Brenner, B. M., and J. L. Troy. 1971. Postglomerular vascular protein concentration: Evidence for a causal role in governing fluid reabsorption and glomerulotubular balance by the renal proximal tubule. *J. Clin. Invest.* **50**:336–349.
303. Brenner, B. M., J. L. Troy, and T. M. Daugharty. 1971. On the mechanism of inhibition of fluid reabsorption by the renal proximal tubule of the volume-expanded rat. *J. Clin. Invest.* **50**:1596–1602.
304. Lassiter, W. E. 1975. Kidney. *Annu. Rev. Physiol.* **37**:371–393.
305. Bartoli, E., J. C. Conger, and L. E. Earley. 1973. Effect of intraluminal flow on proximal tubular reabsorption. *J. Clin. Invest.* **52**:843–849.
306. Richardson, I. W., V. Licko, and E. Bartoli. 1973. The nature of passive flows through tightly folded membranes. *J. Membr. Biol.* **11**:293–308.
307. Alpern, R. J., M. G. Cogan, and F. C. Rector, Jr. 1983. Flow dependence of proximal tubular bicarbonate reabsorption. *Am. J. Physiol.* **245**:F478–F484.
308. Marsh, D. J. 1981. Models of flow and pressure modulating isosmotic reabsorption in mammalian proximal tubules. In: *Physiology of Non-excitable Cells.* J. Salanki, ed. Pergamon Press, Elmsford, N.Y. pp. 47–55.
309. Häberle, D. A., and H. von Baeyer. 1983. Characteristics of glomerulotubular balance. *Am. J. Physiol.* **244**:F355–F366.

CHAPTER 15

The Effects of ADH on Salt and Water Transport in the Mammalian Nephron

The Collecting Duct and Thick Ascending Limb of Henle

Steven C. Hebert and Thomas E. Andreoli

1. Introduction

A cardinal function of the kidney is the separation of salt and water excretion, thus maintaining the constancy of both osmolality and composition of body fluids despite the wide variations in water and solute intake. The processes that accomplish this task are complex and involve the integrated action of virtually all nephron segments, coupled with that of a specialized vascular system (see Chapter 39, Concentrating and Diluting Processes).

This chapter will deal with a specific and fundamental aspect of the concentrating mechanism, i.e., the modulation of salt and water transport by antidiuretic hormone (ADH) in the mammalian medullary thick ascending limb of Henle (mTALH) and the cortical collecting duct (CCT). Sufficient information is currently available to indicate that ADH regulates urinary concentrating ability, independently of salt excretion, by at least two processes: first, ADH increases the osmotic driving force (i.e., interstitial osmolality) for water movement across the collecting duct by stimulating NaCl absorption by the mTALH; and second, the hormone increases dramatically the water permeability of collecting ducts, thus permitting, during antidiuresis, virtually complete equilibration of luminal tubular fluid with a hypertonic medullary interstitium.

Most of the experimental observations cited in this chapter were obtained from studies with the *in vitro* microperfused rabbit CCT and the mouse mTALH, since the effects of ADH are quite dramatic in these nephron segments in these species. In the mouse mTALH, these data will be used to argue that NaCl absorption in this nephron segment is rheogenic and involves a secondary active transport process, and that ADH increases NaCl absorption by increasing the functional number of both electroneutral Na^+–K^+–$2Cl^-$ cotransport units and K^+ conductance units in apical plasma membranes. Similarly, in the rabbit CCT, ADH also affects apical membrane permeability, but in this case almost exclusively to water, by increasing the number of small, water-conductive channels. This latter process may involve the incorporation of "channel-containing" cytosolic vacuoles into apical membranes.

2. Intracellular Mediators of ADH Action

It is well recognized that the effects of ADH on transport processes in renal epithelia are mediated by the intracellular second messenger cAMP. Intracellular cAMP appears to act via phosphorylation of a cell-specific protein kinase, which in turn effects an alteration in transport processes located at the luminal membrane of renal epithelia to augment water transport, in the collecting duct cell, and NaCl transport, in the mTALH cell. The level of cAMP within the cell is reduced, at least in part, through enzymatic cleavage to 5′-AMP by cytosolic phosphodiesterase, a process which serves to terminate hormone action.

Evidence for this chain of events in ADH-responsive epithelia was first provided by Orloff and Handler,[1] who observed that in toad urinary bladder, either cAMP or the phos-

Steven C. Hebert • Division of Nephrology, University of Texas Medical School, Houston, Texas 77225. **Thomas E. Andreoli** • Departments of Internal Medicine, and Physiology and Cell Biology, University of Texas Medical School, Houston, Texas 77225. Present address of S.C.H.: Department of Internal Medicine, Brigham and Women's Hospital, Boston, Massachusetts.

phodiesterase inhibitor theophylline brought about changes in Na^+ and water transport identical to those observed with ADH. This finding has been documented by subsequent work in a number of intact tissues, including isolated rabbit CCT[2] and mouse mTALH.[3]

More recently, Morel and colleagues (see Ref. 4) have performed elegant studies, utilizing individual nephron segments, which have identified the primary loci of vasopressin-stimulated adenylate cyclase to be in the mTALH and along the entire collecting duct. The intimate relation between hormone binding and adenylate cyclase activation was firmly established by Jard *et al.*,[5] who described a close correlation between binding of analogs of lysine vasopressin and adenylate cyclase activation, as well as comparable half-times of lysine vasopressin binding and adenylate cyclase activation. Their work, and that of others,[6] has led to the suggestion that binding of neurohypophyseal hormones to only a small fraction of receptors is necessary to activate sufficient adenylate cyclase for a maximal physiological response; i.e., at usual levels of circulating ADH, "spare" receptors on epithelial cells are unoccupied and cAMP generation rates are less than maximal.

The finding that the bovine renal medulla contains a cAMP-dependent protein kinase which phosphorylates membrane proteins from that tissue has led to the conclusion that cAMP-dependent protein phosphorylation is the next step in the sequence of intracellular events mediating the effects of ADH on renal epithelial transport.[7] Studies in intact renal medullary tissue have shown the rate of activation of protein kinase to be proportional to the concentration of ADH bathing the tissue, and to the concentration of cAMP achieved within the tissue.[8]

The final events in the ADH-activated sequence which alter the transport characteristics of apical membranes are still unknown. However, morphological studies have demonstrated that patches of membrane held in submembrane granules may be added to the apical surface of cells stimulated by ADH.[9] Furthermore, aggregates of intramembranous particles are seen in apical membranes of ADH-responsive epithelia stimulated by hormone and may, in part, be inserted from cytoplasmic vesicles.[9] Treatment of amphibian bladders with colchicine prior to exposure to ADH decreases: the number of microtubules within cells; the number of fusion events between apical membranes and aggregate-containing submembrane vesicles; the number of aggregates appearing in the luminal membrane; and the hydroosmotic response to ADH.[10] The effect is not seen if colchicine is added after tissue stimulation by ADH. These morphological studies,[9,10] considered together, may be used to infer that ADH, working via cAMP and protein kinase, alters transport in hormone-responsive epithelia by causing the microtubule-dependent insertion of specialized membrane units within the apical plasma membranes of these cells. As will be noted subsequently, this notion is also in accord with conclusions derived from physiological studies of the action of ADH on the mammalian CCT, and possibly on the mTALH.

3. The Medullary Thick Ascending Limb

The notion that ADH might regulate urinary concentrating power by modulating the rate of NaCl absorption in the mTALH was set forth originally by Wirz.[11] Subsequently, Morel provided strong support for this view with the demonstration that ADH increased adenylate cyclase activity in isolated mTALH segments.

The initial insights into the transport characteristics of isolated mTALH and cTALH segments were provided by Burg and Green[12] and Rocha and Kokko.[13] Three general features of NaCl absorption by the thick ascending limb, either cTALH or mTALH, emerged from these latter studies. First, net Cl^- absorption proceeded against a transepithelial electrochemical gradient and was associated with a lumen-positive transepithelial voltage (V_e, mV). Both net Cl^- absorption and the transepithelial voltage could be abolished by "loop" diuretics such as furosemide. Second, both net Cl^- absorption and the transport-related transepithelial voltage depended on the activity of basolateral membrane Na^+,K^+-ATPase, an enzyme present in large amounts in TALH.[14–16] Third, the ionic permeability of the TALH was as high as in the proximal nephron, while the water permeability was remarkably low.

A second set of insights into the salt transport characteristics of the TALH was provided by Greger and Schlatter,[17] who noted that Ba^{2+} (an agent known to block K^+ channels in epithelia[18–22] and in excitable tissues[23] reduced the transepithelial electrical conductance (G_e, mS/cm^2); and that net Cl^- absorption depended on the presence of both luminal Na^+ and K^+.[3,17,24,25] These workers therefore deduced that apical membranes of the TALH might contain conductive K^+ channels, and that NaCl uptake from luminal fluids into cells might involve a cotransport process requiring Na^+, K^+, and Cl^-.

3.1. The Mode of NaCl Absorption by the mTALH

Figure 1 presents a general model for net NaCl absorption in the mTALH which reconciles current experimental data on this nephron segment in mammalian species, notably the mouse and the rabbit, and in amphibian diluting segments. There is a general consensus among investigators[26,27] that net transepithelial Cl^- absorption in the mammalian TALH involves a secondary active transport process in which luminal Cl^- entry into cells is mediated by an electroneutral Na^+–K^+–$2Cl^-$ cotransport mechanism. Recent studies assessing either tracer Na^+ and Cl^- uptake,[28] or the binding of the radiolabeled "loop" diuretic [^{3}H]bumetanide by apical membrane vesicles prepared from TALH segments,[29] have confirmed the dependence of Cl^- uptake on both Na^+ and K^+ in this nephron segment.

No measurements have yet been made of the electrochemical gradient for ion cotransport across apical membranes of the mammalian TALH. However, in the *Amphiuma* diluting segment, there is a favorable integrated chemical gradient for entry of neutral Na^+–K^+–$2Cl^-$ units from luminal fluids into cells.[30] The driving force for Cl^- entry across the apical membrane of the mammalian mTALH is probably provided by the Na^+ electrochemical gradient, with the latter maintained by basolateral membrane Na^+,K^+-ATPase. In accord with this notion, maneuvers which inhibit Na^+,K^+-ATPase, such as addition of ouabain to, or removal of K^+ from, peritubular solutions, have been shown to abolish NaCl absorption and the transepithelial voltage in the mTALH.[3,12,13]

Cl^- exit from the cell across the basolateral membrane of the mTALH appears to be primarily conductive.[25,27] This notion derives in part from the observations that net Cl^- absorption accounts for about 90% of the equivalent short-circuit current in both the mouse mTALH[25] and the rabbit cTALH.[31] In addition, the large intracellular negative voltage of −40 to −70 mV (cell with respect to bath) provides a portion of the driving force for conductive transport of Cl^- across the basolateral membrane

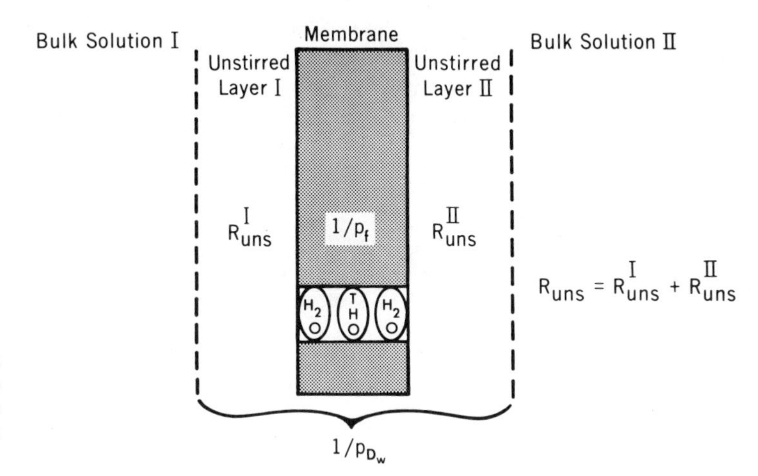

Fig. 1. A model for NaCl absorption and K^+ secretion in the mouse medullary thick ascending limb of Henle. The solid lines denote conservative processes (primary or secondary) and the dashed lines, dissipative processes. From Hebert *et al.*[25]

in these nephron segments.[25–27,32] Furthermore, during net Cl^- absorption, cell Cl^- is above its electrochemical equilibrium concentration in the rabbit cTALH[33] and in *Amphiuma* diluting segments.[30]

Apical membranes of diluting segments, either mammalian[17,25] or amphibian,[30,34,35] also contain Ba^{2+}-sensitive K^+ conductances which account almost entirely for the electrical conductance of apical membranes, i.e., the K^+ transference number across apical membranes is near unity. These K^+ channels exhibit many of the characteristics of K^+ channels in excitable tissues and other epithelia,[18,22,23] such as voltage dependence, concentration dependence, and Ba^{2+}/K^+ competition effects on electrical conductance. These channels also constitute the route for the active K^+ secretion noted in renal tubular diluting segments.[25,36] However, it appears that the majority of K^+ transported from cells to lumen is recycled back into cells via the Na^+–K^+–$2Cl^-$ cotransport process. Thus, the rate of net K^+ secretion constitutes less than 10% of the rate of net Cl^- absorption, while the calculated K^+ current across apical membranes is approximately 60% of the rate of net Cl^- absorption.[25,32]

Since dilution of the urine by the TALH occurs through the net absorption of equal quantities of Na^+ and Cl^-, a stoichiometry of $1Na^+ : 2Cl^-$ for the transcellular transport mechanism requires that one-half of net Na^+ absorption occurs paracellularly. In accord with this view, estimates of the magnitude of the Na^+-permselective shunt conductance in the mouse mTALH[25,37] indicate that the magnitudes of the lumen-positive transepithelial voltage and the calculated Na^+ conductance of the shunt pathway are sufficient to drive a quantity of Na^+ through the paracellular route equal to about half of the rate of net Cl^- absorption.

Finally, the model for NaCl absorption in the TALH shown in Fig. 1 appears to have at least two general consequences. First, about 50% of the rate of total net Na^+ absorption occurs through the paracellular route. In other words, the combination of a lumen-positive transepithelial voltage and a high shunt conductance reduces—with respect to exclusively transcellular, active Na^+ absorption—the metabolic energy expenditure for net Na^+ absorption. Second, regulatory mechanisms in epithelial cells promote the rapid adjustment of the rate of Na^+ entry into cells to equal the rate of Na^+ exit from cells, thus avoiding large and potentially lethal changes in cell volume ("flush-through" effect) when net Na^+ absorption is varied.[38] The "flush-through" effect may be minimized in the mTALH where 50% of net Na^+ absorption proceeds through the paracellular route.

3.2. The Effect of ADH on Net Salt Absorption in the mTALH

It is now well established that ADH increases both the net rate of salt absorption and the spontaneous transepithelial voltage in isolated mouse mTALH segments.[3,39] This stimulating effect of ADH on net salt absorption occurs at peritubular hormone concentrations found in the plasma of mammalian species during ordinary antidiuresis, and cAMP analogs produce the same effect as ADH on mouse mTALH segments.[3,39] Moreover, ADH also increases the rate of salt absorption from the mTALH of homozygous Brattleboro rats having central diabetes insipidus.[40]

3.2.1. Mechanism of the ADH Effect

Coincident with the increase in the net rate of salt absorption and transepithelial voltage in the mouse mTALH, ADH also increases the transepithelial electrical conductance and the rate of net K^+ secretion in that nephron segment.[25,27] Moreover, these ADH effects on net salt transport and on transepithelial electrical conductance are probably linked. The details of these arguments have been presented elsewhere[25,27,32]; a synopsis of the arguments is as follows.

First, the ADH-dependent increase in the transepithelial electrical conductance of the mouse mTALH involves an increase exclusively in the transcellular electrical conductance.[25] The hormone has no effect on the conductance of the paracellular pathway.[3,25]

Second, the primary effect of ADH in increasing transcellular conductance in the mTALH appears to depend on a hormone-mediated increase in the functional number of conductive K^+ channels in apical plasma membranes. This increase in apical membrane K^+ channels also accounts for ADH- or cAMP-mediated increases in the rate of net K^+ secretion.[25,32]

Third, ADH also produces an increase in the Cl^- conductance of basolateral membranes. However, while the ADH-dependent increase in apical K^+ conductance occurs even when net salt absorption is blocked completely with furosemide, the ADH-dependent increase in basolateral Cl^- conductance is abolished by this latter inhibitory agent.[32] This observation has been rationalized by assuming that cell Cl^- activity increases *pari passu* with ADH-mediated increases in the rate of net Cl^- absorption, and that the increase in cell Cl^- activity is responsible for the ADH-dependent increase in basolateral conductance.

Fourth, ADH appears to increase the functional number of Na^+–K^+–$2Cl^-$ cotransport units in apical membranes.[27,32] Obviously, since ADH increases net Cl^- absorption in the mouse mTALH, the rate of Cl^- flux across apical membranes is greater with hormone than without hormone. But as noted above, it is likely that cell Cl^- concentrations rise in the presence of ADH. Consequently, the chemical driving force for electroneutral Na^+–K^+–$2Cl^-$ cotransport (Fig. 1) from lumen to cell may be less in the presence of ADH than in its absence. According to this view, ADH increases the functional number of Na^+–K^+–$2Cl^-$ cotransport units as well as K^+ conductance units in apical plasma membranes.

Finally, as indicated above, the conductance of the apical membrane of the mTALH is negligible for ionic species other than K^+. Thus, the total current across the apical membrane during net Cl^- absorption is equivalent to the K^+ current from cell to lumen through apical K^+ channels; i.e., in the steady-state open-circuit condition, the total current across the apical membrane must equal the total current through the shunt pathway. Accordingly, an ADH-mediated increase in apical membrane K^+ flux from cell to lumen permits an increased Na^+ current through the paracellular pathway, and hence accounts, at least in part, for the ADH-mediated increase in net NaCl absorption. Thus, the K^+ conductance of apical membranes is one of the factors limiting the rate of net transepithelial NaCl absorption.

4. The Collecting Tubule

The second major contribution of ADH to the renal antidiuretic response is to increase strikingly the water permeability of terminal nephron segments—specifically, the CCT, the outer medullary collecting duct, and the papillary collecting duct. The increase in the water permeability of these nephron segments augments osmotic water flow from tubular lumen into a hypertonic medullary interstitium, thus providing for maximal urinary concentrations during antidiuresis. Virtually all information about the effects of ADH on water transport in collecting duct segments, or in other hormone-responsive epithelia, derives from analyses of the effects of ADH on water and solute transport, from assessments of the effects of ADH on the energetic requirements for water and solute transport (i.e., from activation energy measurements), or from studies of the effects of ADH on the morphological characteristics of hormone-responsive epithelia. This section will consider each of these approaches.

4.1. Water and Solute Permeability Measurements

It is well established that ADH increases the water permeability of apical plasma membranes in hormone-responsive epithelia.[41,42] In general, two methods may be used to assess this ADH-mediated increase in water permeability.

In one instance, net water flux is measured when an osmotic pressure gradient exists across the membrane. In accord with the Starling hypothesis, net water flow across the membrane is linearly related to the driving force by P_f (cm/sec^1), the permeability coefficient for net water flow; thus, P_f may be computed from the relation between net water flux and hydrostatic or osmotic pressure.

Table I. The Effect of ADH on Transport Coefficients in the Rabbit Cortical Collecting Tubule; and Transport Coefficients in Synthetic Bilayer Membranes

Preparation	ADH	P_f	P_{D_w}	$P_{D_{urea}}$	$P_f/P_{D_{urea}}$
		(cm/sec^1 × 10^4)			
CCT[a]	–	20	5	0.03	700
	+	186	14	0.02	9900
Synthetic bilayer membranes[b]	–	22	20	0.04	550

[a]Data from Refs. 44–47.
[b]Data from Refs. 43, 48–50.

In the second method, the flux of tracer water, e.g., THO, is measured at zero net volume flow: both solutions bathing a membrane are at the same hydrostatic and/or osmotic pressure and are identical in composition. Tracer water molecules in one solution exchange at random (by diffusion) across the membrane with unlabeled water molecules in the other solution, but there is no net water flux. From Fick's first law of diffusion and the tracer appearance rate in the nonlabeled solution, one may compute P_{D_w} (cm/sec^1), the permeability coefficient for water diffusion across the membrane.

Table I shows the effects of ADH on P_f, P_{D_w} and the permeability coefficient for the small hydrophilic solute urea in the *in vitro* rabbit CCT. The data in Table I indicate that P_f is more than 4-fold higher than P_{D_w} in the absence of ADH and more than 13-fold higher in the presence of the hormone. This hormone-associated increase in the P_f/P_{D_w} ratio occurs because of a greater stimulation in osmotic than diffusional water movement, i.e., P_f increases 9-fold while P_{D_w} increases only 3-fold (Table I).

In spite of the dramatic increase in both diffusional and osmotic water permeability coefficients observed with ADH in the CCT, this epithelium remains virtually impermeable to small hydrophilic nonelectrolytes like urea ($P_{D_{urea}}$, Table I). In other words, the ADH-mediated increase in CCT permeability is highly selective in that, while ADH increases both diffusional and osmotic water permeability, the permeability to small hydrophilic nonelectrolytes having molecular dimensions similar to water remains low.

The remarkable water/solute selectivity induced by ADH may be considered more quantitatively by comparing $P_f/P_{D_{urea}}$ ratios in collecting ducts with and without the hormone. In the absence of ADH, the $P_f/P_{D_{urea}}$ ratio is similar to that of an unmodified lipid bilayer membrane (Table I), but in the presence of this hormone the $P_f/P_{D_{urea}}$ ratio rises dramatically. Thus, the antidiuretic response involves a profound increase in the water permeability of collecting duct epithelium which can discriminate more than 10^3-fold between water and hydrophilic solutes (e.g., urea) having effective radii only slightly larger than that of the water molecule.

4.2. The P_f to P_{D_x} Disparity

In principle, F_f will equal P_{D_w} when net water transport across a given membrane, due to either osmotic or hydrostatic gradients, occurs exclusively by a solubility–diffusion process.[43] Thus, the transport of water across the collecting duct both in the presence and absence of ADH appears more complex than by a simple solubility–diffusion process since P_f is greater than P_{D_w} and the $P_f/P_{D_{urea}}$ ratios far exceed those expected for a simple lipid bilayer (see Table I). Three classes of explanations have been proposed to account for the disparity[41,42] between P_f and P_{D_w} observed in the vast majority of natural membranes.

First, Koefoed-Johnsen and Ussing[51] and Pappenheimer[52,53] suggested that a P_f to P_{D_w} disparity observed in epithelia might result from membranes containing pores sufficiently large to permit laminar or quasilaminar flow during osmosis. According to this view, which may be referred to as the "large pore" hypothesis, a P_f/P_{D_w} ratio will exceed unity because laminar, or Poiseuille, flow varies with r^4 (r = pore radius) while zero-volume flow ^{3}HHO diffusion is proportional to r^2.

However, small hydrophilic species should be entrained with laminar water flow through relatively large pores.[54,55] Thus, solutes that are able to enter relatively large pores (i.e., have reflection coefficients, σ, less than unity) exhibit an acceleration of solute flow in the same direction as solvent flow, and a retardation of solute flow in the opposite direction from solvent flow. This coupling of solvent and solute flow through relatively large pores has been termed the "solvent drag" effect.[54]

While large pores may account for P_f/P_{D_w} ratios greater than unity in porous collodion membranes, in artificial bilayer membranes modified with pore-forming antibiotics, and in certain natural biomembranes, such large pores cannot account for the solute-to-water discrimination ratios observed in the CCT and other ADH-responsive epithelia. For example, calculations based on a large-pore formulation indicate that the ADH-dependent P_f/P_{D_w} ratio of approximately 13 (Table I) in CCT would require an apical membrane pore radius of about 13 Å. Yet, CCT are virtually impermeable to small hydrophilic solutes such as urea ($r \cong 2.2$ Å; see Table I).

A second class of explanations accounting for the disparity between P_f and P_{D_w} assumes that the membrane is homogeneous and that the mode of osmotic water transport across apical plasma membranes is diffusional, but that unstirred layers in series with apical membranes impede THO diffusion at zero volume flow but not net volume flow during osmosis.[56–58] These unstirred layers may be viewed picturesquely as unmixed regions of water adjacent to a membrane which act as constraints to diffusion but not flow. According to this view, the relation between P_f and P_{D_w} is

$$\frac{1}{P_{D_w}} = \frac{1}{P_f} + R_u \qquad (1)$$

where $1/P_{D_w}$ is the resistance to THO diffusion at zero volume flow; $1/P_f$ is the resistance to osmotic volume flow; and R_u represents a series resistance to THO diffusion but not to osmotic volume flow; this relation is shown schematically in Fig. 2. It is generally considered that R_u is referable either to cytosolic diffusion constraints or to unmixed regions in bulk solutions[46,56,59–61] and may be defined as $\beta\Delta x/D^\circ_w$, where Δx is the unstirred layer (or cell layer) thickness, D°_w is the diffusion coefficient for water in water, and β represents a tortuosity factor relating to the geometry of the water pathway.

Finally, if apical plasma membranes contained narrow aqueous channels sufficiently narrow to preclude side-by-side passage of water molecules, water transport would follow single-file kinetics such that $P_f/P_{D_w} = n_w$, where n_w is the number of water molecules in a channel.[62–64] Thus, the relation between P_f and P_{D_w} for a membrane containing narrow channels which is in series with an unstirred layer is[65–67]

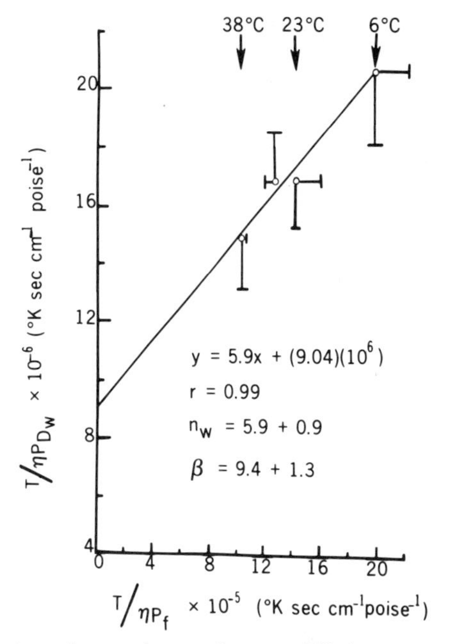

Fig. 2. Series resistance elements for water diffusion across a membrane in series with two resistance elements represented by unstirred layers in solution I and II, denoted as R^{I}_{uns} and R^{II}_{uns}, respectively. For water transport by a solubility–diffusion process, $1/P_f$ represents the diffusion resistance of the membrane to water, and $1/P_{D_w}$ represents the total diffusion resistance from solution I to solution II.

$$\frac{1}{P_{D_w}} = n_w\left(\frac{1}{P_f}\right) + R_u \qquad (2)$$

Equation (2) therefore expresses the combined effects of unstirred layers in series with a membrane, and the single-file effect for water transport within the membrane, on the P_f/P_{D_w} ratio (see Fig. 2).

The relation between P_f and P_{D_w} expressed in Eq. (2) has been tested recently in the rabbit CCT by assessing the temperature dependence of P_f and P_{D_w} in the presence and absence of ADH.[65] Since the unstirred layer resistance is determined in part by the diffusion coefficient for water in water (D°_w), R_u is not temperature independent; however, the Wang relation (ϕ = constant = $\eta\, D^\circ_w/T$) can be used to transport Eq. (2) into a linear form. By combining the Wang relation with Eq. (2) we have

$$\left(\frac{T}{\eta}\cdot\frac{1}{P_{D_w}}\right) = \left(\frac{T}{\eta}\cdot\frac{n_w}{P_f}\right) + \frac{\beta\Delta x}{\phi} \qquad (3)$$

Equation (3) indicates that P_f and P_{D_w} measurements made at varying temperature form a linear relation which permits calculation of n_w from the slope and β from the intercept, by using viscosity values corresponding to the appropriate temperature. The major advantage of this kind of analysis, rather than using temperature-dependent data to calculate activation energies (see below), lies in the fact that the former analysis depends exclusively on permeability data, and as such is not subject to artifacts relating to the effects of multiple diffusion resistances in series and to the effect of temperature-dependent variations in the number of water permeation sites.

Figure 3 shows ADH-dependent P_f and P_{D_w} data measured over the range of 6–38°C in rabbit CCT and plotted ac-

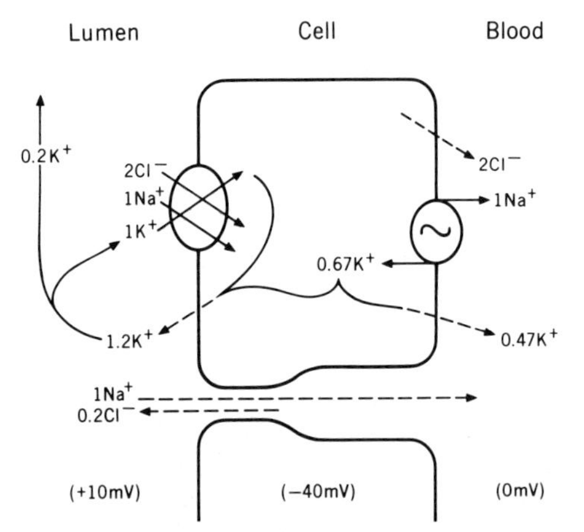

Fig. 3. An analysis of the number of single-file molecules per channel, in apical membranes of the rabbit cortical collecting tubule. P_{D_w} and P_f were measured at 6, 23, and 38°C. η represents the viscosity of water at the absolute temperature, T, for which P_f and P_{D_w} measurements were obtained. From Hebert and Andreoli.[70]

cording to Eq. (3).[65] A close linear relation, as predicted by Eq. (3), was obtained, and yielded a β value computed from the temperature dependence of P_f and P_{D_w} which is consistent with the view that about 50% of the P_f/P_{D_w} disparity in the rabbit CCT is due to unstirred layer effects. Since the geometry of the *in vitro* collecting duct excludes large bulk solution unstirred layers, R_u is most likely related to cytosolic (or cellular) diffusion constraints.[46] Levine *et al.*[60,61] have recently come to a similar conclusion for ADH-stimulated water transport in the toad urinary bladder.

This analysis[65] of the temperature dependence of P_f and P_{D_w} in isolated rabbit CCT (Fig. 3) also has indicated that water transport through apical membranes follows single-file kinetics typical of narrow channels, both in the absence and presence of ADH; i.e., the n_w term in Eq. (3), using the data in Fig. 3, is approximately 6. Analysis of the n_w term, both in the presence and absence of ADH, indicated that n_w was unaltered by the hormone and suggested that ADH does not alter the molecular characteristics of the water channel.[41,42] One may also deduce, given the failure of ADH to increase urea permeability in collecting ducts (Table I), that these channels are sufficiently narrow to preclude significant urea entry. A similar conclusion about apical membrane water permeation routes holds for ADH-sensitive amphibian epithelia.[66,67]

4.3. Comparison of Water Channels in the CCT with Gramicidin A Channels in Lipid Bilayers

Since gramicidin A channels formed in lipid bilayer membranes serve as a paradigm for single-file water diffusion,[66,67] it is instructive to compare the known characteristics of this model channel with those of the proposed water channel in apical membranes of the CCT. These data are presented in Table II.

First, it should be noted that the n_w value computed in the previous section for the narrow apical membrane channels or pores of the rabbit CCT is remarkably similar to that obtained by Rosenberg and Finkelstein[66,67] for gramicidin A channels in lipid bilayers which also exhibit a remarkable degree of water/urea selectivity (Table I). In other words, apical membrane aqueous channels sufficiently narrow to give rise to single-file diffusion of water account for the high water/urea selectivity typical of ADH-responsive epithelia (Table I).

It should be noted, however, there are major differences in the Na^+ *conductances*, G_{Na}, of gramicidin A channels and the narrow channels for water transport in apical membranes of collecting tubules. Using the water permeability of the gramicidin A channel determined by Rosenberg and Finkelstein,[66,67] 9.58×10^{-15} cm³/sec per channel, one requires approximately 2×10^{12} channels/cm² luminal surface area in the CCT to account for a P_f of 186×10^{-4} cm/sec (Table I). Given the Na^+ conductance of the CCT determined by O'Neil and Boulpaep,[68] $0.4 \times 10^{-3}/\Omega$ per cm², the maximal Na^+ conductance of each apical membrane water channel would be $2 \times 10^{-16}/\Omega$ per channel, which is four orders of magnitude less than the Na^+ conductance of gramicidin A channels. Thus, although a narrow channel model serves as a good paradigm for single-file water diffusion, the ion-conducting characteristics of ADH-dependent water channels in apical membranes of CCT appear to be quite different from those of gramicidin A channels.

4.4. Activation Energy Measurements

The notion that water transport through apical membranes of collecting ducts might involve water transport through narrow aqueous channels, and that ADH might increase the number of

Table II. Comparison of the Permeability Properties of Gramicidin A-Treated Lipid Bilayers with the Proposed ADH-Induced Water Channel in Cortical Collecting Tubules[a,b]

	Gramicidin channel	Cortical collecting tubules
P_f	342×10^{-4} cm/sec per mho cm² 9.58×10^{-15} cm³/sec per channel	186×10^{-4} cm/sec[1]
P_{D_w}	65×10^{-4} cm/sec per mho cm² 1.82×10^{-15} cm/sec per channel	14.2×10^{-4} cm/sec[1]
$P_f/P_{D_{urea}}$	$>10^3$	$\simeq 10^4$
G_{Na}	2.8×10^{-12} mho/channel	0.4×10^{-3}
n_w	5–7	$\simeq 6$

[a]Adapted from Ref. 41.
[b]Sources: Gramicidin A: $P_f/P_{D_{urea}}$, Ref. 94; water permeability, Refs. 66, 67; G_{Na}, n_w, Refs. 66, 67. Cortical collecting tubules: $P_f/P_{D_{urea}}$, Table I; G_{Na}, Ref. 68; n_w, Fig. 3.

such channels, is also supported by analyses of the activation energy (E_A, kcal/mole) for water and solute transport across apical membranes of collecting ducts. As a frame of reference, note that the activation energy for water or solute permeation across a pure lipid membrane (termed E_A^w for water or E_A^s for solutes) may be expressed as the sum of two terms: the energy required to break hydrogen bonds between the test species and neighboring water molecules ($E_{A^w}^H$ or $E_{A^s}^H$); and the energy needed for the test species to diffuse through the lipid bilayer core ($E_{A^w}^D$ or $E_{A^s}^D$).[69,70] Cohen[71] computed 1.8 kcal/mole to be the activation energy for hydrogen bond rupture in aqueous solutions for both water and solute molecules. Moreover, in purely lipid membranes of a given composition at the same temperature, $E_{A^w}^D$ and $E_{A^s}^D$ are the same,[69] since water and solutes obviously traverse the same permeation pathway.

In the collecting tubule, the ADH-dependent E_A values for water and moderately lipophilic solute permeation are clearly different, 9–10 and 16.6–19.6 kcal/mole, respectively[70]; the latter data refer to solutes such as butyramide and antipyrine. Now by taking four, three, and two as the number of hydrogen bonds for water, butyramide, and antipyrine, respectively, these E_A^w and E_A^s data yield apparent values of 1.4 and 10.6–15.6 kcal/mole for, respectively, $E_{A^w}^D$ and $E_{A^s}^D$ in apical membranes of CCT. These latter data, i.e., $E_{A^s}^D$ exceeds $E_{A^w}^D$ in rabbit CCT, indicate that the energetic requirements for water transport through apical plasma membranes are considerably less than those for moderately lipophilic species. Accordingly, it is reasonable to conclude that the permeation pathway for water movement through apical plasma membranes involves a specialized permeation pathway, presumably aqueous, rather than the hydrophobic core of apical plasma membranes. This latter conclusion is also supported by the observation (compare above; Table I) that the water/solute permeation ratio in apical plasma membranes is considerably greater than that expected for a simple solubility–diffusion process through the hydrophobic core of apical plasma membranes.

4.5. Morphological Studies

A series of morphological studies of ADH-responsive anuran epithelia (principally toad and frog urinary bladders) has led to the appreciation that a number of ultrastructural changes occur in the apical membranes of granular cells in association with the application of serosal ADH. Chevalier *et al.*[72] were the first to report the aggregation of apical membrane intramembranous particles in frog urinary bladders treated with oxytocin. Kachadorian *et al.*[73] reported similar apical membrane aggregates in toad urinary bladder exposed to ADH.

While extensive structural studies have not yet been performed using mammalian collecting ducts, apical intramembranous particle aggregates have been identified in medullary collecting ducts from rat,[74] and in outer medullary and cortical collecting tubules of rabbits, where the aggregates are confined to the apical membranes of principal cells.[75] The particle aggregates in these mammalian tubules are similar to, but not identical with, those of anuran epithelia.

Although there are no definitive data indicating that these aggregates represent the putative ADH-induced water channels, a large body of evidence has accumulated indicating that, at a minimum, these aggregates are associated with the ADH-induced increase in apical membrane water permeability. Thus, the appearance of aggregates depends on the serosal application of ADH, cAMP and/or *in vivo* dehydration but not hydration, is independent of an imposed osmotic gradient, and can be inhibited by drugs which selectively inhibit the ADH-induced increase in water flow.[9,10,69]

In the absence of ADH stimulation, similar aggregates have been identified in both toad and frog urinary bladders in vacuole membranes beneath the apical membranes of granular cells.[9,76] In the presence of ADH, the number of aggregate-containing vacuoles decreased markedly, and occasionally these vacuoles have been seen fusing with the apical membrane. The frequency of these so-called fusion events, seen on freeze–fracture sections of apical membranes in response to ADH, seems to correlate with the accumulation of aggregates, and has led to the hypothesis that the water permeation sites are "shuttled" from the membranes of these vacuoles to the apical membrane under the influence of ADH.[76] However, it is unknown whether fusion of these aggregate-containing vacuoles with the apical membrane can account entirely for the rather large increase in apical membrane capacitance, presumed to represent an increase in membrane area, seen with application of ADH. Recently, electron micrography has demonstrated that the surface topography of granular cells changes from broad ridgelike villus structures to fine microvillus structures in the presence of ADH.[69]

5. Homology of Hormone Action

The observations presented in the preceding sections may be integrated partially into a general statement about the mechanism of action of ADH. In the mTALH, ADH increases the functional number of apical membrane K^+ conductance units and apical membrane Na^+–K^+–$2Cl^-$ cotransport units (Fig. 1). In the collecting duct, ADH increases the functional number of narrow aqueous channels in apical plasma membranes.[42] And in apical membranes of amphibian epithelia, ADH increases both the functional number of small channels for water transport[60,61] and the functional number of Na^+-conductive channels.[77] Thus, there may exist a general mode of action of ADH in hormone-sensitive epithelia, i.e., to increase the functional number of transport units in apical membranes for those molecular species whose flux is augmented by ADH.

In amphibian epithelia, in the rabbit CCT, and in the mouse mTALH, the ADH-mediated increases in transepithelial transport rates of the target species occur within minutes of hormone application to basolateral membranes.[1–3,27,32,41,42,69] In other words, the ADH-mediated alterations in apical membrane transport processes occur at rates sufficiently rapid to require activation or translocation of existing transport units, i.e., by a recruitment process rather than by *de novo* synthesis.

In this regard, the fusion of subapical vacuoles into intramembranous aggregates in apical membranes correlates reasonably well with the ADH-mediated hydroosmotic response[9,72–76] in amphibian epithelia; however, as indicated above, the explicit relation between apical membrane aggregates and ADH-dependent water channels in apical membranes has not yet been established. Alternatively, Li *et al.*[77] have provided convincing evidence that the natriferic response of toad urinary bladder involves the conversion, or recruitment, of electrically silent Na^+ channels in apical membranes to amiloride-sensitive conductive Na^+ channels in these membranes. It is plausible that similar recruitment mechanisms may underlie ADH-dependent increases in apical membrane transport rates for hormone-targeted molecular species in different ADH-sensitive epithelia.

6. Modulation of the ADH Response

The actions of ADH on NaCl absorption in the TALH, and on water abstraction from the collecting duct, can be modulated by a number of factors. Two of these are particularly pertinent. First, increases in peritubular osmolality rapidly and reversibly inhibit the rate of net NaCl absorption in the mTALH by a mechanism which inhibits transport directly. Second, prostaglandins, particularly those of the E series, inhibit the hydroosmotic action of ADH in the amphibian urinary bladder and the collecting duct, and salt absorption in the mTALH, by mechanisms that reduce ADH-dependent increments in intracellular cAMP levels.

6.1. Peritubular Osmolality

In isolated mouse mTALH segments, increases in peritubular osmolality, produced either with permeant solutes such as urea or with impermeant solutes such as mannitol, rapidly and reversibly inhibit the ADH-stimulated rate of net Cl^- absorption.[78] The increases in peritubular solute concentrations do not affect the dissipative permeability characteristics of the shunt pathway, but rather reduce the rate of conservative transcellular Cl^- absorption. This inhibition of transcellular salt absorption occurs at a locus beyond the generation of cAMP, since supramaximal concentrations of either ADH or cAMP are unable to reverse the bath hypertonicity-mediated reduction in NaCl absorption. Thus, increasing the absolute magnitude of interstitial osmolality provides a feedback signal which reduces, by a mechanism distal to cAMP, the rate of ADH-dependent salt absorption by the mTALH.

6.2. Prostaglandin–ADH Interactions

Considerable evidence points to the importance of locally generated renal prostaglandins in modulating the actions of ADH on renal epithelial transport processes. In the amphibian urinary bladder, Orloff *et al.*[79] found that prostaglandins inhibited the hydroosmotic effect of ADH but did not change the response of the epithelium to cAMP. Based on *in vitro* studies in the isolated rabbit collecting duct, Grantham and Orloff[80] also suggested that prostaglandins might modulate renal urinary concentrating systems by opposing the hydroosmotic effect of ADH, but not cAMP. These authors inferred that prostaglandins acted at a locus proximal to hormone-dependent accumulation of cAMP within the cell, with little or no discernible prostaglandin action on transport events beyond cAMP accumulation.

The major product of prostaglandin synthesis in the renal medulla, PGE_2, appears to be responsible for the majority of the physiological effects on water excretion. In the mammalian CCT, in the medullary interstitial cell, and in the toad urinary bladder, ADH has stimulated the production of PGE_2.[81,82] In these tissues, inhibition of endogenous PGE production with prostaglandin synthetase inhibitors, such as indomethacin or meclofenamate, has increased the rate of Na^+ transport and/or the rate of osmotic water permeation.[82] In both the rabbit CCT and the toad urinary bladder, prostaglandins inhibit the ADH-stimulated accumulation of cAMP within the cell, and exert little or no inhibitory action on transport events beyond the accumulation of cAMP within the cell.[79,80,82] Thus, the ADH effects on transport in these tissues have been modulated by an inhibitor which is synthesized *in situ*, whose production is stimulated by ADH, and whose action is to reduce the ability of ADH to elevate cellular levels of cAMP.

PGE_2 also participates in a local negative feedback system in the renal medulla that modulates the rate of net NaCl absorption by the mTALH. In rat micropuncture experiments, Higashihara *et al.*[83] demonstrated a decrease in NaCl delivery to distal tubule sites, and an increase in papillary NaCl content, following prostaglandin synthesis inhibition; they concluded that prostaglandins might inhibit NaCl transport in the TALH. Complementary studies conducted by Kauker,[84] using micropuncture techniques, and by Stokes,[85] using the isolated, perfused rabbit mTALH, have also been consistent with the notion that PGE_2 inhibits NaCl absorption by the TALH.

Culpepper and Andreoli[86] have examined the interactions between ADH and PGE_2 in the *in vitro* mouse mTALH. In the absence of ADH, PGE_2 had no effect on NaCl absorption either in the mTALH or in the ADH-unresponsive cTALH. However, in the presence of ADH, PGE_2 reduced, in mTALH segments, the ADH-dependent values for transepithelial voltage and net NaCl absorption to ADH-independent values. PGE_2 blocked only the ADH-stimulated components of net NaCl absorption; and the PGE_2-mediated reduction in ADH-dependent NaCl transport could be reversed either by cAMP or by supramaximal concentrations of ADH. Likewise, Torikai and Kurokawa[87] have reported biochemical studies in the mTALH which indicate that PGE_2 has no effect on cellular cAMP concentrations in the absence of ADH, and that PGE_2 markedly inhibits the ADH-dependent stimulation of cytosolic cAMP concentrations. Thus, it appears that PGE_2 reduces the ADH-dependent rise in NaCl absorption in the murine mTALH by blocking an element in the hormone-dependent sequence of cAMP generation.

The explicit molecular locus for the PGE_2-mediated inhibition of cAMP formation is not yet known. However, it has been observed[88] that PGE_2 does not inhibit the component of NaCl transport in the mouse mTALH stimulated by the nonhormonal catalytic subunit activator forskolin, and that PGE_2 does inhibit transport stimulation by cholera toxin, an agent which specifically activates adenylate cyclase at a stimulatory guanine nucleotide-binding subunit. Thus, it is plausible that PGE_2 inhibits the ADH-stimulated generation of cAMP in the mTALH by interaction with a guanine nucleotide-binding subunit.[88] Recently, pertussigen, an agent that stimulates adenylate cyclase by blocking an inhibitory guanine nucleotide-binding subunit, has been utilized to assess the interaction of PTH and PGE_2 on adenylate cyclase activity in the proximal straight tubule.[89] In the presence of pertussigen, PGE_2 failed to reduce the magnitude of the PTH-stimulated adenylate cyclase, suggesting that PGE_2 may inhibit the activity of PTH-stimulated adenylate cyclase in this latter nephron segment by activating an inhibitory guanine nucleotide-binding subunit. Since the interactions of PTH/PGE_2 in the proximal straight tubule and ADH/PGE_2 in the mTALH appear similar, PGE_2 may inhibit ADH-stimulated adenylate cyclase activity in the latter segment by a similar mechanism.

The significance of the interactions between prostaglandins and ADH to *in vivo* renal salt absorption and concentrating ability has been affirmed by a number of studies. Infusions of prostaglandins of the E series into hydrated animals have increased urinary sodium excretion and, in hydropenic dogs, decreased absorption of free water.[82,86] Inhibition of endogenous renal prostaglandin synthesis, either with indomethacin or with meclofenamate, has resulted in antinatriuresis and is an enhanced urinary concentrating ability in response to administration of vasopressin.[90,91] In addition, Ganguli *et al.*[92] demonstrated an increase in medullary NaCl content following prostaglandin synthesis inhibition even in the absence of any discernible change in papillary blood flow.

The synthesis of PGE_2 has been demonstrated in the medullary collecting duct and in interstitial cells. Craven and DeRubertis[93] have shown that PGE_2 synthesis by medullary interstitial cells can be modulated both by ADH and by increases in osmolality produced with urea or NaCl. These agents appear to function in acute experiments by affecting the calcium-dependent acyl hydrolase activity that regulates the availability of arachidonic acid in these cells. Finally, increases in local osmolality in the renal medulla, produced by ADH-mediated increases in NaCl absorption from the mTALH and the consequent enhancement in countercurrent multiplication, might be expected to play a major role in PGE_2 synthesis *in vivo*. For example, hypertonic NaCl stimulates PGE_2 release from medullary cells, and hypertonic urea suppressed this effect as well as the PGE_2 release mediated directly by pharmacological concentrations of ADH.

7. Summary: Integration of ADH Action on Urinary Concentration

The *in vivo* and *in vitro* arguments summarized in this chapter may be integrated into the model for some of the factors which modulate urinary concentrating ability shown in Fig. 4. According to this model, ADH-stimulated NaCl absorption by the mTALH is regulated by two negative feedback loops (depicted by the dashed lines in Fig. 4), each of which is dependent on increases in interstitial osmolality produced by the ADH-mediated enhancement of countercurrent multiplication. During the early stages of antidiuresis, an ADH-mediated increase in NaCl absorption by the mTALH leads to a rapid rise in the interstitial NaCl concentration. This increase in interstitial osmolality stimulates the release of PGE_2 from interstitial cells which, in turn, decreases the rate of ADH-stimulated NaCl absorption. Later during antidiuresis, a rise in medullary interstitial urea concentration would tend to inhibit PGE_2 release from interstitial cells and the ADH-mediated increase in NaCl absorption from the mTALH.[27]

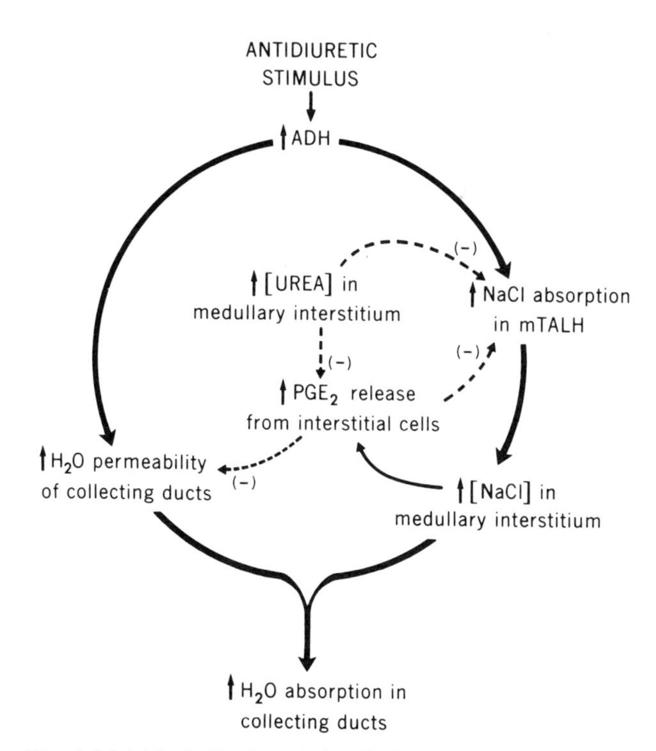

Fig. 4. Model for feedback regulation of urinary concentrating mechanisms. From Hebert and Andreoli.[27]

Thus, the direct inhibition by interstitial hyperosmolality on ADH-dependent NaCl absorption by the mTALH is coupled to PGE_2 production which modulates ADH action through the second messenger cAMP; the net effect is a negative feedback loop on ADH-dependent NaCl addition to the medullary interstitium. A similar negative feedback loop operates at the level of the collecting duct where endogenous PGE_2 production, stimulated by ADH, decreases the ADH-induced increase in water permeability at the level of cellular cAMP accumulation.

ACKNOWLEDGMENTS. We gratefully acknowledge the able secretarial assistance of Ms. Clementine Whitman and Ms. Dot Cowan. This research was supported in part by grants from the National Institutes of Health (5 RO1-AM-25540 and 5 T32-AM-07408), the National Science Foundation (PCM 81-41370), and the American Heart Association (831-294). S.C.H. is an Established Investigator of the American Heart Association.

References

1. Orloff, J., and J. H. Handler. 1962. The similarity of effects of vasopressin, adenosine 3′,5′-monophosphate (cyclic AMP) and theophylline on the toad bladder. *J. Clin. Invest.* **41**:702–709.
2. Grantham, J. J., and M. B. Burg. 1966. Effect of vasopressin and cyclic AMP on permeability of isolated collecting tubules. *Am. J. Physiol.* **211**:255–259.
3. Hebert, S. C., R. M. Culpepper, and T. E. Andreoli. 1981. NaCl transport in mouse medullary thick ascending limbs. I. Functional nephron heterogeneity and ADH-stimulated NaCl cotransport. *Am. J. Physiol.* **241**:F412–F431.
4. Morel, F. 1983. Regulation of kidney functions by hormones: A new approach. *Recent Prog. Horm. Res.* **39**:271–304.
5. Jard, S., C. Roy, T. Barth, R. Rajerison, and J. Bockaert. 1975. Antidiuretic hormone-sensitive kidney adenylate cyclase. *Adv. Cyclic Nucleotide Res.* **5**:31–52.
6. Eggena, P., I. L. Schwartz, and R. Walter. 1970. Threshold and receptor reserve in the action of neurohypophyseal peptides: A study of synergists and antagonists in the hydroosmotic response on the toad urinary bladder. *J. Gen. Physiol.* **56**:250–271.
7. Dousa, T. P., and H. Valtin. 1976. Cellular actions of vasopressin in the mammalian kidney. *Kidney Int.* **10**:4663.
8. Dousa, T. P., L. O. Barnes, and J. K. Kim. 1977. The role of cyclic AMP-dependent protein phosphorylations and microtubules in the cellular action of vasopressin in mammalian kidney. In: *Neurohypophysis*. A. M. Moses and L. Share, eds. Karger, Basel. pp. 220–235.
9. Schwartz, I. L., C. S. Huang, A. J. Fischman, S. K. Marur, and H. R. Wyssbrod. 1981. Current ideas on the sequence of events involved in the hydroosmotic action of antidiuretic hormones. In: *Neurohypophyseal Peptide Hormones and Other Biologically Active Peptides*. D. H. Schlessinger, ed. Elsevier, Amsterdam. pp. 101–110.

10. Muller, J., W. A. Kachadorian, and V. A. DiScala. 1980. Evidence that ADH-stimulated intramembrane particle aggregates are transferred from cytoplasmic to luminal membranes in toad bladder epithelial cells. *J. Cell Biol.* **85**:83–95.
11. Wirz, V. H., B. Hargitay, and W. Kuhn. 1951. Lokalisation des Konzentrierungsprozesses in der Niere durch direkte Kryoskopie. *Helv. Physiol. Pharmacol. Acta* **9**:196–207.
12. Burg, M. B., and N. Green. 1973. Function of the thick ascending limb of Henle's loop. *Am. J. Physiol.* **224**:659–668.
13. Rocha, A. S., and J. P. Kokko. 1973. Sodium chloride and water transport in the medullary thick ascending limb of Henle: Evidence for active chloride transport. *J. Clin. Invest.* **52**:612–623.
14. Schmidt, U., and U. C. Dubach. 1969. Activity of (Na^+, K^+)-stimulated adenosine triphosphatase in the rat nephron. *Pfluegers Arch.* **306**:219–226.
15. Jørgensen, P. L. 1977. The function of (Na^+, K^+)-ATPase in the thick ascending limb of Henle's loop. *Curr. Probl. Clin. Biochem.* **6**:190–199.
16. Katz, A. I., A. Doucet, and F. Morel. 1979. Na-K-ATPase activity along the rabbit, rat and mouse nephron. *Am. J. Physiol.* **237**:F114–F120.
17. Greger, R., and E. Schlatter. 1983. Properties of the lumen membrane of the cortical thick ascending limb of Henle's loop of rabbit kidney. *Pfluegers Arch.* **396**:315–324.
18. O'Neil, R. G. 1983. Voltage-dependent interaction of barium and cesium with the potassium conductance of the cortical collecting duct apical cell membrane. *J. Membr. Biol.* **74**:165–173.
19. van Driessche, W., and W. Zeiske. 1980. Ba^2-induced conductance fluctuations of spontaneously fluctuating K^+ channels in the apical membrane of frog skin (*Rana temporaria*). *J. Membr. Biol.* **56**:31–42.
20. Wills, N. K., W. Zeiske, and W. Van Driessche. 1982. Noise analysis reveals K^+ channel conductance fluctuations in the apical membrane of rabbit colon. *J. Membr. Biol.* **69**:187–197.
21. Zeiske, W., and W. van Driessche. 1983. The interaction of "K^+-like" cations with the apical K^+ channel in frog skin. *J. Membr. Biol.* **76**:57–72.
22. García-Díaz, J. F., W. Nagel, and A. Essig. 1983. Voltage-dependent K conductance at the apical membrane of *Necturus* gallbladder. *Biophys. J.* **43**:269–278.
23. Armstrong, C. M., R. P. Swenson, Jr., and S. R. Taylor. 1982. Block of squid axon K^+ channels by internally and externally applied barium ions. *J. Gen. Physiol.* **80**:663–682.
24. Greger, R., and G. Frömter. 1981. Chloride reabsorption in the rabbit cortical thick ascending limb of the loop of Henle: A sodium dependent process. *Pfluegers Arch.* **390**:38–43.
25. Hebert, S. C., P. A. Friedman, and T. E. Andreoli. 1984. The effects of antidiuretic hormone on cellular conductive pathways in mouse medullary thick ascending limbs of Henle. I. ADH increases transcellular conductance pathways. *J. Membr. Biol.* **80**:201–219.
26. Murer, H., and R. Greger. 1982. Membrane transport in the proximal tubule and thick ascending limb of Henle's loop: Mechanisms and their alterations. *Klin. Wochenschr.* **60**:1103–1113.
27. Hebert, S. C., and T. E. Andreoli. 1984. Control of NaCl transport in the thick ascending limb. *Am. J. Physiol.* **246**:F745–F756.
28. Eveloff, J., and R. Kinne. 1983. Sodium–chloride transport in the medullary thick ascending limb of Henle's loop: Evidence for a sodium–chloride cotransport system in plasma membrane vesicles. *J. Membr. Biol.* **72**:173–181.
29. Forbush, B., and H. C. Palfrey. 1983. [^{3}H]-Bumetanide binding to membranes isolated from dog kidney outer medulla. *J. Biol. Chem.* **258**:11787–11792.
30. Oberleithner, H., W. Guggino, and G. Giebisch. 1982. Mechanism of distal tubular chloride transport in amphiuma kidney. *Am. J. Physiol.* **242**:F331–F339.
31. Greger, R., and E. Schlatter. 1983. Properties of the basolateral membrane of the cortical thick ascending limb of Henle's loop of rabbit kidney—A model for secondary active chloride transport. *Pfluegers Arch.* **396**:325–334.
32. Hebert, S. C., and T. E. Andreoli. 1984. Effects of antidiuretic hormone on cellular conductive pathways in mouse medullary thick ascending limbs of Henle. II. Determinants of the ADH-mediated increases in transepithelial voltage and in net Cl^- absorption. *J. Membr. Biol.* **80**:221–223.
33. Greger, R., H. Oberleithner, E. Schlatter, A. C. Cassola, and C. Weidtke. 1983. Chloride activity in cells of isolated perfused cortical thick ascending limbs of rabbit kidney. *Pfluegers Arch.* **399**:29–34.
34. Guggino, W. B., B. A. Stanton, and G. Giebisch. 1982. Electrical properties of isolated early distal tubule of the amphiuma kidney. *Fed. Proc.* **41**:1597.
35. Oberleithner, H., F. Lang, R. Greger, W. Wang, and G. Giebisch. 1983. Effect of luminal potassium on cellular sodium activity in the early distal tubule of *Amphiuma* kidney. *Pfluegers Arch.* **396**:34–40.
36. Stokes, J. B. 1982. Consequences of potassium recycling in the renal medulla: Effects on ion transport by the medullary thick ascending limb of Henle's loop. *J. Clin. Invest.* **70**:219–229.
37. Hebert, S. C., and T. E. Andreoli. 1984. Kinetic analysis of Ba^{++}-blockade of apical membrane K^+-channels in mouse medullary thick ascending limbs. IX International Congress of Nephrology, P416A.
38. Schultz, S. G. 1981. Homocellular regulatory mechanism in sodium-transporting epithelia: Avoidance of extinction by "flush-through." *Am. J. Physiol.* **241**:F579–F590.
39. Hall, D. A., and D. M. Varney. 1980. Effect of vasopressin on electrical potential difference and chloride transport in mouse medullary thick ascending limb of Henle's loop. *J. Clin. Invest.* **66**:792–802.
40. Work, J., B. Booker, J. A. Schafer, J. Galla, and R. Luke. 1983. *In vivo* and *in vitro* effect of ADH on loop of Henle: Chloride reabsorption in the Brattleboro (DI) rat. *Am. Soc. Nephrol. 16th Ann. Meet.* 185A.
41. Hebert, S. C., J. A. Schafer, and T. E. Andreoli. 1981. The effects of antidiuretic hormone (ADH) on solute and water transport in the mammalian nephron. *J. Membr. Biol.* **58**:1–19.
42. Hebert, S. C., and T. E. Andreoli. 1982. Water permeability of biological membranes: Lessons from antidiuretic hormone-responsive epithelia. *Biochim. Biophys. Acta* **650**:267–280.
43. Cass, A., and A. Finkelstein. 1967. Water permeability of thin lipid membranes. *J. Gen. Physiol.* **50**:1765–1784.
44. Grantham, J. J., and M. B. Burg. 1966. Effect of vasopressin and cyclic AMP on permeability of isolated collecting tubules. *Am. J. Physiol.* **211**:255–259.
45. Grantham, J. J., and J. Orloff. 1968. Effect of prostaglandin E_1 on the permeability response of the isolated collecting tubule to vasopressin, adenosine 3′,5′-monophosphate, and theophylline. *J. Clin. Invest.* **47**:1154–1161.
46. Schafer, J. A., and T. E. Andreoli. 1972. Cellular constraints to diffusion: The effect of antidiuretic hormone on water flows in isolated mammalian collecting ducts. *J. Clin. Invest.* **51**:1264–1278.
47. Schafer, J. A., and T. E. Andreoli. 1972. The effect of antidiuretic hormone on solute flows in mammalian collecting tubules. *J. Clin. Invest.* **51**:1279–1286.
48. Finkelstein, A. 1976. Nature of the water permeability increase induced by antidiuretic hormone (ADH) in toad urinary bladder and related tissues. *J. Gen. Physiol.* **68**:137–143.
49. Gallucci, E., S. Micelli, and C. Lippi. 1971. Nonelectrolyte permeability across thin lipid membranes. *Arch. Int. Physiol. Biochim.* **79**:881–887.
50. Vreeman, H. J. 1966. Permeability of thin phospholipid films. 1. *K. Ned. Akad. Wet. Amsterdam Ser. B* **69**:542–577.
51. Koefoed-Johnsen, V., and H. H. Ussing. 1953. The contributions of diffusion and flow to the passage of D_2O through living membranes. *Acta Physiol. Scand.* **28**:60–76.
52. Pappenheimer, J. R. 1953. Passage of molecules through capillary walls. *Physiol. Rev.* **33**:387–423.

53. Pappenheimer, J. R., E. M. Renkin, and L. M. Boneru. 1951. Filtration diffusion and molecular seiving through peripheral capillary membranes. *Am. J. Physiol.* **167**:13–46.
54. Andersen, B., and H. H. Ussing. 1957. Solvent drag on nonelectrolytes during osmotic flow through isolated toad skin and its response to antidiuretic hormone. *Acta Physiol. Scand.* **39**:228–239.
55. Kedem, O., and A. Katchalsky. 1961. A physical interpretation of the phenomenological coefficients of membrane permeability. *J. Gen. Physiol.* **45**:143–179.
56. Dainty, J. 1963. Water relations of plant cells. *Adv. Bot. Res.* **1**:279–326.
57. Nernst, W. 1904. Theorie der reactionsgeschwindigkeit in heterogenen systemen. *Z. Phys. Chem.* **47**:52–55.
58. Teorell, T. 1936. A method of studying conditions within diffusion layers. *J. Biol. Chem.* **113**:735–748.
59. Holz, R., and A. Finkelstein. 1970. The water and nonelectrolyte permeability induced in thin lipid membranes by the polyene antiobiotics nystatin and amphotericin B. *J. Gen. Physiol.* **56**:125–145.
60. Levine, S. D., M. Jacoby, and A. Finkelstein. 1984. The water permeability of toad urinary bladder. II. The value of $P_f/P_d(w)$ for the antidiuretic hormone-induced water permeation pathway. *J. Gen. Physiol.* **83**:543–561.
61. Levine, S. D., M. Jacoby, and A. Finkelstein. 1984. The water permeability of toad urinary bladder. I. Permeability of barriers in series with the luminal membrane. *J. Gen. Physiol.* **83**:529–541.
62. Dick, D. A. T. 1966. *Cell Water.* Butterworths, London. pp. 102–111.
63. Lea, E. J. A. 1963. Permeation through long narrow pores. *J. Theor. Biol.* **5**:102–107.
64. Levitt, D. G. 1974. A new theory of transport for cell membrane pores. I. General theory and application to red cell. *Biochim. Biophys. Acta* **373**:115–131.
65. Hebert, S. C., and T. E. Andreoli. 1980. Interactions of temperature and ADH on transport in cortical collecting tubules. *Am. J. Physiol.* **238**:F470–F480.
66. Rosenberg, P. A., and A. Finkelstein. 1978. Interaction of ions and water in gramicidin A channels: Streaming potentials across lipid bilayer membranes. *J. Gen. Physiol.* **72**:327–340.
67. Rosenberg, P. A., and A. Finkelstein. 1978. Water permeability of gramicidin A-treated lipid bilayer membranes. *J. Gen. Physiol.* **72**:341–350.
68. O'Neil, R. G., and E. L. Boulpaep. 1979. Effect of amiloride on the apical cell membrane cation channels of sodium-absorbing, potassium-secreting renal epithelium. *J. Membr. Biol.* **50**:365–387.
69. Hebert, S. C., and T. E. Andreoli. 1982. Water movement across the mammalian cortical collecting duct. *Kidney Int.* **22**:526–535.
70. Hebert, S. C., and T. E. Andreoli. 1980. Interactions of temperature and ADH on transport processes in cortical collecting tubules: Evidence for ADH-induced narrow aqueous channels in apical membranes. *Am. J. Physiol.* **238**:F470–F480.
71. Cohen, B.E. 1975. The permeability of liposomes to nonelectrolytes. I. Activation energies for permeation. *J. Membr. Biol.* **20**:205–234.
72. Chevalier *et al.* 1974. Membrane-associated particles: Distribution in frog urinary bladder epithelium at rest and after oxytocin treatment. *Cell Tissue Res.* **152**:129–140.
73. Kachadorian, W. A., J. B. Wade, and V. A. DiScala. 1975. Vasopressin: Induced structural change in toad bladder luminal membranes. *Science* **190**:67–69.
74. Harmanci, M. C., P. Stern, W. A. Kachadorian, H. Valtin, and V. A. DiScala. 1980. Vasopressin and collecting duct intramembranous particle clusters: A dose–response relationship. *Am. J. Physiol.* **239**:F560–F564.
75. Harmanci, M. C., M. Lorenzen, and W. A. Kachadorian. 1982. Vasopressin-induced intramembranous particle aggregates in isolated rabbit collecting duct. *Kidney Int.* **21**:275a.
76. Wade, J. B., D. L. Stetson, and S. A. Lewis. 1981. ADH action: Evidence for a membrane shuttle mechanism. *Ann. N.Y. Acad. Sci.* **372**:106–117.
77. Li, H-YS., L. G. Palmer, I. S. Edelman, and B. Lindeman. 1982. The role of sodium-channel density in the natriferic response of the toad urinary bladder to antidiuretic hormone. *J. Membr. Biol.* **64**:77–89.
78. Hebert, S. C., R. M. Culpepper, and T. E. Andreoli. 1981. NaCl transport in mouse medullary thick ascending limbs. III. Modulation of the ADH effect by peritubular osmolality. *Am. J. Physiol.* **241**:F443–F451.
79. Orloff, J., J. S. Handler, and S. Bergstrom. 1965. Effect of prostaglandin (PGE) on the permeability response of the toad bladder to vasopressin, theophylline and adenosine 3′-5′-monophosphate. *Nature (London)* **205**:397–398.
80. Grantham, J. J., and J. Orloff. 1968. Effect of prostaglandin E_1 on the permeability response of the isolated collecting tubule to vasopressin, adenosine 3′-5′-monophosphate and theophylline. *J. Clin. Invest.* **47**:1154–1161.
81. Handler, J. S. 1981. Vasopressin–prostaglandin interactions in the regulation of epithelial cell permeability to water. *Kidney Int.* **19**:831–838.
82. Beck, T. R., and M. J. Dunn. 1981. The relationship of antidiuretic hormone and renal protaglandins. *Miner. Electrolyte Metab.* **6**:46–59.
83. Higashihara, E., J. B. Stokes, J. P. Kokko, W. B. Campbell, and T. D. DuBose. 1979. Cortical and papillary micropuncture examination of chloride transport in segments of the rat kidney during inhibition of prostaglandin production. *J. Clin. Invest.* **64**:1277–1287.
84. Kauker, M. L. 1977. Prostaglandin E_2 effect from the luminal side on renal tubular ^{22}Na efflux: Tracer microinjection studies. *Proc. Soc. Exp. Biol. Med.* **154**:274–277.
85. Stokes, J. B. 1979. Effect of prostaglandin E_2 on chloride transport across the rabbit thick ascending limb of Henle. *J. Clin. Invest.* **64**:495–502.
86. Culpepper, R. M., and T. E. Andreoli. 1983. Interactions among prostaglandin E_2, antidiuretic hormone, and cyclic adenosine monophosphate in modulating Cl absorption in single mouse medullary thick ascending limbs of Henle. *J. Clin. Invest.* **71**:1588–1601.
87. Torikai, S., and K. Kurokawa. 1983. Effect of PGE_2 on vasopressin-dependent cell cAMP in isolated single segments. *Am. J. Physiol.* **245**:F58–F66.
88. Culpepper, R. M., and T. E. Andreoli. 1984. Prostaglandin E_2 inhibition of vasopressin-stimulated NaCl transport in the mouse medullary thick ascending limb of Henle. *Adv. Prostaglandin Thromboxane Leukotriene Res.* in press.
89. Dominguez, J. H., F. Schuler, T. Brown, T. D. Pitts, and J. B. Puschett. 1984. Pertussigen reverses the inhibition of adenylate cyclase by prostaglandin E_2 in the proximal nephron. *Clin. Res.* **32**:445a.
90. Fejes-Tóth, G., A. Magyer, and J. Walter. 1977. Renal response to vasopressin after inhibition of prostaglandin synthesis. *Am. J. Physiol.* **232**:F416–F423.
91. Berl, T., A. Raz, H. Wald, J. Horowitz, and W. Czaczkes. 1977. Prostaglandin synthesis inhibition and the action of vasopressin: Studies in man and rat. *Am. J. Physiol.* **232**:F529–F537.
92. Ganguli, M., L. Tobin, S. Azar, and M. O'Donnell. 1977. Evidence that prostaglandin synthesis inhibitors increase the concentration of sodium and chloride in rat renal medulla. *Circ. Res. Suppl.* **40**:I135–I139.
93. Craven, P. A., and F. R. DeRubertis. 1981. Effects of vasopressin and urea on Ca^{2+}–calmodulin-dependent renal prostaglandin E. *Am. J. Physiol.* **241**:F649–F658.
94. Finkelstein, A. 1974. Aqueous pores created in thin lipid membranes by the antibiotics nystatin, amphotericin B and gramicidin A: Implications for pores in plasma membranes. In: *Drugs and Transport Processes.* B. A.Callingham, ed. MacMillan, London. pp. 241–250.

CHAPTER 16

Urinary Concentrating and Diluting Processes

Mark A. Knepper and John L. Stephenson

1. Introduction

Body fluid osmolality is normally maintained within narrow bounds through the control of body water balance. Water intake is regulated to some degree by the thirst mechanism.[1] However, the chief means by which body water balance is so precisely maintained is the regulation of renal water excretion.

The predominant signal to the kidney that determines the rate of water excretion is the concentration of the octapeptide vasopressin* in the peripheral plasma. Vasopressin is secreted by specialized nerve endings in the posterior pituitary (neurohypophysis). It is synthesized in the cell bodies of these neurons located in the supraoptic and paraventricular nuclei of the hypothalamus. At near-normal states of extracellular fluid volume and arterial blood pressure, the circulating vasopressin concentration is a direct function of plasma osmolality.[2]

Figure 1 illustrates the whole kidney response to varying rates of vasopressin infusion in rats.[3] Two important observations can be made. First, the kidney is capable of wide variations in water excretion in response to changing levels of circulating vasopressin. Second, these wide variations in urine volume are achieved with little or no change in the rate of solute excretion. This ability to vary water excretion without changing solute excretion is crucial to the independent control of body fluid volume and osmolality. Figure 1 implies that there are reciprocal variations in water excretion and total solute concentration in the urine. When water excretion is low in response to high levels of circulating vasopressin, the urine is normally concentrated to an osmolality much greater than plasma. When water excretion is high in response to reduced levels of circulating vasopressin, the urine is diluted to an osmolality less than plasma.

The ability to concentrate and dilute the urine, and thus to maintain a relative constancy of solute excretion in the face of wide variations in water excretion, comes about as a result of a complex set of interactions among the various epithelial and vascular components of the kidney. Our purpose in the following is to describe and analyze these interactions. Space limitations prevent a comprehensive review. Rather our emphasis will be on general theoretical principles and critical experimental data. Several recent reviews give a comprehensive treatment of experimental studies,[4,5] anatomy,[5,6] and theoretical principles[7] relevant to the urinary concentrating and diluting mechanism.

*Synonym: antidiuretic hormone. The chief form is *arginine* vasopressin in most mammalian species, *lysine* vasopressin in the pig and the hippopotamus.

Mark A. Knepper • National Heart, Lung and Blood Institute, National Institutes of Health, Bethesda, Maryland 20205. **John L. Stephenson** • Department of Physiology, Cornell University Medical College, New York, New York 10021.

2. Renal Structure

2.1. Nephrons

A looped structure is characteristic of mammalian nephrons (Figure 2). This looped configuration brings into close proximity nephron segments that are not in direct axial continuity and allows parallel interactions between segments. Along the length of the nephron there are several types of epithelia, each characteristic of a major portion of the nephron (Fig. 3). These major segments are the proximal tubule, the thin limbs of Henle's loops, the distal tubule (including the thick limbs of Henle's loops), and the collecting duct system. Each is further divided into several subsegments. With the exception of the thin limbs, which we will discuss below, the structure and function of each of these epithelia are described in detail in separate chapters of this volume.

Nephrons may be classified into two populations based on the location of the loop of Henle bends (Fig. 2). Short-looped nephrons bend above the inner–outer medullary border, generally within the inner stripe of the outer medulla. Long-looped

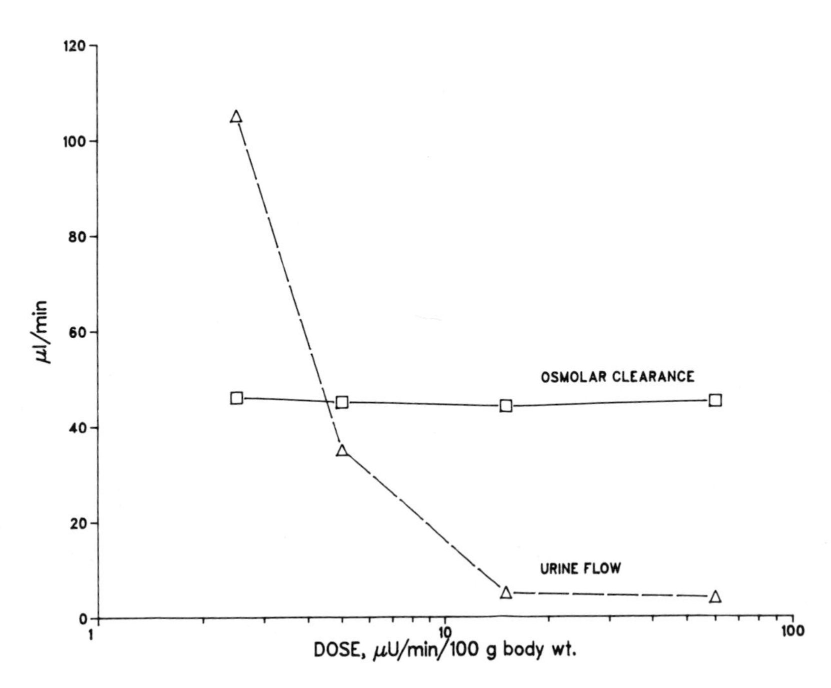

Fig. 1. Steady-state renal response to varying rates of vasopressin infusion in rats. A water load (4% of body weight) was given and maintained throughout the experiments to suppress endogenous vasopressin secretion. Although the urine flow rate was markedly reduced at higher doses, osmolar clearance changed very little. Data from Atherton *et al.*[3]

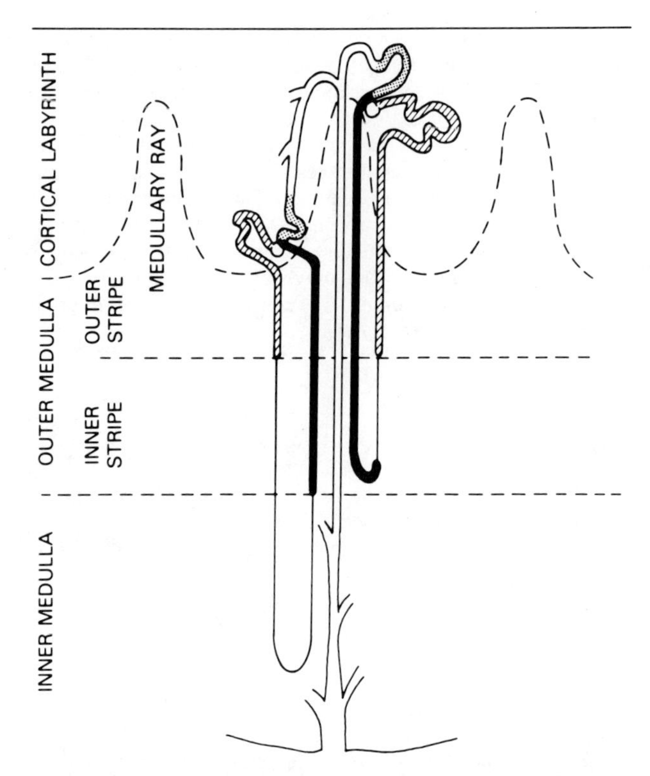

Fig. 2. Mammalian renal structure. Major regions of the kidney are labeled on left. Configurations of a long-looped and a short-looped nephron are shown. The major portions of the nephron are proximal tubules (hatched), thin limbs of Henle's loops (single line), thick ascending limbs of Henle's loops (solid), distal convoluted tubules (stippled), and the collecting duct system (open).

nephrons bend in the inner medulla. The long- and short-looped nephrons converge in the cortical labyrinth into a common collecting duct system.

The inner medullary ascending limbs of long-looped nephrons always have a "thin-limb" morphology. The cells are flat and have few mitochondria. The remaining length of the ascending limb of long-looped nephrons and the entire ascending limb of the short looped nephrons have a "thick-limb" morphology. The cells are considerably taller than those of thin ascending limbs and they are filled with mitochondria. Thus, thin ascending limbs are found only in the inner medulla and thick ascending limbs are found only in the outer medulla and

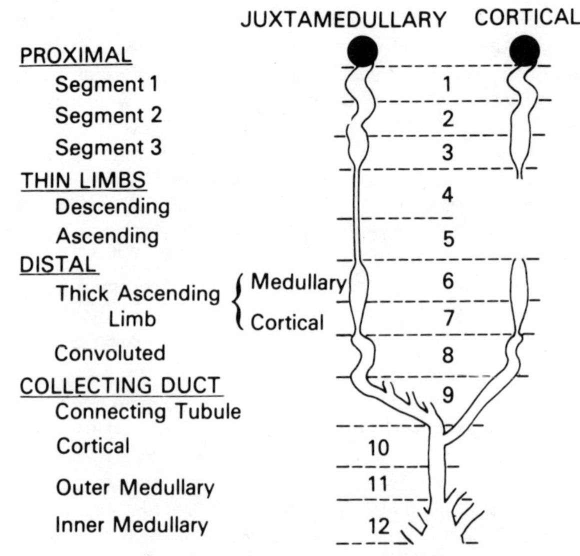

Fig. 3. Nephron segments in the rabbit. From Burg.[8]

medullary rays. There are important functional differences between the two types of ascending limbs which we discuss below.

Thin descending limbs occur in both short- and long-looped nephrons. As with thin ascending limbs, the cells of thin descending limbs are flat and have few mitochondria. Ultrastructural differences between the two are described in detail elsewhere.[5,6,9] Corresponding differences in the transport properties of thin ascending and thin descending limbs have also been described (Section 5). There are morphological variations among descending limbs related to whether they are part of short- or long-looped nephrons and whether they are located in the inner or outer medulla.[5,6,9] Knowledge is lacking, however, as to whether there are corresponding functional differences among thin descending limb segments.

2.2. Regional Organization

The mammalian kidney consists of several distinct regions (Fig. 2). Each of these regions has a characteristic combination of nephron segments and vascular structures in a specific geometric arrangement. The tubular and vascular structures in a given region share a common interstitial space. In general, solute concentrations in the interstitium are determined by the integrated function of the tubules and blood vessels in the region. The luminal composition in a given segment may be altered secondary to transport in a second segment if the transport alters the interstitial composition. The extent to which tubular transport affects interstitial composition depends on the effective rate of blood flow to the region.

Each region can be viewed as possessing regional functions that result from the integrated behavior of the nephron segments and vascular structures that it contains. In the following we describe the structural and functional characteristics of each region.

2.2.1. Inner Medulla and Inner Stripe of Outer Medulla

In both the inner medulla and the inner stripe of the outer medulla, the tubules are arranged in a highly organized parallel manner (Fig. 2). Similarly, the major blood vessels (vasa recta) are arranged in parallel as vascular bundles. This vascular arrangement facilitates countercurrent exchange of solutes between ascending and descending vasa recta,[10] which reduces the effective blood flow to these regions. As a result of the low effective blood flow, tubular transport readily modifies the interstitial composition, and there is a high degree of coupling between nephron segments.

2.2.2. Cortical Labyrinth

In contrast to the medulla, the tubules in the cortical labyrinth are not oriented in a parallel fashion. There is a dense capillary plexus and a high effective blood flow, which provides an efficient sink for substances absorbed from the renal tubules and an efficient source of secreted substances. Because of the high effective blood flow, renal tubular transport probably does not greatly alter the interstitial composition. Consequently, in contrast to the medulla, the functional interaction between adjacent nephron segments is small.

2.2.3. Outer Stripe of Outer Medulla and Medullary Rays

These regions are interposed between the cortical labyrinth and the deep medullary regions. Because the structures in these regions are not accessible to micropuncture measurements, our knowledge of their functional characteristics is limited. Instead, the functions of these regions must be inferred from other types of data (e.g., anatomical studies, *in vitro* perfusion studies, etc.). Both regions contain tubules arranged in parallel as in the deeper regions of the medulla. In contrast to the deep medulla, however, the vasculature is not as highly organized[11] and the effective blood flow may be higher. On the other hand, the capillary plexuses of both the medullary rays and the outer stripe are considerably sparser than that of the cortical labyrinth.[11] Consequently, the effective blood flow to the medullary rays and outer stripe may be intermediate between that of the cortical labyrinth and deep medulla. If so, it is likely that the composition of the interstitial spaces is modified to some extent by tubular transport and that transfer of solutes between nephron segments occurs.

3. Basic Concepts

3.1. The Countercurrent Multiplier Hypothesis

The looped configuration of the nephron and the parallel arrangement of structures in the renal medulla described above (Section 2.2) were well known by the late 19th century. The modern concept of parallel functional interactions among nephron segments arose much later from the theoretical work of Kuhn and colleagues.[12–14] In 1951, Hargitay and Kuhn[13] published the first explicit statement of the countercurrent multiplier hypothesis. They showed that a small concentration difference between bulk solutions could be multiplied many-fold by a hairpin counterflow system to generate an arbitrarily large axial concentration gradient, and suggested that the loop of Henle and collecting duct in the kidney formed such a system.

The prototype Kuhn multiplier is illustrated in Fig. 4. An axial concentration gradient can be induced by either solute transport from ascending flow to descending flow or water trans-

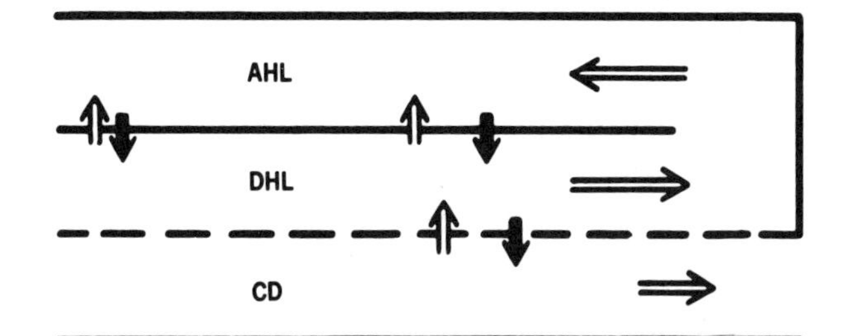

Fig. 4. Kuhn-type countercurrent multiplier. The system consists of loop of Henle and collecting duct. Concentration is built up in the loop by water moving from descending limb (DHL) to ascending limb (AHL) or solute from AHL to DHL. Water and solute move between DHL and collecting duct (CD), so that CD and DHL attain nearly the same osmolality. There is no provision in the model for the transport of water and solute from medullary nephron segments to the interstitium and vasa recta. Open arrows indicate water movement; solid arrows, solute movement.

port from descending to ascending flow. Mass balance requires that the algebraic sum of volume and solute flows in ascending and descending limbs equal the withdrawal at $x = L$ (equivalent to collecting duct flow) where flow toward the loop is counted as positive and away from the loop negative. Thus, we have

$$F_1(x) + F_2(x) = \Delta F(L) = \Delta F$$

and

$$F_1(x)\, C_1(x) + F_2(x)\, C_2(x) = \Delta F(L)\, C(L)$$

where $F_1(x)$ and $C_1(x)$ are volume flow and concentration in descending limb at axial position x, and $F_2(x)$ and $C_2(x)$ are volume flow and concentration in ascending limb. At the bend $C_1(x) = C_2(x) = C(L)$. These two equations combine to give

$$\Delta F\,[C(L) - C_1(x)] = [F_1(x) - \Delta F][C_1(x) - C_2(x)]$$

If there is no withdrawal from the system at $x = L$, $\Delta F = 0$ and everywhere mass balance requires $F_1(x) = -F_2(x)$ and $C_1(x) = C_2(x)$. With withdrawal $F_1(x) > \Delta F > 0$, and

$$C(L) - C_1(x) < [C_1(x) - C_2(x)]/[\Delta F/F_1(x)]$$

where $\Delta F/F_1(x)$ is the fraction of solution entering the descending limb that is withdrawn. $C_1(x) - C_2(x)$ must be less than the maximum equilibrium concentration difference that can be maintained perpendicular to the loop axis, i.e., the "single effect."* Thus, purely by mass balance considerations the axial gradient that can be generated by the system is shown to be limited by the ratio of the single effect to the fractional withdrawal.

Hargitay and Kuhn[13] analyzed in detail a counterflow system in which the concentration difference was generated by a hydrostatic pressure difference driving water without solute from descending to ascending flows. By solving the differential equations describing the system, they showed that the axial gradient generated depended on axial length of the multiplier and the fractional withdrawal. They also pointed out that other mechanisms could be used to generate the concentration difference and, subsequently, Kuhn and Ramel[14] posited active salt transport from ascending to descending flows as the most probable mechanism.

The countercurrent multiplier hypothesis provided predictions which could be tested experimentally in the mammalian kidney. If the theory applies to the kidney, there should be an axial osmolality gradient along the loops of Henle with only a small difference in osmolality between limbs of the loop at a given level. This prediction was confirmed by Wirz *et al.*,[15] who used an ingenious cryoscopic method to estimate the osmolality of tubule fluid in slices along the corticomedullary axis of antidiuretic rat kidneys. Indeed, an axial osmolality gradient was present in the loops (Fig. 5). The theory as applied to the concentrating kidney also requires a third tube, representing the collecting ducts, in which the urine reaches its final concentration by osmotic equilibration with the hypertonic loops. Thus, in the concentrating kidney the osmolality of the collecting duct fluid should parallel that of the loops. This also was confirmed by Wirz *et al.*[15] The osmolality in the large tubes (i.e., collecting ducts) of the medulla increased from cortex to papillary tip and was only slightly lower than that of the loops at a given level.

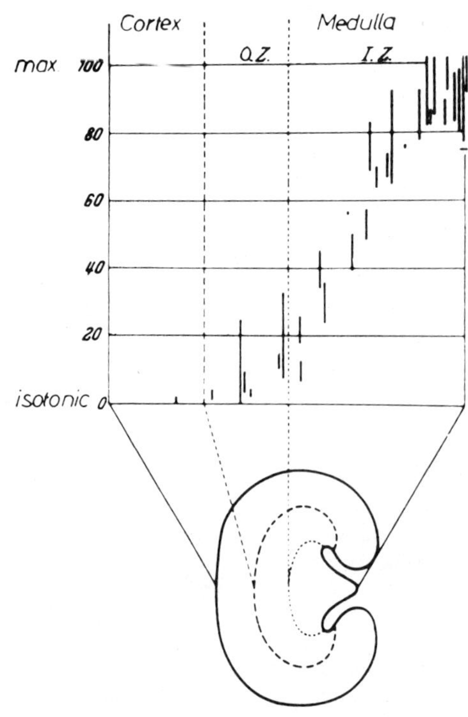

Fig. 5. Relative osmolalities of tubule fluid in slices from kidneys of hydropenic rats. Values are given as percent of maximum. O.Z., outer zone; I.Z., inner zone. From Wirz *et al.*[15]

As originally proposed, the countercurrent multiplier model described a direct interaction between ascending and descending limbs. It was recognized, however, by Hargitay and Kuhn[13] that application to the kidney requires either water or solute transport via the interstitial space. This implied that the medullary osmolality gradient should exist in the interstitial and vascular spaces as well as within the tubules. Consistent with this requirement, plasma collected from the vasa recta of antidiuretic hamsters was found to have an osmolality similar to that of the final urine.[16] Thus, it appeared that the axial osmolality gradient was shared by all the structures of the medulla. That the osmolality of fluid in the loops of Henle, collecting ducts, and vasa recta are all approximately the same at a given level of the inner medulla was confirmed by Gottschalk and Mylle[17] using micropuncture techniques in antidiuretic rats and hamsters.

3.2. The Nature of the Single Effect

As described above, when urine flow in the collecting ducts is taken into account, the countercurrent multiplier model requires an osmolality difference between ascending and descending limbs of the loop of Henle. Micropuncture studies[17–19] demonstrating that luminal fluid in the early distal tubule is hypotonic to plasma are consistent with this requirement. The subsequent demonstration by Lassiter *et al.*[20] of net water ab-

*The term "single effect" can be misleading (even as used by Hargitay and Kuhn). It represents a virtual limiting concentration difference. In general, the actual concentration difference $C_1(x)$ - $C_2(x)$ does not equal the single effect, but lies somewhere between it and zero. A more detailed analysis shows that if the concentration $C_1(x)$ increases from 0 to L, then $C_1(x)$ - $C_2(x)$ decreases from a maximum at $x = 0$ to 0 at $x = L$.

sorption from the loop of Henle showed that the hypotonicity must be generated by the absorption of solute (most likely NaCl) in excess of water rather than by water addition to the loop. This was confirmed by isolated, perfused tubule studies of thick ascending limbs[21,22] which have shown directly that this segment is capable of active NaCl absorption, is impermeable to water, and can decrease the osmolality of the luminal fluid below that of the peritubular environment. These results are supportive of the concept that the generation of a medullary osmolality gradient depends on active NaCl absorption from ascending limbs. It is generally assumed that the descending limb fluid osmotically equilibrates with the hypertonic medullary interstitium by water absorption, solute entry, or both. The mechanism of osmotic equilibration in the descending limb has been the subject of considerable controversy (Section 5).

3.3. The Role of Vasopressin

Although hypotonic in the early distal tubule, the tubule fluid becomes isotonic to plasma by the end of the distal tubule in rats excreting a concentrated urine.[17,19] During water diuresis, the tubule fluid remains hypotonic in the late distal tubule.[19] Vasopressin acts on the late distal tubule to increase the osmotic water permeability. Woodhall and Tisher[23] showed that the portion of the late distal tubule which responds to vasopressin is actually the initial portion of the collecting duct system ("initial collecting tubule"). The cortical collecting duct[24] and the inner medullary collecting duct[25,26] have also been shown to respond to vasopressin with an increase in water permeability. The action of vasopressin to increase water permeability appears to be confined to the collecting duct system.

Surveys of vasopressin-stimulated adenylate cyclase activity along the nephron[27] have revealed that the locus of vasopressin action may include the thin ascending limb in rats and rabbits and the thick ascending limb in rats and mice. A direct effect of vasopressin on transport in the thin ascending limb has not yet been described. However, Hall and Varney[28] demonstrated that vasopressin increases the rate of NaCl absorption in the medullary thick ascending limb of the mouse. This finding was subsequently confirmed by others.[29,30] Vasopressin apparently does not stimulate NaCl absorption in the cortical portion of the mouse thick ascending limb.[30]

3.4. Function of the Vasa Recta

Vascular perfusion of the medulla via the vasa recta is necessary to provide nutrients for cellular metabolic processes. In addition, the vasa recta are an essential part of the counterflow system. In the steady state, solute and water removed from the medulla by the vasa recta and associated capillaries must exactly equal solute and water absorbed from the medullary nephrons. Analysis of the vasa recta system[7] has yielded the following equation:

$$\begin{aligned} T_s &= T_{H_2O}\, C_{DVR}(0) + [F_{DVR}(0) + T_{H_2}O]\, \Delta C \\ &= T_{H_2O} C_{DVR}(0) + T_W \end{aligned}$$

where $F_{DVR}(0)$ and $C_{DVR}(0)$ are the volume flow and concentration of fluid entering the descending vasa recta (DVR), $F_{AVR}(0)$ and $C_{AVR}(0)$ are the volume flow and concentration of fluid leaving in the ascending vasa recta (AVR), and T_{H_2O} and T_s denote water and total solute absorption from the medullary nephrons. The term $C_{AVR}(0) - C_{DVR}(0)$ is denoted ΔC. This is

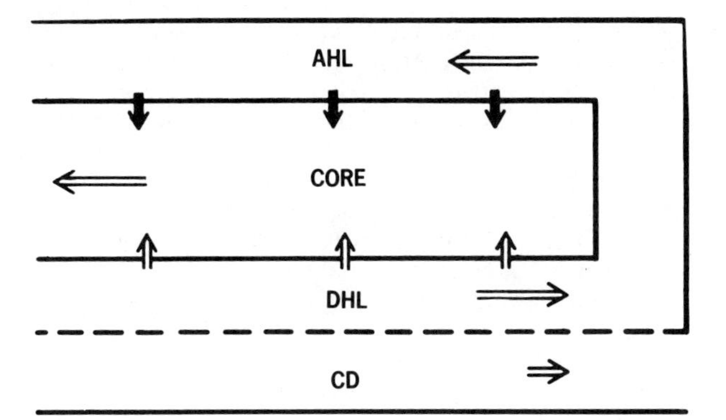

Fig. 6. Fundamentals of the concentrating mechanism. Solute without water from ascending limb increases the osmolality of the vascular interstitial core and extracts water without solute from the descending limb and collecting duct. Solute and water mix and expand in the core. From Stephenson.[31]

the fundamental conservation equation for nephrovascular coupling in the renal medulla. It shows that solute supplied by the nephron and returned to the systemic circulation by the blood vessels can be split into two parts. The first, $T_{H_2O}\, C_{DVR}(0)$, carries with it its isotonic complement of water. This solute, therefore, is utilized in concentrating loop of Henle and collecting duct fluid. The second part, $[F_{DVR}(0) + T_{H_2O}]\, \Delta C$, is due to the concentration difference between ascending and descending vasa recta and is solute that is unavailable for concentration. It is simply lost from the system.

3.5. Central Core Model

At every medullary level we can approximate net axial flow of solute in the combined vasa recta interstitial system[7] by

$$(F_{AVR} + F_{DVR})\, C_I(x) - D \text{ grad } C_I(x) = -\, T_s(x)$$

where $D \cong (F_{DVR})^2[(h_{AVR} + h_{DVR})/h_{AVR}\, h_{DVR}]$ and $T_s(x)$ is net solute uptake by the vasa recta from the nephrons between a given medullary level x and the papillary tip. With suitably defined D, this result applies either to individual solutes or to total concentration. This equation shows that we can approximate the combined vasa recta interstitial space by a single tube, the "central core," closed at the papillary end and open at the cortical medullary junction, where it discharges into the systemic circulation.

The central core model differs from the earlier models in that it incorporates the vasa recta as an essential part of the counterflow system. This allows for any variant of water and solute exchange between nephrons and vasculature. In particular, it permits solute supplied by the ascending limb to extract water from the descending limb and collecting duct and so perform useful osmotic work on the fluid in these segments. This mechanism is illustrated in Fig. 6 for the case of concentration in the descending limb by water extraction alone.

Here we consider a single-solute system with metabolically driven solute transport out of the ascending limb (AHL) with the ascending limb solute permeable, but water impermeable. In contrast, the descending limb (DHL) and the collecting duct (CD) are water permeable, but solute impermeable. Qualitatively, it is clear that if fluid in the core is made hypertonic by solute transport out of AHL, water will be extracted from the downflowing DHL and CD. This water extraction will generate net upward flow in the core. The counterflowing hypertonic solution will continue to extract water from descending Henle's limb and collecting duct until it reaches the corticomedullary junction. Exact analytic solution of the differential equations describing this system are not possible, but analysis of its overall operation is relatively straightforward. If it is assumed that fluid entering the medulla in DHL, CD, and DVR is approximately at plasma concentration and that at the papillary tip

$$C_{DHL}(L) = C_{AHL}(L) \cong C_{DVR}(L) \cong C_{CD}(L) = C(L)$$

analysis of mass conservation yields the result[7] that the concentration ratio generated by the system is

$$r = \frac{C(L)}{C(0)} = \frac{1}{1 - f_T(1 - f_U)(1 - f_W)}$$

where

$$f_T = \frac{T_s}{C(0)[F_{DHL}(0)]}$$

is fractional solute transport out of the ascending limb,

$$f_U = \frac{F_{CD}(0)}{F_{CD}(0) + F_{DHL}(0)} = \frac{F_{CD}(L)}{F_{CD}(L) + F_{DHL}(L)}$$

is fractional urine flow, and

$$f_W = \frac{T_W}{T_s}$$

is fractional dissipation.

This equation was derived under the assumption that concentration in collecting duct and descending Henle's limb is by water extraction alone, but this assumption is not essential. The equation is valid regardless of the mode of equilibration between DHL and/or CD and core.[7]

Dilution of the AHL fluid is given by

$$C(L) - C_{AHL}(0) = \frac{T_s}{F_{DHL}(L)}$$

From this we find the relative hypotonicity of AHL fluid is given by

$$C(0) - C_{AHL}(0) = \frac{T_s}{F_{DHL}(L)} - \frac{T_s - T_w}{F_{DHL}(L) + F_{CD}(L)}$$

The relative hypotonicity increases with both dissipation T_W and withdrawal $F_{CD}(L)$ and vanishes if they both vanish. With finite withdrawal and dissipation, the relative hypotonicity is limited by back-leak which limits T_s and the gradient that can be generated.

3.6. Summary of Basic Concepts

The above-described view of the concentration and dilution of urine can be summarized as follows: Metabolically driven NaCl transport out of the ascending limb system without accompanying water increases the osmolality of the interstitial vascular space, which can be approximated as a single tube (the central core) closed at the papillary end and open at the corticomedullary junction. In part, this solute may be recycled via the descending limb and in part it extracts water from the descending limb and collecting duct, generating a net upward flow in core structures. The net composite flow in the descending limb, core, and collecting duct is downward and numerically equal to the sum of the upward flow in the ascending limb and the final urine flow. In the presence of vasopressin, the ascending limb, collecting duct, and core are approximately osmotically equilibrated and the solute transfer from the ascending limb concentrates the composite downflow and dilutes the ascending limb upflow. The relative hypotonicity of the ascending limb fluid leaving the medulla increases with relative urine flow and dissipative solute loss via the core. It cannot be greater than the maximum equilibrium concentration difference (the single effect) that can be attained between the ascending limb and surrounding bath. In the presence of vasopressin, the dilute fluid leaving the ascending limb regains isotonicity in the distal cortical nephron, losing its water to the cortical interstitium. In the absence of vasopressin, the osmotic water permeability of the collecting duct system is low. The hypotonic fluid generated by active NaCl transport from the thick ascending limbs remains hypotonic as it passes through the collecting duct system resulting in a hypotonic final urine.

3.7. An Unresolved Issue: Concentration in the Inner Medulla

A number of experimental observations indicate that the renal concentrating mechanism may be even more complex than described in the preceding paragraph. Although there is clear evidence for active NaCl transport in the thick ascending limbs, several attempts to detect active NaCl absorption from the thin ascending limb (the inner medullary portion of the ascending limb) have yielded either negative results[25,32,33] or low net absorption rates[34] which are not large enough to have much physiological significance.[35] Thus, the mechanism which generates the axial osmolality gradient in the inner medulla probably differs from that in the outer medulla. Urea and other non-NaCl solutes account for one-half or more of the osmolality of the tissue fluid in the inner medulla and up to 90% of the final urinary osmolality (Fig. 7).[36] Therefore, models of the concentrating mechanism must account for solutes other than NaCl.

These observations necessitate an alternative to the Kuhn–Ramel-type active countercurrent multiplier in the inner medulla. Before considering the possibilities in Section 6, we provide the necessary background by reviewing the characteristics of handling of several osmotically important solutes in the renal medulla (Section 4), and the properties of the thin limbs of the loops of Henle (Section 5).

4. Handling of Individual Solutes in the Medulla

Because water transport in the kidney occurs primarily by osmosis, it is evident that water excretion is highly dependent on the handling of solutes throughout the kidney. Furthermore, because urinary solutes are dissolved in water and the flow of water conveys the solutes along the nephron, renal tubular solute

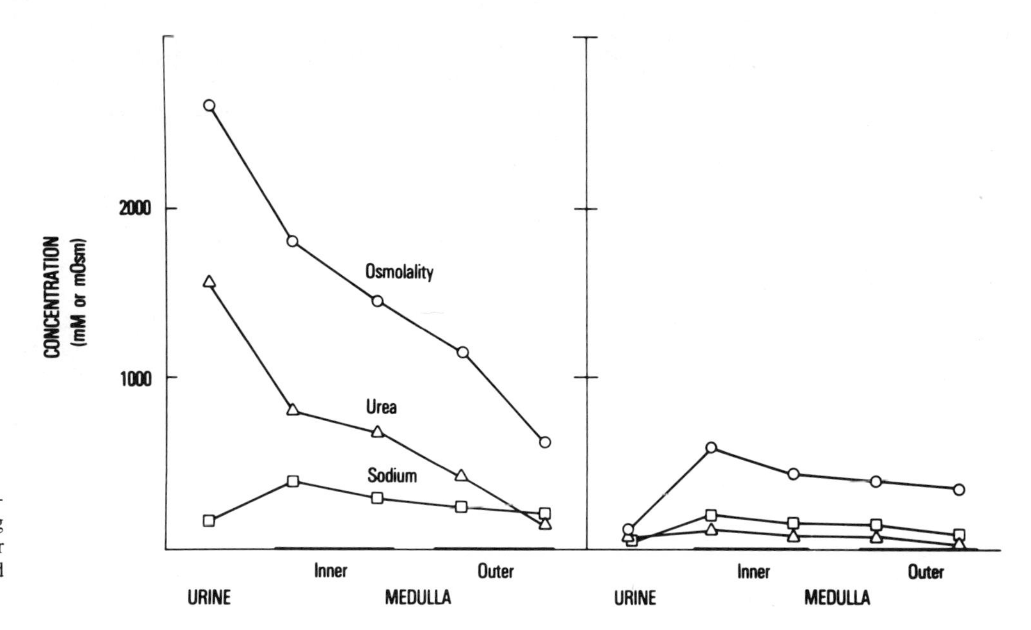

Fig. 7. Composition of renal medulla and urine during antidiuresis (left) and water diuresis (right). Adapted from Hai and Thomas.[36]

transport may be greatly influenced by variations in water absorption. Consequently, analysis of the renal concentrating and diluting mechanism requires consideration of the pattern of solute handling throughout the kidney. A number of solutes (viz. urea, and salts of sodium, potassium, and ammonium) are likely to play important roles by virtue of their accumulation in the final urine or their participation in thc generation of the medullary osmolality gradient. All of these solutes accumulate to some extent in the medullary interstitium (Table I). In the following we describe the renal handling of these solutes with emphasis on their handling in the medulla.

4.1. Urea

In mammals, waste nitrogen is eliminated mainly through urea excretion. Urea generally accounts for 30–60% of the total urinary osmolality.[36,41–43] Therefore, as is true among all of the osmotically important urinary solutes, urea must be highly concentrated in the urine to minimize any reduction in its excretion when the rate of water excretion is reduced during antidiuresis.[44] The amount of water obligated for the excretion of urea is much less than that required for equal osmotic loads of NaCl and other salts.[45,46] In fact, the presence of a high concentration of urea in the urine increases the concentration of nonurea solutes in maximally concentrated urine or, viewed another way, reduces the amount of water needed for excretion of a given amount of nonurea solute. These observations suggest an important role for urea in the concentrating mechanism.

In 1958, using the newly developed microcatheterization technique, Klümper *et al.*[47] observed in nondiuretic hamsters that urea is absorbed in substantial quantities along the inner medullary collecting ducts. Because water absorption occurs at a relatively higher rate, the urea concentration increases toward the papillary tip. These observations have recently been confirmed by micropuncture of the inner medullary collecting ducts.[48] Ullrich and Jarausch[49] observed in antidiuretic dogs that the urea concentration in the tissue water of the inner medulla was nearly as high as in the final urine. Evidently, urea accumulates throughout the inner medulla, not just in the collecting ducts. Analysis of fluid obtained by micropuncture of the vasa recta and loops of Henle directly demonstrated high urea concentrations in these structures.[20,39,50–53]

Berliner *et al.*[10] proposed that urea is concentrated in the inner medullary interstitium as a result of urea absorption from the inner medullary collecting ducts. They suggested that the high urea concentration in the collecting ducts results simply from absorption of water in excess of urea earlier in the nephron. Lassiter *et al.*,[20] using micropuncture techniques, found evidence of substantial entry of urea into the loops of Henle. From these data it was inferred that urea is recycled between the collecting ducts and loops of Henle.[20,54] That is, the urea absorbed from the inner medullary collecting ducts enters the thin limbs of the loops of Henle and from there is conveyed by the flow of the tubular fluid through the distal tubule and back to the collecting ducts. Consistent with this model, Marsh[52] demonstrated by micropuncture in antidiuretic hamsters that urea enters both the thin descending limb and the thin ascending limb of the loop. Urea recycling in the medulla undoubtedly contributes to its high concentration in the collecting ducts and inner medullary interstitium.

The mass flow rate of urea in the long loops of Henle near the loop bends has been shown to be as high or higher than the

Table I. Solutes That Accumulate in the Renal Medulla

Solute	Species	Vasa recta concn/ venous concn[a]
Sodium	Rat	2 (37)
	Hamster	4 (34)
Potassium	Rat	8 (37,38)
Urea	Rat	23 (39)
Ammonium	Rat	90 (40)

[a]Concentrations measured in plasma collected from inner medullary vasa recta by micropuncture. Venous concentrations determined in renal venous or peripheral venous blood. References in parentheses.

filtered load of urea in antidiuretic rodents.[20,39,51–53] This indicates that urea entry occurs either into the proximal tubule or into the thin descending limb. Entry of urea into the inner medullary portion of the descending limb[52] undoubtedly accounts for a major fraction of the net secretion of urea. Part of the urea entry may also take place in the proximal straight tubule where active[55] or passive[56] urea secretion may occur.

The high mass flow rate of urea near the bends of long loops is associated with high urea concentrations at those sites. Values of 280 mM or greater are typical in antidiuretic rodents.[52,53] Based solely on mass balance arguments, the tubular fluid at the end of the thick ascending limbs must be dilute relative to systemic plasma for the kidney to produce a concentrated urine.[57] It has been pointed out that, because the luminal urea concentration near the long-loop bends accounts for an osmolality greater than or equal to plasma, urea absorption would have to occur along the ascending limb if the tubule fluid at the end of the thick ascending limb is to become dilute relative to plasma.[39] Although micropuncture studies have demonstrated net entry of urea into the thin ascending limb,[52] urea absorption may occur in the thick ascending limb. Although the urea permeability of the thick ascending limb is relatively low in the inner stripe of the outer medulla,[26,58] the urea permeability is substantially higher in the outer stripe and medullary rays.[56,58] Presumably, urea absorption from the urea-permeable portion of the thick ascending limb of the long-looped nephrons allows the luminal urea concentration to fall to the extent that the tubule fluid osmolality can be lowered below that of plasma. In this manner, passive urea absorption may contribute to the diluting ability of the thick ascending limb and to the overall concentrating ability of the kidney.

4.2. Sodium

Active transport of sodium fulfills different functions along the length of the nephron.[8] In the proximal tubule, it is linked to the secondary active absorption of glucose, amino acids, bicarbonate, and other substances. As discussed above (Section 3), in the thick ascending limb, active transport of NaCl generates the single effect for countercurrent multiplication in the outer medulla. Sodium is also absorbed from the thin ascending limb[52] but probably primarily by passive diffusion rather than active transport (Section 5). In the collecting duct system, regulation of active transport of sodium is responsible for the fine control of its excretion. The chief regulating factor is thought to be the mineralocorticoid hormone aldosterone.

The inner medullary collecting duct is an especially important site of sodium absorption because of its terminal position along the renal tubule. Despite rapid water absorption during antidiuresis, the luminal sodium concentration decreases along the inner medullary collecting duct to well below plasma levels at the papillary tip, apparently by an active transport process.[59–64] This has been the only demonstrable active transport process in the inner medulla. Therefore, although NaCl transport by the thick ascending limb powers the concentrating mechanism, NaCl itself is not generally highly concentrated in the final urine because of the avid absorptive process in the inner medullary collecting ducts.

4.3. Potassium

Depending primarily on dietary intake of potassium, potassium salts may account for a substantial fraction of the urinary osmolality. The conventional view of renal potassium handling is that it is virtually totally absorbed proximally and that most of the excreted potassium enters the nephron as a result of distal secretion.[65] Micropuncture studies in rats[37,38] have shown a higher mass flow rate of potassium near the bends of the long loops of Henle than is likely to exist at the end of the proximal tubule or descending limb. Potassium concentrations in the vasa recta have been found to be considerably greater than in the systemic plasma (Table I). That is, potassium accumulates in the medulla. It appears likely that potassium is recycled in the medulla as a result of potassium transfer from the collecting ducts, ascending limbs, or both to the descending limbs.[66] It has been proposed that medullary recycling and the associated medullary accumulation of potassium may play an important role in determining its rate of excretion.[66,67]

4.4. Ammonia

Ammonia is an important urinary buffer. Since it is a base with a pH of about 9.0, it is excreted primarily as NH_4^+ (ammonium). Ammonium and its attendant anion (Cl^-) can account for a large fraction of the total urinary osmolality, particularly when the rate of acid excretion is high. The ammonia that is excreted is derived almost totally from its renal synthesis, primarily in the proximal convoluted and proximal straight tubules.[68]

How ammonia is transferred from its chief site of synthesis in the proximal tubules to be concentrated in the final urine is of considerable interest. Buerkert *et al.*[69] demonstrated a reduction in the mass flow rate of ammonia between the late proximal convoluted tubule and the distal tubule of superficial nephrons, i.e., net absorption from the loops of Henle. Tissue slice analysis[70] and measurements of ammonia concentrations in vasa recta plasma[40] (Table I) have shown that ammonia accumulates in the medulla to well above cortical concentrations. In isolated, perfused thick ascending limbs from rats, net absorption of ammonia against a concentration gradient has been observed (D. W. Good, M. A. Knepper, and M. B. Burg, unpublished observations). *In vivo,* such a transport process may generate a single effect for countercurrent multiplication of ammonia analogous to that of the NaCl multiplier and consequently account for ammonia accumulation in the medulla.

Much of the ammonia in the final urine enters the tubule fluid across the collecting duct epithelium.[69,71–74] This occurs predominantly as a result of transport from the loops of Henle via the interstitium rather than from local synthesis by the cells of the collecting ducts.[69] Whether it is the ionic form or the nonionic form that is transported across the collecting duct epithelium has not been established.

5. Properties of the Thin Limbs of Henle's Loops

Knowledge of the permeability properties of the thin limbs of the loops of Henle is critical to a rational analysis of the concentrating mechanism in the inner medulla. Both the thin descending and the thin ascending limbs have been thoroughly studied (Table II).

Most of our knowledge of the properties of the thin descending limb has been obtained from studies of isolated, perfused descending limb segments from the outer medulla of rabbits (Table II). As noted in Section 2.1, there are morphological

Table II. Permeability Properties of the Thin Limbs of Henle's Loop

	Rabbit	Rat	Hamster
Thin descending			
Solute permeabilities (cm/sec × 10^{-5})			
Na^+	1.6 (75)	—	—
	1.9 (76)		
	0.2 (77)		
K^+	2.5 (76)	—	—
Urea	1.5 (78)	13 (25)	—
	1.0 (79)		
Osmotic water permeability (μm/sec)	2315 (75)	516 (25)	5130 (33)
		2139 (33)	
Thin ascending			
Solute permeabilities (cm/sec × 10^{-5})			
Na^+	26 (32)	80 (33)	88 (33)
K^+	—	—[a]	—[a]
Cl^-	117 (32)	184 (33)	196 (33)
Urea	7 (32)	23 (33)	19 (33)
		14 (25)	
Osmotic water permeability (μm/sec)	13 (32)[b]	25 (33)[b]	29 (33)[b]
		41 (25)	

[a] K^+ permeability approximately equal to Na^+ permeability by dilution potentials.[33]

[b] Mean value reported was not significantly different from zero.

variations along the length of the descending limb of the long loops and between descending limbs of short- and long-looped nephrons. We do not yet know whether there are corresponding variations in permeability properties. Based on the appearance of the perfused segments in published photographs, Jamison and Kriz[5] suggested that the perfused segments were from long-looped nephrons. The osmotic water permeability of the thin descending limb was found to be extremely high.* The solute permeabilities have been interpreted as being low.[80] However, Pennell *et al.*[39] pointed out that the measured values of urea permeability are consistent with a significant degree of urea entry *in vivo*. Since the permeabilities of the other solutes are similar, this conclusion can probably be extended to these as well. Nevertheless, the *in vitro* data are compatible with water absorption as the chief mode of osmotic equilibration in the rabbit descending limb. Micropuncture studies in rodents, however, have produced evidence that solute entry may be the chief mode of osmotic equilibration in the descending limbs.[39,51–53] Conceivably, the different conclusions result from the different species studied in the two types of experiments.

The permeability properties of the thin ascending limbs are in marked contrast to those of the descending limb (Table II). The thin ascending limb is virtually impermeable to water. The solute permeabilities are extremely high. The permeability to urea, although high compared to the value in the descending limb, is considerably lower than the values for sodium and chloride.

*Stoner and Roch-Ramel[79] measured a low water permeability in rabbit thin ascending limb (20 mm/sec). As indicated in Table II, several other investigators working with preparations from three different species have found extremely high values. Thus, the preponderance of evidence favors a high osmotic water permeability in this segment.

6. Concentration in the Inner Medulla

6.1. Mass Balance Requirements

In order to describe the concentration process in the inner medulla adequately, it is necessary to consider more than one solute. The basic ideas remain the same as outlined in Section 3.4. Namely, in the steady state there can be no net accumulation of any solute or of water in the volume contained between any two medullary levels, which we will designate by x_1 and x_2. Thus, in the absence of chemical sources

$$\Sigma_i F_{ik}(x_2) - \Sigma_i F_{ik}(x_1) = 0$$

and

$$\Sigma_i F_{iv}(x_2) - \Sigma_i F_{iv}(x_1) = 0$$

where $F_{ik}(x)$ is net axial flow of the kth solute in the ith tube at medullary level x and $F_{iv}(x)$ is net axial volume flow in the ith tube.

As before, we consider flow toward the papilla positive and that toward the cortex negative. The above equations apply to geometries of arbitrary complexity with appropriate summation conventions.[81] In any tube, axial solute flow is the sum of a diffusive and convective term. If we neglect diffusive transport relative to convective, we obtain from the above conservation relations

$$\Sigma_i F_{iv}(x) = \Sigma_i F_{iv}(L) = U$$

$$\Sigma_i F_{iv}(x)\, C_{ik}(x) = \Sigma_i F_{iv}(L)\, C_{ik}(L) = U\, C_k(\mathrm{u})$$

and in summing over all solutes

$$\Sigma_i F_{iv}(x)\, C_{i\mathrm{M}}(x) = \Sigma F_{iv}(L)\, C_{i\mathrm{M}}(L) = UC_M(\mathrm{u})$$

where $C_{ik}(x)$ is the concentration of the kth solute in the ith tube, $C_k(\mathrm{u})$ is the concentration of the kth solute in the urine, $C_{i\mathrm{M}}(x)$ is total solute concentration (idealized osmolality), i.e.,

$$C_{i\mathrm{M}}(x) = \Sigma_k C_{ik}(x)$$

U is total final urine flow exiting the collecting ducts, and $C_{\mathrm{M}}(\mathrm{u})$ is total osmolality of urine emerging from the collecting ducts at the papillary tip. If we define

$$\Delta C_{i\mathrm{M}}(x) = C_{\mathrm{IM}}(x) - C_{i\mathrm{M}}(x)$$

where $C_{\mathrm{IM}}(x)$ is osmolality of the medullary interstitium at level x, the above equations combine to give

$$\Sigma_i F_{iv}(x)\, \Delta\, C_{i\mathrm{M}}(x) = U[C_{\mathrm{M}}(\mathrm{u}) - C_{\mathrm{IM}}(x)]$$

Except for the neglect of diffusional terms relative to convective, the above equation is exact. In the concentrating kidney, $C_{\mathrm{M}}(\mathrm{u}) - C_{\mathrm{IM}}(x) > 0$ for $x < L$ and always $U > 0$. Thus, the right-hand side is positive. This means that the left-hand sum must also be positive. With our convention that flow toward the papilla is positive and away from the papilla negative, hypertonic descending flows and hypotonic ascending flows will contribute positive terms and enhance concentrating ability. Conversely, hypotonic descending flows and hypertonic ascending flows will contribute negative terms and diminish concentrating ability. This mass balance equation, when considered in conjunction with equations for individual tubes, places severe restrictions on possible mechanisms for generating a medullary concentration gradient. For each tube we have the localized conservation requirement

$$\frac{dC_{ik}}{dx} = \frac{J_{iv}C_{ik} - J_{ik}}{F_{iv}}$$

and

$$\frac{dC_{iM}}{dx} = \frac{J_{iv}C_{iM} - J_{iM}}{F_{iv}}$$

where J_{iv} and J_{ik} are the transmural volume and solute fluxes which are determined by the transmural driving forces. These can be conveniently described by the Kedem and Katchalsky phenomenology.[82] A viable model must satisfy both the overall conservation requirement and the local conservation and phenomenological equations. We will examine some of the various mechanisms that have been proposed for generating the inner medullary concentration gradient with respect to both requirements.

6.2. Proposed Models

6.2.1. Active Solute Transport in a Water-Impermeable Segment

Active solute transport out of the thin ascending limb or into the descending limb could generate either a hypotonic ascending limb fluid or a hypertonic descending limb fluid if the segment were also impermeable to water. This would satisfy the mass balance requirement (Section 6.1) for concentration in the inner medulla. Theoretically, active transport *into* the inner medullary collecting duct could satisfy the mass balance requirement. Such a mechanism appears unlikely because of the experimentally observed high water permeability and net solute absorption in the collecting ducts, as well as the low volume flow rate relative to the loop.

A role for active solute secretion into the descending limb in the generation of a single effect for countercurrent multiplication in the medulla was originally proposed by Kuhn and Ramel.[14] Kriz *et al.*[83] also suggested that active solute secretion into the descending limb may contribute to the concentration of solutes in the inner medulla. Evidence for active transport in the thin descending limb, however, has not been found.[75,78] Furthermore, several investigators have found that this segment has a high osmotic water permeability (Table II), not the low value required by the model. The morphological heterogeneity of the thin descending limb (Section 2.1) raises the question of whether the measurements that have been made in descending limbs are descriptive of all thin descending limbs. The recent demonstration of substantial Na^+,K^+-ATPase in the early thin descending limb of the long-looped nephrons of rats[84,85] is consistent with the possibility of active solute transport.

Although several studies[25,32–34] have reported little or no active solute transport in the thin ascending limb, it should be emphasized that these studies do not rule out such a possibility. Certain of the local conditions in the inner medulla *in vivo* (e.g., oxygen tension, osmolality, pH, hormone concentrations, or substrate concentrations) that are possibly critical to the function of the thin ascending limb may not have been reproduced in experiments.

6.2.2. Active Solute Transport with Nonuniform Solute Reflection Coefficients

In the inner medullary collecting duct, the reflection coefficient for urea is substantially lower than that for NaCl.[25,26] Because of its lower reflection coefficient, urea will not be as effective in driving osmotic water flow as will NaCl. The urea concentration is higher in the lumen than in the interstitium[50] and the NaCl is lower in the lumen than in the interstitium (Section 4.2). Under these circumstances, osmotic equilibration across the collecting duct epithelium will be achieved at a luminal osmolality somewhat greater than that of the interstitium at the same level.[86,87] This osmolality difference may satisfy the mass balance requirement (Section 6.1) for concentration in the inner medulla.[87] Because of the asymmetry of reflection coefficients, active NaCl absorption from the inner medullary collecting duct would theoretically result in spontaneous hypotonic fluid absorption and spontaneously increase the luminal osmolality above that of the interstitium. Marsh[88] presented evidence for spontaneous hypotonic fluid absorption from the inner medullary collecting duct in an isolated papilla preparation from rats. Hypotonic fluid absorption can be expected to dilute the interstitium relative to the collecting ducts. Therefore, fluid absorption driven by active NaCl absorption from the inner medullary collecting ducts does not appear to be a feasible means of concentrating the interstitial fluid. However, by increasing collecting duct water absorption, the asymmetry of reflection coefficients may enhance water conservation by the kidney.[86]

6.2.3. Passively Exchanging Single-Solute Models

A detailed analysis of single-solute systems with only passive exchange of solute and water[89] has lead to the conclusion that all concentrations within the system are bounded by the maximum and minimum entering concentrations. This generalization allows the elimination of many models which may satisfy external balance requirements but fail on thermodynamic grounds. For example, with respect to overall mass balance, an osmolality gradient in the inner medulla may be expected to be supported by hypertonicity of the descending limb or descending vasa recta fluid entering the inner medulla. However, it is clear that with only passive exchange and in the absence of transfer of chemical potential energy from another inner medullary structure, the concentration in the descending limb or descending vasa recta cannot rise above the entering concentration.

6.2.4. Passively Exchanging Multisolute Models

The restrictions on concentration by passive exchange do not apply to multisolute systems. Stephenson[57,90] and Kokko and Rector[80] simultaneously proposed a mechanism by which the osmolality in the thin ascending limb can be lowered below that in the interstitium by salt and urea mixing in the interstitial-vascular space. Analysis of the mechanism is facilitated by the central core model shown in Fig. 8. We will suppose that descending limb fluid entering the medulla is relatively rich in salt and collecting duct fluid is relatively rich in urea. The DHL is assumed to be highly permeable to water and relatively impermeable to both salt and urea; AHL is assumed to be essentially

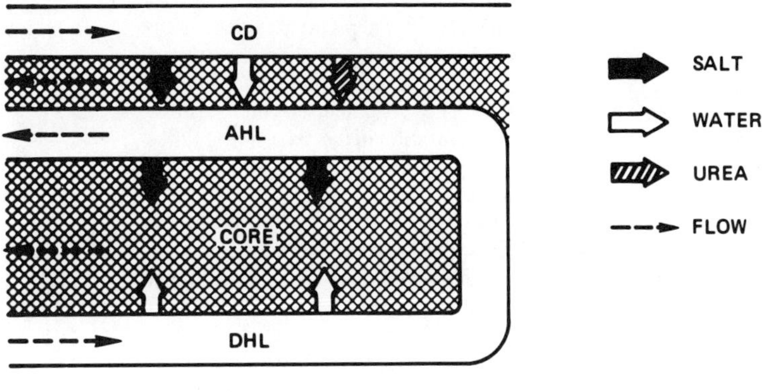

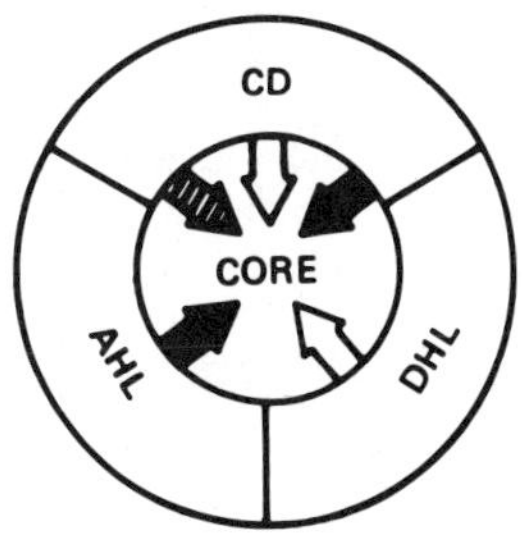

Fig. 8. Increase of concentration in central core model by urea cycling. Urea from the collecting duct (CD) and salt from the ascending Henle's limb (AHL) and CD mix in the core and extract water from the CD and descending Henle's loop (DHL). The combined volume expansion of the core mixture and contraction of the DHL fluid depresses core salt concentration relative to salt concentration at beginning of AHL. The salt concentration difference adds a positive increment to transport of salt out of the AHL by passive diffusion and so enhances concentration.

impermeable to water, highly permeable to salt, and relatively impermeable to urea; CD is permeable to salt, water, and urea. Under these assumptions, total osmolality of DHL, core, and CD are approximately equal, i.e.,

$$[\text{NaCl}]_{\text{DHL}} + [\text{urea}]_{\text{DHL}} = [\text{NaCl}]_{\text{core}} + [\text{urea}]_{\text{core}} = [\text{NaCl}]_{\text{CD}} + [\text{urea}]_{\text{CD}}$$

The concentrations of salt and urea in the core are given by

$$[\text{NaCl}]_{\text{core}} = \frac{\text{net salt entry}}{\text{volume flow up core}}$$

$$[\text{urea}]_{\text{core}} = \frac{\text{net urea entry}}{\text{volume flow up core}}$$

From these relations we find

$$\frac{[\text{NaCl}]_{\text{core}}}{[\text{NaCl}]_{\text{core}} + [\text{urea}]_{\text{core}}} = \frac{\text{salt entry}}{\text{salt entry} + \text{urea entry}}$$

From this equation it is clear that as urea entry into the core increases, the ratio of salt concentration to total osmolality decreases. Under appropriate conditions

$$[\text{NaCl}]_{\text{core}} < [\text{NaCl}]_{\text{DHL}}$$

and

$$[\text{urea}]_{\text{core}} < [\text{urea}]_{\text{CD}}$$

Since at the bend of the loop

$$[\text{NaCl}]_{\text{AHL}} = [\text{NaCl}]_{\text{DHL}}$$

we also have

$$[\text{NaCl}]_{\text{AHL}} > [\text{NaCl}]_{\text{core}}$$

Accordingly, salt can diffuse out of the water-impermeable AHL, lowering its osmolality below that of the core and the osmotically equilibrated DHL and CD and providing the solute needed to generate a medullary concentration gradient. Effectively, urea diffusing into the core from the CD raises its osmolality and extracts water from DHL, this dilutes NaCl in the core and induces salt transport out of AHL, which in turn extracts more water from DHL and CD. Overall, there is mixing and expansion of the salt and urea solution in the core. Part of the free energy lost in the expansion is used to perform useful osmotic work of compression on the solutions in DHL and CD. Overall, the system is dissipative. The energy used to drive it is the metabolic energy used to concentrate salt and urea in the outer medulla and cortex.

Detailed theoretical analysis[90] and computer stimulation[82,91,92] have demonstrated the general feasibility of the passive model. Qualitatively, the permeability properties of the thin limbs appear to be consistent with the passive model (Table II). In addition, direct *in vivo* studies[52] of solute fluxes across the hamster thin ascending limb have demonstrated net absorption of NaCl and net secretion of urea, as predicted by the passive model. As required by the model, the vasa recta urea concentration exceeds that of thc loop of Henle in antidiuretic rodents.[39,52] Furthermore, in antidiuretic rats the vasa recta sodium concentration has been observed to be lower than in the loop fluid.[38,93] Micropuncture results[94] showing a lower osmolality in the thin ascending limb than in the thin descending limb at the same level are also consistent with a single effect in the thin ascending limb. Thus, there is a considerable body of experimental data which is qualitatively consistent with the passive model.

Despite the evidence in favor of the passive model, quantitative considerations have brought into question the importance of the passive model in the generation of an inner medullary osmolality gradient.[35] The arguments center around measurements of the urea concentration in the descending limbs near the bends of the long loops of Henle. These studies[52,53] have shown that although urea concentrations in the luminal fluid are somewhat lower than in the surrounding vasa recta or interstitium, they nevertheless often exceed 280 mM. Previous analyses of the passive mechanism[80,95] have assumed a much lower luminal urea concentration in the descending limbs. A high luminal urea concentration in the descending limb precludes the development of a sufficient lumen-to-bath NaCl difference to drive the passive absorption of a substantial fraction of the NaCl that enters the thin ascending limb. Marsh[96] has pointed out that for the passive model to generate a NaCl gradient in the inner medulla, an absolute requirement is that the urea concentration at the beginning of the thin ascending limb be lower than in the interstitium at the inner–outer medullary junction, a condition which is unlikely to exist. Thus, although the passive model is supported by much experimental data, the luminal urea concentrations near the bends of the long loops of Henle appear to be inconsistent with the model.

7. Summary

Urinary concentration occurs as a result of countercurrent multiplication in the renal medulla. In the outer medulla, active NaCl absorption from the water-impermeable thick ascending limb generates the single effect for the multiplier. However, since experimental evidence for active transport in the thin (inner medullary) ascending limb has not been found, the concentrating mechanism probably differs in the inner medulla. Of the several possible mechanisms that have been proposed for the inner medullary multiplier, none is totally consistent with existing experimental data. The preponderance of evidence, however, favors the passive solute-mixing model of Stephenson[57,89] and Kokko and Rector.[80] Clearly, further theoretical and experimental efforts will be required to fully resolve this issue.

References

1. Fitzsimons, J. T. 1976. The physiological basis of thirst. *Kidney Int.* **10**:3–11.
2. Robertson, G. L., E. A. Mahr, S. Athar, and T. Sinha. 1973. Development and clinical application of a new method for the radioimmunoassay of arginine vasopressin in human plasma. *J. Clin. Invest.* **52**:2340–2352.
3. Atherton, J. C., R. Green, and S. Thomas. 1971. Influence of lysine-vasopressin dosage on the time course of changes in renal tissue and urinary composition in the conscious rat. *J. Physiol. (London)* **213**:291–309.
4. Jamison, R. L. 1981. Urine concentration and dilution: The roles of antidiuretic hormone and urea. In: *The Kidney,* 2nd ed. B. Brenner and F. C. Rector, Jr., eds. Saunders, Philadelphia. pp. 495–550.
5. Jamison, R. L., and W. Kriz. 1982. *Urinary Concentrating Mechanism: Structure and Function.* Oxford University Press, London.
6. Kaissling, B., and W. Kriz. 1979. *Structural Analysis of the Rabbit Kidney.* Springer-Verlag, Berlin.
7. Stephenson, J. L. 1978. Countercurrent transport in the kidney. *Annu. Rev. Biophys. Bioeng.* **7**:15–39.
8. Burg, M. B. 1981. Renal handling of sodium, chloride, water, amino acids, and glucose. In: *The Kidney,* 2nd ed. B. Brenner and F. C. Rector, Jr., eds. Saunders, Philadelphia. pp. 328–370.
9. Schwartz, M. M., and M. A. Venkatachalam. 1974. Structural differences in thin limbs of Henle: Physiological implications. *Kidney Int.* **6**:193–208.
10. Berliner, R. W., N. G. Levinsky, D. G. Davidson, and M. Eden. 1958. Dilution and concentration of the urine and the action of antidiuretic hormone. *Am. J. Med.* **27**:730–744.
11. Rollhäuser, H., W. Kriz, and W. Heinke. 1964. Das Gefässystem der Rattenniere. *Z. Zellforsch. Mikrosk. Anat.* **64**:381–403.
12. Kuhn, W., and K. Ryffel. 1942. Herstellung konzentrierter Lösungen aus verdünnten durch blosse Membranwirkung: Ein Modellversuch zur Funktion der Niere. *Hoppe-Seylers Z. Physiol. Chem.* **276**:145–178.
13. Hargitay, B., and W. Kuhn. 1951. Die Multiplikationsprinzip als Grundlage der Harnkonzentrierung in der Niere. *Z. Elektrochem. Angew. Phys. Chem.* **55**:539–558.
14. Kuhn, W., and A. Ramel. 1959. Activer Salztransport als möglicher (und wahrscheinliche) Einzeleffekt bei der Harnkonzentrierung in der Niere. *Helv. Chim. Acta* **42**:628–660.
15. Wirz, H., B. Hargitay, and W. Kuhn. 1951. Lokalisation des Konzentrierungsprozesses in der Niere durch direkte Kryoskopie. *Helv. Physiol. Pharmacol. Acta* **9**:196–207.
16. Wirz, H. 1953. Der osmotische Druck des Blutes in der Nierenpapille. *Helv. Physiol. Pharmacol. Acta* **11**:20–29.
17. Gottschalk, C. W., and M. Mylle. 1959. Micropuncture study of the mammalian urinary concentrating mechanism: Evidence for the countercurrent hypothesis. *Am. J. Physiol.* **196**:927–936.
18. Walker, A. M., P. A. Bott, J. Oliver, and M. C. MacDowell. 1941. The collection and analysis of fluid from single nephrons of the mammalian kidney. *Am. J. Physiol.* **134**:580–595.
19. Wirz, H. 1956. Der osmotische Druck in den corticalin Tubuli der Rattenniere. *Helv. Physiol. Pharmacol. Acta* **14**:353–362.
20. Lassiter, W. E., C. W. Gottschalk, and M. Mylle. 1961. Micropuncture study of net transtubular movement of water and urea in nondiuretic mammalian kidney. *Am. J. Physiol.* **200**:1139–1146.
21. Burg, M. B., and N. Green. 1973. Function of the thick ascending limb of Henle's loop. *Am. J. Physiol.* **224**:659–668.
22. Rocha, A. S., and J. P. Kokko. 1973. Sodium chloride and water transport in the medullary thick ascending limb of Henle. *J. Clin. Invest.* **52**:612–624.
23. Woodhall, P. B., and C. C. Tisher. 1973. Response of the distal tubule and cortical collecting duct to vasopressin in the rat. *J. Clin. Invest.* **52**:3095–3108.
24. Grantham, J. J., and M. B. Burg. 1966. Effect of vasopressin and cyclic AMP on permeability of isolated collecting tubules. *Am. J. Physiol.* **211**:255–259.
25. Morgan, T., and R. W.Berliner. 1968. Permeability of the loop of Henle, vasa recta, and collecting duct to water, urea, and sodium. *Am. J. Physiol.* **215**:108–115.
26. Rocha, A. S., and J. P. Kokko. 1974. Permeability of medullary nephron segments to urea and water: Effect of vasopressin. *Kidney Int.* **6**:379–387.
27. Morel, F., M. Imbert-Teboul, and D. Chabardes. 1981. Distribution of hormone-dependent adenylate cyclase in the nephron and its physiological significance. *Annu. Rev. Physiol.* **43**:569–581.
28. Hall, D. A., and D. M. Varney. 1980. Effect of vasopressin on electrical potential difference and chloride transport in mouse medullary thick ascending limb of Henle's loop. *J. Clin. Invest.* **66**:792–802.
29. Sasaki, S., and M. Imai. 1980. Effects of vasopressin on water and NaCl transport across the thick ascending limb of Henle's loop of mouse, rat, and rabbit kidneys. *Pfluegers Arch.* **383**:215–221.
30. Hebert, S. C., R. M. Culpepper, and T. E. Andreoli. 1981. NaCl transport in mouse medullary thick ascending limbs. I. Functional nephron heterogeneity and ADH-stimulated NaCl cotransport. *Am. J. Physiol.* **241**:F412–F431.
31. Stephenson, J. L. 1973. Concentrating engines and the kidney. I. Central core model of the renal medulla. *Biophys. J.* **13**:512–545.
32. Imai, M., and J. P. Kokko. 1974. Sodium chloride, urea, and water transport in the thin ascending limb of Henle. *J. Clin. Invest.* **53**:393–402.

33. Imai, M. 1977. Function of the thin ascending limb of Henle of rats and hamsters perfused *in vitro*. *Am. J. Physiol.* **232**:F201–F209.
34. Marsh, D. J., and S. P. Azen. 1975. Mechanism of NaCl reabsorption by hamster thin ascending limbs of Henle's loop. *Am. J. Physiol.* **228**:71–79.
35. Moore, L. C., and D. J. Marsh. 1980. How descending limb of Henle's loop permeability affects hypertonic urine formation. *Am. J. Physiol.* **239**:F57–F71.
36. Hai, M. A., and S. Thomas. 1969. Influence of prehydration on the changes in renal tissue composition induced by water diuresis in the rat. *J. Physiol. (London)* **205**:599–618.
37. Battilana, C. A., D. C. Dobyan, F. B. Lacy, J. Bhattacharya, P. A. Johnson, and R. L. Jamison. 1978. Effect of chronic potassium loading on potassium secretion by the pars recta or descending limb of the juxtamedullary nephron in the rat. *J. Clin. Invest.* **62**:1093–1103.
38. Jamison, R. L., F. B. Lacy, J. P. Pennell, and V. M. Sanjana. 1976. Potassium secretion by the descending limb or pars recta of the juxtamedullary nephron *in vivo*. *Kidney Int.* **9**:323–332.
39. Pennell, J. P., V. Sanjana, N. R. Frey, and R. L. Jamison. 1975. The effect of urea infusion on the urinary concentrating mechanism in protein-depleted rats. *J. Clin. Invest.* **55**:399–409.
40. Stern, L., K. A. Backman, and J. P. Hayslett. 1982. Cortical-medullary gradient for ammonia: Mechanism regulating formation of ammonium in medullary collecting duct. *Kidney Int.* **21**:240a.
41. Valtin, H. 1966. Sequestration of urea and non-urea solutes in renal tissue of rats with hereditary diabetes insipidus: Effect of vasopressin and dehydration on the countercurrent mechanism. *J. Clin. Invest.* **45**:337–345.
42. Schmidt-Nielsen, B., and R. R. Robinson. 1970. Contribution of urea to urinary concentrating ability in the dog. *Am. J. Physiol.* **218**:1363–1369.
43. Gunther, R. A., and L. Rabinowitz. 1980. Urea and renal concentrating ability in the rabbit. *Kidney Int.* **17**:205–222.
44. Shannon, J. A. 1936. Glomerular filtration rate and excretion in relation to urine flow in the dog. *Am. J. Physiol.* **117**:206–225.
45. Gamble, J. L., C. F. McKhann, A. M. Butler, and E. Tuthill. 1934. An economy of water in renal function referable to urea. *Am. J. Physiol.* **109**:139–154.
46. Crawford, J. D., A. P. Doyle, and J. H. Probst. 1959. Service of urea in renal water conservation. *Am. J. Physiol.* **196**:545–548.
47. Klümper, J. D., K. J. Ullrich, and H. H. Hilger. 1958. Das Verhalten des Harnstoff in den Sammelrohren der Säugetierniere. *Pfluegers Arch.* **267**:238–243.
48. Marsh, D. J., and C. M. Martin. 1980. Lack of water or urea movement from pelvic urine to papilla in hydropenic hamsters. *Miner. Electrolyte Metab.* **3**:81–86.
49. Ullrich, K. J., and K. H. Jarausch. 1956. Untersuchungen zum Problem der Harnkonzentrierung und Harnverdünnung. *Pfluegers Arch.* **262**:537–550.
50. Ullrich, K. J., G. Rumrich, and B.Schmidt-Nielsen. 1967. Urea transport in the collecting duct of rats on normal and low protein diet. *Pfluegers Arch.* **295**:147–156.
51. DeRouffignac, C., and F. Morel. 1969. Micropuncture study of water, electrolytes, and urea along the loops of Henle in *Psammomys*. *J. Clin. Invest.* **48**:474–486.
52. Marsh, D. J. 1970. Solute and water flows in thin limbs of Henle's loop in the hamster kidney. *Am. J. Physiol.* **218**:824–831.
53. Pennell, J. P., F. B. Lacy, and R. L. Jamison. 1974. An *in vivo* study of the concentrating process in the descending limb of Henle's loop. *Kidney Int.* **5**:337–347.
54. Ullrich, K. J., K. Kramer, and J. W. Boylan. 1962. Present knowledge of the countercurrent system in the mammalian kidney. In: *Heart, Kidney and Electrolytes*. C. K. Friedberg, ed. Grune & Stratton, New York. pp. 1–37.
55. Kawamura, S., and J. P. Kokko. 1976. Urea secretion by the straight segment of the proximal tubule. *J. Clin. Invest.* **58**:604–612.
56. Knepper, M. A. 1983. Urea transport in nephron segments from medullary rays of rabbits. *Am. J. Physiol.* **244**:F627.
57. Stephenson, J. L. 1972. Concentration of urine in a central core model of the renal counterflow system. *Kidney Int.* **2**:85–94.
58. Knepper, M. A. 1983. Urea transport in isolated, perfused thick ascending limbs and collecting ducts from rats. *Am. J. Physiol.* **245**:F634–F639.
59. Hilger, H. H., J. D. Klümper, and K. J. Ullrich. 1958. Wasserrückresorption und Ionentransport durch die Sammelrohren der Säugetierniere. *Pfluegers Arch.* **267**:218–237.
60. Jamison, R. L. 1970. Micropuncture study of superficial and juxtamedullary nephrons in the rat. *Am. J. Physiol.* **218**:46–55.
61. Diezi, J., P. Michoud, J. Aceves, and G. Giebisch. 1973. Micropuncture study of electrolyte transport across papillary collecting duct of the rat. *Am. J. Physiol.* **224**:623–634.
62. Sonnenberg, H. 1974. Medullary collecting-duct function in antidiuretic and in salt-or water-diuretic rats. *Am. J. Physiol.* **226**:501–506.
63. Stein, J. H., R. W. Osgood, and R. T. Kunau. 1976. Direct measurement of papillary collecting duct sodium transport in the rat: Evidence for heterogeneity of nephron function during Ringer loading. *J. Clin. Invest.* **58**:767–773.
64. Oliver, R.E., D. R. Roy, and R. L. Jamison. 1982. Urinary concentration in the papillary collecting duct of the rat: Role of the ureter. *J. Clin. Invest.* **69**:157–164.
65. Giebisch, G., G. Malnic, and R. W. Berliner. 1981. Renal transport and control of potassium excretion. In: *The Kidney*, 2nd ed. B. Brenner and F. C. Rector, Jr., eds. Saunders, Philadelphia. pp. 408–439.
66. Jamison, R. L., J. Work, and J. A. Schafer. 1982. New pathways for potassium transport in the kidney. *Am. J. Physiol.* **242**:F297–F312.
67. Stokes, J. B. 1982. Consequences of potassium recycling in the renal medulla. *J. Clin. Invest.* **70**:219–229.
68. Good, D. W., and M. B. Burg. 1983. Ammonia production by individual segments of rat nephron. *Kidney Int.* **23**:232a.
69. Buerkert, J., D. Martin, and D. Trigg. 1982. Ammonium handling by superficial and juxtamedullary nephrons in the rat: Evidence for an ammonia shunt between the loop of Henle and the collecting duct. *J. Clin. Invest.* **70**:1–12.
70. Robinson, R. R., and E. E. Owen. 1965. Intrarenal distribution of ammonia during diuresis and antidiuresis. *Am. J. Physiol.* **208**:1129–1134.
71. Ullrich, K. J., H. H. Hilger, and D. J. Klümper. 1958. Sekretion von Ammoniumionen in den Sammelrohren der Säugetierniere. *Pfluegers Arch.* **267**:244–250.
72. Sonnenberg, H., S. Cheema-Dhadli, M. B. Goldstein, B. J. Stinebaugh, D. R. Wilson, and M. L. Halperin. 1981. Ammonia addition into the medullary collecting duct of the rat. *Kidney Int.* **19**:281–287.
73. Sajo, I. M., M. B. Goldstein, H. Sonnenberg, B. J. Stinebaugh, D. R. Wilson, and M. L. Halperin. 1981. Sites of ammonia addition to tubular fluid in rats with chronic metabolic acidosis. *Kidney Int.* **20**:353–358.
74. Graber, M. L., H. H. Bengele, E. Mroz, C. Lechene, and E. A. Alexander. 1981. Acute metabolic acidosis augments collecting duct acidification in rate in the rat. *Am. J. Physiol.* **241**:F669–F676.
75. Kokko, J. P. 1970. Sodium chloride and water transport in the descending limb of Henle. *J. Clin. Invest.* **49**:1838–1846.
76. Rocha, A. S., and J. P. Kokko. 1973. Membrane characteristics regulating potassium transport out of the isolated perfused descending limb of Henle. *Kidney Int.* **4**:326–330.
77. Abramow, M., and L. Orci. 1980. On the "tightness" of the rabbit descending limb of the loop of Henle—Physiological and morphological evidence. *Int. J. Biochem.* **12**:23–27.
78. Kokko, J. P. 1972. Urea transport in the proximal tubule and the descending limb of Henle. *J. Clin. Invest.* **51**:1999–2008.
79. Stoner, L. C., and F. Roch-Ramel. 1979. The effects of pressure on the water permeability of the descending limb of Henle's loops of rabbits. *Pfluegers Arch.* **382**:7–15.
80. Kokko, J. P., and F. C. Rector. 1972. Countercurrent multiplica-

tion system without active transport in inner medulla. *Kidney Int.* **2**:214–223.

81. Stephenson, J. L. 1976. Concentrating engines and the kidney. III. Canonical mass balance equation for multinephron models of the renal medulla. *Biophys. J.* **16**:1273–1286.
82. Stephenson, J. L., R. P. Tewarson, and R. Mejia. 1974. Quantitative analysis of mass and energy balance in non-ideal models of the renal counterflow system. *Proc. Natl. Acad. Sci. USA* **71**:1618–1622.
83. Kriz, W., J. Schnermann, and H. Koepsell. 1972. The position of short and long loops of Henle in the rat kidney. *Z. Anat. Entwicklungsgesch.* **138**:301–319.
84. Ernst, S. A., and J. H. Schreiber. 1981. Ultrastructural localization of Na^+,K^+-ATPase in rat and rabbit kidney medulla. *J. Cell Biol.* **91**:803–813.
85. Garg, L. C., and C. C. Tisher. 1983. Na-K-ATPase activity in thin limbs of rat nephrons. *Kidney Int.* **23**:255a.
86. Sanjana, V. F., C. R. Robertson, and R. L. Jamison. 1976. Water extraction from the inner medullary collecting tubule system: A role for urea. *Kidney Int.* **10**:139–146.
87. Bonventre, J. V., and C. Lechene. 1980. Renal medullary concentrating process: An integrative hypothesis. *Am. J. Physiol.* **239**:F578–F588.
88. Marsh, D. J. 1966. Hypo-osmotic re-absorption due to active salt transport in perfused collecting ducts of the rat renal medulla. *Nature (London)* **210**:1179–1180.
89. Stephenson, J. L. 1966. Concentration in renal counterflow systems. *Biophys. J.* **6**:539–551.
90. Stephenson, J. L. 1973. Concentrating engines and the kidney. II. Multisolute central core systems. *Biophys. J.* **13**:545–567.
91. Mejia, R., and J. L. Stephenson. 1979. Numerical solution of multinephron kidney equations. *J. Comput. Phys.* **32**:235–246.
92. Lory, P., A. Gilg, and M. Horster. 1983. Renal countercurrent system: Role of collecting duct convergence and pelvic urea predicted from a mathematical model. *J. Math. Biol.* **16**:281–304.
93. Johnston, P. A., C. A. Battilana, F. B. Lacy, and R. L. Jamison. 1977. Evidence for a concentration gradient favoring outward movement of sodium from the thin loop of Henle. *J. Clin. Invest.* **59**:234–240.
94. Jamison, R. L., C. M. Bennett, and R. W. Berliner. 1967. Countercurrent multiplication by the thin loops of Henle. *Am. J. Physiol.* **212**:357–366.
95. Stewart, J., and H. Valtin. 1972. Computer simulation of osmotic gradient without active transport in renal inner medulla. *Kidney Int.* **2**:264–270.
96. Andreoli, T. E., R. W. Berliner, J. P. Kokko, and D. J. Marsh. 1978. Questions and replies: Renal mechanisms for urinary concentrating and diluting processes. *Am. J. Physiol.* **235**:F1–F11.

CHAPTER 17

Transport Functions of the Distal Convoluted Tubule

Linda S. Costanzo and Erich E. Windhager

1. Introduction

The distal convoluted tubule (DCT) has traditionally been described as the nephron segment extending from the macula densa to the first confluence with another DCT to form a collecting tubule. Virtually all data on DCT function derive from *in vivo* studies in the rat using micropuncture or microperfusion techniques. Rat DCT are 2.4–2.5 mm in length.[1,2] They can be identified on the kidney surface with light microscopy by their contrast to proximal tubules: the lumina are narrower and the contour more irregular than in proximal tubules. Distal tubular epithelium lacks a brush border and therefore fails to exhibit the light reflex seen in proximal tubules.[3] Eighty percent of the rat DCT is accessible to micropuncture, with only the initial 20% below the kidney surface. At present, the rat provides the most convenient model for studies of DCT function. While the *in vitro* perfused rabbit nephron technique has been applied extensively to study the function of other tubular segments, the technique is not readily applied to the DCT because of its short length in the rabbit.

In contrast to the proximal tubule, where isosmotic reabsorption of more than half of the glomerular filtrate occurs, only small quantities of solute and water are transferred in the DCT under nonisosmotic conditions. The DCT is less permeable to water than the proximal tubule and NaCl reabsorption may not be accompanied by osmotically proportional movement of water. In the absence of antidiuretic hormone (ADH), such hyperosmotic reabsorption results in the formation of hypoosmotic urine.

Although the DCT characteristically transports small quantities of solutes, this segment can establish large electrochemical gradients for electrolytes such as Na^+,[4] Cl^-,[5] K^+,[6] H^+,[7] and Ca^{2+}[8] in the presence of a considerable lumen-negative transepithelial potential difference.[8,9] Because of the large electrochemical gradients observed and the long zonulae occludentes seen in electron micrographs,[10–12] the distal tubular epithelium is characterized as a "tight" epithelium.[13]

The DCT is the only nephron segment in which properties of urine dilution and concentration can be studied simultaneously *in vivo*. Pioneering micropuncture studies[14] showed that the early portion of the DCT constitutes part of the "diluting segment," perpetuating the function of the TALH, whereas the late portion is involved in the kidney's concentration operation. The point along the DCT where concentrating ability appears is related to the pattern of ADH-sensitivity and varies among different species.[1]

Regulation of many excretory functions of the kidney occurs in the DCT. Distal Na^+ reabsorption[15–17] and K^+ secretion[15] are under the influence of mineralocorticoids, the absence of aldosterone resulting in diminished Na^+ reabsorption and K^+ secretion. Administration of aldosterone results in correction of these defects.[17] Osmotic water flow in the DCT depends upon the circulating levels of ADH. In the absence of ADH, the DCT is water-impermeable and tubular fluid does not equilibrate osmotically with plasma, whereas with normal ADH levels, osmotic equilibration is achieved.[18] The final regulation of Ca^{2+} reabsorption is also achieved in the DCT, controlled by parathyroid hormone (PTH) levels.[19,20]

2. Structural Heterogeneity

The micropuncture literature has traditionally regarded the DCT as a single, uniform structure. However, this presumption is not supported by recent anatomical evidence. Instead, the DCT as defined by micropuncturists contains three structurally distinct segments: the "true" DCT, the connecting tubule, and the initial cortical collecting tubule. The "true" DCT is the earliest, post-macula densa segment. The connecting tubule,

Linda S. Costanzo • Department of Physiology and Biophysics, Medical College of Virginia, Richmond, Virginia 23289. **Erich E. Windhager** • Department of Physiology, Cornell University Medical College, New York, New York 10021.

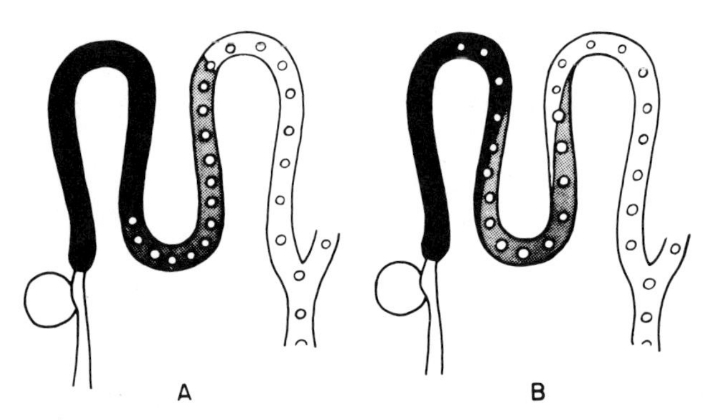

Fig. 1. Patterns of segmentation of superficial distal tubules in different species. From Kaissling.[23]

first described by Schweigger-Seidel in 1865,[21] is the segment interposed between the "true" DCT and the cortical collecting tubule; it is ultrastructurally distinct from the adjacent segments. Whether the embryonic origin of the connecting tubule is from metanephrogenic blastema or ureteral bud is unresolved.[22–24] In some species, including the rat, a portion of the initial cortical collecting tubule is on the kidney surface, is accessible to micropuncture and therefore has been included in studies of the DCT.

There are four cell types distributed among the segments of the DCT. These include DCT cells, connecting tubule cells (CNT cells), intercalated cells (I cells), and principal cells (P cells). In the rabbit and guinea pig, the borders between the DCT (composed exclusively of DCT cells), the connecting tubule (composed of CNT and I cells), and the cortical collecting duct (composed of P and I cells) are distinct.[22] In other species such as rat,[1,25,26] mouse,[26] and human,[27] the cells characteristic of a particular segment intermingle. Figure 1 contrasts the patterns of segmentation in the rabbit (panel A) and the rat (panel B). In rabbits the borders are sharply defined, whereas in rats the transitions are gradual. In the rat, all four cell types are on the kidney surface.

DCT cells are characterized on their basal side by tall platelike processes. These processes amplify the lateral surface area about 20 times over all the basal area of the cell.[28] The luminal surface area is small. There are long intercellular channels between DCT cells separated from the lumen by tight junctions.[29] The lateral cell membranes have a high Na^+,K^+-ATPase activity.[30,31]

In contrast to DCT cells, CNT cells are characterized mainly by amplification of the basal cell membrane.[22] The intercellular junctions between CNT cells have been classified as "tight" but are variable in length.[29] The luminal membrane is smooth and possesses a cell coat. CNT cells have fewer and smaller mitochondria than DCT cells.

P cells are ultrastructurally similar to CNT cells. They are also characterized by basal cell membrane amplification, although to a smaller extent than in CNT cells. The infoldings are restricted to the most basal region of the cells and are situated perpendicular to the basal lamina. The intracellular organelles are located above the basal infoldings so that the infoldings appear as a bright rim under light microscopy.[23] The luminal membrane has short microvilli beneath which is a thick network of microtubules and microfilaments.[22] The tight junctions between P cells are deep. Adjacent P cells are interlocked by small villi, connected by desmosomes. The width of the lateral intercellular space is variable, being dilated in the presence of vasopressin, and contracted in the absence of the hormone.[32,33]

The I cells are present in the connecting tubule, cortical collecting tubule, and DCT of rat, *Psammomys,* human, and mouse.[23] The prominent feature of I cells is amplification of the luminal membrane, with a variable shape in the apical cell pole. Because of their unique appearance, scanning electron microscopy has been used to study I cells.[34] In addition, there is an assortment of round or elongated intracellular vesicles at the apical cell pole. I cells have a dark appearance owing to the extensive endoplasmic reticulum and numerous mitochondria.

Cellular heterogeneity can be used to explain functional differences along the DCT. For example, the late but not the early rat DCT exhibits water absorption in response to a transepithelial osmotic gradient.[1,35] These results are in accord with the biochemical data of Morel and co-workers (see Ref. 36) showing that ADH-sensitive adenylate cyclase is highest in the latter portion of the DCT. The particular cell type(s) of the late DCT which is responsive to ADH is not known. The rat DCT is also sensitive to PTH[20] and contains PTH-sensitive adenylate cyclase.[36] The physiological action of PTH (stimulation of Ca^{2+} transport) is entirely different from that of ADH (stimulation of osmotic water flow). At present, it is not known whether the two hormones produce their effects via a single intracellular messenger, cAMP, by actions on different cell types or not.

3. Transepithelial Net Transport of Solutes and Water

3.1. NaCl

Net reabsorption of Na^+ ions occurs against an electrochemical potential gradient throughout the length of the DCT and is therefore the result of active transport. This conclusion is reached from data on net transport rates of Na^+ during free-flow micropuncture,[37,38] on estimates of transepithelial Na^+ concentration gradients,[4,5,15] and on measurements of transepithelial electrical potential differences.[19,39–41]

In nondiuretic rats, a single DCT reabsorbs 0.295 neq/min, or 6.8% of the filtered load of Na^+.[38] Assuming a DCT length of 2.5 mm and an average luminal diameter of about 20 μm,[41] Na^+ transport proceeds at a rate of 31×10^{-12} eq/sec per mm² as calculated from the free-flow micropuncture data of Khuri *et al.*[38] This rate is in reasonably good agreement with data obtained from *in vivo* microperfusion of rat DCT,[8] where the Na^+ reabsorptive rate was 159 pmoles/min per mm at a delivered Na^+ concentration of 80 mM and perfusion rate of 5 nl/min. When corrected for surface area, the reabsorptive rate in the microperfusion experiments was 42×10^{-12} eq/sec per mm². The rate of reabsorption as determined either by free-flow or pump perfusion techniques corresponds to approximately one-third the rate of Na^+ reabsorption by proximal tubules.

Under control conditions, the rate of fractional Na^+ reabsorption shows a decline along the length of the DCT (Fig. 2), paralleling a simultaneous decrease in the intratubular Na^+ concentration. Tubular fluid/plasma Na^+ concentration, $(TF/P)_{Na}$, averages 0.51 at the extrapolated beginning and 0.22 at the end of the DCT; the corresponding values for reabsorptive rates are 184 and 9×10^{-12} eq/sec per mm², respectively.[38] These results suggest that the rate of transepithelial Na^+ transport is directly related to the intratubular Na^+ concentration. Recent evidence demonstrates conclusively that the early and late portions of the rat DCT have the same transport capacity for Na^+.[35]

Estimates of the ability of the distal tubule to establish a

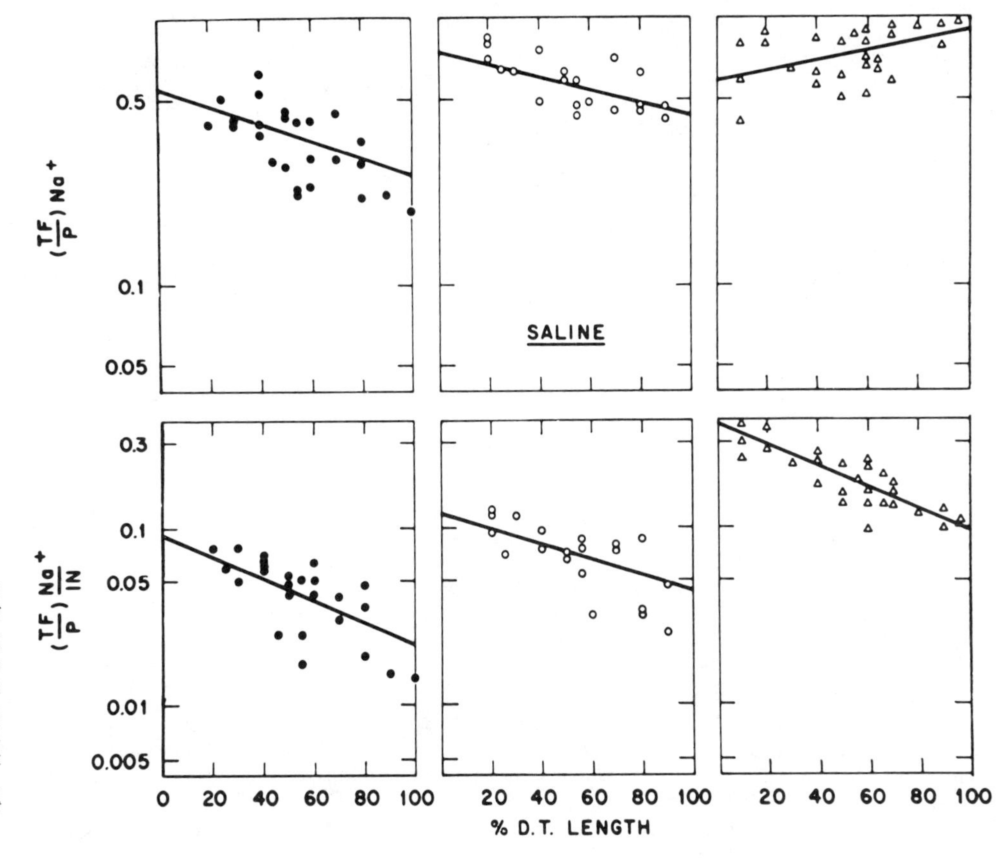

Fig. 2. $(TF/P)_{Na}$ and $(TF/P)_{Na/In}$ as a function of distal tubular length during progressive isotonic saline loading. Left panels: mean distal flow rate, 4.0 nl/min; middle panels: mean distal flow rate, 8.5 nl/min; right panels: mean distal flow rate, 23.0 nl/min. From Khuri *et al.*[38]

limiting Na^+ concentration gradient have been used to evaluate the nature of Na^+ transport.[5,8,15,42] In these stationary microperfusion experiments, a droplet of an electrolyte-free isosmotic solution of raffinose (or other impermeant nonelectrolyte) is deposited into the tubular lumen, separated from the rest of the tubular fluid by columns of oil. The raffinose solution is reaspirated after a measured contact time and analyzed. Initially, Na^+ moves into the luminal fluid until a limiting concentration gradient is reached. At this point, the active reabsorptive outflux is balanced by the passive influx resulting in a state of near-zero net flux of solute and water. The limiting concentration gradient for Na^+ in the DCT is 0.33–0.37.[5,8,15] Concentration ratios as low as 0.10 have been reported for the late DCT.[42] Free-flow Na^+ concentrations approach the equilibrium concentration under nondiuretic conditions, but not at high rates of tubular flow.[38] The limiting Na^+ concentration gradient is reduced in the pressure of poorly permeant anions (such as sulfate) in distal fluid, presumably as a result of increased intraluminal negativity.[4]

Saturation kinetics cannot be demonstrated in direct studies of net Na^+ reabsorption by the DCT.[2,8,38] Costanzo and Windhager,[8] using *in vivo* microperfusion of rat DCT, demonstrated a linear relationship between load and transport rate over the range of delivered Na^+ loads from normal to more than three times normal (Fig. 3). Fractional Na^+ reabsorption for a DCT length of 2.5 mm was 85%. Khuri *et al.* arrived at a similar figure over a 10-fold range of loads using free-flow micropuncture.[38] The capacity for an increased absolute Na^+ reabsorptive rate in response to an increased delivered load provides a "second line" of defense in the maintenance of Na^+ balance during conditions of diminished proximal reabsorption (volume expansion). It also reduces renal Na^+ loss somewhat after administration of diuretics such as furosemide and ethacrynic acid which inhibit NaCl reabsorption in the thick ascending limb of Henle's loop.[43,44]

The decline in the Na^+ reabsorptive rate along the DCT is best explained by load dependence and the observed decrease in the luminal Na^+ concentration. Good and Wright[45] demonstrated that the distal Na^+ reabsorptive rate depends on the luminal concentration. The luminal Na^+ concentration falls as a result of Na^+ transport unaccompanied by water movement in

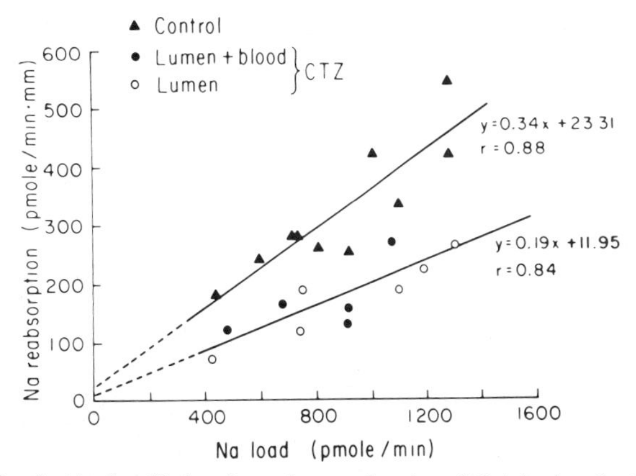

Fig. 3. Absolute Na^+ reabsorption as a function of Na^+ load to distal tubule in *in vivo* microperfusion experiments. Upper line was obtained from control tubules and lower line from CTZ-treated tubules. From Costanzo and Windhager.[8]

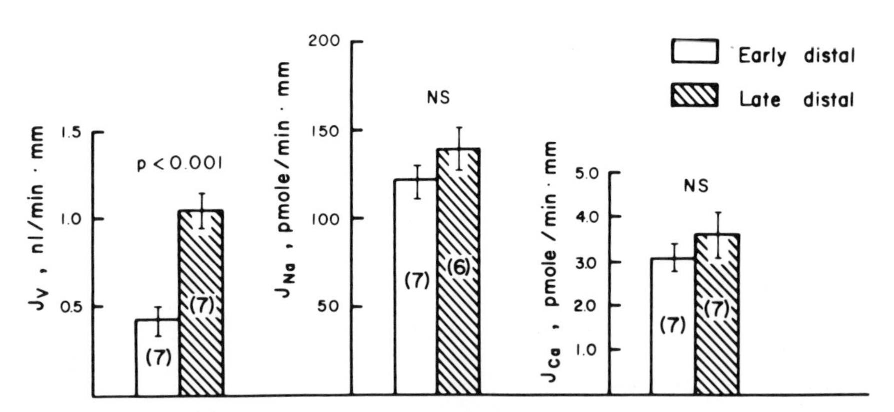

Fig. 4. Comparison of transport rates of fluid, Na^+, and Ca^{2+} in early and late DCT. From Costanzo.[35]

the early DCT.[8,38] Normally, most of the entering Na^+ load is reabsorbed early, so that later parts of the DCT are exposed to a lower intratubular Na^+ concentration or load; hence, there is a decrease in the reabsorptive rate. The declining rate of Na^+ transport along the DCT is not the consequence of an *inherent* difference in transport capacity, as early and late segments of the DCT exhibit identical rates of Na^+ transport when delivered loads are comparable[35] (Fig. 4).

At increased rates of distal Na^+ delivery, the tubular epithelium can drastically increase its rate of reabsorption, consistent with the view that transport is normally unsaturated. The intratubular Na^+ concentration may actually rise along the DCT at very high loads (Fig. 2).[38] The concentration rises because at constant fractional reabsorption, the absolute amount of Na^+ reabsorbed as well as the amount not reabsorbed is increased in response to the increased load. As water is reabsorbed in the late DCT,[1,35] the intraluminal Na^+ concentration rises.

A second factor which contributes to the appearance of high intratubular Na^+ concentrations during saline diuresis concerns the role of urea.[38] Lassiter *et al.*[46,47] and Danielson *et al.*[48] demonstrated that an important fraction of the urea delivered to the collecting ducts reenters the loop of Henle and is concentrated in the DCT, a segment with low urea permeability.[49] These investigators found decreased urea recirculation at high urine flow rates. Since in the rat, late distal tubular fluid equilibrates osmotically with plasma, it is clear that its urea concentration influences the luminal Na^+ concentration. At low flow rates, a significant quantity of urea recirculates, becomes concentrated in the DCT, and permits the development of large transepithelial Na^+ concentration gradients. In contrast, at high rates of flow, smaller amounts of urea recirculate and a larger proportion of the late distal tubular solute content will be contributed by Na^+.

The contribution of different nephron segments to the natriuresis resulting from expansion of the extracellular fluid volume has been studied extensively.[2,38,50,51] Micropuncture data on the DCT not only failed to provide evidence for inhibition of transport in volume expansion, but have shown load-related increases in absolute rates of Na^+ reabsorption.[2,8,38,52,53] Thus, in contrast to proximal tubules[54,55] and collecting ducts,[52,55–57] the distal tubule does not participate in the production of natriuresis resulting from extracellular volume expansion.[58]

Apparently, neither Starling forces nor humoral agents thought to inhibit Na^+ reabsorption in proximal tubules[55] and collecting ducts[52] have a role in regulating salt reabsorption in the distal convolution. The atrial peptide recently purified produces natriuresis largely based upon renal hemodynamic effects.[59] Tubular effects of the natriuretic factor (peptide) have been proposed,[60,61] although the DCT is not thought to be its site of action.[61]

There is little information about the rate of net Na^+ reabsorption by the DCT under conditions of decreased delivery of Na^+ out of the loop of Henle, such as during partial occlusion of the renal artery or vein or during partial occlusion of the ureter. It is known that there is constant fractional Na^+ reabsorption even at loads as low as 0.1 neq/min,[38] approximately one-fourth the normal delivery rate. The limiting concentration gradients for Na^+ during partial renal venous clamping[62] were not different from those in control or saline-expanded animals.[63] Since the establishment of limiting Na^+ concentration gradients depends on active transport and passive backflux of Na^+, it seems likely that neither the activity of the Na^+ pump nor the Na^+ permeability is significantly altered by either renal vein occlusion or volume expansion.

Mineralocorticoids exert at least part of their Na^+-retaining effect in the DCT of the rat, although the major site of action in the rabbit is the cortical collecting tubule.[64,65] The distal epithelium of adrenalectomized rats is unable to lower the intraluminal Na^+ concentration[15] or to reabsorb Na^+[16] to the same extent as that of intact rats. The defect in Na^+ transport can be reversed by aldosterone administration.[66,67] The stimulation of Na^+ transport probably involves the formation of a hormone-induced protein, as actinomycin D blocks the aldosterone-dependent fraction of distal Na^+ reabsorption.[66] With high-resolution, two-dimensional gel electrophoresis and autoradiography, Geheb *et al.* have identified aldosterone-induced proteins of similar isoelectric points and molecular weights in membrane and cytosolic fractions from epithelial cells of toad urinary bladder.[68,69] Both actinomycin D[69] and spironolactone[70] blocked Na^+ transport and aldosterone-induced protein synthesis. On the other hand, amiloride had no effect on the proteins' synthesis, demonstrating that the aldosterone induction of the proteins is not a nonspecific effect relating to changes in Na^+ transport rate.[69] It has been suggested that the aldosterone-induced protein or proteins either increase luminal membrane Na^+ permeability, secondarily increasing the rate of Na^+ pumping, or directly stimulate the active extrusion of Na^+ across the contraluminal membrane. There is evidence from other aldosterone-sensitive epithelia that the primary action of the hormone-induced protein is to increase the Na^+ conductance of the apical membrane.[71–75]

Increases in Na^+,K^+-ATPase activity are well documented following aldosterone administration.[28,76–81] The increase in apical Na^+ conductance discussed above may produce an elevation of intracellular Na^+ concentration which causes secondary increases in Na^+,K^+-ATPase activity in responsive tissues. In the cortical collecting tubule, increases in apical Na^+ conductance[75] and in Na^+,K^+-ATPase do not occur after aldosterone treatment when amiloride is present.[71] In the rabbit, the nephron site of aldosterone-induced increases in Na^+,K^+-ATPase appears to be the connecting tubule and cortical collecting tubule.[28,71,78,80] Kaissling and LeHir have documented that changes in mineralocorticoid-dependent enzyme activity correlate with changes in basolateral membrane surface area, with the true DCT being unresponsive, the connecting tubule having intermediate responsiveness, and the collecting duct being most responsive.[28] The anti-mineralcorticoid, canrenone, blocked these structural changes in CNT and P cells. The pattern of effects of aldosterone in the rat may be explained by the presence of these cell types in the latter portion of the DCT. Thus, the effect of mineralocorticoids is twofold. They may raise the intracellular Na^+ concentration, thereby raising the rate of Na^+ transport. In addition, and possibly secondary to the postulated changes in intracellular Na^+ levels, the activity of the Na^+,K^+-ATPase is increased. It is presently not known what extent increased turnover rate may account for the observed enzyme response.

NaCl transport in the DCT is altered by two classes of diuretics. Thiazide diuretics inhibit Na^+[8,82] and Cl^-[83] reabsorption at an early distal site where salt transport is unaccompanied by water movement. Little is known about the mechanism of action of thiazides, except that the drugs cause increases in distal transepithelial specific resistance without altering the transepithelial potential difference.[8] This observation might imply coincident decreases of Na^+ and Cl^- conductance. The K^+-sparing diuretics have a distal site of action,[84] different from that of thiazides. Amiloride inhibits Na^+ transport only in late segments of the rat DCT, with no effects in the early DCT.[35]

In nondiuretic rats, approximately 6% of the filtered Cl^- is reabsorbed along the DCT.[85] Like Na^+, the concentration of Cl^- in early distal tubular fluid is lower than that in plasma and declines along the distal tubule. The $(TF/P)_{Cl}$ is about 0.50 at the beginning and 0.20 at the end of the distal convolution.[5,14,18,48,85–87] Cl^- reabsorption by the DCT amounts to some 0.23 neq/min, corresponding to a net flux of 2.5×10^{-11} eq/sec per mm^2 or approximately two-thirds of the reabsorptive net flux of Na^+. In the presence of significant concentrations of poorly reabsorbed anions in distal tubular fluid, the Cl^- concentration declines below control levels,[87,88] with the mean $(TF/P)_{Cl}$ being reduced in rats given sodium bicarbonate, sodium sulfate, and acetazolamide.[87]

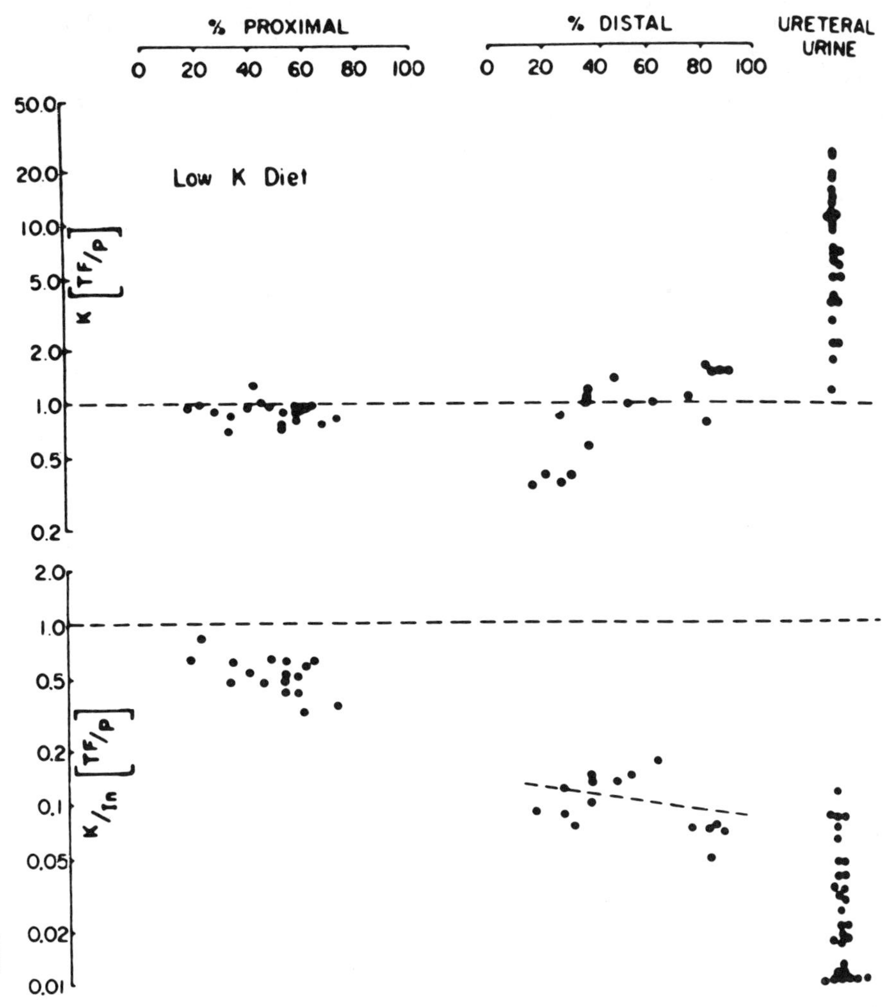

Fig. 5. Summary of $(TF/P)_K$ and $(TF/P)_{K/In}$ along the nephron in control rats. From Malnic *et al.*[6]

3.2. Potassium

Micropuncture and microperfusion studies have amply confirmed Berliner's[89] early hypothesis that the distal nephron is the most important site at which modulation of urinary K^+ excretion occurs. Although urinary K^+ excretion rates may vary by 100-fold, all or nearly all of the filtered K^+ is reabsorbed prior to the distal tubule. Thus, the net secretion of K^+ by the DCT and collecting ducts is the main source of urinary K^+.[90] Only during osmotic diuresis,[42] in massive extracellular volume expansion,[91–93] or after administration of diuretics such as furosemide[84,94] does the delivery of larger than normal loads of K^+ to the distal tubule contribute to kaliuresis.[91,95] In all other experimental conditions studied, particularly in K^+-loaded or -depleted animals,[4] the rate of urinary K^+ excretion depends entirely on the magnitude of net K^+ secretion by the DCT and collecting ducts.

Figures 5–7 provide information on the range of K^+ concentrations and the fraction of filtered K^+ remaining at any point along the nephron and in the final urine.[4] Under nondiuretic conditions (Fig. 5), rats on control diets excrete approximately 20% of the filtered load of K^+. $(TF/P)_K$ increased from 0.2 to approximately 5.0 along the distal tubule. The fraction of filtered K^+ remaining, $(TF/P)_{K/In}$, increased from less than 0.05 at the beginning, to more than 0.2 at the end of the DCT. These data suggest that the average rate of net secretion by a single superficial rat DCT amounts to about 0.03 neq/min or 0.32×10^{-11} eq/sec per mm^2; this is approximately 1/10th of the rate at which Na^+ ions are reabsorbed in this segment. Comparison of the fractional excretion of K^+ at the end of the DCT and in ureteral urine indicates that in rats on a normal dietary potassium intake, collecting ducts contribute little to total K^+ excretion. Of course, such a comparison between data obtained from superficial distal tubules and the final urine may be misleading, since final urine is an admixture of fluid from superficial and deep nephrons. As little is known about K^+ transport by the DCT of juxtamedullary nephrons, some degree of uncertainty attends inferences about the contribution of the collecting ducts to K^+ excretion. *In vitro* microperfusion of cortical and outer medullary collecting ducts from rabbits revealed a secretory K^+ flux in the cortical portion, whereas the medullary portion had a reabsorptive flux.[96] The latter may be related to K^+ recycling in renal medulla.

Net reabsorption of K^+ can be demonstrated in the DCT and collecting duct of rats kept on a low-potassium diet for several weeks (Fig. 6). This is apparent from the decline in

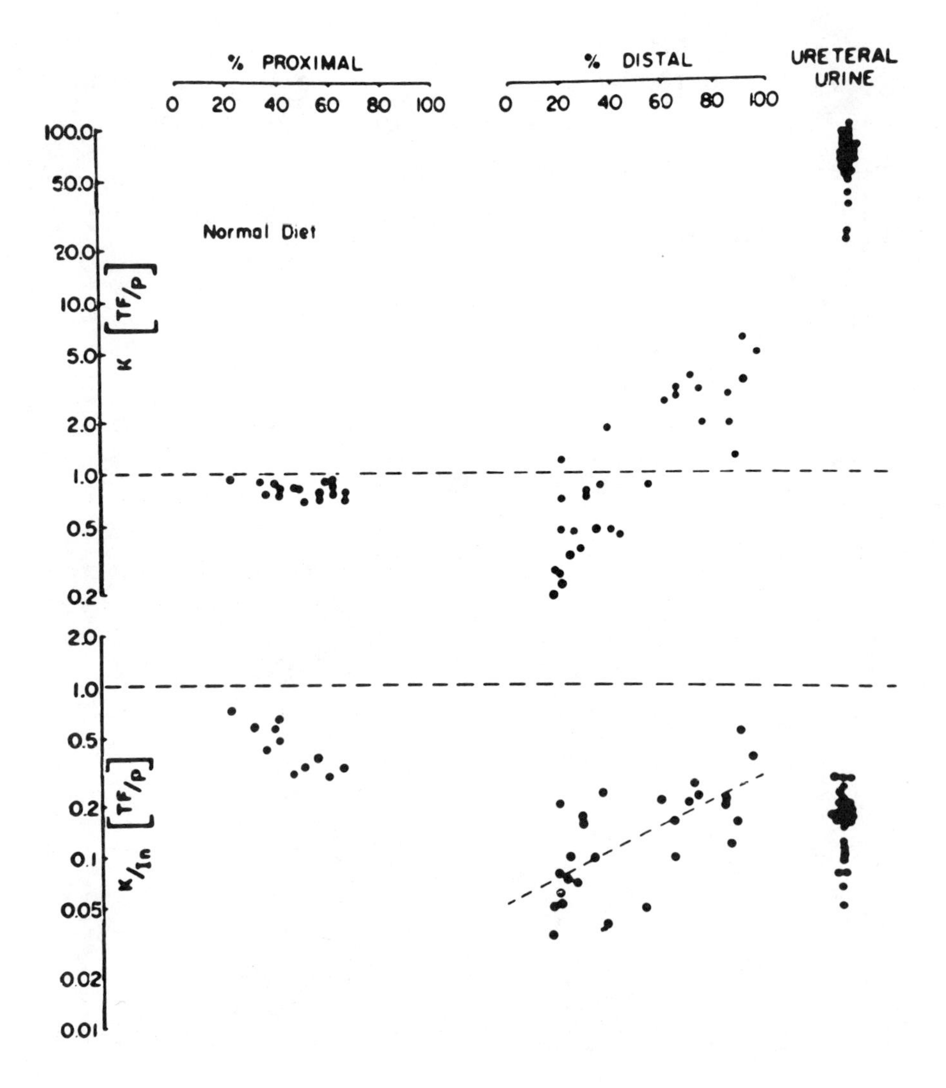

Fig. 6. Summary of $(TF/P)_K$ and $(TF/P)_{K/In}$ along the nephron in rats kept on a low-potassium diet. From Malnic *et al.*[6]

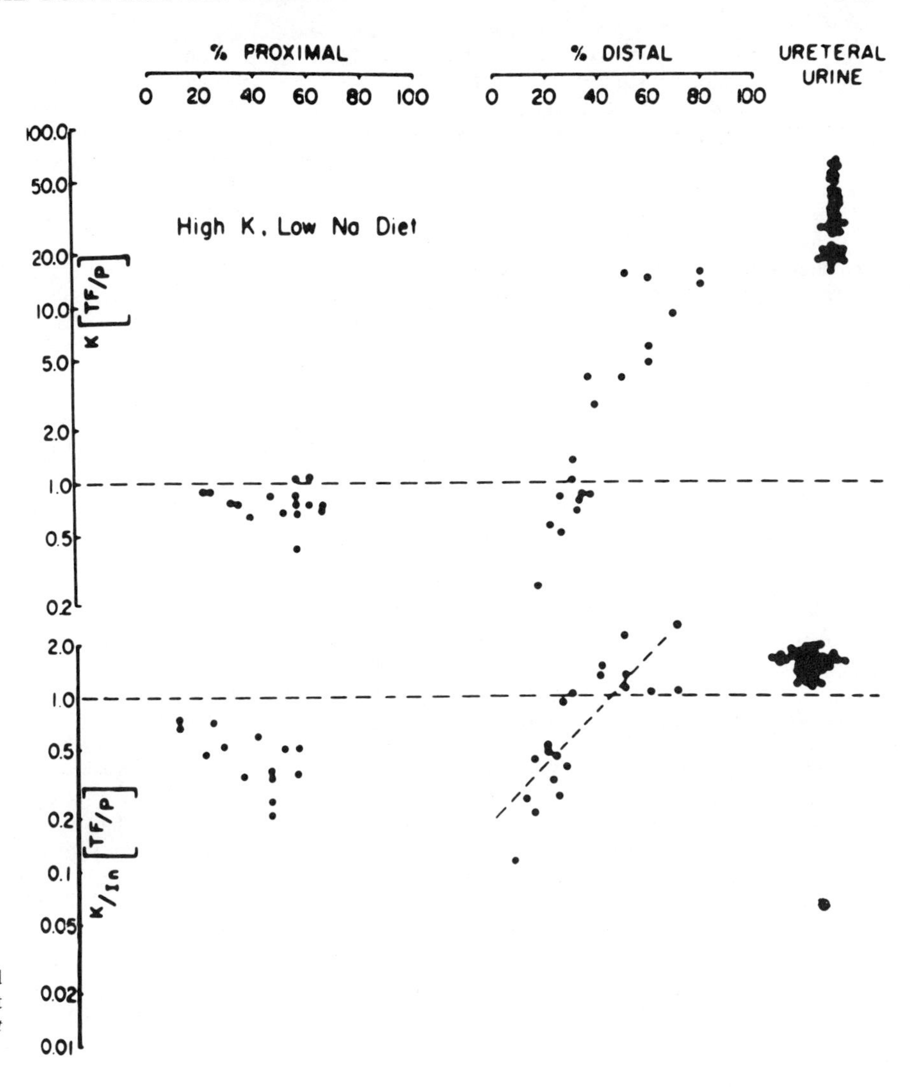

Fig. 7. Summary of $(TF/P)_K$ and $(TF/P)_{K/In}$ along the nephron in rats kept on a high-potassium diet. From Malnic *et al.*[6]

$(TF/P)_{K/In}$ ratios along the DCT, and the observation that the fraction of filtered K^+ in ureteral urine is significantly less than that present at the end of the DCT. On the other hand, distal tubular secretion of K^+ can be increased markedly by feeding rats a high-potassium diet (Fig. 7).

In rodents,[4] dogs,[97,98] and monkeys,[99] K^+ secretion occurs mainly in the DCT. However, in the rabbit, where the DCT is short, the collecting duct is a major secretory site.[100,101] In the rat nephron, sites beyond the DCT contribute to total urinary K^+ excretion in only a few special situations. These include ingestion of a high-potassium diet, where the collecting duct epithelium secretes a significant quantity of K^+.[102] In addition, when rats are pretreated with a low-sodium diet and then acutely challenged with a potassium load, urinary K^+ excretion is augmented by 50%, largely as a result of secretion by the collecting duct.[103] Finally, in the "remnant kidney" model, there is increased net secretion of K^+ in the collecting ducts.[104,105]

Several factors influence the net rate of K^+ secretion by the DCT. Among them are dietary potassium intake, mineralocorticoids, acid–base status, distal tubular flow rate, and ADH. Each of these factors will be discussed in detail.

As discussed above, the distal tubular secretory rate is influenced by changes in potassium intake. With ingestion of potassium salts, the increased K^+ secretory rate is due, in part, to elevation of plasma K^+ concentration.[106] In microperfusion studies on rat DCT *in vivo,* where changes in proximal tubular function were eliminated, dietary potassium supplementation resulted in an increased DCT secretory rate.[107] Potassium-adapted animals had a higher K^+ secretory rate than control animals whether they were infused during experiments with Ringer's solution or challenged with an acute potassium load. A maximal transport rate was achieved when the plasma K^+ concentration was 6 meq/liter. The higher distal secretory rates were associated with higher circulating levels of aldosterone. Plasma aldosterone levels increased as plasma K^+ increased, also reaching a maximum at a plasma K^+ concentration of 6 meq/liter.

In another study,[108] morphometric analysis of microperfused DCT from potassium-adapted rats revealed no structural alterations in early DCT cells. Likewise, short segments of early DCT did not secrete K^+. However, there was a pronounced augmentation in basolateral membrane area in the connecting tubule and initial collecting tubule in potassium-adapted rats as compared to those fed a control diet. The morphological changes were limited to the CNT and P cells, while the I cells present in these segments were unaffected. The functional and morphologial data from this study, taken together, suggest that K^+ is secreted by the CNT and P cells of the accessible rat distal tubule.[109] These observations are in good agreement with those of Kaissling and LeHir[28] in rabbits adapted to a high-potassium diet. Basolateral membrane area of CNT cells doubled after potassium-adaptation and P-cell area quadrupled. In the rabbit, these cells are part of the connecting tubule and cortical collecting duct. Antimineralocorticoid treatment prevented these structural changes. Furthermore, the structural adaptation in response to dietary potassium content correlated well with changes in Na^+,K^+-ATPase activity in the rabbit connecting tubule and cortical collecting duct,[80] supporting the idea that distal K^+ secretion is related to basolateral membrane Na^+,K^+-ATPase activity.[108] Structural changes in I cells have been noted in rat medullary collecting duct following potassium-depletion, suggesting that this cell type may be involved in distal K^+ reabsorption.[110]

Mineralocorticoids have a profound effect on net K^+ secretion in the rat DCT. In marked contrast to normal rats, the tubular fluid K^+ concentration does not increase along the distal tubule of adrenalectomized rats.[111] Stationary microperfusion experiments also indicate that the ability of the late distal tubule to raise the intratubular K^+ concentration is impaired after adrenalectomy.[15] Neither the free flow[111] nor the limiting concentration gradient[15] of early distal tubules is altered in mineralocorticoid deficiency, suggesting that aldosterone increases the net secretory rate of K^+ in the late portion of the DCT. This observation is consistent with recent evidence demonstrating that K^+ secretion by the entire rat distal convolution can be accounted for by the transport in the late portion[108]; K^+ transport does not occur in the early aldosterone-unresponsive DCT. It is likely that the mechanism by which aldosterone increases K^+ secretion includes both an increase in active peritubular uptake of K^+ and an increase in luminal membrane K^+ permeability.[112–114] A recent study in adrenalectomized glucocorticoid-replete rats, treated 24 hr after surgery with physiological doses of aldosterone, demonstrated that the time course of the onset of kaliuresis coincides with that for antinatriuresis, i.e., 60–90 min after aldosterone administration.[67] Both effects were quantitatively related to the Na^+ excretory rate prior to aldosterone treatment. An early study suggested that inhibitors of protein synthesis do not block aldosterone-induced kaliuresis[115]; more recently, Horisberger and Diezi have found that actinomycin D inhibits the antinatriuretic and kaliuretic effects concomitantly, supporting the suggestion that the portions of distal Na^+ and K^+ transport under aldosterone control are related to one another.[116]

K^+ secretion by the DCT is markedly affected by changes in acid–base status.[117] Generally, in alkalosis K^+ secretion is increased, and in acidosis depressed. However, exceptions to this general principle are known. Two reports suggest that metabolic acidosis increases renal K^+ excretion.[118,119] This conflicting information is difficult to evaluate because acid–base disturbances are associated with several alterations which also affect distal K^+ secretion. For example, in acidosis there is increased renal Na^+ and fluid excretion; since increased fluid delivery to the DCT is an important stimulus for K^+ secretion, the kaliuresis observed in some studies during metabolic acidosis may result from the increased distal flow rate, rather than a direct effect of acidosis. This is the most likely explanation for the kaliuresis of diabetic ketoaciduria.[120] On the other hand, metabolic alkalosis produces increased delivery of HCO_3^- and fluid to the DCT, both of which stimulate K^+ secretion.[120] Stanton and Giebisch have performed *in vivo* microperfusion of rat DCT to separately evaluate the effects of plasma and luminal pH on K^+ secretion.[121] They found that metabolic acidosis inhibited and metabolic alkalosis enhanced distal K^+ secretion, when the luminal flow rate and tubular fluid composition were kept constant by microperfusion. Alternatively, when the pH of the luminal perfusate was increased from 6.5 to 8.0 with the systemic pH kept constant, no changes in K^+ secretion were noted. Stanton and Giebisch concluded that alterations in K^+ transport during acid–base disturbances are primarily mediated by the plasma pH. Associated changes in fluid delivery rate but not in luminal HCO_3^- concentration modify K^+ secretion. It is therefore suggested that K^+ uptake at the peritubular membrane and cell K^+ content are regulated by acid–base status. It is known that $^{42}K^+$ uptake across peritubular membranes of rat distal tubules is stimulated by alkalosis.[122]

There are long-standing observations that distal K^+ secretory rate is related to Na^+ excretion and urine and flow rate. A sodium-restricted diet results in decreased Na^+ and K^+ excretion.[6,123] Exctracellular volume expansion,[4,124,125] administration of diuretics,[6,124] release of bilateral ureteral obstruction,[126] and removal of the contralateral kidney[127] enhance Na^+ and K^+ excretion. It is generally held that increasing the rate of delivery of Na^+ and fluid to the DCT increases distal K^+ secretion (Fig. 8).[89,90,120] Because distal Na^+ delivery is the product of early distal Na^+ concentration and flow rate, it has been speculated that either or both of these variables might regulate the distal K^+ secretory rate. In studies employing volume expansion,[38,42,91,95,128] furosemide,[84] contralateral nephrectomy,[50] or microperfusion of the loop of Henle,[2] both flow rate and Na^+ concentration were increased simultaneously. Good and Wright[45] therefore performed experiments using continuous *in vivo* microperfusion of rat DCT where the Na^+ concentration of the luminal perfusate and perfusion rate could be varied independently. Additionally, these two luminal factors could be manipulated without systemic alterations in extracellular volume, plasma, K^+ concentration, mineralocorticoid levels, or acid–base status. When distal flow rate was changed within the range from 4 to 27 nl/min, without altering the perfusate Na^+ concentration, there was an impressive positive relationship between absolute K^+ secretion and perfusion rate. On the other hand, increasing the perfusate Na^+ concentration over the range from 43 to 97 mM did not affect K^+ secretion when the flow rate was kept constant. In these experiments, the rate of net K^+ secretion depended partly on the luminal K^+ concentration, as reduction of the perfusate K^+ concentration to 34 mM actually reversed the direction of net transport. Good and Wright concluded, based on these observations, that increased luminal flow rate may increase K^+ secretion by lowering the luminal K^+ concentration. In fact, at an initial perfusate K^+ concentration of 4 mM, perfused DCT did not elevate the luminal K^+ concentration as much when the flow rate was 26 nl/min as when it was 6 nl/min (Fig. 9). Thus, at high flow rates, distal K^+ secretion proceeds against a smaller transepithelial concentration gradient and net transport is enhanced. In another perfusion study, Good *et al.*[129] concluded that some luminal Na^+ (K_m 10 mM) is necessary to allow K^+ secretion to proceed at the normal rate, and that in the physiologic range, changes in Na^+ concentration do not play a major role in regulating K^+ secretion.

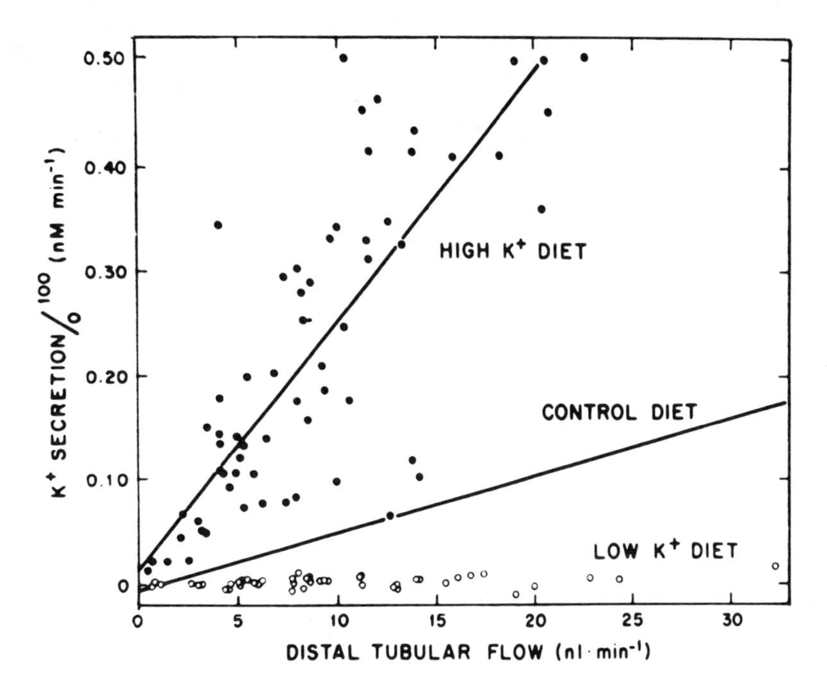

Fig. 8. Plot of absolute rates of distal tubular K^+ secretion as a function of distal tubular flow rate. From Khuri *et al.*[91]

In rats with hereditary hypothalamic diabetes insipidus (DI), ADH stimulates distal K^+ secretion. Field *et al.* microperfused distal tubular segments *in vivo* with isotonic solutions to minimize water absorption and observed significant stimulation of secretory K^+ transport when ADH was infused intravenously.[130] These findings suggest a role for ADH in the prevention of K^+ retention and hyperkalemia during antidiuresis.

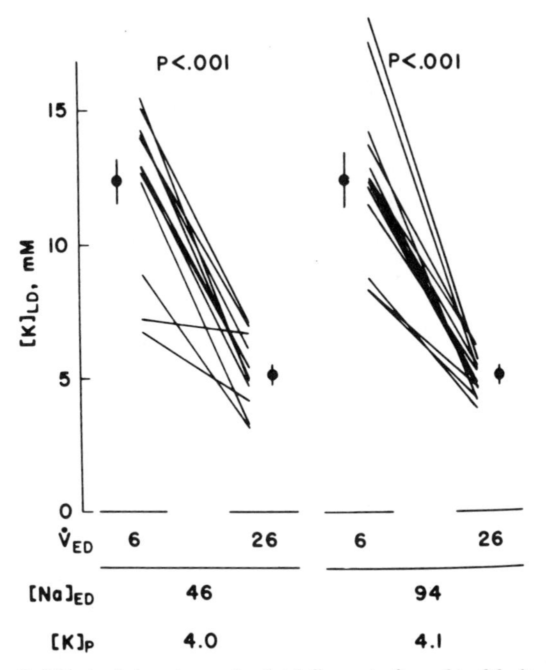

Fig. 9. Effect of changing early distal flow rate from 6 to 26 nl/min with solutions containing 2 mM K^+ on K^+ concentration in fluid collected from late DCT. Lines connect values obtained from different tubules in the same rat. Closed circles give means ±S.E. From Good and Wright.[45]

Other factors which may influence the distal net secretory rate include the transepithelial voltage and the anionic composition of the luminal fluid. The lumen-negative transepithelial potential difference has been considered a driving force for K^+ secretion.[4,42,131,132a] One mechanism of action proposed for K^+-sparing diuretics is based on the diminution of the transepithelial potential difference which these drugs produce.[84,131] Good and Wright[45] have noted that flow-dependent changes in transepithelial voltage are consistent with changes in the luminal K^+ concentration (see above). Furthermore, Velazquez *et al.* [132] observed that replacing all luminal Cl^- with SO_4^{2-} in microperfusion experiments caused increased transepithelial voltage and increased K^+ secretion. However, when low concentrations of amiloride were added to the SO_4^{2-} perfusion solution, the voltage was diminished as expected, but the increase in K^+ secretion was not prevented. Thus, anion substitution-induced changes in K^+ secretory rate cannot be explained entirely on the basis of the coincident changes in transepithelial potential. They suggest that a K^+–Cl^- cotransport process mediating K^+ reabsorption exists in the DCT; if such a process is inhibited by a low luminal Cl^- concentration, then the result would be increased net K^+ secretion.

3.3. Water

In nondiuretic rats, approximately 8% of the filtered water is reabsorbed by the distal convolution in the presence of ADH.[38] About 5.4 nl/min is delivered to the DCT, about 3.0 nl/min leaves the DCT, yielding a net water flux of 0.25 nl/sec per mm^2. In spite of the hypoosmolality of early distal tubular fluid (Fig. 10), the net water flux per unit surface area of the DCT is only about 40% of the value estimated for the isosmotically reabsorbing proximal convoluted tubules.[38] This rel-

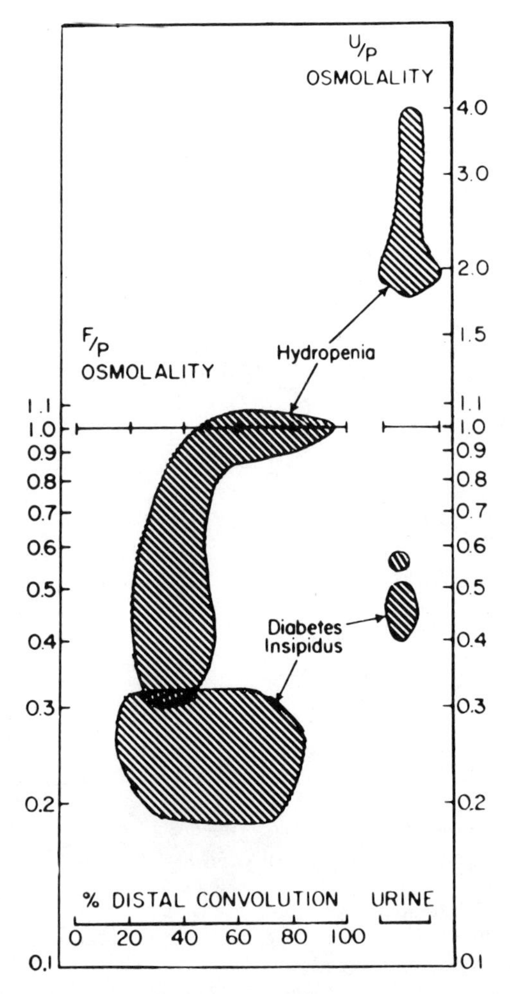

Fig. 10. Comparison of osmolality of distal tubular fluid in hydropenia and in diabetes insipidus in the rat. From Gottschalk.(18)

atively low water permeability has been studied quantitatively. The hydraulic conductance (L_p) of the DCT of rats in antidiuresis is reported to be 3.1–4.9 × 10^{-8} cm^3/cm^2 per sec per cm H_2O.(133–135) A striking decrease in hydraulic conductance was found during water diuresis, the value for L_p being less than 10% of that in antidiuresis, 1.5 × 10^{-9} cm^3/cm^2 per sec per cm H_2O.(134,135) Perfusion of the distal convolution with a solution containing high concentrations of Ca^{2+} (15 meq/liter) reduced L_p to values almost as low as in water diuresis.(136) L_p is increased significantly in adrenalectomized rats and can be restored to normal values by administration of glucocorticoids.(134)

Clearly, values for water flux or hydraulic conductance provide only overall information on DCT function, neglecting cellular heterogeneity. Both structural data(23) and biochemical data on hormone sensitivity(36) stress the interspecies differences in the characteristics of the DCT as well as the marked heterogeneity of the DCT within a given species. These considerations are of particular importance with respect to hydraulic conductance.

Walker *et al.*,(14) Wirz,(137) and Gottschalk and Mylle(138) demonstrated that in rats there is a transition from hypotonic to isotonic tubular fluid along the DCT; this transition occurs at about the midpoint of the DCT (Fig. 10). This observation has been confirmed by Colindres *et al.*(139) using special precautions to avoid retrograde collection of fluid from the collecting ducts. Furthermore, microperfused late segments of rat DCT reabsorb water in response to an imposed osmotic gradient whereas early segments do not (Fig. 4).(35) The early observations clearly provided one of the cornerstones in the development of the countercurrent hypothesis of the concentration and dilution of urine. In dogs(140) and rhesus monkeys,(99) tubular fluid remains hypoosmolar to plasma throughout the length of the DCT.

It is now clear that both the site along the DCT where tubular fluid becomes isosmotic with peritubular blood (e.g., in rat) and the interspecies differences with respect to osmotic equilibration in the DCT are related to ADH sensitivity. Morel and collaborators have performed careful studies to evaluate ADH-sensitive adenylate cyclase activity along the nephron of a number of different species, including rabbit,(141) rat,(36,142) mouse,(143) and human.(144) A summary of their observations is shown in Fig. 11. It is apparent that there are marked differences in arginine vasopressin sensitivity along the nephron and in the corresponding nephron segments from different species. The cortical and medullary collecting ducts are vasopressin-responsive in all species examined. However, the vasopressin-sensitivity of the DCT is variable. Human and rabbit DCT are essentially unresponsive. Mouse and rat DCT have little or no sensitivity in the early portion with high activity in the connecting tubule and initial collecting tubule. While these elegant studies do not differentiate the cell type with which ADH interacts, they do provide a plausible basis for the physiological differences in osmotic water flow observed in numerous studies.(1,35,138,139)

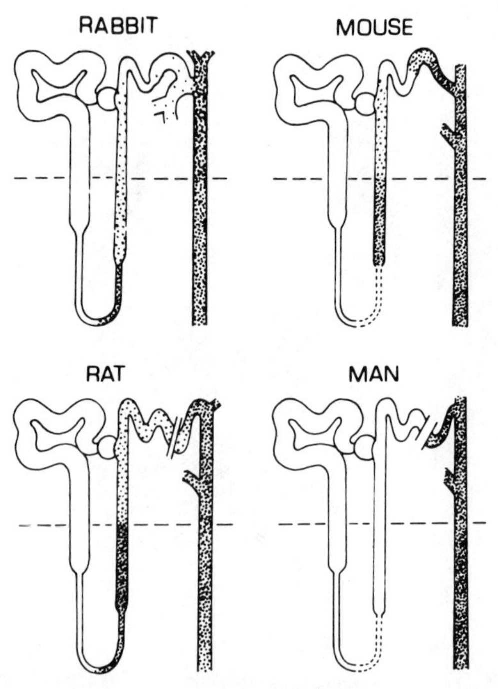

Fig. 11. Localization of vasopressin-sensitive adenylate cyclase in nephrons from different species. The dot density is proportional to increasing adenylate cyclase activity induced by 10^{-6} M arginine vasopressin. From Morel.(36)

Table I. Bicarbonate Reabsorption along the Nephron[a]

Group	Filtered load	Proximal	Loop	Distal	Excretion
Control	25.7	22.1	0.49	2.28	0.83
5% $NaHCO_3$	43.0	24.3	2.70	7.02	8.98
Hypochloremic alkalosis	40.6	24.9	7.35	4.37	3.98
Metabolic acidosis	9.55	8.16	0.51	0.40	0.48
Hyperventilation	18.0	11.0	3.12	2.05	1.83
15% CO_2	27.4	20.0	5.67	1.11	0.62
Diamox	28.5	11.7	0.10	9.30	7.40

[a]Data from Malnic, de Mello-Aires, and Giebisch.[7] All values are in units of μeq/min per kg per ml. GFR.

As in toad bladder,[145,146] vasopressin-induced increases in hydraulic conductance and osmotic water flow are associated with an increase in intramembranous particle clusters in the luminal membrane of rat collecting duct.[147] The frequency and the area occupied by the particle clusters are related to the dose of vasopressin administered, and decreases in frequency and area accompanied decreases in urine osmolarity when vasopressin was discontinued. It has not yet been established whether this specific structural alteration which is related to vasopressin action also occurs in the DCT.

3.4. Hydrogen

The DCT contributes to the overall acid–base balance of the body by reabsorbing a portion of filtered HCO_3^-,[7] by generating titratable acid and by adding ammonia to the tubular fluid.[148,149]

From the older literature,[7] it has been estimated that in normal rats, where the average filtered load of HCO_3^- was 25.7 eq/min per kg per ml GFR, the distal tubule reabsorbed 2.28 eq/min per kg per ml GFR, or 8.9% of the filtered load (Table I). Acute and chronic metabolic alkalosis as well as Diamox administration caused increases in absolute and fractional HCO_3^- reabsorption by the DCT. More recent data of Lucci *et al.*[150] yielded a somewhat different set of conclusions. The absolute rate of HCO_3^- reabsorption was measured during microperfusion of superficial rat DCT where HCO_3^- delivery could be carefully controlled. HCO_3^- reabsorption was assessed from microcalorimetric determinations of total CO_2, rather than estimated from pH measurements. In control hydropenic animals, when DCT were microperfused with a solution containing 25 mM HCO_3^- equilibrated with 15% CO_2, no HCO_3^- reabsorption occurred. Virtually identical results were obtained when the perfusion solution was not preequilibrated with CO_2, but rather the CO_2 tension of the luminal solution was measured *in situ* with pCO_2 microelectrodes. Under the latter experimental conditions, there was net entry of CO_2 into the lumen from surrounding cortical structures and the total CO_2 was corrected for this net addition; again, under this different set of conditions, no HCO_3^- reabsorption was evident. Lucci *et al.* also assessed DCT HCO_3^- transport in various acid–base disturbances. There was no significant reabsorption in combined respiratory acidosis–metabolic alkalosis, pure respiratory acidosis, or acute metabolic acidosis. However, chronic metabolic acidosis exposed significant HCO_3^- reabsorption, 52.6 ± 13.9 pmoles/mm per min. Therefore, the DCT has an adaptive process, occurring over a period of several days, which can respond to acidosis by reabsorbing HCO_3^-. The precise signal for adaptation is not clear, nor is the transport process which is altered. However, the presence of "disequilibrium pH" (see below) during conditions where acidification is stimulated points to the existence of a H^+ secretory mechanism in the DCT.

The DCT is a nephron site where the tubular fluid is more acid than the blood.[7,151–154] Reported values for the *in situ* distal tubular pH range from 6.20 to 6.88.[7,152,155–158]

Early studies suggested that distal tubular fluid is, under normal conditions, characterized by an acid "disequilibrium pH," of between 0.14 and 0.88 pH unit. An acid "disequilibrium pH" is by definition observed when the measured *in situ* pH is lower than the equilibrium pH measured when tubular fluid is collected and allowed to equilibrate at a known pCO_2 *in vitro*. An acid disequilibrium pH was first observed by Rector *et al.* using pH-sensitive glass microelectrodes.[155] Their results have been confirmed by others[7,152] using antimony microelectrodes. The presence of an acid disequilibrium pH has been taken as evidence for H^+ secretion in the DCT in the absence of luminal carbonic anhydrase. Histological evidence supports the absence of this enzyme on the luminal aspect of distal tubular cells,[159] while its presence has been documented in the cytoplasm of I cells and in the basal infoldings of early DCT cells.[160] On the other hand, recent evidence does not support the existence of a disequilibrium pH in distal tubular fluid under normal conditions. DuBose *et al.*[157] have developed an aspiration pH microelectrode which permits measurement of the equilibrium pH *in vivo*. This microelectrode allows tubular fluid to reach chemical equilibrium at the prevailing pCO_2 of renal cortical structures after aspiration of the fluid into a chamber in contact with the environment. The design circumvents the problematic requirement in the earlier measurements of exact knowledge of the pCO_2 of tubular fluid. Thus, DuBose *et al.* proposed to measure *in situ* and equilibrium pH in the same distal tubular segments during free flow *in vivo;* comparison of the two values would provide information about the existence of a disequilibrium pH. Neither control nor HCO_3^--loaded rats exhibited a proximal or distal disequilibrium pH. Intravenous infusion of benzolamide, a carbonic anhydrase inhibitor, revealed disequilibrium in proximal but not distal fluid. However, combined metabolic alkalosis and respiratory acidosis, expected to maximally increase HCO_3^- reabsorption, caused an acid disequilibrium pH in DCT which was reversed by carbonic anhydrase infusion. The discrepancy between these and the older findings has not been entirely resolved. The explanation may reside partly in technical considerations related to electrode construction[154] or in the value assumed for the pCO_2 of distal tubular fluid when determining the equilibrium pH. In view of the disagreement about the exact value for the pCO_2 of cortical structures,[153,161] it seems particularly valuable that DuBose *et al.* have measured the equilibrium pH *in vivo*.

The failure of DuBose *et al.* to demonstrate a significant acid disequilibrium pH in control or HCO_3^--loaded rats can be taken as evidence for a low H^+ secretory rate in the DCT. This conclusion is consistent with the observation that HCO_3^- reabsorption is also low under these experimental conditions.[150] Evidently maximal urinary acidification occurs beyond the superficial DCT. However, Lucci *et al.* did not observe enhancement of HCO_3^- reabsorption during combined respiratory acidosis–metabolic alkalosis, although this setting did produce an acid disequilibrium pH. This apparent discrepancy may be explained by the relatively large change in HCO_3^- concentration required for detection as compared to the small change in H^+ concentration required to produce a noticeable disequilibrium pH. Because an acid, not an alkaline, disequilibrium pH is observed in the DCT, it can be concluded that H^+ secretion is the primary mechanism for HCO_3^- reabsorption. At steady state, the luminal carbonic acid dehydration will equal its rate of formation, which will in turn equal the rate of HCO_3^- reabsorption. Since the uncatalyzed dehydration of carbonic acid is slow,[162] carbonic acid will accumulate in the tubular lumen until the concentration is sufficiently high to drive the reaction at a rate equal to the reabsorptive rate of HCO_3^-. Luminal carbonic acid will not be in equilibrium with CO_2 in plasma or in tubular fluid, and therefore the intratubular pH will be lower than the calculated equilibrium pH.

3.5. Calcium

Ca^{2+} is reabsorbed in all parts of the nephron including the DCT.[163] In the nondiuretic rat, approximately 10–14% of the filtered Ca^{2+} reaches the distal tubule; estimates of reabsorption in the DCT range from 8 to 11% of the filtered load.[8,163] The concentration of Ca^{2+} in tubular fluid relative to that in an ultrafiltrate of plasma is 0.6 at the beginning of the accessible DCT and falls to 0.3 by the end of the distal convolution.[8] This information, taken together with the lumen-negative transepithelial potential difference in this segment,[9] supports an active Ca^{2+} reabsorptive process, occurring against a chemical as well as an electrical gradient. The components of net Ca^{2+} reabsorption have been examined using stationary microperfusion techniques.[8] When droplets of Ca^{2+} and Na^+-free isosmotic raffinose were introduced into the lumina of DCT, the Na^+ concentration quickly rose to the equilibrium value of 55 mM. In contrast, no Ca^{2+} appeared in the luminal raffinose droplets, suggesting that the distal tubular epithelium has a low effective Ca^{2+} permeability. When the aqueous raffinose droplets contained various Ca^{2+} concentrations, then Ca^{2+} disappeared quickly from the luminal fluid at a rate proportional to the initial Ca^{2+} concentration. This concentration dependence of transport suggested that, like Na^+ reabsorption, distal Ca^{2+} reabsorption might be load-dependent. In fact, when DCT were microperfused *in vivo* with solutions simulating early distal tubular fluid, the Ca^{2+} reabsorptive rate was strongly dependent upon the delivered load.[8] There was no evidence of saturation of reabsorption even at loads which were three times the physiological delivery rate. Considering the entire rat DCT, Ca^{2+} reabsorption was 68% of the delivered load over the range examined.

In the kidney as a whole, there is a remarkable parallelism between the excretion of Ca^{2+} and Na^+.[164] In general, maneuvers which alter proximal tubule Na^+ and water absorption cause proportional changes in Ca^{2+} absorption, such that volume contraction increases Ca^{2+} reabsorption, and volume expansion, PTH, or Diamox administration decreases Ca^{2+} reabsorption. TALH has both passive and active Ca^{2+} transport [166,167]; furosemide causes proportionate inhibition of NaCl and Ca^{2+} transport.[19]

In contrast to these more proximal segments, Ca^{2+} transport in the DCT is dissociable from Na^+ transport. (1) For example, diuretics such as thiazides[8,82] and amiloride,[35,168] whose primary natriuretic action is in the DCT, actually enhance Ca^{2+} transport in this segment (see below). (2) PTH and cAMP analogs increase Ca^{2+} reabsorption in rat DCT[20] and rabbit DCT and granular collecting tubule[19] without significant alterations in Na^+ transport. (3) Metabolic acidosis is associated with hypercalciuria.[169,170] This impairment of Ca^{2+} transport is likely to occur in the DCT and is reversed by infusion of HCO_3^-.[171–174] There has been speculation that the load of HCO_3^- to the DCT specifically augments distal Ca^{2+} transport, irrespective of systemic acid–base status.[171,172] (4) Adrenocortical insufficiency is associated with increased Ca^{2+} and decreased Na^+ reabsorption, and this pattern is reversed by mineralocorticoid treatment.[175] Mineralocorticoids have their Na^+-retaining action in the DCT, although it is not yet clear whether their effects on Ca^{2+} transport also occur at this site. (5) PO_4^{3-} depletion causes marked hypercalciuria,[176,177] independent of parathyroid status.[176] The defect in Ca^{2+} reabsorption occurs in the distal nephron and is reversible by PO_4^{3-} repletion.[176,177] Thus, the DCT is not only a site where regulation of Ca^{2+} reabsorption appears to be separate from that of Na^+ reabsorption, but it is also a site where Ca^{2+} and Na^+ transport may be inversely related. Studies with diuretics have provided some information about the mechanisms of distal Ca^{2+} transport. In particular, it is now known that the sites for the hypocalciuric actions of thiazide diuretics and amiloride are the same as for their natriuretic actions.

The Ca^{2+}-retaining activity of thiazides is confined to the early portion of the rat DCT as is their inhibitory action on Na^+ transport.[82] Likewise, amiloride, whose natriuretic activity is limited to the late DCT, increases Ca^{2+} transport at the same site.[35] These observations support earlier speculation that in the DCT, a portion of Ca^{2+} reabsorption is linked to Na^+ reabsorption in an inverse fashion.[164] In fact, *in vivo* microperfusion studies with amiloride in the late DCT[35] showed a strong correlation between enhancement of Ca^{2+} transport and the magnitude of inhibition of Na^+ transport (Fig. 12). One possible interpretation of this observation is that the increased Ca^{2+} transport is secondary to decreased Na^+ transport. Walser proposed that such a relationship in the DCT might result from an effect of distal diuretics on the transepithelial potential difference, with diuretic-induced decreases in potential causing decreased back-flux and thus increased net flux of Ca^{2+}.[164] This explanation is unlikely in view of the extremely low Ca^{2+} permeability of the DCT.[8] However, it may be that amiloride increases net Ca^{2+} flux secondary to reduction of luminal membrane Na^+ conductance.[178–180] The expected decrease in intracellular Na^+ activity might increase the rate of Ca^{2+}/Na^+ exchange in the basolateral membrane, thus increasing Ca^{2+} transport; such an exchange process has now been documented for various epithelia.[181–184] Alternatively, decreased luminal Na^+ conductance might hyperpolarize the luminal membrane, increasing Ca^{2+} entry into

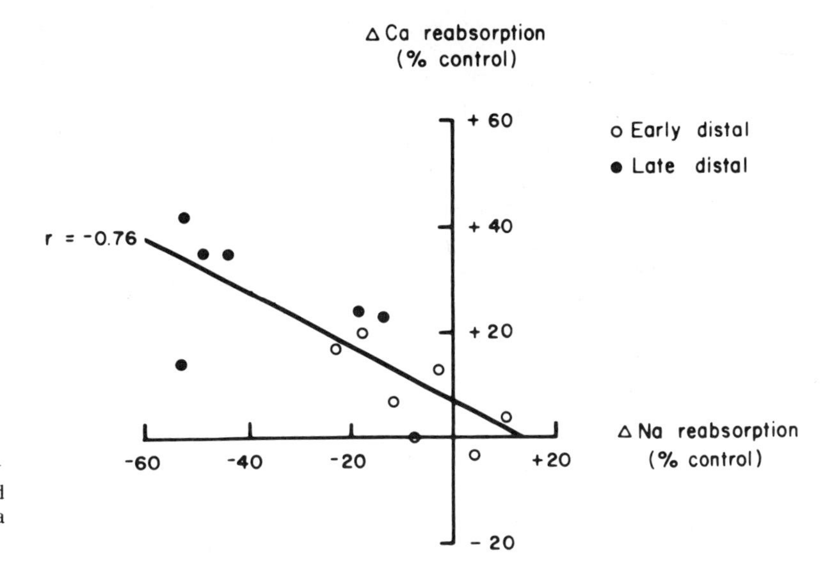

Fig. 12. Amiloride-induced change of Ca^{2+} reabsorption as a function of amiloride-induced change in Na^+ reabsorption. Each point represents a single tubule. From Costanzo.[35]

DCT cells and transepithelial Ca^{2+} movement. It is not possible to distinguish between these two explanations without measurements of individual membrane voltages and cytosolic Na^+ and Ca^{2+} activities. It is not yet known whether similar mechanisms may account for the inverse relationships between Ca^{2+} and Na^+ transport seen with thiazide diuretics and adrenalectomy.

3.6. Magnesium

Under normal conditions, approximately 5–10% of the filtered load of Mg^{2+} is reabsorbed between the early distal tubule and the final urine of the dog[185] and rat.[186,187] Estimates of fractional reabsorption by the DCT *per se* range from 2 to 5% of the filtered load.[188–192] In control animals, the concentration of Mg^{2+} in distal tubular fluid is less than in an ultrafiltrate of plasma,[185–187] indicating active reabsorption. The transport of Mg^{2+} within the DCT is load-dependent,[192] although the transport process operates normally close to capacity. Thus, when the delivered Mg^{2+} load was increased in microperfused rat DCT, fractional reabsorption fell strikingly.[192] These observations contrast with those for Na^+ and Ca^{2+} transport where reabsorption is unsaturated.[8,38]

Under certain conditions,[193,194] net addition of Mg^{2+} to tubular fluid between the early distal tubule and the final urine has been observed. LeGrimellec *et al.*[194] found that urinary Mg^{2+} excretion exceeded the filtered load by 20% during acute Mg^{2+} loading in rats. One interpretation of these observations is that the superficial DCT and collecting tubules secrete Mg^{2+}. Alternatively, Mg^{2+} reabsorption in deep nephrons may be far less than in superficial nephrons, resulting in a higher amount of Mg^{2+} delivered to the final urine than is delivered from the superficial DCT. It is possible that a medullary gradient for Mg^{2+} is established during Mg^{2+} loading, which then provides a favorable driving force for Mg^{2+} diffusion into collecting duct fluid. Such a process could also result in medullary recycling of Mg^{2+} into descending limbs of long Henle's loops. At present, neither the corticopapillary Mg^{2+} gradient nor the permeability of the collecting ducts to Mg^{2+} has been established. However, Brunnette *et al.*[187] and Carney *et al.*[188] could not detect reabsorption or secretion in terminal nephron segments of rats after acute elevation of the plasma Mg^{2+} concentration.

Available data suggest that the DCT is not a major site of hormonal regulation of Mg^{2+} reabsorption.[195] PTH causes renal Mg^{2+} retention,[196,197] although the hypocalcemia attending hormone administration has caused magnesuria in some studies.[198,199] Direct examination of renal tubular effects of PTH with micropuncture techniques has revealed small effects in the loop of Henle and distal tubule.[191,192,200] In the golden hamster, a species with particular sensitivity to PTH, about 20% of the reabsorption of filtered Mg^{2+} is under hormonal control, the site of action being the loop of Henle.[189] PTH-sensitive Mg^{2+} transport in the isolated rabbit TALH[201] supports the observations in golden hamster. Distal diuretics, such as thiazides, have little or no effect on Mg^{2+} excretion.[202]

3.7. Phosphate

A number of studies demonstrate a small but significant reabsorption of filtered PO_4^{3-} between the early distal tubule and the final urine[203–205] which is unmasked by parathyroidectomy[204,206–208] and inhibited by PTH.[204,208] While these results suggest a role for the distal nephron in the regulation of renal PO_4^{3-} reabsorption, a few studies have not detected distal PO_4^{3-} reabsorption. For example, in stationary microperfusion experiments, no PO_4^{3-} was reabsorbed from distal segments of parathyroid-intact or thyroparathyroidectomized rats.[209] At the present time, the extent and regulation of distal PO_4^{3-} reabsorption remain controversial.

4. Electrophysiological Considerations

4.1. Transepithelial Voltage

In contrast to the proximal tubule where each major segment is characterized by approximately the same transepithelial electrical potential difference (PD), voltages across the wall of the DCT vary along its length. Wright[9] was the first to show that the early distal tubule is characterized by a PD of 9–12 mV, lumen negative, whereas in the latter half of the DCT, the PD exceeds −45 mV.

In isolated mammalian distal tubule preparations, so far only obtained from rabbit kidney, the initial segment (DCT_a), immediately adjacent to the macula densa, and inaccessible to surface micropuncture, can be studied directly. In this nephron segment, transepithelial voltages ranging from +7 to −40 mV have been observed.[19,65,210] With this preparation, Gross *et al.*[65] found an average of −40 mV and observed that ouabain in the peritubular bath, or amiloride in the luminal perfusate, diminished these lumen-negative voltages. Such results may be expected when Na^+ transport from lumen to cell occurs via amiloride-sensitive ion channels, in series with Na^+ extrusion via an ouabain-sensitive Na^+,K^+-ATPase located within the basolateral cell membrane. Significantly, addition of vasopressin to the contraluminal bath failed to produce significant changes in transepithelial voltages in these perfused, isolated DCT, in contrast to marked effects on the cortical collecting tubule.[65] Other studies from the same laboratory indicate that the initial segment is insensitive to both mineralocorticoids and the K^+-sparing diuretic triamterine.[210] In one study, [19] lumen-positive PD were found. It is not known at present whether this difference in polarity of the voltage was due to the fact that the tubules had been obtained from juxtamedullary nephrons, or due to uncertainties in localization of tubule segments when lumen-negative potentials were observed.

In amphibian kidneys, the so-called "diluting segment"[211] resembles the thick ascending limb and the initial segment of the distal tubule of mammals. It is located between the point where the tubule lies adjacent to the glomerulus and the approximate midpoint to the confluence with a collecting duct. The transepithelial voltage in the "diluting segment" averages about +14 mV.[211]

In the adjacent portion of the DCT (DCT_b), which corresponds to the "early" DCT of micropuncture studies, transepithelial voltages of +7 mV[39,131,212–214] to −20 mV[9,40,41,212,215–217,218a] have been observed. All of these measurements were obtained in rats *in vivo*. Magnitude and polarity of the voltage apparently depend on the type of microelectrodes used, and the technique of voltage recording. This controversy is presently unresolved, but a definitive answer may eventually be obtained when DCT_b segments of rat kidneys can be isolated and perfused *in vitro*.

The third subdivision (DCT_g) of the DCT, referred to as the connecting tubule, is generally located beneath the kidney surface. In isolated, perfused DCT_g preparations obtained from rabbit kidneys, lumen-negative voltages ranging from −13 to −30 mV have been observed.[19,219]

The most distally located segment of the DCT (DCT_1), the so-called "late" DCT of micropuncture studies, corresponds to the initial portion of the cortical collecting tubule. Lumen-negative transepithelial voltages in this nephron segment in rats range from −20 to −50 mV[4,9,39–41,103,104,131,132a,212–218] and are augmented by mineralocorticoids.[103,114,214,222] In dogs, a PD of −43 mV,[223] in *Amphiuma* −45 mV,[224] and in *Necturus* an average value of −35 mV[225,226] have been reported. Late distal tubules of *Necturus,*[211] frog,[211] salamander,[211] and *Amphiuma*[224] have transepithelial PD of −30 to −40 mV. Jentsch *et al.*[227] found lumen-positive voltages in distal tubules of *Amphiuma*. When lumen-positive and -negative transepithelial PD were detectable in the same animal, the negative voltages were localized in late distal segments.[227] Lumen-positive segments were not sensitive to amiloride.[228] The lumen-positive voltage of early distal tubules of *Amphiuma* is abolished, however, by luminal application of furosemide or perfusion with either Na^+- or Cl^--free fluid.[229]

The interior of distal tubule cells is about 70 mV negative with respect to peritubular fluid.[212] The technical difficulties in recording stable intracellular PD are formidable, and explain why many studies report mostly transient membrane potentials. Nevertheless, the peritubular membrane potential of the DCT_b of mammalian kidney was reported to range from −55 to −65 mV.[103,218] Somewhat higher values were found in the DCT_g, ranging from −65 to −90 mV.[9,40,103,114,132a,215] At present, the reason for such an apparent difference of basolateral voltages in early and late DCT is not clear and might, in fact, be artifactual. In view of the considerable overlap in recorded basolateral membrane voltages in early and late DCT, it appears that the much larger differences in transepithelial potentials along the DCT are due to differences in the luminal membrane potential. A progressive depolarization of the luminal cell border along the length of the DCT could account for the concurrent increase in the lumen-negative transepithelial voltage. This view is consistent with the demonstration of amiloride-sensitive Na^+ transport in the late, but not in the early DCT.[35]

Ion-substitution experiments have provided some insight into a number of permeability and transport properties of the DCT. The magnitude of the transepithelial PD is sensitive to the luminal Na^+ concentration. For instance, when luminal Na^+ was reduced from 150 to less than 15 mM by substitution with choline,[4,9,40,215] Tris,[9] or mannitol,[40] the transepithelial voltage decreased from −40 to −50 mV to some −8 mV. In more recent microperfusion studies of the late DCT by Good *et al.*,[129,220] the transepithelial voltage was approximately constant at −35 mV at luminal Na^+ concentrations higher than 35 mM but voltages declined steeply when Na^+ concentrations were reduced to 10 or 3 mM. On the other hand, Hayslett *et al.* [216] found that the voltage was more sensitive to luminal Na^+ concentration changes in the range between 30 and 150 mM, and less sensitive at concentrations below 30 mM. Such differences in results may be explained by undetectable artifacts but, more likely, may be caused by different permeability properties in different groups of rats, possibly reflecting the specific endocrine (steroid) status of the animals under study. An additional note of caution is indicated because of the contribution to the transepithelial voltages of shunt currents due to rheogenic cellular pumps.[216] In fact, as pointed out by Hayslett *et al.*,[216] unknown changes in pump rates in response to ion substitutions in luminal perfusates, make an analysis of the permselectivity of the luminal cell membrane on the basis of transepithelial voltages of mammalian distal tubules, at best, a semiquantitative undertaking.

Cytosolic activities of Na^+ (a_{Na}^i), K^+ (a_K^i) measured with ion-selective microelectrodes are not different in early and late DCT cells, averaging 16 mM[216] and 47 mM,[218] respectively. a_K^i is reduced in dietary potassium depletion, acidosis,[218] and after adrenalectomy.[114] Alkalosis or potassium loading leads to increased levels of a_K^i in distal tubule cells of the rat.[218] Measurements of cytosolic Cl^- activity (a_{Cl}^i), also obtained with ion-selective microelectrodes in rat DCT_1 cells, indicate an average value of 42 mM,[221] i.e., higher than predicted from passive distribution. As pointed out in a recent review by Koeppen and Giebisch,[212] the reported values for a_{Na}^i and a_{Cl}^i are likely to be overestimates and those of a_K^i underestimates since most of the electrode recordings were only transient in nature because of the large size of the double-barreled electrodes employed.

Studies on amphibian DCT[228,230] indicate that changes in luminal Na^+ concentration manifest themselves as electrical potential changes, not across the luminal, but rather across the peritubular cell membrane. In *Amphiuma* tubules,[228] an increase in luminal Na^+ concentration from 10 to 100 mM reversibly increased the peritubular PD. In fact, the magnitude of the basolateral membrane voltage was found to be a saturable function of the luminal Na^+ concentration. Either luminal amiloride or peritubular ouabain abolished this hyperpolarization of the peritubular membrane voltage. The most logical explanation for these results is that a rise in luminal Na^+ concentration leads to elevated intracellular concentrations of Na^+, which in turn enhance the activity of the Na^+,K^+ exchange pump; due to a coupling ratio of Na^+/K^+ in excess of 1, a hyperpolarization of the basolateral cell membrane results. Thus, the Na^+,K^+ pump contributes directly to the generation of the peritubular cell membrane potential.

Several electrophysiological studies[224,230] have shown that the peritubular cell membrane of *Amphiuma* distal tubules has a high K^+ permeability. This conclusion rests mainly on the observation that stepwise increases in peritubular K^+ concentration result in a progressive decline in basolateral membrane voltage. In other studies,[229] electrochemical equilibrium for Cl^- across the basolateral membrane has been demonstrated after addition of furosemide to the tubule lumen. Luminal cell membranes of distal tubule cells are characterized by a high degree of K^+ permeability,[40,100,215,231,232a] a membrane property that is critical for generating positive transepithelial voltages in early distal tubules of *Amphiuma* where cotransport of Na^+, K^+, and Cl^- ions have been observed. The luminal K^+ permeability permits recirculation of K^+ from cell interior to tubule lumen, thus hyperpolarizing the cell with respect to tubular fluid, increasing the transepithelial positive voltage and augmenting the electrical driving force for passive Na^+ reabsorption via intercellular pathways. During adaptation to a high-potassium diet, the K^+ conductance of the luminal cell membrane of early distal *Amphiuma* tubules is increased, an event that promotes net secretion of K^+ into the tubule lumen. The luminal membrane K^+ conductance is decreased by lowering the pH of the peritubular fluid,[232] thus reducing K^+ secretion in states of acidosis.

Measurements of transepithelial resistance during ion-substitution experiments have demonstrated that the high permeability of the DCT to K^+ is responsible for the relatively high transepithelial electrical conductance of this tubular segment. On the other hand, the K^+ permeability of the collecting duct epithelium is much lower, a property essential for restricting the loss of K^+ from the collecting duct lumen after its secretion into the distal tubule.[233] Studies by Malnic and Giebisch[40] and Wright[234] have shown that the transepithelial conductance of K^+ exceeds that of Na^+, while the latter greatly exceeds the Cl^- conductance.

The transepithelial PD is influenced by metabolic acidosis,[87,152] metabolic alkalosis,[87,152] adrenalectomy,[113,114] aldosterone,[114] cycloheximide,[113] actinomycin D,[114] ouabain,[230] acetazolamide,[87] and amiloride.[113] Sodium depletion or a high-potassium diet leads to increased PD in the DCT.[4,54] The relative partial conductance of K^+ is increased by potassium loading,[235] decreased by potassium depletion,[234] and diminished by adrenalectomy,[113] but can be restored to normal values by adminstration of aldosterone.[114]

5. Mechanisms of Transport

Central to nearly all considerations of ion transport across renal tubular epithelium, including the DCT, is the action of a Na^+,K^+ exchange pump which is capable of maintaining a low Na^+ and high K^+ concentration within the cytosol. Transfer of ions across the epithelium is then accomplished by a characteristic polar arrangement of membrane permeabilities (via ion channels or carrier-proteins) in the luminal and peritubular cell membranes whereby the electrochemical driving forces can act upon ions only in such a way that ion fluxes of specific magnitude and direction are uniquely determined.

Mechanisms of transport in the DCT encompass some processes which are probably shared by all cell types present in amphibian and mammalian distal tubules. Examples are the basolateral Na^+,K^+ pump and K^+ permeabilities of luminal cell membranes. Other transport processes are apparently specific for mammalian DCT cells, such as the apical cotransport of Na^+ and Cl^- described by Velazquez and Wright.[236] Alternatively, a given transport process present in the amphibian DCT may be absent in the mammalian DCT but appear instead in a neighboring segment. For example, the early distal tubule of *Amphiuma*[228,232a] resembles the TALH of the mammalian loop of Henle to a greater extent than the mammalian DCT.[236,237] Cell heterogeneity within the DCT of a given species must also be recognized. For instance, in rat nephrons, DCT_b cells do not secrete K^+, whereas avid K^+ secretion is attributable to the CNT cell, and the P cell of the connecting tubule and the initial collecting tubule.[108] Active reabsorption of K^+ may only occur in I cells of the DCT.[107] Therefore, in contrast to other nephron segments where functional heterogeneity is less pronounced, it is not possible to provide a simple comprehensive model of ion transport for an idealized DCT. Nevertheless, with these caveats in mind, a number of generalizations regarding mechanisms of ion transport by the DCT may be advanced.

Net reabsorption of Na^+ in the DCT is ultimately achieved by the rheogenic operation of a Na^+,K^+ pump[228] which probably extrudes 3 Na^+ in exchange for 2 K^+, mediated by an ATPase located within the basolateral cell membrane.[238,239] In addition to extrusion of Na^+, the active accumulation of K^+ by the cell, against an electrochemical potential gradient, is achieved by the same transport system. This has been demonstrated by measurements of a cytosolic K^+ activity gradient and the basolateral membrane voltage using ion-selective microelectrodes in distal cells of the amphibian kidney: the electrical PD failed to account for the measured difference in extra- and intracellular K^+ activities.[235] By lowering the cytosolic Na^+ activity to some 12 mM,[240] the Na^+,K^+ pump provides the chemical driving force for passive entry of Na^+ into the cell across the luminal membrane. Luminal entry and peritubular extrusion are transport steps arranged in series, which result in net reabsorptive movement of Na^+ from lumen to peritubular fluid.

The mechanism of Na^+ entry across the luminal cell membrane is fairly well defined in the "late" DCT of the mammalian kidney. In this segment, Na^+ reabsorption is inhibited by luminal application of about 10^{-6} to 10^{-5} M amiloride,[19,84,210,211,241] i.e., a dose generally recognized to inhibit Na^+-selective ion channels in epithelia (apparent $K_i = 10^{-8}$ to 10^7 M in tight epithelia),[242] as compared to a K_i of 3×10^{-5} M for inhibition of Na^+/H^+ exchange.[243] Luminal Na^+ entry into cells of the rat DCT may also be mediated by coupled, carrier-mediated NaCl transport.[132,236,244] This tentative conclusion rests on the observation that net Na^+ reabsorption more than doubled with only minor changes in transepithelial voltage, when Cl^- was substituted for SO_4^{2-} in luminal perfusion fluid. Velazquez *et al.*[132] have refrained from assigning this mode of transfer to a specific subsection of the DCT. On the other hand, coupled Na^+–Cl^-–K^+ cotransport is well documented in the early portion of the amphibian distal tubule.[232]

In analogy to the cortical collecting tubule,[245–247] the rate of luminal Na^+ entry may be influenced by the level of intracellular Ca^{2+} activity, which in turn may depend, in part, upon a 3 Na^+/2 Ca^{2+} exchange process located within the basolateral cell membrane. In the isolated, perfused cortical collecting tubule, experimental maneuvers thought to increase cytosolic Ca^{2+} activity lead to a decrease in the rate of Na^+ entry across the luminal cell boundary.[245–247] These data are consistent with the existence of a negative feedback process linking the rate of active Na^+ extrusion in the peritubular membrane with the rate of Na^+ entry across the luminal cell border in which cytosolic Ca^{2+} activity acts as a regulatory factor.[181,248] Cytosolic Ca^{2+} activity would be increased by reduced Na^+/Ca^{2+} exchange across the peritubular cell membrane whenever the electrochemical potential gradient for Na^+ across this membrane is reduced, i.e., when a^i_{Na} is increased or extracellular Na^+ activity (a^o_{Na}) or basolateral membrane voltage is decreased. Thus, inhibition of the Na^+ pump by ouabain or increased luminal Na^+ concentration are examples for a condition in which first a^i_{Na}, then a^i_{Ca} are increased and the luminal rate of Na^+ entry subsequently reduced. Cytosolic Ca^{2+} levels may, in addition, govern the rate of luminal K^+ conductance[248] so that rates of luminal Na^+ entry not only lead to negative feedback inhibition of luminal Na^+ conductance but also to increased K^+ secretion by enhancement of the K^+ conductance of this membrane. Schultz[249] has proposed that a^i_{Ca} may also activate a K^+ conductance in the basolateral membrane such that the rate of passive K^+ leakage out of the cell across the peritubular membrane is kept in step with the rate of K^+ uptake by the Na^+,K^+-ATPase within the same plasma membrane.

The Na^+,K^+ pump plays a similar role in net secretion of K^+ by distal tubular epithelium as it does in net reabsorption of Na^+ in this nephron segment. Although direct evidence for active K^+ uptake via the Na^+,K^+ pump across the basolateral cell membrane of mammalian DCT cells is still missing, some evidence in support of this view has been obtained in isolated cortical collecting tubules of the rabbit.[250] Addition of ouabain caused initially a rapid decrease in voltage across the basolateral membrane, followed by a more gradual and sustained depolarization across this boundary. This result is best explained by inhibition of the Na^+,K^+ pump during the initial rapid depolarization and a progressive dissipation of the K^+ gradient across the basolateral membrane during the later phase. The latter interpretation rests on assuming a high K^+ conductance of the peritubular membrane which permits some back-leakage of K^+ out of cell into the contraluminal space under control conditions. This back-leakage contributes as a K^+ diffusion potential to the total magnitude of the basolateral voltage.

Active K^+ uptake by the Na^+,K^+ pump provides the cytosolic K^+ activity necessary to drive K^+ across the luminal cell membrane whereby net secretion of this ion is brought about.[241,250] This view is supported by kinetic studies[122] of unidirectional K^+ fluxes that led de Mello-Aires *et al.* to propose a model of distal tubular K^+ transfer which provides some insight into the mechanisms of regulation. Figure 13 is a summary of the kinetic parameters of distal tubular K^+ transport in control, potassium-loaded, potassium-deprived, and alkalotic animals. As compared to control conditions, the rate coefficients of luminal K^+ exchange (k_{21} and k_{12}) were not significantly affected, whereas the flux from cell to lumen was diminished in low-potassium animals and stimulated in high-potassium and alkalotic animals. Stimulation of tubular K^+ secretion was associated with an increase of S_2, the amount of K^+ labeled within cells. This transport pool increased as a consequence of a rise in the flux component of K^+ uptake from the peritubular environment. On the other hand, during reduced net secretion in low-potassium animals, the uptake of K^+ across the peritubular cell membrane was depressed and a reduction of the cellular transport pool occurred. Changes in transport pool probably reflect changes in intracellular K^+ activity, since studies with K^+-sensitive microelectrodes[218] as well as peritubular potential measurements[234] support the view that cellular K^+ concentrations in distal tubules are elevated in potassium-loaded and alkalotic animals but reduced in potassium-depleted and acidotic animals. Other studies provide evidence that the activity of the Na^+,K^+ pump is also stimulated by adrenal steroids.[251,252]

The luminal membrane serves as the preferential exit route for K^+ and thus allows net secretion of this ion to occur.[250] In the mammal, there are presently no data available on the magnitude of the K^+ permeability of luminal as compared to contraluminal cell membranes. However, considering the much smaller membrane voltage (cell negative) at the luminal than the peritubular cell border in the late DCT (some −40 mV vs. −70 mV), it is evident that in terms of electrical driving force, K^+ diffusion into the tubule lumen rather than into the peritubular fluid is favored. In the cortical collecting tubule, the magnitude of the luminal voltage is determined by the relative contribution of a lumen-directed K^+ diffusion potential, opposed and hence reduced by a cell-directed Na^+ diffusion potential.[241,250] Amiloride-sensitive Na^+ channels and Ba^{2+}-sensitive K^+

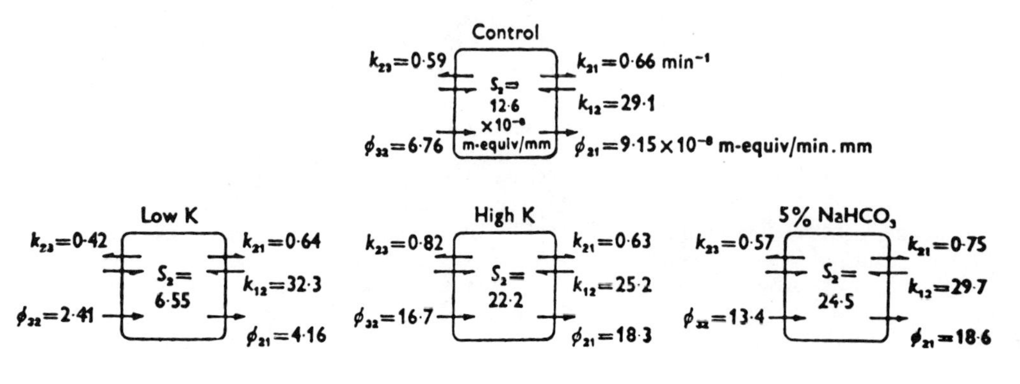

Fig. 13. Summary of main kinetic parameters of distal tubular K^+ transfer in control, potassium-loaded, alkalotic, and potassium-deprived animals. From de Mello-Aires *et al.*[122]

channels are the molecular basis of these diffusion potentials.[250] In fact, using patch-clamp techniques a Ca^{2+}-activated, Ba^{2+}-sensitive K^+ channel could be demonstrated directly in the luminal membrane of rabbit collecting tubules.[248]

Electrodiffusion of K^+ across the luminal cell membrane, however, may not be the only mode by which K^+ crosses from cell to lumen during net secretion. In studies on rats, electroneutral cotransport of KCl as a secretory pathway in K^+ secretion has recently been proposed to complement the channel-mediated diffusive pathway.[237]

There is indirect evidence for the existence of an active reabsorptive K^+ pump in the luminal membrane.[4,54,253,254] In potassium depletion, distal tubular secretion of K^+ disappears despite the existence of a lumen-negative transepithelial voltage,[4,84] and luminal steady-state K^+ concentrations are usually,[4] although not in all studies,[255] below the level predicted for electrochemical equilibrium. Also, administration of ouabain to rats[84,256] causes an increase in the luminal K^+ concentration, an effect that cannot be explained by inhibition of the peritubular uptake mechanism; rather it is likely that ouabain inhibits a luminal K^+ reabsorptive pump.

The fact that active transepithelial net reabsorption of K^+ has never been demonstrated in free-flow[4] or tubule perfusion[108] experiments does not preclude the existence of active K^+ transport from lumen to cytosol, since opposing K^+ pumps (basolateral and luminal) might cancel each other in terms of transepithelial net movement. Active reabsorptive K^+ transport is thought to occur in potassium depletion in the apical membrane of I cells of the medullary collecting duct because of the associated marked increase in luminal membrane area which is found exclusively in I cells.[110]

Cl^- reabsorption by the mammalian DCT has been categorized as an active transport process, particularly in its "early" portion.[40] This conclusion was based on the small transepithelial voltages, either lumen-negative[9,41] or slightly lumen-positive,[39,131,213,214] and Cl^- concentrations in tubular fluid significantly less than in plasma.[18] However, it is likely that this transfer mechanism is not primarily active but is ultimately driven by the Na^+,K^+ pump in the peritubular cell membrane. Such secondary active transport of Cl^- can be envisioned if a neutral Na^+–Cl^- cotransporter is located within the luminal cell membrane. The low intracellular Na^+ concentration, produced by the peritubular Na^+,K^+ pump, may then provide the energy for Cl^- entry across the luminal cell membrane against an electrochemical potential gradient. This mode of luminal Cl^- transfer has been suggested by Wright and his collaborators.[132,236,244] It is inhibited by furosemide and chlorothiazide, but not by bumetanide.[236,244] Intracellular Cl^- activity can be raised by this mode of transfer to levels higher than the electrochemical equilibrium across the peritubular cell membrane, allowing passive movement of Cl^- out of the cell into the contraluminal fluid compartment. The mechanism of Cl^- movement across the peritubular membrane may, at least in part, consist of cotransport of KCl.[132]

Whereas much of the current views on Cl^- transport in mammalian DCT remains speculative, there is a large body of pertinent experimental results available on the DCT of *Amphiuma*.[229,240] In the early *Amphiuma* DCT, direct measurements of Cl^- activity and membrane voltages have shown that Cl^- transport across the luminal cell membrane occurs against an electrochemical potential gradient. The intracellular Cl^- activity of 11.5 mM under control conditions is higher than the equilibrium value of 3.1 mM calculated from the measured luminal membrane voltage of 82.2 mV. As expected, when cotransport of Na^+ and Cl^- occurs, cytosolic Cl^- activity decreases (to 3.3 mM) approaching electrochemical equilibrium when furosemide is added to the luminal perfusate.[229] Subsequent studies from the same laboratories[235] have shown that the apical cotransport for Na^+ and Cl^- is actually mediated by a triple-ion carrier for Na^+, 2 Cl^-, and K^+. This conclusion rests on the observations that the luminal entry of Na^+ depends on the presence of K^+ in the perfusate[232a] and that net reabsorption of K^+, Cl^-, and Na^+ is markedly inhibited by luminal application of furosemide. In fact, in control animals the direction of net K^+ transport is reversed from reabsorption to secretion after administration of furosemide or by deletion of Cl^- from the luminal perfusate. These authors[235] concluded that the net K^+ secretion "unmasked" by inhibition of apical Na^+–K^+–Cl^- cotransport involves an electrodiffusive pathway for K^+ within the luminal cell membrane of early distal *Amphiuma* tubules.

The stoichiometry of luminal cotransport of Na^+, K^+, and Cl^- in the early distal tubule of *Amphiuma* is still unknown.[228] In the mammalian cortical TALH, coupling ratios of 1 Na^+ : 1 K^+ : 2 Cl^- have been proposed by Greger *et al.*[257] Ion transport in this mammalian nephron segment exhibits many similarities to those in the early distal *Amphiuma* tubule and it would be tempting to assume that the same electroneutral carrier molecule mediates Na^+, K^+, and Cl^- transport in the luminal membrane of both mammalian loops of Henle and amphibian DCT. This view is favored by Oberleithner *et al.*[258] whereas Sackin *et al.*[259] consider a positively charged cotransport carrier-ion complex (1 Na^+ : 1 K^+ : 1 Cl^-) as the more likely transport mode. The observation that furosemide leads to peritubular hyperpolarization does not help to distinguish between neutral and positively charged carrier transport.[228] In the case of a neutral carrier, the reduction in a_{Cl}^i found after furosemide administration might increase a peritubular Cl^- diffusion potential and thus hyperpolarize the basolateral cell membrane. Electrochemical equilibrium of Cl^- across this cell membrane has indeed been observed in furosemide-treated distal tubules of *Amphiuma*.[229] On the other hand, inhibition of transport via a positively charged carrier would also hyperpolarize the luminal and, secondarily, the peritubular membrane by decreasing the current flowing across these membrane resistances. Further studies are needed to resolve this problem.

There is no direct information regarding the mode of H^+ secretion by the DCT. Transepithelial net movement of H^+ against their electrochemical gradient occurs and therefore qualifies as active transport.[7] However, it is not known whether a primary active transport process, driven directly by ATP, is responsible for net secretion. If an ATP-driven H^+ pump can be demonstrated, the question of specificity of cell types remains to be answered.

Finally, there is some evidence that the transepithelial ion permeability of the DCT is influenced by the magnitude of the prevailing osmotic pressure difference between luminal and peritubular fluid.[41] The effect of changes in osmolality on the properties of other epithelia is well known,[260–262] but the functional significance of this effect is not understood. In the distal nephron, however, it appears likely that the changes in osmolality of tubular fluid, which normally occur along the distal convolution, participate in an intraepithelial feedback mechanism, influencing both the efficiency of net solute transport in the diluting segment and the osmotic equilibration which occurs in the late DCT and collecting ducts in the presence of ADH.

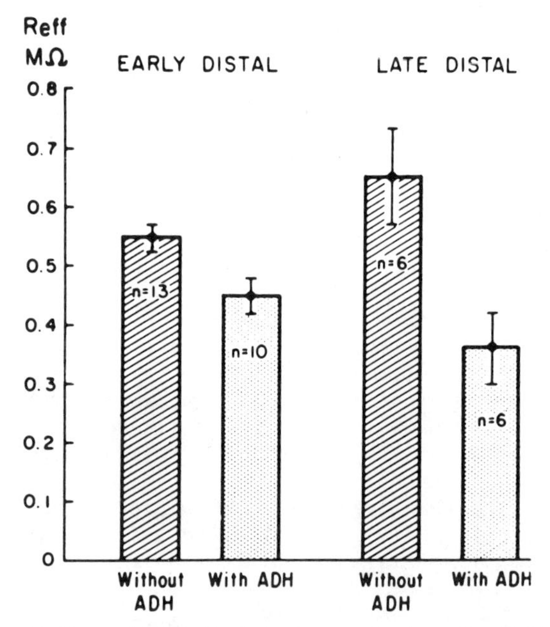

Fig. 14. Transepithelial effective resistence of early and late DCT in diabetes insipidus rats before and after administration of vasopressin. From de Bermudez and Windhager.(41)

Evidence in favor of such a hypothesis consists of the observation that raising luminal osmolality decreases the transepithelial electrical resistance, whereas increasing the osmolality of the peritubular fluid results in a marked elevation of resistance.(41) Similar conclusions have been reached simultaneously and independently by Schafer *et al.*(263,264) in collecting ducts. Osmotic gradients, rather than absolute osmolalities, determine the change in transepithelial conductance.(41)

Therefore, the finding that the electrical resistance across the tubular wall decreases as a function of length along the distal tubule in antidiuretic rats(41) can be explained in terms of this osmotic effect. The decline in electrical resistance along the DCT seen in free flow was predicted from experiments in which the tubular lumen was perfused with solutions of known osmolalities. This hypothesis is also supported by the observation that the decline in transepithelial resistance in the late DCT does not occur in rats with congenital diabetes insipidus (Fig. 14). Measurements of changes in transepithelial PD in the DCT during alterations in the osmotic environment suggest that the change in ion permeability involves intercellular, rather than transcellular, ion pathways.(41)

6. Summary

The location of the DCT on the kidney surface provides a unique opportunity to study the renal diluting and concentrating mechanism in a single nephron segment *in vivo*. The early distal tubule is similar in structure to the ascending limb of Henle's loop, and is a site at which NaCl is reabsorbed without an osmotically proportionate amount of water. The late distal tubule resembles collecting tubules and is the site at which osmotic equilibration occurs in the presence of ADH. The hydraulic conductance of the late distal tubule is under the influence of ADH, although the location along the nephron where this sensitivity appears is quite variable.

The distal tubular epithelium can maintain steep concentration gradients for Na^+, K^+, Cl^-, H^+, and Ca^{2+}. Na^+ reabsorption occurs against an electrochemical gradient, and depends on the Na^+ concentration in tubular fluid. There is a strong relationship between load and reabsorption consistent with the view that late distal reabsorptive sites are normally unsaturated. Mineralocorticoids exert their Na^+-retaining action in the latter part of the DCT. The distal tubule does not contribute to the natriuresis of volume expansion and Starling forces apparently have little influence on net reabsorption. There is evidence that active Na^+ extrusion from cells is an electrogenic process.

The distal tubule is the site of regulation of K^+ excretion. Net reabsorption of K^+ is seen when body K^+ stores are low, whereas net secretion occurs when stores are high. Alkalosis increases and acidosis depresses net secretion; these effects are probably mediated by a pH effect on peritubular uptake of K^+ and on K^+ permeability of the luminal cell membrane. Aldosterone increases peritubular uptake and net secretion. At low K^+ concentrations in tubular fluid, the rate of K^+ secretion is strongly dependent on flow rate, so most diuretic states are associated with kaliuresis. K^+ secretion consists of active peritubular uptake, with passive diffusion from the cell into the lumen down the existing electrochemical gradient. An active reabsorptive mechanism also exists on the luminal membrane. Cl^- is reabsorbed by secondary active transport in the early part of the DCT.

Active H^+ secretion participates in the reabsorption of HCO_3^- and the acidification of the urine. The rate of H^+ secretion depends on the buffer load reaching the distal tubule.

Ca^{2+} is actively reabsorbed by the DCT, but its net reabsorption does not always parallel net Na^+ reabsorption as in other parts of the nephron.

The distal transepithelial PD is not constant: early distal segments have small potentials, whereas the late DCT has large, lumen-negative potentials. The peritubular membrane potential is at least accounted for in part by electrogenic Na^+ transport. Both luminal and peritubular membranes have high K^+ conductances. The transepithelial resistance declines along the DCT, possibly as a result of the increasing osmolality of tubular fluid. Transepithelial osmotic gradients influence the conductance of intracellular ion pathways.

ACKNOWLEDGMENTS. Work in the authors' laboratories was supported by grants from the National Institutes of Health (AM-28358 and AM-11489). L.S.C. is an Established Investigator of the American Heart Association.

References

1. Woodhall, P. B., and C. C. Tisher. 1973. Response of the distal tubule and cortical collecting duct to vasopressin in the rat. *J. Clin. Invest.* **52**:3095–3108.
2. Morgan, T., and R. W. Berliner. 1969. A study by continuous microperfusion of water and electrolyte movements in the loop of Henle and distal tubule of the rat. *Nephron* **6**:388–405.
3. Gottschalk, C. W., and M. Mylle. 1957. Micropuncture study of pressures in proximal and distal tubules and peritubular capillaries of the rat kidney during osmotic diuresis. *Am. J. Physiol.* **189**:323–328.
4. Malnic, G., R. M. Klose, and G. Giebisch. 1966. Micropuncture study of distal tubular potassium and sodium transport in rat nephron. *Am. J. Physiol.* **211**:529–547.
5. Kashgarian, M. H., H. Stockle, C. W. Gottschalk, and K. J.

Ullrich. 1963. Transtubular electrochemical potentials of sodium and chloride in proximal and distal renal tubules of rats during antidiuresis and water diuresis (diabetes insipidus). *Pfluegers Arch.* **277**:89–106.
6. Malnic, G., R. M. Klose, and G. Giebisch. 1964. Micropuncture study of renal potassium excretion in the rat. *Am. J. Physiol.* **206**:674–686.
7. Malnic, G., M. de Mello-Aires, and G. Giebisch. 1972. Micropuncture study of renal tubular hydrogen ion transport in the rat. *Am. J. Physiol.* **222**:147–158.
8. Costanzo, L. S., and E. E. Windhager. 1978. Calcium and sodium transport by the distal convoluted tubule. *Am. J. Physiol.* **235**:F492–F506.
9. Wright, F. S. 1971. Increasing magnitude of electrical potential along the renal distal tubule. *Am. J. Physiol.* **220**:624–638.
10. Farquhar, M. G., and G. E. Palade. 1963. Functional complexes in various epithelia. *J. Cell Biol.* **17**:375–412.
11. Tisher, C. C., and W. E. Yarger. 1973. Lanthanum permeability of the tight junction (zonula occludens) in the renal tubule of the rat. *Kidney Int.* **3**:238–250.
12. Martinez-Palomo, A., and D. Erlij. 1973. The distribution of lanthanum in different tubular segments of the rat kidney. *Pfluegers Arch.* **343**:267–272.
13. Erlij, D. 1976. Solute transport across isolated epithelia. *Kidney Int.* **9**:76–87.
14. Walker, A., P. Bott, J. Oliver, and M. MacDowell. 1941. The collection and analysis of fluid from single nephrons of the mammalian kidney. *Am. J. Physiol.* **134**:580–595.
15. Hierholzer, K., M. Wiederholt, H. Holzgreve, G. Giebisch, R. M. Klose, and E. E. Windhager. 1965. Micropuncture study of renal transtubular concentration gradients of sodium and potassium in adrenalectomized rats. *Pfluegers Arch.* **285**:193–210.
16. Hierholzer, K., M. Wiederholt, and H. Stolte. 1966. Hemmung der Natriumresorption in proximalen und distalen konvolut andrenalektomierter Ratten. *Pfluegers Arch.* **291**:43–63.
17. Hierholzer, K., and H. Stolte. 1969. The proximal and distal tubular action of adrenal steroids on sodium reabsorption. *Nephron* **6**:188–204.
18. Gottschalk, C. W. 1961. Micropuncture studies of tubular function in mammalian kidney. *Physiologist* **4**:35.
19. Shareghi, G. R., and L. C. Stoner. 1978. Calcium transport across segments of the rabbit distal nephron *in vitro*. *Am. J. Physiol.* **235**:F367–F375.
20. Costanzo, L. S., and E. E. Windhager. 1980. Effects of PTH, ADH and cyclic AMP on distal tubular Ca and Na reabsorption. *Am. J. Physiol.* **239**:F478–F485.
21. Schweigger-Seidel, F. 1865. *Die Nieren des Menschen und der Saugetiere in inhrem feineren Baue.* Halle.
22. Kaissling, B., and W. Kriz. 1979. Structural analysis of the rabbit kidney. *Adv. Anat. Embryol. Cell. Biol.* **56**:1–123.
23. Kaissling, B. 1982. Structural aspects of adaptive changes in renal electrolyte excretion. *Am. J. Physiol.* **243**:F211–F226.
24. Oliver, J. 1968. *Nephrons and Kidneys: A Quantitative Study of Developmental and Evolutionary Mammalian Renal Architecture.* Harper and Row, New York.
25. Crayen, M., and W. Thoenes. 1975. Architektur und Cytologischer Aufbau des distalen Tubulus in der Rattenniere. *Fortschr. Zool.* **23**:279–288.
26. Kriz, W., B. Kaissling, and M. Psczolla. 1978. Morphological characterization of the cells in Henle's loop and the distal tubule. In: *New Aspects of Renal Function.* H. G. Vogel and K. J. Ullrich, eds. Excerpta Medica, Amsterdam. pp. 67–78.
27. Tisher, C. C., R. E. Bulger, and B. F. Trump. 1968. Human renal ultrastructure. III. The distal tubule in healthy individuals. *Lab. Invest.* **18**:655–668.
28. Kaissling, B., and M. LeHir. 1982. Distal tubular segments of the rabbit kidney after adaptation to altered Na and K intake. I. Structural changes. *Cell Tissue Res.* **224**:469–492.
29. Schiller, A., R. Taugner, and B. Roesinger. 1978. Vergleichende Morphologie der zonulae occludentes am Nierentubulus. *Verh. Anat. Ges.* **72**:229–234.
30. Schmidt, U., and U. C. DuBach. 1971. Na-K-ATPase in the rat nephron related to sodium transport: Results with quantitative histochemistry. In: *Recent Advances in Quantitative Histochemistry and Cytochemistry.* Huber, Bern. pp. 320–344.
31. Schmidt, U., and M. Horster. 1977. Na-K-activated ATPase: Maturation in rabbit nephron segments. *Am. J. Physiol.* **233**:F55–F60.
32. Ganote, C. E., J. J. Grantham, H. L. Moses, M. B. Burg, and J. Orloff. 1968. Ultrastructural studies of vasopressin effect on isolated perfused renal collecting tubules of the rabbit. *J. Cell Biol.* **36**:355–367.
33. Tisher, C. C., R. E. Bulger, and H. Valtin. 1971. Morphology of renal medulla in water diuresis and vasopressin-induced anti-diuresis. *Am. J. Physiol.* **220**:87–94.
34. Hagege, J., M. Gage, and G. Richet. 1974. Scanning of the apical pole of distal tubular cells under differing acid–base conditions. *Kidney Int.* **5**:137–146.
35. Costanzo, L. S. 1984. Comparison of Ca and Na transport in early and late rat distal tubules: Amiloride effect. *Am. J. Physiol.* **246**:F937–F945.
36. Morel, F. 1981. Sites of hormone action in the mammalian nephron. *Am. J. Physiol.* **240**:F159–F164.
37. Giebisch, G., and E. E. Windhager. 1964. Renal tubular transfer of sodium chloride and potassium. *Am. J. Med.* **36**:643–669.
38. Khuri, R. N., M. Wiederholt, N. Strieder, and G. Giebisch. 1975. Effects of graded solute diuresis on renal tubular sodium transport in the rat. *Am. J. Physiol.* **228**:1262–1268.
39. Barratt, L. J., F. C. Rector, Jr., J. P. Kokko, C. C. Tisher, and D. W. Seldin. 1975. Transepithelial potential difference profile of the distal tubule of the rat kidney. *Kidney Int.* **8**:368–375.
40. Malnic, G., and G. Giebisch. 1972. Some electrical properties of distal tubular epithelium in the rat. *Am. J. Physiol.* **223**:797–808.
41. de Bermudez, L., and E. E. Windhager. 1975. Osmotically induced changes in electrical resistance of distal tubules of rat kidney. *Am. J. Physiol.* **229**:1536–1546.
42. Malnic, G., R. M. Klose, and G. Giebisch. 1966. Microperfusion study of distal tubular potassium and sodium transfer in rat kidney. *Am. J. Physiol.* **211**:548–559.
43. Burg, M. 1976. Tubular chloride transport and the mode of action of some diuretics. *Kidney Int.* **9**:189–197.
44. Schlatter, E., R. Gregor, and C. Weidtke. 1983. Effect of "high ceiling" diuretics on active salt transport in the cortical thick ascending limb of Henle's loop of rabbit kidney. *Pfluegers Arch.* **396**:210–217.
45. Good, D. W., and F. S. Wright. 1979. Luminal influences on potassium secretion: Sodium concentration and fluid flow rate. *Am. J. Physiol.* **236**:F192–F205.
46. Lassiter, W. E., C. W. Gottschalk, and M. Mylle. 1961. Micropuncture study of net transtubular movement of water and urea in nondiuretic mammalian kidney. *Am. J. Physiol.* **200**:1139–1146.
47. Lassiter, W. E., C. W. Gottschalk, and M. Mylle. 1964. Micropuncture study of net transtubular movement of water and urea in rat kidney during saline diuresis. *Am. J. Physiol.* **206**:669–673.
48. Danielson, R. A., B. Schmidt-Nielsen, and C. Hohberger. 1970. Micropuncture study of the regulation of urea excretion by the collecting ducts in rats on high and low protein diets. In: *Urea and the Kidney.* B. Schmidt-Nielsen, ed. Excerpta Medica, Amsterdam. pp. 375–384.
49. Capek, K., R. Rumrich, and K. J. Ullrich. 1966. Harnstoffpermeabilitat der corticalen Tubulusabschnitte von Ratten in Antidiurese and Wasserdiurese. *Pfluegers Arch.* **290**:237–249.
50. Diezi, J., P. Michoud, A. Grandchamp, and G. Giebisch. 1976. Effects of nephrectomy on renal salt and water transport in the remaining kidney. *Kidney Int.* **10**:450–462.
51. Knox, F. G., and J. Gasser. 1974. Altered distal sodium reabsorption in volume expansion. *Mayo Clin. Proc.* **49**:775–781.
52. Stein, J. H., and H. J. Reineck. 1974. Role of the collecting duct in the regulation of excretion of sodium and other electrolytes. *Kidney Int.* **6**:1–9.
53. Kunau, R. 1972. Changes in Na^+ reabsorption in the loop of

Henle and distal convolution of the rat nephron following minimal and marked increases in Na delivery. *Clin. Res.* **20**:762.

54. Giebisch, G., and E. E. Windhager. 1973. Electrolyte transport across renal tubular membranes. In: *Handbook of Physiology*, Section 8. J. Orloff and R. W. Berliner, eds. American Physiological Society, Washington, D.C. pp. 315–376.
55. Windhager, E. E., and G. Giebisch. 1976. Proximal sodium and fluid transport. *Kidney Int.* **9**:121–133.
56. Stein, J. H., R. W. Osgood, S. Boonjarern, and T. F. Ferris. 1973. A comparison of the segmental analysis of sodium reabsorption during Ringer's and hyperoncotic albumin infusion in the rat. *J. Clin. Invest.* **52**:2313–2323.
57. Sonnenberg, H. 1973. Proximal and distal tubular function in salt-deprived and salt-loaded deoxycorticosterone acetate-escape rats. *J. Clin. Invest.* **52**:263–272.
58. Osgood, R. W., H. J. Reineck, and J. H. Stein. 1978. Further studies on segmented sodium transport in the rat kidney during expansion of the extracellular fluid volume. *J. Clin. Invest.* **62**:311–320.
59. Camargo, M. J. F., H. D. Kleinert, S. A. Atlas, J. E. Sealey, J. H. Laragh, and T. Maack. 1984. Ca-dependent hemodynamic and natriuretic effects of atrial extract in isolated rat kidney. *Am. J. Physiol.* **246**:F447–F456.
60. Sonnenberg, H., W. A. Copples, A. J. DeBold, and A. T. Veress. 1982. Intrarenal localization of the natriuretic effect of cardiac atrial extract. *Can. J. Physiol. Pharmacol.* **60**:1149–1152.
61. Briggs, J. P., B. Steipe, G. Schubert, and J. Schnermann. 1982. Micropuncture studies of the renal effects of atrial natriuretic substance. *Pfluegers Arch.* **395**:271–276.
62. Anagnostopoulos, T., M. U. Kinney, and E. E. Windhager. 1971. Salt and water reabsorption by short loops of Henle during renal vein constriction. *Am. J. Physiol.* **220**:1060–1066.
63. Landwehr, D. M., R. M. Klose, and G. Giebisch. 1967. Renal tubular sodium and water reabsorption in the isotonic sodium chloride-loaded rat. *Am. J. Physiol.* **212**:1327–1333.
64. Schwartz, G. J., and M. B. Burg. 1978. Mineralocorticoid effects on cation transport by cortical collecting tubules *in vitro*. *Am. J. Physiol.* **235**:F576–F585.
65. Gross, J. B., M. Imai, and J. P. Kokko. 1975. A functional comparison of the cortical collecting tubule and the distal convoluted tubule. *J. Clin. Invest.* **55**:1284–1294.
66. Wiederholt, M. 1966. Mikropunktionsuntersuchungen am proximalen und distalen Konvolut der Rattenniere uber den Einfluss von Actinomycin D auf den mineralocorticoidabhangigen Na-Transport. *Pfluegers Arch.* **292**:334–342.
67. Horisberger, J. D., and J. Diezi. 1983. Effects of mineralocorticoids on Na and K excretion in the adrenalectomized rat. *Am. J. Physiol.* **245**:F89–F99.
68. Geheb, M., E. Hercker, I. Singer, and M. Cox. 1981. Subcellular localization of aldosterone-induced proteins in toad urinary bladders. *Biochim. Biophys. Acta* **641**:422–426.
69. Geheb, M., G. Huber, E. Hercker, and M. Cox. 1981. Aldosterone-induced proteins in toad urinary bladders. *J. Biol. Chem.* **256**:11716–11723.
70. Geheb, M., R. Alvis, E. Hercker, and M. Cox. 1983. Mineralocorticoid-specificity of aldosterone-induced protein synthesis in giant toad (*Bufo marinus*) urinary bladders. *Biochem. J.* **214**:29–35.
71. Petty, K. J., J. P. Kokko, and D. Marver. 1981. Secondary effect of aldosterone on Na-K ATPase activity in the rabbit cortical collecting tubule. *J. Clin. Invest.* **68**:1514–1521.
72. Perkins, F. M., and J. S. Handler. 1981. Transport properties of toad kidney epithelia in culture. *Am. J. Physiol.* **241**:C154–C159.
73. Palmer, L. G., J. H. Y. Li, B. Lindemann, and I. S. Edelman. 1981. Aldosterone control of the density of sodium channels in the toad urinary bladder. *J. Membr. Biol.* **69**:91–102.
74. Garty, H., and I. S. Edelman. 1983. Amiloride-sensitive trypsinization of apical sodium channels. *J. Gen. Physiol.* **81**:785–803.
75. Sariban-Sohraby, S., M. B. Burg, and R. J. Turner. 1983. Apical sodium uptake in toad kidney epithelial cell line A6. *Am. J. Physiol.* **245**:C167–C171.
76. Schmidt, U., J. Schmid, H. Schmid, and U. C. Dubach. 1975. Sodium and potassium-activated ATPase: A possible target of aldosterone. *J. Clin. Invest.* **55**:655–660.
77. Horster, M., H. Schmid, and U. Schmidt. 1980. Aldosterone *in vitro* restores nephron Na-K ATPase of distal segments from adrenalectomized rabbits. *Pfluegers Arch.* **384**:203–206.
78. Garg, L. C., M. A. Knepper, and M. B. Burg. 1981. Mineralocorticoid effects on Na-K ATPase in individual nephron segments. *Am. J. Physiol.* **240**:F536–F544.
79. Doucet, A., and A. I. Katz. 1981. Mineralocorticoid receptors along the nephron: ^3H aldosterone binding in rabbit tubules. *Am. J. Physiol.* **241**:F605–F611.
80. Le Hir, M., B. Kaissling, and U. C. Dubach. 1982. Distal tubular segments of the rabbit kidney after adaptation to altered Na and K intake. II. Changes in Na-K ATPase activity. *Cell Tissue Res.* **224**:493–504.
81. El Mernissi, G., and A. Doucet. 1983. Short-term effects of aldosterone and dexamethasone on Na-K ATPase along the rabbit nephron. *Pfluegers Arch.* **399**:147–151.
82. Costanzo, L. S. 1985. Localization of diuretic action in microperfused rat distal convoluted tubules: Ca and Na transport. *Am. J. Physiol.* in press
83. Kunau, R. T., D. R. Weller, and H. L. Webb. 1975. Clarification of the site of action of chlorothiazide in the rat nephron. *J. Clin. Invest.* **56**:401–407.
84. Duarte, C. G., F. Chomety, and G. Giebisch. 1971. Effect of amiloride, ouabain, and furosemide on distal tubular function in the rat. *Am. J. Physiol.* **221**:632–639.
85. Gottschalk, C. W. 1962–1963. Renal tubular function: Lessons from micropuncture. *Harvey Lect.* **58**:99–123.
86. Windhager, E. E., and G. Giebisch. 1961. Micropuncture study of renal tubular transfer of sodium chloride in the rat. *Am. J. Physiol.* **200**:581–590.
87. Malnic, G., M. de Mello-Aires, and F. Viera. 1970. Chloride excretion in single nephrons of rat kidney during alterations of acid–base equilibrium. *Am. J. Physiol.* **218**:20–26.
88. Rector, F. C., and J. R. Clapp. 1962. Evidence for active chloride reabsorption in the distal renal tubule of the rat. *J. Clin. Invest.* **41**:101–107.
89. Berliner, R. W. 1961. Renal mechanisms for potassium secretion. *Harvey Lect.* **55**:141–171.
90. Wright, F. S., and G. Giebisch. 1978. Renal potassium transport: Contributions of individual nephron segments and populations. *Am. J. Physiol.* **235**:F515–F527.
91. Khuri, R. N., M. Wiederholt, N. Strieder, and G. Giebisch. 1975. The effect of flow rate and potassium intake on distal tubular potassium transfer. *Am. J. Physiol.* **228**:1249–1261.
92. Brenner, B. M., and R. W. Berliner. 1969. Relationship between extracellular volume and fluid reabsorption by the rat kidney. *Am. J. Physiol.* **217**:6–12.
93. Cortney, M. D., M. Mylle, W. E. Lassiter, and C. W. Gottschalk. 1965. Renal transport of water, solute and PAH in rats loaded with isotonic saline. *Am. J. Physiol.* **209**:1199–1205.
94. Morgan, T., M. Tadokoro, D. Margin, and R. W. Berliner. 1970. Effect of furosemide on Na^+ and K^+ transport studied by microperfusion of the rat nephron. *Am. J. Physiol.* **218**:292–297.
95. Kunau, R. T., Jr., H. L. Webb, and S. C. Borman. 1974. Characteristics of the relationship between the flow rate of tubular fluid and potassium transport in the distal tubule of the rat. *J. Clin. Invest.* **54**:1488–1495.
96. Stokes, J. B. 1982. Na and K transport across the cortical and outer medullary collecting tubule of the rabbit: Evidence for diffusion across the outer medullary portion. *Am. J. Physiol.* **242**:F514–F520.
97. Watson, J. F. 1966. Potassium reabsorption in the proximal tubule of the dog nephron. *J. Clin. Invest.* **45**:1341–1348.

98. Bennett, C. M., J. R. Clapp, and R. W. Berliner. 1967. Micropuncture study of the proximal and distal tubule in the dog. *Am. J. Physiol.* **213**:1254–1262.
99. Bennett, C. M., B. M. Brenner, and R. W. Berliner. 1968. Micropuncture study of nephron function in the Rhesus monkey. *J. Clin. Invest.* **47**:203–216.
100. Grantham, J. J., M. B. Burg, and J. Orloff. 1970. The nature of transtubular Na and K transport in isolated rabbit renal collecting tubules. *J. Clin. Invest.* **49**:1815–1826.
101. Stokes, J. B. 1981. Potassium secretion by the cortical collecting tubule: Relation to sodium absorption, luminal sodium concentration and transepithelial voltage. *Am. J. Physiol.* **241**:F395–F402.
102. Giebisch, G. 1971. Renal potassium excretion. In: *The Kidney: Morphology, Biochemistry, Physiology,* Volume 3. C. Rouiller and A. Muller, eds. Academic Press, New York. pp. 329–382.
103. Wright, F. S., N. Strieder, N. B. Fowler, and G. Giebisch. 1971. Potassium secretion by the distal tubule after potassium adaptation. *Am. J. Physiol.* **221**:437–448.
104. Bank, N., and H. S. Aynedjian. 1973. A micropuncture study of potassium excretion by the remnant kidney. *J. Clin. Invest.* **52**:1480–1490.
105. Finklestein, F. O., and J. P. Hayslett. 1974. Role of medullary structures in the functional adaptation of renal insufficiency. *Kidney Int.* **6**:419–425.
106. Orloff, J., and D. G. Davidson. 1959. The mechanism of potassium excretion in the chicken. *J. Clin. Invest.* **38**:21–30.
107. Stanton, B. A., and G. H. Giebisch. 1982. Potassium transport by the renal distal tubule: Effects of potassium loading. *Am. J. Physiol.* **243**:F487–F493.
108. Stanton, B. A., D. Biemesderfer, J. B. Wade, and G. Giebisch. 1981. Structural and functional study of the rat distal nephron: Effects of potassium adaptation and depletion. *Kidney Int.* **19**:36–48.
109. Field, M. J., R. W. Berliner, and G. H. Giebisch. 1985. Regulation of renal potassium metabolism. In: *Clinical Disorders of Fluid and Electrolyte Metabolism.* 4th ed. M. M. Maxwell, C. R. Kleeman, and R. G. Narens, eds. McGraw-Hill, New York, in press.
110. Stetson, D. L., J. B. Wade, and G. Giebisch. 1980. Morphologic alterations in the rat medullary collecting duct following potassium depletion. *Kidney Int.* **17**:45–56.
111. Cortney, M. A. 1969. Renal tubular transfer of water and electrolytes in adrenalectomized rats. *Am. J. Physiol.* **216**:589–598.
112. Wiederholt, M., C. Behn, W. Schoormans, and L. Hansen. 1972. Effect of aldosterone on sodium potassium transport in the kidney. *J. Steroid Biochem.* **3**:151–159.
113. Wiederholt, M., W. Schoormans, F. Fischer, and C. Behn. 1973. Mechanism of action of aldosterone on potassium transfer in the rat kidney. *Pfluegers Arch.* **345**:159–178.
114. Wiederholt, M., S. K. Aguilian, and R. N. Khuri. 1974. Intracellular potassium in the distal tubule of the adrenalectomized and aldosterone treated rat. *Pfluegers Arch.* **347**:117–123.
115. Fimognari, G. M., D. D. Fanestil, and I. S. Edelman. 1967. Induction of RNA and protein synthesis in the action of aldosterone. *Am. J. Physiol.* **213**:954–962.
116. Horisberger, J. D., and J. Diezi. 1984. Inhibition of aldosterone-induced anti-natriuresis and kaliuresis by actinomycin D. *Am. J. Physiol.* **246**:F201–F204.
117. Malnic, G., M. de Mello-Aires, and G. Giebisch. 1971. Potassium transport across renal distal tubules during acid–base disturbances. *Am. J. Physiol.* **221**:1192–1208.
118. Scott, D., and G. H. McIntosh. 1975. Changes in blood composition and in urinary mineral acid excretion in the pig in response to acute acid–base disturbances. *Q. J. Exp. Physiol.* **60**:131–140.
119. Rostand, S., and J. Watkins. 1977. Response of the isolated rat kidney to metabolic and respiratory acidosis. *Am. J. Physiol.* **233**: F82–F88.
120. Giebisch, G., G. Malnic, and R. W. Berliner. 1981. Renal transport and control of potassium excretion. In: *The Kidney,* 2nd ed. B. M. Brenner and F. C. Rector, eds. pp. 408–439.
121. Stanton, B. A., and G. Giebisch. 1982. Effects of pH on potassium transport by renal distal tubule. *Am. J. Physiol.* **242**: F544–F551.
122. de Mello-Aires, M., G. Giebisch, and G. Malnic. 1973. Kinetics of potassium transport across single distal tubules of rat kidney. *J. Physiol. (London)* **232**:47–70.
123. Peterson, L. N., and F. S. Wright. 1977. Effect of sodium intake on renal potassium excretion. *Am. J. Physiol.* **233**:F225–F234.
124. Dirks, J. H., and J. F. Seely. 1970. Effect of saline infusions and furosemide on the dog distal nephron. *Am. J. Physiol.* **219**:114–121.
125. Seely, J. F., and J. H. Dirks. 1969. Micropuncture study of hypertonic mannitol diuresis in the proximal and distal tubule of the dog kidney. *J. Clin. Invest.* **48**:2330–2340.
126. McDougal, W. S., and F. S. Wright. 1972. Defect in proximal and distal sodium transport in post-obstructive diuresis. *Kidney Int.* **2**:304–317.
127. Peters, G. 1963. Compensatory adaptation of renal functions in the unanesthetized rat. *Am. J. Physiol.* **205**:1042–1048.
128. Reineck, H. J., R. W. Osgood, T. F. Ferris, and J. H. Stein. 1975. Potassium transport in the distal tubule and collecting duct of the rat. *Am. J. Physiol.* **229**:1403–1409.
129. Good, D. W., H. Velazquez, and F. S. Wright. 1984. Luminal influences on potassium secretion: Low sodium concentration. *Am. J. Physiol.* **246**:F609–F619.
130. Field, M. J., B. A. Stanton, and G. H. Giebisch. 1984. Influence of ADH on renal potassium handling: A micropuncture and microperfusion study. *Kidney Int.* **25**:502–511.
131. Barnatt, L. J. 1976. The effect of amiloride on the transepithelial potential difference of the distal tubule of the rat kidney. *Pfluegers Arch.* **361**:251–254.
132. Velazquez, H., F. S. Wright, and D. W. Good. 1982. Luminal influences on potassium secretion: Chloride replacement with sulfate. *Am. J. Physiol.* **242**:F46–F55.
132a. Garcia-Filho, E., G. Malnic, and G. Giebisch. 1982. Effects of changes in electrical potential difference on tubular potassium transport. *Am. J. Physiol.* **238**:F235–F246.
133. Ullrich, K. J., G. Rumrich, and G. Fuchs. 1964. Wasserpermeabilitat und transtubularer Wasserfluss cortikaler Nephronabschnitte bei verschiedenen Diuresezustanden. *Pfluegers Arch.* **280**:99–119.
134. Ullrich, K. J. 1973. Permeability characteristics of the mammalian nephron. In: *Handbook of Physiology,* Section 8. J. Orloff and R. W. Berliner, eds. American Physiological Society, Washington, D.C. pp. 377–398.
135. Stolte, H., J. P. Brecht, M. Wiederholt, and K. Hierholzer. 1968. Einfluss von Adrenalektomie und Glucocorticoiden auf die Wasserpermeabilitat cortikaler Nephronabschnitte der Rattenniere. *Pfluegers Arch.* **299**:99–127.
136. Lassiter, W. E., A. Frick, G. Rumrich, and K. J. Ullrich. 1965. Influence of ionic calcium on the water permeability of proximal and distal tubules in the rat kidney. *Pflugers Arch.* **285**:90–95.
137. Wirz, H. 1956. Der osmotische Druck in den corticalen Tubuli der Rattenniere. *Helv. Physiol. Pharmacol. Acta* **14**:353–362.
138. Gottschalk, C. W., and M. Mylle. 1959. Micropuncture study of the mammalian urinary concentrating mechanism: Evidence for the countercurrent hypothesis. *Am. J. Physiol.* **196**:927–936.
139. Colindres, R. E., R. Kramp, M. E. Allison, and C. W. Gottschalk. 1977. Hydrodynamic alterations during distal tubular fluid collections in the rat kidney. *Am. J. Physiol.* **232**:F497–F505.
140. Clapp, J. R., and R. R. Robinson. 1966. Osmolality of distal tubular fluid in the dog. *J. Clin. Invest.* **45**:1847–1853.
141. Imbert, M., D. Chabardes, M. Montegut, A. Clique, and F. Morel. 1975. Vasopressin dependent adenylate cyclase in single segments of rabbit kidney tubule. *Pfluegers Arch.* **357**:173–186.
142. Imbert-Teboul, M., D. Chabardes, M. Montegut, A. Clique, and

F. Morel. 1978. Vasopressin-dependent adenylate cyclase activities in the rat kidney medulla: Evidence for two separate sites of action. *Endocrinology* **102**:1254–1261.
143. Chabardes, D., M. Imbert-Teboul, M. Gagnon-Brunette, and F. Morel. 1978. Different hormonal target sites along the mouse and rabbit nephrons. In: *Biochemical Nephrology*. W. G. Guder and U. Schmidt, eds. Huber, Bern. pp. 447–454.
144. Chabardes, D., M. Gagnon-Brunette, M. Imbert-Teboul, O. Gontcharevskaia, M. Montegut, A. Clique, and F. Morel. 1980. Adenylate cyclase responsiveness to hormones in various portions of the human nephron. *J. Clin. Invest.* **65**:439–448.
145. Kachadorian, W. A., J. B. Wade, and V. A. DiScala. 1975. Vasopressin: Induced structural change in toad bladder luminal membrane. *Science* **190**:67–69.
146. Kachadorian, W. A., S. D. Levine, J. B. Wade, V. A. DiScala, and R. M. Hays. 1977. Relationship of aggregated intramembranous particles to water permeability in vasopressin-treated toad urinary bladder. *J. Clin. Invest.* **59**:576–581.
147. Harmanci, M. C., P. Stern, W. A. Kachadorian, H. Valtin, and V. A. DiScala. 1980. Vasopressin and collecting duct intramembranous particle clusters: A dose–response relationship. *Am. J. Physiol.* **239**:F560–F564.
148. Glabman, S., R. M. Klose, and G. Giebisch. 1963. Micropuncture study of ammonia excretion in the rat. *Am. J. Physiol.* **205**:127–132.
149. Hayes, C. P., J. S. Mayson, E. E. Owen, and R. R. Robinson. 1964. A micropuncture evaluation of renal ammonia excretion in the rat. *Am. J. Physiol.* **207**:77–83.
150. Lucci, M. S., L. R. Pucacco, N. W. Carter, and T. D. DuBose. 1982. Evaluation of bicarbonate transport in rat distal tubule: Effects of acid–base status. *Am. J. Physiol.* **243**:F335–F341.
151. Gottschalk, C. W., W. E. Lassiter, and M. Mylle. 1960. Localization of urine acidification in the mammalian kidney. *Am. J. Physiol.* **198**:581–585.
152. Vierra, F. L., and G. Malnic. 1968. Hydrogen ion secretion by rat renal cortical tubules as studied by an antimony microelectrode. *Am. J. Physiol.* **214**:710–718.
153. DuBose, T. D., L. R. Pucacco, M. S. Lucci, and N. W. Carter. 1979. Micropuncture determination of pH, PCO_2 and total CO_2 concentration in accessible structures of the rat renal cortex. *J. Clin. Invest.* **64**:476–482.
154. DuBose, T. D. 1983. Application of the disequilibrium pH method to investigate the mechanism of urinary acidification. *Am. J. Physiol.* **245**:F535–F544.
155. Rector, F. C., N. W. Carter, and D. W. Seldin. 1965. The mechanism of bicarbonate reabsorption in the proximal and distal tubules of the kidney. *J. Clin. Invest.* **44**:278–290.
156. Warnock, D. G., and F. C. Rector. 1981. Renal acidification mechanisms. In: *The Kidney,* 2nd ed. B. M. Brenner and F. C. Rector, eds. Saunders, Philadelphia. pp. 440–494.
157. DuBose, T. D., L. R. Pucacco, and N. W. Carter. 1981. Determination of disequilibrium pH in the rat kidney *in vivo:* Evidence for hydrogen secretion. *Am. J. Physiol.* **240**:F138–F146.
158. Karlmark, B., P. Jaeger, and G. Giebisch. 1983. Luminal buffer transport in rat cortical tubule: Relationship to potassium metabolism. *Am. J. Physiol.* **245**:F584–F592.
159. Lonnerholm, G. 1971. Histochemical demonstration of carbonic anhydrase activity in the rat kidney. *Acta Physiol. Scand.* **81**:433–439.
160. Lonnerholm, G., and Y. Ridderstrale. 1980. Intracellular distribution of carbonic anhydrase in the rat kidney. *Kidney Int.* **17**:162–174.
161. Sohtell, M., and B. Karlmark. 1976. *In vivo* micropuncture PCO_2 measurements. *Pfluegers Arch.* **363**:179–180.
162. Maren, T. H. 1967. Carbonic anhydrase: Chemistry, physiology and inhibition. *Physiol. Rev.* **47**:595–781.
163. Lassiter, W. E., C. W. Gottschalk, and M. Mylle. 1963. Micropuncture study of renal tubular reabsorption of calcium in normal rodents. *Am. J. Physiol.* **204**:771–775.
164. Walser, M. 1971. Calcium–sodium interdependence in renal transport. In: *Renal Pharmacology*. Appleton, New York. pp. 21–41.
165. Sutten, R. A. L., and J. H. Dirks. 1981. Renal handling of calcium, phosphate and magnesium. In: *The Kidney,* 2nd ed. B. M. Brenner and F. C. Rector, eds. Saunders, Philadelphia. pp. 551–618.
166. Bourdeau, J. E., and M. B. Burg. 1979. Voltage dependence of calcium transport in the thick ascending limb of Henle's loop. *Am. J. Physiol.* **236**:F357–F364.
167. Suki, W. N., D. Rouse, R. C. K. Ng, and J. P. Kokko. 1980. Calcium transport in the thick ascending limb of Henle. *J. Clin. Invest.* **66**:1004–1009.
168. Costanzo, L. S., and I. M. Weiner. 1976. Relationship between clearances of Ca and Na: Effect of distal diuretics and PTH. *Am. J. Physiol.* **230**:67–73.
169. Lemann, J., J. R. Litzow, and E. J. Lennon. 1967. Studies of the mechanism by which chronic metabolic acidosis augments urinary calcium excretion in man. *J. Clin. Invest.* **46**:1318–1328.
170. Sutton, R. A. L., N. L. M. Wong, and J. H. Dirks. 1979. Effects of metabolic acidosis and alkalosis on Na and Ca transport in the dog kidney. *Kidney Int.* **15**:520–533.
171. Sutton, R. A. L., N. L. M. Wong, and J. H. Dirks. 1976. Renal tubular Na and Ca reabsorption: Dissociation by maneuvers which increase bicarbonate excretion. *Clin. Res.* **24**:413a.
172. Peraino, R. A., and W. N. Suki. 1980. Urine HCO_3^- augments renal Ca absorption independent of systemic acid–base changes. *Am. J. Physiol.* **238**:F394–F398.
173. Peraino, R. A., W. N. Suki, and B. J. Stinebaugh. 1983. Renal excretion of calcium and magnesium during correction of metabolic acidosis by bicarbonate infusion in the dog. *Miner. Electrolyte Metab.* **3**:87–93.
174. Marone, C. C., N. L. M. Wong, R. A. L. Sutton, and J. H. Dirks. 1983. Effects of metabolic alkalosis on calcium excretion in the conscious dog. *J. Lab. Clin. Med.* **101**:264–273.
175. Robinson, B. H. S., E. B. Marsh, J. W. Duckett, and M. Walser. 1962. Adrenocortical modification of the interdependence of calcium and sodium reabsorption in the kidney. *J. Clin. Invest.* **41**:1394.
176. Goldfarb, S., G. R. Westby, M. Goldberg, and Z. S. Agus. 1977. Renal tubular effects of chronic phosphate depletion. *J. Clin. Invest.* **59**:770–779.
177. Wong, N. L. M., G. A. Quamme, T. J. O'Callaghan, R. A. L. Sutton, and J. H. Dirks. 1980. Renal tubular transport in phosphate depletion: A micropuncture study. *Can. J. Physiol. Pharmacol.* **58**:1063–1071.
178. Biber, T. U. L. 1971. Effect of changes in transepithelial transport on the uptake of sodium across the outer surface of the frog skin. *J. Gen. Physiol.* **58**:131–144.
179. Nagel, W., and A. Dorge. 1970. Effect of amiloride on sodium transport of frog skin: Action on intracellular sodium content. *Plfuegers Arch.* **317**:84–92.
180. O'Neil, R. G., and E. L. Boulpaep. 1979. Effect of amiloride on the apical cell membrane cation channels of a sodium-absorbing, potassium-secreting renal epithelium. *J. Membr. Biol.* **50**:365–387.
181. Taylor, A., and E. E. Windhager. 1979. Possible roles of cytosolic calcium and Na–Ca exchange in regulation of transepithelial sodium transport. *Am. J. Physiol.* **236**:F505–F512.
182. Friedman, P. A., J. F. Figueiredo, T. Maack, and E. E. Windhager. 1981. Sodium–calcium interactions in the renal proximal convoluted tubule of the rabbit. *Am. J. Physiol.* **240**:F558–F568.
183. Frindt, G., E. E. Windhager, and A. Taylor. 1982. Hydroosmotic response of collecting tubules to ADH or cAMP at reduced peritubular sodium. *Am. J. Physiol.* **243**:F503–F513.
184. Lorenzen, M., C. O. Lee, and E. E. Windhager. 1984. Cytosolic Ca^{+2} and Na^{+} activities in perfused proximal tubules of *Necturus* kidney. *Am. J. Physiol.* **247**:F93–F102.
185. Wen, S. F., R. L. Evanson, and J. H. Dirks. 1970. Micropuncture study of renal magnesium transport in proximal and distal tubule of the dog. *Am. J. Physiol.* **219**:570–576.

186. LeGrimellec, C. L., N. Roinel, and F. Morel. 1973. Simultaneous Mg, Ca, P, K, Na and Cl analysis in rat tubular fluid during perfusion of either inulin or ferrocyanide. *Pfluegers Arch.* **340**:181–196.
187. Brunnette, M., H. Vigneault, and S. Carriere. 1974. Micropuncture study of magnesium transport along the nephron in the young rat. *Am. J. Physiol.* **227**:891–896.
188. Carney, S. L., N. L. M. Wong, G. A. Quamme, and J. H. Dirks. 1980. Effect of magnesium deficiency on renal magnesium and calcium transport in the rat. *J. Clin. Invest.* **65**:180–188.
189. Harris, C. A., M. A. Burnatowska, J. F. Seely, R. A. L. Sutton, G. A. Quamme, and J. H. Dirks. 1979. Effects of parathyroid hormone on electrolyte transport in the hamster nephron. *Am. J. Physiol.* **236**:342–348.
190. Quamme, G. A. 1980. Effect of calcitonin on calcium and magnesium transport in the rat nephron. *Am. J. Physiol.* **238**:573–578.
191. Quamme, G. A. 1981. Effect of furosemide on calcium and magnesium transport in the rat nephron. *Am. J. Physiol.* **241**:340–347.
192. Quamme, G. A., and J. H. Dirks. 1980. Effect of intraluminal and contraluminal magnesium on magnesium and calcium transfer in the rat nephron. *Am. J. Physiol.* **238**:187–198.
193. Wen, S. F., N. L. M. Wong, and J. H. Dirks. 1971. Evidence for renal magnesium secretion during magnesium infusion in the dog. *Am. J. Physiol.* **220**:33–37.
194. LeGrimellec, C., N. Roinel, and F. Morel. 1973. Simultaneous Mg, Ca, P, K, Na and Cl analysis in rat tubular fluid. II. During acute Mg plasma loading. *Pfluegers Arch.* **340**:197–210.
195. Quamme, G. A., and J. H. Dirks. 1983. Renal magnesium transport. *Rev. Physiol. Biochem. Pharmacol.* **97**:69–110.
196. Massry, S. G., J. W. Coburn, and C. R. Kleeman. 1969. Renal handling of magnesium in the dog. *Am. J. Physiol.* **216**:1460–1467.
197. MacIntyre, I. 1967. Magnesium metabolism. *Adv. Intern. Med.* **13**:143–154.
198. Wacker, W. E. C., and A. F. Parisi. 1968. Magnesium metabolism. *N. Engl. J. Med.* **278**:658–776.
199. King, R. G., and S. W. Stanbury. 1970. Magnesium metabolism in primary hyperparathyroidism. *Clin. Sci.* **39**:281–303.
200. Kuntziger, H., C. Amiel, N. Roinel, and F. Morel. 1974. Effects of parathyroidectomy and cyclic AMP on renal transport of phosphate, calcium and magnesium. *Am. J. Physiol.* **227**:905–911.
201. Shareghi, G. R., and Z. S. Agus. 1982. Magnesium transport in the cortical thick ascending limb of Henle's loop of the rabbit. *J. Clin. Invest.* **69**:759–769.
202. Wong, N. L. M., G. A. Quamme, and J. H. Dirks. 1982. Effect of chlorothiazide on renal calcium and magnesium handling in the hamster. *Can. J. Physiol. Pharmacol.* **60**:1160–1165.
203. LeGrimellec, C., N. Roinel, and F. Morel. 1974. Simultaneous Mg, Ca, P, K and Cl analysis in rat tubular fluid. IV. During acute phosphate plasma loading. *Pfluegers Arch.* **346**:189–204.
204. Amiel, C., H. Kuntziger, and G. Richet. 1970. Micropuncture study of handling of phosphate by proximal and distal nephron in normal and parathyroidectomized rat: Evidence for distal reabsorption. *Pfluegers Arch.* **317**:93–109.
205. Poujeol, P., D. Chabardes, N. Roinel, and C. DeRouffinac. 1976. Influence of extracellular fluid volume expansion on magnesium, calcium and phosphate handling along the rat nephron. *Pfluegers Arch.* **365**:203–211.
206. Beck, L. H., and M. Goldberg. 1974. Mechanism of the blunted phosphaturia in saline-loaded thyroidectomized dogs. *Kidney Int.* **6**:18–23.
207. Knox, F. G., and C. Lechene. 1975. Distal site of action of parathyroid hormone on phosphate reabsorption. *Am. J. Physiol.* **229**:1556–1560.
208. Pastoriza-Munoz, E., R. E. Colindres, W. E. Lassiter, and C. Lechene. 1978. Effect of parathyroid hormone on phosphate reabsorption in rat distal convolution. *Am. J. Physiol.* **235**:F321–F330.
209. Lang, F., R. Gregor, G. Marchand, and F. G. Knox. 1976. Stationary microperfusion study of phosphate reabsorption in proximal and distal nephron segments. *Pfluegers Arch.* **368**:45–48.
210. Gross, J. B., and J. P. Kokko. 1977. Effects of aldosterone and potassium-sparing diuretics on electrical potential differences across the distal nephron. *J. Clin. Invest.* **59**:82–89.
211. Stoner, L. C. 1977. Isolated, perfused amphibian renal tubules: The diluting segment. *Am. J. Physiol.* **233**:F438–F444.
212. Koeppen, B. M., and G. Giebisch. 1983. Electrophysiology of mammalian renal tubules: Inferences from intracellular microelectrode studies. *Annu. Rev. Physiol.* **45**:497–517.
213. Allen, G. G., and L. J. Barratt. 1981. Electrophysiology of the early distal tubule: Further observations of electrode techniques. *Kidney Int.* **19**:24–35.
214. Allen, G. G., and L. J. Barratt. 1981. Effect of aldosterone on the transepithelial potential difference of the rat distal tubule. *Kidney Int.* **19**:678–686.
215. Giebisch, G., G. Malnic, R. M. Klose, and E. E. Windhager. 1966. Effect of ionic substitutions on distal potential differences in rat kidney. *Am. J. Physiol.* **211**:560–568.
216. Hayslett, J. P., E. L. Boulpaep, and G. H. Giebisch. 1978. Factors influencing transepithelial potential difference in mammalian distal tubule. *Am. J. Physiol.* **234**:F182–F191.
217. Hayslett, J. P., E. L. Boulpaep, M. Kashgarian, and G. H. Giebisch. 1977. Electrical characteristics of the mammalian distal tubule: Comparison of Ling–Gerard and macroelectrodes. *Kidney Int.* **12**:324–331.
218. Khuri, R. N., S. K. Agulian, and K. Kallognlian. 1972. Intracellular potassium in cells of the distal tubule. *Pfluegers Arch.* **335**:297–308.
218a. Temple-Smith, P., L. Costanzo, and E. E. Windhager. 1977. Reexamination of transepithelial potential difference in distal convoluted tubules of the rat. *Electrophysiology of the Nephron.* T. Anagnostopoulos, ed. Inserm Editions, Paris. pp. 115–124.
219. Imai, M. 1979. The connecting tubule: A functional subdivision of the rabbit distal nephron segments. *Kidney Int.* **15**:346–356.
220. Good, D. W., and F. S. Wright. 1980. Luminal influences on potassium secretion: Transepithelial voltage. *Am. J. Physiol.* **239**: F289–F298.
221. Khuri, R. N., S. K. Agulian, and K. Bogharian. 1974. Electrochemical potential of chloride in distal renal tubule of the rat. *Am. J. Physiol.* **227**:1352–1355.
222. Wiederholt, M., W. Schoormans, L. Hanson, and C. Behn. 1974. Sodium conductance changes by aldosterone in the rat kidney. *Pfluegers Arch.* **348**:155–165.
223. Boulpaep, E. L., and J. F. Seely. 1971. Electrophysiology of proximal and distal tubules in the autoperfused dog kidney. *Am. J. Physiol.* **221**:1084–1096.
224. Sullivan, J. 1968. Electrical potential differences across distal renal tubules of *Amphiuma*. *Am. J. Physiol.* **214**:1096–1103.
225. Maude, D. L., I. Shehadeh, and A. K. Solomon. 1966. Sodium and water transport in single perfused distal tubules of *Necturus* kidney. *Am. J. Physiol.* **211**:1043–1049.
226. Windhager, E. E., and G. Giebisch. 1965. Electrophysiology of the nephron. *Physiol. Rev.* **45**:214–244.
227. Jentsch, T., M. Koch, A. Krolik, and M. Wiederhold. 1982. Calcium transport in the distal tubule of the *Amphiuma* kidney. *Pfluegers Arch.* **392**(Suppl.):R15.
228. Cohen, B., G. Giebisch, L. L. Hansen, U. Teuscher, and M. Wiederholt. 1984. Relationship between peritubular membrane potential and net fluid reabsorption in the distal renal tubule of *Amphiuma*. *J. Physiol. (London)* **348**:115–134.
229. Oberleithner, H., W. Guggino, and G. Giebisch. 1982. Mechanism of distal tubular chloride transport in *Amphiuma* kidney. *Am. J. Physiol.* **242**:F331–F339.
230. Wiederholt, M., and G. Giebisch. 1974. Some electrophysiological properties of the distal tubule of *Amphiuma* kidney. *Fed. Proc.* **33**:387.
231. Oberleithner, H., W. Guggino, and G. Giebisch. 1981. The cellular mechanism of potassium adaptation in the distal amphibian nephron. *J. Physiol. (London)* **318**:55P–56P.

232. Stanton, B. A., W. B. Guggino, and G. Giebisch. 1982. Acidification of the basolateral solution reduces potassium conductance of the apical membrane. *Fed. Proc.* **41**:1006.

232a. Oberleithner, H., F. Lang, R. Gregor, W. Wang, and G. Giebisch. 1983. Effect of luminal potassium on cellular sodium activity in the early distal tubule of *Amphiuma* kidney. *Pfluegers Arch.* **396**:34–40.

233. Burg, M., and L. Stoner. 1974. Sodium transport in the distal nephron. *Fed. Proc.* **33**:31–36.

234. Wright, F. S. 1971. Alterations in electrical potential and ionic conductance of renal distal tubule cells in potassium adaptation. *Proc. Int. Union Physiol. Sci.* **9**:609.

235. Oberleithner, H., W. Guggino, and G. Giebisch. 1983. The effect of furosemide on luminal sodium, chloride and potassium transport in the early distal tubule of *Amphiuma* kidney: Effects of potassium adaptation. *Pfluegers Arch.* **396**:27–33.

236. Velazquez, H. E., and F. S. Wright. 1984. Sodium, chloride and potassium transport by the distal nephron: Effect of bumetanide and chlorothiazide. *Kidney Int.* **25**:319a.

237. Ellison, D. H., H. E. Velazquez, and F. S. Wright. 1984. Effects of barium and chloride on net and unidirectional potassium fluxes across distal tubules. *Proc. 9th Int. Congr. Nephrol.* p. 411A.

238. Schmidt, U., and I. C. Dubach. 1969. Activity of (Na + K)-stimulated adenosine triphosphatase in the rat nephron. *Pfluegers Arch.* **306**:219–227.

239. Ernst, S. A. 1975. Transport ATPase cytochemistry: Ultrastructural localization of potassium-dependent and potassium-independent phosphatase activities in rat kidney cortex. *J. Cell Biol.* **66**:586–608.

240. Oberleithner, H., F. Lang, W. Wang, and G. Giebisch. 1982. Effects of inhibition of chloride transport on intracellular sodium activity in distal amphibian nephron. *Pfluegers Arch.* **394**:55–60.

241. O'Neil, R. G., and S. C. Sansom. 1984. Characterization of apical membrane Na and K conductances of cortical collecting duct using microelectrode techniques. *Am. J. Physiol.* **247**:F14–F24.

242. Cuthbert, A. W., and W. K. Shum. 1975. Effects of vasopressin and aldosterone on amiloride binding in toad bladder epithelial cells. *Proc. R. Soc. London Ser. B* **189**:543–575.

243. Kinsella, J. L., and P. S. Aronson. 1981. Amiloride inhibition of the Na–H exchanger in renal microvillus membrane vesicles. *Am. J. Physiol.* **241**:F371–F379.

244. Velazquez, H., and F. S. Wright. 1983. Distal tubular pathways for sodium, chloride and potassium transport assessed by diuretics. *Kidney Int.* **23**:269a.

245. Frindt, G., and E. E. Windhager. 1983. Effect of quinidine, low peritubular Na or Ca on Na transport in isolated perfused rabbit cortical collecting tubules. *Fed. Proc.* **42**:305a.

246. Frindt, G., and E. E. Windhager. 1984. Transepithelial electrical resistance (R_t) of cortical collecting tubules at reduced peritubular Na concentration. *Proc. 9th Int. Congr. Nephrol.* p. 413A.

247. Windhager, E. E., and G. Frindt. 1984. Role of cytosolic calcium in renal tubular transport. In: *Proceedings of the IXth International Congress of Nephrology—Nephrology Today.* Springer-Verlag, Berlin, in press.

248. Taylor, A., and E. E. Windhager. 1983. Regulatory role of intracellular calcium ions in epithelial Na transport. *Annu. Rev. Physiol.* **45**:519–532.

249. Schultz, S. G. 1981. Homocellular regulatory mechanisms in sodium-transporting epithelia: Avoidance of extinction by "flush-through." *Am. J. Physiol.* **241**:F579–F590.

250. Koeppen, B. M., B. A. Biagi, and G. H. Giebisch. 1983. Intracellular microelectrode characterization of the rabbit cortical collecting duct. *Am. J. Physiol.* **244**:F35–F47.

251. Stanton, B., A. Janzen, T. Klein-Robbenhaar, J. Wade, G. Giebisch, and R. DeFronzo. 1983. Role of physiological levels of aldosterone in regulation of distal tubule morphology and potassium transport. *Kidney Int.* **23**:267a.

252. Wade, J. B., R. G. O'Neil, J. L. Pryor, and E. L. Boulpaep. 1979. Modulation of cell membrane area in renal collecting tubules by corticosteroid hormones. *J. Cell Biol.* **81**:439–445.

253. Giebisch, G. 1978. Renal potassium transport. In: *Membrane Transport in Biology*, Volume IVA. G. Giebisch, D. C. Tosteson, and H. H. Ussing, eds. Springer-Verlag, Berlin. pp. 215–298.

254. Stanton, B., and G. Giebisch. 1981. Mechanism of urinary potassium excretion. *Min. Electrolyte Metab.* **5**:100–120.

255. Jones, S. M., and J. P. Hayslett. 1983. Demonstration of active potassium secretion in the late distal tubule. *Am. J. Physiol.* **245**:F83–F88.

256. Strieder, N., R. N. Khuri, M. Wiederholt, and G. Giebisch. 1974. Studies on the renal action of ouabain in the rat: Effects in the non-diuretic state. *Pfluegers Arch.* **349**:91–107.

257. Greger, R., E. Schlatter, and F. Lang. 1983. Evidence for electroneutral sodium chloride cotransport in the cortical thick ascending limb of Henle's loop of rabbit kidney. *Pfluegers Arch.* **396**:308–314.

258. Oberleithner, H., G. Giebisch, F. Lang, and W. Wang. 1982. Cellular mechanism of the furosemide sensitive transport system in the kidney. *Klin. Wochenschr.* **60**:1173–1179.

259. Sackin, H., N. Morgunov, and E. L. Boulpaep. 1982. Electrical potentials and luminal membrane ion transport in the amphibian renal diluting segment. *Fed. Proc.* **41**:1495.

260. Ussing, H. H., and E. E. Windhager. 1964. Nature of shunt path and active sodium transport path through frog skin epithelium. *Acta Physiol. Scand.* **61**:484–504.

261. Franz, T. J., W. R. Galey, and J. T. VanBruggen. 1968. Further observations on asymmetrical solute movement across membranes. *J. Gen. Physiol.* **51**:1–12.

262. DiBona, I. R., and M. D. Civan. 1973. Pathways for movement of ions and water across toad urinary bladder. I. Anatomical site of transepithelial shunt pathways. *J. Membr. Biol.* **12**:101–122.

263. Schafer, J. A., C. S. Patlak, and T. E. Andreoli. 1974. Osmosis in cortical collecting tubules: A theoretical and experimental analysis of the osmotic transient phenomenon. *J. Gen. Physiol.* **64**:201–237.

264. Shafer, J. A., S. L. Troutman, and T. E. Andreoli. 1974. Osmosis in cortical collecting tubules: ADH-independent osmotic flow rectification. *J. Gen. Physiol.* **64**:228–240.

CHAPTER 18

The Respiratory Epithelium

Michael J. Welsh

1. Introduction

The respiratory epithelium forms a continuous layer of cells that separates air from liquid throughout the lung. The integrity and function of the epithelium are a critical requirement for effective gas exchange, the uptake of O_2 from the environment and elimination of CO_2 from the organism. Transfer of gases between the air and the blood requires two anatomical structures: the conducting airways, which distribute the inspired air within the lungs, and the alveoli, which are the site of O_2 and CO_2 diffusion between the gas phase and the pulmonary capillary blood. The epithelia in both regions serve as passive barriers between gas and fluid phases and perform active ion transport functions. In the airways, the epithelium can actively secrete Cl^- or absorb Na^+; ion transport is an important determinant of the quantity and composition of the respiratory tract fluid, an essential component of mucociliary clearance. In the alveoli, the epithelium actively absorbs Na^+; Na^+ absorption is a major factor that maintains a fluid-free alveolus. In the fetal lung, the epithelium actively secretes Cl^-; fluid secretion is a major requirement for normal pulmonary growth and development.

In this chapter, I will describe the transport functions of the airway epithelium, including the trachea and bronchi, the alveolar epithelium, and the fetal lung. However, I will not discuss how submucosal glands produce mucus nor how alveolar cells secrete surfactant. The reader may consult several reviews[1–3] for this information.

2. The Tracheal Epithelium

The pulmonary airways are covered by a thin layer of fluid composed of two layers: the superficial mucus and a periciliary fluid layer that lies beneath the mucus.[1,4–6] Both layers are essential for mucociliary clearance, a major pulmonary defense mechanism. The viscoelastic, gellike mucus traps particulate material and is propelled up the airways by cilia. The periciliary fluid provides a less viscous, "watery" environment, one that allows the cilia to beat effectively and couples the superficial layer of mucus to the movement of the underlying cilia. The volume and composition of the periciliary fluid and the hydration of the mucus appear to be determined by the airway epithelium's active ion transport and passive permeability properties.

Olver *et al.*,[7] studying the posterior membranous portion of canine tracheal epithelium, were the first to find that airway epithelium could actively transport ions. Since then, most studies of the control and mechanisms of airway epithelial ion transport have been performed with dog trachea. Therefore, in the following discussion, all references to tracheal epithelium will refer to the posterior membranous portion of canine trachea.

2.1. Morphologic Considerations

Figure 1 is a low-power electron micrograph of tracheal epithelium,[8] a pseudostratified columnar epithelium composed of three major cell types: ciliated, goblet, and basal. Ciliated epithelial cells and goblet cells (in a ratio of approximately 4 to 1) extend from the lumen or mucosal surface to the basement membrane or submucosal surface. Ciliated cells are 40 to 60 μm in length and 3 to 5 μm in width at their apices; cilia, interspersed with microvilli, cover the apical surface, while the basolateral membrane displays a moderate degree of interdigitation with adjacent cells. Junctional complexes consisting of zonula occludens, zonula adhaerens, and desmosomes connect the cells at their apices. The epithelium also contains basal cells that lie near the submucosal surface and are thought to be the precursors of ciliated and goblet cells. These histologic features resemble those reported in the trachea of other species[9] and those observed in other airway epithelia.[10,11] Tracheal epithelium also contains a number of mucus-secreting submucosal glands. In canine trachea, submucosal glands are primarily located in the cartilaginous portion of the trachea, rather than in the posterior membranous portion.

Michael J. Welsh • Laboratory of Epithelial Transport and Pulmonary Division, Department of Internal Medicine, University of Iowa College of Medicine, Iowa City, Iowa 52242.

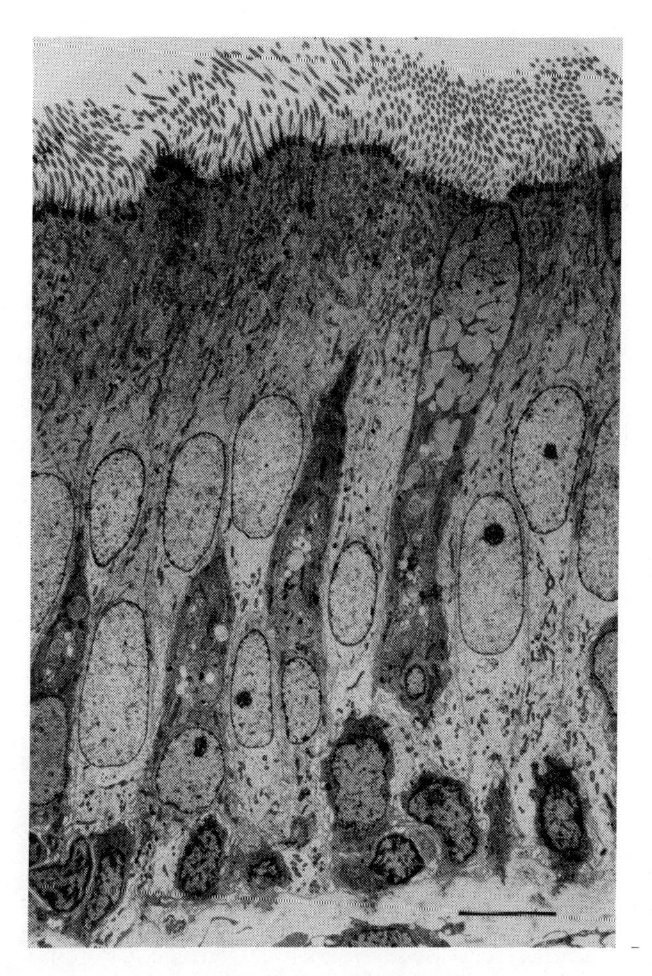

Fig. 1. Low-power electron micrograph of tracheal epithelium, revealing three cell types: ciliated, goblet, and basal. Three goblet cells are shown in which the apical portion is outside the plane of the section: they can be identified by dense cytoplasm and immature secretory granules. The cilia and shorter microvilli are apparent at the apical surface of ciliated cells. Bar = 10 μm. From Widdicombe *et al.*[8] with permission.

2.2. General Properties of Ion Transport

In vitro canine tracheal epithelium generates a transepithelial electrical potential difference (Ψ_t) of 10 to 30 mV (with reference to the mucosal solution),[7,12] a value in excellent agreement with that measured *in vivo*.[13,14] The transepithelial electrical resistance (R_t) ranges from 150 to 700 Ω-cm². When Ψ_t is clamped to zero, the short-circuit current (I_{sc}) is entirely accounted for by secretion of Cl^- from the submucosal to the mucosal surface and by absorption of Na^+ from the mucosal to the submucosal surface. Under spontaneously transporting short-circuit conditions, with no agents added to the bathing solutions, the net flux of both ions varies from animal to animal, but usually Cl^- is secreted at a rate 1½ to 4 times the rate of Na^+ absorption.

Intracellular microelectrode studies indicate that the Cl^- secretory process and the Na^+ absorptive process reside within the same cell type.[15] This feature gives the epithelium the ability to precisely control the direction and volume of transepithelial fluid transport. In this regard, the Cl^- secretory process underlies the capacity of the trachea for net fluid secretion.[16]

Cl^- secretion by tracheal epithelium is an electrogenic transport process, as indicated by a direct relation between the rate of Cl^- secretion and the transepithelial conductance.[7,12,15] Cl^- secretion requires Na^+ in the submucosal bathing solution[12,17] and depends upon the Na^+,K^+ pump. If ouabain is added to or K^+ removed from the submucosal bathing solution, the basolateral Na^+,K^+-ATPase is inhibited, and thus Cl^- secretion is also inhibited.[12,17,18]

2.3. Control of Cl^- Secretion

A variety of neurohumoral and pharmacologic agents regulate the rate of Cl^- secretion in tracheal epithelium. These agents fall into three categories, as shown in Table I: (1) those that stimulate electrogenic Cl^- secretion, (2) those that inhibit electrogenic Cl^- secretion, and (3) those that induce neutral NaCl secretion.

The first group of agents, those that stimulate both the rate of Cl^- and the short-circuit current, includes cAMP.[19] Moreover, theophylline, β-adrenergic agonists, and prostaglandins of the E series all increase the cAMP content of the surface epithelial cells and induce Cl^- secretion.[20,21] In addition, Lazarus *et al.*,[22,23] using immunocytochemical techniques, found that both β-adrenergic agonists and prostaglandin (PG) E_1 increase the immunoreactive cAMP staining of ciliated epithelial cells without inducing goblet cell staining. Together, these observations clearly implicate cAMP as a physiologic mediator of electrogenic Cl^- secretion. However, there is no direct correlation between the intracellular cAMP concentration and the rate of Cl^- secretion[20]; perhaps other intracellular mediators exert a parallel control over the secretory processes or perhaps the cAMP content of a specific cellular compartment is critical in determining the transport rate.

The Ca^{2+} ionophore A23187 also stimulates maximal electrogenic Cl^- secretion[24]; addition of theophylline or cAMP to A23187-treated tissues produces no further increase in the rate of Cl^- secretion. The stimulation produced by A23187 requires Ca^{2+} in the bathing media[24] but does not increase

Table I. Agents That Mediate Chloride Secretion in Canine Tracheal Epithelium[a]

Agent	cAMP	Ref.
Stimulate electrogenic Cl^- secretion		
β-Adrenergic agonists	↑	20, 26, 43
Theophylline, cAMP	↑	12, 16, 19, 20
Prostaglandin E	↑	20, 21
Substance P	?	44
Ca^{2+} ionophore A23187	–	20, 24
Prostaglandin $F_{2\alpha}$	–	20, 21
Inhibit electrogenic Cl^- secretion		
Indomethacin and other prostaglandin synthesis inhibitors	↓	20, 21
Stimulate neutral NaCl secretion		
Histamine	–	20, 25
Acetylcholine	–	20, 27
α-Adrenergic agonists	–	20, 26

[a]The effect of each agent on the cAMP content of the surface cells is indicated (↑, increase; ↓, decrease; –, no change; ?, unknown).

cellular levels of cAMP.[20] These findings indicate that intracellular Ca^{2+} may also mediate the rate of Cl^- secretion.

$PGF_{2\alpha}$ also increases the rate of electrogenic Cl^- secretion, but it does not increase intracellular cAMP levels.[20,21] At this time, the intracellular mediator responsible for $PGF_{2\alpha}$-induced Cl^- secretion is unknown. Perhaps $PGF_{2\alpha}$ increases intracellular Ca^{2+} or increases cAMP but to a level below the threshold of detection; the latter suggestion is consistent with the lack of correlation between cellular cAMP levels and the rate of Cl^- secretion.

The second group of agents (Table I), those that inhibit Cl^- secretion, includes the prostaglandin synthesis inhibitor indomethacin and other cyclooxygenase inhibitors. Indomethacin reduces the rate of spontaneous Cl^- secretion, inhibits PGE_2 production by the epithelium, and decreases intracellular levels of cAMP.[20,21] Furthermore, indomethacin does not alter the Cl^- secretory response that follows the addition of exogenous prostaglandins or other secretagogues. Thus, it appears that cyclooxygenase inhibitors do not directly interfere with the secretory process, but act on the epithelial cells at a step preceding cAMP generation. These findings suggest that prostaglandins are physiologically important mediators determining the spontaneous rate of Cl^- secretion in tracheal epithelium. Because indomethacin suppresses the baseline rate of secretion, but does not inhibit the response to secretagogues, it has also been used to minimize basal Cl^- secretion and thus magnify the response to subsequent addition of exogenous secretagogues.[15]

The third group of agents in Table I elicit nearly equivalent increases in the rates of Cl^- and Na^+ secretion with minimal changes in the short-circuit current. Furthermore, none of the agents in this group alters the cAMP content of the surface epithelial cells. Thus, the secretion produced by histamine,[25] α-adrenergic agonists,[26] and acetylcholine[27] appears to result from electrically neutral secretion from the submucosal glands. In this regard, these agents have been observed to directly stimulate mucus secretion by submucosal glands.[28–30] Thus, the extracellular mediators that control Cl^- secretion from the surface epithelium are different from those that control mucus secretion from the submucosal glands.

2.4. Mechanism of Ion Transport in Canine Tracheal Epithelium

2.4.1. A Cellular Model of Ion Transport

Cl^- secretion by canine tracheal epithelium conforms to a cellular model of secretion that is also thought to operate in other secretory epithelia including shark rectal gland,[31,32] corneal epithelium,[33,34] large and small mammalian intestine,[35–37] opercular epithelium of teleosts,[38] and gastric mucosa.[39] In this section, I will first briefly review this model, since it currently provides the best explanation of the experimental observations. Then I will discuss the mechanisms of ion transport within the framework of this model. Even though I present the model before the discussion of the experimental results, I do not intend to indicate that the model is completely proven; rather, the model is a means of focusing and organizing the ensuing discussion.

Figure 2 is a cellular model of ion transport in tracheal epithelium. The cell secretes Cl^- from the submucosal surface to the mucosal surface and absorbs Na^+ in the opposite direction; both processes are electrogenic. The main features of the model are:

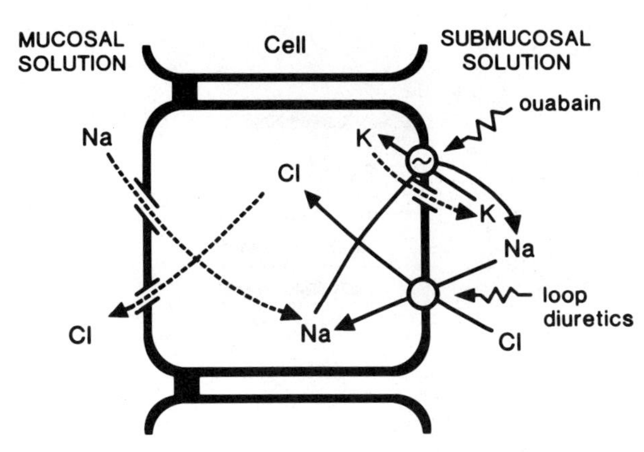

Fig. 2. Cellular model of ion transport in canine tracheal epithelium. See text for details.

1. Cl^- enters the cell across the basolateral membrane via an electrically neutral NaCl cotransport process. Because Cl^- is coupled to Na^+, the movement of Na^+ "down" a favorable electrochemical gradient across the basolateral membrane drives Cl^- "uphill" against its electrochemical gradient. Thereby, Cl^- is accumulated within the cell at an activity greater than predicted for electrochemical equilibrium with the energy provided indirectly by the Na^+ electrochemical gradient.
2. Na^+, which enters the cell at the basolateral membrane coupled to Cl^-, exits back across the basolateral membrane via the Na^+,K^+ATPase. This enzyme keeps intracellular Na^+ activity low and thus supplies the nonconjugate energy for transepithelial Cl^- secretion. K^+, which enters the cell in exchange for Na^+ on the Na^+,K^+ pump, exits passively across the K^+-permeable basolateral membrane.
3. Cl^- leaves the cell passively, moving down a favorable electrochemical gradient across a Cl^- conductive apical cell membrane. Secretagogues appear to regulate the permeability of the apical membrane to Cl^-, thus controlling the rate of Cl^- secretion.
4. The apical membrane also contains a Na^+ permeability. Na^+ may enter the cell passively at the apical membrane driven by a favorable electrochemical gradient.
5. Accordingly, during secretion, transepithelial current flow has several components. At the apical membrane, current is carried by conductive Cl^- exit and, to a lesser degree, Na^+ entry, while at the basolateral membrane, current is carried by conductive K^+ exit and Na^+ exit via the Na^+,K^+ pump.

2.4.2. The Cl^- Entry Step

Transepithelial Cl^- secretion requires the presence of Na^+ in the submucosal bathing solution.[12,17] This observation is explained by several studies indicating that Cl^- enters the cell coupled to Na^+. The results of unilateral ion substitution experiments[15] indicate that Cl^- enters the cell via an electrically neutral transport process. This conclusion rests on two observations. First, when either sulfate or gluconate replaces Cl^- in the submucosal bathing solution, the electrical potential difference across the basolateral membrane (Ψ_b) does

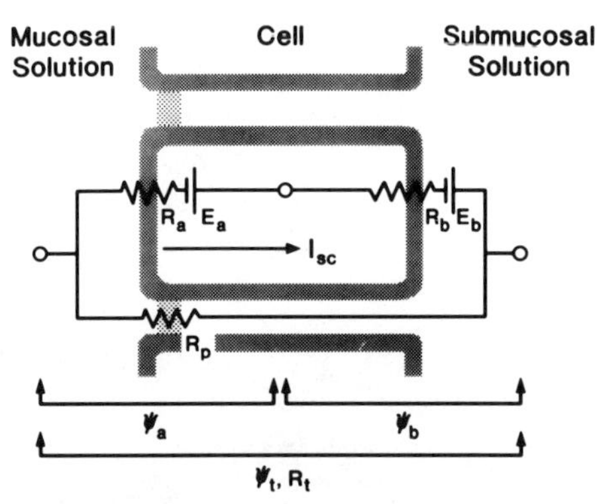

Fig. 3. Equivalent electrical circuit model of ion transport by canine tracheal epithelium. R_a, R_b, R_p, and R_t refer to the resistances of the apical and basolateral membranes, the paracellular pathway, and transepithelial resistance, respectively. E_a and E_b represent the electromotive force across the apical and basolateral membranes. Ψ_a, Ψ_b, and Ψ_t refer to the electrical potential difference across the apical membrane, the basolateral membrane, and the epithelium, respectively. I_{sc} refers to the short-circuit current.

Table II. Effect of Furosemide on the Electrical Properties and Equivalent Circuit Parameters of Canine Tracheal Epithelium[a]

	I_{sc} (μA/cm^2)	R_t (Ω-cm^2)	Ψ_a (mV)	R_a/R_b
Control	140	245	−60	3.2
	±22	±31	±3	±1.0
Furosemide	56*	322*	−65*	4.1
	±10	±43	±2	±1.0
	R_b (Ω-cm^2)	R_a (Ω-cm^2)	E_b (mV)	E_a (mV)
Control	95	313	72	−21
	±16	±53	±3	±4
Furosemide	115	508*	71	−41*
	±26	±81	±2	±4

[a]Values represent the mean ± S.E.M. from eight tissues. Epinephrine (10^{-6} M, submucosal solution) was present to stimulate Cl^- secretion during both the control condition and following addition of furosemide (10^{-3} M, submucosal solution). Adapted from Welsh.[42]
*Statistically significant difference from the control value, $p < 0.05$.

not change. This finding indicates that the chemical potential difference for Cl^- plays little or no part in determining Ψ_b. Second, when Cl^- is replaced in the submucosal bathing solution by one of these impermeant, nontransported ions, the membrane resistance ratio (R_a/R_b) increases (where R_a and R_b refer to the resistances of the apical and basolateral membrane, respectively). If Cl^- movement at the basolateral membrane were electrically conductive, R_b would increase and R_a/R_b would decrease under these conditions. The opposite finding, that R_a/R_b increases, is consistent with a known apical Cl^- conductance; when submucosal Cl^- is removed, intracellular Cl^- will decrease, thus increasing R_a.

Further insight into the mechanism of ion transport across the individual cell membranes has come from the application of an equivalent electrical circuit analysis to the transport processes of tracheal epithelium.[40] The model, shown in Fig. 3, is similar to that used in a variety of Na^+-absorbing epithelia.[41] The flow of ions is represented as the short-circuit current (I_{sc}), electrochemical driving forces (or zero current potentials) as electromotive forces or batteries (E), and ion permeabilities as electrical resistances (R). Each cell membrane is represented as a resistance and electromotive force in series. The cellular pathway is composed of the apical and basolateral membranes in series. The paracellular pathway is represented as a resistance in parallel with the cellular pathway. A value for each of the circuit parameters can be obtained during steady-state conditions from an analysis of the acute electrical response produced by addition of either epinephrine or amiloride.[40]

The equivalent electrical circuit model was used to analyze furosemide's effect in tracheal epithelium. When added to the submucosal bathing solution, furosemide specifically inhibits Cl^- secretion but does not alter the rate of Na^+ absorption.[42] Table II shows the effect of submucosal furosemide (10^{-3} M) on the transepithelial and intracellular electrical properties and the equivalent resistances and EMFs at the two cell membranes. Furosemide decreases the I_{sc} (indicating an inhibition of Cl^- secretion), increases R_t slightly, and hyperpolarizes the electrical potential difference across the apical membrane (Ψ_a). Despite these changes, furosemide does not alter R_b. These observations suggest that furosemide does not inhibit an electrically conductive Cl^- entry step, nor does it appear to alter other basolateral membrane ionic conductances.

Submucosal furosemide also decreases the electromotive force across the apical membrane (E_a). Since, under stimulated conditions, the apical membrane is predominately Cl^- permeable,[15] the chemical potential difference for Cl^- across the apical membrane largely determines E_a.[42] Thus, the intracellular Cl^- activity (a_c^{Cl}) can be estimated from the Nernst equation:

$$E_a = \frac{RT}{zF} \ln \left(\frac{a_m^{Cl}}{a_c^{Cl}} \right) \tag{1}$$

where a_m^{Cl} refers to the Cl^- activity in the mucosal solution and R, T, z, and F have their usual meanings. The value of E_a shown in Table II indicates that under baseline conditions, a_c^{Cl} is 41 mM. However, following the addition of furosemide to the submucosal solution, E_a becomes more negative, indicating a decrease in a_c^{Cl} to 20 mM. These estimates of a_c^{Cl} agree with values of a_c^{Cl} measured directly with intracellular Cl^--selective microelectrodes[45]; following submucosal addition of bumetanide (10^{-4} M), a loop diuretic structurally and functionally related to furosemide, a_c^{Cl} decreases. These results indicate that furosemide inhibits electrically neutral entry of Cl^- and thereby decreases a_c^{Cl}. Furosemide does not alter the basolateral membrane EMF (E_b), a finding which also suggests that Cl^- transport is electrically neutral. Finally, although R_a did increase following furosemide, the small magnitude of the increase indicates that furosemide did not substantially inhibit Cl^- transport by a decrease in apical membrane Cl^- permeability.

Table III shows the thermodynamic activity of intracellular Cl^- (a_c^{Cl}) measured with Cl^--selective and conventional microelectrodes under short-circuit conditions. When the rate of Cl^- secretion is minimal (indomethacin added to the mucosal bathing solution), a_c^{Cl} is 37 mM.[45] This value is 3.7 times greater than the activity expected for an equilibrium distribution

Table III. Intracellular Chloride Activities (a_c^{Cl}) in Nonsecreting and Secreting Tissues[a]

	I_{sc} (μA/cm^{-2})	R_t (Ω-cm^2)	Ψ_a (mV)	a_c^{Cl} (mM)	a_{eq}^{Cl} (mM)	$\Delta\tilde{\mu}_a^{Cl}$ (mV)
Nonsecreting	30	584	−60	37	10	35
	±7	±71	±2	±3	±1	±3
Secreting	115*	252*	−46*	39	17*	22*
	±11	±21	±2	±3	±2	±2

[a]Values represent the mean ± S.E.M. from seven epithelia. a_{eq}^{Cl} refers to the intracellular Cl^- activity expected for an equilibrium distribution based on the intracellular voltage. $\Delta\tilde{\mu}_a^{Cl}$ is the electrochemical gradient for Cl^- across the apical membrane. Indomethacin (10^{-6} M) was added to the mucosal solution to minimize the rate of Cl^- secretion under "nonsecreting" conditions and remained in the mucosal solution following stimulation of secretion with epinephrine (10^{-6} M, submucosal solution).
*Statistically significant difference between "Nonsecreting" and "Secreting" conditions, $p < 0.005$.

Table IV. Intracellular Chloride Activities (a_c^{Cl}) in the Absence of Extracellular Sodium[a]

	I_{sc} (μA/cm^2)	R_t (Ω-cm^2)	Ψ_a (mV)	a_c^{Cl} (mM)	a_{eq}^{Cl} (mM)	$\Delta\tilde{\mu}_a^{Cl}$ (mV)
Na^+-free Ringers	32	322	−51	17	14	−4
	±19	±39	±3	±2	±2	±2
Na^+ Ringer's	127*	270*	−53	34*	14	−26*
	±36	±26	±6	±2	±4	±6

[a]Values represent the mean ± S.E.M. from five epithelia. Measurements were made following incubation in Na^+-free, choline Ringer's solution for 1½ to 2 hr and after the Na^+-free solution was replaced with Na^+-containing solution. Epinephrine (10^{-6} M) was present in the submucosal bathing solution during both experimental conditions.
*Statistically significant difference between the two periods, $p < 0.05$.

across the two cell membranes based on the intracellular voltage. The electrochemical gradient for Cl^- across the basolateral membrane, $\Delta\tilde{\mu}_b^{Cl}$ (which is exactly equal and opposite to the electrochemical gradient across the apical membrane under short-circuit conditions), was calculated, in mV, from the relation

$$\Delta\tilde{\mu}_b^{Cl} = \Psi_b + \frac{RT}{zF}\ln\left(\frac{a_m^{Cl}}{a_c^{Cl}}\right) \quad (2)$$

The data indicate that Cl^- is accumulated against an electrochemical gradient of 35 mV. As Table III shows, when the secretagogue epinephrine is added to the submucosal solution, I_{sc} increases, R_t decreases, and Ψ_a depolarizes. Despite the stimulation of Cl^- secretion, a_c^{Cl} is not significantly altered, at 39 mM, and Cl^- is still accumulated intracellularly against a $\Delta\tilde{\mu}_b^{Cl}$ of 22 mV. These observations indicate that addition of secretagogue enhances the rate of Cl^- entry; there are equivalent increases in the rate of Cl^- exit and the rate of Cl^- entry. This raises the possibility that some intracellular mediator may regulate the Cl^- entry process. One reasonable candidate to regulate the Cl^- entry mechanism is intracellular cAMP, since it is known to increase with stimulation of secretion. Furthermore, cAMP is known to mediate electrically neutral Cl^- entry mechanisms in other epithelia, shark rectal gland,[46] and flounder intestine.[47]

When choline replaces Na^+ in the bathing solutions, a_c^{Cl} decreases (Table IV). In the absence of Na^+, a_c^{Cl} is only 17 mM, a value not substantially different from the Cl^- activity of 14 mM expected for an equilibrium distribution across the apical membrane. The subsequent change to Na^+-containing Ringer's increases a_c^{Cl} to 34 mM. These data establish that Cl^- accumulation is Na^+ dependent; removing Na^+ inhibits Cl^- entry at the basolateral membrane and allows Cl^- to distribute passively across a Cl^- conductive apical membrane.

These observations, taken together, indicate that Cl^- is accumulated across the basolateral membrane, against its electrochemical gradient, via a Na^+-dependent, electrically neutral transport process that is inhibited by loop diuretics. Certainly the Na^+ electrochemical gradient across the basolateral membrane is substantially greater than the $\Delta\mu_b^{-Cl}$ of 22 mV. If the intracellular Na^+ concentration is 20 mM, as determined chemically in isolated tracheal epithelial cells,[8] then the electrochemical gradient for Na^+ would be 98 mV, more than four times greater than the electrochemical gradient for Cl^- and more than sufficient to energize the "uphill" movement of Cl^- into the cell. In this way, the energy for "uphill" Cl^- entry is derived from the Na^+ electrochemical gradient, and ultimately, from the basolateral Na^+,K^+-ATPase, which maintains the Na^+ gradient; thus, no *direct* link between Cl^- entry and metabolic energy is required.

While these considerations provide compelling evidence for Na^+-coupled Cl^- entry, the identity of the transport process is not known with certainty. There are several possible neutral Cl^- transport processes that might mediate Cl^- entry: (1) NaCl cotransport, as shown in Fig. 2; (2) cotransport of 2 Cl^-, 1 Na^+, and 1 K^+, a process found in Ehrlich ascites cells[48]; or (3) parallel countertransport of Na^+–H^+ and Cl^-–HCO_3^-, proposed as the mechanism in rabbit ileum.[49] The third alternative seems unlikely since neither removing HCO_3^- from the bathing solution[12] nor adding substituted stilbenes (which inhibit Cl^-/HCO_3^- exchange in other cells)[50] inhibits Cl^- secretion in tracheal epithelium. There is currently no information that allows one to choose between the first two hypotheses. In either case, the Na^+ electrochemical gradient would be sufficient to energize Cl^- uptake.

2.4.3. The Cl^- Exit Step

The apical membrane of tracheal epithelium is Cl^- conductive. The magnitude of the Cl^- conductance varies, depending on the presence or absence of secretagogues and the intracellular cAMP levels.[15,40,51]

Under nonsecreting conditions (indomethacin 10^{-6} M, mucosal solution), the apical membrane possesses little if any Cl^- conductance. Two observations support this conclusion. First, when nontransported anions replace Cl^- in the bathing solution of indomethacin-treated tissues, neither the cellular electrical potential profile nor the membrane resistance ratio, R_a/R_b,[15] is substantially altered. Second, under nonsecreting conditions, the chemical potential difference for Na^+, as well as Cl^-, determine E_a,[40] indicating that the magnitude of the Cl^- conductance is usually less than that of the Na^+ conductance.

On the other hand, following addition of secretagogue, the apical membrane becomes predominantly Cl^- conductive. This conclusion is warranted by the findings that: first, Ψ_a and R_a/R_b are dependent upon the Cl^- concentration of the mucosal bathing solution[15]; and second, the chemical gradient for Cl^- across the apical membrane determines E_a.[40,45] Na^+ and K^+

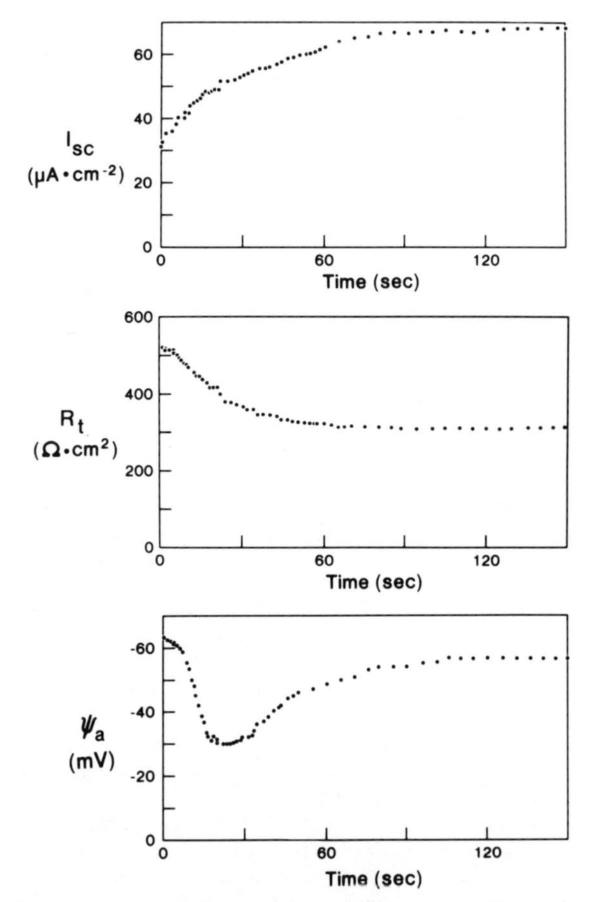

Fig. 4. Effect of epinephrine (10^{-6} M, submucosal solution) on the short-circuit current (I_{sc}), transepithelial resistance (R_t), and electrical potential difference across the apical membrane (Ψ_a). Values represent the data obtained from one tissue. Time-zero indicates the onset of the response to epinephrine (10^{-6} M) added to the submucosal solution. All values were stable prior to the onset of stimulation, so that time-zero represents the prestimulation steady state. Indomethacin (10^{-6} M, mucosal solution) was present throughout to minimize the baseline rate of Cl^- secretion in the absence of secretagogue, thus maximizing the changes observed with addition of epinephrine. Adapted from Welsh *et al.*[40]

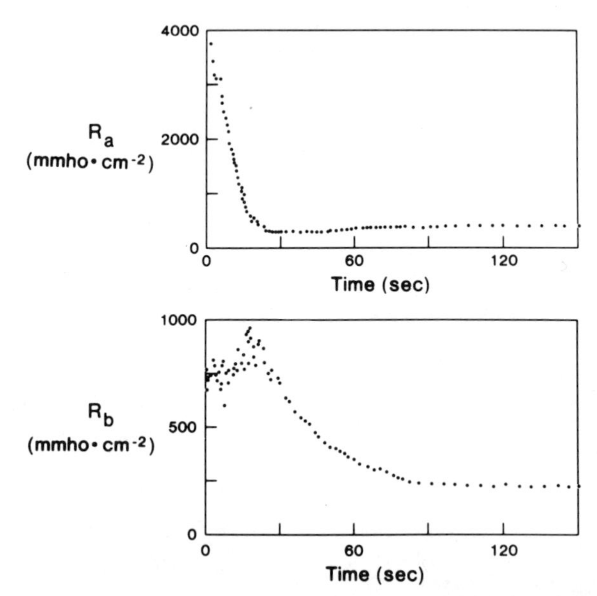

Fig. 5. Effect of epinephrine on the apical (R_a) and basolateral (R_b) membrane resistance. The data were obtained from the same tissue shown in Fig. 4.

substitution studies indicate that the apical membrane contains a small apical Na^+ conductance but no appreciable K^+ conductance under either nonsecreting or secreting conditions.[40,52,53]

Figures 4 and 5 illustrate the time course of the changes in electrical properties when epinephrine stimulates Cl^- secretion.[40] I_{sc} increases and R_t decreases. In contrast, the electrical potential difference across the apical membrane, Ψ_a (and Ψ_b which is equal and opposite to Ψ_a under short-circuit conditions), undergoes a biphasic response; first depolarizing and then repolarizing. These acute changes are partly explained by the changes in membrane resistances shown in Fig. 5. Following addition of epinephrine, apical membrane resistance decreases rapidly to approximately 1/10th the initial value and then stabilizes, 20–30 sec following the onset of stimulation. As a result of the decrease in R_a (and increase in Cl^- conductance), Ψ_a depolarizes, approaching the value of the Cl^- chemical gradient across the apical membrane. These findings suggest that the first effect of cAMP-mediated secretagogues is an increase in the apical Cl^- permeability.

In the steady state, before stimulation, the Thevenin equivalent electromotive force across the apical membrane, E_a, varies in magnitude and is frequently positive, up to a value of +60 mV. After stimulation, E_a uniformly decreases to a value in the range of −20 to −30 mV. These values of E_a reflect the known transport functions of the epithelium. Under nonsecreting (indomethacin-treated) conditions, the apical membrane is primarily Na^+ conductive and the Na^+ chemical potential difference across the apical membrane is an important determinant of E_a. On the other hand, following stimulation of secretion, the apical membrane is primarily Cl^- conductive and thus the Cl^- chemical potential difference across the apical membrane determines E_a [see Eq. (1)].

As shown in Table III, $\Delta\tilde{\mu}_a^{Cl}$ favors Cl^- exit under both secreting and nonsecreting conditions. In fact, $\Delta\tilde{\mu}_a^{Cl}$ is larger under nonsecreting conditions, but when the apical membrane is relatively Cl^- impermeable, the rate of Cl^- secretion is low. With addition of secretagogue, the apical membrane becomes Cl^- permeable allowing Cl^- to exit passively with its electrochemical gradient. This raises the question of whether $\Delta\tilde{\mu}_a^{Cl}$ is sufficient to drive passive Cl^- exit. If apical Cl^- exit is diffusional, an estimate of the required Cl^- conductance can be obtained from the rate of Cl^- secretion (estimated from the epinephrine-induced increase in I_{sc}) and the value of $\Delta\tilde{\mu}_a^{Cl}$; for diffusional exit, the apical Cl^- conductance would have to be 3.9 mS/cm^2. This value is in excellent agreement with total apical conductances in the range of 4 mS/cm^2 obtained from an equivalent circuit analysis, suggesting that the electrochemical driving force for Cl^- is sufficient to drive Cl^- exit.

2.4.4. The Basolateral Membrane Na^+,K^+ Pump and K^+ Conductance

Tracheal epithelium contains a Na^+,K^+-ATPase[54] that has been localized to the basolateral membrane by auto-

Table V. Intracellular Potassium Activities (a_c^K) in Nonsecreting and Secreting Tissues[a]

	I_{sc} (μA/cm^2)	R_t (Ω-cm^2)	Ψ_a (mV)	a_c^K (mM)	a_{eq}^K (mM)	$\Delta\tilde{\mu}_b^K$ (mV)
Nonsecreting	71	330	−58	69	34	19
	±9	±54	±2	±5	±3	±1
Secreting	151*	225*	−54*	70	28*	24*
	±23	±30	±2	±5	±2	±2

[a]Values represent the mean ± S.E.M. from eight epithelia that were studied under nonsecreting conditions (indomethacin 10^{-6} M, submucosal solution) and during the steady state following stimulation of secretion (epinephrine 10^{-6} M, submucosal solution). Indomethacin remained in the mucosal solution following addition of epinephrine. a_{eq}^K refers to the intracellular K^+ activity expected for an equilibrium distribution based on the electrical potential difference across the basolateral membrane. $\Delta\tilde{\mu}_b^K$ is the electrochemical gradient for K^+ across the basolateral membrane.
*Statistically significant difference between the two condtiions, $p < 0.05$.

Table VI. Effect of Barium on Electrical Properties and Membrane Resistances[a]

	I_{sc} (μA/cm^2)	R_t (Ω-cm^2)	Ψ_a (mV)	R_a (Ω-cm^2)	R_b (Ω-cm^2)
Control	134	280	−51	241	146
	±28	±27	±4	±56	±28
Barium	65*	352*	−28*	238	302*
	±8	±38	±4	±57	±41

[a]Values represent the mean ± S.E.M. from six tissues. Epinephrine (10^{-6} M) was present in the submucosal solution to stimulate Cl^- secretion during both the "Control" period and following addition of barium (2 mM) to the submucosal solution.
*Statistically significant difference between the two periods, $p < 0.05$.

radiographic ouabain-binding.[18] Transepithelial Cl^- transport requires the Na^+,K^+-ATPase: Cl^- secretion ceases when the Na^+,K^+-ATPase is inhibited either by adding ouabain to or removing K^+ from the submucosal bathing solution.[12,17] The Na^+,K^+-ATPase functions in the epithelium in a mode similar to that observed in other cells, extruding Na^+ from the cell and accumulating K^+. As a result, the intracellular Na^+ concentration is low and the K^+ activity is high. In isolated cells of tracheal epithelium, intracellular Na^+ concentration measured chemically is 20 mM and K^+, 150 mM.[8] Table V shows the intracellular K^+ activity (a_c^K) measured with K^+-selective microelectrodes under short-circuit conditions. The intracellular K^+ activity of 70 mM indicates that K^+ is actively accumulated across the basolateral membrane against an adverse electrochemical gradient ($\Delta\tilde{\mu}_b^K$) of 19 mV under nonsecreting conditions and 24 mV under secreting conditions. Smith and Frizzell[53] obtained similar values in tracheal epithelium under open-circuit conditions.

K^+ entering the cell on the Na^+,K^+ pump must be recycled across the basolateral membrane, since the rate of transepithelial K^+ secretion is minimal. Several observations indicate that K^+ exits passively across a basolateral membrane that is predominantly K^+ conductive:

1. When the submucosal K^+ concentration is increased, the basolateral membrane depolarizes; in contrast, neither Na^+ nor Cl^- substitutions alter Ψ_b.[15,52,53]
2. The Na^+,K^+ pump does not appreciably contribute to the basolateral membrane conductance. When ouabain is added to secreting tissues, neither R_t nor the membrane resistance ratio, R_a/R_b, decreases; however, ouabain does decrease I_{sc} and depolarize Ψ_a.[52]
3. K^+ accumulates intracellularly at an activity greater than predicted for electrochemical equilibrium (Table V),[53] providing a favorable electrochemical driving force for passive K^+ exit.
4. Ba^{2+}, which blocks the K^+ conductance of a variety of epithelial and nonepithelial cells,[55–58] also blocks a basolateral K^+ conductance in tracheal epithelium.[59] Table VI shows that when Ba^{2+} (2 mM) is added to the submucosal bathing solution, R_b doubles, Ψ_a and Ψ_b depolarize, but R_a remains stable. As a result, I_{sc} decreases and R_t increases. Moreover, submucosal Ba^{2+} limits the magnitude of the depolarization produced by elevation of the submucosal K^+ concentration.

These observations provide considerable evidence that K^+ is accumulated in the cell via the activity of the Na^+,K^+ pump and then recycled "down" a favorable electrochemical gradient across a K^+-permeable basolateral membrane.

As Figs. 4 and 5 show, epinephrine first decreases R_a and depolarizes Ψ_a. However, R_b then decreases to one-third to one-fourth its initial value and the intracellular voltage partially repolarizes.[15,40] The data presented above clearly indicate that R_b decreases because basolateral membrane K^+ conductance increases. This increase in K^+ conductivity has two important consequences.[40,52]

First, an increase in K^+ conductance will prevent large transport-related alterations in a_c^K. Consider the following scenario. As the rate of Cl^- secretion increases, the rate of Na^+ entry into the cell (via the Na^+-coupled Cl^- entry process) will increase; as a result, the rate of Na^+ extrusion (via the Na^+,K^+ pump) will increase. Given a constant pump stoichiometry, the rate of K^+ entry into the cell will also increase. If K^+ is to recycle back across the basolateral membrane, either the driving force for K^+ must increase or the basolateral membrane K^+ permeability must increase. In fact, a_c^K remains constant following stimulation of secretion (Table V).[53] At the same time, the electrochemical driving force for K^+ exit does increase, owing to the depolarization of the intracellular voltage. However, the increase (25%) is by itself insufficient to account for the required increase in the rate of K^+ exit without an increase in K^+ permeability (assuming a relatively constant stoichiometry for the Na^+,K^+ pump and Na^+-coupled Cl^- entry). Thus, it is the increased K^+ permeability which appears to be the major factor enhancing the rate of K^+ exit in tracheal epithelium (Fig. 5).

Second, the increase in basolateral K^+ permeability prevents a sustained depolarization of intracellular voltage. This can be appreciated by examining the time course of I_{sc}, Ψ_a, R_a, and R_b (Figs. 4 and 5). The initial decrease in R_a depolarizes Ψ_a, as Ψ_a approaches the value of E_a, the Cl^- chemical gradient. Subsequently, the decrease in R_b repolarizes Ψ_a, shifting intracellular voltage toward E_b and away from E_a. A negative Ψ_a is critical for Cl^- exit, since Cl^- leaves the cell against its chemical gradient, driven by an electrical gradient.[45] In fact, if a secondary repolarization did not occur, there would be little net secretion of Cl^-.

The importance of a negative Ψ_a in maintaining Cl^- secretion is demonstrated by two other findings. First, Fig. 6 shows

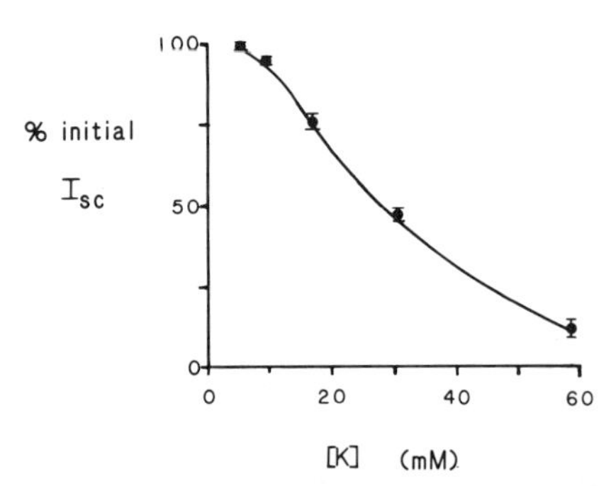

Fig. 6. Effect of increasing K^+ concentration on the short-circuit current (I_{sc}). Na^+ concentration was maintained constant at 75 mM and K^+ was substituted for tetramethylammonium in random order. Epinephrine (10^{-6} M, submucosal solution) and amiloride (10^{-4} M, mucosal solution) were present throughout, so that I_{sc} primarily reflects the rate of Cl^- secretion. Values represent the mean ± S.E.M. of data obtained in six tissues.

that increasing the bathing solution K^+ concentration, which depolarizes the intracellular voltage, decreases the rate of Cl^- secretion.[52] Second, submucosal Ba^{2+}, which also depolarizes intracellular voltage (Table VI),[59] decreases the short-circuit current and the rate of Cl^- secretion.

Clearly, control of the basolateral membrane K^+ permeability is an important regulatory mechanism, one that protects the intracellular ionic content and maintains the rate of Cl^- secretion in tracheal epithelium. In fact, similar regulatory mechanisms may be present in other transporting epithelia. Schultz[60] recently summarized the evidence that mechanisms exist which regulate the basolateral K^+ permeability in Na^+-absorbing epithelia. Two possible explanations have been considered for the direct relation between the rate of transepithelial transport and the K^+ permeability.[60]

The first possibility is that the K^+ leak is an inherent property of the Na^+,K^+ pump[60,61]; however, this does not appear to be the case in tracheal epithelium. One reason is that submucosal furosemide inhibits Cl^- secretion but does not alter R_b (Table II). Since furosemide probably decreases the activity of the Na^+ pump by decreasing the rate of Na^+ entry, the failure of furosemide to alter R_b weighs against direct coupling. More direct evidence in this regard is the finding that submucosal ouabain, a specific inhibitor of the Na^+,K^+ pump, inhibits ion transport without altering R_b in secreting tissues.[52]

A second possibility is that some intracellular mediator regulates the K^+ conductance. The observation that intracellular Ca^{2+} mediates K^+ permeabilities in a variety of cells[62] suggests that Ca^{2+} might also mediate the K^+ permeability in tracheal epithelium. Although the Ca^{2+} ionophore does stimulate Cl^- secretion and appears to produce a decrease in R_b similar to that observed with epinephrine, there is as yet no direct evidence for a Ca^{2+}-activated K^+ permeability in tracheal epithelium.

2.4.5. The Apical Membrane Na^+ Entry Process

Na^+ absorption by tracheal epithelium is an electrogenic process that conforms to the classic Koefoed-Johnsen and Ussing model.[63] Mucosal amiloride inhibits both Na^+ absorption and the I_{sc}.[64] Amiloride also increases R_a and hyperpolarizes Ψ_a[40], changes identical to those observed in a variety of Na^+-absorbing epithelia.

Although a variety of neurohumoral agents mediate the rate of Cl^- secretion (Table I), no endogenous substance has yet been found that stimulates the rate of Na^+ absorption. While secretagogues have been observed to inhibit the rate of Na^+ absorption, the effect is often small and variable.[64] In addition, it is not clear whether the decrease in Na^+ absorption is caused by a decrease in apical Na^+ permeability or by a decrease in the electrochemical driving force for Na^+ entry. With regard to the latter suggestion, the small depolarization of ψ_a accompanying stimulation of secretion may decrease the electrochemical gradient for Na^+ entry; it is unknown whether the rate of Cl^- secretion influences intracellular Na^+ activity.

2.4.6. Similarities between Cellular Mechanisms of Ion Transport in Cl^--Secreting and Na^+-Absorbing Epithelia

There are several interesting similarities between the cellular mechanisms of electrogenic ion transport in Cl^--secreting epithelia, such as canine trachea, and Na^+-absorbing epithelia, such as amphibian skin, urinary bladder, mammalian colon, etc. First, the rate of transepithelial transport is regulated, at least in the short term, by an ion-specific apical membrane conductance. Second, ion movement across the apical membrane is passive; both Cl^- exit and Na^+ entry are driven by favorable electrochemical gradients. Third, a basolateral membrane Na^+,K^+-ATPase couples metabolic energy conversion to ion transport. Fourth, as the rate of transepithelial transport increases, the basolateral membrane K^+ permeability increases, thus preventing large changes in cellular ionic content.

The similarities between the mechanisms of secretion and absorption are of particular interest in tracheal epithelium; there the same cell appears to contain the machinery for both Cl^- secretion and Na^+ absorption, with the balance between the two regulated by the neurohumoral environment. These similarities may also have implications for the embryologic development of transport processes. For example, in mammalian intestine the crypt cells, which contain active Cl^- secretory processes, are the precursors of the surface epithelial cells, which contain active Na^+ absorptive mechanisms.[41,65]

3. The Bronchial Epithelium

3.1. Morphologic Considerations

Bronchial epithelium is histologically similar to tracheal epithelium[11,66]; the epithelium is primarily composed of ciliated columnar epithelial cells, goblet cells, and basal cells, in decreasing order of frequency. Although other cells are present in the airways,[10] they are much less frequent.

Although tight junctions connect the epithelial cells at all levels of the airway,[67,68] freeze–fracture techniques reveal morphologic differences in different regions.[69] In tracheal epithelium, tight junctions consist of multiple parallel, sparsely interconnected fibrils, whereas in the bronchial epithelium, the fibrils are fewer and more highly interconnected. Furthermore, the junctional complexes joining ciliated cells differ from those joining mucus and serous cells; junctions of ciliated cells have a greater number of fibrils but fewer interconnections than those of mucus and serous cells. The relation between these morphologic characteristics and the permeability properties of the epithelium is unclear. However, it is of interest that the structural heterogeneity is matched by a heterogeneity in ionic permeability, i.e., transepithelial resistance is greater in tracheal than in bronchial epithelium.[65]

3.2. Ion Transport

The ion transport properties of the airway epithelia are not uniform throughout the lung. As Table VII summarizes, the transepithelial electrical properties and ion fluxes vary from the trachea to the main stem bronchi to the more distal fourth- to sixth-generation bronchi.[66,70] When measured both *in vitro*[66] and *in vivo*,[13] ψ_t decreases as one moves from the proximal to the distal airways owing to a decrease in R_t in more distal portions of the airway. While the I_{sc} does not exhibit marked regional variations, the ion fluxes which account for the I_{sc} do. In tracheal epithelium, the I_{sc} is primarily attributed to Cl^- secretion with a lesser rate of Na^+ absorption. On the other hand, in bronchial epithelium, the I_{sc} is attributed to Na^+ absorption. A low rate of K^+ secretion has also been observed in bronchial epithelium but there is no net movement of Cl^- under short-circuit conditions.

Canine bronchial epithelium absorbs Na^+ via an electrogenic process that appears to conform to the Koefoed-Johnsen and Ussing model.[63] Two observations support this deduction. The apical membrane probably contains an electrically conductive Na^+ entry process, since mucosal amiloride inhibits Na^+ absorption and increases transepithelial resistance.[66] On the other hand, the basolateral membrane probably contains the Na^+,K^+ pump, since submucosal ouabain inhibits Na^+ absorption.

Although a variety of neurohumoral agents have been examined, none has yet been found that either elicits electrogenic Cl^- secretion or increases Na^+ absorption. Acetylcholine has been observed to increase the unidirectional flux of Cl^- from the submucosal to the mucosal solution and to decrease the net rate of Na^+ absorption, by increasing the unidirectional Na^+ flux from the submucosal to the mucosal solution.[70] However, these changes were not paralleled by changes in the transepithelial electrical properties. Thus, acetylcholine probably induces neutral NaCl secretion by submucosal glands, as is the case in tracheal epithelium.

While agents that stimulate bronchial Na^+ absorption have not been found, the indoleamines serotonin, melatonin, and harmaline have been found to inhibit bronchial Na^+ transport in the baboon.[71] Serotonin inhibits Na^+ absorption with characteristics similar to those observed with amiloride inhibition: it is only effective from the mucosal surface; it is reversible; it increases transepithelial resistance; and it does not have synergistic effects with amiloride. Since serotonin-producing enterochromaffin cells are located throughout the airways, it has been suggested that indoleamines may mediate Na^+ absorption. However, some doubt about the physiologic role of the indoleamines is raised by the large doses required (0.4 mM inhibits 50% of the amiloride-sensitive I_{sc}) and the requirement for luminal application.

4. The Alveolar Epithelium

The alveolus is the principal site of gas exchange. Both oxygen and carbon dioxide diffuse across the alveolar epithelial cells and the capillary endothelial cells, driven by their concentration gradients. The major barrier between the gas in the alveolus and the fluid in the vascular interstitial space is the alveolar epithelium. This epithelium plays a major role in keeping the alveolus fluid-free and air-filled; it also controls the ionic composition of the aqueous subphase, the thin layer of fluid that underlies the monolayer of surface-active material lining the alveoli.[72] In turn, the active and passive transport properties of the epithelium allow it to perform these functions.

Because the pulmonary architecture is complicated, it has been difficult to elucidate the functional properties of the alveolar epithelium. As a result, several experimental approaches have been tried. The following discussion is organized according to these approaches.

4.1. Morphologic Considerations

The alveolar epithelium is composed of two cell types: membranous type I cells and granular type II cells.[73,74] Type I pneumocytes are broad, thin (0.1 to 0.3 μm) cells that cover approximately 97% of the alveolar surface. Figure 7 shows a section of a type I cell lining the alveolar surface; the cell is directly adjacent to the pulmonary capillary endothelium. The air–blood barrier is quite thin, consisting of epithelial cell, interstitial space, and endothelium; thus, the distance for gas diffusion is small. Figure 7 also shows a cuboidal type II cell; although they are slightly more numerous than type I cells, they occupy less than 3% of the alveolar surface. Whereas type I cells have relatively few intracellular organelles, type II cells contain multivesicular bodies, multilamellar bodies, mitochondria, peroxisomes, rough endoplasmic reticulum, and a prominent Golgi apparatus. Furthermore, type I cells have a smooth surface whereas type II cells have blunt microvilli projecting from their surface. Type II cells produce the surface-active material that coats the alveoli.[75,76]

Table VII. Regional Differences in the Ion Transport Properties of Canine Airway Epithelium[a]

	Trachea	Main stem bronchus	Bronchi
Φ_t (mV)	10 to 40	9 to 15	5 to 12
R_t (Ω-cm^2)	200 to 700	140 to 200	100 to 140
I_{sc} (μA/cm^2)	30 to 100	30 to 100	30 to 100
Net J^{Cl}_{sm} (μeq/cm^2 per hr)	−1.0 to −3.5	0.5 to −1.5	−0.5 to +0.5
Net J^{Na}_{ms} (μeq/cm^2 per hr)	0.5 to 1.5	0.8 to 1.5	1.5 to 3

[a]Derived from Refs. 7, 12, 17, 66, 70. J^{Na} and J^{Cl} refer to the net fluxes of Na^+ and Cl^-, respectively. A flux from mucosal to submucosal solution is considered positive and a flux from submucosa to mucosa is negative, in accord with the direction of current flow. The values of Ψ_t are identical to those obtained *in vivo*.[13]

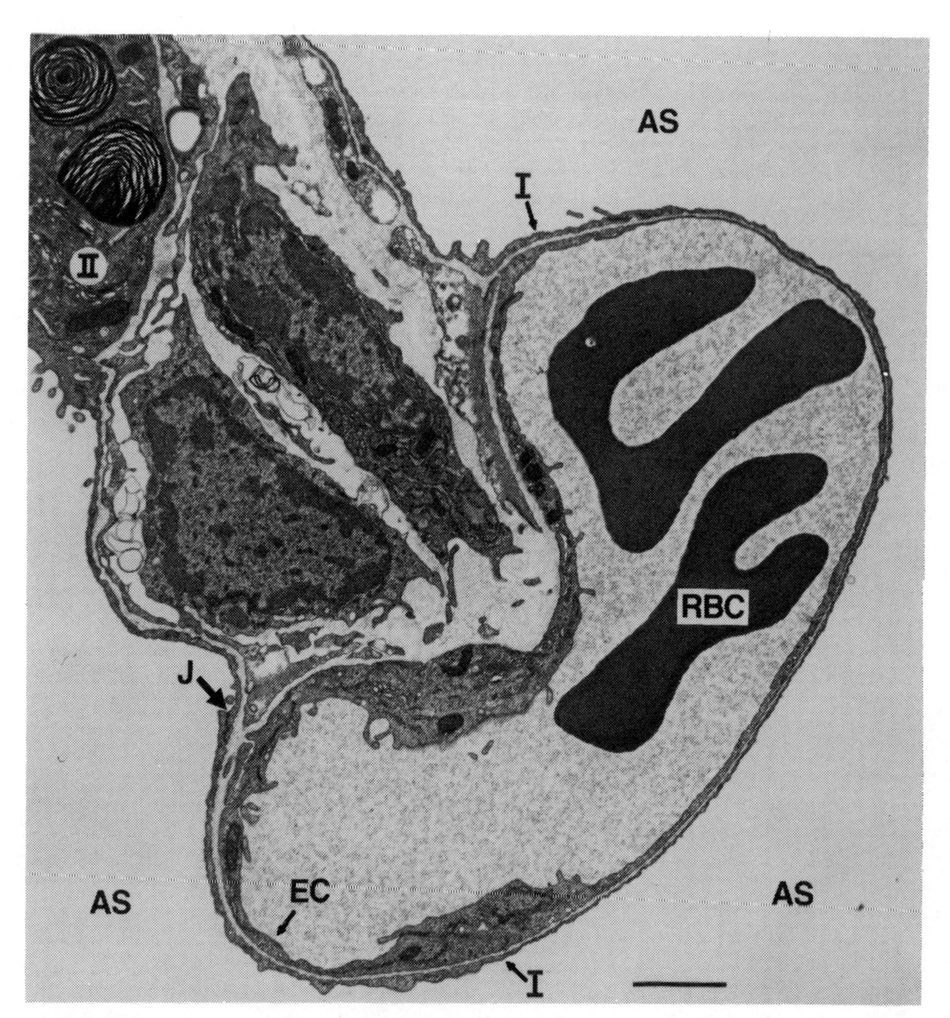

Fig. 7. Electron micrograph of rat alveolar capillary membrane. The pulmonary capillary containing a red blood cell (RBC) is lined by endothelial cells (EC). The capillary endothelium is adjacent to a thin type I pneumocyte (I). The type I epithelial cells are joined by tight junctions (J). Also shown is a portion of a type II epithelial cell (II) which contains a lamellar body, the source of pulmonary surfactant. AS, alveolar space. Bar = 2 μm. From Schneeberger[74] with permission.

Freeze–fracture studies reveal that the tight junctions between epithelial cells consist of three to six continuous, highly interconnected strands that completely encircle the cells.[77] In contrast, tight junctions between capillary endothelial cells consist of one to three partly interconnected rows of intramembranous particles which show areas of apparent discontinuity. These structural observations suggest that the alveolar epithelium provides the main permeability barrier of the alveolar capillary membrane, a conclusion supported by studies employing intravenously injected enzymatic tracers. Following injection of cytochrome *C* (12,000 daltons), an electron-dense reaction product is observed in endothelial cell junctions and in the basement membrane between the alveolar epithelium and the capillary endothelium[78]; however, it does not cross the alveolar epithelium. Following injection of a larger tracer, horseradish peroxidase (40,000 daltons), the reaction product is confined to the capillary lumen.[79] When vascular volume is expanded, and presumably pulmonary vascular pressure is increased, the tracer passes through the endothelial wall but fails to penetrate the alveolar epithelium.[79,80]

4.2. Studies of Alveolar Permeability in Whole Lung Preparations

Whole lung preparations have frequently been studied to determine alveolar epithelial permeability. The alveolar air space is usually filled with fluid and then the movement of radiolabeled substances between the alveolar space and the vascular space is measured. Such studies demonstrate a high permeability to water, a relatively low permeability to small water-soluble molecules, and nearly complete impermeability to macromolecules.[81–84] Based on the permeabilities of urea, glucose, and sucrose, Taylor and Gaar[82] calculated an "equivalent pore radius" of 0.6 to 1.0 nm for the alveolar epithelium. Theodore *et al.*[83] examined the permeability of the solutes sucrose, inulin, and dextran (60,000 to 90,000 daltons) and suggested that the membrane is not homogeneous, but that it may also contain a few "larger pores" in the range of 8 nm. Despite the problems inherent in such measurements in a whole lung preparation, the results indicate that the pulmonary epithelium provides a continuous barrier that has nonelectrolyte permeabilities similar to those found in a variety of other epithelia.

Evidence that the alveolar epithelium actively transports ions was obtained by Matthay *et al.* in the lungs of anesthetized sheep.[85] They infused either a Ringer's solution or an isosmotic serum solution into one lobe of the lung and measured lymph flow and the clearance of water and protein from the lung. Fluid was absorbed under both conditions, but protein from the serum instillate was concentrated by 50% after 4 hr in the alveoli. Fluid absorption cannot be accounted for by passive forces, the hydrostatic and oncotic pressures across the alveoli. On the other hand, there is evidence for active transport: amiloride added to the instilled serum significantly decreased the absorption of fluid and prevented the concentration of alveolar fluid protein.[86]

There is one drawback to studies performed in whole lung: transport processes cannot be unequivocally localized to the alveolar epithelium. While the large surface area of the alveolar epithelium makes its involvement likely, ion transport, particularly Na^+ absorption, by large and small airways may also make a substantial contribution to the results.

4.3. Ion Transport by Amphibian Alveolar Epithelium

Because the anatomy of mammalian lung is complex, several workers have turned to the hollow amphibian lung as a model of the alveolar epithelium. The frog lung consists of a large hollow sac with trabecular folds. Furthermore, it is lined by cells that have some of the characteristics of type I and type II pneumocytes.

In vitro bullfrog lung generates a spontaneous transepithelial electrical potential difference, ψ_t, of approximately +15 and +20 mV (with reference to the mucosal solution).[87–89] The transepithelial electrical resistance is in the range of 1700 Ω-cm^2,[89] placing it in the category of "tight" epithelia. When ψ_t is clamped to zero, the epithelium produces a short-circuit current of 10–20 μA/cm^2 by secreting Cl^- from the submucosal to the luminal surface.[88] Calculation of an "equivalent pore radius" from the measured transepithelial fluxes of nonelectrolytes suggests a pore radius of 0.8 to 0.9 nm.[89] These results indicate that the alveolar epithelium of the frog is a relatively impermeable, high-resistance epithelium. The demonstration of active ion transport also suggests that the epithelium may control the ionic composition of fluid in the alveolus.

4.4. Cultured Monolayers of Alveolar Epithelial Cells

Again, because of the complex anatomy, other researchers have examined the transport properties of cultured alveolar epithelial cells.[90,91] Such studies have focused on cultured monolayers of type II epithelial cells from rat lungs. In primary tissue cultures, type II cells maintain: (1) their histologic characteristics; (2) their polarity, with microvilli on the apical surface (facing the culture medium); (3) their biochemical properties, including the ability to secrete surfactant; and (4) their ability to form typical-appearing tight junctions between cells.[91,93]

Type II cells grown as monolayers in plastic culture dishes form domes, a formation also observed with a variety of other cultured epithelial cells[94–96] and attributed to active cellular transport of solute from the culture medium to the space beneath the cell monolayer. The transport results in a localized accumulation of fluid at an increased hydrostatic pressure beneath the monolayer, i.e., a dome.

Cultured type II cells appear to form domes by active Na^+ transport. When metabolic activity is inhibited by either dinitrophenol, potassium cyanide, or incubation at 4°C, domes do not form.[97] Similarly, both amiloride, which blocks an apical Na^+ conductance, and ouabain, a specific inhibitor of the Na^+,K^+-ATPase, cause domes to disappear and prevent further formation.[98]

More direct evidence for active Na^+ absorption by cultured type II cells comes from the study of monolayers of cells grown on collagen-coated Millipore filters mounted in "Ussing chambers." Monolayers develop a transepithelial electrical potential difference, ψ_t, of approximately 1 mV (mucosal solution reference) and an R_t of 200–250 Ω-cm^2. Amiloride abolishes ψ_t when it is added to the apical side of the epithelium and ouabain inhibites ψ_t most effectively when added to the basal side. Furthermore, Na^+ absorption appears to be under hormonal control: β-adrenergic agonists increase both the ψ_t measured on monolayers grown on Millipore filters and the rate of dome formation in monolayers grown on plastic.[91]

5. The Fetal Lung

In contrast to the adult lung, the fetal lung is a fluid-filled organ. The lung fluid is a major determinant of normal fetal lung development; when normal fluid dynamics are disrupted, lung maturation is altered. On one hand, chronic drainage of liquid from the fetal lamb lung retards normal lung growth, prevents the usual increase in lung weight, and inhibits alveolar development.[99] On the other hand, ligation of the trachea, which prevents drainage of secreted fluid, increases lung tissue weight, enhances the growth of future gas-exchanging areas, and produces thinner, better-defined alveolar walls than those of control animals. Similar observations have been made in infants with congenital atresia of the bronchi and trachea.[100,101]

Most studies investigating fetal fluid movement have used the fetal lamb as the model. Therefore, unless otherwise noted, the ensuing discussion of fetal fluid transport will refer to the sheep. (Full-term gestation is 130–140 days.)

5.1. Morphologic Considerations

Early in gestation (from 39 to 74 days), the lung consists of branching, tubular structures lined by a single layer of columnar epithelium.[102] With lung growth, the airways and future alveolar spaces begin to develop and there is a shift from columnar to cuboidal epithelium at 74–94 days. After 95 days, the future air spaces develop a flattened, single layer of cuboidal epithelial cells.

To secrete fluid and to maintain ionic concentration differences between lung liquid and plasma, the lung requires a continuous sheet of epithelial cells. Indeed, by 69 days of gestation, the pulmonary epithelium is both structurally and functionally intact. Morphologically, freeze–fracture electron microscopy shows well-developed, continuous tight junctions[102,103] as early as 39 days of gestation; from 39 days of gestation through 121 days, the number of junctional strands increases slightly. Physiologically, the airway and alveolar epithelia are relatively impermeable to lipid-insoluble nonelectrolytes by 69 days of gestation.[103,104]

5.2. Fluid Secretion by the Fetal Lung

Fluid production by the fetal lung has the following characteristics:

1. Fetal lung secretes fluid[103,105] at a rate that increases with gestational age, reaching rates of 2–5 ml/hr per kg body wt. The volume of fetal fluid is 30–40 ml/kg body wt.
2. The composition of fetal lung fluid differs from that of plasma and amniotic fluid, as shown in Table VIII.[105–107] Both Cl^- and K^+ concentrations are greater in lung fluid than in plasma or amniotic fluid, although the Cl^- concentration

Table VIII. Composition of Fetal Lung Liquid, Plasma, and Amniotic Fluid[a]

Source	Na^+ (mM)	K^+ (mM)	Ca^{2+} (mM)	Cl^- (mM)	PO_4^{3-} (mM)	HCO_3^- (mM)	pH (units)	Protein (g/100 ml)
Lung liquid	150	6.3	0.8	157	0	2.8	6.3	0.06
Plasma	150	4.8	3.3	107	2.3	24	7.4	4.5
Amniotic fluid	113	7.6	1.6	87	3.2	19	7.0	0.1

[a]Derived from Refs. 105–107, 114.

difference is quantitatively greater. In contrast, the Na^+ concentration is the same in lung fluid as in plasma, while the Ca^{2+} and HCO_3^- concentrations are lower in lung fluid than in plasma or amniotic fluid. This ionic composition suggests Cl^- secretion.

3. The fetal lung generates an electrical potential difference of −1 to −10 mV (mean −4 mV) (lumen with reference to plasma).[108]
4. When unidirectional, radioisotope fluxes are measured between fetal lung liquid and plasma, the Cl^- flux-ratio (flux into the lung/out of the lung) is 1.6, compared to a flux-ratio of 0.6 predicted for a passive distribution.[108] K^+ also enters lung fluid against an electrochemical gradient but is quantitatively less important; Na^+ and Ca^{2+} appear to be passively distributed.
5. When added to lung liquid, cyanide abolishes secretion and net absorption of fluid occurs.[108] Absorption presumably results from a favorable osmotic pressure gradient, due to the protein concentration difference between lung fluid and plasma.

These observations provide compelling evidence that active secretion of Cl^- against its electrochemical gradient is responsible for fetal lung fluid production; Na^+ may enter the lung passively, with a favorable electrochemical gradient. Furthermore, the rate of secretion may be hormonally controlled: addition of prolactin to lung fluid (2.5 μg/ml) triples the rate of secretion.[109]

5.3. Absorption of Fetal Lung Liquid

At birth the lung is transformed from a fluid-filled to an air-filled organ; fluid secretion gives way to fluid absorption as liquid is cleared from the lung. In fact, a slowing of the rate of spontaneous fluid production begins 2 days before delivery.[110]

The prenatal decrease in fluid secretion is not related to fetal arterial blood gas tensions, arterial pH, blood hematocrit, or plasma proteins; nor is it mediated through the vagal or cholinergic pathways.[110] Instead, epinephrine regulates the rate at which the lung absorbs fluid in the term fetus. Several observations support this conclusion. First, epinephrine (infused intravenously at rates that result in plasma epinephrine concentrations comparable to those obtained spontaneously during labor and delivery) reversibly inhibits fluid secretion and, in fact, results in fluid absorption.[111] Second, inhibition of secretion results from a β-adrenergic effect; infusion of norepinephrine fails to inhibit secretion, and β-adrenergic blockade with propranolol prevents the inhibition produced by epinephrine.[111] Third, the gestational age of the lamb determines the response to β-adrenergic agonists[111]; with increasing fetal maturity, infusion of isoproterenol produces a progressively greater inhibition of secretion and, as term is approached, isoproterenol infusion induces fluid absorption. Fourth, fluid secretion and plasma concentrations of epinephrine are inversely related during labor[112]; epinephrine concentrations rise nearly 100-fold during the 900 min before birth while fluid secretion (7 ml/min) reverses to fluid absorption (30 ml/min). Finally, similar effects are seen in the rabbit fetus; the β-adrenergic agonist isoxsuprine, administered 3 days prior to term, decreases the quantity of pulmonary fluid.[113] Taken together, these findings provide convincing evidence that epinephrine mediates the physiologic reversal in the direction of fluid transport at birth.

There are two possible mechanisms for the absorption of lung fluid during late term and labor. First, net absorption might be a passive process, resulting from a protein osmotic gradient of nearly 20 mm Hg (see Table VIII). Thus, with active secretion inhibited, the fluid could be absorbed passively. Such a reversal of flow has been observed when cyanide inhibits active transport.[108] Second, the lung may actively absorb fluid. Supporting this suggestion, the electrical potential difference across the lung increases both during infusion of epinephrine (from −2.7 mV to −4.0 mV) and with gestational age.[112] Furthermore, amiloride added to fetal lung liquid not only prevents absorption of fluid produced by epinephrine infusion, but also reverses the direction of fluid movement from absorption to secretion.[112]

While these observations clearly indicate that fetal lung actively transports ions, the site of transport has not been localized. It has been assumed that fluid is produced by the alveolar epithelium since it constitutes a large interface between the lung liquid and the interstitial fluid.[108] However, the adult airway epithelium both secretes and absorbs salt and water and the adult alveolar epithelium absorbs fluid; thus, it is possible that the airways play a major role in fetal lung fluid dynamics.

6. Summary

The principal function of the lung is gas exchange. A system of epithelia—the conducting airways and alveoli—perform this function; the airways deliver gas to the alveoli where the diffusional exchange of O_2 and CO_2 occurs. These epithelia, the boundary separating air and liquid throughout the lung, have both active ion transport and passive barrier functions.

In the conducting airways, ion transport plays an important role in controlling the quantity and composition of the respiratory tract fluid. The respiratory tract fluid, in turn, is required for effective pulmonary mucociliary clearance, a lung defense mechanism that removes inhaled particulate material. Ion transport by the airway epithelia varies depending upon the location. Trachea and proximal airway epithelia have a moderately high transepithelial electrical resistance, and Cl^- secretion is the predominant active transport process. In contrast, the bronchial epithelium has a lower transepithelial electrical resistance, and Na^+ absorption is the major active transport process.

Tracheal epithelium secretes Cl^- via an electrogenic transport process. The rate of secretion is regulated by a variety of neurohumoral agents; the effect of secretagogues is regulated, in turn, by intracellular levels of cAMP and Ca^{2+}. In canine tracheal epithelium, Cl^- secretion conforms to a cellular model that appears to apply to several other Cl^--secreting epithelia. Cl^- enters the cell across the basolateral membrane via an electrically neutral process: Cl^- is accumulated intracellularly against an adverse electrochemical gradient, its entry coupled to and energized by the movement of Na^+ down a favorable electrochemical gradient. As a result of the intracellular accumulation, Cl^- may exit passively down a favorable electrochemical gradient across a Cl^--permeable apical membrane. The basolateral membrane also contains the Na^+, K^+-ATPase which indirectly supplies the energy for transport: the enzyme maintains the Na^+ gradient by extruding Na^+ from the cell. The Na^+,K^+-ATPase also pumps K^+ into the cells against its electrochemical gradient; K^+ then recycles across the basolateral membrane, exiting passively via a K^+ conductance. The K^+ permeability of the basolateral membrane increases when secretagogues increase the rate of Cl^- secretion; as a result, the intracellular K^+ activity is constant and the intracellular voltage remains negative.

In the alveolus the epithelium contains active ion transport processes that help keep the alveolus fluid-free and that may control the composition of the thin layer of fluid which lines the alveoli. In mammals the epithelium absorbs Na^+ while in amphibians it secretes Cl^-.

In the fetus, fluid secretion by the respiratory epithelium plays a central role in the normal development of the lung. Fluid secretion results from active Cl^- transport into the lung lumen. At birth, the lung is transformed from a fluid-filled to an air-filled organ; epinephrine mediates this conversion by decreasing the rate of Cl^- secretion and stimulating Na^+ absorption.

ACKNOWLEDGMENTS. Research cited herein from the author's laboratory was supported by grants from the NIH (NHLBI) and Cystic Fibrosis Foundation. The author is an established investigator of the American Heart Association. I wish to thank Drs. E. E. Schneeberger, J. H. Widdicombe, and C. B. Basbaum for generously providing electron micrographs and Barbara Yerkes for a critical reading of the manuscript.

List of Symbols

ψ_t Transepithelial electrical potential difference; submucosal solution with respect to mucosal solution, in mV

ψ_a Electrical potential difference across the apical membrane; cell interior with respect to mucosal solution, in mV

ψ_b Electrical potential difference across the basolateral membrane; submucosal solution with respect to cell interior, in mV

ψ_i Electrical potential difference recorded by an ion-selective microelectrode, in mV

I, I_{sc} Current; I_{sc} represents the short-circuit current or the current required to clamp ψ_t to 0, in $\mu A/cm^2$

R Electrical resistance, in Ω-cm^2

G Electrical conductance; ($G = 1/R$), in mS/cm^2

f_R Fractional resistance of the apical membrane; $= \Delta\psi_a/\Delta\psi_b = R_a/(R_a + R_b)$

α Membrane resistance ratio, R_a/R_b

J Flux; measured with radiolabeled tracers, in $\mu eq/cm^2$ per hr

$\Delta\tilde{\mu}^i$ Electrochemical potential difference for an ion (i), in mV

a^i Activity of an ion (i), in mM

Subscripts

a Apical membrane
b Basolateral membrane
t Transepithelial
p Paracellular
m Mucosal bathing solution
s Submucosal bathing solution
c Cell interior

References

1. Nadel, J. A., B. Davis, and R. J. Phipps. 1979. Control of mucus secretion and ion transport in airways. *Annu. Rev. Physiol.* **41**:369–381.
2. *Ciba Foundation Symposium* No. 54. 1978. *Respiratory Tract Mucus.* Elsevier, Amsterdam.
3. King, R. J. 1982. Pulmonary surfactant. *J. Appl. Physiol.* **53**(1):1–8.
4. Wanner, A. 1977. Clinical aspects of mucociliary transport. *Am. Rev. Respir. Dis.* **116**:73–125.
5. Yoneda, K. 1976. Mucous blanket of rat bronchus: An ultrastructural study. *Am. Rev. Respir. Dis.* **114**:837–842.
6. Yoneda, K. 1976. Mucous blanket of rat bronchus. *Am. Rev. Respir. Dis.* **114**:837–842.
7. Olver, R. E., B. Davis, M. G. Marin, and J. A. Nadel. 1975. Active transport of Na^+ and Cl^- across the canine tracheal epithelium *in vitro. Am. Rev. Respir. Dis.* **112**:811–815.
8. Widdicombe, J. H., C. B. Basbaum, and E. Highland. 1981. Ion contents and other properties of isolated cells from dog tracheal epithelium. *Am. J. Physiol.* **241**:C184–C192.
9. Rhodin, J. A. G. 1966. Ultrastructure and function of the human tracheal mucosa. *Am. Rev. Respir. Dis.* **93**:1–15.
10. Breeze, R. G., and E. B. Wheeldon. 1977. The cells of the pulmonary airways. *Am. Rev. Respir. Dis.* **116**:705–777.
11. Frasca, J. M., O. Auerbach, V. R. Parks, and J. D. Jamieson. 1968. Electron microscopic observations of the bronchial epithelium of dogs. *Exp. Mol. Pathol.* **9**:363–379.
12. Al-Bazzaz, F. J., and Q. Al-Awqati. 1979. Interaction between sodium and chloride transport in canine tracheal mucosa. *J. Appl. Physiol.* **46**:111–119.
13. Boucher, R. C., P. A. Bromberg, and J. T. Gatzy. 1980. Airway transepithelial electric potential *in vivo:* Species and regional differences. *J. Appl. Physiol.* **48**:169–176.
14. Welsh, M. J. 1982. Cigarette smoke inhibition of ion transport in canine tracheal epithelium. *J. Clin. Invest.* **71**:1614–1623.
15. Welsh, M. J., P. L. Smith, and R. A. Frizzell. 1982. Chloride secretion by canine tracheal epithelium. II. The cellular electrical potential profile. *J. Membr. Biol.* **70**:227–238.
16. Welsh, M. J., J. H. Widdicombe, and J. A. Nadel. 1980. Fluid transport across the canine tracheal epithelium. *J. Appl. Physiol.* **45**:905–909.
17. Widdicombe, J. H., I. F. Ueki, I. Bruderman, and J. A. Nadel. 1979. The effects of sodium substitution and ouabain on ion transport by dog tracheal epithelium. *Am. Rev. Respir. Dis.* **120**:385–392.
18. Widdicombe, J. H., C. B. Basbaum, and J. Y. Yee. 1979. Localization of Na pumps in the tracheal epithelium of the dog. *J. Cell Biol.* **82**:380–390.
19. Al-Bazzaz, F. J. 1981. Role of cyclic AMP in regulation of chloride secretion by canine tracheal mucosa. *Am. Rev. Respir. Dis.* **123**:295–298.

20. Smith, P. L., M. J. Welsh, J. S. Stoff, and R. A. Frizzell. 1982. Chloride secretion by canine tracheal epithelium. I. Role of intracellular cAMP levels. *J. Membr. Biol.* **70**:217–226.
21. Al-Bazzaz, F., V. P. Yadava, and C. Westenfelder. 1981. Modification of Na and Cl transport in canine tracheal mucosa by prostaglandins. *Am. J. Physiol.* **240**:F101–F105.
22. Lazarus, S. C., C. B. Basbaum, and W. M. Gold. 1982. Cellular localization of cyclic AMP in the trachea of dog, cat, and ferret. *Am. Rev. Respir. Dis.* **125**:244.
23. Lazarus, S. C., C. B. Basbaum, and W. M. Gold. 1982. Effect of prostaglandin E_1 on intracellular cyclic AMP in cat and dog trachea. *Clin. Res.* **30**:434a.
24. Al-Bazzaz, F., and T. Jayaram. 1981. Ion transport by canine tracheal mucosa: Effect of elevation of cellular calcium. *Exp. Lung Res.* **2**:121–130.
25. Marin, M. G., B. Davis, and J. A. Nadel. 1977. Effect of histamine on electrical and ion transport properties of tracheal epithelium. *J. Appl. Physiol.* **42**:735–738.
26. Al-Bazzaz, F. J., and E. Cheng. 1979. Effect of catecholamines on ion transport in dog tracheal epithelium. *J. Appl. Physiol.* **47**:397–403.
27. Marin, M. G., B. Davis, and J. A. Nadel. 1976. Effect of acetylcholine on Cl^- and Na^+ fluxes across dog tracheal epithelium *in vitro*. *Am. J. Physiol.* **231**:1546–1549.
28. Ueki, I., V. F. German, and J. A. Nadel. 1980. Micropipette measurement of airway submucosal gland secretion. *Am. Rev. Respir. Dis.* **121**:351–357.
29. Phipps, R. J., J. A. Nadel, and B. Davis. 1980. Effect of alpha-adrenergic stimulation on mucus secretion and on ion transport in cat trachea *in vitro*. *Am. Rev. Respir. Dis.* **121**:359–365.
30. Nadel, J. A., and B. Davis. 1980. Parasympathetic and sympathetic regulation of secretion from submucosal glands in airways. *Fed. Proc.* **39**:3075–3079.
31. Silva, P., J. Stoff, M. Field, L. Fine, J. N. Forrest, and F. H. Epstein. 1977. Mechanism of active chloride secretion by shark rectal gland: Role of Na-K-ATPase in chloride transport. *Am. J. Physiol.* **233**:F298–F306.
32. Welsh, M. J., P. L. Smith, and R. A. Frizzell. 1983. Intracellular chloride activities in the isolated perfused shark rectal gland. *Am. J. Physiol.* **245**:F640–F644.
33. Nagel, W., and P. Reinach. 1980. Mechanism of stimulation by epinephrine of active transepithelial Cl transport in isolated frog cornea. *J. Membr. Biol.* **56**:73–79.
34. Candia, O. A., H. F. Schoen, L. Low, and S. M. Podos. 1981. Chloride transport inhibition by piretanide and MK-196 in bullfrog corneal epithelium. *Am. J. Physiol.* **240**:F25–F29.
35. Field, M. 1979. Intracellular mediators of secretion in the small intestine. In: *Mechanisms of Intestinal Secretion*. H. J. Binder, ed. Liss, New York. pp. 83–91.
36. Frizzell, R. A., M. Field, and S. G. Schultz. 1979. Sodium-coupled chloride transport by epithelial tissues. *Am. J. Physiol.* **236**:F1–F8.
37. Frizzell, R. A., and K. Heintze. 1979. Electrogenic chloride secretion by mammalian colon. In: *Mechanisms of Intestinal Secretion*. H. J. Binder, ed. Liss, New York. pp. 101–110.
38. Degnan, K. J., K. J. Karnaky, and J. A. Zadunaisky. 1977. Active chloride transport in the *in vitro* opercular skin of a teleost (*Fundulus heteroclitus*), a gill-like epithelium rich in chloride cells. *J. Physiol. (London)* **271**:155–191.
39. Machen, T. E., and W. L. McLennan. 1980. Na^+-dependent H^+ and Cl^- transport in *in vitro* frog gastric mucosa. *Am. J. Physiol.* **238**:G403–G413.
40. Welsh, M. J., P. L. Smith, and R. A. Frizzell. 1983. Chloride secretion by canine tracheal epithelium. III. Membrane resistances and electromotive forces. *J. Membr. Biol.* **71**:209–218.
41. Schultz, S. G., R. A. Frizzell, and H. N. Nellans. 1977. Active sodium transport and the electrophysiology of rabbit colon. *J. Membr. Biol.* **33**:351–384.
42. Welsh, M. J. 1983. Inhibition of chloride secretion by furosemide in canine tracheal epithelium. *J. Membr. Biol.* **71**:219–226.
43. Davis, B., M. B. Marin, J. W. Yee, and J. A. Nadel. 1979. Effect of terbutaline on movement of Cl^- and Na^+ across the trachea of the dog *in vitro*. *Am. Rev. Respir. Dis.* **120**:547–552.
44. Al-Bazzaz, F. J., and J. Kelsey. 1982. Effect of substance P on ion transport by tracheal mucosa. *Am. Rev. Respir. Dis.* **125**:243.
45. Welsh, M. J. 1982. Intracellular chloride activities in canine tracheal epithelium: Direct evidence for sodium-coupled intracellular chloride accumulation in a chloride secreting epithelium. *J. Clin. Invest.* **71**:1392–1401.
46. Silva, P., K. Spokes, J. Epstein, and F. H. Epstein. 1979. Action of cAMP and theophylline on ion movement during ouabain inhibition of Na-K-ATPase in perfused rectal gland of *Squalus acanthias*. *Bull. Mt. Desert Isl. Biol. Lab.* **19**:70–72.
47. Frizzell, R. A., P. L. Smith, E. Vosburgh, and M. Field. 1979. Coupled sodium–chloride influx across brush border of flounder intestine. *J. Membr. Biol.* **46**:27–39.
48. Geck, P., C. Pietrzyk, B. C. Burckhardt, B. Pfeiffer, and E. Heinz. 1980. Electrically silent cotransport of Na^+, K^+ and Cl^- in Ehrlich cells. *Biochim. Biophys. Acta* **600**:432–447.
49. Liedtke, C. M., and U. Hopfer. 1982. Mechanism of Cl^- translocation across small intestinal brush-border membrane. I. Absence of Na^+–Cl^- cotransport. *Am. J. Physiol.* **242**:G263–G271.
50. Cabantchik, Z. I., and A. Rothstein. 1974. Membrane proteins related to anion permeability of human red blood cells. *J. Membr. Biol.* **15**:207–226.
51. Shorofsky, S., M. Field, and H. Fozzard. 1983. Electrophysiology of Cl secretion in canine trachea. *Membr. Biol.* **72**:105–115.
52. Welsh, M. J. 1983. Evidence for a basolateral membrane potassium conductance in canine tracheal epithelium. *Am. J. Physiol.* **244**:C377–C384.
53. Smith, P. L., and R. A. Frizzell. 1982. Changes in intracellular K activities after stimulation of Cl secretion in canine tracheal epithelium. *Chest* **81**:5S.
54. Westenfelder, C., W. R. Earnest, and F. J. Al-Bazzaz. 1980. Characterization of Na-K-ATPase in dog tracheal epithelium: Enzymatic and ion transport measurements. *J. Appl. Physiol.* **48**:1008–1019.
55. Eaton, D. C., and M. S. Brodwick. 1980. Effects of barium on the potassium conductance of squid axon. *J. Gen. Physiol.* **75**:727–750.
56. Kirk, K. L., D. R. Halm, and D. C. Dawson. 1980. Active sodium transport by turtle colon via an electrogenic Na-K exchange pump. *Nature (London)* **287**:237–239.
57. Standen, N. B., and P. R. Stanfield. 1978. A potential- and time-dependent blockade of inward rectification in frog skeletal muscle fibres by barium and strontium ions. *J. Physiol. (London)* **280**:169–191.
58. van Driessche, W. V., and W. Zeiske. 1980. Ba^{2+}-induced conductance fluctuations of spontaneously fluctuating K^+ channels in the apical membrane of frog skin (*Rana temporaria*). *J. Membr. Biol.* **56**:31–42.
59. Welsh, M. J. 1983. Barium inhibition of basolateral membrane potassium conductance in tracheal epithelium. *Am. J. Physiol.* **244**:F639–F645.
60. Schultz, S. G. 1981. Homocellular regulatory mechanisms in sodium-transporting epithelia: Avoidance of extinction by "flush-through." *Am. J. Physiol.* **241**:F579–F590.
61. Higgins, J. T., B. Gebler, and E. Frömter. 1977. Electrical properties of amphibian urinary bladder. II. The cell potential profile in *Necturus maculosa*. *Pfluegers Arch.* **371**:87–97.
62. Gárdos, G. 1959. The role of calcium in the potassium permeability of human erythrocytes. *Acta Physiol. Acad. Sci. Hung.* **15**:121–125.
63. Koefoed-Johnsen, V., and H. H. Ussing. 1958. The nature of the frog skin potential. *Acta Physiol. Scand.* **42**:298–308.
64. Widdicombe, J. H., and M. J. Welsh. 1980. Ion transport by dog tracheal epithelium. *Fed. Proc.* **39**:3062–3066.
65. Welsh, M. J., P. L. Smith, M. Fromm, and R. A. Frizzell. 1982.

Crypts are the site of intestinal fluid and electrolyte secretion. *Science* **218**:1219–1221.

66. Boucher, R. C., M. J. Stutts, and J. T. Gatzy. 1981. Regional differences in bioelectric properties and ion flow in excised canine airways. *J. Appl. Physiol.* **51**:706–714.
67. Marin, M. L., B. P. Lane, R. E. Gordon, and E. Drummond. 1979. Ultrastructure of rat tracheal epithelium. *Lung* **156**:223–236.
68. Inoue, S., and J. C. Hogg. 1977. Freeze–etch study of the tracheal epithelium of normal guinea pigs with particular reference to intercellular junctions. *J. Ultrastruct. Res.* **61**:89–99.
69. Schneeberger, E. E. 1980. Heterogeneity of tight junction morphology in extrapulmonary and intrapulmonary airways of the rat. *Anat. Rec.* **198**:193–208.
70. Boucher, R. C., and J. T. Gatzy. 1982. Regional effects of autonomic agents on ion transport across excised canine airways. *J. Appl. Physiol.* **52**:893–901.
71. Legris, G. J., P. C. Will, and U. Hopfer. 1982. Inhibition of amiloride-sensitive sodium conductance by indoleamines. *Proc. Natl. Acad. Sci. USA* **79**:2046–2050.
72. Nielson, D. W., J. Goerke, and J. A. Clements. 1981. Alveolar subphase pH in the lungs of anesthetized rabbits. *Proc. Natl. Acad. Sci. USA* **78**:7119–7123.
73. Haies, D. M., J. Gil, and E. R. Weibel. 1981. Morphometric study of rat lung cells. I. Numerical and dimensional characteristics of parenchymal cell population. *Am. Rev. Respir. Dis.* **123**:533–541.
74. Schneeberger, E. E. 1978. Structural basis for some permeability properties of the air–blood barrier. *Fed. Proc.* **37**:2471–2478.
75. King, R. J. 1979. Utilization of alveolar epithelial type II cells for the study of pulmonary surfactant. *Fed. Proc.* **38**:2637–2643.
76. Mason, R. J., and J. Nellenbogen. 1982. Synthesis of saturated phosphatidylcholine and phosphatidylglycerol by freshly isolated rat alveolar type II cells. *Fed. Proc.* **41**:1600.
77. Schneeberger, E. E., and M. J. Karnovsky. 1976. Substructure of intercellular junctions in freeze-fractured alveolar–capillary membranes of mouse lung. *Circ. Res.* **38**:404–411.
78. Schneeberger, E. E. 1976. Ultrastructural basis for alveolar–capillary permeability to protein. *Ciba Found. Symp.* **38**:3–28.
79. Schneeberger, E. E., and M. J. Karnovsky. 1971. The influence of intravascular fluid volume on the permeability of newborn and adult mouse lungs to ultrastructural protein tracers. *J. Cell Biol.* **49**:319–334.
80. Schneeberger-Keeley, E. E., and M. J. Karnovsky. 1968. The ultrastructural basis of alveolar–capillary membrane permeability to peroxidase used as a tracer. *J. Cell Biol.* **37**:781–793.
81. Taylor, A. E., A. C. Guyton, and V. S. Bishop. 1965. Permeability of the alveolar membrane to solutes. *Circ. Res.* **16**:353–362.
82. Taylor, A. E., and K. A. Gaar. 1970. Estimation of equivalent pore radii of pulmonary capillary and alveolar membranes. *Am. J. Physiol.* **218**:1133–1140.
83. Theodore, J., E. D. Robin, R. Gaudio, and J. Acevedo. 1975. Transalveolar transport of large polar solutes (sucrose, inulin, and dextran). *Am. J. Physiol.* **229**:989–996.
84. Egan, E. A., R. M. Nelson, and R. E. Olver. 1976. Lung inflation and alveolar permeability to non-electrolytes in the adult sheep *in vivo*. *J. Physiol.* (*London*) **260**:409–424.
85. Matthay, M. A., C. C. Landolt, and N. C. Staub. 1982. Differential liquid and protein clearance from the alveoli of anesthetized sheep. *J. Appl. Physiol.* **53**:96–104.
86. Matthay, M. A., J. H. Widdicombe, and N. C. Staub. 1982. Clearance of alveolar fluid in sheep may involve an active ion transport process. *Fed. Proc.* **41**:1244.
87. Gatzy, J. T. 1967. Bioelectric properties of the isolated amphibian lung. *Am. J. Physiol.* **213**:425–431.
88. Gatzy, J. T. 1975. Ion transport across the excised bullfrog lung. *Am. J. Physiol.* **228**:1162–1171.
89. Crandall, E. D., and K. J. Kim. 1981. Transport of water and solutes across bullfrog alveolar epithelium. *J. Appl. Physiol.* **50**:1263–1271.
90. Goodman, B. E., and E. D. Crandall. 1982. Dome formation in primary cultured monolayers of alveolar epithelial cells. *Am. J. Physiol.* **243**:C96–C100.
91. Mason, R. J., M. C. Williams, J. H. Widdicombe, M. J. Sanders, D. S. Misfeldt, and L. C. Berry. 1982. Transepithelial transport by pulmonary alveolar type-II cells in primary culture. *Proc. Natl. Acad. Sci. USA* **79**:6033–6037.
92. Mason, R. J., and M. C. Williams. 1981. Pulmonary alveolar epithelial cells form domes in primary culture. *Am. Rev. Respir. Dis.* **123**:216.
93. Dobbs, L. G., R. J. Mason, M. C. Williams, B. J. Benson, and K. Sueishi. 1982. Secretion of surfactant by primary cultures of alveolar type II cells isolated from rats. *Am. Rev. Respir. Dis.* **125**:205.
94. Misfeldt, D. S., S. T. Hamamoto, and D. R. Pitelka. 1976. Transepithelial transport in cell culture. *Proc. Natl. Acad. Sci. USA* **73**:1212–1216.
95. Cereijido, M., E. S. Robbins, W. J. Dolan, C. A. Rotunno, and D. D. Sabatini. 1978. Polarized monolayers formed by epithelial cells on a permeable and translucent support. *J. Cell Biol.* **77**:853–880.
96. Handler, J. S., F. M. Perkins, and J. P. Johnson. 1980. Studies of renal cell function using cell culture techniques. *Am. J. Physiol.* **238**:F1–F9.
97. Goodman, B. E., R. S. Fleischer, and E. D. Crandall. 1982. Effects of metabolic inhibitors on dome formation by cultured alveolar epithelial cells. *Fed. Proc.* **41**:1245.
98. Goodman, B. E., R. S. Fleischer, and E. D. Crandall. 1982. Evidence for sodium transport by cultured alveolar epithelial cells. *Am. Rev. Respir. Dis.* **125**:278.
99. Alcorn, D., T. M. Adamson, T. F. Lambert, J. E. Maloney, B. C. Ritchie, and P. M. Robinson. 1977. Morphological effects of chronic tracheal ligation and drainage in the fetal lamb lung. *J. Anat.* **123**:649–660.
100. Griscom, N. T., G. B. C. Harris, M. E. B. Wohl, G. F. Vawter, and A. J. Eraklis. 1969. Fluid-filled lung due to airway obstruction in the newborn. *Pediatrics* **43**:383–390.
101. Potter, E. L., and G. P. Bohlender. 1941. Intrauterine respiration in relation to development of the fetal lung. *Am. J. Obstet. Gynecol.* **42**:14–22.
102. Schneeberger, E. E., D. V. Walters, and R. E. Olver. 1978. Development of intercellular junctions in the pulmonary epithelium of the foetal lamb. *J. Cell Sci.* **32**:307–324.
103. Olver, R. E., E. E. Schneeberger, and D. V. Walters. 1981. Epithelial solute permeability, ion transport and tight junction morphology in the developing lung of the fetal lamb. *J. Physiol.* (*London*) **315**:395–412.
104. Normand, I. C. S., R. E. Olver, E. O. R. Reynolds, and L. B. Strang. 1971. Permeability of lung capillaries and alveoli to non-electrolytes in the foetal lamb. *J. Physiol.* (*London*) **219**:303–330.
105. Mescher, E. J., A. C. G. Platzker, P. L. Ballard, J. A. Kitterman, J. A. Clements, and W. H. Tooley. 1975. Ontogeny of tracheal fluid, pulmonary surfactant, and plasma corticoids in the fetal lamb. *J. Appl. Physiol.* **39**:1017–1021.
106. Adamson, T. M., R. D. H. Boyd, H. S. Platt, and L. B. Strang. 1969. Composition of alveolar liquid in the foetal lamb. *J. Physiol.* (*London*) **204**:159–168.
107. Humphreys, P. W., I. C. S. Normand, E. O. R. Reynolds, and L. B. Strang. 1967. Pulmonary lymph flow and the uptake of liquid from the lungs of the lamb at the start of breathing. *J. Physiol.* (*London*) **193**:1–29.
108. Olver, R. E., and L. B. Strang. 1974. Ion fluxes across the pulmonary epithelium and the secretion of lung liquid in the foetal lamb. *J. Physiol.* (*London*) **241**:327–357.
109. Perks, A. M., and S. Cassin. 1982. The effects of arginine vasopressin and other factors on the production of lung fluid in fetal goats. *Chest* **81**:63S–65S.
110. Kitterman, J. A., P. L. Ballard, J. A. Clements, E. J. Mescher,

and W. H. Tooley. 1979. Tracheal fluid in fetal lambs: Spontaneous decrease prior to birth. *J. Appl. Physiol.* **47**:985–989.

111. Walters, D. V., and R. E. Olver. 1978. The role of catecholamines in lung liquid absorption at birth. *Pediatr. Res.* **12**:239–242.
112. Walters, D. V., C. A. Ramsden, M. J. Brown, R. E. Olver, and L. B. Strang. 1982. Fetal lung liquid absorption during epinephrine infusion and spontaneous labor in the lamb. *Chest* **81**:65S–66S.
113. Enhorning, G., D. Chamberlain, C. Contreras, R. Burgoyne, and B. Robertson. 1977. Isoxsuprine-induced release of pulmonary surfactant in the rabbit fetus. *Am. J. Obstet. Gynecol.* **129**:197–202.
114. Adams, F. H., A. J. Moss, and L. Fagan. 1963. The tracheal fluid in the fetal lamb. *Biol. Neonate* **5**:151–158.

Index

A23187, and tubuloglomerular feedback mechanism, 272–273
Absorption
of fat, lipids involved during, 213–214
of fetal lung liquid, 378
intestinal, *see* Intestinal absorption
of lipids, *see* Lipid absorption
by mTALH, of sodium chloride, 318–319
effect of ADH on, 319–320
Accommodation, in cardiac sodium channel excitability, 91
Acetylcholine (ACh)
and cardiac electrical activity, 101
gating and, 109
potentiation with oxyntic cells, 154–155
storage in synaptic vesicles, 33
Acetylcholine receptor (AChR)
aggregation at synapses of, 117–118
amino acid sequence of, 108–109
associated molecules and, 118
biogenesis of, 114–115
biosynthesis and regulation of, 109
developmental regulation of, 117
and dose–response relationship, 119
glycosylation of subunits in, 115
in intracellular compartment, 115
junctional versus extrajunctional, 118
monoclonal antibodies and, 113–114
nicotinic, 109–112
immunological approaches to study of, 112–114
in rough endoplasmic reticulum, 115
subunit maturation and assembly in, 115–117
Acetylcholine receptor (AChR) channel
and ACh molecule cooperation, 119–120
cations and, 109
permeation, 124
selectivity, 123–124
closed, 123
defined, 108
open, 122
as pentameric integral membrane protein, 108
size and shape of, 108
ACh, *see* Acetylcholine (ACh)
AChR, *see* Acetylcholine receptor (AChR)
Acid–base status, and intestinal electrolyte transport control, 188–189
Acid extrusion, across luminal cell membrane, 301–302
Acidification
of intracellular organelles, 25–35
urinary, 32
Acid interior, evidence for, 25
Action potential
cardiac, *see* Cardiac action potentials
repolarization of, 96–97
of skeletal muscle, 67, 68
Activation, of cardiac calcium channels, 94
Activation energy, effect of ADH on, 322–323
Active transport
and chemiosmotic hypothesis, 6
energetics of, 5–10
and β-galactoside transport system, 10–19
membrane vesicles and, 3–5
proton electrochemical gradient and, 9–10
urinary concentration and, models of, 338
ADH, *see* Antidiuretic hormone (ADH)
Adrenergic agents, and intestinal absorption, 197–198
β-Adrenergic modulation
of calcium channels, 99–100
of cardiac activity, 99
of outward potassium currents, 100–101
of pacemaker current, 100
Agonists, of cardiac calcium channels, 95
Airway epithelium, 367–379
ion transport regional differences in, 375
Albumin–ligand–membrane interactions, organic anions and, 238
Alkaline secretion, gastric mucosa and, 168–169
Alveolar epithelium, 375
cell cultures from, 377
ion transport in, 377
morphology of, 375–376
and permeability studies, 376–377
Alveolus, 375
Ambystoma, and proximal tubule osmotic gradient, 305–307
Amiloride, and calcium transport in DCT, 354
Amiloride-sensitive sodium entry, in intestinal absorption, 181
Amines, biogenic storage of, 32–33
Amino acid
in AChR, 108–109
in bile, 243
transepithelial potentials and, 290–292
transport, across basolateral membrane, 183
γ-Amino-butyric acid (GABA), 125–126
Ammonia, in renal medulla, 336
Amphiuma
bicarbonate ion and
absorption, 185–186
secretion, 185
electrical potential difference in, 356
potassium permeability in, 356
transport mechanisms in DCT of, 359–360
Angiotensin, and intestinal absorption, 199
Anion reflection coefficients, versus sodium chloride diffusion, 307–308
Anions, *see also* Organic anions
canalicular bile acid secretion and, 235
and hydrogen secretion, 164–165
unidentified, canalicular bile acid secretion and, 235
Antagonists, organic, in cardiac calcium channels, 95
Antiarrhythmic drugs, in cardiac sodium channel excitability, 91–92
Antibodies
in β-galactoside transport system, 17
monoclonal, *see* Monoclonal antibodies (mAbs)
polyclonal, 17
Antidiuretic hormone (ADH)
action mechanism of, 323
activation energy and, 322–323
and channel effects compared, 322
hepatocyte bile formation and, 237
intracellular mediators of, 317–318
net salt absorption and, 319–320
peritubular osmolality and, 324

Antidiuretic hormone (ADH) (*cont.*)
and prostaglandin interactions, 324–325
-responsive anuran epithelia, 323
sensitivity in DCT, 352
and urinary concentration, 325
water and solute permeability and, 320–322
Apical membranes, *see also* Brush border membrane (BBM) vesicles
ATPase activity and hydrogen transport in, 158–159, 160
and basolateral membranes, transport interactions between, 299–300
and intestinal sodium and chloride absorption, 177–181
basolateral interactions in, 181–182
neutral ion co- and countertransport across, 140–141
permeability and conductance of, 162–164
potassium channels in, 141
sodium entry across, 136–137
organic solutes and, 139–140
sodium entry process, in tracheal epithelium, 374
sodium permeability regulation for, 137–139
Aplysia, synaptic transmission in, 126
Arachidonic acid metabolites, in intestinal electrolyte transport control, 190–191
Arrhythmias, slow responses in, 95
Arterioles
afferent and efferent, 254–255
resistance in, glomerular filtration rate and, 268–269
ATP
and cardiac potassium channel inhibition, 97
and intestinal secretion, 197
ATPase
in apical membrane vesicles, 158–159
and canalicular bile acid secretion, 234–235
identity of, 34–35
and peritubular sodium extrusion, 298–299

BADF, *see* Bile acid-dependent flow (BADF)
BAIF, *see* Bile acid-independent flow (BAIF)
B. alkalophilus, TMG transport and, 10
Barium, effect on electrical properties of tracheal epithelium, 373
Basement membrane, of glomerular capillary wall, 257
Bases
distribution of, 39–40
effects of external application, 42–44
Base transport, basolateral pathways for, 303
physiological roles for, 303
Basolateral membranes
amino acid transport across, 183
and apical membranes, transport interactions between, 299–300
of epithelial cells, 141–143
and intestinal sodium and chloride absorption, 182–183
Basolateral membranes (*cont.*)
and intestinal sodium and chloride absorption (*cont.*)
apical membrane interaction in, 181–182
pathways for base transport, 303
physiological roles of, 303
permeability and conductance of, 162–164
of tracheal epithelium, sodium–potassium pump in, 372–374
BBM vesicles, *see* Brush border membrane (BBM) vesicles
BC, *see* Bile caniculus (BC)
Bicarbonate ion
and fluid movement across proximal tubular epithelium, 303–304
reabsorption of, in DCT, 353–354
transport, 184–186
Bile
composition of, 226
defined, 225
flow of, 230–232
formation sites for, 230–231
hepatocyte, physiological modifiers of formation of, 236–238
lipid excretion in, 240–241
miscellaneous substances found in, 243
proteins in, 242–243
secretory functions of, 226
structural determinants of, 225, 226–229
vitamins in, 243
Bile acid-dependent flow (BADF), 231–232
components of, 232
Bile acid-independent flow (BAIF), 231–232
Bile acid micelles, in intestinal lipid absorption, 221–222
Bile acids
canalicular excretion and secretion of, 233–235
hepatic uptake of, 233
Bile acid transport, 232
intracellular, 233
Bile caniculus (BC), 225, 226
and canal of Hering, 229
Bile duct
bile secretory function and, 229
function of, 243–244
Bile duct columnar epithelium, 227
Bile ductules, 230–231
Bile salt, secretion and, 231–232
Biliary lipid secretion, mechanisms of, 241
Biliary permeability, to cations and anions, 235
Biliary protein secretion, mechanisms of, 242–243
Binding
acetylcholine, gating and, 109
of bungarotoxin, 116
mAbs sites for, 113
Binding proteins, organic anions and
intracellular, 239
membrane, 239
Biochemical buffering, 42
Biogenic amine storage, 32–33
Biological membranes, *see* Membranes
BK, *see* Bradykinin (BK)
Blood flow, and intestinal electrolyte transport control, 189
Body fluid osmolality, 329
Bombesin, and intestinal secretion, 196
Bowman's space, and hydrostatic pressure gradient, 259
Bradykinin (BK), 190–191
Bromoacetylcholine, in AChR structure–function correlations, 111
Bromosulfophthalein (BSP), 238, 239
Bronchial epithelium
ion transport in, 375
morphology of, 374–375
Brush border membrane (BBM) vesicles, *see also* Luminal brush border membrane
ATPase activity and H^+ transport in, 158–159, 160
ion transport by, 160–161
BSP, *see* Bromosulfophthalein (BSP)
Buffering
biochemical, 42
definitions of, 40–41
organellar, 42
physicochemical, 41–42
Bungarotoxin
binding, and AChR maturation and assembly, 116
blotting approach, in AChR structure–function correlations, 111–112

Calcium
absorption, in DCT, 354–355
hepatocyte bile formation and, 236
and muscle contraction activation, 73–78
oxyntic cells and, 155
release in skeletal muscle, 69–70
and transient inward currents, 98–99
Calcium channels
and cardiac electrical activity, 89
action potential recordings and, 92–93
β-adrenergic modulation of, 99–100
cholinergic modulation of, 101
and contractile activation, 96
drug effects on, 95
voltage clamp studies and, 93–95
in skeletal muscle, 67
Calcium current, and contractile activation, 96
Calcium transport
between apical and basolateral cell membranes, 299–300
sarcoplasmic reticulum membranes and, 75
cAMP, *see* Cyclic AMP (cAMP)
Canal of Hering, 229
Canalicular bile acid, 233–235
Canaliculi
organic anion excretion from, 240
in oxyntic cells, 152
Capillary
glomerular, *see* Glomerular capillaries
models in glomerular dynamics
multiple, 266
single, 264–265
Carbonylcyanide *m*-chlorophenylhydrazone (CCCP), 4
Carbonylcyanide *p*-trifluoromethoxyphenylhydrazone (FCCP), 26–27
Cardiac action potentials
and calcium channels, 92–93
purposes of, 85

Cardiac action potentials (*cont.*)
in sodium channel excitability, 89–90
specialized functions of, 89
Cardiac arrhythmias, slow responses in, 95
Cardiac impulse, spread of, 87–88
normal sequence of, 88
Carrier-mediated uptake, of organic anions, 238–239
Carrier protein, *see lac* carrier protein
Catecholamines, and GFR, 275–276
Cations
and AChR channel, 109
selectivity and permeation in, 123–124
canalicular bile acid secretion and, 235
and hydrogen secretion, 164
organic, secretion of, 240
CCCP, *see* Carbonylcyanide *m*-chlorophenylhydrazone (CCCP)
CCK, *see* Cholecystokinin (CCK)
CCT, *see* Cortical collecting tubule (CCT)
Cells
connecting tubule (CNT), 344
distal convoluted tubule (DCT), 344
epithelial, *see* Epithelial cells
glandular, conductance of, 162
Goormaghtigh, 255
intercalated (I), 344
juxtaglomerular (JG), 254–255
mesangial, 257
principal (P), 344
surface, *see* Surface cells
Cell-to-cell communication, in heart, 85–87
Cellular buffering processes, 40–42
Central core model, of urinary concentration and dilution, 333–334
Channels
AChR, *see* Acetylcholine receptor (AChR) channel
ADH effects compared, 322
closed, 123
defined, 107–108
glutamate-activated, 124–125
ionic, *see* Ionic channels, 53–63
ligand-gated, 107
potassium, *see* Potassium channels
voltage-gated, 107
Chemical coupling, in heart, 88
Chemical excitability, 54
Chemiosmotic hypothesis, 5–7
Chloride
absorption by epithelial cells, model for, 140–141
and fluid movement across proximal tubular epithelium, 303–304
secretion, *see* Chloride secretion
transport, *see* Chloride transport
Chloride/bicarbonate ion exchange, in ion-transport systems, 48
Chloride entry step, in tracheal epithelium, 369–371
Chloride exit step
in intestinal absorption, 182–183
in tracheal epithelium, 371–372
Chloride secretion
by epithelial cells, model for, 143–144
nonacidic, in gastric mucosa, 166–167
in tracheal epithelium, 368–369
Chloride transport
gastric mucosa and, 165–167
Chloride transport (*cont.*)
in intestine, 175–177
with potassium, in basolateral membranes, 143
sodium-coupled, across luminal cell membrane, 296–298
Cholecystokinin (CCK), 197
Cholesterol, absorption in intestine, 219–221
Cholinergic modulation
of ACh-sensitive potassium channels, 101
of calcium channels, 101
of cardiac activity, 101
Cholinergic muscarinic agents, and intestinal secretion, 195–196
Chromaffin granules
and biogenic amine storage, 32–33
and hydrogen diffusion potential, 28
and hydrogen pump, 28–29
Chromophore, endogenous, pH measurement in, 27
Closed channels, and single-channel measurements, 123
CNT cells, *see* Connecting tubule (CNT) cells
Colloid osmotic pressure, in glomerular filtration rate, 258–259
Concentration, urinary, *see* Urinary concentration
Conduction
along T system, 69
in cardiac sodium channel excitability, 90
of cell membranes to ions, 162–163
fluctuation analysis of, *see* Fluctuation analysis
hydraulic, and filtration coefficient, 261–262
neutral exchangers and, 167
pathways, in gastric mucosa, 169
of potassium, in tracheal epithelium, 372–374
single-channel, AChR channel and, 122
of surface cells versus glandular cells, 162
in tubular membrane, 66–68
Conjugation, hepatic, 239–240
Connecting tubule (CNT) cells, 344
Connexons, 85–86
Contractile activation, calcium current and, 96
Corpuscle, glomerular, 253
Cortical collecting tubule (CCT), 320–323
Cortical labyrinth, 331
Cortisol, hepatocyte bile formation and, 237
Countercurrent multiplier hypothesis, 331–332
Coupling
in heart
chemical, 88
electrical, 87–88
in oxyntic cells, 153–155
solute–solvent, and intercellular shunt pathway, 304–309
Currents, *see also* Gating currents
endplate, and dose–response relationship, 119
inward, and pacemaker activity, 97–99
ionic, 62
macroscopic, 120–122
outward potassium, β-adrenergic modulation of, 100–101
Currents (*cont.*)
pacemaker, 97–98
β-adrenergic modulation of, 100
Cyclic AMP (cAMP)
and calcium transport in DCT, 354
hepatocyte bile formation and, 236
as second messenger for histamine action, 154
Cyclic nucleotides, in regulation of ion transport, 141
Cytoplasmic buffering, mechanisms of, 41–42

DAD, *see* Diaminodurene (DAD)
DCCD, *see* *N,N′*-Dicyclohexylcarbodiimide (DCCD)
DCT, *see* Distal convoluted tubule (DCT)
Deprotonation, in β-galactoside transport system, 15
Desmarestia, pH measurement in, 27
Dextran, in molecular sieving studies, 263
Diabetes insipidus
DCT and, 360
hypothalamic, *see* Hypothalamic diabetes insipidus
Diaminodurene (DAD), 4
Dibutyryl cAMP, hepatocyte bile formation and, 236
N,N′-Dicyclohexylcarbodiimide (DCCD), 4
8-(*N,N*-Diethylamino)-octyl-3,4,5-trimethoxybenzoate (TMB-8), 273
Diffusion
ADH effects on CCT, 320
exchange, in chloride transport, 166
water, series resistance elements for, 321
Dilution, urinary, *see* Urinary dilution
Distal convoluted tubule (DCT)
cells of, 344
electrophysiological considerations for, 355–357
ion transport in
bicarbonate, 353–354
calcium, 354–355
functions, 343–360
magnesium, 355
mechanisms, 357–360
phosphate, 355
potassium, 348–351
sodium chloride, 344–347
water, 351–353
segmentation patterns in, 344
structural heterogeneity of, 343–344
versus proximal tubule, 343
Diurnal rhythm, hepatocyte bile formation and, 237–238
Dopamine, and intestinal absorption, 198
Dose–response, AChR and, 119–120
Dual countertransport model, 140
Dyes, pH-sensitive, 40

EAMG, *see* Experimental autoimmune myasthenia gravis (EAMG)
Effective filtration pressure (EFP)
in glomerular filtration rate, 260–261
in single capillary model, 264
EFP, *see* Effective filtration pressure (EFP)
Electrical coupling, in heart, 87–88
Electrical excitability, 54–55
Electrical potential

Electrical potential (*cont.*)
difference (PD), 355–357
profile, of surface and oxyntic cells, 163
Electrogenic hypothesis, of hydrogen secretion, 165
Electrolytes, secretion model for hepatocytes, 235–236
Electrolyte transport, intestinal
hormonal control of, 189–190
kinins and arachidonic acid metabolites in, 190–191
models of, 175–177
neural and neuroendocrine control of, 191–199
systemic factors in, 188–189
Electrophorus, nicotinic AChR in, 109–112
Endocrine–paracrine system, of intestine, 193
Endocytosis, receptor-mediated, 31–32
Endogenous chromophore, pH measurement in, 27
Endothelium, of glomerular capillary wall, 257
Endplate currents (EPCs), 119
Energetics
of active transport, 5–10
in ion-transport systems, 45–46, 47
Enkephalin, and intestinal absorption, 198–199
ENS, *see* Enteric nervous system (ENS)
Enteric nervous system (ENS), 192–193
neuroactive agents of
absorptive, 196
secretory, 195
Entry processes, in intestinal absorption
sodium-coupled
chloride, 177–179
glucose, 179–180
other, 180
uncoupled and amiloride-sensitive sodium ion, 181
Enzymes
proteolytic, in β-galactoside transport system, 17
purification of, ATPase identity and, 35
EPCs, *see* Endplate currents (EPCs)
Epinephrine, electrical properties in tracheal epithelium and, 372
Epithelial cells
alveolar, cultured monolayers of, 377
chloride secretion by, 143–144
gastric, *see* Gastric epithelial cells
ion transport and conductance pathways in, 169
potassium absorption and secretion in, 186–187
sodium- and chloride-absorbing, 136–143
Epithelial ion transport, cellular models of, 135–136
Epithelial water permeability, in proximal tubule, 307
Epithelium, *see* Airway epithelium; Alveolar epithelium; Bronchial epithelium; Tracheal epithelium
Equilibrium kinetic studies, 121
Equivalent electrical circuit model of ion transport, 370
Escherichia coli, active transport in, 3–19
Estrogens, hepatocyte bile formation and, 237
N-Ethylmaleimide (NEM), 18–19
Exchange diffusion, in chloride transport, 166
Exchangers, in chloride transport, 167
Excitability
chemical, 54
electrical, 54–55
Excitable tissues, of heart, 85–102
Excitation, in T system, 69–73
Excitatory–inhibitory system, of gut innervation, 191
Excretion
canalicular bile acid, 233–234
of neutral compounds, 240
Exit processes, in intestinal absorption
chloride, 182–183
and glucose and amino acid transport, 183
sodium ion, 182
Exocytosis, canalicular bile acid secretion and, 235
Experimental autoimmune myasthenia gravis (EAMG), 112–113
External pH, and proton electrochemical gradient, 8

FAD, *see* Flavin-adenine dinucleotide (FAD)
Fat absorption, lipids involved during, 213–214
Fatty acids
absorption in intestine, 219–221
short-chain, 186
FCCP, *see* Carbonylcyanide *p*-trifluoromethoxyphenyl-hydrazone (FCCP)
FD, *see* Fluorescein-labeled dextran (FD)
Ferrocyanide technique, and intrarenal distribution of GFR, 276–277
Fetal lung, 377
fluid secretion by, 377–378
liquid, absorption of, 378
morphology of, 377
FF, *see* Filtration fraction term (FF)
Fick's law of diffusion, and ADH effects on CCT, 320
Ficoll, in molecular sieving studies, 263
Filtration, *see also* Glomerular filtration rate (GFR)
hydrostatic, 235
osmotic, 234–235
Filtration coefficient, in glomerular filtration rate, *see* Glomerular filtration coefficient
Filtration equilibrium, in glomerular dynamics, 265–266
Filtration fraction term (FF), 258
Flavin-adenine dinucleotide (FAD), 4
Fluctuation analysis, 61
Fluid secretion, by fetal lung, 377–378
Fluorescein-labeled dextran (FD), 27
Fluorescent polymers, pH measurement in, 27
Fluorescent weak bases, pH measurement in, 27
Flux
lactose transport and, 14–15
of neutral weak acid, 42
and anionic, conjugate weak base, 42–43
of neutral weak base, 43
Flux (*cont.*)
of neutral weak base (*cont.*)
and cationic, conjugate weak acid, 43–44
Furosemide, effect in tracheal epithelium, 370

GABA, *see* γ-Amino-butyric acid (GABA)
β-Galactoside transport system, 11
immunological reagents and, 17–18
and *lac* carrier protein
purification of, 11–12
secondary structure model for, 16, 17
mechanistic studies in, 14–15, 17
mutagenesis in, 18–19
polypeptide requirement in, 12–14
proteoliposomes and, 12, 13
subunit structure in, 18
Gallbladder bile, *see* Bile acids
Gas exchange, in respiratory epithelium, 367–379
Gastric epithelial cells
histology of, 151–152
and ultrastructure of oxyntic cell, 152–153
Gastric inhibitory peptide (GIP), 197
Gastric microsomal vesicles, ion transport by, 157–158
Gastric mucosa
hydrochloric acid secretion in, 155–156
ion transport by, 151–170
electrophysiological and tracer flux studies of, 161–169
membrane changes during stimulation, 163–164
membrane fractionation in, 156–161
and parallel exchanger model, 168
Gastrin
and intestinal secretion, 197
potentiation with oxyntic cells, 154–155
Gating, *see also* Gating currents
and acetylcholine binding, 109
in AChR channels
complementary methods in, 120–121
kinetics of, 120
macroscopic measurements in, 121–122
single-channel measurements in, 122–123
in ionic channels, 55
equilibrium behavior of, 56
kinetic behavior of, 56–57
Gating currents
in ionic channels, 57
in potassium channels, 62–63
in sodium channels, 59–61
GFR, *see* Glomerular filtration rate (GFR)
GIP, *see* Gastric inhibitory peptide (GIP)
Glandular cells, conductance of, 162
Glomerular capillaries
permeability of, to macromolecules, 262–263
wall of, 257
Glomerular corpuscle, ontogeny of, 253
Glomerular dynamics, 264–266
Glomerular filtration coefficient
factors influencing, 267
and hydraulic conductivity, 261–262
Glomerular filtration rate (GFR)
colloid osmotic pressure in, 258–259
effective filtration pressure in, 260–261
forces governing, 258–259

Glomerular filtration rate (GFR) (*cont.*)
and glomerular capillary permeability, 262–263
hydraulic conductivity and filtration coefficient in, 261–262
hydrostatic pressure gradients in, 259–260
intrarenal distribution of, 276–277
in kidneys, 253–277
mathematical models for, 264–266
physiological regulation of, 267–276
single nephron, *see* Single nephron glomerular filtration rate (SNGFR)
Glomerular plasma flow, in glomerular dynamics, 265–266
Glomerulus
microvasculature of, 254–255
structural characteristics of, 256
structural innervation of, 257–258
Glucagon
hepatocyte bile formation and, 237
and intestinal secretion, 197
Glucocorticoids, in intestinal electrolyte transport control, 190
Glucose
transepithelial potentials and, 290–292
transport across basolateral membrane, 183
Glutamate-activated channels, 124–125
Goldman–Hodgkin–Katz equation, resting potential and, 53–54
Golgi complex, AChR subunit assembly in, 117
Goormaghtigh cells, 255
Gramicidin A channels, 322
Granules
chromaffin, *see* Chromaffin granules
neurohypophyseal, 29–30
platelet, and biogenic amine storage, 32–33
platelet dense, 29
Growth factors, in ion-transport systems, 47–48
Gut function, chemical messengers and, 193
Gut innervation, 191

Halobacterium halobium, TMG transport and, 10
Heart
adrenergic effects on, 99–101
cell-to-cell communication in, 85–87
chemical coupling in, 88
cholinergic effects on, 101
electrical coupling and impulse spread in, 87–88
excitable tissues of, 85–102
impulse generation in, *see* Cardiac impulse, spread of
multicellular structure of, 85–88
regional differences in electrical activity, 88–89
ionic basis and function of, 89
Hemoglobin–haptoglobin complexes, biliary protein secretion and, 242
Henle, loops of, 329–331
thick ascending, 318
sodium chloride absorption by, 318–319
thin limbs of, properties of, 336–337
Hepatic bile, *see* Bile
Hepatic conjugation, organic anions and, 239–240
Hepatic transport, bile secretion and, 225–244
Hepatic uptake, of bile acids, 233
Hepatocellular electrolyte secretion, 230–234
Hepatocellular water secretion, 230–234
Hepatocyte bile, physiological modifiers of formation of, 236–238
Hepatocytes
and bile formation, 230–231
lobular gradient and, 229
water and electrolyte secretion model, 235–236
Histamine
and intestinal secretion, 197
oxyntic cells and, 154
Homo-oligomers, AChR maturation and assembly and, 117
Hormones
antidiuretic, *see* Antidiuretic hormone (ADH)
in bile, 243
in intestinal electrolyte transport control, 189–190
in ion-transport systems, 47–48
steroids in bile and, 243
thyroid, hepatocyte bile formation and, 236–237
5HT, and intestinal secretion, 197
Hydraulic conductivity, and filtration coefficient, 261–262
Hydrocarbon continuum, nonpolar lipids in, 223
Hydrochloric acid secretion, in gastric mucosa, 155–156
Hydrocortisone, hepatocyte bile formation and, 237
Hydrogen
diffusion potential for, 27–28
in β-galactoside transport system, 15
ion pump, acidification by, 28–31
potassium-ATPase, in gastric mucosa, 157
and substrate stoichiometry, 10
Hydrogen secretion
anion dependence of, 164–165
cation dependence of, 164
electrogenic hypothesis for, 165
luminal pathways for, 301–302
Hydrogen transport
in apical membrane vesicles, 158–159, 160
gastric mucosa and, 164–165
Hydrostatic filtration, for canalicular bile acid secretion, 234
Hydrostatic pressure gradient, in glomerular filtration rate, 259–260
Hyperosmolality, in proximal convoluted tubules, 306
Hypothalamic diabetes insipidus
and potassium secretion, 351
transepithelial effective resistance of DCT in, 360
Hypotonicity, luminal, in proximal tubule, 306

IBMX, *see* 3-Isobutyl-1-methylxanthine (IBMX)
I cells, *see* Intercalated (I) cells
IgA, *see* Immunoglobulin A (IgA)
Immunoglobulin A (IgA), 242
Immunological reagents, as probes in β-galactoside transport system, 17–18
Impulse, cardiac, *see* Cardiac impulse, spread of
Inactivation, of cardiac calcium channels, 94
Inhibitor studies, of ATPase identity, 34–35
Inner medulla, 331
urinary concentration in, 334
mass balance requirements for, 337–338
proposed models for, 338–340
Inside-out (ISO) vesicles, 4
Insulin
hepatocyte bile formation and, 237
in intestinal electrolyte transport control, 190
Intercalated (I) cells, 344
Intercellular shunt pathway
of proximal tubule, 292–293
and solvent–solute coupling, 304–309
Internephron heterogeneity, 287–288
Intestinal absorption
agents stimulating, 197–199
of bicarbonate ion, 184–185
of cholesterol, 219–221
of fatty acid, 219–221
of potassium, 187
of sodium and chloride, 177–183
Intestinal microvillus membrane barrier, lipid absorption and, 216–217
Intestinal mucosa, lipid uptake into, 213–223
Intestinal secretion
agents stimulating, 195–197
bicarbonate ion, 185–186
potassium, 186–187
of sodium and chloride, 183
intracellular mediators of, 183
mechanisms of, 184
Intestinal unstirred water layer barrier, lipid absorption and, 217–219
Intestine
barriers to lipid absorption in, 214–216
bicarbonate ion transport in, 184–186
electrical resistance and shunt permselectivity of, 187–188
and electrolyte transport
control in, 188–199
models for, 175–177
fatty acid and cholesterol absorption in, 219–221
and lipid uptake during fat absorption, 213–214
potassium transport in, 186–187
short-chain fatty acid absorption in, 186
sodium and chloride in
absorption of, 177–183
secretion of, 183–184
Intracellular binding proteins, organic anions and, 239
Intracellular chloride, activities in tracheal epithelium, 371
Intracellular organelles
acidification of, 25–35
bile secretory function and, 225, 226–228
Intracellular pH
measurement of, 39–40
regulation of, 39–48

Intracellular potassium, activities in tracheal epithelium, 373
Intracellular vacuoles, and hydrogen pump, 30–31
Intramembranous charge movements, in skeletal muscle, 70–72
Intrinsic autonomic nervous system, intestinal, 192–193
Intrinsic control mechanisms, and glomerular filtration rate, 269
Inward currents
  calcium-activated transient, 98–99
  and pacemaker activity, 97–99
Ionic channels, 55
  and cardiac electrical activity, 89
  between T system and sarcoplasmic reticulum, 72
  two-state model for, 55–57
    extended, 57
Ionic currents, in potassium channels, 62
Ionophores, effect on proton electrochemical gradient, 8–9
Ion pumps, 293
Ions
  bicarbonate, *see* Bicarbonate ion
  conductance of cell membranes to, 162–163
  permeability of, in proteoliposomes, 12
  potassium, and cardiac sodium channel excitability, 91
  in skeletal muscle, 65–78
Ion transport
  in airway epithelium, regional differences in, 375
  in alveolar epithelium, 377
  by apical membrane vesicles, 160–161
  across apical membranes, 140–141
  in bronchial epithelium, 375
  epithelial, 135–136
  by gastric microsomal vesicles, 157–158
  gastric mucosa and, 151–170
  in intestine, 175–199
  systems identification in, 44–48
  in tracheal epithelium, 368
    cellular model of, 369
    and chloride entry step, 369–371
    and chloride exit step, 371–372
    equivalent electrical circuit model of, 370
    secretion and absorption compared, 374
    and sodium entry process, 374
    and sodium–potassium pump, 372–374
3-Isobutyl-1-methylxanthine (IBMX), 273

JG cells, *see* Juxtaglomerular (JG) cells
Junctions, tight versus leaky, in gastric mucosa, 162
Juxtaglomerular apparatus–macula densa complex, 255–257
Juxtaglomerular (JG) cells, 254–255
Juxtamedullary proximal tubule, versus superficial proximal tubule, 287–288

Kallikrein system, and intestinal electrolyte transport control, 190, 191
Kidney
  glomerular filtration rate regulation in, 253–277
Kidney (*cont.*)
  and nephron classification, 329–331
  regional organization of, 331
  species comparison of, 253, 254
Kinins, in intestinal electrolyte transport control, 190–191
KJU, *see* Koefoed–Johnsen–Ussing (KJU) model
Koefoed–Johnsen–Ussing (KJU) model, 135–136
Kuhn-type countercurrent multiplier, 331

lac carrier protein
  proteoliposomes reconstituted with, 12, 13
  purification of, 11–12
  turnover numbers for, 13–14
D-Lactate dehydrogenase (D-LDH), 6
Lactose transport, polypeptide requirement for, 12–14
Landis–Pappenheimer relationship, and glomerular filtration rate, 258
"Large pore" hypothesis, 321
D-LDH, *see* D-Lactate dehydrogenase (D-LDH)
LDS, *see* Lithium dodecyl sulfate (LDS)
Leukotrienes, in intestinal electrolyte transport control, 191
Ligand-gated channels, 124
  GABA-activated, 125–126
  glutamate-activated, 124–125
  muscarinic ACh-activated, 126
  synaptic transmission and, 126
Lipid absorption
  barriers in intestine to, 214–216
  and bile acid micelles, 221–222
  and intestinal microvillus membrane barrier, 216–217
  into intestinal mucosa, 213–223
  and intestinal unstirred water layer barrier, 217–219
  process of, 223
Lipids
  and biliary secretion mechanism, 241
  composition and source in bile, 240–241
  in fat absorption, 213–214
  nonpolar, 222–223
Lipid uptake, *see* Lipid absorption
Lithium dodecyl sulfate (LDS), 111
Liver, surface membrane domains of, 227, 228
Lobular gradient, bile secretory function and, 229
Luminal brush border membrane
  and peritubular potassium permeability, 300
  and transepithelial potentials, 290–292
Luminal cell membrane
  acid extrusion across, 301–302
  and pathways for hydrogen secretion, 301–302
  sodium and calcium transport in, 299–300
  sodium entry across, 296–298
    in DCT, 357–358
Luminal hypotonicity, in proximal tubule, 306
Lung, fetal, *see* Fetal lung
Lutoids, and vacuolar solute storage, 34
Lysosomal protein degradation, 31
Lysosomes
Lysosomes (*cont.*)
  and hydrogen diffusion potential, 28
  and hydrogen pump, 30
  pH measurement in, 26

mAbs, *see* Monoclonal antibodies (mAbs)
Macromolecules, permeability of glomerular capillaries to, 262–263
Macroscopic current methods, 121–122
  versus single-channel recording, 120–121
Magnesium, reabsorption in DCT, 355
Main immunogenic region (MIR), 113
4-(*N*-Maleimido)-α-benzyltrimethylammonium iodide (MTA), 111–112
Mechanistic studies, in β-galactoside transport system, 14–15, 17
Medulla
  individual solute handling in, 334–335
    ammonia, 336
    potassium, 336
    sodium, 336
    urea, 335–336
  inner and outer, 331
  and medullary rays, 331
Medullary rays, 331
Medullary thick ascending limb of Henle, *see* Henle, loops of
Membrane binding proteins, organic anions and, 239
Membrane fractionation, in gastric mucosa, 156–161
Membrane junctions, tight versus leaky, 162
Membrane polarity, bile secretory function and, 225, 226–228
Membrane potential
  activation and restoration of force and, 70
  and chemical excitability, 54
  determination of, 7
  and electrical excitability, 54–55
  substrate accumulation and, 9
Membrane recycling hypothesis, 153
Membranes
  apical, *see* Apical membranes
  basolateral, *see* Basolateral membranes
  brush border, *see* Brush border membrane (BBM) vesicles
  conductance in, *see* Conduction
  excitable, ionic channels in, 53–63
  gastric, changes during stimulation, 163–164
  luminal brush border, *see* Luminal brush border membrane
  luminal cell, *see* Luminal cell membrane
  microvillus, lipid absorption and, 216–217
  peritubular cell, 357
  of terminal cisternae, electrical properties of, 72–73
  tubular, 66–68
Membrane vesicles, *see* Vesicles
Mesangial cells, of glomerular capillary wall, 257
Metabolic acidosis, and calcium transport in DCT, 354
Metals, in bile, 243
Methylamine
  in acidic vesicle, 26
  and pH measurement techniques, 26–27
1-Methyl-3-isobutylxanthine (MIX), 236

Methyl-1-thio-β-D-galactopyranoside (TMG), 10
MF, *see* Myosin filament (MF)
Micellar sink hypothesis, 233–234
Micelles, bile acid, lipid absorption and, 221–222
Microelectrodes, pH-sensitive, 40
Microperfusion studies, and potassium transport in DCT, 348–350
Micropuncture
 GFR control and, 276
 and potassium transport in DCT, 348
Microrheological network model, in glomerular dynamics analysis, 266
Microsomal enzyme inducers, hepatocyte bile formation and, 237
Microvillus membrane, lipid absorption and, 216–217
Mineralocorticoids
 DCT and
  and potassium secretion, 350
  and sodium retention, 346
 in intestinal electrolyte transport control, 189–190
MIR, *see* Main immunogenic region (MIR)
Mitochondria, in oxyntic cells, 152
MIX, *see* 1-Methyl-3-isobutylxanthine (MIX)
Molecular sieving, and glomerular filtration rate, 262–263
Monoclonal antibodies (mAbs)
 and AChR organization, 113–114
 in β-galactoside transport system, 17
Monosaccharides, in bile, 243
Motilin, and intestinal secretion, 197
MTA, *see* 4-(*N*-Maleimido)-α-benzyltrimethylammonium iodide (MTA)
Multiple capillary models, in glomerular dynamics, 266
Muscarine ACh-activated channels, 126
Muscle
 contraction of, cellular calcium movements in
  background to, 73–75
  sarcoplasmic reticulum membranes and, 75
  in single fibers, 75–78
 skeletal, ion movements in, 65–78
Mutagenesis, in β-galactoside transport system, 18–19
Myasthenia gravis, AChR antibody and, 112–113
Myofilaments, in skeletal muscle, 73
Myosin filament (MF), 73

NAP-taurine, *see* *N*-(4-Azido-2-nitrophenyl)-2-aminoethylsulfonate (NAP-taurine)
NDS-TEMPO, *see* *N*-4-(2,2,6,6-Tetramethyl-1-oxyl)piperidinyl-*N*′-4(4′-nitro-2,2′-stilbenedisulfonic acid) thiourea (NDS-TEMPO)
NE, *see* Norepinephrine (NE)
Necturus
 electrical potential difference in, 356
 luminal cell membrane in, 296
 membrane potentials of, 166
Necturus (*cont.*)
 potassium chloride cotransport in, 143
 and proximal tubule osmotic gradient, 305–307
NEM, *see* *N*-Ethylmaleimide (NEM)
Nephrons
 bicarbonate reabsorption along, 353
 classification of, 329–331
Nerves, renal, and GFR, 275–276
Neurogenic factors, hepatocyte bile formation and, 237
Neurohumoral agents, in intestinal electrolyte transport control, 195
Neurohypophyseal granules, and hydrogen pump, 29–30
Neuromodulation, in intestinal electrolyte transport control, 193–195
Neuromuscular junction (NMJ), 54
Neurotensin, and intestinal secretion, 197
Neurotransmitters, and intestinal electrolyte transport control, 193–195
Neutral compounds, excretion of, 240
Nexus, 85–86
 membrane permeability in, 87
Nicotinic agents, and intestinal absorption, 199
Nigericin, substrate accumulation and, 9–10
NMJ, *see* Neuromuscular junction (NMJ)
NMR, *see* Nuclear magnetic resonance (NMR)
Nodal tissue, slow responses in, 95–96
Nonequilibrium kinetic studies, 121
Nonpolar lipids, 222–223
Norepinephrine (NE), 276
Nuclear magnetic resonance (NMR), 27, 40

Oligonucleotide-directed, site specific mutagenesis, 18–19
Open channels, and single-channel measurements, 122
Optical probes, and pH measurement techniques, 27
Organellar buffering, 42
Organic anions, 238
 albumin–ligand–membrane interactions and, 238
 and binding proteins
  intracellular, 239
  membrane, 239
 canalicular excretion of, 240
 carrier-mediated uptake of, 238–239
 hepatic conjugation and, 239–240
 solute transport, 238–240
Organic antagonists, of cardiac calcium channels, 95
Organic cations, excretion of, 240
Organic solutes, and sodium entry across apical membrane, 139–140
Osmolality
 body fluid, 329
 of distal tubular fluid, disease states compared, 352
 hyper-, in proximal convoluted tubules, 306
 peritubular, ADH response and, 324
Osmosis
 and canalicular bile, 230
 in proximal convoluted tubules, 306
Osmotic filtration, canalicular bile acid secretion and, 234–235
Osmotic gradient, site at proximal tubule, 305–307
Osmotic pressure, colloid, 258–259
Outer medulla, 331
Outward current, potassium, β-adrenergic modulation of, 100–101
Oxyntic cells
 electrical potential profile of, 163
 ion transport and conductance pathways in, 169
 stimulus–secretion coupling in, 153–155
 ultrastructure of, 152–153

Pacemaker activity, and inward currents, 97–99
Pacemaker current, 97–98
 β-adrenergic modulation of, 100
Paracellular shunt pathway, canalicular bile secretion and, 235
Parallel exchanger model, in gastric mucosa, 168
Parathyroid hormone (PTH), 354
Passively exchanging models, of urinary concentration
 multisolute, 338–340
 single-solute, 338
Passive permeability
 and intercellular shunt path, 292–293
 in proximal convoluted tubule, 289–290
 in proximal straight tubule, 293
 and sodium cotransport with sugars and amino acids, 290–292
P cells, *see* Principal (P) cells
PD, *see* Electrical potential, difference (PD)
Peptides, in bile, 243
Peritubular cell membrane, and amphibian DCT studies, 357
Peritubular osmolality, ADH response and, 324
Peritubular potassium permeability, 300
Peritubular sodium/calcium exchange, 299–300
Peritubular sodium extrusion, 298–299
Permeability
 alveolar, 376–377
 effect of ADH on, 320–322
 epithelial water, in proximal tubule, 307
 passive, *see* Passive permeability
 peritubular potassium, 300
 properties of thin limbs of Henle's loops, 336–337
Permeation, of AChR cations, 124
PGs, *see* Prostaglandins (PGs)
pH, *see also* Acidification; pH gradient
 buffering processes and, 40–42
 in DCT, 353–354
 external, and proton electrochemical gradient, 8
 in β-galactoside transport system, 15
 intracellular, *see* Intracellular pH
 measurement techniques for, 26–27
 regulation of, in ion-transport systems, 44–48
 -sensitive dyes, 40
 -sensitive microelectrodes, 40
 transmembrane difference

pH (*cont.*)
transmembrane difference (*cont.*)
generation of, 27–31
uses of, 31–34
Phenazine methosulfate (PMS), 4
pH gradient
determination of, 7–8
substrate accumulation and, 9
Phosphate, reabsorption in DCT, 355
Phosphoenolpyruvate–phosphotransferase system (PTS), 4
Physicochemical buffering, 41–42
Plasma proteins, concentration in glomerular dynamics, 265–266
Platelet granules
and biogenic amine storage, 32–33
dense, and hydrogen pump, 29
PMS, *see* Phenazine methosulfate (PMS)
Polyclonal antibodies, in β-galactoside transport system, 17
Polypeptides, lactose transport and, 12–14
Polyvinylpyrrolidone (PVP), 263
Porphyrins, in bile, 243
Potassium
and active transport in RSO vesicles, 5
and canalicular bile acid secretion, 234–235
in cardiac sodium channel excitability, 91
conductance in tracheal epithelium, 372–374
leak, in basolateral membranes, 142–143
and peritubular sodium extrusion, 298–299
permeability of peritubular cell membrane in *Amphiuma*, 357
in renal medulla, 336
transfer in DCT, kinetic parameters of, 358
Potassium channels, 61–62
in apical membranes, 141
and cardiac electrical activity, 89
cholinergic modulation of, 101
delayed rectifier current and, 96–97
and early repolarization, 97
and inhibition by ATP, 97
inwardly rectifying, 96
gating currents in, 62–63
ionic currents in, 62
in skeletal muscle, 68
Potassium chloride, cotransport, in basolateral membranes, 143
Potassium secretion, hypothalamic diabetes insipidus and, 351
Potassium transport
in DCT, 348
in gastric mucosa, 167–168
in intestine, 186–187
Potentials
action, *see* Action potential; Cardiac action potentials
electrical, *see* Electrical potential
hydrogen diffusion, 28
membrane, *see* Membrane potential
resting, *see* Resting potential
transepithelial, *see* Transepithelial potentials
Principal (P) cells, 344
Probes
immunological reagents as, 17–18
optical, and pH measurement techniques, 27
Prostaglandins (PGs)
and ADH interactions, 324–325
hepatocyte bile formation and, 237
in intestinal electrolyte transport control, 191
renal, effect on GFR, 274–275
Prostanoids, in intestinal electrolyte transport control, 191
Protein kinase, oxyntic cell regulation via, 155
Proteins, *see also* Binding proteins, organic anions and
in bile, 242–243
lysosomal degradation of, 31
pentameric integral membrane, 108
plasma, in glomerular dynamics, 265–266
Proteoliposomes, and *lac* carrier protein, 12, 13
Proton electrochemical gradient
and active transport relationship, 9–10
effect of ionophores on, 8–9
external pH and, 8
Proximal convoluted tubule
fluid transport model in, 304
osmotic gradient in, 306
passive permeabilities in, 289–290
transepithelial potential differences in, 288–289
transport along, 287
Proximal nephron, general properties of, 285–286; *see also* Proximal tubule
Proximal sodium transport, active and passive components of, 304
Proximal straight tubule
osmotic gradient in, 306
transepithelial potentials across, 293
transport along, 287
Proximal tubule, *see also* Proximal convoluted tubule; Proximal straight tubule
and base exit across basolateral membrane, 302–303
epithelial fluid movement in, 303–304
fluid movement across
anion-driven, 303–304
anion reflection coefficients versus sodium chloride diffusion, 307–308
route of, 308–309
osmotic gradient site at, 305–307
and peritubular sodium extrusion, 298–299
and proximal sodium transport components, 304
and solvent–solute coupling, 304–309
stop-flow pressure, in glomerular filtration rate, 259–260
superficial versus juxtamedullary, 287–288
and transepithelial sodium transport, 293–295
cell model for, 295
and transport between apical and basolateral cell membranes, 299–300
transport functions distribution along, 286–288
versus distal convoluted tubule, 343
Psammomys, I cells in, 344
PTH, *see* Parathyroid hormone (PTH)
PTS, *see* Phosphoenolpyruvate–phosphotransferase system (PTS)
Purkinje fibers, in cardiac sodium channel excitability, 89–91
PVP, *see* Polyvinylpyrrolidone (PVP)

QNB, *see* [^{3}H]-Quinuclidinyl benzylate (QNB)
[^{3}H]-Quinuclidinyl benzylate (QNB), 196

Receptor-mediated endocytosis, 31–32
Refractory period, in cardiac sodium channel excitability, 90–91
Renal medulla, outer, 331; *see also* Inner medulla; Medulla
Renal nerves, and GFR, 275–276
Renal prostaglandins, and GFR, 274–275
Renin–angiotensin system, and GFR, 274–275
Repolarization, of action potential, 96–97
Respiratory epithelium, *see* Airway epithelium
Resting potential
calculation for, 53–54
cardiac potassium channels and, 96–97
Reticulum, *see* Sarcoplasmic reticulum (SR)
Right-side-out (RSO) vesicles, 4

SA, *see* Stimulation-associated (SA) vesicles
Salmonella typhimurium, PTS in, 4
Salt transport, effect of ADH on, 317–325
Sarcoplasmic reticulum (SR)
calcium transport mechanism and, 75
ultrastructure of, 65–66
Saturation, kinetics of, DCT and, 345
SCFAs, *see* Short-chain fatty acids (SCFAs)
SDS, *see* Sodium dodecyl sulfate (SDS)
Secondary structure model, for *lac* carrier protein, 17
Secretin
and bile flow, 244
in intestinal electrolyte transport control, 190
Secretion
alkaline, in gastric mucosa, 168–169
of bile, and hepatic transport, 225–244
bile salt-dependent and independent, 231–232
biliary lipid, 240–241
biliary protein, 242–243
canalicular bile acid, 234–235
electrolyte, model for hepatocytes, 235–236
by fetal lung, 377–378
hepatocellular water and electrolyte, 230–234
hydrogen, *see* Hydrogen secretion
intestinal, *see* Intestinal secretion
potassium, hypothalamic diabetes insipidus and, 351
Secretory granules, pH measurement in, 26–27
Selectivity, of cardiac calcium channels, 95
Short-chain fatty acids (SCFAs), 186
Shunt pathways
intercellular
of proximal tubule, 292–293
and solute–solvent coupling, 304–309
paracellular, canalicular bile secretion and, 235

Shunt pathways (*cont.*)
and water transport, 187–188
Single capillary model, in glomerular dynamics, 264–265
Single-channel recording
closed channels and, 123
conductance and, 122
open channels and, 122
in sodium channels, 61
versus macroscopic current methods, 120–121
Single effect, urinary concentration and, 332–333
Single nephron glomerular filtration rate (SNGFR), 269–271
intrarenal distribution of, 276–277
and stop-flow pressure relationship, 272
Skeletal muscle, ion movements in, 65–78; *see also* Muscle
Slow responses
in arrhythmias, 95
and cardiac calcium channels, 92–95
in nodal tissue, 95–96
SNGFR, *see* Single nephron glomerular filtration rate (SNGFR)
Sodium
absorption by epithelial cells, models for, 136–143
and apical membrane, *see* Apical membranes
and canalicular bile acid secretion, 234–235
luminal cell membrane and, 296–298
peritubular extrusion, 298–299
in renal medulla, 336
transepithelial potentials and, 290–292
Sodium/bicarbonate–chloride/hydrogen exchange
chemical dependence in, 45
energetics in, 45–46
in ion-transport systems, 45–46
pH dependence in, 45
pharmacology of, 46
Sodium–calcium countertransport, in basolateral membranes, 143
Sodium channels, 58–59
and cardiac electrical activity, 89
action potential recordings and, 89–90
consequences of, 90–92
and voltage clamp studies, 90
fluctuation analysis and single-channel recordings in, 61
gating currents in, 59–61
in skeletal muscle, 67–68
Sodium chloride
absorption, by mTALH, 318–319
effect of ADH on, 319–320
reabsorption by DCT, 344–347
versus anion reflection coefficients, 307–308
Sodium–chloride cotransport, in gastric mucosa, 167
Sodium chloride transport, peritubular, 299
Sodium dodecyl sulfate (SDS), 111
Sodium exit, in intestinal absorption, 182
Sodium/hydrogen exchange
energetics in, 47
hormones and growth factors and, 47–48
in ion-transport systems, 46–48
pH dependence in, 46–47
Sodium/hydrogen exchange (*cont.*)
pharmacology of, 47
sodium involvement in, 47
Sodium, potassium-ATPase, in gastric mucosa, 156–157
Sodium–potassium pump
in basolateral membranes, 142
in DCT, 358
in tracheal epithelium, 372–374
Sodium transport
between apical and basolateral cell membranes, 299–300
in gastric mucosa, 167–168
in intestine, 175–176
membrane vesicles and, 10, 11
proximal, cell model for, 295
transepithelial, evidence for, 293–295
Solutes
handling in renal medulla, 334–336
transport in DCT, 344–351, 353–355
transport models, in urinary concentration, 338
vacuolar storage of, 33–34
Solute–solvent coupling, intercellular shunt pathway role and, 304–309
Solvent drag, ADH and, 321
Somatostatin
and bile flow, 244
and intestinal absorption, 198
SR, *see* Sarcoplasmic reticulum (SR)
Staphylococcus aureus, and mAbs binding sites, 113
Starling filtration–reabsorption principle, 258
Starling forces, and sodium reabsorption in proximal tubules, 346
Steroids, in bile, 243
Stimulation-associated (SA) vesicles, 158–159, 160
Substance P, and intestinal secretion, 197
Substrate accumulation
effect of valinomycin and nigericin on, 9–10
effect on pH gradient and membrane potential, 9
Substrate stoichiometry, and hydrogen ion, 10
Sucrose, and vacuolar solute storage, 33–34
Superficial proximal tubule, versus juxtamedullary proximal tubule, 287–288
Surface cells
conductance of, 162
electrical potential profile of, 163
nonacidic chloride secretion by, 166–167
Synaptic transmission, 126
Synaptic vesicles, and biogenic amine storage, 33

TC, *see* Terminal cisternae (TC)
Temperature, hepatocyte bile formation and, 237
Terminal cisternae (TC)
electrical properties of membranes of, 72–73
in tubular system, 66
Tetramethylphenylenediamine (TMPD), 4
Tetrodotoxin (TTX)
in cardiac sodium channel excitability, 89
and conduction in T system, 69
Thevenin equivalent electromotive force, and chloride exit step, 372
Thiazides, and calcium transport in DCT, 354–355
Thyroid hormones, hepatocyte bile formation and, 236–237
Tight junctions
bile secretory function and, 228, 229
permeability and conductance of, 162–164
Tissue resistance, in gastric mucosa, 163–164
TMB-8, *see* 8-(*N*,*N*-Diethylamino)-octyl-3,4,5-trimethoxybenzoate (TMB-8)
TMG, *see* Methyl-1-thio-β-D-galactopyranoside (TMG)
TMPD, *see* Tetramethylphenylenediamine (TMPD)
Torpedo, nicotinic AChR in, 109–112
and monoclonal antibodies, 113–114
Tracheal epithelium, 367
chloride secretion control in, 368–369
ion transport in, 368
cellular model of, 369
and chloride entry step, 369–371
and chloride exit step, 371–372
and secretion and absorption mechanisms compared, 374
and sodium entry process, 374
and sodium–potassium pump, 372–374
morphology of, 367, 368
Transepithelial effective resistence, of DCT, in diabetes insipidus, 360
Transepithelial electrical potential difference (PD), of DCT, 355–357
Transepithelial potentials
and intercellular shunt path, 292–293
across proximal convoluted tubule, 288–289
across proximal straight tubule, 293
and sodium cotransport with sugars and amino acids, 290–292
Transepithelial voltage, in DCT, 355–357
Transglomerular pressure, in glomerular dynamics, 265–266
Transmembrane pH difference, *see also* pH gradient
generation of, 27–31
uses of, 31–34
Transmitter-activated channels, 126–127
Transport
active, *see* Active transport
active solute, models of urinary concentration, 338
along proximal tubule, 286–288
amino acid, across basolateral membrane, 183
between apical and basolateral cell membranes, 299–300
base, and basolateral pathways, 302–303
bicarbonate ion, 184
bile acid, 232
intracellular, 233
calcium, *see* Calcium transport
chloride, *see* Chloride transport
electrolyte, *see* Electrolyte transport, intestinal
functions of the DCT, 343–360
hepatic, bile secretion and, 225–244
hydrogen, *see* Hydrogen transport

Transport (*cont.*)
  intestinal secretory, mechanisms of, 184
  lactose, 12–14
  organic anion solute, 238–240
  potassium, *see* Potassium transport
  salt, effect of ADH on, 317–325
  short-chain fatty acid, 186
  sodium, *see* Sodium transport; Sodium–potassium pump
  sodium-dependent, membrane vesicles and, 10
  transepithelial, in gastric mucosa, 165–166
  of water, *see* Water transport
Transport coefficients, effect of ADH on, 320
Transverse tubular system (T system)
  excitation conductance along, 69
  and transmission at triad level, 69–73
  ultrastructure of, 66
Triglyceride lipolysis, nonpolar lipids and, 223
T system, *see* Transverse tubular system (T system)
TTFB, *see* 4,5,6,7-Tetrachloro-2-trifluoromethylbenzimidazole (TTFB)
TTX, *see* Tetrodotoxin (TTX)
Tubular membrane, electrical properties of, 66–68
Tubular oxyntic gland, 152
Tubular system
  membrane electrical properties in, 66–68
  in skeletal muscle fibers, 65–66
  transverse, *see* Transverse tubular system (T system)
Tubuloglomerular feedback, and glomerular filtration rate, 269–274
Tubulovesicles (TV)
  membrane recycling hypothesis and, 153
  in oxyntic cells, 152
TV, *see* Tubulovesicles (TV)
Two-state model, of ionic channels, 55–57
  extended, 57–58

Unstirred layers, barrier in intestine, 217–219
Urea
  cycling in central core model, 338–339
  in renal medulla, 335–336
Urinary acidification, 32
Urinary concentration
  ADH action and, 325
  central core model of, 333–334
  countercurrent multiplier hypothesis of, 331–332
  in inner medulla, 334
    mass balance requirements for, 337–338
    proposed models for, 338–340
  and nature of the single effect, 332–333
  processes of, 329–340
  role of vasopressin in, 333
  vasa recta function in, 333
Urinary dilution
  central core model of, 333–334
  and countercurrent multiplier hypothesis, 331–332
  and nature of the single effect, 332–333
  processes of, 329–340
  role of vasopressin in, 333
  vasa recta function in, 333

Vacuoles
  and hydrogen pump, 30–31
  solute storage in, 33–34
Valinomycin, substrate accumulation and, 9–10
Vasa recta, function in urinary concentration and dilution, 333
Vascular resistance, and glomerular filtration rate, 268–269
Vasoactive intestinal polypeptide (VIP), 196
Vasopressin, role in urinary concentration, 333
Vesicles, *see also* Tubulovesicles (TV)
  acidic, methylamine in, 26
  and active transport, 3–5
Vesicles (*cont.*)
  BBM, *see* Brush border membrane (BBM) vesicles
  gastric microsomal, ion transport by, 157–158
  inside-out, 4
  right side-out, 4
  stimulation-associated, 158–159, 160
  synaptic, and biogenic amine storage, 33
VIP, *see* Vasoactive intestinal polypeptide (VIP)
Vitamins, in bile, 243
Volatile fatty acids, 186
Voltage clamp analysis
  and cardiac calcium channels, 93–94
    activation in, 94
    inactivation in, 94
    properties of, 94
    and selectivity, 95
  in cardiac sodium channel excitability, 90

Water
  permeability of proximal tubular epithelium, 307
  secretion model for hepatocytes, 235–236
Water channels, ADH-induced, versus gramicidin A channels, 322
Water diffusion, series resistance elements for, 321
Water transport
  in DCT, 351–353
  effect of ADH on, 317–325
  in intestine, 175–199
    models of, 175–177
  shunt pathways and, 187–188
Weak acids
  distribution of, 39–40
  effects of external application, 42–44

Xenobiotics, in bile, 243

Yeast vacuoles, solute storage in, 34

Zonulae occludens, *see* Tight junctions